Dychweler y llyfr hwn erb... dyddia...
...y dyc...wel isod,
neu pan elwir amdan... ga... y ...lyfrgellydd.
Dirwyir y darllenydd ... dychwel y llyfr
y dyddiad hwn.

This book m... returned on or before the
date shown ... or when required by the
Librarian.
F... will be imposed for the late return of
this book.

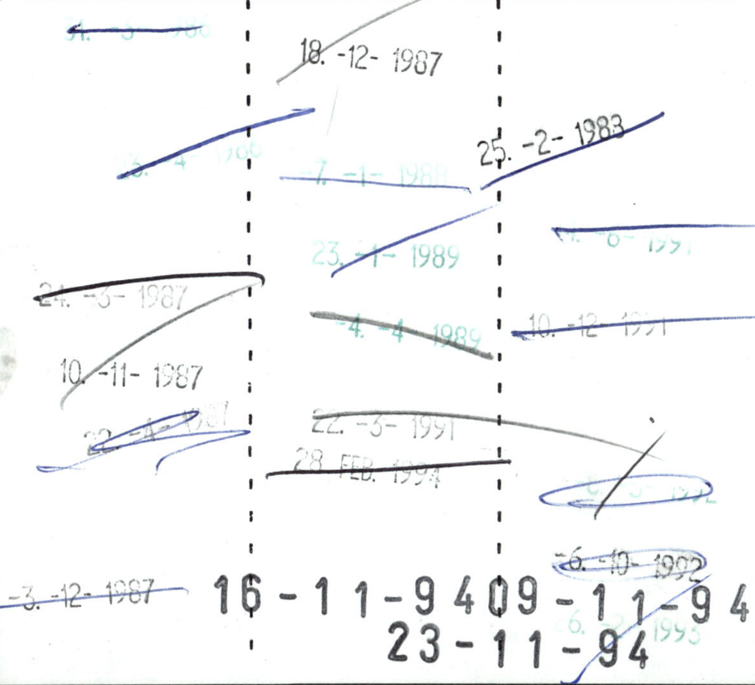

COMPREHENSIVE INSECT PHYSIOLOGY BIOCHEMISTRY AND PHARMACOLOGY

Volume 3

INTEGUMENT, RESPIRATION AND CIRCULATION

Rd 1 B

CIP VOL 3-A

COMPREHENSIVE INSECT PHYSIOLOGY BIOCHEMISTRY AND PHARMACOLOGY

Volume 3

INTEGUMENT, RESPIRATION AND CIRCULATION

Executive Editors

G. A. KERKUT
Department of Neurophysiology, University of Southampton, UK

L. I. GILBERT
Department of Biology, University of North Carolina, USA

PERGAMON PRESS

OXFORD · NEW YORK · TORONTO · SYDNEY · PARIS · FRANKFURT

852677

UK	Pergamon Press Ltd., Headington Hill Hall, Oxford OX3 0BW, England
USA	Pergamon Press Inc., Maxwell House, Fairview Park, Elmsford, New York 10523, USA
CANADA	Pergamon Press Canada Ltd., Suite 104, 150 Consumers Road, Willowdale, Ontario M2J 1P9, Canada
AUSTRALIA	Pergamon Press (Aust.) Pty. Ltd., P.O. Box 544. Potts Point, N.S.W. 2011, Australia
FRANCE	Pergamon Press SARL, 24 rue des Ecoles, 75240 Paris, Cedex 05, France
FEDERAL REPUBLIC OF GERMANY	Pergamon Press GmbH, Hammerweg 6, D-6242 Kronberg-Taunus, Federal Republic of Germany

Copyright © 1985 Pergamon Press Ltd.

All Rights Reserved. No part of this publication may be reproduced, stored in a retrieval system or transmitted in any form or by any means, electronic, electrostatic, magnetic tape, mechanical, photocopying, recording or otherwise, without permission in writing from the publishers.

First edition 1985

Library of Congress Cataloging in Publication Data

Main entry under title:
Comprehensive insect physiology, biochemistry, and pharmacology.

Contents: v. 1. Embryogenesis and reproduction.
1. Insects—Physiology—Collected works. I. Kerkut, G. A.
II. Gilbert, Lawrence I. (Lawrence Irwin), 1929–

QL495.C64 1984 595.7′01 83–25743

British Library Cataloguing in Publication Data

Comprehensive insect physiology, biochemistry and pharmacology.
1. Insects
I. Kerkut, G. A. II. Gilbert, Lawrence I.
595.7 QL463

ISBN 0–08–030804–X (volume 3)
ISBN 0–08–026850–1 (set)

SCIENCE
QL495. C55

Filmset by Filmtype Services Ltd., Scarborough
Printed in Great Britain by A. Wheaton & Co. Ltd., Exeter

Contents

Foreword

Aristotle was enchanted by the phenomenon of insect metamorphosis and the early microscopists such as Robert Hook, Marcello Malpighi, Anton van Leeuwenhoek, René de Réaumur and Pieter Lyonet were fascinated by the structure and function of the different parts of insects and made some of the first important contributions to our knowledge of insect physiology. More detailed functional studies were made by Borelli in his book "De Motu Animalium", published in 1680, and his interpretation of insect walking patterns remained in our textbooks until 1955.

In general, the 18th and 19th century research workers were more concerned with the morphology and classification of insects, though physiologists such as Claude Bernard and naturalists such as John Lubbock and Henri Fabre were always interested in the functional analysis of insects.

One of the milestones in the study of insect physiology was the publication by Wigglesworth of his small book on insect physiology in 1934. This was stimulated by an appreciation of the way in which studies on the basic physiology of insects were necessary before one could understand and ultimately control the activity of insect pests of man and crops.

Wigglesworth initially studied medicine and then carried out research at the London School of Hygiene and Tropical Medicine. His innate gift for planning simple but fundamental experiments on *Rhodnius* led rapidly to an increase in our knowledge about moulting, the control of larval and adult stages, and provided the foundation for insect endocrinology. Furthermore he inspired a group of co-workers who later played a key role in the application of modern techniques to solve the problems of insect physiology and biochemistry.

Wigglesworth's "Insect Physiology" was followed by a more detailed and full-sized textbook, "Principles of Insect Physiology", which was published in 1947 and is now in its 7th edition (1972).

The three-volume edition of "Physiology of Insecta", a multi-authored work edited by Morris Rockstein, was published in 1964 and a new edition in six volumes followed in 1973.

The study of insect biochemistry developed more slowly, partly because there was no special distinction between physiology and biochemistry; the investigator just used the methods available for his studies. David Keilin started his studies working on insects: "From 1919 onwards I had been actively engaged in the study of the anatomy of the respiratory system, respiratory adaptation and respiration of dipterous larvae and pupae. Among the vast amount of material I was investigating, special attention was given to the larvae of *Gasterophilus intestinalis*". For these studies Keilin developed a method for the spectroscopic analysis of respiratory pigments of insect pupae under the microscope, which ultimately led to the discovery of the cytochromes.

The pteridines were discovered in insect pigments, and the one gene–one enzyme hypothesis of Beadle and Tatum, which was the cornerstone of molecular biology, was a result of biochemical and genetic analysis of *Drosophila*.

The rapid expansion of biochemistry after 1945 led to many more workers studying insect biochemistry and the first textbook on the subject by Darcy Gilmore was published in 1961. This was followed by the multi-authored "Biochemistry of Insects", edited by Morris Rockstein, in 1978.

The first evidence that a steroid hormone acts at the level of the gene came from the studies of Clever and Karlson in the 1960s on the puffing by the polytene chromosomes of *Chironomus*.

Though insect physiologists and biochemists initially published their papers in journals such as *Biological Bulletin*, *Journal of Biological Chemistry*, *Biochemical Journal*, *Journal of Physiology*, *Journal of Experimental Zoology*, *Journal of Experimental Biology*, *Roux' Archiv für Entwicklungsmechanik*, and *Zeitschrift für vergleichende Physiologie*, the great expansion of insect physiology and biochemistry from 1945 onwards led to the establishment of journals and other periodicals specialising in insects, such as the *Journal of Insect Physiology*, *Insect Biochemistry*, *Annual Review of Entomology*, and *Advances in Insect Physiology*.

It is also fitting to mention the work of other pioneers in the study of insect physiology and biochemistry, such as Autrum, Bounhiol, Bodenstein, Butenandt, Chadwick, Dethier, Fraenkel, Fukuda, Joly, Lees, Karlson, Kopec, Piepho, Richards, Roeder, Berta Scharrer, Snodgrass and Williams; these and many others laid the foundations of the subject and all following research workers have stood on the shoulders of these giants.

In July 1980 a meeting was held at Pergamon Press in Oxford to discuss the possibility of publishing a series of volumes on insect physiology, biochemistry and pharmacology. The idea was to produce 12 volumes that would provide an up-to-date summary and orientation on the physiology, biochemistry, pharmacology, behaviour and control of insects that would be of value to research workers, teachers and students. The volumes should provide the reader with the classical background to the literature and include all the important basic material. In addition, special attention would be given to the literature from 1950 to the present day. Emphasis would be given to illustrations, graphs, EM pictures and tabular summaries of data.

We were asked to act as Executive Editors and by December 1980 we had produced a 27-page booklet giving details of the aims and objectives of the project, details of the proposed volumes and chapters, suggested plans within the chapters, abbreviations, preparation of diagrams and tables, and journal citations to ensure uniformity of presentation as far as possible. This booklet was sent to authors of the chapters and their comments invited. By the middle of 1981 most of the chapters had been assigned to authors and the project was under way. The details of the volumes and the chapters they contain are given on the following pages so that the reader can see the contents of each of the other volumes.

In addition, there is a final volume, Volume 13, which is the Index Volume. Although each volume will contain its own subject index, species index and author index, Volume 13 will contain the combined subject, species and author indexes for all 12 text volumes so that any material in these volumes can be rapidly located.

All references in the volumes are given with full titles of papers, journal, volume, and first and last pages. The references to the authors in the text are given with their initials so that it is clear that the text refers to D. Smith and not, say, to A. Smith. There are more than 50,000 references to the literature, more than 10,000 species of insect referred to, and all should be readily found in the 12 different volumes.

There are 240 authors of the 200 chapters in the volumes and they have produced a series of very readable, up-to-date, and critical summaries of the literature. In addition, they have considered the problems associated with their subject, indicated the present state of the subject and suggested its developmental pathway over the next decade.

We are very grateful to our colleagues for the efficient way that they have met the challenge and the deadlines in spite of their many other commitments.

This series of volumes will be very useful to libraries, but an important case can be made that the books should be considered as research instruments. A set of volumes should also be available in the laboratory for constant reference. They will provide the research worker with an account of the literature and will always be instantly available for consultation. For this reason they should be considered as research equipment equally important as microscopes, oscilloscopes or spectrophotometers.

The volumes should save research workers many weeks of time each year in that not only will they provide an awareness of the literature and the background, and so save valuable research time, but the full index to authors, subject and species, and the full literature references, should also make it much easier to write reports and papers on their own new research work.

It is hoped that these volumes will do much to strengthen the case for insects as a source of research material, not only because insects are important medical and agricultural pests (over 200 million people at present have malaria: insects eat or destroy about 20% of planted food crops), but also because in many cases insects are the ideal unique research material for studying and solving fundamental biological problems.

G. A. Kerkut
Southampton

L. I. Gilbert
Chapel Hill

Preface to Volume 3

The main topics of Volume 3 in this masterwork concern integument, respiration and circulation. These three seemingly separate subjects represent precisely those areas of physiology most different from vertebrates, in fact, one might say opposite of vertebrates.

Our skeletons are internal, the insects are hung from an exoskeleton; our circulatory system is enclosed, the insects have an open system; we have immunoglobulins, the insects defend themselves without them; we have hemoglobin, the insects have evolved a more efficient heme protein, but only on rare occasions. Despite these obvious differences, work on insect physiology has often led to knowledge of general import, such as the discovery of the cytochromes.

We realize from this that in spite of major differences in physiology, so much of the underlying biochemistry, the cell–cell interactions, and the overall genetic direction and control are startlingly similar in all animals. To find these fundamental underlying principles, one often must grapple with the technical problems of small amounts of material and limitations in analytical chemistry.

Historically, the study of the composition of the cuticle has been aided by new analytical procedures. The basic principles of the cuticular tanning process have not changed much in 40 years, yet we know considerably more now about the final structures. We are just beginning to learn about cuticular lipid and its role in the physiology of the cuticle, but questions of the water proofing of cuticle, its control and maintenance still need to be investigated. In fact, the important questions of how chemicals impinge on the cuticle and penetrate to the interior of the insect have never been satisfactorily described, and what is worse, there does not seem to be a continuing interest or study of this vital subject.

Studies on the hemolymph as a tissue have taken traditionally several directions. The most obvious are the classification of hemocytes, the least obvious concern is the biochemistry of the hemolymph. After many years of attempts at classifying insect hemocytes, the present volume presents the beginnings of some order from the confusion. Surprisingly the function of insect hemocytes are now being compared with similar functions in vertebrate blood cells. Again, one is struck by a fundamental similarity between animals.

Some of the most exciting advances mentioned in the present volume are those related to the subject of cell recognition. The detection of foreign matter, particularly the detection of invading organisms, often boils down to cell surface chemistry. The principles underlying defensive mechanisms will undoubtedly be found to be similar for many animals.

I am particularly pleased to see the chapter on chemistry and physiology of the hemolymph. The relationship between the hemolymph and the organs and cells of the hemocoel has not been given the attention it deserves. After years of measurements of hemolymph composition and a growing appreciation of the role of proteins as carriers, we are still a long way from finding a reasonable saline solution, and we are hindered by a lack of understanding of the role of many of the simpler hemolymph elements. A feedback link between energy metabolism and general respiration needs to be found and its physiology understood better.

The use of immunological techniques, particularly immunohistochemistry, has been applied to many subjects in insect physiology and related fields. It is obvious that these newer, powerful tools will make a great impact on many of the subjects presented here.

The ability to detect extremely small amounts of hormones, proteins and other substances in the hemolymph should eliminate some of the problems of working with smaller insects. Advances often come as discrete jumps and the use of immunological techniques is one such jump whose time has come.

T. A. Miller
Riverside, California

Contributors to Volume 3

Andersen, S. O.
Angust Krogh Institute of Biological Chemistry A, 13 Universitetsparken, 2100 Copenhagen, Denmark

Ashhurst, D. E.
Department of Anatomy, St George's Hospital Medical School, University of London, Cranmer Terrace, Tooting, London SW17 0RE, UK

Blomquist, G. J.
Department of Biochemistry, School of Medical Sciences, University of Nevada, Reno, NV 89557, USA

Boman, H. G.
Department of Microbiology, University of Stockholm, S-10691 Stockholm, Sweden

Collins, J. V.
Department of Biology, Dalhousie University, Halifax, Nova Scotia, B3H 4J1, Canada

Crossley, A. C.
School of Biological Sciences, Zoology Building, University of Sydney, Sydney, New South Wales 2006, Australia

Dean, R. L.
Department of Zoology, University of Western Ontario, London, Ontario, N6A 5B7, Canada

Dillwith, J. W.
Department of Entomology, University of Missouri, Columbia, MO 65211, USA

Dziadik-Turner, C.
Department of Biochemistry, Kansas University Medical Center, Kansas City, KS 66103, USA

Götz, P.
Institut für Allgemeine Zoologie, der Freien Universität Berlin, Konigs-Louise-Str. 1-3, D-1000 Berlin 33, Federal Republic of Germany

Gupta, A. P.
Department of Entomology and Economic Zoology, Cook College, Rutgers University, The State University of New Jersey, PO Box 231, New Brunswick, NJ 08903, USA

Hepburn, H. R.
Department of Physiology, Medical School, University of the Witwatersrand, 1 Jan Smuts Avenue, Johannesburg 2001, Republic of South Africa

Keeley, L. L.
Department of Entomology, Texas A & M University, College Station, TX 77843, USA

Koga, D.
Department of Agricultural Chemistry, Yamaguchi University, Yamaguchi 753, Japan

Kramer, K. J
US Grain Marketing Research Laboratory, US Department of Agriculture, 1515 College Avenue, Manhattan, KS 66506, USA

Locke, M.
Department of Zoology, University of Western Ontario, London, Ontario, N6A 5B7, Canada

Mill, P. J.
Department of Pure and Applied Zoology, Baines Wing, University of Leeds, Leeds LS2 9JT, UK

Miller, T. A.
Department of Entomology, Division of Toxicology and Physiology, University of California, Riverside, CA 92521, USA

Mullins, D. E.
Department of Entomology, Virginia Polytechnic Institute and State University, Blacksburg, VA 24061, USA

Contents of All Volumes

Volume 13 — CUMULATIVE INDEXES

1 Structure of the Integument

H. R. HEPBURN

University of the Witwatersrand, Johannesburg, Republic of South Africa

Dedicated to K. M. Rudall.

1 INTRODUCTION

The bones, cartilage and tendons of the vertebrate principally serve an endoskeletal function. The integument of insects, on the other hand, serves so many more purposes: it is the interface between a living animal and an environment from which both beneficial and harmful influences come. Thus, besides serving as an exoskeleton, the integument is a rich sensory depot which must perceive stimuli in many different forms.

The integument is very far from being an inert substance. While it consists of a cuticle and an epidermis, neither of these layers has any physiological integrity without the other. The integument is a collection and an historical record of synthetic activities performed by the animal. But this record is not static as a rune, the kinds and qualities of layers come and go, especially during metamorphosis. It is hardly surprising that, with great structural changes occurring from instar to instar, equally great physical and chemical changes also occur. It is the purpose of this chapter to document some of the striking and important properties of the integument and to make inferences as to their significance in the lives of "... these narrow Engines ...".

2 EPICUTICLE

2.1 General Properties

The epicuticle is an extraordinarily complex and very thin composite structure that overlies the chitin-bearing procuticle of insects. An epicuticle generally covers the entire external surface of an insect but there are a few exceptions associated with special functions, the most notable being the surfaces of some chemoreceptive sensillae first described by Slifer, E. (1961). While the fore- and hindguts of insects are also covered with an epicuticle, the midgut, which is endodermal in origin, lacks an epicuticular covering (Bertram, D. and Bird, R., 1961).

The functions of the epicuticle, shown experimentally or inferred, are many and varied and in most cases it is not yet clear how precisely function is related to structure. The epicuticle makes important contributions to water homeostasis both in the uptake of water through the integument of terrestrial insects such as the flea *Xenopsylla brasiliensis* (Edney, E., 1947) or in the aquatic alderfly, *Sialis lutaria*, (Shaw, J., 1955) as well as through resistance to water loss first studied in any experimental detail in the cockroach *Periplaneta americana* by Ramsay, J. (1935). The epicuticle is thought to set limits on the dimensions an exoskeleton can attain during intrastadial growth of holometabolous larvae such as has been documented for the skipper, *Calpodes ethlius* (Locke, M., 1961), and the silkworms *Bombyx mori* (Hackman, R., 1976). Similarly, the extent to which a recently ecdysed cuticle may expand, or that of a sap- or blood-sucking insect may distend, during engorgement are thought to reside in the mechanical properties of the epicuticle in each particular case (Bennet-Clark, H., 1963; Wigglesworth, V., 1959).

Additional functions for the epicuticle have

been inferred, but the structural bases for such functions have remained completely elusive to date. For example, the epicuticle must be selectively permeable to the transport of moulting fluid precursors into the apolysial space (necessary for the partial digestion of the outer-lying presumptive exuvial cuticle) and to the return of the endocuticular digests into the body proper of the pharate instar. At the same time, the pharate epicuticle must itself be protected from chemical dissolution by moulting fluid enzymes.

Although we have no direct quantitative measurements of the mechanical properties of the epicuticle it has been inferred from its bending and fracture characteristics that it is in some ways analogous to a lacquer, stronger in compression than in tension (Joffe, I. *et al.*, 1975). The epicuticle may also serve as a chemical reservoir, either as a depot for the storage of metabolic waste products or for the brief retention of defensive secretions (Filshie, B. and Waterhouse, D., 1969) and possibly even for juvenile hormone (Wigglesworth, V., 1970).

Turning to the structure of the epicuticle, this subject is made all the more difficult because of arguments over homology that have been in progress for about 150 years. The nomenclatural problems remind us of Alice: "... What's the point of their having names if they don't come when you call them?" Nonetheless, a generalized view of the epicuticle has been developing over the last several years and can be summarized as follows. It consists of four principal layers each of which is of unique composition and properties (Fig. 1). There is an outermost cement layer that is subtended by a wax layer. Below the wax layer there is an outer epicuticle, and below that an inner epicuticle. Unlike the layered subdivisions of the chitin-bearing procuticle, the layers of the epicuticle are synthesized in a temporal sequence that is the reverse of that in which they are anatomically situated in a fully differentiated cuticle. The inner and outer epicuticles are formed before apolysis and thus define the reality of the pharate cuticle (Jenkin, P. and Hinton, H., 1966) while the cement and wax layers follow apolysis (Locke, M., 1966).

2.2 Cement layer

The cement layer is generally regarded as the result of the combined secretory products of specialized dermal glands (specialized epidermal cells) and thought to consist of proteins and lipids that have been stabilized by various polyphenolic substances (Wigglesworth, V., 1933; 1976; Beament, J., 1955). The distribution and relative thickness of the cement layer varies enormously in the same or in different insects. For example, most of the exoskeleton of *Rhodnius prolixus* is covered by a cement layer, but certain regions of the abdominal tergites actually lack a cement layer (Wigglesworth, V., 1947) and some caterpillars are said to be only partially covered with a cement layer as in *Diataraxia oleracea* (Way, M., 1950). The cement layer may be completely absent as from the honeybee (Locke, M., 1961).

It has very often been suggested that the properties of the cement layer are roughly analogous to those of commercial shellac, the latter being the secretory products of the lac insect, *Laccifer lacca.* Its formation has been compared with that of the cockroach ootheca (as, incidentally, has the tanning of procuticle been compared with oothecal tanning) which is secreted by separate collaterial glands and the combined products of secretion after

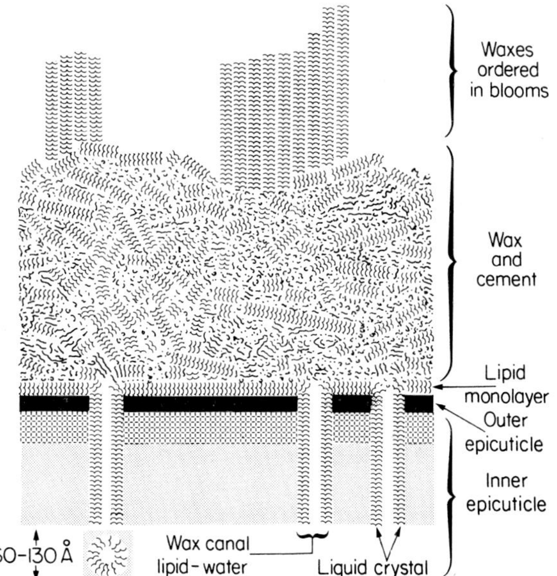

FIG. 1. Schematic diagram of an insect epicuticle. In this instance the most complex situation known is illustrated. Thus, wax is associated with the cement layer and wax also occurs in a bloom above the cement layer (cf. Fig. 2). (Adapted from Locke, M., 1974.)

mixing (Pryor, M., 1940a,b; Wigglesworth, V., 1948).

The formation of the cement layer has been principally studied by Wigglesworth who has recently summarized its morphogenesis and properties in the sucking bug *Rhodnius prolixus*. On ecdysis, the new teneral cuticle is hydrophobic but shortly after eclosion, secretions from the specialized dermal glands either penetrate or cover the cement layer so that it becomes hydrophilic. Some hours later the same structure becomes hydrophobic again, possibly because it has become stabilized chemically through processes analogous to tanning and/or because the cement layer itself becomes impregnated with waxy material which is hydrophobic in nature (Wigglesworth, V., 1976). The cement layer is thought to function primarily by protecting the subtending wax layer against possible abrasive or impact damage. A comparative study of cuticles with and without cement layers would contribute much to our very limited knowledge of this epicuticular layer.

2.3 Wax layer

The wax layer is situated just below the cement layer in insects where the latter is in fact present. The composition of insect waxes has been investigated over the last few decades, the earlier work being summarized by Gilmour, D. (1960) and that of the interim by Jackson, L. and Blomquist, G. (1976; see Blomquist and Dillwith, this volume). Generalizations about cuticular waxes are fraught with difficulty because of variations in solvent extraction procedures and conditions. Indeed, composition has recently been shown to vary biologically also, and variations in lipid class occur between individuals of the same species, as well as at different times of development (Gilby, A., 1980; Warthen, J. *et al.*, 1981).

In general, the composition of insect cuticular waxes (Table 1) is dominated by hydrocarbon molecules varying in length from 12 to 31 carbons and by esters of even-numbered fatty acids and alcohols. The physiological significance of these compounds may well lie in their physical attributes: saturated and esterified fatty acids are associated with hardness in commercial waxes (Warth, A., 1956) while unsaturated analogues tend to be soft and even greasy. The waxes are mainly long-chain saturated alcohols esterified with acids. However, owing to the extremely labile nature of lipids, the precise composition of epicuticular wax for any insect must be regarded as tentative.

The structural arrangement of wax in the cuticle is not a simple matter, nor can we describe a discrete wax layer. At least four different variations of arrangement have been described to date. A lipid monolayer was described by Beament, J. (1945) and this layer can be seen immediately above the outer epicuticle. A more complex arrangement based on studies of several species has been visualized by Locke, M. (1974) who shows that in some insects wax may also be combined with the cement layer (Fig. 1) as well as extending above the surface of the cement layer in the form of wax blooms commonly

Table 1: Major classes of insect cuticular lipids

Fraction	*Pteronarcys*[1] *californica*		*Anabrus*[2] *simplex*	*Periplaneta*[3] *australasie*, *Periplaneta brunnea*, *Periplaneta fulginosa*	*Lucilia*[4] *cuprina*		*Cynthia*[5] *ricini*	*Tenebrio*[6] *molitor*	*Melanoplus*[7] *sanguinipes*
	Larva	Adult			Pupa	Adult			
Hydrocarbon	3	12	48-58	90	33	62	5	10	60
Wax esters	1	4	9-11	—	—	—	16	13	28
Triglyceride	78	7	—	7	25	16	—	—	1
Free fatty acids	12	49	15-18	2	16	7	Tr	5	6
Aliphatic alcohols	—	—	2- 3	—	—	—	70	58	2
Sterols	1	18	—	1	2	—	Tr	1	1
Others	5	10	15-18	—	15	24	8	13	—

[1] Armold, M. *et al.*, 1969; [2] Baker, G. *et al.*, 1960; [3] Jackson, L., 1970; [4] Goodrich, B., 1970; [5] Bowers, W. and Thompson, M., 1965; [6] Bursell, E. and Clements, A., 1967; [7] Soliday, C. *et al.*, 1974.

associated with Heteroptera and Homoptera and also very many beetles (Fig. 2). Finally a structural role for lipid in the procuticle has been intimated by Wigglesworth, V. (1975a,b).

The lipids of the wax layer are probably secreted by the epidermis just prior to ecdysis. Generally, when lipid is initially released at the surface of an insect it is in a mobile liquid phase but is transformed into wax at ecdysis. They are then transported through the pore canal system to the somewhat finer and arborescing wax canals. Locke, M. (1961) has concluded that the final wax product is actually synthesized in the wax canal filament where esterases modify final wax precursors that have come from the epidermis. His conclusions were based on studies of several insects such as the honeybee, wax-moth, the skipper *Calpodes ethlius* and the beetle *Tenebrio molitor*. Against this, it has been claimed that there is a wax layer on the caterpillar *Diataraxia* (Way, M., 1950) which is said to lack pore canals, and in *Calpodes ethlius* (Locke, M., 1960a,b). Clearly it would be of interest to reinvestigate many of these allegedly unusual cases with the considerably more sophisticated techniques presently available.

The physiological significance of the insect wax layer has been best documented in terms of its relationship to waterproofing. The relationship between cuticular transpiration and temperature change has been studied in some detail in the case of the butterfly *Pieris* (Beament, J., 1959). For this animal there is an initial critical transition temperature at which there is a sudden increase in cuticular water permeability followed by a plateau region of little change and again another transition temperature at which permeability further increases, as shown in Fig. 3.

At present it remains uncertain whether or not the observed changes in cuticular permeability are due to either failure or changes in the lipid phase that might be related to lipid melting points, or whether there might be shifts in structure toward

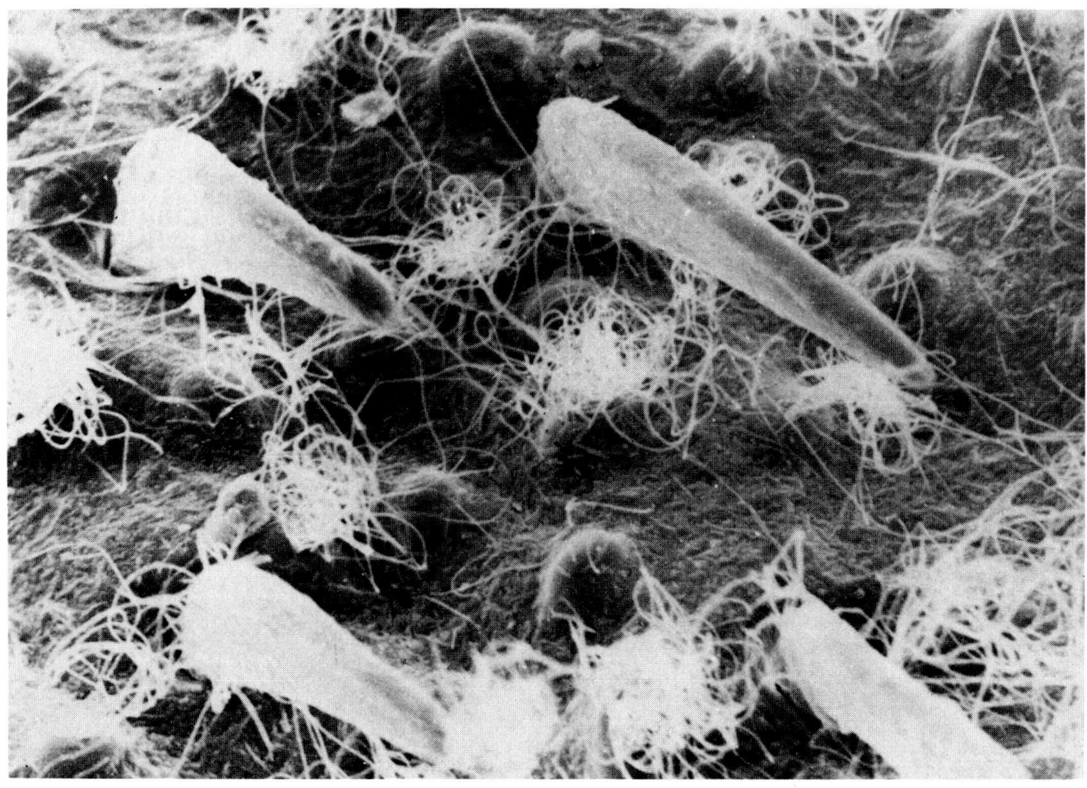

FIG. 2. SEM of the ephemeral wax bloom filaments of the Namib desert dune beetle, *Cardiosis fairmairei*. (Courtesy of Dr E. McClain, Gobabeb, Namibia.)

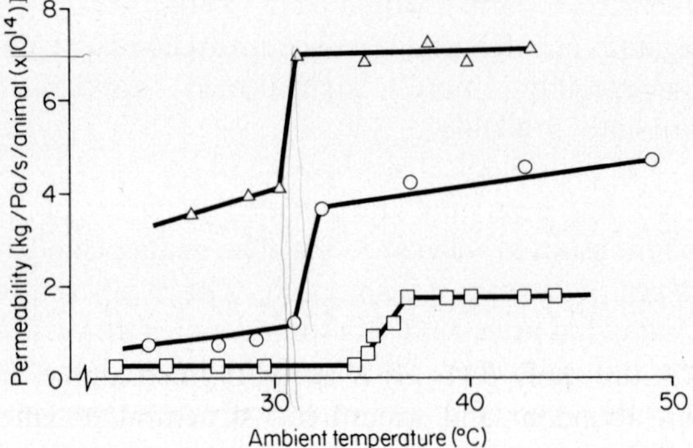

FIG. 3. Cuticular water permeability as related to changes in temperature. Squares denote pupae of *Tenebrio molitor*, circles, the larvae of *Rhodnius prolixus*, the triangles, pupae of the butterfly *Pieris rapae*. (After Beament, J., 1959.)

liquid crystal phases as suggested by Beament, J. (1959). Recently, a number of branched lipids have been reported from several different desert beetles, *Eleodes, Cryptoglossa* and *Centrioptera*, (Hadley, N., 1977, 1978) but it is not clear how these compounds relate to permeability *vis-à-vis* lipid structure.

Numerous biological observations as to changes in wax layer properties have been recorded in recent times. For example, diapausing pupae of the tomato sphinx, *Manduca*, produce three times more epicuticular wax than do non-diapausing pupae, and the inference is that changes in epicuticular wax synthesis could be mediated by neuroendocrine events.

2.4 Outer epicuticle

The outer epicuticle (often designated as cuticulin) appears thus far to be of universal occurrence as an epicuticular layer of the exoskeleton, as well as occurring in the linings of tracheae and serving as the insertion layer for the tonofibrillae of muscles (Richards, A., 1951). It has been described by Locke, M. (1965, 1966) as a trilaminar layer of about 12–18 nm in thickness. Its chemistry and composition remain unknown but its resistance to degradation suggests properties similar to polythene-like polymers and to quinone-tanned protein (Locke, M., 1976).

The outer epicuticle is the first-formed layer of the pharate cuticle. It is derived from specialized regions of the apical membrane of the epidermal cells, namely the plasma membrane plaques (Fig. 4). During development discrete plates appear above the membrane and gradually form a continuous layer through accretion (Filshie, B. and Waterhouse, D., 1969; Locke, M., 1976; vol. 2). Shortly after the outer epicuticle has been formed, small pores of about 3 nm diameter can be detected in this layer and it has been suggested that these pores allow the egress of apolysial droplets into the apolysial space for the partial digestion of the presumptive exuvial cuticle and for the transport of the products of digestion into the body of the pharate instar.

In view of the critically important functional roles commonly ascribed to the outer epicuticle, whatever it may be called, it is surprising that the composition of this material has so seldom come under the scrutiny of those of chemical bent. Because of its position in the cuticle, it has important contributions as a selective permeability barrier in the metabolism of old cuticle, particularly during the course of pharate cuticle development. The outer epicuticle apparently limits the extent to which intrastadial growth, through stretching (Hackman, R., 1976) or intussusception (Locke, M., 1974) can occur. Similarly, it is thought to define the final dimensions to which the exoskeleton will ultimately expand following an ecdysis, and thus give a basis for the observations variously known as "Brooks' rule", "Dyar's rule" or "Przibram's law".

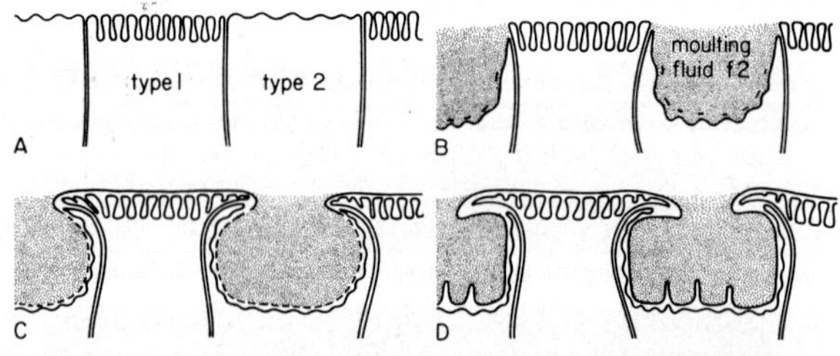

FIG. 4. Diagram summarizing formation of the outer epicuticle in the evaporative surface area of the stink bug, *Nezara viridula*. Stage **A** represents differentiation of epidermal cells prior to secretion of apolysial fluid. The plasma membranes of the two cells are continuous. **B** shows the apolysial fluid associated with a type-2 cell. The forming outer epicuticle occurs in small patches above the microvilli. In **C**, outer epicuticle formation proceeds at a greater rate over type-1 cells laterally displacing the forming cuticle. Finally, **D**, the expanding outer epicuticle, causes surface folding of the sheeted material. (After Filshie, B. and Waterhouse, D., 1969.)

The intricate and often complex surface patterns of insects, specialized areas such as diffraction gratings, plastrons, strigils and the like, are all formed at the same time as the outer epicuticle, so that its morphogenetic significance is thought to be great (Locke, M., 1974). In this respect, the "chicken or the egg" question arises: to what extent is the outer epicuticle itself the determinant of final exoskeleton shape *vis-à-vis* the crenulated microvilli of the apical plasma membrane plaques which actually secrete the products of the outer epicuticle.

2.5 Inner epicuticle

The inner epicuticle is an optically isotropic (Weis-Fogh, T., 1970) layer thought to consist of polymerized lipoprotein stabilized by quinones, a conclusion reached in early studies by Dennell, R. (1946). It is thought to be a product of secretion of the epidermis to which contributions from the oenocytes have been added (Wigglesworth, V., 1947). It has also been postulated that the inner epicuticle serves as the reservoir for extracellular enzymes associated with wound repair (Wigglesworth, V., 1937).

3 PROCUTICLE

3.1 General properties

The procuticle of the insect integument has been classically understood, and without exception, to be the chitin-bearing part of the cuticle proper. While the chitin-free epicuticle has been the subject of several interesting controversies, particularly with reference to permeability properties, differing interpretations of various aspects of the procuticle have led to most intriguing arguments of a fundamental nature. Studies of the principal components of procuticle, chitin and protein have not developed apace, and we know considerably more about the molecular dimensions and configurations of chitin than we do of protein. Even very fundamental problems, such as how exactly are the chitins and proteins structurally situated in a piece of procuticle, remain unresolved. This section will, perforce, not be a particularly integrated one with respect to protein and chitin but will trace the development of that information which we do have, and will sometimes suggest functional relationships where possible.

3.2 Chitins

3.2.1 Molecular structures and conformation

Chitin, poly-β-(1→4)-*N*-acetyl-D-glucosamine, is an abundant and widespread structural polymer akin to the celluloses and is synthesized by fungi, protozoans and most of the protostomian invertebrates (Jeuniaux, C., 1971; Kramer *et al.*, this volume). In insects it occurs naturally in crystalline form but it is crystallographically heterogeneous, and three polymorphs which differ both in the packing and polarities of adjacent chains have thus far been described (Rudall, K., 1963).

α-Chitin is the most prevalent of the chitin polymorphs and a detailed structure for it was proposed by Carlström, D. (1957). He confirmed the orthorhombic unit cell of earlier crystallographers, defined its dimensions and space group symmetry (adjacent chains are anti-parallel) and showed that the crystallites arise from hydrogen-bonded sheets of chains. The Carlström structure has been the subject of much refinement over the last 20 years and a recent treatment by Blackwell's group (Minke, R. and Blackwell, J., 1978) appears to have overcome some of the "noise" problems inherent in the 1957 model.

Thus, α-chitin is currently understood to consist of an orthorhombic unit cell with the following dimensions: 4.74 Å in the crystallographic *a*-axis, 18.86 Å in the *b*-axis and 10.32 Å in the *c*-, or fibre, axis. The unit cell (Fig. 5) consists of disaccharide sections of two anti-parallel chains bonded in two ways. There is an intramolecular hydrogen bond (O3′–H ··· O5) and successive chains are connected by C=O ··· H—N bonds, thus forming sheets along the *a*-axis. The CH_2OH side-chains adopt very different conformations on each side of the two chains within the unit cell structure: the CH_2OH group on one side forms an intramolecular hydrogen bond with the carbonyl group of the next residue (O6–H ··· 07′); the CH_2OH group of the other chain forms an intermolecular or intersheet bond (O6–H ··· O7′) as shown in Fig. 6. Different

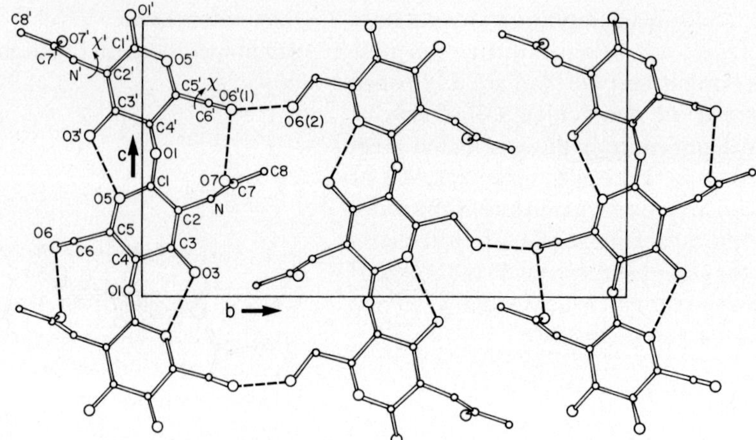

FIG. 5. Proposed structure for α-chitin showing points of rotation (χ′ and χ) about the side-chains. Basic model of Carlström, D. (1957) as refined by Minke, R. and Blackwell, J. (1978). *bc* projection. The CH_2OH in upper left corner forms an O6′(1)-H … O7 intramolecular bond and the CH_2OH of the centre chain forms an O6(2)-H … O6′ intermolecular bond. (After Blackwell, J. *et al.*, 1978.)

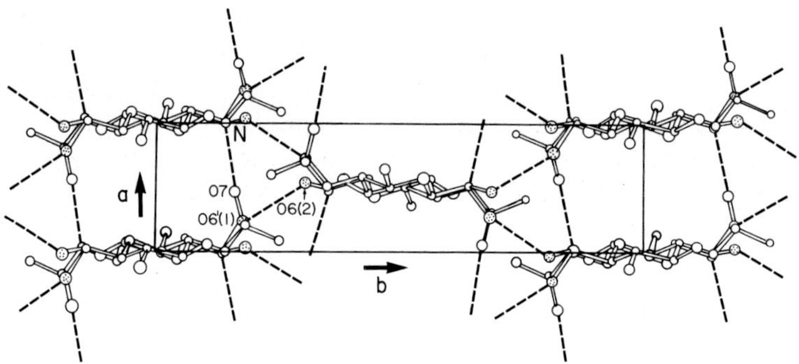

FIG. 6. *ab* projection of α-chitin showing the C7=O7 … H—N and the O6(2)—H … O6′ intermolecular bonds. (After Blackwell, J. *et al.*, 1978.)

amide groups also participate in hydrogen bonding. In Blackwell's refinement there is a statistical mixture of the CH_2OH bonding groups (Blackwell, J. *et al.*, 1978). Finally, the basic α-chitin chain is sterically rigid except for rotational freedom about the side-chains (Fig. 5) and this facilitates transconformational changes that can be induced chemically and mechanically (Rudall, K. and Kenchington, W., 1973).

β-Chitin was first distinguished from α-chitin by Lotmar, W. and Picken, L. (1950) in diffraction studies of the squid pen (*Loligo*). It must be considered as rare among insects because it has thus far been identified only in the cocoons of the figwort weevils, *Cionus* and *Cleopus*, and that of the cerylonid beetle, *Murmidius* (Rudall, K. and Kenchington, W., 1973). The structure of β-chitin has also been the subject of refinement studies through the years and has recently been characterized in some detail (Gardner, K. and Blackwell, J., 1975).

β-Chitin consists of a monoclinic unit cell with the following dimensions: 4.85 Å in the crystallographic *a*-axis, 9.26 Å in the *b*-axis and 10.38 Å in the *c*- or fibre axis. The unit cell (Fig. 7) thus contains but a single repeating disaccharide (the *b*-axial dimension of β-chitin being only about one-half that of α-chitin). The chains are arranged in parallel so that the sense is opposite that of α-chitin. The chain conformation is the same as that of α-chitin and sheets of β-chitin result from

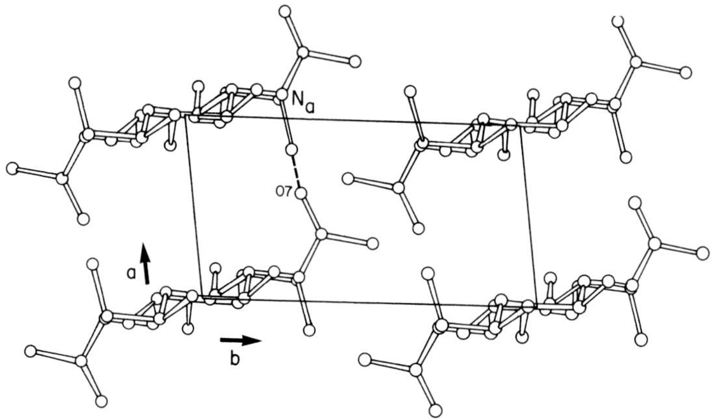

FIG. 7. Proposed structure for anhydrous β-chitin in *ab* projection showing the C7=O7 ··· H—N intermolecular bond. (After Blackwell, J. *et al.*, 1978.)

C=O ··· H—N hydrogen bonds. CH_2OH groups are hydrogen bonded to the carbonyl group of the next chain along the *a*-axis. There is no intermolecular hydrogen bonding between sheets, and this provides a basis for the ease of solvation in this polymorph (Blackwell, J. *et al.*, 1978).

γ-Chitin was discovered and partially characterized by the indefatigable crystallographer Rudall, K. (1962b, 1963) in studies of squid (*Loligo*) stomach. In insects this polymorph is known from the cocoons of the weevils *Prionomerus* and *Rhynchaenus*, the leaf beetle *Donacia*, the spider beetles *Ptinus* and *Gibbium* and the great dung beetle *Pacylomera* (Rudall, K., 1976) as well as from the peritrophic membranes of the cockroach *Blaberus, Schistocerca*, the caterpillar *Antheraea pernyi* and the sawfly *Phymatocera* (Rudall, K. and Kenchington, W., 1973). It is of interest to note that these same membranes in honeybees and other aculeate Hymenoptera consist only of α-chitin.

The unit cell dimensions of γ-chitin are regarded as tentative (awaiting additional confirmatory data) and are suggested to be as follows: ~4.7 Å in the crystallographic *a*-axis, ~29 Å in the *b*-axis and ~10.3 Å in the *c*- or fibre axis. It will be noted that the *b*-axial dimension of γ-chitin is about three times that of β-chitin and this leads to the suggestion that the γ-chitin molecule folds back on itself, as diagrammatically shown in Fig. 8. γ-Chitin is similar to β-chitin in its propensity for hydration and solvation (Rudall, K., 1963). The ease with which both β- and γ-chitin form hydrates mitigates against their usefulness as skeletal fibres. The fact that α-chitin is considerably less susceptible to hydration effects (Hepburn, H. and Chandler, H., 1978) may also be *one* of the reasons why most insects must moult in order to appreciably increase their surface dimensions.

All three polymorphs of chitin, though variable in degree of crystallinity, occur in insects as microfibrils and are classified as viscoelastic polymers (Hepburn, H. and Chandler, H., 1978). That is, when macroscopic samples of all of these chitins are deformed by tensile loading their mechanical properties such as stiffness, strength and extensibility vary within and among polymorphs with differing rates of deformation. Precise interpretation of the observed mechanical behaviour is made difficult because of differences in chitin samples with respect to crystallinity, orientation and microfibrillar packing in actually testable macrofibres.

Nonetheless, mechanochemical studies of chitin polymorphs indicate some important properties that ultimately derive from the structure of the

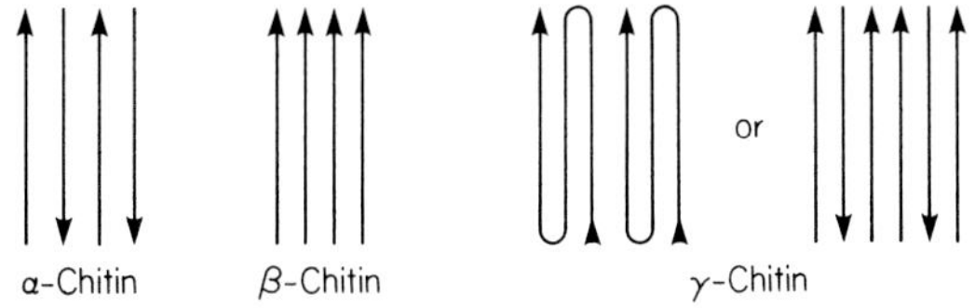

FIG. 8. Schematic representation of chitin chains in the unit cells of α-, β- and γ-chitin. (After Rudall, K., 1963.)

crystallites concerned. Owing to the inter-chain bonding characteristics of α-chitin *vis-à-vis* β- and γ-chitin, γ-chitin appears to deform viscoelastically by neighbouring chains slipping past one another while the deformation characteristics of β- and γ-chitin are consistent with molecular conformational changes (Rudall, K., 1963; Hepburn, H. and Chandler, H., 1978). While the tertiary structure of γ-chitin remains equivocal, mechanochemical evidence is consistent with a ↑∪∩↑-shaped unit cell rather than the ↑↓↑-shaped cell of Fig. 8. One can envisage the native conformation of γ-chitin as crudely analogous to the cross-β conformation described from the eggstalk silks of *Chrysopa* (Parker, K. and Rudall, K., 1957; Geddes, A. *et al.*, 1968).

Chitin crystals always occur in insects as collections of microfibrils, and measurements from many different preparations show that these microfibrils range from about 25 Å to 30 Å in diameter (Rudall, K., 1967; Neville, A., 1975). Microfibrils are in turn always associated with protein in a chitin–protein complex, be it cuticle, cocoon or peritrophic membrane (Rudall, K. and Kenchington, W., 1973). However, the insect exoskeleton is thought to contain only complexed α-chitin in microfibrils consisting on average of between 18 and 21 chains passing through a crystallite at any particular level, implying 3 sheets containing 6 or 7 chains (Rudall, K., 1963; Rudall, K. and Kenchington, W., 1973; Neville, A., 1975).

The structure of chitin has been classically defined on the basis of diffraction data obtained after chemical purification of a chitin-bearing sample, hence the "pure chitin lattice". However, it has long been known that X-ray diffraction patterns of intact cuticles are "modified" owing to the presence of protein (Fraenkel, G. and Rudall, K., 1947). Rudall has extended these observations over the years and has defined two basic chitin–protein complexes: a 31 Å axial repeat system in which diffraction spacings are associated with three chitobiose molecules (or 6 GlcNAc residues) and a 41 Å axial repeat system based on four chitobiose (8 GlcNAc, residues) (Rudall, K. and Kenchington, W., 1973; Rudall, K., 1976).

The origins and functional significance of these two repeating systems are not at all clear. It could be that in the 31 Å system, for example, every sixth residue differs from the preceding ones, a substitution of glucosamine for *N*-acetylglucosamine. To date, the 31 Å system has been observed in all of a wide variety of bees, wasps and beetles that have been examined; the 41 Å system occurring alone is recorded from the cranefly *Tipula* and a praying mantis while the occurrence of both 31 Å and 41 Å repeats in the same animal are documented for the moth *Attacus atlas*, the locusts *Schistocerca* and *Anacridium* and the cockroach *Blaberus*.

3.2.2 Problems in Chitin Microfibrillar Arrangements

One of the most intriguing, yet not entirely resolved, problems in integumental biology is that of chitin fibre architecture, the regular (?) geometry of fibres in three-dimensional space. This is of great importance because the procuticle is a secretory product of the epithelium and questions arise as to whether the resulting structures represent self-assembly crystallization phenomena, or whether the observed geometry is in fact both produced and controlled by the synthesizing epidermal cells.

The more immediate problem, which has been extensively studied — particularly in crustaceans — since the time of Schmidt, W. (1924), has been to resolve the structural arrangement of chitin microfibrils as they occur in the intact cuticles of insects and other arthropods (cf. Richards, A., 1951; Neville, A., 1975 and Hepburn, H., 1976). The attendant arguments and evidence advanced thus far have proved as contentious and stimulating for those interested in cuticle as has the honeybee dance language controversy for students of apiculture and animal behaviour (Wells, P. and Wenner, A., 1973; Gould, J., 1976).

Following the introduction of electron microscopy and the subsequent publication of literally hundreds of both transmission and scanning electron photomicrographs of cuticle, we must still consider the well-known photomicrograph illustrating the very familar parabolic patterning of chitin microfibrils first described from insect cuticle by Locke, M. (1960a). How ought these patterns to be understood? In a major extension of the ideas of Schmidt, W. (1924), Bouligand, Y. (1965) proposed a geometrical interpretation to explain the parabolic patterns such as illustrated in Fig. 9 and commonly encountered in arthropod cuticles.

The Bouligand model basically asserts that chitin microfibrils occur in sheets within the cuticle. Within any given sheet all of the microfibrils are parallel to one another and, often, are also parallel to the surface plane of the piece of cuticle within which they occur. (However, there are numerous exceptions to this latter condition where microfibrils curve in arcs with reference to the surface plane of cuticle, and this fact provided the basis for the early model of Locke, M., 1960a.) Neighbouring sheets of microfibrils are rotated with respect to one another and always in the same direction of rotation. The *angle* of rotation may be variable or fixed in different cases (Neville, A., 1975). Any collection of sheets that has rotated through 180° constitutes a lamella (Fig. 10). The *apparent* thickness of a lamella is susceptible to change depending upon the degree of obliquity with which the piece of cuticle was cut on the ultra-microtome (Filshie, B. and Smith, D., 1980). In this view, parabolic patterns of microfibrils (Fig. 9) arise as optical artefacts, the genesis of which is apparent from a consideration of the Moiré pattern shown in Fig. 11.

We have recapitulated the *ideal* case in the above model. However, there are a number of exceptions and anomalies which have not yet been satisfactorily explained in terms of Bouligand's helicoidal model. The most important of these is the occurrence of what seem to be unequivocal vertically disposed microfibrils, not to be confused with pore canal filaments. To overcome this difficulty Weis-Fogh, T. (1970) proposed a different model in which the microfibrils are continuously secreted and the sheets of microfibrils gradually change position vertically. In this way, a microfibril present in one region of a helicoid turns slightly and can therefore contribute to more than one apparent sheet of microfibrils (Fig. 12). This is usually termed the "screw-carpet" model, and is analogous to crystal structure in some synthetic plastics. The general applicability of this model has never really been explored to any extent. It is certainly worthy of further consideration.

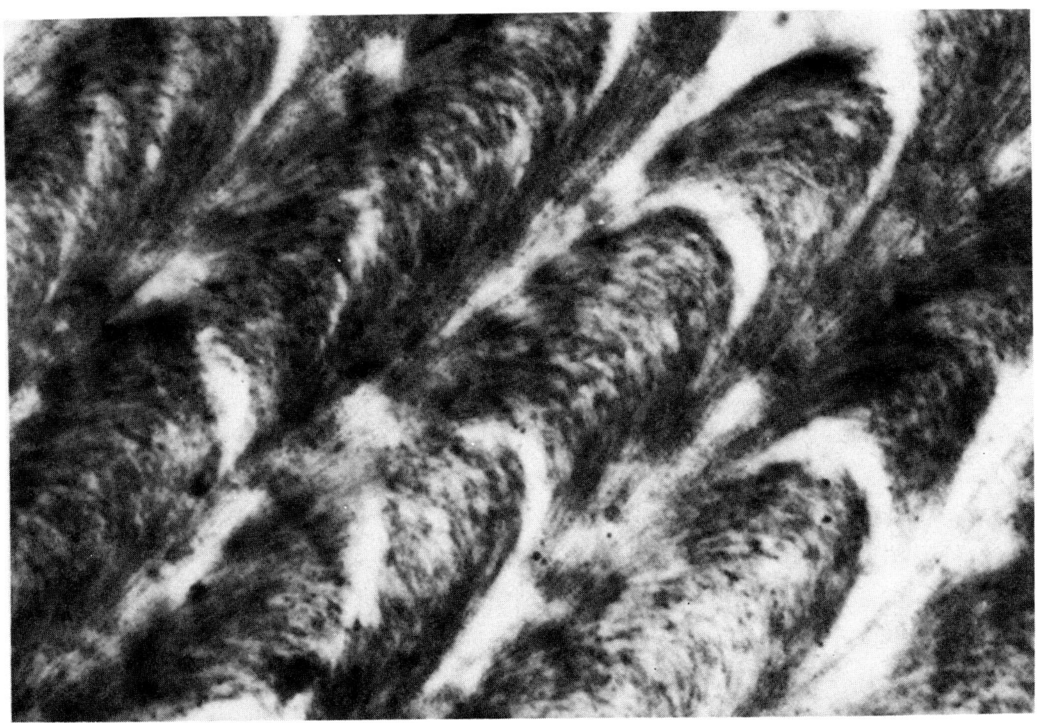

FIG. 9. A: Transmission electron photomicrograph of a section of the cuticle of the African honeybee (*Apis mellifera adansonii* L.) worker abdominal sternite showing the characteristic and artefactual parabolic patterns associated with chitin microfibrils arranged in a helicoidal array (× 16,000).

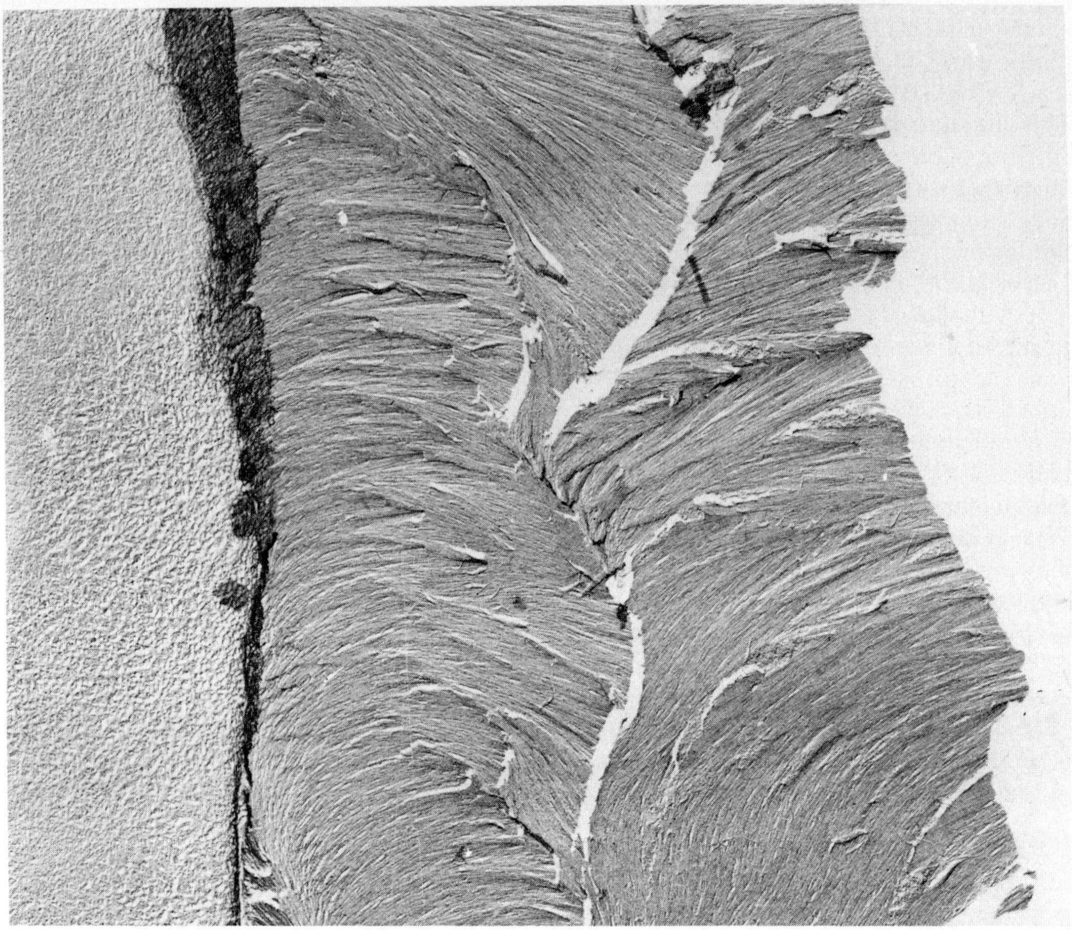

FIG. 9. **B**: Freeze-fracture preparation of **A**, (× 32,000). (**B**, courtesy A. G. Richards.)

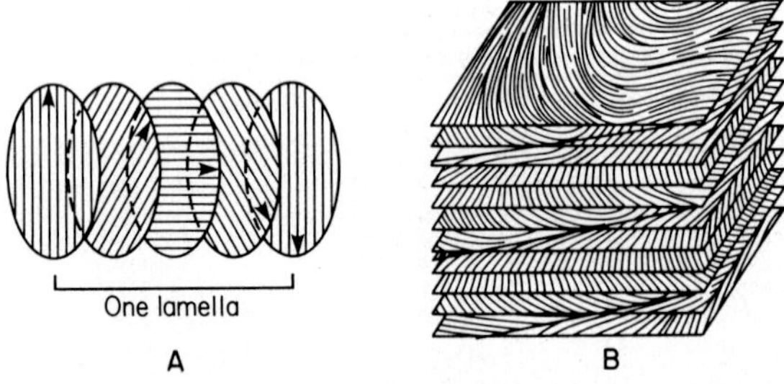

FIG. 10. **A**: Diagrammatic representation of only a few sheets of chitin microfibrils rotating through 180° and defining a single lamella. The arrows indicate the orientation of the microfibrils within each of the sheets. **B**: A variation of **A** in which the chitin microfibrils of any one sheet are curved instead of being parallel.

FIG. 11. Moiré pattern used by Bouligand, Y. (1965) to demonstrate how the parabolic pattern of arcs arise in sections of cuticle viewed electron microscopically.

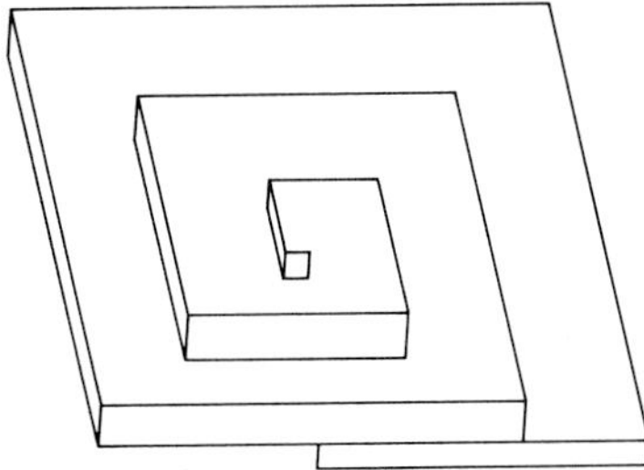

FIG. 12. Diagrammatic representation of "screw-carpet" model of Weis-Fogh, T. (1970) to account for the occurrence of vertically orientated chitin microfibrils in otherwise helicoidally arranged sheets of microfibrils.

There has been considerable doubt expressed as to the reality of helicoidal arrays in many cases. A number of arguments have been advanced against the Bouligand model, and that of Weis-Fogh by implication, as follows. Both helicoidal models preclude the possibility of separate lamellae or laminae of microfibrils, yet evidence (Fig. 13) for the existence of discrete layers has been provided from numerous decapod crustaceans (Dennell, R., 1973; Mutvei, H., 1974; Dalingwater, J., 1975a). Geometrical considerations of both the Bouligand and Weis-Fogh models require that the artefactual laminae of microfibrils actually spiral upwards, so that what are only *apparent* laminae would be out of register somewhere around four faces of a truncated pyramid. Dalingwater, J. (1975b) has published a photograph of *Cancer* cuticle in which continuity of laminae around the angles of the pyramid is clearly discernible. In the helicoidal hypothesis there must be a discontinuity at one of the four pyramidal corners; in Dalingwater's photograph there is not a discontinuity.

The helicoidal model of Bouligand fails to satisfactorily explain the occurrence of vertical microfibrils in cuticle (Fig. 14) yet these have been variously reported from both crustaceans (Mutvei, H., 1974; Dalingwater, J., 1975a) and from insects (Hepburn, H. and Ball, A., 1973; Hepburn, H. and Chandler, H., 1976). Finally, the helicoidal models have not been able to successfully accommodate

FIG. 13. Scanning electron photomicrograph of an oblique fracture of the chela of the crayfish *Austropotamobius* illustrating what appears to be a discrete lamina (or lamella) in the cuticle of this animal. (From Dalingwater, J., 1975a.)

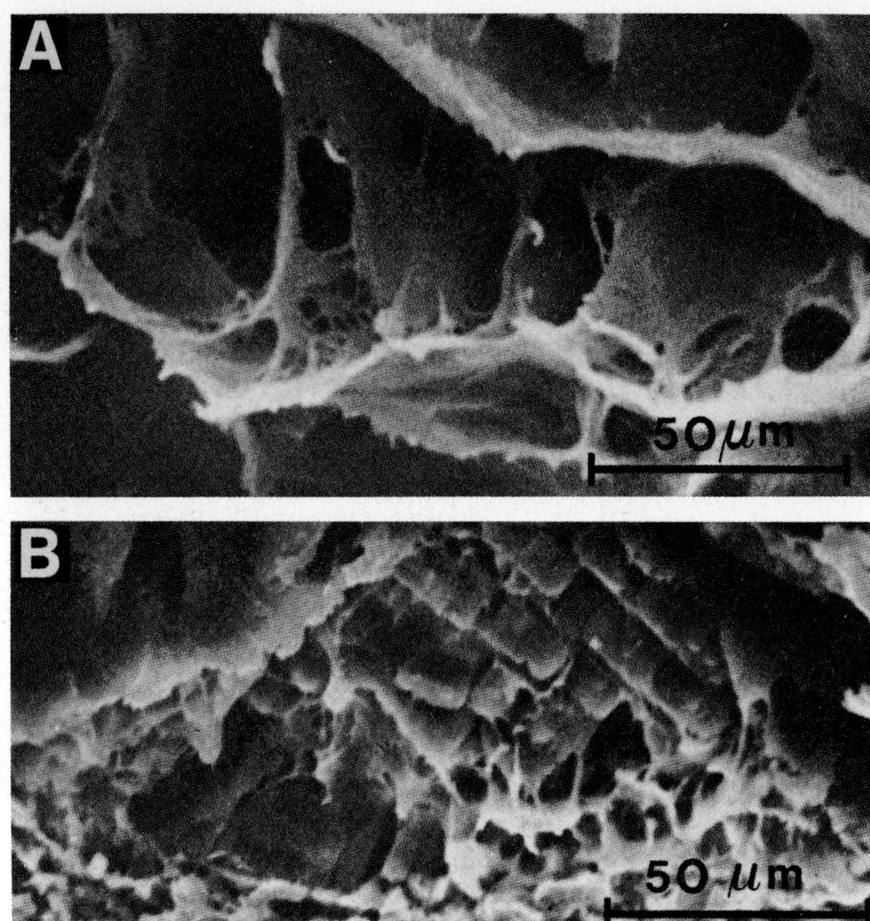

FIG. 14. Scanning electron photomicrographs of fractured (**A**) beetle, *Pachynoda sinuata*, and (**B**) locust (*Locusta migratoria*), intersegmental membrane showing vertically arranged chitin fibres.

spiralling bow-curves that occur in cuticular bumps and corneal lens cuticle (Fig. 15) of various insects (Chu, H. *et al.*, 1975; Neville, A., 1975). This latter microfibrillar pattern forms the basis of the mathematical arguments of Gordon, H. and Winfree, A. (1978) that the Bouligand model is not universally applicable to insect cuticles.

Most recently, Filshie, B. and Smith, D. (1980) have contributed to the resolution of this fine structural puzzle in arthropod cuticle with their studies of a crayfish, *Panulirus*. Essentially, they suggest that the reality of vertically arranged structures in the cuticle which *appear* to be microfibrils can be explained by including regular rows of particles (possibly proteinaceous) oriented perpendicularly to the chitin microfibrils (Fig. 16). But Filshie, B. and Smith, D. (1980) state that their vertical components are certainly not the same as those claimed by Dennell, R. (1973); Mutvei, H. (1974); and Dalingwater, J. (1975a); nor do they agree with the latter three authors that interlamellar (or interlaminar) microfibrils form an integral part of the cuticle.

In any event, we are left with two observations on the arrangement of chitin microfibrils in cuticle. The first of these is that there are no known or even suggested cases in which a random array of microfibrils (such as glass fibre in man-made chopped strand mat) occurs. The second observation is that we do have a basic working model, whatever its shortcomings, for the distribution of chitin microfibrils in insect procuticle as proposed by

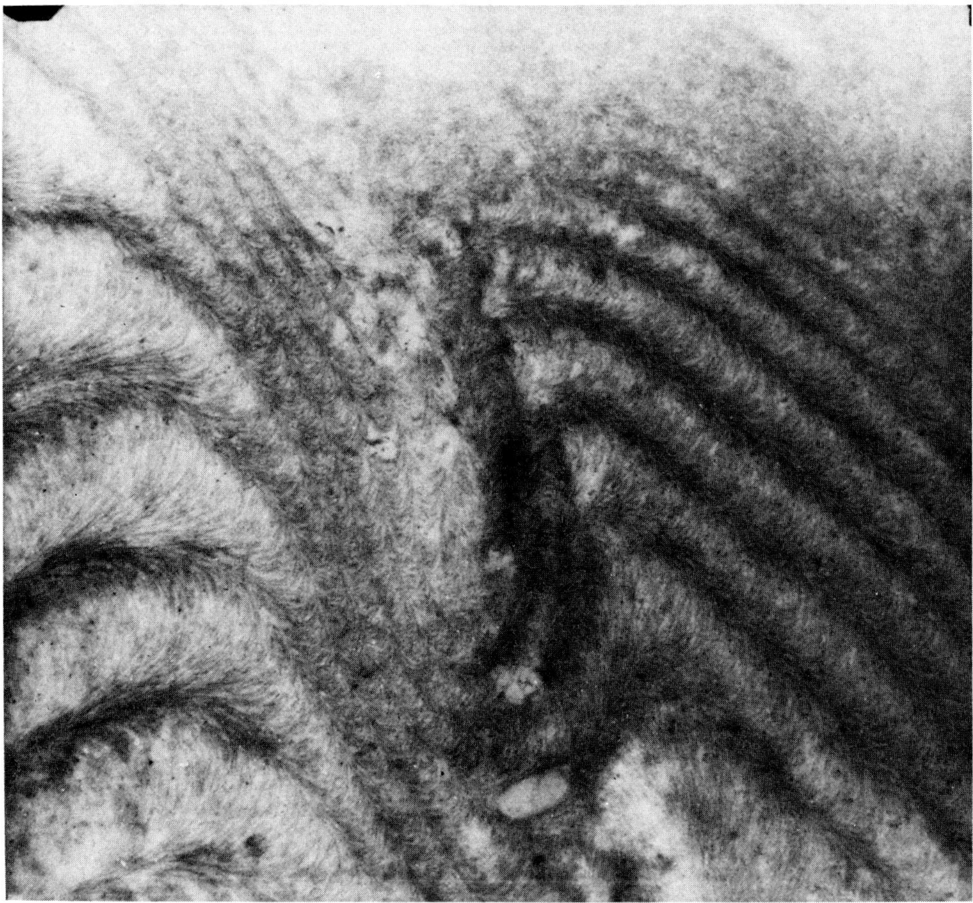

FIG. 15. Transmission electron photomicrograph of the corneal cuticle of *Boreus californicus* illustrating double spiral of bow-curves said by Gordon, H. and Winfree, A. (1978) to be inconsistent with the Bouligand model.

Neville, A. and Luke, B. (1969). Bearing in mind reservations discussed above, and that chitin microfibrillar architecture is not the same thing as whole cuticular structure, the model of Neville and Luke is certainly of heuristic value.

The "two-system model" of cuticle architecture advanced by Neville, A. and Luke, B. (1969) avers that:

(1) sheets of chitin microfibrils are helicoidally arranged in the Bouligand manner in the protein matrix and/or
(2) neighbouring sheets of microfibrils may be oriented in the same direction and so form a "preferred" layer. (Preferred is not used here in the usual physical way to mean statistically dominant, but in an absolute sense.)

Neville, A. and Luke, B. (1969) claim that most exocuticles are helicoidal throughout their thickness (exhibit artefactual parabolic lamellae) with the few exceptions of some locusts, mantids and beetles, and that various permutations occur in the endocuticle (neither exocuticle or endocuticle are clearly defined by these authors). The following patterns have been described from the endocuticle of insects and are summarized by Neville, A. (1975):

(1) Entirely helicoidal microfibrillar sheets have been reported for Lepidoptera, Diptera, Coleoptera, Collembola, Diplura and Thysanura.
(2) Entirely preferred or unidirectional sheets occur in the femoral apodeme of a grasshopper, *Eutropidacris*.

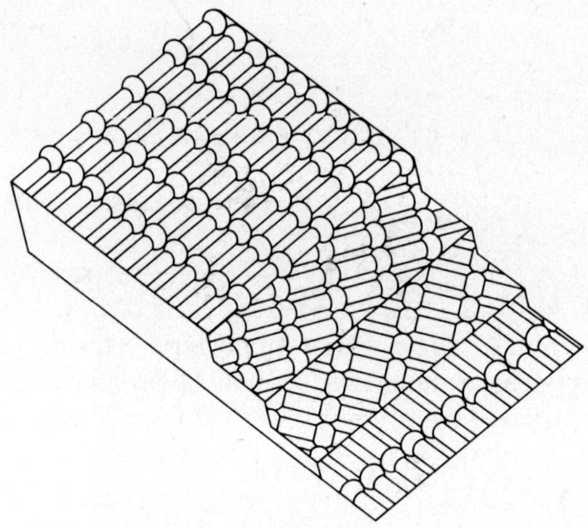

FIG. 16. Model proposed by Filshie, B. and Smith, D. (1980) to explain the occurrence of apparent vertical microfibrils in the inner endocuticle of the crayfish, *Panulirus*. They suggest that the spheres are not chitinous. After Filshie, B. and Smith, D. (1980).

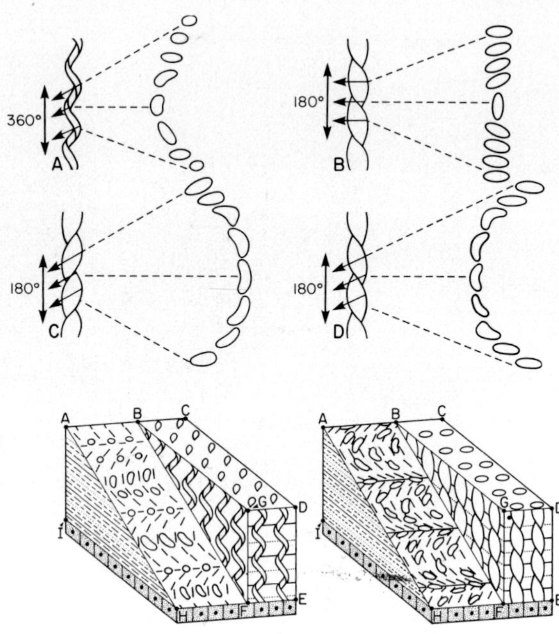

FIG. 17. Pore canal model of Neville, A. *et al.* (1969) developed to relate pore canal structure to helicoidally arranged chitin microfibrils as proposed by Bouligand, Y. (1965). In the surface plane (top right of diagram) the shapes and directions of pore canals are the same but change with the changing direction of the helicoid microfibril sheets.

(3) Alternating helicoidal and preferred layers in which all of the preferred layers lie in the same orientation, occur in Orthoptera, Cursoria and Mantodea; the alternation of layers is controlled by circadian clocks (Neville, A., 1965).
(4) Alternating helicoidal and preferred layers in which the preferred layers change direction with respect to one another (analogous to orthogonal plywoods) are found in Hemiptera, Odonata, some Coleoptera, Phasmida and some Mecoptera.

Interestingly, this proposed classification of Neville, A. and Luke, B. (1969) can be regarded as a by-product of investigations on the structures of pore canals. Light microscopic investigations had led early workers (Drach, P., 1939; Wigglesworth, V., 1948) to infer that pore canals seemed to be helicoidal, and electron microscopic studies led to the same interpretation (Richards, A. and Anderson, T., 1942; Locke, M., 1957). Neville, A. *et al.* (1969) were able to show that a pore canal is a tube of elliptical cross-section which is twisted regularly along its length so as to form a twisted ribbon. Comparisons of beetles and waterbugs, on the one hand, and honeybees and horseshoes crabs on the other, showed that in the former, plywood type of cuticle, the ends of the ellipses were in the axis of the chitin sheet while in the latter helicoidal cases the pore canals twist as do the planes of the helicoids (Fig. 17).

In applying the "chicken-or-the-egg" question to pore canals versus helicoidal fibre sheets, there are several examples of cuticle from which pore canals are absent but the chitin microfibrils are helicoidally arranged (Locke, M., 1960a; Caveney — cited in Neville, A. 1975). While this kind of evidence is admittedly circumstantial, it seems reasonable to infer that pore canal shape results from the forces associated with crystallization of the forming cuticle and not vice-versa (Neville, A., 1975).

3.2.3 Biosynthesis and Degradaton of Chitin

While a complete description of the synthesis of chitin monomers and their eventual polymerization into growing chains has not been made for any organism, substantial advances have been made in parallel studies on fungal and insect material (Kramer *et al.*, this volume). In the first instance, sufficient data have been collected using electron microscopical and radiochemical techniques to

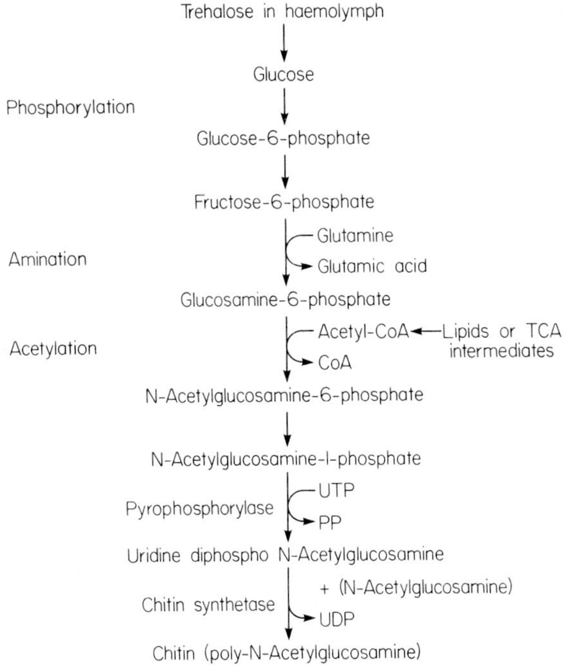

FIG. 18. Pathway for the biosynthesis of chitin. (Based on various sources.)

establish that chitin is in fact synthesized by the insect epidermis (Candy, D. and Kilby, B., 1962; Jaworski, E. *et al.*, 1963). The pathway for the synthesis of chitin in fungi and in insects as presently understood is briefly indicated in Fig. 18.

It is generally assumed that active chitin monomers are secreted into the apolysial spaces of insects where they are then polymerized and added to the growing chain by the enzyme chitin synthetase which is attached to the plasma membrane in some unknown manner. The eventual resolution of this point is crucial to an understanding of chitin microfibrillar architecture, because it is this last step in the synthetic pathway that holds the key as to whether chitin microfibrils are formed by self-assembly mechanisms or by cellular organelles.

Some light is shed on this problem in recent studies of fungal preparations by Bartnicki-Garcia, S. *et al.* (1978). In fungi, chitin synthetase is very probably produced as an inactive zymogen in the endoplasmic reticulum of a synthetically active epithelium. The proenzyme is contained in cytoplasmic vesicles, termed chitosomes, in which it is transported to the plasma membrane. The zymogen is then activated by proteases on the cell surface which release the chitin molecules and so eventually polymerize the chitin monomers. This account is somewhat hypothetical because there is no unequivocal proof that chitosomes actually occur in living cells. Chitosomes could be artefacts of the ultracentrifuge.

As far as insects are concerned, it remains to be shown that chitin is indeed secreted into the apolysial space in monomeric form. *N*-acetylglucosamine residues may be presumed to be transferred from UDPAG to the forming chain. If chitin synthetase is located in the plasma membrane, the UDPAG could well come from inside the cell, with the chains being polymerized in the extracellular apolysial space; but confirmation is required. As we have said, the chemical synthesis of chitin appears to be the same in fungi and in insects. However, the fine physiological controls of biosynthesis will certainly differ since we know that chitin synthesis in insects is controlled by 20-hydroxyecdysone (Agui, N. *et al.* 1969; Riddiford, vol. 8) and is also pH- and temperature-dependent (Hettick, B. and Bade, M., 1978).

The degradation of chitin is a normal catabolic process associated with growth and moulting in insects (Jeuniaux, C., 1971; Bade, M. and Stinson, A., 1978). Because the exuvium generally consists of epicuticle and exocuticle, most of the chitin which is digested and resorbed during moulting must be endocuticular in origin. It is digested by the enzymes chitinase and chitobiase but recent work (Kimura, S., 1976; Bade, M., 1978) suggests that we are dealing with a collection of as yet unpurified enzymes.

Just as the synthesis of chitin is a cyclic process (Porter, C. and Jaworski, E., 1965; Surholt, B. and Zebe, E., 1972), the elaboration of chitinases is also periodic. In the silkworm, *Bombyx*, the epidermis of the intermoult caterpillar lacks chitinases, and chitinolytic activity can be detected only at the time of apolysial (moulting) fluid secretion (Jeuniaux, C., 1963). There are other physiological variations on chitinase activity. For example, in diapausing pupae of *Hyalophora cecropia* the epidermis secretes an enzymically inactive gel which becomes activated when the gel becomes fluid toward the end of diapause (Passonneau, J. and Williams, C., 1953) while in non-diapausing Lepidoptera the apolysial fluid is active on secretion (Jeuniaux, C., 1963). The difference in diapausing and non-diapausing forms may be related in turn to basic differences in salt mobilization, since the activation of the chitinolytic proenzymes is calcium ion dependent in the latter (Bade, M., 1978).

3.3 Proteins of cuticles and allied structures

3.3.1 Heterogeneity of cuticular proteins

Proteins constitute the bulk of insect cuticle on the basis of dry mass measurements (Richards, A., 1951), yet we have far less knowledge about the structure and varieties of proteins in the cuticle than we do of chitin (see Silvert, vol. 2). To date, not a single primary structural analysis has been reported for an oligomeric fragment, much less an entire cuticular protein. There is, however, a steadily growing literature on protein bands obtained by various electrophoretic and chromatographic techniques and using different solvents to separate protein "species" according to their bond characteristics. For example, proteins can be classified as those which are water-soluble (these used to be termed "arthropodin" but the term means very little today) and others which are held in the cuticle by van der Waal's forces, hydrogen bonding or covalent bonding. The sequential extraction and electrophoretic characterization of proteins in this way has come to be known as the "Hackman series" after Hackman, R. (1972).

The sequential extraction of cuticular proteins is of direct importance to understanding protein structure within the cuticle, because the products of the extraction sequence indicate to what extent a given fraction might interact with other components such as chitin or proteins in the intact cuticle (Andersen, S., 1979; Hackman, R., 1972). But caution must be introduced in the interpretation of proteins studied in this way. Different solvents will affect different bonds in different ways, since not all hydrogen bonds, for example, are of the same strength; nor are given bonds necessarily equally accessible to a solvent. Because of the extreme sensitivity of proteins to variations in analytical technique, it is very likely that many of the bands obtained are but fragments of whole proteins so that a protein banding profile might well contain numerous oligomers (Rudall, K., 1976).

Protein bands vary enormously, and heterogeneity of bands occurs within the same animal over the course of its life cycle, as in the waxmoth *Galleria mellonella* (Srivastava, R., 1970) or not, as in *Schistocerca gregaria* (Andersen, S., 1973). Protein variations certainly occur in different kinds of cuticle taken from the same animal, e.g. solid and intersegmental cuticle in *Hyalophora cecropia* (Willis, J., 1970); but, interestingly enough, there are no apparent differences in the cuticular protein bands of physogastrous and non-physogastrous termite queens (Bordereau, C. and Andersen, S., 1978).

The significance of protein heterogeneity to structure and function remains an open question. Nonetheless, an origin for such heterogeneity has been proposed by Hackman, R. (1972, 1976) taking a leaf from genetic variations in insect population studies. He suggests that natural insect populations (not to mention that "laboratory artefact" *Drosophila*) are polymorphic at thousands of structural gene loci and that every individual in a large out-breeding population is likely to have a unique protein profile (Silvert, vol. 2). The argument naturally follows that any character, genotypic or phenotypic, need only be non-lethal or neutral in terms of natural selection forces to be retained in a population. Indeed Jukes, T. (1980) has recently provided a compelling argument that evolution can be viewed as the retention of silent nucleotides with respect to structural protein composition.

Hackman, R. (1976) has noted that there is a simple relationship in amino acid coding based on the second letter of RNA, and that there is a high probability of interchange between the first and second letters of the code (substitutions of this kind are well known and have been recently summarized by Jukes, T., 1980). Hackman, R. (1976) has compared the amino acid residues of tanned and untanned cuticles and grouped them according to the second letter of their genetic code origins (Table 2). In this way it can be seen that what at first appears to be a series of entirely unique cuticles can actually be resolved into very similar profiles indeed. Thus, untanned cuticles constitute a group similar in amino acid composition and one that is markedly different from both tanned cuticles and from resilin. We can suggest that it may be the *kinds*

Table 2: Amino acid composition of insect cuticles and cuticular proteins grouped according to the second letter of the genetic code (residues amino acid/1000 total residues)

Species	Second letter of the genetic code				Reference
	U	C	A	G	
	Phe,Leu,Ile, Met,Val	Ser,Pro, Thr,Ala	Tyr,His,Glx, Asx,Lys	Cys,Trp, Gly,Arg	
Agrianome spinicollis					
larval cuticle water-soluble protein	226	358	285	131	Hackman, R. (1974)
Anoplognathus montanus					
larval cuticle	231	283	307	179	Hackman, R. (1976)
Anabrus simplex					
abdominal intersegmental connectives	218	282	325	175	DeHaas, B. *et al*,. (1957)
Bombyx mori					
larval cuticle	185	362	310	142	Hackman, R. and Goldberg, M. (1971)
Lucilia cuprina					
larval cuticle	196	303	338	162	Hackman, R. and Goldberg, M. (1971)
Musca fergusoni					
larval cuticle	178	318	328	176	Gilbey, A. and McKellar, J. (1970)
Pison spinolae					
larval cuticle	248	294	335	123	Hackman, R. (1976)
Xylophasia monoglypha					
abdominal cuticle	179	338	331	152	Hunt, S. (1971)
resilin	92	298	184	426	Andersen, S. (1971)
Rhodnius prolixus					
abdominal cuticular protein	219	446	156	179	Hackman, R. (1975)
Periplaneta americana					
adult cuticle	199	435	236	129	Hackman, R. and Goldberg, M. (1971)

of proteins rather than particular proteins which are important to structure, a generalization consistent with the effects of residue composition on molecular behaviour (discussed below; see sections 3.3.2 and 3.3.5).

Before discussing the properties of amino acids and particular kinds of proteins we should mention the possible origins of cuticular proteins. There is still uncertainty as to whether proteins occurring in the cuticle are synthesized only by the epidermis and/or whether they might be produced somewhere else in the body and transported to the cuticle through the haemolymph. While it is often tacitly assumed that cuticular proteins are synthesized by the epidermis there is close similarity between some of the proteins in the haemolymph and cuticle in the moth *Manduca sexta* (Koeppe, J. and Gilbert, L., 1973) and in *Locusta migratoria* (Phillips, D. and Loughton, B., 1976), which implies the possibility of transport. Similarly, radiolabelled proteins injected into label-free insects can subsequently be recovered from the cuticle (Locke, M., 1969; Koeppe, J. and Gilbert, L., 1974; Geiger, J. *et al.*, 1977).

3.3.2 Amino acids

Differences in the amino acid compositions of different kinds of cuticles from different insects of varying age are well established in the literature and have been discussed in considerable detail (Hackman, R., 1975, 1976; Andersen, S., 1979; Chen, vol. 10) (cf. Table 3). Rather than retabulate this vast amount of data, we rather enquire as to the relationships between amino acid composition and properties versus structure and function in intact cuticles. A few generalizations can be made in this regard.

The protein fractions of untanned cuticles from holometabolous larvae are usually rich in generally hydrophobic residues with bulky side chains (Hackman, R., 1971). Such residues are associated with loose molecular packing in structural protein chemistry (Warwicker, J., 1960; Fraser, R. and Macrae, T., 1973). Thus, the resulting packing is

*Table 3: Amino acid composition of various insect cuticles**

Species	Residues with small side-chains	Highly polar residues	Proline	Alanine	Cuticle type
Coleoptera					
Agrianome spinicollis[1]	32	43	10	–	arthrodial/caterpillar
Tenebrio molitor[1]	40	36	9	?	solid
Tenebrio molitor[2]	42	36	7	?	solid
Tenebrio molitor[3]	42	27	9	?	solid
Pachynoda epphipiata[1]	37	45	8	–	arthrodial/caterpillar
Pachynoda epphipiata[2]	38	42	10	trace	solid
Pachynoda epphipiata[3]	50	29	7	trace	solid
Xylotrupes gideon[3]	47	25	8	1	solid
Dictyoptera					
Periplaneta americana[3]	43	17	10	25	solid
Diptera					
Lucilia cuprina[1]	29	43	9	–	arthrodial/caterpillar
Lucilia cuprina[4]	28	46	9	9	arthrodial/caterpillar
Calliphora augur[4]	28	47	9	–	arthrodial/caterpillar
Drosophila melanogaster[4]	26	47	5	1	arthrodial/caterpillar
Musca domestica[4]	29	45	6	8	arthrodial/caterpillar
Hemiptera					
Rhodnius prolixus[1]	53	9	8	–	arthrodial/caterpillar
Lepidoptera					
Bombyx mori[1]	31	47	10	–	arthrodial/caterpillar
Bombyx mori[2]	21	38	10	14	arthrodial/caterpillar
Sphinx ligustri[1]	28	51	9	?	arthrodial/caterpillar
Sphinx ligustri[2]	38	41	12	?	solid
Sphinx ligustri[3]	44	39	8	?	solid
Hyalophora cecropia[1]	29	50	10	?	arthrodial/caterpillar
Aglais urticae[1]	29	52	9	?	arthrodial/caterpillar
Xylophasia monoglypha[3]	40	38	7	+	solid
Odonata					
Aeshna juncea[3]	64	20	7	–	resilin
Orthoptera					
Schistocerca gregaria[3]	63	20	7	–	transitional

* The data given as residues of animo acids/100 residues.
[1] = larvae, [2] = pupae, [3] = adults, [4] = puparia. Compiled from various sources and where more than one analysis is available for a given case, averaged values are used.

less crystalline and this property imparts greater flexibility to larger structures where such residues occur. There are also several suggestive mechanical correlates: cuticles that are untanned but rich in bulky side-chain residues are often tough materials, but are of relatively low strength and stiffness (Hepburn, H. and Joffe, I., 1976). Similarly, loose packing in a cuticle coincides with increased creep, and this property could indeed facilitate intrastadial growth in caterpillars (Hackman, R., 1976). A corollary to this is that cuticular protein hydrolysates in which there is a preponderance of residues with small side-chains implies closer chain packing and a corresponding increase in stiffness and strength.

On average, the smaller amino acid residues predominate in the hydrolysates of sclerotized cuticle. There is also a correlation between the relative abundance of non-polar residues and relative cuticular hardness (Welinder, B., 1975). Similarly the number of free amino groups available for substitution decreases as sclerotization progresses.

Table 4: Hydrophobicity indices of various insect cuticles according to their amino acid composition

Species	HI	Reference
Coleoptera		
Pachynoda epphipiata		
larva	4	Andersen, S. (1975)
pupa	32	Andersen, S. (1975)
imago	16	Andersen, S. (1975)
Tenebrio molitor		
larva	30	Andersen, S. *et al.* (1973)
pupa	32	Andersen, S. *et al.* (1973)
imago	34	Andersen, S. *et al.* (1973)
Diptera		
Lucilia cuprina		
larva	−4	Hackman, R. and Goldberg, M. (1976)
puparium	−13	Hackman, R. and Goldberg, M. (1976)
Calliphora vicina		
larva	−4	Hackman, R. and Goldberg, M. (1976)
Calliphora augur		
puparium	−24	Hackman, R. and Goldberg, M. (1971)
Lepidoptera		
Bombyx mori		
larva	9	Hackman, R. and Goldberg, M. (1971)
Sphinx ligustri		
larva	−19	Palm, in Andersen, S. (1979)
pupa	27	Palm, in Andersen, S. (1979)
imago	27	Palm, in Andersen, S. (1979)
Dictyoptera		
Periplaneta americana		
imago	41	Hackman, R. and Goldberg, M. (1971)
Isoptera		
Macrotermes bellicosus		
physogastric queen, pleural membrane	−38	Bordereau, C. and Andersen, S. (1978)
neotergites	1	Bordereau, C. and Andersen, S. (1978)
Hemiptera		
Rhodnius prolixus		
fifth-instar larva, abdomen	59	Hackman, R. (1975)
Orthoptera		
Locusta migratoria		
imago, pronotum	55	Hackman, R. and Goldberg, M. (1976)
Schistocerca gregaria		
fifth-instar larva, femur	51	Andersen, S. (1973)
imago, femur	52	Andersen, S. (1971)
tibia	51	Andersen, S. (1971)
ventral thorax	53	Andersen, S. (1971)
abdominal terga	36	Andersen, S. (1971)
abdominal intersegmental membrane	−5	Andersen, S. (1971)
membrane of coxal cavity	−10	Andersen, S. (1971)
wing hinge	−12	Andersen, S. (1971)

Returning to the scheme of Hackman, as outlined in Table 2, one observes the relationships that group A contains the hydrophilic amino acid residues and that sclerotized cuticles are consistently low in such residues.

Andersen, S. (1979) recently proposed some very interesting extensions of these ideas, and points out that "hydrophobic and hydrophilic" are boundary terms defining widely ranging behaviour of residues with respect to hydrophobicity (see Andersen, this volume). Taking calculated values for the energy changes associated with the transfer of a given residue from water to ethanol, a hydrophobicity index of amino acids can be obtained. Hydrophobicity indices for whole cuticles are obtained by extrapolation from the relative percentage contributions of different residues. While some simplifying assumptions have been made (for example, the acid amide residues may be either amides or free acids) a meaningful indication of the relative strengths of hydrophobic interactions between protein molecules is still obtained.

It can be seen from Table 4 that there is a spectrum of values from + 60 to − 40, and that there is no sharp division of cuticle types along the way. It is of interest to note that stiffness and strength of cuticles also cover a spectrum (Hepburn, H. and Joffe, I., 1976) and that very stiff cuticles have a high positive index, those of low strength and stiffness a more negative index (Table 4). Citing the example given by Andersen, S. (1979), unsclerotized cuticles from *Bombyx mori* and *Locusta migratoria* both appear to lack covalent bonding in the protein matrix so that when samples of these cuticles are stretched, resistance to displacement will derive principally from the chitin microfibrils. This implies that the elastic modulus (stiffness) will indicate an approximate measure of the strength of the reactions and these coincide with hydrophobicity indices of Table 4. One can conclude that a general consistency appears between the mechanical properties spectrum on the one hand, and both hydrophobicity index as well as the grouping of amino acid residues according to the second letter of the genetic code (Table 2) (Hackman, R., 1976; Hepburn, H. and Joffe, I., 1976).

3.3.3 THE PROTEINS OF EGGS

Insect oothecae consist primarily of proteins. They occur widely amongst the Cursoria, Mantodea and Orthoptera and are also secreted to a lesser extent by Coleoptera and Lepidoptera. The ootheca of the cockroach, *Blatta orientalis* is crystallographically related to the extended β-fibroins and is central to the history of tanning in insects in that they were used in what is considered to be the first modern chemical analysis of tanning by Pryor, M. (1940a,b). Oothecae are usually secreted from paired collateral glands of the abdomen as proteins in solution which then crystallize as the secreted materials come together outside of the animal.

The proteins of the praying mantis ootheca are also secreted in liquid form, but the final product consists of a series of highly oriented ribbons (Rudall, K., 1956). The ootheca of the tortoise beetle, *Aspidomorpha*, has recently been studied both by diffraction and electron microscopic techniques (Atkins, E. *et al.*, 1966) and found to consist of highly ordered tactoid bodies which contain a microfibrillar structure. The conformational affinities of *Aspidomorpha* oothecal protein are uncertain; however it has mechanical similarities to feather keratin (Hepburn, H. and Sinclair, G., 1978). Little more can be said about oothecal protein structures other than that they are rich in the polar residues glycine and tyrosine, and rather low in cysteine and methionine (Pau, R. *et al.*, 1971; Kramer, K. *et al.*, 1973). These last observations, *vis-a-vis* solid cuticle, would make comparative hardness indices and hydrophobicity values for oothecal material extremely interesting, to say the least.

The insect egg shell is a multiple-laminated structure synthesized by the epithelial cells of the ovarian follicles and is, like the cuticles of insects, a physical record of the activities of the cells that produce it (Telfer, W. and Smith, D., 1970). The egg consists of a chorion, vitelline membrane and oolemma (see vol. 1). The chorion has generally been considered to be the principal structural layer of the egg, conferring toughness and rigidity to the whole (Beament, J., 1946). More recently it has been suggested that the vitelline membrane also confers structural stability to eggshells, on the grounds that dechorionated eggs maintain their shape (Furneaux, P. and MacKay, A., 1976).

The principal structural protein of the insect egg is termed chorionin and this material has been analysed chemically, microscopically and crystallographically for many orders of insects (Furneaux, P. and MacKay, A., 1976). The chorionins of numerous species have been fractionated and, following the tradition in keratin chemistry, have been classified as type I or type II chorionin proteins depending upon either a high or a low sulphur content. Similarly, the amino acid compositions of various species of eggshells have been reported (Table 5). The same difficulties apply to the interpretation of egg proteins as to cuticular proteins:

Table 5: Amino acid composition of protein fractions from insect eggshells

Species	*Bombyx mori*[2]			*Antheraea pernyi*[2]		*Sympetrum infuscatum*[3]			*Tenebrio molitor*[4]			*Rhodnius prolixus*[5]		*Oryctes rhinoceros*[5]	*Gryllus mitratus*[6]	
Fraction[1]	I	II	III	I	III	I	II	III	I	II	III	I	II		I	III
Cys	1	3	14	5	13	1	3	58	46	44	32	16	25	41	6	19
His	4	0	6	0	4	0	0	71	9	12	10	38	32	18	12	22
Arg	23	28	14	22	18	6	6	3	38	23	10	27	32	20	18	27
Asp	39	42	53	36	44	41	22	17	81	73	16	31	53	69	49	36
Thr	32	17	28	28	27	33	23	6	41	27	22	54	41	32	10	21
Ser	31	39	37	38	47	80	56	11	48	71	113	12	32	69	237[7]	50
Glu	36	30	64	43	45	100	11	30	60	81	33	33	70	70	105	35
Pro	42	20	56	36	30	60	72	120	62	87	169	49	59	67	13	107
Gly	361	302	101	324	137	152	340	245	68	61	48	226	142	79	21	66
Ala	83	27	46	127	60	73	15	158	63	50	137	80	34	61	16	70
½Cys[6]	57	256	5	59	7	180	218	0	29	36	0	49	147	35	12	2
Val	60	44	33	62	38	21	8	9	65	55	29	73	82	68	77	31
Met	0	3	7	3	8	1	6	17	8	9	10	0	1	3	trace	6
Ile	32	18	21	38	24	14	13	5	48	43	12	40	11	39	37	14
Leu	64	12	39	69	45	46	39	7	75	69	15	12	26	50	22	30
Tyr	79	13	12	73	25	31	34	103	35	21	27	84	20	43	6	21
Phe	19	10	14	14	14	0	0	3	31	26	4	22	14	19	6	20
Try	18	7	4	7	9	5	9	trace	2	1	0				17	12
Dityrosine						0	0	4	0	0	2.1				0	0.6
Trityrosine						0	0	1	0	0	0.7				0	0.2
Amide	32	30	76	29	62	121	19	16	68	78	40	61	118	70	34	35

Values expressed as μmol per 100 mg dry sample. [1] I and II refer to soluble chorionins which differ in cystine content. III refers to the insoluble residue after extraction of soluble chorionins. [2] Kawasaki, H. *et al.* (1971b); [3] Kawasaki, H. *et al.* (1974); [4] Kawasaki, H. *et al.* (1975); [5] Furneaux, P. and MacKay, A. (1976); [6] Kawasaki, H. *et al.* (1971a); [7] as *O*-phosphorylated serines.

degradation by chemical analytical techniques probably produced collections of oligomers so that we cannot state unequivocally whether "chorionin" is a protein or a collection of proteins.

The eggshell consists of an intrachorionic space near the inner margin of the chorion, and is accessible to the external atmosphere via aeropyles which traverse the chorion (Margaritis, vol. 1). It has been suggested by Hinton, H. (1969) that the position of the air space shortens the path of oxygen to the developing embryo, and that water loss is reduced through the limited cross-sectional area of the aeropyle. However, physiological measurements in support of this conjecture are wanting. There is considerable variation in the surface properties of eggs and the distribution of aeropyles in insect eggs (Hinton, H., 1980). This is somewhat surprising in view of the chemical similarity of egg chorionins (Smith, D. *et al.*, 1971).

Recently electron microscopic studies summarizing data from many different insects as to the structure of the chorion were published by Furneaux, P. and MacKay, A. (1976). For beetle chorions that reflect a highly ordered type of structure these authors have suggested a cross-ply structure of fibres embedded in a more amorphous matrix substance, presumably also of protein (Fig. 19). They envisage that each of the fibre bundles consists of four microfibrils of about 40 Å diameter. For other eggs such as those of the moth *Hyalophora cecropia*, Smith, D. *et al.* (1971) have reported that the microfibrils are arranged in helicoidal lamellae.

The vitelline membrane appears to be both more variable, and less understood, than is the chorion. It ranges from negligible as a percentage of the total egg in the grasshopper, *Melanoplus differentialis*, (Slifer, E. and Sekhon, S., 1963) to nearly four-fifths of eggshell mass in the dragonfly, *Sympetrum infuscatum* (Kawasaki, H. *et al.*, 1974). The vitelline membrane is the site of the principal wax layer in eggshells and the wax is formed after the bulk of the shell has been synthesized (Beament, J., 1946). A putative arrangement (Fig. 20) for the vitelline membrane from serial sections has been prepared by Furneaux, P. and MacKay, A. (1976). Despite the recent appearance of H. Hinton's large work on many aspects of insect eggs (1980), the subject must be regarded as in its infancy.

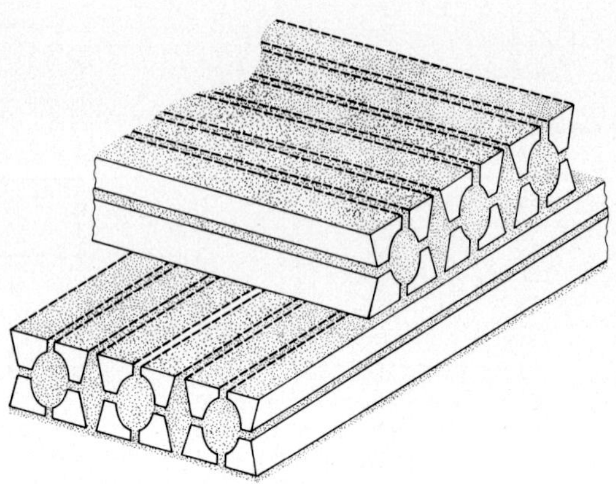

FIG. 19. Cross-ply arrangement of fibres and impregnating matrix (stipling) between fibres as reconstructed for the chorion of the beetle, *Oryctes*. (After Furneaux, P. and MacKay, A., 1976.)

3.3.4 RESILIN

The discovery of resilin by Weis-Fogh, T. (1960), and its subsequent characterization by that author and his colleagues, represents what must certainly be regarded as an intellectual *tour-de-force* in the first rank of insect structural studies. It was initially characterized from the prealar arm and wing-hinge ligaments of the locust, *Schistocerca gregaria*, and from the pleuro-subalar muscle tendon of the dragonfly, *Aeshna grandis*. The recoverable mechanical deformability of these preparations, in contrast to sclerites, led to the distinction between "rubber-like" cuticle and "solid' cuticle.

Pure resilin is both colourless and transparent, and reveals no evidence of structure in electron micrographic preparations (Elliott, G. *et al.* 1965). It is optically and mechanically isotropic but does exhibit strain birefringence in the direction of deformation. When solvated it can be distended reversibly over 200% of its initial length, held to the stretched length by drying out, and when resolvated and released will return to its original rest position. Its principal physical properties are its long-range elasticity, elastic efficiency (which is an index of recoverable energy obtained when deformation is discontinued), and lack of either creep or stress relaxation (Weis-Fogh, T., 1961).

Weis-Fogh, T. predicted (1960, 1961) that resilin ought to show molecular structural similarities to botanical rubbers in which there is a three-dimensional network of chains nearly free of one another, thermally agitated, randomly kinked and fixed within a network by only a very few stable cross-links. Andersen, S. (1963, 1964, 1966) subsequently characterized these necessary covalent cross-links as new fluorescent amino acids, dityrosine and tertyrosine, which ultimately derive from tyrosine. Amino acid analyses of several preparations from different species indicate quite a close overall similarity in composition (Table 6).

On the basis of composition, resilin differs markedly from other structural proteins in that it is rich in glycine, hydrophilic residues, and proline but completely lacks hydroxyproline, tryptophan or the sulphur-bearing residues (Andersen, S., 1971). Resilin remains unsequenced but a very likely candidate for future sequencing studies of insect proteins. Though its biosynthesis also remains to be established, the deposition of resilin is certainly linked to the median neurosecretory cells of the brain, and it is also known that these same cells secrete bursicon which is associated with the deposition of endocuticle in *Sarcophaga bullata* (Fogal, W. and Fraenkel, G., 1969; Reynolds, vol. 8) and in *Locusta migratoria* (Vincent, J., 1971) so it may be that resilin synthesis and deposition are likewise under direct or even indirect control of bursicon.

In insects, resilin functions as a mechanical spring in tension, compression and in bending members. Resilin (or resilins) probably occur in all insects. It

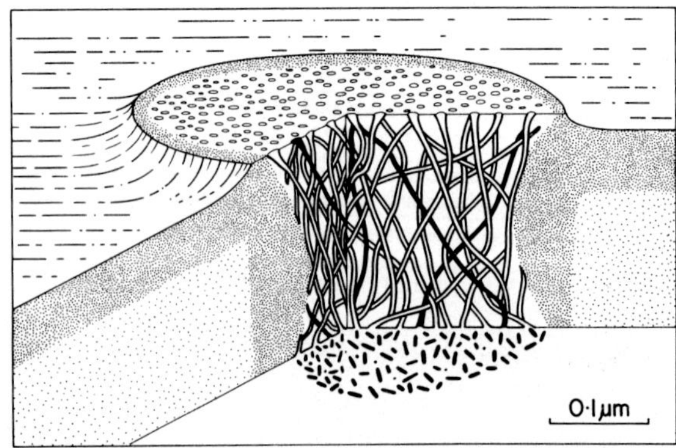

FIG. 20. Reconstruction by Furneaux, P. and MacKay, A. (1976) of the vitelline membrane of the beetle, *Dermestes* spp. Twisted and intertwining tubules open to the surface. These tubules are surrounded by an electron-dense (stippling) "sleeve". It is suggested that this system is responsible for wax transport in the egg shell. (After Furneaux, P. and MacKay, A., 1976.)

Table 6: Amino acid composition of various resilins (residues of amino acid per 100 residues)

	Schistocerca gregaria					*Aeshna juncea*	*Oryctes rhinoceros*
	Wing hinges		Prealar arms		Clypeo-labral spring	flight tendon	abdominal spring
	a	b	a	b	b	b	b
Aspartic acid	9.9	10.2	10.8	11.3	7.5	9.4	22.6
Threonine	3.1	2.9	3.0	3.1	3.4	2.0	4.1
Serine	7.9	7.7	8.1	7.5	8.9	12.8	4.9
Glutamic acid	4.6	4.3	4.8	3.8	6.4	4.2	5.9
Proline	7.6	7.5	7.9	7.7	7.2	7.5	8.9
Glycine	39.1	41.1	36.7	39.7	31.4	42.2	33.5
Alanine	11.2	11.2	10.6	11.2	13.5	7.0	4.6
Valine	2.6	2.3	3.2	2.5	3.1	1.2	2.5
Cystine	—	—	—	—	—	—	—
Methionine	—	—	—	—	—	—	—
Isoleucine	1.7	1.3	1.7	1.4	1.9	0.9	2.6
Leucine	2.3	2.2	2.4	2.8	5.2	3.0	3.9
Tyrosine	2.5	2.1	3.1	1.2	1.9	1.2	0.5
Phenylalanine	2.6	2.5	2.4	2.4	2.8	2.1	2.2
Lysine	0.5	0.4	—	0.5	0.9	0.8	0.5
Histidine	0.9	0.8	5.3	1.1	0.9	1.2	0.5
Arginine	3.5	3.6	—	3.8	5.1	4.6	2.9
Sum of polar residues	32.9	32.0	35.1	32.3	35.0	36.2	41.9

a = Bailey, K. and Weis-Fogh, T. (1961); b = Anderson, S. (1971).

functions as a spring in tension in the flight system of dragonflies (Weis-Fogh, T., 1960); in compression in the jumping fleas (Bennet-Clark, H. and Lucey, E., 1967) and the recoil of the moth and butterfly proboscis (Hepburn, H., 1971) and bending (tension and compression) in the wing hinges of locusts (Andersen, S. and Weis-Fogh, T., 1964). Several other examples can be found in Neville, A. (1975).

3.3.5 Silks

(*a*) *Structure* Protein silks are elaborated by virtually every order of insects in which they have been sought. In addition to the familiar cocoon formed by spun fibre of circular cross-section, protein silks also occur in ribbons and sheets and are used in myriad ways: for anchoring dragonfly eggs to rocks in fast streams or as egg stalks in lacewing flies, as food-traps for caddis-flies and tunnels for web-spinners, and in very many other ways. Aside from these special uses in some life stage of an insect, silk holds importance for the study of the integument because it is produced by an epithelium (if not always the epidermis) and because it is vastly more accessible to crystallographic and chemical study than are proteins occurring *in situ* in a cuticle. Silk affords us much insight into the behaviour of structural proteins, and as to how behaviour can be related to molecular structure.

There is another point of interest in silks. It will be seen that silks are relatively distensible in fibroin form. However it has very often been suggested that one of the main "reasons" why an insect must moult is owing to the inextensible nature of α-chitin. It should be pointed out that all of the structural proteins are relatively distensible compared with chitin, but this is not so when they are tanned in a cuticle. It may well be that tanning is required to impart stability to the structural protein, and this could be a "reason" why insects moult.

The discovery of transconformational changes in keratin established that a protein polymer could exist in at least two different configurations, a folded α and an unfolded parallel β form (Astbury, W. and Woods, H., 1933). This landmark in molecular biology was given important extension by Rudall, K. (1946) who discovered that epidermal protein from the cow's lip could be thermally supercontracted into yet another stable configuration, the cross-β form. Rudall also established experimentally the relationships for the KMEF (keratin, myosin, epidermin and fibrinogen) group of structural proteins.

All three of these protein configurations were subsequently discovered and characterized from naturally occurring insect material. Thus, parallel-β fibroin in Tussah silk from the caterpillar *Antheraea pernyi* was described by Marsh, R. *et al.* (1955); the α-helical configuration was found in the honeybee cocoon silk (Rudall, K., 1962a) and the cross-β pattern in the egg stalk of the green lacewing flies (*Chrysopa*) by Parker, K. and Rudall, K. (1957).

The basic crystal structures for these three kinds of configurations are shown in Fig. 21. The β-structure (as reconfirmed by Geddes, A. *et al.*, 1968) consists of highly oriented fibres in which the crystallites are arranged with their long axes in the same direction as the fibre axis. The ribbon-like crystallites are linked by short bends so that a continuously folded polypeptide chain results. These chains lie perpendicular to the ribbon surface and parallel to the long axis of the ribbon. The structure of the α-helical pattern characteristic of honeybee silks, as reconfirmed by Atkins, E. (1967), is seen as a crystal structure in which four helices coil upon one another, much like threads in a piece of cotton, and parallel to the fibre axis. The β-pleated sheet configuration is the well-known form in which the crystallographic *b*-axis is coincident with the fibre axis.

The distinctness of structure in the different configurations as well as the behaviour of such fibroins, greatly depends upon the secondary structure of the protein. That is, the structure depends upon the hydrogen bonding topologies, and the transconformations which can occur between them require the disruption of the secondary protein structure through changes in the positions of the crystallites with respect to their position and arrangement within the fibre (Harris, W. *et al.* 1977).

(b) Problems of crystallinity At the time of writing the amino acid compositions of several hundred different insect silks have been described, many of which have been tabulated in the review of Lucas, F. and Rudall, K. (1968a,b). In the quest to relate amino acid composition of cuticular proteins to structure of cuticles and to functional properties, relationships involving "kinds" of amino acids and their relationships to water emerge; this is analogous to the hydrophobicity index with respect to mechanical properties suggested by Andersen, S. (1959; see also section 3.3.2).

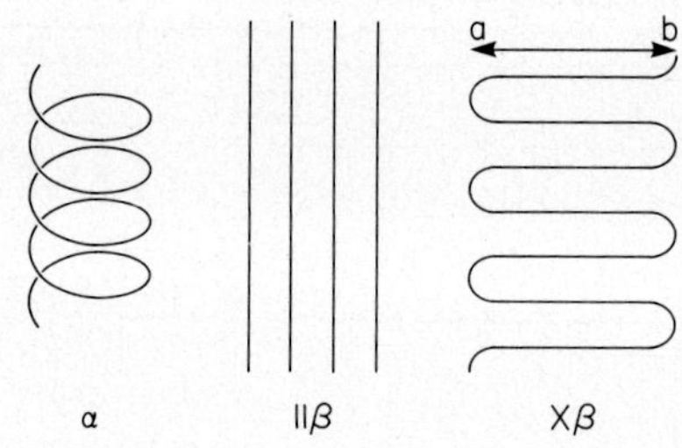

FIG. 21. Highly diagrammatic representations of fibroin crystal structures. (After Rudall, K., 1962b.)

It is well known that the degree of crystallinity, the ratio of ordered or crystalline to disordered or amorphous regions, greatly affects the bulk material properties of a polymer, especially in terms of the relative availability of chemically reactive sites. Moreover, the average crystalline content can decrease with increasing fibre diameter (Iizuka, E., 1963). Some estimate of the degree of crystallinity of fibroins can be made from a consideration of amino acid composition (Warwicker, J., 1960). Thus, amino acids with long side-chains tend to occur in the amorphous regions, and those with short side-chains in the crystalline regions when the unit cell in the *b* direction is small. Since only glycine, alanine, threonine and serine occur in highly crystalline fractions of *Bombyx* fibroin (Drucker, B. *et al.*, 1953) these specific residues are taken as short side-chain standards for purposes of comparison.

Lucas, F. *et al.* (1960) used this approach to estimate the degree of crystallinity for fibroins of *Anaphe, Antheraea* and *Bombyx*, and we have similarly calculated short side-chain–long side-chain (SC/LC) ratios for *Apis, Chrysopa* and *Galleria* from the data of Lucas, F. and Rudall, K. (1968a). These ratios are given in Table 7, inspection of which immediately establishes honeybee silk as being anomalous because this silk has a very low glycine content yet is known to be relatively crystalline (Lucas, F. and Rudall, K., 1968b; Atkins, E., 1967). By the same token, it is also possible for long side-chain residues to occur in crystalline regions (e.g. braconid and spider fibroins; Warwicker, J., 1960). A possibly more useful approach for estimating relative crystallinity is suggested by studies of

Table 7: Relative crystallinity of α, cross-β, and parallel-β fibroins assessed in terms of elastic moduli and amino acid residues

Specimens	(SC/LC)*	(E_{wet}/E_{dry})†
Parallel-β		
Anaphe	23.4[1]	0.81[5]
Bombyx	6.6[1]	0.64[5]
Antheraea	3.9[1]	0.32[5]
Galleria	3.0[2]	0.18[5]
Cross-β		
Chrysopa	10.9[3]	0.29[5]
α		
Apis	~1.5[4]	0.53[5]

* = based on number of amino acid residues estimated per 1000 residues.
† = elastic modulus in MPa.
LC = amino acid residues with long side-chains.
SC = amino acid residues with short side-chains.
Data sources: [1]Lucas, F. *et al.* (1955); [2]Lucas, F. *et al.* (1960); [3]Lucas, F. *et al.* (1957); [4]Lucas, F. and Rudall, K. (1968a); [5]Hepburn, H. *et al.* (1979).

cellulose bast fibres, reported by Wainwright, S. *et al.* (1976).

The argument from cellulose states that the degree of crystallinity will be reflected in the sensitivity of the material to solution effects. For example, water can penetrate amorphous regions in a capillary manner, thereby diminishing the interactions between crystallites, or alternatively, actually compete for potential hydrogen-bonding sites within the fibre (Hepburn, H. and Joffe, I., 1976; Wainwright, S. *et al.* 1976). In the work on cellulose materials it was assumed that the effect of hydration was to loosen the interaction between neighbouring crystalline regions, so reducing stiffness. It was further assumed that the modulus of the dry material approached that of crystalline cellulose, as a result of which the ratio of modulus wet to modulus dry provided an approximate index of the degree of crystallinity, a ratio of 1 indicating complete, and smaller values progressively less, crystallinity. When the six fibroins of Table 7 are examined for hydration sensitivity, expressed as the ratio of elastic modulus of wet to that of dry fibre, entirely different estimates of crystallinity are obtained. Indeed, a value of 0.53 for honeybee silk shows that this fibre is rather crystalline and not dissimilar to *Bombyx*, results that are consistent with crystallographic estimates for the same materials (Lucas, F. and Rudall, K., 1968a).

The importance of the insect silks in the study of fibrous proteins is that they provide relatively simple preparations for investigations of conformational changes. We have seen in the tensile stress–strain curves of the waxmoth silk, and of the other parallel-β fibroins, what can largely be described as linear behaviour in the dry state (Hepburn, H. *et al.* 1979). Differences within this group are related to variations in crystallinity as estimated from ratios of wet to dry moduli, and to a lesser extent the SC/LC ratios. Honeybee silk, on the other hand, is α-helical in nature and its ability to undergo transconformational changes is well documented (Rudall, K., 1962a; Lucas, F. and Rudall, K., 1968a). Thus highly extended specimens of honeybee silk provide diffraction patterns, the main features of which are practically identical to those of *Bombyx* or *Galleria*. The same is true of highly extended *Chrysopa* silk.

If we assume from the diffractional data of Parker, K. and Rudall, K. (1957); Rudall, K. (1962); Atkins, E. (1967); and Geddes, A. *et al.* (1968) that extensions of the required magnitude do result in diagrams that are indistinguishable in their main features, then it could be expected that similar extensometric relationships would occur in measured deformations of these configurations. Silks tested in distilled water closely approximate the expected behaviour, as shown in Fig. 22. Here, all three configurations have been plotted to the same scale. However, the origins of the honeybee and waxmoth curves have been zero-shifted to the right along the abscissa in order to compensate for intrinsic differences in total extensibility. If, for example, the α fibroin were a cross-β fibroin it would require an additional 0.68 strain to achieve a total strain of 2.09 (about 700%). Similarly, the parallel-β form is zero-shifted to the right by a strain of 1.4 with respect to the cross-β form, and by 0.68 with respect to the α form.

The results obtained by shifting the α-helical and parallel-β silks with respect to the cross-β lacewing silk extensibility data obtained on experiment agree within 10% of the deformations that are required to distort similarly accurately scaled molecular models. This observation enables us to make reasonable estimates of the relationship between fibre crystal structure and the measured mechanical properties of the three crystal systems of the silk fibroins. The deformation of the parallel-β fibroins

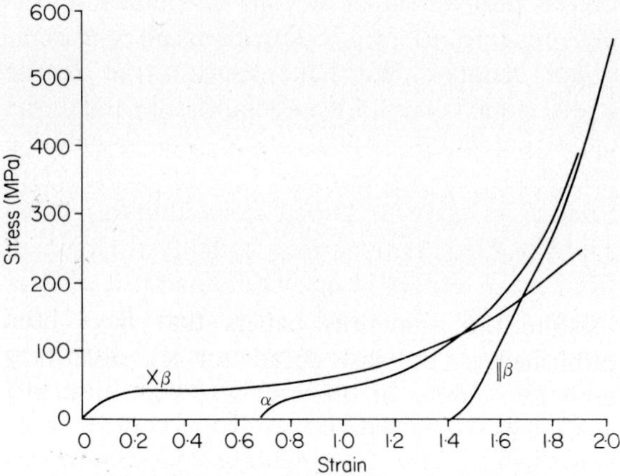

FIG. 22. Superposition of tensile stress–strain curves of honeybee α-helical, waxmoth parallel-β and lacewing fly cross-β silks. The honeybee and lacewing fly curves exhibit low strain yield points associated with transconformational change. Both the honeybee and the waxmoth curves have been zero-shifted to the right to illustrate the goodness of fit for the cross-$\beta \rightarrow$ parallel-β and $\alpha \rightarrow$ parallel-β transconformations.

is largely limited to the direct stretching of the peptide backbone, and hence stiffness reflects the deformation resistance of the primary bonding with total extensibility (including initial "slack") being further affected by the extent of bulky side-chain residues present.

Reference to Fig. 22 further shows that the α and cross-β silks share similarity in that both exhibit a very low strain yield, a long region of transformation with finally increasing stiffness. The initial deformation behaviour of these kinds of fibroins involves the stretching and then breaking of the secondary hydrogen-bonding lattice of the crystals followed by gradual unfolding of the molecules. The unfolding phase is the low stress, plateau area which is the greater in the cross-β silk because it is more folded than the α silk. Once the unfolding has been completed the continued stretching of the fibre is essentially a process wherein the now parallel-β form has appeared, and the deformation process becomes one of stretching the peptide backbone as occurs in the parallel-β silks.

Thus it appears that the mechanical properties of the three crystallographic forms of fibroin are related both to molecular structure and to transconformational changes, and are also completely consistent with crystallographic observations. Extrapolation back to cuticular proteins is difficult at this time, since we have but a single report on the conformation of larval cuticular proteins in a fly *Calliphora vicina* (Hackman, R. and Goldberg, M., 1979).

3.4 Other constituents

3.4.1 WATER

Water has an enormous effect on the chemical properties of structures and for the freedom (or lack) of movement within them. Innumerable, but seldom studied physical properties of biological matter will vary in association with water: specific heat capacity, dielectric properties, viscosity, density distribution of water molecules around other molecules, dissociation characteristics and, of the greatest importance to the structure and performance of insect cuticles, the hydrogen bonds between liquid water and structural molecules. All of these things will also vary with temperature. The significance of water to integuments is supported in experimental measurements; however, interpretations of observed behaviour largely come from inferferences on the physics of water (Pauling, L., 1957, 1960).

While we have no information on the molecular configurations of water in cuticles, some generalizations can nevertheless be made. Water content of cuticle varies greatly; for example, from 15% in the femora of *Locusta* (Fraenkel, G. and Rudall, K., 1947) to 55% in the larval cuticle of *Sarcophaga* (Dennell, R., 1946). By extrapolation from mechanical behaviour it has been suggested that water can be present through either capillary or molecular sorption (Hepburn, H. and Joffe, I., 1976).

Physical properties such as stiffness and strength of chitins, whole cuticles and silk fibroins increase on dehydration while distensibility decreases (Hepburn, H. and Joffe, I., 1976; Hepburn, H. *et al.*, 1979). All three of these materials exhibit anisotropy in swelling and the direction of size increase is generally normal rather than parallel to the axes of the structural fibre. The same is true for sclerites because the fibres in the procuticle generally run parallel to the surface (Fraenkel, G. and Rudall, K., 1947).

The long-range elasticity of proteins such as resilin and fibroin are solvation-dependent (Weis-Fogh, T., 1960; Rudall, K., 1946) as is the plasticization of the abdominal cuticle of sucking bugs like *Rhodnius* (Bennet-Clark, H., 1962). Conversely, it is well documented that increasing density and cuticular hardening go hand-in-hand with dehydration (Fraenkel, G. and Rudall, K., 1947); but one must guard against thinking that dehydration is sufficient to account for hardening. This important distinction was established over three decades ago in studies of pupariation by *Sarcophaga* (Fraenkel, G. and Rudall, K., 1947) and most recently reiterated in studies of *Locusta migratoria* by Andersen, S. (1981).

Thus, skeletal rigidity is largely achieved through tanning or hardening processes, and partial dehydration of the procuticle is a natural consequence of stabilization (Andersen, this volume). Great skeletal flexibility, as found in caterpillars and arthrodial membranes, is associated with a high water content and perhaps relatively few covalently bound proteins in the matrix.

3.4.2 Pigments

Whatever the significance of pigments in insect biology may be, at the outset it is important to state that there is no correlation between the biochemical events that lead to hardening and those which lead to darkening (Dennell, R. and Malek, S., 1955). That very dark cuticles are often hard is merely coincidental. Although the term "melanization" has been in use for many decades to describe "darkening" the actual chemical recognition of melanin is recent indeed: melanin as such has been described as conjugated with protein in the puparium of the fly *Lucilia cuprina* (Hackman, R., 1967) or as a discrete layer of material immediately below the epicuticle in *Sarcophaga bullata* (Fogal, W. and Fraenkel, G., 1969).

There are numerous external factors which apparently influence the distribution of melanins in cuticles. For example, melanin content may vary with temperature, as in *Sarcophaga*, population density of animals in a particular place, as in the locusts *Schistocerca* and *Locusta*, and in certain genetic mutants of *Drosophila*. It is likely that melanin formation is mediated by bursicon (cf. vols 7 and 8). Black pigments also occur as granules in some insects, the larvae of the butterfly *Papilio* and in the toad bugs *Gelastocoris*. Virtually all of the myriad pigments which colour insects occur as cytoplasmic granules in the epidermis, fat body, haemolymph or other organs beneath the cuticle (Kayser, vol. 10).

3.4.3 Lipids

Despite the numerous papers that have been published in the last decade or so, describing cuticular lipids of insects, a severe difficulty surrounds the interpretation of such information. This problem was recently put in perspective by Gilby, A. (1980) who noted that very little is actually known about the location of lipids within cuticles. Thus, it is tacitly assumed that the lipids discussed in section 2 are in fact epicuticular in origin.

Few studies have distinguished the cuticular compartments from which lipids have come. Among these, Locke and his colleagues (Locke, M., 1965, 1974) have documented the transport of waxes or their precursors to the cuticular surface via pore canals and even finer, arborescing wax canals in the caterpillar, *Calpodes ethlius*. As for the procuticle, the possible significance of lipids to matrix stabilization was first mooted by Wigglesworth, V. (1947) and he has only very recently touched upon this problem experimentally.

Combining histochemical and electron microscopical techniques in the study of *Rhodnius prolixus*, Wigglesworth, V. (1970, 1975a) was able to show that the lamellate endocuticle of this bug exhibits the parabolic pattern so common to insect cuticle (Fig. 9). In addition, he has described a series of lipid laminae of the same spacing and superimposed upon the chitin–protein lamellae. These lipid laminae are quite distinct from pore canals, the walls of which also contain dense concentrations of lipid. Wigglesworth has postulated that this procuticular lipid fraction might initiate or control the deposition of the helicoidal lamellae if formation of the latter is a crystallization process as envisaged by Bouligand, Y. (1972).

These very few observations suggest many possibilities, the most important of which is that lipids might contribute to the stabilization of the procuticle by being incorporated in some of the polymerization reactions that occur there. Indeed, the suggestions (Rudall, K. and Kenchington, W., 1971; Wigglesworth, V., 1976), that polymerization of lipids can be viewed as analogous to the polymerization of hydrocarbons or of isoprenoid units into materials akin to nylon or polyethylene hold exciting possibilities for the future.

3.4.4 Inorganic salts

The occurrence of any appreciable amounts of inorganic salts in the cuticles of insects can be regarded as relatively rare. Calcareous deposits in the form of calcium carbonate have been recorded from the external surfaces of the larvae of some Stratiomyidae and Psychodidae (Richards, A., 1951) and from the procuticles of some Lepidoptera and Hymenoptera (Weaver, A., 1958). Such salts as calcium phosphate have also been recorded from the puparium of *Rhagoletis cerasi* (Richards, A., 1951) and from the puparium of *Musca autumnalis* (Fraenkel, G. and Hsiao, C., 1967) and *Musca fergusoni* (Gilby, A. and McKellar, J., 1976) and are thought to derive from the Malpighian tubules of the larval instar which forms the pupa, a process which appears to be under control of the moulting hormone.

The eggs of several genera of phasmids have been variously reported to contain calcium salts as carbonate, citrate and oxalate (Hackman, R. and Goldberg, M., 1960; Parker, K. and Rudall, K., 1955). How, if at all, these salts are structurally integrated in the cuticles or eggshells of insects is not known. As to their functions we can only make inferences from Crustacea wherein calcification results in compressive strengthening of the skeleton. Perhaps this property would be desirable for fly pupae that occur in unstable substrates. Because inorganic salts are so rare in insects *vis-à-vis* Crustacea this suggests some very fundamental differences in the metabolism of these two related groups, and one worthy of further detailed investigation.

4 SOLID CUTICLES

4.1 General properties

A random sample of cuticles taken from the various developmental stages of an insect's life history will show great variation. Indeed, the physical and chemical properties of such structures vary both horizontally and vertically within an animal, and all of these properties will, in turn, be seen to change over time. For example, the tough and distensible body cuticle of the cuckoo larva of Chrysididae, in due course, is replaced by the extremely brittle and indistensible sclerotized plates of the adult wasp. Over time, cuticles vary in state of hydration, crystallinity, stiffness and strength, toughness and brittleness as well as in electrical, thermal and optical properties.

Such cuticles superimposed on structural variation are far from being physiologically stable or inert. Electrical, permeability and metabolic properties of cuticles are subject to periodic modification by the epidermis. The metabolic changes, especially with reference to metamorphosis, are enormous. Likewise, the myriad properties of the integument can be affected by the general health or sex of the animal, for example, physogastry, as well as in response to feeding, the cone-nosed sucking bug *Rhodnius prolixus* being a classical example of the latter. The structure of the cuticle can also change in response to injury, infection by parasites and parasitoids, and even daily changes in weather.

Cuticles exhibit a temporal and spatial spectrum of properties. Nevertheless, areas of cuticles can be grouped on functional grounds into the so-called "solid" cuticle typical of sclerites, apodemes, phragmata, mandibles and claws; those which consist of nothing more than resilin and covering epithelium such as the pleuro-subalar muscle tendon of the dragon fly *Aeshna cyanea* (Weis-Fogh, T., 1960); the transitional cuticles such as found in the clypeolabral spring and distal part of abdominal tergites in the grasshopper *Schistocerca gregaria* (Weis-Fogh, T., 1960); and, finally the thin arthrodial membranous cuticle of all insects as well as the cuticles of caterpillars and most other holometabolous larvae.

4.2 Classification schemes

The bulk of insect cuticle is very largely occupied by procuticle and consequently considerations of possible subdivisions of procuticle have tended to assume predominance in classification schemes for the insect integument proposed through the years. This is not at all surprising since the all-important epicuticle has remained intractable for so long, and because there have been relatively few extensive studies of the epidermis in conjunction with the cuticle as such. In the literature of this century there have been three more or less distinct ways of classifying the enormous variation in cuticle first extensively catalogued by light microscopists in the 19th century (Richards, A., 1951).

These schemes can reasonably be described in terms of the investigative approaches used in studying cuticles. For example, there is a "structural" and a "histological" way of describing integuments. Each method describes different aspects of cuticular composition and arrangement of constituents.

The structural school of classification gradually developed in the last century (Richards, A., 1951) and is well represented by the classical work of Wigglesworth, V. (1933) which distinguished cuticle into chitinous and non-chitinous regions and many others in more recent years. Where definitions of terminology are not given in papers dealing with cuticle, it is usually safe to infer that the authors have a "structural" view of the cuticle in mind. The characteristics of cuticles seen in this way have been described by Neville, A. (1975). In this view, exocuticles are often coloured brown or black on tanning and the chitin microfibrils of this region are arranged helicoidally and of a smaller pitch than are the microfibrils of the endocuticle. The chitin microfibrils of exocuticle are said to be more highly crystalline and better separated from the protein matrix than are the microfibrils of the post-ecdysial endocuticle. The endocuticle, apart from the above, consists of well-defined lamellae which are digested and resorbed during moulting. The various permutations of chitin architecture and the occurrence of growth layers are said to occur in the endocuticle. In this scheme, mesocuticle is regarded as impregnated with lipid but not tanned, and as a transitory stage through which presumptive exocuticle passes.

The "histological" scheme for subdividing procuticle is based on the colour reactions (or lack of reaction) of integumental material to Mallory's Triple Connective Tissue stains and was extensively developed by Richards, A. (1967). Viewing cuticles in this way, one is given a qualitative indication of the extent of tanning and hence indirectly

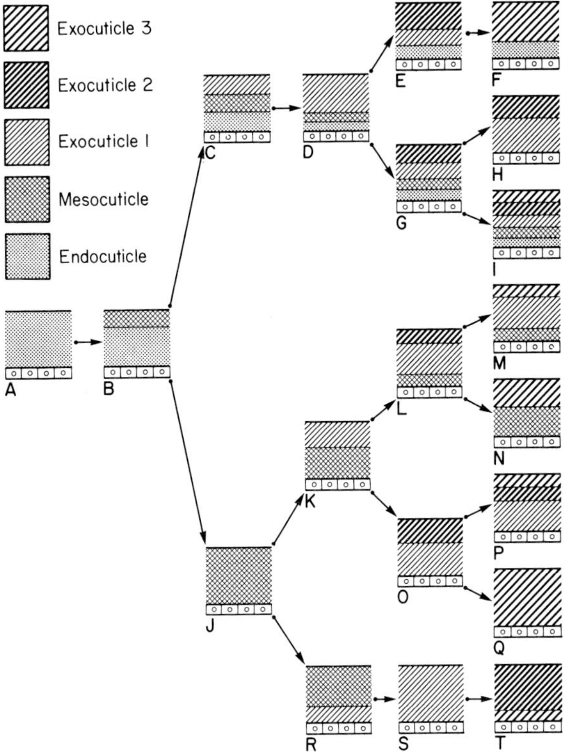

FIG. 23. A. Richards's (1967) scheme for the various routes and stages associated with the histological differentiation of the chitin-bearing procuticles of insects. The sequences A to F, A to I, A to K and A to T have been documented in developmental studies. Examples of all end-points have also been observed; however the sequence positions of H, M, N, P and Q have been assigned to presumed developmental routes.

something of the mechanical properties depending on the presence of histologically defined regions. One is also given an indication of the sclerotization route (not biochemical pathway), and the status of cuticle for a given insect can be determined with reference to the life cycle of the whole anatomical region of the animal as outlined in Fig. 23.

Both the structural approach – which mainly describes the arrangements of chitin microfibrils in procuticle — and the histological view — which mainly describes the extent to which the cuticle is tanned in a qualitative way and so protein matrix — provide much information about a cuticle, but because they are both qualitative in nature fine discriminations between similar kinds of cuticle are not readily made. A very tentative attempt was made recently to combine the useful features of

Table 8: Classification of arthropod cuticles according to mechanical, histological and structural criteria

Specimen	Proposed classification	Histological classification	Structural classification
Thoracic tendon of dragonfly, *Aeshna*	I*	mesocuticle	none
Intersegmental membrane of locust abdomen, *Locusta*	II†	endocuticle	preferred
Abdominal terga of silkworm, *Bombyx*	III*	endocuticle	helicoidal
Carapace of *Limulus*			
(a) outer fraction	IVa*	exocuticle '2' & 3	helicoidal
(b) inner fraction	IVa*	exocuticle 1	preferred
Wing membrane of butterfly, *Danaus*	IVb*	?	helicoidal
Gena of locust, *Locusta*	(IVa > IVb)*	exocuticle endocuticle	helicoidal/ preferred
Larval abdominal terga of *Agenius zebra*	(IVa < IVb)*	endocuticle	helicoidal
Elytron of adult *Agenius zebra*	(IVa > IVb)	exocuticle, endocuticle	pseudo- orthogonal
Normal abdominal larval terga of *Rhodnius*	(IVa > IVb)†	exocuticle, endocuticle	pseudo- orthogonal
Plasticized abdominal larval terga of *Rhodnius*	III†	endocuticle?	pseudo- orthogonal
Teneral adult, locust femur, *Locusta*	(IVa < IVb)*	endocuticle	helicoidal/ preferred
6-hr-old locust femur, *Locusta*	(IVa > IVb)*	exocuticle, endocuticle	helicoidal/ preferred
1-month-old locust femur, *Locusta*	(IVa• IVb)*	exocuticle, mesocuticle, endocuticle	helicoidal/ preferred

From Hepburn, H. and Chandler, H. (1975).
*, isotropic; †, anisotropic.

both schemes and to superimpose upon them measurements of the mechanical properties of very different kinds of cuticle as outlined in Table 8 (Hepburn, H. and Chandler, H., 1975).

Thus, there are four main types of cuticle which, on the basis of their tensile mechanical hysteresial behaviour, can be conveniently described as those which (I) exhibit long-range rubbery elasticity; (II) cycle-soften; (III) cycle-harden independently of matrix; and (IV) cycle-harden dependent on the matrix (matrix-dependence tends to inhibit cycle-hardening). Types III and IV are easily separable: in III there are no significant decreases in the elastic modulus, relative stiffness, breaking force and strain between fresh test specimens and those from which the matrix has been removed by hydrolysis, while in IV there is a several-fold reduction in modulus, relative stiffness, force at breaking and a concomitant increase in breaking strain on hydrolysis. Type IV cuticles include a wide spectrum of properties, the end-points of which are sharply definable as follows: IVa cuticles which are characterized by a heavily tanned brittle matrix, a high relative stiffness, elastic modulus and breaking stress and low breaking strain; and IVb cuticles in which there is a highly plastically-deformable matrix, a low relative stiffness, elastic modulus but large breaking strains. Fractographically IVa cuticles are brittle and show little evidence of the role of the chitin phase, while IVb forms commonly display the influence of the fibre phase.

The majority of cuticles fall somewhere in between the end-points of IVa and IVb, so that they are blends of these two sub-types and they are analogous to case-hardened engineering materials. In destructive tensile tests of such materials, the outer hardened (tanned) region behaves and fails like a IVa material, while the inner, relatively untanned region behaves like a IVb material. Such differences are readily apparent on fractographic examination of such a material. As the ratio of these two sub-types varies in a given cuticle, so does the extent of material compromise. Thus, when a relatively large fraction of brittle to plastic matrix occurs, the simple notation IVa > IVb can be used, and conversely, IVa < IVb when there is a larger fraction in the reverse direction. In such a relative scale more or less equal fractional amounts can be denoted simply as IVa ≃ IVb. Similarly, in a IVb-like structure where the matrix is not sufficiently stabilized to check the expression of chitin fibre arrangement, surface planar isotropy (*) or anisotropy (†) can be indicated by superscription, for example, IVa* > IVb* for an isotropic specimen such as locust femoral cuticle which has a brittle outer layer and an isotropic plastically deformable inner layer. Several examples have been classified in this way, and their synonymies with other classifications are given in Table 8. This table also shows that the histological classification is useful in qualitatively distinguishing cuticles of types III and IV; that is, those which show well-defined differences in sclerotization, while the structural classification provides a limited amount of information on the probable planar isotropy of a given piece of cuticle.

The classification of cuticles on the basis of mechanical properties as suggested here appears to overcome many of the limitations of other schemes. For example, in type IVb cuticles, actual chitin architecture is not so important as is the fact that the absence of a stabilized matrix allows participation in the deformation process (whether of intra-stadial growth or mechanical hysteresis) of chitin–chitin cross-links and the formation of mechanical cross-links while, in IVa cuticles, increased matrix stabilization inhibits the potential contribution of these cross-links. This scheme allows for the fact that cuticles are not only metabolically active but are also mechanically dynamic.

Whatever usefulness this latter approach may or may not have in approximating the functional properties of different kinds of cuticles, its general application in cuticular studies seems to be greatly restricted by the extremely specialized nature of experimental measurements required to study the cuticles (Hepburn, H. and Chandler, H., 1980).

4.3 Composite cuticles

4.3.1 General properties of solid cuticles

The term "solid cuticle" was first proposed by Jensen, M. and Weis-Fogh, T. (1962) to describe sclerotized and hardened pieces of insect exoskeleton. This term has the convenience of an engineering description in that some functional quality is implied *vis-à-vis* a term such as "sclerite", which has special meaning in morphology.

These kinds of cuticles break in a brittle manner and the fracture faces of broken specimens reveal that the cuticles are mechanically dominated by a rigid and heavily tanned protein matrix (Hepburn, H. and Joffe, I., 1976). At the other end of the spectrum are those cuticles, which we shall call "plastic cuticles", in which the proteinaceous matrix exhibits plastic flow when specimens of the material are stretched to breaking. These materials are of relatively low stiffness, low elastic modulus and breaking stress, and show a relatively large breaking strain. These sorts of cuticles demonstrate the importance of the chitin microfibrils to the mechanical integrity of the whole structure when fractured specimens are examined (Hepburn, H. and Joffe, I., 1976).

Solid cuticles can exhibit either planar isotropy or planar anisotropy, depending upon the spatial arrangement of chitin microfibrils and the extent to which the embedding medium, i.e. the protein matrix, is tanned. By way of example, stiff–brittle solid cuticle (as determined by physical measurement) occurs in very many beetle elytra (Krzelj, S., 1969), thoracic pleura of locusts (Jensen, M. and Weis-Fogh, T., 1962), and the tanned abdominal sclerites of honeybees (Thompson, P. and Hepburn, H., 1978). Plastic cuticles occupy the other end of the spectrum and can be illustrated by the abdominal tergites and sternites of dragonflies and the membranous regions of butterfly wings (Hepburn H. and Joffe, I., 1976).

The great majority of solid cuticles are intermediate between these two extremities, mainly because they are differentiated into exocuticle and endocuticle. In other words, stiff and brittle cuticles closely correspond to what in the histological classification scheme is called "exocuticle", and the plastic cuticles correspond to "endocuticle" and are illustrated in Fig. 24.

Changes in solid cuticle can be seen if one traces the physical, chemical and histological developmental events over time. Such studies are extremely few in number and information is available from a single paurometabolous insect, the locust *Locusta migratoria migratorioides*, and a

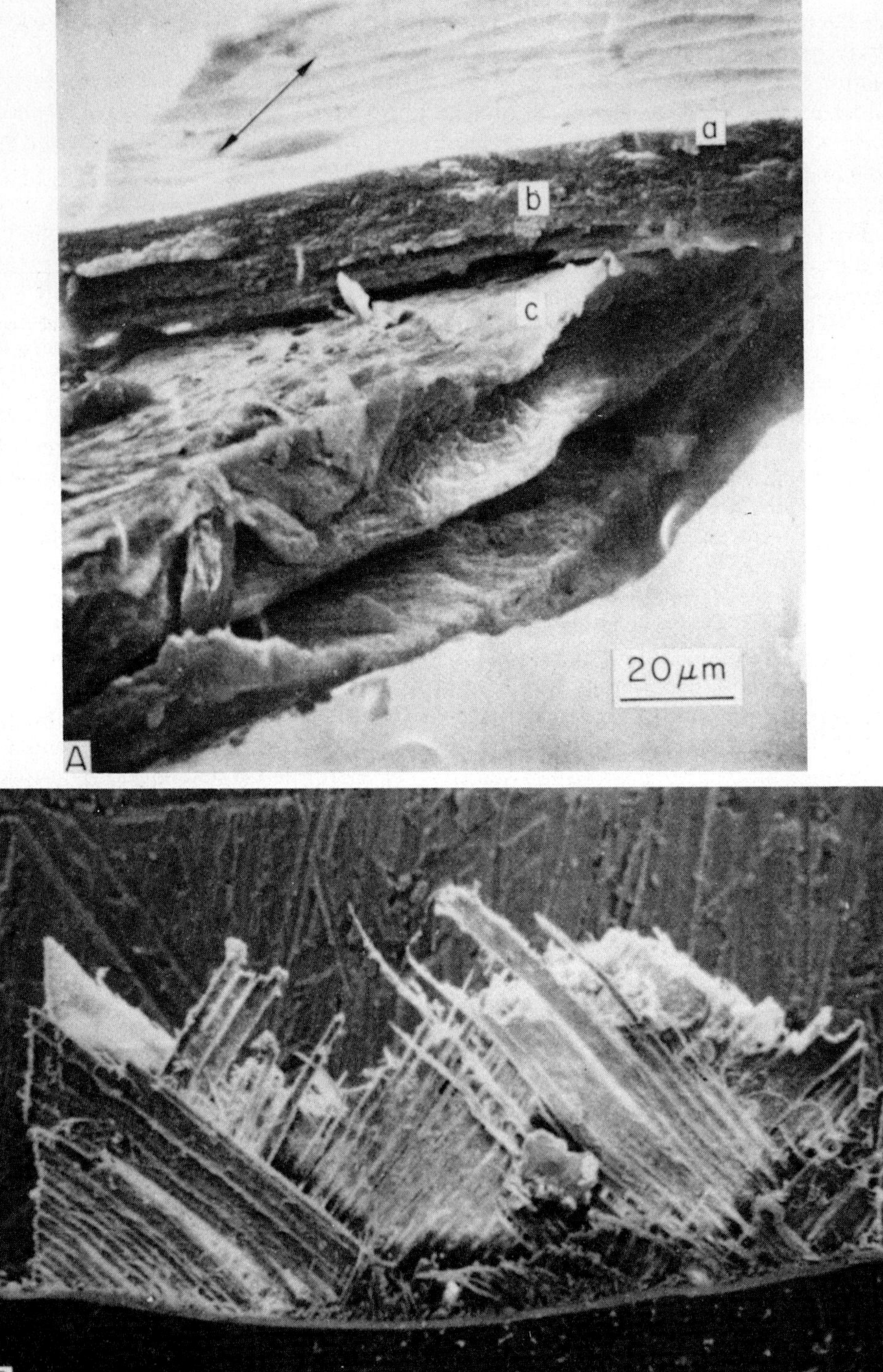

FIG. 24. **A**: Scanning electron photomicrograph of the fractured femoral cuticle of the locust, *Locusta migratoria*, broken in tension (arrow denotes direction of load application). a = epicuticle, b = exocuticle and c = endocuticle. From Hepburn, H. and Joffe, I. (1974); **B** as in **A**, but for the beetle *Pachynoda sinuata*.

single holometabolous one, the African honeybee *Apis mellifera adansonii.*

The relative changes in the amounts of NaOH non-extractable protein, stiffness (elastic modulus), and reactions to histological stains as a function of time during the development of the third abdominal tergite of the pharate and post-pharate adult honeybee are shown in Fig. 25 and similar parameters for the larval and adult locust hind femur are shown in Fig. 26. It is obvious that the trends are the same in both cases. Stiffness is highly correlated with NaOH non-extractable protein (an index of cross-linking in the matrix; Andersen, S. and Barrett, F., 1971) so that as the amount of non-extractable protein increases (relative to total mass) so the stiffness increases. The introduction of additional cross-links in the matrix renders it increasingly insoluble (and the recovery of cross-link derivatives following drastic hydrolysis is correlated with the amounts of non-extractable protein). These relationships hold despite the fact that different tanning systems occur in the examples. The locust is β-sclerotized (Andersen, S., 1970, 1974a,b) and the honeybee mainly quinone-tanned (Andersen, S. *et al.*, 1981).

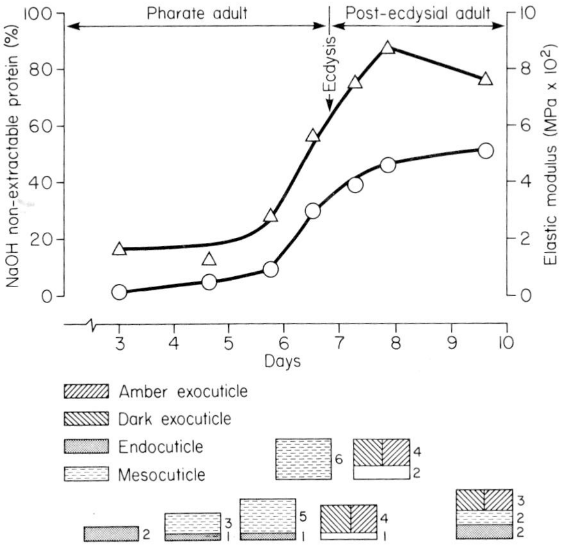

FIG. 25. Changes in the amounts of NaOH non-extractable protein (circles) and elastic modulus (triangles) during pharate and post-ecdysial development of the third abdominal tergite of the honeybee, *Apis mellifera adansonii.* Cuticular staining patterns (following Richards, A., 1967) at appropriate assay intervals are given below the time scale, and the thickness of the cuticle for that stage indicated (micrometres). From Thompson, P. and Hepburn, H. (1978).

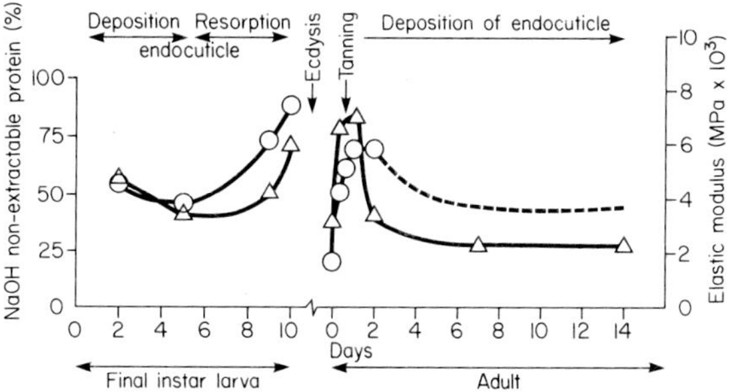

FIG. 26. Changes in NaOH non-extractable protein (circles) and elastic modulus (triangles) in the hind leg femoral cuticle of the locust, *Locusta migratoria*, that occur during larval and adult development. (From Hepburn, H. and Joffe, I., 1974.)

Therefore the generalization emerges that the precise nature of cross-link formation is not so important as is the actual extent of cross-linking.

That the fully formed adult locust femur and honeybee abdominal tergite are less stiff than earlier stages (Figs 25 and 26) is only superficially anomalous because of the deposition of *more* endocuticle in both locust and honeybee *after* the exocuticle has been tanned, and the fact that the stiffness depends upon the *total* cross-sectional area of the specimen. If, however, one compensates for thickness and replots that data as *relative* stiffness the appropriate "intuitively" correct results are obtained (Figs 27 and 28). The relative stiffness is also highly correlated with NaOH non-extractable protein. A dip in the real stiffness curves will occur when the rate of sclerotization falls below that of endocuticle deposition (as in the 1-week-old adult locust and post-ecdysial honeybee). However, it is only the *relative* amount of non-extractable protein that has decreased in this case since the *absolute* amount cannot decrease. The *relative* stiffness ($N\,mm^{-1}$) is directly related to the absolute amount of insoluble (tanned) matrix, hence *total* number of cross-links. For real stiffness ($N\,mm^{-2}$) it is the *relative* number of cross-links that is reflected when two different pieces of cuticle are compared. The trends for ultimate tensile strength are the same as those of stiffness while the breaking force (force to break the material irrespective of its thickness) follows that of the *relative stiffness*.

The total extensibility of the cuticles (breaking strain) is highly negatively correlated with the stiffness and non-extractability. This is because, as the extent of sclerotization proceeds, the ability of

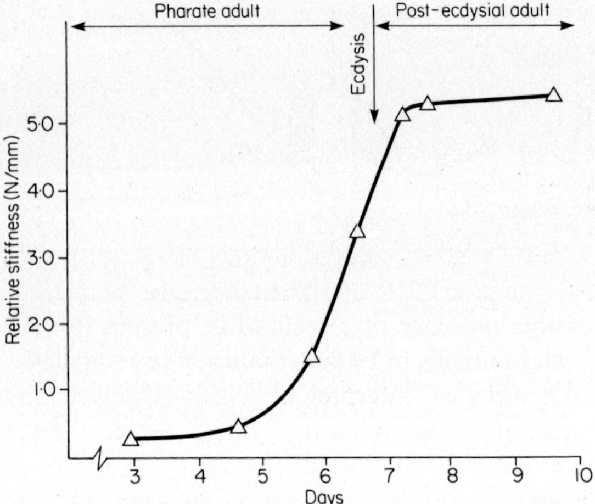

FIG. 27. The relative stiffness of the cuticle of the third abdominal tergite of the honeybee, *Apis mellifera adansonii*, during pharate and post-ecdysial development. Based on the elastic modulus curve in Fig. 25 but replotted to compensate for changes in thickness. (From Thompson, P. and Hepburn, H., 1978.)

proteins to slip past one another or to be stretched is greatly reduced owing to the cross-linking of the matrix. Where the matrix is not cross-linked to any extent the material can flow plastically; where it is well-tanned it fails in a brittle manner (Fig. 24).

Returning briefly to cuticle classification, it can be seen that locust and honeybee solid cuticle vary enormously over time, and we can ask how these changes relate to histological staining reactions. Figure 25 shows that there is indeed a *qualitative* relationship between staining and the physico-chemical properties: day 3 honeybee cuticle is blue-staining endocuticle, highly extractable and soft and flexible but neither stiff nor strong, suggesting little (if any) covalent cross-linking. This is further supported by the fall in stiffness and extractability beginning on day 8 when blue endocuticle appears again. At 6½ days the cuticle is entirely red-staining mesocuticle and its extractability has decreased and stiffness has increased (cross-links have been introduced). The appearance of non-staining exocuticle leads to further rises in stiffness and even less extractability, so that it appears that mesocuticle does represent an intermediate stage on its way to becoming exocuticle. Note, however, that histological exocuticle can vary greatly in stiffness (an order of magnitude difference exists between the elytron of the meloid, *Ecapatoma*, and the mesosternum of the scarab, *Pachynoda*; Hepburn, H. and Joffe, I., 1976) so that although a qualitative relationship exists between stiffness, non-extractability and staining reactions, molecular rationales for the relationship remain equivocal. Some of the physical properties of solid cuticles are summarized graphically in Fig. 29.

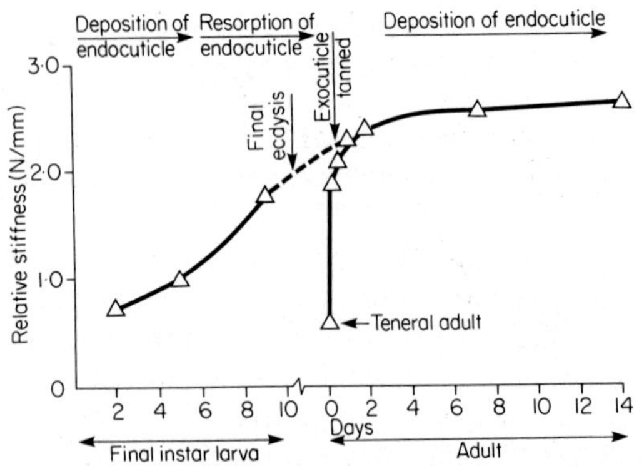

FIG. 28. The relative stiffness of the hind leg femoral cuticle of the locust, *Locusta migratoria* during larval and adult development. Based on curve in Fig. 26 but replotted to compensate for changes in cuticular thickness. (From Hepburn, H. and Joffe, I., 1976.)

4.3.2 FUNCTIONAL COMPOSITES

The extensive research on insect cuticle over the last 50 years has been very largely dominated by

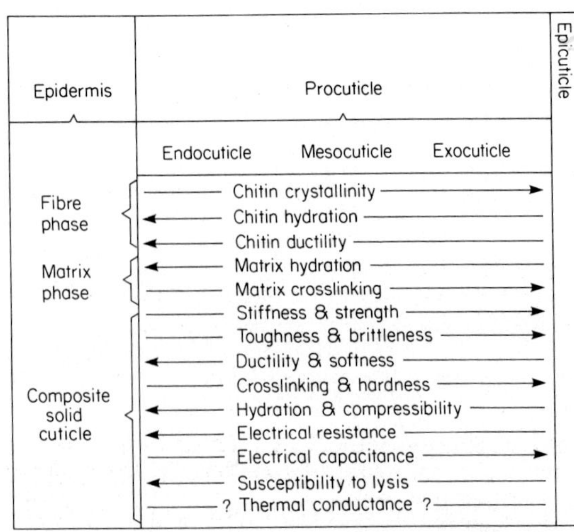

FIG. 29. A summary of the direction of change in some of the physical properties of insect procuticle during the course of differentiation and sclerotization.

fractionation studies of cuticular material into various components and the subsequent detailed analysis of very small parts, for example the study of lipids or X-ray diffraction of the chitin micelle. While a great deal has been learned in this way about minute aspects of structure and composition and sometimes, function, the literature clearly reflects the fact that cuticle biologists have been unable to see the forest for the trees. A major exception is to be seen in the approach and in the experiments of Torkel Weis-Fogh, who tried to resolve some of the difficulties associated with the energetics of flapping flight (noted by Boettiger, E. and Furshpan, E., 1952; Sotavolta, O., 1952; Pringle, J., 1957) by detailed analyses of the mechanical properties of cuticle.

Weis-Fogh's studies of elasticity in arthropod locomotion were briefly introduced in 1958 and subsequently developed in two major classical papers on the functional properties of whole cuticles (Weis-Fogh, T., 1960; Jensen, M. and Weis-Fogh, T., 1962). The 1960 work established that the elastic properties of the pre-alar arm and wing-hinge elements of the locusts, *Schistocerca*, and a flight-related tendon of the dragonfly, *Aeshna*, derive from the protein rubber, resilin (see section 3.3.4). Before proceeding to further analyses of insect flight, Jensen, M. and Weis-Fogh, T. (1962) proposed a working hypothesis for cuticle consisting of a laminate of four components (excluding epidermis and epicuticle):

(1) a system of chitin lamellae parallel to the surface, not necessarily continuous but of considerable tensile strength;
(2) a rubber-like proteinaceous glue between lamellae and cross-linked to differing extents but poor in tyrosine and not tanned;
(3) additional protein which when tanned is rich in tyrosine and resin-like in nature; and finally,
(4) water.

They introduced the additional constraint that the laminae are of greater than molecular dimensions to avoid the kind of molecular interlocking proposed by Fraenkel, G. and Rudall, K. (1940, 1947) which would result in very hard and brittle materials. The working hypothesis of Jensen and Weis-Fogh has proved, and continues to prove, useful as a conceptual framework for cuticular analysis. The arrangements of chitin microfibrils have in general been found to be parallel to the surface plane of the cuticle (literature summarized in Neville, A., 1975 and Hepburn, H., 1976) and of considerable tensile strength (Hepburn, H. and Joffe, I., 1976). The tanned protein matrix has been very well established (summarized by Andersen, S., 1979) as has the importance of water (Hepburn, H. and Joffe, I., 1976; Hepburn, H. and Chandler, H., 1976). The possible presence of a resilin-like protein in hard cuticles remains to be systematically investigated.

Attempts to interpret cuticular structure and modification in terms of function have been remarkably few during the last 20 years; however, a few examples point to a very modest beginning in this direction pioneered by Weis-Fogh and his colleagues. Hepburn, H. (1971) reported some simple investigations on the mechanisms which might

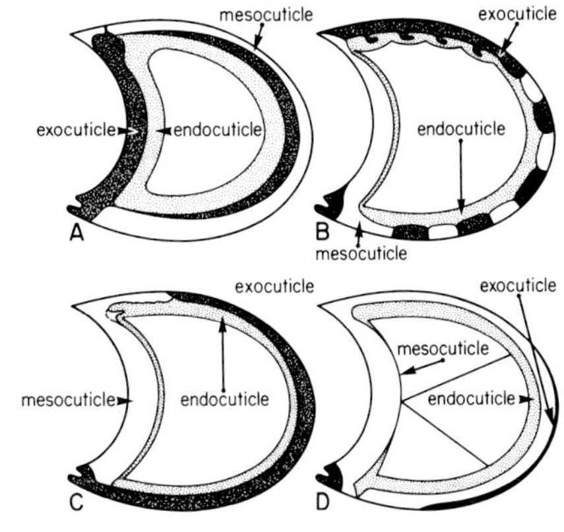

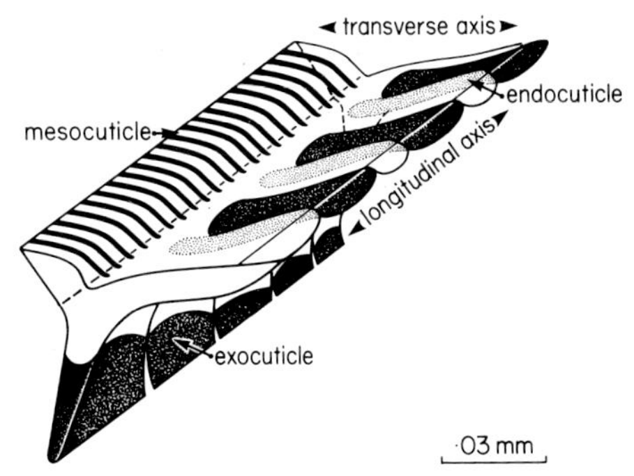

FIG. 30. Distribution of kinds of cuticle in the transverse planes of the galeae of moths and butterflies. (**A**) *Agrotis exclamationis*; (**B**) *Deilephila elpenor*; (**C**) *Danaus plexippus*; (**D**) *Pieris brassicae*. Distribution of cuticle types in the longitudinal axis of a portion of the galea of the moth *Deilephila elpenor* (lower drawing). (From Hepburn, H., 1971.)

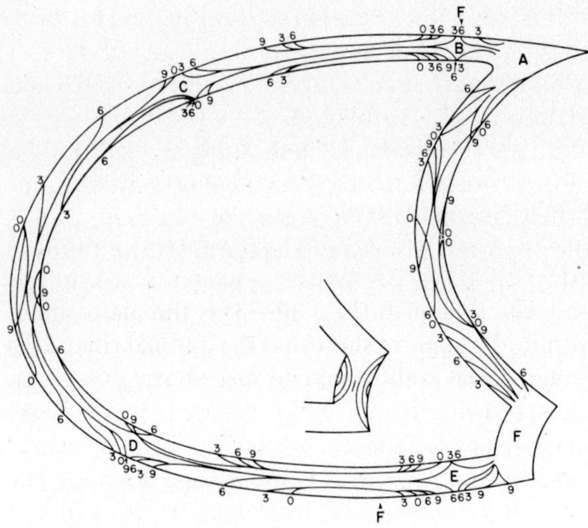

FIG. 31. Distribution map of the isoclinic and isochromatic fringes obtained from a compression-loaded Perspex model of a butterfly or moth galeal cross-section. B to E indicate areas of principal stress concentration, A and F indicate areas of maximum shear stress. (From Hepburn, H., 1971.)

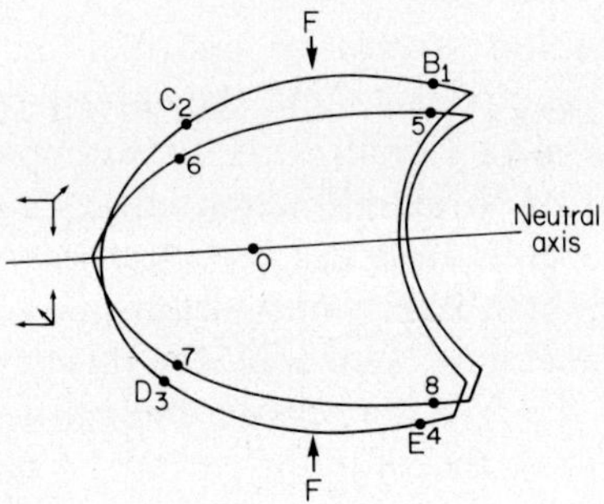

FIG. 32. Relative changes in position (strain) observed by different regions of a Perspex model of a galeal cross-section. (From Hepburn, H., 1971.)

allow for the extension and the recoil of the proboscis of moths and butterflies. Aside from important contributions from the investing musculature and internal galeal hydrostatic pressure to proboscis manoeuvrability previously established by Eastham, L. and Eassa, Y. (1955), the problem of how structure of the cuticle related to required functions of the proboscis remained.

Hepburn and colleagues approached the problem by coupling histological observations of cuticular types found in galeal cross-sections with results of photoelastic analyses of Perspex models of the galeae. Serial sections of proboscides showed variations in the kinds of cuticle present transversely and longitudinally (Fig. 30). While the cuticle types were then defined histologically as exocuticle, mesocuticle and endocuticle, we now know, by extrapolation from subsequent mechanical measurements on cuticles, that these correspond to very stiff and brittle material, intermediate, and soft cuticles respectively (section 4.3.1). In addition, a resilin-based spring in the dorsal bar was identified.

The photoelastic analyses involved the mechanical loading of Perspex models of galeal cross-sections in a manner to simulate muscle contractions. The isoclinic and isochromatic fringes seen in the specimens placed between crossed polarizing lenses that were rotated through fixed angles were then recorded. In this way a map of the distribution and direction of fringes which relate to the principal stresses can be constructed as shown in Fig. 31. One can then extrapolate the model data back to the histologically observed cross-sections and attempt to rationalize the cuticle types present in terms of functional requirements of the coiling or uncoiling proboscis. Thus, the tensile and compressive stresses on the specimen are greatest at B, C, D, and E (Fig. 32).

Functionally, these four areas must absorb or transfer fluctuating compressive and tensile stresses. In the model the stresses are fixed at a particular loading; but in the real galea the stresses fluctuate as the proboscis coils or uncoils. The regions A and F are subjected to maximum shear stress due to bending, and produce the isochromatics typical of this kind of stress. When the models fatigue, breaks always occur in these two regions, usually at F.

By drawing the lumen of the model, at rest and while stressed in the loading frame, and superimposing the images, the physical displacement of individual points anywhere in the model can be determined in reference to the neutral axis, an intermediate layer which is neither compressed nor stretched when the proboscis bends. By marking the areas of principal stress, the behaviour of these four regions could be shown (Figs 31 and 32). The areas which are subjected to greatest anteroposterior compressive displacement are the CD axis (20%) and the BE axis (15%). Lateral displacement at C (the angle 2-0-6) = 12°, B (angle 1-0-5) = 6°, E (angle 4-0-8) = 6°, and D (angle 3-0-7) = 0°. Movements are in the directions of the lines indicated above and below the neutral axis. The areas of

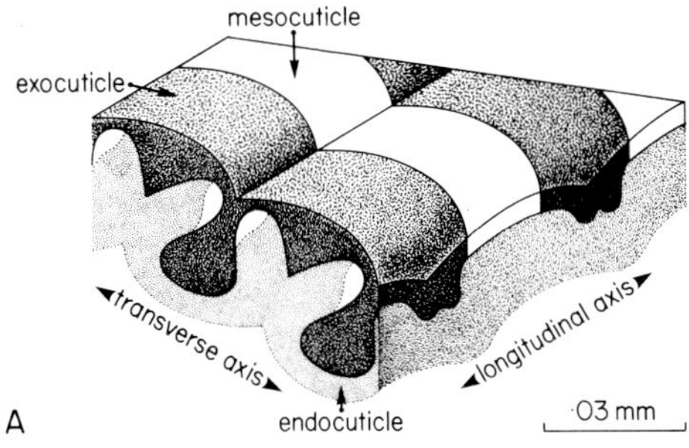

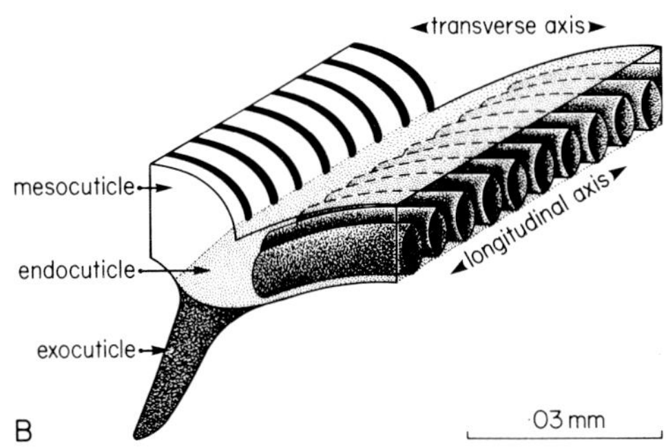

FIG. 33. **A**: Reconstruction from serial sections of the outer galeal wall of the moth, *Deilephila elpenor* showing reinforcing rods of exocuticle surrounded by softer and distensible patches of endocuticle. **B**: Dorsal bar region of *Pieris brassicae*. (From Hepburn, H., 1971.)

principal stress are the same as those obtained directly by distortion. The third axis, which corresponds to force directions in the sagittal plane, is not considered for the two-dimensional model.

The deformations just described accompany extension of the proboscis in the sagittal plane. The amount of local cross-sectional change is the lineal distance between the origin and insertion points of the individual galeal muscles. This distance decreases from the base to the apex of the proboscis while muscle declination increases. Uncoiling is a summation of many local cross-sectional effects. In applying these model findings to the prototype, specific cuticular modifications are seen to coincide with the principal stresses and the regions of maximum shear stress (Figs 30–33).

The regions of maximum shear stress are treated in the following ways. The ventral linkage is reinforced mesally by the consistently heavily sclerotized and rigid exocuticular linkage hook. The transition of cuticle layers on the lumen side from exocuticle to either endocuticle or mesocuticle results in a flexible cuticular padding against shear in this direction. At the dorsal linkage where shear is less, there is usually a transition of cuticle types through the shear zone from a harder, less flexible to a softer, more flexible kind of cuticle. In a working proboscis the ventral linkage hooks are fixed and rigid; the dorsal hooks freely slide over each other.

There are a variety of modifications in the galea coinciding with areas of principal stress in the models. Region B (Figs 30–33) is the main site of resilin. The figures show that the internal ribs of the outer galeal wall terminate just above this point, thus allowing for flexibility between the rib bases and the dorsal linkage hooks. The opposite ventral point (E) is variously modified by thinning out the exocuticle, folding it, indenting it, or eliminating it (Figs 30–33). The reactions to stress at regions C and D usually involve transferral of stress by thinning or thickening of the cuticle on either side of them. Sphingid moths have solved the problem by a complicated alternation of triangular wedges and pyriform lobes of exocuticle connected by thin exocuticular rods (Fig. 33). These rods are supported by a softer mesocuticular padding. In the butterflies *Pieris brassicae* and *Papilio demoleus* the stress at C and E is displaced by two septa (Fig. 30) which may act as guy-ropes during proboscis extension (Eastham, L. and Eassa, Y., 1955).

The arthropod cuticle has considerably more modifications than horizontal and vertical variation of cuticular types, grossly seen as sclerotized and arthrodial cuticles. The cuticle must also allow the ingress of environmental information through numerous sensory devices such as chemoreceptors (many of which lack an epicuticle — Slifer, E., 1961) and mechanoreceptors. These latter organs have been the subject of relatively intense research in the last few years, with attention having been paid to specific problems such as how deformation of the cuticle can be exploited by the mechanoreceptor, how the receptor cuticle is modified for function, and how deformation of cuticle leads to nervous discharge from the input signal to the central nervous system.

Barth and colleagues have investigated cuticles and Perspex models of mechanoreceptors, the so-called lyriform organs of spiders, and coupled these observations to experiments on the stimulus-conducting properties of cuticular slit sense organs

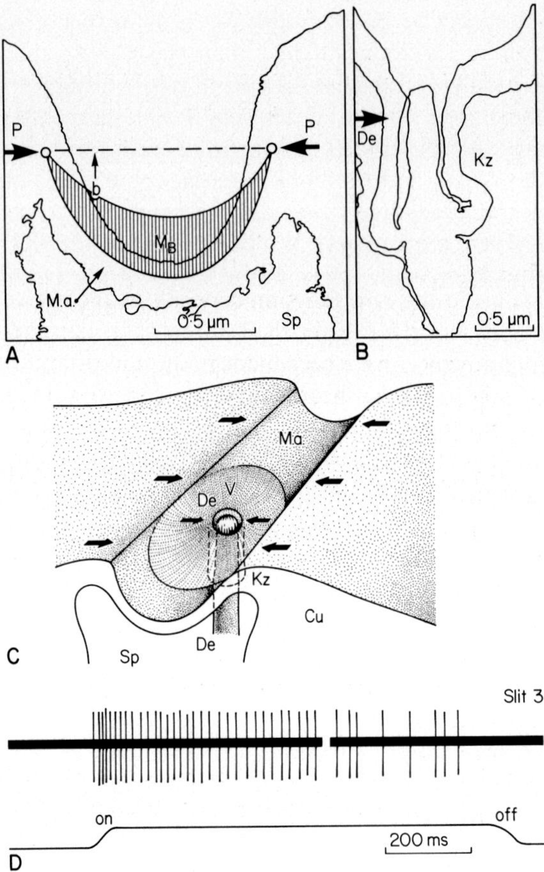

FIG. 34. Mechanoreceptor of a lyriform slit sense organ of the spider *Cupiennius salei*. **A** shows the covering membrane (Ma), the moment of bending (Mb) resulting from compression of slit (P). **B** indicates the covering membrane's coupling cylinder (Kz) into which the end of the dendrite projects (De). **C** schematic representation of directions of slit displacement. **D** neurophysiological recording from stimulated (compressed) slit sense organ. (From Barth, F., 1972b, 1973.)

(Barth, F. 1972a, 1976; Barth, F. and Pickelmann, P., 1975). He constructed Perspex discs of different degrees of similarity to slits naturally occurring in spiders, and loaded these models in different modes (tension, compression and shear). These experiments showed that deformation of the slits in the models is greatest when the applied load is normal to the long axis of the slits.

The sensory slit in the spider is actually trough-like (Fig. 34) and if the Perspex model is slightly modified to simulate the cuticular thickening around the natural slit then any deformations related to stresses parallel to the slit are reduced, and those normal to the axis of the slit enhanced. Likewise, deformation of a slit is greatest in the middle region and decreases at both ends. The extent of deformability, given the same load, increases with slit length. Extending these model observations back to cuticular structure, Barth observed that the lamellae of the exocuticle are arranged very closely to stress trajectories observable around a simple notch in a bar that is uniformly loaded in tension (Fig. 35).

These observations were subsequently extended in a far more elaborate series of experiments using seven slits in the Perspex disc, thus simulating more exactly the real structure found in the spiders' legs. In this study it was discovered that the single slit is inadequate as a model and that a collection of slits behaves differently. For example, the extent of deformation of a given slit varies both with slit length and whether or not the slit occupies a peripheral position within the group. Moreover, different areas of the same slit vary in the extent to which they deform in response to direction of an

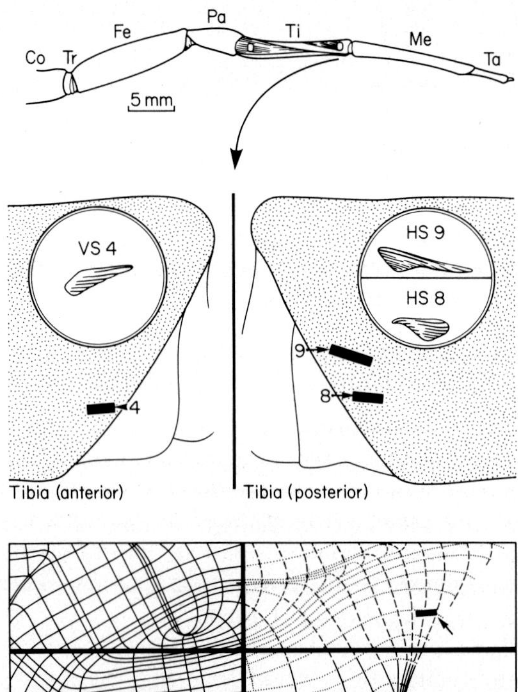

FIG. 35. Distribution of principal stresses as a trajectory map of spider's leg segment. ---- pressure, —— tension, ···· no stress, a lyriform organ indicated by arrow. (From Barth, F., 1976.)

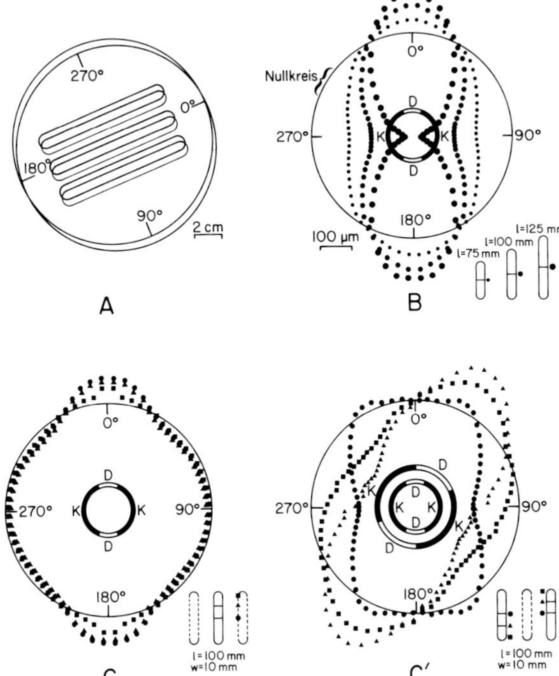

FIG. 36. Slit deformation behaviour of mechanoreceptor model. **A**: Simple model for loading studies. **B**: Deformation of single model slits varying in length in different directions. The small circle indicates load directions for compression (K) and dilatation (D). **C**: deformation of median slit in a group of three slits. **C′**: deformation of peripheral slit in a group of three slits. (After Barth, F., 1976.)

applied load (Fig. 36). Indeed Barth found that in most cases the receptor dendrite attaches to the cuticle at the point where the greatest deformation occurs with the least force.

In extensions of this work, Barth examined photoelastic models of tibiae and showed that the topographical features of the leg coincide with pressure stress distribution in the models so that compressive forces act perpendicular to both model and to real spider. Clearly, the structure of the cuticle is interpretable in terms of model stresses. The physiological significance is that the architectural arrangement of lyriform slits in the tibial cuticle allows for a large range of stimuli to be perceived at a specific receptor site (and a marked stimulus intensity fractionation by different slits in the same mechanoreceptor group).

Studies of the kind conducted by Hepburn, H. (1971) on the lepidopteran proboscis and Barth (Barth, F., 1972a; Barth, F. and Pickelmann, P., 1975) on the lyriform slit sense organs of spiders essentially employed optical techniques and model analysis to rationalize cuticle structure in terms of function. Much earlier work, notably that of Pringle, J. (1948) on the gyroscopic mechanism of the fly haltere and other works in that tradition, were mainly concerned with elucidating function. The recent trend has been to exploit what is known of the physical (mechanical) and structural properties of cuticle and how function depends on and derives from them. Thus, over the last few years considerable advances have been made in the understanding of how structure, hence physical properties, relate to function in the campaniform sensilla of the cricket, *Gryllus bimaculatus*, by Gnatzy, and his colleagues (Gnatzy, W. and Romer, F., 1980; Gnatzy, W. and Tautz, J., 1980) and in the cockroach (Moran, D. *et al.*, 1971; Chapman, K. *et al.*, 1973), and in the apodemes of locusts (Bennet-Clark, H., 1976).

4.4 Arthrodial membrane and caterpillar-like cuticles

We refer to "arthrodial membranes" as a general term to denote those soft and highly flexible regions of cuticle around which sclerites articulate, setae deflect in sockets, and body segments move (thus avoiding use of the term "intersegmental membrane" which has restricted meaning in morphology — Snodgrass, R., 1935). These kinds of cuticles are grouped together because of their very evident mechanical similarities and reactions to histological stains (Wigglesworth, V., 1956; Richards, A., 1967). Unfortunately, general familiarity with these materials has been on the decline since freeze-drying techniques have supplanted the old-school method of "blowing" caterpillars and maggots.

The groundwork for understanding these cuticles stems from the studies of Fraenkel, G. and Rudall, K. (1940, 1947) on the physical and chemical consequences of pupariation in the flies *Calliphora erythrocephala* and *Sarcophaga faculata*. Fraenkel and Rudall were able to demonstrate several important aspects of structure here: the principal being that there is relatively little crystalline order in the cuticles and that the chitin crystallites can be oriented when suitably loaded in compression or tension, a property lost on completion of pupariation.

Similarly, the already hydrated larval cuticle is capable of considerable swelling on solvation in aqueous media (Rudall, K., 1963). The protein fraction is largely soluble in water and this observation is the basis for the term "arthropodin" introduced by Fraenkel, G. and Rudall, K. (1947) in contrast to stabilized protein formerly termed "sclerotin" by Pryor, M. (1940a). Thus the interaction between chitin fibre and protein matrix is a very loose one and is dramatically illustrated in deformation and fractographic studies of such cuticles. Taken to destruction in a tensile manner, these cuticles are visco-plastic. There is a "necking" of the specimens as in mild steel and the material literally flows in response to continued extension (Hepburn, H. and Chandler, H., 1976).

Arthrodial and caterpillar-like cuticles are materials of very low stiffness, great distensibility and reasonable toughness. There is, however, remarkable variation in the extent to which such cuticles can be extended. A most exceptional example is the arthrodial membrane cuticle in the abdomen of the female locust, *Locusta migratoria*. This material is capable of nearly 2000% distension (Vincent, J. and Wood, S., 1972) which is nearly three times greater than can be obtained from cross-β silks. Part of the explanation certainly lies in the fact that the chitin fibres are arranged normal to the longitudinal axis of the animal, and one can presume that the protein matrix is not completely bound in itself or to the chitin fraction.

The relatively large distensibility of caterpillar cuticle seems to have been exploited in intra-stadial growth by these animals. In such cases the epicuticle primarily unfolds rather than stretches but the underlying procuticle is slowly stretched in the course of a stadium, as evidenced by the continuous thinning of the lamellae (effected by hydrostatic pressure?) in the procuticle of the silkworm *Bombyx mori* (Hackman, R., 1975, 1976). Under the appropriate circumstances, the abdominal cuticle can also behave much like arthrodial cuticle (cf. Reynolds, vol. 8).

Arthrodial membranous cuticles, be they the general integumental exoskeleton of fly larvae or of caterpillars or proper intersegmental cuticle, have been studied by Rudall, K. (1963) and found to be basically indistinguishable on the basis of the X-ray diffraction characteristics. Despite the fact that these kinds of cuticles are a *sine qua non* of arthropody, they remain very little studied and require extensive further investigation. It is interesting to note that serious study of tanning by Pryor, M. (1940a,b) began in the 1939-45 war, when there was considerable applied interest in the stability of laminated aircraft structures. In much the same way, perhaps the physics of plastic fabrics now used in gossamer gliders can be exploited to study the very thin and very tough fabrics that make up caterpillars and arthrodial membranes.

4.5 Cuticular processes

There is a vast literature, largely taxonomic, describing and cataloguing the distribution and kinds of invaginated and evaginated processes associated with various parts of the integument. Basic classifications of such protuberances in the past have centred round the nature of innervation supplying the process. Snodgrass, R. (1935) emphasized what he regarded as the important distinction: whether a cuticular process was served by a single or multiple innervation or whether or not innervated at all. More recently, Richards, A. and Richards, P. (1979) have attempted to bring some order to the chaos, and have proposed a classification based on the ontogenesis for such processes as follows: multicellular in which the epidermal cells underlying the process appear similar to neighbouring epidermal cells, and others for which the epidermis is specifically differentiated; unicellular processes and subcellular ones where there is more than one projection per cell.

Multicellular processes lacking a differentiated and specialized epidermis may be movable owing to an unsclerotized basal region as in claws or leg spurs (Fig. 37), and can be extended to include eversible osmeteria of caterpillars or defensive glands in chrysomelid beetle larvae or eversible pheromonal dispersal glands such as the Nasanov glands of honeybees or those of male Mecoptera.

Multicellular processes having a well-differentiated and specialized epidermis include bristles, chaetae, hairs or setae and usually function in chemosensory or tactile ways. These are the trichoid sensilla and related structures (Dethier, V., 1963). These processes arise from three epidermal cells that are specialized: a trichogen cell which secretes the cuticle of the process, a tormogen cell responsible for formation of the setal socket and associated cuticle, and finally the third cell is differentiated into a bipolar primary sense cell with peripheral dendrites and an axonal connection to the central nervous system. While tactile setae are usually covered with solid and continuous cuticle, that of chemoreceptive setae is porous, the pores arising during secretion of the cuticle by the trichogen cell (Ernst, K., 1969, 1972).

Cuticular processes arising from single epidermal cells lack sockets and sense cells, and the term "acanthae" has been proposed for these now known to be widespread processes (Richards, A., 1965). These sclerotized processes occur in the foreguts of the adults of many orders of insects, in the reproductive system of moths (Richards, A. and Richards, P., 1979) and as tenent hairs on the tarsi of flies (Whitten, J., 1969a,b) or the combs of fleas. Again, such very disparate examples are, whatever their functions, put into perspective when considered in terms of the ontogenetic origin: processes arising from a single epidermal cell lacking innervation.

Finally there are microtrichia which are very small hairs, and there are several to many of these arising as projections from a single epidermal cell. These are widespread on the surfaces of insects, particularly on the abdomen.

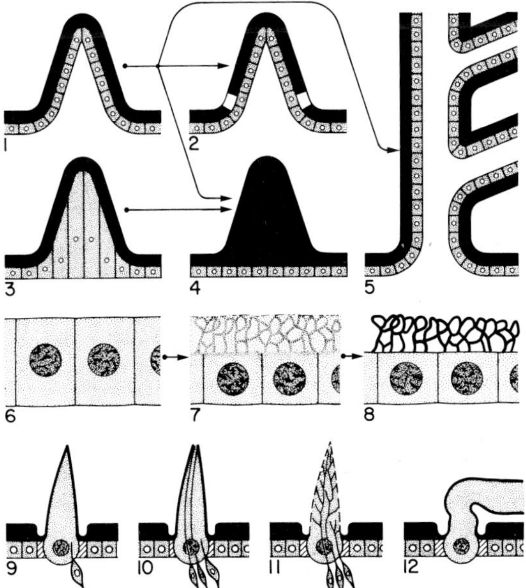

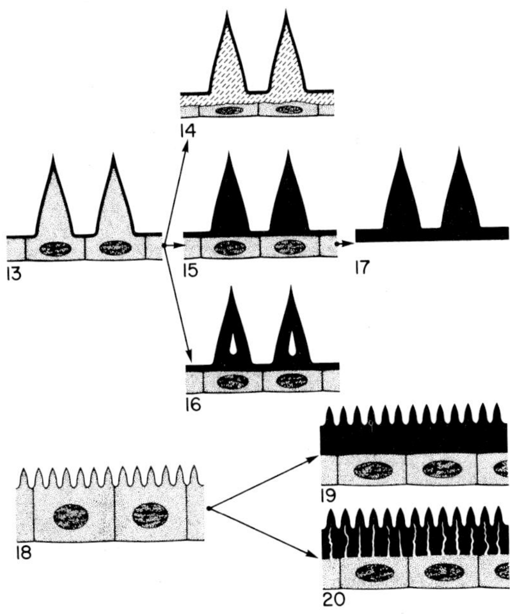

Fig. 37. Diagrams of multicellular projections. Cuticle solid black, cytoplasm stippled. **1**: A simple multicellular spine. **2**: Modification of a simple spine with membranous ring at base, **3,4**: Unusual solid spine termed Brunner's organ, **5**: Projection such as an antenna. **6–8**: Developmental sequence for formation of the "felt chamber" of larval spiracular discs of higher Diptera. **9–12**: Major types of protruding trichoid sensilla. Tormogen cells are lined rather than stippled. **13–17**: Sequences for development of acanthae — **14**: Cell processes withdrawn from lumens of acanthae and space filled with acid mucopolysaccharide (cross-hatched) as in *Panorpa*; **15**: cell processes withdrawn, lumens filled with cuticle and processes underlaid with cuticle as in "cornuti" in reproductive system of moths; **16**: cell processes cut off by secretion of underlying cuticle as in proventriculi of certain staphylinid beetles; **17**: cell processes withdrawn, lumens filled and underlaid with cuticle, then secreting cells phagocytized by haemocytes as in tenent hairs of flies. **18**: Early stage with cytoplasmic filaments extending into each microtrichium. **19**: Solidified microtrichia underlaid by cuticle. **20**: Microtrichia with relationship to pore canals. (From Richards, A. and Richards, P., 1979.)

4.6 Peritrophic membranes

Peritrophic membranes are secreted by the midgut epithelium, consist of chitin and protein, and occur in most insects, immature and adult, whatever their food preferences might be, solid or liquid (Richards, A. and Richards, P., 1977). In addition to a role in alimentation, they may also be used in the special construction of beetle cocoons (Streng, R., 1969; Kenchington, W., 1976). Whether peritrophic membranes are examples of "true" integument or not simply depends upon definition.

Two general types of peritrophic membranes (PM) were described by Wigglesworth, V. (1930): multilayered PM secreted by the surface of the midgut and single-layered PM arising from the anterior end of the midgut. This classification has become refined during intervening years with application of electron microscopy, so that Peters, W. (1968, 1969) recognizes three types based on the arrangement of chitin fibres within a PM: a reticulate or orthogonal arrangement, a hexagonal system, and finally a random arrangement of chitin fibres. If a random arrangement does in fact occur it is probably unique in insects where there are no other known cases of randomly arranged fibres in the integumentary system.

5 MOULTING, APOLYSIS AND ECDYSIS

5.1 General

The terms "moulting" and "ecdysis" have often been used interchangeably since the time of Snodgrass, and additional confusion was introduced when Snodgrass, R. (1935) restricted the term "moulting" to mean the separation of the old cuticle from a new cuticle formed beneath it, and held "ecdysis" to be the actual shedding of the presumptive exuvial cuticle. Modifications to these usages have been introduced by various authors during the ensuing decades.

Over the last 20 years or so use of varying terminology to describe the events associated with metamorphosis has culminated in a somewhat bitter but very edifying controversy between Wigglesworth, V. (1973) and Hinton, H. (1973). The two essays in question are well worth the reading, if for no other reason, because they point to extreme variations in both the timing of events as well as the actual sequences of structural change that occur during metamorphosis. For example, one can encounter in the very same animal regions which are in the larval stage and yet others in the pupal stage, as in the honeybee (Thompson, P., 1978).

In the present chapter we will follow the refinements suggested by Zacharuk, R. (1976) as follows. "Moulting" shall refer to all of the events and processes leading up to a shedding of the old cuticle. "Apolysis" means the separation of the epidermis from the old cuticle (this is as was originally proposed by Jenkin, P. and Hinton, H., 1966). As such, apolysis is but one aspect of moulting. Finally, "ecdysis" refers to the shedding of the cuticle. Ecdysis is mediated by the so-called eclosion hormone, currently thought to be of universal occurrence and controlling all ecdyses in insects (Truman, J. *et al.*, 1981; Truman, vol. 8). "Ecdysis" would appear to be synonymous with the term "eclosion".

Added to the above, we have the terms "moult", meaning the time period and events occurring from apolysis to ecdysis; "post-moult" from ecdysis to completion of bulk tanning; and, "intermoult" from the end of the post-moult to the next apolysis.

5.2 Events of metamorphosis

Mindful of the very large number of exceptions and special or peculiar cases, one can nevertheless describe a general outline of the events associated with moulting and ecdysis in insects as follows:

(1) apolysis along a usually anteroposterior gradient;
(2) mitotic division of the epidermal cells as well as the expansion of epidermal cell surface area and volume;
(3) the secretion of moulting fluid;
(4) formation of the outer epicuticle of the newly forming pharate cuticle and surface determination of same;
(5) secretion and formation of the new inner epicuticle of the forming pharate cuticle;
(6) activation of the enzymes of moulting fluid and the lysis and presumed reabsorption of the endocuticle of the old cuticle;
(7) deposition of the presumptive pharate exocuticle;
(8) ecdysis;
(9) expansion of the newly ecdysed cuticle;
(10) onset of tanning (which sometimes occurs before ecdysis);
(11) secretion of the endocuticle;
(12) secretion of wax;
(13) continuous deposition and partial tanning of newly formed endocuticle;
(14) formation of an apolysial membrane for the next moult where one occurs.

These are the kinds of events that occur in insects. Not all of them need occur in any one animal, nor need the sequence be exactly the same from animal to animal. Apolysis appears to be mediated, at least in part, by ecdysteroids or juvenile hormone (Riddiford, L., 1976; Agui, N., 1977; Riddiford, vol. 8; Locke, vol. 2).

While much has been made of developmental processes proceeding along anteroposterior gradients (Lawrence, P., 1970) and while it is readily observable in many insects chosen at random that such gradients are very real, it has nevertheless been reported that, in the moth *Ephestia kuhniella*, moulting (= apolysis?) commences in the thorax and extends thence to head and abdomen (Köhler, W., 1932). Coupled with the most recent discovery that eclosion hormone may come from a different anatomical source in different developmental stages of the same animal (Truman, J. *et al.*, 1981), detailed comparative developmental studies would be highly desirable.

Cell division of the epidermis is usually correlated with the onset of apolysis in insects as in the wax-moth *Galleria mellonella* (Barbier, R., 1971) and cockroach *Blattella germanica* (Kunkel, J., 1975). Alternatively mitotic activity may not be at all an apparent event associated with apolysis, as suggested for some Diptera by Richards, A. (1951).

Apolysis itself is not extremely well known, and at present it seems that it may progress in slightly different ways. In their study of the silkworm, Passonneau, J. and Williams, C. (1953) suggested that the forming apolysial space occurs at the level of the subcuticle. Dense apolysial droplets appear under the subcuticle and within the endocuticle just prior to the formation of the apolysial space. Alternatively, Noble-Nesbitt, J. (1963) reported the secretion in *Podura aquatica* of a foam-like substance between the plasma membrane and the old cuticle and in the process so forming the apolysial (= ecdysial) droplets which enter the apolysial space by exocytosis of the plasma membrane. The release of apolysial droplets apparently may precede apolysis in some cases, for example *Drosophila melanogaster* (Mitchell, H. *et al.*, 1971) or the butterfly *Calpodes* (Locke, M. and Krishnan, N., 1973) or follow apolysis as in *Hyalophora cecropia* (Passonneau, J. and Williams, C., 1953).

There is general agreement that the apolysial space becomes larger after the secretion of the moulting fluid, and that part of this larger space is derived from the partial disgestion of the old endocuticle. Since the apolysial droplets are very closely associated with the breakdown of the old endocuticle, it has long been inferred that these droplets are collections of cuticulolytic enzymes. While exact studies of apolysis are in their infancy, Zacharuk, R. (1976) has emphasized that the physical separation of old cuticle from the epidermis (apolysis) is a process entirely separate from that of enzymolysis of the old endocuticle.

The very thin apolysial (= ecdysial, exuvial, moulting) membrane first described by Passonneau, J. and Williams, C. (1953) from *H. cecropia* is appressed to, but not connected with, the newly forming pharate cuticle and seems to be present in most kinds of insects. However, recent ultrastructural studies of certain Diptera (Filshie, B., 1970; Mitchell, H. *et al.*, 1971) indicate that some flies lack the apolysial membrane. The combined evidence to date indicates that the apolysial membrane is derived from the innermost lamellae of the old endocuticle because such membranes are always associated with cuticles in which there is enzymolysis of the endocuticle, and are absent from cuticles where there is no significant lysis of endocuticle.

The stability of the apolysial membrane has been credited to the occurrence of lipoprotein complexes added to the cuticle prior to apolysis in the beetle *Tenebrio molitor* (Delachambre, J., 1967). In any event, the physiological significance of the apolysial membrane remains an enigma. It occurs in some insects and is absent from others. Following apolysis and the secretion of the enzymes responsible for lysis of the endocuticle into the apolysial space, the old cuticle may be substantially digested or apparently not at all (Zacharuk, R., 1976). The fate of those portions of endocuticle which are lysed is unknown. It could be that such materials are recycled in new cuticular synthesis but unequivocal supporting evidence is still wanting.

Decades of study have so often made the cuticle seem to be a physically inert and intractable substance that it is gratifying to record information as to the intense cycles of activity associated with renewal of the insect skeleton, work on the butterfly *Calpodes* particularly associated with Locke and colleagues (Locke, vol. 2).

The main sequence of pharate cuticle synthesis is shown in Fig. 38 and is briefly summarized as follows. In this insect the first indication of moulting is the production and release of apolysial (= ecdysial) droplets which become dispersed amongst endocuticular lamellae. Prior to lysis of the old cuticle the epidermis switches over to the secretion of the outer epicuticle. Once the outer epicuticle has become tanned the proenzymes of the apolysial droplets are activated and enzymolysis of the old endocuticle commences. The epidermis then switches to the production of the inner epicuticle. Finally the epidermis switches to the production of a procuticle and to the stabilization of cuticle through various tanning pathways.

Rather than present a pastiche of fragmentary observations from various species, we summarize M. Locke's (1976) elegant essay in which the secretion and synthesis of the numerous components of the pharate cuticle are rationalized as the alternation of two different forms of epidermal secretory activity. These concern the plasma membrane plaques associated with the elaboration of the outer epicuticle and procuticle on the one hand, and the Golgi complex whose secretory vesicles are associated with the inner epicuticle and the apolysial droplets (Locke, M., 1967). The Golgi complex serves the additional roles of enzyme supply to multivesicular bodies and these contribute to the cyclical renewal of the apical plasma membrane and other components of the cuticle.

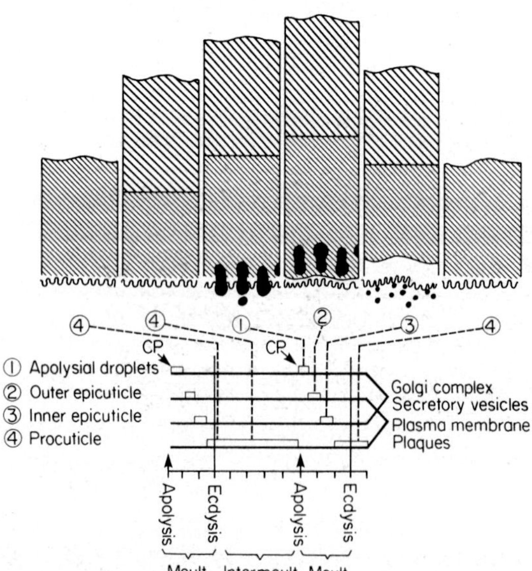

FIG. 38. Schematic representation of the events associated with apolysis and ecdysis in insects. There is an alternation in the activities of the plasma membrane plaques and the Golgi complex. CP indicates the critical period for prothoracic gland activity, after which the sequences of moulting will proceed independently of the prothoracic glands. (Slightly modified after Locke, M., 1976.)

In a series of studies Locke has shown that although numerous proteins of the cuticle may have very different functions (inner and outer epicuticles, apolysial droplets, phenolases and peroxidases), the origins of these proteins, whenever they have been investigated, are derived from complex vesicles of the Golgi apparatus (Locke, M. 1969a,b; Locke, M. and Krishnan, N., 1973). The formation and appearance of the apolysial droplets indicates the beginning of the moult period. (Fig. 39).

The production of the outer epicuticle represents a shift in epidermal secretory activity from that of Golgi complex vesicle formation to one of plasma membrane plaque activity. The outer epicuticle arises from specialized areas of the cell membrane, the so-called plasma membrane plaques. Discrete plates of outer epicuticle gradually increase in size until they fuse and the epidermal surface becomes completely covered by the outer epicuticle (Fig. 4) (Locke, M., 1976). However, it is not clear whether the plasma membrane plaques serve as physical templates for outer epicuticle formation or whether they transport precursor material to the site of chemical assembly. On completion of the outer epicuticle the surface of the cell becomes separated from the pharate cuticle, the secretory vesicles from the Golgi complex are released between microvilli to form the inner epicuticle. Finally epidermal activity changes once more and the plasma membrane plaques (which have remained intact but quiescent during inner epicuticle formation) come to the fore in the formation of procuticular fibrous cuticle, possibly the chitin microfibrils themselves (Locke, M., 1976).

During the moulting process many events naturally follow on the apolysis of a given region — the synthesis of the new and partial degradation of the old cuticle. The fate of the old cuticle during moulting has probably been the least studied aspect of metamorphosis and investigations of cast exuvia are extremely rare. The extent to which the old cuticle is digested is variable. In the larvae of the click beetle, *Ctenicera*, about one-half of the old

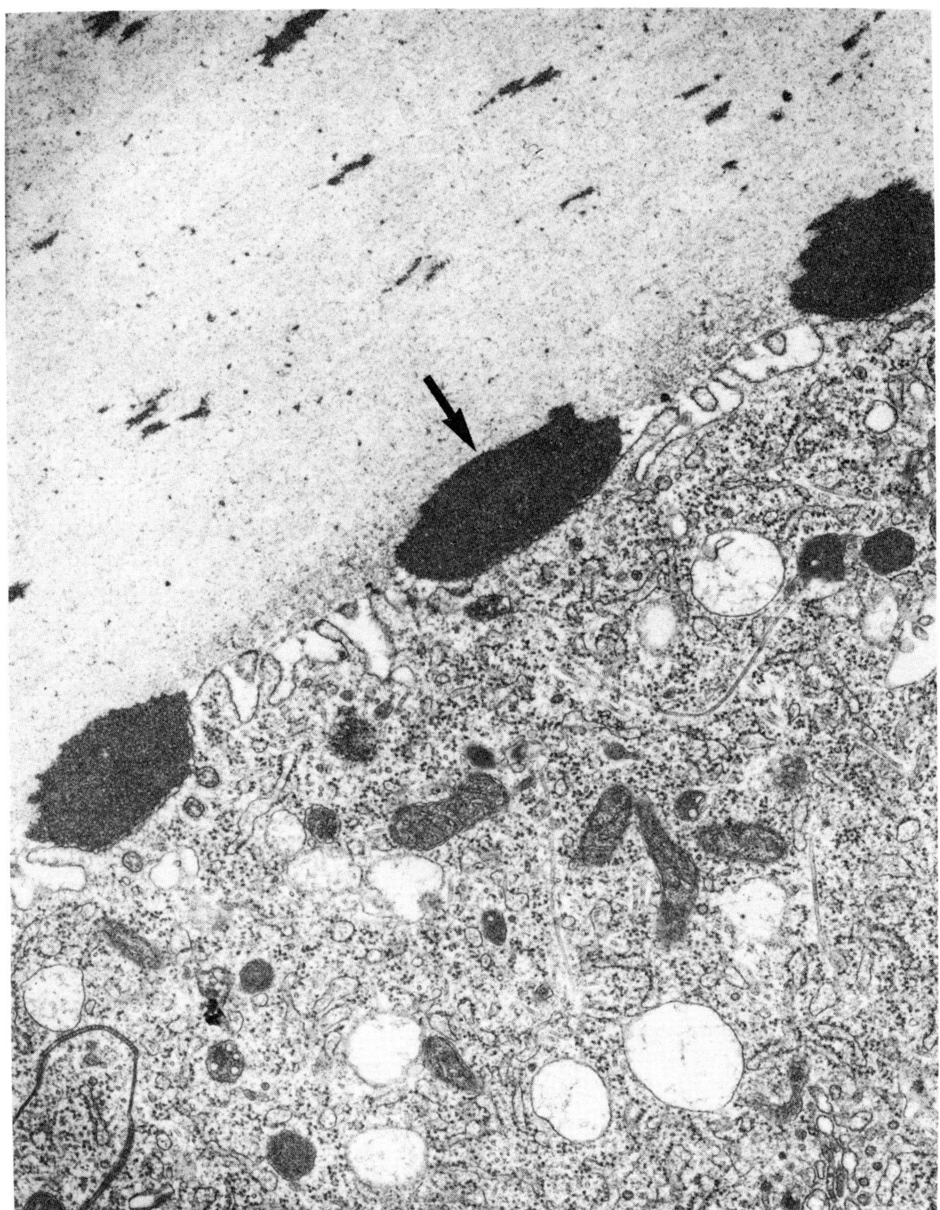

FIG. 39. Electron photomicrograph showing secretion of ecdysial droplets (arrowed). This signals the onset of the moulting cycle. (Courtesy M. Locke.)

cuticle is degraded, including material that would histologically be defined as endocuticle and the last secreted layers of mesocuticle (Zacharuk, R., 1972). The situation is somewhat different in flies. In *Drosophila* substantial amounts of procuticle are degraded (Mitchell, H. *et al.*, 1971) while in *Lucilia cuprina* apparently none of the endocuticular lamellae are broken down completely (Filshie, B., 1970). Similarly, the fates of the cuticle digests have not been unequivocally determined. It has been suggested that they are reabsorbed through the body surface before ecdysis (Wigglesworth, V., 1973) or that they are drunk by the insect at ecdysis (Wachter, S., 1930; Zacharuk, R., 1972).

It would appear that there are no changes in the epicuticle of the presumptive exuvial cuticle during moulting (Zacharuk, R., 1976). Other structures of direct concern involving the cuticle during moulting are the tracheae and tonofibrillae or muscle attachments. J. Noble-Nesbitt's (1963) ultrastructural studies of *Podura aquatica* indicate that tonofibrillae are resistant to moulting (apolysial) fluid enzymes and that they maintain their attachments to the old cuticle until ecdysis is reached, or nearly so.

6 EPIDERMIS

6.1 General

The epidermis usually forms a continuous sheet of single-layered cells beneath the cuticle as well as covering skeletal invaginations such as apodemes, fore- and hindguts and the inner surfaces of evaginated structures such as wings, gills, styli, spurs, etc. (An exception is *Oncopeltus* which has two layers of cells below the procuticle — Richards, personal communication; Neville, A., 1975). In some insects the number of epidermal cells appears to remain constant; instead of dividing mitotically they increase in size as the whole larva grows as in cyclorrhaphan flies and in many Hymenoptera (Locke, M., 1974). In yet other insects, such as the nematoceran flies, the epidermal cells both divide mitotically with larval growth but also increase in size (Hinton, H., 1961). Although the epidermis is differentiated tissue, it nevertheless performs many of the kinds of functions common to other cells of other organ systems — metabolism for self-maintenance, the production of organelles and chemical synthesis and secretion.

There are other interesting, if not unique, properties of the epidermis to consider. For example, the population density of cells in the epidermis is regulated through control of mitosis (Wigglesworth, V., 1937) with possible contributions from programmed cell death (Lockshin, R. and Beaulaton, J., 1974). Extensive studies of polarity and pattern, mainly in the cone-nosed bug *Rhodnius prolixus* (Wigglesworth, V., 1959; Lawrence, P., 1970), indicate that epidermal cells recognize both their position and orientation within the segmental gradient of the animal. While the basis of this ability remains to be more fully documented, it appears to partially derive from the transfer of information between cells that are electrochemically polarized and structurally asymmetrical with respect to junctional membrane contacts between cells (Loewenstein, W. and Kanno, Y., 1964; Caveney, S., 1976, vol. 2). Epidermal cells are likewise capable of site detachment and amoeboid-like movement in the repair of cuticular wounds (Wigglesworth, V., 1937, 1959). Other kinds of movement vary from the changing of shape in the control of lens aperture to movements towards tracheoles under conditions of hypoxia (Wigglesworth, V., 1937, 1959). More obviously, the epidermal cells contribute much to ionic homeostasis, the synthesis of waxes, greases and the numerous instances of exocrine secretions such as pheromones and compounds used for defence.

The considerable versatility of epidermal cells is modulated by a variety of both extrinsic and intrinsic factors. The former include the numerous humoral factors and hormones that percolate through the haemocoel (see volumes 7 and 8), the population size of cells, physiological and health status of neighbours, wounds, disease and parasites, intercellular interactions, etc. The effects of water, temperature and humidity on the cellular milieu is well known, as also are trophic factors and changes associated with metamorphosis. Intrinsic factors are similarly numerous and include structural asymmetry of the individual cells, hence variations in membrane permeability. Age, health, nutritional status and ionic conditions also affect both activeness and potential responsiveness of individual epidermal cells.

6.2 Intercellular communication

6.2.1 IONIC COUPLING

Since the wonderful observations of H. V. Wilson (1907) that mechanically isolated sponge cells will spontaneously coalesce, all sorts of biologists have been concerned with the means by which animal cells "recognize" one another and communicate. In more modern times, precisely how is functional coordination achieved in any tissue? The insect epidermis is a more restricted case because it is almost invariably a single cell thick; but of great interest because of the many different functions simultaneously performed by different areas of the same tissue. For example, given apparently excess amounts of both enzymes and the appropriate substrates, different integumental regions of the locust, *Schistocerca gregaria*, produce overlying cuticles which vary in thickness, extent of tanning, and hence stiffness (Andersen, S. 1974a,b); or old honeybees may return to the synthesis of wax if colony demand requires such services. Similarly, Wichard, W. and Komnick, H. (1971) have shown that the epidermal cells of the gills of mayflies are specialized so that some are involved in chloride transport; others are not.

Recent electrophysiological studies by Caveney and his colleagues have opened an entirely new and very promising line of research in understanding how cells communicate (see Caveney, vol. 2). The basic observation comes from the studies of Loewenstein, W. and Kanno, Y. (1964) who showed that the epithelial cells of *Drosophila* salivary glands are electrically coupled. In extensions of this work, the following observations emerge.

Each cell in the larval abdominal sternites of the beetle, *Tenebrio molitor*, is defined by a plasma membrane which has four specialized regions: an apical plasma membrane, junctional membrane, a lateral non-contact membrane and the basal plasma membrane (Fig. 40). Adjacent epidermal cells are attached to one another via the specialized junctional membrane region of the cells. This junctional contact is further differentiated into septate and gap junctions, and it is thought that the gap junction is the most likely channel through which cells can communicate (Caveney, S. and Podgorski, C., 1975).

Caveney, S. (1976) has recently suggested that, because adjacent cells are coupled ionically when in membrane contact, the epidermis can be visualized as a functional, if not anatomical, syncytium. A further property of this arrangement is that electrical resistance between contacting regions is relatively low *vis-à-vis* non-contact regions, so that any movement of ions from one cell to another is an anisotropic one. Because of the relatively high resistance between two cells not in junctional membrane contact, there is likely to be little leakage of ions into such extracellular spaces (Caveney, S., 1976).

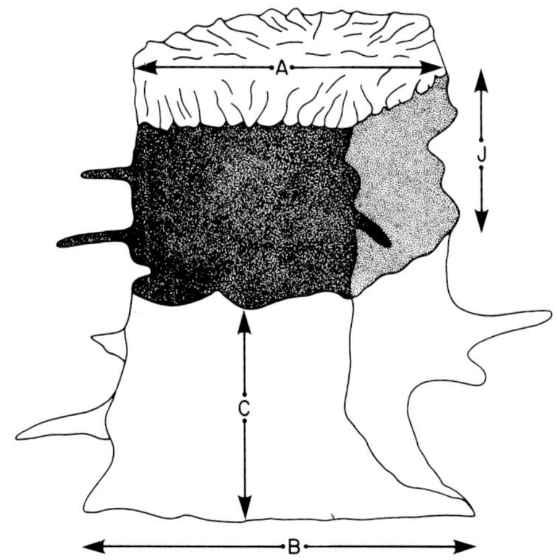

FIG. 40. Surface of an epidermal cell as defined by its differentiated plasma membranes. The junctional membrane area is stippled (J) while the other three areas are non-junctional and include the apical plasma membrane (A), the lateral non-contact membrane (C) and the basal membrane (B). (After Caveney, S. and Podgorski, C., 1975.)

The extent to which a series of cells might exert an influence is limited by the distances between them, since the magnitude of the potentials is directly related to the distance between stimulating and recording electrodes. Nonetheless, the general resistance of these pathways is low, as evidenced by the fact that the movement of ions over about 25 cells has been recorded (Caveney, S., 1974).

In a compelling series of measurements, Caveney, S. (1976) has shown that ionic conduction very likely fluctuates during the pupal-adult metamorphosis of *Tenebrio molitor*, as evidenced by great changes in the specific resistance of non-junctional membrane, resistance of the intercellular pathway, and in membrane potential (Fig. 41). Moreover, Hax, W. *et al.* (1974) have shown that both 20-hydroxyecdysone and cAMP increase the resistance between membranes in contact, and that juvenile hormone decreases the resistance between membrane regions not in contact. The electrical resistance of non-junctional membranes is greatly affected by changing concentrations of potassium while junctional contact regions are less susceptible. Finally, the ionic flow between junctional membranes can be reduced if calcium concentrations in the epidermal cytoplasm are increased (Caveney, S., 1976).

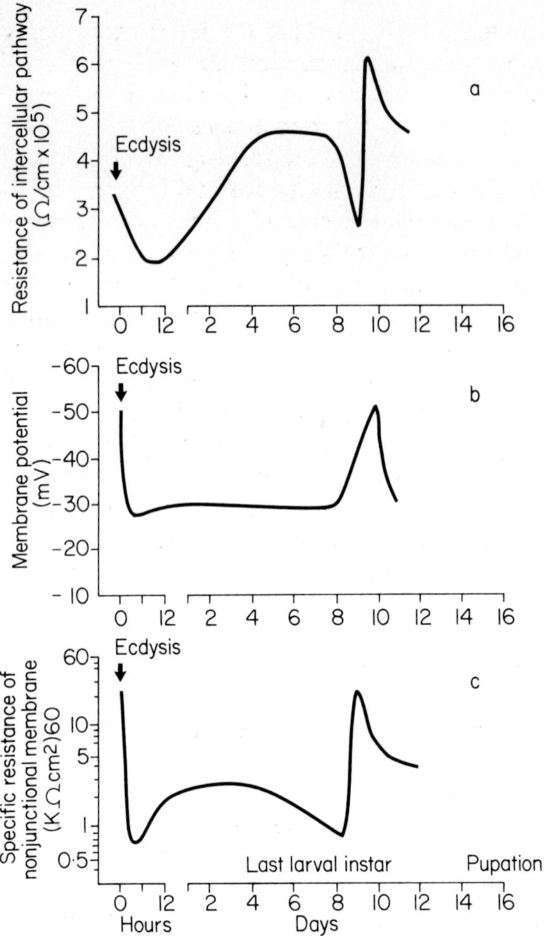

FIG. 41. Changes in the ionic properties of the developing epidermis of *Tenebrio molitor*. (From Caveney, S., 1976.)

Recalling the wound experiments of Wigglesworth, V. (1937) one might wonder what would happen to the normal function of this ionic network given some perturbation to the system: given the death of one cell in a series of connected epidermal cells, might large areas break down? This problem is apparently overcome by a general shut-down of connecting membrane regions, a response mediated, at least in part, by mitochondrial and endoplasmic reticular stores of calcium (Caveney, S., 1976).

The above observations show that charged particles pass through neighbouring cells and these electrical signals could well provide a means of communication. Future research in this area might well shed light on the regulation of growth and morphogenesis generally: polarity of cuticular patterns, establishment of segmental gradients or the arrangements of chitin microfibrils.

The transfer of electrical and chemical signals in the surface plane of cuticle can be taken as unequivocal. In addition to these two dimensions it is very probable that cuticle is also a piezoelectric material (Scheie, P., 1980). If so, this would mean that local deformations of the cuticle would likewise produce electrical signals and provide a third dimension of information for the underlying epidermis. A possible role for cuticle to make electrical contributions is also indicated, if indirectly, in A. Richards's (1957) observations that the conductance properties of *Sarcophaga* and *Blaberus* cuticles are reduced if the epidermis is removed from the test preparation. One can envisage piezoelectric transduction as an important part of a three-dimensional electric grid in an intact integument . . . but we are left only speculation in the absence of careful measurement.

6.2.2 SEGMENTAL GRADIENTS

The possible transfer of information — electrical, chemical, etc. — between cells within a tissue has been most extensively investigated using transplantation and grafting techniques (Piepho, H., 1955a,b; Wigglesworth, V., 1959). Such studies on the insect integument have been greatly facilitated by the fact that insect material is largely free of the incompatibility and rejection problems so commonly encountered in mammals. Thus, while we find that it is readily possible to perform interspecific transplants, we are left with this anomaly: in *Rhodnius prolixus* (principal experimental animal for such studies) there is a positional incompatibility problem in that grafted material is far more easily accepted within the same lateral plane than in the longitudinal axis of the insect (Locke, M., 1959, 1967).

In Locke's experiments (1959) on *Rhodnius* he discovered that pieces of larval abdominal sternites taken from one individual and placed in exactly the same place in the recipient as they had been in the donor would be completely acceptable and result in a normal cuticular pattern. However, when he rotated the donor cuticle through 180° before inserting it in the recipient the resulting transplant will

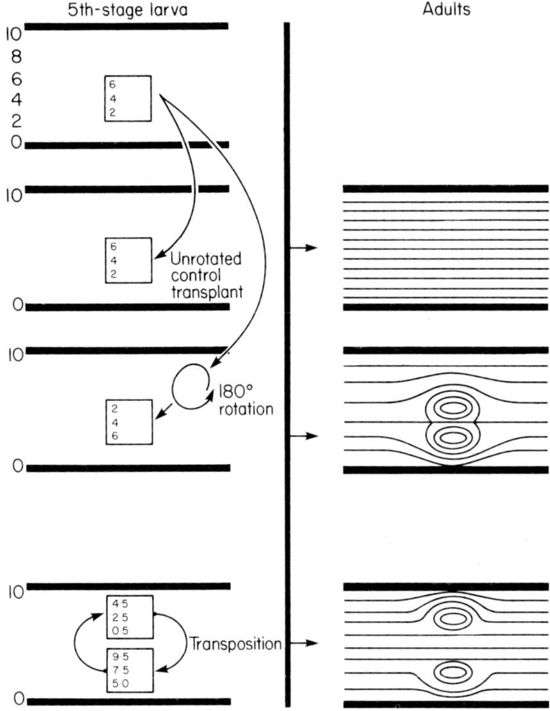

FIG. 42. A summary of Locke's experiments on *Rhodnius prolixus*. The numbers (0 to 10) indicate the relative position of points on the cuticle with respect to the longitudinal axis of the insect. Stylized ripple patterns that develop in the adult are to the right of the figure. (From Locke, M., 1959.)

take, but marked discontinuities in the cuticular pattern occur. These observations (Fig. 42) imply that smooth continuity of surface pattern requires continuity of similar epidermal contacts. In a slightly different experiment, intersegmental membranous cuticle was transplanted into segmental cuticle and again discontinuities of surface were observed (Locke, M., 1960a; Lawrence, P., 1966). Finally, polarity of pattern is clearly maintained in epidermal cells, as evidence by cuticular pattern following transposition transplants (Fig. 42).

These collective observations certainly suggest that epidermal cells have, receive and send information as to their whereabouts. If pattern and polarity were determined to any great extent by extrinsic humoral factors quite different results could have been expected. What is most gratifying is that the transfer of information between epidermal cells in the transplantation experiments is entirely consistent with the polarized and anisotropic distribution of junctional membrane complexes described by Caveney, S. (1976). Like the latter experiments, transplantation studies have not yet incorporated possible piezoelectric contributions impinging on the epidermis in the third dimension.

The extensive and appealing experiments of Wigglesworth, Piepho, Locke, and Lawrence clearly point to the existence of a repeating segmental gradient. Cells are "alike" side to side but not with respect to the longitudinal axis of an insect. The characteristic variant is the recognition of position within a linear axial order. Pieces of integument that are displaced from this order react to form a graft pattern that is predictable on the basis of their origin and final placement in a transplantation experiment (Lawrence, P., 1970). This reconfirms the interpretation that epidermal cells have what might be called "neighbourhood characteristics", and that each neighbourhood in turn has its own electrical grid system.

6.2.3 TRACHEAL GROWTH

Tracheae are in general structurally similar to exoskeletal cuticle, having an epicuticle, procuticle and surrounding epidermal epithelium (Richards, A., 1951). Our principal interest in tracheae concerns what seem to be unique aspects of tracheal formation and growth. It has long been known that continuity of the epithelium is necessary for the growth of tracheae; even so, growth proceeds in a peculiar way. In a paurometabolous insect such as *Rhodnius prolixus*, the diameter of a trachea increases at each moult and the rate of increase of the main tracheae is equal to the rate of increase in terminal branches of the tracheal system (Locke, M., 1958).

Control of tracheal diameter is usually attributed to the synthesizing epidermis but it would be of interest to know how the epidermal cells "decide" how large the diameter ought to be. Ryerse, J. and Locke, M. (1978) have added to this the observations that tracheal cuticle deposition can be induced *in vitro* with 20-hydroxyecdysone, but this hormone does not facilitate increase in tracheal diameter. Similarly, how are the relative rates of growth coordinated to produce the observed geometrical relationships? The growth of tracheae begins with the secretion of cuticle near to the spiracles and then proceeds into the haemocoel.

It would be of interest to determine whether or

not the distribution of junctional and non-junctional plasma membrane regions of tracheal cells is similar to that of "normal" epidermal cells, and to establish whether or not ionic coupling of tracheal cells reflects the direction of growth in these structures.

6.3 Growth and size

An insect cuticle can change in pattern, form and size during the course of metamorphosis. Such changes are inevitably related to major events in the epidermis. Among these, there are major variations in the rates and kinds of lytic and synthetic activities, cell turnover through ageing and death (Lockshin, vol. 2) and mitoses. Most of these activities are greatly affected by the distribution, kinds and amounts of hormones in a given region of the body. For example, juvenile hormone both directs the kinds of synthesis performed by certain cells as well as affects the rate of cell death (see vol. 8 of this series). Moreover, it has also been suggested that there is an inverse relationship between the levels of moulting hormone and cAMP, and a positive correlation between bursicon activity in the haemolymph and epidermal cAMP with the implication that cAMP might mediate bursicon production and release (Delachambre, J. *et al.*, 1979). Against this, it might also be the case that moulting hormone regulates cAMP levels by depressing them, and bursicon by elevating them. So, it is conceivable that bursicon might mediate cAMP production and not vice-versa. This problem remains unresolved.

Cell turnover would seem to be a very pertinent but unstudied aspect of growth and size (but see Kato, Y. and Oba, T., 1977). Since the ultimate size of any instar must eventually depend upon the amount of material synthesized by the underlying epidermis, the ratio of functional to non-functional cells ought to be considered. Such information would be of considerable interest. Another problem attending the study of growth is how to measure growth in insects. Growth of the exoskeleton (increase in surface area) can be restricted to post-ecdysial expansion prior to hardening or can be intra-stadial in nature — intussusceptive (Locke, M., 1974) or creep (Hackman, R., 1976). Mitotic activity in the epidermis is initially associated with a relative reduction in the amount of nuclear material as the volume of the cells increase in size before telophase.

Even though apolysis, for example, proceeds along an anteroposterior gradient, hence the epidermal cells of the head region are functioning differently to those in the abdomen, it has recently been shown that even adjacent cells in the abdomen of the moth, *Hyalophora cecropia*, exhibit differences in organelle content and staining intensity so that even small populations of epidermal cells are not homogeneous (Sedlak, B. and Gilbert, L., 1976; Wielgus, J. and Gilbert, L., 1978).

Also recently some of the intracellular changes that occur during growth have been documented, mainly for *Calpodes ethlius* (Locke, M., 1976; Dean, R. *et al.*, 1980; Locke, vol. 2). Cuticle deposition and composition are controlled by dynamic changes in the secretion of synthesized products as well as by the uptake of various molecules. Synthesis is cyclical in nature, and this is related to the turnover of the apical plasma membranes (Locke, M. and Huie, P., 1979). The turnover of these membrane plaques is apparently related to moult/intermoult cycles and not to the kind of cuticle being secreted. The cyclical formation and involution of membrane plaques is synchronous with the hormonal control of moulting. The dissolution of plaques is effected by lytic enzymes elaborated in the Golgi complex and carried in lysosomes to multivesicular bodies. The latter digest fragments of the plasma membrane. The Golgi complex is also the source of apolysial droplets which digest portions of the old cuticle after apolysis. Thus, the apical plasma membrane is in a dynamic state of cyclical formation through the addition of new plaques and loss through endocytosis.

The cyclical behaviour of plaque formation and dissolution are aspects of epidermal function that are affected by hormones (Dean, R. *et al.*, 1980). Indeed, both epidermal cell morphology and cuticle production are influenced both by 20-hydroxy-ecdysone and juvenile hormone (Sedlak, B. and Gilbert, L., 1976). In *Manduca sexta* apolysis is associated with high ecdysteroid titres and larval ecdysis with high juvenile hormone titres (Sedlak, B. and Gilbert, L., 1979). Whitten, J. (1969a,b) has found polytene chromosomes in the pupal epidermis of *Sarcophaga bullata* specifically related to the synthesis of the entire dorsum of the pulvillus. These observations are quite encouraging as they hold the possibility of locating the precise region of the genome responsible for the synthesis of individual constituents of the cuticle. (For greater details of growth see volumes 2, 7 and 8 of this series.)

7 BASEMENT MEMBRANE

In most insects there is a layer of material immediately below and connected to the overlying epidermis (Richards, A., 1951). This layer of the integument is undoubtedly the one about which we know least (see Ashhurst, this volume). On the basis of electron micrographs it would appear that the basement membrane consists of collagen fibres in *Locusta migratoria* (Rinterknecht, E. and Levi, P., 1966) and in a beetle, *Hydrocyrius columbiae* (Neville, A., 1975). The origins and formation of the basement membrane in *Rhodnius prolixus* have been studied electron microscopically by Wigglesworth, V. (1973) who concluded that it derives from plasmatocytes in the haemolymph.

8 CONCLUDING REMARKS

The incredible variation to be found in the integuments of insects ultimately derives from simple principles. Extracting these principles from the integument is, of course, our main task. As I see the literature on the insect integument, there are many interesting and important problems to be resolved. For example, the general applicability of the Bouligand model to insect cuticles is in doubt and more work is needed here. We know nothing of the molecular conformations of cuticular proteins and such knowledge is essential to an understanding of the physical chemistry of tanning. We know virtually nothing of the significance of the basement membrane or of procuticular lipids. We do not know to what extent formation of cuticle is a phenomenon of crystallization or of cell-directed synthesis and assembly. While exciting discoveries have been made concerning the electrophysiology of the epidermis, this work is in an embryonic phase and sorely in need of greater efforts. Is piezoelectric transmission a viable route by which information can enter the epidermis? How is resilin synthesized? Those concerned with insect pest management will want to know the mode of action of insecticides inhibiting chitin synthesis. What we have learned to date holds more for titillation rather than education of the intellect.

REFERENCES

AGUI, N. (1977). Time studies of ecdysone action on *in vitro* apolysis of *Chilo suppressalis* integument. *J. Insect Physiol. 23*, 837–842.

AGUI, N., YAGI, S. and FUKAYA, M. (1969). Induction of moulting of cultivated integument taken from diapausing rice stem borer larva in the presence of ecdysterone. *Appl. Ent. Zool. 4*, 156–157.

ANDERSEN, S. O. (1963). Characterization of a new type of cross-linkage in resilin, a rubber-like protein. *Biochem. Biophys. Acta. 69*, 249–262.

ANDERSEN, S. O. (1964). The cross-links in resilin identified as dityrosine and trityrosine. *Biochem. Biophys. Acta 93*, 213–215.

ANDERSEN, S. O. (1966). Covalent cross-links in a structural protein, resilin. *Acta Physiol. Scand. 66*, (Suppl. 263), 1–81.

ANDERSEN, S. O. (1970). Isolation of arterenone (2-amino-3′,4′-dihydroxyacetophenone) from hydrolysates of sclerotized insect cuticle. *J. Insect Physiol. 16*, 1951–1959.

ANDERSEN, S. O. (1971). Resilin. In *Extracellular and Supporting Structures*. Edited by M. Florkin and E. H. Stotz. *Comprehensive Biochemistry*. Vol. 26c, pages 633–657.

ANDERSEN, S. O. (1973). Comparison between the sclerotization of adult and larval cuticle in *Schistocerca gregaria. J. Insect Physiol. 19*, 1603–1614.

ANDERSEN, S. O. (1974a). Cuticular sclerotization in larval and adult locusts, *Schistocerca gregaria. J. Insect Physiol. 20*, 1537–1552.

ANDERSEN, S. O. (1974b). Evidence for two mechanisms of sclerotization in insect cuticle. *Nature 251*, 507–508.

ANDERSEN, S. O. (1975). Cuticular sclerotization in the beetles *Pachynoda epphipiata* and *Tenebrio molitor. J. Insect Physiol. 21*, 1225–1232.

ANDERSEN, S. O. (1979). Biochemistry of insect cuticle. *Ann. Rev. Ent. 24*, 29–61.

ANDERSEN, S. O. (1981). The stabilization of locust cuticle. *J. Insect Physiol. 27*, 393–396.

ANDERSEN, S. O. and BARRETT, F. M. (1971). The isolation of ketocatechols from insect cuticle and their possible role in sclerotization. *J. Insect Physiol. 17*, 69–83.

ANDERSEN, S. O. and WEIS-FOGH, T. (1964). Resilin, a rubber-like protein in arthropod cuticle. *Adv. Insect Physiol. 2*, 1–65.

ANDERSEN, S. O., CHASE, A. M. and WILLIS, J. H. (1973). The amino-acid composition of cuticles from *Tenebrio molitor* with special reference to the action of juvenile hormone. *Insect Biochem. 3*, 171–180.

ANDERSEN, S. O., THOMPSON, P. R. and HEPBURN, H. R. (1981). Cuticular sclerotization in the honeybee (*Apis mellifera adansonii*). *J. Comp. Physiol. 145*, 17–20.

ARMOLD, M. T., BLOMQUIST, G. J. and JACKSON, L. L. (1969). Cuticular lipids of insects. III. The surface lipids of the aquatic and terrestrial life forms of the big stonefly *Pteronarcys californica* (Newport). *Comp. Biochem. Physiol. 31*, 685–692.

ASTBURY, W. T. and WOODS, H. J. (1933). X-ray studies of the structure of hair, wool and related fibres — II. The molecular structure and elastic properties of hair keratin. *Phil. Trans. Roy. Soc. (A), 232*, 333–394.

ATKINS, E. D. T. (1967). A four-strand coiled-coil model for some insect fibrous proteins. *J. Mol. Biol. 24*, 139–141.

ATKINS, E. D. T., FLOWER, N. E. and KENCHINGTON, W. (1966). Studies on the oothecal protein of the tortoise beetle, *Aspidomorpha. J. Roy. Mic. Sci. 86*, Pt. 2. 123–135.

BADE, M. L. (1978). Enzymatic breakdown of the chitin component in insect cuticle during moult. In *Proceedings of the First International Conference of Chitin/Chitosan*. Edited by R. R. A. Muzzarelli and E. R. Pariser. Pages 472–482. Massachusetts Institute of Technology, Cambridge, Mass.

BADE, M. L. and STINSON, A. (1978). Digestion of cuticle chitin during the moult of *Manduca sexta*. (Lepidoptera: Sphingidae). *Insect Biochem. 9*, 221–231.

BAILEY, K. and WEIS-FOGH, T. (1961). Amino acid composition of a new rubber-like protein, resilin. *Biochim. Biophys. Acta 48*, 452–459.

BAKER, G., PEPPER, J. H., JOHNSON, L. H. and HASTINGS, E. (1960). Estimation of the composition of the cuticular wax of the mormon cricket, *Anabrus simplex*. *J. Insect Physiol. 5*, 47–60.

BARBIER, R. (1971). Recherches sur la morphogenèse tegumentaire d'un insecte holométabole, *Galleria mellonella* L. (Lepidoptère Pyralidae). Thèses de Docteur Es-Sciences Naturelles, Université de Rennes, U.E.R. des Sciences Biologiques. Série C, Ordre 122, Série 42. (original not seen).

BARTH, F. G. (1972a). Die Physiologie der Spaltsinnesorgane I. Modellversuche zur Rolle des cuticularen Spaltes beim Reiztransport. *J. Comp. Physiol. 78*, 315–336.

BARTH, F. G. (1972b). Die Physiologie der Spaltsinnesorgane. II. Funktionelle Morphologie eines Mechanorezeptors. *J. Comp. Physiol. 81*, 159–186.

BARTH, F. G. (1973). Bauprinzipien und adäquater Reiz bei einem Mechanorezeptor. *Verh. Dt. Zool. Ges. 66*, 25–30.

BARTH, F. G. (1976). Sensory information from strains in the exoskeleton. In *The Insect Integument*. Edited by H. R. Hepburn. Pages 445–473. Elsevier, Amsterdam.

BARTH, F. G. and PICKELMANN, P. (1975). Lyriform slit sense organs. Modelling an arthropod mechano-receptor. *J. Comp. Physiol. 103*, 39–54.

BARTNICKI-GARCIA, S., BRACKER, C. E. and RUIZ-HERRERA, J. (1978). Synthesis of chitin microfibrils in vitro by chitin synthetase particles, chitosomes, isolated from *Mucor rouxii*. In *Proceedings of the First International Conference on Chitin/Chitosan*. Edited by R. R. A. Muzzarelli and E. R. Pariser. Pages 450–463. Massachusetts Institute of Technology, Cambridge, Mass.

BEAMENT, J. W. L. (1945). The cuticular lipids of insects. *J. Exp. Biol. 21*, 115–131.

BEAMENT, J. W. L. (1946). The formation and structure of the chorion of the egg in an hemipteran, *Rhodnius prolixus*. *Quart. J. Mic. Sci. 87*, 393–439.

BEAMENT, J. W. L. (1955). Wax secretion in the cockroach. *J. Exp. Biol. 32*, 514–538.

BEAMENT, J. W. L. (1959). The waterproofing mechanism of arthropods. The effect of temperature on cuticle permeability in terrestrial insects and ticks. *J. Exp. Biol. 36*, 391–442.

BENNET-CLARK, H. C. (1962). Active control of the mechanical properties of insect endocuticle. *J. Insect Physiol. 8*, 627–633.

BENNET-CLARK, H. C. (1963). The relation between epicuticular folding and the subsequent size of an insect. *J. Insect Physiol. 9*, 43–46.

BENNET-CLARK, H. C. (1976). Energy storage in jumping insects. In *The Insect Integument*. Edited by H. R. Hepburn. Pages 421–443. Elsevier, Amsterdam.

BENNET-CLARK, H. C. and LUCEY, E. C. A. (1967). The jump of the flea: a study of the energetics and a model of the mechanism. *J. Exp. Biol. 47*, 59–76.

BERTRAM, D. S. and BIRD, R. G. (1961). Studies on mosquito-borne viruses in their vectors. I. The normal fine structure of the mid-gut epithelium of adult female *Aedes aegypti* (L.) and the functional significance of its modification following a blood meal. *Trans. Roy. Soc. Trop. Med. Hyg. 55*, 404–423.

BLACKWELL, J., MINKE, R. and GARDNER, K. H. (1978). Determination of the structures of α- and β-chitins by X-ray diffraction. In *Proceedings of the First International Conference on Chitin/Chitosan*. Edited by R. R. A. Muzzarelli and E. R. Pariser. Pages 108–123. Massachusetts Institute of Technology, Cambridge, Mass.

BOETTIGER, E. G. and FURSHPAN, E. (1952). The mechanics of flight movements in Diptera. *Biol. Bull., Woods Hole. 102*, 200–211.

BORDEREAU, C. and ANDERSEN, S. O. (1978). Structural cuticular proteins in termite queens. *Comp. Biochem. Physiol. 60B*, 251–256.

BOULIGAND, Y. (1965). Sur une architecture torsadée répandue dans de nombreuses cuticles d'Arthropodes. *C.R. Hebd. Séanc. Acad. Sci., Paris. 261*, 3665–3668.

BOULIGAND, Y. (1972). Twisted fibrous arrangements in biological materials and cholesteric mesophases. *Tissue Cell 4*, 189–217.

BOWERS, W. S. and THOMPSON, M. J. (1965). Identification of the major constituents of the crystalline powder covering the larval cuticle of *Samia cynthia ricini*. *J. Insect Physiol. 11*, 1003–1011.

BURSELL, E. and CLEMENTS, A. N. (1967). The cuticular lipids of the larva *Tenebrio molitor*. L. (Coleoptera). *J. Insect Physiol. 13*, 1671–1678.

CANDY, D. and KILBY, B. (1962). Studies on chitin synthesis in the desert locust. *J. Exp. Biol. 39*, 129–140.

CARLSTRÖM, D. (1957). The crystal structure of α-chitin (poly-N-acetyl-D-glucosamine). *J. Biophys. Biochem. Cytol. 3*, 669–683.

CAVENEY, S. (1974). Intercellular communication in a positional field: Movement of small ions between epidermal cells. *Dev. Biol. 40*, 311–322.

CAVENEY, S. (1976). The insect epidermis: a functional syncytium. In *The Insect Integument*. Edited by H. R. Hepburn. Pages 259–274. Elsevier, Amsterdam.

CAVENEY, S. and PODGORSKI, C. (1975). Intercellular communication in a positional field. Ultrastructural correlates and tracer analysis of communication between insect epidermal cells. *Tissue Cell. 7*, 559–574.

CHAPMAN, K. M., DUCKROW, R. B. and MORAN, D. T. (1973). Form and role of deformation in excitation of an insect mechanoreceptor. *Nature. 244*, 453–454.

CHU, H., NORRIS, D. M. and CARLSON, S. D. (1975). Ultrastructure of the compound eye of the diploid female beetle, *Xyleborus ferrugineus*. *Cell Tiss. Res. 165*, 23–26.

DALINGWATER, J. E. (1975a). SEM observations on the cuticle of some decapod crustaceans. *Zool. J. Linn. Soc. 56*, 327–330.

DALINGWATER, J. E. (1975b). The reality of arthropod cuticular laminae. *Cell. Tiss. Res. 163*, 411–413.

DEAN, R. L., BOLLENBACHER, W. E., LOCKE, M., SMITH, S. L. and GILBERT, L. I. (1980). Haemolymph ecdysteroid levels and cellular events in the intermoult/moult sequence of *Calpodes ethlius*. *J. Insect Physiol. 26*, 267–280.

DEHAAS, B. W., JOHNSON, L. H., PEPPER, J. H., HASTINGS, E. and BAKER, G. L. (1957). Proteins of Mormon cricket exoskeleton. 1. Amino acid composition of (a) pronotal shields, (b) abdominal tergites, (c) abdominal intersegmental connectives. *Physiol. Zool. 30*, 121–127.

DELACHAMBRE, J. (1967). Origine et nature de la membrane exuviale chez la nymph de *Tenebrio molitor* L. (Coleoptera). *Z. Mikr. Anat. 81*, 114–134.

DELACHAMBRE, J., DELBECQUE, J. P., PROVANSAL, A., GRILLOT, J. P., DE REGGI, M. L. and CAILLA, H. L. (1979). Total and epidermal cyclic AMP levels related to the variations of ecdysteroids and bursicon during the metamorphosis of the mealworm *Tenebrio molitor* L. *Insect Biochem. 9*, 95–99.

DENNELL, R. (1946). A study of an insect cuticle: the larval cuticle of *Sarcophaga faculata* Pand. (Diptera). *Proc. Roy. Soc. B.* 133, 348–373.

DENNELL, R. (1973). The structure of the cuticle of the shore-crab *Carcinus maenas* (L.). *Zool. J. Linn. Soc. 52*, 159–163.

DENNELL, R. and MALEK, S. (1955). The cuticle of the cockroach *Periplaneta americana* III. The hardening of the cuticle: impregnation preparatory to phenolic tanning. *Proc. Roy. Soc. B. 143*, 414–426.

DETHIER, V. G. (1963). *The Physiology of Insect Senses*. Methuen, London.

DRACH, P. (1939). Mue et cycle d'intermues chez les Crustacés décapodes. *Ann. Inst. Océan. 19* (3), 103–392.

DRUCKER, B., HAINSWORTH, R. and SMITH, S. G. (1953). *Shirley Inst. Mem. 26*, 191 (cited from Lucas, F. *et al.*, 1955 — original not seen).

EASTHAM, L. E. S. and EASSA, Y. E. E. (1955). The feeding mechanism of the butterfly *Pieris brassicae* L. *Phil. Trans. Roy. Soc. B 245*, 137–169.

EDNEY, E. B. (1947). Laboratory studies on the bionomics of the rat fleas, *Xenopsylla brasiliensis* Baker and *X. cheopis* Roths. II. Water relations during the cocoon period. *Bull. Ent. Res. 38*, 263–280.

ELLIOTT, G. F., HUXLEY, A. F. and WEIS-FOGH, T. (1965). On the structure of resilin. *J. Mol. Biol. 13*, 791–795.

ERNST, K. D. (1969). Die Feinstruktur von Riechsensillen auf der Antenne des Aaskäfers *Necrophorus*. *Z. Mikr. Anat. 94*, 72–102.

ERNST, K. D. (1972). Die Ontogenie der basiconischen Riechsensillien auf der Antenne von *Necrophorus*. *Z. Mikr. Anat.* 129, 217–236.

FILSHIE, B. K. (1970). The fine structure and deposition of the larval cuticle of the sheep blowfly *Lucilia cuprina*. *Tissue Cell 2*, 479–498.

2 Sclerotization and Tanning of the Cuticle

SVEND OLAV ANDERSEN

Angust Krogh Institute, Copenhagen, Denmark

1 INTRODUCTION

Tanning and sclerotization are terms which, when used for insect cuticle, refer to the processes by which the cuticular structure is made stiffer, less soluble, more resistant to degradation, and often darker. As a rule this transformation occurs after ecdysis, but some cuticular structures are sclerotized before ecdysis, and the dipteran puparium is the last instar larval cuticle which is sclerotized to form a protective covering within which pupal and adult development can occur. The terms tanning and sclerotization are occasionally used with a slightly different meaning (to describe whether the hardened structure has a tanned color or not), but in this chapter I shall use the two terms synonymously.

There has been an active interest in cuticle sclerotization for many years, as is evident from the many reviews which have appeared (Wigglesworth, V., 1957; Dennell, R., 1958a; Richards, A., 1958, 1967, 1978; Pryor, M., 1962; Brunet, P., 1963, 1980; Cottrell, C., 1964; Hackman, R., 1971, 1974; Andersen, S., 1976, 1977, 1979a). Several theories have been proposed to account for the changes occurring during sclerotization, and there is still no general agreement on what happens in the cuticle during the process. In the following I shall outline the main features of the controversy, and of the available evidence.

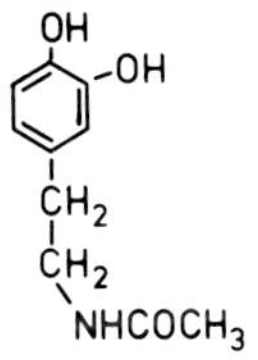

I 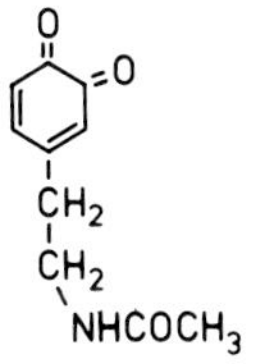 II

The generally accepted theory is based upon Pryor's suggestion (Pryor, M., 1940a,b, 1962) that cuticular proteins are cross-linked by reaction with quinones, which change the cuticle into a hard, dark, insoluble material. Some of the details in Pryor's original scheme have been modified, but the main features are still accepted. According to the current version of the scheme (Hackman, R., 1974) a metabolic product from tyrosine, the *ortho*-diphenol *N*-acetyldopamine (I), is secreted by the epidermal cells into the cuticle; in the epicuticle it meets a diphenol-oxidase, which oxidizes it to the corresponding quinone (II). The quinone diffuses into the bulk of the cuticle, where it reacts spontaneously with available amino groups (and sulfhydryl groups, if any). When a quinone reacts with an amino group the nitrogen atom is linked to the aromatic ring by a covalent bond, and the quinone is reduced to a catechol residue. The catechol groups can be non-enzymatically reoxidized to the quinone stage by means of another free quinone molecule and react with a second amino group. If this amino group is located on a second protein molecule a cross-link is established. The protein-bound quinone can also react with more quinones, thereby forming polymers, which may be able to bridge large distances between reactive groups in the proteins. Depending upon the relative proportions of diphenolic material secreted into the cuticle, and the amount of reactive groups, the material incorporated may end up in the reduced or the oxidized state. Test tube experiments (Mason, H. 1955) have indicated that when proteins are in excess, the bound material will be present in the reduced catecholic state and be lightly colored, whereas if quinones are formed in excess, the bound material will be in the oxidized state, and the product will be darkly colored. According to a suggestion by Hackman, R. (1959) the difference between dark and lightly colored, tanned cuticle could be due to the amount of quinone which is formed. The assumed reaction sequence of quinone tanning is shown in Scheme I.

The possibility that an alternative sclerotization mechanism may exist in insect cuticle was suggested by Andersen, S. and Barrett, F., in 1971. The suggestion has been further elaborated by Andersen and has recently been reviewed (Andersen, S., 1979a). To account for the structure of some degradation products obtained from sclerotized cuticle, it was suggested that the tanning agent, *N*-acetyldopamine, is not always oxidized to a quinone, but that the side-chain can be activated to give a reactive intermediate, which would link the cuticular proteins together. No suggestions were made at that time concerning the structure of the intermediate, but it was suggested that the carbon atom adjacent to the aromatic ring (the β-carbon atom) was involved in the reaction. Recent evidence has made it more likely that both the α- and the β-carbon atoms are involved (Andersen, S. *et al.*, 1980), but it seems reasonable to retain the term β-sclerotization for this mechanism. The structure of the hypothetical cross-link in its simplest form would accordingly be as shown (Structure III). It has been suggested (Andersen, S., 1974b) that the color of sclerotized cuticle can be used as an indication of the type of tanning which has occurred: β-sclerotization giving lightly colored cuticles, and quinone tanning resulting in dark brown cuticles.

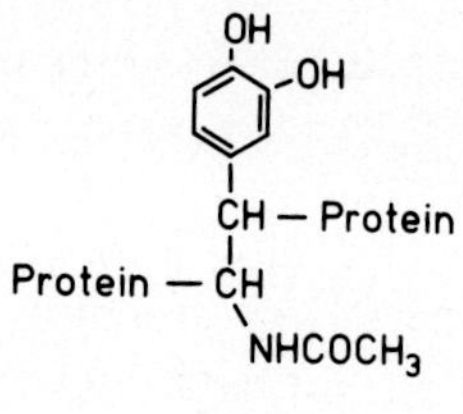

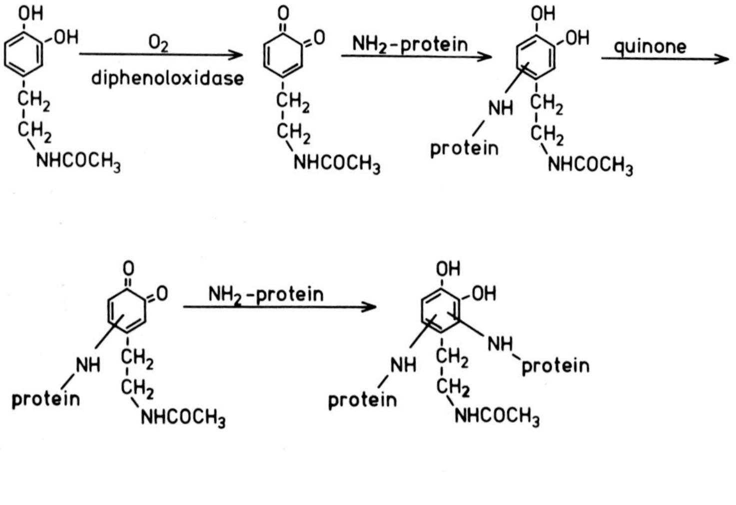

Scheme I

The main differences between the two schemes (quinone tanning and β-sclerotization) involve the exact structure of the intermediates and the detailed structure of the eventual cross-links. Cross-links have never been purified from sclerotized cuticle and consequently the structure has not been established, so it is necessary to rely upon more or less circumstantial evidence when attempting to decide the relative merits of the alternative sclerotization schemes. Since the two schemes are not mutually exclusive, oxidation products generated by both pathways may well be formed simultaneously in the cuticle, thereby giving rise to a mixed product where several types of cross-links and polymers are present.

It has been suggested that the stiffening of cuticle, which is part of the natural sclerotization, is caused by dehydration of the material (Vincent, J. and Hillerton, J., 1979). Biological materials in general become stiffer when dried, and this also applies to cuticle. According to the suggestion the phenolic material, which is incorporated into the cuticle, facilitates the removal of water from the structure. Cross-links between cuticular proteins would not be necessary to explain the observed changes in properties of the material.

Both the appearance and the properties of cuticle vary widely between different insect species and different parts of the same animal. There is no reason to believe that the same stabilization mechanism has been used in all cases, and generalizations based upon results obtained with a single species may be dangerous. A number of different cuticular types will have to be analyzed in detail to determine the mechanisms involved in sclerotization, and how the reactions are controlled to give the optimal degree of sclerotization. Most of the results and ideas presented in this chapter were obtained from experiments utilizing very few species, and it will be interesting to see how far they will be valid.

2 TANNING AGENTS

Tanning agents are the compounds which during sclerotization are secreted by the epidermal cells into the cuticle, where they are acted upon by enzymes to make them reactive and able to connect to proteins and other components. According to the definition compounds deposited unchanged in the cuticle are not to be considered as tanning agents, although they may influence the properties of the structure. This could be the case for uric acid (Caveney, S., 1971) and 3,4-dihydroxyphenyl acetic acid (Andersen, S., 1975), which can be extracted unmodified from cuticle of some Coleoptera.

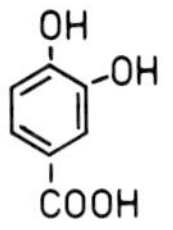

OH, OH, COOH — IV

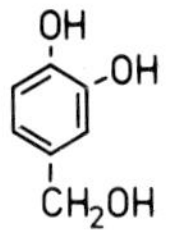

OH, OH, CH_2OH — V

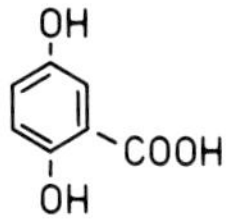

OH, COOH, OH — VI

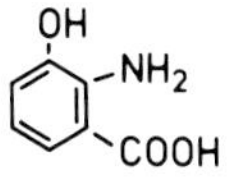

OH, NH_2, COOH — VII

N-acetyldopamine (I) is so far the only compound for which there is convincing evidence that it functions as a tanning agent for insect cuticles. Other phenolic compounds are used for tanning of non-cuticular materials, such as 3,4-dihydroxybenzoic acid (IV) and 3,4-dihydroxybenzyl alcohol (V) for tanning of cockroach egg-cases (Brunet, P. and Kent, P., 1955; Pau, R. and Acheson, R., 1968). Gentisic acid (VI) and 2-amino-3-hydroxybenzoic acid (VII) are used for stabilizing the silks of the silk moths *Antheraea pernyi* and *Hyalophora cecropia*, respectively (Brunet, P. and Coles, B., 1974). Even these species, which use unconventional phenolic compounds for tanning the secreted materials, apparently use *N*-acetyldopamine for stabilizing their cuticle.

The importance of *N*-acetyldopamine in cuticle stabilization was first established in 1962 (Karlson, P. and Sekeris, C., 1962; Sekeris, C. and Karlson, P., 1962; Sekeris, C. and Fragoulis, E., vol. 8). It was shown that insects can synthesize the compound, and that it is readily incorporated into cuticle during sclerotization. After it has been incorporated, *N*-acetyldopamine cannot be recovered in significant amount from sclerotized cuticle. *N*-acetyldopamine is synthesized from the amino acid tyrosine according to the reaction sequence shown

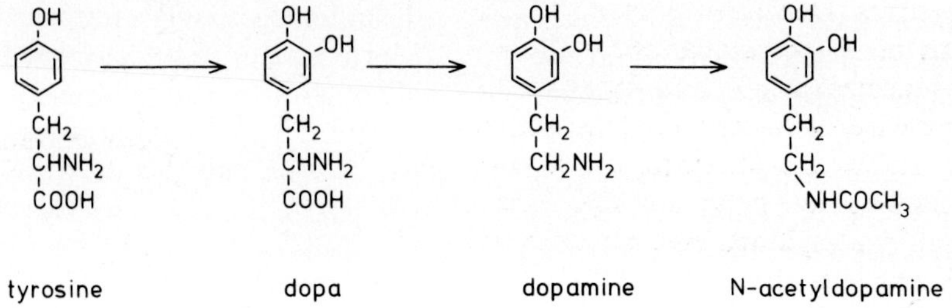

tyrosine dopa dopamine N-acetyldopamine

Scheme II

(Scheme II) (Sekeris, C. and Karlson, P., 1962). There is no agreement on where the transformation takes place; both hemocytes, oenocytes, and epidermal cells have been reported as being involved. It is possible that the various steps in the sequence occur in different places, so that intermediates have to be transported from one cell type to another, or maybe several cell types are capable of independently synthesizing *N*-acetyldopamine. The first step is the hydroxylation of tyrosine to dihydroxyphenylalanine (dopa, VIII). The enzyme responsible for the hydroxylation has not been identified with certainty. It has repeatedly been suggested that the phenoloxidase, which is present as an inactive proenzyme in the hemolymph, might be responsible for the hydroxylation (Pryor, M., 1940b; Richards, A., 1951; Karlson, P. and Wecker, E., 1955; Karlson, P. and Schweiger, A., 1961; Brunet, P., 1963, 1980; Mitchell, H. *et al.*, 1967; Preston, J. and Taylor, R., 1970; Pau, R. and Kelly, C., 1975). When activated, the enzyme will form dopa from tyrosine, but it readily oxidizes dopa to the corresponding quinone, thereby giving rise to melanin. Since bursicon-stimulated hemocytes will produce dopa from tyrosine (Mills, R. and Whitehead, D., 1970; Post, L., 1972), it is possible that it is an intracellular enzyme which is involved.

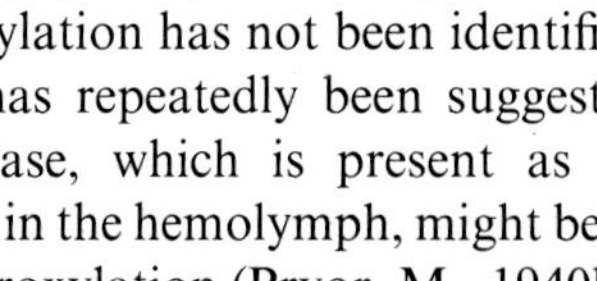

VIII IX

Dopa can be decarboxylated to dopamine (IX) by means of dopa-decarboxylase, an enzyme which has been demonstrated in several insect species (Sekeris, C., 1963; Lunan, K. and Mitchell, H., 1969; Chen, T. and Hodgetts, R., 1976). Mutants of *Drosophila melanogaster*, defective in the production of the enzyme, have been obtained (Wright, T. *et al.*, 1976). The cuticle of such mutants has also been affected, confirming that dopamine is an essential intermediate for proper tanning of the cuticle. Inhibitors of dopa-decarboxylase will block normal hardening and darkening of flesh-fly puparia (Bodnaryk, R., 1971b). In blowfly larvae the enzyme is present in the epidermal cells and is induced by the molting hormone 20-hydroxyecdysone (Fragoulis, E. and Sekeris, C., 1975), but it has also been reported to be present in hemocytes (Whitehead, D., 1970).

An enzyme acetylating dopamine to *N*-acetyldopamine by means of acetyl-coenzyme A has been demonstrated in several insect species, such as *Calliphora, Drosophila,* and *Tenebrio* (Karlson, P. and Ammon, H., 1963; Sekeris, C. and Herlich, P., 1966; Maranda, B. and Hodgetts, R., 1977). All enzymes needed for the transformation of tyrosine to *N*-acetyldopamine have thus been demonstrated in insects, but there is still uncertainty concerning their exact location in the animals, and how their activity is regulated.

All available evidence indicates that *N*-acetyldopamine is of widespread importance as a tanning agent in insects. It should, however, be remembered that *N*-acetyldopamine can serve other functions in the life of insects; dopamine acetyltransferase has been found in significant amounts in insect nervous systems (Dewhurst, S. *et al.*, 1972), and the synthesis of *N*-acetyldopamine in locust ganglia has been reported (Mir, A. and Vaughan, P., 1981). So the presence of *N*-acetyldopamine in a given insect is not sufficient evidence for the suggestion that it is used for cuticular tanning. Compounds other than *N*-acetyldopamine may also be utilized as tanning agents.

The amino acid β-alanine is incorporated in many types of cuticle during sclerotization (Brunet, P., 1980), and is apparently of importance for obtaining the optimal properties of the sclerotized material. The role of β-alanine has been investigated in most detail in puparium formation in *Sarcophaga bullata* and *Drosophila*, but little is known about how it is incorporated. It becomes chemically bound to the cuticular structure in such a way that its amino group is free to react with various blocking agents (Bodnaryk, R. and Levenbook, L., 1969), and it may be placed as N-terminal residue on the proteins (Bodnaryk, R., 1971a). β-alanyltyrosine (X) is present in the hemolymph of *Sarcophaga bullata* species before pupariation, but disappears when sclerotization is initiated (Levenbook, L. *et al.*, 1969). The dipeptide is not incorporated directly; it is first hydrolyzed to free β-alanine and tyrosine, then both amino acids are incorporated. Tyrosine is presumably first transformed to dopamine, since dopa-decarboxylase inhibitors prevent its incorporation (Bodnaryk, R., 1971b). Such inhibitors also prevent the incorporation of β-alanine, indicating that β-alanine can only function in sclerotization when it is accompanied by dopamine or a dopamine derivative. The suggestion is supported by the observation that injection of dopamine, and to a certain extent also *N*-acetyldopamine, stimulated incorporation of β-alanine in larvae where the formation of dopamine is blocked (Bodnaryk, R., 1971b). It has been proposed that the actual tanning agent could be *N*-β-alanyldopamine (XI) (Andersen, S. 1977), but the evidence obtained so far is inconclusive.

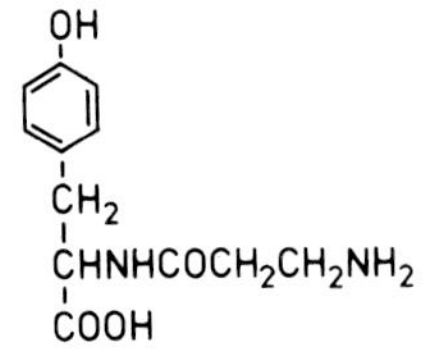

It is generally assumed that tanning agents are diphenols of low molecular weight, which are secreted into the cuticle, there to react with proteins. Evidence has been published (Koeppe, J. and Gilbert, L., 1974) indicating that dopamine or some derivative of dopamine can be linked to a hemolymph protein in pupae of *Manduca sexta*,

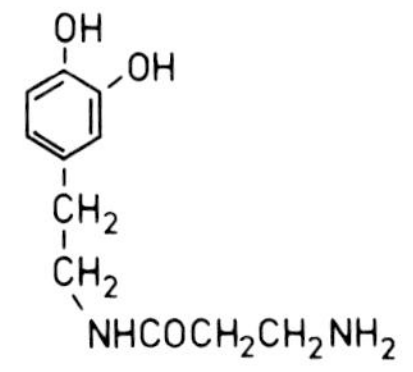

OH
(benzene ring)
CH_2
$CHNHCOCH_2CH_2NH_2$
COOH

X

OH, OH
(benzene ring)
CH_2
CH_2
$NHCOCH_2CH_2NH_2$

XI

and that this phenol–protein complex is transported through the epidermal cells to be incorporated into the cuticle. Strictly speaking, the phenol–protein complex would thus be a tanning agent in this species. More information on the nature of the complex is needed.

3 ENZYMES INVOLVED IN TANNING

For practical reasons it is convenient to divide the enzymes concerned with tanning into two groups: those involved in the pathway leading to synthesis of *N*-acetyldopamine and other possible tanning agents, and those involved in the actual tanning mechanism. All the enzymes belonging to the former group occur either intracellularly or in the hemolymph and have been discussed above, while the actual tanning enzymes occur extracellularly in the cuticle and are concerned with the further transformation of the tanning agents. The cuticular enzymes, which can use *N*-acetyldopamine as substrate, are the most interesting, since this compound is of dominating importance. Our knowledge of cuticular enzymes is still limited, and some of the enzyme activities, which have been described in cuticle, may be involved in reactions other than those concerning with tanning.

The presence of enzymes catalyzing the oxidation of phenols has been demonstrated in many types of cuticle, both histochemically and by quantitative measurements. Such phenoloxidases appear to be of two types: one type will oxidize both monophenols and *ortho*-diphenols and resembles the tyrosinase described in mushrooms; the other type will oxidize *ortho*- and *para*-diphenols but not mono-phenols, and it resembles the laccase obtained from some fungi and higher plants.

Characterizing cuticular enzymes is difficult; it is not possible to determine the specific activity of the enzymes as long as they are inside the cuticular structure, and there is considerable risk of modifying the enzymes when they are solubilized. Often it can be a problem to get enzyme activity out of the cuticle, where it may be bound to the cuticular structure in some unknown way. Extracts from whole insects are of little use when studying the cuticular enzymes, since insect hemolymph is rich in prophenoloxidase, which can easily be transformed into the active enzyme. The activation can be caused by limited proteolysis (Ashida, M. *et al.*, 1974), or by lipid-like substances (Heyneman, R. and Vercauteren, R., 1964). In *Drosophila* it has been shown that activation of the proenzyme is the result of a cascade of reactions (Seybold, W. *et al.*, 1975), which is started when animals are damaged. The activated enzymes from insect hemolymph tend to aggregate readily and can easily be adsorbed to cellular debris or to fragments of cuticle. Only investigations, using isolated and purified cuticles, can therefore be assumed to give reliable information on the enzymes involved in sclerotization. That the cuticular diphenoloxidases are not identical with those from hemolymph was shown by Hackman, R. and Goldberg, M. (1967) who, from dipteran larval cuticle, obtained extracts containing both mono- and diphenoloxidase activity, and the activities differed in several respects from the activities obtained from hemolymph. Ohnishi, E. (1954) showed that *Drosophila virilis* larval cuticle contains an inextractable enzyme catalyzing the oxidation of various diphenols and of dimethyl-*p*-phenylenediamine, which is not a substrate for the hemolymph enzyme. The enzyme has been characterized in more detail by Yamazaki, H. (1969), who used powdered, acetone-dried cuticle from fruit-fly white puparia. The cuticular powder has a wide substrate specificity. It will oxidize *ortho*- and *para*-diphenols and aromatic diamines, but has negligible activity towards tyrosine. The activity is unusually thermostable, and it is not inhibited by a number of compounds, such as thiourea, which inhibits the hemolymph enzyme. The activity resisted all attempts at solubilization, but a similar enzymatic activity could be obtained in soluble form by trypsin digestion of cuticle from silkwork pupae (*Bombyx mori*), collected during larval–pupal ecdysis, when the cuticle is still unhardened (Yamazaki, H., 1972). As soon as the cuticle has become sclerotized it is impossible to obtain the enzyme by trypsin treatment, indicating that it becomes sclerotized together with the structural proteins in the cuticle. There is a drastic increase in enzymatic activity in the intact cuticle during the first hours after ecdysis, indicating either activation of a proenzyme or deposition of fresh enzyme. The enzyme activity drops again to reach a low level at about 1 day after ecdysis. The oxidase released by trypsin treatment was purified, and a nearly homogeneous preparation was obtained and partly characterized. It was concluded that presumably it is a copper protein, and it was classified as a laccase. Enzymes resembling that obtained by Yamazaki from silkworm pupal cuticle have since been obtained from other sources, such as unsclerotized locust cuticle and blowfly larval cuticle. If the ability to catalyze the oxidation of *para*-diphenols is used as criterion for the presence of laccases, this group of enzymes is of widespread occurrence in cuticles of insects (unpublished observations).

Laccases have been purified after trypsin digestion of unsclerotized cuticle from the locusts, *Schistocerca gregaria* (Andersen, S., 1978) and *Locusta migratoria*. From blowfly (*Calliphora vicina*) larval cuticle two different laccases have been obtained both with and without trypsin treatment (Barrett, F. and Andersen, S., 1981). Since some insect cuticles may contain proteolytic activity (Dohke, K., 1973; Ashida, M. *et al.*, 1974), the possibility cannot be ruled out that the enzyme is released by proteolysis from blowfly larval cuticle, even when no external trypsin is added. From *Drosophila* larval cuticle (wandering stage) small amounts of oxidase activity have been extracted without addition of trypsin (unpublished observation). The substrate specificity corresponds to that of the other enzymes mentioned, but it has not been obtained in sufficient quantities for purification and characterization.

Electrophoresis of purified preparations in

polyacrylamide gels under denaturing conditions gave three bands having molecular weights between 62,000 and 70,000 in the case of silkworm pupal cuticle (Yamazaki, H., 1972), and two bands with molecular weights between 90,000 and 100,000 from *S. gregaria* cuticle (Andersen, S., 1978). The major laccase-like enzyme from blowfly larval cuticle gave only a single band with molecular weight about 90,000 (Barrett, F. and Andersen, S., 1981). The multiple bands obtained from two species could be due to limited proteolysis during release of the enzymes by trypsin digestion, but it may also indicate the presence of isoenzymes.

The insect laccases have a rather broad substrate specificity, as a number of substituted catechols and hydroquinones are readily oxidized, whereas neither mono-phenols nor *meta*-diphenols are oxidized. Compounds carrying negative groups on the side-chain, such as protocatechuic acid and dopa, are rather poor substrates, whereas a positive charge has little effect on the rate of oxidation. *N*-acetyldopamine is a good substrate for the enzymes, although not the best. The K_m value for *N*-acetyldopamine was determined to be 0.53 mM for the blowfly larval enzyme and 1.3 mM for the enzyme from locust cuticle. The enzymes are inhibited by cyanide, fluoride, and azide ions, whereas complexing agents such as *o*-phenanthroline, dipyridyl, and phenylthiourea have little effect.

The pH optima of the laccases are in the acidic range, varying from about pH 4.5 for the major blowfly enzyme to pH 5.5 for the silkworm enzyme. It is doubtful whether the cuticular pH will ever be as low as this, and such low pH optima need not have a functional importance. There is a tendency for the pH-optima to be slightly higher for intact cuticle than for the solubilized enzymes.

There is no evidence that diphenoloxidases other than those described are present in locust or silkworm pupal cuticle. From the larval blowfly cuticle we have obtained two diphenoloxidases different from the major laccase enzyme described above. One is also of the laccase type, but has a pH optimum at a slightly alkaline reaction and differs from the major laccase in its substrate specificity.

The other enzyme is not a laccase, but resembles the hemolymph diphenoloxidase so much that it seems likely that it is the same enzyme which occurs both in cuticle and in hemolymph. The enzyme belongs to the tyrosinase type, as it catalyzes the hydroxylation of tyrosine to dopa and the oxidation of *o*-diphenols to *o*-quinones. It has no activity towards *p*-diphenols, nor towards aromatic diamines. The enzyme is strongly inhibited by thiourea and phenylthiourea, it has a pH optimum near neutrality, and is less thermostable than the cuticular enzymes of the laccase type. The enzyme is apparently strongly aggregated in the extracts, since it is excluded from 5% polyacrylamide during electrophoresis (Barrett, F. and Andersen, S., 1981). Such aggregation has been analyzed in more detail for the hemolymph diphenoloxidase, where it has been shown that the enzyme is present as an unaggregated, inactive precursor, which after activation tends to form large, ill-defined aggregates (Munn, E. and Bufton, S., 1973). There is some evidence that the cuticular tyrosinase-type enzyme in blowfly larvae also occurs as an inactive precursor, which is activated by proteolysis when the cuticle is damaged. At least it has been found that when pieces of blowfly cuticle are cleaned at pH 8.5, they contain high tyrosinase activity when tested soon after cleaning. If they are cleaned at pH 5 the activity is low, but increases rapidly after transferring the samples to pH 8.5. This increase can be prevented by the presence of trypsin inhibitors (Barrett and Andersen, unpublished observations). The activity of the laccase-type enzymes is unaffected by the pH during preparation and by the presence of trypsin inhibitors, indicating that the activities of the various enzymes in blowfly larval cuticle are controlled in different ways.

Only a limited number of insect cuticles has been investigated with respect to the types of diphenoloxidases they contain, but it appears that the laccase type of enzyme is present in cuticles destined to be sclerotized, both when sclerotization occurs soon after ecdysis, as in locusts and silkworm pupae, and when it occurs at a later stage, as in puparium formation in Diptera. The tyrosinase type of diphenoloxidase has been found in soft cuticles which are never sclerotized, such as larval cuticle of the silkworm, *Hyalophora cecropia* (unpublished observation), and in blowfly larval cuticle, which has to function first as a soft cuticle during larval life and then as a hard, sclerotized cuticle in the puparium. The blowfly larval cuticle is the most complex example we have encountered with regard to enzymatic complexity. The correlation between presence of tyrosinase-like enzyme and soft texture of the cuticle has led to the suggestion that this type of phenoloxidase is not involved in sclerotization, but in healing of damaged cuticle (Barrett, F. and Andersen, S., 1981). According to this suggestion the laccase type of diphenoloxidase should be the enzyme involved in sclerotization.

Peroxidases are another group of enzymes which might be involved in oxidizing diphenols to reactive intermediates. It has been suggested they may be responsible for catalyzing the cross-linking of the cuticular, rubber-like protein, resilin, by oxidizing protein-bound tyrosine residues to free radicals (Andersen, S., 1966; Coles, G., 1966). Peroxidases may also be involved in sclerotization of hard, cuticular structures (Locke, M., 1969). Peroxidase activity has been demonstrated both in the epidermal cells and in cuticle, but the enzyme has not been purified and characterized from insects. Since catalase and other heme-containing proteins can have slight peroxidatic activity, and catalase has been found present in some cuticles, it is necessary to purify and characterize the enzyme to ascertain whether it should be classified as peroxidase or catalase. The substrates which are commonly used to demonstrate peroxidase histochemically, such as diaminobenzidine and benzidine, are also oxidized to colored products by the cuticular laccases. Accordingly a positive reaction with these substrates will not prove the presence of peroxidases.

Although it is well established that the diphenol, *N*-acetyldopamine, is the tanning agent in many cuticles, enzymes other than diphenoloxidases might also be involved in activating the compound. Some years ago evidence was obtained that locust cuticle contains an enzyme activity catalyzing the release of tritium specifically located on the carbon atoms of the side-chain of *N*-acetyldopamine (Andersen, S., 1972). The enzyme is firmly bound to the cuticular structure, and its properties have only been characterized in intact cuticle, since it has not been possible to obtain active extracts. The relationship between this activity and the diphenoloxidase present in cuticle is ambiguous, since it has only been possible to compare the two activities in intact pieces of cuticle (Andersen, S., 1979b), and such comparisons are made complicated by penetration and diffusion problems. The comparison showed that the activity, which is responsible for the release of tritium from the side-chain of *N*-acetyldopamine, resembles the laccase activity. The two activities have the same high thermostability, the same pH stability, nearly the same pH optimum and inhibitor specificity. The small differences which were observed could be explained as diffusion artefacts due to the differences in assay conditions. It was observed that the trypsin-solubilized laccase would release tritium from the aromatic ring of labelled *N*-acetyldopamine, while tritium located on the side-chain was not released to any significant extent. This observation can be explained by assuming that an enzyme different from the laccase is responsible for releasing side-chain-located tritium from *N*-acetyldopamine, when intact pieces of cuticle are used. An alternative explanation is that a single enzyme is responsible for both activities in the intact cuticle, and that trypsin digestion modifies the enzyme by limited proteolysis, thereby destroying the side-chain-directed activity, while leaving the laccase activity intact.

The one-enzyme hypothesis was supported by the observation that *ortho*- and *para*-diphenols, which are good substrates for the laccase, will competitively inhibit the release of tritium from the side-chain of *N*-acetyldopamine when intact cuticle is used as enzyme source. Moreover, the best substrates for the laccase were the most effective inhibitors of tritium release. This relationship can be expected, if it is the same active site which binds both *N*-acetyldopamine and the various other diphenols.

Results obtained by characterizing the reaction products, when *N*-acetyldopamine is used as substrate, indicate that intact cuticle contains enzyme activities which are different from the purified laccase. When intact locust cuticle acts upon *N*-acetyldopamine the main product is a dimeric compound for which the structure XII was first suggested (Andersen, S., 1972). On the basis of NMR studies the structure has been revised, and the correct formula is XIIIa and/or XIIIb (Andersen, S. *et al.*, 1980). It is probably a mixture of isomers which is formed.

Another dimeric compound has recently been isolated from the incubation medium after reacting *N*-acetyldopamine with intact locust cuticle. It is characterized by having a double bond in one of the side-chains, as shown in structure XIV

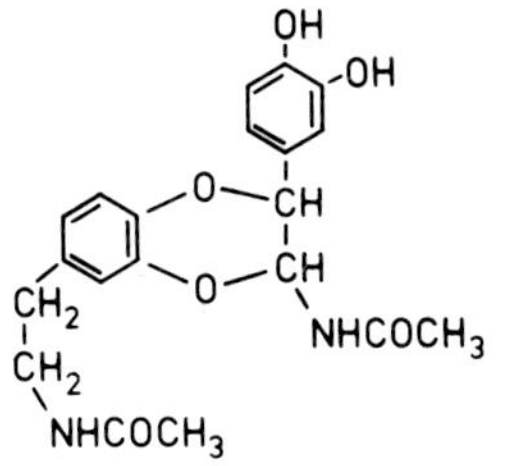

XIIIa

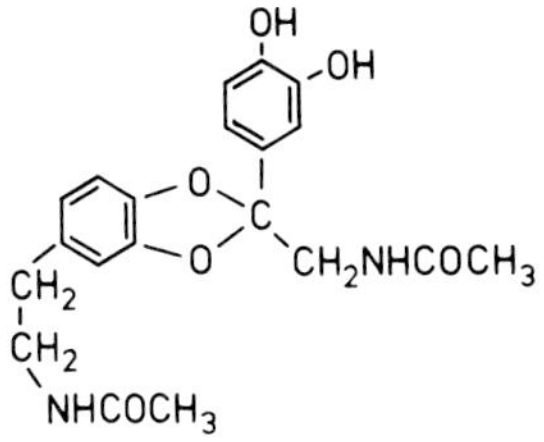

XII

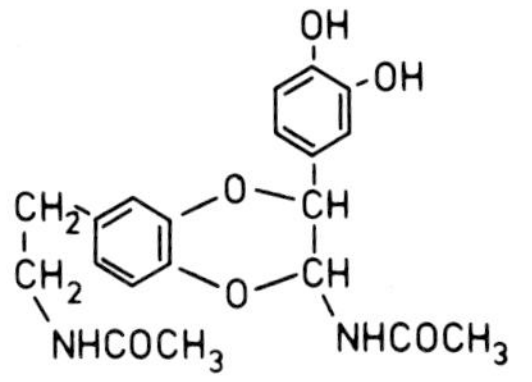

XIIIb

XIV

XV

(Andersen, S. and Roepstorff, P., 1982). It is formed in much smaller amounts than compound II.

None of these dimeric compounds are formed when purified locust laccase is incubated with *N*-acetyldopamine. The fact that they are formed when intact cuticle is used as the enzyme source can be explained by assuming the presence in cuticle of an enzyme that will introduce a double bond into the side-chain of *N*-acetyldopamine. The presence of such an enzyme can be demonstrated by adding some amino acids or small peptides to the incubation medium. They will partly inhibit the laccase, and in their presence a metabolite of *N*-acetyldopamine accumulates in the medium. It has been isolated and identified as dehydro-*N*-acetyldopamine (XV) (Andersen, S. and Roepstorff, P., 1982). It is a far better substrate for the laccase than *N*-acetyldopamine, and it is accordingly consumed as rapidly as it is formed if the laccase is not inhibited.

The "desaturase" activity, responsible for the formation of the double bond, has not yet been characterized in detail, nor has it been solubilized, but it can be assumed that it corresponds to the enzyme activity responsible for the release of tritium from the side-chain of *N*-acetyldopamine. However, the true identity of the various enzymatic activities in insect cuticle can only be ascertained when all the activities have been rendered soluble, purified, and characterized in detail.

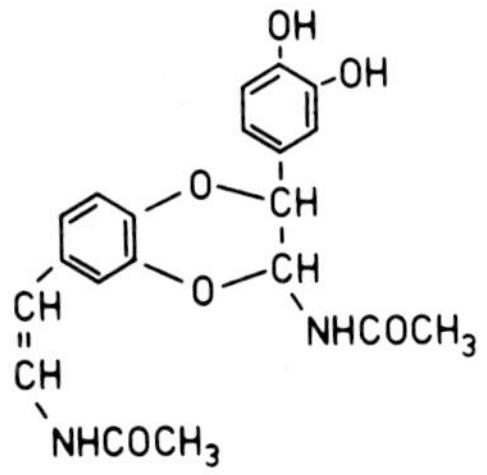

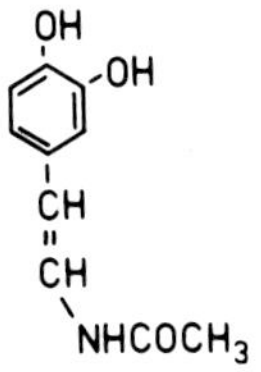

4 THE CHEMISTRY OF TANNING

Several attempts have been made to obtain insight into the chemical reactions going on during sclerotization by studying the fragments which can be obtained after degradation of fully sclerotized cuticle. Such studies are necessary, but can be misleading, as artefacts are easily produced during degradation. Dennell, R. (1958b) used alkaline hydrolysis, under reducing conditions, of blowfly puparia to liberate what he identified as a substituted aminohydroquinone. The identification was mainly based upon one-dimensional paper chromatography and color reactions without having appropriate standards available, and the results cannot be considered reliable.

Acid hydrolysis of sclerotized cuticles has been used to liberate substituted catechols, among which ketocatechols (XVI–XIX) are the dominating species (Andersen, S. and Barrett, F., 1971; Andersen, S. and Roepstorff, P., 1978). The type of ketocatechol obtained depends upon the exact conditions of hydrolysis, and it has been argued that they are degradation products of a common precursor in the cuticle (Andersen, S., 1971). However,

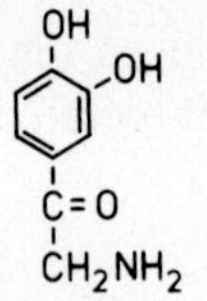

OH, OH — C=O — CH_2NH_2

XVI

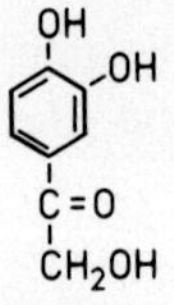

OH, OH — C=O — CH_2OH

XVII

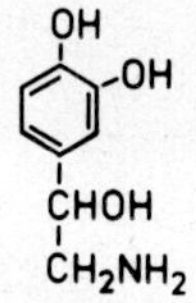

OH, OH — CHOH — CH_2NH_2

XX

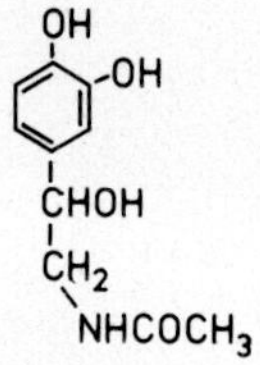

OH, OH — CHOH — CH_2 — $NHCOCH_3$

XXI

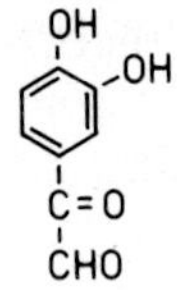

OH, OH — C=O — CHO

XVIII

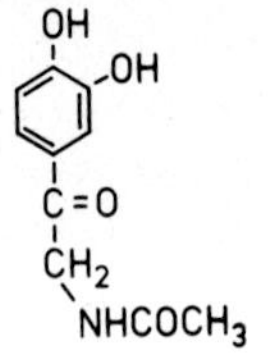

OH, OH — C=O — CH_2 — $NHCOCH_3$

XIX

OH, OH — CHO

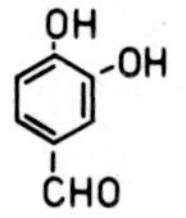

XXII

small amounts of ketocatechols have also been obtained after extraction of cuticle under non-hydrolytic conditions (Atkinson, P. *et al.*, 1973), and it was suggested that they are present as such in the cuticle before extraction, and that their function might be to act as antioxidants and not to be involved in protein stabilization via cross-linking.

Digestion with proteolytic enzymes has also been used to liberate ketocatechols and other catechols from sclerotized cuticle. Catechols obtained in this way have not been linked to the cuticular structure, but may have been trapped inside the cuticular network. The ketocatechols, which can be obtained in this way, only account for a small fraction of what can be released by acid hydrolysis. The free ketocatechols are thus only minor components in sclerotized cuticle: they may have a function as antioxidants, or they may be unimportant byproducts formed during sclerotization. The following catechols have been obtained from proteolytic digests of locust sclerotized cuticle: dopamine (IX), *N*-acetyldopamine (I), arterenol (XX), *N*-acetylarterenol (XXI), hydroxyketocatechol (XVIII), protocatechualdehyde (XXII), protocatechuic acid (IV), and several dimeric compounds, including compounds XIII and XIV (Andersen, S. and Roepstorff, P., 1982, and unpublished observations).

When exposed to acid hydrolysis the dimeric compounds are degraded, and the moiety with two free phenolic groups is recovered as a ketocatechol. A fraction of the ketocatechols obtained after acid hydrolysis of cuticle will be derived from the dimeric compounds present in the cuticle. The fraction is, however, of quantitatively minor importance since the dimer, which is present in largest amounts, can account for less than 1% of the total amount of ketocatechols which are obtained. Higher oligomers are presumably also present in sclerotized cuticle and will give ketocatechols on acid hydrolysis, but so far we have only indirect evidence for their existence.

During proteolytic digestion of sclerotized cuticle most of the catechol-containing material is released as a mixture of compounds which, although they are not ketocatechols, give ketocatechols by acid hydrolysis (Andersen, S., 1970; Andersen, S. and Roepstorff, P., 1982). During hydrolysis with acid a number of free amino acids are also formed. Despite several attempts to fractionate the proteolytic digest it has not been possible to obtain a pure compound for identification; neither has it been possible to separate the moiety which gives amino acids on hydrolysis from that which gives ketocatechols. It appears that the catechols are chemically linked to peptide material, but the nature of the links is still unknown. Since the same ketocatechols are obtained after acid hydrolysis of the dimeric compounds and of the catechol-containing peptide material, it can be assumed that links from both the α- and β-carbon atom in the side-chain of *N*-acetyldopamine are involved in both cases. A significant fraction of the phenolic groups from tyrosine residues and of the ε-amino groups from lysine residues in the cuticular proteins is not available for substitution in sclerotized cuticle (Andersen, S., 1972), and these groups may be involved in the linkage. The protein-linked catechols are presumably present in polymeric form, and a reliable estimate of the degree of polymerization is needed.

There are several compounds among those obtained by mild acid hydrolysis which are not easily accounted for, such as arterenol and 3,4-dihydroxyphenylglycol (XXIII), catechols with a secondary alcohol group in the β-position of the side-chain (Andersen, S. and Roepstorff, P., 1978). They may be reduction products from the corresponding ketones, but there is no supporting evidence that such a reduction takes place during hydrolysis. The compounds might also be derived from material linked to proteins with a single, hydrolyzable bond, as presumably the keto-derivatives are formed by hydrolyzing two such bonds. More difficult to explain is the isolation of a ketocatechol acetyl ester (XXIV) after mild acetic acid hydrolysis of sclerotized cuticle (Andersen, S. and Roepstorff, P., 1978). The acetyl group is not introduced during hydrolysis, since it is present also when formic acid or propionic acid is used for hydrolysis. The compound is derived from *N*-acetyldopamine incorporated into the cuticle, but how the acetamide group is transformed into an acetyl ester is so far inexplicable.

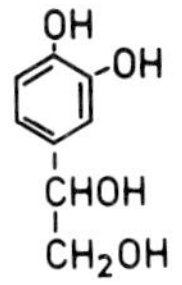

OH, -OH, CHOH, CH_2OH

XXIII

OH, -OH, C=O, CH_2, $OCOCH_3$

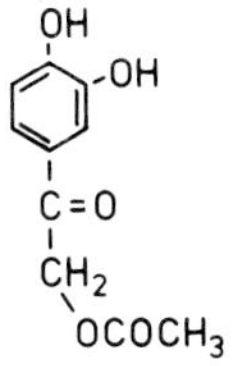

XXIV

Ketocatechols can be obtained from nearly all sclerotized cuticles from insects. Their amounts can correspond to more than 10% of the cuticular dry weight (Barrett, F., 1977), and values down to 0.2% have been obtained for locust cornea and intersegmental membranes (Andersen, S., 1974a). Ketocatechols can be obtained from cuticles sclerotized before ecdysis (Andersen, S. *et al.*, 1981) as well as from those sclerotized after ecdysis. Cuticles which only sclerotize after ecdysis will not yield ketocatechols until the process of sclerotization is initiated. Ketocatechols have not been obtained from sclerotized cuticle from non-insect arthropods or from apterygote insects (Thysanura, Collembola, Diplura) (unpublished observations), indicating that the ability to fly and the ability to produce ketocatechol-yielding material have been developed in parallel during evolution.

It was mentioned above that β-alanine plays a role in the sclerotization of some cuticles. Acid hydrolysis releases the amino acid in significant quantities from sclerotized dipteran puparia, but not from the still unsclerotized, white puparium (Bodnaryk, R. and Levenbook, L., 1969; Bodnaryk, R., 1971a). It has, to my knowledge, not been possible to obtain β-alanyl derivatives by limited degradation of sclerotized puparia. The bonds between β-alanyl residues and the cuticular material are very labile towards acid treatment (Bodnaryk, R., 1971b). If the suggestion that β-alanyldopamine is a tanning agent in fruit-fly puparia is correct, and if it is incorporated by the same pathway as *N*-acetyldopamine, it should be possible to obtain β-alanyl-substituted ketocatechols by very mild acid treatment of *Drosophila* puparia. We have searched for such β-alanyl derivatives without success. Degradation studies have so far been of no help in investigating the role of β-alanine in sclerotization.

5 PROPOSAL FOR A SCLEROTIZATION MECHANISM

A revised scheme for the chemical reactions occurring during sclerotization has recently been proposed (Andersen, S. and Roepstorff, P., 1982). It is based upon the results obtained from degradation studies of sclerotized cuticle and from characterization of enzyme activities in cuticle. The scheme incorporates aspects of both the quinone tanning scheme and of the β-sclerotization scheme, and it appears to be able to account for most of the observations made on various sclerotized cuticles.

The scheme (Scheme III) suggests that cuticles which are able to sclerotize contain two different enzyme activities, both of which can use *N*-acetyldopamine as substrate. One of the enzymes is a diphenoloxidase which, in the presence of oxygen, will oxidize *N*-acetyldopamine to the corresponding

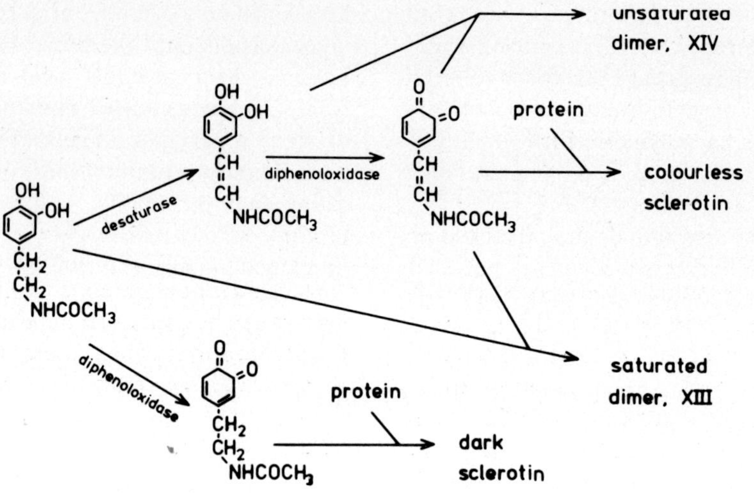

Scheme III

quinone. The enzyme is supposed to be identical with the laccases which have been isolated from various insect cuticles. The quinone can react with the cuticular proteins as assumed in the classical quinone tanning scheme (Scheme I), and the result is darkly colored sclerotins. The other enzyme, a desaturase, can remove hydrogen from the side-chain of *N*-acetyldopamine, thereby introducing a double bond in the side-chain. The unsaturated compound, dehydro-*N*-acetyldopamine (XV), is a substrate for the same diphenoloxidase (laccase) as that which oxidizes *N*-acetyldopamine to a quinone. The resulting unsaturated quinone (XXV) possesses an extended system of conjugated double bonds, and it can be predicted to be a highly reactive molecule. It reacts with the *ortho*-diphenols available (*N*-acetyldopamine and dehydro-*N*-acetyldopamine) to form substituted benzodioxins (XIII and XIV) as well as with the cuticular proteins. The cuticle is thereby sclerotized, and the end-product is a colorless material.

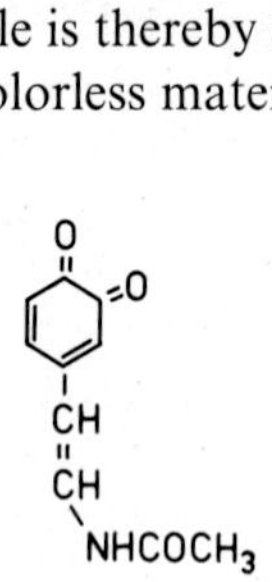

XXV

The main features of the scheme are:

(1) quinone tanning and β-sclerotization occur at the same time and use the same primary substrate, *N*-acetyldopamine;
(2) quinone tanning can occur without simultaneous β-sclerotization, whereas β-sclerotization cannot occur without some quinone tanning going on at the time;
(3) the relative importance of the two pathways depends both upon the relative activities of the two enzymes and upon the affinities of the two substrates, *N*-acetyldopamine and dehydro-*N*-acetyldopamine, for the diphenoloxidase;
(4) since the unsaturated quinone is two oxidative steps removed from *N*-acetyldopamine, it can react spontaneously with two groups and thereby form a cross-link, and no protein-bound intermediates, which need reoxidation, are formed as assumed in the quinone tanning scheme.

Evidence is available for all the steps suggested in the scheme, but not all the evidence is equally conclusive. The diphenoloxidase (laccase) has been purified from three different types of cuticle, and it is active *in vitro* towards both *N*-acetyldopamine and dehydro-*N*-acetyldopamine. The desaturase has not been purified and characterized, but its activity has been demonstrated in locust femur and fruit-fly puparial cuticle. Oxidation of dehydro-*N*-acetyldopamine *in vitro* leads to the formation of the unsaturated dimer (XIV), and when *N*-acetyldopamine is added to the incubation medium the saturated dimer (XIII) is also formed. If dehydro-*N*-acetyldopamine is oxidized by the diphenoloxidase in the presence of foreign proteins some of the substrate is coupled to the proteins, but it is not yet known what type of bonds is formed. If unsclerotized locust cuticle is incubated with dehydro-*N*-acetyldopamine some of the compound is incorporated into the cuticle; the cuticle remains colorless, it becomes stiffer, and the proteins become insoluble. After acid hydrolysis of the cuticle the usual types of ketocatechols can be recovered (Andersen, S. and Roepstorff, P., 1982).

The most serious weakness in the scheme is that the linkages between the derivatives of *N*-acetyldopamine and the proteins are unknown, despite many attempts and efforts to identify them. There is good evidence that the linkages exist, but they have proved highly elusive. The two most important problems in cuticle sclerotization today are: What is the precise nature of the interprotein links?, and How is sclerotization controlled?

6 CONTROL OF SCLEROTIZATION

It is necessary that sclerotization occurs in the right place, at the right time, and to the right degree. Too little or too much sclerotization might prove fatal to the insect, but it is not known how precise the control has to be; it will probably vary from species to species. It seems reasonable to assume that a precise control of the degree of sclerotization will not be essential for the dipteran puparium, as long as it is sufficient to protect the animal developing inside the puparium. As the wings of flying insects are exposed to strong and varying bending and twisting forces, they will need mechanical strength, some resilience, and extreme lightness. Too little sclerotization could make them too soft, and too much sclerotization could make them too brittle and heavy. However, little experimental evidence bearing on this problem is available, due to difficulties in obtaining reliable quantitative measurements of mechanical properties and of the degree of sclerotization of selected samples of cuticle.

No reliable methods exist for determining the total degree of sclerotization of cuticles. It would be valuable if we were able to determine the number of cross-links between macromolecules in a piece of cuticle, or to determine the total amount of diphenolic material incorporated into the cuticle. The yield of neutral ketocatechols after hydrolysis of cuticle samples in 1 M HCl can be used as an approximate measure for the amount of *N*-acetyldopamine incorporated via the β-sclerotization pathway, but we cannot determine how much is incorporated via quinone tanning. Ketocatechol analyses can be used as an indication for the degree of sclerotization, when there is evidence from enzyme analyses that β-sclerotization is the dominating pathway during sclerotization.

The enzymatic characteristics of locust femur cuticle indicate that about 90% of the total sclerotization is due to β-sclerotization, and the yield of ketocatechols probably gives a reasonably good estimate of the total sclerotization in this sample. The method has been used both to follow the progress of sclerotization and to obtain an impression of the variability in sclerotization of femur cuticle among mature individuals of *Locusta migratoria* grown together in the same cage. The ketocatechol yield varied from a minimum value of 0.44% to a maximum value of 2.30% of the femur dry weight; the mean value for 58 samples was 1.26%, and the standard deviation was 0.54%. The mean dry weight of the femur cuticles was 11.17 ± 0.98 mg, and the heaviest samples weighed 1.5 times more than the lightest (unpublished observations). The ketocatechol yield can thus vary with a factor of five, although the animals were reared together under controlled conditions. It is not known whether a similar variability is present in natural populations. Neither is it known whether the differences in yield of ketocatechols correlate with other properties of cuticle. If the ketocatechol yield is a reliable estimate of the degree of sclerotization these preliminary measurements indicate that there is no very precise control of the degree to which locust femur cuticle is sclerotized.

The amount of enzyme activity, measured as the release of tritium from the side-chain of *N*-acetyldopamine, has been determined in cuticle from various body parts of locusts. The values were compared to the yield of ketocatechols from corresponding parts from animals analyzed either 1 day or 10–12 days after ecdysis (Andersen, S., 1974a). No correlation was found between enzyme activity and the yield of ketocatechols, which indicates that it is not the amount of enzyme which determines to what degree the cuticle is sclerotized, and that the limiting factor must be looked for elsewhere.

Injection of dopamine, generally labelled with tritium, into adult locusts about 2–6 h after ecdysis, resulted in incorporation of much of the label into the cuticle. The distribution of label between samples of cuticle from various body parts corresponded closely to the degree of sclerotization measured as ketocatechol yield from the corresponding parts of control animals (Andersen, S., 1974a). The mandibles were the only exception; their incorporation of labelled dopamine was nearly twice that expected from the ketocatechol yield. The mandibles, in contrast to the other samples of locust cuticle, are dark brown, and they will liberate tritium both from the side-chain and from the ring positions in *N*-acetyldopamine, indicating that both quinone tanning and β-sclerotization occur in this sample (Andersen, S., 1974b). The results indicate that a single dose of dopamine is distributed into various parts of cuticle corresponding to the rate at which they are sclerotized via the β-sclerotization pathway. It therefore seems likely that it will be the uptake of dopamine or *N*-acetyldopamine in the epidermal cells which will be the rate-limiting step, and that the epidermal cells from various parts of the integument will compete for the available dopamine. The cuticle above the epidermal cells with the greatest ability to take up dopamine/*N*-acetyldopamine will accordingly be the areas which will become most heavily sclerotized.

The degree of sclerotization will also depend upon the total amount of sclerotizing agent synthesized in the animal. It was mentioned above that *N*-acetyldopamine is mainly synthesized from tyrosine via dopa and dopamine, and that there is no general agreement on where this synthesis takes place. Both hemocytes, oenocytes, and epidermal cells have been suggested to play a role. It appears that the synthesis is under hormonal control, and there is suggestive evidence that both ecdysteroids and bursicon can be involved in initiating sclerotization (vol. 8). Ecdysteroids induce sclerotization in ligated abdomens of blowfly larvae, and one of the direct effects of 20-hydroxyecdysone in this system is the induction of *de novo* synthesis of the enzyme dopa-decarboxylase in the epidermal cells (Fragoulis, E. and Sekeris, C., 1975). To my knowledge there is no evidence that this enzyme is involved in the regulation of sclerotization in systems other than puparium formation in dipteran species.

In several instances it has been possible to induce cuticle synthesis and tanning *in vitro* by the addition of 20-hydroxyecdysone to the medium (Marks, E. and Sowa, B., 1976; Mitsui, T. and Riddiford, L. 1976). Thus it appears that the epidermis of some insects is equipped with the complete system for synthesizing the tanning agent and can be activated by appropriate hormone treatment. Marks, E. and Sowa, B. (1976) report that cockroach epidermis can produce cuticle *in vitro*, when stimulated by 20-hydroxyecdysone, although the cuticle appears incomplete and it is not sclerotized. They suggest that some controlling factor, for instance bursicon, is missing in the *in vitro* system.

Bursicon was described as a hormone necessary for cuticle tanning and endocuticle formation of flies and other insects (Fraenkel, G. and Hsiao, C., 1965; Fogal, W. and Fraenkel, G., 1969; Zdarek, J., Reynolds, S., vol. 8). There is some evidence that the hormone can stimulate the uptake of tyrosine into the hemocytes and thereby initiate its conversion to *N*-acetyldopamine (Mills, R. and Whitehead, D., 1970; Post, L., 1972). Bursicon also appears to be able to stimulate the epidermal cells directly, as demonstrated by a conspicuous rise in epidermal cyclic AMP when the level of bursicon activity is increased in the hemolymph of *Tenebrio molitor* (Delachambre, J. *et al.*, 1979).

From the available evidence it can be concluded that the synthesis of the tanning agent is hormonally controlled, and that at least 20-hydroxyecdysone and bursicon may be involved. The mechanisms of the hormonal controls are poorly understood, and the possibility that further hormones are involved cannot be excluded.

The ultrastructure of insect epidermal cells has been studied in detail, and much information has been obtained on the structural changes in the cells during the molting cycle (Locke, M., vol. 2). We lack a correspondingly detailed study of the biochemical changes in the epidermis during the same period, so that the ultrastructure can be correlated with function on the molecular level.

It appears that the post-ecdysial sclerotization of

insect cuticle is controlled on two levels: a central, hormonal regulation of the onset of sclerotization and a local control of the extent and type of sclerotization. In some insects parts of the cuticle are sclerotized before ecdysis, while the animal is in the pharate state, and in the honeybee it has been shown that this pre-ecdysial sclerotization utilizes the same chemical reactions as post-ecdysial sclerotization (Andersen, S. *et al.*, 1981). Nothing is known about how the pre-ecdysial sclerotization is regulated, but it is possible that it is initiated by a rise in the level of ecdysteroids.

ACKNOWLEDGEMENTS

My thanks are due to the NOVO foundation, the Carlsberg foundation, and the Danish Science Research Council for the support which has made this work possible.

REFERENCES

ANDERSEN, S. O. (1966). Covalent cross-links in a structural protein, resilin. *Acta Physiol. Scand. 66*, Suppl. 263, 1–81.

ANDERSEN, S. O. (1970). Isolation of arterenone (2-amino-3′,4′-dihydroxyacetophenone) from hydrolysates of sclerotized insect cuticle. *J. Insect Physiol. 16*, 1951–1959.

ANDERSEN, S. O. (1971). Phenolic compounds isolated from insect hard cuticle and their relationship to the sclerotization process. *Insect Biochem. 1*, 157–170.

ANDERSEN, S. O. (1972). An enzyme from locust cuticle involved in the formation of cross-links from *N*-acetyldopamine. *J. Insect Physiol. 18*, 527–540.

ANDERSEN, S. O. (1974a). Cuticular sclerotization in larval and adult locusts, *Schistocerca gregaria*. *J. Insect Physiol. 20*, 1537–1552.

ANDERSEN, S. O. (1974b). Evidence for two mechanisms of sclerotization in insect cuticle. *Nature 251*, 507–508.

ANDERSEN, S. O. (1975). Cuticular sclerotization in the beetles, *Pachynoda epphipiata* and *Tenebrio molitor*. *J. Insect Physiol. 21*, 1225–1232.

ANDERSEN, S. O. (1976). Cuticular Enzymes and Sclerotization in Insects. In *The Insect Integument*. Edited by H. R. Hepburn. Pages 121–144. Elsevier, Amsterdam, Oxford and New York.

ANDERSEN, S. O. (1977). Arthropod cuticles: their composition, properties and functions. *Symp. Zool. Soc. London 39*, 7–32.

ANDERSEN, S. O. (1978). Characterization of a trypsin-solubilized phenoloxidase from locust cuticle. *Insect Biochem. 8*, 143–148.

ANDERSEN, S. O. (1979a). Biochemistry of insect cuticle. *Ann. Rev. Ent. 24*, 29–61.

ANDERSEN, S. O. (1979b). Characterization of the sclerotization enzyme(s) in locust cuticle. *Insect Biochem. 9*, 233–239.

ANDERSEN, S. O. and BARRETT, F. M. (1971). The isolation of ketocatechols from insect cuticle and their possible role in sclerotization. *J. Insect Physiol. 17*, 69–83.

ANDERSEN, S. O. and ROEPSTORFF, P. (1978). Phenolic compounds released by mild acid hydrolysis from sclerotized cuticle: purification, structure, and possible origin from cross-links. *Insect Biochem. 8*, 99–104.

ANDERSEN, S. O. and ROEPSTORFF, P. (1982). Sclerotization of insect cuticle. III. An unsaturated derivative of *N*-acetyldopamine and its role in sclerotization. *Insect Biochem. 12*, 269–276.

ANDERSEN, S. O., JACOBSEN, J. P. and ROEPSTORFF, P. (1980). Studies of the sclerotization of insect cuticle. The structure of a dimeric product formed by incubation of *N*-acetyldopamine with locust cuticle. *Tetrahedron 36*, 3249–3252.

ANDERSEN, S. O., THOMPSON, P. R. and HEPBURN, H. R. (1981). Cuticular sclerotization in the honeybee (*Apis mellifera adansonii*). *J. Comp. Physiol. 145*, 17–20.

ASHIDA, M., DOHKE, K. and OHNISHI, E. (1974). Activation of prephenoloxidase. III. Release of a peptide from prephenoloxidase by the activating enzyme. *Biochem. Biophys. Res. Commun. 57*, 1089–1095.

ATKINSON, P. W., BROWN, W. V. and GILBY, A. R. (1973). Phenolic compounds from insect cuticle. Identification of some lipid antioxidants. *Insect Biochem. 3*, 309–315.

BARRETT, F. M. (1977). Recovery of ketocatechols from exuviae of last instar larvae of the cicada, *Tibicen pruinosa*. *Insect Biochem. 7*, 209–214.

BARRETT, F. M. and ANDERSEN, S. O. (1981). Phenoloxidases in larval cuticle of the blowfly, *Calliphora vicina*. *Insect Biochem. 11*, 17–23.

BODNARYK, R. P. (1971a). *N*-terminal β-alanine in the puparium of the fly *Sarcophaga bullata*: evidence from kinetic studies of its release by partial acid hydrolysis. *Insect Biochem. 1*, 228–236.

BODNARYK, R. P. (1971b). Studies on the incorporation of β-alanine into the puparium of the fly, *Sarcophaga bullata*. *J. Insect Physiol. 17*, 1201–1210.

BODNARYK, R. P. and LEVENBOOK, L. (1969). The role of β-alanyl-L-tyrosine (sarcophagine) in puparium formation in the fleshfly *Sarcophaga bullata*. *Comp. Biochem. Physiol. 30*, 909–921.

BRUNET, P. C. J. (1963). Tyrosine metabolism in insects. *Ann. N.Y. Acad. Sci. 100*, 1020–1034.

BRUNET, P. C. J. (1980). The metabolism of the aromatic amino acids concerned in the cross-linking of insect cuticle. *Insect Biochem. 10*, 467–500.

BRUNET, P. C. J. and COLES, B. C. (1974). Tanned silks. *Proc. Roy. Soc. B. 187*, 133–170.

BRUNET, P. C. J. and KENT, P. W. (1955). Observations on the mechanism of a tanning reaction in *Periplaneta* and *Blatta*. *Proc. Roy. Soc. B. 144*, 259–274.

CAVENEY, S. (1971). Cuticle reflectivity and optical activity in scarab beetles: the role of uric acid. *Proc. Roy. Soc. B. 178*, 205–225.

CHEN, T. and HODGETTS, R. B. (1976). Some properties of dopa decarboxylase from *Sarcophaga bullata*. *Comp. Biochem. Physiol. 53B*, 415–418.

COLES, G. C. (1966). Studies on resilin biosynthesis. *J. Insect Physiol. 12*, 679–691.

COTTRELL, C. B. (1964). Insect ecdysis with particular emphasis on cuticular hardening and darkening. *Adv. Insect Physiol. 2*, 175–218.

DELACHAMBRE, J., DELBECQUE, J. P., PROVANSAL, A., GRILLOT, J. P., DE REGGI, M. L. and CAILLA, H. L. (1979). Total and epidermal cyclic AMP levels related to the variations of ecdysteroids and bursicon during the metamorphosis of the mealworm *Tenebrio molitor* L. *Insect Biochem. 9*, 95–99.

DENNELL, R. (1958a). The hardening of insect cuticles. *Biol. Rev. 33*, 178–196.

DENNELL, R. (1958b). The amino acid metabolism of a developing insect cuticle: the larval cuticle and puparium of *Calliphora vomitoria*. II. The non-specific hydroxylation of aromatic amino acids and the production of polyphenols by the cuticle. *Proc. Roy. Soc. B. 148*, 280–284.

DEWHURST, S. A., CROKER, S. G., IKEDA, K. and MCCAMAN, R. E. (1972). Metabolism of biogenic amines in *Drosophila* nervous tissue. *Comp. Biochem. Physiol. 43B*, 975–981.

DOHKE, K. (1973). Studies on prephenoloxidase-activating enzyme from cuticle of the silkworm *Bombyx mori*. II. Purification and characterization of the enzyme. *Arch. Biochem. Biophys. 157*, 210–221.

FOGAL, W. and FRAENKEL, G. (1969). The role of bursicon in melanization and endocuticle formation in the adult fleshfly *Sarcophaga bullata*. *J. Insect Physiol. 15*, 1235–1247.

FRAENKEL, G. and HSIAO, C. (1965). Bursicon, a hormone which mediates tanning of the cuticle in the adult fly and other insects. *J. Insect Physiol. 11*, 513–556.

FRAGOULIS, E. G. and SEKERIS, C. E. (1975). Induction of dopa (3,4-dihydroxyphenylalanine) decarboxylase in blowfly integument by ecdysone. *Biochem. J. 146*, 121–126.

HACKMAN, R. H. (1959). Biochemistry of the insect cuticle. *Proc. 4th Int. Congr. Biochem. Vienna 1958 12*, 48–62.

HACKMAN, R. H. (1971). The integument of Arthropoda. In *Chemical Zoology*. Edited by M. Florkin and B. T. Scheer. Vol. 6, pages 1–62. Academic Press, New York and London.

HACKMAN, R. H. (1974). Chemistry of the insect cuticle. In The *Physiology of Insecta*. Edited by M. Rockstein. Vol. 6, pages 215–270. Academic Press, New York and London.

HACKMAN, R. H. and GOLDBERG, M. (1967). The *o*-diphenoloxidases of fly larvae. *J. Insect Physiol. 13*, 531–544.

HEYNEMAN, R. A. and VERCAUTEREN, R. E. (1964). Activation of the latent phenoloxidase of *Tenebrio molitor* (Ins. Coleoptera). *Enzymologia 28*, 85–88.

KARLSON, P. and AMMON, H. (1963). Zum Tyrosinstoffwechsel der Insekten. XI. Biogenese und Schicksal der Acetylgruppe des *N*-Acetyldopamins. *Z. Physiol. Chem. 330*, 161–168.

KARLSON, P. and SCHWEIGER, A. (1961). Zum Tyrosinstoffwechsel der Insekten. IV. Das Phenoloxydase-System von *Calliphora* und seine Beeinflussing durch das Hormon Ecdyson. *Z. Physiol. Chem. 323*, 199–210.

KARLSON, P. and SEKERIS, C. E. (1962). *N*-acetyldopamine as sclerotizing agent of the insect cuticle. *Nature 195*, 183–184.

KARLSON, P. and WECKER, E. (1955). Die Tyrosinaseaktivität während der Pupariumbildung von *Calliphora erythrocephala*. *Z. Physiol. Chem. 300*, 42–48.

KOEPPE, J. K. and GILBERT, L. I. (1974). Metabolism and protein transport of a possible pupal cuticle tanning agent in *Manduca sexta*. *J. Insect Physiol. 20*, 981–992.

LEVENBOOK, L., BODNARYK, R. P. and SPANDE, T. F. (1969). β-Alanyl-L-tyrosine. Chemical synthesis, properties and occurrence in larvae of the fleshfly *Sarcophaga bullata* Parker. *Biochem. J. 113*, 837–841.

LOCKE, M. (1969). The localization of a peroxidase associated with hard cuticle formation in an insect, *Calpodes ethlius* Stoll, Lepidoptera, Hesperiidae. *Tissue Cell 1*, 555–574.

LUNAN, K. D. and MITCHELL, H. K. (1969). The metabolism of tyrosine-O-phosphate in *Drosophila*. *Archiv. Biochem. Biophys. 132*, 450–456.

MARANDA, B. and HODGETTS, R. (1977). A characterization of dopamine acetyltransferase in *Drosophila melanogaster*. *Insect Biochem. 7*, 33–43.

MARKS, E. P. and SOWA, B. A. (1976). Cuticle Formation in vitro. In *The Insect Integument*. Edited by H. R. Hepburn, pages 330–375. Elsevier, Amsterdam, Oxford and New York.

MASON, H. S. (1955). Comparative biochemistry of the phenolase complex. *Adv. Enzymol. 16*, 105–183.

MILLS, R. R. and WHITEHEAD, D. L. (1970). Hormonal control of tanning in the American cockroach: changes in blood cell permeability during ecdysis. *J. Insect Physiol. 16*, 331–340.

MIR, A. K. and VAUGHAN, P. F. T. (1981). Biosynthesis of *N*-acetyldopamine and *N*-acetyloctopamine by *Schistocerca gregaria* nervous tissue. *J. Neurochem. 36*, 441–446.

MITCHELL, H. K., WEBER, U. M. and SCHAAR, G. (1967). Phenol oxidase characteristics in mutants of *Drosophila melanogaster*. *Genetics 57*, 357–368.

MITSUI, T. and RIDDIFORD, L. M. (1976). Pupal cuticle formation by *Manduca sexta* epidermis in vitro: Patterns of ecdysone sensitivity. *Devel. Biol. 54*, 172–186.

MUNN, E. A. and BUFTON, S. F. (1973). Purification and properties of a phenoloxidase from the blowfly *Calliphora erythrocephala*. *Eur. J. Biochem. 35*, 3–10.

OHNISHI, E. (1954). Tyrosinase in *Drosophila virilis*. *Annotnes Zool. Jap. 27*, 33–39.

PAU, R. N. and ACHESON, R. M. (1968). The identification of 3-hydroxy-4-O-β-D-glucosidobenzyl alcohol in the left colleterial gland of *Blaberus discoidalis*. *Biochim. Biophys. Acta 158*, 206–211.

PAU, R. N. and KELLY, C. (1975). The hydroxylation of tyrosine by an enzyme from third-instar larvae of the blowfly *Calliphora erythrocephala*. *Biochem. J. 147*, 565–573.

POST, L. C. (1972). Bursicon: its effect on tyrosine permeation into insect haemocytes. *Biochim. Biophys. Acta 290*, 424–428.

PRESTON, J. W. and TAYLOR, R. L. (1970). Observations on the phenoloxidase system in the haemolymph of the cockroach *Leucophaea maderae*. *J. Insect Physiol. 16*, 1729–1744.

PRYOR, M. G. M. (1940a). On the hardening of the ootheca of *Blatta orientalis*. *Proc. Roy. Soc. B 128*, 378–393.

PRYOR, M. G. M. (1940b). On the hardening of the cuticle of insects. *Proc. Roy. Soc. B 128*, 393–407.

PRYOR, M. G. M. (1962). Sclerotization. In *Comparative Biochemistry*. Edited by M. Florkin and H. S. Mason. Vol. 4, part B, pages 371–396. Academic Press, New York.

RICHARDS, A. G. (1951). *The Integument of Arthropods*. University of Minnesota Press, Minneapolis.

RICHARDS, A. G. (1958). The cuticle of arthropods. *Ergeb. Biol. 20*, 1–26.

RICHARDS, A. G. (1967). Sclerotization and the localization of brown and black colors in insects. *Zool. Jahrb. Anat. 84*, 25–62.

RICHARDS, A. G. (1978). The chemistry of insect cuticle. In *Biochemistry of Insects*. Edited by M. Rockstein. Pages 205–232. Academic Press, New York, San Francisco and London.

SEKERIS, C. E. (1963). Zum Tyrosinstoffwechsel der Insekten. XII. Reinigung, Eigenschaften und Substratspecifität der Dopa-Decarboxylase. *Z. Physiol. Chem. 332*, 70–78.

SEKERIS, C. E. and HERLICH, P. (1966). Zum Tyrosinstoffwechsel der Insekten. XVII. Der Tyrosinstoffwechsel von *Tenebrio molitor* und *Drosophila melanogaster*. *Z. Physiol. Chem. 344*, 267–275.

SEKERIS, C. E. and KARLSON, P. (1962). Zum Tyrosinstoffwechsel der Insekten. VII. Der katabolische Abbau des Tyrosins und die Biogenese der Sklerotisierungsubstanz, *N*-Acetyldopamine. *Biochim. Biophys. Acta 62*, 103–113.

SEYBOLD, W. D., MELTZER, P. S. and MITCHELL, H. K. (1975). Phenoloxidase activation in *Drosophila*. A cascade of reactions. *Biochem. Genet. 13*, 85–108.

VINCENT, J. F. V. and HILLERTON, J. E. (1979). The tanning of insect cuticle — a critical review and a revised mechanism. *J. Insect Physiol. 25*, 653–658.

WHITEHEAD, D. L. (1970). L-dopa-decarboxylase in the haemocytes of diptera. *FEBS Lett. 7*, 263–266.

WIGGLESWORTH, V. B. (1957). The physiology of insect cuticle. *Ann. Rev. Ent. 2*, 37–54.

WRIGHT, T. R. F., BEWLEY, G. C. and SHERALD, A. F. (1976). The genetics of dopa decarboxylase in *Drosophila melanogaster*. II. Isolation and characterization of dopa-decarboxylase-deficient mutants and their relationship to the α-methyl-dopa hypersensitive mutants. *Genetics 84*, 287–310.

YAMAZAKI, H. I. (1969). The cuticular phenoloxidase in *Drosophila virilis*. *J. Insect Physiol. 15*, 2203–2211.

YAMAZAKI, H. I. (1972). Cuticular phenoloxidase from the silkworm *Bombyx mori*: properties, solubilization, and purification. *Insect Biochem. 2*, 431–444.

3 Chitin Metabolism in Insects*

KARL J. KRAMER

U.S. Grain Marketing Research Laboratory and Kansas State University, USA

CAROL DZIADIK-TURNER

Kansas University Medical Center, Kansas City, Kansas, USA

and

DAIZO KOGA

Yamaguchi University, Japan

1 INTRODUCTION

In the course of its evolution the insect has developed a cuticle composed primarily of chitin and cross-linked protein (sclerotin) that enables it to cope with the complexities of its environment, providing for growth, mobility, protection and communication. The former component of cuticle, chitin, is the simplest of the glycosaminoglycans and is a $\beta(1\rightarrow4)$ homopolymer of *N*-acetylglucosamine $(C_8H_{13}O_5N)_{n\gg1}$. This polysaccharide also occurs in nature combined with other materials in algae, fungi and marine invertebrates. In cuticle and

* Contribution No. 82–387-B-j, Department of Biochemistry, Kansas Agricultural Experiment Station, Manhattan, Kansas 66506. Cooperative investigation between ARS, USDA, and the Kansas Agricultural Experiment Station. Mention of a proprietary product in this paper does not imply approval by the USDA to the exclusion of other products that may also be suitable. Supported in part by USDA competitive research grants 3001–0217 and 82–1–1055 to KJK.

peritrophic membrane it is thought to exist as long microfibrils strongly bound to protein and possibly to pigments. The morphogenesis of the chitin–protein complex into cuticle is one of the many mysteries in developmental biology and can be understood adequately only by comprehending both chitin and protein assembly, whether self or directed, as well as the synthesis and catabolism of these components. This chapter is concerned with the metabolism of chitin, which is not understood in sufficient detail to justify making broad generalizations. Insects make great use of chitin in structural tissues, in which respect they differ markedly from higher animals. We think that some of these pathways are relatively unique in the animal kingdom and thus they may become targets for precisely designed biochemical pesticides. It is no doubt apparent that the potential for exploitation of chitinous materials in insects is very great.

The metabolism of chitin in micro-organisms is better understood than in other life forms. In recent years research on chitin in insects has intensified because of the serendipitous discovery of insecticides which apparently block the formation of chitin. Many laboratories are now characterizing in more precise biochemical terms how chitin is synthesized, degraded and assembled into supramolecular structures in many different organisms. Consequently, it is difficult to make a comprehensive review of chitin metabolism which does not soon become outdated. The following is an effort to summarize the precarious state of our knowledge in this field of research involving insects.

2 CUTICLE STRUCTURE AND COMPOSITION

Insect cuticle is composed of chitin and protein plus several other components (lipids, salts, pigments, etc.) which contribute only a few percent to its total dry weight. Chitin can vary from as little as several percent to as high as 60% of the total dry weight (Richards, A., 1978). This structural polysaccharide is therefore of utmost importance to the growth and development of insects. Before focusing our discussion on chitin alone, it is worthwhile to review cuticle structure in general and the interaction of chitin with the other components.

Research investigating arthropod cuticle has increased in recent years, and the topic has been approached in a variety of ways. These studies use physical, chemical, kinetic and immunological techniques to answer important and pressing questions in molecular biology, e.g. control of metamorphosis, the mechanism of hormone induction, structure/function relationships of macromolecules and receptor sites, chromosomal puffing, supramolecular assembly and insect control. Reviews have been published discussing various aspects of cuticle biology, and include several comprehensive texts, one citing work done before 1950 (Richards, A., 1951) and two citing more recent papers (Neville, A., 1975; Hepburn, H., 1976). Shorter reviews covering selected topics include those by Wigglesworth, V. (1948, 1957, 1970); Hackman, R. (1947a); Locke, M. (1974); Richards, A. (1978) and Andersen, S. (1979 and this volume). Hepburn, H., discusses integument structure elsewhere in this volume.

The cuticle of insects cannot be viewed as a static structure as it changes almost continuously throughout the life of the animal. Structure and composition differ from species to species, and even differ from one part of an animal to another. An idealized overview will be given, though, to formulate a simple model. The terminology has changed over the years, making a review of the literature confusing, but the generally accepted terms are those suggested by Weis-Fogh, T. (1970).

The basic structure of the insect integument comprises a layer of epithelial cells covered by an overlaying cuticle which the cells have secreted (Fig. 1; see M. Locke, vol. 2). The cuticle is multilayered, each layer having been sequentially deposited by the cells, the outermost being deposited first. The horizontal layers are traversed by vertical pore canals, through which cell secretions can be transferred to the surface of the cuticle.

The cuticle can be divided into several sections: the epicuticle, exocuticle, mesocuticle and endocuticle, the last three making up the procuticle. The former covers the whole of the insect except for chemoreceptors, midgut and the ends of gland cells. Less than 1 μm in depth (Richards, A., 1978), it is fairly uniform throughout the animal (Neville, A., 1975). The general opinion is that the epicuticle contains no chitin (Andersen, S., 1979) although some authors have disagreed (Krishnan, G. *et al.*, 1955,

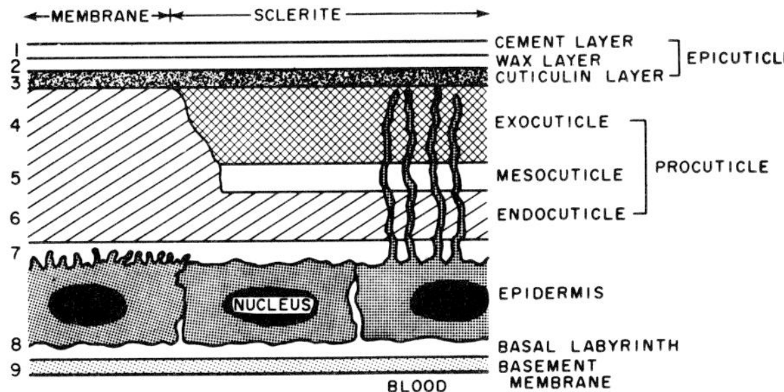

FIG. 1. Idealized cuticle structure. A diagrammatic section through the integument of an insect showing two types of cuticle: membrane and sclerite. The procuticle layer(s) are the chitin-containing portion of the cuticle. The horizontal layers are transversed by the pore canals which provide a route from the epidermal cells to the epicuticle.

Krishnan, G., 1956). The outermost layer of epicuticle, not always present, is the cement layer. Composed of tanned protein and lipid, its probable role is to protect the wax layer just beneath it. Studies where the wax was damaged or dissolved have demonstrated that the wax layer is essential for maintenance of the insect's water content (Ramsay, J., 1935; King, G., 1944; Wigglesworth, V., 1944, 1945; G. Blomquist and J. Dillwith, this volume).

The procuticle is that portion of the cuticle which contains chitin. The properties of the procuticle differ widely from one part of the animal to another (Weis-Fogh, T., 1970) and most probably depend on the particular combination of chitin and proteins present (Neville, A., 1975).

The exo-, endo- and mesocuticular layers were initially described on the basis of histological staining. When exposed to Mallory's triple strain the exocuticle stained blue, the mesocuticle red and the endocuticle was refractile to the stain (Richards, A., 1951). The exocuticle is secreted before ecdysis and appears in the electron microscope to have chitin fibrils in what has been called a "highly crystalline pattern". The proteins are tanned and this layer is not dissolved at ecdysis. The mesocuticle is less organized and untanned. The endocuticle, which is next to the epidermal cells, appears striated in electron micrographs (Wigglesworth, V., 1970; Neville, A., 1975). The striations are growth layers that are visible because of a change in the orientation of the chitin microfibrils (Neville, A., 1967a,b, 1975). The endocuticle is resorbed from one instar to the next and in starvation can be used as a nutrient source (Neville, A., 1975).

The evolutionary selection of an exoskeleton composed of crystalline chitin and cross-linked protein has dictated a pattern of discontinuous growth for insects who must shed the exoskeleton and replace it by a larger "shell" periodically (Weis-Fogh, T., 1970; Jeauniaux, C., 1971). In general terms a new cuticle partly forms within the old cuticle, the inner portion of the old cuticle is dissolved and resorbed, and then the outermost layer (exuvium) of the old cuticle is shed. In many larval insects, exuviae are consumed by the new instar so that many components are recycled.

Morphological studies during insect development have contributed to the understanding of the individual components of cuticle and their dynamics during the molting process (Locke, M., 1966, 1970; Zacharuk, R., 1972, 1976; Kunkel, J., 1975, Kayser-Wegmann, I., 1976; Sedlak, B. and Gilbert, L., 1979; Price, J. and Holdich, O., 1980; Dorn, A. and Hoffmann, P., 1981; Wolfgang, W. and Riddiford, L., 1981). The chitin microfibrils as visualized by the electron microscope seem to originate directly at the epidermal cell border, reinforcing the idea that the enzyme chitin synthetase is located in the cell membrane (Weis-Fogh, T., 1970). Chitin is known to be produced from precursors located in the epidermal cells (Coles, G., 1966; Condoulis, W. and Locke, M., 1966), but the origin for cuticle proteins is less clear (Locke, M., 1969; Locke, M. and Krishnan, N., 1971; Fox, F. *et al.*, 1972; Koeppe, J. and Gilbert, L., 1973; Phillips, D. and Loughton, G., 1976; Geiger, J. *et al.*, 1977).

The cuticle proteins of interest here are those associated with chitin. They can be classified into two groups:

(1) those which bind chitin covalently or non-covalently and contribute to the supramolecular assembly and disassembly of cuticle; and
(2) those which are directly responsible for the synthesis and degradation of chitin.

Interactions between chitin and cuticle proteins can be weak and non-specific or strong and involve covalent linkages. For example, cuticular proteins from *Calliphora* larvae bind crab chitin non-covalently and non-specifically (Hackman, R. and Goldberg, M., 1978). There have been several reports of amino acid or protein residues in chitin preparations (Hackman, R., 1960; Hunt, S., 1970) and some covalently bound chitin protein mucopolysaccharides have been fractionated (Brine, C. and Austin, P., 1981b). Poly-*N*-acetylglucosamine from sarcophagid puparial cases remains covalently conjugated with peptides even after extensive chemical and enzymatic degradation, as well as chromatographic separation (peptidochitodextrins, Lipke, H. and Geoghegan, T., 1971a; Lipke, H. and Strout, V., 1972).

The differences in solubility properties between chitin and protein sometimes make it possible to carry out their chemical characterization separately. Unlike most structural macromolecules, many of the cuticle proteins are water-soluble, at least until they become sclerotized (see S. Andersen, this volume). The following remarks will deal with the chemical properties of cuticular proteins. Chitin will be addressed in the next section.

To date the study of cuticle has not resulted in a systematic picture of the structure, synthesis or degradation of its protein components, although some generalizations can be made. The most common method of studying cuticle protein is to subject the cuticle tissue to a series of extraction buffers (Hackman, R. and Goldberg, M., 1958; Mills, R. *et al.*, 1967; Srivastava, R., 1970; Sharma, S. and Pant, R., 1973) and then to characterize the components of each extraction, in recent years relying on data from gel electrophoresis (D. Silvert, vol. 2). Investigations of the properties of proteins extracted from insect cuticle have been conducted primarily by Richards, A. (1951); Andersen, S. (1971); Andersen, S. and Weis-Fogh, T. (1964) and Hackman, R. (1959, 1967, 1971, 1972, 1974a,b, 1975, 1976); Hackman, R. and Goldberg, M. (1958, 1967, 1971, 1976, 1977, 1978, 1979). Attempts also have been made to study proteins in different layers of cuticle separately (Welinder, B., 1975).

Cuticular proteins tend to have low isoelectric points (pH 3–6), relatively low molecular weights (10–80 K, especially <20 K) and they contain a rather high percentage of amino acids with hydrophobic side chains (Richards, A., 1978). More recently structural proteins from sarcophagid larval exoskeleton have been partially characterized as part of a study on cuticle sclerotization (Lipke, H. *et al.*, 1981).

Lipids represent only a small percentage of the total dry weight of cuticle, but their role as a water barrier between the insect and its environment is crucial for survival (Wigglesworth, V., 1944, 1945, 1948; G. Blomquist and J. Dillwith, this volume). No reports of chitin having an interaction with lipid are known, although chitin–lipid complexes are indeed very possible, at least non-covalent ones. Cuticular lipids are extremely heterogeneous and do not possess the degree of similarity apparent among cuticle proteins (Richards, A., 1951; Howard, R. and Blomquist, G., 1982).

3 CHITIN STRUCTURE AND OCCURRENCE

Chitin was first described in 1811 by Henri Beaconnot, a member of the French Academy of Science. In the course of his research on mushrooms he treated specimens with warm, dilute alkali and isolated chitin, a product he termed "fungine". Odier, A. (1823) found the same substance in insect cuticle, which he renamed "chitin" from the Greek word for tunic or envelope. The chemical composition of chitin (glucosamine and acetic acid) was first described by Ledderhose, G. (1878) and confirmed by Gilson, E. (1894). In spite of this information, the idea that chitin could be defined as a polymer remained controversial for many years. Many believed it to be a mixture of glucosamine and an unspecified nitrogen compound. Chitin was found to be a polysaccharide by Purchase, E. and Braun, C., in 1946. More recently there has been renewed interest in chitin chemistry, its commercial preparation and its laboratory uses, prompting several reviews and conferences (e.g. Kent, P., 1964; Brimacombe, J. and Webber, J., 1964; Hunt, S., 1970; Muzzarelli, R., 1977; Muzzarelli, R. and Pariser, E., 1978; Austin, P. *et al.*, 1981).

Chitin is widely distributed in animals and represents the skeletal polysaccharide of several phyla, i.e. Arthropoda, Annelida, Mollusca and Coelenterata (Jeuniaux, C., 1963; Kent, P., 1964; Neville, A., 1975). In several groups of fungi, chitin replaces cellulose as the structural polysaccharide. In insects chitin has been found in the body wall, gut lining, cuticle and muscle attachment points in all species tested (Richards, A., 1951).

Chitin is a polymer of *N*-acetylglucosamine joined in $\beta(1\text{–}4)$ glycosidic links (Fig. 2). Its

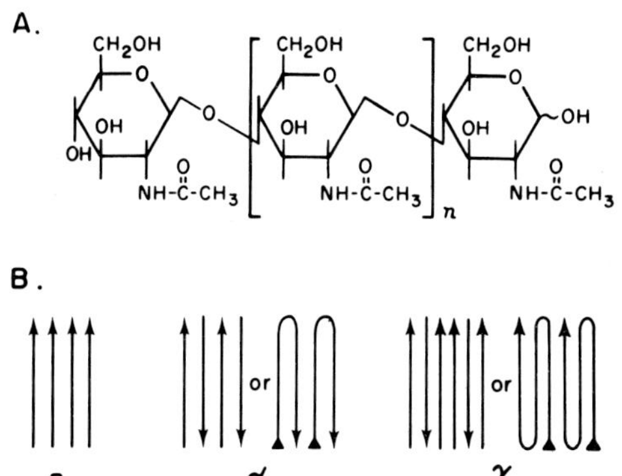

FIG. 2. Chitin structural components. (**A**) $\beta(1\rightarrow4)$ linked 2-acetamido-2-deoxy-D-glucose, the repeating chitobiose unit of a chitin chain. (**B**) The molecular interpretation of the arrangement of chitin chains in the unit cells of three crystallographic types of chitin. The forms are termed, left to right: β (parallel chains), α (antiparallel chains) and γ (two chains up for every chain down). All three types are normally associated with protein.

structure may be considered a substituted cellulose with an *N*-acetyl moiety replacing the OH on carbon-2 found in cellulose. Because of their chemical similarities, research on chitin and cellulose has often been intertwined. The completely *N*-acetylated chitin polymer may only be an idealized structure, however, as studies with lobster chitin show the acetyl group is missing on some residues (Giles, C. *et al.*, 1958). The free amino group may serve as a nucleophilic attachment site for other components in cuticle. *In situ* insect chitin is present as microfibrils and is surrounded by cuticular proteins. Depending on the type of cuticle, the chitin/protein matrix contains different relative amounts of each component. Highly elastic cuticle has relatively low levels of chitin (Andersen, S., 1979). The fibrilar arrangement suggests that only the outermost chitin chains in any group are available for bonding with cuticle protein. Some researchers contend, however, that chitin in cuticle is saturated with protein and that non-covalent binding between the two is non-specific (Hackman, R. and Goldberg, M., 1978). The chitin microfibrils in cuticle are organized in any of several patterns which have been studied in detail with the electron microscope (Neville, A., 1975). The microfibril model replaces an earlier model of cuticle structure in which chitin and protein were described as monomolecular sheets occurring alternately (Neville, A. *et al.*, 1976).

Early attempts to describe and identify chitin were based on its solubility properties, i.e. it being the cuticular component insoluble in hot, dilute alkali. By the early 1950s several color reactions were used to identify chitin in biological specimens or histological sections (Richards, A., 1951). The chitosan/iodine color test uses alkali to deacetylate chitin, forming chitosan; chitosan plus iodine, in the presence of acid, turns violet. The diphenol–iodine–zinc chloride test also gives a blue/violet color in the presence of chitin. The Schiff polysaccharide reaction has been used to indicate chitin, but this non-specific reaction cannot distinguish chitin from other polysaccharides. Calcofluor, a dye that seems to stain chitin selectively in yeast cell walls, has been used to study patterns of chitin synthesis by fluorescent microscopy (Cabib, E. and Bowers, B., 1975; Sloat, B. and Pringle, J., 1978; Sloat, B. *et al.*, 1981). Chitin has also been localized in cell walls with electron microscopy of colloidal gold particles with wheat germ agglutinin or chitinase (Molano, J. *et al.*, 1980). More recently, fluorescein-labeled isothiocyanate conjugated with wheat germ agglutinin or with chitinase has been used to locate chitin *in situ* (Benjaminson, M., 1969; Dempster, A. and McFarlane, J., 1981; Galun, M. *et al.*, 1981). Methods for estimating the amount of chitin in biological samples have been reviewed elsewhere (Hackman, R., 1974a; Muzzarelli, R., 1977). Hackman, R. and Goldberg, M. (1981) have reported a colorimetric assay useful for determining microgram amounts of chitin present in arthropod cuticles. The assay detects 2,5-anhydrohexoses produced from deacetylated chitin. Color is developed with 3-methyl-2-benzothiazolone hydrochloride and $FeCl_3$, and is stable for 90 min.

The preparation of chitin has been a constant problem in experimental work due to its limited solubility in aqueous systems. Chitin is soluble only in caustic solvents such as hot, concentrated mineral acids which may lead to covalent bond rearrangements (Muzzarelli, R., 1976). Procedures for the preparation of chitin have been published and reviewed elsewhere (Muzzarelli, R., 1977). The usual source for chitin is crab or shrimp shells. Most preparative procedures include long incubations in concentrated mineral acids or hydrogen peroxide solutions, exposing the sample to oxidizing conditions and to pH and temperature extremes. These harsh treatments are known to deacetylate chitin and to cause fragmentation of the long chains. Colloidal chitin can be prepared by dissolving chitin in cold, concentrated hydrochloric acid and then filtering it quickly into ice cold water (Berger, L. and Reynolds, D., 1958; Lingappa, Y. and Lockwood, J., 1962). This method has problems, because the product may not resemble native chitin. Oligomers of chitin have been prepared from native material for use in studying protein–saccharide interactions (see section 5.2). Hackett, C. and Chen, K. (1978) report a method that extracts chitin from ascospores of a fungus (*Sordria brevicollis*) and from eggs of a nematode (*Ascaris lumbricoides*) at neutral pH using an ozone–peroxide–EDTA solvent. Non-aqueous solvents such as hexafluoroacetates or chloroalcohols mixed with mineral acids can also extract chitin (Capozza, R., 1975; Austin, P., 1975a,b). A chitin–protein complex was prepared from *Heliothis zea* integument by boiling in 1% sodium dodecylsulfate for 2–4 h (Smith, R. *et al.*, 1981). All of these procedures can be applied to insect cuticle for the preparation of insect chitin.

Researchers have used various physical techniques to investigate the architecture of chitin polysaccharide, including electron microscopy (e.g. Hall, L. *et al.*, 1981; Bal, A. *et al.*, 1981), nuclear magnetic resonance (Saito, H. *et al.*, 1981a,b) and electron diffraction (Gemperle, A. *et al.*, 1982). The most successful description of the supramolecular structure, though, has come from X-ray diffraction studies. Descriptive models have been developed, but these must not be interpreted as representing chitin in general. Instead, they represent only those parts of a specimen that show a high degree of order. Examples of tissue that are especially suited to X-ray study, because of the minimal amounts of foreign material present, are diatom fibers (*Thalassiosira fluviatilis*) and cuttlefish pen (*Loligo*).

Deproteinization of the samples sharpens the pattern somewhat, but raises questions about the "nativeness" of the specimen under study (Rudall, K., 1976). Chitin chains are helical and contain two residues per turn. Three polymorphic forms of chitin have been described, differing in the arrangement of their polymer chains (Fig. 2). The α-chitin structure was proposed by Carlstrom, D. (1957), 1962) and is an antiparallel sheet. The hydrogen bonding pattern in α-chitin has been determined recently (Minke, R. and Blackwell, J., 1978). In the β-structure the chitin chains are oriented in parallel fashion (Lotmar, W. and Picken, L., 1950; Blackwell, J. *et al.*, 1965, 1967; Blackwell, J., 1969). The γ-structure is arranged in sets of three chains: two parallel and one anti-parallel (Rudall, K., 1962, 1963). α-Chitin occurs in arthropod cuticles and is the most common arrangement; β- and γ-chitins occur in cocoons (Kenchington, W., 1971). α-Chitin is the most stable form and β- and γ-chitins can change to the α-form after various treatments (Rudall, K. and Kenchington, W., 1973).

Recently a new model has been proposed for chitin–protein complexes based on an X-ray diffraction study of ichneumon fly (*Megarhyssa*) ovipositor (Blackwell, J. and Weih, M., 1980). The data support a model of a supercoil of 15–30 chitin chains as the core of a microfibril that is surrounded by a protein sheath, giving a total fibril diameter of 7.25 nm. The protein is composed of subunits spaced around the chitin core in a 6_1 helix (Fig. 3). The chitin–protein fibrils are arranged in a hexagonal network and exhibit a fiber repeat of *ca.* 15.3 nm, or a multiple thereof, possibly due to additional supercoiling. The fly ovipositor is thus seen as a highly oriented structure in which both chitin and protein components are ordered at the molecular level, making X-ray diffraction patterns discernible. The structure of chitin–protein complexes in cuticle may not be as regular as in this specialized organ, as cuticle proteins are known to differ depending on the type and age of the tissue (Hunt, S., 1970; Neville, A., 1975). Cuticle structure

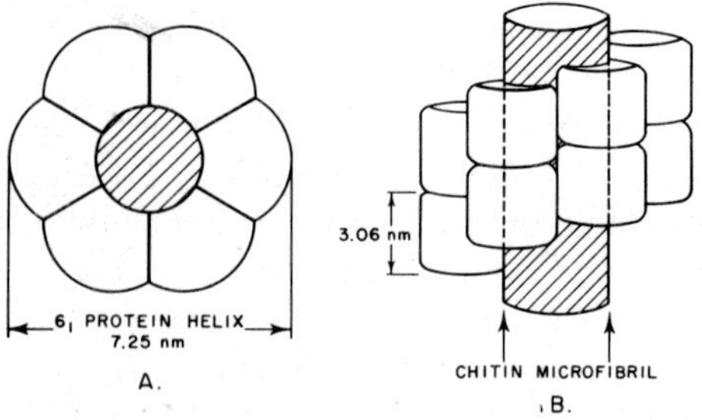

FIG. 3. Chitin/protein complex of α-chitin. The arrangement of protein subunits around the chitin core in a microfibril, based on X-ray fiber diagrams of ichneumon fly ovipositor (*Megarhyssa*). The fibril is viewed (**A**) perpendicular to the fiber axis and (**B**) along the fiber axis. The protein subunits are represented as globular structures that repeat in 3.06 nm along the axis. Adapted from Blackwell, J. and Weih, M. (1980) and used with permission.

may therefore prove to be too complex to be elucidated by the X-ray diffraction technique, and other physical techniques that have the capability of analyzing less ordered chitinous structures are needed. Other types of studies have corroborated the protein-wrapped-around-chitin structure. For example, it was found necessary to use the sequential action of protease followed by chitinase to disrupt the larval cuticle of corn earworm (*Heliothis zea*) (Smith, R. *et al.*, 1981).

4 CHITIN SYNTHESIS

Enzymatic processes that control the synthesis and breakdown of chitin play important roles in the growth, differentiation, nutrition, and behavior of insects. Researchers are slowly coming to understand how nature manipulates chitin in some organisms, in particular yeasts and fungi. Other species, including insects, are less well understood.

Chitin synthesis takes place via a metabolic pathway of eight or more steps converting hemolymph trehalose into the cuticle carbohydrate component (Fig. 4). The pathway was first outlined using cell-free extracts of the mold *Neurospora crassa* (Glaser, L. and Brown, D., 1957). Since then studies using insect tissues have elucidated the same general sequence, in addition to identifying a lipid intermediate, dolichyl diphosphate-*N*-acetylglucosamine, that may also be involved. For example, the epidermal cells of the desert locust (*Schistocerca gregaria*) contain all of the enzymes necessary to convert trehalose or glucose to uridine-diphospho-*N*-acetylglucosamine (Candy, D. and Kilby, B., 1962); similar results were obtained using cultured imaginal wing disks from *Drosophila* (Fristrom, J., 1968). A whole-animal preparation of the Southern armyworm (*Prodenia*) has been used to assess the activity of chitin synthetase (UDP-2-acetamido-2-deoxy-D-glucose: chitin 4-β-acetamidodeoxy-D-glucoyltransferase) at various developmental stages (Porter, C. and Jaworski, E., 1965). Synthetase activity was maximal late in the last instar and during the first 24 h of the pupal stage. More recently, an assay system using freshly dissected cuticle tissue has been described (Vardanis, A., 1976, 1979). Which of these enzymes is rate-limiting for chitin synthesis is unknown.

Most of the detailed knowledge of the chitin synthetase enzyme has been obtained from studies of yeast and fungal systems (Ruiz-Herrera, J. and Bartnicki-Garcia, S., 1974; Ruiz-Herrera, J. *et al.*, 1977; Braun, P. and Calderone, R., 1979a,b). Some of the many reports on chitin synthetase are listed in Tables 1 and 2. Chitin synthetase zymogen has been isolated from yeast in an almost completely inactive form (Cabib, E., 1972; Cabib, E. and Ulane, R., 1973). It is a very large, multisubunit protein that is activated by a protease (activating factor, AF) which appears to be identical to the proteinase B and tryptophan synthetase inactivase II identified in other yeast studies (Lenney, J., 1956; Hata, T. *et al.*, 1967; Cabib, E. and Ulane, R., 1973; Hasilik, A. and Holzer, H., 1973; Hasilik, A., 1974; Lenney, J., 1975; Ulane, R. and Cabib, E., 1976). AF appears to be a serine protease and is found in vesicles within the yeast cell. The action of AF can be completely blocked by the presence of an inhibitor (AFI) which has been purified (Ulane, R. and Cabib, E., 1974). AFI is a heat-stable protein with an apparent molecular weight of 8.5×10^3 daltons. AFI apparently binds to and directly inhibits AF since AFI has no effect on preactivated chitin synthetase (Fig. 5). Whether a similar activating mechanism occurs with insect chitin synthetase is unknown. Certain inhibitors may actually be hydrolytic enzymes. An "inhibitor" was isolated from fungal cytosol that turned out to be an endochitinase (Lopez-Romero, E. *et al.*, 1982).

Yeast chitin synthetase zymogen is uniformly distributed on the plasma membrane (Duran, A. *et al.*, 1979). The primary septum that develops between the mother and daughter cells is composed almost exclusively of chitin and is synthesized by local activation of the synthetase zymogen, presumably by directing AF-containing vesicles to appropriate sites on the membrane. This may be a general mechanism for morphogenic processes requiring growth in a specific direction and could be an important consideration for insect cuticle. The epidermal cell secretes and assembles components of cuticle on only one of its surfaces, resulting in precise spatial regulation of cuticle chitin synthesis.

In crustacea the relative rates of incorporation of glucose and *N*-acetylglucosamine are similar to those in locusts; glucose uptake is maximal after ecdysis and uptake of *N*-acetylglucosamine is

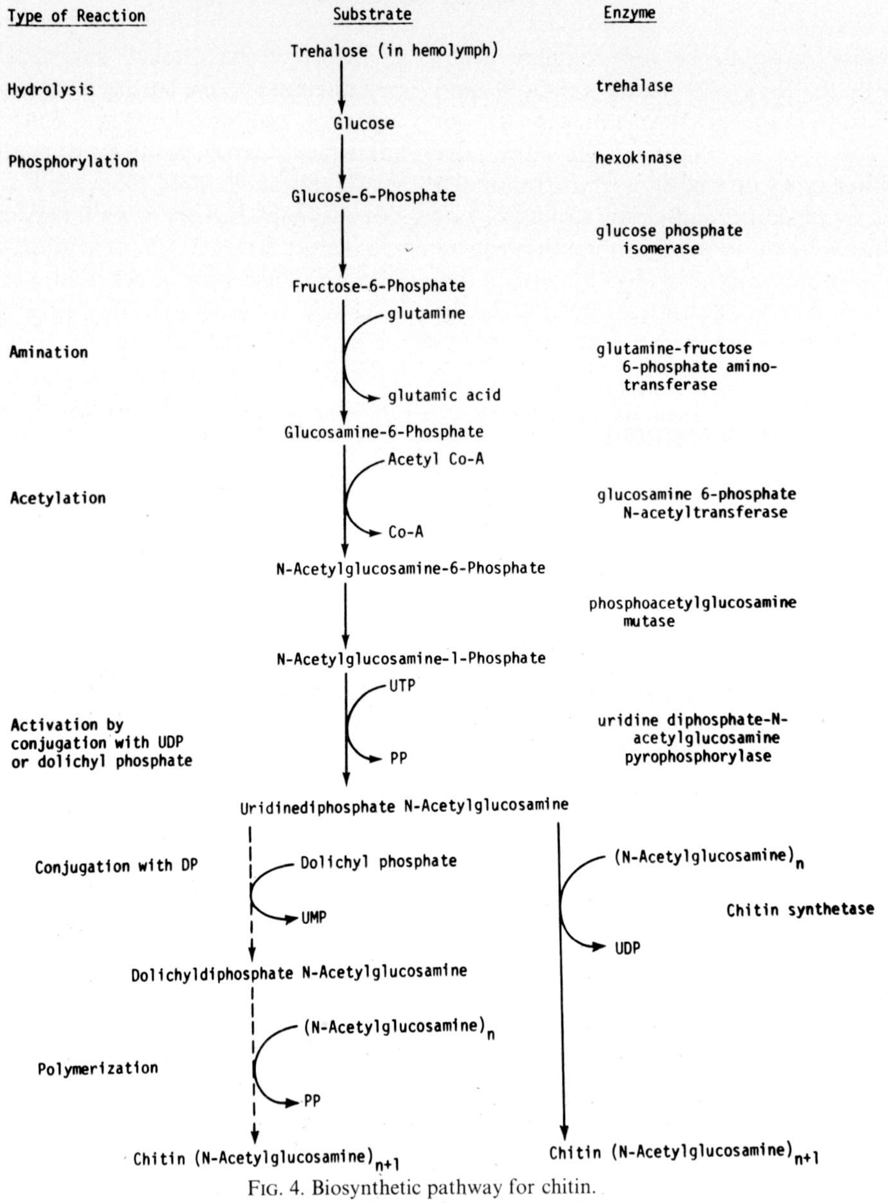

Type of Reaction
Substrate
Enzyme
Trehalose (in hemolymph)
Hydrolysis
trehalase
Glucose
Phosphorylation
hexokinase
Glucose-6-Phosphate
glucose phosphate isomerase
Fructose-6-Phosphate
glutamine
Amination
glutamine-fructose 6-phosphate amino-transferase
glutamic acid
Glucosamine-6-Phosphate
Acetyl Co-A
Acetylation
glucosamine 6-phosphate N-acetyltransferase
Co-A
N-Acetylglucosamine-6-Phosphate
phosphoacetylglucosamine mutase
N-Acetylglucosamine-1-Phosphate
UTP
Activation by conjugation with UDP or dolichyl phosphate
uridine diphosphate-N-acetylglucosamine pyrophosphorylase
PP
Uridinediphosphate N-Acetylglucosamine
Conjugation with DP
Dolichyl phosphate
UMP
(N-Acetylglucosamine)n
Chitin synthetase
UDP
Dolichyldiphosphate N-Acetylglucosamine
(N-Acetylglucosamine)n
Polymerization
PP
Chitin (N-Acetylglucosamine)n+1
Chitin (N-Acetylglucosamine)n+1

FIG. 4. Biosynthetic pathway for chitin.

Chitin Synthetase Zymogen
I
Activating Factor
Activating Factor · I
Chitin Synthetase
(NAG)n + UDP-N-acetylglucosamine
(NAG)n+1

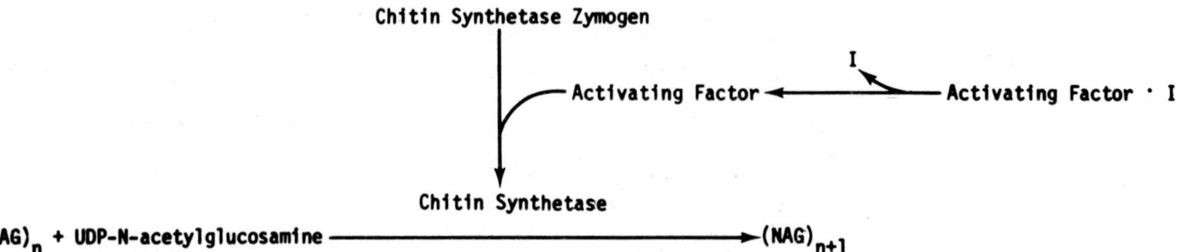

FIG. 5. Activation of chitin synthetase zymogen.

Table 1: Chitin synthetase in unicellular organisms

Organism	Comments	Reference
Blastocladiella emersonii (watermold)	Sensitivity of first hexosamine pathway-specific enzyme to end-product inhibition is *not* constant throughout life cycle.	Selitrennikoff, C. *et al.* 1980
	Vesicles derived from gamma particles resemble chitosomes and synthesized chitin when supplied with UDP-GlcNAc.	Mills, G. and Cantino, E., 1981
Candida albicans (dimorphic fungus)	Cytoplasmic component isolated that inhibited activation of chitin synthetase; MW $\cong$ 400 K; activity increased by treatment with phosphatidyl serine.	Braun, P. and Calderone, R., 1979
	Chitin synthetase activity in two polyene-resistant, ergosterol-deficient strains higher than parent strain. Chitin synthetase zymogen from mutant also more susceptible to trypsin digestion.	Pesti, M. *et al.* 1981
Mucor racemosus (yeast)	Chitin and chitosan syntheses in yeast and mycelial morphology compared and found to be three-fold higher in mycelial cells; modification of newly synthesized chitin occurs in both forms.	Domek, D. and Borgia, P., 1981
Mucor rouxii (yeast)	Cell wall microfibrils synthesized by soluble chitin synthetase in absence of cell membrane.	Ruiz-Herrera, J. and Bartnicki-Garcia, S., 1974
	Chitin synthetase particles (chitosomes) and microfibrils studied with EM; regard chitosomes as cytoplasmic conveyors of synthetase to cell surface.	Bracker, C. *et al.* 1976
	Chitin synthetase purified 120-fold; chitosomes with MW $> 7 \times 10^6$; required proteolytic activation.	Ruiz-Herrera, J. *et al.* 1977
	Amphotericin-B and nystatin tested as inhibitors of chitin synthetase; Am-B $K_i = 0.13$ mM, non-competitive.	Rast, D. and Bartnicki-Garcia, S., 1981
	Inhibition study with nucleoside–peptide antibiotics. Polyoxin A $K_i = 0.6\,\mu$M; Nikkomycin X $K_i = 0.5\,\mu$M; Nikkomycin Z $K_i = 3.5\,\mu$M, all competitive; Polyoxin A does *not* exhibit antifungal activity, nikkomycins do.	Muller, H. *et al.* 1981
	Digitonin stimulates chitin synthetase at low concentrations and inhibits at high concentrations; also causes release of 16S chitosome subunits with chitin synthetase activity.	Ruiz-Herrera, J. *et al.* 1980
Neurospora crassa (fungus)	Tunicamycin inhibited chitin synthetase in competitive manner with $K_i = 480\,\mu$M.	Selitrennikoff, C., 1979
Phycomyces (grease mold)	Benzoylphenylurea insecticides may act on activation of chitin synthetase zymogen and not directly on synthetase enzyme.	Leighton, T. *et al.* 1981
Rhodotorula glutinis (yeast)	pH change from 4.5 to 2 stimulates chitin synthesis; no proportional value in between decreases chitin content or increases total hexosamines.	Berthe, M. *et al.* 1981
Saccharomyces (yeast)	Yeast primary septum development sketched from EM studies.	Cabib, E., 1976
	Chitin synthetase zymogen distributed over yeast cell membrane; scar formed from local activation of zymogen.	Duran, A. *et al.* 1979
Saccharomyces carlsbergensis (yeast)	Chitin synthetase pH_{opt} 6.2; K_m for UDP-GlcNAc 0.6–0.9 mM, polyoxin A a potent inhibitor.	Keller, F. and Cabib, E., 1971
	Localization of chitin in bud scars studied by EM, before and after action of specific enzymes.	Cabib, E. and Bowers, B., 1971
	Purification of protein inhibitor of chitin synthetase activating factor, MW $\cong$ 8.5 K.	Ulane, R. and Cabib, E., 1974
Saccharomyces cerevisiae (yeast)	Heat-stable protein inhibitor of activating factor characterized.	Cabib, E. and Ulane, R., 1973
	Purification of protein inhibitor of chitin synthetase activating factor.	Ulane, R. and Cabib, E., 1974
	Chitin synthetase activating factor purified with affinity chromatography; MW $\cong$ 44 K; $pH_{opt} = 6.5–7$; serine protease inhibited by PMSF.	Ulane, R. and Cabib, E., 1976
	Assay method for chitin synthetase suitable for kinetic experiments described; mechanism is random or ping-pong.	Fahnrich, M. and Ahlers, J., 1981
	Temperature-sensitive mutant able to grow but cannot form buds. Defective gene product appears to be involved in selection of budding site and formation of chitin.	Sloat, B. *et al.* 1981

Table 2: Chitin synthetase studies in invertebrates

Organism	Tissue	Comments	Reference
Artemia salina (brine shrimp)	whole larvae	Chitin synthetase pH_{opt} 7, endogeneous acceptor and Mg^{2+} required; inhibited by UDP, diflubenzuron; not inhibited by polyoxin D.	Horst, M., 1981
Bombyx mori (silkmoth)	whole larvae	Radioactive tracer study after injection of [^{14}C]glucose. Glycogen synthesized early in feeding period was degraded and used as energy source and precursor in chitin synthesis.	Kimura, S., 1974b
Galleria mellonela (greater wax moth)	gut and peritrophic membrane	Mg^{2+} required for synthetase activity; inhibited by polyoxin D and UDP.	Cohen, E. and Casida, J., 1980a
	epidermis	*In vitro* incorporation of GlcNAc followed; chitin synthetase preceded by elevated ecdysone titer.	Ferkovich, S. *et al.* 1981
Hemigrapsus nudus (purple shore crab)	integument	Chitin synthetase activity increases in premolt and early postmolt, $K_m = 1.7 \times 10^{-4}$ M for UDP GlcNAc.	Hohnke, L., 1971
Leucophaea maderae (cockroach)	leg regenerates	Sensitive quantitative assay for inhibition of chitin synthetase using leg regenerates; $ID_{50} = 6.11 \times 10^{-10}$ M for diflubenzuron and 7.53×10^{-7} M for polyoxin D.	Sowa, B. and Marks, E., 1975
Locusta migratoria (grasshopper)	whole animal & abdominal integument	Incorporation of ^{14}C precursors of chitin followed; prior to imaginal ecdysis chitin precursors originated from old cuticle.	Surholt, B., 1975a, b
Manduca sexta (tobacco hornworm)	fifth stadium larvae	Diflubenzuron ($ID_{50} = 1.1 \times 10^{-9}$ M) inhibited glucose and glucosamine incorporation into chitin; decrease in cuticle production achieved when fed or applied topically.	Mitsui, T. *et al.* 1980
	cuticular disks	A short-term culturing method is described for analysis of chitin biosynthesis in larval cuticle. Incorporation of [^{14}C]GlcNAc followed.	Hettick, B. and Bade, M., 1978
Melanoplus sanguinipes (grasshopper)	abdominal integument	*In vitro* assay system for chitin synthesis labeled glucose, GlcNAc or UDP-GlcNAc as substrate; linear rate for 2 h.	Vardanis, A., 1976
	integument	Results consistent with chitin synthesis pathway in fungus; chitin synthetase accessible to substrate only from cell exterior.	Vardanis, A., 1979
Orconectes obscurus (crayfish)	whole animal	Incorporation of [^{14}C]glucose into chitin studied; chitin synthesis occurs in all stages of molt cycle; rate of chitin synthesis increases in premolt and peaks after molt.	Hornung, D. and Stevenson, J., 1971
Orconectes sanborni (crayfish)	whole animal	[^{3}H]glucose and [^{14}C]GlcNAc incorporated into chitin.	Hettick, B., 1976
	epidermis, midgut gland and muscle extracts	Early enzymes in chitin synthesis pathway, and not late enzymes, increase in premolt causing accumulation of UDP-GlcNAc and GlcNAc-6-P; late enzymes increase in postmolt.	Stevenson, J. and Hettick, B., 1980
Pieris brassicae (cabbageworm)	fifth stadium larvae	Polyoxin D or diflubenzuron injected and cuticle examined; total inhibition of cuticle growth and impairment of chitin synthesis noted.	Gijswijt, M. *et al.* 1979
Plodia interpunctella (Indian meal moth)	whole larvae and pupal epidermis	*In vitro* incorporation of GlcNAc followed; peaks of chitin synthesis at beginning of last-instar prepupae and in white pupae; exposure to ecdysone stimulated chitin synthesis.	Ferkovich, S. *et al.* 1981
Stomoxys calcitrans (stable fly)	pupae	Incorporation of radiolabeled GlcNAc in pupae studied; peaks at 1 and 4 days post-pupation, coinciding with production of ecdysial membrane and imaginal cuticle, respectively.	Mayer, R. *et al.* 1979

Table 2: Chitin synthetase studies in invertebrates — Continued.

Organism	Tissue	Comments	Reference
	pupae	Chitin synthetase isolated from pupal homogenates; pH_{opt} 6.5; divalent cations not required; $K_m = 31.7$ μM and $V_m = 135$ pmol h^{-1} mg^{-1} for UDP-GlcNAc. GlcNAc, UTP, UDP, glycerine and *N,N′*-diacetylchitobiose inhibit enzyme; no inhibition by diflubenzuron.	Mayer, R. *et al.* 1980a
	pupal epidermis	Assay system for studying chitin synthesis described; suitable substrates are radiolabeled glucose, glucosamine, fructose and GlcNAc; no MH required.	Mayer, R. *et al.* 1980b
	whole animal	Tunicamycin and five insect growth regulators tested on pupal chitin synthesis; no inhibition noted.	Mayer, R. *et al.* 1981
Tetranychus urticae (two-spotted spider mite)	oocytes and hypodermis	EM study; role of cellular structures in chitin synthesis; nikkomycin used as inhibitor. Two separate systems described, one in oocyte and other in hypodermal membrane.	Mothes, U. and Seitz, K., 1981
Triatoma infestans (blood-sucking bug)	whole animal and crude homogenates	Maximum incorporation of precursors into chitin occurs just after ecdysis; *N*-acetylglucosaminyl-phospholipid involved in chitin synthesis; tunicamycin inhibited conversion of GlcNAc to chitin; dolichyl diphosphate *N*-acetylglucosamine suggested as a chitin precursor.	Quesada-Allue, L. *et al.* 1976; Quesada-Allue, L., 1982
Tribolium castaneum (red flour beetle)	larvae	Chitin synthetase requires Mg^{2+}; inhibited by polyoxin D, UDP and nikkomycin.	Cohen, E. and Casida, J., 1980a,b
	gut	Chitin synthetase not inhibited by diflubenzuron and SIR8514 in *in vitro* assay.	Cohen, E. and Casida, J., 1980b
Tenebrio molitor (yellow mealworm)	gut and peritrophic membrane	Mg^{2+} required for chitin synthetase activity; inhibited by polyoxin D and UDP.	Cohen, E. and Casida, J., 1980a

maximal before ecdysis (Stevenson, J., 1972). Some researchers believe that in crustaceans glycogen also serves as a source of precursors for chitin synthesis (Hohnke, L., 1971). Study of chitin precursors in crayfish has suggested that early enzymes in the chitin synthetic pathway increase in activity during the premolt period, causing a build-up of *N*-acetylglucosamine-6-phosphate and UDP-*N*-acetylglucosamine. The activity of later enzymes increases during the post-molt period, causing a rise in overall chitin synthesis (Stevenson, J. and Hettick, B., 1980).

Cell-free biosynthesis of crustacean chitin has been achieved using a microsomal preparation from the brine shrimp *Artemia salina* (Horst, M., 1981). The enzyme preparation was able to catalyze the transfer of UDP-*N*-acetylglucosamine to an endogenous acceptor. The product was identified as chitin since it was degraded by purified chitinase and resisted extraction with alkali and urea. The synthetase appears to be located in the microsomal membrane.

Less information is known about the enzymes catalyzing chitin synthesis in insects because few, if any, of the enzymes and their effectors have been purified. The chitin synthetase of insects, like that in yeasts, appears to be an integral part of the cell structure (e.g. Cohen, E. and Casida, J., 1980a). Even slight damage to epidermal cells leads to a shut-down of chitin synthesis (Surholt, B., 1975a). Incorporation studies done with radiolabeled chitin precursors show that insect chitin synthesis can be regulated at several points. Before ecdysis, *N*-acetylglucosamine is used for the synthesis of new chitin, presumably because this substrate is produced by chitinase(s) hydrolyzing old cuticle (Waterhouse, D. *et al.*, 1961). After ecdysis, preformed *N*-acetylglucosamine is not readily available; glucose uptake is increased and the control point appears to be the amination of fructose-6-phosphate (Surholt, B., 1975b). This is in agreement with the finding of Fristrom, J. (1968) that glucosamine and *N*-acetylglucosamine are sequestered in two separate metabolic pools. A detailed characterization of purified enzymes in the synthetic pathway for insect chitin remains to be done.

5 CHITIN DEGRADATION

β-Glucosaminidases are among the most important carbohydrate-splitting enzymes. They catalyze the hydrolysis of β-glucosamine linkages and their substrates can be oligosaccharides, polysaccharides, glycoproteins, mucopolysaccharides or mucolipids. One type of glucosaminidase is the β-*N*-acetylglucosaminidases. In many animals they are important lysosomal enzymes whose deficiency leads to cellular pathology and clinical disorders such as Tay-Sach's and Sandhoff's diseases (Neufeld, E. *et al.*, 1975). In insects β-*N*-acetylglucosaminidases are also molting enzymes whose excess or deficiency may lead to developmental arrest. The substrates for the insect enzymes are chitin, the primary structural polysaccharide in cuticle and peritrophic membrane, and chitin oligosaccharides. The catabolism of chitin is catalyzed by two specialized β-glucosaminidases, endochitinase [poly(1,4-β-(2-acetamido-2-deoxy-D-glucoside)) glycanohydrolase] and exochitinase [oligo(1,4-β-(2-acetamido-2-deoxy-D-glucoside) glycanohydrolase].

Some of the many reports of endochitinases, exochitinases and hexosaminidases in invertebrates are listed, alphabetically by source, in Table 3, along with description of the enzymes' characteristics. These specific citations are not meant to be complete and all-inclusive. In many of these reports it is hard to ascertain if hexosaminidase or true chitinase activity was being studied, often due to the conditions chosen for the enzyme assay. A substrate (artificial) may have been used which masks the true specificity such that the *in vivo* function was not clearly defined. Hexosaminidase (chitinase) was found in many animals in different organs/locations, including integument, molting fluid, hemolymph, saliva, digestive juice, midgut, silk gland, fat body and testis; often it was present in separate organs within the same animal. In a few cases activity was clearly shown to belong to multiple enzymes (Spindler, K., 1976; Kimura, S., 1977; Zielkowski, R. and Spindler, K., 1978; Mommsen, T., 1980; Dziadik-Turner, C. *et al.*, 1981; Koga, D. *et al.*, 1982a,b). Many of the enzymes parallel the larval–pupal–adult transformation and therefore probably play some role in metamorphosis. Others do not fluctuate as such, and function in different capacities such as digestion. Midgut chitinase may hydrolyze food sources containing chitin; it may also have lysozomal enzyme functions such as glycoprotein and glycolipid processing.

5.1 Molting fluid

Molting fluid digests the old endocuticle, thus weakening the rest of the cuticle in preparation for its mechanical rupture at the time of ecdysis. After detachment and retraction of the epidermis from the overlying cuticle (apolysis), the molting space formed in some insects is filled with a colorless and viscous secretion, the molting gel. Soon thereafter the epidermis begins to secrete a new cuticle which thickens as a result of sequential deposition of layers, and hardens as a result of sclerotization. Also, a marked decrease in the viscosity of molting gel occurs, forming the molting fluid. At this time, digestion of the old cuticle begins and, within a time period of several hours to a few days, the endocuticle is completely degraded. The molting fluid is then gradually resorbed and only the thin sclerotized outer layers of old cuticle remain. These are mechanically ruptured at ecdysis and discarded as the exuviae.

Many of the organic and inorganic components of molting fluid in two silkmoths and the tobacco hornworm have been determined (Katzenellenbogen, B. and Kafatos, F., 1970: Jungreis, A., 1973, 1974, 1978a,b). Larval–pupal molting fluid of *Hyalophora cecropia* is a hyperosmotic salt solution relative to hemolymph with potassium representing the principal cationic and osmotic component. Bicarbonate is the principal anionic component. Pupal–adult molting fluid from *H. cecropia* is similar except that it is hypo-osmotic relative to hemolymph (Passonneau, J. and Williams, C., 1953) while larval–pupal fluid from *Manduca sexta* is iso- or slightly hypo-osmotic. The pH of the molting fluids characterized varies from 7.2 to 8.0 and appears to depend on the method of anesthesia used to quiet the animals. Carbon dioxide anesthesia results in a more alkaline pH (pH 7.7 versus pH 7.2

Table 3: Occurrence of exo- and endo-chitinases (hexosaminidases) in invertebrates

Source	Activity[1]	Comments	Reference
Alphitobius diaperinus (mealworm)	exo- or endochitinase	Activity present in digestive juice of larvae but not in adults.	Saxena, S. and Sarin, K., 1972
Anthrenus verbasci (carpet beetle)	exo- or endochitinase	No chitinase activity in digestive tract.	Jeuniaux, C., 1955b
Aphaenogaster treatae and *Apterostigma dentigerum* (ants)	exo- or endochitinase	Activity in midgut > rectum.	Martin, M. *et al.* 1976
Atta cephalotes (ant)	exo- or endochitinase	Activity in midgut ≈ rectum.	Martin, M. *et al.* 1973
Atta columbia (ant)	exo- or endochitinase	Activity in rectum > midgut.	Martin, M. *et al.* 1973
Blattidae sp. (cockroaches)	exo- or endochitinase	Activity in salivary gland and alimentary canal of 12 species	Balan, J. and Fisk, F., 1974; Masih, S., 1973
Balanus amphitrite (barnacle)	exo- or endochitinase	20-Hydroxyecdysone stimulated enzyme in mantle epidermis.	Freeman, J., 1980
Bombyx mori (silkmoth)	exo- or endochitinase	Activity in exuviae.	Hamamura, Y. and Kanehara, Y., 1940; Hamamura, Y. *et al.* 1954
	exo- or endochitinase	Activity in molting fluid, $pH_{opt} = 5.4$.	Jeuniaux, C. and Amanieu, M., 1955
	exo- or endochitinase	Both activities low in integument until after third ecdysis when five-fold increase occurred. Activity in soluble fraction.	Kimura, S., 1973a
	exo-β-*N*-acetylglucosaminidase	20-Hydroxyecdysone injected into abdomen increased exochitinase 5-fold.	Kimura, S., 1973b
	exo-β-*N*-acetylglucosaminidase	Distinct enzymes found in molting fluid and hemolymph. $MW_{app} \approx 1.5 \times 10^5$. $pH_{opt} \approx 5.0–5.8$. K_m's for phenyl-*N*-acetyl-β-D-glucosaminide ≈ 2–3 mM. $HgCl_2$ inhibits both enzymes.	Kimura, S., 1974
	exo-β-*N*-acetylglucosaminidase and endochitinase	Two endochitinases and three β-*N*-acetylglucosaminidases in molting fluid separated by chromatography. Endo enzyme more stable than β-*N*-acetylglucosaminidase.	Kimura, S., 1976a
	exo-β-*N*-acetylglucosaminidase	Larval hemolymph enzyme purified. Glycoprotein composed of 2 subunits of $MW_{app} = 6.1 \times 10^4$.	Kimura, S., 1976b
	exo-β-*N*-acetylglucosaminidase	Larval hemolymph and molting fluid enzyme compared. Enzymes are distinct immunologically and kinetically. Molting fluid enzyme also detected in integument, posterior silk gland, midgut, fat body, and testis. Hemolymph enzyme found in midgut, fat body and testis.	Kimura, S. 1977
	β-*N*-acetylglucosaminidase and endochitinase	Activity in alimentary canal peaked during larval–pupal transformation. MW_{app} for both enzymes $> 6.7 \times 10^4$.	Kimura, S., 1981a
	β-*N*-acetylglucosaminidase	Enzyme in hemolymph perhaps controlled by gene affording association of subunits into active complex.	Kimura, S., 1981b
Carcinus maenas (crab)	endochitinase	Activity detected in gut.	Lunt, M. and Kent, P., 1960
Chamelea gallina (bivalve)	exo-β-*N*-acetylglucosaminidase	Exoenzyme purified from digestive gland.	Perez, N. and Cabezas, J., 1977

Table 3: Occurrence of exo- and endo-chitinases (hexosaminidases) in invertebrates — Continued

Source	Activity[1]	Comments	Reference
Coptotermes lacteus (termite)	exo-β-*N*-acetylglucosaminidase and endochitinase	Activity in extracts of whole bodies	Tracey, M. and Youatt, G., 1958; Waterhouse, D. *et al.* 1961
Cupiennius salei (spider)	exo-β-*N*-acetylglucosaminidase and endochitinase	Two different enzymes purified from digestive fluid. Endochitinase $MW_{app} = 4.8 \times 10^4$ and $pH_{opt} = 7.2$. Exo-β-*N*-acetylhexosaminidase $MW_{app} = 1.1 \times 10^5$, $pH_{opt} = 5.4$, K_m for p*N*p βGlcNAc = 0.35 mM, for $\beta GlcNAc_2 = 0.93$ mM, k_i for GlcNAc = 12 mM, K_i for GlcNAc lactone = 24 μM.	Mommsen, T., 1980
Cyphomyrmex costatus and *Cyphomyrmex rimosus* (ants)	exo- or endochitinase	Activity in midgut > rectum.	Martin, M. *et al.* 1973
Dolichoderus taschenbergi (ant)	exo- or endochitinase	Activity in midgut > rectum.	Martin, M. *et al.* 1976
Drosophila hydei (fruit-fly)	exo- hexosaminidase and endochitinase	Two different enzymes in integument. Hexosaminidase also in gut and fat body. pH_{opt} for both enzymes = 5.8. $MW_{app} = 4 \times 10^4$ and $> 1 \times 10^5$ for endoenzyme and hexosaminidase, respectively. Hexosaminidase $K_m = 57$ mM for p*N*pβGlcNAc, specific activity = 20 nmol min^{-1} mg^{-1}. Endo: $K_m = 5$ mg ml^{-1} for [^{14}C]chitin.	Spindler, V., 1976
Drosophila melanogaster (fruit-fly)	endochitinase	Activity in whole body just before each ecdysis. Probably more than one enzyme.	Winicur, S. and Mitchell, H., 1974
Eciton burchelli, *Ectatomma ruidum*, *Formica dakotensis*, *Formica pallidefulva* and *Formica ulkei* (ants)	endo- or exochitinase	Rectal activity > midgut	Martin, M., *et al.* 1976
Helicella ericetorum (snail)	endochitinase	Enzyme purified from digestive gland. $pH_{opt} = 4.5$. $MW_{app} = 1.3 \times 10^5$. p*I* = 4.83. K_m for p*N*p-substrates ≈ 200–400 μM. $V_m = 10$–30 μmol min^{-1} mg^{-1}. Hg^{2+}, Fe^{3+} inhibit.	Calvo, P. *et al.* 1978
	exo-β-*N*-acetylglucosaminidase and endochitinase	Enzyme from digestive gland. $K_m = 420$ μM and V_m = 30 μmol min^{-1} mg^{-1} for p*N*pβGlcNAc; K_m = 190 μM and $V_m = 8.6$ μmol min^{-1} mg^{-1} for p*N*pGalNAc. Enzyme shows endo-activity towards natural substrates with a low level of exo-activity. Mannose inhibits cleavage of p*N*p-substrates.	Cabezas, J. *et al.* 1981
Helix peliomphala (snail)	exochitinase	Enzyme purified from digestive juice. pH_{opt} = 5.4, $\lambda_{max} = 281$ mm. Substrate specificity and metal ion effects studied.	Kimura, S. *et al*, 1965, 1966a,b
Helix pomatia (snail)	exo-β-*N*-acetylglucosaminidase and endochitinase	β-*N*-Acetylglucosaminidase separated by gel filtration from two endochitinases in digestive juice. Exoenzyme $MW_{app} = 1.6 \times 10^5$, p*I* = 4.95, pH_{opt} = 3.4. Endoenzymes $MW_{app} = 1.3 \times 10^4$ and 2.6×10^4 and $pH_{opt} = 4.2$.	Lundblad, G. *et al.* 1974a,b; Woollen, J. *et al.* 1961; Zechmeister, L. *et al.* 1939
Hirudo medicinalis (leech)	exo-β-*N*-acetylglucosaminidase	Activity present in head.	Woollen, J. *et al.* 1961
Homarus americanus (lobster)	exo-β-*N*-acetylglucosaminidase	Activity present in digestive juice.	Kuhn, R. and Tiedeman, H., 1954
Hyalophora cecropia (silkmoth)	exochitinase	Activity detected in molting fluid.	Passonneau, J. and Williams, C., 1953
Locusta migratoria (locust)	exohexosaminidase and endochitinase	Two separate enzymes in supernatant fraction from integument. pH_{opt} between 4 and 5. Hexosaminidase K_m for p*N*pβGlcNAc = 5 mM.	Zielkowski, R. and Spindler, K., 1978.

Table 3: Occurrence of exo- and endo-chitinases (hexosaminidases) in invertebrates — Continued

Source	Activity[1]	Comments	Reference
Lycoperdon perlatum (puff ball)	exo-β-*N*-acetylglucosaminidase	Activity in whole body extract. $pH_{opt} = 3.7–4.8$. K_m for β-phenyl βGlcNAc = 280 μM.	Powning, R. and Irzykiewicz, H., 1964
Manduca sexta (hornworm)	exo- or endochitinase	Activity localized in old cuticle.	Bade, M., 1974
	exo- or endochitinase	Activity rises sharply about 30 h before pupal ecdysis; it disappears abruptly 12 h later.	Bade, M., 1975
	exo- or endochitinase	Chitinase is allosteric enzyme.	Bade, M. and Stinson, A., 1979, 1981c
	exo- or endo chitinase	Assay system with soluble enzyme(s) from molting fluid and insoluble substrate. Data suggest chitinase contains several subunits (MW = 20–145 K) and acts with processive behavior on chitin chains.	Bade, M. and Stinson, A., 1981a,b
	exochitinase	Three enzymes (I, II, III) in pupal hemolymph. $MW_{app} = 6.1 \times 10^4$. I present in molting fluid and integument. II present in larval hemolymph. I and III much more active toward chitin oligosaccharides. K_m for I and chitobiose = 0.1 mM, $k_{cat} = 250\ s^{-1}$.	Dziadik-Turner, C. *et al.* 1981; Koga, D. *et al.* 1982
	endochitinase	Three enzymes (I, II, III) in pharate pupal integument. $MW_{app} = 7.5 \times 10^4$, 6.2×10^4 and 5×10^5, respectively. I present in molting fluid. K_m for I and glycol chitin = 0.15 mg ml^{-1}, $k_{cat} = 2\ s^{-1}$.	Koga, D. *et al.* 1983
Marine invertebrates (~ 65 species)	exo-β-*N*-acetylglucosaminidase	Digestive systems showed highest activities.	Molotsov, N. and Vafina, M., 1972
Myrmicocrypta ednaella (ant)	exo- or endochitinase	Activity in midgut > rectum.	Martin, M. *et al.* 1973
Mytilus edulis (mussel)	exo-β-*N*-acetylglucosaminidase	Two forms isolated from hepatopancreas.	Sanchez-Mozo, P. *et al.* 1977
Nasutitermes sp. (termite)	exo- or endochitinase	Activity present in whole-body extract.	Tracey, M. and Youatt, G., 1958
Oryctes rhinoceros (beetle)	exo-β-*N*-acetylglucosaminidase	Activity in integument derived from fungus *Metarrhizium anisopliae*.	Ratault, C. and Vey, A. 1977
Patella vulgaris (limpet)	exo-β-*N*-acetylglucosaminidase	Enzyme detected in digestive gland.	Findlay, J. *et al.* 1958; Woollen, J. *et al.* 1961; Woollen, J. *et al.* 1961
Periplaneta americana (cockroach)	exo- or endochitinase	Activity detected in blood, digestive juice, cuticle and saliva. pH_{opt} 5.4–6.0.	Waterhouse, D. *et al.* 1961; Waterhouse, D. and McKellar, J., 1961
		Multiple enzymes detected in gut, blood and cuticle. pH_{opt} 4.6–5.1.	Powning, R. and Irzykiewicz, H., 1963, 1964
		Multiple enzymes produced by embryonic cell line and hemolymph.	Landureau, J. and Jolles, P., 1970; Bernier, I. *et al.* 1974
Schistocerca gregaria (locust)	exo-β-*N*-acetylglucosaminidase	Activity detected in crop fluid.	Woollen, J. *et al.* 1961
Sericomyrmex amabalis (ant)	exo- or endochitinase	Activity in midgut $\approx$ rectum.	Martin, M. *et al.* 1973
Sitophilus oryzae (rice weevil)	exo-β-*N*-acetylglucosaminidase	Exoenzyme in ovary and whole-body extracts. Derived from weevil and symbiont. $MW_{app} = 1.5 \times 10^5$, $pH_{opt} = 4.7$. K_m for p*N*pβGlcNAc = 0.2–0.8 mM.	Nardon, P. *et al.* 1978
Spodoptera eridania (Southern armyworm)	exo-β-*N*-acetylglucosaminidase	Activity in lysosomal and light mitrochondrial fractions prepared from midgut.	Young, R., 1979

Table 3: Occurrence of exo- and endo-chitinases (hexosaminidases) in invertebrates — Continued

Source	Activity[1]	Comments	Reference
Stomoxys calcitrans (stable fly)	exo-β-*N*-acetylglucos-aminidase	Soluble and particulate (lysosomal) activities detected. 2–3-fold activity changes during post-pupation period. $pH_{opt} = 4.5$–5.5	Deloach, J. and Mayer, R., 1979
	endochitinase	Single enzyme isolated from prepupal whole body. $pH_{opt}=5$, $pI=4.85$, $MW_{app}=4.8 \times 10^4$.	Chen, A. *et al.* 1982
Tegenaria atrica (spider)	exo-β-*N*-acetylglucos-aminidase and endochitinase	One endo- and two exoenzymes identified in digestive juice. K_m for endo = 0.9 mg chitin ml^{-1}. K_m for exo = 0.6 mM (p*N*pβGlcNAc).	Mommsen, T., 1978
Tenebrio molitor (mealworm)	exo-β-*N*-acetylglucos-aminidase and endochitinase	Activity detected in extract of cast skins.	Jeuniaux, C., 1955b
Turbo cornutus (gastropod)	exo-β-*N*-acetylglucos-aminidase	Two isozymes purified from liver. $MW_{app} \approx 10^5$. $pH_{opt} = 4.0$. K_m for p*N*p substrate = 3 mM.	Yeung, K. *et al.* 1979

[1] Activity listed according to substrate used in enzyme assay. Chitinase activity listed where chitin or chitin derivative (glycol chitin, chitin oligosaccharide) was used. Glucosaminidase shown where nitrophenylated derivative was utilized.

for molting fluid from non-anesthetized animals). Molting fluid contains proteins, polypeptides and amino acids but, interestingly, no sugars.

The idea that molting fluid contained enzymes capable of digesting cuticle was postulated early in this century (Plotnikov, 1904, cited by Passonneau, J. and Williams, C., 1953; Tower, W., 1906). In view of the chitin–protein nature of endocuticle, it is not surprising that both chitinases and proteinases are present (Passonneau, J. and Williams, C., 1953; Jeuniaux, C. and Amanieu, M., 1955; Katzenellenbogen, B. and Kafatos, F., 1970; Bade, M. and Shoukimas, J., 1974). Two proteinases are present in fluid from *Antheraea polyphemus* pharate adults (Katzenellenbogen, B. and Kafatos, F., 1971a). A single exochitinase and several endochitinases occur in molting fluid from *Manduca sexta* pharate pupae (Dziadik-Turner, C. *et al.*, 1981; Koga, D. *et al.*, 1982, 1983). Esterases are also present, including carboxylesterases and arylesterases. Two esterase fractions were separated from the molting fluid of *A. polyphemus* (Katzenellenbogen, B. and Kafatos, F., 1971b) while three were identified in *M. sexta* (Mai, M. and Kramer, K., 1983). Molting gel lacks proteolytic activity but does contain an inactive form of proteinase (proenzyme) that can be activated (Katzenellenbogen, B. and Kafatos, F., 1971c). Analogously, it may contain an inactive form of chitinase (proenzyme?) that becomes activated in the molting fluid. Tyrosinase may also be present (Katzenellenbogen, B. and Kafatos, F., 1970).

5.2 Chitinase assays

There are many substrates available for assaying chitinolytic enzymes (Table 4). The ideal substrate, of course, is chitin itself. However, chitin is insoluble and preparations frequently contain undefined chemical constituents such as amino acids, peptides or proteins (Brine, C. and Austin, P., 1981b). Thus, it is very difficult to obtain a substrate that truly reflects the native structure of chitin in cuticle and at the same time provides enough sensitivity so that low levels of enzymes are not overlooked.

Cuticle is a highly organized chitin–protein–lipid supramolecular structure. Studies on enzymes acting on artificial or modified substrates are difficult to interpret in physiological terms because the kinetic results *in vitro* may demonstrate only a part of the enzyme's true specificity and catalytic efficiency. Modified chitin and pseudochitin substrates suffer from such shortcomings, although some pseudosubstrates offer the advantages of increased sensitivity and ease of assay. Many assay methods have been devised for detecting chitinase activity, none of which have the hoped-for

Table 4: Substrates available for chitinolytic enzyme assays

Substrate	Endochitinase	Exochitinase	Comment
Chitin, glycol chitin, Chitin Red®, ^{3}H-chitin chitin containing tissue	+	±	Substrates and products poorly defined. Generally these are very poor substrates for exoenzyme.
$\beta(1\rightarrow4)$GlcNAc oligosaccharides, dNp$\beta(1\rightarrow4)$GlcNAc$_4$	+	+	Well defined chemically. The best choice.
pNp$\beta(1\rightarrow4)$GlcNAc, fluorescent conjugates of $\beta(1\rightarrow4)$GlcNAc, $\beta(1\rightarrow4)$GlcNAc$_2$	–	+	Well defined chemically but will not react with endoenzyme.

combination of ease, accuracy and nativeness of cuticle chitin structure. Several of these procedures will be described.

Endochitinase is measured primarily on large polymeric substrates. Colloidal chitin has been prepared from crustaceans (Berger, L. and Reynolds, D., 1958), insects (Bade, M. and Stinson, A., 1981b) and fungi (Hackett, C. and Chen, K., 1978) and it can be utilized as a chitinase substrate in several ways. The decrease in turbidity of the substrate can be used as a direct measurement of enzymatic activity (Jeuniaux, C., 1966; Powning, R. and Irzykiewicz, H., 1965) or the amount of *N*-acetylglucosamine released can be measured colorimetrically after enzyme inactivation (Morgan, W. and Elson, L., 1934; Aminoff, D. *et al.*, 1952; Reissig, J. *et al.*, 1955; Benson, R., 1975; Rouleau, M., 1980). The latter result is not reliable because it actually measures the activity of exochitinase whose substrates (small oligosaccharides) are the products of endochitinase catalysis. Chromogenic assay of the *N*-acetylglucosamine released from deproteinized cuticle squares has been used to assess chitinase activity (Bade, M., 1975). Although these assays have the advantage of more "nativeness of substrate", they are insoluble substrates whose chemical structure is very poorly defined. The catalysis then, is a heterogeneous one, and such experiments are unable to characterize precisely the enzyme's specificity because controlled kinetic studies are practically impossible. Some chitinase enzyme sources are devoid of exochitinase such that no *N*-acetylglucosamine release occurs. To circumvent this potential problem, C. Jeuniaux's (1961, 1966) method of assay includes the addition of exochitinase either as lobster serum or as a reconstituted pure enzyme.

Glycol chitin, a commercially available modified chitin that is soluble in aqueous solutions at physiological pH, has been used as a substrate for chitinase. Its activity can be detected by repeatedly testing the viscosity of the substrate/enzyme mixture (Ohtakara, A., 1961; Lundblad, G. and Hultin, E., 1966). This method can be used to distinguish exo- and endo-chitinase activities as the endoenzyme will cause a much more rapid decrease in the viscosity of the substrate solution (Ohtakara, A., 1963, 1964). However, performing the viscometric assay is tedious. The use of potassium ferriferrocyanide as an oxidizing agent to measure the generation of reducing end groups colorimetrically has made the glycol chitin assay for endochitinase easier to carry out (Imoto, T. and Yagishita, K., 1971; Koga, D. *et al.*, 1983). The degree of glycolation of chitin determines the solubility and susceptibility to enzymatic attack (Yamada, H. and Imoto, T., 1981). Glycolation is done using ethylene oxide or 2-chloroethanol, the latter being preferred because of the difficulty of preparing the former reagent. Deacetylated glycol chitin is produced under the conditions of chitin glycolation, and it is necessary to reacetylate the product with acetic anhydride.

Two chitinase assays take advantage of the solubility of *N*-acetylglucosamine oligomer/monomer products which are prepared from the insoluble chitin. One uses tritiated chitin, either prepared *in vivo* (Bade, M., 1974; Bade, M. and Stinson, A., 1981b) or by reacetylating chitosan with tritiated acetic anhydride (Molano, J. *et al.*, 1977). Cleavage of the labeled substrate is detected as radioactivity soluble in 10% trichloroacetic acid. The second assay uses Chitin Red® (Calbiochem) as substrate, a preparation of γ-chitin coupled to a red chromophore. Chitinase activity produces a red-colored supernatant.

Chitin oligomers have been used as substrates in

kinetic studies of non-specific chitinases such as lysozyme to measure both endo- and exochitinase activities (Rupley, J. and Gates, V., 1967; Banerjee, S. *et al.*, 1973; Masaki, A. *et al.*, 1981). The chitin oligomers ranging in size from disaccharide to heptasaccharide are prepared by hydrolyzing crab or shrimp shells with concentrated hydrochloric acid and separating the oligomers with either charcoal-celite or gel permeation chromatography (Rupley, J., 1964; Raferty, M. *et al.*, 1969; Tsai, C., 1970). As a precaution, chitin oligosaccharides should be treated with an acylating agent such as acetic anhydride to ensure that no chitosan oligomers are generated. Endo- or exo-cleavage patterns can be determined by following the reaction time-course. *N*-Acetylglucosamine will be an initial product only if the exoenzyme is present. The reaction products are detected by these same chromatographic methods, paper chromatography (Wadstrom, T., 1971; Tsai, C., 1970), thin-layer chromatography (Powning, R. and Irzykiewicz, H., 1967a) or high-performance liquid chromatography (van Eikeren, P. and McLaughlin, H., 1977; Dziadik-Turner, C. *et al.*, 1981; Turner, C. *et al.*, 1981; Koga, D. *et al.*, 1982, 1983; Blumberg, K. *et al.*, 1982; Fukamizo, T. and Hayashi, K., 1982). Radiolabeled oligosaccharides or oligomers oxidized at the non-reducing end by periodate treatment may also be used (Molano, J. *et al.*, 1979; Hirano, S. and Yagi, Y., 1980, 1981). If a need for glucosamine or chitosan oligomer analysis arises, they can be identified and quantified using a conventional amino acid analyzer and either ninhydrin (Chang, J. and Hash, J., 1979), *o*-phthaldialdehyde (Carroll, S. and Nelson, D., 1979), or fluorescamine (Chen, A. and Mayer, R., 1981) for derivatization.

An unusual chromophoric oligosaccharide substrate, 3,4-dinitrophenyl-tetra-*N*-acetyl-β-chitotetraoside, distinguishes endo- from exoenzyme activity (Ballardie, F. and Capon, B., 1972; Otaki, N. and Kimura, M., 1975; Koga, D. *et al.*, 1982, 1983). Exochitinase will release dinitrophenol either immediately or following a lag phase after incubation with substrate, depending on whether hydrolysis proceeds from the reducing or non-reducing end of the oligosaccharide. Endochitinase does not release dinitrophenol, but it will generate smaller dinitrophenylated sugar products.

Exochitinase activity can be assayed using chitin pseudosubstrates containing *N*-acetylglucosamine and a chromophore (e.g. *p*-nitrophenol) or fluorescent component (e.g. 4-methylumbelliferone). The increase in absorbance or fluorescence can be followed continuously at a discrete wavelength through the course of the reaction (Ford, J. *et al.*, 1973; Shulman, M. *et al.*, 1980). These substrates are commercially available and the assays are sensitive, quick and relatively inexpensive. Structurally related compounds are also available that can be used to study in detail structure–activity relationships and the mechanism of catalysis (Jones, C. and Kosman, D., 1980; Dziadik-Turner, C. *et al.*, 1981; Koga, D. *et al.*, 1982).

5.3 Exochitinases [exo-β(1$\rightarrow$4)-oligo-*N*-acetylglucosaminidases] and Endochitinases [endo-β(1$\rightarrow$4)-poly-*N*-acetylglucosaminidases]

Invertebrate enzymes that hydrolyze substrates which contain *N*-acetylglucosamine as part of their structure have been studied for many years. Table 3 lists some of the reports on these carbohydrases, along with comments about their properties. Although the importance of chitinolytic enzymes in the growth and development of arthropods has been recognized for many years, little work has been published on the properties of purified enzymes. Most of the detailed knowledge of the chitinolytic enzymes has come from studies of fungal and yeast enzymes (Leaback, D., 1970). In many cases concerning insects there is ambiguity about the kind of activity measured (exo- or endo-) and the homogeneity of the enzymes. Primarily, heterogeneous enzyme preparations and/or poorly defined substrates were used. Several of the more definitive studies of chitinolytic enzymes will be described here, including those on the silkworm, *Bombyx mori*; the tobacco hornworm, *Manduca sexta*; the fruitfly, *Drosophila hydei*; the migratory locust, *Locusta migratoria* and the hunting spider, *Cupiennius salei* (Table 3).

Forty years ago chitinase activity was noted in cast exuviae and in molting fluid of *Bombyx mori* (Hamamura, Y. and Kanehara, Y., 1940; Hamamura, Y. *et al.*, 1954; Jeuniaux, C. and Amanieu, M., 1955; Jeuniaux, C., 1955a). Subsequently, Kimura, S. (1973a,b, 1974a,b,

1976a,b, 1977, 1981a,b) has conducted very elegant studies of the chitinolytic enzymes in silkworm tissues during larval and pupal development. Endo- and exo-β-*N*-acetylglucosaminidases were detected in integument with the exo-activity increasing several-fold after injections of 20-hydroxyecdysone. Molting fluid contained three exo- and two endoenzymes that were separated by chromatographic methods. Similar enzymes were found in midgut, fat body, testis and silk gland. Larval hemolymph contained an antigenically distinct exo-β-*N*-acetylglucosaminidase and no endo-enzyme. Molting fluid and integument contained endochitinase and chitobiase, the latter cleaving only β-GlcNAc$_2$ and not chitin (Kimura, S., 1982). Larger oligosaccharides were not tested.

Another insect whose chitinolytic enzymes has been studied extensively is *M. sexta*. Bade, M. and Stinson, A. (1978a,b) have proposed that hornworm chitinase hydrolyzes cuticle chitin after proteases in molting fluid unmask the chitin from bound proteins. Smith, R. *et al.* (1981) report that digestion of chitin–protein ghosts derived from 1% sodium dodeyl sulfate-boiled extracts of *Heliothis zea* larvae occurred using a combination of a proteolytic enzyme followed by chitinase whose source was not revealed. Conversely, Lipke, H. and Geoghegan, T. (1971b) found that native chitin in *Sarcophaga bullata* was not made more available to chitinase by prior treatment with proteases. Bade and her colleagues also suggested that molting fluid contains a single chitinase which exhibits positive cooperativity and acts via a processive mechanism, cleaving off successive GlcNAc monomers from the same chitin chain without diffusing from one polymer to another (Bade, M. and Stinson, A., 1979, 1981a,b,c). In all of these studies, crude molting fluid and insoluble substrates were used. Since multiple chitinolytic enzymes are known to be present in molting fluid (see Table 3), it is difficult to interpret the kinetic data derived from such a heterogeneous catalysis. These conclusions must be regarded with caution until studies with purified enzymes and better substrates are conducted.

Kramer and colleagues have used purified enzymes and well-defined soluble substrates to determine the kinetic behavior of the chitinolytic enzymes from the tobacco hornworm (Dziadik-Turner, C. *et al.*, 1981; Koga, D. *et al.*, 1982, 1983). Their studies, and those of other laboratories (see Table 3), suggest that two classes of enzymes, exochitinase and endochitinase, degrade cuticle chitin in insects. One exo-β-*N*-acetylglucosaminidase (ExI) occurred in molting fluid and pharate pupal integument and a second exoenzyme (ExII) was present in larval hemolymph. Both enzymes were found in pupal hemolymph. Either enzyme could hydrolyze oligosaccharide substrates following an exo-cleavage pattern beginning at the non-reducing end. Their physical, chemical and kinetic properties were very similar with one major exception. ExI was more efficient at hydrolyzing β-GlcNAc oligosaccharides. ExII was at least two orders of magnitude less active (Table 5). The two enzymes were unrelated immunologically. The tissue localizations, together with the kinetic behavior, demonstrated that the former is a true exochitinase and that the latter is an exo-β-*N*-acetylhexosaminidase with a "sluggish" exochitinase activity. ExI degraded substrates as large as hexamer (and probably larger ones) to monomer which can be recycled *in vivo* and utilized as a precursor for the chitin formed in cuticle of the next developmental stage.

Table 5: Kinetic parameters for exochitinase (EI) and exo-β-N-acetylhexosaminidase (EII) (from Koga, D. et al. *1982a).*

	k_{cat} (s^{-1})		K_m (M × 10^3)		k_{cat}/K_m ($s^{-1} M^{-1} \times 10^{-4}$)	
Substrate*	EI	EII	EI	EII	EI	EII
p*Np*βGlcNAc	392±30	285.0±6	0.247±0.028	0.14±0.01	158	205
βGlcNAc$_2$	265±43	3.0±1.3	0.128±0.039	0.55±0.31	207	0.57
βGlcNAc$_3$	157±17	3.5±0.2	0.084±0.020	0.68±0.05	185	0.50
βGlcNAc$_4$	93±5	3.7±1.8	0.051±0.009	0.73±0.47	183	0.50
βGlcNAc$_5$	88±5	5.0±0.8	0.058±0.009	0.96±0.22	150	0.52
βGlcNAc$_6$	77±7	4.5±0.5	0.063±0.013	0.87±0.12	121	0.53

*for EI: E_0 = 0.34 nM, S_0 = 0.065–0.390 mM, pH 5.6, substrate inhibition observed at [S] ≫ 0.2 mM.
For EII: E_0 = 1.89 nM, S_0 = 0.065–0.78 mM, pH 5.6

pH rate studies of *M. sexta* exochitinase suggested that two ionizable groups in the active site with apparent pK_a values of 3.8 and 8.1 are important for binding and hydrolysis, respectively (Koga, D. *et al.*, 1982a). The identities of these residues are unknown, but speculations can be made based upon an examination of another β-*N*-acetylglucosaminidase, hen's eggwhite lysozyme (Imoto, T. *et al.*, 1972). Lysozyme has an acidic group, aspartic acid 101 ($pK_a = 4.5$), involved in a binding contact with substrate in subsite A that must be ionized to interact (Fig. 6). Exochitinase may have a similar acidic group with a slightly lower p*K* in its active site (denoted as A^- in Fig. 6). Two other carboxylic acid side-chains participate in acetal bond hydrolysis in lysozyme (glutamic acid 35 and aspartic acid 52), but apparently only one other group in exochitinase. There appears to be no counterpart in exochitinase to lysozyme's aspartic acid 52, which stabilizes a carbonium ion or covalently displaces the leaving group in the lysozyme mechanism. There is a protonated group with a relatively high p*K* in exochitinase (denoted as BH in Fig. 6) that may participate as a proton donor analogous to the role played by glutamic acid 35 in lysozyme. Whereas lysozyme exhibited a rather narrow pH optimum (pH 4–6), *M. sexta* exochitinase had a pH optimum extending into the alkaline range (pH 4–8) which is due to the basic group, BH, in the active site. Candidate residues for this basic group include histidine and lysine. A similar pH-activity profile was found for another exo-β-*N*-acetylhexosaminidase from a fungus, *Aspergillus niger* (Jones, C. and Kosman, D., 1980).

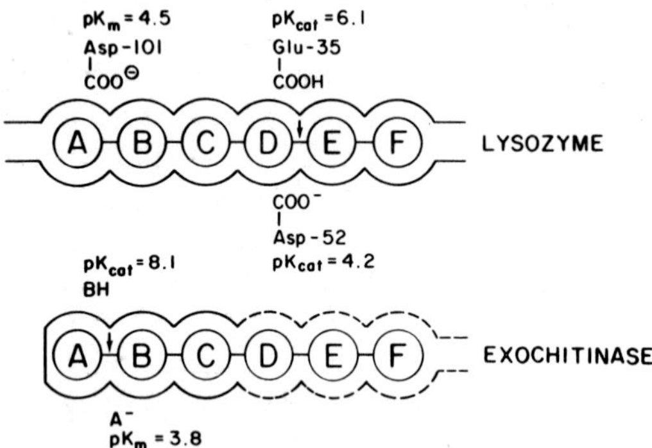

FIG. 6. A comparison of the proposed active site of *M. sexta* exochitinase with that of hen's eggwhite lysozyme. $\beta(1\rightarrow4)$ GlcNAc hexasaccharide denoted by A — F with A being the non-reducing end.

X-ray crystallographic studies have shown that the active site of lysozyme is an open cleft with six subsites (A–F, Fig. 6, Imoto, T. *et al.*, 1972). The binding site of *Manduca sexta* exochitinase may also have subsites, probably only three or four forming a close-ended cleft (A–D, Fig. 6) since trimer through hexamer had similar K_m values (Table 5). Exochitinase exhibited substrate and product inhibition effects *in vitro*, indicating that a large active site exists for both productive and non-productive modes of binding. Whether these inhibition phenomena happen *in vivo* is unknown.

The substrates for exochitinase are generated from cuticle chitin by endo-β-acetylglucosaminidase or endochitinase. Three endo-β-*N*-acetylglucosaminidases were purified from the integument of *M. sexta* pharate pupae (Koga, D. *et al.*, 1983). These enzymes are glycoproteins that have unique molecular weights, 7.5×10^4 for E_NI, 6.2×10^4 for E_NII and 5.0×10^4 for E_NIII, and also contain neutral hexoses and hexosamine. Molting fluid contained mostly E_NI, and smaller amounts of E_NII and E_NIII. The latter was immunologically related to E_NII but not to E_NI. Analysis of initial velocity experiments and products formed from $\beta(1\rightarrow4)$ GlcNAc oligosaccaride and glycol chitin substrates revealed that the longer substrates were preferred and that all three enzymes behaved primarily as endochitinases (Table 6). Oligosaccharides rapidly disappeared until only disaccharide and trisaccharide remained, after which a small amount of free *N*-acetylglucosamine was produced from trisaccharide very slowly. Thus, the enzymes exhibited a "sluggish" exochitinase activity, but only with chitotriose as substrate. No activity was expressed toward p*N*pβGlcNAc, βGlcNAc$_2$ or *Micrococcus lysodeikticus* cell walls. The endochitinases were also susceptible to substrate inhibition. The active site of these enzymes is probably a large open cleft, much like that of lysozyme, which accommodates both productive and non-productive binding of substrate.

Endochitinolytic enzymes from various sources differ widely in their physical and kinetic properties. Insect endochitinases occur in several tissues and sizes. Mammalian sera contain enzymes that are about the same size, $5–6 \times 10^4$ (Lundblad, G. *et al.*, 1974a, 1979), while plant endochitinases are smaller, 3×10^4 (Molano, J. *et al.*, 1979). Whether the

Table 6: Rates and cleavage patterns of substrate hydrolysis by Manduca sexta *endochitinases I, II and III (from Koga, D.* et al. *1983).*

Substate	Velocity (p mol s^{-1})*			Cleavage pattern
	I	II	III	
$\beta(1\rightarrow4)GlcNAc_2$	0	0	0	x_1-x_2 ↓ ↓
$\beta(1\rightarrow4)GlcNAc_3$	0.62 ± 0.08 [5]† (51)‡	1.21 ± 0.15 [4] (100)	0.89 ± 0.12 [6] (73)	$x_1-x_2-x_3$
$\beta(1\rightarrow4)GlcNAc_4$	1.26 ± 0.01 [10] (40)	3.18 ± 0.15 [11] (100)	1.90 ± 0.01 [12] (60)	↓ $x_1-x_2-x_3-x_4$
$\beta(1\rightarrow4)GlcNAc_5$	2.01 ± 0.05 [17] (53)	3.81 ± 0.23 [13] (100)	3.21 ± 0.04 [21] (84)	↓ ↓ $x_1-x_2-x_3-x_4-x_5$
Glycol chitin	12.02 ± 1.54 [100] (42)	28.53 [100] (100)	15.58 ± 3.74 [100] (55)	

* $E_0 = 18 \times 10^{-9}$ M; oligosaccharide concentration $(S_0) = 94 \times 10^{-6}$ M; glycol chitin concentration $(S_0) = 0.02$–0.1 mg ml^{-1}; 50 mM sodium phosphate, pH 5.3, 25°C.

† Numbers in brackets give relative rate of hydrolysis of all substrates by one enzyme.

‡ Numbers in parenthesis give relative rate of hydrolysis of one substrate by all enzymes.

mammalian and plant enzymes are true chitinases *in vivo* is uncertain. The pH optimum appears to depend on the enzyme source and substrate used. Low optima (~pH 2) were observed with yeast, goat and bovine endochitinases when glycol chitin or radiolabeled chitin was the substrate (Lundblad, G. *et al.*, 1974, 1979; Correa, J. *et al.*, 1982). However, the latter two enzymes showed maximum activity with colloidal chitin at the same pH as that found for the *M. sexta* enzymes (pH 6). Fungal and plant endochitinases also had pH optima of 6 (Berger, L. and Reynolds, D., 1958; Molano, J. *et al.*, 1979). Bade, M. and Stinson, A. (1981b) reported that chitinolytic activity in *M. sexta* molting fluid was maximum at a slightly higher pH, 7, but that it was more reproducible at pH 6.5. Raw molting fluid and insoluble chitin were used in their experiments, however. Similar K_m values (mg ml^{-1}) are exhibited for glycol chitin by mammalian and insect enzymes.

A proposed mechanism of chitin catabolism by the two types of chitinolytic enzymes in *M. sexta* is shown in Fig. 7. After endochitinase (E_N) hydrolyzes chitin ($\beta GlcNAc_{n>1}$) to yield oligosaccharides ($\beta GlcNAc_n$), the latter may bind to exochitinase (E_x) to give a productive complex $(E_x.\beta GlcNAc_n)_p$ where the substrate is susceptible to the action of the catalytic site. At higher concentrations additional substrate molecules may also bind either enzyme such that none of the bonds are properly exposed to the catalytic amino acid functional groups and non-productive complexes are formed ($[E_x.(\eta+1)\beta GlcNAc_n]_{np}$). Compounds may inhibit E_x in a competitive pattern where more than one inhibitor may bind ($E_x{\cdot}I_n$) depending on the inhibitor's structure and concentration. The oligosaccharides may be sequentially degraded by exochitinase from the non-reducing end to yield monosaccharide ($n\beta$ GlcNAc). This kind of mechanism is not unusual for enzymes that modify polymeric and oligomeric substrates, and models for such hydrolyses have been proposed that envisage substrate interacting with the entire active site (productive complex) or only parts of it (non-productive complex) (Hutny, J. and Ugorski, M., 1981; Klesov, A. *et al.*, 1981; Thoma, J. and Crook, C., 1982).

Chitinolytic enzymes have also been partially characterized from integument of *Drosophila hydei* (Spindler, K., 1976) and *Locusta migratoria* (Zielkowski, R. and Spindler, K., 1978). Endochitinase and exo-β-*N*-acetylhexosaminidase from *D. hydei* exhibited pH optimum between 5.5 and 6.2 and temperature optima of 35° and 50°, respectively. Three endoenzymes and one

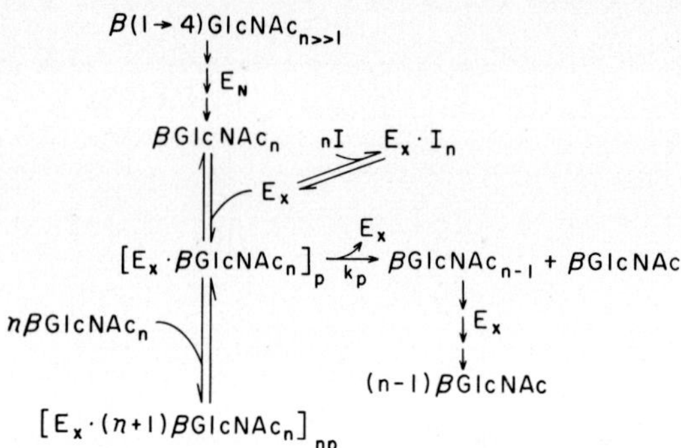

FIG. 7. Proposed pathway for the catalysis and inhibition of *M. sexta* exochitinase. Key to the symbols: E_N, endochitinase; E_X, exochitinase; $\beta(1\rightarrow4)$ $GlcNAc_{n>>1}$, chitin; β $GlcNAc_n$, chitin oligosaccharides; $[E_X \cdot \beta GlcNAc_n]_p$, productive enzyme substrate complex; $[E_X \cdot (\eta + 1)\beta GlcNAc_n]_{np}$, nonproductive complex; I, inhibitor; βGlcNAc, *N*-acetylglucosamine.

exoenzyme fraction were partially purified by gel filtration, which is identical to the number of enzymes found in *M. sexta* integument (Dziadik-Turner, C. *et al.*, 1981; Koga, D. *et al.*, 1982, 1983). Exo activity was also detected in *Drosophila* gut and fat body. The pH optimum for *Locusta* integument endochitinase and exo-β-*N*-acetylhexosaminidase was between 4 and 5. Temperature optima were 50° and 70°, respectively. Both endo- and exo-activities reached maximum levels during the latter part of the molting cycle and declined before ecdysis.

Some of the most active hydrolytic enzymes in the digestive fluid are chitinolytic ones (Martin, M. *et al.*, 1973, 1976; Mommsen, T., 1978, 1980). These probably digest chitin and other carbohydrate-containing food sources. Two of the better-characterized digestive chitinases were obtained from the digestive juice of the hunting spider, *Cupiennius salei* (Mommsen, T., 1980). These are similar to the integument enzymes found in insects. An endochitinase ($MW_{app} = 4.8 \times 10^4$) degraded chitin to disaccharide and trisaccharide. Exochitinase ($MW_{app} = 1 \times 10^5$) hydrolyzed the oligosaccharides to monomer.

The reason for multiple chitinolytic enzymes in insects is unknown. The enzymes may act in a concerted mechanism or individually. In mammalian fibroblasts, *N*-acetylhexosaminidases are synthesized in the form of precursors of larger molecular weight and are shortened later to the size of chains found in tissues (Hasilik, A. and Neufeld, E., 1980a,b). This processing and additional post-translational modification (phosphorylation) allow recognition for incorporation into lysosomes (Gustafson, G. and Milner, L., 1980; Waheed, A. *et al.*, 1981; Knecht, D. and Dimond, R., 1981). Whether these modifications occur with insect enzymes is unknown.

5.4 Lysozymes

In addition to its bacteriolytic action, lysozyme (mucopeptide *N*-acetylmuramolylhydrolase) also cleaves chitin substrates (Imoto, T. *et al.*, 1972). This enzyme may be mistaken for a chitinase and it is imperative to assay suspected enzymes for lytic activity (muramidase) towards bacterial cell walls such as *Micrococcus lysodeikticus* (Neuberger, A. and Wilson, B., 1967). Chitinase does not exhibit such activity.

Several insect lysozymes, as well as some other invertebrate ones, have been characterized (Table 7). They are generally small basic proteins (approx. M.W. $1.5–2 \times 10^4$) with properties (heat stability, pH and ionic strength optima) similar to those of vertebrate lysozymes. Tissue sources include eggs, gut and hemolymph. Insect lysozymes are probably antibacterial in function because enzymatic activity increases dramatically when the animal is challenged with bacteria.

5.5 Insect cell lines

Several insect cells exhibit chitinolytic enzyme activity (Table 8). The first reported were two cell lines from *Periplaneta americana* whose spent media contained colloidal chitin degrading activity (Landureau, J. and Jolles, P., 1970) and also hydrolyzed bacterial cell walls. It was subsequently shown by chromatographic methods that lytic and chitinolytic activities were distinct (Bernier, I. *et al.*, 1974). β-*N*-Acetylhexosaminidase activity was detected in culture media from cells derived from *Culex quinquefaciatus*, *Drosophila melanogaster* and *Culex fumiterana*, but not from *Trichoplusia ni* or *Aedes pseudoscutelaris* (Turner, C. *et al.*, 1981; Koga *et al.*, unpublished). The *Culex* medium hydrolyzed βGlcNAc oligosaccharides in a manner that demonstrated the presence of exo- and endo-chitinases.

Table 7: Occurrence of lysozyme in invertebrates

Species	Tissue source	Comments	Reference
Asterias rubens (starfish)	hemolymph	Enzyme with MW_{app} 1.55×10^4 purified. Ser is N-terminal residue.	Jolles, J. and Jolles, P., 1975
Bombyx mori (silkmoth)	hemolymph	Activity increased 4-fold after injection of *M. lysodeikticus* cells. Enzyme with $MW_{app} = 1.65 \times 10^4$ purified.	Powning, R. and Davidson, W., 1973
	hemolymph	Enzyme with $MW_{app} = 1.5 \times 10^4$ purified. Antibody elicited in rabbit does not cross react with *Galleria* lysozyme.	Croizier, G. and Croizier, L., 1978
Ceratitis capitata (Mediterranean fruit fly)	eggs	Enzyme purified with MW_{app} 2.3×10^4. $pH_{opt} = 6.5$. Gly is N-terminal residue.	Fernandez-Sousa, J. *et al.* 1977
Chlamys opercularis (bivalve)	crystalline style and digestive gland	—	McHenery, J. *et al.* 1979
Crassostrea virginica (oyster)	hemolymph and mantle mucus	—	McDade, J. and Tripp, M., 1967a,b
	hemolymph	Enzyme present in serum and leucocytes.	Rodrick, G. and Cheng, T., 1974
	hemolymph	Acidic protein.	Feng, S., 1974
Galleria mellonella (wax moth)	hemolymph	Low activity increased 9-fold after injection of *M. lysodeikticus* into prepupae. Enzyme with $MW_{app} = 1.47 \times 10^4$ purified.	Powning, R. and Davidson, W., 1973
	hemolymph	Enzyme with $MW_{app} = 1.5 \times 10^4$ purified. Antibody elicited in rabbit does not cross react with *Bombyx* lysozyme.	Croizier, G. and Croizier, L., 1978
Modiolus modiolus (bivalve)	crystalline style and digestive gland	—	McHenery, J. *et al.* 1979
Mya arenaria (soft shelled clam)	hemolymph	$pH_{opt} = 5$.	Cheng, G. and Rodick, T., 1974
	gills and digestive gland	—	McHenery, J. *et al.* 1979
Mytilus edulis (mussel)	hemolymph	—	Hardy, S. *et al.* 1976
	digestive gland and crystalline style	$pH_{opt} = 6.4$.	McHenery, J. *et al.* 1978, 1979
Nephthys hombergii (annelid)	whole body	Enzyme with $MW_{app} = 1.3 \times 10^4$ purified. Lys is N-terminal. $pH_{opt} = 6.0–6.5$.	Perin, J. and Jolles, P., 1972
Ornithodoros lahorensis and *Ornithodoros tholozani* (ticks)	whole body and egg	Enzymes with $MW_{app} = 1.38 \times 10^4$ purified.	Podboronov, V. *et al.* 1978
Periplaneta americana (American cockroach)	gut and hemolymph	Optimum activity at pH = 6.5–7 and 0.1 ionic strength.	Powning, R. and Irzykiewicz, H., 1963
Spodoptera eridania (armyworm)	hemolymph and hemocytes	Activity increased by injection of various foreign materials.	Anderson, R. and Cook, M., 1979
Tellina tenuis (mussel)	digestive gland and crystalline style	—	McHenery, J. *et al.* 1979 Jolles, P. and Zuili, S., 1960

Table 8: Chitinolytic enzymes associated with insect cell lines.

Species	Tissue source	Comments	Reference
Choristoneura fumiferana (spruce budworm)	embryonic cell line	p*N*p*β*GlcNAc and glycol chitin hydrolyzed by medium.	Koga, D. *et al.* unpublished
Culex quinquefaciatus (house mosquito)	ovary	Both endo- and exoenzymes present in medium. $MW_{app} = 6 \times 10^4$. K_m for p*N*p*β*GlcNAc = 70 μM.	Turner, C. *et al.* 1981
Drosophila melanogaster (fruit-fly)	embryo	Hexosaminidase activity in spent medium.	Turner, C. *et al.* 1981
Periplaneta americana (cockroach)	embryonic cell lines E Pa and hemocyte line H Pa 33.	Colloidal chitin and cell wall hydrolytic activities present.	Landureau, J. and Jolles, P., 1970
		Chitinolytic activity separated from lytic activity.	Bernier, I. *et al.* 1974

6 REGULATION OF CHITIN METABOLISM

The complex structure of insect cuticle, and the difficulty in obtaining stable, purified preparations of insect chitinase and chitin synthetase, have complicated the study of control mechanisms in chitin metabolism. Because chitin is rarely found without surrounding protein, the study of chitin synthesis and degradation is of necessity linked to the study of the whole cuticular structure. Some of the papers cited here will, therefore, discuss events in whole cuticle as a way of observing changes in chitin suprastructure. The recent developments in genetic studies will soon make the investigation of individual cuticle components technically feasible (O'Brien, S. and MacIntyre, R., 1978; Fristrom, J. *et al.*, 1978; Synder, M. *et al.*, 1981; Hirsh, J. and Davidson, N., 1981; Kimura, S., 1981b; Silvert, D., vol. 2; Berger, E. and Ireland, R., vol. 2). Studies concerning chitin metabolism divide themselves into two categories: those investigating hormonal and developmental modes of control and those investigating other than hormonal regulators, such as pesticides and insect growth regulators. These modes of control will be considered separately.

Each insect molt is a response to the hormonal and enzymatic milieu present at the time previous to and concurrent with the event. Hormonal control of insect growth and development has been reviewed extensively (Wigglesworth, V., 1970; Wyatt, G., 1972; Gilbert, L. and King, D., 1973; Willis, J., 1974; Riddiford, L., 1980; Riddiford, L. and Truman, J., 1978; Highnam, K., 1978; Sridhara, S. *et al.*, 1978; Mordue, W. and Stone, J., 1979; Gilbert, L. *et al.*, 1980; Stone, J. and Mordue, L., 1980; see volumes 7 and 8). A variety of hormones, many of them peptides and still uncharacterized, control the homeostatic events in the life of an insect. Research of many years has shown, though, that the cyclical process of growth and molting is controlled primarily by discrete quantities of three or four hormones which are secreted at critical time points in development. They are prothoracicotrophic hormone (PTTH), molting hormone (MH), juvenile hormone (JH), and eclosion hormone (EH). These are discussed in detail in other chapters in this series. Table 9 lists some of the studies of hormonal effects on cuticle and chitin metabolism.

The effects of MH and JH on chitin biosynthesis have been studied in various species using metabolite incorporation experiments. In both *Triatoma infestans* and *Locusta migratoria* maximum incorporation of radiolabeled precursors into chitin occurred after ecdysis (Surholt, B., 1975a, Quesada Allue, L. *et al.*, 1976). Incorporation peaks have also been noted post-pupation (Mayer, R. *et al.*, 1979). Administration of 20-hydroxyecdysone causes a decrease in the rate of [^{14}C]glucose incorporation into chitin in *Bombyx mori*, followed by increased incorporation until ecdysis (Kimura, S., 1973b). Exposure of *Plodia interpunctella* larval epidermis to 20-hydroxyecdysone stimulated chitin synthesis, but only after a 24 h delay following exposure to the hormone (Ferkovich, S. *et al.*, 1981); this is consistent with the effects of MH observed in crayfish

(Armstrong, P. and Stevenson, J., 1979). Apparently a preincubation period is required before cells become competent to respond to hormone (Riddiford, L., vol. 8). Studies have also been done in which the change in integument enzyme activity patterns were correlated with the molt cycle (Hohnke, L., 1971; Surholt, B., 1976).

The desire to study the developmental events of epidermal cells in a system in which exogenous effects could be better controlled led investigators to try tissue explant techniques. Successes were few, however, until the knowledge of insect biochemistry had increased (reviewed by Marks, E., 1970). Gradually it was learned that a source of molting hormone was necessary to induce synthesis of cuticular components and cuticle formation (Marks, E. and Leopold, R., 1970, 1971) and that 20-hydroxyecdysone was the active form of molting hormone (e.g. Oberlander, H., 1976). Studies have investigated the effect of hormone concentration, duration and interruption of dose. Marks, E. (1972), using cultured cockroach leg explants, was able to show that cuticle deposition could be induced up to five times in the same explant when the molting hormone dose was repeated at discrete intervals. Other studies (Riddiford, L., 1975) looked at the role of JH and found that in *Manduca sexta* formation of larval cuticle depended on exposure to JH, at the proper titer, along with 20-hydroxyecdysone. If JH was absent, pupal cuticle was synthesized. The timing of cuticular events in cockroach leg regenerates has been studied by time-lapse photomicrography (Marks, E., 1973). It has been noted that the deposition of cuticle in the Indian meal moth lasts only 2 days, but that the epidermal cells have a 4-day preparation period before this, during which protein is synthesized (Oberlander, H., 1976). Data concerning the control of chitin biosynthesis are consistent with the hypothesis that ecdysteroids promote the synthesis of RNA and protein followed by a period of ecdysteroid independent protein and chitin syntheses (Oberlander, H. *et al.*, 1980).

The authenticity of cuticle synthesized *in vitro* has been tested in a number of ways. The product varies considerably from system to system and photography from various experiments is displayed in Marks, E. and Sowa, B. (1976). Chitin formed *in vitro* was tested for histochemically, with chitinase conjugated fluorescent stain, and was found in cultured cockroach leg regenerates (Marks, E. and Leopold, R., 1971). In another study radiolabeled *N*-acetylglucosamine was taken up by 20-hydroxyecdysone stimulated leg regenerates (Marks, E. and Sowa, B., 1974). A synthesis and digestion study done on cultured wing discs showed that radiolabeled D-glucosamine was taken up into the epidermal and cuticular regions; chitinase treatment resulted in release of labeled material into the medium (Oberlander, H. and Leach, C., 1975).

Other radiolabeled incorporation studies have shown that wing tissue from *Antheraea polyphemus* (silkmoth) can be induced by MH to specifically synthesize epidermal cell proteins (Ruh, M. *et al.*, 1974). Cockroach embryos cultured *in vitro* with radiolabeled precursors and MH show that the localization of the synthesized cuticular products is a function of the stage of development of the epidermal cells (Bulliere, F., 1977). The many studies done on invertebrate tissue culture have been reviewed elsewhere (Maramorosch, K., 1976; Kurstak, E. and Maramorosch, K., 1976; Kurstak, E. *et al.*, 1980).

7 INSECT GROWTH REGULATORS

Chitin is an ideal target for pesticide development because of its specialized occurrence in nature (Kent, P., 1964; Muzzarelli, R., 1977) and its critical role at each stage of morphogenesis. An understanding of chitin enzymology may help to develop a new class of insect growth regulators that interfere with chitin metabolism (volumes 11 and 12). Either chitin excess or deficiency during any morphogenetic cycle can produce deleterious and lethal effects. Table 10 lists some of the published reports of the effects of pesticides on chitin/cuticle metabolism.

Natural enzyme inhibitors, especially inhibitors of hydrolases, are found in many plants, animals, bacteria and fungi. Some of these have been useful in the treatment of carbohydrate-dependent metabolic disorders such as diabetes mellitus (Lednicer, D. and Babcock, J., 1962). Hydrolases, including chitinase, found in plants and microorganisms, may act as defense mechanisms (Abeles, F. *et al.*, 1970; Morrissey, R. *et al.*, 1976; Powning, R. and Davidson, W., 1979). Natural products that

Table 9: Hormonal and developmental control of cuticle/chitin metabolism

Organism	Tissue	Comments	Reference
Aedes aegypti (yellow fever mosquito)	embryonic cell culture	MH prevents typical monolayer formation; MH inhibits uptake of U, T and leucine; MH causes *de novo* synthesis of several proteins.	Lanir, N. and Cohen, E., 1978
Antheraea euclypti (euclypti moth)	ovarian cell culture	No uptake of MH noted; no metabolic products of MH detected; JH inhibited RNA and protein synthesis.	Cohen, E. and Gilbert, L., 1972
Antheraea polyphemus (polyphemus moth)	pupal wing	MH *in vitro* induces synthesis of epidermal cell proteins.	Ruh, M. *et al.* 1974
Balanus amphitrite (barnacle)	epidermal explants	MH stimulated and molt inhibiting hormone reduced amount of chitinolytic activity *in vitro*.	Freeman, J., 1980
Blaberus discoidalis (cockroach)	embryos in culture	MH added to culture hastens peaks of RNA and chitin synthesis compared to control cultures.	Bulliere, F., 1977
Bombyx mori (silkworm)	abdominal body walls	Increase in chitinase activity in isolated abdomens 9 h after ecdysterone injection; increase not blocked by actinomycin D or puromycin.	Kimura, S., 1973b
	isolated larval abdomen	Light and electron microscopic study on effects of MH on epidermal cells; within 12 h of MH injection inner endocuticle digested; new epicuticle laid down within 18 h.	Kimura, S. *et al.* 1974
	whole animal	Homozygotes isolated with high and low chitinase activities in hemolymph; genetic study.	Kimura, S., 1981b
Calliphora vicina (blowfly)	whole animal fat body	MH causes *de novo* polyA-RNA synthesis; inducibility correlated to specific developmental stages.	Scheller, K. and Karlson, P., 1977
Calpodes ethlius (larger canna leaf-roller)	larval tracheae explants	Use increase in diameter of cultured tracheae as index of growth; MH causes cuticle deposition without growth.	Ryerse, J. and Locke, M., 1978
	whole animal (larvae and pupae)	Correlate hemolymph MH titers with cellular events in molt/intermolt sequence; four peaks of MH seen in last instar with associated structural changes.	Dean, R. *et al.* 1980
Culex molestus (mosquito)	ovarian cell culture	Effects of JH and JH mimic on cell growth; JH inhibits RNA and DNA synthesis.	Himeno, M. *et al.* 1979
Drosophila melanogaster (fruit-fly)	embryonic cell culture	New culture medium described.	Echalier, G. and Ohanessian, A., 1970
	whole animal	Chitinase activity appears just before larval ecdysis; data suggest multiple enzymes present.	Winicur, S. and Mitchell, H., 1974
	whole animal	Temperature sensitive mutant used to generate colonies of larvae with differing MH titers; effect on fat body protein and translatable mRNA studied.	Lepesant, J. *et al.* 1978
	Kc and Ca cell lines	MH induces β-galactosidase activity; enzyme properties similar to *E. coli* enzyme.	Best-Belpomme, M. *et al.* 1978
Gecarcinus lateralis (land crab)	whole adult	Metabolic fate of ecdysone examined in animals with and without stimulus of regenerated limbs.	McCarthy, J., 1980
Hyalophora cecropia (silkworm)	pupal wings	JH inhibits cuticle formation *in vitro*; postulates JH acts through effect on polyamine synthesis.	Willis, J., 1981
	whole adult	Inject MH and JH; blebbing of apical membrane within 1 h; epidermal cells develop rough endoplasmic reticulum.	Sedlack, B. and Gilbert, L., 1975
Leucophaea maderae (cockroach)	leg regenerates	Deposition of cuticle induced *in vitro* with MH.	Marks, E. and Leopold, R., 1970
	leg regenerates	Two sheaths deposited in leg regenerates; 1st has no chitin and can be induced by a number of substances; 2nd induced only by MH, contains chitin and resembles cuticle.	Marks, E. and Leopold, R., 1971
	leg regenerates	Time–dosage study of MH effect on cuticular deposition; concentration of MH and length of exposure make *ca.* equal contributions to effect of dose.	Marks, E., 1972

Table 9: Hormonal and developmental control of cuticle/chitin metabolism — Continued

Organism	Tissue	Comments	Reference
Locusta migratoria (migratory locust)	integument of abdomen and wings	Enzymes of intermediary metabolism measured 24 h before, 24 h after, and 12 days after ecdysis.	Surholt, B., 1976
Manduca sexta (tobacco hornworm)	abdominal cuticle	Pulsed pattern of chitinase in cuticle; activity peaks before ecdysis.	Bade, M., 1975
	integument explants	Time–dose reciprocity for various regions of abdominal epidermis studied; distinguish MH titer from duration of exposure.	Mitsui, T. and Riddiford, L., 1976
	abdominal epidermis (fifth stadium larvae and pharate pupae)	Light microscope study of abdominal epidermis; looks at intermolt cuticle secretion; notes intracellular spaces during last instar.	Wielgus, J. and Gilbert, L., 1970
	whole larvae	Transient apolysis of abdominal epidermis noted with 1st surge of MH in final feeding stage.	Riddiford, L. and Curtis, A., 1978
	epidermis (fifth stadium)	Day-by-day observation of epidermal cell structure; correlate changes with hormone titers.	Sedlak, B. and Gilbert, L., 1979
	cell culture (embryo)	Ecdysone must be converted to 20-hydroxyecdysone before cells can undergo morphological response to hormone.	Marks, E. and Holman, G., 1979
	whole animal	Examine actions and interactions of brain, CA and PG in post-embryonic development; draw attention to shifting role of JH in control of MH secretion, metabolism and action.	Safranek, L. *et al.* 1980
	whole animal	Re-examines role of brain (PTTH) in ecdysteroid-dependent transformations.	Safranek, L. and Williams, C., 1980
	whole animal	Change of commitment from larval to pupal gene products studied; DNA synthesis not essential, RNA and protein synthesis are necessary for change to occur.	Riddiford, L., 1981
Oncopeltus fasciatus (milkweed bug)	whole larvae	JH inhibits cuticle formation in last instar larvae *in vivo*.	Willis, J., 1981
Orconectes obscurus (crayfish)	whole animal	[^{14}C]GlcNAc incorporation monitored 1 h and 24 h after ecdysterone injection; increase in incorporation noted in intermolt and early premolt after 24 h hormone treatment.	Armstrong, P. and Stevenson, J., 1979
Plodia interpunctella (Indian meal moth)	imaginal discs	MH stimulates cuticle synthesis; increasing MH causes increased chitin synthesis; protein synthesis necessary for chitin synthesis.	Oberlander, H. *et al.* 1978
	imaginal discs (wing)	MH causes 2-fold increase in protein synthesis after 2 h incubation; maximum rate 16 h after MH treatment; no qualitative difference in electrophoretic pattern of proteins.	Oberlander, H. and Leach, C., 1978
	imaginal discs	Chitin synthesis 8 h after 24 h exposure to MH; RNA and protein synthesis continue in hormone-contact period; synthesis of protein, not RNA, required in posthormone culture period.	Oberlander, H. *et al.* 1980
	cultured wing discs	Chitin synthesis dependent on protein synthesis in hormone incubation phase and in post-stimulus lag period.	Ferkovich, S. *et al.* 1980
Trichoplusia ni (cabbage looper)	ovarian cell line	No uptake of MH noted; no metabolic products of MH detected; JH inhibited RNA and protein synthesis.	Cohen, E. and Gilbert, L., 1972
—	—	Review of literature published 1915 to 1970.	Marks, E., 1970

MH = molting hormone; JH = juvenile hormone; CA = corpus allatum; PG = prothoracic gland.

Table 10: *Effects of pesticides on chitin/cuticle synthesis*

Organism	Tissue	Comments	Reference
Aedes aegypti (yellow fever mosquito)	whole larvae	TH6040 and methoprene applied to shallow prairie pools; bioassay with *A. aegypti* to determine long-term persistence.	Madder, D. and Lockhart, W., 1980
	larval cell line	Glycosylation of cell proteins studied; detected decrease of lipid-bound oligosaccharides with tunicamycin.	Butters, T. and Hughes, R., 1980
Boarmia bistortata (blueberry moth)	whole animal	TH6040 caused no histological change in internal tissues. Endocuticle not properly attached to epicuticle.	Salama, H. *et al.* 1976
Calliphora erythrocephala (blowfly)	whole adult and primary culture	Structural changes seen in adult peritrophic membrane formed during exposure to TH6040 and Captan.	Becker, B., 1978
Cephalosporium (fungi)	—	Readily degraded TH6040; metabolites are borderline mutagens.	Senferer, S. *et al.* 1979
Chrotogonus trachypterus (locust)	integument	Effect of TH6040 and penfluron on quantity of chitin, protein and lipid in integument.	Saxena, S. and Kumar, V., 1981
Culex pipiens (northern house mosquito)	ovarian cell line	Effect of drugs and pesticides on growth and respiration of cultured cell studied; comparisons made with mouse cells.	Yoshida, M. *et al.* 1979
Fusarium (fungi)	—	Readily degraded TH6040; metabolites are borderline mutagens.	Senferer, S. *et al.* 1979
Leptinotarsa decemlineata (Colorado potato beetle)	whole adult	Effect of TH6040 on chitin formation in the elytra studied.	Grosscurt, A., 1978
Leucophaea maderae (cockroach)	leg regenerates	MH-dependent *in vitro* chitin synthesis confirmed and quantitative bioassay described; synthesis inhibited by polyoxin D ($I_{50} = 7.53 \times 10^{-7}$ M) and TH6040 ($I_{50} = 6.11 \times 10^{-10}$ M)	Sowa, B. and Marks, E., 1975
	cockroach leg regenerates	Triazine compounds CGA-19235 and CGA-72662 inhibited chitin synthesis *in vitro*.	Miller, A. *et al.* 1980
Locusta migratoria (locust)	whole adult	TH6040 caused decrease in chitin and protein of peritrophic membrane.	Clarke, L. *et al.* 1977
Lymentria monacha (nun moth)	whole animal	TH6040 did not affect development of testes or spermatogenesis; epicuticle and endocuticle not properly attached.	Salama, H. *et al.* 1976
Manduca sexta (tobacco hornworm)	fifth stadium larval	TH6040 inhibited cuticle production when fed or topically applied.	Mitsui, T. *et al.* 1980
	abdominal segments	TH6040 inhibited endocuticle deposition and MH-induced pupal cuticle synthesis; inhibited Glc and $GlcNH_2$ incorporation into chitin ($ID_{50} = 1.1 \times 10^{-9}$ M)	Mitsui, T. *et al.* 1980
Musca domestica (house fly)	whole larvae	Mechanism of resistance to TH6040 studied in selected strains; resistant strains showed decreased penetration of insecticide and increased excretion of metabolites.	Pimprikar, G. and Georghiou, G., 1979
	larval body wall	*In vitro* assay for chitin synthesis described; TH6040, DU 19111 and polyoxin D caused inhibition with chitin synthetase as the target enzyme.	Van Eck, W., 1979
	whole animal	TH6040 decreased chitin in cuticle and increased cuticle chitinase activity to 180% at 1 ppm and 240% at 2.5 ppm.	Ishaaya, I. and Casida, J., 1974
Neurospora crassa (fungi)	—	Tunicamycin a linear competitive inhibitor of chitin synthetase $K_i = 480\ \mu$M.	Selitrennikoff, C., 1979
Oncopeltus fasciatus (milkweed bug)	whole animal (fifth instar nymph)	Structure/activity study with 24 benzolyphenyl ureas; good correlation between toxicity and potency as *in vivo* inhibitor of chitin synthetase.	Hajjar, N. and Casida, J., 1978

Table 10: Effects of pesticides on chitin/cuticle synthesis — Continued

Organism	Tissue	Comments	Reference
Penicillium (fungi)	—	Readily degraded TH6040; metabolites are borderline mutagens.	Senferer, S. *et al.* 1979
Pieris brassicae (cabbage worm)	whole larvae	Polyoxin D and benzoylphenyl-ureas inhibited chitin synthesis; measured by decrease in [^{14}C]glucose uptake and microscopic cuticle thickness.	Gijswijt, M. *et al.* 1979
	whole animal	Neither TH6040 nor DU19111 affected chitinase activity *in vivo* or *in vitro*; both inhibited chitin synthesis in larvae cuticle.	Deul, D. *et al.* 1978
Plodia interpunctella (Indian meal moth)	wing imaginal discs	Effects of JH and pesticides on MH-induced cuticle deposition; cytochalasin B and polyoxin D but not TH6040 inhibited chitin synthesis.	Oberlander, H. and Leach, C., 1975
Plorthetria dispar (gypsy moth)	whole animal	TH6040 caused 33% decrease in chitin content of integument; no change in protein content.	Salama, H. *et al.* 1976
	whole animal	TH6040 caused increase in hemolymph glucose and cuticle glycogen.	Baeumler, W. and Salama, H., 1976
Rhodotorula (yeast)	—	Readily degraded TH6040; metabolites are borderline mutagens.	Senferer, S. *et al.* 1979
Rattus norvegicus (rat)	C6 glial cells	Characterized uptake, subcellular location and metabolism of [^{14}C]TH6040; no significant metabolism of TH6040 observed, no inhibition of complex carbohydrate synthesis.	Bishai, W. and Stoolmiller, A., 1979
Schistocerca gregaria (locust)	whole larvae and adults	TH6040 caused decreased chitin in cuticle and increased tendency to fracture; structural coherence and optical properties of affected cuticle depend on protein.	Ker, R., 1977
Stomoxys calcitrans (face fly)	pupal imaginal epidermal tissue	Assay system for screening chitin synthesis inhibitors described.	Mayer, R. *et al.* 1980b
Tribolium castaneum (red flour beetle)	whole animal	TH6040 caused dose dependent decrease in activity of trehalase, invertase and amylase *in vivo*; no inhibition *in vitro*.	Ishaaya, I. and Ascher, K., 1971
Swiss mice	whole adult	TH6040 did not inhibit mammalian hexosamine transferases as monitored by *in vivo* incorporation of labeled precursors into hyaluronic acid and condroitin sulfate.	Bentley, J. *et al.* 1979
NONE	processed chitin	Interaction of chitin and chitosan with pesticides studied; uptake of 2,4-D by chitosan and of Dicamba by both polymers.	Kemp, M. and Wightman, W., 1981

are used as fungal antibiotics have been described that interfere with chitin deposition (Corbett, J., 1974). The polyoxins, structural analogs of UDP-*N*-acetylglucosamine, have been studied in detail and their mode of action defined as a competitive inhibitor of chitin synthetase (Misato, T. *et al.*, 1979).

Although some fungicides and antibiotics have insecticidal activity in the laboratory (Leighton, T. *et al.*, 1981), they have not been used as agricultural chemicals. However, since 1970 a group of insecticides has been developed that specifically interferes with chitin metabolism (reviewed by Verloop, A. and Ferrell, D., 1977; A. Chen and R. Mayer, vol. 12). The best-studied of these is Diflubenzuron® TH6040, [*N*-(4-chloropheny)-*N*′-(2,6-difluorobenzoyl)-urea], a substituted urea compound that apparently disrupts production of chitin in newly formed cuticle so that the insect has difficulty in molting and consequently dies (Mulder, R. and Gijswijt, M., 1973; Wellinga, K. *et al.*, 1973; McGregor, H. and Kramer, K., 1976; Kramer, K. and McGregor, H., 1979). Histological studies note a variety of structural changes after the administration of benzoylureas (Ker, R., 1970; Clarke, L. *et al.*, 1977; Grosscurt, A., 1978), including an inability to incorporate [^{3}H]glucose into endocuticle (Post, L. *et al.*, 1974), and an absence of normal

Table 11: Inhibitors of chitin synthesis and degradation

Compound	Comment	Reference
Chitin synthesis		
I. Dinitrophenols and benzoylphenylureas	Twenty-four compounds examined in structure/function study in *Oncopeltus fasciatus*. Toxicity correlated with chitin synthetase inhibition.	Hajjar, N. and Casida, J., 1978, 1979
TH 6040	$ID_{50} = 6.11 \times 10^{-10}$ M in cockroach leg regenerate assay.	Sowa, B. and Marks, E., 1975
	No inhibition of chitin synthesis in *Plodia interpunctella* tissue culture assay.	Oberlander, H. and Leach, C., 1975
	33% decrease in cuticle chitin; no change in level of cuticle protein.	Salama, H. *et al.* 1976
	More effect seen in first instar than in fourth instar of *Tribolium castaneum*; decrease in trehalase, invertase and amylase activities.	Ishaaya, I. and Ascher, K., 1971
	Total inhibition of cuticular growth in *Pieris brassicae* larvae.	Gijswijt, M. *et al.* 1979
	Decrease in cuticle chitin in *Schistocerca gregaria.*	Ker, R., 1977
	Decrease in cuticle chitin and protein in peritrophic membrane of *Locusta migratoria.*	Clarke, L. *et al.* 1977
	Cuticle formation inhibited *in vivo* and *in vitro* in *Calliphora erythrocephala.*	Becker, B., 1978
	Inhibition of chitin formation and change in elytra penetrability by TH 6040 show same kinetics.	Grosscurt, A., 1978
	Virtually complete inhibition of chitin synthesis 15 min after application to *Pieris brassicae.*	Deul, D. *et al.* 1978
	In vitro system for measuring chitin synthesis in housefly larvae; accumulation of UDP-GlcNAc seen in presence of TH 6040.	Van Eck, W., 1979
	Inhibited whole cuticle growth and chitin synthesis.	Gijswijt, M. *et al.* 1979
	Inhibited cuticle formation in *Manduca sexta* when given orally or topically applied.	Mitsui, T. *et al.* 1980
	Screening assay in *Stomoxys calcitrans* described. $ID_{50} = 52$ nM *in vitro.*	Mayer, R. *et al.* 1980b
	Did not inhibit chitin synthetase preparation from *S. calcitrans in vitro.*	Mayer, R. *et al.* 1980a
	Insect growth regulator without effect on chitin synthetase in aqueous, cell-free assay.	Leighton, T. *et al.* 1981; Mayer, R. *et al.* 1981
	Inhibited purified chitin synthetase from brine shrimp *in vitro.*	Horst, M., 1981
	$ID_{50} = 1.1 \times 10^{-9}$ M in *Manduca sexta.*	Mitsui, T. *et al.* 1980
	20 μg per nymph reduces cuticle chitin level in *Chrotogonus trachypterus*; no effect on cuticle protein(s).	Saxena, S. and Kumar, V., 1981
DU 19111	Inhibited chitin synthesis in *Pieris brassicae.*	Deul, D. *et al.* 1978
	Inhibited chitin synthetase; accumulation of UDP-GlcNAc in metabolite incorporation study.	Van Eck, W., 1979
	Insect growth regulator without effect on chitin synthetase in aqueous, cell free assay.	Leighton, T. *et al.* 1981 Mayer, R. *et al.* 1981
SIR8514	Insect growth regulator without effect on chitin synthetase in aqueous, cell free assay.	Leighton, T. *et al.* 1981; Mayer, R. *et al.* 1981
	$ID_{50} = 440$ nM in *S. calcitrans* epidermal chitin assay *in vitro.*	Mayer, R. *et al.* 1980b
SIR6874	Insect growth regulator without inhibitory effect on chitin synthetase in aqueous, cell free assay.	Leighton, T. *et al.* 1981; Mayer, R. *et al.* 1981
Penfluron	20 μg per nymph greatly reduced cuticle chitin in *C. trachypterus.*	Saxena, S. and Kumar, V., 1981
EL494	$ID_{50} = 8.6 \mu$M when assayed in *Stomoxys calcitrans* epidermal tissue assay.	Mayer, R. *et al.* 1980b
II. Nucleoside analogs		
Polyoxin A	Competitive inhibitor of chitin synthetase; $K_i = 0.6 \mu$M when assayed with UDP-GlcNAc.	Muller, H. *et al.* 1981
	$K_i = 5 \times 10^{-7}$ M, competitive *vs.* UDP-GlcNAc.	Keller, F. and Cabib, E., 1971

Table 11: Inhibitors of chitin synthesis and degradation — Continued

Compound	Comment	Reference
Polyoxin A-M	Competitive inhibitors for UDP-GlcNAc in chitin synthetase reaction *in vitro*.	Misato, T. *et al.* 1979
Polyoxin D	$ID_{50} = 13\,\mu M$ in *S. calcitrans* epidermal chitin synthesis assay.	Mayer, R. *et al.* 1980b
	$ID_{50} = 7.53 \times 10^{-7}$ M in cockroach leg regenerate assay.	Sowa, B. and Marks, E., 1975
	Inhibited chitin synthesis in *Plodia interpunctella* wing disk assay.	Oberlander, H. and Leach, C., 1975
	Inhibited chitin synthesis *in vivo* in *Pieris brassicae*.	Gijswijt, M. *et al.* 1979
	Accumulation of UDP-GlcNAc in housefly chitin synthesis assay *in vitro*.	Van Eck, W., 1979
	Inhibited gut chitin synthetase from *Tribolium* when assayed as microsomal fraction.	Cohen, E. and Casida, J., 1980a,b
Tunicamycin	Blocked synthesis of dolichyl diphosphate *N*-acetylglucosamine in *Triatoma infestans*.	Quesada-Allue, L. *et al.* 1976; Quesada-Allue, L., 1982
	Competitive inhibitor of *N. crassa* chitin synthetase, $K_i = 480\,\mu M$ with respect to UDP-*N*-GlcNAc.	Selitrennikoff, C., 1979; Mayer, R. *et al.* 1981
Cytocalasin B	Inhibited chitin synthesis in *Plodia interpunctella* wing disk assay.	Oberlander, H. and Leach, C., 1975
Nikkomycin	Inhibits *Tribolium* gut chitin synthetase *in vitro*.	Cohen, E. and Casida, J., 1980a,b
Nikkomycin X	$K_i = 0.5\,\mu M$, competitive *vs.* UDP-GlcNAc.	Muller, H. *et al.* 1981
Nikkomycin Z	$K_i = 0.5\,\mu M$, competitive *vs.* UDP-GlcNAc.	Muller, H. *et al.* 1981
III. Polyene Antibiotics		
Nystatin Amphotericin B Filipin Pimaricin	Inhibitory at levels $\geq 10\,\mu g\,ml^{-1}$ when tested on purified chitin synthetase.	Rast, D. and Bartnicki-Garcia, S., 1981
IV. Sulfenimides		
Captan	Inhibited peritrophic membrane formation *in vivo* and *in vitro* in *Calliphora erythrocephala*.	Becker, B., 1978
Captafol	Inhibited *Tribolium* gut chitin synthetase *in vitro*.	Cohen, E. and Casida, J., 1980a,b
V. Organophosphate		
Kitazin P	Inhibited chitin synthetase from *P. oryzae* non-competitive with UDP-GlcNAc.	Misato, T. *et al.* 1979
	Insect growth regulator without inhibitory effect on chitin synthetase in aqueous, cell-free assay.	Leighton, T. *et al.* 1981
VI. Chloro hydrocarbons		
Dichlorofluanid	Inhibited *Tribolium* gut chitin synthetase *in vitro*.	Cohen, E. and Casida, J., 1980a,b
VII. Triazines		
CGA-1235	Insect growth regulator without inhibitory effect on chitin synthetase in aqueous, cell-free assay.	Mayer, R. *et al.* 1981
CGA-72662	Inhibited chitin synthesis in cultured cockroach leg regenerate system.	Miller, A. *et al.* 1981
VIII. Miscellaneous compounds		
Heat-stable cytoplasmic factor	Obtained from yeast cells and hyphae of *Candida albicans*.	Braun, P. and Calderone, R., 1979a,b
UDP, UTP	Inhibited chitin synthetase from *Tribolium* gut assayed as microsomal fraction from larvae.	Cohen, E. and Casida, J., 1980a,b
	Inhibited purified chitin synthetase from *Stomoxys calcitrans in vitro*.	Mayer, R. *et al.* 1980a
UDP	Inhibited purified chitin synthetase from brine shrimp.	Horst, M., 1981
CDP, CTP	Inhibited chitin synthetase from *Tribolium* gut assayed as microsomal fraction from larvae.	Cohen, E. and Casida, J., 1980a,b

Table 11: *Inhibitors of chitin synthesis and degradation* — Continued

Compound	Comment	Reference
Glycine $\beta(1,4)$ $GlcNAc_2$	Inhibited purified chitin synthetase from *Stomoxys calcitrans in vitro*.	Mayer, R. *et al.* 1980a
Chitin degradation		
I. Metal ions		
Hg^{2+}, Fe^{3+}	*In vitro* inhibition of exochitinase.	Kimura, S., 1974a; Calvo, P. *et al.* 1978; Dziadik-Turner, C. *et al.* 1981
II. Sugar analogs		
GlcNAc, GalNAc, lactones, *N*-acyl derivatives	*In vitro* inhibition of exochitinase.	Mommsen, T., 1980; Dziadik-Turner, C. *et al.* 1981; Koga, D. *et al.* 1982

endocuticle deposition (Mitsui, T. *et al.,* 1980). Diflubenzuron® is commercially available and is currently used to control certain forest insects.

The exact mode of action of this and other antichitin compounds is unknown (Kramer, K., 1976, 1979) although it has been shown that both chitin synthesis (Marks, E. and Sowa, B., 1974; Deul, D. *et al.,* 1978) and degradation (Ishaaya, I. and Casida, J., 1974; Yu, S. and Terriere, L., 1975) are affected in insects. A structure/function study done on a group of 24 benzoylureas in the milkweed bug (*Oncopeltus fasciatus*) showed good correlation between toxicity and inhibition of chitin synthesis without alteration in the metabolism of ecdysone and 20-hydroxyecdysone (Hajjar, N. and Casida, J., 1979). Because inhibition of chitin synthesis occurs first, this phenomenon rather than chitin degradation appears to be the predominant mechanism leading to insect death.

Unlike the polyoxins, the benzoylphenylureas may not act directly on chitin synthetase. Recent experiments show that these compounds are unable to inhibit chitin synthetase in a cell-free system (Mayer, R. *et al.,* 1981; Leighton, T. *et al.,* 1981) but they do act effectively in the cockroach leg regenerate assay (Leighton, T. *et al.,* 1981) and a partially purified plasma membrane preparation (Horst, M., 1980). It has been proposed that these compounds act as serine protease inhibitors, preventing the conversion of chitin synthetase zymogen to active enzyme (Leighton, T. *et al.,* 1981). However, no direct effect on a protease has been demonstrated. In nearly every inhibition experiment it has not been determined if chitin is the only carbohydrate affected. Levels of other metabolites, such as glycoproteins and mucolipids may also be decreased.

To date no cytotoxic effects of TH6040 have been noted in higher animals, nor has it been demonstrated to affect the synthesis of complex carbohydrates (Bentley, J. *et al.,* 1979; Bishai, W. and Stoolmiller, A., 1979). However, toxic effects at 10–30 parts per billion have been noted in brine shrimp (Cunningham, P., 1976; Horst, M., 1981) and some resistant strains of housefly have already developed (Pimprikar, G. and Georghiou, G., 1979).

Because of the critical protective function of cuticle, chitinases have not been overlooked as insecticides and antimicrobial agents, although no insecticide is known presently that specifically interferes with chitin degradation. A preparation of β-1,3 glucanase and chitinase has been tested and found effective as a treatment for rice blight (Muzzarelli, R., 1977). The insecticidal effect of *Bacillus thuringiensis* spores and toxin against the spruce budworm is increased when applied with the addition of chitinase (vol. 12). The combined preparations show an improved mortality rate and better foliage protection; the chitinase apparently facilitates entry of the pathogen by hydrolyzing the chitinous layer of the gut (Smirnoff, W., 1971, 1974). Various preparations of *B. thuringiensis* and chitinase have been field-tested (e.g. Smirnoff, W. *et al.,* 1973a,b; Morris, D., 1976; Dimond, J., 1972). Chitinase itself has been found to be both toxic and non-toxic to insects (Smirnoff, W., 1974; Lysenko, O., 1976; Fast, P., 1978).

A number of chemicals have been tested for inhibition of chitin synthesis and degradation. These are listed in Table 11 for reference purposes, and will not be dealt with in detail here (see chapter 12 in this volume for some details). Suffice it to say that this number is not very large, especially in regard to chitinase. Pesticide chemists have put forth relatively little effort to develop compounds directed against chitin metabolism. We have only begun to understand the enzymatic mechanisms involved in the synthesis and degradation of chitin. Hopefully, these results will stimulate greater effort to synthesize and test compounds that may perturb chitin biochemistry.

REFERENCES

ABELES, F. B., BOSSHART, R. P., FORRENCE, L. E. and HABIG, W. H. (1970). Preparation and purification of glucanase and chitinase from bean leaves, *Plant Physiol. 47*, 129–134.

AMINOFF, D., MORGAN, W. T. J. and WATKINS, W. M. (1952). Studies in immunochemistry. II. The action of dilute alkali on the N-acetylhexosamines and the specific blood-group mucoids. *Biochem. J. 51*, 379–389.

ANDERSEN, S. O. (1971). Resilin. In *Comprehensive Biochemistry*. Edited by M. Florkin and E. H. Stotz, *26C*, 633–657. Elsevier, Amsterdam.

ANDERSEN, S. O. (1979). Biochemistry of insect cuticle. *Ann. Rev. Ent. 24*, 29–61.

ANDERSEN, S. O. and WEIS-FOGH, T. (1964). Resilin. A rubberlike protein in arthropod cuticle. *Adv. Insect Physiol. 2*, 1–65.

ANDERSON, R. S. and COOK, M. L. (1979). Induction of lysozyme like activity in the hemolymph and hemocytes of an insect, *Spodoptera eridania*. *J. Invert, Path, 33*, 197–203.

ARMSTRONG, P. W. and STEVENSON, J. R. (1979). The effect of ecdysterone on ^{14}C-N-acetylglucosamine incorporation into chitin in the crayfish *Orconectes obscurus* during the molt cycle. *Comp. Biochem. Physiol. 63B*, 63–65.

AUSTIN, P. R. (1975a). Purification of chitin. *U.S. Patent 3*, 878, 377.

AUSTIN, P. R. (1975b). Solvents and purification of chitin. *U.S. Patent 3*, 892, 731.

AUSTIN, P. R., BRINE, C. J., CASTLE, J. E. and ZIKAKIS, J. P. (1981). Chitin: new facets of research. *Science 212*, 749–753.

BADE, M. L. (1974). Localization of molting chitinase in insect cuticle. *Biochim. Biophys. Acta 372*, 474–477.

BADE, M. L. (1975). Time pattern of appearance and disappearance of active molting chitinase in *Manduca* cuticle. The endogenous activity. *FEBS Lett. 51*, 161–163.

BADE, M. L. and SHOUKIMAS, J. J. (1974). Neutral metal chelator-sensitive protease in insect molting fluid. *J. Insect Physiol. 20*, 281–290.

BADE, M. L. and STINSON, A. (1978a). Activation of old cuticle chitin as a substrate for chitinase in molt of *Manduca sexta*. *Biochem. Biophys. Res. Commun. 84*, 381–388.

BADE, M. L. and STINSON, A. (1978b). Digestion of cuticle chitin during the moult of *Manduca sexta* (Lepidoptera: Sphingidae). *Insect Biochem. 9*, 221–231.

BADE, M. L. and STINSON, A. (1979). Molting fluid chitinase: a homotropic allosteric enzyme. *Biochem. Biophys. Res. Commun. 87*, 349–353.

BADE, M. L. and STINSON, A. (1981a). Biochemistry of insect differentiation. A system for studying the mechanism of chitinase activity *in vitro*. *Arch. Biochem. Biophys. 206*, 213–221.

BADE, M. L. and STINSON, A. (1981b). Biochemistry of insect differentiation. Requirements for high *in vitro* moulting fluid chitinase activity. *Insect Biochem. 11*, 599–604.

BADE, M. L. and STINSON, A. (1981c). Chitin and chitinase: a kinetic model. *J. Theor. Biol. 93*, 697–700.

BAEUMLER, W. and SALAMA, H. S. (1976). Some biochemical changes induced by Dimilin in the gypsy moth *Porthretia dispar* (L.). *Z. Angew. Entomol. 81*, 304–310.

BAL, A. K., MURPHY, A. M. and HAMPSON, M. C. (1981). Ultrastructure and chemical analysis of the resting sporangium wall of *Synchytrium endobioticum*. *Canad. J. Plant Path. 3*, 83–89.

BALAN, J. S. and FISK, F. W. (1974). A comparative study of digestive chitinases in cockroaches. *Proc. North Central Branch Ent. Soc. Amer. 29*, 149.

BALLARDIE, F. W. and CAPON, B. (1972). 3,4-Dinitrophenyltetra-N-acetyl-β-chitotetraoside, a good chromophoric substrate for hen's egg-white lysozyme. *Chem. Commun.* 828–829.

BANERJEE, S. K., KREGAR, I., TURK, V. and RUPLEY, J. A. (1973). Lysozyme-catalyzed reaction of the N-acetylglucosamine hexasaccharide. Dependence of rate on pH. *J. Biol. Chem. 248*, 4786–4792.

BECKER, B. (1978). Effects of 20-hydroxyecdysone, juvenile hormone, Dimilin and captan on *in vitro* synthesis of peritrophic membranes in *Calliphora erythrocephala*. *J. Insect Physiol. 24*, 699–705.

BENJAMINSON, M. A. (1969). Conjugates of chitinase with fluorescein isothiocyanate or lissamine rhodamine as specific stains for chitin *in situ*. *Stain Technol. 44*, 27–31.

BENSON, R. L. (1975). Assay for 2-amino-2-deoxy-D-glucose in the presence of other primary amines. *Carbohyd. Res. 42*, 192–196.

BENTLEY, J., WEBER, G. and GOULD, D. (1979). The effect of Diflubenzuron feeding of glycosaminoglycan and sulfhemoglobin biosynthesis in mice. *Pest. Biochem. Physiol. 10*, 162–167.

BERGER, L. R. and REYNOLDS, D. M. (1958). The chitinase system of a strain of *Streptomyces griseus*. *Biochim. Biphys. Acta 29*, 522–534.

BERNIER, I., LANDUREAU, J.-C., GRELLET, P. and JOLLÈS, P. (1974). Characterization of chitinases from haemolymph and cell cultures of cockroach (*Periplaneta americana*). *Comp. Biochem. Physiol. 47B*, 41–44.

BERTHE, M. C., CHARPENTIER, C., LEMATRE, J. and BONALY, R. (1981). Glucosamine and chitin accumulation in cell walls of the yeast *Rhodotorula glutinis* CBS3044. Influence of culture conditions. *Biochem. Biophys. Res. Commun. 100*, 1504–1514.

BEST-BELPOMME, M., COURGEON, A. M. and RAMBACH, A. (1978). β-Galactosidase is induced by hormone in *Drosophila melanogaster* cell cultures. *Proc. Natl. Acad. Sci. USA 75*, 6102–6106.

BISHAI, W. R. and STOOLMILLER, A. C. (1979). Uptake of Diflubenzuron (*N*-[[(4-chloro-phenyl)amino]carbonyl]-2,6-difluorobenzamide) by rat C6 glial cells *in vitro*. *Pest. Biochem. Physiol. 11*, 258–266.

BLACKWELL, J. (1969). Structure of β-chitin or parallel chain systems of poly-β(1→4)-*N*-acetyl-*D*-glucosamine. *Biopolymers 7*, 281–298.

BLACKWELL, J. and WEIH, M. A. (1980). Structure of chitin-protein complexes: ovipositor of the Ichneumon fly *Megarhyssa*. *J. Mol. Biol. 137*, 49–60.

BLACKWELL, J., PARKER, K. D. and RUDALL, K. M. (1965). Chitin in pogonophore tubes. *F. Mar. Biol. Ass. 45*,659–661.

BLACKWELL, J., PARKER, K. D. and RUDALL, K. M. (1967).Chitin fibers of the diatoms *Thalassiosira fluviatilis* and *Cyelotella cryptica*. *J. Mol. Biol. 28*, 383–385.

BLUMBERG, K., LINIERE, F., PUSTILNIK, L. and BUSH, C. A. (1982). Fractionation of oligosaccharides containing *N*-acetyl amino sugars by reverse phase high pressure liquid chromatography. *Anal. Biochem. 119*, 407–412.

BRACKER, C. E., RUIZ-HERRERA, J. and BARTNICKI-GARCIA, S. (1976). Structure and transformation of chitin synthetase particles (chitosomes) during microfibril synthesis *in vitro*. *Proc. Natl. Acad. Sci. USA 73*, 4570–4574.

BRAUN, P. C. and CALDERONE, R. A. (1979a). Regulation and solubilization of *Candida albicans* chitin synthetase. *J. Bact. 140*, 666–670.

BRAUN, P. C. and CALDERONE, R. A. (1979b). Proteolytic regulation of chitin synthesis in yeast and hyphal forms of *Candida albicans*. DHEW Publ. (NIH) (U.S.), NIH-79-1591, Ltd, Proteolysis Microorg., pp.135–8.

BRIMACOMBE, J. S. and WEBBER, J. M. (1964). Chitin. In *Mucopolysaccharides: Chemical Structure, Distribution and Isolation*. Pages 18–42. Amsterdam, Elsevier.

BRINE, C. J. and AUSTIN, P. R. (1981a). Chitin variability with species and method of preparation. *Comp. Biochem. Physiol. 69B*, 283–286.

BRINE, C. J. and AUSTIN, P. R. (1981b). Chitin isolates: species variation in residual amino acids. *Comp. Biochem. Physiol. 70B*, 173–178.

Bullière, F. (1977). Effect of moulting hormone on RNA and cuticle synthesis in the epidermis of cockroach embryos cultured *in vitro*. *J. Insect Physiol. 23*, 393–401.

Butters, T. D. and Hughes, R. C. (1980). Effects of tunicamycin on insect cells. *Biochem. Soc. Trans. 8*, 170–171.

Cabezas, J. A., Reglero, A., DePedro, A., Diez, T. and Calvo, P. (1981). Hydrolysis of natural and synthetic substrates by α-L-fucosidase, β-D-glucuronidase and β-*N*-acetylhexosaminidase purified from molluscs. *Int. J. Biochem. 13*, 389–393.

Cabib, E. (1972). Chitin synthetase system from yeast. In *Methods in Enzymology*. Edited by Ginsberg. Vol. 28, pp.572–580. Academic Press, New York.

Cabib, E. (1976). Yeast primary septum — (chitin synthetase) — a journey into three-dimensional biochemistry. *Trends Biol. Sci. 1*, 275–277.

Cabib, E. and Bowers, B. (1971). Chitin and yeast budding: localization of chitin in yeast bud scars. *J. Biol. Chem. 246*, 152–159.

Cabib, E. and Bowers, B. (1975). Timing and function of chitin synthesis in yeast. *J. Bact. 124*, 1586–1593.

Cabib, E. and Ulane, R. (1973). Chitin synthetase activating factor from yeast, a protease. *Biochem. Biophys. Res. Commun. 50*, 186–191.

Calvo, P., Reglero, A. and Cabezas, J. (1978). Purification and properties of β-*N*-acetylhexaosaminidase from the mollusc *Helicella ericetorum* Müller. *Biochem. J. 175*, 743–750.

Candy, D. J. and Kilby, B. A. (1962). Studies on chitin synthesis in the desert locust. *J. Exp. Biol. 39*, 129–140.

Capozza, R. C. (1975). Enzymically decomposable biodegradable pharmaceutical carrier. German Patent 2, 505, 305.

Carlström, D. (1957). The crystal structure of α-chitin (poly-N-acetyl-D-glucosamine). *J. Biophys. Biochem. Cytol. 3*, 669–683.

Carlström, D. (1962). The polysaccharide chain of chitin. *Biochim. Biophys. Acta 59*, 361–364.

Carroll, S. F. and Nelson, D. R. (1979). Fluorometric quantitation of amino sugars in the picomole range. *Anal. Biochem. 98*, 190–197.

Chang, J. J. and Hash, J. H. (1979). The use of an amino acid analyzer for the rapid identification and quantitative determination of chitosan oligosaccharides. *Anal. Biochem. 95*, 563–567.

Chen, A. C. and Mayer, R. T. (1981). Fluorescamine as a tool for amino sugar analysis. *J. Chromatogr. 207*, 445–448.

Chen, A. C., Mayer, R. T. and DeLoach, J. R. (1982). Purification and characterization of chitinase from the stable fly, *Stomoxys calcitrans*. *Arch. Biochem. Biophys. 216*, 314–321.

Cheng, T. C. and Rodrick, G. E. (1974). Identification and characterization of lysozyme from the hemolymph of the soft-shelled clam *Mya arenaria*. *Biol. Bull. 147*, 311–320.

Clarke, L., Temple, G. H. R. and Vincent, J. F. V. (1977). Effects of a chitin inhibitor — Dimilin on production of peritrophic membrane in the locust, *Locusta migratoria*. *J. Insect Physiol. 23*, 241–246.

Cohen, E. and Casida, J. E. (1980a). Properties of *Tribolium* gut chitin synthetase. *Pest. Biochem. Physiol. 13*, 121–128.

Cohen, E. and Casida, J. E. (1980b). Inhibition of *Tribolium* gut chitin synthetase. *Pest. Biochem. Physiol. 13*, 129–136.

Cohen, E. and Gilbert, L. I. (1972). Metabolic and hormonal studies on two insect cell lines. *J. Insect Physiol. 18*, 1061–1076.

Coles, G. C. (1966). Studies on resilin biosynthesis. *J. Insect Physiol. 12*, 679–691.

Condoulis, W. V. and Locke, M. (1966). The deposition of endocuticle in an insect, *Calpodes ethlius* Stoll (Lepidoptera: Hesperiidae). *J. Insect Physiol. 12*, 311–323.

Corbett, J. R. (1974). *The Biochemical Mode of Action of Pesticides*. Academic Press, New York.

Correa, J. U., Elango, N., Polacheck, I. and Cabib, E. (1982). Endochitinase, a mannan-associated enzyme from *Saccharomyces cerevisiae*. *J. Biol. Chem. 257*, 1392–1397.

Croizier, G. and Croizier, L. (1978). Purification and immunological comparison of two insect lysozymes. *C.R. Acad. Sci. Paris 286D*, 469–472.

Cunningham, P. A. (1976). Effects of Dimilin (TH6040) on reproduction in the brine shrimp, *Artemia salina*. *Environ. Ent. 5*, 701–706.

Dean, R. L., Bollenbacher, W. E., Locke, M., Smith. S. L. and Gilbert, L. I. (1980). Haemolymph ecdysteroid levels and cellular events in the intermoult/moult sequence of *Calpodes ethlius*. *J. Insect Physiol. 26*, 267–280.

DeLoach, J. R. and Mayer, R. T. (1979). The pupal instar of *Stomoxys calcitrans*: developmental changes in acid phosphatase, cytochrome oxidase and lysosomal glycosidases. *Insect Biochem. 9*, 653–659.

Dempster, A. and McFarlane, J. E. (1981). Anomalies in the staining of chitin by the fluorescent chitinase technique. *Stain Technol. 56*, 50–51.

Deul, D., DeJong, B. and Kortenbach, J. (1978). Inhibition of chitin synthetase by two 1-(2,6-disubstituted benzoyl)-3-phenylurea insecticides. *Pest. Biochem. Physiol. 8*, 98–105.

Dimond, J. B. (1972). Demonstration of *Bacillus thuringiensis* plus enzyme chitinase against spruce budworm in Maine. Part I. Efficacy. Misc. Report No. 144, Maine Agricul. Expt. Station.

Domek, D. B. and Borgia, P. T. (1981). Changes in the rate of chitin-plus-chitosan synthesis accompany morphogenesis of *Mucor racemosus*. *J. Bact. 146*, 945–951.

Dorn, A. and Hoffmann, P. (1981). The "embryonic moults" of the milkweed bug as seen by the S.E.M. *Tissue Cell 13*, 461–473.

Duran, A., Cabib, E. and Bowers, B. (1979). Chitin synthetase distribution on the yeast plasma membrane. *Science 203*, 363–365.

Dziadik-Turner, C., Koga, D. and Kramer, K. J. (1981). Secretion of exo- and endo-β-*N*-acetylglucosaminidases by insect cell lines. *Insect Biochem. 11*, 215–219.

Dziadik-Turner, C., Koga, D., Mai, M. S. and Kramer, K. J. (1981). Purification and characterization of two β-*N*-acetylhexosaminidases from the tobacco hornworm, *Manduca sexta* (L.) (Lepidoptera: Sphingidae). *Arch. Biochem. Biophys. 212*, 546–560.

Echalier, G. and Ohanessian, A. (1970). *In vitro* culture of *Drosophila melanogaster* embryonic cells. *In Vitro* 6, 162–172.

Fahnrich, M. and Ahlers, J. (1981). Improved assay and mechanism of the reaction catalyzed by the chitin synthetase from *Saccharomyces cerevisiae*. *Eur. J. Biochem. 121*, 113–118.

Fast, P. G. (1978). Laboratory bioassays of mixtures of *Bacillus thuringiensis* and chitinase. *Canad. Ent. 110*, 201–203.

Feng, S. Y. (1974). Lysozymelike activities in the hemolymph of *Crassostrea virginica*. *Contemp. Topics Immunobiol. 4*, 225–231.

Ferkovich, S. M., Oberlander, H., Leach, C. E. and VanEssen, F. (1980). Hormonal control of chitin biosynthesis in imaginal discs. In *Invertebrate Systems* in vitro: *Fifth International Conference on Invertebrate Tissue Culture*. Pages 209–216. Edited by E. Kurstak, K. Maramorosch and A. Dubendorfer. Elsevier/North Holland, Amsterdam.

Ferkovich, S. M., Oberlander, H. and Leach, C. E. (1981). Chitin synthesis in larval and pupal epidermis of the Indian meal moth, *Plodia interpunctella* (Hubner), and the greater wax moth, *Galleria mellonella* (L.). *J. Insect Physiol. 27*, 509–514.

Fernandez-Sousa, J. M., Gavilanes, J. G., Munico, A. M., Perez-Aranda, A. and Rodriguez, R. (1977). Lysozyme from the insect *Ceratitis capitata* eggs. *Eur. J. Biochem. 72*, 25–33.

Findlay, J., Levvy, G. A. and Marsh, C. A. (1958). Inhibition of glycosidases by aldonolactones of corresponding configuration. 2. Inhibitors of β-*N*-acetylglucosaminidase. *Biochem. J. 69*, 467–476.

Ford, J. R., Nunley, J. A., Li, Y., Chambers, R. P. and Cohen, W. (1973). A continuously monitored spectrophotometric assay of glycosidases with nitrophenyl glycosidases. *Anal. Biochem. 54*, 120–128.

Fox, F. R., Seed, J. R. and Mills, R. R. (1972). Cuticle sclerotization by the American cockroach: immunological evidence for the incorporation of blood proteins into the cuticle. *J. Insect Physiol. 18*, 2065–2070.

Freeman, J. A. (1980). Hormonal control of chitinolytic activity in the integument of *Balanus amphitrite, in vitro*. *Comp. Biochem. Physiol. 65A*, 13–17.

Fristrom, J. W. (1968). Hexosamine metabolism in imaginal disks of *Drosophila melanogaster*. *J. Insect Physiol. 14*, 729–740.

Fristrom, J. W., Hill, R. J. and Watt, F. (1978). The procuticle of *Drosophila*: Heterogeneity of urea-soluble proteins. *Biochemistry 17*, 3917–3924.

Fukamizo, T. and Hayashi, K. (1982). Separation and mutarotation of chitooligosaccharides. *J. Biochem. Tokyo 91*, 619–626.

Galun, M., Malki, D. and Galun, E. (1981). Visualization of chitin-wall formation in hyphal tips and anastomoses of *Diplodia natalensis* by fluorescein-conjugated wheat germ agglutinin and ^{3}H-*N*-acetyl-D-glucosamine. *Arch. Microbiol. 130*, 105–110.

GEIGER, J. G., KROLAK, J. M. and MILLS, R. R. (1977). Possible involvement of cockroach haemocytes in the storage and synthesis of cuticle proteins. *J. Insect Physiol. 23*, 227–230.

GEMPERLE, A., HOLAN, Z. and POKORNÝ, V. (1982). The glucan-chitin complex in *Saccharomyces cerevisiae*. IV. The electron diffraction of crustacean and yeast cell wall chitin. *Biopolymers 21*, 1–16.

GIJSWIJT, M. J., DEUL, D. H. and DEJONG, B. J. (1979). Inhibition of chitin synthesis by benzoyl-phenylurea insecticides. III. Similarity in action in *Pieris brassicae* (L.) with Polyoxin D. *Pest. Biochem. Physiol. 12*, 87–94.

GILBERT, L. I. and KING, D. S. (1973). Physiology of growth and development. In *The Physiology of Insecta*. Edited by M. Rockstein. Vol. 1, pp.249–370. Academic Press, New York.

GILBERT, L. I., BOLLENBACHER, W. E. and GRANGER, N. A. (1980). Insect endocrinology: regulation of endocrine glands, hormone titer and hormone metabolism. *Ann. Rev. Physiol. 42*, 493–510.

GILES, C. H., HASSAN, A. S. A., LAIDLOW, M. and SUBRAMANIAN, R. V. R. (1958). Adsorption at organic surfaces. III. Some observations on the constitution of chitin and on its adsorption of inorganic and organic acids from aqueous solution. *J. Soc. Dyers Colour 74*, 647–654.

GILSON, E. (1894). Recherces chimiques sur la membrane cellulaire des champignons. *Bull. Soc. Chim. Paris 3*, 619–687.

GLASER, L. and BROWN, D. H. (1957). The synthesis of chitin in cell-free extracts of *Neurospora crassa*. *J. Biol. Chem. 228*, 729–742.

GROSSCURT, A. C. (1978). Effects of diflubenzuron on mechanical penetrability, chitin formation, and structure of the elytra of *Lepinotarsa decemlineata*. *J. Insect Physiol. 24*, 827–831.

GUSTAFSON, G. L. and MILNER, L. A. (1980). Immunological relationship between β-*N*-acetylglucosaminidase and proteinase I from *Dictyostelium discoideum*. *Biochem. Biophys. Res. Commun. 94*, 1439–1444.

HACKETT, C. J. and CHEN, K. C. (1978). Quantitative isolation of native chitin from resistant structures of *Sordaria* and *Ascaris* species. *Anal. Biochem. 89*, 487–500.

HACKMAN, R. H. (1959). Biochemistry of the insect cuticle. *Proc. 4th Int. Congr. Biochem. Vienna 12*, 48–62.

HACKMAN, R. H. (1960). Studies on chitin. 4. The occurrence of complexes in which chitin and protein are covalently linked. *Aus. J. Biol. Sci. 13*, 568–577.

HACKMAN, R. H. (1967). Melanin in an insect, *Lucilia cuprina* (Wied). *Nature 216*, 163.

HACKMAN, R. H. (1971). Distribution of cystine in a blowfly larval cuticle and stabilization of the cuticle by disulphide bonds. *J. Insect Physiol. 17*, 1065–1071.

HACKMAN, R. H. (1972). Gel electrophoresis and Sephadex thin layer studies of proteins from an insect cuticle, *Agrianome spinicollis*. (Coleoptera) *Insect Biochem. 2*, 235–242.

HACKMAN, R. H. (1974a). Chemistry of the insect cuticle. In *Physiology of Insecta*. Edited by M. Rockstein. Vol. 6, pp.215–270. Academic Press, New York.

HACKMAN, R. H. (1974b). The soluble cuticular proteins from three arthropod species: *Scylla serrata*, *Boophilus microplus* and *Agrianome spinicollis*. *Comp. Biochem. Physiol. 49B*, 457–64.

HACKMAN, R. H. (1975). Expanding abdominal cuticle in the bug *Rhodnius* and the tick *Boophilus*. *J. Insect Physiol. 21*, 1613–1623.

HACKMAN, R. H. (1976). The interactions of cuticular proteins and some comments on their adaptation to function. In *The Insect Integument*. Edited by H. R. Hepburn. Pages 107–120. Elsevier, New York.

HACKMAN, R. H. and GOLDBERG, M. (1958). Proteins of the larval cuticle *Agrianome spinicollis* (Col.). *J. Insect Physiol. 2*, 221–231.

HACKMAN, R. H. and GOLDBERG, M. (1967). The *o*-diphenoloxidases of fly larvae. *J. Insect Physiol. 13*, 531–544.

HACKMAN, R. H. and GOLDBERG, M. (1971). Studies on the hardening and darkening of insect cuticles. *J. Insect Physiol. 17*, 335–347.

HACKMAN, R. H. and GOLDBERG, M. (1976). Comparative chemistry of arthropod cuticular proteins. *Comp. Biochem. Physiol. 55B*, 201–206.

HACKMAN, R. H. and GOLDBERG, M. (1977). Molecular crosslinks in cuticles. *Insect Biochem. 7*, 175–184.

HACKMAN, R. H. and GOLDBERG, M. (1978). The non-covalent binding of two insect cuticular proteins by a chitin. *Insect Biochem. 8*, 353–357.

HACKMAN, R. H. and GOLDBERG, M. (1979). Some conformational studies of larval cuticular protein from *Calliphora vicina*. *Insect Biochem. 9*, 557–561.

HACKMAN, R. H. and GOLDBERG, M. (1981). A method for determinations of microgram amounts of chitin in arthropod cuticles. *Anal. Biochem. 110*, 277–280.

HAJJAR, N. P. and CASIDA, J. E. (1978). Insecticidal benzoylphenyl ureas: structure–activity relationships as chitin synthesis inhibitors. *Science 200*, 1499–1500.

HAJJAR, N. P. and CASIDA, J. E. (1979). Structure–activity relationships of benzoylphenyl ureas as toxicants and chitin synthesis inhibitors in *Oncopeltus fasciatus*. *Pest. Biochem. Physiol. 11*, 33–45.

HALL, L. D., YALPANI, M. and YALPANI, N. (1981). Ultrastructure of chitosan and some gel-forming, branched-chain chitosan derivatives. *Biopolymers 20*, 1413–1419.

HAMAMURA, Y. and KANEHARA, Y. (1940). Enzymatic studies on exuvial fluid of silkworm, *Bombyx mori*. II. Chitinase. *J. Agric. Chem. Soc. Japan 16*, 907–909.

HAMAMURA, Y., IIDA, S., OTUKA, M., KANEHARA, Y. and ITO, S. (1954). The enzymatic study of exuvial fluid of silkmoth, *Bombyx mori*. L. *Bull. Fac. Tex. Fibres, Kyoto Univ. 1*, 127–130.

HARDY, S. W., FLETCHER, T. C. and GERRIE, L. M. (1976). Factors in haemolymph of the mussel *Mytilus edulis* L., of possible significance as defence mechanisms. *Biochem. Soc. Trans. 4*, 473–475.

HASILIK, A. (1974). Inactivation of chitin synthase in *Saccharomyces cerevisiae*. *Arch. Microbiol. 101*, 295–301.

HASILIK, A. and HOLZER, H. (1973). Participation of the tryptophan synthase inactivating system from yeast in the inactivation of chitin synthase. *Biochem. Biophys. Res. Commun. 53*, 552–559.

HASILIK, A. and NEUFELD, E. F. (1980a). Biosynthesis of lysosomal enzymes in fibroblasts. Synthesis as precursors of higher molecular weight. *J. Biol. Chem. 255*, 4937–4945.

HASILIK, A. and NEUFELD, E. F. (1980b). Biosynthesis of lysosomal enzymes in fibroblasts. Phosphorylation of mannose residues. *J. Biol. Chem. 255*, 4946–4950.

HATA, T., HAYASHI, R. and DOI, E. (1967). Purification of yeast proteinases. Part I. Fractionation and some properties of the proteinases. *Agric. Biol. Chem. 31*, 150–159.

HEPBURN, H. R. (1976). *The Insect Integument*. Elsevier, Amsterdam.

HETTICK, B. P. (1976). A quantitative determination of chitin intermediate pools during the incorporation of ^{14}C-*N*-acetyl-glucosamine and ^{3}H-glucose in the crayfish, *Orconectes sanborni*. Ph.D. dissertation, Kent State University, Kent, Ohio.

HETTICK, B. P. and BADE, M. L. (1978). Chitin synthesis in culture. *Proceedings of the First International Conference on Chitin/Chitosan*. Edited by R. Muzzarelli and E. Pariser. Pages 464–471. Massachusetts Institute of Technology Sea Grant Report No. MITS 78–7.

HIGHNAM, K. C. (1978). Insect hormones. In *Topics in Hormone Chemistry*. Vol. I. Edited by W. Butt. Pages 78–79. John Wiley, New York.

HIMENO, M., TAKAHASHI, J. and KOMANO, T. (1979). Effect of juvenile hormone on macromolecular synthesis of an insect cell line. *Agric. Biol. Chem. 43*, 1285–1292.

HIRANO, S. and YAGI, Y. (1980). Effects of the N-substitution and physical form of chitosan and the modification of its nonreducing end groups on the rates of hydrolysis by chitinase from *Streptomyces griseus*. *Agric. Biol. Chem. 44*, 963–964.

HIRANO, S. and YAGI, Y. (1981). Periodate oxidation of the non-reducing end-groups of substrates increases the rates of enzymic hydrolyses by chitinase and by lysozyme. *Carbohyd. Res. 92*, 319–322.

HIRSH, J. and DAVISON, N. (1981). Isolation and characterization of the dopa decarboxylase gene of *Drosophila melanogaster*. *Mol. Cell. Biol. 1*, 475–485.

HOHNKE, L. A. (1971). Enzymes of chitin metabolism in the decapod, *Hemiograpsus nudus*. *Comp. Biochem. Physiol. 40B*, 757–779.

HORNUNG, D. E. and STEVENSON, J. R. (1971). Changes in the rate of chitin synthesis during the crayfish molting cycle. *Comp. Biochem. Physiol. 40B*, 341–346.

HORST, M. N. (1980). The biosynthesis of crustacean chitin by a microsomal enzyme from larval brine shrimp. *Fed. Proc. 39*, 1634.

HORST, M. N. (1981). The biosynthesis of crustacean chitin by a microsomal enzyme from larval brine shrimp. *J. Biol. Chem. 256*, 1412–1419.

HOWARD, R. W. and BLOMQUIST, G. J. (1982). Chemical ecology and biochemistry of insect hydrocarbons. *Ann. Rev. Ent. 27*, 149–172.

HUNT, S. (1970). Polysaccharide-protein complexes in invertebrates. In *Chitin*. Chapter 8, pp.129–145. Academic Press, New York.

HUTNY, J. and UGORSKI, M. (1981). Kinetics of hog pancreas α-amylase, development of the multiple attack model. *Arch. Biochem. Biophys. 206*, 29–42.

IMOTO, T. and YAGISHITA, K. (1971). A simple activity measurement of lysozyme. *Agric. Biol. Chem. 35*, 1154–1156.

IMOTO, T., JOHNSON, L. N., NORTH, A. C. T., PHILLIPS, D. C. and RUPLEY, J. A. (1972). Vertebrate enzymes. In *The Enzymes*. Edited by P. D. Boyer. Vol. 7, pp.665–868. Academic Press, New York.

ISHAAYA, I. and ASCHER, K. R. S. (1971). Effect of Diflubenzuron on growth and carbohydrate hydrolases of *Tribolium castaneium*. *Phytoparasitica 5*, 149–158.

ISHAAYA, I. and CASIDA, J. E. (1974). Dietary TH6040 alters composition and enzyme activity of housefly larval cuticle. *Pest. Biocehm. Physiol. 4*, 484–490.

JEUNIAUX, C. (1955a). Properties chitinolytiques des estraits aqueux de exuvies larvaires, prenymphales et numphales de *Tenebrio molitor*. *Arch. Int. Physiol. Biochem. 63*, 114–120.

JEUNIAUX, C. (1955b). Inability of varied-carpet beetle larvae (*Anthrenus verbasci*) to digest chitin. *Nature, Lond. 176*, 1129–1130.

JEUNIAUX, C. (1961). Chitinase: an addition to the list of hydrolases in the digestive tract of vertebrates. *Nature, Lond. 192*, 135–136.

JEUNIAUX, C. (1963). *Chitine et chitinolyse, un chapitre de la biologie moleculaire*. Edited by P. Masson, Paris.

JEUNIAUX, C. (1966). Chitinases. In *Methods in Enzymology*. Edited by E. R. Neufeld and V. Ginsberg. Vol. 8, pp.644–650. Academic Press, New York.

JEUNIAUX, C. (1971). On some biochemical aspects of regressive evolution in animals. In *Biochemical Evolution and the Origin of Life*. Edited by E. Schoffeniels. Pages 204–313. North-Holland, Amsterdam.

JEUNIAUX, C. and AMANIEU, M. (1955). Chitinolytic properties of the exuvial fluids of the silkworm. *Experientia 11*, 195–196.

JOLLES, J. and JOLLES, P. (1975). The lysozyme from *Asterias rubens*. *Eur. J. Biochem. 54*, 19–23.

JOLLES, P. and ZUILI, S. (1960). Purification et étude comparée de nouveaux lysozymes. Extraits du poumon de poule et de *Nephthys hombergi*. *Biochim. Biophys. Acta 39*, 212–217.

JONES, C. S. and KOSMAN, D. J. (1980). Purification, properties, kinetics, and mechanism of β-*N*-acetylglucosamidase from *Aspergillus niger*. *J. Biol. Chem. 255*, 11861–11869.

JUNGREIS, A. M. (1973). Formation and composition of moulting fluid in the silkmoth *Hyalophora cecropia*. *Amer. Zoo. 13*, 270A.

JUNGREIS, A. M. (1974). Physiology and composition of molting fluid and midgut lumenal contents in the silkmoth *Hyalophora cecropia*. *J. Comp. Physiol. 88*, 113–127.

JUNGREIS, A. M. (1978a). The composition of larval-pupal moulting fluid in the tobacco hornworm, *Manduca sexta*. *J. Insect Physiol. 24*, 65–73.

JUNGREIS, A. M. (1978b). Physiology of moulting in insects. *Adv. Insect Physiol. 14*, 109–183.

KATZENELLENBOGEN, B. S. and KAFATOS, F. C. (1970). Some properties of silkmoth moulting gel and moulting fluid. *J. Insect Physiol. 16*, 2241–2256.

KATZENELLENBOGEN, B. S. and KAFATOS, F. C. (1971a). Proteinases of silkmoth moulting fluid: physical and catalytic properties. *J. Insect Physiol. 17*, 775–800.

KATZENELLENBOGEN, B. S. and KAFATOS, F. C. (1971b). General esterases of silkmoth moulting fluid: preliminary characterization. *J. Insect Physiol. 17*, 1139–1151.

KATZENELLENBOGEN, B. S. and KAFATOS, F. C. (1971c). Inactive proteinases in silkmoth moulting gel. *J. Insect Physiol. 17*, 823–832.

KAYSER-WEGMANN, I. (1976). Ultrastructural differences between larval and pupal cuticles of *Pieris brassicae* (Lepidoptera). *Protoplasma 90*, 319–331.

KELLER, F. A. and CABIB, E. (1971). Chitin and yeast budding: properties of chitin synthetase from *Saccharomyces carlsbergensis*. *J. Biol. Chem. 246*, 160–166.

KEMP, M. V. and WIGHTMAN, W. (1981). Interaction of 2,4-D and Dicamba with chitin and chitosan. *Virginia J. Sci. 32*, 34–37.

KENCHINGTON, W. R. (1971). Variety in structure and production of insect silks. In *Structural Macromolecules in Arthropods*. Shell Centre, London.

KENT, P. W. (1964). Chitin. In *Comparative Biochemistry*. Edited by M. Florkin and H. W. Mason, Vol. 7, pp.93–136.

KER, R. F. (1977). Investigation of locust cuticle using the insecticide diflubenzuron. *J. Insect Physiol. 23*, 39–48.

KIMURA, S. (1973a). Chitinolytic enzymes in the larval development of the silkworm, *Bombyx mori* L. *Appl. Ent. Zool. 8*, 234–236.

KIMURA, S. (1973b). The control of chitinase activity by ecdysterone in larvae of *Bombyx mori*. *J. Insect Physiol. 19*, 115–123.

KIMURA, S. (1974a). The β-*N*-acetylglucosaminidases of *Bombyx mori* L. *Comp. Biochem. Physiol. 49B*, 345–351.

KIMURA, S. (1974b). On the metabolic fate of the carbohydrates during larval-larval transformation in the silkworm, *Bombyx mori* L., with special reference to the lytic and synthetic systems of polysaccharides. *Jap. J. Appl. Ent. Zool. 18*, 183–188.

KIMURA, S (1976a). The chitinase system in the cuticle of the silkworm *Bombyx mori*. *Insect Biochem. 6*, 479–482.

KIMURA, S. (1976b). Insect hemolymph exo-β-*N*-acetylglucosaminidase from *Bombyx mori*. *Biochim. Biophys. Acta 446*, 399–406.

KIMURA, S. (1977). Exo-β-*N*-acetylglucosaminidase and chitobiase in *Bombyx mori*. *Insect Biochem. 7*, 237–245.

KIMURA, S. (1981a). The occurrence of chitinase in the alimentary canal of the silkworm, *Bombyx mori*. *J. Sericult. Sci. Japan 50*, 101–108.

KIMURA, S. (1981b). Genetics of insect hemolymph β-*N*-acetylglucosaminidase in the silkworm, *Bombyx mori*. *Biochem. Genet. 19*, 1–14.

KIMURA, S. (1982). Properties of chitobiase and β-N-acetylglucosaminidase from *Bombyx mori* L. (Lepidoptera: Bombycidae) with special reference to adsorption on chitin. *Appl. Ent. Zool. 17*, 20–31.

KIMURA, A., KURAMOTO, M., OSUMI, S., IMAOKA, I., IZITI, K., ISONO, S., OKAMOTO, T., SUGIMURA, S., TIKAMORI, K., OSUMI, J., OSUMI, T., YOSHINAKA, M., OBAYASHI, I., ISHIMURA, Y. and OKAZAKI, Y. (1965). Studies of the enzyme from the digestive tract of *Helix peliomphala*. Report I. Purification and physico-chemical properties of the chitinase from the digestive tract of *Helix peliomphala*. *J. Faculty of Medicine, Tokushima Univ. 1965*, 122–130.

KIMURA, A., KURAMOTO, M., OSUMI, S., IMAOKA, I., IZITI, K., ISONO, S., OKAMOTO, T., SUGIMURA, S., TIKAMORI, K., OSUMI, J., OSUMI, T., YOSHINAKA, M., OBAYASHI, I., ISHIMURA, Y. and OKAZAKI, Y. (1966a). Report II. Substrate specificity of chitinase from snail digestive tract. *J. Faculty of Medicine, Tokushima Univ. 1966*, 679–683.

KIMURA, A., KURAMOTO, M., OSUMI, S., IMAOKA, I., IZITI, K., ISONO, S., OKAMOTO, T., SUGIMURA, S., TIKAMORI, K., OSUMI, J., OSUMI, T., YOSHINAKA, M., OBAYASHI, I., ISHIMURA, Y. and OKAZAKI, Y. (1966b). Report III. Effect of metal ions on chitinase from snail digestive tract. *J. Faculty of Medicine, Tokushima Univ. 1966*, 684–687.

KIMURA, S., AKAI, H. and KOBAYASHI, M. (1974). Light and electron microscopic studies of the effects of ecdysterone upon the epidermal cells of the larva of the silkworm, *Bombyx mori*. *Cell. Diff. 3*, 259–265.

KING, G. (1944). Permeability of keratin membranes. *Nature, Lond. 154*, 575–576.

KLESOV, A. A., PARBUZIN, V. S. and RABINOVICH, M. L. (1981). Enzymatic conversion of polymers. Causes of apparent inhibition by products in enzymatic degradation of polymeric substrates. *Biokhimiya 46*, 1840–1846.

KNECHT, D. A. and DIMOND, R. L. (1981). Lysosomal enzymes possess a common antigenic determinant in the cellular slime mold, *Dictyostelium discoideum*. *J. Biol. Chem. 256*, 3564–3575.

KOEPPE, J. K. and GILBERT, L. I. (1973). Immunochemical evidence for the transport of haemolymph protein into the cuticle of *Manduca sexta*. *J. Insect Physiol. 19*, 615–624.

KOGA, D., DZIADIK-TURNER, C., MAI, M. S. and KRAMER, K. J. (1982). Kinetics and mechanism of exochitinase and β-*N*-acetylhexosaminidase from the tobacco hornworm, *Manduca sexta* L. (Lepidoptera: Sphingidae). *Insect Biochem. 12*, 493–499.

KOGA, D., JILKA, J. and KRAMER, K. J. (1983). Insect endochitinases; glycoproteins from molting fluid, integument and pupal hemolyph. *Insect Biochem. 13*, 295–305.

KRAMER, K. J. (1976). Selective biochemical approaches to insect control. In *Proceedings of the Joint United States-Japan Seminar on Stored Product Insects*. Edited by R. B. Mills, L. A. Bulla and S. Utida. Pages 226–243. Kansas State University.

KRAMER, K. J. (1979). Research in biochemistry and the action of insect growth regulators. In *Proceedings: Prevention and Control of Insects in Stored Food Products*. Edited by R. Strong, R. B. Mills and L. A. Bulla. Pages 269–276. Kansas State University.

KRAMER, K. J. and MCGREGOR, H. E. (1979). Activity of seven chitin synthesis inhibitors against development of stored product insects. *Environ. Ent. 8*, 274–276.

KRISHNAN, G. (1956). The nature and composition of the epicuticle of some arthropods. *Physiol. Zool. 29*, 324–337.

KRISHNAN, G., RAMACHANDRAN, G. N. and SANTANAM, M. S. (1955). Occurrence of chitin in the epicuticle of an arachnid *Palamneus swammerdami*. *Nature, Lond. 176*, 557–558.

KUHN, R. and TIEDEMAN, H. (1954). *N*-Acetyl-β-hexosaminidase in *Homerus*. *Chem. Ber. 87*, 1141–1144.

KUNKEL, J. G. (1975). Cockroach molting. I. Temporal organization of events during molting cycle of *Blatella germanica* (L.). *Biol. Bull. 148*, 259–273.

KURSTAK, E. and MARAMOROSCH, K. (eds) (1976). *Invertebrate Tissue Culture: applications in medicine, biology and agriculture*. Academic Press, New York.

KURSTAK, E., MARAMOROSCH, K. and DUBENDORFER, A. (eds) (1980). *Invertebrate systems in vitro: Fifth International Conference on invertebrate tissue culture*. Elsevier/North-Holland Biomedical Press, Amsterdam.

LANDUREAU, J. C. and JOLLÈS, P. (1970). Lytic enzyme produced *in vitro* by insect cells: lysozyme or chitinase? *Nature, Lond. 225*, 968–969.

LANIR, N. and COHEN, E. (1978). Studies on the effect of the moulting hormone in a mosquito cell line. *J. Insect Physiol. 24*, 613–621.

LEABACK, D. H. (1970). The metabolic hydrolysis of hexosaminidase linkages. In *Metabolic Conjugation and Metabolic Hydrolysis*. Edited by W. H. Fishman. Vol. II, pp.443–457. Academic Press, New York.

LEDDERHOSE, G. (1878). Uber chitin und zein spaltung produkte. *Hoppe-Seyler's Z. Physiol. Chem. 2*, 213–227.

LEDNICER, D. and BABCOCK, J. C. (1962). Preparation and reactions of steroidal α-aminonitriles. *J. Org. Chem. 27*, 2541–2544.

LEIGHTON, T., MARKS, E. and LEIGHTON, F. (1981). Pesticides: insecticides and fungicides are chitin synthesis inhibitors. *Science 213*, 905–907.

LENNEY, J. F. (1956). A study of two yeast proteinases. *J. Biol. Chem. 221*, 919–930.

LENNEY, J. F. (1975). Three yeast proteins that specifically inhibit yeast proteases A, B and C. *J. Bact. 122*, 1265–1273.

LEPESANT, J. A., KEJZLAROVA-LEPESANT, J. and GAREN, A. (1978). Ecdysone-inducible functions of larval fat bodies in *Drosophila*. *Proc. Natl. Acad. Sci., USA, 75*, 5570–5574.

LINGAPPA, Y. and LOCKWOOD, J. L. (1962). Chitin media for selective isolation and culture of actinomycetes. *Phytopathology 52*, 317–323.

LIPKE, H. and GEOGHEGAN, T. (1971a). The composition of peptidochitodextrins from Sacrophagid puparial cases. *Biochem. J. 125*, 703–716.

LIPKE, H. and GEOGHEGAN, T. (1971b). Enzymolysis of sclerotized cuticle from *Periplaneta americana* and *Sarcophaga bullata*. *J. Insect Physiol. 17*, 415–425.

LIPKE, H. and STROUT, V. (1972). Peptidochitodextrins of sarcophagid cuticle. II. Studies on the linkage region. *Israel J. Ent. 7*, 117–128.

LIPKE, H., STROUT, K., HENZEL, W. and SUGUMARAN, M. (1981). Structural proteins of Sarcophagid larval exoskeleton. *J. Biol. Chem. 256*, 4241–4246.

LOCKE, M. (1966). The structure and formation of the cuticulin layer in the epicuticle of an insect, *Calpodes ethlius* (Lepidoptera: Hesperiide). *J. Morph. 118*, 461–494.

LOCKE, M. (1969). The structure of an epidermal cell during the development of the protein epicuticle of the uptake of molting fluid in an insect. *J. Morph. 127*, 7–40.

LOCKE, M. (1970). The molt/intermolt cycle in the epidermis and other tissues of an insect *Calpodes ethlius* (Lepidoptera: Hesperiidae). *Tissue Cell 2*, 197–223.

LOCKE, M. (1974). The structure and formation of the integument in insects. In *The Physiology of Insecta*. Edited by M. Rockstein. Vol. VI, Chap. 2, p.123–213. Academic Press, New York.

LOCKE, M. and KRISHNAN, N. (1971). The distribution of phenoloxidases and polyphenols during cuticle formation. *Tissue Cell 3*, 103–126.

LOPEZ-ROMERO, E., RUIZ-HERRERA, J. and BARTNICKI-GARCIA, S. (1982). The inhibitory protein of chitin synthetase from *Mucor rouxii* is a chitinase *Biochim. Biophys. Acta 702*, 233–236.

LOTMAR, W. and PICKEN, L. E. R. (1950). A new crystallographic modification of chitin and its distribution. *Experientia 6*, 58–59.

LUNDBLAD, G. and HULTIN, E. (1966). Human serum lysozyme (muramidase). I. Viscosimetric determination with glycol chitin and purification by selective adsorption. *Scand. J. Clin. Lab. Invest. 18*, 201–208.

LUNDBLAD, G., HEDERSTEDT, B., LIND, J. and STEBY, M. (1974). Chitinase in goat serum. Preliminary purification and characterization. *Eur. J. Biochem. 46*, 367–376.

LUNDBLAD, G., ELANDER, M. and LIND, J. (1976). Chitinase and β-*N*-acetylglucosaminidase in the digestive juice of *Helix pomatia*. *Acta Chem. Scand. B30*, 889–894.

LUNDBLAD, G., ELANDER, M., LIND, J. and SLETTENGREN, K. (1979). Bovine serum chitinase. *Eur. J. Biochem. 100*, 455–460.

LUNT, M. R. and KENT, P. W. (1960). Chitinase system from *Carcinus maenas*. *Biochim. Biophys. Acta 44*, 371–373.

LYSENKO, O. (1976). Chitinase of *Serratia maccescens* and its toxicity to insects. *J. Invert. Path. 27*, 385–386.

MADDER, D. J. and LOCKHART, W. L. (1980). Studies on the dissipation of diflubenzuron and methoprene from shallow prairie pools. *Canad. Ent. 112*, 173–177.

MAI, M. S. and KRAMER, K. J. (1983). Comparative biochemistry of esterases in pharate pupal molting fluid from the tobacco hornworm, *Manduca sexta*. *Comp. Biochem. Physiol. 74B*, 769–773.

MARAMOROSCH, K. (ed) (1976). *Invertebrate Tissue Culture: research applications*. Academic Press, New York.

MARKS, E. P. (1970). The action of hormones in insect cell and organ cultures. *Gen. Comp. Edocr. 15*, 289–302.

MARKS, E. P. (1972). Effects of ecdysterone on the deposition of cockroach cuticle *in vitro*. *Biol. Bull, 142*, 293–301.

MARKS, E. P. (1973). Induction of molting in insect organ cultures. In *Proc. 3rd Int. Colloq. Invertebrate Tissue Culture*. Edited by J. Rehacek, D. Blaskovic and F. Hink. Pages 221–232. Slovak Acad. Sci., Bratislava.

MARKS, E. P. and HOLMAN, G. M. (1979). Ecdysone action on insect cell lines. *In Vitro 15*, 300–307.

MARKS, E. P. and LEOPOLD, R. A. (1970). Cockroach leg regeneration: effects of ecdysterone *in vitro*. *Science, 167*, 61–62.

MARKS, E. P. and LEOPOLD, R. A. (1971). Deposition of cuticular substances *in vitro* by leg regenerates from the cockroach, *Leucophaea maderae*. (F.) *Biol. Bull. 140*, 73–83.

MARKS, E. P. and SOWA, B. (1974). An *in vitro* model system for the production of insect cuticle. In *Mechanism of Pesticide Action*. Edited by G. K. Kohn. ACS Symp. Series No. 2, pp.145–155. Amer. Chem. Soc., Washington, DC.

MARKS, E. P. and SOWA, B. A. (1976). Cuticle formation *in vitro*. p. In *The Insect Integument*. Edited by H. R. Hepburn. Pages 339–358. Elsevier, New York.

MARTIN, M. M., GIESELMANN, M. J. and MARTIN, J. S. (1973). Rectal enzymes of attine ants, α-amylase and chitinase. *J. Insect Physiol. 19*, 1409–1416.

MARTIN, M. M., GIESELMANN, M. J. and MARTIN, J. S. (1976). Presence of chitinase in digestive fluid of ants. *Comp. Biochem. Physiol. 53A*, 331–332.

MASAKI, A., FUKAMIZO, T., OTAKARA, A., TORIKATA, T., HAYASHI, K. and IMOTO, T. (1981). Lysozyme-catalyzed reaction of chitooligosaccharides. *J. Biochem. Tokyo 90*, 527–533.

MASIH, S. (1973). A comparative study of carbohydrates in four species of cockroaches (Dictyoptera: Blattaria). *Proc. Ind. Natl. Sci. Acad. 39B*, 598–603.

MAYER, R. T., MEOLA, S. M., COPPAGE, D. L. and DELOACH, J. R. (1979). The pupal instar of *Stomoxys calcitrans*: cuticle deposition and chitin synthesis. *J. Insect Physiol. 25*, 677–683.

MAYER, R. T., CHEN, A. C. and DELOACH, J. R. (1980a). Characterization of a chitin synthetase from the stable fly, *Stomoxys calcitrans* (L.). *Insect Biochem. 10*, 549–556.

MAYER, R. T., MEOLA, S. M., COPPAGE, D. L. and DELOACH, J. R. (1980b). Utilization of imaginal tissues from pupae of the stable fly for the study of chitin synthesis and screening of chitin synthesis inhibitors. *J. Econ. Ent. 73*, 76–80.

MAYER, R. T., CHEN, A. C. and DELOACH, J. R. (1981). Chitin synthesis inhibiting insect growth regulators do not inhibit chitin synthetase. *Experientia 37*, 337–338.

McCarthy, J. F. (1980). Ecdysone metabolism and the interruption of proecdysis in land crab, *Gecarcinas lateralis*. *Biol. Bull.* *158*, 91–102.

McDade, J. E. and Tripp, M. R. (1967a). Lysozyme in the hemolymph of the oyster, *Crassostrea virginica*. *J. Invert. Path.* *9*, 531–535.

McDade, J. E. and Tripp, M. R. (1967b). Lysozyme in oyster mantle mucus. *J. Invert. Path.* *9*, 581–582.

McGregor, H. E. and Kramer, K. J. (1976). Activity of Dimilin® (TH6040) against Coleoptera in stored wheat and corn. *J. Econ. Ent.* *69*, 479–480.

McHenery, J. G., Allen, J. A. and Birkbeck, T. H. (1978). The lysozyme-like activity of the common mussel, *Mytelius edulis*. *Soc. Gen. Microbiol. Proc.* *5*, 107.

McHenery, J. G., Birkbeck, T. H. and Allen, J. A. (1979). The occurrence of lysozyme in marine bivalves. *Comp. Biochem. Physiol.* *63B*, 25–28.

Miller, A. L., Kress, B. C., Lewis, L., Stein, R. and Kinnon, C. (1980). Effect of tunicamycin and cycloheximide on the secretion of acid hydrolases from I-cell cultured fibroblasts. *Biochem. J.* *186*, 971–975.

Mills, G. L. and Cantino, E. C. (1981). Chitosome-like vesicles from gamma particles of *Blastocladiella emersonii* synthesize chitin. *Arch. Microbiol.* *130*, 72–77.

Mills, R. R., Greenslade, F. C., Fox, F. R. and Nielsen, D. J. (1967). Purification of KCl-soluble proteins from the cuticle of *Acheta domesticus* (L.). *Comp. Biochem. Physiol.* *22*, 327–332.

Minke, R. and Blackwell, J. (1978). The structure of α-chitin. *J. Molec. Biol.* *120*, 167–181.

Misato, T., Kakiki, K. and Hori, M. (1979). Chitin as a target for pesticide action: progress and prospect. In *Advances in Pesticide Science*. Edited by H. Geissbuhler. Pages 458–464. Pergamon Press, New York.

Mitsui, T. and Riddiford, L. M. (1976). Pupal cuticle formation by *Manduca sexta* epidermis *in vitro*: Patterns of ecdysone sensitivity. *Devel. Biol.* *54*, 172–186.

Mitsui, T., Nobusawa, C., Fukami, J., Collins, J. and Riddiford, L. M. (1980). Inhibition of chitin synthesis by diflubenzuron in *Manduca* larvae. *J. Pest. Sci.* *5*, 335–341.

Molano, J., Duran, A. and Cabib, E. (1977). A rapid and sensitive assay for chitinase using ^{3}H-chitin. *Anal. Biochem.* *83*, 648–656.

Molano, J., Polacheck, I., Duran, A. and Cabib, E. (1979). An endochitinase from wheat germ. Activity on nascent and preformed chitin. *J. Biol. Chem.* *254*, 4901–4907.

Molano, J., Bowers, B. and Cabib, E. (1980). Distribution of chitin in the yeast cell wall: an ultrastructural and chemical study. *J. Cell Biol.* *85*, 199–212.

Molotsov, N. V. and Vafina, M. G. (1972). Distribution of β-*N*-acetylglucosaminidase in marine invertebrates. *Comp. Biochem. Physiol.* *41B*, 113–120.

Mommsen, T. P. (1978). Digestive enzymes of a spider (*Tegenaria atrica* Koch). II. Carbohydrases. *Comp. Biochem. Physiol.* *60A*, 371–375.

Mommsen, T. P. (1980). Chitinase and β-*N*-acetylglucosaminidase from the digestive fluid of the spider *Cupiennius salei*. *Biochim. Biophys. Acta* *612*, 361–372.

Mordue, W. and Stone, J. V. (1979). Insect hormones. In *Hormones and Evolution*. Edited by E. J. W. Barrington. Vol. 1, pp.215–271. Academic Press, New York.

Morgan, W. T. J. and Elson, L. A. A. (1934). CXXVII. A colorimetric method for the determination of *N*-acetylglucosamine and *N*-acetylchondrosamine. *Biochem. J.* *28*, 988–995.

Morris, O. N. (1976). A two year study of *Bacillus thuringiensis*–chitinase combinations in spruce budworm control. *Canad. Ent.* *108*, 225–233.

Morrissey, R., Dugan, E. and Koths, J. (1976). Chitinase production by *Arthobacter sp.* lysing cells of *Fusarium roseum*. *Soil Biol. Biochem.* *8*, 23–28.

Mothes, U. and Seitz, K. A. (1981). A possible pathway of chitin synthesis as revealed by electron microscopy in *Tetranychus urticae* (Acari; Tetranychidae). *Cell. Tiss. Res.* *214*, 443–448.

Mulder, R. and Gijswijt, M. J. (1973). The laboratory evaluation of the two promising new insecticides which interfere with cuticle deposition. *Pest. Sci.* *4*, 737–745.

Muller, H., Furter, R., Zahner, H. and Rast, D. M. (1981). Metabolic products of microorganisms. 203. Inhibition of chitosomal chitin synthetase and growth of *Mucor rouxii* by nikkomycin Z, nikkomycin X and polyoxin A: a comparison. *Arch. Microbiol.* *130*, 195–197.

Muzzarelli, R. A. A. (1976). Biochemical modification of chitin. In *The Insect Integument*. Edited by H. R. Hepburn. Pages 63–87. Elsevier, Amsterdam.

Muzzarelli, R. A. A. (1977). *Chitin*. Pergamon Press, New York.

Muzzarelli, R. A. A. and Pariser, E. R. (1978). *Proceedings of the First International Conference on Chitin/Chitosan*. Massachusetts Institute of Technology Sea Grant Report No. MITSG 78–7.

Nardon, P., Wicker, C., Grenier, A. A. and Laviolette, P. (1978). Contrôle de la proliferation des symbiotes par l'hôte: étude préliminaire de l'exo-β-*N*-acétylglucosaminidase chez le curculionide *Sitophilus oryzae*. *C.R. Acad. Sci. Paris.* *287D*, 1157–1160.

Neuberger, A. and Wilson, B. M. (1967). Inhibition of lysozyme by derivatives of D-glucosamine. *Biochim. Biophys. Acta* *147*, 473–486.

Neufeld, E. F., Lim, T. W. and Shapiro, L. J. (1975). Inherited disorders of lysosomal metabolism. *Ann. Rev. Biochem.* *44*, 357–376.

Neville, A. C. (1967a). Daily growth layers in animals and plants. *Biol. Rev.* *42*, 421–441.

Neville, A. C. (1967b). Chitin orientation in cuticle and its control. *Adv. Insect Physiol.* *4*, 213–286.

Neville, A. C. (1975). *Biology of the Arthropod Cuticle*. Springer-Verlag, New York.

Neville, A. C., Parry, D. A. D. and Woodhead-Galloway, J. (1976). The chitin crystallite in arthropod cuticle. *J. Cell Sci.* *21*, 73–82.

Oberlander, H. (1976). Hormonal contol of growth and differentiation of tissues cultured *in vitro*. *In vitro* *12*, 225–235.

Oberlander, H. and Leach, C. (1975). Inhibition of chitin synthesis in *Plodia interpunctella*. In *Proc. 1st Int. Congr. Stored Products Entomol.*, pp.651–655.

Oberlander, H. and Leach, C. E. (1978). Protein synthesis in imaginal disks of *Plodia interpunctella* during development *in vivo* and *in vitro*. *In Vitro* *14*, 723–727.

Oberlander, H., Ferkovich, S. M., Van Essen, F. and Leach, C. E. (1978). Chitin biosynthesis in imaginal disks cultured *in vitro*. *Wilhelm Roux's Arch. Dev. Biol.* *185*, 95–98.

Oberlander, H., Ferkovich, S., Leach, E. and Van Essen, F. (1980). Inhibition of chitin biosynthesis in cultured imaginal discs. Effects of alpha-amanitin, actinomycin-D, cycloheximide and puromycin. *Wilhelm Roux's Arch. Dev. Biol.* *188*, 81–86.

O'Brien, S. J. and MacIntyre, R. J. (1978). Genetics and biochemistry of enzymes and specific proteins of *Drosophila*. In *The Genetics and Biology of Drosophila*. Edited by M. Ashburner, and T. Wright. Vol. 2a, 395–551. Academic Press, New York.

Odier, A. (1823). Memoire sur la composition chimique des parties cornees des insects. *Mem. Soc. Hist. Nat. Paris* *1*, 29–42.

Ohtakara, A. (1961). Studies on the chitinolytic enzymes of black-koji mold. 1. Viscometric determination of chitinase activity by application of chitin glycol as a new substrate. *Agric. Biol. Chem.* *25*, 50–54.

Ohtakara, A. (1963). Studies on the chitinolytic enzymes of the black-koji mold. 5. Participation of two different enzymes in the discomposition of glycochitin to the constituent aminosugar. *Agric. Biol. Chem.* *27*, 454–460.

Ohtakara, A. (1964). Chitinolytic enzymes of black-koji mold. 6. Isolation of some properties of *N*-acetyl-β-glucosaminidase. *Agric. Biol. Chem.* *28*, 745–751.

Otaki, N. and Kimura, M. (1975). Studies on lysozymes. VI. Application of a new colormetric assay on lysozymes. *Ind. Health.* *13*, 23–29.

Passonneau, J. V. and Williams, C. M. (1953). Moulting fluid of cecropia silkworm. *J. Exp. Biol.* *30*, 545–560.

Perez, N. and Cabezas, J. A. (1977). β-*N*-Acetylhexosaminidase du mollusque *Chamelea gallina* L. *Biochimie* *59*, 729–733.

Périn, J. P. and Jolles, P. (1972). The lysozyme from *Nephthys hombergi* (annelid). *Biochim. Biophys. Acta* *263*, 683–689.

Pesti, M., Campbell, J. M. and Peberdy, J. F. (1981). Alteration of ergosterol content and chitin synthase activity in *Candida albicans*. *Curr. Micro.* *5*, 187–190.

Phillips, D. R. and Loughton, G. (1976). Cuticle protein in *Locusta migratoria*. *Comp. Biochem. Biophys.* *55B*, 129–135.

Pimprikar, G. D. and Georghiou, G. P. (1979). Mechanisms of resistance to Diflubenzuron in the house fly, *Musca domestica* (L.). *Pest. Biochem. Physiol.* *12*, 10–22.

Podboronov, V. M., Revina, T. A., Grokhoyskaya, I. M., Zhurovleva, T. P. and Finnik, V. P. (1978). Some biochemical properties of lysozymes of argasid ticks. *Medit. Parazitol. Parazitar. Bol.* *47*, 96–99.

PORTER, C. A. and JAWORSKI, E. G. (1965). Biosynthesis of chitin during various stages in the metamorphosis of *Prodenia eridania*. *J. Insect Physiol. 11*, 1151–1160.

POST, L. C., DEJONG, B. J. and VINCENT, W. R. (1974). 1-(2,6-Disubstituted benzoyl)-3-phenylurea insecticides: inhibitors of chitin synthesis. *Pest Biochem. Physiol. 4*, 473–483.

POWNING, R. F. and DAVIDSON, W. J. (1973). Studies on insect bacteriolytic enzymes I. Lysozyme in haemolymph of *Galleria mellonella* and *Bombyx mori*. *Comp. Biochem. Physiol. 45B*, 669–686.

POWNING, R. F. and DAVIDSON, W. J. (1979). Studies on insect bacteriolytic enzymes. III. Lytic activities in some plant materials of possible benefit to insects. *Comp. Biochem. Physiol. 63B*, 199–206.

POWNING, R. F. and IRZYKIEWICZ, H. (1963). A chitinase from the gut of cockroach *Periplaneta americana*. *Nature Lond. 200*, 1128.

POWNING, R. F. and IRZYKIEWICZ, H. (1964). β-Acetylglucosaminidase in the cockroach (*Periplaneta americana*) and in the puff-ball (*Lycoperdon perlatum*). *Comp. Biochem. Physiol. 12*, 405–415.

POWNING, R. F. and IRZYKIEWICZ, H. (1965). Studies on the chitinase system in bean and other seeds. *Comp. Biochem. Physiol. 14*, 127–133.

POWNING, R. F. and IRZYKIEWICZ, H. (1967). Separation of chitin oligosaccharides by thin-layer chromatography. *J. Chromatogr. 29*, 115–119.

PRICE, J. B. and HOLDICH, O. M. (1980). An untrastructural study of the integument during the moult cycle of the woodlouse, *Oniscus asellus* (Crustacea: Isopoda). *Zoomorphologie 95*, 250–263.

PURCHASE, E. R. and BRAUN, C. E. (1946). D-Glucosamine hydrochloride. *Org. Synth. 26*, 36–37.

QUESADA-ALLUE, L. A. (1982). The inhibition of insect chitin synthesis by tunicamycin. *Biochem. Biophys. Res. Commun. 105*, 312–319.

QUESADA, ALLUE, L. A., MARECHAL, L. R. and BELOCOPITOW, E. (1976). Chitin synthesis in *Triatoma infestans* and other insects. *Acta Physiol. Latinoamer. 26*, 349–363.

RAFERTY, M. A., RAND-MEIR, T., DAHLQUIST, F. W., PARSONS, S. M., BORDERS, C. L., WOLCOTT, R. G., BERANEK, W. and JAO, L. (1969). Separation of glycosaminoglycan saccharide and glycoside mixtures by gel filtration. *Anal. Biochem. 30*, 427–435.

RAMSAY, J. A. (1935). The evaporation of water from the cockroach. *J. Exp. Biol. 12*, 373–383.

RAST, D. M. and BARTNICKI-GARCIA, S. (1981). Effects of amphotericin B., nystatin and other polyene antibiotics on chitin synthase. *Proc. Natl. Acad. Sci., USA 78*, 1233–1236.

RATAULT, C. and VEY, A. (1977). Production d'esterases et de *N*-acetyl-β-D-glucosaminidase dan le tegument du Coleoptere *Oryctes rhinoceros* par le champignon entomopathogens *Metarrhizium anisopliae*. *Entomophaga 22*, 289–294.

REISSIG, J. L., STROMINGER, J. K. and LELOIR, L. F. (1955). A modified colorimetric method for the estimation of *N*-acetylamino sugars. *J. Biol. Chem. 217*, 956–966.

RICHARDS, A. G. (1951). *The Integument of Arthropods*, University of Minnesota Press, Minneapolis.

RICHARDS, A. G. (1978). The Chemistry of Insect Cuticle. In *Biochemistry of Insects*. Edited by M. Rockstein. Pages 205–232. Academic Press, New York.

RIDDIFORD, L. M. (1976). Juvenile hormone control of epidermal commitment *in vivo* and *in vitro*. In *The Juvenile Hormones*. Edited by L. I. Gilbert. Pages 198–219. Plenum Press, New York.

RIDDIFORD, L. M. (1980). Insect endocrinology: action of hormones at the cellular level. *Ann. Rev. Physiol. 42*, 511–528.

RIDDIFORD, L. M. (1981). Hormonal control of epidermal cell development. *Amer. Zool. 21*, 751–762.

RIDDIFORD, L. M. and CURTIS, A. T. (1978). Hormonal control of epidermal detachment during the final feeding stage of *Manduca sexta*. *J. Insect Physiol. 24*, 561–568.

RIDDIFORD, L. M. and TRUMAN, J. W. (1978). Biochemistry of insect hormones and insect growth regulators. In *Biochemistry of Insects*. Edited by M. Rockstein. Pages 308–357. Academic Press, New York.

RODRICK, G. E. and CHENG, T. C. (1974). Kinetic properties of lysozyme from the hemolymph of *Crassostrea virginica*. *J. Invert. Path. 24*, 41–48.

ROULEAU, M. (1980). Effects of cations, sugars, detergents, sulfhydryl compounds and cryoprotective agents on the colorimetric determination of *N*-acetylglucosamine by the method of Reissig. *Anal. Biochem. 103*, 144–151.

RUDALL, K. M. (1962). Silk and other cocoon proteins. In: *Comparative Biochemisty*. Edited by M. Florkin and H. S. Mason. Vol. 4, pp.397–433. Academic Press, New York.

RUDALL, K. M. (1963). The chitin/protein complexes of insect cuticles. *Adv. Insect Physiol. 1*, 257–313.

RUDALL, K. M. (1976). Molecular structure in arthropod cuticles. In *The Insect Integument*. Edited by H. Hepburn. Pages 21–41. Elsevier, Amsterdam.

RUDALL, K. M. and KENCHINGTON, W. (1973). The chitin system. *Biol. Rev. 48*, 597–636.

RUH, M. F., RUH, T. S., DEWERT, W. and DUENAS, V. (1974). Ecdysterone-induced protein synthesis *in vitro*. *J. Insect Physiol. 20*, 1729–1736.

RUIZ-HERRERA, J. and BARTNICKI-GARCIA, S. (1974). Synthesis of cell wall microfibrils *in vitro* by a "soluble" chitin synthetase from *Mucor rouxii*. *Science 186*, 357–358.

RUIZ-HERRERA, J., LOPEZ-ROMERO, E., and BARTNICKI-GARCIA, S. (1977). Properties of chitin synthetase in isolated chitosomes from yeast cells of *Mucor rouxii*. *J. Biol. Chem. 252*, 3338–3343.

RUIZ-HERRERA, J., BARTNICKI-GARCIA, S. and BRACKER, C. E. (1980). Dissociation of chitosomes by digestion into 16S subunits with chitin synthetase activity. *Biochim. Biophys. Acta 629*, 201–216.

RUPLEY, J. A. (1964). The hydrolysis of chitin by concentrated hydrochloric acid, and the rapid preparation of low molecular weight substrates for lysozyme. *Biochim. Biophys. Acta 83*, 245–255.

RUPLEY, J. A. and GATES, V. (1967). Studies on the enzymic activity of lysozyme, II. The hydrolysis and transfer reactions of *N*-acetylglucosamine oligosaccharides. *Proc. Natl. Acad. Sci. USA 57*, 496–510.

RYERSE, J. S., and LOCKE, M. (1978). Ecdysterone-mediated cuticle deposition and the control of growth in insect tracheae. *J. Insect Physiol. 24*, 541–550.

SAFRANEK, L. and WILLIAMS, C. M. (1980). Studies of the prothoracicotropic hormone in the tobacco hornworm, *Manduca sexta*. *Biol. Bull. 158*: 141–153.

SAFRANEK, L., CYMBOROWSKI, B. and WILLIAMS, C. M. (1980). Effects of juvenile hormone on ecdysone-dependent development in the tobacco hornworm, *Manduca sexta*. *Biol. Bull. 158*, 248–256.

SAITO, H., TABETA, R. and HIRANO, S. (1981a). Conformation of chitin and *N*-acyl chitosans in solid state as revealed by ^{13}C cross polarization/magic angle spinning (CP/MAS) NMR spectroscopy. *Chem. Lett.* 1479–1482.

SAITO, H., MAMIZUKA, T., TABETA, R. and HIRANO, S. (1981b). High resolution ^{13}C NMR spectra of chitin oligomers in aqueous solution. *Chem. Lett.* 1483–1484.

SALAMA, H. S., MOTAGALLY, Z. A. and SKATULLA, U. (1976). On the mode of action of Dimilin as a molting inhibitor in some lepidopterous insects. *Z. Angew. Ent. 80*, 396–407.

SANCHEZ-MOZO, P., FREIRE RAMMA, M., VAZQUEZ PERMAS, R. and RUIZ AMIL, M. (1977). Purification and properties of two enzymatic forms of β-*N*-acetylglucosaminidase from *Mytilus edulis* (L.) hepatopancreas. *Comp. Biochem. Physiol. 58B*, 29–34.

SAXENA, S. C. and KUMAR, V. (1981). Effect of Difluron and Penfluron on integumentary chitin, protein and lipid of *Chrotogonus trachypterus* (Orthoptera: Acrididae). *Ind. J. Exp. Biol. 19*, 669–670.

SAXENA, S. C. and SARIN, K. (1972). Chitinase in the alimentary tract of the lesser mealworm, *Alphitobius diaperinus*. *Appl. Ent. Zool. 7*, 94.

SCHELLER, K. and KARLSON, P. (1977). Synthesis of poly(A) containing RNA induced by ecdysterone in fat body cells of *Calliphora vicina*. *J. Insect Physiol. 23*, 435–440.

SEDLAK, B. and GILBERT, L. (1975). Hormonal control of insect epidermal cell activities and ultrastructural analysis. *Trans. Amer. Mic. Soc. 94*, 480–500.

SEDLAK, B. J. and GILBERT, L. I. (1979). Correlations between epidermal cell structure and endogenous hormone titers during the fifth larval instar of the tobacco hornworm, *Manduca sexta*. *Tissue Cell 11*, 643–653.

SELITRENNIKOFF, C. D. (1979). Competitive inhibition of *Neurospora crassa* chitin synthetase activity by tunicamycin. *Arch. Biochem. Biophys. 195*, 243–244.

SELITRENNIKOFF, C. P., DALLEY, N. E. and SONNEBORN, D. R. (1980). Regulation of the hexosamine biosynthetic pathway in the water mold in *Blastocladiella emersonii*: sensitivity to endproduct inhibition is dependent upon the life cycle phase. *Proc. Natl. Acad. Sci. USA 77*, 5998–6002.

SENFERER, S., BRAYMER, H. and DUNN, J. (1979). Metabolism of Diflubenzuron by soil microorganisms and mutagenicity of the metabolites. *Pest Biochem. Physiol. 10*, 174–180.

SHARMA, S. C. and PANT, R. (1973). Larval and pupal cuticular proteins of *Philosamia rieini. Ind. J. Exp. Biol. 11*, 349–351.

SHULMAN, M. L., KULSHIN, V. A., KHORLIN, A. Y. (1980). A continuous fluorimetric assay for glycosidase activity: human *N*-acetyl-*β*-D-hexosamindiase. *Anal. Biochem. 101*, 342–348.

SLOAT, B. F. and PRINGLE, J. R. (1978). A mutant of yeast defective in cellular morphogenesis. *Science 200*, 1171–1173.

SLOAT, B. F., ADAMS, A. and PRINGLE, J. R. (1981). Roles of the CDC24 gene product in morphogenesis during the *Saccharomyces cerevisiae* cell cycle. *J. Cell Biol. 89*, 395–405.

SMIRNOFF, W. A. (1971). Effect of chitinase on the action of *Bacillus thuringiensis. Canad. Ent. 103*, 1829–1831.

SMIRNOFF, W. A. (1974). The symptoms of infection by *Bacillus thuringiensis* and chitinase formulation in larvae of *Choristoneura fumiferana. J. Invert. Path. 23*, 397–399.

SMIRNOFF, W. A., RANDALL, A. P., MARTINEAU, R., HALLIBURTON, W. and JUNEAU, A. (1973a). Field test for effectiveness of chitinase additive to *Bacillus thuringiensis* Berliner against *Choristoneura fumiferana. Canad. J. For. Res. 3*, 228–236.

SMIRNOFF, W. A., FETTES, J. J. and DESAULNIENS, R. (1973b). Aerial spraying of a *Bacillus thuringiensis*-chitinase formulation for control of the spruce budworm. *Canad. Ent. 105*, 1535–1544.

SMITH, R. J., PEKRUL, S. and GRULA, E. A. (1981). Requirement for sequential enzymatic activities for penetration of the integument of the corn earworm (*Heliothis zea*). *J. Invert. Path. 38*, 335–344.

SNYDER, M., HIRSH, J. and DAVIDSON, N. (1981). The cuticle genes of *Drosophila*: a developmentally regulated gene cluster. *Cell 25*, 165–177.

SOWA, B. A. and MARKS, E. P. (1975). An *in vitro* system for the quantitative measurement of chitin synthesis in the cockroach inhibition by TH6040 and polyoxin D. *Insect Biochem. 5*, 855–859.

SPINDLER, K. D. (1976). Initial characterization of chitinase and chitobiase from the integument of *Drosophila hydei. Insect Biochem. 6*, 663–667.

SRIDHARA, S., NOWOCK, J. and GILBERT, L. I. (1978). Biochemical endocrinology of insect growth and development. In *Biochemistry and Mode of Action of Hormones II*, Vol. 20. Edited by H. V. Rickenberg. Pages 133–188. University Park Press, Baltimore.

SRIVASTAVA, R. P. (1970). Electrophoretic behavior of cuticular proteins of different developmental stages of *Galleria mellonella. J. Insect Physiol. 16*, 2345–2351.

STEVENSON, J. R. (1972). Changing activities of the crustacean epidermis during the molting cycle. *Amer. Zool. 12*, 373–380.

STEVENSON, J. R. and HETTICK, B. P. (1980). Metabolism of chitin precursors by crayfish tissues during chitin synthesis. *J. Exp. Zool. 214*, 37–48.

STONE, J. V. and MORDUE, W. (1980). Isolation of insect neuropeptides. *Insect Biochem. 10*, 229–239.

SURHOLT, B. (1975a). Studies *in vivo* and *in vitro* on chitin synthesis during larval–adult moulting cycle of the migratory locust, *Locusta migratoria* L. *J. Comp. Physiol. 102*, 135–147.

SURHOLT, B. (1975b). Formation of glucosamine-6-phosphate in chitin synthesis during ecdysis of the migratory locust, *Locusta migratoria. Insect Biochem. 5*, 585–593.

SURHOLT, B. (1976). Changes in enzyme patterns in the integument of *Locusta migratoria* during formation of adult cuticle. *Insect Biochem. 6*, 79–83.

THOMA, J. A. and CROOK, C. (1982). Subsite mapping of enzymes. Double inhibition studies. *Eur. J. Biochem. 122*, 613–618.

TOWER, W. L. (1906). Observations on the changes in the hypodermis and cuticula of coleoptera during ecdysis. *Biol. Bull. 10*, 176–192.

TRACEY, M. V. and YOUATT, G. (1958). Cellulase and chitinase in two species of Australian termites. *Enzymologia 19*, 70–72.

TSAI, C. S. (1970). Determination of degree of polymerization of N-acetyl chitooligoses by chromatographic methods. *Anal. Biochem. 36*, 114–122.

ULANE, R. E. and CABIB, E. (1974). The activating system of chitin synthetase from *Saccharomyces cerevisiae*. Purfication and properties of an inhibitor of the activating factor. *J. Biol. Chem. 249*, 3418–3422.

ULANE, R. E. and CABIB, E. (1976). The activating system of chitin synthetase from *Saccharomyces cerevisiae. J. Biol. Chem. 251*, 3367–3374.

VAN ECK, W. H. (1979). Mode of action of two benzoylphenyl ureas as inhibitors of chitin synthesis in insects. *Insect Biochem. 9*, 295–300.

VAN EIKEREN, P. and MCLAUGHLIN, H. (1977). Analysis of the lysozyme-catalyzed hydrolysis and transglycosylation of *N*-acetyl-D-glucosamine oligomers by high-pressure liquid chromatography. *Anal. Biochem. 77*, 513–522.

VARDANIS, A. (1976). *In vitro* assay system for chitin synthesis in insect tissue. *Life Sci. 19*, 1949–1956.

VARDANIS, A. (1979). Characteristics of the chitin-synthesizing system of insect tissue. *Biochim. Biophys. Acta 588*, 142–147.

VERLOOP, A. and FERRELL, D. (1977). Benzoylphenyl ureas- a new group of larvicides interfering with chitin deposition. In *Pesticide Chemistry in the 20th Century*. ACS Symposium Series No. 37. Edited by J. Plimmer. Pages 237–270. Amer. Chem. Soc., Washington, DC.

WADSTROM, T. (1971). Chitinase activity and substrate specificity of endo-*β*-*N*-acetylglucosaminidase of *Staph. aureus*, Strain M18. *Acta Chem. Scand. 25*, 1807–1812.

WAHEED, A., HASILIK, A. and VON FIGURA, K. (1981). Processing of the phosphorylated recognition marker in lysosomal enzymes. Characterization and partial purification of a microsomal *α*-*N*-acetylglucosaminyl phosphodiesterase. *J. Biol. Chem. 256*, 5717–5721.

WATERHOUSE, D. F. and MCKELLAR, J. W. (1961). The distribution of chitinase activity in the body of the American cockroach. *J. Insect Physiol. 6*, 185–195.

WATERHOUSE, D. F., HACKMAN, R. H. and MCKELLAR, J. W. (1961). An investigation of chitinase activity in cockroach and termite extracts. *J. Insect Physiol. 6*, 96–112.

WEIS-FOGH, T. (1970). Structure and formation of insect cuticle. In *Insect Ultrastructure*. Edited by A. C. Neville. Pages 165–185. Blackwell, Oxford.

WELINDER, B. S. (1975). The crustacean cuticle. 3. Composition of individual layers in *Cancer pagurus* cuticle. *Comp. Biochem. Physiol. 52A*, 659–663.

WELLINGA, K., MULDER, R. and VAN DAALEN, J. J. (1973). Synthesis and laboratory evaluation of 1-(2,6-disubstituted benzoyl)-3-phenylureas, a new class of insecticides. I. 1-(2,6-dichloro benzoyl)-3-phenylureas. *J. Agric. Food Chem. 21*, 348–354.

WIELGUS, J. J. and GILBERT, L. I. (1978). Epidermal cell development and control of cuticle deposition during last larval instar of *Manduca sexta. J. Insect Physiol. 24*, 629–637.

WIGGLESWORTH, V. B. (1944). Action of inert dusts on insects. *Nature, Lond. 153*, 493–494.

WIGGLESWORTH, V. B. (1945). Transpiration through the cuticle of insects. *J. Exp. Biol. 21*, 97–114.

WIGGLESWORTH, V. B. (1948). The insect cuticle. *Biol. Rev. 23*, 408–451.

WIGGLESWORTH, V. B. (1957). The physiology of insect cuticle. *Ann. Rev. Ent. 2*, 37–54.

WIGGLESWORTH, V. B. (1970). *Insect Hormones*. Oliver & Boyd, Edinburgh.

WILLIS, J. H. (1974). Morphogenetic action of insect hormones. *Ann. Rev. Ent. 19*, 97–116.

WILLIS, J. H. (1981). Juvenile hormone: the status of "status quo." *Amer. Zool. 21*, 763–773.

WINICUR, S. and MITCHELL, H. K. (1974). Chitinase activity during *Drosophila* development. *J. Insect Physiol. 20*, 1795–1805.

WOLFGANG, W. J. and RIDDIFORD, L. M. (1981). Cuticular morphogenesis during continuous growth of the final instar larva of a moth. *Tissue Cell 13*, 757–772.

WOOLLEN, J. W., WALKER, P. G. and HEYWORTH, R. (1961). Studies on glucosaminidase. VI. *N*-Acetyl-*β*-glucosaminidase and *N*-acetyl-*β*-galactosaminidase activities of a variety of enzyme preparations. *Biochem. J. 79*, 294–298.

WYATT, G. R. (1972). Insect hormones. In *Biochemical Actions of Hormones*. Edited by G. Litwack. Pages 386–490. Academic Press, New York.

YAMADA, H. and IMOTO, T. (1981). A convenient synthesis of glycochitin, a substrate of lysozyme. *Carbohydrate Res. 92*, 160–162.

YEUNG, K. K., OWEN, A. J. and DAIN, J. A. (1979). Purification and properties of two isoenzymes of β-N-acetylhexosaminidase from *Turbo cornutus*. *Comp. Biochem. Physiol. 63B*, 329–334.

YOSHIDA, M., ONAKA, M., FUJITA, T. and NAKAJIMA, M. (1979). Inhibitory effects of pesticides on growth and respiration of cultured cells. *Pest. Biochem. Physiol. 10*, 313–321.

YOUNG, R. G. (1979). Midgut lysosomal phosphatases and carbohydrases of the last larval instar of the southern armyworm, *Spodoptera eridania*. *Ann. Ent. Soc. Amer. 72*, 193–196.

YU, S. J. and TERRIERE, L. C. (1975). Activities of hormone metabolizing enzymes in houseflies treated with some substituted urea growth regulators. *Life Sci. 17*, 619–626.

ZACHARUK, R. Y. (1972). Fine structure of the cuticle, epidermis, and fat body of larval elateridae (Coleoptera) and changes associated with molting. *Canad. J. Zool. 50*, 1463–1488.

ZACHARUK, R. Y. (1976). Structural changes of the cuticle associated with molting. In *The Insect Integument*. Edited by H. R. Hepburn. Chap. 16, pp.299–321. Elsevier, New York.

ZECHMEISTER, L., TOTH, G. and VAJDA, E. (1939). Chromatographic adsorption of chitinase. *Naturwissenschaften 27*, 367–378.

ZIELKOWSKI, R. and SPINDLER, K. D. (1978). Chitinase and chitobiase from the integument of *Locusta migratoria*: Characterization and titer during the fifth larval instar. *Insect Biochem. 8*, 67–71.

4 Cuticular Lipids

GARY J. BLOMQUIST

University of Nevada, Reno, Nevada, USA

and

JACK W. DILLWITH

University of Missouri, Columbia, Missouri, USA

INTRODUCTION

The cuticular lipids of insects consist of aliphatic material which is present on the outer layer of the integument. These waxes serve the critical function of restricting water loss to prevent desiccation in most terrestrial insects, and also serve as sex attractants and aphrodisiacs, as species and caste recognition cues, and as kairomones in certain insect groups. The cuticular lipids affect the absorption of insecticides and other chemicals from the environment and have been suggested to serve as a barrier to the penetration of micro-organisms. Some insects, such as the scale insects and honeybees, produce much larger amounts of waxy material for protection and for raising brood and storing honey.

All insects which have been carefully studied possess hydrocarbons in their cuticular lipids, and they are the most abundant class of compounds in many species. Other types of waxes commonly found in cuticular extracts include wax esters, ketones, epoxides, secondary alcohols, primary alcohols, free fatty acids, sterols, sterol esters and triacylglycerols. In this context, wax is used in its broad sense to include a variety of lipids rather than in the strict chemical definition.

Cuticular lipids are synthesized by cells which are located near the integument or in the peripheral fat body, probably oenocytes. The biosynthetic pathways for the major cuticular components, including hydrocarbons, wax esters and secondary alcohols, have been established. Prior to deposition, the lipids must pass through the epidermal cells and then through the cuticle to reach the surface of the insect (Fig. 1). The pore canals have been implicated in the process by which cuticular lipids cross the cuticle (Locke, M., 1965; Hepburn, H., this volume). The thickness of the cuticular lipid layer appears to vary considerably among different species, and values from a few molecules thick to over half a micrometer have been reported.

Various aspects of insect cuticular lipids have recently been reviewed. Discussions focusing on aspects of the chemistry and biochemistry of cuticular lipids have been presented by Blomquist, G. and Jackson, L. (1979); Nelson, D. (1978); Lockey, K. (1980a); Jackson, L. and Blomquist, G. (1976a); Blomquist, G. and de Renobales, M. (1982) and Jackson, L. and Baker, G. (1970). Analytical techniques used to characterize insect waxes have been reviewed by Jackson, L. and Armold, M. (1977) and Gilby, A. (1980a). The role of cuticular hydrocarbons in chemical communication has been dealt with by Howard, R. and Blomquist, G. (1982), and discussions of the role of cuticular lipids in limiting water loss have recently been presented by Hadley, N. (1980a, 1981); Gilby, A. (1980b); Machin, J. (1980); Edney, E. (1977) and Ebeling, W. (1974). This chapter summarizes the functions attributed to insect cuticular lipids, outlines analytical procedures used in characterizing wax components, reviews the types of chemicals commonly found in cuticular lipids, discusses the biosynthesis of cuticular lipids and reviews the physiological and endocrine regulation of cuticular lipid formation.

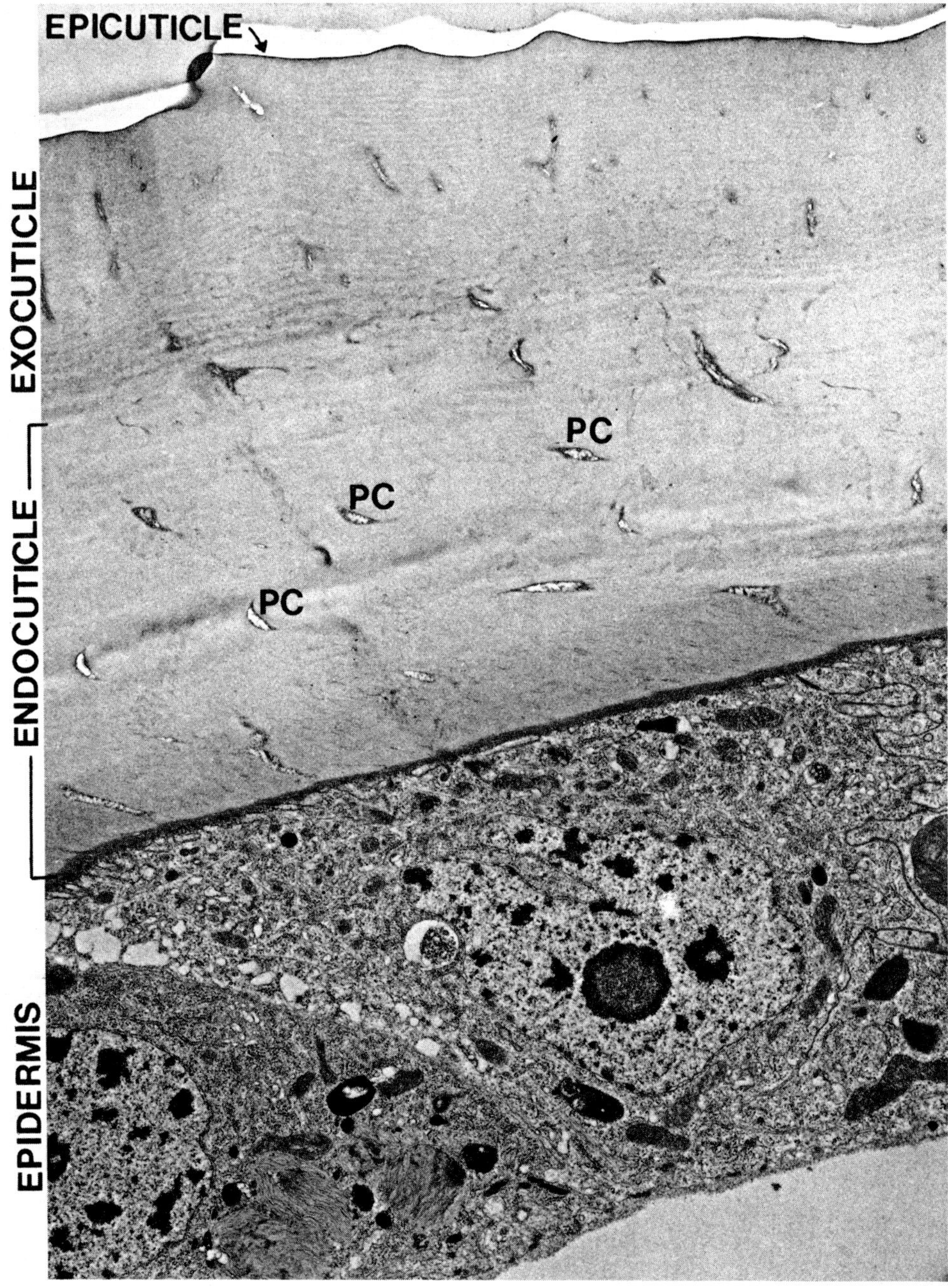

FIG. 1. Electron micrograph of the integument of adult cabbage looper, *Trichoplusia ni*, female abdominal tergite. The epidermis, endocuticle, exocuticle and epicuticle are marked. PC is pore canal (× 12,200). (Courtesy of Dr Jean Percy, Canadian Forestry Service, Sault Ste. Marie, Ontario, Canada.)

1 FUNCTIONS OF CUTICULAR LIPIDS

1.1 Prevent desiccation

Perhaps the most critical of the functions attributed to insect cuticular lipids is the formation of a waterproof layer which prevents a lethal rate of desiccation. Many investigators have measured the transpiration of water from insects, and these studies have been reviewed periodically (Hadley, N., 1980a, 1981; Gilby, A., 1980b; Machin, J., 1980; Wharton, G. and Richards, A., 1978; Edney, E., 1977; Ebeling, W., 1974 and Beament, J., 1976, 1964, 1967). The evidence is in agreement that the cuticle is permeable to water vapors, and that the cuticular lipids play a major role in reducing transpiration.

Major lines of evidence which support this view are:

(1) Cuticles from which the surface lipids are removed with organic solvents are relatively permeable to water. Vegetable oils, lecithin, and a series of wetting agents and detergents affect the permeability of the cuticle to water.
(2) The waterproofing observed with intact insects is closely duplicated when extracted cuticular lipids are deposited on collodium membranes or intact wing membranes (Beament, J., 1945).
(3) Rapid desiccation occurs as a result of scratching the outer surface of the cuticle with abrasive dusts, and this has been used in insect control. Adsorption of the lipids onto the dust may also play a role in increasing transpiration (Ebeling, W., 1974).
(4) If the abraded insect is kept in a moist atmosphere to prevent desiccation, the lipid layer is restored, and along with it, the resistance to desiccation returns.
(5) Positive correlations have been shown between the quantity of extracted cuticular wax and the measured or potential transpiration. In addition, a greater amount of cuticular lipid is often present when there is a greater need for water conservation (Hadley, N., 1980a, 1981; Bell, R. *et al.*, 1975).
(6) Insects which inhabit desiccating environments have a surface-wax composition that provides maximum impermeability to water vapor (Hadley, N., 1980a, 1981).
(7) The transpiration rate from an insect is found to increase rather abruptly at a temperature that corresponds to the transition point or change of phase point of the lipids on the cuticle of the particular species (Beament, J., 1976).

This latter point has become quite controversial. The discovery that the permeability of insect cuticle underwent abrupt increases at specific temperatures prompted Beament, J. (1945) and Wigglesworth, V. (1945) to propose that cuticular permeability is controlled by a monolayer of polar lipids oriented at an angle of 65° to the cuticle surface. It was proposed that at the transition temperature (Tc), sufficient thermal energy is available to disrupt this ordered state, leading to increased transpiration. Another model proposed to account for the permeability changes at Tc includes M. Locke's (1965) suggestion that the molecules in the wax canals undergo a phase change at Tc.

Machin, J. (1980) and Hadley, N. (1981) have summarized the recent evidence which argues against the oriented monolayer model proposed by Wigglesworth and Beament. The major lines of evidence against the model are:

(1) The cuticular lipids of most insects studied to date are not composed primarily of aliphatic alcohols and acids, but rather contain large amounts of hydrocarbon;
(2) Forced area curve data for cockroach cuticular lipid (Lockey, K., 1976) indicated that a monolayer was absent and that the molecules were only weakly attracted to the water surface;
(3) Toolson, E. *et al.* (1979), using spin-labeled molecules, showed that there was not a preferred orientation on the cuticle, and furthermore showed that the phase transitions take place with lipids associated with the surface rather than the pore canals;
(4) Toolson, E. (1978), using mathematical arguments, has shown that the abrupt permeability increases which occur at Tc when water loss is plotted against temperature are in reality artifacts created by incorrectly dividing the observed rate of water loss by the saturation deficit to correct for thermally induced changes in the diffusion gradient.

Using Toolson's method, Machin, J. (1980) calculated data for a number of insects, and suggests that the presence or absence of temperature transitions depends upon whether the insect has a "greasy" or "waxy" cuticle. A generally accepted model for the waterproofing observed by cuticular lipids awaits development.

1.2 Effect on entry of insecticides and micro-organisms

The cuticular lipids play a role in controlling the entry of insecticides and other chemicals from the environment (Brooks, G., 1976). Olson, W. (1970) suggests that the cuticular lipid layer constitutes a significant barrier to the penetration of lipoidal solutes through the integument. He concludes that partitioning of lipid-soluble insecticides such as DDT from the epicuticular lipid into lower phases is the rate-limiting step in penetration. Disruption of the cuticular lipids increases the rate of solute penetration into the cockroach (Olson, W., 1970). Other workers (Szeicz, F. *et al.*, 1973) consider the underlying proteinacious part of the cuticle as the main barrier to insecticide penetration, and consider the surface lipid as facilitating the lateral movement of topically applied insecticides. Gerolt, P. (1969) has obtained evidence contradicting the widely held view that contact insecticides reach the site of action by penetrating through the integument. He suggests that topically applied dieldrin accumulates in the cuticular lipids, spreads laterally, and then reaches the site of action via the tracheal system.

The protection from micro-organisms provided by the cuticular lipids may be due to the non-wetting nature they provide the insect's surface, making it difficult for the organisms to establish themselves. It is also possible that the cuticular lipids provide chemical protection by being toxic to micro-organisms. Koidsumi, K. (1957) reported that the cuticular lipids of the larvae of *Bombyx mori* L. and *Chilo simplex* Butler were capable of restricting the growth of pathogenic fungi *in vitro,* and indicated that the antifungal compounds were short-chain saturated fatty acids. However, no definitive structural identification of these compounds was made.

1.3 Chemical communication

In addition to their critical role in preventing desiccation, insect cuticular lipids also serve in chemical communication. Major roles assigned to cuticular hydrocarbons include serving as sex pheromones, as species and caste recognition cues in several social insects, as thermoregulatory pheromones and as kairomonal cues for parasites. Table 1 lists the species reported to use cuticular components in chemical communication and lists the types of components involved.

1.3.1 Sex pheromones

Cuticular hydrocarbons have been widely implicated as dipteran sex pheromones. Rogoff, W. *et al.* (1964) reported that benzene extracts of female *Musca domestica* elicited both attraction and excitation mating behavior patterns in males. Carlson, D. *et al.* (1971) reported that (Z)-9-tricosene was a "sex attractant" of *Musca domestica* and that it could be extracted from the bodies of females or from their fecal deposits.

A low level of response to (Z)-9-tricosene, both in olfactometer and pseudofly bioassays, prompted several workers to re-examine the female housefly's surface lipid chemistry, and led to the identification of over 100 isomeric methylalkanes (Nelson, D. *et al.,* 1981), (Z)-9,10-epoxytricosane, (Z)-14-tricosene-10-one, and a C_{36} wax ester (Uebel, E. *et al.*, 1978a). The methylalkanes are terminally and internally branched monomethylalkanes and 3,X-, 4,X- and internally branched dimethylalkanes. Uebel, E. *et al.* (1976) showed that addition of branched alkanes of 28–30 carbons to a pseudofly treated with (Z)-9-tricosene enhanced the mating strike activity of the males, although the alkanes by themselves had no activity. This stimulatory effect of branched alkanes was confirmed by Rogoff, W. *et al.* (1980). In addition, Uebel, E. *et al.* (1978a) showed that the epoxide and ketone both released strike activity by the males in the absence of (Z)-9-tricosene. The ester showed no biological activity. Combinations of all active and synergistic components still failed, however, to match a live virgin female as a target for a mating strike. Richter, I. *et al.* (1976) proposed that (Z)-9-tricosene may serve as a releaser mechanism for responsiveness to optical cues, leading to both mating strikes and aggregation.

Considerable effort has also been made at elucidating the importance of cuticular lipids as pheromones for the face-fly *Musca autumnalis*. Chaudhury, M. *et al.* (1972, 1973, 1974) suggested the presence of a sex pheromone in this species from olfactometer bioassay data. Uebel, E. *et al.* (1975c) and Sonnet, P. *et al.* (1975) isolated and characterized several cuticular hydrocarbons which elicited strike behavior by males in a modified pseudofly bioassay. Active components included (Z)-14-nonacosene, (Z)-13-nonacosene and (Z)-13-heptacosene. These olefinic components were found in approximately equal quantities on both males and females. Males, however, contain greater quantities of heptacosane and nonacosane than do females, and the authors suggest that these alkanes attenuate the effects of the olefins. The overall hydrocarbon differences are however slight and a closer examination of this species likely to be fruitful.

An even more complex situation exists regarding the sex pheromone complex of the stable-fly, *Stomoxys calcitrans* (L). Muhammed, S. *et al.* (1975) reported that the female produced both "sex attractant" and "mating stimulation" pheromones. Uebel, E. *et al.* (1975b) showed that (Z)-9-hentriacontene, (Z)-9-tritriacontene and methyl branched hentria- and tritriacontenes incited males to attempt copulation, as did a series of monomethyl- and dimethylalkanes. Later work by Sonnet, P. *et al.* (1977) on the saturated components showed that 15-methyl- and 15,19-dimethyltritriacontanes were most active. Harris, R. *et al.* (1976) showed that increasing amounts of these hydrocarbons were produced from emergence until the female was 3 days old. Meola, R. *et al.* (1977) reported that this species required a blood-meal for sex pheromone production and successful sexual behavior. Sonnet, P. *et al.* (1979) further characterized the female pheromone complex which induces male copulatory behavior. The major methyl branched olefins were identified as 13-methyl-1-hentriacontene and 13-methyl-1-tritriacontene. The most active preparation was a combination of the normal alkenes, methylalkenes, and mono- and dimethylalkanes.

Three species of *Fannia* have been examined for sex pheromones using the pseudofly bioassay. (Z)-9-Pentacosene constitutes the major cuticular hydrocarbon component of female *Fannia canicularis*, and was reported to stimulate the male to initiate copulatory attempts (Uebel, E. *et al.*, 1975a, 1978b, 1977). The mating stimulants from female *Fannia pusio* are normal mono-olefins of 31 and 33 carbons (Uebel, E. *et al.*, 1978c). Bioassay data indicate that (Z)-11-hentriacontene is the most active component. Essentially the same mixture of alkenes is found on females of *Fannia femoralis*. For both species alkanes are also present, and Uebel, E. *et al.* (1978a) showed that addition of the alkanes of *F. femoralis* to its C_{31} olefins increased mating strike frequency.

Mackley, J. *et al.* (1981) and Bolton, H. *et al.* (1980) investigated the possibility that the horn-fly (*Haematobia irritans*) courtship behavior involved chemical components. Alkenes, including (Z)-9-tricosene, (Z)-5-tricosene, (Z)-9-pentacosene and (Z)-9-heptacosene were identified from both sexes. Males had greater quantities of the (Z)-9-tricosene than did females, but the females contained greater quantities of all other olefins. Each alkane was individually active in releasing strike behavior, but only mixtures elicited the remaining steps of the courtship.

The fruit-fly *Drosophila melanogaster* also utilizes cuticular hydrocarbons as part of its sex pheromone. Newly emerged males and females have an almost identical hydrocarbon profile (Antony, C. and Jallon, J., 1981). A sexual dimorphism in long-chain hydrocarbons occurs starting on the second day after eclosion, and males lose sexual attractiveness (Antony, C. and Jallon, J., 1981). An aphrodisiac pheromone begins accumulating on females 1 day after eclosion (Venard, R. and Jallon, J., 1980). This hydrocarbon pheromone gives a dose-dependent wing vibration response in males.

Initial attraction of male tsetse flies (*Glossina* spp.) to females is known to involve visual cues (Swynnerton, C., 1936) with no evidence presented for olfactory sex attractants (Dean, J. *et al.*, 1969; Turner, D., 1971). Good evidence does exist, however, for the presence of a contact sex pheromone in *Glossina morsitans* which elicits copulatory behavior by the males (Langley, P. *et al.*, 1975). Similar evidence exists for *Glossina palpalis palpalis* and *Glossina pallidipes* (Carlson, personal communication). Carlson, D. *et al.* (1978) found that male behavior associated with copulation was elicited only by three high-molecular weight hydrocarbons. The most active hydrocarbon was

Table 1: Cuticular Lipids in Chemical Communication

Process/species	Components involved	References
Sex pheromone		
Musca domestica	(Z)-9-tricosene, (Z)-9, 10-epoxytricosane	Carlson, D. *et al.*, 1971
	(Z)-14-tricosene-10-one, methylalkanes	Uebel, E. *et al.*, 1976
Musca autumnalis	(Z)-14-nonacosene, (Z)-13-nonacosene,	Uebel, E. *et al.*, 1975c
	(Z)-13-heptacosene	Sonnet, P. *et al.*, 1975
Stomoxys calcitrans	(Z)-9-hentriacontene, (Z)-9-tritriacontene, methylalkanes	Uebel, E. *et al.*, 1975b
	15-methyl and 15, 19-dimethyltritriacontanes	Sonnet, P. *et al.*, 1977
	13-methyl-1-hentriacontene, 13-methyl-1-tritriacontene	Sonnet, P. *et al.*, 1979
Fannia canicularis	(Z)-9-pentacosene + alkanes	Uebel, E. *et al.*, 1978c
Fannia pusio	(Z)-11-hentriacontene + alkanes	Uebel, E. *et al.*, 1978c
Fannia femoralis	(Z)-11-hentriacontene + alkanes	Uebel, E. *et al.*, 1978c
Haematobia irritans	(Z)-5-tricosene, (Z)-9-pentacosene, (Z)-9-heptacosene	Bolton, H. *et al.*, 1980
Glossina morsitans	15, 19, 23-trimethyheptatriacontane, 15, 19- and	Langley, P. *et al.*, 1975
	17, 21-dimethylheptatriacontane	Carlson, D. *et al.*, 1978
Glossina sp. (seven other species)	methylalkanes	Huyton, P. *et al.*, 1980
Lucilia cuprina	cuticular lipids (unidentified)	Emmens, R. 1981
Pikonema alaskensis	9, 19-alkadienes of C_{29}, C_{31}, C_{33}, C_{35} and C_{37}	Bartelt, R. *et al.*, 1982
Lycoriella mali	*n*-heptadecane	Kostelc, J. *et al.*, 1975, 1980
Colias eurytheme	13-methylheptacosane and esters	Grula, J. *et al.*, 1980
Copulation release pheromone		
Callosobruchus chinensis	mono- and dimethylalkanes of C_{26}–C_{33} and (E)-3, 7-dimethyl-2-octene-1, 8-dioic acid	Tanaka, K. *et al.*, 1981
Species and caste recognition		
Zooptermopsis angusticollis	hydrocarbon composition	Blomquist, G. *et al.*, 1979a
Reticulitermes flavipes	hydrocarbon composition	Howard, R. *et al.*, 1978
Reticulitermes virginicus	hydrocarbon composition	Howard, R. *et al.*, 1982
Trichopsenius frosti	hydrocarbon composition	Howard, R. *et al.*, 1980a
Trichopsenius depressus	hydrocarbon composition	Howard *et al.*, unpublished
Philotermes howardi	hydrocarbon composition	Howard *et al.*, unpublished
Xenitusa hexagonalis	hydrocarbon composition	Howard *et al.*, unpublished
Kairomones		
Heliothis zea (attract parasites)	13-methylhentriacontane	Jones, R. *et al.*, 1971
Heliothis virescens (attract parasites)	11-methylhentriacontane, 16-methyl-dotriacontane, 13-methyltriacontane	Vinson, S. *et al.*, 1975
Thermoregulatory		
Vespa crabro	(Z)-9-pentacosene	Veith, H. and Koeniger, N., 1978
Oviposition markers		
Cardiochiles nigriceps	nC_{23}–nC_{33}	Vinson and Williams, unpublished data

identified as 15,19,23-trimethylheptatriacontane, with 15,19-dimethylheptatriacontane and 17,21-dimethylheptatriacontane being less active. Female *G. morsitans* was shown to contain about 4.2 μg of the trimethylalkane and 1.2 μg each of the dimethylalkanes. With all three compounds, male copulatory behavior was released only on physical contact with a surrogate female containing the chemicals (Carlson, D. *et al.*, 1978). Huyton, P. *et al.* (1980) showed that solvent-extracted dead males baited with the trimethylalkanes gave an ED_{95} of 2 μg. No synergistic effects were found with the dimethylalkanes, and Huyton, P. *et al.* (1980) concluded that the trimethyl compound is the sex pheromone of this species. The natural pheromone appears on the pharate adult female about 2 days before emergence from the puparium and is present throughout her life. Male responsiveness to pheromone baited decoys increases to a maximum 3–4 days after emergence. These authors suggest that another possible function of this pheromone may be to ensure that the union of male and female is maintained throughout the lengthy copulation period which is required for successful ejaculation and spermatophore transfer (Pollock, J., 1970).

Huyton, P. *et al.* (1980) conducted interspecific mating tests with seven species of *Glossina* to evaluate the specificity of the contact pheromone isolated from *G. morsitans*. Their tests with either live or dead females as targets indicated that the responses of the males are largely the consequence of differences in composition of the female surface cuticular hydrocarbons and not of female behavior.

In addition to Diptera, other insects use long-chain hydrocarbons as sex pheromones. Several sibling species in the genus *Holomelina* (tiger moths) use 2-methylheptadecane, as do females of *Pyrrharctia isabella* (Roeloffs, W. and Cardé, R., 1971). Conner, W. *et al.* (1980) noted that the female arctiid moth *Utetheisa ornatrix* emits (Z,Z,Z)-3,6,9-heneicosatriene plus small quantities of an unidentified C_{21} tetraene as her sex pheromone. These hydrocarbons appear gland-specific and are probably not cuticular components.

Two sulfur butterflies, *Colias eurytheme* and *Colias philodice* (Pieridae) coexist over much of North America and originally were thought to interbreed. Reproductive isolation is based upon female response to species-specific visual and chemical signals (Silberglied, R. and Taylor, O., 1978). Males of *C. philodice* contain on their wings three major esters (*n*-hexylmyristate, *n*-hexylpalmitate, and *n*-hexylstearate), a series of *n*-alkanes with odd carbon numbers (C_{23} to C_{29}), and small quantities of 13-methylheptacosane. Males of *C. eurytheme* contain on their wings the same *n*-alkanes, no esters, and a large amount of 13-methylheptacosane (Grula, J. *et al.*, 1980). The esters and 13-methylheptacosane all have significant electrophysiological activity. Preliminary behavioral data suggest that the esters are the important species recognition cues and that the branched hydrocarbon functions as an aphrodisiac. Grula, J. and Taylor, O. (1979) studied the inheritance of male pheromone production by gas chromatographic analysis of wing extracts of various genotypes derived from forced interspecific crosses. The X-chromosome carried most of the gene(s) controlling production of 13-methylheptacosane. Expression of this compound in hybrids displayed a codominant pattern, and was not influenced by diet.

Cuticular hydrocarbon components are part of the sex pheromone of the yellowheaded spruce sawfly, *Pikonema alaskensis* Rohwer (Bartelt, R. *et al.*, 1982). A series of (Z,Z)-9,19-alkadienes of 29, 31, 33, 35 and 37 carbons are present on the surface of the female and possess pheromone activity.

Long-chain methyl branched alkanes are components of the copulation release pheromone, erectin, of the azuki bean weevil, *Callosobruchus chinensis* L. (Tanaka, K. *et al.*, 1981). 3-Methyl-, internally branched mono-methyl- and dimethylalkanes of C_{26}–C_{35} along with a dicarboxylic acid comprise the copulation release pheromone, which is distinct from the sex attractant.

1.3.2 Species and Caste Recognition Cues

Social insects are well known for their ability to recognize conspecifics (Wilson, E., 1971), and to distinguish the caste and sex of the individuals they interact with. Because many social insects live either wholly in the soil or distributed between soil and wood (especially subterranean termites and many ants), much of their sensory perception must be either olfactory or tactile. The nests and galleries of these insects are often tightly closed, low-volume systems. Such systems place a premium on semiochemicals which are complex enough to have a high information content, but which are of low enough volatility to minimize sensory habituation. Howard, R. *et al.* (1978) and Blomquist, G. *et al.* (1979a) suggested that the cuticular hydrocarbons of termites might serve as semiochemical cues for caste and species recognition. They found that *Reticulitermes flavipes* (Rhinotermitidae) and *Zooptermopsis angusticollis* (Hodotermitidae) possess drastically different hydrocarbon profiles, and both of these differ markedly from *Nasuititermes exitiosus* (Nasuititermitidae) (Moore, B., 1969). In addition, R. Howard and co-workers completely characterized the cuticular hydrocarbons of *R. virginicus* (Howard, R. *et al.*, 1982) and partially characterized by GLC the hydrocarbons of four other species of *Reticulitermes,* as well as those of *Coptotermes formosanus*, *Marginitermes hubbardi*, *Incistitermes minor*, *Pterotermes occidentis*, and *Tenuirostitermes tenuirostris* (unpublished data). In every case, the hydrocarbon composition is unique. Furthermore, examination of the hydrocarbon compositions by caste indicates caste-specific ratios of hydrocarbon components. Preliminary bioassay data have confirmed the role of cuticular hydrocarbons as species recognition cues for *Reticulitermes virginicus* and *Reticulitermes flavipes* (Howard, R. *et al.,* 1982).

Additional evidence that cuticular hydrocarbons serve as species recognition cues comes from the finding that the highly integrated, host-specific termitophilous beetle *Trichopsenius frosti* has an identical cuticular hydrocarbon profile as its host termite *R. flavipes* (Howard, R. *et al.,* 1980a). In addition, the cuticular hydrocarbons of three species of termitophilous beetles (representing two subfamilies of Staphylinidae) associated with *R. virginicus* are identical to those of that termite (Howard, unpublished data). Biosynthetic studies (Howard, R. *et al.,* 1980a) showed that *T. frosti* produces its own hydrocarbons, rather than procuring them in some manner from its termite host.

The scarab beetle *Myrmecaphodius excavaticollis* associated with *Solenopsis* spp. ("fire ants") has a cuticular hydrocarbon composition that loosely mimics that of its current ant host. The mechanism by which the beetles achieve this is unknown. The cuticular hydrocarbons of two *Solenopsis* species (*Solenopsis invicta* and *Solenopsis richteri*) have been thoroughly characterized (Nelson, D. *et al.,* 1980) and those of two other species (*Solenopsis geminata* and *Solenopsis xyloni*) characterized by GLC (Vander Meer, personal communication). All four species have unique hydrocarbon profiles, suggesting that, like termites, ants may use these chemicals as species recognition cues.

1.3.3 Thermoregulatory pheromones

Wasps and hornets are known to actively heat their brood combs, especially those cells containing pupae (Ishay, J., 1972a,b). They do this by positioning their respiratory spiracles over the silken dome of the pupal cell and alternately retracting and extending their abdominal segments up to 180 times per minute. Ishay showed by preliminary bioassays that it was the pupae themselves that triggered this brooding behavior, and that the stimulus involved was that of a "volatile" chemical which could be extracted with alcohol from the pupae. Veith, H. and Koeniger, N. (1978) identified the pheromone as (Z)-9-pentacosene.

1.3.4 Kairomonal cues for parasites

Parasites are such ubiquitous components of practically every ecosytem that it is not surprising that some of them have managed to evolve to utilize cuticular hydrocarbon cues produced by their hosts as primary cues in locating their hosts. Hymenopteran parasites of *Heliothis* spp. exhibit intensified searching behavior for their hosts when exposed to either frass or scales of the ovipositing female moth. Jones, R. *et al.* (1971) showed that 13-methylhentriacontane was the major component of the hemolymph, cuticle, and frass of larvae *Heliothis zea* which was responsible for intense searching behavior by the larval parasite *Microplitis croceipes*. Vinson, S. *et al.* (1975) similarly showed that 11-methylhentriacontane, 16-methyldotriacontane and 13-methyltritriacontane, produced in the mandibular glands of *Heliothis virescens* and deposited on the leaf surface by feeding larvae, stimulate *Cardiochiles nigriceps* to search the immediate vicinity more efficiently. Jones, R. *et al.* (1973) reported that an egg parasite of *H. zea* (*Trichogramma evanescens*) was stimulated by extracts of scales left by the ovipositing *H. zea* female, and that the most active factor in these extracts was *n*-tricosane. Lewis and co-workers have since conducted extensive laboratory and field investigations on the role and practical utilization of *n*-tricosane in manipulating population levels of *H. zea* (Lewis, W. *et al.,* 1975a,b, 1976, 1977, 1979; Nordlund, D. *et al.,* 1976, 1977a,b). Much of their findings regarding the behavioral sequence of host selection by parasites, the chemistry of the kairomones involved, their role in the host finding process, and the basis for their potential employment in pest management programs has recently been reviewed (Lewis, W. *et al.,* 1976).

1.4 Wax of honeybees, scale insects

In addition to a thin layer of cuticular lipid, some insects secrete large amounts of wax. These include the honeybees, which use the wax as structural material to build the honeycomb, and the scale insects, which use the wax for protection. The scale insects are so called because in many species the female is protected by a scale or shield consisting of a mixture of wax and cast skins. The amount of lipid covering the scale insects is much larger than that usually seen on the epicuticle of most insects; however, its chief function is still most likely to protect the insects against desiccation (Hackman, R., 1951). The adult females of scale insects are degenerate and are attached or fastened to the host plant by the mouthparts. It is probably because they are immobile that the scale insects form their characteristic protective coat, which also protects them from insect predators and the weather.

2 ANALYTICAL PROCEDURES

2.1 Extraction and separation

The analytical techniques used in insect wax chemistry have been reviewed comprehensively by Jackson, L. and Armold, M. (1977) and by Gilby, A. (1980a). The cuticular lipids from whole insects or cast exuviae can be conveniently extracted by immersion in organic solvents such as hexane or chloroform for short periods of time. Usually a few seconds to 15 min is recommended for hexane, and shorter extraction times for chloroform. Chloroform is more penetrating and the chances of extracting internal lipids increase as the extraction time increases. Chloroform–methanol mixtures and diethylether are not commonly used except for very short extraction periods due to their interaction with water and the increased possibility of extracting internal lipids. Most often one immersion does not extract all of the cuticular lipid; two or more short immersions are usually better than one long immersion.

After extraction, the major lipid classes can usually be separated by column chromatography (Nelson, D. and Sukkestad, D., 1970; Blomquist, G. *et al.,* 1972) and/or thin-layer chromatography (TLC) (Jackson, L. *et al.,* 1974). The use of silver nitrate-impregnated TLC plates for unsaturated components and molecular sieve for branched components are often very helpful. Gas–liquid chromatography (GLC) is routinely used to separate and quantify the components of each class of lipid. Infrared spectrophotometry, nuclear magnetic resonance spectrometry, mass spectrometry (MS) and combined GLC–MS, including chemical ionization (CI)–MS techniques have proven very useful in structure determination. GLC–MS is the principal tool used in obtaining structure assignments for hydrocarbons.

2.2 Structure determination of hydrocarbons

n-Alkanes can be identified by their retention times on GLC and inclusion in 5 Å molecular sieve. Structure assignments can be confirmed by GLC–MS.

The methyl branched alkanes have retention times on GLC which are somewhat less than an *n*-alkane of the same total carbons. Monomethylalkanes with methyl branches on carbons 2, 3 or 4 elute about 0.3 or 0.4 carbon units in front of *n*-alkanes of the same total carbon number on packed columns. Internally branched monomethylalkanes elute about 0.6 to 0.7 carbon units in front of corresponding *n*-alkanes. Internally branched dimethylalkanes elute about 1.3 to 1.4 carbon units before corresponding *n*-alkanes, and dimethlyalkanes with one methyl branch near the end and one in the middle (3,X–, 4,X–) elute about 1 carbon unit in front of corresponding *n*-alkanes.

When subjected to mass spectrometry each type of hydrocarbon yields a diagnostic spectrum. Discussions on the interpretation of spectra from insect hydrocarbons can be found in papers by Jackson, L. and Blomquist, G. (1976a), Nelson, D. and Sukkestad, D. (1970), Nelson, D. *et al.* (1972), Nelson, D. (1978), and Pomonis, J. *et al.* (1978). Recently, the application of chemical ionization–mass spectrometry to insect hydrocarbons has proven useful (Howard, R. *et al.,* 1980b).

Most of the alkenes characterized from insects to date are *cis*/(Z). This assignment is usually based on the presence of a peak at 730 cm^{-1} and the absence of a peak at 970 cm^{-1} (indicative of *trans* or E double bonds) in the infrared spectrum. Z and E isomers can be separated by silver nitrate-impregnated TLC. Oxidative cleavage followed by GLC or oxidation of the alkene to a diol with osmium tetroxide followed by silylation and mass spectrometry have been successfully used to determine the position of double bonds. More recently, methoxymercuration–demercuration followed by mass spectral analysis has been successfully applied to alkenes (Blomquist, G. *et al.*, 1980c). This technique is especially advantageous with mixtures of components, as it eliminates the need for isolating each component by preparative GLC.

2.3 Characterization of wax esters, fatty alcohols and sterols

Wax esters are usually characterized by saponification and characterization of the component fatty acids and fatty alcohols. However, some information can be obtained about the total number of carbon atoms in the wax ester chains by GLC on a high-temperature silicone column. Since GLC only gives peaks according to total number of carbons in the wax ester there may be a mixture of various alcohols esterified to a mixture of fatty acids in any one peak. In order to isolate the alcohols and fatty acids, the wax esters are usually either saponified or transesterified. Saponification has the advantage that the alcohols can be extracted with an organic solvent from the alkaline saponification mixture and then extraction of the acidified solution will yield the fatty acids. Transesterification provides the methyl esters of the fatty acids which are easily separated from the alcohols by TLC or column chromatography and then the fatty acid methyl esters and fatty alcohols can be readily characterized by GLC (Jackson, L. and Armold, M., 1977). Free primary alcohols, secondary alcohols and sterols can usually be separated from the other lipid components by TLC or column chromatography (Soliday, C. *et al.*, 1974). The purified fractions of the various alcohols are then submitted to GLC and GLC–MS for further characterization.

Secondary alcohol wax esters require some special techniques since they do not saponify or transesterify well under normal conditions. Secondary alcohol wax esters have been characterized by GLC of the total fraction to determine the range of chain lengths. This step is followed by $LiAlH_4$ reduction of the ester linkage to yield secondary alcohols and primary alcohols. TLC separation of primary and secondary alcohols followed by GLC or GLC–MS can be used to characterize the components of the secondary alcohol wax esters (Blomquist, G. *et al.*, 1972).

2.4 Characterization of free fatty acids and acylglycerols

The free fatty acid fraction is generally characterized by methylation followed by GLC on a polyester column which separates the methylesters according to chain length and number of double bonds. The triacylglycerols, diacylglycerols and monoacylglycerols can be separated by TLC or column chromatography and then each fraction can be analyzed by GLC on high-temperature silicone columns, or they can be simply saponified and methylated or transesterified and the resulting methylesters can be subjected to GLC analysis (Jackson, L. and Armold, M., 1977).

3 STRUCTURE AND COMPOSITION OF CUTICULAR LIPIDS

The cuticular lipids of many insect species consist of complex mixtures of various types of hydrocarbons and other components. The structures of the major types of insect cuticular lipids are presented in Fig. 2.

3.1 Hydrocarbons

Hydrocarbons have been the most extensively studied cuticular components and frequently are the only class characterized. They are common components of surface extracts and are present on all insects which have been carefully examined. The insect surface lipids of some species are $> 90\%$ hydrocarbon (Jackson, L. and Blomquist, G., 1976a) whereas hydrocarbons account for only 0.5% of the surface lipid of the tobacco budworm pupae (Coudron, T. and Nelson, D., 1978). Insect surface lipid hydrocarbons are usually a mixture of components which includes *n*-alkanes, *n*-alkenes, terminally branched monomethylalkanes, internally branched monomethylalkanes, dimethylalkanes, trimethylalkanes and others. The types of hydrocarbons present in over 80 species are

n-alkanes	$CH_3-(CH_2)_x-CH_3$
n-alkenes	$CH_3-(CH_2)_x-CH{=}CH-(CH_2)_y-CH_3$
2-methylalkanes	$CH_3-CH(CH_3)-(CH_2)_x-CH_3$
3-methylalkanes	$CH_3-CH_2-CH(CH_3)-(CH_2)_y-CH_3$
internally branched monomethylakanes	$CH_3-(CH_2)_x-CH(CH_3)-(CH_2)_y-CH_3$
dimethylalkanes	$CH_3-(CH_2)_x-CH(CH_3)-(CH_2)_y-CH(CH_3)-(CH_2)_z-CH_3$
trimethylalkanes	$CH_3-(CH_2)_y-[CH(CH_3)-(CH_2)_y]_3-(CH_2)_z-CH_3$
primary alcohol wax esters	$CH_3-(CH_2)_x-O-C(=O)-(CH_2)_y-CH_3$
secondary alcohol wax esters	$CH_3-(CH_2)_x-C(H)[(CH_2)_y-CH_3]-O-C(=O)-(CH_2)_z-CH_3$
epoxides	$CH_3-(CH_2)_x-CH(-O-)CH-(CH_2)_y-CH_3$
ketones	$CH_3-(CH_2)_x-C(=O)-(CH_2)_y-CH_3$
secondary alcohols	$CH_3-(CH_2)_x-CH(OH)-(CH_2)_y-CH_3$
primary alcohols	$CH_3-(CH_2)_x-CH_2-OH$
free fatty acids	$CH_3-(CH_2)_x-C(=O)-OH$
sterols (cholesterol)	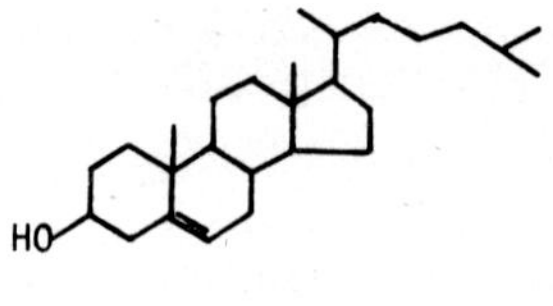

FIG. 2. Structures of common insect cuticular components.

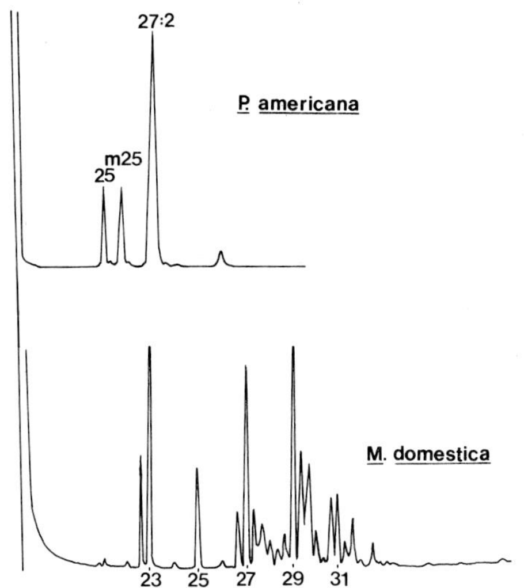

FIG. 3. GLC traces of the cuticular hydrocarbons of *Periplaneta americana* (Jackson, L., 1972) and *Musca domestica* (Nelson, D. *et al.*, 1981). The components in the top trace are 25(*n*-pentacosane), m25(3-methylpentacosane), and 27 : 2 ([Z,Z]-6,9-heptacosadiene). The numbers below the bottom trace refer to the carbon numbers of *n*-alkanes.

presented in Table 2. The hydrocarbons of some insects have a relatively simple composition, such as is present in *Periplaneta americana* (Fig. 3A) (Jackson, L., 1972). Other insects have exceedingly complex mixtures, such as the female housefly (Fig. 3B), which has 134 methyl branched components (Nelson, D. *et al.*, 1981).

The complexity of insect cuticular hydrocarbons has led to the suggestion that hydrocarbon composition might be used as a taxonomic character (Lockey, K., 1976; Jackson, L. and Blomquist, G., 1976a). However, only a relatively small number of species have been examined and from the data presented in Table 2 there do not appear to be obvious correlations of hydrocarbon compositions within insect groups. In addition, the composition of hydrocarbons often varies tremendously among different developmental forms, between sexes, and at different times of the year. In general, most investigators have concentrated on one developmental stage. These observations indicate that the distribution pattern of hydrocarbon components is unlikely to be a useful discriminating factor in the near future.

3.2 *n*-Alkanes

n-Alkanes occur in all the insect surface lipids so far investigated (Table 2). The *n*-alkanes of insect surface lipids are generally in the range C_{21}–C_{36} with alkanes having an odd number of carbon atoms predominating. The complexity of the *n*-alkane mixtures varies considerably. For example, *n*-pentacosane comprises 85% of the *n*-alkane fraction of *P. americana* (Jackson, L., 1972), but *n*-alkanes from C_{17} to C_{34} with both odd and even numbers of carbon atoms are major components of the hydrocarbons of pupal tobacco budworms, *Heliothis virescens* (Coudron, T. and Nelson, D., 1978). *n*-Alkanes are not usually found alone, but only *n*-alkanes were reported from the hydrocarbon fraction from cast skins of the beetle, *Tenebrio molitor* (Bursell, E. and Clements, A., 1967). Analysis of the hydrocarbons of seven scale insects showed that they were entirely *n*-alkanes of C_{25}–C_{35}, with odd-numbered carbon chain-lengths predominating (Faurot-Bouchet, E. and Michel, G., 1964, 1965).

3.3 Terminally branched monomethylalkanes

2-Methylalkanes and/or 3-methylalkanes are prevalent in insect surface lipids. Like the *n*-alkanes, the terminally branched monomethylalkanes range from simple compositions such as in *P. americana* where only 3-methylpentacosane is present (Jackson, L., 1972) to complex mixtures of 3-methylalkanes, 2-methylalkanes or mixtures of 2- and 3-methylalkanes (Blomquist, G. and Jackson, L., 1979). The 2-methylalkanes are somewhat unique in that components with both odd- and even-numbered carbon chains are present in substantial amounts. This appears to reflect their biosynthesis, which involves the carbon skeleton of either valine or leucine (Blailock, T. *et al.*, 1976). Many of the insects studied to date that have internally branched monomethylalkanes also have 3-methylalkanes (Table 2), which may reflect a similar biosynthetic origin (Blomquist, G. *et al.*, 1975a; Blomquist, G. and Kearney, G., 1976). The majority of the 3-methylalkanes have an odd-numbered carbon chain (even number of total carbons).

Table 2: Occurrence of hydrocarbons in insect surface lipids

Species				Branched alkanes					Reference
		Unsaturated		Monomethylalkanes					
				Terminally branched					
	n-Alkanes	Alkenes	Alkadienes	2-Me	3-Me	Internally branched	Dimethyl-Alkanes	Trimethyl-Alkanes	
Plecoptera									
Pteronarcys californica	+	+			+	+			Armold, M. *et al.*, 1969
Orthoptera									
Anabrus simplex	+				+	+	+		Jackson, L. and Blomquist, G., 1976b
Acheta domesticus	+	+	+	+		+	+		Hutchins, R. and Martin, M., 1968; Blomquist, G. *et al.*, 1976
Gryllus pennsylvanicus	+	+	+	+		+	+		Blomquist, G. *et al.*, 1976; Warthen, J. and Uebel, E., 1980a
Nemobius fasciatus	+		+	+					Blomquist, G. *et al.*, 1971; Warthen, J. and Uebel, E., 1980a
Melanoplus sanguinipes	+				+	+	+		Soliday, C. *et al.*, 1974
Melanoplus packardii	+				+	+	+		Soliday, C. *et al.*, 1974
Melanoplus bivittatus	+			+		+	+		Jackson, L., 1981
Melanoplus femurrubrum	+			+		+	+		Jackson, L., 1981
Melanoplus dawsoni	+			+		+	+		Jackson, L., 1981
Schistocerca vaga	+					+	+	+	Nelson, D. and Sukkestad, D., 1975
Schistocerca gregaria	+					+	+		Lockey, K., 1976
Schistocerca americana	+					+	+	+	Jackson, L., 1982
Locusta migratoria	+				+	+	+		Lockey, K., 1976
Dictyoptera									
Periplaneta americana	+	+	+		+				Baker, G. *et al.*, 1963; Jackson, L., 1972; Beatty, I. and Gilby, A., 1969
Periplaneta japonica	+	+	+		+	+			Jackson, L., 1972
Periplaneta australasiae	+	+			+	+			Jackson, L., 1970
Periplaneta brunnea	+				+	+			Jackson, L., 1970
Periplaneta fuliginosa	+	+			+	+			Jackson, L., 1970
Leucophaea maderae	+				+	+			Tartivita, K. and Jackson, L., 1970
Blatta orientalis	+				+	+			Tartivita, K. and Jackson, L., 1970
Phasmatodea									
Diapheromera femorata	+				+	+	+		Warthen, J. *et al.*, 1981
Diptera									
Sarcophaga bullata	+	+	+		+	+			Jackson, L. *et al.*, 1974
Phormia regina	+	+	+	+					Louloudes, S. *et al.*, 1962
Cochliomyia hominivorax	+	+	+	+					Louloudes, S. *et al.*, 1962
Calliphora vicina	+	+	+	+					Louloudes, S. *et al.*, 1962

Musca domestica	+	+		+	+	+	+		Nelson, D. *et al.*, 1981; Uebel, E. *et al.*, 1976
Musca autumnalis	+	+							Uebel, E. *et al.*, 1975c
Stomoxys calcitrans	+	+				+	+		Uebel, E. *et al.*, 1975b
Fannia canicularis	+	+							Uebel, E. *et al.*, 1977
Fannia pusio	+	+							Uebel, E. *et al.*, 1978c
Fannia femoralis	+	+							Uebel, E. *et al.*, 1978c
Glossina morsitans	+						+	+	Carlson, D. *et al.*, 1978
Lycoriella mali	+								Kostelc, J. *et al.*, 1975
Drosophila melanogaster	+	+	+	+					Jackson, L. *et al.*, 1981
Hymenoptera									
Atta colombica	+	+						+	Martin, M. and MacConnell, J., 1970
Atta sexdens	+							+	Martin, M. and MacConnell, J., 1970
Atta ceptalotes isthmicola	+							+	Martin, M. and MacConnell, J., 1970
Myremecia gulosa	+			+	+	+			Cavill, G. *et al.*, 1970
Iridomyrex humilis	+	+			+	+			Cavill, G. and Houghton, E., 1973
Camponotus intrepidus	+	+			+	+			Brophy, J. *et al.*, 1973
Formica nigricans	+	+	+			+			Bergstrom, G. and Lofquist, J., 1973
Formica rufa	+	+	+			+			Bergstrom, G. and Lofquist, J., 1973
Formica polyctena	+	+	+			+			Bergstrom, G. and Lofquist, J., 1973
Solenopsis invicta	+				+	+	+		Lok, J. *et al.*, 1975; Nelson, D. *et al.*, 1980
Solenopsis richteri	+				+	+	+		Lok, J. *et al.*, 1975; Nelson, D. *et al.*, 1980
Pogonomyrmex rugosus	+				+	+	+		Regnier, F. *et al.*, 1973
Pogonomyrmex barbatus	+				+	+	+		Regnier, F. *et al.*, 1973
Apis mellifera	+	+	+			+			Blomquist, G. *et al.*, 1980c
Bombus appositus	+	+							Hadley, N. *et al.*, 1981
Bombus occidentalis	+	+				+			Hadley, N. *et al.*, 1981
Nomia bakeri	+								Hadley, N. *et al.*, 1981
Bembix pruinosa	+	+			+	+	+		Hadley, N. *et al.*, 1981
Coleoptera									
Curculio caryae	+	+	+			+			Mody, N. *et al.*, 1975
Phyllobius maculicornis	+			+		+	+		Jacob, J., 1977
Chrysomela sp.	+			+					Jacob, J., 1977
Cassida sp.	+			+					Jacob, J., 1977
Donacia sp.	+			+					Jacob, J., 1977
Anthonomus grandis	+	+				+			Hedin, P. *et al.*, 1972
Tribolium confusum	+				+	+	+		Lockey, K., 1978a
Tribolium castaneum	+			+	+	+	+		Baker, J. *et al.*, 1978; Lockey, K., 1978a
Eleodes armata	+			+	+	+	+		Jackson, L. *et al.*, 1980
Cryptoglossa verrucosa	+				+	+			Hadley, N., 1978
Centrioptera muricata	+				+	+			Hadley, N., 1978
Centrioptera variolosa	+				+	+			Hadley, N., 1978
Pelecyphorus adversus	+				+	+			Hadley, N., 1978

Table 2: Occurrence of hydrocarbons in insect surface lipids (Continued)

Species	*n*-Alkanes	Unsaturated: Alkenes	Unsaturated: Alkadienes	Branched alkanes: Monomethylalkanes: Terminally branched: 2-Me	Branched alkanes: Monomethylalkanes: Terminally branched: 3-Me	Branched alkanes: Monomethylalkanes: Internally branched	Branched alkanes: Dimethyl-Alkanes	Branched alkanes: Trimethyl-Alkanes	Reference
Tenebrio molitor	+	+			+	+	+		Lockey, K., 1978b
Tenebrio obscurus	+	+			+	+	+		Lockey, K., 1978b
Attagenus megatoma	+	+			+	+	+		Baker, J., 1978
Lasioderma serricorne	+				+	+	+		Baker, J. *et al.*, 1979b
Cylindrinotus laevioctos triatus	+				+	+	+	+	Lockey, K., 1981
Phylan gibbus	+				+	+	+	+	Lockey, K., 1981
Ceutorrhynchus assimilis	+			+		+	+		Richter, I. and Krain, H., 1980
Blaps mucronta	+				+	+	+		Lockey, K., 1980
Alphitophagus bifasciatus	+	+		+	+	+	+		Lockey, K., 1979
Alphitobius diaperinus	+			+	+	+	+		Lockey, K., 1979
Rhagonycha fulva	+	+	+						Jacob, J., 1978
Cantharis livida	+	+							Jacob, J., 1978
Canthans livida	+	+							Jacob, J., 1978
Popillia japonica	+			+	+	+	+		Nelson, D. *et al.*, 1977; Bennett, G. *et al.*, 1972
Trichopsenius frosti	+	+	+	+	+	+	+		Howard, R. *et al.*, 1980a
Trichopsenius depressus	+	+	+	+	+	+	+		Howard, McDaniel and Blomquist, unpublished
Philotermes howardi	+	+	+	+	+	+	+		Howard, McDaniel and Blomquist, unpublished
Xenitusa hexagonalis	+	+	+	+	+	+	+		Howard, McDaniel and Blomquist, unpublished
Callosobruchus maculatus	+				+	+	+		Baker, J. and Nelson, D., 1981
Isoptera									
Zootermopsis angusticollis	+				+	+	+		Blomquist, G. *et al.*, 1979a
Reticulitermes flavipes	+	+	+	+	+	+	+		Howard, R. *et al.*, 1978
Reticulitermes virginicus	+	+	+	+	+	+	+		Howard, R. *et al.*, 1982
Lepidoptera									
Heliothis virescens	+			+	+	+	+		Coudron, T. and Nelson, D., 1978
Heliothis zea	+					+			Jones, R. *et al.*, 1971
Manduca sexta	+	+	+			+	+		Nelson, D. *et al.*, 1971; Coudron, T. and Nelson, D., 1981; Nelson, D. and Sukkestad, D., 1970
Homoptera									
Acyrthosiphon pisum	+								Stransky, K. *et al.*, 1973
Diceroprocta apache	+				+	+			Hadley, N., 1980b

3.4 Internally branched mono-, di- and trimethylalkanes

A very comprehensive review (Nelson, D., 1978) covers the occurrence and analysis of internally methyl-branched hydrocarbons. The majority of the internally branched monomethylalkanes have the methyl branch located on an odd-numbered carbon atom which often is carbon number 5, 7, 9, 11, 13, 15 or 17. The general range is from 20 to 40 carbons. Monomethylalkanes with the methyl branch on an even-numbered carbon atom are usually reported mixed with odd-numbered carbon isomers and as minor components.

Most of the long-chain internally branched dimethylalkanes have an isoprenoid type spacing, although they are not derived from isoprenoid units. Common isomers are 9,13-, 11,15-, 13,17-, 15,19- and 17,21-. Methyl branches on positions 11,21- have been reported (Nelson, D. *et al.,* 1977) and an unusual symmetrical component, 5,17-dimethylheneicosane, has been reported from the termite *Zooptermopsis angusticollis* (Blomquist, G. *et al.,* 1979a). Other types of dimethylalkanes include a 3,X- and 4,X-dimethylalkane series from the housefly (Nelson, D. *et al.,* 1981) and cigarette beetle (Baker, J. *et al.*, 1978), 11,19-dimethylalkanes in the tobacco budworm (Coudron, T. and Nelson, D., 1978), 11,12-dimethylalkanes in the cigarette beetle (Baker, J. *et al.,* 1978) and 13,15-dimethylalkane in the fire ant (Nelson, D. *et al.,* 1980). As careful analyses of dimethylalkanes are made on more organisms, it appears that the methyl groups can be positioned almost anywhere on the chain. Isomers with methyl groups on odd-numbered carbons are more prevalent than isomers with methyl branches on even-numbered carbons. Internally branched trimethylalkanes with isoprenoid spacing of the methyl branches have been reported from *Manduca sexta* (Nelson, D. and Sukkestad, D., 1970; Nelson, D. *et al.*, 1972), *Schistocerca vaga* (Nelson, D. and Sukkestad, D., 1975) and the tsetse fly *Glossina morsitans* (Carlson, D. *et al.,* 1978).

Two homologous series of trimethylalkanes have been observed as major constituents of the saturated hydrocarbons from ants (Martin, M. and MacConnell, J., 1970). 3,7,11-Trimethylalkanes of C_{34}, C_{36} and C_{38} and 4,8,12-trimethylalkanes of C_{35}, C_{37} and C_{39} are the major constituents of the surface lipid-saturated hydrocarbons of the ant, *Atta columbia.* In *Atta sexdens*, both 3,7,11- and 4,8,12-trimethylalkanes are present as major components, but in *Atta cephalotes isthmicola*, only the 3,7,11-trimethyl series is present and accounts for only about 15% of the alkanes (Martin, M. and MacConnell, J., 1970).

3.5 Alkenes

n-Alkenes, with one, two, or three double bonds, have been characterized from about one-half of the insects examined to date (Table 2). The chain length of cuticular alkenes usually ranges from C_{20} to C_{37}, with odd-numbered chain lengths predominating. The positions of the double bonds can be almost anywhere in the chain. Alkenes with the double bond in the 1-position are present in the red flour beetle, *Tribolium castaneum* (Baker, J. *et al.*, 1978). Two and 3-alkenes, as well as a series of alkenes with double bonds in unspecified positions, ranging from C_{20} to C_{31}, have been reported in the pecan weevil *Curculio caryae* (Mody, N. *et al.,* 1975). A number of Dipteran species possess *n*-alkenes of 23 to 31 carbons which function in chemical communication (Table 1). The double bonds are usually either in the 7 or 9 positions. In three cockroaches, *Periplaneta australiasia*, *Periplaneta fuliginosa* (Jackson, L., 1970) and *Periplaneta japonica* (Jackson, L., 1972), (Z)-9-alkenes from C_{27} to C_{31} were observed. All castes of the termites *Reticulitermes flavipes* (Howard, R. *et al.*, 1978) and *Reticulitermes virginicus* (Howard, R. *et al.,* 1982) possess (Z)-9-pentacosene as a major cuticular component as do four species of termitophilous beetles associated with them (Howard, R. *et al.,* unpublished data). Alkenes with double bond positions at 8 and 10 have been reported from the honeybee, *Apis mellifera* (Blomquist, G. *et al.,* 1980b). The face-fly, *Musca autumnalis*, has a complex series of 6-, 8-, 10-, 11- and 12-pentacosene, 10-, 11-, 12- and 13-heptacosene, and 10-, 11-, 12-, 13-, and 14-nonacosene (Uebel, E. *et al.,* 1975c).

Polyunsaturated hydrocarbons are not as common insect surface lipid components and are often minor components. However, the major hydrocarbon component of the surface lipids of the American cockroach, *Periplaneta americana*, is (Z,Z)-6,9-heptacosadiene (Baker, G. *et al.,* 1963).

Although most of the other cockroaches studied have methylalkanes as the major components, *P. japonica* hydrocarbons are 17% (Z,Z)-6,9-nonacosadiene and lesser quantities of (Z,Z)-6,9-heptacosadiene, -octacosadiene, -tricontadiene and hentriacontadiene (Jackson, L., 1972). Cuticular dienes having chain lengths of C_{29}–C_{37} with double bonds in the (Z,Z)-9,19-positions are sex pheromone components of the female yellowheaded spruce sawfly, *Pikonema alaskensis* (Bartelt, R. *et al.*, 1982).

4 OXYGENATED COMPONENTS

The non-hydrocarbon components of insect surface lipids have not been extensively studied in many insects, but it appears that wax esters seldom predominate except in beeswax. Free alcohols, free fatty acids and lesser quantities of other components are common. Some insects have appreciable quantities of unusual lipids and some of those will be discussed below.

4.1 Wax esters and sterol esters

Wax esters are not as common in insect cuticular lipids, nor have they been as widely studied, as hydrocarbons. Wax esters from different insects vary in composition, although most of them are composed of long-chain normal primary alcohols and acids (Jackson, L. and Baker, G., 1970). The cuticular lipids of both the adult and naiad *Pteronarcys californica* contain saturated and unsaturated C_{30} to C_{40} wax esters but the adults have more of the longer-chain wax esters (Armold, M. *et al.*, 1969). The pea aphid, *Acyrthosiphon pisum*, contains C_{32} to C_{54} aliphatic wax esters composed of C_{12} to C_{34} normal saturated primary alcohols and acids (Stransky, K. *et al.*, 1973). Sterol esters are also reported in the cuticular lipids of some insects (Armold, M. *et al.*, 1969; Bursell, E. and Clements, A., 1967) with the principal sterol being cholesterol.

Novel wax esters were identified in the cuticular lipids of the grasshoppers *Melanoplus sanguinipes* and *Melanoplus packardii* (Blomquist, G. *et al.*, 1972). These wax esters, comprising 18 and 28% respectively of the cuticular lipids of these two insects, consist of saturated fatty acids esterified to secondary alcohols. The fatty acids range from C_{12} to C_{22}. The secondary alcohols range from C_{21} to C_{27} with the hydroxyl group near the center of the carbon chain. In *M. sanguinipes,* 59% of the secondary alcohols consist of the C_{23} compounds in which the major isomer is tricosan-11-ol, with smaller amounts of tricosan-12-ol and tricosan-10-ol. A number of other species of *Melanoplus* also contain these unusual wax esters (Jackson, L., 1981; Warthen, J. and Uebel, E., 1980b), and to date they have not been reported from other insects.

The primary alcohols found in the cuticular lipids of insects are usually of even-numbered carbon atoms and range from C_{22} to C_{34}. Seventy percent of the cuticular lipid of the larva of the eri silkworm, *Samia cynthia ricini* (Bowers, W. and Thompson, M., 1965) is *n*-triacontanol and *n*-octacosanol. The alcohols from *Eleodes armata* (Hadley, N., 1977) are primarily saturated C_{20}–C_{34} alcohols with the suggestion of branched and/or unsaturated alcohols containing 30–40 carbon atoms. Over 90% of the wax on *Eriocampa ovata* larvae is primary alcohol of C_{20}–C_{32}, with hexacosanol comprising about 73% of the fraction (Percy, and Blomquist, unpublished data). In *M. sanguinipes* and *M. packardii* the primary alcohols are C_{24}–C_{30} and comprise only a small percentage of the surface lipids (Soliday, C. *et al.*, 1974). Trace amounts of free secondary alcohols are also present from *M. sanguinipes* and *M. packardii.* Half of the surface lipid extract of larvae of *Tenebrio molitor* is pentacosan-8,9-diol with a melting point of 115° (Bursell, E. and Clements, A., 1967). A distinction between insects with "soft" surface lipids in which the proportion of hydrocarbons is high, and those with "hard" surface lipids in which long-chain alcohols (and diols) appear to predominate, is suggested.

4.2 Fatty acids

The C_{14}–C_{20} fatty acids with an even number of carbon atoms predominate. In most cases saturated and unsaturated fatty acids have been observed. However, the surface lipid fatty acids of *Lucilia cuprina* are all saturated (Goodrich, B., 1970). Branched fatty acids are not usually present in insect waxes. Free α-hydroxy acids have been observed from the pea aphid, *Acyrthosiphon pisum* (Stransky, K. *et al.*, 1973).

4.3 Other non-hydrocarbon cuticular components

The three major non-hydrocarbon components of the surface lipids of female houseflies are (Z)-14-tricosene-10-one, (Z)-9,10-epoxytricosane and 9-hexadecenyl-9-octadecenoate (Uebel, E. *et al.*, 1978a). The first two compounds may be involved in the sex pheromone complex. The surface lipid of male *Drosophila melanogaster* contain (Z)-vaccenyl acetate (Jackson, L. *et al.*, 1981) as a major non-hydrocarbon component. The (Z)-vaccenyl acetate is found in the ejaculatory bulb of adult males and is transferred to females during mating (Brieger, G. and Butterworth, F., 1970). Heneicosan-8-ol acetate was identified as a major (27%) non-hydrocarbon constituent of 5-day-old male little houseflies, *Fannia canicularis* (Uebel, E. *et al.*, 1977). Triacylglycerols are present in some cuticular extracts, but usually as minor amounts. Their presence in extracts of cast skins would suggest that the triacylglycerols are true cuticular components, and not simply contaminants from internal lipids.

4.4 Beeswax — surface lipids and comb wax

The composition of the surface lipid of the honeybee, *Apis mellifera*, is quantitatively different from that of the comb wax. The major component of the surface lipid is hydrocarbon (58%), followed by monoesters (23%), diesters (9%), triesters (2%), free fatty acids and more polar lipids (8%) (Blomquist, G. *et al.*, 1980b). The comb wax composition is hydrocarbon (13–17%) (Tulloch, A., 1971; Stransky, K. *et al.*, 1971), wax monoester (31–35%), diesters (10–14%), triesters (3%), hydroxymonoesters (3–6%) and hydroxypolyesters, free primary alcohols, diols, and acid monoesters (7–10%) (Tulloch, A., 1970, 1971). The structures of the major types of esters found in beeswax are shown in Fig. 4).

5 BIOSYNTHESIS OF CUTICULAR LIPIDS

Most of the work on the biosynthesis of insect cuticular lipids has focused on the hydrocarbons,

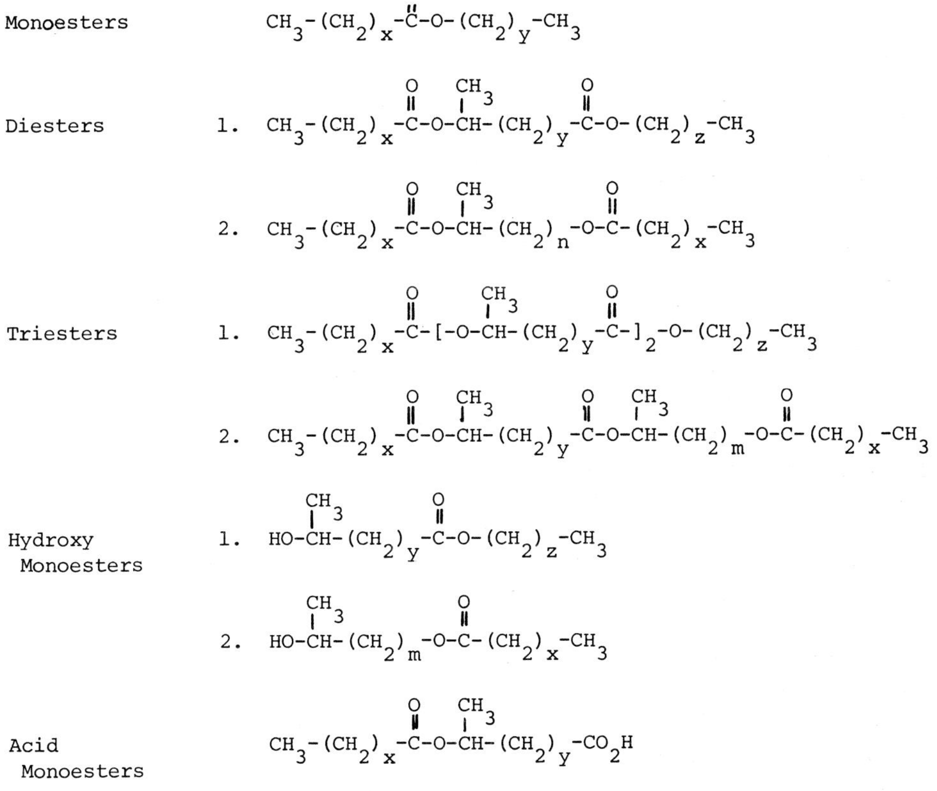

Monoesters		$CH_3\text{-}(CH_2)_x\text{-}\overset{O}{\overset{\parallel}{C}}\text{-}O\text{-}(CH_2)_y\text{-}CH_3$
Diesters	1.	$CH_3\text{-}(CH_2)_x\text{-}\overset{O}{\overset{\parallel}{C}}\text{-}O\text{-}\overset{CH_3}{\overset{\vert}{C}}H\text{-}(CH_2)_y\text{-}\overset{O}{\overset{\parallel}{C}}\text{-}O\text{-}(CH_2)_z\text{-}CH_3$
	2.	$CH_3\text{-}(CH_2)_x\text{-}\overset{O}{\overset{\parallel}{C}}\text{-}O\text{-}\overset{CH_3}{\overset{\vert}{C}}H\text{-}(CH_2)_n\text{-}O\text{-}\overset{O}{\overset{\parallel}{C}}\text{-}(CH_2)_x\text{-}CH_3$
Triesters	1.	$CH_3\text{-}(CH_2)_x\text{-}\overset{O}{\overset{\parallel}{C}}\text{-}[\text{-}O\text{-}\overset{CH_3}{\overset{\vert}{C}}H\text{-}(CH_2)_y\text{-}\overset{O}{\overset{\parallel}{C}}\text{-}]_2\text{-}O\text{-}(CH_2)_z\text{-}CH_3$
	2.	$CH_3\text{-}(CH_2)_x\text{-}\overset{O}{\overset{\parallel}{C}}\text{-}O\text{-}\overset{CH_3}{\overset{\vert}{C}}H\text{-}(CH_2)_y\text{-}\overset{O}{\overset{\parallel}{C}}\text{-}O\text{-}\overset{CH_3}{\overset{\vert}{C}}H\text{-}(CH_2)_m\text{-}O\text{-}\overset{O}{\overset{\parallel}{C}}\text{-}(CH_2)_x\text{-}CH_3$
Hydroxy Monoesters	1.	$HO\text{-}\overset{CH_3}{\overset{\vert}{C}}H\text{-}(CH_2)_y\text{-}\overset{O}{\overset{\parallel}{C}}\text{-}O\text{-}(CH_2)_z\text{-}CH_3$
	2.	$HO\text{-}\overset{CH_3}{\overset{\vert}{C}}H\text{-}(CH_2)_m\text{-}O\text{-}\overset{O}{\overset{\parallel}{C}}\text{-}(CH_2)_x\text{-}CH_3$
Acid Monoesters		$CH_3\text{-}(CH_2)_x\text{-}\overset{O}{\overset{\parallel}{C}}\text{-}O\text{-}\overset{CH_3}{\overset{\vert}{C}}H\text{-}(CH_2)_y\text{-}CO_2H$

FIG. 4. Structures of the major ester components of beeswax.

with fewer studies on the biosynthesis of the oxygenated components reported. The interest in hydrocarbons probably reflects the large amount of this fraction in surface lipid extracts, the ease in which they can be extracted and quantified and the predominance of unique methyl branched components.

5.1 Hydrocarbons

Considerable progress has been made in the past few years on understanding the biosynthesis of insect hydrocarbons. *In vivo* studies in the 1960s established that labeled acetate was readily incorporated into insect hydrocarbons (Lamb, N. and Monroe, R., 1968; Lambremont, E. *et al.*, 1966; Louloudes, S. *et al.*, 1961; Nelson, D., 1969; Robbins, W. *et al.*, 1960; Vroman, H. *et al.*, 1965; Young, R., 1963). Later studies with specific radiolabeled precursors and careful analysis of metabolic products have established the biosynthetic pathways for the most common hydrocarbon components. Experiments using carbon-13 labeled precursors have confirmed some of the conclusions based on radiochemical data. Most experiments to date have used either *in vivo* or tissue slice systems, and only recently have studies with cell-free preparations been initiated. This section will review the information available on the site of hydrocarbon synthesis and the biosynthetic pathways for normal, unsaturated and methyl branched hydrocarbons.

5.1.1 Dietary contribution

In a study comparing the cuticular hydrocarbons of the tobacco hornworm to its dietary hydrocarbons, Nelson, D. *et al.* (1971) presented data indicating that a portion of the dietary alkanes was incorporated into the cuticular lipids. A similar conclusion was reached by Blomquist, G. and Jackson, L. (1973a) using radiolabeled alkanes fed to a grasshopper. However, in both of these cases the contribution of dietary alkanes to the insect's complement of hydrocarbon appeared small, and the insect biosynthesized the majority of its hydrocarbon components. In a study on the source of the hydrocarbons which serve as kairomones of the corn earworm, Hendry, L. *et al.* (1976) concluded that the diet was the major source of the kairomone (*n*-tricosane). However, they did not examine the ability of the corn earworm to produce *n*-tricosane *de novo*. In as much as many other insects are known to do so readily (Blomquist, G. and Jackson, L., 1979), it would seem wise to check this point before excluding the possibility.

5.1.2 Site of synthesis

The most probable site(s) for cuticular hydrocarbon biosynthesis are cells associated with the epidermal layer or the peripheral fat body. Nelson, D. (1969) demonstrated that acetate and palmitate were incorporated *in vitro* into hydrocarbons by the integument of *Periplaneta americana* and *Manduca sexta*, whereas the fat body did not efficiently incorporate either of these precursors into hydrocarbon. Similarly, studies (Armold, M. and Regnier, F., 1975b; Blomquist, G. and Kearney, G., 1976; Dillwith, J. *et al.*, 1981; Jackson, L. and Baker, G., 1970) have demonstrated that hydrocarbon production in *Periplaneta americana, Sarcophaga bullata* and *Musca domestica* occurs in epidermal related cells. Diehl, P. (1973, 1975) separated the oenocyte-rich peripheral fat body from the central yellow fat body tissue in the locust, *Schistocerca gregaria*, and observed the highest rate of hydrocarbon synthesis in the oenocyte-rich peripheral fat body. In *Tenebrio molitor*, Romer, F. (1980) demonstrated that isolated oenocytes efficiently and specifically incorporated [1-^{14}C]acetate into hydrocarbon.

5.1.3 Transport

Chino and co-workers have recently shown that the diacylglycerol-carrying lipoprotein (DGLP) (or lipophorin) (Chino, H. *et al.*, 1981a) from the American cockroach and a locust contain large amounts of hydrocarbon (Chino, H. *et al.*, 1981b; Chino, H. and Kitazawa, K., 1981; H. Chino, vol. 10). Approximately one-fourth of the total lipids associated with lipophorin were hydrocarbon which were similar in composition to those found on the surface of the insects. They postulate that one function of lipophorin is to carry hydrocarbon from the site of synthesis to the site of deposition.

5.1.4 Biosynthetic Pathways

The two most considered metabolic routes for hydrocarbon biosynthesis involve condensation–reduction and elongation–decarboxylation pathways (Fig. 5). In the bacterium, *Sarcinia lutea*, evidence has been presented favoring a modification of the condensation–reduction pathway (Albro, P., 1976), in which two acyl groups condense head-to-head with the loss of a carboxyl group, and the molecule is subsequently reduced. In contrast, Kolattukudy and co-workers have presented convincing evidence that in plants hydrocarbon biosynthesis occurs by the elongation of fatty acids followed by a reductive decarboxylation (Kolattukudy, P., 1980). Both indirect and direct evidence favors the elongation–decarboxylation pathway in insects.

(a) *Elongation–decarboxylation pathway: indirect evidence.* Long-chain fatty acids of the same carbon range as the common insect hydrocarbons are not found in most insect cuticular lipid extracts. However, a comparison of the structures of the primary and secondary alcohols with *n*-alkanes lends circumstantial evidence favoring the elongation–decarboxylation pathway for hydrocarbon biosynthesis in insects. The structural relationships between the primary alcohols and *n*-alkanes in several insects (Soliday, C. *et al.*, 1974; Stransky, K. *et al.*, 1973) suggest the possibility of a similar precursor. The primary

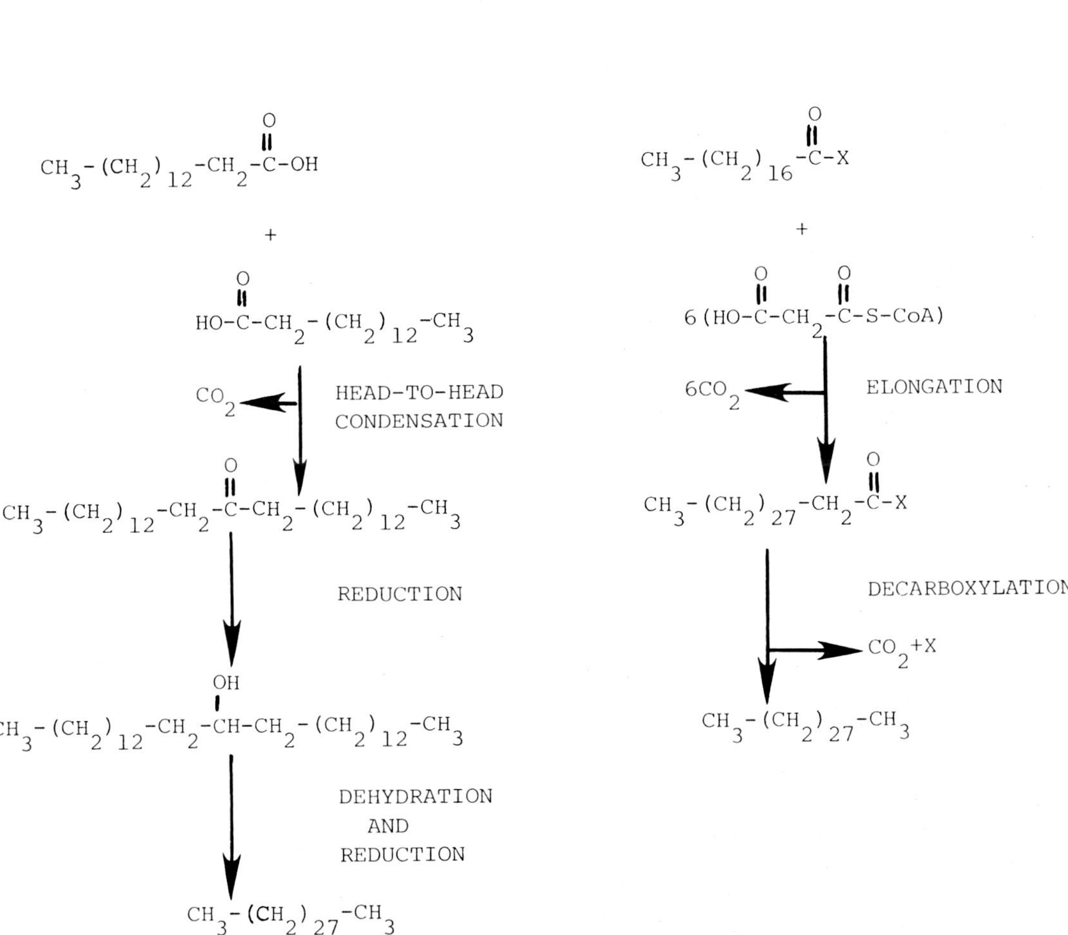

Fig. 5. The two most considered pathways for hydrocarbon biosynthesis. **A:** Head-to-head condensation pathway; **B:** elongation–decarboxylation pathway.

alcohols and *n*-alkanes from insects are often of a similar chain-length range (C_{22}–C_{34}). The *n*-alkanes consist primarily of odd-numbered carbon chain components and the primary alcohols are of even-numbered carbon chains. Lambremont, E. (1972) has demonstrated the reduction of palmitic and stearic acid to the corresponding fatty alcohols and the conversion of these fatty alcohols to fatty acids in the tobacco hornworm. It is likely that a similar reduction of very long-chain fatty acids to fatty alcohols is involved in the biosynthesis of cuticular primary alcohols in insects. Similarly, one can envisage the decarboxylation of very long-chain acids to *n*-alkanes one carbon unit shorter. Since very long-chain fatty acids are not commonly found in insect cuticular lipids, it appears that they must be efficiently decarboxylated to alkanes or reduced to the corresponding fatty alcohol. Lack of substantial amounts of very long-chain *n*- or branched fatty acids appears characteristic of most insect cuticular lipids.

Over one-half of the secondary alcohols in the grasshopper *Melanoplus sanguinipes* would have to arise from odd-chain fatty acids for a direct condensation mechanism to be operative in their biosynthesis. Tricosan-11-ol is the major isomer comprising over a third of the secondary alcohols in this insect (Blomquist, G. *et al.*, 1972). The fatty acids 11 : 0 and 13 : 0 would be required for a direct condensation pathway of biosynthesis to occur. Odd-chain fatty acids have not been extensively studied in insects, but unless insects possess an unusually active odd-chain fatty acid synthesizing system, the structures of the secondary alcohols make it appear unlikely that they arise directly from a condensation pathway.

Several lines of evidence from radioisotope tracer studies also argue against a condensation–reduction pathway for hydrocarbon biosynthesis in insects. In the migratory grasshopper, *M. sanguinipes,* the proposed intermediates of the condensation pathway, symmetrical ketones and secondary alcohols of 23, 27 and 31 carbons, were not converted to hydrocarbons (Blomquist, G. and Jackson, L., 1973b). Furthermore, the unidirectional metabolism of alkanes to secondary alcohols in this insect does not support the condensation pathway for hydrocarbon biosynthesis. Similar studies in the cockroach, *Periplaneta fuliginosa*, showed that a symmetrical C_{23} secondary alcohol and ketone were not reduced to *n*-tricosane, the major *n*-alkane in this insect (Major, M. and Blomquist, G., 1978). Since the possibility of a non-symmetrical condensation of a C_{10} with a C_{14} or a C_8 with a C_{16} fatty acid was conceivable for the biosynthesis of *n*-tricosane, the incorporation of 7-, 8-, 10-, 11- and 12-tricosanols into hydrocarbon was also studied. None of these possible intermediates were incorporated into hydrocarbon in *P. fuliginosa* (Major, M. and Blomquist, G., 1978).

(b) *Decarboxylation of long-chain fatty acids to n-alkanes.* Direct evidence in favor of the elongation–decarboxylation pathway in insects was obtained by demonstrating the direct decarboxylation of long-chain fatty acids to *n*-alkanes in several cockroaches (Major, M. and Blomquist, G., 1978) and in the termite *Zootermopsis angusticollis* (Chu, A. and Blomquist, G., 1980b). *In vivo* studies showed that [R-^{3}H]hexacosanoic acid was converted to *n*-pentacosane in the American cockroach, and that [R-^{3}H]tetracosanoic acid was converted to *n*-tricosane in *Periplaneta fuliginosa* (Major, M. and Blomquist, G., 1978). The decarboxylation of high specific activity [15,16-^{3}H]tetracosanoic acid in the termite *Z. angusticollis* was demonstrated both *in vivo* and *in vitro*. An example of the type of data obtained is presented in Fig. 6. Whereas [1-^{14}C]acetate labeled all the cuticular hydrocarbon components (Fig. 6A), [R-^{3}H]tetracosanoic acid labeled *n*-tricosane exclusively both *in vivo* (Fig. 6B) and *in vitro* (Fig. 6C). α-Hydroxytetracosanoic acid was incorporated into hydrocarbon more efficiently than was tetracosanoic acid, suggesting that the first step in the reductive decarboxylation involves α-hydroxylation. Most of the decarboxylase activity was located in the microsomal fraction, and was stimulated two-fold by the addition of ascorbic acid (Chu, A. and Blomquist, G., 1980b).

Although all the evidence to date favors an elongation–decarboxylation pathway, alternative pathways cannot be ruled out. There exists no direct evidence demonstrating the elongation of 16 or 18 carbon fatty acids to the putative precursors of hydrocarbons, and very long-chain fatty acids and methyl branched fatty acids are not usually reported in insects. This suggests that if the elongation–decarboxylation pathway does occur, the elongation and decarboxylation reactions must

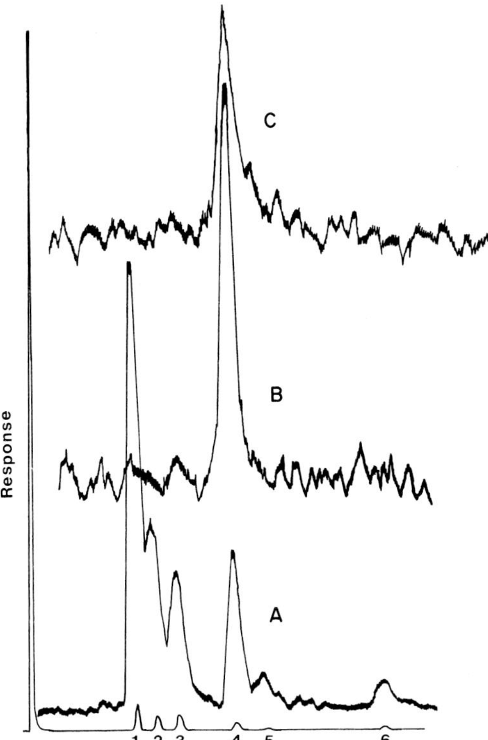

FIG. 6. Radio–GLC traces of the distribution of [1-^{14}C]acetate *in vivo* (**A**) and [15,16-^{3}H]tetracosanoic acid *in vivo* (**B**) and *in vitro* (**C**) into the major cuticular hydrocarbons of *Zootermopsis angusticollis*. The bottom trace is a mass trace. Components are identified as (1) *n*-heneicosane, (2) 5-methylheneicosane, (3) 5,17-dimethylheneicosane, (4) *n*-tricosane, (5) 5-methyltricosane, and (6) *n*-pentacosane. (Chu, A. and Blomquist, G., 1980b.)

be tightly coupled, as significant amounts of intermediates are not released.

The decarboxylation of long-chain acids cannot be demonstrated in all insects. For example, the female housefly, which synthesizes *n*-tricosane, did not decarboxylate [15,16-^{3}H]tetracosanoic acid *in vivo*. Rather, most of the radioactivity from this acid was recovered in 16 and 18 carbon fatty acids after a 2 h incubation, suggesting a very efficient chain shortening mechanism (Dillwith and Blomquist, unpublished observations).

(c) *Biosynthesis of alkenes.* A comparison of the data on the incorporation of [1-^{14}C]acetate, [1-^{14}C]stearate, [1-^{14}C]oleate, and [9,10-^{3}H]oleate into hydrocarbon by the housefly, *Musca domestica*, suggests a pathway in which oleic acid is elongated and decarboxylated to form the alkenes, including the sex pheromone component, (Z)-9-tricosene (Dillwith, J. *et al.*, 1981).

The major cuticular hydrocarbon component of the American cockroach is (Z,Z)-6,9-heptacosadiene, comprising 70% of the total hydrocarbons (Jackson, L., 1972). Labeled acetate was incorporated about equally into the saturated and diene hydrocarbons, whereas [1-^{14}C]linoleic acid was preferentially incorporated into the diene fraction (Jackson, L. and Baker, G., 1970). Recent work showed that [9,10-^{3}H]oleate was incorporated almost exclusively into the diene, with less than 6% of the radiolabel being recovered in the saturated fraction (Dwyer, L. *et al.,* 1981b). [^{13}C]NMR experiments demonstrated that [2-^{13}C]acetate labeled carbons 25 and 27 but not carbon 3 of the C_{27} alkadiene. In addition, ozonolysis of the diene labeled from [1-^{14}C]acetate followed by radio–GLC analysis showed that carbons 1–6 were not labeled. These data indicate that linoleate from the diet or synthesized *de novo* (Dwyer, L. and Blomquist, G., 1981) is elongated by the addition of acetate units and is then decarboxylated (Dwyer, L. *et al.,* 1981b).

(d) *Methyl branched alkanes.* Methyl branched alkanes often comprise a significant portion of cuticular hydrocarbon mixtures (Nelson, D., 1978) and have been shown to serve as both pheromones and kairomones in many insects (Howard, R. and Blomquist, G., 1982). Consequently, the origin of the methyl branch units has received considerable attention.

2-Methylalkanes. The biosynthesis of the 2-methylalkanes has been investigated in the ground cricket, *Nemobius fasciatus*, and in the field cricket, *Gryllus pennsylvanicus* (Blailock, T. *et al.,* 1976). The data suggest that the carbon skeleton of valine and leucine serve as the precursors to the even and odd-numbered carbon chain 2-methylalkanes, respectively. 2-Methylalkanes of 23 and 25 carbons comprise 20% of the cuticular hydrocarbons of *N. fasciatus* with the other 80% *n*-alkanes (Blomquist, G. *et al.,* 1976). Of the labeled acetate that was incorporated into hydrocarbon, 28% was recovered in the 2-methylalkanes and the remainder in the *n*-alkanes. In contrast, valine was preferentially incorporated into 2-methylalkanes, with 98% of the radio-activity that was incorporated into hydrocarbon recovered in the 2-methylalkane fraction. Likewise, isobutyric acid preferentially labeled the 2-methylalkanes. This observation suggests that the cricket converts valine to isobutyric acid which is then incorporated into the even-chain 2-methylalkanes during the initial stages of chain elongation (Fig. 7). Both odd- and even-chain 2-methylalkanes are present in the cuticular hydrocarbons of *G. pennsylvanicus*. Labeled valine, leucine, isobutyric acid and isovaleric acid were preferentially incorporated into the branched alkanes, suggesting that leucine was converted to isovaleric acid which served as the precursor to the odd-chain 2 methylalkanes (Fig. 7) (Blailock, T. *et al.*, 1976).

3-Methylalkanes. Two pathways have been suggested for the biosynthesis of 3-methylalkanes. In plants (Kolattukudy, P., 1980) and micro-organisms (Albro, P., 1976) isoleucine is converted to 2-methylbutyric acid which is then incorporated into 3-methylalkanes during the initial stages of chain synthesis. In contrast, considerable evidence has accumulated that in several cockroaches a methylmalonyl derivative serves as the precursor to the branching methyl group of 3-methylalkanes (Blomquist, G. and Kearney, G., 1976; Blomquist, G. *et al.*, 1975a; Conrad, C. and Jackson, L., 1971). Isoleucine was not efficiently incorporated into 3-methylpentacosane in *Periplaneta americana* (Blomquist, G. *et al.*, 1975a).

A remarkable specificity occurs in the biosynthesis of the branched alkanes of *P. americana*, in that 3-methylpentacosane is the only component present. The lack of methyl branched fatty acids in this insect had led to the suggestion that the methylmalonyl–CoA branching unit was added as the penultimate unit (Conrad, C. and Jackson, L., 1971; Blomquist, G. *et al.*, 1975a). [^{13}C]NMR experiments were used to determine if the branching unit was added as the second unit or near the end of the biosynthetic process. A knowledge of the position of the labeled carbon from [1-^{13}C]propionate in 3-methylpentacosane would allow a determination of whether propionate was incorporated in early (Fig. 8A1) or late (Fig. 8A2) stages of the elongation process. If carbon 2 was enriched, the methyl group would have been incorporated during the penultimate step of chain elongation. If carbon 4 was enriched, the methyl group would have been inserted as the second unit in chain elongation (Fig. 8A). The data from [^{13}C]NMR experiments showed that the labeled carbon from [1-^{13}C]propionate was incorporated exclusively in the 4 position (Fig. 8B), demonstrating that the methyl branch unit is added as the second unit in the elongation process (Fig. 8) (Dwyer, L. *et al.*, 1981a).

Internally branched alkanes. Methyl groups on the internal carbon atoms of alkane chains could originate either during the elongation process by substitution of a methylmalonyl–CoA in place of a

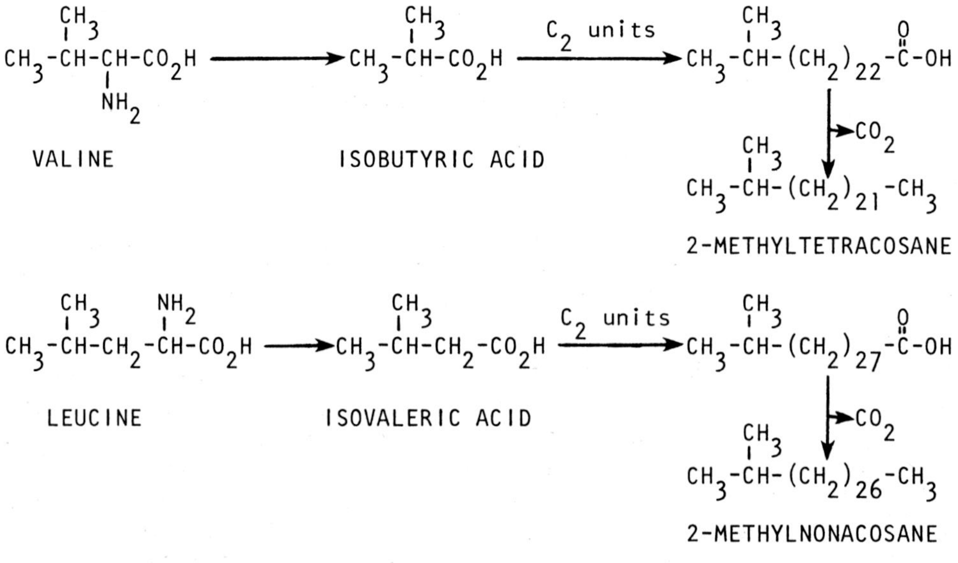

CH_3-CH(CH_3)-CH(NH_2)-CO_2H → CH_3-CH(CH_3)-CO_2H —(C_2 units)→ CH_3-CH(CH_3)-$(CH_2)_{22}$-C(=O)-OH —(−CO_2)→ CH_3-CH(CH_3)-$(CH_2)_{21}$-CH_3

VALINE — ISOBUTYRIC ACID — 2-METHYLTETRACOSANE

CH_3-CH(CH_3)-CH_2-CH(NH_2)-CO_2H → CH_3-CH(CH_3)-CH_2-CO_2H —(C_2 units)→ CH_3-CH(CH_3)-$(CH_2)_{27}$-C(=O)-OH —(−CO_2)→ CH_3-CH(CH_3)-$(CH_2)_{26}$-CH_3

LEUCINE — ISOVALERIC ACID — 2-METHYLNONACOSANE

FIG. 7. Proposed biosynthetic pathways for 2-methylalkanes in insects. (Blailock, T. *et al.*, 1976.)

A

1. $C\text{-}C(=O)\text{-}X + C\text{-}{}^{*}C(=O)\text{-}X$ (C branch on the first C) $\longrightarrow C\text{-}C\text{-}C(C)\text{-}{}^{*}C(=O)\text{-}X \xrightarrow{11\ (2:0),\ -CO_2} C_1\text{-}C_2\text{-}C_3(C)\text{-}{}^{*}C_4\text{-}C_{5\text{-}25}$

2. $C\text{-}C_{20}\text{-}C(=O)\text{-}X + C\text{-}{}^{*}C(=O)\text{-}X$ (C branch on the first C) $\longrightarrow C\text{-}C_{20}\text{-}C\text{-}C(C)\text{-}{}^{*}C(=O)\text{-}X \xrightarrow{2:0,\ -CO_2} C\text{-}C_X\text{-}C_4\text{-}C_3(C)\text{-}{}^{*}C_2\text{-}C_1$

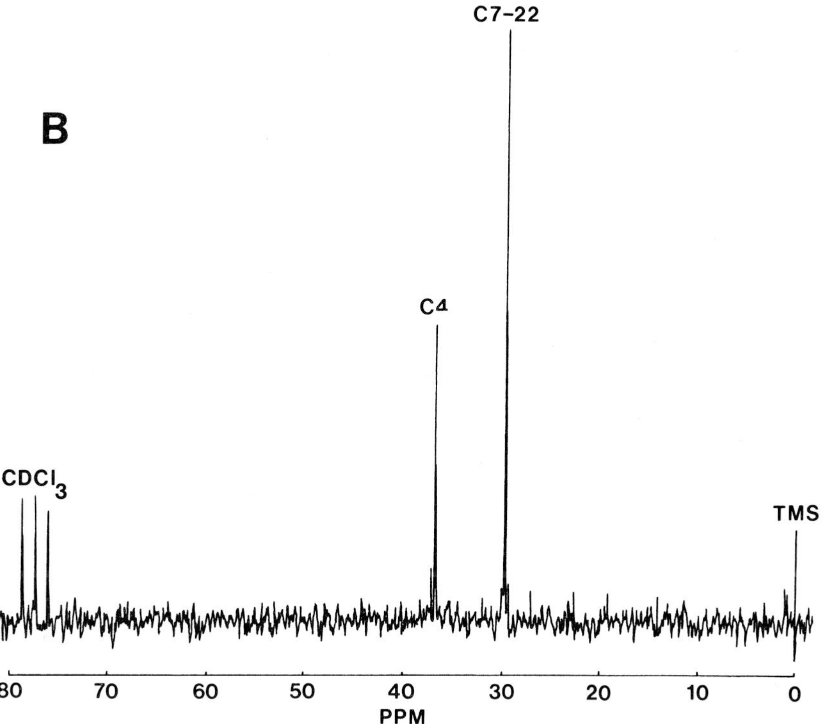

FIG. 8. **A:** Incorporation of [1-^{13}C]propionate into 3-methylpentacosane in *Periplaneta americana* in early (A1) or late (A2) stages of chain elongation; **B:** [^{13}C]NMR spectrum of 3-methylpentacosane enriched from [1-^{13}C]propionate. Only carbon-4 is enriched, indicating that pathway (A1) in which propionate is incorporated as the second unit is operative. (Dwyer, L. *et al.*, 1981a.)

malonyl–CoA, or by the methylation of a preformed chain. In algae, the methylation of a preformed chain by the methyl group of *S*-adenosylmethionine is involved in the biosynthesis of 7- and 8-methylheptadecanes (Fehler, S. and Light, R., 1970; Han, J. *et al.*, 1969). The variety of positions substituted (5, 7, 9, 11, 13 or 15) and the occurrence of dimethylalkanes with methyl branches separated by 1, 3, 7, 9 or 11 methylene groups favors the methylmalonyl–CoA hypothesis in insects.

Studies on the biosynthesis of 13-methylpentacosane in *Periplaneta fuliginosa* showed that [1-^{14}C]propionate preferentially labeled this methyl branched alkane (Blomquist, G. and Kearney, G., 1976). Similarly, [1-^{14}C]propionate was incorporated almost exclusively into the mono- and dimethylalkanes in the termite *Zootermopsis angusticollis* (Blomquist, G. *et al.*, 1979b) and the housefly, *Musca domestica* (Dillwith, J. *et al.*, 1981). [Methyl-^{14}C]Methionine did not serve as an efficient precursor to the branched alkanes in any of these insects. These data suggest that propionate, as a methylmalonyl derivative, is incorporated in place of malonyl–CoA during chain elongation.

Although propionate is readily incorporated into the methylalkanes in a variety of insects, it may not be the endogenous precursor to the methylmalonyl–CoA used in branched hydrocarbon synthesis. Studies on the biosynthesis of branched alkanes in the termite, *Z. angusticollis,* suggested that succinate is a precursor to the methylmalonyl derivative which is the methyl branched precursor (Chu, A. and Blomquist, G., 1980a).

Definitive evidence that succinate is a source of

$$^{-}O_2C{-}^{13}CH_2{-}^{13}CH_2{-}CO_2^{-} \longrightarrow {}^{-}O_2C{-}^{13}CH(^{13}CH_3){-}\underset{\underset{O}{\|}}{C}{-}X \longrightarrow CH_3{-}CH_2{-}CH_2{-}CH_2{-}^{13}CH(^{13}CH_3){-}CH_2{-}(CH_2)_x{-}H$$

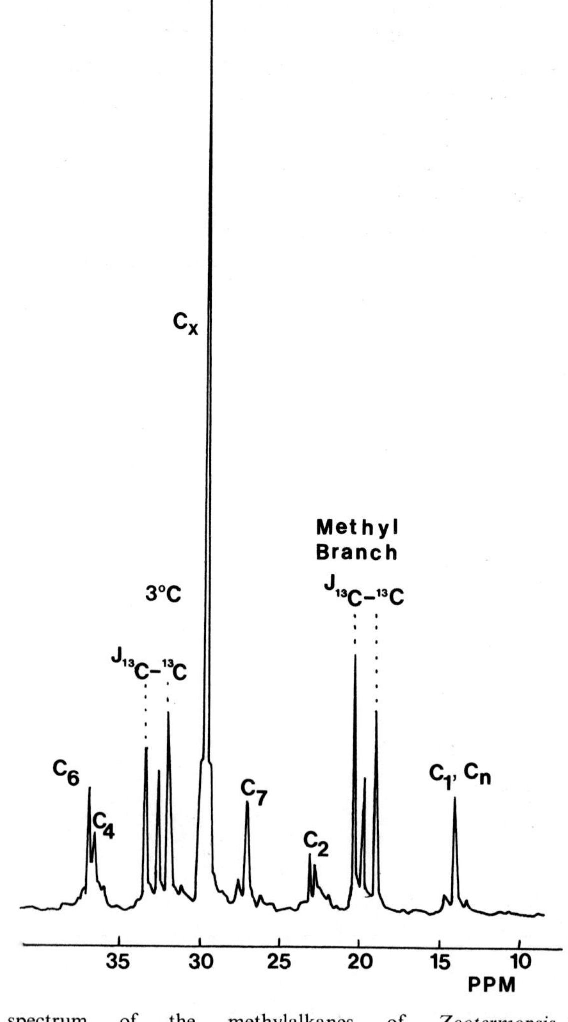

FIG. 9. [^{13}C]NMR spectrum of the methylalkanes of *Zootermopsis angusticollis* enriched from [2,3-^{13}C]succinate. The carbon–carbon coupling shows that the labeled carbons are incorporated intact into the branching methyl carbon and the tertiary carbon. (Blomquist, G. *et al.*, 1980a.)

the branching methyl unit of methylalkanes was obtained by using carbon-13 NMR to examine the incorporation of [2,3-^{13}C]succinate into methylalkanes in *Z. angusticollis*. The data showed that carbons 2 and 3 of succinate became the methyl branch and tertiary carbon in branched alkanes (Fig. 9). Furthermore, the carbon–carbon coupling of adjacent labeled carbons showed that the bond between carbons 2 and 3 of succinate remains intact (Blomquist, G. *et al.*, 1980a).

An alternative source of methylmalonyl–CoA has been demonstrated in the housefly. In this insect there was no detectable incorporation of the label from [2,3-^{13}C]succinate into the branching methyl group or tertiary carbon. In contrast, the labeled carbons from D,L[3,4,5-$^{13}C_3$]valine were incorporated intact (as determined by ^{13}C–^{13}C coupling) into the branching methyl carbon, tertiary carbon, and carbon adjacent to the tertiary carbon (Fig. 10). This suggests that valine, and perhaps other amino acids that are metabolized via methylmalonyl–CoA, serve as a precursor to the methyl branch unit (Dillwith, Nelson, Pomonis, Nelson and Blomquist, unpublished observations).

Sodium [1-^{13}C]propionate was readily and specifically incorporated into the methylalkanes of the housefly, with an approximate 50-fold enrichment in the carbon adjacent to the tertiary carbon observed (Dillwith, Nelson, Pomonis, Nelson and Blomquist, unpublished observations). Mass spectral analysis of the [^{13}C]-enriched alkanes showed that for most components the branching unit was inserted toward the beginning of the elongation process rather than toward the end. Figure 11 summarizes our current understanding of methylalkane synthesis in the housefly, and is likely applicable to other insects.

The choice of precursors to methylmalonyl–CoA may be related to diet. The low amounts of protein in a termite's diet makes succinate, which can be derived from carbohydrate, a logical choice. In contrast, the housefly has a much higher intake of protein, and apparently utilizes selected amino acids to form methylmalonyl–CoA. The precursor to the methyl branching group in most insects is not known.

In addition to being incorporated as the methyl branching carbon, the labeled carbons from

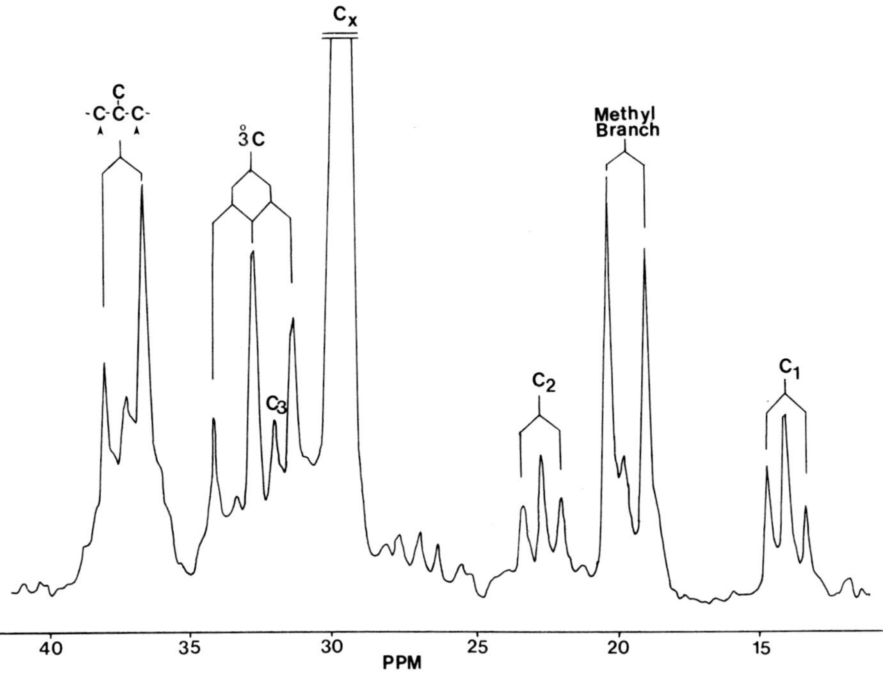

FIG. 10. [^{13}C]NMR of the methylalkanes of the female housefly, *Musca domestica*, enriched from D,L-[3,4,5-$^{13}C_3$]valine. The carbon–carbon coupling shows that the three labeled carbons are incorporated intact into the methyl branching carbon, the tertiary carbon, and carbon adjacent to the tertiary carbon. (Dillwith *et al.*, unpublished.)

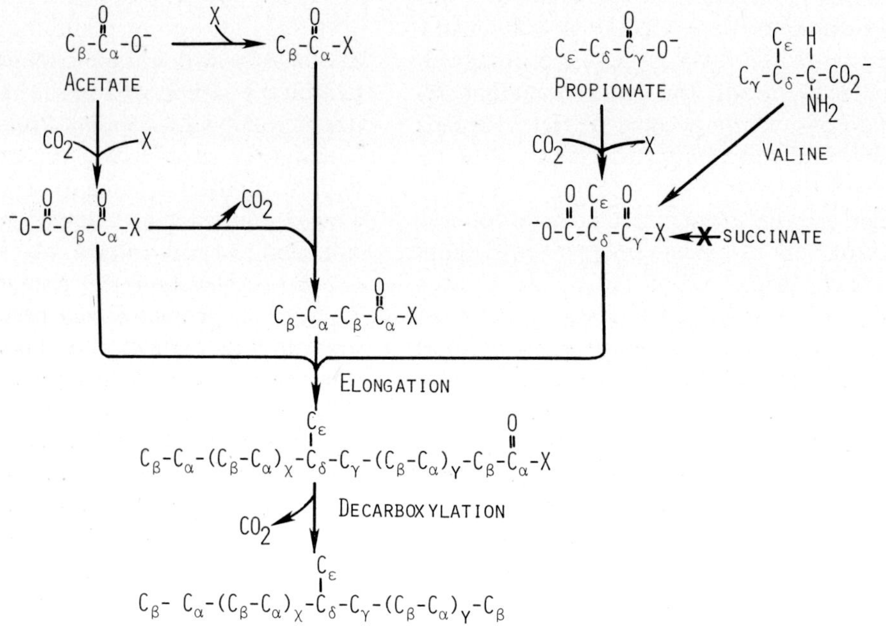

FIG. 11. Proposed biosynthetic pathway of the methylalkanes of the female housefly, *Musca domestica*. The origin of each carbon has been determined by [^{13}C]NMR spectroscopy.

[3-^{13}C]propionate and [methyl-^{13}C]methylmalonate also labeled the even-numbered carbons in the alkanes and alkenes. Likewise, [2-^{13}C] propionate labeled the odd-numbered carbons of both the alkanes and alkenes. These data suggest that propionate is converted to an acetyl derivative, with carbon 3 of propionate converted to the carboxyl carbon of acetate, and carbon 2 of propionate converted to the methyl carbon of acetate. The occurrence of this pathway for propionate metabolism in the housefly, which does not utilize the methylmalonyl–CoA mutase reaction, prompted an examination of the housefly and other insects for vitamin B_{12} levels. The housefly contained undetectable levels of vitamin B_{12} (Wakayama, Dillwith and Blomquist, unpublished observations), which was consistent with its inability to use succinate as a precursor to methylmalonyl–CoA and the presence of enzymes which directly convert propionate to an acetate derivative. In contrast, termite *Z. angusticollis* contained extremely high levels of vitamin B_{12}, which was consistent with its ability to convert succinate to methylmalonate.

5.2 Biosynthesis of non-hydrocarbon cuticular components

The biosynthesis of the non-hydrocarbon components of insect cuticular wax has not been extensively studied. It is postulated that the long-chain fatty acids found in cuticular extracts arise from the elongation of 16 and 18 carbon fatty acids by the addition of two carbon units. Likewise, it is assumed that very long-chain primary alcohols arise from the reduction of the corresponding fatty acid, but direct evidence on these processes is not available.

5.2.1 SECONDARY ALCOHOL WAX ESTERS

Labeled *n*-alkanes included in the diet or administered to the surface of the grasshopper, *Melanoplus sanguinipes*, were metabolized to secondary alcohols and esterified (Blomquist, G. and Jackson, L., 1973a,b). Chain-length specificity was evident in that the shorter-chain length *n*-alkanes (C_{21}, C_{23} and C_{25}) were metabolized more readily than the longer-chain (C_{27} and C_{29}) compounds (Blomquist, G. and Jackson, L., 1973b). This chain-length specificity appears to explain the distribution of secondary alcohols and *n*-alkanes in this insect. The major naturally occurring secondary alcohols are $C_{23} > C_{25} > C_{21}$, whereas the principal *n*-alkanes are C_{27} and C_{29} in this organism (Soliday, C. *et al.*, 1974).

The nature of the reaction which introduces an oxygen into long-chain alkanes was investigated in *M. sanguinipes* with $^{18}O_2$ and $H_2{}^{18}O$. Two possible mechanisms had been suggested for this reaction. A dehydrogenation followed by hydration could occur, or alternatively, a mixed-function oxidase type enzyme could incorporate one atom of oxygen from O_2 (Fig. 12). Incubation of *M. sanguinipes* with $^{18}O_2$ or $H_2{}^{18}O$ followed by mass spectrometry of the isolated secondary alcohol showed that molecular oxygen serves as the oxygen donor (Blomquist, G. *et al.*, 1975b). The ^{18}O from $^{18}O_2$ was incorporated into secondary alcohols. This observation suggests that a mixed-function oxidase type enzyme is involved in secondary alcohol biosynthesis.

The esterification of labeled secondary alcohols occurs readily in *M. sanguinipes,* with a fairly evident chain-length specificity. The shorter chain secondary alcohols C_{21}, C_{23}, C_{25} and C_{27} were esterified more readily than the C_{31} compound (Blomquist, G. and Jackson, L., 1973b). The esterification reaction was very rapid and efficient. Greater than 80% of exogenous C_{23} secondary alcohol was esterified in 18 h, and almost all the administered *n*-alkane which was hydroxylated appeared in the form of a secondary alcohol wax ester (Blomquist, G. and Jackson, L., 1973b). In accord with this observation, only trace amounts of free secondary alcohol were present in the cuticular extract of this insect (Soliday, C. *et al.*, 1974).

5.2.2 Primary alcohol wax esters

Labeled acetate, palmitate and tetracosanol injected beneath the cuticle of the honeybee were incorporated into wax monoester. Acetate and palmitate labeled both the alcohol and acid moieties of the monoester, whereas tetracosanol was incorporated almost exclusively into the alcohol moiety (Blomquist, G. *et al.*, 1980b).

The esterification of primary alcohols was studied with a microsomal preparation obtained from the abdomens of insects not actively producing comb wax. Microsomal preparations (105,000 ***g*** pellet) gave about a 2-fold higher specific activity than did a 10,000 ***g*** pellet or a 105,000 ***g*** supernatant. When a microsomal preparation was incubated in the presence of labeled tetracosanol and palmitoyl-CoA, the distribution of products showed that most of the primary alcohol incorporated into the esters was recovered in the monoester fraction, with lesser amounts in the di- and triester fractions (Blomquist, G. and Ries, M., 1979).

The rate of wax monoester synthesis from [1-^{14}C]palmitate was stimulated by CoA, ATP and $MgCl_2$ and the addition of palmitoyl-CoA resulted in a 5-fold increase in monoester synthesis from labeled tetracosanol, indicating that the acyl donor group is activated in the form of an acyl-CoA. Palmitoyl-CoA stimulated monoester formation from labeled tetracosanol and increased amounts of wax ester were formed with increasing amounts of palmitoyl-CoA up to 10^{-4} M, after which it inhibited, presumably due to a detergent effect (Blomquist, G. and Ries, M., 1979).

The chain-length specificity of the primary alcohols incorporated into monoester was examined, and the results showed that labeled alcohols of 16, 18, 24 and 28 carbons were readily

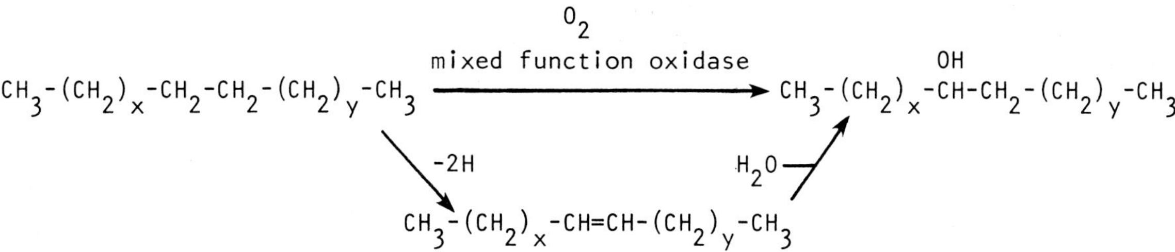

FIG. 12. Proposed pathway for the biosynthesis of secondary alcohols in *Melanoplus sanguinipes*. Stable isotope studies have demonstrated that the oxygen in the hydroxyl group arises from molecular oxygen. (Blomquist, G. *et al.*, 1975b.)

incorporated into monoester with the shorter-chain 16 and 18 carbon alcohols incorporated at higher rates (Blomquist, G. and Ries, M., 1979; Lambremont, E. and Wykle, R., 1979). The naturally occurring primary alcohols are of chain lengths C_{24}–C_{32}. This suggests that the chain-length specificity of the alcohols present in monoester does not reside in the esterification step, but rather in the reduction of fatty acids to alcohols.

The synthesis of wax esters in honeybees actively producing comb wax was examined by Lambremont, E. and Wykle, R. (1979). The post-mitochondrial supernatant fraction readily esterified tetracosanol to palmitoyl-CoA and the characterization of this process suggests that the synthesis of wax esters by the specialized wax glands of the honeybee is similar to that of the epidermal tissue.

5.2.3 Epoxides and ketones

The biosynthesis of (Z)-9,10-epoxytricosane and 14-tricosen-10-one was examined in the housefly, *Musca domestica*. Both of these components, which are part of the female sex pheromone, were formed from [9,10-^{3}H](Z)-9-tricosene. Both male and female insects were able to convert (Z)-9-tricosene to the corresponding epoxide and ketone with the male possessing higher activity (Blomquist and Dillwith, unpublished data). The male housefly does not possess either of these 2 oxygenated derivatives in its cuticular lipids (Uebel, E. *et al.*, 1978a) which is apparently due to its inability to synthesize (Z)-9-tricosene (Dillwith, J. *et al.*, 1981).

6 PHYSIOLOGICAL CONSIDERATIONS

6.1 Development and environment

The variation in insect surface lipids, as affected by developmental and environmental factors, has been studied in a few insect species. The amount and composition of cuticular lipids often changes as an insect goes through developmental stages, particularly if the insect's environment undergoes major changes. For example, the amount and the distribution of lipid classes and the composition of components within a class are quite different in the aquatic naiad form of the big stonefly, *Pteronarcys californica*, compared with the terresterial adult (Armold, M. *et al.*, 1969). It appears that in this insect the surface lipid composition varies with life stage, depending upon the need for water conservation. Similarly, the cuticular hydrocarbons vary both quantitatively and qualitatively throughout the life cycle of the flesh-fly, *Sarcophaga bullata*, and the quantity of cuticular hydrocarbon correlates well with the water conservation needs of the insect (Armold, M. and Regnier, F., 1975a). Marked differences were noted in the cuticular lipids of immature and adult forms of the cigarette beetle, *Lasioderma serricorne* (Baker, J. *et al.*, 1979b), the black carpet beetle, *Attagenus megatoma* (Baker, J. *et al.*, 1979b), two species of imported fire-ants, *Solenopsis invicta* and *Solenopsis richteri* (Lok, J. *et al.*, 1975) and the desert cicada, *Diceroprocta apache* (Hadley, N., 1980b). There were marked increases in either the percentage of hydrocarbons among the cuticular lipid classes, the total amount of hydrocarbon, the composition of hydrocarbons, or all three. In the desert cicada there was approximately five times more hydrocarbon in the adult than in the nymphal exuvium, and this difference correlated with the increased transpiration potential experienced by the adult. Qualitative and quantitative differences in the cuticular lipids of different life forms of the sheep blowfly, *Lucilia cuprina* (Goodrich, B., 1970) and pecan weevils, *Curculio caryae* (Mody, N. *et al.*, 1975) may also be related to the need to reduce cuticular transpiration in the adult form.

The cuticular lipids of insects also vary in amount and composition depending upon the season. For example, the desert tenebrionid beetle, *Eleodes armata*, possessed more hydrocarbons and a higher percentage of long-chain components in summer than they did in winter (Hadley, N., 1977). Similar results were obtained when winter-collected beetles were acclimated to 35°, suggesting that this change in hydrocarbon is an adaptation to an altered environment. Again, the changes appear to be related to the water conservation needs of the insect.

Diapausing pupae of the tobacco hornworm, *Manduca sexta* (Bell, R. *et al.*, 1975) secrete three times as much surface wax as do non-diapausing pupae. The extra thickness of the wax layer apparently protects the insect from desiccation and the authors speculate that the deposition of additional wax may result from hormonal changes accompanying entry into diapause.

6.2 Endocrine regulation

Very little is known about the general area of the control of hydrocarbon biosynthesis in insects. Work by Armold, M. and Regnier, F. (1975b), however, suggests that 20-hydroxyecdysone stimulates hydrocarbon formation. They found that in the development of *S. bullata* there are two periods of rapid accumulation of hydrocarbons. The most rapid accumulation is during the 4-day period preceding the pupal–adult ecdysis ($>20\ \mu g\ day^{-1}\ insect^{-1}$). An earlier accumulation ($5–8\ \mu g\ day^{-1}\ insect^{-1}$) occurs during pupariation and the 3-day period following pupariation. At pupariation, exogenous 20-hydroxyecdysone was shown to stimulate hydrocarbon biosynthesis from [^{3}H]acetate at a rate of 1.3 times the rate of controls 10.5 h after hormone administration and at a rate of 3.3 times the rate of controls 24 h after hormone administration. In an *in vitro* study they showed that integuments of 20-hydroxyecdysone-treated insects incorporated acetate into hydrocarbons 9.8 times better than integuments of control insects (Armold, M. and Regnier, F., 1975b). This shows that cuticular hydrocarbon biosynthesis in *S. bullata* occurs in the integument, and that the locus of regulation is also present in the integument.

Dramatic changes in the composition of the cuticular lipids of the housefly, *Musca domestica*, occur with age in the adult insect (see R. Sohal, vol. 10). Newly emerged males and females have almost identical cuticular compositions. Dramatic changes take place in the female between days 2 and 3. (Z)-9-Tricosene, (Z)-9,10-epoxytricosane, and (Z)-14-tricosen-10-one first appear at this time, and increasing amounts of these components and methylalkanes accumulate in the female as the insect ages (Dillwith, Nelson and Blomquist, unpublished observations). These components, which are part of the housefly sex pheromone, are absent, or present in very small amounts, in the male. Because of the apparent similarity in timing of the beginning of pheromone production and ovarian maturation, the production of the pheromone components was correlated to ovarian development.

Production of the cuticular sex pheromone components of the female housefly, including (Z)-9-tricosene (muscalure), (Z)-9,10-epoxytricosane, (Z)-14-tricosen-10-one, and methyl branched alkanes, correlated with ovarian development. No pheromone was produced by females with ovaries in pre-vitellogenesis, production was initiated in insects during early vitellogenesis, and larger amounts of pheromone were found in females as the ovarian follicles matured (Dillwith, Adams and Blomquist, unpublished observations).

Females ovariectomized shortly after adult emergence did not produce pheromone. Reimplanting pre-vitellogenic ovaries into ovariectomized females and allowing them to reach early vitellogenic stages restored pheromone production (Fig. 13). Implanting ovaries into male insects resulted in altering the composition of their cuticular lipids such that they then produced (Z)-9-tricosene (muscalure), the C_{23} epoxide and ketone, and large amounts of methylalkanes (Blomquist, Adams and Dillwith, unpublished data). Injection of 20-hydroxyecdysone into ovariectomized females or into newly emerged males caused the same alteration in cuticular lipids as did implanting of ovaries, and production of pheromone components was initiated. The major cuticular component produced by ovariectomized females and control males is (Z)-9-heptacosene. The presence of maturing ovaries or injection of 20-hydroxyecdysone causes a change in the chain length of the alkenes such that (Z)-9-tricosene is produced, and increases the amount of methylalkanes synthesized. Adult male and female insects both possess the enzyme(s) which convert (Z)-9-tricosene to the corresponding epoxide and ketone at all ages, and the production of these oxygenated derivatives occurs when substrate ((Z)-9-tricosene) is present (Blomquist, Adams and Dillwith, unpublished data).

6.3 Biosynthesis of beeswax: effect of season, age

In addition to producing a thin layer of surface lipids, some insects, such as the honeybee, produce a much larger amount of wax for the honeycomb. The biosynthesis of beeswax has been investigated by Piek, T. (1963) and Young, R. (1963), who

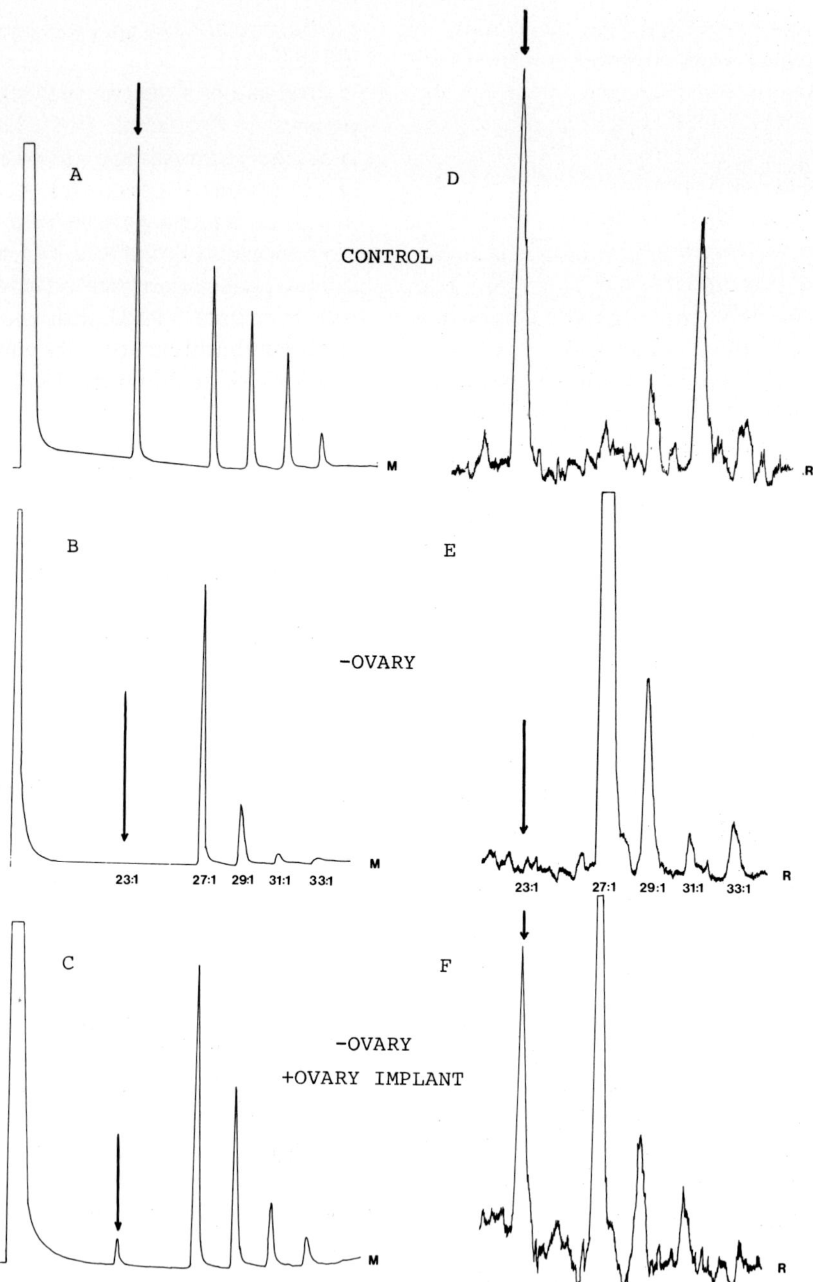

FIG. 13. The effect of ovariectomy and re-implanting ovaries on alkene synthesis in the female housefly, *Musca domestica*. **A-C** are mass traces and **D-F** are radioactivity traces. Ovaries were removed within 12 h after adult emergence. Ovaries from 24 h-old donor insects were implanted into 4-day-old ovariectomized females. Six-day-old insects were injected with [1-^{14}C]acetate, sacrificed after 2 h, and the alkenes isolated and analyzed by radio–GLC (Dillwith, Adams and Blomquist, unpublished.).

showed that labeled acetate was incorporated into the hydrocarbon, ester and acid fractions of beeswax. The biosynthesis of wax by the honeybee was studied in winter and summer. It was observed that in winter, bees not only incorporated acetate into hexane-extractable lipids at much lower levels than in summer, but also demonstrated a 4-fold higher incorporation of acetate into hydrocarbon than into monoester (Blomquist, G. *et al.*, 1980b). In contrast, in summer, when many of the bees were actively producing comb wax, acetate was incorporated at about the same rates into monoester and into hydrocarbon. A comparison of the cuticular and comb wax of the honeybee showed that the major component in the cuticular wax was hydrocarbon, which comprised about 58% of this wax. Monoester accounted for 23%. In contrast, hydrocarbon and monoester account for 14 and 35% respectively, of the comb wax of the honeybee (Tulloch, A., 1971). Data from studies on the incorporation of labeled acetate into wax components by insects not producing comb wax showed that it was incorporated into the various wax fractions in about the same proportion as the amount of each component. In bees actively producing comb wax, a lower rate of incorporation into hydrocarbon and a higher rate of incorporation into monoester was observed.

It is generally accepted that the activities which bees engage in (i.e. cell cleaning, feeding larvae, building activities, gathering nectar and pollen, etc.) are age-related phenomena, although there is great flexibility in the age–activity relationship. The comb wax of bees is secreted by wax glands located on the ventral abdomen and the wax glands are best developed and most productive in bees 12–18 days old (Gary, N., 1975). Bees from 13 to 17 days old incorporated acetate about equally into both the hydrocarbon and monoester fractions. From age 19 to 31 days there was an increase in the incorporation of acetate into hydrocarbon and a decreased incorporation into monoester, so that by day 31 over 71% of the label that was incorporated into wax was present in the hydrocarbon fraction, and less than 10% in monoester. Thus, it appears that from day 13 to 17, both cuticular and comb wax are synthesized, whereas in older bees, only cuticular wax is being formed (Blomquist, G. *et al.*, 1980b).

To further explore this phenomenon, the incorporation of labeled acetate into fat body, thorax, and both dorsal and ventral integument tissue from the abdomens of bees was studied. The ventral abdominal integument tissue contained wax glands. An analysis of the data showed that fat body tissue did not efficiently incorporate labeled acetate into either hydrocarbon or monoester. Both the thorax and the dorsal abdominal integument tissue incorporated acetate into wax. These tissues incorporated radioactivity at about a 4- to 5-fold higher rate into hydrocarbon than into monoester. In contrast, the ventral abdominal tissue, which contained the wax glands, incorporated radioactivity about equally into both hydrocarbon and monoester (Blomquist, G. *et al.*, 1980b). These data demonstrate that the site of wax synthesis in the honeybee is associated with the integument and points out the difference in the wax produced by the glands (comb wax) compared with the cuticular wax.

7 CONCLUDING REMARKS

The development of microanalytical techniques, particularly GLC–MS, has allowed rapid advances in our understanding of the chemistry of insect cuticular lipids. However, our knowledge of the chemical composition of insect cuticular lipids is still limited to representative species of only seven orders, and in many cases is restricted to one lipid class — the hydrocarbons. Although it has been known for decades that a major function of insect cuticular lipids is to restrict water loss, a generally accepted model for the observed waterproofing is not available. Indeed, the recent data which argue against the long-held view of an "oriented monolayer" should result in a renewed interest in gaining an understanding of the waterproofing properties of cuticular lipids. The recognition that cuticular lipid components function in chemical communication in certain species has resulted in a great deal of interest to determine how general this phenomenon is, and to examine the types of roles cuticular lipids play. The general metabolic pathways have been determined for some of the cuticular lipid components, and information is becoming available on the endocrine regulation of cuticular lipid formation in a few species. However, these studies have only begun to explore the processes by which cuticular lipid components are formed and regulated and many intriguing questions remain. A strong case can be made for the joining of forces between behavioral ecologists, physiologists, morphologists, biochemists and chemists to better our understanding of insect cuticular lipids.

ACKNOWLEDGEMENTS

The work from G.J.B.'s laboratory was supported by the Science and Education Administration of the US Department of Agriculture under grant 7801064 from the Competitive Research Grants Office and National Science Foundation grants PCM76-21694 and PCM-8118305. Contribution of the Nevada Agricultural Experiment Station, Journal Series No. 556. The authors express their thanks to colleagues who provided unpublished data for inclusion in this chapter. We thank Drs Neil Hadley, Dennis Nelson, and Larry Jackson for their helpful suggestions in the preparation of this manuscript. We thank Dr. Jean Percy for providing the micrograph in Fig. 1.

REFERENCES

ALBRO, P. W. (1976). Bacterial waxes. In *Chemistry and Biochemistry of Natural Waxes*. Edited by P. E. Kolattukudy. Pages 419–445. Elsevier, Amsterdam.

ANTONY, C. and JALLON, J.-M. (1981). Evolution des hydrocarbures comportalement actifs de *Drosophila melanogaster* au cours de la maturation sexuelle. *C.R. Acad. Sci. Paris 292,* 239–242.

ARMOLD, M. T. and REGNIER, F. E. (1975a). A developmental study of the cuticular hydrocarbons of *Sarcophaga bullata. J. Insect Physiol. 21,* 1827–1833.

ARMOLD, M. T. and REGNIER, F. E. (1975b). Stimulation of hydrocarbon biosynthesis by ecdysterone in the flesh fly *Sarcophaga bullata. J. Insect Physiol. 21,* 1581–1586.

ARMOLD, M. T., BLOMQUIST, G. J. and JACKSON, L. L. (1969). Cuticular lipids of insects. III. The surface lipids of the aquatic and terrestrial life forms of the big stone fly. *Pteronarcys californica* Newport. *Comp. Biochem. Physiol. 31,* 685–692.

BAKER, G. L., VROMAN, H. E. and PADMORE, J. (1963). Hydrocarbons of the American cockroach. *Biochem. Biophys. Res. Commun. 13,* 360–365.

BAKER, J. E. (1978). Cuticular lipids of larvae of *Attagenus megatoma. Insect Biochem. 8,* 287–292.

BAKER, J. E. and NELSON, D. R. (1981). Cuticular hydrocarbons of adults of the cowpea weevil, *Callosobruchus maculatus. J. Chem. Ecol. 7,* 175–182.

BAKER, J. E., NELSON, D. R. and FATLAND, C. L. (1979a). Developmental changes in cuticular lipids of the black carpet beetle, *Attagenus megatoma. Insect Biochem. 9,* 335–339.

BAKER, J. E., SUKKESTAD, D. R., NELSON, D. R. and FATLAND, C. L. (1979b). Cuticular lipids of larvae and adults of the cigarette beetle, *Lasioderma serricorne. Insect Biochem. 9,* 603–611.

BAKER, J. E., SUKKESTAD, D. R., WOO, S. M. and NELSON, D. R. (1977). Cuticular hydrocarbons of *Tribolium castaneum*: Effects of the food additive tricalcium phosphate. *Insect Biochem. 8,* 159–167.

BARTELT, R. J., JONES, R. L. and KULMAN, H. M. (1982). Hydrocarbon components of the yellowheaded spruce sawfly sex pheromone: A series of (Z,Z)-9,19 dienes. *J. Chem. Ecol. 8,* 95–114.

BEAMENT, J. W. L. (1945). The cuticular lipids of insects. *J. Exp. Biol. 21,* 115–131.

BEAMENT, J. W. L. (1964). The active transport and passive movement of water in insects. *Adv. Insect Physiol. 2,* 67–129.

BEAMENT, J. W. L. (1967). Lipid layers and membrane models. In *Insects and Physiology*. Edited by J. W. L. Beament and J. E. Treherne. Pages 303–313. Oliver & Boyd, London.

BEAMENT, J. W. L. (1976). The ecology of cuticle. In *The Insect Integument*. Edited by H. R. Hepburn. Pages 359–374. Elsevier, Amsterdam.

BEATTY, I. M. and GILBY, A. R. (1969). The major hydrocarbon of a cockroach cuticular wax. *Naturwissenschaften 25,* 373.

BELL, R. A., NELSON, D. R., BORG, T. K. and CARDWELL, D. L. (1975). Wax secretion in non-diapausing and diapausing pupae of the tobacco hornworm, *Manduca sexta. J. Insect Physiol. 21,* 1725–1729.

BENNETT, G. A., KLEIMAN, R. and SHOTWELL, O. L. (1972). Hydrocarbons in haemolymph from healthy and diseased Japanese beetle larvae. *J. Insect Physiol. 18,* 1343–1350.

BERGSTROM, G. and LOFQVIST, J. (1973). Chemical congruence of the complex odoriferous secretions from Dufour's gland in three species of ants in the genus Formica. *J. Insect Physiol. 19,* 877–907.

BLAILOCK, T. T., BLOMQUIST, G. J. and JACKSON, L. L. (1976). Biosynthesis of 2-methylalkanes in the crickets *Nemobius fasciatus* and *Gryllus pennsylvanicus. Biochem. Biophys. Res. Commun. 68,* 841–849.

BLOMQUIST, G. J., BLAILOCK, T. T., SCHEETZ, R. W. and JACKSON, L. L. (1976). Cuticular lipids of insects — VII — Cuticular hydrocarbons of the crickets *Acheta domesticus, Gryllus pennsylvanicus* and *Nemobius fasciatus. Comp. Biochem. Physiol. 54B,* 381–386.

BLOMQUIST, G. J., CHU, A. J., NELSON, J. H. and POMONIS, J. G. (1980a). Incorporation of [2,3-^{13}C]succinate into methyl branched alkanes in a termite. *Arch. Biochem. Biophys. 204,* 648–650.

BLOMQUIST, G. J., CHU, A. J. and REMALEY, S. (1980b). Biosynthesis of wax in the honeybee, *Apis mellifera* L. *Insect Biochem. 10,* 313–321.

BLOMQUIST, G. J. and DE RENOBALES, M. (1983). Biosynthesis of insect cuticular hydrocarbons: Applications of carbon-13 NMR spectroscopy. In *Metabolic Aspects of Lipid Nutrition in Insects*. Edited by T. E. Mittler and R. Dadd. Pages 203–222. Westrien Press, Boulder, Colorado.

BLOMQUIST, G. J., HOWARD, R. W. and MCDANIEL, C. A. (1979a). Structures of the cuticular hydrocarbons of the termite *Zootermopsis angusticollis* (Hagen). *Insect Biochem. 9,* 365–370.

BLOMQUIST, G. J., HOWARD, R. W. and MCDANIEL, C. A. (1979b). Biosynthesis of the cuticular hydrocarbons of the termite *Zootermopsis angusticollis* (Hagen). Incorporation of propionate into dimethylalkanes. *Insect Biochem. 9,* 371–374.

BLOMQUIST, G. J., HOWARD, R. W., MCDANIEL, C. A., REMALEY, S., DWYER, L. A. and NELSON, D. R. (1980c). Application of methoxymercuration–demercuration followed by mass spectrometry as a convenient microanalytical technique for double-bond location in insect-derived alkenes. *J. Chem. Ecol. 6,* 257–269.

BLOMQUIST, G. J. and JACKSON, L. L. (1973a). Incorporation of labelled dietary *n*-alkanes into cuticular lipids of the grasshopper *Melanoplus sanguinipes. J. Insect Physiol. 19,* 1639–1647.

BLOMQUIST, G. J. and JACKSON, L. L. (1973b). Hydroxylation of *n*-alkanes to secondary alcohols and their esterification in the grasshopper *Melanoplus sanguinipes. Biochem. Biophys. Res. Commun. 53,* 703–708.

BLOMQUIST, G. J. and JACKSON, L. L. (1979). Chemistry and biochemistry of insect waxes. *Prog. Lipid Res. 17,*319–345.

BLOMQUIST, G. J. and KEARNEY, G. P. (1976). Biosynthesis of the internally branched monomethylalkanes in the cockroach, *Periplaneta fuliginosa. Arch. Biochem. Biophys. 173,* 546–553.

BLOMQUIST, G. J., MAJOR, M. A. and LOK, J. B. (1975a). Biosynthesis of 3-methylpentacosane in the cockroach *Periplaneta americana. Biochem. Biophys. Res. Commun. 64,* 43–50.

BLOMQUIST, G. J., MCCAIN, D. C. and JACKSON, L. L. (1975b). Incorporation of oxygen-18 into secondary alcohols of the grasshopper *Melanoplus sanguinipes. Lipids 10,* 303–306.

BLOMQUIST, G. J. and RIES, M. K. (1979). The enzymatic synthesis of wax esters by a microsomal preparation from the honeybee *Apis mellifera L. Insect Biochem. 9,* 183–188.

BLOMQUIST, G. J., SOLIDAY, C. L., BYERS, B. A., BRAKKE, J. W. and JACKSON, L. L. (1972). Cuticular lipids of insects: V. Cuticular wax esters of secondary alcohols from the grasshoppers *Melanoplus packardii* and *Melanoplus sanguinipes. Lipids 7,* 356–362 (1972).

BOLTON, H. T., BUTLER, J. F. and CARLSON, D. A. (1980). A mating stimulant pheromone of the horn fly, *Haematobia irritans* (L.): Demonstration of biological activity in separated cuticular components. *J. Chem. Ecol. 6,* 951–964.

BOWERS, W. S. and THOMPSON, M. J. (1965). Identification of the major constituents of the crystalline powder covering the larval cuticle of *Samia cynthia ricini* (Jones). *J. Insect Physiol. 11,* 1003–1011.

BRIEGER, G. and BUTTERWORTH, F. M. (1970). *Drosophila melanogaster*: Identity of male lipid in reproductive system. *Science, Wash. 167*, 1212.

BROOKS, G. T. (1976). Penetration and distribution of insecticides. In *Insecticide Biochemistry and Physiology*. Edited by C. F. Wilkinson. Pages 3–58. Plenum Press, New York and London.

BROPHY, J. J., CAVILL, G. W. K. and SHANNON, J. S. (1973). Venom and Dufour's gland secretions in an Australian species of *Camponotus*. *J. Insect Physiol. 19*, 1791–1798.

BURSELL, E. and CLEMENTS, A. N. (1967). The cuticular lipids of the larva of *Tenebrio molitor* L. (Coleoptera). *J. Insect Physiol. 13*, 1671–1678.

CARLSON, D. A., LANGLEY, P. A. and HUYTON, P. (1978). Sex pheromone of the tsetse fly: isolation, identification and synthesis of contact aphrodisiacs. *Science 201*, 750–753.

CARLSON, D. A., MAYER, M. S., SILHACEK, D. L., JAMES, J. D., BEROZA, M. and BIERL, B. A. (1971). Sex attractant pheromone of the housefly: isolation, identification and synthesis. *Science 174*, 76–77.

CAVILL, G. W. K., CLARK, D. V., HOWDEN, M. E. H. and WYLLIE, S. G. (1970). Hydrocarbon and other lipid constituents of the bull ant, *Myrmecia gulosa*. *J. Insect Physiol. 16*, 1721–1728.

CAVILL, G. W. K. and HOUGHTON, E. (1973). Hydrocarbon constituents of the Argentine ant, *Iridomyrmex humilis*. *Aust. J. Chem. 26*, 1131–1135.

CHAUDHURY, M. F. B. and BALL, H. J. (1973). The effect of age, nutritional factors, and gonadal development on the mating behavior of the face fly, *Musca autumnalis*. *J. Insect Physiol. 19*, 57–64.

CHAUDHURY, M. F. B. and BALL, H. J. (1974). Effect of age and time of day on sex attraction and mating of the face fly, *Musca autumnalis*. *J. Insect Physiol. 20*, 2079–2085.

CHAUDHURY, M. F. B., BALL, H. J. and JONES, C. M. (1972). A sex pheromone of the female face fly, *Musca autumnalis*, and its role in sexual behavior. *Ann. Ent. Soc. Amer. 65*, 607–612.

CHINO, H., DOWNER, R. G. H., WYATT, G. R. and GILBERT, L. I. (1981a). Lipophorins, a major class of lipoproteins of insect haemolymph. *Insect Biochem. 11*, 491.

CHINO, H., KATASE, H., DOWNER, R. G. H. and TAKAHASHI, K. (1981b). Diacylglycerol-carrying lipoprotein of hemoplymph of the American cockroach: Purification, characterization and function. *J. Lipid Res. 22*, 7–15.

CHINO, H. and KITAZAWA, K. (1981). Diacylglycerol-carrying lipoprotein of hemolymph of the locust and some insects. *J. Lipid Res. 22*, 1042–1052.

CHU, A. J. and BLOMQUIST, G. J. (1980a). Biosynthesis of hydrocarbons in insects: Succinate is a precursor of the methyl branched alkanes. *Archiv. Biochem. Biophys. 201*, 304–312.

CHU, A. J. and BLOMQUIST, G. J. (1980b). Decarboxylation of tetracosanoic acid to *n*-tricosane in the termite *Zootermopsis angusticollis*. *Comp. Biochem. Physiol. 66B*, 313–317.

CONNER, W. E., EISNER, T., VANDER MEER, R. K., GUERRERO, A., GHIRINGELLI, D. and MEINWALD, J. (1980). Sex attractant of an arctiid moth (*Utetheisa ornatrix*): A pulsed chemical signal. *Behav. Ecol. Sociobiol. 7*, 55–63.

CONRAD, C. W. and JACKSON, L. L. (1971). Hydrocarbon biosynthesis in *Periplaneta americana*. *J. Insect Physiol. 17*, 1907–1916.

COUDRON, T. A. and NELSON, D. R. (1978). Hydrocarbons in the surface lipids of pupal tobacco budworms, *Heliothis virescens*. *Insect Biochem. 8*, 59–66.

COUDRON, T. A. and NELSON, D. R. (1981). Characterization and distribution of the hydrocarbons found in diapausing pupae tissues of the tobacco hornworm, *Manduca sexta* (L.). *J. Lipid Res. 22*, 103–112.

DEAN, J. W., CLEMENTS, S. A. and PAGET, J. (1969). Observations on sex attraction and mating behaviour of the tsetse fly, *Glossina morsitans orientalis* Vanderplank. *Bull. Ent. Res. 59*, 355–365.

DIEHL, P. A. (1973). Paraffin synthesis in the oenocytes of the desert locust. *Nature, Lond. 243*, 468–470.

DIEHL, P. A. (1975). Synthesis and release of hydrocarbons by the oenocytes of the desert locust, *Schistocerca gregaria*. *J. Insect Physiol. 21*, 1237–1246.

DILLWITH, J. W., BLOMQUIST, G. J. and NELSON, D. R. (1981). Biosynthesis of the hydrocarbon components of the sex pheromone of the housefly, *Musca domestica*. *Insect Biochem. 11*, 247–254.

DILLWITH, J. W. and BLOMQUIST, G. J. (1982). Site of sex pheromone biosynthesis in the female housefly, *Musca domestica* L. *Experientia 38*, 471–473.

DWYER, L. A. and BLOMQUIST, G. J. (1981). Biosynthesis of linoleic acid in the American cockroach. *Prog. Lipid Res. 20*, 215–218.

DWYER, L. A., BLOMQUIST, G. J., NELSON, J. H. and POMONIS, J. G. (1981a). A ^{13}C-NMR study of the biosynthesis of 3-methylpentacosane in the American cockroach. *Biochim. Biophys. Acta 663*, 536–544.

DWYER, L. A., DE RENOBALES, M. and BLOMQUIST, G. J. (1981b). Biosynthesis of (Z,Z)-6,9-heptacosadiene in the American cockroach. *Lipids 16*, 810–814.

EBELING, W. (1974). Permeability of insect cuticle. In *The Physiology of Insecta*. Vol. VI. Edited by M. Rockstein. 2nd edn. Pages 271–343. Academic Press, New York and London.

EDNEY, E. B. (1977). *Water Balance in Land Arthropods*. (*Zoophysiology and Ecology*, vol. 9). Springer-Verlag, Berlin, Heidelberg and New York.

EMMENS, R. L. (1981). Evidence for an attractant in cuticular lipids of female *Lucilia cuprina* (Weid.), Australian sheep blowfly. *J. Chem. Ecol. 7*, 529–541.

FAUROT-BOUCHET, E. and MICHEL, G. (1964). Composition of insect waxes. I. Waxes of exotic coccidae: *Gascardia madagascariensis*, *Coccus ceriferus* and *Tachardia Lacca*. *J. Amer. Oil Chem. Soc. 41*, 418–421.

FAUROT-BOUCHET, E. and MICHEL, G. (1965). Composition des cires d'insects. II. Cires des cochenilles *Ceroplates rusci*, *Icerya purchasi*, *Pulvinaria flacifera* et *Quadraspidiotus perniciosus*. *Bull. Soc. Chim. Biol. 47*, 93–97.

FEHLER, S. W. G. and LIGHT, R. J. (1970). Biosynthesis of hydrocarbons in *Anabaena variabilis*. Incorporation of [methyl-^{14}C]- and [methyl-$^{2}H_3$]methionine into 7- and 8-methylheptadecanes. *Biochemistry 9*, 418–422.

GARY, N. E. (1975). Activities and behavior of honeybees. In *The Hive and The Honeybee*. Edited by Dadant and Son. Pages 185–264. Dadant & Sons, Hamilton, Illinois.

GEROLT, P. (1969). Mode of entry of contact insecticides. *J. Insect Physiol. 15*, 563–580.

GILBY, A. R. (1980a). Chemical methods (Lipids). In *Cuticle Techniques in Arthorpods*. Edited by T. A. Miller, Pages 217–252. Springer-Verlag, New York.

GILBY, A. R. (1980b). Transpiration, temperature and lipids in insect cuticle. *Adv. Insect Physiol. 15*, 1–33.

GOODRICH, B. S. (1970). Cuticular lipids of adults and puparia of the Australian Sheep Blowfly, *Lucilia cuprina* (Weid.). *J. Lipid Res. 11*, 1–6.

GRULA, J. W., MCCHESNEY, J. D. and TAYLOR, O. R., JR. (1980). Aphrodisiac pheromones of the sulfur butterflies *Colias eurytheme* and *C. philodice* (Lepidoptera, Pieridae). *J. Chem. Ecol. 6*, 241–256.

GRULA, J. W. and TAYLOR, O. R., JR. (1979). The inheritance of pheromone production in the sulfur butterflies *Colias eurytheme* and *C. philodice*. *Heredity 42*, 359–371.

HACKMAN, R. H. (1951). The chemical composition of the wax of the white wax scale, *Ceroplastes destructor* (Newstead). *Archiv. Biochem. Biophys. 33*, 150–154.

HADLEY, N. F. (1977). Epicuticular lipids of the desert tenebrionid beetle, *Eleodes armata*: seasonal and acclimatory effects on composition. *Insect Biochem. 7*, 277–283.

HADLEY, N. F. (1978). Cuticular permeability of desert tenebrionid beetles: correlations with epicuticular hydrocarbon composition. *Insect Biochem. 8*, 17–22.

HADLEY, N. F. (1980a). Surface waxes and integumentary permeability. *Amer. Sci. 68*, 546–553.

HADLEY, N. F. (1980b). Cuticular lipids of adults and nymphal exuviae of the desert cicada, *Diceroprocta apache* (Homoptera, Cicadidae). *Comp. Biochem. Physiol. 65B*, 549–553.

HADLEY, N. F. (1981). Cuticular lipids of terrestrial plants and arthropods: A comparison of their structure, composition and waterproofing function. *Biol. Rev. 56*, 23–47.

HADLEY, N. F., BLOMQUIST, G. J. and LANHAM, U. N. (1981). Cuticular hydrocarbons of four species of Colorado *hymenoptera*. *Insect Biochem. 11*, 173–177.

HAN, J., CHAN, W. H. S. and CALVIN, M. J. (1969). Biosynthesis of alkanes in *Nostoc muscorum*. *J. Amer. Chem. Soc. 91*, 5156–5159.

HARRIS, R. L., OEHLER, D. D. and BERRY, I. L. (1976). Sex pheromone of the stable fly: Affect on cuticular hydrocarbons of age, sex, species and mating. *Environ. Ent. 5*, 973–977.

HEDIN, P. A., THOMPSON, A. C., GUELDNER, R. C. and MINYARD, J. P. (1972). Volatile constituents of the boll weevil. *J. Insect Physiol. 18*, 79–86.

HENDRY, L. B., WICHMANN, J. K., HINDENLANG, D. M., WEAVER, K. M. and KORZENIOWSKI, S. H. (1976). Plants — the origin of kairomones utilized by parasitoids of phytophagus insects? *J. Chem. Ecol. 2*, 271–283.

HOWARD, R. W. and BLOMQUIST, G. J. (1982). Chemical ecology and biochemistry of insect hydrocarbons. *Ann. Rev. Ent. 27*, 149–172.

HOWARD, R. W., MCDANIEL, C. A. and BLOMQUIST, G. J. (1978). Cuticular hydrocarbons of the Eastern subterranean termite, *Reticulitermes flavipes* (Kollar) (Isoptera: Rhinotermitidae). *J. Chem. Ecol. 4*, 233–245.

HOWARD, R. W., MCDANIEL, C. A. and BLOMQUIST, G. J. (1980a). Chemical mimicry as an integrating mechanism: Cuticular hydrocarbons of a termitophile and its host. *Science 210*, 431–433.

HOWARD, R. W., MCDANIEL, C. A., NELSON, D. R. and BLOMQUIST, G. J. (1980b). Chemical ionization mass spectrometry application to insect-derived cuticular alkanes. *J. Chem. Ecol. 6*, 609–623.

HOWARD, R. W., MCDANIEL, C. A., NELSON, D. R., BLOMQUIST, G. J., GELBAUM, L. T. and ZALKOW, L. H. (1982). Cuticular hydrocarbons as possible species- and caste-recognition cues in *Reticulitermes* sp. *J. Chem. Ecol. 8*, 1227–1239.

HUTCHINS, R. F. N. and MARTIN, M. M. (1968). The lipids of the common house cricket, *Acheta domesticus* L. II. Hydrocarbons. *Lipids 3*, 250–255.

HUYTON, P. M., LANGLEY, P. A., CARLSON, D. A. and SCHWARTZ, M. (1980). Specificity of contact sex pheromones in tsetse flies *Glossina sp. Physiol. Ent. 5*, 253–264.

HUYTON, P. M., LANGLEY, P. A., CARLSON, D. A. and COATES, T. W. (1980). The role of sex pheromones in initiation of copulatory behavior by male tsetse flies, *Glossina morsitans morsitans*. *Physiol. Ent. 5*, 243–252.

ISHAY, J. (1972a). Thermoregulatory pheromones in wasps. *Experientia 28*, 1185–1187.

ISHAY, J. (1972b). Thermoregulatory pheromones in wasps. *Israel J. Med. Sci. 8*, 1773.

JACKSON, L. L. (1970). Cuticular lipids of insects: II. Hydrocarbons of the cockroaches *Periplaneta australasiae*, *Periplaneta brunnea* and *Periplaneta fuliginosa*. *Lipids 5*, 38–41.

JACKSON, L. L. (1972). Cuticular lipids of insects IV. Hydrocarbons of the cockroaches *Periplaneta japonica* and *Periplaneta americana* compared to other cockroach hydrocarbons. *Comp. Biochem. Physiol. 41B*, 331–336.

JACKSON, L. L. (1981). Cuticular lipids of insects. IX. Surface lipids of the grasshoppers *Melanoplus bivittatus*, *Melanoplus femurrubrum* and *Melanoplus dawsoni*. *Comp. Biochem. Physiol. 70B*, 441–445.

JACKSON, L. L. (1982). Cuticular lipids of insects-X. Normal and branched alkanes from the surface of the grasshopper *Schistocerca americana*. *Comp. Biochem. Physiol. B. 71B*, 739–742.

JACKSON, L. L. and ARMOLD, M. T. (1977). Insect lipid analysis. In *Analytical Biochemistry of Insects*. Edited by R. B. Turner. Pages 171–206. Elsevier, Amsterdam.

JACKSON, L. L., ARMOLD, M. T. and BLOMQUIST, G. J. (1981). Surface lipids of *Drosophila melanogaster*: Comparison of the lipids from female and male wild type and sex-linked yellow mutant. *Insect Biochem. 11*, 87–91.

JACKSON, L. L., ARMOLD, M. T. and REGNIER, F. E. (1974). Cuticular lipids of adult fleshflies, *Sarcophaga bullata*. *Insect Biochem. 4*, 369–379.

JACKSON, L. L. and BAKER, G. L. (1970). Cuticular lipids of insects. *Lipids 5*, 239–246.

JACKSON, L. L. and BLOMQUIST, G. J. (1976a). Insect waxes. In *Chemistry and Biochemistry of Natural Waxes*. Edited by P. E. Kolattukudy. Pages 201–233. Elsevier, Amsterdam.

JACKSON, L. L. and BLOMQUIST, G. J. (1976b). Cuticular lipids of insects: VIII. Alkanes of the mormon cricket *Anabrus simplex*. *Lipids 11*, 77–79.

JACKSON, L. L., HADLEY, N. F. and BLOMQUIST, G. J. (1980). Epicuticular lipids of the desert tenebrionid beetle, *Eleodes armata*: identification of the branched hydrocarbons. *Insect Biochem. 10*, 399–402.

JACOB, J. (1977). The cuticular lipids of the beetle, *Phyllobius maculicornis*. *Hoppe-Seyler's Z. Physiol. Chem. 358*, 1375–1377.

JACOB, J. (1978). Sex-dependent composition of cuticular lipids from the beetle *Rhagonycha fulva*. *Hoppe-Seyler's Z. Physiol. Chem. 359*, 653–656.

JONES, R. L., LEWIS, W. J., BEROZA, M., BIERL, B. A. and SPARKS, A. N. (1973). Host seeking stimulants (kairomones) for the egg parasite, *Trichogramma evanescens*. *Environ. Ent. 2*, 593–596.

JONES, R. L., LEWIS, W. J., BOWMAN, M. C., BEROZA, M. and BIERL, B. A. (1971). Host seeking stimulant for parasite of corn earworm: Isolation, identification and synthesis. *Science 173*, 842–843.

KOIDSUMI, K. (1957). Antifungal action of cuticular lipids in insects. *J. Insect Physiol. 1*, 40–51.

KOLATTUKUDY, P. E. (1980). Cutin, suberin and waxes. In *The Biochemistry of Plants: A Comprehensive Treatise*. Vol. 4. Edited by P. K. Stumpf and E. V. Conn. Pages 571–645. Academic Press, New York.

KOSTELC, J. G., GIRARD, J. E. and HENDRY, L. B. (1980). Isolation and identification of a sex attractant of a mushroom-infesting sciarid fly. *J. Chem. Ecol. 6*, 1–11.

KOSTELC, J. G., HENDRY, L. B. and SNETSINGER, R. (1975). A sex pheromone complex of the mushroom-infesting sciarid fly, *Lycoriella mali* Fitch. *J. New York Ent. Soc. 83*, 255–256.

LAMB, N. J. and MONROE, R. E. (1968). Lipids synthesis from acetate-^{14}C by the cereal leaf beetle, *Oulema malanopus*. *Ann. Ent. Soc. Amer. 61*, 1164–1166.

LAMBREMONT, E. N. (1972). Lipid metabolism of insects: Interconversion of fatty acids and fatty alcohols. *Insect Biochem. 2*, 197–202.

LAMBREMONT, E. N., BAUMGARNER, J. W. and BENNETT, A. F. (1966). Lipid biosynthesis in the boll weevil (*Anthonomus grandis* Boheman) (Coleoptera: Curculionidae): Distribution of radioactivity in the principal lipid classes synthesized from C^{14}-1-acetate. *Comp. Biochem. Physiol. 19*, 417–429.

LAMBREMONT, E. N. and WYKLE, R. L. (1979). Wax synthesis by an enzyme system from the honey bee. *Comp. Biochem. Physiol. 63B*, 131–135.

LANGLEY, P. A., PIMLEY, R. W. and CARLSON, D. A. (1975). Sex recognition pheromone in tsetse fly *Glossina morsitans*. *Nature 254*, 51–53.

LEWIS, W. J., BEEVERS, M., NORDLUND, D. A., GROSS, H. R. JR. and HAGEN, K. S. (1979). Kairomones and their use for management of entomophagous insects. IX. Investigations of various kairomone-treatment patterns for *Trichogramma* spp. *J. Chem. Ecol. 5*, 673–680.

LEWIS, W. J., JONES, R. L., GROSS, H. R., JR. and NORDLUND, D. A. (1976). The role of kairomones and other behavioral chemicals in host finding by parasitic insects. *Behav. Biol. 16*, 267–289.

LEWIS, W. J., JONES, R. L., NORDLUND, D. A. and GROSS, H. R., JR. (1975a). Kairomones and their use for management of entomophagous insects: II. Mechanisms causing increase in rate of parasitization by *Trichogramma* spp. *J. Chem. Ecol. 1*, 349–360.

LEWIS, W. J., JONES, R. L., NORDLUND, D. A. and SPARKS, A. N. (1975b). Kairomones and their use for management of entomophagus insects: I. Evaluation for increasing rates of parasitization by *Trichogramma* spp. in the field. *J. Chem. Ecol. 1*, 343–347.

LEWIS, W. J., NORDLUND, D. A., GROSS, H. R., JR., JONES, R. L. and JONES, S. L. (1977). Kairomones and their use for management of entomophagous insects. V. Moth scales as a stimulus for predation of *Heliothis zea* (Boddie) eggs by *Chrysopa carnea* Stephens larvae. *J. Chem. Ecol. 3*, 483–487.

LOCKE, M. (1965). Permeability of insect cuticle to water and lipids. *Science 147*, 295–298.

LOCKEY, K. H. (1976). Cuticular hydrocarbons of *Locusta*, *Schistocerca* and *Periplaneta*, and their role in waterproofing. *Insect Biochem. 6*, 457–472.

LOCKEY, K. H. (1978a). Hydrocarbons of adult *Tribolium castaneum* Hbst. and *Tribolium confusum* Duv. (Coleoptera: Tenebrionidae). *Comp. Biochem. Physiol. 61B*, 401–407.

LOCKEY, K. H. (1978b). The adult cuticular hydrocarbons of *Tenebrio molitor* L. and *Tenebrio obscurus* F. (Coleoptera: Tenebrionidae). *Insect Biochem. 8*, 237–250.

LOCKEY, K. H. (1979). Cuticular hydrocarbons of adult *Alphitophagus bifasciatus* (Say.) and *Alphitobius diaperinus* (Panz.) (Coleoptera: Tenebrionidae). *Comp. Biochem. Physiol. 64B,* 47–56.

LOCKEY, K. H. (1980a). Insect cuticular hydrocarbons. *Comp. Biochem. Physiol. 65B,* 457–462.

LOCKEY, K. H. (1980b). Cuticular hydrocarbons of adult *Blaps mucronata* Latreille (Coleoptera: tenebrionidae). *Comp. Biochem. Physiol. 67B,* 33–40.

LOCKEY, K. H. (1981). Cuticular hydrocarbons of adult *Cylindrinotus laevioctostriatus* (Goeze) and *Phylan gibbus* (Fabricius) (Coleoptera: Tenebrionidae). *Insect Biochem. 11,* 549–561.

LOK, J. B., CUPP, E. W. and BLOMQUIST, G. J. (1975). Cuticular lipids of the imported fire ants, *Solenopsis invicta* and *richteri. Insect Biochem. 5,* 821–829.

LOULOUDES, S. J., CHAMBERS, D. L., MOYER, D. B. and STARKEY, J. H. (1962). The hydrocarbons of adult houseflies. *Ann. Ent. Soc. Amer. 55,* 442–448.

LOULOUDES, S. J., KAPLANIS, J. N., ROBBINS, W. E. and MONROE, R. E. (1961). Lipogenesis from C^{14}-acetate by the American cockroach. *Ann. Ent. Soc. Amer. 54,* 99–103.

MACHIN, J. (1980). Cuticle water relations: Towards a new cuticle waterproofing model. In *Insect Biology in the Future.* Edited by M. Locke and D. S. Smith. Pages 79–103. Academic Press, New York.

MACKLEY, J. W., CARLSON, D. A. and BUTLER, J. F. (1981). Identification of the cuticular hydrocarbons of the hornfly and assays for attraction. *J. Chem. Ecol. 7,* 669–683.

MAJOR, M. A. and BLOMQUIST, G. J. (1978). Biosynthesis of hydrocarbon in insects: Decarboxylation of long chain acids to *n*-alkanes in *Periplaneta. Lipids 13,* 323–328.

MARTIN, M. M. and MACCONNELL, J. G. (1970). The alkanes of the ant, *Atta colombica. Tetrahedron 26,* 307–319.

MEOLA, R. W., HARRIS, R. L., MEOLA, S. M. and OEHLER, D. D. (1977). Dietary-induced secretion of sex pheromone and development of sexual behavior in the stable fly. *Environ. Ent. 6,* 895–897.

MODY, N. V., HEDIN, P. A., NEEL, W. W. and MILES, D. H. (1975). Hydrocarbons from males, females and larvae of pecan weevil. *Curculio caryae. Lipids 10,* 117–119.

MOORE, B. P. (1969). Biochemical studies in termites. In *Biology of Termites.* Edited by K. Krishna and F. M. Weesner. Vol. 1, pages.407–432. Academic Press, New York.

MUHAMMED, S., BUTLER, J. F. and CARLSON, D. A. (1975). Stable fly sex attractant and mating pheromones found in female body hydrocarbons. *J. Chem. Ecol. 1,* 387–398.

NELSON, D. R. (1969). Hydrocarbon synthesis in the American cockroach. *Nature Lond. 221,* 854–855.

NELSON, D. R. (1978). Long-chain methyl-branched hydrocarbons: Occurrence, biosynthesis and function. *Adv. Insect Physiol. 13,* 1–33.

NELSON, D. R., DILLWITH, J. W. and BLOMQUIST, G. J. (1981). Cuticular hydrocarbons of the housefly, *Musca domestica. Insect Biochem. 11,* 187–197.

NELSON, D. R., FATLAND, C. L. and CARDWELL, D. L. (1977). Long-chain methylalkanes from haemolymph of larvae of Japanese beetles, *Popillia japonica. Insect Biochem. 7,* 439–446.

NELSON, D. R., FATLAND, C. L., HOWARD, R. W., MCDANIEL, C. A. and BLOMQUIST, G. J. (1980). Re-analysis of the cuticular methylalkanes of *Solenopsis invicta* and *S. richteri. Insect Biochem. 10,* 409–418.

NELSON, D. R. and SUKKESTAD, D. R. (1970). Normal and branched aliphatic hydrocarbons from the eggs of the tobacco hornworm. *Biochemistry 9,* 4601–4611.

NELSON, D. R. and SUKKESTAD, D. R. (1975). Normal and branched alkanes from cast skins of the grasshopper *Schistocerca vaga* (Scudder). *J. Lipid Res. 16,* 12–18.

NELSON, D. R., SUKKESTAD, D. R. and TERRANOVA, A. C. (1971). Hydrocarbon composition of the integument, fat body, hemolymph, and diet of the tobacco hornworm. *Life Sci. 10,* 411–419.

NELSON, D. R., SUKKESTAD, D. R. and ZAYLSKIE, R. G. (1972). Mass spectra of methyl-branched hydrocarbons from eggs of the tobacco hornworm. *J. Lipid Res. 13,* 413–421.

NORDLUND, D. A., LEWIS, W. J., JONES, R. L. and GROSS, H. R., JR. (1976). Kairomones and their use for management of entomophagous insects. IV. Effect of kairomones on productivity and longevity of *Trichogramma pretiosum* Riley (Hymenoptera: Trichogrammatidae). *J. Chem. Ecol. 2,* 67–72.

NORDLUND, D. A., LEWIS, W. J., JONES, R. L., GROSS, H. R., JR. and HAGEN, K. S. (1977a). Kairomones and their use for management of entomophagous insects. VI. An examination of the kairomones for the predator *Chrysopa carnea* Stephens at the oviposition sites of *Heliothis zea* (Boddie). *J. Chem. Ecol. 3,* 507–511.

NORDLUND, D. A., LEWIS, W. J., TODD, J. W. and CHALFANT, R. B. (1977b). Kairomones and their use for management of entomophagous insect. VII. The involvement of various stimuli in the differential response of *Trichogramma pretiosum* Riley to two suitable hosts. *J. Chem. Ecol. 3,* 513–518.

OLSON, W. P. (1970). Penetration of ^{14}C-DDT into and through the cockroach integument. *Comp. Biochem. Physiol. 35,* 273–282.

PIEK, T. (1964). Synthesis of wax in the honeybee (*Apis mellifera* L.). *J. Insect Physiol. 10,* 563–572.

POLLOCK, J. N. (1970). Sperm transfer by spermatophores in *Glossina austeni* Newstead. *Nature 225,* 1063–1064.

POMONIS, J. G., FATLAND, C. F., NELSON, D. R. and ZAYLSKIE, R. G. (1978). Insect hydrocarbons. Corroboration of structure by synthesis and mass spectrometry of mono- and dimethylalkanes. *J. Chem. Ecol. 4,* 27–39.

REGNIER, F. E., NIEH, M. and HÖLLDOBLER, B. (1973). The volatile Dufour's gland components of the harvester ants *Pogonomyrmex rugosus* and *P. barbatus. J. Insect Physiol. 19,* 981–992.

RICHTER, I. and KRAIN, H. (1980). Cuticular lipid constituents of cabbage seedpod weevils and host plant oviposition sites as potential pheromones. *Lipids 15,* 580–586.

RICHTER, I., KRAIN, H. and MANGOLD, H. K. (1976). Long-chain (Z)-9-alkenes are "psychedelics" to houseflies with regard to visually stimulated sex attraction and aggregation. *Experientia 32,* 186–188.

ROBBINS, W. E., KAPLANIS, J. N., LOULOUDES, S. J. and MONROE, R. E. (1960). Utilization of 1-C^{14}-acetate in lipid synthesis by adult houseflies. *Ann. Ent. Soc. Amer. 53,* 128–129.

ROELOFFS, W. L. and CARDÉ, R. T. (1971). Hydrocarbon sex pheromone in tiger moths (Arctiidae). *Science 171,* 684–686.

ROGOFF, W. M., BELTZ, A. D., JOHNSON, J. D. and PLAPP, F. W. (1964). A sex pheromone in the housefly *Musca domestica* L. *J. Insect Physiol. 10,* 239–246.

ROGOFF, W. M., GRETZ, G. H., SONNET, P. F. and SCHWARTZ, M. (1980). Responses of male house flies to muscalure and to combinations of hydrocarbons with and without muscalure. *Environ. Ent. 9,* 605–606.

ROMER, F. (1980). Histochemical and biochemical investigations concerning the function of larval oenocytes of *Tenebrio molitor* L. (Coleoptera, Insecta). *Histochemistry 69,* 69–84.

SILBERGLIED, R. E. and TAYLOR, O. R. (1978). Ultraviolet reflection and its behavioral role in the courtship of the sulfur butterflies *Colias eurytheme* and *C. philodice* (Lepidoptera, Pieridae). *Behav. Ecol. Sociobiol. 3,* 203–242.

SOLIDAY, C. L., BLOMQUIST, G. J. and JACKSON, L. L. (1974). Cuticular lipids of insects. VI. Cuticular lipids of the grasshoppers, *Melanoplus sanguinipes* and *Melanoplus packardii. J. Lipid Res. 15,* 399–405.

SONNET, P. E., UEBEL, E. C., HARRIS, R. L. and MILLER, R. W. (1977). Sex pheromone of the stable fly: Evaluation of methyl- and 1,5-dimethylalkanes as mating stimulants. *J. Chem. Ecol. 3,* 245–249.

SONNET, P. E., UEBEL, E. C., LUSBY, W. R., SCHWARTZ, M. and MILLER, R. W. (1979). Sex pheromone of the stable fly: identification, synthesis and evaluation of alkenes from female stable flies. *J. Chem. Ecol. 5,* 353–361.

SONNET, P. E., UEBEL, E. C. and MILLER, R. W. (1975). Sex pheromones of the face fly and compounds influencing pheromone activity. *Environ. Ent. 4,* 761–764.

STRANSKY, K., STREIBL, M. and KUBELKA, V. (1971). On natural waxes. XIX. Complex esters of the wax of the honey bee (*Apis mellifera* L.). *Coll. Czech. Chem. Commun.* (English edn.) *36,* 2267–2280.

STRANSKY, K., UBIK, K., HOLMAN, J. and STREIBL, M. (1973). Chemical composition of compounds produced by the pea aphid *Acyrthosiphon pisum* (Harris): Pentane extract of surface lipids. *Coll. Czech. Chem. Commun. 38,* 770–780.

SWYNNERTON, C. F. M. (1936). The tsetse flies of East Africa. A first study of their ecology, with a view to their control. *Trans. R. Ent. Soc. London 84,* 1–579.

SZEICZ, F. M., PLAPP, F. W., JR. and VINSON, S. B. (1973). Tobacco budworm: Penetration of several insecticides into the larva. *J. Econ. Ent. 66,* 9–15.

Tanaka, K., Ohsawa, K., Honda, H. and Yamamoto, I. (1981). Copulation release pheromone, erectin, from the azuki bean weevil (*Callosobruchus chinensis* L.). *J. Pest. Sci. 6*, 75–82.

Tartivita, K. and Jackson, L. L. (1970). Cuticular lipids of insects: I. Hydrocarbons of *Leucophaea maderae* and *Blatta orientalis*. *Lipids 5*, 35–37.

Toolson, E. C. (1978). Diffusion of water through the arthropod cuticle: Thermodynamic consideration of the transition phenomenon. *J. Thermol. Biol. 3*, 69–73.

Toolson, E. C., White, T. R. and Glaunsinger, W. S. (1979). Electron paramagnetic resonance spectroscopy of spin-labeled cuticule of *Centruroides sculpturatus* (Scorpiones: Buthidae). Correlation with thermal effects on cuticular permeability. *J. Insect Physiol. 25*, 271–275.

Tulloch, A. P. (1970). The composition of beeswax and other waxes secreted by insects. *Lipids 5*, 1–12.

Tulloch, A. P. (1971). Beeswax: Structure of the esters and their component hydroxyacids and diols. *Chem. Phys. Lipids. 6*, 235–265.

Turner, D. A. (1971). Olfactory perception of live hosts and carbon dioxide by the tsetse fly *Glossina morsitans orientalis* Vanderplank. *Bull. Ent. Res. 61*, 75–96.

Uebel, E. C., Menzer, R. E., Sonnet, P. E. and Miller, R. W. (1975a). Identification of the copulatory sex pheromone of the little housefly, *Fannia canicularis* (L.) (Diptera; Muscidae). *J. New York Ent. Soc. 83*, 258–259.

Uebel, E. C., Schwartz, M., Lusby, W. R., Miller, R. W. and Sonnet, P. E. (1978a). Cuticular non-hydrocarbons of the female housefly and their evaluation as mating stimulants. *Lloydia 41*, 63–67.

Uebel, E. C., Schwartz, M., Miller, R. W. and Menzer, R. E. (1978b). Mating stimulant pheromone and cuticular lipid constituents of *Fannia femoralis* (Stein) (Diptera: Muscidae). *J. Chem. Ecol. 4*, 83–93.

Uebel, E. C., Schwartz, M., Sonnet, P. E., Miller, R. W. and Menzer, R. E. (1978c). Evaluations of the mating stimulant pheromones of *Fannia canicularis, F. pusio* and *F. femoralis* as attractants. *Florida Ent. 61*, 139–143.

Uebel, E. C., Sonnet, P. E., Bierl, B. A. and Miller, R. W. (1975b). Sex pheromone of the stable fly: Isolation and preliminary identification of compounds that induce mating strike behavior. *J. Chem. Ecol. 1*, 377–385.

Uebel, E. C., Sonnet, P. E., Menzer, R. E., Miller, R. W. and Lusby, W. R. (1977). Mating-stimulant pheromone and cuticular lipid constituents of the little housefly, *Fannia canicularis* (L.). *J. Chem. Ecol. 3*, 269–278.

Uebel, E. C., Sonnet, P. E. and Miller, R. W. (1976). Housefly sex pheromone: Enhancement of mating strike activity by combination of (Z)-9-tricosene with branched saturated hydrocarbons. *J. Econ. Ent. 5*, 905–908.

Uebel, E. C., Sonnet, P. E., Miller, R. W. and Beroza, M. (1975c). Sex pheromone of the face fly, *Musca autumnalis* De Geer (Diptera: Muscidae). *J. Chem. Ecol. 1*, 195–202.

Veith, H. J. and Koeniger, N. (1978). Identifizierung von cis-9-pentasocen als auslöser für das wärmen der brut bei der hornisse. *Naturwissenschaften 65*, 263.

Venard, R. and Jallon, J.-M. (1980). Evidence for an aphrodisiac pheromone of female *Drosophila*. *Experientia 36*, 211–212.

Vinson, S. B., Jones, R. L., Sonnet, P. E., Bierl, B. A. and Beroza, M. (1975). Isolation, identification and synthesis of host-seeking stimulants for *Cardiochiles nigriceps*, a parasitoid of tobacco budworm. *Entomol. Exp. Appl. 18*, 443–450.

Vroman, H. E., Kaplanis, J. N. and Robbins, W. E. (1965). Effect of allatectomy on lipid biosynthesis and turnover in the female cockroach, *Periplaneta americana*. *J. Insect Physiol. 11*, 897–904.

Warthen, J. D., Jr. and Uebel, E. C. (1980a). Major unsaturated cuticular hydrocarbons of the field crickets, *Gryllus pennsylvanicus* and *Nemobius fasciatus*. *Lipids 15*, 601–603.

Warthen, J. D., Jr. and Uebel, E. C. (1980b). Differences in the amounts of two major cuticular esters of *Melanoplus differentialis* males, females and nymphs. *Acrida 9*, 101–106.

Warthen, J. D., Jr. and Uebel, E. C. (1980c). Comparison of the unsaturated cuticular hydrocarbons of male and female house crickets, *Acheta domesticus* (L.) (Orthoptera: Gryllidae). *Insect Biochem. 10*, 435–439.

Warthen, J. D., Jr., Uebel, E. C., Lusby, W. R. and Adler, V. E. (1981). The cuticular lipids of the walking stick, *Diapheromera femorata* (Say). *Insect Biochem. 11*, 467–472.

Wharton, G. W. and Richards, A. G. (1978). Water vapor exchange kinetics in insects and acarines. *Ann. Rev. Ent. 23*, 309–328.

Wigglesworth, V. B. (1945). Transpiration through the cuticle of insects. *J. Exp. Biol. 21*, 97–114.

Wilson, E. O. (1971). *The Insect Societies*. Harvard University Press, Cambridge, Mass.

Young, R. G. (1963). The biosynthesis of beeswax. *Life Sci. 2*, 676–679.

5 Structure of the Fat Body

R. L. DEAN, and MICHAEL LOCKE

University of Western Ontario, London, Ontario, Canada,

and

J. V. COLLINS

Dalhousie University, Halifax, Nova Scotia, Canada

"... all observation must be for or against some view if it is to be of any service."
(Attributed to C. Darwin by I. Stone in *The Origin*, 1981.)

"The fat body hangs in its bathing medium, the haemolymph, growing and dividing like a child's guide to cell biology." (M. Locke, 1980a.)

"The fat body of insects is a rather ungrateful material for cytological studies."
(A. B. Dutkowski, 1974.)

INTRODUCTION

In biology it is axiomatic that structure reflects function. Fat body is so pleiomorphic and physiologically versatile a tissue that it would be impossible to describe its structure in the singular as our title suggests. Indeed, Wyatt, G. (1980) has stated that the functional diversity of fat body cells may be unequalled by any other metazoan cell type. An understanding of the morphology of the fat body is therefore not accomplished by a simple description of structure, as it changes during development. The appropriate physiology has yet to be discovered for much of the structure that has already been described (see L. Keeley, this volume). The fat body varies in structure between organisms, between stages, according to its internal location and by differentiation from cell to cell. It changes its structure cyclically in relation to growth in the intermoult/moult cycle, in relation to changes in commitment at metamorphosis, in adaptation to stress, nutrition and metabolic demand and in relation to reproduction. In this account we point out how the predictable cyclical activities make the fat body useful for studying several problems of cell biology in addition to those relating directly to insect physiology.

(a) Glossary of cell biology terms

The fat body is increasingly being recognized as a tissue suitable for studying aspects of cell biology that are difficult to investigate with vertebrate cells. Fat body development is cyclical and sequential compared to liver, which is in a dynamic equilibrium (Locke, M., 1980b). The fat body is thus convenient for observing evanescent processes such as autophagy or mitochondrial division, and it is the only material that has been used to study the Golgi complex beads.

Autophagic vacuole (*AV*) A 2° lysosome derived from isolation bodies and 1° lysosomes. Synonyms: cytolysome, cytosegresome.

Confronting cisternae RER apposed to the plasma membrane or to peroxisomes at localized regions free of ribosomes.

Cortex Cytoplasm below the plasma membrane containing microfilaments and other cytoskeletal elements to the exclusion of ER and ribosomes.

Dense body Descriptive term for many inclusions that have been insufficiently studied for satisfactory characterization.

Endoplasmic reticulum (*ER*) Membrane-bound compartments continuous with the nuclear envelope that connect with Golgi complexes (GC) through transition vesicles (tv). All fat body ER bears ribosomes (RER). Smooth ER is abundant in oenocytes but absent from fat body which differs from liver in this respect.

Golgi complex (*GC*) Fat body GCs consist of rings of GC beads on the RER where transition vesicles arise, three or four saccules together with secretory vesicles (SVs) and 1° lysosomes at the secretory face.

Golgi complex beads (*GCb*) 10 nm particles arranged in rings around the transition vesicles on the smooth face of RER where GCs arise.

Heterophagic vacuole Membrane-bound vacuole containing extracellular material scheduled for lysis. Derived from endocytosed material and 1° lysosomes, i.e. a 2° lysosome.

Instar A developmental stage characterized by an integument outwardly unchanged between ecdyses, but with a fat body that varies with age in the stadium.

Intermoult/moult cycle The developmental sequence which repeats itself from stadium to stadium. In the fat body it begins with cell preparation for intermoult syntheses (such as perioxisome formation, RER and GC formation, mitochondrial growth and division), is followed by intermoult syntheses (such as lipid, glycogen,-haemolymph protein), and concludes with ecdysial events (such as tyrosine mobilization) and preparations for the next stadium.

Isolation body (*IB*) The fragment of cytoplasm isolated by the paired isolation membranes. It becomes an autophagic vacuole after fusing with a 1° lysosome. Several IBs often fuse together.

Isolation membranes The paired membranes of the envelope which invests organelles scheduled for lysis.

Lymph Fluid in intercellular spaces below the basal lamina through which it has filtered from the haemolymph.

Multivesicular body (*MVB*) A structure bound by a single unit membrane containing material carried to it by pinocytosis vesicles and 1° lysosomes. MVBs have inner vesicles arising by invagination at their surface and are concerned in membrane turnover and protein digestion.

Peroxisome A particle with a single unit membrane enclosing dense contents and a core, and which always contains catalase and oxidases. Synonym: microbody.

Pinocytic vesicle Microvesicle arising at the plasma membrane surface that is involved in conveying membrane and protein into the cell.

Plasma membrane reticular system (*PMRS*) The system of plasma membrane infolds forming an extra inner surface in many large insect cells.

Primary lysosome (*1° L*) Microvesicles from GC carrying lytic enzymes or proenzymes.

Protein granule (*PG*) Membrane-bound vesicle storing protein that has been pinocytosed from the haemolymph. The protein is often crystalline. PGs are often composite structures, having fused with AVs.

Provacuoles Plasma membrane derived compartments that fuse together to form tyrosine storage vacuoles.

Residual body Membrane-bound inclusion characterized by undigested residues (membrane fragments or whorls, myelin figures, ferritin-like particles, etc.)

Secondary lysosome (*2° L*) Membrane-bound vesicle resulting from fusion of primary lysosome with vacuole carrying material to be digested.

Secretory vesicle (*SV*) Membrane-bound vesicle of protein destined for secretion outside the cell by exocytosis. Synonyms: secretory granules, zymogen granules.

Stadium The period from one ecdysis to the next, during which the fat body changes in the characteristic intermoult/moult sequence of development.

Transition vesicle (*tv*) Microvesicle between the RER and the outer saccule of the Golgi complex. They bud off through the centre of the GCb rings.

Tyrosine vacuole Large vacuole storing tyrosine in larval fat body cells.

Urate granule Membrane-bound structure at metamorphosis storing urate and uric acid sometimes with fibrous and vesicular contents. Of uncertain origin.

Urate vacuole Large vacuole storing urate in cockroach urate cells. Structurally similar to large urate granules in Lepidoptera.

Vacuole Large membrane-bound structure which in the fat body may store either tyrosine or urate. Synonym: watery vacuole.

Vitellogenesis Synthesis of vitellogenin by fat body.

(b) Abbreviations

AV	Autophagic vacuole
B	Bacteria
BPM	Bacterial plasma membrane
BL	Basal lamina
cv	Coated vesicle
D	Desmosomes
E	Envelope of Gram-negative bacterium
G	Glycogen
S	GC saccule
GC	Golgi complex
GCb	Golgi complex beads
H	Haemocoel
HP	Haemolymph proteins
HD	Hemidesmosome
L	Lipid droplet
LP	Lymph proteins
Ls	Lymph space
mf	Microfilament
mt	Microtubule
M	Mitochondria
MVB	Multivesicular bodies
N	Nucleus
P	Peroxisomes

PMRS	Plasma membrane reticular system
PM	Plasma membrane
PG	Protein storage granules
PV	Provacuole
RER	Rough endoplasmic reticulum
SV	Secretory vesicle
tv	transition vesicle
UG	Urate storage granules
UV	Urate vacuole
C	Urate vacuole core
V	Vacuole
VM	Vacuole membrane
1° L	1° lysosomes

(c) Preservation of the fat body

The fat body is an ungrateful kind of cell for structural studies as Dutkowski, A. (1974) pointed out. Published electron micrographs attest to the difficulty of fixation. There are many reasons for this. Fat body usually contains different kinds of vacuoles and compartments that require different conditions for optimal fixation of the natural shape. It may contain phagic vacuoles, that if broken, liberate hydrolytic enzymes into the cell. It nearly always contains large areas of lipid and glycogen that lack the mechanical stability given to regions of cytoplasm containing a cytoskeleton. Lastly, it is exposed in the haemocoel, its natural shape being partly due to tension on the basal lamina strands that connect it to other tissues. The conditions appropriate for good fixation are similar to those needed for the epidermis (Locke, M. and Huie, P., 1980a). The following details are often neglected:

(1) Fixation should be by inflation, preferably after some haemolymph has been removed to allow replacement without mechanical distortion. Fat body should not be dissected in saline prior to fixation.
(2) The initial fixation should be in a small volume, since large volumes leach out cell components. This is best accomplished by leaving an injected insect for several minutes before opening it up.
(3) The pH should be kept on the alkaline side of 7.4 since fixation is much slower close to neutrality.
(4) Fixation and other steps prior to in-block staining should be on ice, since cold discourages the solution of components before they can be fixed.
(5) The fat body should be carried through to embedding with its natural shape protected within a ring of integument.

Good preservation is particularly important for distinguishing between the partial autolysis that is part of metamorphic cell reprogramming and cell death preceding replacement by stem cells (section 3.3.1). Poorly preserved cells undergoing organelle autophagy may sometimes have been misinterpreted as dying larval cells.

1 STRUCTURE

1.1 The diversity of tissue arrangement and gross morphology in different orders

The fat body is typically located in two body regions that reflect its embryonic origin: close to or surrounding the gut, and adjacent to the integument (Buys, K., 1924). In most orders, segmentally arranged somites with a central coelomic cavity appear in the embryo. As the coelomic cavity becomes continuous with the haemocoel the splanchnic mesoderm of the somite grows around the gut and some cells differentiate to form perivisceral fat body. The somatic mesoderm of the somite contributes cells which form the peripheral fat body next to the integument (Anderson, D., 1972a,b). These layers are not always discrete: continuity between perivisceral and peripheral fat body is frequently encountered and the initial segmental arrangement is typically lost in later development.

The structure and arrangement of fat body is characteristically constant within a species, and members of the same order usually show the same general plan (Buys, K., 1924). Descriptions of fat body arrangements and gross morphology are included in Table 1. Whatever the precise layout of the tissue, a constant feature is that a very large area is exposed to the haemolymph which allows rapid metabolite exchange.

1.2 Basic cell structure: trophocytes

In some orders, only one type of cell is found in the fat body: the trophocyte or adipocyte. In other

Table 1: Gross morphology of fat body tissue

Order and species	Observations	Reference*
Coleoptera		
Leptinotarsa decemlineata	Tissue consists of a few small lobes, some next to gonads and some near integument. Peripheral lobes consist of an oenocyte surrounded by trophocytes. (A)†	5
Dictyoptera		
Periplaneta americana	Flat lobes, mycetocytes centrally placed and more or less surrounded by	1
Diploptera punctata	urate cells. Trophocytes tend to occupy edge of lobe. (A)	11
Diptera		
Calliphora stygia	(L) Flat sheets, one cell thick in paired anterior and dorsolateral lobes. Posterior lobe is one cell thick, highly fenestrated and folded. Cell number about 11,500 throughout larval life. Progressive histolysis from anterior to posterior. (L)	14
Sarcophaga argyrostoma	Cell separation starting at anterior is completed within 48 h of puparium formation. (L)	7
Calliphora erythrocephala	Oenocytes among fat body trophocytes. (A)	13
Glossina austeni	Oenocytes among fat body trophocytes. (A)	15
Sarcophaga bullata	Oenocytes among fat body trophocytes. (A)	12
Dacus tryoni	Oenocytes among fat body trophocytes. (L,A)	6
Musca domestica	Oenocytes among fat body trophocytes. (A)	3
Ephemeroptera		
Heptagenia sp.	Thin irregular sheets broadening to form irregular pads around the	2
Chirotonetes sp.	spiracles. (N)	
Hemiptera		
Rhodnius prolixus	Abdominal tissue is lace-like, predominantly a single cell layer but thicker at lateral margins. Spaces in tissue reduced by increasing cell number in successive instars. (N).	16
Lepidoptera		
Pontia rapae	Perivisceral tissue in longitudinal ribbons which may be much folded and	2
Phlegethontius quinquemaculata	cut. This tissue attached in places to smaller quantity of tissue adjacent to	
Melitta satyriniformis	integument. (L)	
Calpodes ethlius	Sheets of cells one layer thick growing to two layers thick in each stadium. (L)	10
	During metamorphosis larval tissue disaggregates into contiguous clumps of cells that reorganize into nodular clumps surrounding tracheoles or developing gonads. (A)	8
Orthoptera		
Schistocerca gregaria	Throughout abdomen and thorax.	4
Locusta migratoria	Peripheral trophocytes interspersed with oenocytes firmly attached to integument, central loose network of anastomozing	9
Melanoplus femur-rubrum	lobes in haemocoel that is frequently continuous with peripheral	2
Orchelimum sp.	tissue. (N,A)	
Gryllus abbreviatus		
Trichoptera		
Hydropsyche sp.	Many leaf-like bladders, some forming separate masses while others are connected at their bases by tissue lying close to body wall. (L)	2

* 1, Bodenstein, D. (1953); 2, Buys, K. (1924); 3, Clark, M. and Dahm, P. (1973); 4, Coupland, R. (1957); 5, de Loof, A. and Lagasse, A. (1970); 6, Evans, J. (1967); 7, Fraenkel, G. and Hsiao, C. (1968); 8, Larsen, S. (1976); 9, Lauverjat, S. (1977); 10, Locke, M. and Collins, J. (1965, 1968); 11, Stay, B. and Clark, J. (1971); 12, Stoppie, P. *et al.* (1981); 13, Thomsen, E. and Thomsen, M. (1974); 14, Thomson, J. (1973); 15, Tobe, S. *et al.* (1973); 16, Wigglesworth, V. (1967a).

† (A), adult; (L), larva; (N), nymph.

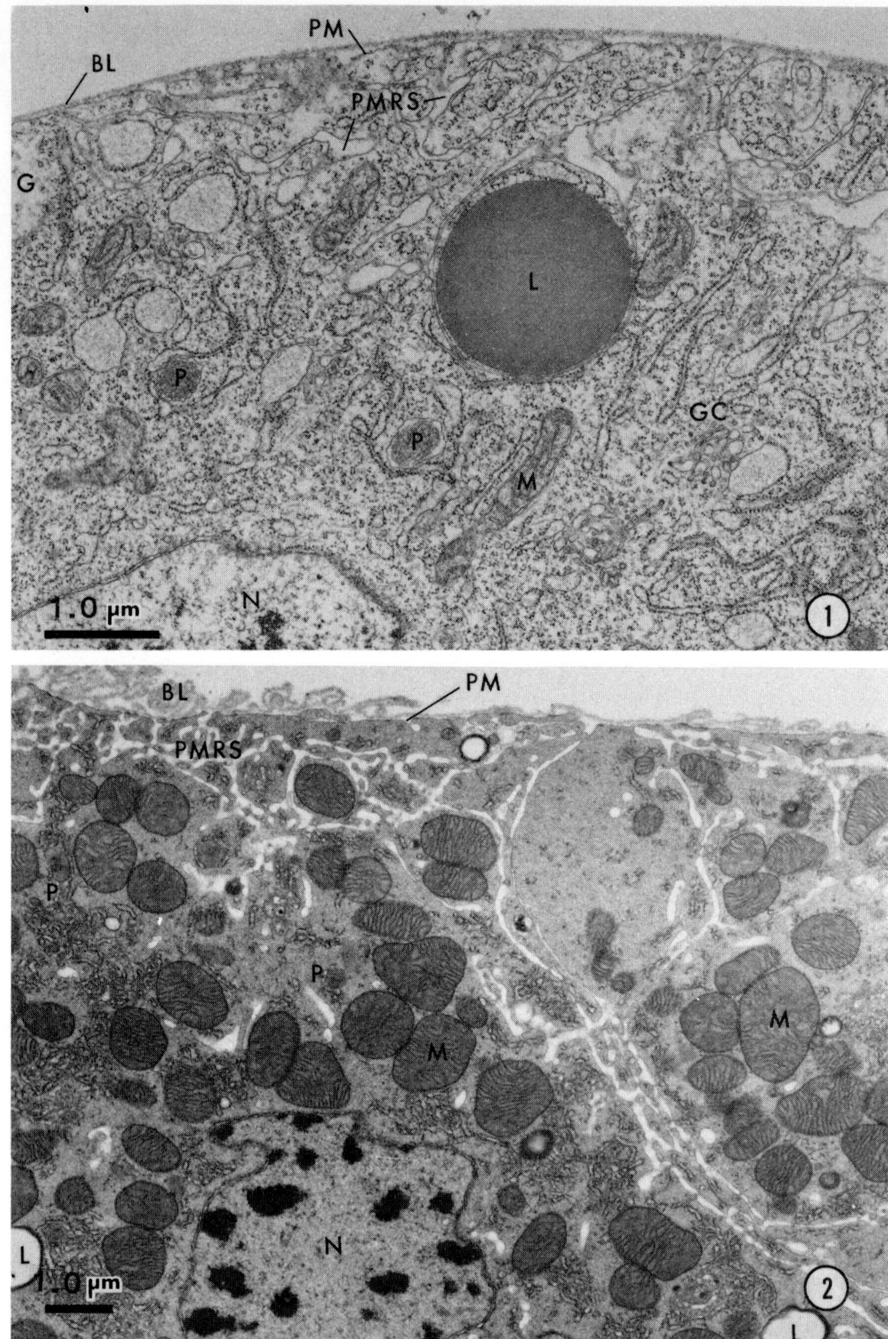

FIG. 1–3. The structure of generalized trophocytes. FIG. 1. *Calpodes* fifth-stage larva during intermoult preparation. The cytoplasm contains many free polysomes, elongate mitochondria (M) and perioxisomes (P) forming next to confronting RER cistern. The plasma membrane reticular system (PMRS) consists of irregular inward extensions of the plasma membrane (PM). A variable portion of the cell is composed of lipid droplets (L) and glycogen (G). (×21,000.)

FIG. 2. *Calpodes* adult. The PMRS is well developed. Mitochondria are spherical and ribosomes are mainly on RER. Peroxisomes are present. (×12,000.)

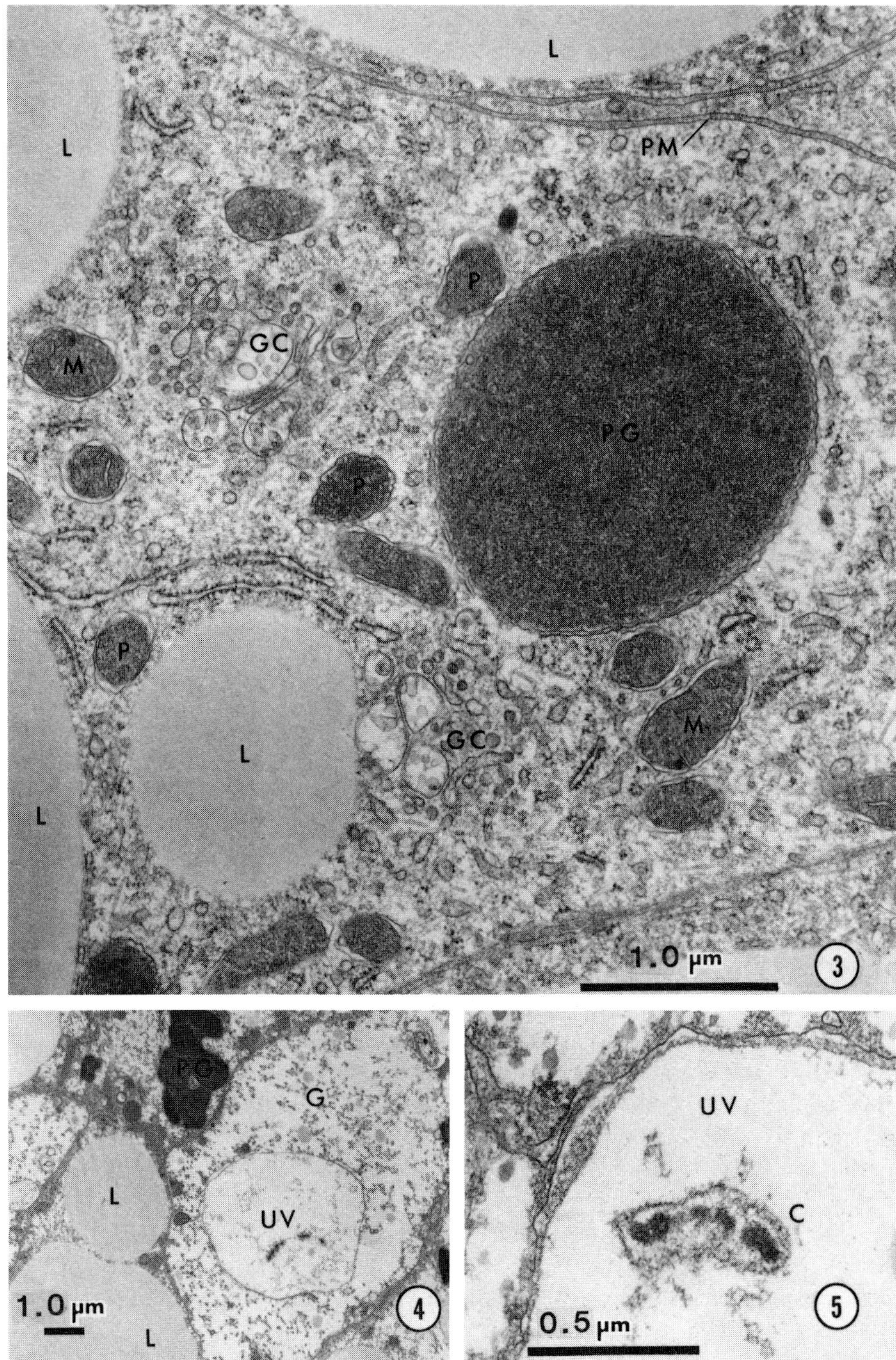

FIG. 3. *Periplaneta* fifth-stage nymph. Very large lipid droplets (L) almost isolate small islands of cytoplasm containing protein storage granules (PG), Golgi complexes (GC), peroxisomes (P) and mitochondria (M). The narrow spaces between cells contain protein. (× 35,000.)

FIGS 4 and 5. A specialized fat body cell. The urocyte from adult *Periplaneta*. A few very empty appearing cells contain large urate filled vacuoles (UV). The urate vacuoles contain a characteristic core (C). (Fig. 4 × 7000; Fig. 5 × 61,000.)

orders trophocytes are most numerous but the fat body is an aggregate of two or three cell types (Table 1). During embryogenesis and early postembryonic development trophocytes have a rounded nucleus which may be polyploid, and the cytoplasm contains few inclusions. Thereafter, trophocyte morphology varies with nutritional state, with the time in the intermoult/moult cycle and also with the developmental stage of the insect (Figs 1–3 and 28). At times of great metabolic activity much rough endoplasmic reticulum (RER) and numerous Golgi complexes (GC) occur together with abundant mitochondria and peroxisomes. This contrasts sharply with the start of non-feeding periods when the fat body may be laden with stores of protein, lipid and glycogen.

In some species, for example *Calpodes ethlius*, the trophocytes appear to differ from each other only in size. Peripheral cells are smaller than those in the perivisceral ribbons, which probably only reflects differences in ploidy. Their activities are similar and more or less synchronous. In other species the trophocytes show regional specialization (see section 2.2.5).

1.2.1 Trophocyte shape and the intercellular environment

The form of the fat body as a tissue depends upon three factors: the basal lamina, junctions, and the shape of each cell determined by its cytoskeleton.

(a) *The basal lamina.* Except for the haemocytes (A. Gupta, this volume), all tissues in an insect are separated from the haemolymph by a basal lamina (D. Ashhurst, this volume) — including even those tissues such as fat body, epidermis, ovarioles, wing discs and pericardial cells that traffic in haemolymph proteins. Of these, the fat body stands out as a tissue that must both take up and secrete proteins through the basal lamina. The appearance of the basal lamina (Fig. 9) suggests that it may serve as a barrier to some haemolymph components such as cells and large proteins. The basal lamina around Malpighian tubules, for example, only allows the smallest particles of colloidal gold to penetrate as far as the cell surface (Oschman, J. and Berridge, M., 1971). For this reason we propose to refer to the spaces around fat body cells and below the basal lamina as lymph spaces, since the fluid in them may be filtered from the haemolymph bathing the tissue. A property of the fat body basal lamina must be permeability or selective permeability to haemolymph proteins in both directions. There is nothing in the appearance of the basal lamina that gives a clue how this may be managed. The texture may be loose enough to allow smaller proteins to pass through, but some of the larger polymers may well be excluded. The main larval storage proteins synthesized by the fat body in *Calpodes* have molecular weights of 720, 580, and 470 K daltons (Locke, J. *et al.*, 1982) or more (D. M. Webster, personal communication) and are polymers of six or more peptides with molecular weights of 70–80 K daltons (L. Levenbook, vol. 10). Negative staining shows the smallest of these large polypeptides to be 11–12 nm in diameter (Fig. 10) making it unlikely that they can traverse the basal lamina unimpeded. If these large polymers are excluded then it may be that they pass through the basal lamina as monomers with which the polymers are in equilibrium.

The basal lamina around tissues such as nerves is layered, the number of layers increasing from stadium to stadium as though they have been added on at each stage. The fat body (and other tissues that grow extensively) on the other hand, have single-layered basal laminae without extra layers in later stages. Since the basal lamina does not get thinner, material must presumably be added to it as the fat body grows. Why then is it not layered like that around nerves?

The absence of layering is probably the result of involution of the old stage basal lamina and the secretion of a new one at each moult. Immediately after the third to fourth ecdysis in *Calpodes*, the basal lamina is endocytosed in large vacuoles within which it disappears, presumably as a result of digestion. The basal lamina is held to the plasma membrane by hemidesmosomes. These concentrate in patches that become vacuoles, dragging the basal lamina into the vacuole lumen with them (Figs 11 and 12). This phagocytosis of the basal lamina occurs both at the outer surface and between cells.

The relationship between the basal lamina and the fat body is a complicated one, for it is both at the haemolymph face of the tissue and threaded through the interior between the cells in reticular strands. The strands are particularly obvious at the interior meeting places where three or four cells may

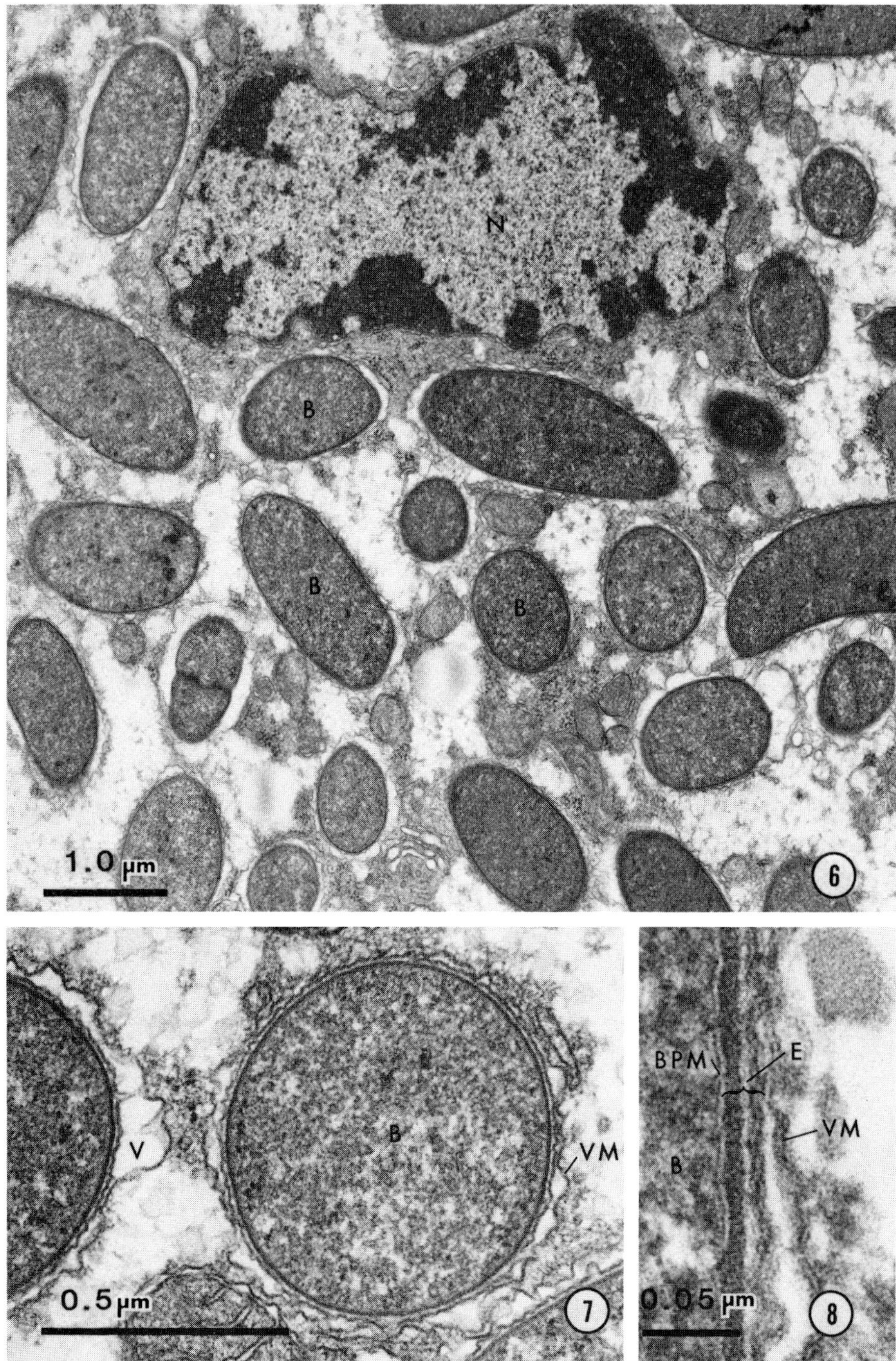

FIGS 6–8. A specialized cell from adult *Periplaneta* fat body — the mycetocyte. Mycetocytes have little of the regular structure of trophocytes but are filled with vacuoles (V) containing Gram-negative bacteria (B). The envelope (E) of Gram-negative bacteria is a complicated structure consisting of a cell wall and an outer unit membrane-like layer. The bacterial plasma membrane (BPM) is often difficult to see. All bacteria are separated from the host cell by a vacuole membrane (VM). (Fig. 6 × 22,000; Fig. 7 × 88,000; Fig. 8 × 350,000.)

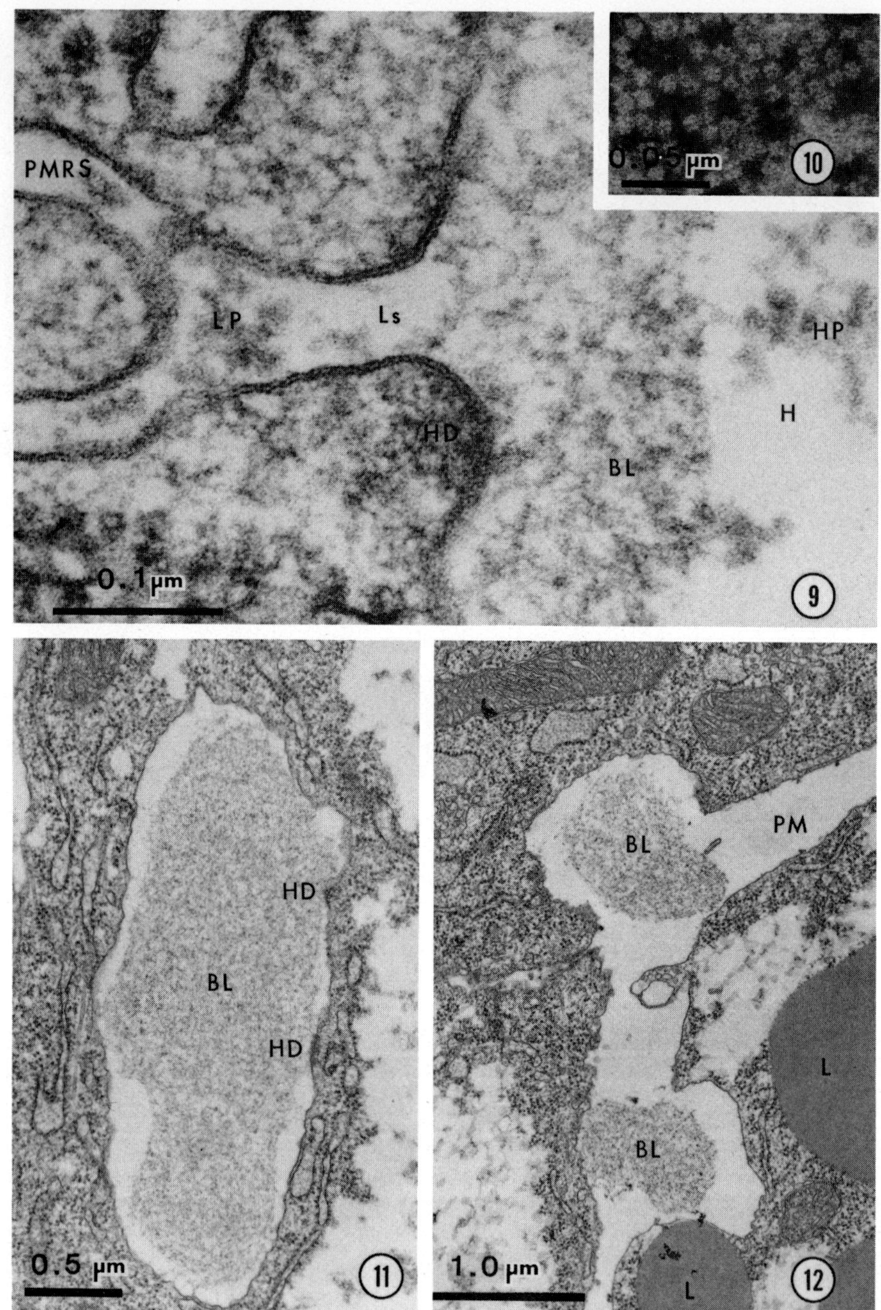

FIGS 9–16. The basal lamina (BL) (see also Figs 1, 2, 19, 20, 21, 37, 38 and 65). FIGS 9 and 10. The basal lamina in relation to entrances to the PMRS. The basal lamina is attached to the surface of fat body cells through hemidesmosomes (HD). The texture of the basal lamina is probably dense enough to filter some haemocoel (H) components such as large proteins (HP) from the lymph spaces (Ls) between cells and in the PMRS. Figure 10 negatively stained 470K dalton haemolymph storage protein to the same scale. (Courtesy D. Webster.) (Figs 9 and 10 × 310,000.)

FIGS 11 and 12. Basal lamina involution in an early fourth-stage *Calpodes* larva. After ecdysis the old basal lamina is gathered by its hemidesmosomes (HD) into plasma membrane pockets which separate as vacuoles within which the basal lamina disappears. (Fig. 11 × 35,000; Fig. 12 × 28,000.)

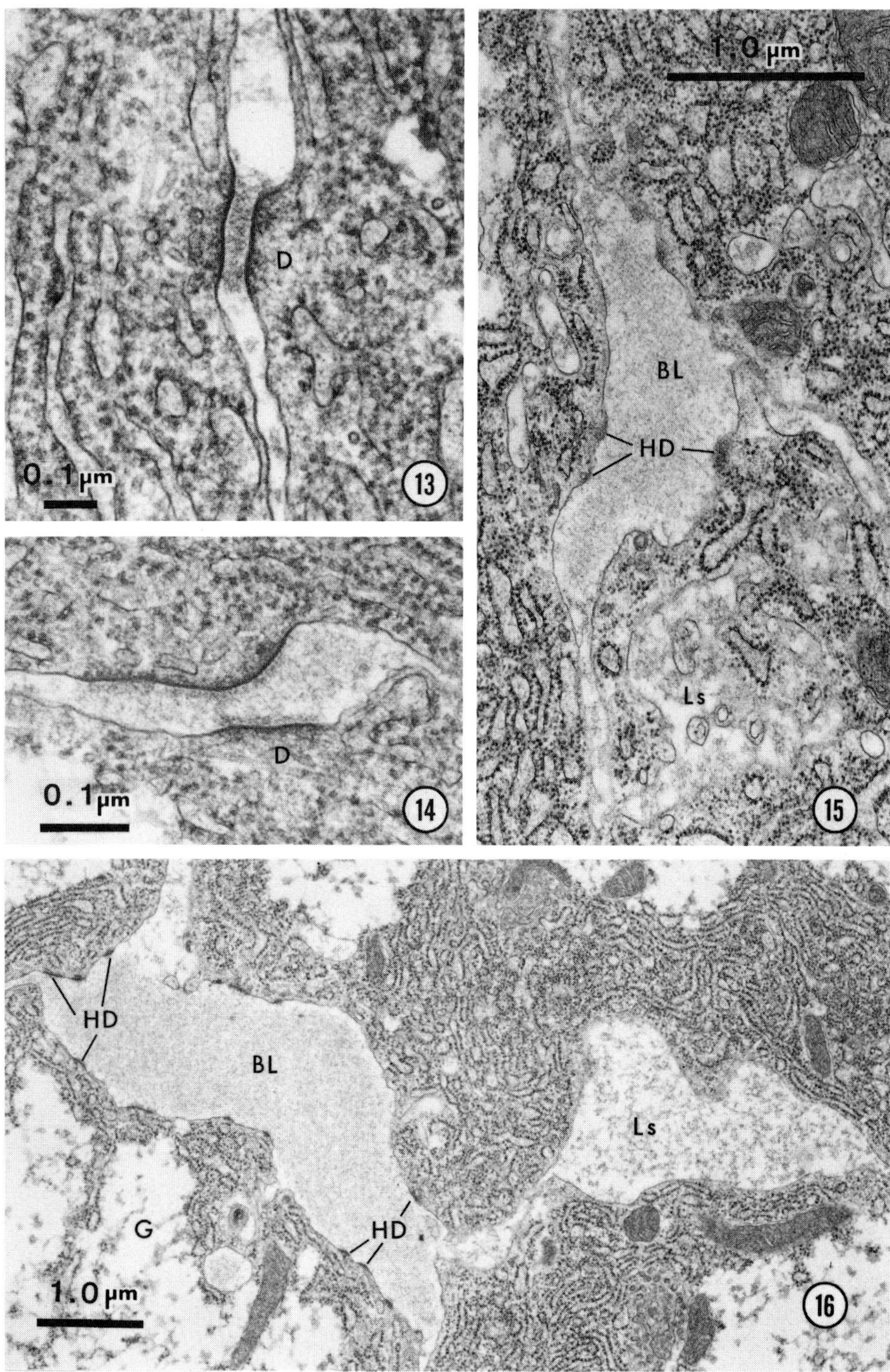

FIGS 13–16. Hemidesmosomes, desmosomes, the internal basal lamina and lymph spaces. There is a continuous intergradation between the desmosomes (D) that attach fat body cells to one another like spot welds, and the hemidesmosomes (HD) attaching cells to the internal basal lamina strands. There are also discrete internal lymph spaces (Ls). (Fig. 13 × 95,000; Fig. 14 × 160,000; Fig. 15 × 36,000; Fig. 16 × 19,000.)

join together. There is a continuous gradation between desmosomes joining cells together directly, paired hemidesmosomes abutting the basal lamina, unpaired hemidesmosomes of adjacent cells attached to the same basal lamina and single separate hemidesmosomes (Figs 13–16). The close association between the basal lamina and desmosomes suggests that the fat body has a direct role in basal lamina secretion. It is difficult to see how haemocytes, that may be the source of basal laminae, could be involved in joining desmosomes with basal lamina deep within the fat body tissue. No haemocyte associations with desmosomes have been observed.

The fat body is strung to other tissues by strands of basal lamina that often have a core of elastic fibres (Locke, M. and Huie, P., 1972).

(b) *Junctions.* Junctions between fat body cells are desmosomes and hemidesmosomes (Figs 13–16), gap junctions (Fig. 18) and occasionally rather long-lived mid bodies (Fig. 17). The electrical coupling of *Calpodes* larval fat body cells varies with stage in the intermoult/moult cycle. Although most of the coupling is presumably due to the gap junctions, it is relatively easy to observe the mid bodies remaining after cell division during intermoult preparation. Mid bodies must therefore remain intact for many hours before cell separation is complete.

(c) *Cell shape and the cytoskeleton.* Fat body cells are not simple polygons strung in strips, or one or two layered sheets as they are often depicted, but have differentiated outer surfaces and an inner relationship with basal lamina strands and lymph spaces. There are two categories of outer surface differentiation dependent upon the cytoskeleton: surface processes, and infolds that give rise to the plasma membrane reticular system (PMRS). Within the fat body there are transport routes along the intercellular surfaces leading to lymph spaces (Fig. 16).

Surface processes. In *Calpodes* and other insects the epidermis forms basal extensions or feet whose elongation and shortening is related to changes in shape of the integument (Locke, M. and Huie, P., 1981a,b). Fat body cells in *Calpodes* and *Periplaneta* develop similar surface processes. In *Calpodes* fifth-stage larvae they arise at the end of the intermoult preparation after the formation of a bilayered strip of tissue, at the same time as they develop in the epidermis. They extend over neighbouring cells forming stubby interdigitations that increase the surface area between the surfaces of adjacent cells.

Infolds and the PMRS. Fat body cells are like oenocytes (Locke, M., 1969) and other giant cells in having a subsurface PMRS. Early in the fourth, fifth and imaginal stadia of *Calpodes* (Locke, M. and Collins, J., 1968), *Sarcophaga* (Stoppie, P. *et al.*, 1981), *Calliphora* (Thomsen, E. and Thomsen, M., 1974), and *Leptinotarsa* (Dortland, J. and Esch, T., 1979; de Loof, A. and Lagasse, A., 1970) the fat body cells have little surface differentiation. The surface also lacks specialization after periods of starvation. During intermoult preparation the plasma membrane folds in from the surface creating interconnected channels. These infolds may have confronting ER cisternae. In the fourth stage *Calpodes* provacuoles arise from these infolds as well as from the external surface, (see section 3.2.1c tyrosine storage vacuoles). The irregular infolds mature into a precise meshwork of extracellular spaces forming a subsurface connected by infolds to the surface, the plasma membrane reticular system (PMRS) (Figs 19 and 20). The subsurface is parallel to the surface and separated from it by a constant distance of 1–1.5 μm. The spaces of the PMRS also have rather constant dimensions with membrane to membrane separations of 100 to 150 nm in both the subsurface and its connections. The gateways between the outside and the intercellular environment also have constant dimensions but we have been unable to see a perforate diaphragm of the kind that occurs in pericardial cells (A. Crossley, this volume) and between the podia of kidney cells. The PMRS may extend laterally around the fat body cells (Fig. 24) and even as far as the lymph spaces.

These surface differentiations can be related to development of the cytoskeleton. All over the surface directly below the plasma membrane, bundles of microfilaments and microtubules form a cortex upon which the desmosomes and hemidesmosomes can insert (Fig. 21). Microfilaments and microtubules also form a meshwork around the PMRS. The cytoskeleton is particularly obvious around the lymph spaces, as though it is concerned with their maintenance. Filaments may also be involved. Fat body cells in both *Calpodes* and *Periplaneta* occasionally contain skeins and whorls

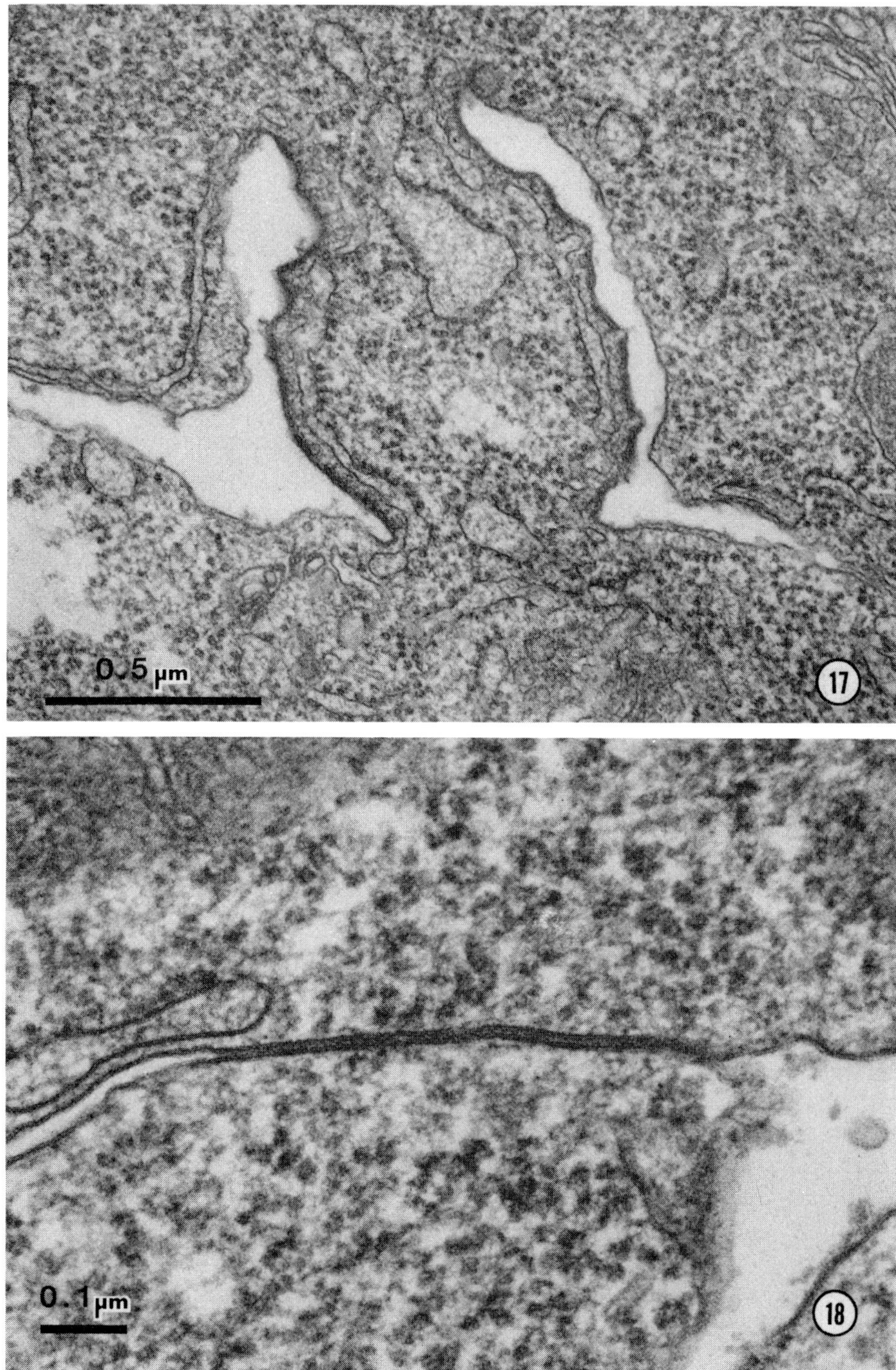

FIGS 17 and 18. Junctions allowing communication between fat body cells in fifth-instar *Calpodes* larvae.

FIG. 17. Mid bodies have a life span of many hours and survive long after the cells appear separate by light microscopy. (× 76,000.)

FIG. 18. Gap junctions are present between fat body cells except during pupal metamorphic rearrangement. (× 150,000).

of filaments. In tissue culture much of the surface differentiation is lost, and balls of filaments appear in the cytoplasm.

Lymph spaces. The picture of the fat body which emerges is not one of cells packed tightly into a solid tissue but of cells having a peripheral sponge structure, separated by lymph spaces, and linked to one another or to the basal lamina by (hemi)desmosomes. The lymph spaces naturally contain material resembling fixed protein (Fig. 16). These lymph spaces are not merely accumulations of exocytosed protein synthesized by the fat body. Horseradish peroxidase injected into the haemocoel appears in the lymph spaces, which must therefore receive material from the haemolymph as well as contribute material to it. This poses the question of the way that the lymph spaces communicate with the haemocoel and what their functions may be.

Intercellular surface transport routes. In cockroach (Figs 22 and 23) and *Calpodes* (Figs 24 and 25) intermoult fat body there are evenly spaced surfaces between all the cells. The desmosomes serve not only to hold the cells together at particular points but also to maintain a rather constant separation between cells. The separation between cells is similar to the separation between plasma membrane faces in the PMRS and is not the standard 20 nm separation commonly seen in epithelia. Fat body cells are not bonded together but bonded apart. In many places these intersurface compartments and the nearby PMRS contain packets of fixed protein presumably en route to and from the sites of pinocytosis and exocytosis that occur even deep within the tissue. Golgi complexes, MVBs and protein storage granules are all found squeezed between the interstices of lipid droplets close to plasma membranes but with little cytoplasmic communication with the rest of the cell. The structure suggests that fat body cells both need and use the spaces between them for the transport of protein. The mechanism for protein movement through these inter-surface compartments which are only 100 nm wide for distances as great at 100 μm can scarcely be unaided diffusion.

This description of the fat body as clusters of cells surrounded by lymph with moving packets of protein between them confirms the appropriateness of the term applied to them — trophocytes (see L. Keeley, this volume).

1.3 Cell diversity

1.3.1 Chromatocytes

The integument of aquatic larvae of the Simuliidae and Thaumaleidae is transparent. Their colour pattern is determined by the distribution of pigment-laden cells lying beneath the epidermal basal lamina. These chromatocytes are flat and so thin that the nucleus causes a distinct bulge in the centre of the cell. Chromatocytes are connected in sheaths, one cell thick, which surround ribbons of trophocytes but remain separated from them by a basal lamina. The cells increase in number through successive larval instars but maintain the same relative position. At metamorphosis, changes in the insect colour pattern depend on mass migrations and redistribution of these cells. Although chromatocytes have a distinctive appearance they are probably modified trophocytes. Like trophocytes they accumulate lipids during larval growth and these reserves are depleted during adult development (Hinton, H., 1958, 1959).

1.3.2 Urocytes

Although uric acid is excreted into the lumen of Malpighian tubules in insects it is also commonly found stored in internal tissues, particularly the fat body (see also section 3.1.3). This feature is common in cockroaches where uric acid is stored thoughout life and may comprise as much as 10% of the dry weight (Bursell, E., 1967). Urates do not accumulate in trophocytes or mycetocytes (section 1.3.3) of the cockroach fat body but in cells specialized for urate storage, the urocytes (Figs 4 and 5). The cytoplasm of urocytes has little ER and few mitochondria, and consists mainly of very large vacuoles containing crystalloid spherules of uric acid or urate that are birefringent under plane polarized light (Cochran, D. *et al.*, 1979). Such uric acid storage in vacuoles has often been called storage excretion, but the ease with which the uric acid may be mobilized in fat body deposits suggests that it may serve as a reservoir of nitrogen. An early study by Bodenstein, D. (1953) showed that, in *Periplaneta americana*, urates disappeared from the fat body when the corpora cardiaca were removed. Corpora cardiaca replacement restored urates to the urocytes, suggesting humoral control of protein or urate metabolism. However, the situation is complex because urate is redeposited in urocytes of cardiectomized insects during starvation. Uric acid is deposited in the fat body of *Periplaneta* maintained on a high-protein diet. The deposits are absent in the roaches switched to a pure dextrin diet (Haydak, M., 1953). In a cytological study, Cochran, D. and coworkers (1979) also showed that the quantity of stored urate is related to dietary nitrogen levels. In addition, these authors demonstrated that the urate spherule is not a homogeneous structure but contains an irregular electron-dense core (Fig. 5). A role for this structure in urate deposition or mobilization is suggested but is significance awaits further analysis. Distinct urocytes are not restricted to the Dictyoptera; they occur also in the fat body of Collembola and Thysanura (Wigglesworth, V., 1972).

Urate storage vacuoles of urocytes resemble the urate storage granules of lepidopteran fat body (section 3.1.3) except that they are very much larger. Both may have plasma membrane-derived vacuole membranes, that in the Lepidoptera are derived from pinocytic vesicles. The urocyte vacuole membranes have the thickness of plasma membranes rather than the ER (McDermid, H. and Locke, M., 1982) as do the tyrosine storage vacuoles (section 2.2.1c).

1.3.3 Mycetocytes

Many insect species possess intracellular microorganisms contained within specialized cells, the mycetocytes (Figs 6–8). It has long been assumed that this relationship is symbiotic for the following reasons. The non-pathogenic micro-organisms are present in each individual, they are required for normal development in the absence of dietary supplements, and elaborate mechanisms have evolved for transmitting them from one generation to the next. A general discussion of symbiosis lies outside the scope of this chapter but a review by Houk, E. and Griffiths, G. (1980) lists references to earlier reviews. In cockroaches, some Hemiptera such as the rice planthopper, *Laodelphax striatellus* (Noda, H., 1977), and the primitive termite *Mastotermes* (Wigglesworth, V., 1972), the mycetocytes are scattered singly throughout the fat body, forming an integral part of the tissue. In cockroaches the mycetocytes are characteristically located near the centre of the flattened fat body lobes (Bodenstein, D., 1953). The cells are usually large with a multilobed nucleus but the volume of cytoplasm is reduced since it is often densely packed with symbiotes (Fig. 6). The symbiotes occupy their own compartment surrounded by host cell membrane. Each symbiote is bounded by a plasma membrane and a complex envelope consisting of a cell wall of peptidoglycan and the structured outer layer, similar in appearance to a unit membrane, that is characteristic of Gram-negative bacteria (Fig. 8). In *Blattella germanica*, mycetocytes differentiate even in aposymbiotic embryos; that is, in embryos developed free of symbiotes. The "empty mycetocytes" have large nuclei but only small amounts of fibrous appearing cytoplasm (Brooks, M. and Richards, A., 1955). These authors also reported that normal mycetocytes undergo mitosis in the second half of each larval stadium. In the absence of symbiotes the cells remain small and do not divide. If the initial symbiote population is kept artificially small, the mycetocytes may grow to unusually large sizes without cell division, even though there appears to be no inhibition of symbiote replication. Mycetocytes frequently contain areas of stored glycogen but lipid droplets are rare and protein granules absent (Wüest, J., 1978). Wüest also observed that the quantity of glycogen seemed unrelated to phases of the reproductive cycle, in marked contrast to the fluctuations seen in trophocytes.

Although mycetocytes may be derived from fat body cells, in other insects they are located in specialized organs called mycetomes that are separate from the fat body. They may be connected with the gut, gonads or fat body (Wigglesworth, V., 1972). An attempt has been made to show a symbiotic relationship between host and microorganism in the pea aphid, *Acyrthosiphon pisum*, where the mycetome is a distinct abdominal organ lying dorsal to the gut among the fat body lobes (Griffiths, G. and Beck, S., 1973). Large central cells of the mycetome containing one kind of microorganism are partially surrounded by flattened sheath cells containing a second type. Electron microscope autoradiography coupled with the incorporation of [^{3}H]mevalonate, suggested that both types of micro-organism synthesize cholesterol. The

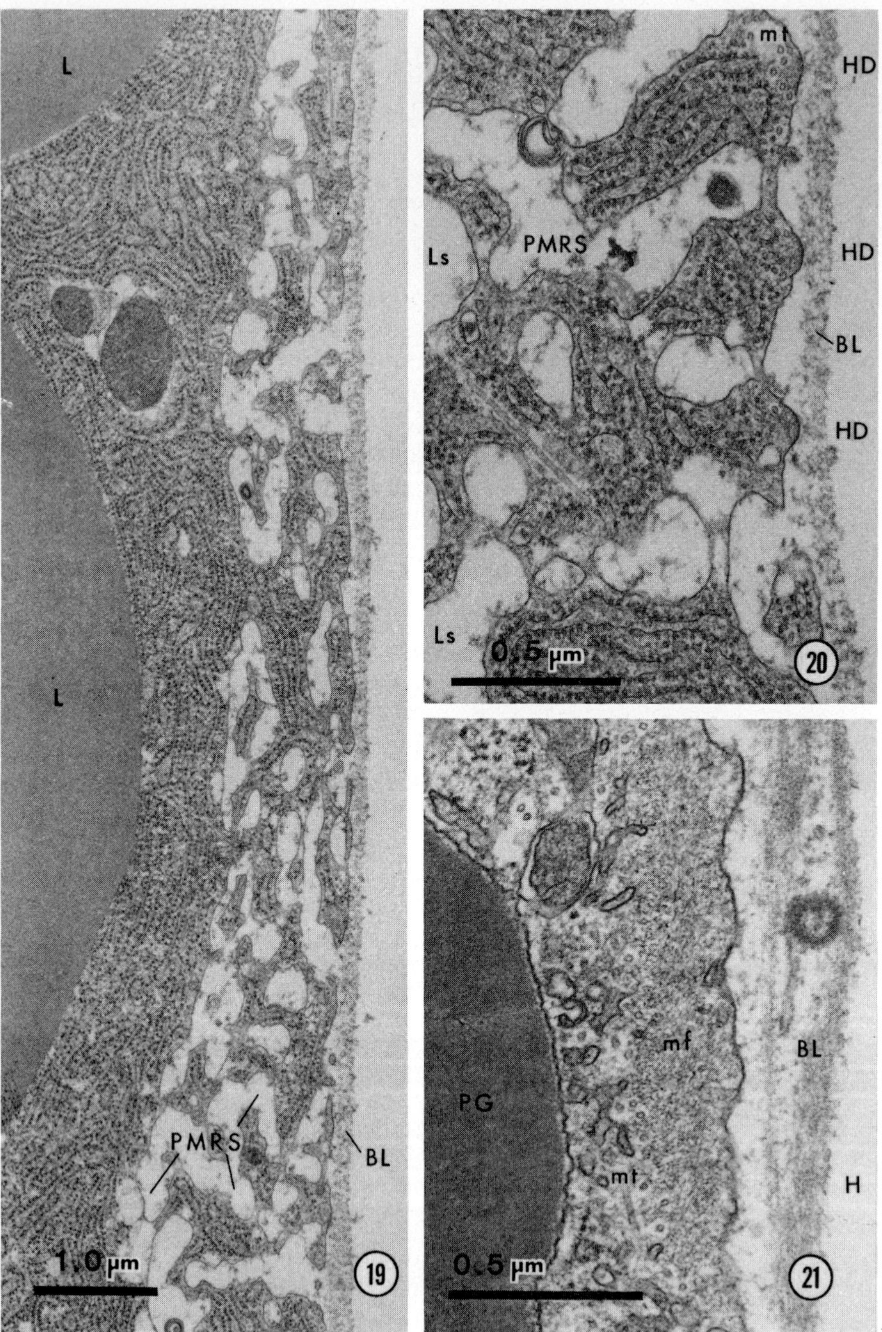

FIGS 19–25. External and internal surfaces and lymph spaces (LS) of the plasma membrane reticular system (PMRS).

FIGS 19 and 20. The plasma membrane surfaces of larval *Calpodes* fat body cells involved in intermoult syntheses fold inwards in a network, the PMRS. The PM is attached to the basal lamina of the external surface by hemidesmosomes (HD) that are associated with microtubular (mt) and microfilamentous (mf) elements of the cytoskeleton. (Fig. 19 ×24,000; Fig. 20 ×62,000.)

FIG. 21. The cortex of fat body cells is often a dense feltwork of microfilaments and microtubules as is this *Periplaneta* adult. (×72,000.)

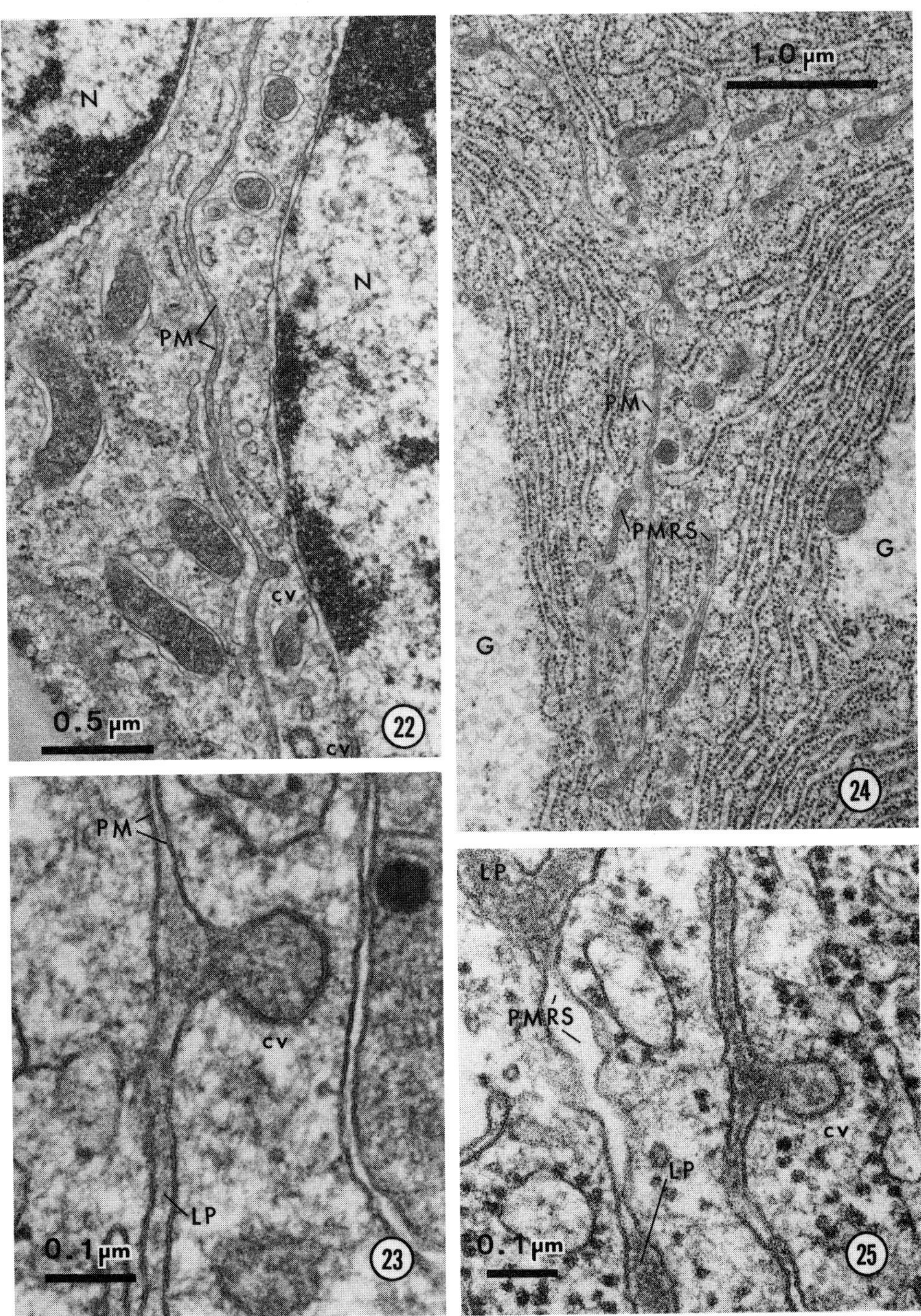

FIGS 22–25. Intercellular surface transport routes. FIGS 22 and 23. Internal surfaces between trophocytes have coated vesicles (cv) and are often filled with lymph proteins (LP) as in these *Periplaneta* cells. (Fig. 22 ×43,000; Fig. 23 ×170,000.)

FIGS 24 and 25. Internal surfaces may connect to protein filled elements of the PMRS as in these intermoult *Calpodes* larvae. (Fig. 24 ×28,000; Fig. 25 ×140,000.)

majority of silver grains were at first associated with the symbiote membranes but the frequency of grains over surrounding mycetocyte tissue increased with increasing incubation time. Perhaps, then, symbiotes export cholesterol to the host cell. Griffiths, G. and Beck, S. (1975, 1977) observed a vesicular system that may participate in this type of transport.

1.3.4 Oenocytes

Oenocytes are cells of ectodermal origin that are usually associated with the epidermis. In many species they also extend inward into the fat body where they are interspersed among the trophocytes (Wigglesworth, V., 1972). Oenocytes may be associated with the peripheral layer of fat body as well as with the perivisceral tissue. This distribution occurs in *Schistocerca gregaria*, with the peripheral location being more frequent (Coupland, R., 1957). The association persists throughout all developmental stages including the mature adult. In old adults the oenocytes may contain yellow pigment and uric acid crystals. In adults of *Leptinotarsa*, oenocytes are found in fat body just below the integument at a density of up to two or three cells per fat body lobe (Dortland, J. and Esch, T., 1979). The structure of these oenocytes is similar to that of the permanent sub-epidermal oenocytes described by Locke, M. (1969). They contain a large nucleus, few mitochondria and extensively developed tubular ER lacking ribosomes. As in trophocytes there may be small fields of glycogen (de Loof, A. and Lagasse, A., 1970). Trophocytes and oenocytes are intimately associated in *Leptinotarsa*. Although gap junctions have not been demonstrated, the plasma membranes of the trophocytes are as closely associated with those of the oenocytes as with each other. Oenocytes have also been described in the fat body of some Diptera (Table 1). In *Sarcophaga bullata*, Stoppie, P. and co-workers (1981) observed that oenocytes and trophocytes were always separated by a basal lamina and no gap junctions were seen (cf. Fig. 1, Stoppie, P. *et al.*, 1981 and Fig. 4, de Loof, A. and Lagasse, A., 1970). Evans, J. (1967) has examined the fate of larval oenocytes associated with trophocytes at metamorphosis in *Dacus tryoni*. Both cell types undergo cytolysis when the larval fat body disintegrates at metamorphosis (section 3.3), and a new population of imaginal oenocytes differentiates in the adult fat body. The significance of this close and persistent relationship between oenocytes and trophocytes is not known. It is all the more interesting since it is not universal among insects. Romer, F. and co-workers (1974) have shown that isolated oenocytes of *Tenebrio molitor* synthesize ecdysteroids *in vitro*. Since fat body is known to convert ecdysone to 20-hydroxyecdysone (S. Smith or W. E. Bollenbacher and N. Granger, vol. 7) and to respond to the latter, it is of interest to know more about the interrelations of steroid metabolism between oenocytes and fat body.

2 DEVELOPMENT

2.1 Polyploidy, polyteny and nucleolar activity

In insects, tissue growth involving DNA synthesis and cell enlargement rather than cell division is common (G. Richards, vol. 2). It is found in tissues which must rapidly synthesize large amounts of a limited number of peptides. In fat body cells this growth pattern often precedes the synthesis of storage proteins during the last larval stadia of holometabola or vitellogenin synthesis in adult females.

Larval nuclear development. In most insect orders, except the Diptera, this type of nuclear growth is accomplished by endomitosis. In the fat body of both *Rhodnius prolixus* (Wigglesworth, V., 1967b) and *Calpodes*, the chromosomes replicate and the centromeres separate, but the nuclear membrane remains intact. For example in *Rhodnius*, Wigglesworth, V. (1976b) reported that fat body cells are predominantly tetraploid at hatching, although a few octaploid cells are found. In fully nourished insects this condition persists throughout life although occasional 16*n* and 32*n* nuclei occur. Higher levels of ploidy, which result from cell and nuclear fusion, are found only after extreme starvation. In the early larval life of *Calpodes* the cells are tetraploid, and cell number increases by mitosis. At the beginning of the fifth stadium a period of DNA synthesis coupled with amitosis results in an increase both in ploidy and in cell number. Amitosis of the cells perpendicular to the plane of the fat body ribbon produces a tissue which is two cells thick. During the remainder of the stadium, the cells are predominantly octaploid although $16n$ and $32n$ cells are seen (K. Nair and M. Locke, unpublished observations).

Polyteny is a special case of polyploidy in which the chromosomes undergo repeated cycles of replication without separation of the resulting strands. Polyteny, which is characteristic of the Diptera, is usually studied in tissues other than fat body, notably the salivary gland. An exception is the detailed report on the development of polyteny in the fat body of *Calliphora stygia* (Thomson, J., 1973, 1975). *Calliphora* larvae hatching on day 0 ecdyse to the second instar 1 day later and to the third instar in the early half of day 3. At hatching, the fat body chromatin, which includes a highly condensed heterochromatin body, surrounds a prominent nucleolus. During the next 2 days the nuclei enlarge rapidly, and by the time the larva weighs ~2.5 mg, separate, irregular and elongate chromosomes can be resolved. At ~48 h after hatching polytene chromosomes can be recognized. The chromosomes contain short segments showing the characteristic banding pattern of polyteny. Between these are segments of diffuse, unbanded chromatin. The condensed heterochromatin body persists but is less conspicuous than at hatching. It apparently does not replicate as much as the rest of the genome. As growth continues with further cycles of replication, the chromosomes shorten and progressively occupy less of the nuclear volume. From the middle of the second stadium to the end of feeding on day 6 the rate of fat body protein synthesis is maximal. During this period further replication results in increasing chromosome width and banded segments continue to contract so that individual bands lie more closely together. Chromosomal replication ceases at the end of the feeding period. The banded sections of the chromosomes are thought to represent groups of reversibly inactivated loci while the intervening dispersed sections are analogous to the interphase chromatin of non-polytene nuclei. The segments which are dispersed during early larval life presumably include the loci involved in large-scale transcription prior to, or during, the peak period of protein synthesis.

Thomson has also described a relationship between developing polytene chromosomes and proliferating ribonucleoprotein (RNP) particles. During the first two stadia there is active RNA synthesis. The nucleoli enlarge and, by the time the period of maximum protein synthesis begins, there are multiple ribonucleoprotein particles of more or less uniform size. During the period of maximal protein synthesis these particles are associated with chromosomal segments other than the nucleolar organizers. Thomson suggests that the particles may participate in some aspects of processing, stabilizing or transport of gene products from the associated, transcriptionally active, chromosome segments. Further, this physical association may be a manifestation of a mechanism which couples mRNA synthesis with the synthesis of specific ribosomal populations (Thomson, J., 1973, 1975). This relationship between RNP particles and specific segments of polytene chromosomes can be visualized because the latter are so large. There is, however, no reason to suppose that a similar relationship does not exist in polyploid nuclei. The RNP particles near polytene chromosomes are presumably the same as the perichromatin granules that are believed to be packaged messenger RNA in other cells. From about 30 h into the last larval stadium of *Calpodes* the number of fat body perinucleolar ribosomal precursor particles and perichromatin granules increases, remains constant for about 30 h and then declines just prior to the intermoult period of massive protein synthesis (Figs 26 and 27) (Dean, R. *et al.*, 1980). This increase in the number of RNP particles correlates with increased [^{3}H]uridine incorporation (Locke, M., 1970) and the appearance of large numbers of ribosomes in the cytoplasm (Locke, M. and Collins, J., 1968).

Adult nuclear development. The fat body cells of *Locusta migratoria* females are mostly tetraploid at ecdysis to the adult. These cells become octaploid and may even attain higher levels of ploidy at the start of vitellogenin synthesis. The ploidy values reported do not meet the predicted values for a doubling of DNA content during increasing ploidy, so that the interesting question of whether all the genome is replicated remains open. Allatectomy prevents the increase in ploidy but the application of a juvenile hormone (JH) analogue will induce it in allatectomized individuals. This JH-dependent

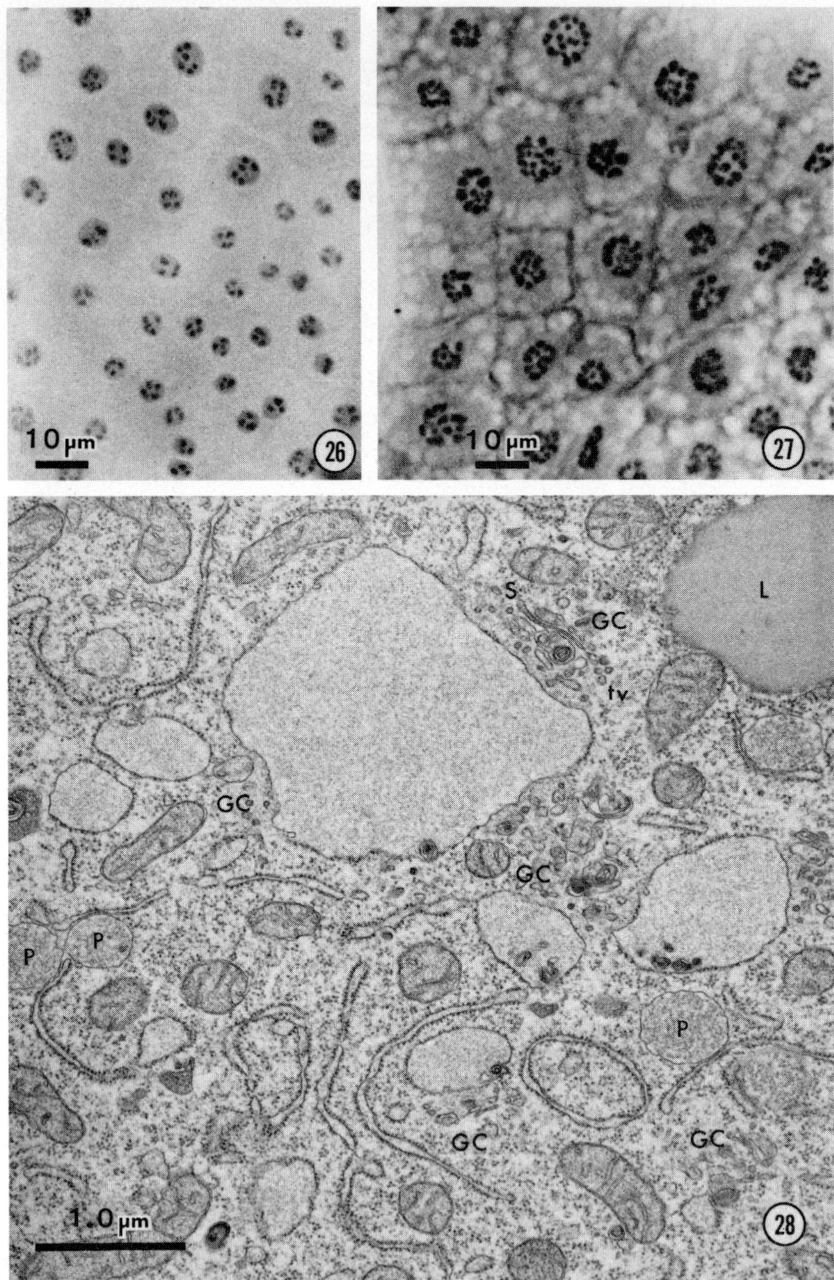

FIGS 26–30. *Calpodes* fat body in development. FIGS 26 and 27. Changes in the nucleoli and in polyploidy during intermoult preparation in fifth-instar larvae. Shortly after ecdysis to the fifth stage (Fig. 26) the nuclei clearly vary in their ploidy. Later in intermoult preparation (Fig. 27) all the cells are more polyploid but in addition the nucleoli are multilobed. Electron microscopy at this time shows numerous ribosomal precursor granules (Locke, M. and Huie, P., 1980b) and auto-radiography shows elevated uridine uptake (Locke, M., 1970). (×1000.)

FIG. 28. The structure of the fat body during intermoult preparation. Fifth instar larva shortly after ecdysis. There are many free polysomes. Some RER is distended, particularly in the cisternae destined to become the forming face of a Golgi complex (GC). Flattened RER cisternae confront the newly formed and growing peroxisomes (P). The newly formed GCs are limited to a smooth face of RER, a few transition vesicles (tvs) and a very few irregular saccules (S). (×29,000.)

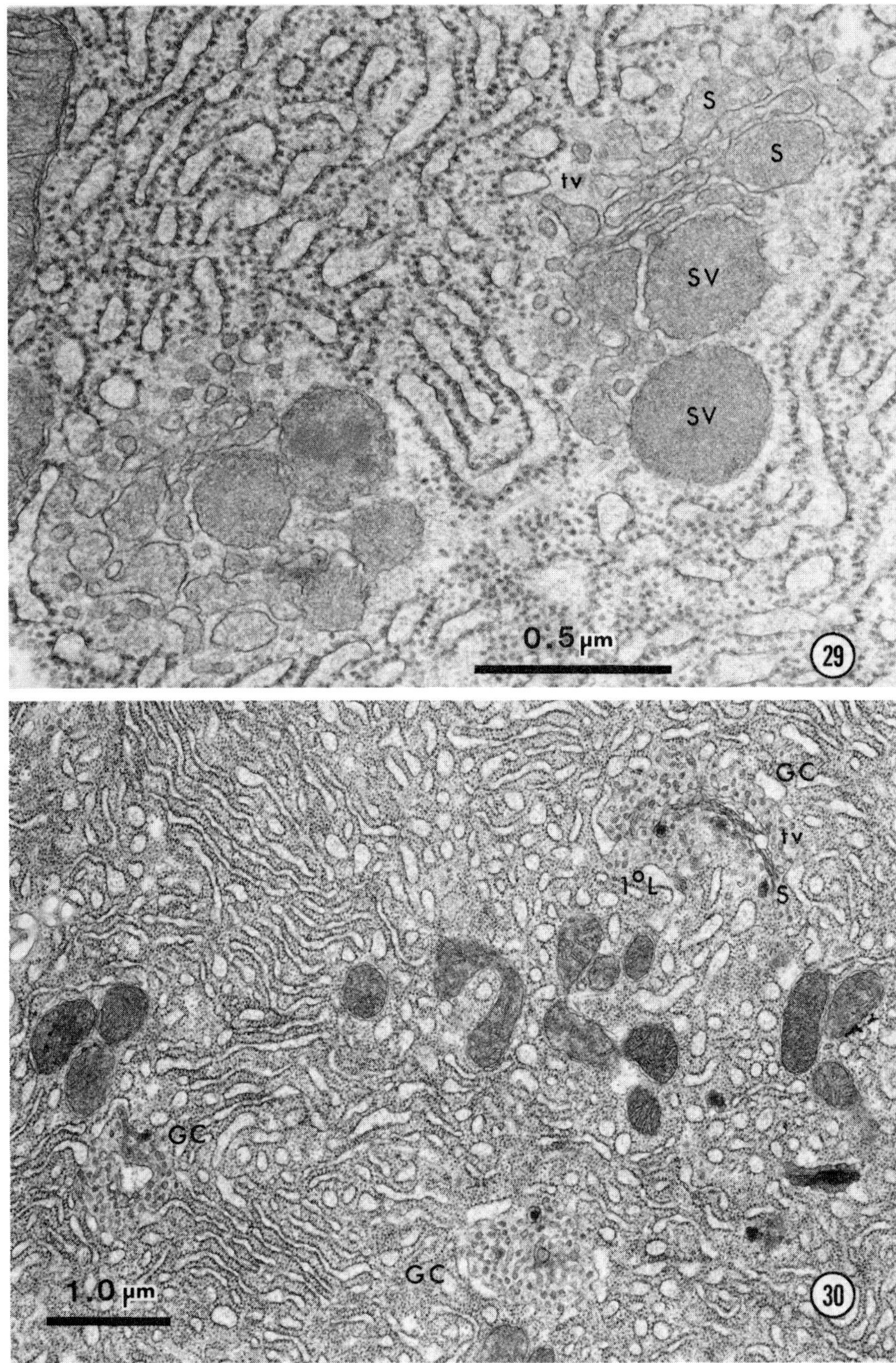

FIG. 29. During intermoult protein synthesis the main part of the fat body is composed of RER and well-developed Golgi complexes (GC) like these with many transition vesicles (tv), three or four saccules (S) and several secretory vesicles (SVs).

FIG. 30. Shortly after the second critical period the fat body stops secreting haemolymph proteins and the larva begins to pupate. The RER cisternae distend with material (Fig. 32) and the GCs change the form of the secretory face, losing secretory vesicles but forming many microvesicles that are presumably 1° lysosomes.

polyploidy is thought to provide multiple copies of the genome to expedite the selective transcription of large amounts of mRNA specific for vitellogenin synthesis (Nair, K. *et al.*, 1981; Irvine, D. and Brasch, K., 1981; J. Koeppe *et al.*, vol. 8).

Although polyploidy is characteristic of the fat body cells of larval Diptera the nuclei of both young and old imaginal fat body cells of *Drosophila melanogaster* are the same size. They do not contain the thick, banded chromosomes seen in larvae (Butterworth, F. *et al.*, 1965).

2.2 Synthesis, secretion and storage — structure in relation to function

2.2.1 LARVAL SYNTHESES

(a) *Preparation for synthesis.* The fat body of the last larval instar has been the target of most studies on cell preparation for larval syntheses for two reasons. First, as the larva approaches its maximum weight the fat body weight increases proportionately and thus provides a convenient source of material for biochemical analysis. Secondly, most investigations have been primarily interested in describing the preparation for metamorphosis.

The synthetic activity of the fat body reflects the cyclical nature of growth and development. In *Calpodes* fat body (Locke, M., 1970, 1980a,b), as in the epidermis, (Locke, M. and Huie, P., 1980b), there are intermoult/moult cycles of activity. In the fifth stadium both intermoult and moult are preceded by phases of preparation (Locke, M. and Collins, J., 1968). The first is a period of preparation for larval syntheses that begins shortly after ecdysis at the fourth to fifth moult. This is followed by a phase of massive larval syntheses which ends with preparation for metamorphosis. A few hours after ecdysis the cells are small, with little cytoplasm, in which the mitochondria appear conspicuous. There is little RER and GCs are small with only a few transition vesicles and an incomplete stack of saccules lacking secretory vesicles (SV) (Fig. 28). During the first $2\frac{1}{2}$ days after ecdysis the cells become basophilic with many free ribosomes, an increasing amount of RER and larger GCs with the complete complement of three or four saccules. Lipid droplets arise and enlarge slightly as small areas of glycogen appear. This preparative phase is characterized by RNA synthesis. Autoradiography showed that incorporation of [^{3}H]uridine into cytoplasmic RNA increased until 66 h after ecdysis, just before the main period of protein synthesis (Locke, M., 1970). Incorporation of ^{32}P into ribosomal RNA of *Philosamia cynthia* was demonstrated by Takahashi, S. (1966) in early fifth-stage larvae in what was considered to be the main phase of fifth-instar rRNA synthesis. RNA synthesis during the preparative phase includes ribosomal and messenger RNA.

The accumulation of ribosomes and early stages of RER proliferation coincide with other syntheses for cell growth and metabolism. In *Calpodes*, preparation for fifth-stage larval syntheses includes the loss of fourth-stage peroxisomes and the formation of a new fifth-stage generation of peroxisomes from the RER (section 2.2.4b) (Locke, M. and McMahon, J., 1971). Peroxisomes are synthesized early in the stadium before the main secretory phase begins, and function through the rest of the intermoult.

The ultrastructural events associated with preparation for syntheses may be induced by nutritional changes. When fourth- or fifth-instar *Calpodes* larvae are starved for even a few hours, the cells undergo a resorptive phase during which lipid droplets are depleted, mitochondria fuse and mitochondria and RER are destroyed by autophagy. Refeeding is followed by a phase of mitochondrial division, completed within 4 h, and proliferation of RER (Figs 37 and 38). In *Rhodnius*, mitochondrial destruction was observed after prolonged starvation of fourth instars, and refeeding resulted in synthesis of ribonucleoprotein, proliferation of RER, and mitochondrial recovery and division (Wigglesworth, V., 1967a, 1982). Bosquet, G. (1979) showed that upon refeeding starved *Bombyx mori* larvae, the polysome level in the fat body increased 2.5-fold, thus dramatically increasing the protein synthetic capacity of the cells. There are many reports of protein synthesis preceded by RNA synthesis, often in relation to hormonal stimulation (Table 2; L. Riddiford, vol. 8). Although few of these are cytological studies, the available data are consistent with the general principle that elaboration of the RER and GC are prerequisities for massive synthesis and secretion.

Table 2: Cell preparation for cycles of syntheses

Order and species	Observations	Reference*
Coleoptera		
Leptinotarsa decemlineata	Proliferation of RER, 1–2 days post-emergence. (A)†	7
Tenebrio molitor	RNA synthesis during the first 10 days shows cyclical pattern similar to protein synthesis. (A)	15
Dictyoptera		
Leucophaea maderae	Vitellogenin specific polysomes appear early in adult. ER–ribosome associations develop. (A)	8, 9
Nauphoeta cinerea	RER proliferates 3–7 days post-emergence. (A)	22
Diptera		
Aedes aegypti	RER proliferation 8 h after blood meal. (A)	2
Calliphora erythrocephala	RER, mitochondria, free ribosomes increase in third instar. (L) RER proliferation, 1–2 days post-emergence. (A)	6 18
Calliphora stygia	Microsomal RNA synthesis increases after crustecdysone treatment in third instar. (L)	14
Glossina austeni	RNA synthesis at emergence precedes protein synthesis. (A)	19
Hemiptera		
Rhodnius prolixus	Starved fourth instars, refed, nucleolus enlarges, cytoplasmic RNA increases. (L)	21
Lepidoptera		
Antheraea pernyi	RNA synthesis increases early in adult development, and after injury to pupa. (P)	1
Bombyx mori	Starved fifth instars, refed, polysome level increased. (L)	4
Calpodes ethlius	RNA synthesis, RER proliferation mitochondrial division. (L) RER, mitochondria increase shortly after emergence. (A)	13, 20 11
Hyalophora cecropia	RNA synthesis increases in early adult development and after injury to pupa. (P)	23
Philosamia cynthia ricini	Ribosomal RNA synthesis mainly in fifth instar. Uridine incorporation is highest on day 3 of fourth instar. (L)	3,17
Orthoptera		
Gryllus bimaculatus	RER proliferation 3–4 days post emergence. (A)	10
Locusta migratoria	Maximal uridine incorporation by day 8 of adult life. Polysomes specific for vitellogenin appear, RER and ribosomes increase. (A).	5, 12, 16

* 1, Barth, R. *et al.* (1964); 2, Behan, M. and Hagedorn, H. (1978); 3, Berry, S. *et al.* (1967); 4, Bosquet, G. (1979); 5, Couble, P. *et al.* (1979); 6, de Priester, W. and van der Molen, L. (1979); 7, Dortland, J. and Esch, T. (1979); 8, Engelmann, F. (1977); 9, Engelmann, F. and Barajas, L. (1975); 10, Favard-Sereno, C. (1973); 11, Larsen, S. (1970, 1976); 12, Lauverjat, S. (1977); 13, Locke, M. (1970, 1980a); 14, Neufeld, G. *et al.* (1968); 15, Pemrick, S. and Butz, A. (1970); 16, Reid, P. and Chen, T. (1981); 17, Takahashi, S. (1966); 18, Thomsen, E. and Thomsen, M. (1974); 19, Tobe, S. and Davey, K. (1974a); 20, Tychsen, C. (1978); 21, Wigglesworth, V. (1967a); 22, Wüest, J. (1978); 23, Wyatt, G. and Linzen, B. (1965).

† (A), adult; (L), larva; (P), pupa.

(b) *Phases of larval synthesis and secretion.* The switch at E (ecdysis) + 66 h from the phase of cell preparation to larval synthesis in *Calpodes* is characterized by an increase in the amount of RER and the number and size of GC secretory vesicles (SV, Fig. 29) together with increases in the rate of lipid and glycogen accumulation (Locke, M. and Collins, J., 1968; Locke, M., 1970, 1980a,b). The build-up of secretory protein in the intermoult is in the haemolymph, not the fat body. SVs are quickly secreted, suggesting that the control of exocytosis is not separable at this time from SV formation. These fat body SVs are the principal source of the large variety of haemolymph proteins secreted at all stages (reviewed by Wyatt, G. and Pan, M., 1978). In particular during the last larval intermoult they give rise to the haemolymph storage proteins. Storage proteins may also be synthesized in earlier stadia (Tojo, S. *et al.*, 1981; L. Levenbook, vol. 10). They have been isolated and characterized for several species and are generally

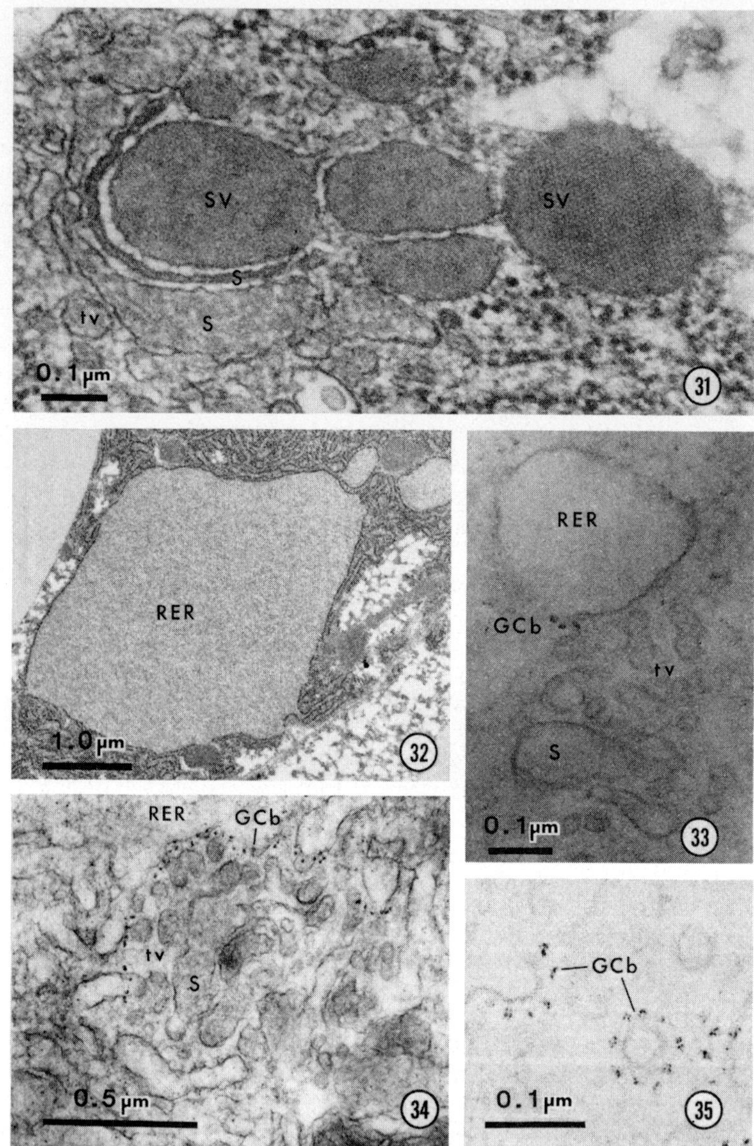

FIGS 31–35. Golgi complexes in development (see also Figs 28–30). FIG. 31. An intermoult GC showing a characteristic crystalline pattern of a secretory vesicle (SV). (× 140,000.)

FIG. 32. Distension of the rough endoplasmic reticulum (RER) (see also Figs 1 and 28). The immediate response to moulting hormone, both naturally and experimentally, is for the RER to distend as the GCs stop transporting secretory protein (Fig. 30). Fifth-stage *Calpodes* larva shortly before the beginning of pupation. (× 19,000.)

FIGS 33–35. The GC beads (GCb) and GC development. Preparations stained only with bismuth to show the GC beads.

FIG. 33. A newly formed GC similar to those in Fig. 28. At this stage the smallest GCs may have only a single ring of beads. (× 130,000.)

FIGS 34 and 35. The beads in mid-instar GCs similar to Figs 29 and 31. There are many rings on the smooth face of the RER where transition vesicles arise. (Fig. 34 × 66,000; from Locke, M. and Huie, P., 1976b; Fig. 35 × 210,000, from Locke, M. and Huie, P., 1976a.)

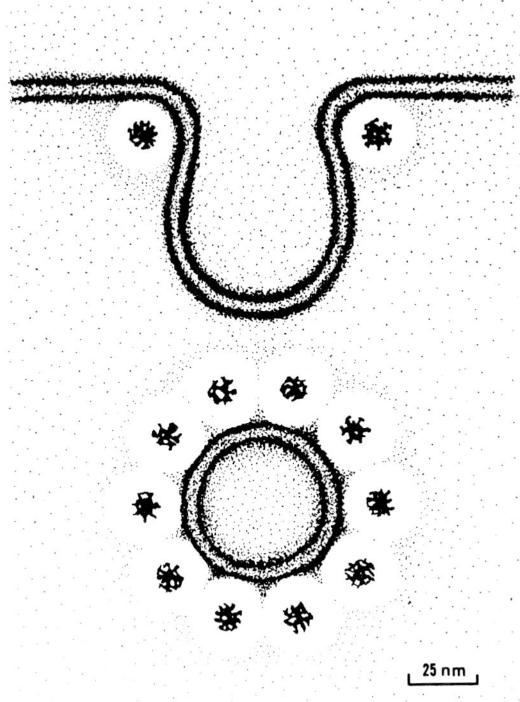

FIG. 36. The arrangement of the GC beads around transition vesicles. The beads are arranged in rings just above the smooth face of the RER where transition vesicles arise to form the Golgi complex. Transition vesicles pop through the rings which they leave behind them as they go on to become part of the outer saccule. (From Locke, M. and Huie, P., 1967a.)

similar in size, composition and among the Diptera, immunological cross-reactivity has been observed (Wyatt, G., 1980). These storage proteins secreted into the haemolymph are later sequestered and stored in the fat body at metamorphosis (section 3.1). No larval function for these proteins has been identified other than accumulation of reserves. In *Chironomus*, the fat body of the last instar secretes haemoglobins (Bergtrom, G. and Robinson, J., 1977; Schin, K. *et al.*, 1977) which are regarded as primarily storage proteins although they may also function in oxygen transport (Thomson, J., 1975). Other haemolymph proteins secreted by the larval fat body include lipoproteins such as the diglyceride carrier, lipophorin (Pattnaik, N. *et al.*, 1979; Chino, vol. 10), JH carrier protein (Nowock, J. *et al.*, 1975; W. Goodman and E. Chang, vol. 7) and JH esterases (B. Hammock, vol 7). In *Trichoplusia ni* (Sparks, T. *et al.*, 1979) and in *Manduca sexta* (Vince, R. and Gilbert, L., 1977) there is a small peak of JH esterase activity immediately before the larval–pupal ecdysis. The fat body evidently retains its secretory ability at the time of metamorphosis, even when it is no longer primarily engaged in secretion.

Although there are no striking cytological correlates to this variety of syntheses for secretion, it is evident that not all products are secreted all the time, and the overall complement of secreted proteins depends upon the ambient controlling factors. Observations on *Calpodes* suggest that the changing pattern of some syntheses may be viewed ultrastructurally. Secretory vesicles with crystalline cores have been observed late in the fifth stadium but not at other ages (Fig. 31). If these patterns are specific for vacuole contents, this may provide a marker for identifying proteins secreted at different times.

The end of the main secretory phase is characterized by a switch in GC structure with loss of SVs and an increase in 1° lysosomes (Fig. 30). This is followed by autophagy of peroxisomes and mitochondria, and finally by the isolation and destruction of most of the RER in autophagic vacuoles (section 3.2). This regression of the synthetic apparatus is characteristic of the last larval stage and is controlled hormonally (M. Locke, vol. 2). Cycles of regression may occur in earlier stages but are less obvious than in a metamorphic moult.

(c) *Storage and mobilization of reserves*

Protein, lipid, glycogen. The fat body is not only a source of secreted products but is also a principal storage organ. It may contain deposits of lipid, glycogen and protein. Protein storage granules are mainly associated with metamorphosis but they also occur in relation to gestation in adult females of the viviparous cockroach, *Diploptera punctata* (Stay, B. and Clark, J., 1971) and in relation to diapause in larvae of *Diatraea* (Brown, J. and Chippendale, G., 1977) and adults of *Leptinotarsa decemlineata* (de Loof, A. and Lagasse, A., 1970). Protein storage granules also occur in nymphal stages of *Periplaneta americana* (Fig. 3). Storage of lipid and glycogen occurs at all ages, and reflects nutritional and metabolic requirements. Mobilization of stored material occurs in relation to moulting, adult syntheses, and starvation. Wigglesworth,

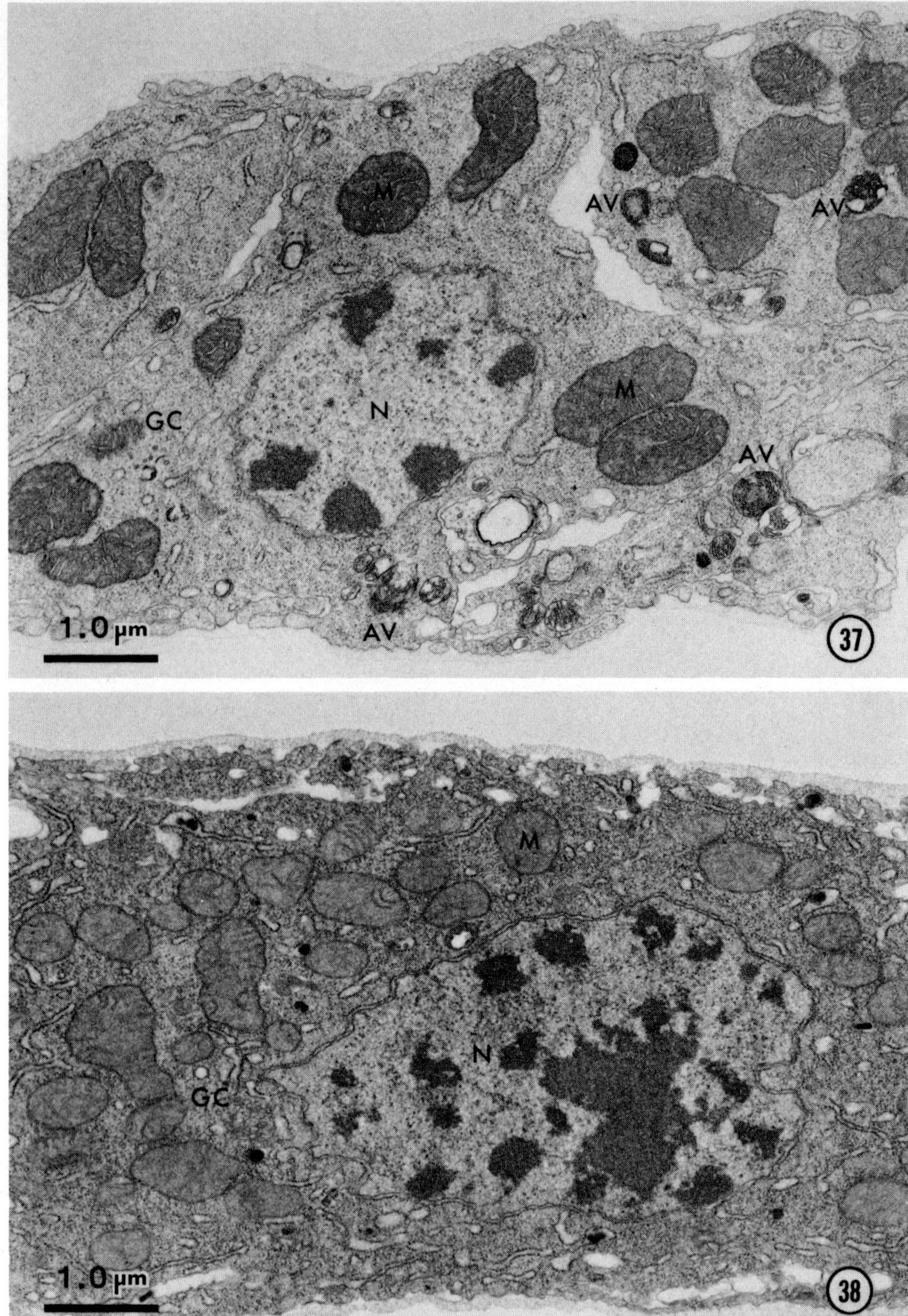

FIGS 37 and 38. The initiation of development after starvation by feeding in fifth-stage *Calpodes* larvae.

FIG. 37. After 24 h of starvation following ecdysis many mitochondria have fused to make large spherical structures. There is little RER and the GCs are small with few transition vesicles (tvs). There is some autophagy (AV = autophagic vacuole). (×20,000.)

FIG. 38. Only 4 h after feeding the mitochondria have divided back to normal dimensions. There is more RER and the enlarged GCs have many tvs, but peroxisomes have yet to reform or lipid or glycogen to be deposited. (×20,000.)

V. (1942) observed that glycogen was depleted in *Rhodnius* during synthesis of chitinous cuticle, whereas proteins were mobilized during deposition of cuticular proteins. During starvation all reserves were mobilized, and a fully starved *Rhodnius* was depleted not only of reserves but also of mitochondria which are destroyed by autophagy (Wigglesworth, V., 1967a). Starvation has similar effects on *Blaberus* (Walker, P., 1965) and on *Calpodes* (Figs 37 and 38). Refeeding after starvation stimulates a burst of syntheses, organelles such as RER and mitochondria are quickly restored, followed by the formation of lipid droplets and other reserves. Although the biochemistry of lipid and glycogen synthesis and breakdown are well understood, little is known of the ultrastructural events during their deposition. The appearance and disappearance of these reserves have been frequently remarked upon (e.g. in *Leptinotarsa*, Labour, G., 1974), but little attention has been given to their genesis.

Tyrosine storage vacuoles. At certain times the fat body contains large fluid-filled vacuoles (Figs 39, 56 and 57). They were first described in live *Chironomus* larvae by Voinov, V. (1927) who saw small vacuoles grow and fuse to form single large vacuoles in the centre of each cell. Wigglesworth, V. (1942) found vacuoles in newly ecdysed *Aedes* larvae and also in *Rhodnius* that had been fed after prolonged starvation (Wigglesworth, V., 1967a). According to Wigglesworth, V. (1982) the watery vacuoles in *Rhodnius* fat body are derived from enlarged autophagic vacuoles that have lost their contents except for a rim of lipofuscin deposited around the inner membrane surface. The relation of these watery vacuoles to urate storage vacuoles (section 1.3.2) or tyrosine storage vacuoles is therefore uncertain. In the fat body of *Leptinotarsa* larvae, vacuoles arise during the first stadium and decline at ecdysis. They then grow during the next three stadia before decreasing in size in the pupa. They are absent by the third day of adult life (Labour, G., 1970, 1974; de Loof, A., 1972; Dortland, J. and Esch, T., 1979). Vacuoles also occur in the fat body of *Carausius morosus* (Rutschke, E. and Brozio, F., 1975) and of *Calliphora* larvae (de Priester, W. and van der Molen, L., 1979). These large, intermittent structures have received surprisingly little attention. At ecdysis to the fourth instar in *Calpodes*, the vacuoles originating in the third instar occupy about 40% of the fat body volume (McDermid, H. and Locke, M., 1983) (Fig. 42). These vacuoles disappear within the next 6–12 h and a new population of fourth-instar vacuoles forms between 18 and 24 h after ecdysis. Brief (12 h) post-ecdysial starvation synchronizes their formation from the beginning of feeding (the technique can also be used to synchronize other kinds of cell behaviour such as mitochondrial division). The peak period for new vacuole formation occurs between 6 and 8 h after feeding. Surface infolds of plasma membrane (Figs 40 and 41) form provacuoles which fuse to form vacuoles. By the further fusion and growth of these vacuoles each cell comes to contain a single giant vacuole 20–40 nm in diameter. They reach a maximum size at about 24 h before the end of the fourth stadium and then disappear abruptly in a 12 h period which centres on ecdysis to the fifth instar. In the fifth stadium the vacuoles are never as abundant as in the fourth, although they become a little more common prior to pupation. Vacuoles are not found during the pupal stadium or at the pupal–adult moult (Larsen, W., 1976). The vacuoles are, therefore, a conspicuously larval, intermoult–moult-related phenomenon.

The vacuoles contain tyrosine and changes in the proportion of cell volume occupied by vacuoles correspond to changes in tyrosine content (Fig. 42). The decline in fat body tyrosine levels and disappearance of the vacuoles at about the time of the fourth to fifth instar ecdysis correspond to an increase in haemolymph tyrosine concentration (Fig. 43). Thus, fat body vacuoles release their stored tyrosine into the haemocoel during moulting to give a maximal concentration immediately after ecdysis. The disappearance of tyrosine from the haemolymph (Fig. 43) coincides with a need for phenolic precursors by the epidermis for cuticular tanning. Similar fat body vacuoles occur in *Leptinotarsa*, *Phormia*, *Manduca*, *Hyalophora* and *Chironomus*, suggesting that the storage of phenolic precursors, particularly tyrosine, in vacuoles may be a generality in insects.

2.2.2 Adult Syntheses

(a) *The fat body at emergence.* The fat body of the adult is primarily a site of storage and synthesis of molecules which are mobilized for reproduction

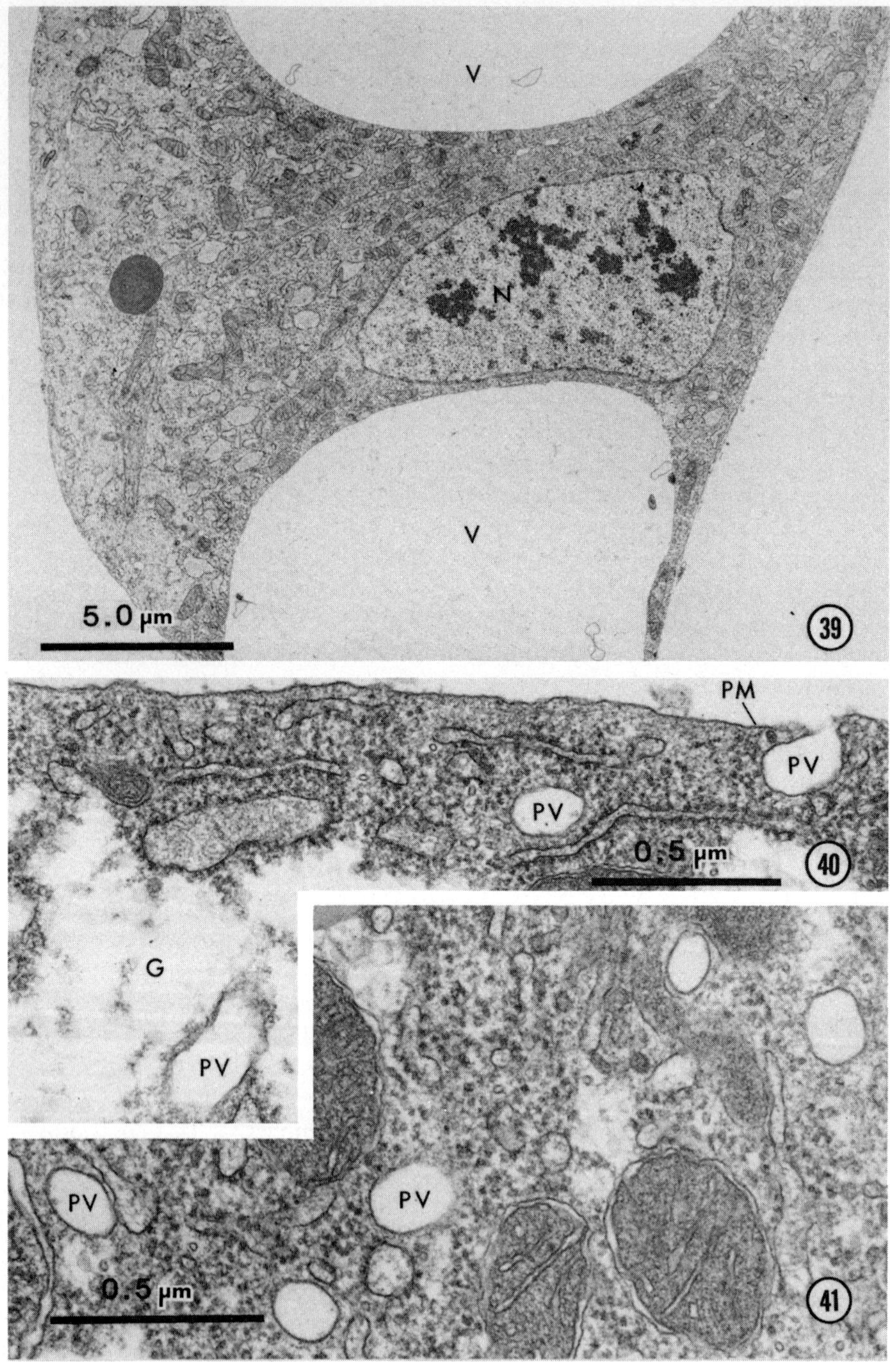

FIGS 39–43. Tyrosine storage vacuoles and their formation (see also Figs 56 and 57). FIG. 39. The fat body from mid-fourth-stage *Calpodes* larvae has many very large watery vacuoles (V) that come to constitute half of the volume of the cell (compare with Figs 42 and 43). (×7000.)

FIGS 40 and 41. At the beginning of the fourth stadium, plasma membrane infolds separate from the surface as provacuoles (PV) that fuse to make the large vacuoles seen in later stages. (Fig. 40 ×67,000; Fig. 41 ×77,000.)

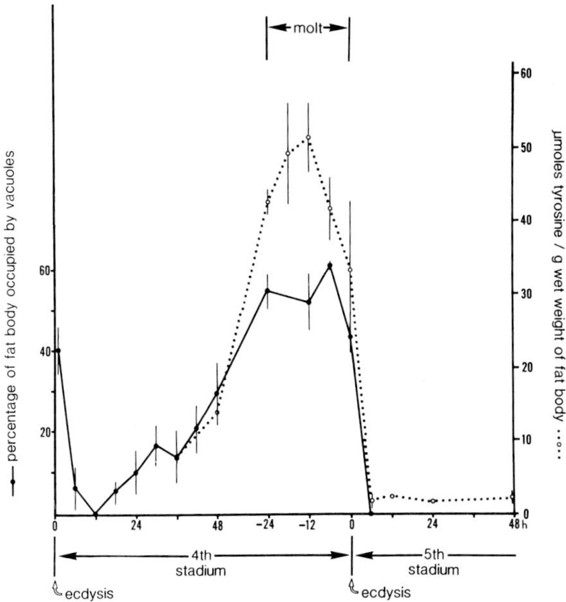

FIG. 42. Changes in the proportion of the fat body of *Calpodes* occupied by vacuoles during the fourth and early 5th stadia (——) compared with the free tyrosine in the fat body (·····). The size of the vacuoles and the tyrosine content of the fat body go hand in hand. (From McDermid, H. and Locke, M., 1983.)

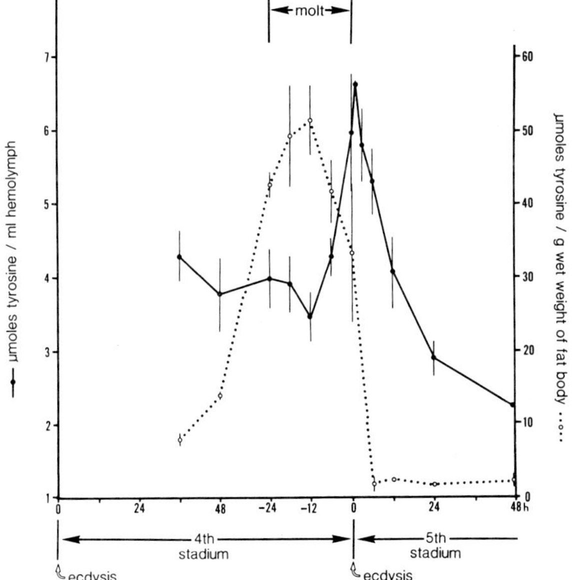

FIG. 43. The correlation between loss of tyrosine from the fat body and its elevation in the haemolymph during the fourth and fifth stadia of *Calpodes*. As the vacuoles decline in the fat body so tyrosine is briefly elevated in the haemolymph. (From McDermid, H. and Locke, M., 1983.)

and flight. Ultrastructural studies of adult fat body have been predominantly concerned with development in the female, especially in relation to vitellogenin synthesis and secretion. At emergence, the cells have three notable features: large lipid droplets, extensive glycogen deposits and the residue of protein storage granules laid down in the premetamorphic stadium (Table 3). The extent to which protein granules persist varies with species and sex. In most species in which the juvenile fat body is reorganized to form the adult tissue (see section 3.3) the storage granules persist throughout the pupal stage and are slowly depleted early in adult life. In silkmoths, development of the adult fat body coincides with ovarian maturation so that mobilization of the stored proteins and other preparative changes actually occur in the pupal stadium (Bhakthan, N. and Gilbert, L., 1972). In other Lepidoptera the granules persist for several days in the adult (Larsen, W., 1972), while in cockroaches they may be retained throughout adult life though their numbers fluctuate with oviposition cycles (Stay, B. and Clark, J., 1971). In the higher Diptera, the new tissue that develops after the larval fat body degenerates contains protein granules. Neither their origin nor fate have been described, although at some stage they contain hydrolytic enzymes (Butterworth, F. *et al.*, 1972; Thomsen, E. and Thomsen, M., 1974).

There is a marked sexual dimorphism in fat body ultrastructure that reflects the differing role of trophocytes in the reproductive process and sometimes in flight behaviour. Male adult fat body is characterized by relatively large lipid and glycogen stores and little ER (Odhiambo, T., 1967; Lauverjat, S., 1977; Stoppie, P. *et al.*, 1981). The cytoplasm is generally sparse and distributed between the large lipid droplets and in the perinuclear region. The large lipid store in male *Hyalophora* has been correlated with their energy requirements for very active flying in mate seeking (Bhakthan, N. and Gilbert, L., 1972). The overall impression is that the male tissue is not involved in massive synthesis of protein for export.

(b) *Preparation for syntheses.* The ultrastructure of adult female tissue shows phases of activity which are remarkably similar to those in larval fat body (section 2.2.1a). In the larva the phases are intermoult preparation, intermoult larval syntheses and cell

Table 3: Ultrastructure of fat body of pharate and newly emerged adults

Order and species	Description	Reference*
Coleoptera		
Leptinotarsa decemlineata	Cells contain glycogen, lipid droplets, protein granules, vacuoles. Little RER. (A)†	4
Dictyoptera		
Nauphoeta cinerea	Mitochondria, lipid, glycogen, protein granules abundant. RER, Golgi complexes sparse. (A)	13
Diptera		
Aedes aegypti	Nuclear chromatin condensed, large nucleoli present. Mitochondria, lipid, glycogen abundant. Some protein granules, free ribosomes. RER sparse. (A)	1, 8
Calliphora erythrocephala	Nucleus spheroid, large nucleolus. Mitochondria, glycogen, lipid, dense bodies, free ribosomes, present. RER sparse. (A)	12
Dacus tryoni	Mitochondria, lipid, glycogen evident. RER slightly distended. (PA)	6
Sarcophaga bullata	Lipid, glycogen abundant. Mitochondria, RER sparse. (A)	11
Lepidoptera		
Calpodes ethlius	Cells show steady loss of lipid, MVBs absent at E−40 h, appear at E−30 h, peroxisomes appear at E−24 h, RER sparse. (PA) Autophagic vacuoles, protein storage granules with crystalline cores, MVBs peroxisomes, lipid droplets, dividing mitochondria. RER increases by E+24 h. (A)	9
Galleria mellonella	Autophagic vacuoles, lipid droplets present, protein storage granules partly degraded. (PA)	5
Hyalophora cecropia	Lipid, glycogen, present. Protein storage granules with crystalline cores steadily degenerate. (PA) Mitochondria, lipid, glycogen abundant. Free ribosomes, protein granules present in females, absent in males. (A)	2
Orthoptera		
Gryllus bimaculatus	Mitochondria, lipid droplets abundant, autophagic vacuoles present, RER, Golgi complexes sparse.	7
Locusta migratoria	Glycogen, lipid droplets, mitochondria often associated with lipid droplets, RER sparse. (A)	3
	Nuclei highly indented, large nucleolus, lipid, glycogen, protein granules (labelled lysosomes) present. (A)	10

* 1, Behan, M. and Hagedorn, H. (1978); 2, Bhakthan, N. and Gilbert, L. (1972); 3, Couble, P. *et al.* (1979); 4, Dortland, J. and Esch, T. (1979); 5, Dutkowski, A. (1974); 6, Evans, J. (1967); 7, Favard-Sereno, C. (1973); 8, Kan, S.-P. and Ho, B.-P. (1972); 9, Larsen, S. (1976); 10, Lauverjat, S. (1977); 11, Stoppie, P. *et al.* (1981); 12, Thomsen, E. and Thomsen, M. (1974); 13, Wüest, J. (1978).
† (A), adult; (PA), pharate adult; MVB, multivesicular body.

remodelling for metamorphosis. In the adult these correspond to the previtellogenic phase, vitellogenin (Vg) synthesis and secretion, followed by the recovery or lysosomal phase.

The cells prepare for synthesis in the previtellogenic phase when the RER and GCs, poorly developed at emergence, quickly become abundant as the lipid and glycogen decrease (Table 2). In *Locusta*, RER proliferation coincides with increases in total tissue RNA, uridine incorporation into RNA and the appearance of a class of polysomes specific for Vg synthesis (Chen, T. *et al.*, 1976, 1979; Reid, P. and Chen, T., 1981).

(c) *Vitellogenin synthesis and secretion.* The vitellogenic phase is characterized by well-developed RER and GCs. The plasma membrane may be highly invaginated, and, in some species, e.g. *Leptinotarsa decemlineata* (Dortland, J. and Esch, T., 1979) and *Calliphora erythrocephala* (Thomsen, E. and Thomsen, M., 1974; Jensen, P. *et al.*, 1981), there are relatively large secretory vesicles which, in the latter, have been shown to contain Vg. In *Locusta migratoria*, Vg was localized in RER cisternae, in GC vesicles and extracellularly in plasma membrane invaginations (Couble, P. *et al.*, 1979). The vitellogenic phase is closely linked in time with the development of

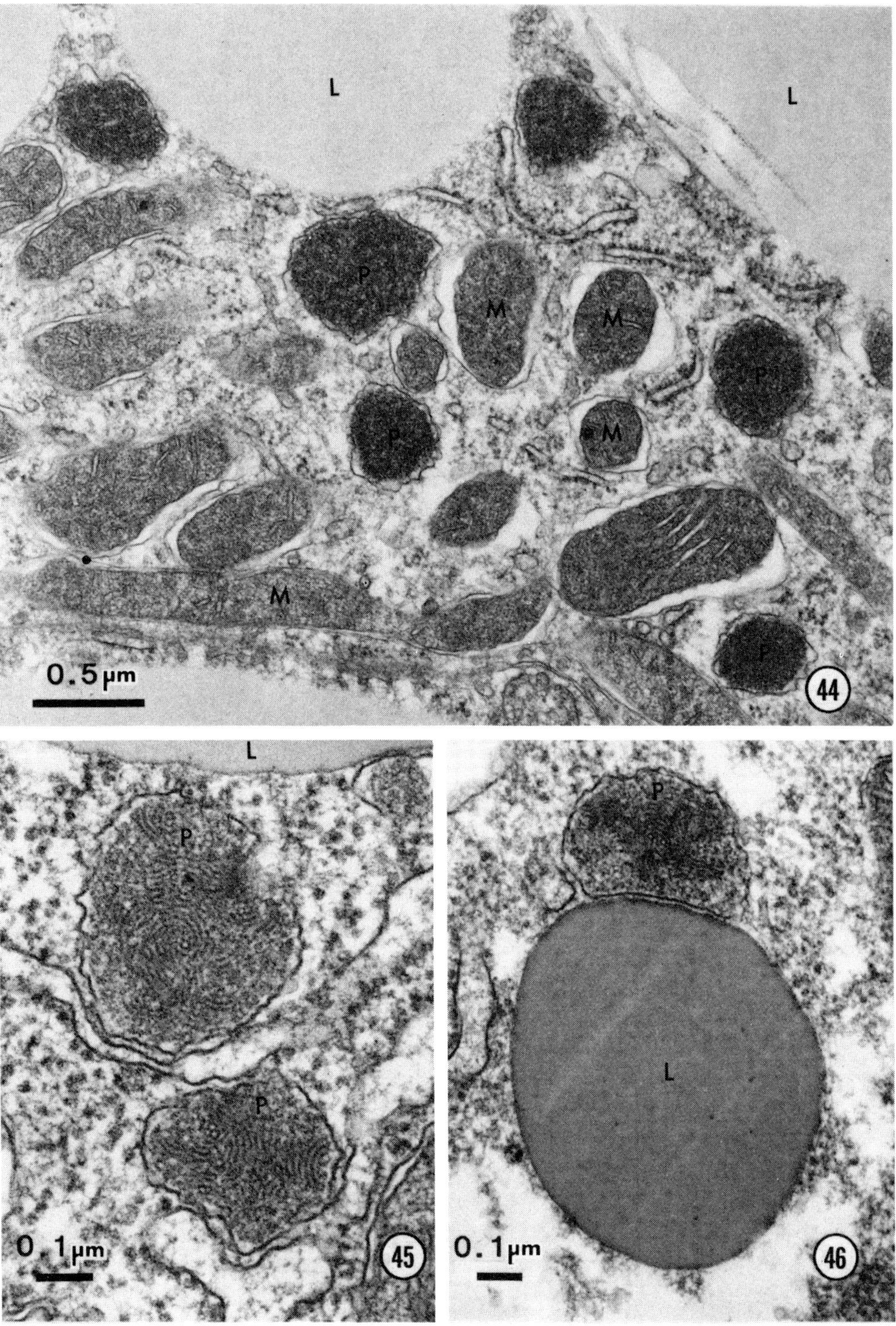

FIGS 44–46. Peroxisomes. FIG. 44. Peroxisomes in *Periplaneta* trophocytes. (×43,000.)

FIG. 45. The characteristic whorled pattern present in *Calpodes* peroxisomes. The core is the dense homogeneous mass usually indented on one side. These growing peroxisomes have a close relation with confronting cisternae of RER. (×96,000.)

FIG. 46. Peroxisomes are often closely applied to newly forming lipid droplets. (×82,000.)

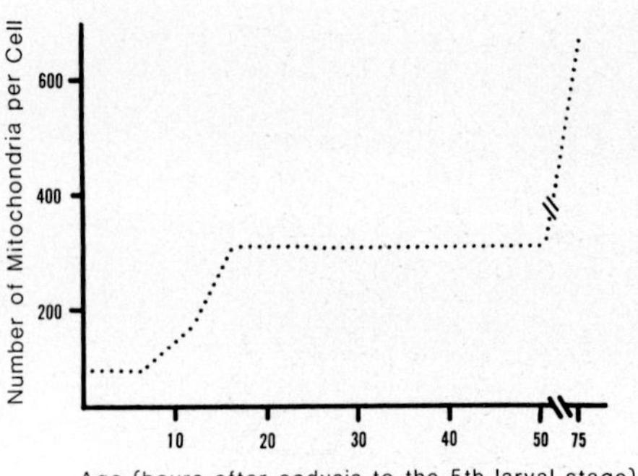

FIGS 47–55. The growth, division and differentiation of fat body mitochondria. FIG. 47. The number of mitochondria per cell during intermoult preparation at the beginning of the fifth larval stage of *Calpodes*. The level is initially low as a result of cell division. It then builds up by mitochondrial division (Figs 52 and 53) as the cells become more polyploid. (After Tychsen, C., 1978.)

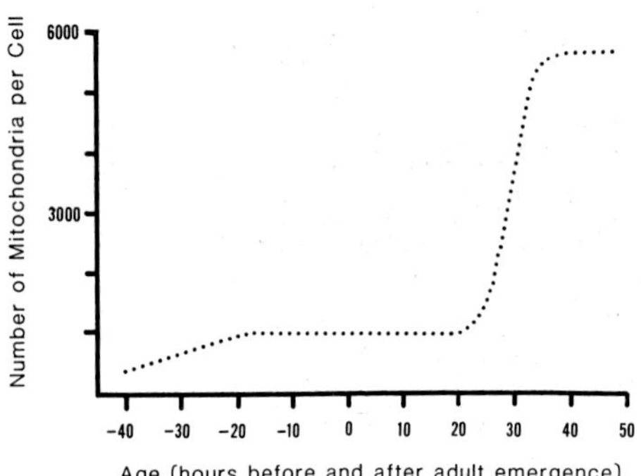

FIG. 48. The number of mitochondria per cell at the beginning of adult life in *Calpodes*. The number of mitochondria per cell is initially low because of prepupal autophagy (Fig. 67). Mitochondria grow and divide (Figs. 49 and 55) to repopulate the cells shortly after emergence. (After Larsen, S., 1970.)

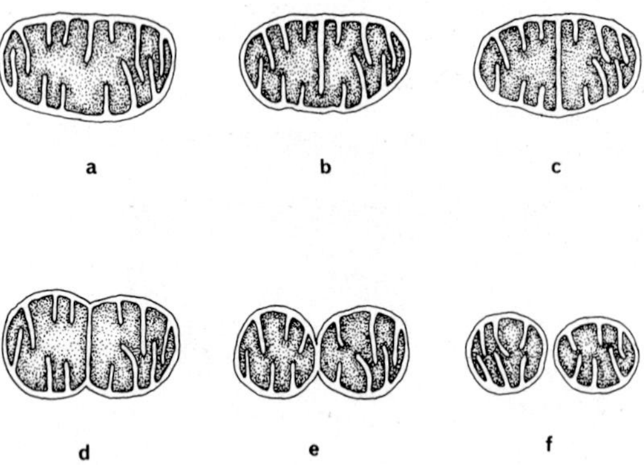

Fig. 49. The sequence of mitochondrial division in newly emerged adult *Calpodes* deduced from profiles (Fig. 55). (From Larsen, S., 1970.)

terminal oocytes in the ovary and ends with oocyte maturation.

(d) *Post-synthetic recovery phase.* The extent to which fat body cells are reorganized after termination of Vg synthesis varies. Some work clearly shows repeating cycles of activity. Rinterknecht, E. and Roussel, J.-P. (1978) reported a reduction in RER and an increase in free ribosomes. In *Locusta*, Couble, P. and co-workers (1979) described a distinct cycle associated with vitellogenesis which ends with the appearance of lysosomes. The content of the lysosomes is not clear, but bulk lysis of ribosomes seems unlikely since they persist for later Vg synthesis (Chen, T. and Wyatt, G., 1981). Lauverjat, S. (1977), on the other hand, reported that after ovarian maturation the fat body is modified in a way that does not reflect the cyclic nature of vitellogenesis and subsequent egg-laying. Lysosomal structures were continuously present in older females. These observations may be reconciled by the variable extent to which the fat body is engaged in the cyclical synthesis of Vg and the continuous synthesis of other proteins. Haemolymph proteins such as lipophorin are also secreted during ovarian cycles of development but not in an overtly cyclic manner (Gellissen, G. and Wyatt, G., 1981). Cellular change related to vitellogenesis is much more obviously cyclic in the Diptera where the fat body, as a consequence of the isolation and destruction of RER, reverts to a morphology similar to that seen in the previtellogenic stage (Thomsen, E. and Thomsen, M., 1974; Behan, M. and Hagedorn, H., 1978; Stoppie, P. *et al.*, 1981).

(e) *Factors affecting cell structure related to vitellogenesis.* Vitellogenin synthesis is controlled by more than one humoral factor (J. Koeppe *et al.*, H. Hagedorn, vol. 8). The influence of gland ablation and replacement or hormone therapy therefore varies between species. Some vitellogenin synthesis is clearly JH-dependent (reviewed by Engelmann, F., 1979). In *Locusta*, allatectomy inhibits Vg synthesis and the associated cytological changes, including those in the nucleus (section 2.1). Indeed, abnormal lipid accumulation rather than depletion follows corpora allata removal. Cell structure is rapidly returned to normal by implantation of active corpora allata (Lauverjat, S., 1977) or by the injection or topical application of JH or its analogues (Rinterknecht, E. and Roussel, J.-P., 1978; Couble, P. *et al.*, 1979). In

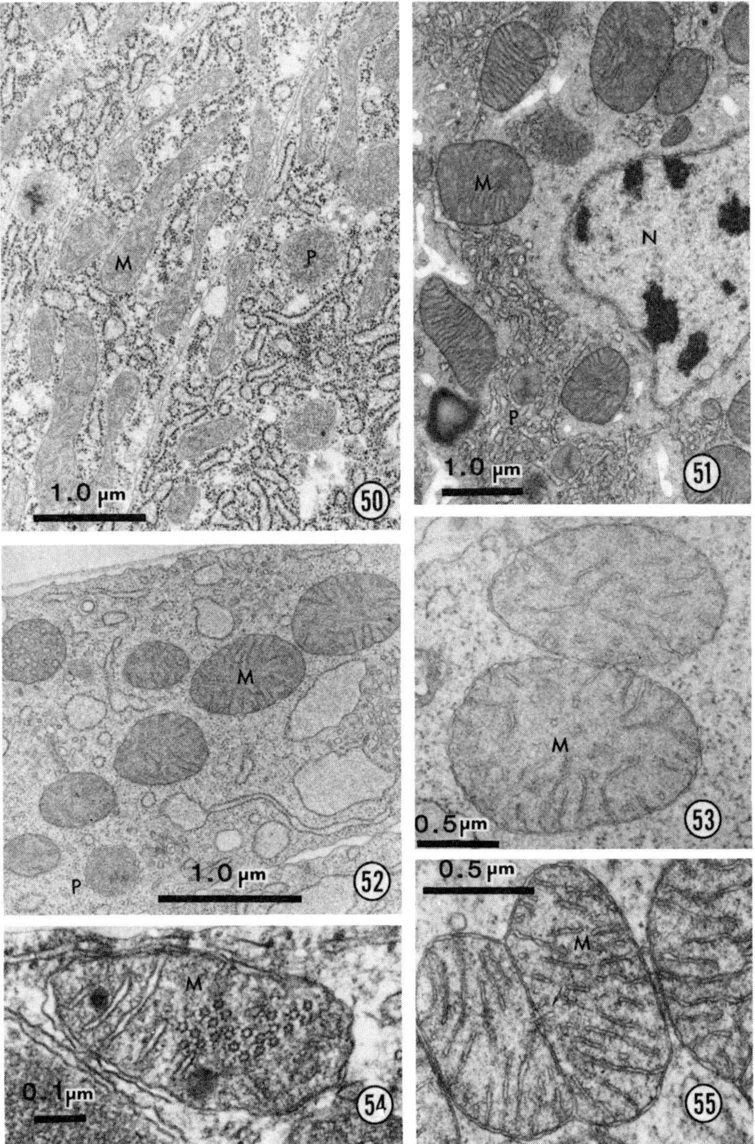

FIG. 50. The shape of larval mitochondria (mid-fifth-instar larval *Calpodes*). The mitochondria are rod-shaped. (×24,000.)

FIG. 51. The shape of adult mitochondria (6-day-old *Calpodes* adult). The mitochondria are oval or spherical. (×17,000.)

FIGS 52 and 53. Shape change and division in larval mitochondria.

FIG. 52. About a day after ecdysis to the fifth-stage, *Calpodes* larval mitochondria round-up. (×30,000.)

FIG. 53. They then divide after which they return to their elongate form (Fig. 50). (×36,000.)

FIG. 54. Differentiation of tubular cristae. Not all fat body mitochondria have lamellate cristae. Cristae may sometimes be tubular as is this *Periplaneta* fat body. (×110,000.)

FIG. 55. Mitochondrial division in recently emerged adult *Calpodes* fat body. (×48,000; from Larsen, S., 1970.)

Leucophaea, JH stimulated the incorporation of labelled choline and orthophosphate into microsomal membranes *in vivo* and *in vitro* (Della-Cioppa, G. and Engelmann, F., 1980). This membrane synthesis was temporally coupled to synthesis of Vg. Although no ultrastructural evidence was presented, the results are consistent with the observations of Couble, P. and co-workers (1979) on the JH-dependent increase in RER in *Locusta*. Since vitellogenesis in *Locusta* is JH-dependent it is not surprising that ovary removal does not prevent the development of fat body structure appropriate for protein synthesis. In fact, Lauverjat, S. (1977) reported that this process was slightly accelerated when compared to controls. Further, the peripheral fat body cells, which Lauverjat claims do not normally undergo differentiation for large-scale protein synthesis, develop the same appearance as the periovarian cells after ovariectomy. Chen, T. and Wyatt, G. (1981) have shown that in *Locusta* Vg continues to be produced and accumulates to abnormally high levels in the haemolymph in ovariectomized females. Ovariectomy of *Leucophaea maderae* results in accumulation of Vg in the haemolymph and fat body (Engelmann, F., 1978) and in *Leptinotarsa* also results in an increase in haemolymph Vg concentration (de Loof, A. and de Wilde, J., 1970). Surprisingly, in *Leptinotarsa* this surgery results in hypertrophy of the peripheral fat body cell type without any apparent increase in their protein synthetic machinery.

In the Diptera, Vg synthesis may be influenced by ecdysteroid secreted by the ovary because ovariectomy has a profound effect on vitellogenesis (H. Hagedorn, vol. 8). In *Calliphora*, ovariectomy does not prevent the formation of a well-developed protein synthetic apparatus in the fat body, but there are no secretory vesicles in the GCs. Injection of 20-hydroxyecdysone counteracts the effect of ovariectomy (Thomsen, E. and Thomsen, M., 1978). Immunocytochemical studies have shown that Vg is not present in the GCs and secretory vesicles of ovariectomized flies, but can be localized at these sites in controls and in adults injected with ecdysteroid after ovariectomy (Thomsen, E. *et al.*, 1980). In *Aedes aegypti* the role of ecdysteroids is more controversial. Behan, M. and Hagedorn, H. (1978) observed that in newly emerged adults the fat body cells show no signs of massive protein synthesis and secretion until several hours after a blood meal. The changes in ultrastructure, and the stimulation of Vg synthesis, correlate well with the known sequence of hormonal events that include secretion of the egg development neurosecretory hormone by the brain, and ecdysone secretion by the ovaries after the blood meal. However, Lea, A. (1982) has concluded that ovarian ecdysone does not stimulate vitellogenesis by the fat body even in blood-fed animals. On the other hand, Stoppie, P. and co-workers (1981) report that fat body in male *Sarcophaga bullata*, which does not normally generate large amounts of RER or plasma membrane infolding, can be induced to do so by the ingestion of moulting hormone that also induces subsequent Vg synthesis in males (Huybrechts, R. and de Loof, A., 1977). In *Anopheles stephensi* also, 20-hydroxyecdysone stimulates Vg synthesis (Redfern, C., 1982). In *Aedes* and *Calliphora* it is not clear what stimulates the development of cytoplasmic components necessary for large-scale Vg synthesis. Thomsen, E. and Thomsen, M. (1978) concluded that in *Calliphora* a factor from the medial neurosecretory cells is involved, although the evidence was not presented. The role of JH must also be questioned since the fat body must be exposed to JH before it becomes competent to respond to ovarian ecdysteroids (Flanagan, T. and Hagedorn, H., 1977). However, the cells show no evidence of major ultrastructural change following the natural period of JH exposure except for a slight alteration in the appearance of the RER (Behan, M. and Hagedorn, H., 1978).

One question which has so far remained unanswered is whether the oenocytes scattered among the trophocytes in the Diptera secrete ecdysteroids which might have a role in the control of ultrastructure. Since males of *Aedes*, and females after ovariectomy, contain about one-third the amount of ecdysone of normal-fed females (Engelmann, F., 1979), there must be another source of this hormone, possibly oenocytes (section 1.2.2d). The conflicting results on *Aedes*, and on species other than Diptera, point to the need for the same careful assay of ecdysteroid titres and tissues of origin in adults as have been done on earlier stages.

There is some confusion in the literature as to whether the fat body in some species is regionally specialized for vitellogenesis. In *Locusta*, Lauverjat, S.

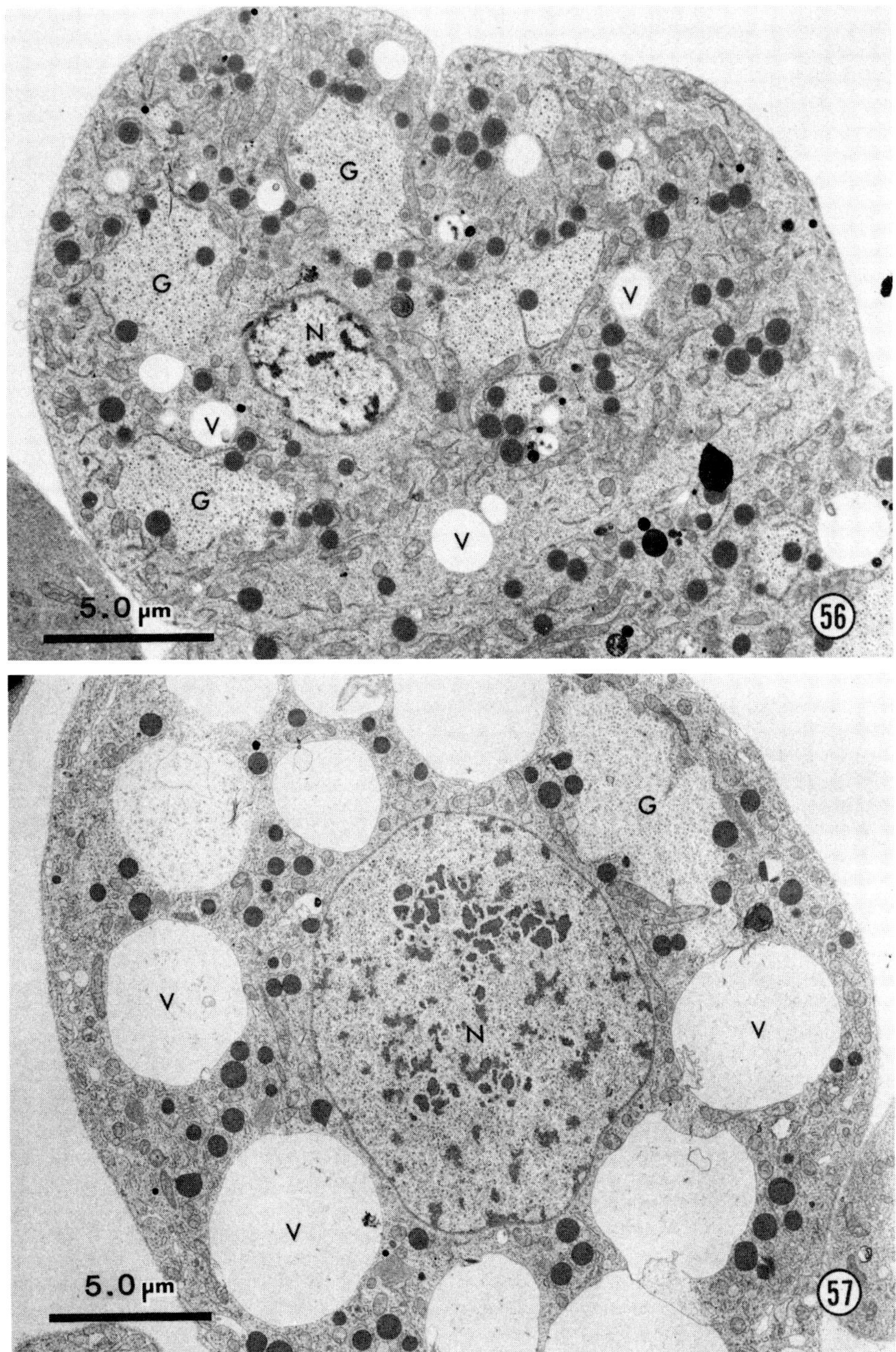

FIGS 56 and 57. Regional differentiation of the fat body. The fat body of *Leptinotarsa* is regionally differentiated with respect to vacuole development. The central fat body (Fig. 56) has smaller vacuoles than that at the periphery (Fig. 57). From McDermid, H. and Locke, M., 1983). (Fig. 56 ×6300; Fig. 57 ×5900.)

(1977) claims that the peripheral fat body does not develop the necessary apparatus for Vg synthesis. In contrast, Couble, P. and co-workers (1979) claim no essential differences in fat body development in different body regions, and clearly state that peripheral fat body can secrete Vg *in vitro*. Similarly, de Loof, A. and Lagasse, A. (1970) suggest that peripheral fat body in *Leptinotarsa decemlineata* has an inappropriate structure for large-scale protein synthesis and yet, after ovariectomy which results in Vg accumulation in the haemolymph, only fat body cells having the structure of peripheral cells are found. It seems important that structural observation should not be over-interpreted. If fat body tissue can be shown to produce Vg *in vitro*, then structural arguments about how much RER or other cytoplasmic elements need be present for its synthesis become pointless.

2.2.3 Other adult syntheses

Since most studies of synthesis by adult fat body have been concerned with vitellogenesis, there is little information on other synthetic activities at this stage. There is evidence, however, that the adult fat body secretes a variety of haemolymph proteins. Indeed, even pupal fat body can be induced to secrete some of the same proteins that are made in larval and adult stages. Kramer, S. (1978) showed that increased JH esterase activity in adults of *Leptinotarsa* was stimulated by treatment with JH. Inhibition by actinomycin and puromycin suggested that the increased activity was due to synthesis. Esterase activity was also induced in pupae of *Hyalophora gloveri* (Whitmore, D. *et al.*, 1974). Lipophorin, a diglyceride carrier protein, has been isolated from the adult fat body of *Locusta* by Gellissen, G. and Wyatt, G. (1981), who showed that its production does not follow the same cyclical pattern as vitellogenesis. In adults of *Glossina austeni* labelled amino acids were incorporated into proteins secreted during larval development *in utero* (Tobe, S. and Davey, K., 1974b, 1975). Protein granules were seen in the fat body during the first pregnancy (Tobe, S. *et al.*, 1973), but neither their formation nor fate was described.

Structural changes related to adult diapause. In *Leptinotarsa*, females reared on short-day regimes enter diapause soon after adult ecdysis, whereas long-day females begin to oviposit within 5 days and continue to produce 50–60 eggs daily (D. Denlinger, vol. 8). During the first 2 days after adult ecdysis the fat body of pre-diapause and non-diapause females show no differences. At ecdysis both contain large vacuoles (section 2.2.1c) and numerous protein granules, some of which may contain urate. The cytoplasmic matrix is poorly developed. The protein granules decline rapidly in number and are absent by the second day. The vacuoles decrease and are missing by day 3 when differences in fat body structure between the two adult types become evident. In long-day females RER proliferates below the plasma membrane and adjacent to the nucleus, while increasing quantities of lipid appear between these two regions. The plasma membrane invaginates deeply and large numbers of mitochondria appear. In contrast, short-day females develop some RER and store much lipid after day 2, resulting in cell enlargement. On the 6th day of adult life protein granules begin to accumulate again and fat body cells swell to occupy most of the abdomen. By 10 days after adult ecdysis mixed autophagic/heterophagic vacuoles appear (section 3.1) and the beetles enter diapause 2 days later (Dortland, J. and Esch, T., 1979). These observations correlate well with data on protein synthesis (Dortland, J., 1978). The fat body of both long- and short-day females synthesizes two vitellogenins and three diapause-related proteins. Long-day females produce little diapause protein and large amounts of Vg, which are rapidly sequestered by the ovary. Short-day females produce small amounts of Vg and larger amounts of the three diapause proteins, two of which are returned to the fat body for storage. The autophagy, which starts before diapause inception, reduces much of the RER and many mitochondria to stored products for use in post-diapause development. The sequestered diapause proteins are thought to act as a nutrient store for diapause itself.

Fat body in allatectomized long-day females develops in the same manner as in untreated short-day females. This supports the contention that Vg synthesis is stimulated by JH, and that the production of protein storage granules is induced by its absence (de Loof, A. and Lagasse, A., 1970).

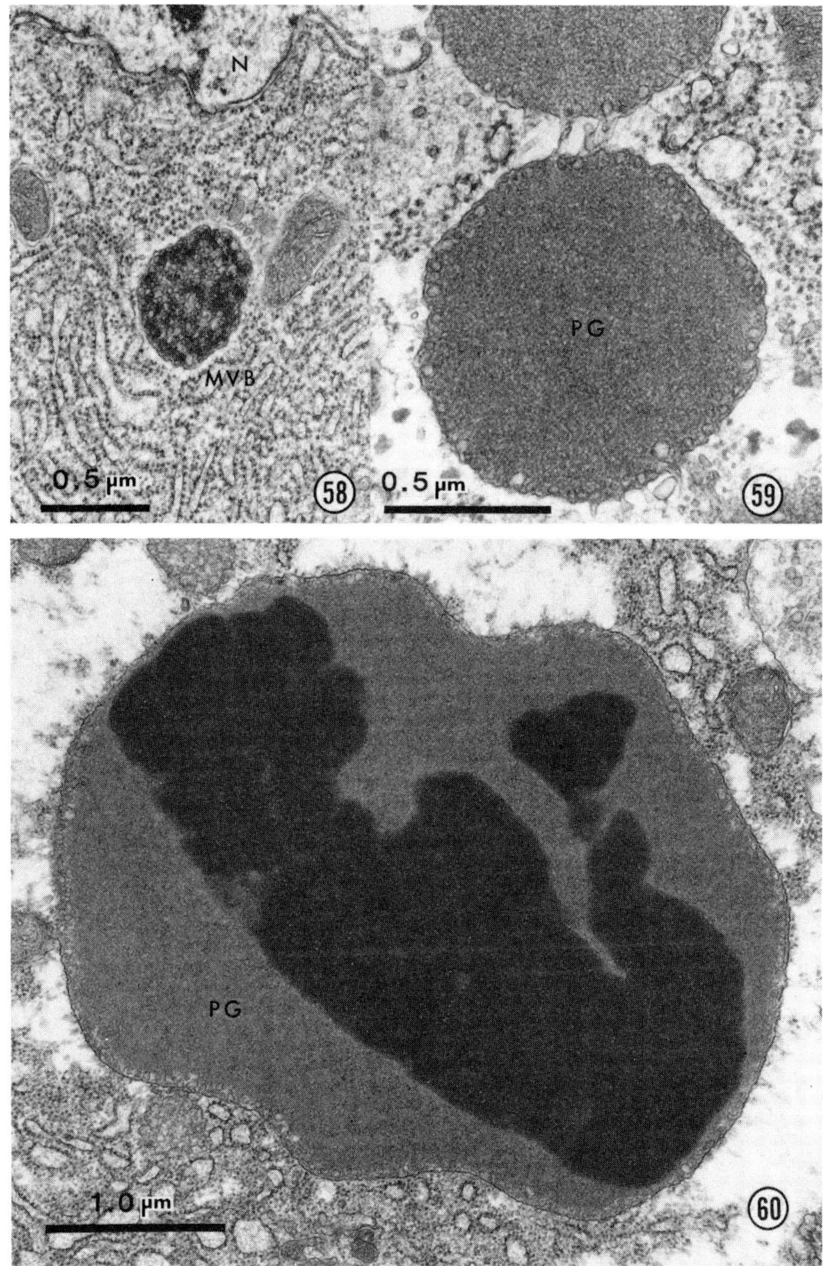

FIGS 58–60. The switch from heterophagy in multivesicular bodies (MVBs) to sequestration and storage in protein storage granules (PGs) in *Calpodes* (Fig. 67). FIG. 58. Intermoult MVBs. Throughout the intermoult in *Calpodes*, but particularly when intermoult preparation is complete, the fat body pinocytoses protein into perinucler MVBs. (×43,000.)

FIG. 59. PG formation. Pupation begins with a phase of autophagy after which protein in MVBs accumulates to make storage granules. (×63,000.)

FIG. 60. Crystallization of protein in storage granules. As PGs accumulate more and more protein they also condense as it crystallizes out. MVBs and PGs both internalize membrane brought to them in pinocytosis vesicles. (×35,000.)

2.2.4 Cyclical changes in cell organelles

(a) *Golgi complex.* The Golgi complexes (GCs) of insects are structurally less complex than those found in vertebrates or plants. In insects the GC at maximal development typically consists of a smooth face of RER bearings rings of GC beads and forming transition vesicles, an outer saccule, two or three inner saccules and according to the state of activity, secretory vesicles, 1° lysosomes and isolation envelope precursor vesicles (Figs 29, 30 and 31). The GC acts as a distribution centre for membranes from the RER and the materials they enclose. As a consequence, the appearance of GC profiles varies with the kind and amount of synthetic products passing through them. The route that these materials take through the GC depends on their fate. There are at least three major processes. These are the processing and packaging of proteins in secretory vesicles for export to the haemolymph (Fig. 31), the production of 1° lysosomes which contribute lytic enzymes to both autophagic and heterophagic vacuoles (Fig. 30), and the formation of isolation membranes for autophagy (Fig. 67). Isolation membranes are probably formed from vesicles derived from the outer saccule of the GC because both show pronounced osmiophilia (Locke, M. and Sykes, A., 1975) and stain with lead (McClintock, J. and Locke, M., 1982). The isolation membranes themselves do not carry the lytic enzymes required for organelle or protein breakdown. These enzymes are transported in primary lysosomes arising from the inner saccules of the GC, as suggested by the demonstration of acid phosphatase activity in both structures. In *Calpodes*, large secretory vesicles containing protein for export form from the inner GC saccules. Large-scale protein secretion does not always correlate with the presence of large secretory vesicles. For example, vitellogenesis in *Calliphora vicina* coincides with the presence of large compound secretory vesicles in the cytoplasm (Thomsen, E. *et al.*, 1980), whereas in *Locusta* no Vg transport between the GC and the extracellular space could be found (Couble, P. *et al.*, 1979). The difference may be due to the rate of movement of SVs from the GC to the surface and the coupling to exocytosis. Control of the rate of exocytosis may be separate from the control of the rate of protein synthesis and, in *Locusta migratoria*, protein may be exocytosed rapidly.

A unique staining property of arthropod GCs has permitted the discovery in fat body cells of a subcellular structure found in all GCs. The smooth surface of the RER near GC elements has bead-like particles arranged in rings at the base of forming transition vesicles. In arthropods, but not in other phyla, these GC beads stain specifically with bismuth salts (Figs 33–35) (Locke, M. and Huie, P., 1977). They have a diameter of 10–12 nm and are separated from each other and the membrane by a clear halo, giving them a centre-to-centre spacing of 27 nm and a centre-to-membrane spacing of 14 nm (Fig. 36) (Locke, M. and Huie, P., 1976a).

Transport of secretory proteins and membrane between RER and GC saccules is accomplished by the transition vesicles in an alternating fission–fusion process. This transport is sensitive to inhibitors of oxidative phosphorylation which block the energy-dependent step at the forming face of the GC, between the smooth surface of the RER and the transition vesicles. Brodie, D. (1981) studied the effect of a variety of inhibitors of protein synthesis and oxidative phosphorylation, and the effect of ionophores on bead ring integrity. He found that bead ring integrity is independent of protein synthesis and of late GC processing of secretory protein. Of the nine experimental treatments used, only those that lowered intracellular ATP levels altered ring morphology. When ATP levels were reduced the beads clumped together and lost their ring structure, although no change in centre-to-centre spacing or centre-to-membrane distance occurred. The concentration of beads into a close-packed sheet may be caused by a loss of membrane, which reduces the area of the forming face of the GC. This study suggests that bead ring integrity is ATP-dependent. Further, the block to transition vesicle formation concomitant with the collapse of bead rings suggests that formation of transition vesicles is dependent upon the integrity of the rings. The beads may therefore be the structural correlate of the energy-dependent gating mechanism controlling transport in the ER–GC transition region described by Jamieson, J. and Palade, G. (1968).

The size of the GC varies with the secretory activity of the cell. At times of low secretory activity the GCs may be barely recognizable, which perhaps accounts for their seeming absence in some older

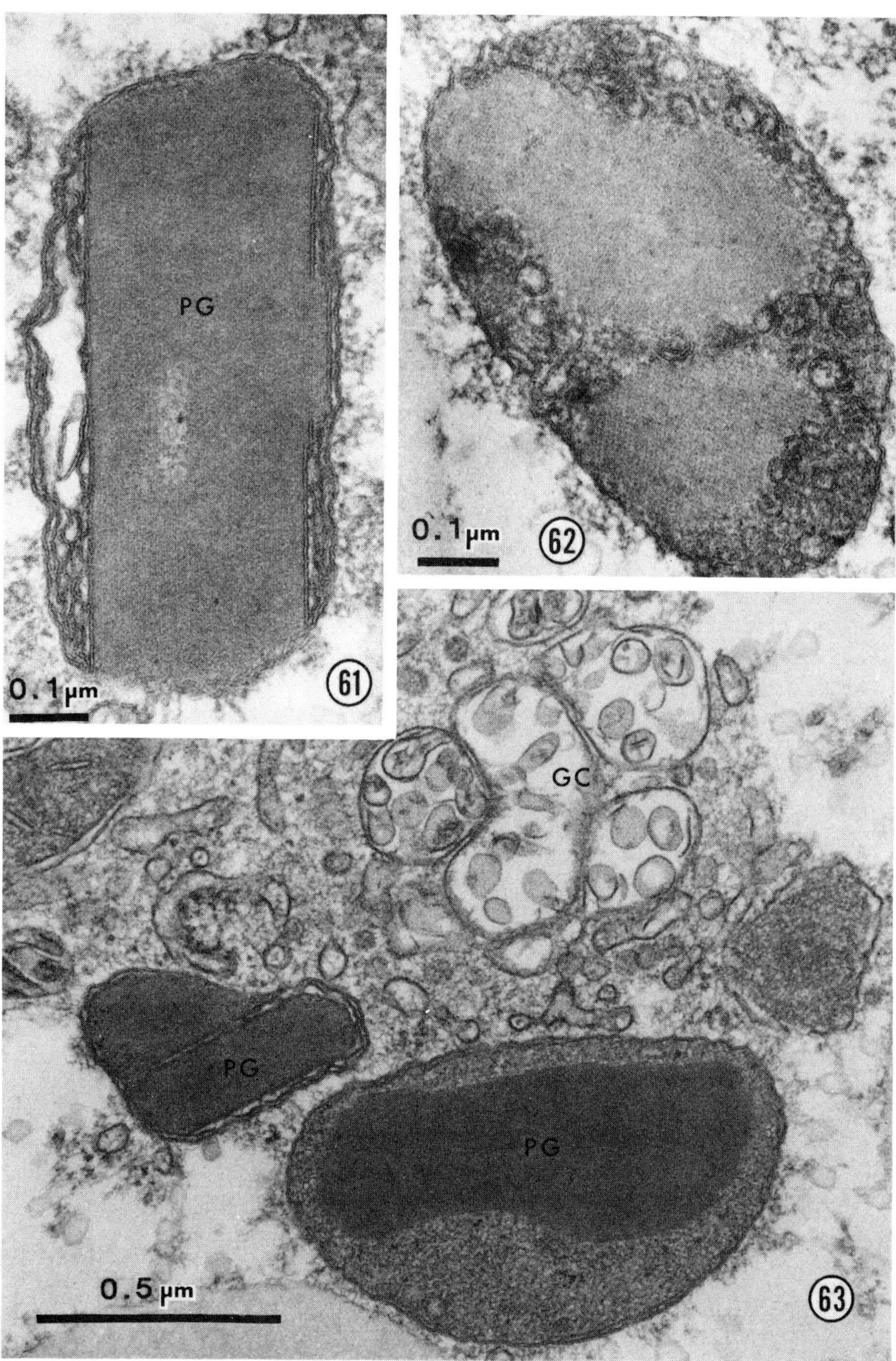

FIGS 61–63. Protein storage in *Periplaneta* (see also Fig. 3). Both nymphs and adults sequester protein in storage granules which show a continuous range in structure from MVBs to giant stored protein crystals. Many PGs seem to have extra internal membranes (Figs 61 and 63). These may be derived from the internalized surface membrane (Fig. 62) but other origins for these PGs are possible. (Fig. 61 × 140,000; Fig. 62 × 140,000; Fig. 63 × 86,000.)

ultrastructural studies. In *Calpodes* no stage can be identified when the GC beads are absent. The minimal structure is one or a few rings of beads adjacent to a smooth surface of ER, as found during GC formation at the beginning of intermoult preparation early in the last larval stadium in *Calpodes ethlius* (Fig. 33). The GC beads are prime movers in the differentiation of GCs from the RER (Locke, M., 1980a).

(b) *Peroxisomes.* In *Calpodes*, mature peroxisomes are flattened with a depression on one side. A dense core of coiled tubules is connected to the limiting membrane at the depression by a finely granular mass (Fig. 45). Peroxisomes contain catalase and urate oxidase (see section 3.1.3). In the fourth larval stadium the peroxisomes decrease in size before ecdysis to the fifth stage. During the day following ecdysis all the old perioxisomes atrophy while a new generation appears. Peroxisomes form as diverticula of the RER, and grow by the addition of material from confronting RER cisternae. The mature peroxisomes become progressively smaller during the phase of intense fat body syntheses and are isolated and lysed before pupal ecdysis (see section 3.2) (Locke, M. and McMahon, J., 1971). Pupal fat body contains no perioxisomes until about 24 h prior to adult emergence, when they arise and grow in the same manner as in larvae (Larsen, W., 1976). Although adult peroxisomes contain a coiled core it is never as well developed as that in larvae. The presence and degree of complexity of the core depends on fixation temperature. Coiled structures present after fixation at 0–4° are not seen after fixation at room temperature, suggesting that they are protein liquid crystals (Larsen, W., 1976). In both larvae and adults there is often an association between peroxisomes and lipid droplets (Fig. 46). Peroxisomes occur in the fat body of other insects such as *Periplaneta* (Fig. 44) as well as in other tissues, but they have been little studied.

(c) *Mitochondria.* As a consequence of mitosis at the end of the fourth stadium in *Calpodes* there are fewer mitochondria per cell after ecdysis to the fifth stage. However, during a 10 h period between E+6 and E+16 h the number of mitochondria per cell increases from about 100 to 300. Detailed stereological analysis showed that the increase in number results from mitochondrial division which follows a period of rapid growth and shape change (Tychsen, C., 1968). A further round of replication at E+17 h increases the number of mitochondria to ~700 per cell (Fig. 47). Thus, during the period of greatest fat body synthetic activity (E+66 h to E+156 h) a large population of mitochondria is present to supply metabolic needs. After this period, most larval mitochondria are isolated and destroyed (Locke, M. and Collins, J., 1965) (see section 3.2) so that few remain during the pupal stadium. At emergence mitochondrial number rises again to culminate in a dramatic 6-fold increase between 22 and 34 h after adult eclosion (Fig. 48) (Larsen, W., 1970). Mitochondrial division has a characteristic morphological sequence both in the larva and adult. It begins as one crista grows across the mitochondrial matrix and fuses with the opposite inner membrane, thereby separating the matrix into separate parts. A constriction forms at the site of the transecting crista which tightens to separate the mitochondrion in two (Figs 49 and 55). The mitochondria of larval fat body, including those which persist into the early pupal stage, are characteristically rod-shaped, while those of late pupae and adults tend to be spherical or oval (Figs 50 and 51). In larval fat body the rod-shaped mitochondria (Fig. 50) first assume a spherical form (Fig. 52) before dividing (Fig. 53). A brief study of *Calliphora erythrocephala* showed differences in mitochondrial morphology in feeding and post-feeding larvae. During feeding, mitochondrial profiles were irregular with a faint matrix; they later became smaller with a more regular outline and dense matrix (Marx, R., 1971). A morphometric study of pre-metamorphic changes in the fat body of *Calliphora* showed a similar pattern. At the start of the last larval stage, fat body mitochondria were irregular and elongate but as the end of feeding approached they became more rounded or oval. The absolute area of mitochondria increased throughout this period but dropped by 50% when feeding ceased, presumably as a result of a round of autophagy occurring at this time (de Priester, W. and van der Molen, L., 1979).

The change in form during development suggests a kind of mitochondrial differentiation. There may also be morphological differentiation of the cristae, which can sometimes be tubular rather than lamellate (Fig. 54). Do these differences reflect changed functions and enzyme compositions? For

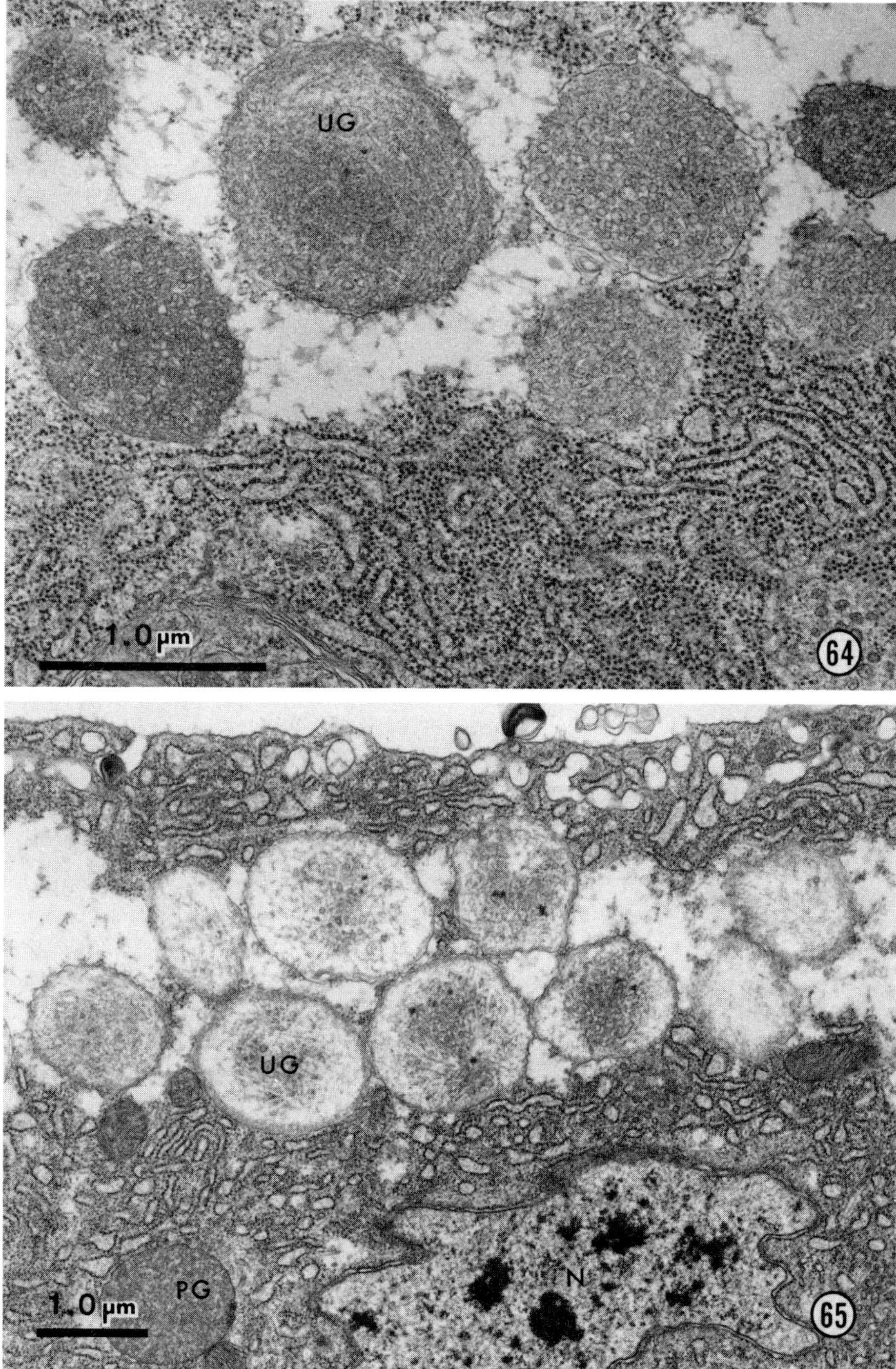

FIG. 64 and 65. Urate storage granules in *Calpodes*. The loss of peroxisomes containing urate oxidase at pupation (Fig. 67) is followed by the appearance of MVB-like vacuoles with fibrous contents that store urate (Fig. 66). Since urate is usually not preserved, they eventually appear like empty vacuoles, smaller but otherwise similar to the urate vacuoles of *Periplaneta* (Figs 4 and 5). (Fig. 64 ×40,000; Fig. 65 ×20,000.)

fat body mitochondria these questions remain to be answered. An investigation on the fat body of *Aldrichina grahami* attempted to correlate developmental changes in mitochondrial size and morphology with the activity of three enzymes from mitochondrial fractions (Tsuyama, S. and Miura, K., 1979). During feeding, the mitochondria showed a greater average diameter in thin sections, were more electron-dense, and were claimed to be more numerous than in older larvae. In post-feeding larvae both large and small mitochondria were found. Early in the stadium, the fraction of large mitochondria accounted for almost all of the activity of three mitochondrial enzymes. After feeding ceased most of this activity was associated with the cell fraction containing small mitochondria.

Little is known about factors controlling mitochondrial morphology. In *Rhodnius*, starvation results in mitochondrial fusion so that each cell comes to contain only a few giant mitochondria (Wigglesworth, V., 1967a). After feeding, the giant mitochondria divide to produce many normal organelles. In *Calpodes* (Locke, M., 1980a) giant mitochondria also result from even a brief starvation (Fig. 37). These divide to become normal again within 4 h of feeding (Fig. 38). Clearly, then, nutrition can influence mitochondrial size and division. It is less certain that it influences mitochondrial growth, DNA replication and division during normal development. Both starvation and a ligature placed posterior to the prothoracic glands prevent any mitochondrial genesis in the first 26 h of the last larval stadium (Tychsen, C., 1978). It is difficult to separate the effects of starvation from possible influences of anterior endocrine centres. However, the coincidence between the timing of the two phases of mitochondrial replication and two significant elevations in haemolymph ecdysteroid titre early in the last larval stadium of *Calpodes ethlius* (Dean, R. *et al.*, 1980) suggests a proximal endocrine influence worth further study.

2.2.5 Regional differences in fat body structure and function

There is considerable evidence that trophocytes, at least in some insects, show regional diversity of structure and function. Differences have been observed at all stages of development and in several taxonomic groups. Several authors have described ultrastructural and functional differences between the peripheral and visceral regions of larval and adult tissue, and have claimed that there are differential regional responses to a variety of conditions including different hormonal milieux. The evidence suggests that in several species the peripheral tissue is primarily a site of lipid synthesis and storage, while the visceral tissue is specialized for protein synthesis and storage, and for heterophagy associated with storage. However, this dimorphism is not consistent, as the case of *Chironomus* shows.

The clearest case for regional specialization has been made in the Diptera where there are two kinds of differentiation. The first is a functional differentiation of the peripheral and visceral tissues. Schin, K. and co-workers (1977) studied the timing and localization of haemoglobin (Hb) synthesis in fourth-instar larvae of *Chironomus thummi*. Incorporation of [^{14}C]aminolevulinic acid, and localization of peroxidatic activity of Hb by the diamino benzidine reaction, showed that Hb synthesis occurred in the peripheral (sub-epidermal) tissue. Visceral tissue was principally involved in pinocytosis of haemolymph proteins, including injected ferritin, and at times, the autophagy of cell organelles.

A second kind of tissue variation observed in the Diptera is an anterior–posterior gradient of development which occurs at metamorphosis. In the last larval stage, fat body cells sequester and store blood proteins in granules. In *Drosphila* the formation of storage granules begins in the anterior region of the fat body and proceeds posteriorly (Tysell, B. and Butterworth, F., 1978). A similar gradient has been described for the cell separation that precedes histolysis in *Sarcophaga* (Whitten, J., 1962; Fraenkel, G. and Hsiao, C., 1968), *Phormia* and *Drosophila* (Whitten, J., 1962). The pattern of lysosomal activity and autolysis is similar in *Calliphora* (van Pelt-Verkuil, E., 1978; Thomsen, E. and Thomsen, M., 1974; Kinnear, J. and Thomson, J., 1975).

Regional differences in the rate of tissue autolysis have been described in other groups, though not with an anterior–posterior gradient (R. Lockshin, vol. 2). In *Leptinotarsa*, the peripheral and some visceral cells are lysed, whereas the dorsolateral

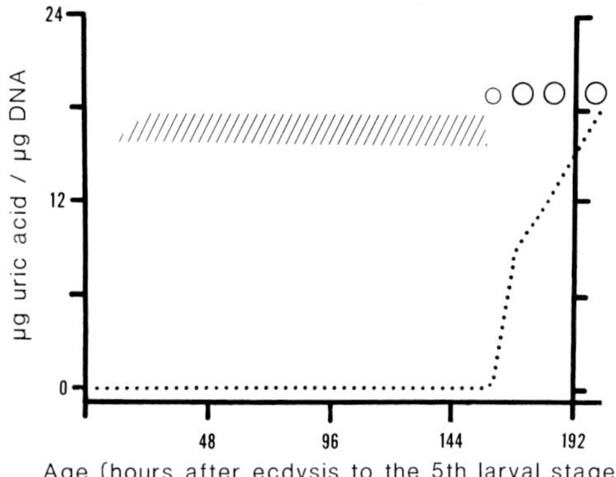

FIG. 66. The relation between the disappearance of peroxisomes (/////), the accumulation of urate in the fat body (·····) (Dobson and Collins, unpublished) and the accumulation of urate storage granules (○ ○) in *Calpodes*.

group of visceral cells persist into the adult stage and are thought to give rise to the adult fat body (Labour, G., 1974). Waku, Y. and Sumimoto, K. (1969) concluded that there is differential autolysis of the visceral cells of *Bombyx mori*.

Gradients of autolysis and other cytological changes occur in response to such physiological conditions as starvation. In *Rhodnius* nymphs, prolonged starvation results in increasing polyploidy after nuclear fusion (Wigglesworth, V., 1967b). Cells first become depleted of glycogen and lipid and then destroy their mitochondria by autophagy. These cytological changes occur in a wave from posterior to anterior within the abdomen, and are evident last of all at the sides of the anterior half of the abdomen (Wigglesworth, V., 1967a,b).

Experiments on the control of increase in ploidy in *Calpodes* suggest that some factor stimulating nuclear replication is transmitted from cell to cell. Fat body cells attached to epidermal wounds may increase ploidy enormously compared to cells further away. Presumably the gap junctions (Fig. 18) are involved in all such examples of linearly graded differentiation.

Functional differences between fat body cells from different parts of the animal have been reported in relation to metamorphosis and adult syntheses. In a study of RNA synthesis during diapause and adult development in silkmoths, Berry, S. and co-workers (1967), observed that the perigonadal fat body incorporated uridine at a much higher rate than the rest of the fat body. In *Locusta* Lauverjat, S. (1977) characterized the peripheral fat body of adult females as having a male phenotype. These cells are not ultrastructurally differentiated for protein synthesis and secretion, and show no cylical activity. Visceral fat body, especially near the ovaries and gut, has well-developed RER and GC, the abundance of which varies cyclically with secretion of Vg during ovarian maturation cycles. As observed by Lauverjat, S. (1977), the cells of these two regions showed different responses to ovariectomy and allatectomy. Visceral cells lost their protein secretory ultrastructure and assumed the appearance of the peripheral cells that store lipids. These observations on regional differentiation have not been confirmed by others working with *Locusta*, possibly because of differences in the timing of their observations or other variables in the studies. The two cell types described by Lauverjat for normal females were observed by Rinterknecht, E. and Roussel, J.-P. (1978) in the visceral fat body of allatectomized females but Couble, P. and co-workers (1979) saw no ultrastructural differences between peripheral and visceral tissue in normal or allatectomized females. In *Leptinotarsa* also there is no agreement on the number of functional cell types, some authors recognize two (de Loof, A. and Lagasse, A., 1970; Dortland, J. and Esch, T., 1979) while others propose three (Labour, G., 1974). The evidence is also contradictory as to the principal site of Vg synthesis and secretion; the visceral tissue is the candidate favoured by de Loof, A. and Lagasse, A. (1970), whereas Dortland, J. and Esch, T. (1979) found evidence for protein synthesis and secretion, as well as lipid storage, in the peripheral tissue. Labour, G. (1974) showed that peripheral fat body cells contain numerous large vacuoles (see section 2.2.1c), whereas in the perivisceral tissue the vacuoles are much smaller (Figs 56 and 57).

The many allusions to topographical and functional differences in the fat body of the species described above reveal an unresolved problem. Although the descriptions provided by each author are suggestive, they are not entirely convincing, largely because clarification of these distinctions has not been a principal objective of any of these studies, except those involving Diptera. The evidence from this order provides convincing evidence for unequal competence of cells in

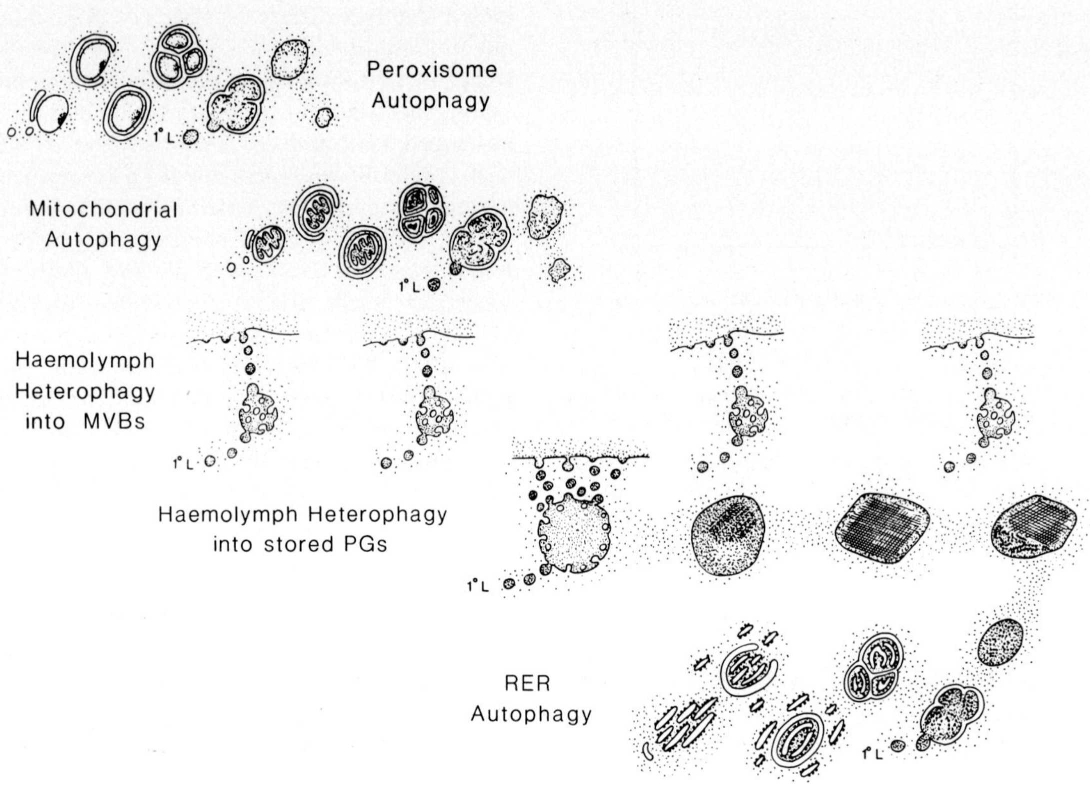

FIG. 67. Autophagy and heterophagy during the preparation for pupation in *Calpodes* in the 36 h prior to pupation. The events shown here are based on work in Locke, M. and Collins, J., 1965, 1968, 1980; Locke, M. and McMahon, J., 1971; Locke, M. and Sykes, A., 1975. These events begin with the elevation of haemolymph ecdysteroid (Dean, R. *et al.*, 1980).

different regions to respond to particular developmental and physiological stimuli. In *Drosophila*, the antero-posterior gradient of protein granule formation is partly controlled by the intrinsic capacity of cells to respond to their milieu (Tysell, B. and Butterworth, F., 1978). Such variations may be genetically determined, as is the formation of the autofluorescent kynurenine granules in the mature larval fat body of *Drosophila melanogaster* (Rizki, T. and Rizki, R., 1962; Rizki, T., 1964). The evidence suggests that even the incorporation of precursors for specific syntheses may be regionally controlled by different genes. During milk production by the uterine gland in adult *Glossina*, fat body cells secrete lipids which are incorporated into the milk. Langley, P. and Bursell, E. (1980) have shown that the number of cells active in synthesis at any time is variable, and speculate that there may be mosaics of differentially competent cells within the fat body. This might explain the observations of uneven incorporation of labelled precursors during vitellogenesis in *Nauphoeta cinerea* (Wüest, J., 1978), the variable activation of cells after injection of JH into allatectomized adults of *Locusta migratoria* (Rinterknecht, E. and Roussel, J.-P., 1978), and the variations in change of ultrastructure from a synthetic to an autolytic appearance during metamorphosis of the fat body in *Galleria mellonella* (Dutkowski, A., 1974).

It is difficult to evaluate all of the ultrastructural reports critically, and to synthesize the information presented. In most cases the descriptions of regional or temporal variation have not been clearly illustrated by micrographs. In addition, with few exceptions, no attempt was made to standardize the regions from which tissue was taken for study. Thus temporal variations could have been mistaken for regional ones. Systematic analysis is required before a more thorough synthesis can be attempted, and the question of the existence and significance of some of these putative differences must remain open.

3 METAMORPHOSIS

Metamorphosis in the fat body involves a series of events that result in significant changes in cell and tissue structure. Several characteristic processes have been described in the Holometabola, though not all of these occur in every species. The main events are: heterophagy and storage of haemolymph proteins; autophagy of some cell organelles; cell separation followed by tissue and cell remodelling; and histolysis of the larval tissue and replacement by a new adult tissue.

3.1 Heterophagy and storage of haemolymph proteins

Larval fat body of many species takes up proteins from the haemolymph by pinocytosis. The proteins enter in small vesicles which fuse with each other to form larger granules. In *Calpodes*, pinocytosis occurs throughout the fifth stadium. During the feeding stage the proteins are hydrolysed in multivesicular bodies (MVBs), which are a constant feature of all or most cell types because of their role in membrane turnover (Fig. 58). During metamorphosis there is a switch from lysis to storage, and the sequestered protein is stored in large membrane-bound protein granules (Locke, M. and Collins, J., 1968, 1980) (Figs 59 and 60). Pinocytosis of haemolymph proteins into granules occurs during the pre-metamorphic phase of last instars of *Calliphora erythrocephala* (Collins, J., 1967) and *Drosophila* (Butterworth, F. *et al.*, 1979). Although these granules were not identified ultrastructurally, they are probably MVBs since in other structural and developmental characteristics these cells resemble those of *Calpodes*. Tobe, S. and Loughton, B. (1969a,b) failed to find MVBs in fat body of fifth-instar *Locusta*, even though there was significant uptake of labelled blood proteins. Coupland, R. (1957) did not see protein granules in fat body of first to fifth instars of *Schistocerca gregaria*. However, Lauverjat, S. (1977) showed crystalline protein granules in the fat body of newly emerged females of *Locusta*. The absence of protein granules in fifth instars may reflect a very rapid turnover of pinocytosed proteins at this stage, and a much later switch from lysis to storage. In *Periplaneta* both recently emerged fifth-stage nymphs and adults have crystalline granules of stored protein (Figs 61–63).

The appearance of storage granules varies with larval and pupal age. They become increasingly crystalline as the pupal stage progresses, and in some species they are almost completely crystalline at adult emergence. They may also appear heterogeneous in their contents (Fig. 67) as the granules of pinocytosed proteins fuse with autophagic vacuoles containing mitochondria and RER (section 3.2). Such mixed granules may also develop crystalline regions. Crystallization may have particular significance in relation to the number and kind of haemolymph storage proteins that have been pinocytosed. Does crystallization imply that only one protein species survives in a storage granule, for example?

3.1.1 Selective uptake of haemolymph proteins

The formation of storage granules from sequestered haemolymph proteins was described in *Calpodes* by Locke, M. and Collins, J. (1968). Although the original demonstration of uptake involved the use of foreign proteins injected into the haemolymph, the process is normally selective and only specific proteins are sequestered (Table 4). Tojo, S. and co-workers (1978) isolated the storage granules from last-instar larvae and pupae of *Hyalophora cecropia*, and identified their contents as the two major protein fractions of the haemolymph of the last instar. Storage proteins have also been isolated from pupal fat body of *Bombyx mori* (Tojo, S. *et al.*, 1980), and are presumed to be stored in the protein granules of that tissue. The existence of storage proteins and their sequestration in granules in fat body are widespread phenomena (Table 4), but little is known of the cellular basis of selective uptake. Most storage proteins are thought to be hexamers (Fig. 10) with molecular weights near 500 K daltons (Kramer, S. *et al.*, 1980) though they may be as large as 900 K daltons (D. Webster,

Table 4: Selective uptake and storage of haemolymph proteins at metamorphosis

Order and species	Method*	Observations	Reference†
Coleoptera			
Leptinotarsa decemlineata	EM	Storage granules, fused autophagic vacuoles and storage granules.	15
Diptera			
Calliphora erythrocephala	EM	Storage granules, fused autophagic vacuoles and storage granules.	9, 22, 23
Calliphora stygia	EP	Uptake and storage of calliphorin (protein C) from haemolymph.	21
Drosophila melanogaster	LM	Storage granules in last instar larvae.	2
	LM,Ce	Storage granules isolated.	25
	EM	Storage granules, autophagic vacuoles.	11
	IP,EP,EC,EM	Storage protein, drosophilin, isolated.	24
Lepidoptera			
Bombyx mori	Ce,Ch,EP,ID,IE IE,EM	Storage granules, autophagic vacuoles, fused storage granule and autophagic vacuoles.	27
	EM,LM	Storage (albuminoid) granules, urate granules, autophagic vacuoles.	28
Calpodes ethlius	Ch,EP,Fl,ID	Uptake of 3 antigens 470 K, 580 K, 720 K, M.W.	5, 30
	EP,Fl	Uptake of two proteins, 580 K, 720 K.	18
	EM,LM	Storage granules, autophagic vacuoles, fused storage granules and autophagic vacuoles.	6, 16, 19
Diatraea grandiosella	EP	Storage of three haemolymph proteins.	3
Ephestia kuhniella	Ce,EC,EP,IE, EM,LM	Storage of two haemolymph proteins. Storage granules.	8
Galleria mellonella	EP,ID	Uptake of three haemolymph antigens.	7
	EM	Storage granules, autophagic vacuoles.	10
Hyalophora cecropia	Ce,EP,ID,EM LM	Storage granules, containing two haemolymph proteins, urate granules.	26
	EM	Storage granules.	1
Malacosoma americanum	ES,ID	Several haemolymph antigens in late larval and pupal fat body.	20
Manduca sexta	Ch,Cm,EP ID,IE	Haemolymph protein, manducin occurs in pupal fat body.	14
Philosamia cynthia ricini	EM	Storage granules.	13
	EM	Storage granules, autophagic vacuoles.	29
Pieris brassicae	EP	Two haemolymph proteins stored.	4
Protoparce quinquemaculata	ES,ID	Uptake of two haemolymph antigens.	12
Orthoptera			
Locusta migratoria	EM	Storage granules (lysosomes) at imaginal ecdysis.	17

* Methods: Ce, centrifugation; Ch, chromatography; Cm, chemical analysis; EC, electrophoresis in cellulose acetate; EP, electrophoresis in polyacrylamide; ES, electrophoresis in starch; ID, immunodiffusion; IE, immunoelectrophoresis; IP, immunoprecipitation; IF, isoelectric focusing; EM, electron microscopy; LM, light microscopy; Fl, fluorography.

† 1, Bhakthan, N. and Gilbert, L. (1972); 2, Butterworth, F. *et al.* (1965); 3, Chippendale, G. (1970); 4, Chippendale, G. and Kilby, B. (1969); 5, Collins, J. (1974, 1975a); 6, Collins, J. (1979): 7, Collins, J. and Downe, A. (1970); 8, Cölln, K. (1973); 9, de Priester, W. and van der Molen, L. (1979); 10, Dutkowski, A. (1974); 11, von Gaudecker, B. (1963); 12, Hudson, A. (1966); 13, Ishizaki, H. (1965); 14, Kramer, S. *et al.* (1980); 15, Labour, G. (1974); 16, Larsen, S. (1976); 17, Lauverjat, S. (1977); 18, Locke, M. *et al.* (1982); 19, Locke, M and Collins, J. (1968); 20, Loughton, B. and West, A. (1965); 21, Martin, M. *et al.* (1971); 22, van Pelt-Verkuil, E. and Dirkx, C. (1979); 23, Price, G. (1969); 24, Roberts, D. *et al.* (1977); 25, Thomasson, W. and Mitchell, H. (1972); 26, Tojo, S. *et al.* (1978); 27, Tojo, S. *et al.* (1980); 28, Waku, Y. and Sumimoto, K. (1969); 29, Walker, P. (1966); 30, Webster, D. (1982).

personal communication). It is not known whether they are taken up as monomers, nor is it obvious how particles the size of the polymer would cross the basal lamina to enter the cells by selective endocytosis. Studies on the molecular mechanisms of selective uptake have been impeded by the difficulty of isolating pure storage proteins from haemolymph and fat body. These problems have been overcome for several species of Lepidoptera, e.g. *Manduca*, *Bombyx*, and *Hyalophora*. Kramer, S. and co-workers (1980) have evidence that the chromatographic, electrophoretic and immunological properties, as well as the amino acid composition, of manducin are the same in the larval haemolymph and the pupal fat body of *Manduca sexta*. This suggests that there are no significant molecular changes during uptake that might mediate the selective process, as does the removal of sialic acid in vertebrate liver (Ashwell, G. and Morell, A., 1974).

3.1.2 Sex differences in protein storage

The amount of protein stored in granules in the fat body may vary with sex. In *Calpodes* pupae the proportion of cell area occupied by granules was greater in females than in males (Locke, M. and Collins, J., 1968). Striking differences between sexes were reported in *Bombyx* by Waku, Y. and Sumimoto, K. (1969). These were quantified by Tojo, S. and co-workers (1980) who reported that in female pupae 60% of the total fat body protein was in storage granules compared with only 20% in the male. Similar sex differences were seen in *Hyalophora cecropia* pupae (Tojo, S. *et al.*, 1978), and Bhakthan, N. and Gilbert, L. (1972) found no protein granules in adult males. Male silkmoth pupae also differ from females in containing significantly more urate/protein granules (section 3.1.3).

3.1.3 Urate storage

Uric acid is stored in the fat body of many species (Wigglesworth, V., 1972) in membrane-bound compartments varying in size from the giant vacuoles of urocytes (section 1.3.2) to the small MVB-like granules of Lepidoptera. In *Hyalophora cecropia* it begins to accumulate after cessation of spinning, and correlates with an increasing accumulation of granules showing a characteristic fibrous internal structure. These granules, which are much more numerous in males, have been isolated and shown to contain 75% uric acid and about 24% protein (Tojo, S. *et al.*, 1978). Unfortunately the granule fraction was too hard to section but its apparent homogeneity under the microscope led these authors to suggest that the protein is part of the granule and not a contaminant. Granules with a similar fibrous matrix are evident in micrographs of fat body from other silkmoths (Mori, T. *et al.*, 1970; Waku, Y. and Sumimoto, K., 1969; Walker, P., 1966). In *Calpodes*, granules with a fibrous matrix (Figs 64 and 65) develop at a time when uric acid accumulates in the fat body (Fig. 66). The development of these granules overlaps the switch from the formation of lysosomal MVBs to the formation of protein storage granules at the start of metamorphosis. This also coincides with the loss of urate oxidase activity by the fat body when the peroxisomes are destroyed (Locke, M. and McMahon, J., 1971) (sections 3.2 and 2.2.4b). The stored urate in the granules presumably arises from the metabolism of nucleic acids (Cochran, D., 1975) that accompanies the end of massive protein syntheses and the autophagy of the RER (section 3.2). There may also be uptake of uric acid from the haemolymph as observed in *Manduca* by Buckner, J. and Caldwell, J. (1980). The origin of the protein component of the urate granules may also be extracellular as urates may be transported as protein-bound complexes (Cochran, D., 1975). Tojo, S. and co-workers (1978) concluded that the urate granules contain proteins distinct from the proteins in the storage granules. In *Calpodes*, vacuoles having a fibrous matrix show a developmental sequence that suggests an origin from a MVB-like structure (Fig. 64). Initially there are MVBs with dense contents of a slightly fibrous appearance. Later stages show less dense, more fibrous contents, and still later, the vacuoles appear almost empty except for the accumulated fibres. These fibrous vacuoles were identified by Locke, M. and Collins, J. (1968) as residual bodies since they may also contain acid phosphatase. The development sequence described above suggests that urate granules may arise partly by endocytosis, with endogenous uric acid being added to the vacuole of endocytosed protein.

Larsen, S. (1976) showed that in late pupal *Calpodes* MVBs may develop into granules with a fibrous matrix. When thorotrast, a non-degradable tracer, was injected into the haemolymph it was first localized in pinocytic vesicles, and later in what appeared to be residual bodies (probably urate granules). Urate granules in the late pupa of *Calpodes* may therefore be vacuoles containing haemolymph proteins pinocytosed at different times, together with uric acid and lysosomal enzymes. Isolation of the granules, and analysis of their contents, should clarify their relationship to residual bodies, and their role in the changes in nitrogen metabolism from larva to pupa.

3.2 Autophagy of cell organelles

Early in metamorphosis the fat body switches from activities concerned with synthesis and secretion to the storage and reutilization of reserves during the pupal and adult stages. This change involves removal of some organelles and their subsequent replacement. Studies on *Calpodes* have shown that components to be digested are first isolated from the cytoplasm (Locke, M. and Collins, J., 1965; Locke, M. and McMahon, J., 1971) and fuse with 1° lysosomes to digest the contents (Locke, M. and Sykes, A., 1975). In this way all peroxisomes, and most of the mitochondria and endoplasmic reticulum, are destroyed in autophagic vacuoles within which these organelles are clearly identifiable soon after their isolation. As lysis proceeds, organelles lose their integrity and gradually degenerate into dense amorphous granules. This organelle-specific sequence of destruction (Locke, M. and Collins, J., 1980) shows that cells have a mechanism for selecting particular cell components for autophagy. In fat body at least, organelle turnover is not random. Autophagic vacuoles may also fuse with each other and with protein storage granules to form structures of heterogeneous appearance and origin (Fig. 67). This has been reported for many species, as shown in Table 4. Autophagy results in complete or partial destruction of organelles, preparatory to a change in cellular activity. Although some RER persists, the cells are principally engaged in storing protein, lipid, glycogen, and in some species, urate.

3.3 Formation of the adult fat body

3.3.1 Cell and tissue reorganization

Two mechanisms have been described for the origin of the adult fat body. It may develop by reorganization of the larval tissue, or differentiate from stem cells carried in the pupa. Both mechanisms have been described among the Diptera. In *Aedes aegypti*, larval tissue is thought to persist in the adult (Trager, W., 1937), whereas Evans, A. (1935) and many subsequent authors have claimed that in the cyclorrapha there is complete destruction of the larval tissue. The process begins with separation of the larval cells from each other, and the breakdown of the basal lamina (Whitten, J., 1962). The cells appear to die early in the pupal stage and by the third day of adult life the larval tissue is no longer recognizable. Butterworth, F. (1972) chronicled the disappearance of larval tissue in *Drosphila*, but the process was not illustrated. In *Dacus tryoni*, Evans, J. (1967) has described the origin of imaginal fat body from groups of cells that appear among imaginal oenocytes. They were said to arise from the new adult epidermis that develops from groups of stem cells carried in the pupa (Anderson, D., 1964). These accounts of the fate of larval fat body and the origin of the adult tissue in the cyclorrapha deserve further examination. Rizki, T. (1978) has described the arrangement of larval cells in the larva and early adult of *Drosophila melanogaster*. At hatching the number of larval fat body cells is fixed and there is no further cell division. Following pupariation the fat body is reorganized so that the cells occupy new positions in the pupa and adult. The abdominal cavity, thorax and head capsule of the adult contain larval cells which float out freely into the bathing medium during dissection. These cells are frequently associated with haemocytes, loosely attached to their surfaces. Observations on a red-pigmented mutant have cast some doubt on the claim that the larval fat body is completely replaced by a new adult tissue. Rizki described the localization of red-pigmented larval cells in the head capsule of newly emerged flies of the *tu-W-rc* mutant. Later, pigmented elements were observed in the adult tissue, but he was unable to distinguish whether these were transformed larval cells, or adult cells that had incorporated larval elements. This raises the question of whether, and to what extent, larval cells contribute to the formation of adult tissue. The view that larval fat body cells die in the adult has not been supported by microscopical evidence. Several authors (for example Butterworth, F. (1972); Thomsen, E. and Thomsen, M. (1974); and Evans, J. (1967)) refer to the disappearance of the larval cells in the adult, but there are no illustrations of the process.

The origin of the adult tissue is equally uncertain. Evans, J. (1967) concluded that in *Dacus* it develops from the new epidermis. This view conflicts with the embryonic origin of the larval fat body from mesoderm. Although Anderson, D. (1964) identified the imaginal discs from which the adult epidermis develops he gave no indication of mesodermal rudiments from which fat body might develop. It is difficult to see how the adult tissue could arise from the epidermal discs, unless the concepts of classical embryology have no relevance to this group of insects. The reorganization of the larval tissue in the pupa and adult, the association with haemocytes, the lack of microscopical evidence of cell death, and the absence of an identified stem cell population, all indicate that not enough is known about the metamorphosis of the fat body in the cyclorrapha. Although there are clear descriptions of the beginning of the metamorphic process there is little ultrastructural evidence for later stages. We know of no evidence on the fate of larval fat body cells, or the origin of the imaginal cells, which could not be interpreted as an extreme example of cell and tissue remodelling.

In other orders the remodelling of larval cells into adult tissue has been described in greater detail. However, because of the difficulties experienced by some investigators in obtaining adequate fixation of the tissue, there are some confusing conclusions in the literature. Larsen, S. (1976) described the process in *Calpodes*, identifying three characteristic stages. As in the Diptera, the cells lose their intercellular connections shortly before pupation, but they remain as a tissue enclosed by a basal lamina. Later, they become clumped around tracheoles, losing their sheet-like arrangement by 2 days before adult ecdysis. The reorganized cells re-establish intercellular connections, as shown by the lanthanum permeation of gap junctions. These changes in cell and tissue reorganization do not involve cell death or cell division. Haemocytes may be involved in the rearrangment of cells. Walters, D. (1969) observed that during metamorphosis in silkmoths, cells which had lost contact with each other and the basal lamina were rearranged initially in clumps, but were later arrayed along tracheae within a reformed basal lamina. In cultures of disaggregated fat body cells, haemocytes pulled cells together to form new aggregates (Walters, D. and Williams, C., 1966). A similar role for haemocytes in tissue reorganization during metamorphosis was proposed (Walters, D., 1969). Some authors have interpreted the changed relationships between fat body cells, and their content of autophagic vacuoles, as evidence for cell death and disintegration of the larval tissues (for example, Ishizaki, H., 1965; Waku, Y. and Sumimoto, K., 1969). In *Leptinotarsa*, Labour, G. (1974) observed that extensive autophagy was not uniform throughout the fat body, and he concluded that only some parts of the fat body undergo autophagy and survive in the adult, whereas other areas disappear completely. However, Krishnakumaran, A. and co-workers (1967) recovered intact pieces of larval fat body, transplanted into silkmoth pupae, from the newly emerged adults. This result supports the ultrastructural evidence for the persistence of most larval cells into the adult. What remains to be clarified is the generality of cell death or its absence during metamorphosis of larval tissue (R. Lockshin, vol. 2).

3.3.2 Cell remodelling

The transformation of larval fat body into the adult tissue begins during the last larval stadium, with the precisely executed autophagy of cell organelles. In *Calpodes* the process involves the isolation and lysis of peroxisomes, mitochondria and RER, in that order, with the RER being cut off in fragments of about mitochondrial size (Locke, M. and Collins, J., 1965, 1980; Locke, M., 1980a,b). In a morphometric study, Larsen, S. (1976) showed that the cells remain depleted of the lysed organelles throughout most of the pupal stage. Regeneration of RER and peroxisomes begins in the pharate adult and continues after emergence. Mitochondria increase in number following a phase of division (Larsen, S., 1970) 22–34 h after emergence. The new generation of mitochondria are morphologically distinct from the larval form (Figs 50 and 51). The new generation of peroxisomes is also different (Larsen, S., 1976).

A remarkable feature of the remodelled tissue in *Calpodes* is the regeneration of organelles during the same period as a resurgence in lytic activity in the cells. The period immediately before and after ecdysis (E−24 to E+36 h) includes a phase of heterophagy of haemolymph proteins as shown by the uptake of injected tracers into multivesicular bodies. During this period, lysis of the stored protein granules also begins, and is gradually continued as vitellogenesis begins. The timing of the disappearance of the storage granules that were laid down in the larva is variable, depending upon such factors as the start of vitellogenesis. In silkmoths vitellogenesis occurs during the pharate adult stage and lysis of protein granules occurs during reorganization of the tissue (Bhakthan, N. and Gilbert, L., 1972). Partial autolysis of cell organelles at metamorphosis is an extreme form of a process that occurs at other stages, particularly in relation to phases of synthetic activity. In *Calpodes* there are small bursts of autophagy in relation to the fourth to fifth moult, and at about the time of pupal commitment in the fifth stage. After adult emergence, some species undergo cycles of vitellogenesis which terminate with phases of autolysis (section 2.2.2b). Autophagy and heterophagy may be intrinsic properties of fat body cells while the degree to which they occur may be influenced by the hormonal milieu.

4 CONTROL OF FAT BODY STRUCTURE

4.1 Control of protein heterophagy and storage

Although larval fat body cells synthesize and secrete haemolymph proteins throughout most of the feeding stage, in most insects there is little or no storage of protein at this time. Haemolymph proteins that are pinocytosed are lysed in MVBs (Fig. 58). In ligated larvae of *Calpodes ethlius* (Collins, M., 1969) this limited heterophagy has been shown to be independent of ecdysone from the prothoracic glands. Small protein granules of about the same size as the MVBs of *Calpodes* were formed in larval fat body of *Drosophila* transplanted into adults (Butterworth, F. and Bodenstein, D., 1967), and *in vitro* (Thomasson, W. and Mitchell, H., 1972; Butterworth, F. *et al.*, 1979). In *Calpodes* the lysis of pinocytosed proteins ceases abruptly and the cells switch to sequestration and storage of hemolymph proteins early in the phase of preparation for pupation (Table 4, Figs 58–60). The switch is stimulated by the rising ecdysteroid titre that initiates the larval to pupal moult (Collins, J., 1969; Dean, R. *et al.*, 1980). This control of storage granule formation has been demonstrated in *Drosophila, in vivo* and *in vitro* (Table 5). Early in the third stadium (E + 24 h) small protein granules occupy 0.5% of the cell area, while at E + 48 h this has increased to 23% (Butterworth, F. *et al.*, 1965). The increase in protein granules can be correlated with the ecdysteroid titres for *Drosophila* which are 39.0 and 119.0 ng g^{-1} wet wt respectively at these times (Borst, D. *et al.*, 1974). Sass, M. and Kovacs, J. (1977) observed premature induction of protein granule formation by 20-hydroxyecdysone in the penultimate larval stage of *Mamestra brassicae*. However, Martin, M. and co-workers (1971) concluded that in third instars of *Calliphora stygia* massive uptake of haemolymph proteins is not stimulated by 20-hydroxycedysone since it begins before the rise in ecdysteroid titre that stimulates pupation (Shaaya, E. and Karlson, P., 1965). The question of 20-hydroxyecdysone stimulation in *Calliphora* must remain open, however, since Koolman, J. (1978) used radioimmune assay to demonstrate a small but significant peak of ecdysteroid during the feeding period (day 4) preceding the main peak on day 8.

Juvenile hormone has also been implicated in the control of storage granule formation. In *Bombyx mori*, allatectomy of larvae in the penultimate stage results in premature granule formation (Tojo, S. *et al.*, 1981). Injection of JH analogue into allatectomized penultimate larvae reduced the amount of precocious granule formation in a dose-dependent manner. These results were interpreted as showing that uptake of storage proteins is induced by ecdysteroids after a decline in JH levels. Observations on the role of JH are not, however, entirely consistent. The fat body structure of diapausing *Diatraea grandiosella* larvae reflects its primary storage function and low synthetic activity. It greatly resembles the fat body in non-diapausing

Table 5: Hormonal control of heterophagy, protein storage, autophagy and histolysis

Order and species	Method*	Hormone*	Effect	Reference†
Diptera				
Calliphora erythrocephala	EM,TR	20-OH Ec	Increased acid phosphatase Autophagic vacuoles formed. (L)*	8, 9
Calliphora vicina	EM	20-OH Ec	Autophagic vacuoles formed. (A)	13
Drosophila melanogaster	LM, Tr, Vt	20-OH Ec	Protein storage, lysosomal granules Protein storage. (L)	1, 2, 3 12, 15
	LM	+JH	Histolysis of larval fat body in adult. (L)	10
Lepidoptera				
Bombyx mori	EM	20-OH Ec, (−JH)	Protein storage granules, autophagic vacuoles formed. (L)	14
Calpodes ethlius	LM, EM, Vt	20-OH Ec	Protein storage granules, autophagic vacuoles formed.	4, 7
	Cm		Acid protease increased. (L)	5
Ephestia kuhniella	EM	20-OH Ec	Protein storage granules formed. (L)	6
Mamestra brassicae	EM, Vt	20-OH Ec, (−JH)	Protein storage granules, autophagic vacuoles formed. (L)	11

* 20-OH Ec, 20-hydroxyecdysone; +JH, juvenile hormone present; −JH, juvenile hormone diminished or absent; Cm, chemical assay; EM, electron microscopy; LM, light microscopy; Tr, tissue transplanted; Vt, *in vitro*; (L), larva; (A), adult.

† 1, Butterworth, F. *et al.* (1965); 2, Butterworth, F. *et al.* (1972); 3, Butterworth, F. *et al.* (1979); 4, Collins, J. (1969); 5, Collins, J. (1979); 6, Cölln, K. (1973); 7, Dean, R. (1978); 8, van Pelt-Verkuil, E. (1979); 9, van Pelt-Verkuil, E. and Dirkx, C. (1979); 10, Postlethwait, J. and Jones, G. (1978); 11, Sass, M. and Kovacs, J. (1975, 1977, 1980); 12, Thomasson, W. and Mitchell, H. (1972); 13, Thomsen, E. and Thomsen, M. (1978); 14, Tojo, S. *et al.* (1981); 15, Tysell, B. and Butterworth, F. (1978).

individuals at pupal ecdysis (Brown, J. and Chippendale, G., 1977). Although preparation for diapause takes longer than preparation for pupation, both follow the same pattern. They are characterized by the loss of RER and mitochondria by autophagy (section 3.2) and the accumulation of protein granules. Since pupation can be prevented and diapause induced experimentally by a single application of JH to the larva, it appears that in this species JH permits, or does not inhibit, the formation of protein granules.

4.2 Control of autophagy and histolysis

The precisely timed beginning of autophagy during metamorphosis and cell remodelling has been correlated with changes in the haemolymph ecdysteroid titre. The isolation and lysis of peroxisomes and mitochondria coincide with increased hormone levels in *Calpodes* (Locke, M. and Collins, J., 1965; Locke, M. and McMahon, J., 1971; Dean, R. *et al.*, 1980). Dean, R. (1978) has shown that the presence of 20-hydroxyecdysone is a sufficient stimulus for induction of autophagy in fat body of fifth-stage *Calpodes* larvae *in vitro*; there is no autophagy in hormone-free medium. However, autophagy once induced does not require the continuing presence of the hormone. This shows that fat body cells are intrinsically capable of sustaining autophagic activity. Why does autophagy not occur in response to the elevated ecdysteroid titre which precedes each larval–larval moult? In penultimate-stage larvae of *Mamestra*, ecdysteroids induce fat body autophagy when injected during the last day, but not during the preceding 3 days of the stadium. A reduction in the JH titre occurs during the penultimate larval stadium so that JH-inhibition of ecdysteroid-induced autophagy might be invoked to explain the result (Sass, M. and Kovacs, J., 1977). However, when these authors incubated fat body of different ages in the presence of ecdysone, but no JH, only cells taken from larvae on the day preceding fourth- to fifth-instar ecdysis underwent autophagy. It was suggested that the cells require a period of time to gain competence to respond to ecdysteroid after exposure to high JH levels. The change in competence of the fat body may be similar to the change in commitment of the epidermis in response to the commitment peak of ecdysteroid in *Manduca* (Riddiford, L., 1978; vol. 8) and *Calpodes* (Dean, R. *et al.*, 1980).

The control of fat body cell lysis in *Drosophila*

is less well understood. Of the larval cells which survive metamorphosis in wild-type flies, about 95% disappear by the end of the second day of adult life. Surviving larval cells in flies homozygous for the ap^4 mutant fare better. In ap^4 flies the larval cells disappear more slowly during the first 2 days of adult life and their number remains unchanged over the next 3 days. Transplantation experiments with fat body from newly ecdysed adults showed that ap^4 cells survive better because their environment induces less histolysis and not because of any intrinsic cellular property. Also, in both wild-type and ap^4 flies the internal environment becomes decreasingly potent in inducing lysis with increasing age (Butterworth, F., 1972). There are anterior–posterior gradients of heterophagic activity (Tysell, B. and Butterworth, F., 1978) and cell separation (Whitten, J., 1962) in *Drosophila*, and regional differences in acid phosphatase activity in *Calliphora erythrocephala* (van Pelt-Verkuil, E., 1980). These observations support the view that fat body cells intrinsically control their ability to respond to the milieu that triggers granule formation and autolysis.

In an earlier study Butterworth, F. and Bodenstein, D. (1967) performed transplantation experiments to discover the fate of late second-instar fat body cells in the adult environment. These cells showed no histolysis in normal adult males but when implanted together with ring glands from late third-instar larvae massive degeneration was inferred because the cells of the transplant could not be recognized. When second-instar cells were implanted into recently emerged females, cytological changes were also induced but were less complete and later than those seen in males with ring gland implants. Implants into ovariectomized females showed no degeneration, whereas those in males containing a transplanted ovary showed some histolysis. Since higher dipteran larval ring glands secrete ecdysone (Bollenbacher, W. *et al.*, 1976) and so do dipteran ovaries (section 2.2.2e; H. Hagedorn, vol. 8), these results can be explained by ecdysteroid induced autophagy of fat body cells. Butterworth, F. (1973) later reported that fat body cells from mid and late third-instar larvae transplanted into adult males did eventually begin to disappear, but not nearly as rapidly as the cells of the host which had undergone metamorphosis. He suggested that rapid lysis must depend on additional preparative changes which make the cells responsive to the lytic environment in the adult. These studies are all complicated by the difficulty of quantifying cell degeneration and death discussed, for example, by Butterworth, F. (1972). 20-Hydroxyecdysone is also implicated in the induction of autophagy in *Calliphora*. Precocious autophagy and storage granule formation is induced in the fat body of younger larvae when transplanted into older larvae (van Pelt-Verkuil, E. and Dirkx, C., 1979) and by 20-hydroxyecdysone injection into premetamorphic larvae (van Pelt-Verkuil, E., 1979). The problem is complicated by the observation of Day, M. (1943) who found that ring gland extirpation prevented the histolysis of larval fat body in adult *Lucilia*, and concluded that the corpora allata are required. Postlethwait, J. and Jones, G. (1978) concluded that the ring gland factor is JH, since fat body histolysis is inhibited in isolated abdomens of *Drosophila* but can be stimulated by the application of a JH analogue. Perhaps a less equivocal result on the factors inducing autophagy in dipteran fat body will be found through *in vitro* studies where direct hormone action can be observed.

The respective roles of ecdysteroids and JH are not understood. In *Calliphora* there is increased activity of lysosomal enzymes in fat body at metamorphosis and this increase can be induced by 20-hydroxyecdysone (van Pelt-Verkuil, E., 1979). In *Calpodes ethlius* increased activity of acid hydrolases is correlated with the increased haemolymph ecdysteroid titre (Collins, J., 1975b, 1979; Dean, R. *et al.*, 1980). These increases in lytic enzymes do not by themselves ensure degradative activity, for in *Calpodes* and other insects, destruction of organelles in autophagic vacuoles occurs side by side with the storage of haemolymph proteins in granules. Hydrolase activity remains relatively high thoughout the puparial stage of *Drosophila melanogaster* without affecting the granules (Butterworth, F. *et al.*, 1972) and through the pupal stage of *Calpodes* (J. Collins, unpublished observations).

It has been suggested that RNA synthesis 9–12 h before increased hormone titre is a prerequisite for induction of autophagy by 20-hydroxyecdysone in *Mamestra* (Sass, M. and Kovacs, J., 1980). Similarly, Postlethwait, J. and Jones, G. (1978) have proposed that protein synthesis is necessary before JH-stimulated histolysis of larval fat body of *Drosophila* will occur. The nature of such protein(s) is unknown.

ACKNOWLEDGEMENTS

We are grateful to Mr P. Huie and Ms H. Kirk for technical assistance, and to Mrs J. Sexsmith for word-processing. The work was supported by Natural Sciences Engineering Research Council grants A6607 to M. Locke, and A6470 to J. V. Collins.

REFERENCES

Anderson, D. T. (1964). The larval development of *Dacus tryoni* (Frogg.) (Diptera: Trypetidae). II. Development of imaginal rudiments other than the principal discs. *Aust. J. Zool. 12*, 1–8.

Anderson, D. T. (1972a). The development of hemimetabolous insects. In *Developmental Systems: Insects* Vol. 1. Edited by S. J. Counce and C. H. Waddington. Pages 85–163. Academic Press, London.

Anderson, D. T. (1972b). The development of holometabolous insects. In *Developmental Systems: Insects*. Vol. 1. Edited by S. J. Counce and C. H. Waddington. Pages 165–242. Academic Press, London.

Ashwell, G. and Morell, A. G. (1974). The role of surface carbohydrates in the hepatic recognition and transport of circulating glycoproteins. *Adv. Enzymol. 41*, 99–128.

Barth, R. H., Bunyard, P. P. and Hamilton, T. H. (1964). RNA metabolism in pupae of the oak silkworm, *Antherea pernyi*. The effects of diapause, development and injury. *Proc. Natl. Acad. Sci. U.S.A. 52*, 1572–1580.

Behan, M. and Hagedorn, H. H. (1978). Ultrastructural changes in the fat body of adult female *Aedes aegypti* in relationship to vitellogenin synthesis. *Cell. Tiss. Res. 186*, 499–506.

Bergtrom, G. and Robinson, J. M. (1977). Ultrastructural localization of the site of hemoglobin synthesis in *Chironomus thummi* (Diptera). *J. Ultrastruct. Res. 60*, 395–405.

Berry, S. J., Krishnakumaran, A., Oberlander, H. and Schneiderman, H. A. (1967). Effects of hormones and injury on RNA synthesis in saturniid moths. *J. Insect Physiol. 13*, 1511–1537.

Bhakthan, N. M. G. and Gilbert, L. I. (1972). Studies on the cytophysiology of the fat body of the American silk-moth. *Z. Zellforsch. 124*, 433–444.

Bodenstein, D. (1953). Studies on the humoral mechanisms in growth and metamorphosis of the cockroach, *Periplaneta americana*. III. Humoral effects on metabolism. *J. Exp. Zool. 124*, 105–115.

Bollenbacher, W. E., Goodman, W., Vedeckis, W. V. and Gilbert, L. I. (1976). The *in vitro* synthesis and secretion of α-ecdysone by the ring glands of the fly, *Sarcophaga bullata*. *Steroids 27*, 309–324.

Borst, D. W., Bollenbacher, W. E., O'Connor, J. D., King, D. S. and Fristrom, J. W. (1974). Ecdysone levels during metamorphosis of *Drosophila melanogaster*. *Devel. Biol. 39*, 308–316.

Bosquet, G. (1979). Occurrence of an active regulatory mechanism of protein synthesis during starvation and refeeding in *Bombyx mori* larvae. *Biochimie 61*, 165–170.

Brodie, D. A. (1981). Bead rings at the endoplasmic reticulum-Golgi complex boundary: morphological changes accompanying inhibition of intracellular transport of secretory proteins in arthropod fat body tissue. *J. Cell. Biol. 90*, 92–100.

Brooks, M. and Richards, A. G. (1955). Intracellular symbiosis in cockroaches. II. Mitotic division of mycetocytes. *Science 122*, 242.

Brown, J. J. and Chippendale, G. M. (1977). Ultrastructure and respiration of the fat body of diapausing and non-diapausing larvae of the corn borer, *Diatraea grandiosella*. *J. Insect Physiol. 23*, 1135–1142.

Buckner, J. S. and Caldwell, J. M. (1980). Uric acid levels during last larval instar of *Manduca sexta*, an abrupt transition from excretion to storage in fat body. *J. Insect Physiol. 26*, 27–32.

Bursell, E. (1967). The excretion of nitrogen in insects. *Adv. Insect Physiol. 4*, 33–67.

Butterworth, F. M. (1972). Adipose tissue of *Drosophila melanogaster*. V. Genetic and experimental studies of an extrinsic influence on the rate of cell death in the larval fat body. *Devel. Biol. 28*, 311–325.

Butterworth, F. M. (1973). Adipose tissue of *Drosophila melanogaster*. VI. Non-susceptibility of the immature larval fat body to the lytic environment of the young adult. *Wilhelm Roux Arch. EntwMech 172*, 263–270.

Butterworth, F. M. and Bodenstein, D. (1967). Adipose tissue of *Drosophila melanogaster*. II. The effect of the adult internal environment on growth, protein deposition, and histolysis in the larval fat body. *J. Exp. Zool. 164*, 251–266.

Butterworth, F. M., Bodenstein, D. and King, R. C. (1965). Adipose tissue of *Drosophila melanogaster*. I. An experimental study of larval fat body. *J. Exp. Zool. 158*, 141–154.

Butterworth, F. M., Dene, H. and Wachnicki, D. (1972). Acid phosphatase in the protein granules of the larval fat body. *Drosoph. Inf. Serv. 49*, 107–108.

Butterworth, F. M., Tysell, B. and Waclawski, I. (1979). The effect of 20-hydroxyecdysone and protein on granule formation in the *in vitro* cultured fat body of *Drosophila*. *J. Insect Physiol. 25*, 855–860.

Buys, K. S. (1924). Adipose tissue in insects. *J. Morph. 38*, 485–527.

Chen, T. T., Couble, P., Abu-Hakima, R. and Wyatt, G. R. (1979). Juvenile hormone-controlled vitellogenin synthesis in *Locusta migratoria* fat body. *Devel. Biol. 69*, 59–72.

Chen, T. T., Couble, P., De Lucca, F. L. and Wyatt, G. R. (1976). Juvenile hormone control of vitellogenin synthesis in *Locusta migratoria*. In *The Juvenile Hormones*. Edited by L. I. Gilbert. Pages 505–529. Plenum Press, New York.

Chen, T. T. and Wyatt, G. R. (1981). Juvenile hormone control of vitellogenin synthesis in *Locusta migratoria*. *Scientific Papers of the Institute of Organic and Physical Chemistry of Wroclaw Technical University*, No. 22, Conference 7, pp. 535–566.

Chippendale, G. M. (1970). Metamorphic changes in fat body proteins of the Southwestern corn borer, *Diatraea grandiosella*. *J. Insect Physiol. 16*, 1057–1068.

Chippendale G. M. and Kilby, B. A. (1969). Relationship between the proteins of the haemolymph and fat body during development of *Pieris brassicae*. *J. Insect Physiol. 15*, 905–926.

Clark, M. K. and Dahm, P. A. (1973). Phenobarbital-induced, membrane-like scrolls in the oenocytes of *Musca domestica* Linnaeus. *J. Cell Biol. 56*, 870–875.

Cochran, D. G. (1975). Excretion in insects. In *Insect Biochemistry and Function*. Edited by D. J. Candy and B. A. Kilby. Pages 177–282. Chapman & Hall, London.

Cochran, D. G., Mullins, D. E. and Mullins, K. J. (1979). Cytological changes in the fat body of the American cockroach, *Periplaneta americana*, in relation to dietary nitrogen levels. *Ann. Ent. Soc. Amer. 72*, 197–207.

Collins, J. V. (1967). The formation of protein granules in the fat body of an insect, *Calpodes ethlius* Stoll. Lepidoptera Hesperiidae. Ph.D. Thesis. Western Reserve University.

Collins, J. V. (1969). The hormonal control of fat body development in *Calpodes ethlius*. (Lepidoptera, Hesperiidae). *J. Insect Physiol. 15*, 341–352.

Collins, J. V. (1974). Hormonal control of protein sequestration in the fat body of *Calpodes ethlius* Stoll. *Canad. J. Zool. 52*, 639–642.

Collins, J. V. (1975a). Secretion and uptake of ^{14}C proteins by fat body of *Calpodes ethlius* Stoll. (Lepidoptera, Hesperiidae). *Differentiation 3*, 143–148.

Collins, J. V. (1975b). Soluble acid phosphatases of the hemolymph and fat body of *Calpodes ethlius* (Stoll.) and the control of protein storage by fat body. *Canad. J. Zool. 53*, 480–489.

Collins, J. V. (1979). Acid hydrolase activity and the control of autophagy and heterophagy in larval fat body of *Calpodes ethlius* Stoll. (Lepidoptera). *Comp. Biochem. Physiol. 62B*, 317–324.

COLLINS, J. V. and DOWNE, A. E. R. (1970). Selective accumulation of haemolymph proteins by the fat body of *Galleria mellonella*. *J. Insect Physiol.* *16*, 1697–1708.

CÖLLN, K. (1973). Uber die Metamorphose der Proteinspektren von Hamolymphe und Fettkorper bei *Ephestia kuhniella*. *Z. Wilhelm Roux Arch. EntwMech.* *172*, 231–257.

COUBLE, P., CHEN, T. T. and WYATT, G. R. (1979). Juvenile hormone-controlled vitellogenin synthesis in *Locusta migratoria* fat body: cytological development. *J. Insect Physiol.* *25*, 327–337.

COUPLAND, R. E. (1957). Observations on the normal histology and histochemistry of the fat body of the locust (*Schistocerca gregaria*). *J. Exp. Biol.* *34*, 290–296.

DAY, M. F. (1943). The function of the corpus allatum in muscoid Diptera. *Biol. Bull.* *84*, 127–140.

DEAN, R. L. (1978). The induction of autophagy in isolated insect fat body by β-ecdysone. *J. Insect Physiol.* *24*, 439–447.

DEAN, R. L., BOLLENBACHER, W. E., LOCKE, M., SMITH, S. L. and GILBERT, L. I. (1980). Haemolymph ecdysteroid levels and cellular events in the intermoult/moult sequence of *Calpodes ethlius*. *J. Insect Physiol.* *26*, 267–280.

DELLA-CIOPPA, G. and ENGELMANN, F. (1980). Juvenile hormone-stimulated proliferation of edoplasmic reticulum in fat body cells of a vitellogenic insect, *Leucophaea maderae* (Blattaria). *Biochem. Biophys. Res. Commun.* *93*, 825–832.

DORTLAND, J. F. (1978). Synthesis of vitellogenins and diapause proteins by the fat body of *Leptinotarsa* as a function of photoperiod. *Physiol. Ent.* *3*, 281–288.

DORTLAND, J. F. and ESCH, T. H. (1979). A fine structural survey of the development of the adult fat body of *Leptinotarsa decemlineata*. *Cell Tiss. Res.* *201*, 423–430.

DUTKOWSKI, A. B. (1974). Fat body of *Galleria mellonella* during metamorphosis. Cytochemical and ultrastructural studies. *Folia Histochem. Cytochem.* *12*, 269–280.

ENGELMANN, F. (1977). Undegraded vitellogenin polysomes from female insect fat bodies. *Biochem. Biophys. Res. Commun.* *78*, 641–647.

ENGELMANN, F. (1978). Synthesis of vitellogenin after long-term ovariectomy in a cockroach. *Insect Biochem.* *8*, 149–154.

ENGELMANN, F. (1979). Insect vitellogenin: identification, biosynthesis, and role in vitellogenesis. *Adv. Insect Physiol.* *14*, 49–108.

ENGELMANN, F. and BARAJAS, L. (1975). Ribosome–membrane association in fat body tissue from reproductively active females of *Leucophaea maderae*. *Exp. Cell Res.* *92*, 102–110.

EVANS, A. (1935). Some notes on the biology and physiology of the sheep blowfly, *Lucilia sericata*, Merg. *Bull. Ent. Res.* *26*, 115–122.

EVANS, J. J. T. (1967). Development and ultrastructure of the fat body cells and oenocytes of the Queensland fruit fly, *Dacus tryoni* (Frogg.). *Z. Mikr. Ant.* *81*, 49–61.

FAVARD-SERENO, C. (1973). Terminal differentiation of the adipose tissue in relation to vitellogenesis in the cricket. *Mol. Biol. Rep.* *1*, 179–186.

FLANAGAN, T. R. and HAGEDORN, H. H. (1977). Vitellogenin synthesis in the mosquito: the role of juvenile hormone in the development of responsiveness to ecdysone. *Physiol. Ent.* *2*, 173–178.

FRAENKEL, G. and HSIAO, C. (1968). Morphological and endocrinological aspects of pupal diapause in a fleshfly, *Sarcophaga argyrostoma*. *J. Insect Physiol.* *14*, 707–718.

GAUDECKER, B. VON (1963). Uber den Formwechsel einiger Zellorganelle bei der Bildung der Reservestoffe im Fettkorper von *Drosophila*-Larven. *Z. Mikr. Ant.* *61*, 56–95.

GELLISSEN, G. and WYATT, G. R. (1981). Production of lipophorin in the fat body of adult *Locusta migratoria*: comparison with vitellogenin. *Canad. J. Biochem.* *59*, 648–654.

GRIFFITHS, G. W. and BECK, S. D. (1973). Intracellular symbiotes of the pea aphid, *Acyrthosiphon pisum*. *J. Insect Physiol.* *19*, 75–84.

GRIFFITHS, G. W. and BECK, S. D. (1975). Ultrastructure of pea aphid mycetocytes: evidence for symbiote secretion. *Cell Tiss. Res.* *159*, 351–367.

GRIFFITHS, G. W. and BECK, S. D. (1977). *In vivo* sterol biosynthesis by pea aphid symbiotes as determined by digitonin and electron microscope autoradiography. *Cell Tiss. Res.* *176*, 179–190.

HAYDAK, M. H. (1953). Influence of the protein level of the diet on the longevity of cockroaches. *Ann. Ent. Soc. Amer.* *46*, 547–560.

HINTON, H. E. (1958). On the nature and metamorphosis of the colour pattern of *Thaumalea* (Diptera, Thaumaleidae). *J. Insect Physiol.* *2*, 249–260.

HINTON, H. E. (1959). Function of chromatocytes in the Simuliidae, with notes on their behaviour at the pupal-adult stage moult. *Quart. J. Mic. Sci.* *100*, 65–71.

HOUK, E. J. and GRIFFITHS, G. W. (1980). Intracellular symbiotes of the Homoptera. *Ann. Rev. Ent.* *25*, 161–187.

HUDSON, A. (1966). Proteins in the haemolymph and other tissues of the developing tomato hornworm, *Protoparce quinquemaculata* Haworth. *Canad. J. Zool.* *44*, 541–555.

HUYBRECHTS, R. and DE LOOF, A. (1977). Induction of vitellogenin synthesis in male *Sarcophaga bullata* by ecdysterone. *J. Insect Physiol.* *23*, 1359–1362.

IRVINE, D. J. and BRASCH, K. (1981). The influence of juvenile hormone on polyploidy and vitellogenesis in the fat body of *Locusta migratoria*. *Gen. Comp. Endocr.* *45*, 91–99.

ISHIZAKI, H. (1965). Electron microscopic study of changes in the subcellular organization during metamorphosis of the fat body cell of *Philosamia cynthia ricini* (Lepidoptera). *J. Insect Physiol.* *11*, 845–855.

JAMIESON, J. D. and PALADE, G. E. (1968). Intracellular transport of secretory proteins in the pancreatic exocrine cell. IV. Metabolic requirements. *J. Cell Biol.* *39*, 589–603.

JENSEN, P. V., HANSEN, B. L., HANSEN, G. N. and THOMSEN, E. (1981). Vitellogenin and vitellin from the blowfly *Calliphora vicina*: occurrence, purification and antigenic characterization. *Insect Biochem.* *11*, 129–135.

KAN, S.-P. and HO, B.-C. (1972). Development of *Breinlia sergenti* (Dipetalonematidae) in the fat body of mosquitoes. Part 2: Ultrastructural changes in the fat body. *J. Med. Ent.* *9*, 255–261.

KINNEAR, J. F. and THOMSON, J. A. (1975). Nature, origin and fate of major haemolymph proteins in *Calliphora*. *Insect Biochem.* *5*, 531–552.

KOOLMAN, J. (1978). Ecdysone oxidase in insects. *Hoppe-Seyler's Zeit. Physiol. Chem.* *359*, 1315–1321.

KRAMER, S. J. (1978). Regulation of the activity of JH-specific esterases in the Colorado potato beetle, *Leptinotarsa decemlineata*. *J. Insect Physiol.* *24*, 743–747.

KRAMER, S. J., MUNDALL, E. C. and LAW, J. H. (1980). Purification and properties of manducin, an amino acid storage protein of the haemolymph of larval and pupal *Manduca sexta*. *Insect Biochem.* *10*, 279–288.

KRISHNAKUMARAN, A., BERRY, S. J., OBERLANDER, H. and SCHNEIDERMAN, H. A. (1967). Nucleic acid synthesis during insect development — II. Control of DNA synthesis in the cecropia silkworm and other saturnid moths. *J. Insect Physiol.* *13*, 1–57.

LABOUR, G. (1970). Etude histologique et histochimique du corps adipeux an cours du developpement post-embryonnaire du Doryphore. *Arch. Anat. Micr.* *59*, 235–262.

LABOUR, G. (1974). Etude ultrastructurale de l'evolution du tissu adipeux au cours de developpement larvaire et nymphale chez le Doryphore. *Ann. Soc. Ent. Fr. (N.S.)* *10*, 943–958.

LANGLEY, P. A. and BURSELL, E. (1980). Role of fat body and uterine gland in milk synthesis by adult female *Glossina morsitans*. *Insect Biochem.* *10*, 11–17.

LARSEN, W. J. (1970). Genesis of mitochondria in insect fat body. *J. Cell Biol.* *47*, 373–383.

LARSEN, W. J. (1976). Cell remodeling in the fat body of an insect. *Tissue Cell* *8*, 73–92.

LAUVERJAT, S. (1977). L'evolution post-imaginale du tissu adipeux femelle de *Locusta migratoria* et son controle endocrine. *Gen. Comp. Endocr.* *33*, 13–34.

LEA, A. O. (1982). Artifactual stimulation of vitellogenesis in *Aedes aegypti* by 20-hydroxyecdysone. *J. Insect Physiol.* *28*, 173–176.

LOCKE, J., MCDERMID, H., BRAC, T. and ATKINSON, B. G. (1982). Developmental changes in the synthesis of haemolymph polypeptides and their sequestration by the prepupal fat body in *Calpodes ethlius* Stoll (Lepidoptera, Hesperiidae). *Insect Biochem.* *12*, 431–440.

LOCKE, M. (1969). The ultrastructure of the oenocytes in the molt/intermolt cycle of an insect. *Tissue Cell* *1*, 103–154.

LOCKE, M. (1970). The molt/intermolt cycle in the epidermis and other tissues of an insect *Calpodes ethlius* (Lepidoptera, Hesperiidae). *Tissue Cell* *2*, 197–223.

LOCKE, M. (1980a). The cell biology of fat body development. In *Insect Biology in the Future*. Edited by M. Locke and D. S. Smith. Pages 227–252. Academic Press, New York.

LOCKE, M. (1980b). Cell structure during insect metamorphosis. In *Metamorphosis: A Problem in Developmental Biology*. Edited by E. Frieden and L. I. Gilbert. Pages 75–103. Plenum Press, New York.

LOCKE, M. and COLLINS, J. V. (1965). The structure and formation of protein granules in the fat body of an insect. *J. Cell Biol. 26*, 857–884.

LOCKE, M. and COLLINS, J. V. (1968). Protein uptake into multivesicular bodies and storage granules in the fat body of an insect. *J. Cell Biol. 36*, 453–483.

LOCKE, M. and COLLINS, J. V. (1980). Organelle turnover in insect metamorphosis. In *Pathologic Aspects of Cell Membranes*, Vol. 2. Edited by B. F. Trump and A. Arstila. Pages 223–248. Academic Press, New York.

LOCKE, M. and HUIE, P. (1972). The fiber components of insect connective tissue. *Tissue Cell 4*, 601–612.

LOCKE, M. and HUIE, P. (1976a). The beads in the Golgi complex-endoplasmic reticulum region. *J. Cell Biol. 70*, 384–394.

LOCKE, M. and HUIE, P. (1976b). Vertebrate Golgi complexes have beads in a similar position to those found in arthropods. *Tissue Cell 8*, 739–743.

LOCKE, M. and HUIE, P. (1977). Bismuth staining of the Golgi complex is a characteristic arthropod feature lacking in *Peripatus*. *Nature 270*, 341–343.

LOCKE, M. and HUIE, P. (1980a). Ultrastructure methods in cuticle research. In *Cuticle Techniques in Arthropods*. Edited by T. A. Miller. Pages 91–144. Springer-Verlag, Berlin.

LOCKE, M. and HUIE, P. (1980b). The nucleolus during epidermal development in an insect. *Tissue Cell 12*, 175–194.

LOCKE, M. and HUIE, P. (1981a). Epidermal feet in insect morphogenesis. *Nature 293*, 733–735.

LOCKE, M. and HUIE, P. (1981b). Epidermal feet in pupal segment morphogenesis. *Tissue Cell 13*, 787–803.

LOCKE, M. and MCMAHON, J. T. (1971). The origin and fate of microbodies in the fat body on an insect. *J. Cell Biol. 48*, 61–78.

LOCKE, M. and SYKES, A. K. (1975). The role of the Golgi complex in the isolation and digestion of organelles. *Tissue Cell 7*, 14, 3–158.

LOOF, A. DE (1972). Diapause phenomena in non-diapausing last instar larvae, pupae and pharate adults of the Colorado beetle. *J. Insect Physiol. 18*, 1039–1047.

LOOF, A. DE and DE WILDE, J. (1970). the relation between haemolymph proteins and vitellogenesis in the Colorado beetle, *Leptinotarsa decemlineata*. *J. Insect Physiol. 16*, 157–169.

LOOF, A. DE and LAGASSE, A. (1970). Juvenile hormone and the ultrastructural properties of the fat body of the adult Colorado beetle, *Leptinotarsa decemlineata* Say. *Z. Mikr. Ant. 106*, 439–450.

LOUGHTON, B. G. and WEST, A. S. (1965). The development and distribution of haemolymph proteins in Lepidoptera. *J. Insect Physiol. 11*, 919–932.

MARTIN, M. D., KINNEAR, J. F. and THOMSON, J. A. (1971). Developmental changes in the late larva of *Calliphora stygia*. IV. Uptake of plasma protein by the fat body. *Aust. J. Biol. Sci. 24*, 291–299.

MARX, R. (1971). Morphological changes in fat body mitochondria during larval development of the blowfly, *Calliphora erythrocephala* Meigen. *Cytobiologie 3*, 417–420.

MCCLINTOCK, J. and LOCKE, M. (1982). Lead staining in the Golgi complex. *Tissue Cell 14*, 541–544

MCDERMID, H. and LOCKE, M. (1983). Tyrosine storage vacuoles in insect fat body. *Tissue Cell 15*, 137–158

MORI, T., AKAI, H. and KOBAYASHI, M. (1970). Ultrastructural changes of the fat body in the silkworm during postembryonic development. *J. Seric. Sci. Japn. 39*, 51–61.

NAIR, K. K., CHEN, T. T. and WYATT, G. R. (1981). Juvenile hormone-stimulated polyploidy in adult locust fat body. *Devel. Biol. 81*, 356–360.

NEUFELD, G. J., THOMSON, J. A. and HORN, D. S. (1968). Short-term effects of crustecdysone (20-hydroxyecdysone) on protein and RNA synthesis in third instar larvae of *Calliphora*. *J. Insect Physiol. 14*, 789–804.

NODA, H. (1977). Histological and histochemical observations of intracellular yeast-like symbiotes in the fat body of the smaller brown planthopper, *Laodelphax striatellus* (Homoptera, Delphacidae). *Appl. Ent. Zool. 12*, 134–141.

NOWOCK, J., GOODMAN, W., BOLLENBACHER, W. E. and GILBERT, L. I. (1975). Synthesis of juvenile hormone binding proteins by the fat body of *Manduca sexta*. *Gen. Comp. Endocrinol. 27*, 230–239.

ODHIAMBO, T. R. (1967). The fine structure and histochemistry of the fat body in the locust, *Schistocerca gregaria*. *J. Cell Sci. 2*, 235–242.

OSCHMAN, J. L. and BERRIDGE, M. J. (1971). The structural basis of fluid secretion. *Fed. Proc. 30*, 49–56.

PATTNAIK, N. M., MUNDALL, E. C., TRAMBUSTI, B. G., LAW, J. H. and KEZDY, F. J. (1979). Isolation and characterization of a larval lipoprotein from the haemolymph of *Manduca sexta*. *Comp. Biochem. Physiol. 63B*, 469–476.

PELT-VERKUIL, E., VAN (1978). Increase in acid phosphatase activity in the fat body during larval and pharate pupal development in *Calliphora erythrocephala*. *J. Insect Physiol. 24*, 375–382.

PELT-VERKUIL, E., VAN (1979). Hormone mediated induction of acid phosphatase activity in the fat body of *Calliphora erythrocephala* prio to metamorphosis. *J. Insect Physiol. 25*, 965–973.

PELT-VERKUIL, E., VAN (1980). The induction of lysosomal enzyme activity in the fat body of *Calliphora erythrocephala*: changes in the internal environment. *J. Insect Physiol. 26*, 91–101.

PELT-VERKUIL, E., VAN and DIRKX, C. (1979). Transplantation experiments in the fat body of *Calliphora*: A morphometric study of induced ultrastructural changes. *Cell Tiss. Res. 203*, 267–281.

PEMRICK, S. M. and BUTZ, A. (1970). RNA synthesis of the fat body of adult *Tenebrio molitor*. *J. Insect Physiol. 16*, 1171–1177.

POSTLETHWAIT, J. H. and JONES, G. J. (1978). Endocrine control of larval fat body histolysis in normal and mutant *Drosophila melanogaster*. *J. Exp. Zool. 203*, 207–214.

PRICE, G. M. (1969). Protein synthesis and nucleic acid metabolism in the fat body of the larva of the blowfly, *Calliphora erythrocephala*. *J. Insect Physiol. 15*, 931–944.

PREISTER, W. DE and VAN DER MOLEN, L. G. (1979). Premetamorphic changes in the ultrastructure of *Calliphora* fat cells. *Cell Tiss. Res. 198*, 79–93.

REDFERN, C. P. F. (1982). 20-Hydroxy-ecdysone and ovarian development in *Anopheles stephensi*. *J. Insect Physiol. 28*, 97–109.

REID, P. C. and CHEN, T. T. (1981). Juvenile hormone controlled vitellogenin synthesis in the fat body of the locust (*Locusta migratoria*): isolation and characterization of vitellogenin polysomes and their induction *in vivo*. *Insect Biochem. 11*, 297–305.

RIDDIFORD, L. M. (1978). Ecdysone-induced change in cellular commitment of the epidermis of the tobacco hornworm, *Manduca sexta*, at the initiation of metamorphosis. *Gen. Comp. Endocr. 34*, 438–446.

RINTERKNECHT, E. and ROUSSEL, J.-P. (1978). Modification ultrastructurales induites par les hormones juveniles dans le cellules du corps gras de *Locusta migratoria*. L. *Bull. Soc. Zool. Fr. 103*, 359–366.

RIZKI, T. M. (1964). Mutant genes regulating the inducibility of kynurenine synthesis. *J. Cell Biol. 21*, 203–211.

RIZKI, T. M. (1978). Fat body. In *The Genetics and Biology of* Drosophila. Vol. 2b, pp. 561–601. Academic Press, London.

RIZKI, T. M. and RIZKI, R. M. (1962). Cytodifferentiation in the rosy mutant of *Drosophila melanogaster*. *J. Cell Biol. 12*, 149–157.

ROBERTS, D. B., WOLFE, J. and AKAM, M. E. (1977). The developmental profiles of two major haemolymph proteins from *Drosophila melanogaster*. *J. Insect Physiol. 23*, 871–878.

ROMER, F., EMMERICH, H. and NOWOCK, J. (1974). Biosynthesis of ecdysones in isolated prothoracic glands and oenocytes of *Tenebrio molitor in vitro*. *J. Insect Physiol. 20*, 1975–1987.

RUTSCHKE, E. and VON U BROZIO, F. (1975). Untersuchungen zur Ultrastruktur des Fettkorpers von Stabschreken bei *Carausius morosus* deren Veranderung nach Verabreichung subletaler Dosen von Toxaphen. *Zool. Jb. Anat. 94*, 296–318.

SASS, M. and KOVACS, J. (1975). Ecdysterone and an analogue of juvenile hormone on the autophagy in the cells of fat body of *Mamestra brassicae*. *Acta Biol. Acad. Sci. Hung. 26*, 189–196.

SASS, M. and KOVACS, J. (1977). The effect of ecdysone on the fat body cells of the penultimate larvae of *Mamestra brassicae*. *Cell Tiss. Res. 180*, 403–409.

SASS, M. and KOVACS, J. (1980). The effects of actinomycin D, cycloheximide and puromycin on the 20-hydroxyecdysone induced autophagocytosis in larval fat body cells of *Pieris brassicae*. *J. Insect Physiol. 26*, 569–577.

SCHIN, K., LAUFER, H. and CARR, E. (1977). Cytochemical and electrophoretic studies of haemoglobin synthesis in the fat body of a midge, *Chironomus thummi. J. Insect Physiol. 23*, 1233–1242.

SHAAYA, E. and KARLSON, P. (1965). Der ecdysontiter Wahrend der Insektenentwicklung, II. Die postembryonale entwicklung der schmeissfliege *Calliphora erythrocephala* Meig. *J. Insect Physiol. 11*, 65–69.

SPARKS, T. C., WILLIS, W. S., SHOREY, H. H. and HAMMOCK, B. D. (1979). Haemolymph juvenile hormone esterase activity in synchronous last instar larvae of the cabbage looper, *Trichoplusia ni. J. Insect Physiol. 25*, 125–132.

STAY, B. and CLARK, J. K. (1971). Fluctuation of protein granules in the fat body of the viviparous cockroach. *Diploptera punctata*, during the reproduction cycle. *J. Insect Physiol. 17*, 1747–1762.

STOPPIE, P., BRIERS, T., HUYBRECHTS, R. and DE LOOF, A. (1981). Moulting hormone, juvenile hormone and the ultrastructure of the fat body of adult *Sarcophaga bullata* (Diptera). *Cell Tiss. Res. 221*, 233–244.

TAKAHASHI, S. (1966). Studies on ribonucleic acid in the fat body of *Philosamia cynthia ricini* Donovan (Lepidoptera) during development. *J. Insect Physiol. 12*, 789–801.

THOMASSON, W. A. and MITCHELL, H. K. (1972). Hormonal control of protein granule accumulation in fat bodies of *Drosophila melanogaster* larvae. *J. Insect Physiol. 18*, 1885–1899.

THOMSEN, E. and THOMSEN, M. (1974). Fine structure of the fat body of the female of *Calliphora erythrocephala* during the first egg maturation cycle. *Cell Tiss. Res. 152*, 193–217.

THOMSEN, E. and THOMSEN, M. (1978). Production of specific-protein secretion granules by fat body cells of the blowfly, *Calliphora erythrocephala. Cell Tiss. Res. 193*, 25–33.

THOMSEN, E., HANSEN, B. L., HANSEN, G. N. and JENSEN, P. V. (1980). Ultrastructural immunocytochemical localization of vitellogenin in the fat body of the blowfly, *Calliphora vicina* Rob.-Desv. (*erythrocephala* Meig.) by use of the unlabeled antibody-enzyme method. *Cell Tiss. Res. 208*, 445–455.

THOMSON, J. A. (1973). Patterns of gene activity in larval tissues of the blowfly *Calliphora*. In *The Biochemistry of Gene Expression in Higher Organisms*. Edited by J. K. Pollack and J. W. Lee. Pages 320–332. Australia and New Zealand Book Company, Sydney.

THOMSON, J. A. (1975). Major patterns of gene activity during development in holometabolous insects. *Adv. Insect Physiol. 11*, 321–398.

TOBE, S. S. and DAVEY, K. G. (1974a). Nucleic acid synthesis during the reproductive cycle of *Glossina austeni. Insect Biochem. 4*, 215–223.

TOBE, S. S. and DAVEY, K. G. (1974b). Autoradiographic study of protein synthesis in abdominal tissues of *Glossina austeni. Tissue Cell 6*, 255–268.

TOBE, S. S. and DAVEY, K. G. (1975). Synthesis and turnover of haemolymph proteins during the reproductive cycle of *Glossina austeni. Canad. J. Zool. 53*, 614–629.

TOBE, S. S., DAVEY, K. G. and HUEBNER, E. (1973). Nutrient transfer during the reproductive cycle in *Glossina austeni* Newst. I. Histology and histochemistry of the milk gland, fat body and oenocytes. *Tissue Cell 5*, 633–650.

TOBE, S. S. and LOUGHTON, B. G. (1969a). An autoradiographic study of haemolymph protein uptake by tissues of the fifth instar locust. *J. Insect Physiol. 15*, 1331–1346.

TOBE, S. S. and LOUGHTON, B. G. (1969b). An investigation of haemolymph protein economy during the fifth instar of *Locusta migratoria migratorioides. J. Insect Physiol. 15*, 1659–1672.

TOJO, S., BETCHAKU, T., ZICCARDI, V. J. and WYATT, G. R. (1978). Fat body protein granules and storage proteins in the silkmoth, *Hyalophora cecropia. J. Cell Biol. 78*, 823–838.

TOJO, S., NAGATA, M. and KOBAYASHI, M. (1980). Storage proteins in the silkworm, *Bombyx mori. Insect Biochem. 10*, 289–303.

TOJO, S., KIGUCHI, K. and KIMURA, S. (1981). Hormonal control of storage protein synthesis and uptake by the fat body in the silkworm, *Bombyx mori. J. Insect Physiol. 27*, 491–497.

TRAGER, W. (1937). Cell size in relation to the growth and metamorphosis of the mosquito *Aedes aegypti. J. Exp. Zool. 76*, 467–489.

TSUYAMA, S. and MIURA, K. (1979). Structural and enzymic changes in fat body mitochondria of the blowfly, *Aldrichina grahami*, during larval growth. *Insect Biochem. 9*, 435–442.

TYCHSEN, C. M. (1978). Mitochondrial generation in the fat body cells of *Calpodes ethlius* Stoll. (Lepidoptera: Hesperiidae). M.Sc. thesis, University of Western Ontario.

TYSELL, B. and BUTTERWORTH, F. M. (1978). Different rate of protein granule formation in the larval fat body of *Drosophila melanogaster. J. Insect Physiol. 24*, 201–206.

VINCE, R. K. and GILBERT, L. I. (1977). Juvenile hormone esterase activity in precisely timed last instar larvae and pharate pupae of *Manduca sexta. Insect Biochem. 7*, 115–120.

VOINOV, V. (1927). Sur l'existence d'un tissu mesenchymateux vacuolaire dans les larves de Chironomus. *C.R. Soc. Biol. 96*, 1015–1017.

WAKU, Y. and SUMIMOTO, K. (1969). Light and electron microscopical study of the fat body cells in the metamorphosing silkworm. *Bull. Fac. Text. Fibers, Kyoto Univ. Ind. Arts Text. Fibers*, 5, 256–287.

WALKER, P. A. (1965). The structure of the fat body in normal and starved cockroaches as seen with the electron microscope. *J. Insect Physiol. 11*, 1625–1631.

WALKER, P. A. (1966). An electron microscope study of the fat body of the moth *Philosamia* during growth and metamorphosis. *J. Insect Physiol. 12*, 1009–1018.

WALTERS, D. R. (1969). Reaggregation of insect cells *in vitro* I. Adhesive properties of dissociated fat body cells from developing Saturniid moths. *Biol. Bull. 137*, 217–227.

WALTERS, D. R. and WILLIAMS, C. M. (1966). Reaggregation of insect cells as studied by a new method of tissue and organ culture. *Science 154*, 516–517.

WEBSTER, D. M. (1982). The major haemolymph proteins of an insect. M.Sc. thesis, University of Western Ontario.

WHITMORE, D., GILBERT, L. I. and ITTYCHERIAH, P. I. (1974). The origin of hemolymph carboxylesterases induced by the insect juvenile hormone. *Molec. Cell. Endocr. 1*, 37–54.

WHITTEN, J. M. (1962). Breakdown and formation of connective tissue in the pupal stage of an insect. *Quart. J. Mic. Sci. 103*, 359–367.

WIGGLESWORTH, V. B. (1942). The storage of protein, fat, glycogen and uric acid in the fat body and other tissues of mosquito larvae. *J. Exp. Biol. 19*, 56–77.

WIGGLESWORTH, V. B. (1967a). Cytological changes in the fat body of *Rhodnius* during starvation, feeding and oxygen want. *J. Cell Sci. 2*, 243–256.

WIGGLESWORTH, V. B. (1967b). Polyploidy and nuclear fusion in the fat body of *Rhodnius* (Hemiptera). *J. Cell Sci. 2*, 603–616.

WIGGLESWORTH, V. B. (1972). *The Principles of Insect Physiology*, 7th edn. Chapman & Hall, London.

WIGGLESWORTH, V. B. (1982). Fine structural changes in the fat body cells of *Rhodnius* (Hemiptera) during extreme starvation and recovery. *J. Cell Sci. 53*, 337–346.

WÜEST, J. (1978). Histological and cytological studies on the fat body of the cockroach *Nauphoeta cinerea* during the first reproductive cycle. *Cell Tiss. Res. 188*, 481–490.

WYATT, G R. (1980). The fat body as a protein factory. In *Insect Biology in the Future*. Edited by M. Locke and D. S. Smith. Pages 201–225. Academic Press, New York.

WYATT, G. R. and LINZEN, B. (1965). The metabolism of ribonucleic acid in cecropia silkmoth pupae in diapause, during development and after injury. *Biochim. Biophys. Acta 103*, 588–600.

WYATT, G. R. and PAN, M. L. (1978). Insect plasma proteins. *Ann. Rev. Biochem. 47*, 779–817.

6 Physiology and Biochemistry of the Fat Body

LARRY L. KEELEY

Texas A&M University, College Station, Texas, USA

INTRODUCTION

The fat body is the principal tissue for intermediary metabolism in insects. While appearing superficially to be an adipose storage tissue, the fat body is in reality a tissue of considerable metabolic activity and is the main source for the hemolymph proteins, lipids and carbohydrates that serve as precursors for metabolism in other tissues. The fat body undergoes growth and development along with the other insect tissues and its functions change in accordance with the developmental stage of the insect (see M. Locke, vol. 2). Because of its changing metabolic role and its integral position in maintaining metabolic homeostasis, the fat body is a target tissue for hormones (see Keeley, L., 1978a) and serves with increasing frequency as a model for examining endocrine regulations of metabolism (Kunkel, J., 1981).

The fat body has a broad metabolic role and the present review will attempt to discuss those areas that are receiving frequent research attention. Although this review seeks to be comprehensive, it concerns itself primarily with studies published during the nearly two decades that have elapsed since the last general review of fat body biochemistry (Kilby, B., 1963). For reviews of the literature prior to the last 20 years, the interested reader is referred to Kilby, B., 1963 and 1965. The present discussion will focus primarily on unique fat body functions and their contribution to the overall physiology of the insect, and will not emphasize general metabolic processes such as glycolysis, the TCA cycle, oxidation of fatty acids, etc. that are common to most animals (see vol. 10).

It was deemed essential to be selective for the topics and articles for this review. Undoubtedly some items of interest were omitted because they were overlooked or their involvement with the fat body was not clearly evident. In the latter case, examples are studies that use whole-body homogenates for which the results can be attributed mainly to the fat body. The quantity of literature dealing specifically with the fat body makes unnecessary the inclusion of studies such as these, that use imprecise methods.

The subject-matter of this chapter focuses mainly on the biochemistry of the fat body with only minimal references to structure since an entire chapter in this series reviews fat body structure (M. Locke, this vol.). Nevertheless, structural organization and biochemical function are complementary so that some redundancy is necessary and inevitable.

1 FAT BODY ORGANIZATION

The insect fat body is structurally heterogeneous. The fat body arrangement is described variously by Buys, K. (1924) as: pads, layers, sheets, masses, bladders, ribbons or bead-like. Its color is usually white, or either pale yellow or green. It is generally the most obvious organ in the insect abdomen where the largest deposits of fat body are located. However, in addition to the large abdominal deposits the fat body is also scattered throughout the remainder of the body where it surrounds other organs including the brain, central nervous system, gut, muscles and gonads. This diffuse distribution is probably an adaptation to the open hemocoel. Metabolites are transported to insect tissues, in part, by diffusion along hemolymph concentration gradients. By being widely distributed, the fat body reduces the diffusion distances between itself, as the source tissue for the metabolites, and the peripheral tissues where the metabolites are used.

The fat body exhibits regional differentiation. Peripheral or subcuticular fat body is found near the body wall outside the musculature (Buys, K., 1924; Coupland, R., 1957), and central or visceral fat body surrounds the alimentary canal. The two regions may have different functions. The visceral tissue appears more biosynthetically active while the peripheral tissue stains more heavily for glycogen and appears to have a storage role (deLoof, A. and Lagasse, A., 1970; personal observations). The two areas of fat body are easily distinguished in *Blaberus discoidalis* cockroaches since subcuticular fat body lacks the symbiote-containing cells (mycetocytes) that are common in the lobes of the visceral fat body (Fig. 1).

Regional biochemical differences are demonstrated by localized deposits of eye pigments in the fat body of *Drosophila melanogaster* fruit-flies. Kynurenine deposits form in the anterior fat body of larvae but not in the posterior tissue (Rizki, T., 1961). During adult metamorphosis, these pigment-containing cells localize in the future head–thorax region. In the *Ore-R* strain of *D. melanogaster* only the anterior fat body makes kynurenine, but tryptophan feeding induces tryptophan pyrrolase activity in other parts of the tissue so that other cells also begin to produce kynurenine (Rizki, T., 1964). The number of responding cells is related to the amount and duration of tryptophan feeding and to larval age. These results indicate the presence of a tryptophan pyrrolase gene that is inactive in most fat body cells but is inducible at differing levels of sensitivity by the presence of the substrate. In a related situation, wild-type *D. melanogaster* show localization of the eye pigment isoxanthopterin in only the posterior fat body (Rizki, T. and Rizki, R., 1962). The regional formation of isoxanthopterin results from substrate localization since the

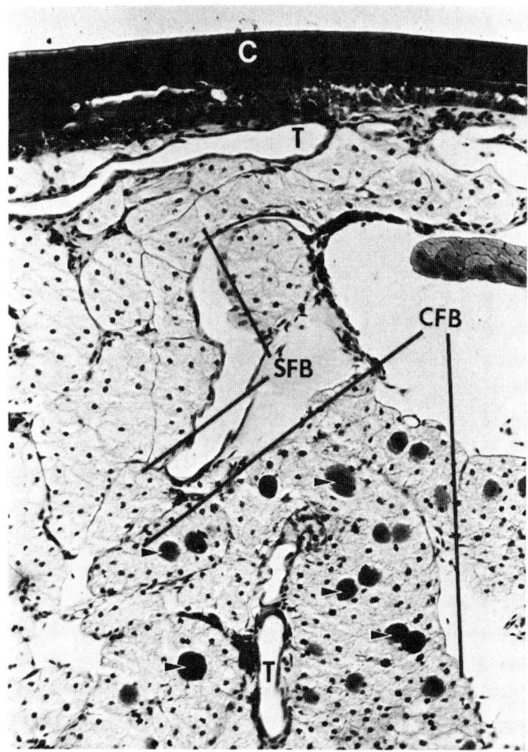

FIG. 1. Localization of mycetocytes within the central fat body relative to the subcuticular fat body in 4-day-old adult male *Blaberus discoidalis* (× 102). The sections were stained with hematoxylin and eosin. C = cuticle; SFB = subcuticular fat body; CFB = central fat body; T = trachea. Arrows indicate several representative mycetocytes.

biosynthetic enzyme for isoxanthopterin is ubiquitous throughout the tissue. Further regional differentiation is observed in the fat body of *D. melanogaster* larvae where posterior cells store greater concentrations of protein granules than do anterior cells (Tysell, B. and Butterworth, M., 1978). Finally, in the bloodworm, *Chironomus thummi*, hemoglobin is synthesized by subcuticular fat body, whereas the central tissue appears to function for metabolite storage (Schin, K. *et al.*, 1977).

The predominant fat body cell is the adipocyte. The older literature uses the term "trophocyte" which seems to this author to be a misnomer. Trophocyte suggests a "nourishment or food" cell. The lipoidal appearance of the fat body may suggest a nutrient-storage function, but we know now that the fat body is more than a storage depot. Both trophocyte and adipocyte are used in the current literature. Therefore, in keeping with the vertebrate terminology for cells from fatty tissues, I propose that the term adipocyte be used in the future in place of the inexact "trophocyte" terminology.

The fat body is composed of either two or three cell types. The predominant metabolic-storage cells are the adipocytes which resemble vertebrate brown fat in that they contain multiple fat droplets rather than the single droplet found in vertebrate white fat (Afzelius, B., 1970). Furthermore, filamentous mitochondria, free ribosomes and glycogen are frequent in adipocytes of both the fat body and brown fat. The second common cell type is the urocyte that sequesters uric acid for storage–excretion. A third cell type that is found in the fat body of some insect species is the mycetocyte; a cell that contains various types of symbiotes, especially bacteria-like organisms. Hemocytes are scattered coincidentally between the lobes of the fat body, and oenocytes, adipocyte-like blood cells of obscure function, are often found in conjunction with fat body lobes. Finally, Malpighian tubules ramify between the fat body lobes. These latter three components constitute cellular contaminants that are virtually impossible to prevent or remove when isolating the fat body. Nevertheless, they are usually a minor component in the tissue and their contributions to metabolism are negligible.

The fat body is structurally organized to provide maximal exposure to the hemolymph. As the principal metabolic–storage tissue in an organism with an open, diffusion-type circulatory system, the fat body is well-suited for both absorbing and releasing metabolites. Observations on the fat body arrangement within an insect show a juxtaposition with the midgut so that assimilated food is in near, or direct, contact with the fat body for efficient uptake.

Cells are arranged two or three layers thick within a fat body lobe. The metabolically active adipocytes are outermost in the lobe with urocytes situated within the lobe (Fig. 2). If mycetocytes are present they are usually central in the lobe and surrounded by urocytes. The more exterior position of the adipocytes gives them maximal exposure to the hemolymph for metabolite exchange. Urocytes, by contrast, are degenerate cells that lack cytoplasmic structures such as ribosomes, mitochondria or endoplasmic reticulum. Presumably, urocytes have low metabolic activity and are mainly storage-excretion depots for uric acid, a nitrogenous end

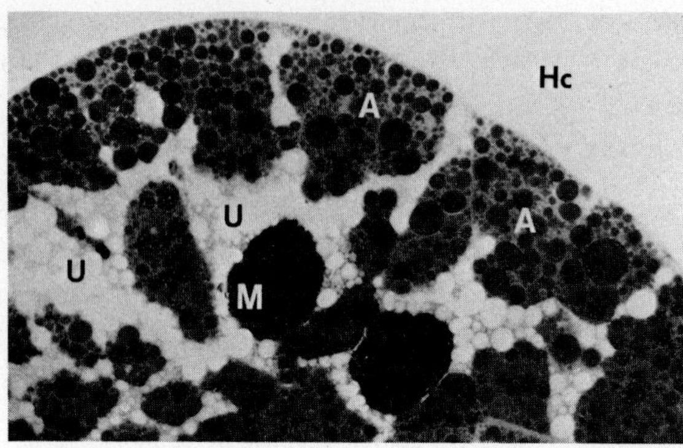

FIG. 2. Distribution of cell types in a typical lobe of the fat body from adult male *Blaberus discoidalis* (× 258). Fat body was fixed in 3% glutaraldehyde–3% acrolein in 0.1 M cacodylate buffer (pH 7.4); post-fixed in 1% osmium tetroxide in the same buffer and embedded in araldite-epon for thin sectioning. The sections were stained with toluidine blue for light microscopy. A = adipocyte; U = urocyte; M = mycetocyte containing bacteroids, Hc = Hemocoel.

product in many insect species or a storage form for nitrogen. The spatial arrangement between urocytes and mycetocytes suggests that they interact. It is reported that bacteroids convert uric acid to either TCA cycle intermediates (Donnellan, J. and Kilby, B., 1967) or to nitrogenous products used for growth (Pierre, L., 1964).

In summary, it is obvious that the structure of the fat body is highly adapted to the unique organization and physiology of the insect. Whereas the fat body appears initially as a diffuse and amorphous storage tissue, closer scrutiny suggests its amorphic distribution throughout the insect body is really an adaptation to the open circulatory–diffusion system. Furthermore, the fat body has regions of specialization. The tissue located in the center of the body, nearest the intestine, appears most metabolically active, while the subcuticular fat body may have a greater storage function. The individual fat body lobes are organized so that metabolic cells are assured maximum exposure to the circulatory system, while cells involved in storage and the recovery of storage products are more secluded within the lobe.

2 METAMORPHOSIS

There are two aspects for consideration in fat body metamorphosis. The first concerns structural reorganization of the tissue while the second concerns biochemical changes. The biochemical changes are discussed throughout the remainder of this chapter as aspects of metabolism rather than in this section as aspects of metamorphosis.

The fat body undergoes three types of imaginal metamorphosis. In the first type, adult adipocytes are derived by structural rearrangement of the cytoplasm within existing larval adipocytes. In the second type, adult fat body forms by lysis of larval adipocytes followed by replacement with precursor imaginal cells. Finally, the fat body of the Colorado potato beetle, *Leptinotarsa decemlineata* forms from a combination of the first two types.

In some insect species, larval adipocytes persist into the adult where their cytoplasm reorganizes itself for adult functions. This has been reported for fat bodies from mosquitoes (Clements, A., 1963), Lepidoptera (Larsen, W., 1976) and cockroaches (Keeley, L., 1981).

Lepidoptera usually show fat body metamorphosis by cytoplasmic reorganization. Ishizaki, H. (1965) suggested that cytolysis and regrowth may occur in adipocytes of the saturniid silkmoth, *Samia cynthia ricini*, with release of cytoplasmic contents into the hemolymph during adult metamorphosis. Furthermore, [^{3}H]thymidine incorporates into fat body DNA during adult metamorphosis in several saturniid species, including *S. cynthia ricini*, (Krishnakumaran, A. *et al.*, 1967) which suggests tissue regrowth by mitosis. However, Krishnakumaran, A. *et al.* (1967) question whether the adipocytes actually disintegrate as suggested by Ishizaki. They find, by histology, that the adipocytes separate and discharge their contents, hence the fat body appears "mushy" during metamorphosis. But in fact, the adipocyte cell membranes remain intact and the tissue later reorganizes into a definitive structure. Dissociation of the adipocytes appears ecdysteroid regulated and occurs *in vitro* in a dose-dependent relationship with 20-hydroxyecdysone (20-OH ecdysone) (Oberlander, H., 1976).

A definitive experiment was performed with the lepidopteran fat body to confirm whether the tissue persisted and reorganized or histolyzed and regrew. Larval fat body DNA was labeled with [^{3}H]thymidine in the silkmoth *Hyalophora cecropia*. After pupation, a piece of prelabeled tissue was transplanted into an unlabeled early pharate adult.

The intact, labeled fat body transplant was located later in the resulting adult and showed adult characters (Krishnakumaran, A. *et al.*, 1967). These studies confirm that the larval adipocytes of Lepidoptera persist through metamorphosis and differentiate into adult adipocytes.

Cytolysis and regrowth is illustrated clearly in higher Diptera. The larval fat body disintegrates in an anterior to posterior direction shortly after pupariation in the cyclorrhaphous Diptera (Fraenkel, G. and Hsiao, C., 1968). Whitten, J. (1962, 1964) reported that disintegration of the fat body results from loss of the surrounding basement membrane, but that although the adipocytes are free, they remain *in situ*. The freed larval adipocytes constitute the "pupal" fat body (Weissmann, R., 1963). Most of the pupal adipocytes undergo cytolysis about the third day of pupariation (Fraenkel, G. and Hsiao, C., 1968); however, some persist into the adult where they eventually undergo cytolysis and disappear (Weissmann, R., 1963; Evans, J., 1967). Adult adipocytes arise either from imaginal precursor cells located under the pupal integument (Evans, J., 1967) or from embryonic cells in the posterior of the larva (Weissmann, R., 1963). Hemocytes are also proposed as precursors for imaginal adipocytes (Whitten, J., 1964; Evans, J., 1967).

The imaginal environment causes histolysis of those larval adipocytes that persist into adults. Approximately 1000 larval adipocytes survive into the adult in *Drosophila melanogaster* (Butterworth, F., 1972). Normally, 90% of these cells disappear within 2 days of adult emergence and are totally gone by 4 days. However, larval adipocytes survive for longer times when transplanted into older flies. If the larval cells are pre-exposed to the environment of young flies for 1 day before transplantation into older flies, then the larval cells degenerate as quickly as if kept continuously in the lytic environment of young flies.

Endocrine factors appear to regulate the lytic effect. For example, larval adipocytes degenerate rapidly in young *Musca domestica* house-flies when vitellogenic ovaries are present, but either ovariectomy or allatectomy of the young flies causes persistence of the larval adipocytes (Adams, T. and Nelson, D., 1969). Furthermore, treatments with juvenile hormone (JH) analog increase by 3-fold the cytolysis rate of persistent larval adipocytes in abdomens isolated from adult *D. melanogaster* (Postlethwait, J. and Jones, G., 1978). Cycloheximide administration blocks the lytic action of JH on larval adipocytes and indicates that protein synthesis facilitates the lytic process. Histolysis of larval adipocytes occurs only after protein body formation and in the presence of either the ring gland, corpora cardiaca–allata complex or functioning ovaries (Butterworth, F. and Bodenstein, D., 1967).

In combination with our present knowledge of the hormone balance in adult Diptera and several past references to fat body histolysis, it can be suggested that ecdysteroids stimulate the lysis of larval adipocytes by the formation of lysosomal protein granules. Acid phosphatase is present in protein granules of *Calliphora erythrocephala* as shown by dissociation of the granules at pH > 8.6 (van Pelt-Verkuil, E. *et al.*, 1979). Furthermore, both acid phosphatase activity and protein granules appear in the fat body of *C. erythrocephala* larvae after treatment with 20-OH ecdysone (van Pelt-Verkuil, E., 1979; dePriester, W. *et al.*, 1979). Hence, these adipocyte protein granules are lysosomal in origin and function, and they form in response to increasing ecdysteroid titer at the onset of metamorphosis. It is now known that ovaries secrete ecdysone in adult mosquitoes (Hagedorn, H. *et al.*, 1975; Hagedorn, vol. 8). Therefore, the larval ring gland and the active adult ovary may be sources for ecdysone so that the presence of either structure in an adult fly makes the environment lytic for persistent larval adipocytes. It is not clear why allatectomy causes persistence of larval adipocytes or why JH treatment induces the lytic milieu. Possibly, in the adult fly, JH acts in conjunction with a brain neurohormone to stimulate the ovaries to secrete ecdysone similar to the case in mosquitoes (Hanaoka, K. and Hagedorn, H., 1980).

A third type of fat body metamorphosis combines both cytolysis–regrowth and cytoplasmic restructuring. Some parts of the pupal fat body undergo lysis in *Leptinotarsa decemlineata* while other pupal adipocytes persist into the adult and show cytoplasmic reorganization (Dortland, J. and Esch, Th., 1979).

3 FAT BODY BIOCHEMISTRY AND ITS REGULATION

3.1 Basal metabolism

Basal metabolism is measured by the rate of oxygen consumption and is a function of the rate of energy production which reflects the total biosynthetic level in the tissue. Protein synthesis is the most energy-demanding process in the cell so that oxygen use is mainly a function of the rate for protein synthesis.

Oxidative metabolism occurs in the fat body of insects via the TCA cycle and electron transport just as it does in other animals. However, it should be noted that respiratory enzyme activities often fluctuate both with the developmental stage of the insect and within any given instar. For example, fat body cytochrome *c* oxidase activity is high at the start of the fourth nymphal instar in *Schistocerca gregaria*, declines by 97% by the end of the instar, then increases sharply at the adult molt and is high in young adults (Hearfield, A. and Kilby, B., 1958). There is now little reason to question the integrity of the complete complement of usual mitochondrial enzymes in the insect fat body and with activities and substrate preferences comparable to those reported for rat liver mitochondria (Keeley, L., 1971).

The fat body is a tissue rich in lipid stores, but carbohydrate is most frequently its main energy source. A respiratory quotient (RQ) of 0.94 is determined for adult fat bodies from *Leucophaea maderae* cockroaches (Wiens, A. and Gilbert, L., 1965). However, in adult male *S. gregaria* the fat body has an RQ of 0.64–0.72, indicating that lipid is the main substrate for energy in this insect (Walker, P. *et al.*, 1970). Addition of a corpora cardiaca extract to *in vitro* fat bodies from adult female *L. maderae* increases the fat body O_2 consumption (Lüscher, M. and Leuthold, R., 1965; Wiens, A. and Gilbert, L., 1965, 1967a). This increase in O_2 uptake occurs because of a 30–35% increase in the rate of oxidation of lipid instead of carbohydrate (Wiens, A. and Gilbert, L., 1965, 1967a). The shift from the use of carbohydrate to lipid for energy production is a compensation for the increase in rate of conversion of carbohydrate substrates to hemolymph trehalose that occurs in response to the presence of a hypertrehalosemic factor in the corpora cardiaca extracts (see J. Steele, vol. 8).

It is expected that hormones affect the basal metabolic rate of the fat body since it is the principal biosynthetic tissue of the insect and should respond to metabolic demands associated with endocrine regulations of growth and development. Lüscher, M. (1968) reported a 50% decrease in fat body respiration, a 70% decrease in fat body protein synthesis and arrested oocyte growth after decapitation of adult female *Nauphoeta cinerea* cockroaches. Corpora allata implants reverse these effects, suggesting that JH is the main hormone regulating all of these processes. The effect of JH on fat body respiration in female insects is related to the JH-dependent production of yolk precursors by the fat body for vitellogenesis. In addition to the JH effect, a small but significant increase in respiration also occurs after brain–corpora cardiaca implants, suggesting some additional neuroendocrine stimulation to the basal metabolism (Lüscher, M., 1968). Corpora cardiaca extracts stimulate respiration by *in vitro* fat bodies from adult female *L. maderae* if the initial fat body respiration is low, but are inhibitory if the initial respiration is high (Engelmann, F. and Müller, H., 1966; Müller, H. and Engelmann, F., 1968). Furthermore, fat body respiration is reduced by about 40% in adult male *Blaberus discoidalis* 30 days after extirpation of the corpora cardiaca–corpora allata complex (Keeley, L. and Friedman, S., 1967). This reduction in fat body basal metabolism is corpora cardiaca-related, and the effect is manifested in fat body mitochondria at the level of electron transport activity (Keeley, L., 1971). This latter result suggests neuroendocrine regulation for the energy-generating capacity in fat body mitochondria.

Therefore, fat body respiration responds to two endocrine regulations. The corpora allata effect fat body respiration associated with protein synthesis for ovarian development (Lüscher, M., 1968). In addition, neuroendocrine factors increase fat body respiration by affecting the types of substrates used and the intrinsic capacity of mitochondria for electron transport. The corpora allata effect accelerates basal metabolism through respiratory control by ADP generated from endergonic, biosynthetic processes, while the neuroendocrine effect determines the exergonic capability within the adipocytes.

New mitochondria form and show synchronous development in imaginal adipocytes shortly after adult emergence. This mitochondriogenesis raises the exergonic capability of the fat body by increasing its number of functional mitochondrial units. Adult adipocytes form in *Calpodes ethlius* by cytoplasmic reorganization of larval adipocytes, and larval mitochondria are destroyed before pupation by engulfment into autophagic vacuoles during the cytoplasmic degeneration associated with metamorphosis (Larsen, W., 1976). Pupal adipocytes contain few mitochondria, but new mitochondria form in the young adult. During the first 22 h of adult life, mitochondrial remnants from the pupal adipocytes increase in volume (Larsen, W., 1970). From 22 to 34 h the mitochondria decrease in volume but increase in number. During this latter period the inner membrane frequently transects the mitochondria into two inner compartments, which suggests mitochondrial division. These observations indicate that new adipocyte mitochondria form in adults by enlargement and division of existing mitochondria carried over from the larva through the pupal stage. In the cockroach *B. discoidalis*, adipocyte reorganization occurs during the first 4 days of adult life (Mannix, J. and Keeley, L., 1980; Keeley, L., 1981), and fat body mitochondria increase their respiratory capacity by 3-fold during the first 10 days of adult life (Keeley, L., 1972). The respiratory increase is associated with elevated heme synthesis between 4 and 6 days of adult age (Keeley, L., 1978b) and an increase in mitochondrial cytochromes (Keeley, L, 1977). The increase in heme synthesis is dependent on a "cytochromogenic hormone" from the brain–corpora cardiaca complex (Hayes, T. and Keeley, L., 1981).

Mitochondriogenesis appears to be an integral aspect of adipocyte metamorphosis. Presumably the fat body converts from a mainly nutrient storage function in immature insects into a more biosynthetically active adult tissue. This is especially true for the adult female fat body which must produce large quantities of protein and lipids for transfer to ovaries for vitellogenesis. The maturation of new adipocyte mitochondria insures an adequate exergonic capacity to fulfill the increased endergonic demands of vitellogenesis, and appears neuroendocrine regulated at the level of the cytochromes.

3.2 Nucleic acids

3.2.1 DNA

Fat body DNA can either remain constant or vary depending on the insect species and its developmental stage. Although adipocyte mitosis is observed in the *C. ethlius* fat body at the end of the penultimate larval instar, DNA synthesis occurs without further mitosis during the last larval instar (Locke, M., 1970). This increase in adipocyte DNA without subsequent cell division is indicative of polyploidy and endomitosis (Locke, M., 1970; Dean, R. *et al.*, 1980). Incorporation of [^{3}H]-thymidine shows evidence for DNA synthesis during larval life in the fat body of saturniid silkmoths (Krishnakumaran, A. *et al.*, 1967). However, in *Calliphora erythrocephala*, the fat body DNA remains constant at about 0.016 μg DNA per fat body throughout larval life (Price, G., 1969). In *Hyalophora cecropia* the fat body DNA content did not change during adult metamorphosis, although the DNA content doubled in the wing epidermis during this time (Linzen, B. and Wyatt, G., 1964). In a comprehensive study of saturniid silkmoths, Krishnakumaran, A. *et al.* (1967) reported that fat body DNA is synthesized during adult metamorphosis in some species but not at the time of the larval–pupal transformation.

Hormones may influence the ability of adipocytes to synthesize DNA at various stages of growth. Exposure to JH induces DNA synthesis in pupal adipocytes of saturniid silkmoths during a stationary pupal–pupal molt if the cells are also exposed to ecdysteroids (Krishnakumaran, A. *et al.*, 1967). However, adipocytes do not synthesize DNA in pupal abdomens grafted to adult male silkmoth abdomens where the endogenous JH titer is naturally high and ecdysteroids are absent. This suggests that the ecdysteroids are the hormones that facilitate DNA synthesis in the pupal fat body. In adult insects it is generally accepted that fat body DNA does not increase (Krishnakumaran, A. *et al.*, 1967). However, in young adult *Locusta migratoria*, adipocyte DNA doubles during the first previtellogenic period (Chen, T. *et al.*, 1979). This doubling of adipocyte DNA in adult locusts is presumably an early maturational response by the fat body to JH, since this increase in DNA content does not occur after allatectomy. DNA synthesis may be needed for fat body maturation preparatory to the later production of yolk constituents. Polyploidy in adipocytes of adult female locusts may enhance the capacity of the adipocytes to produce yolk precursors since the increase in DNA copies and their accompanying genes permit the production of increased amounts of biosynthetic enzymes. These findings suggest that the hormones that regulate adipocyte DNA synthesis change between pupal and adult life as the tissue switches from growth to reproductive functions.

DNA was characterized from adipocytes of diapausing pupae and pharate adults of *H. cecropia* by isopycnic ultracentrifugation (Burke, D. *et al.*, 1976). The DNA patterns remained similar regardless of the developmental state of the tissue source. A major DNA band occurred at an equilibrium density of 1.693 with two minor satellite DNA bands comprising 6% and 3% of the DNA and located at densities of 1.705 and 1.709, respectively. These densities indicate G+C amounts of 34% for the major band and 46% and 50%, respectively, for the 1.705 and 1.709 bands. Renaturation kinetics showed 30% redundant sequences in the genome and 70% unique sequences.

Fat body DNA increases in pharate pupae of the silkworm *Bombyx mori*, but not by new DNA synthesis (Chinzei, Y. and Tojo, S., 1972). Administration of [^{3}H]thymidine fails to label accumulating adipocyte DNA and suggests that the new DNA originates by degradation and transport to the fat body from another tissue. The silk glands degenerate at this time in pharate *B. mori* pupae and their DNA disappears. This suggests the silk glands are the source for the accumulating fat body DNA (Chinzei, Y. and Tojo, S., 1972). Analysis of the pupal adipocyte DNA in *Bombyx mori* separates three types: α, β_1 and β_2 (Chinzei, Y., 1974). All three types have a G + C content of 38% although the α-DNA is several times larger in molecular weight than the β-DNA. No β-DNA occurs in larval adipocytes, but it predominates in pupal cells. Tissues that degenerate, such as the pupal silk gland or midgut, contain β-DNA; whereas the larval silk glands and developing tissues, such as the integument, muscles and gonads, contain only α-DNA. Although β-DNA increases in adipocytes during early pupal life, only α-DNA becomes labeled by [^{3}H]thymidine (Chinzei, Y., 1975). This indicates that the fat body β-DNA has already formed before its appearance in the adipocytes. Comparisons of DNA labeling times and specific activities between various tissues indicate further that the β-DNA found in the pupal fat body forms from the degradation and transfer of larval silk gland DNA (Chinzei, Y., 1975). The β-DNA is located in the cytoplasm of the adipocytes and is ultimately transferred to the ovaries during vitellogenesis for use as a source of nucleic acids for embryogenesis.

3.2.2 RNA

Fat body RNA/DNA ratios provide a relative measure of protein synthetic capacities but not necessarily of synthetic activities. Price, G. (1969) reported exceedingly high ratios of 60 to 125 for RNA/DNA in fat bodies of *C. erythrocephala* larvae. These ratios are more than 50 times greater than those generally reported for vertebrate tissues. However, other reports on fat body RNA/DNA ratios are not so extreme. The fat body of adult male *Blaberus discoidalis* shows an increase in RNA/DNA from a value of 3.5 at adult emergence to a value of 7 by day 6 (Mannix, J. and Keeley, L., 1980). RNA/DNA ratios in the fat body of adult *Leptinotarsa decemlineata* range from 6 to 15 for females but are only about 4.0 for males (Dortland, J. and deKort, C., 1978). In contrast to the adult situation, fat body RNA/DNA ratios decrease from a high of 8 in early fifth-instar larvae of *Philosamia cynthia ricini* to values of 2 to 3 at pupation due to the loss of rRNA from the microsomal fraction (Takahashi, S., 1966).

Developmental changes in fat body RNA are illustrated in *Calliphora erythrocephala*. Adipocytes have the highest RNA content during early larval life with a decline as pupariation approaches (Price, G., 1965, 1969). Examination of the fine structure of the adipocytes shows that ribosomes are abundant in young larvae but disappear before pupariation (Price, G., 1969). Protein synthesis follows this same pattern (Price, G., 1969). Measurements of RNA content per fat body at the time of pupariation suggest that the RNA content is still high even though the ribosomes are degraded. As much as half of the larval RNA is fat body RNA, and most of the adipocyte RNA in young larvae is ribosomal with as much as 25 mg of ribosomes per gram of fat body (Sridhara, S. and Levenbook, L., 1974). This amount of ribosomes is eight times greater than that found in vertebrate liver. During the last 3 days of larval life, 85–90% of these ribosomes are destroyed in association with an increase in degraded 4S-7S, insoluble RNA (Sridhara, S. and Levenbook, L., 1974). The degraded adipocyte RNA is not replaced and forms an inert, insoluble "pellet RNA" after centrifugation at 20,000 ***g*** for 10 min (Protzel, A. *et al.*, 1976). This inert RNA is formed exclusively from ribosomes and is excreted in the meconia after adult emergence. Pharate adult cells synthesize 90% of their ribosomal RNA anew during metamorphosis (Protzel, A. *et al.*, 1976).

In contrast to Diptera, little RNA synthesis occurs in the fat body of Lepidoptera at the onset of adult metamorphosis. A general survey of RNA synthesis in the tissues of several silkmoth species shows that some fat body RNA synthesis occurs during the early half of pharate adult development but declines during the last half (Berry, S. *et al.*, 1967). No measurable differences occur in fat body RNA/DNA ratios between diapause termination and the onset of metamorphosis in *Hyalophora cecropia* pupae (Linzen, B. and Wyatt, G., 1964); however, fat body RNA synthesis increases by 3-fold during the first few days of metamorphosis, then ceases by day 3 (Wyatt, G. and Linzen, B., 1965). In *Antheraea pernyi*, fat body rRNA forms before the onset of visible adult development and is one of the earliest events of diapause termination (Barth, R. *et al.*, 1964). However, after the third day no further rRNA accumulates in the fat body of the pharate adult even though [^{14}C]uridine incorporation still occurs. This suggests that the rate of rRNA synthesis is approximately equal to its rate of breakdown so that no net increase in fat body rRNA occurs at this time. These studies do not identify whether new mRNA is produced at the onset of metamorphosis or whether pre-existing mRNA transcripts survive dormancy and are translated on newly formed ribosomes. The status of fat body RNA and its synthesis in response to hormones at diapause termination may be a useful model for elucidating fundamental endocrine effects on cell metabolism and differentiation. The dormant adipocyte is at a minimal metabolic level, and primary endocrine effects should be observable relative to a minimum of "background" metabolism.

3.2.3 RNA POLYMERASE AND HORMONE EFFECTS

RNA formation is an early fat body response after exposure to either ecdysteroids or JH. Either hormone stimulates RNA synthesis in adipocyte nuclei isolated from *C. erythrocephala* larvae (Congote, L. *et al.*, 1969), but the simultaneous presence of both hormones inhibits their individual effects. Hybridization of adipocyte RNA and DNA increases after exposure of the adipocytes to either hormone (Congote, L. *et al.*, 1970); however, DNA hybridization to RNA produced by either one of the hormones is blocked if the DNA is pretreated with RNA produced in response to the other hormone. This suggests that both hormones regulate common genes in the fat body but are antagonistic for the regulations when present together.

Ecdysteroids, in particular, stimulate RNA synthesis in adipocytes. In *A. pernyi* larvae, fat body RNA synthesis increases by 9-fold within 90–120 min after ecdysone injection and returns to its original base level by 270 min (Sahota, T. and Mansingh, A., 1970). Fat body RNA synthesis increases by over 5-fold, 8 h after 20-hydroxyecdysone (20-OH-ecdysone) treatment of adult female *Gromphadorhina portenosa* cockroaches along with a concomitant 3-fold increase in protein synthesis (Bar-Zev, A. and Kaulenas, M., 1975). Injections of 20-OH-ecdysone stimulate the production of poly(A) RNA in adipocytes of *Calliphora vicina* larvae (Scheller, K. and Karlson, P., 1977a) which indicates hormone induction of new messenger RNA transcripts. Both ecdysone and 20-OH-ecdysone stimulate fat body RNA synthesis with 20-OH-ecdysone having the greater effect, but only when endogenous ecdysteroid titers are low (Scheller, K. and Karlson, P., 1977b). Presumably, hormone treatment when endogenous ecdysteroid titers are high fails to stimulate RNA synthesis above its already high base level. DNA–RNA hybrid competition studies show that RNA from untreated control larvae is not competitive with RNA from hormone-treated larvae, and indicates that new types of RNA appear in response to the ecdysteroids. Both mRNA and rRNA increase in response to ecdysteroid treatment with the majority being rRNA (Scheller, K. and Karlson, P., 1977b). Similar studies with larvae of *Sarcophaga peregrina* indicate that injections of 20-OH-ecdysone preferentially activate RNA polymerase I in adipocyte nuclei for rRNA synthesis (Natori, S., 1976). These results indicate that, although part of the effect of ecdysteroids is on expression of new genomic information, a major response is to make the adipocyte competent for increased protein synthesis by enlarging its ribosomal pool.

There is considerable interest in the RNA polymerases of insect tissues because of their possible involvement in endocrine regulation of growth and metamorphosis. Activation of RNA polymerase may be an essential component in hormonal regulation of specific genes for cell growth and differentiation. Ecdysteroids and JH both appear to act on transcription, hence RNA polymerase activities may be significantly influenced by these hormones. The majority of fat body RNA polymerase is sensitive to the inhibitor α-amanitin during larval life in the tobacco hornworm, *Manduca sexta*, which suggests that the predominant larval enzyme is RNA polymerase II for mRNA production (Sridhara, S. and Gilbert, L., 1975). At the time of wandering, when ecdysteroids appear, RNA polymerase activity declines in association with a change in the enzyme from a soluble to an insoluble form. The insoluble RNA polymerase continues into adult life, but even though the enzyme does not disappear, it is doubtful that it is still functional for gene transcription. The conversion from soluble to insoluble RNA polymerase coincident with the burst of ecdysteroid titer that occurs at the onset of wandering suggests that the structural form of RNA polymerase is endocrine regulated. Sridhara, S. and Gilbert, L. (1978) isolated RNA polymerase from the fat body and found two bands sensitive to α-amanitin and stimulated by Mn^{2+}. They suggest that the two bands represent multiple forms of the enzyme, and speculate that the major band is the free enzyme and the minor band a bound form. Fat body RNA polymerases I, II and III are also described for larvae of the corn earworm, *Heliothis zea* (West, T. *et al.*, 1980). Characterization by DEAE–Sephadex chromatography of the *H. zea* fat body RNA polymerases resolves the usual three major bands but also shows two additional minor bands with RNA polymerase activity (Grula, M. and Weaver, R., 1981).

Most of the studies to date have been inconclusive with respect to relationships between fat body RNA polymerases, hormones and regulations of growth and metamorphosis. However, as the fat body enzymes become better characterized and an optimal assay system developed, the fat body could prove valuable as a model for elucidating how hormones regulate genetic expression. This is particularly true for fat body proteins, such as vitellogenin, which are well-characterized and produced in large amounts at specific times in the life cycle in response to hormone stimulation.

3.3 Proteins and protein synthesis

The fat body is the principal synthetic source for hemolymph proteins. Shigematsu, H. (1958) first demonstrated that soluble proteins released *in vitro* from the fat body of *Bombyx mori* are electrophoretically similar to hemolymph proteins, and the rate of fat body protein secretion *in vitro* matches the rate of hemolymph protein formation *in vivo*. The fat body produces many unique and physiologically significant proteins such as vitellogenins for oocyte maturation; stage-specific, amino acid-storage proteins such as calliphorin, drosophilin and manducin (L. Levenbook, vol. 10); diapause proteins and even hemoglobin in the chironomid bloodworms. The fat body is the source for lipid-binding proteins, and the JH-binding protein that protects circulating JH from degradation. Conversely, the fat body produces the specific esterase that degrades JH. Finally, the fat body has a unique role in protein storage and conservation by forming large proteinaceous storage granules that degrade and serve as a source for proteins and amino acids during metamorphosis. Various aspects of these proteins and their relationships to the fat body were reviewed by Chen, P. (1966); Price, G. (1973); Wyatt, G. and Pan, M. (1978); and Keeley, L. (1978a).

3.3.1 Endocrine effects

Neurohormones appear to increase the general protein synthetic capacity in the fat body. The neuroendocrine effects are especially apparent in female adult insects during yolk precursor production. In female adult insects the vitellogenic proteins comprise a major portion of the proteins synthesized by the fat body and their formation is JH-regulated. Nevertheless, the corpora cardiaca also affect fat body protein synthesis in female adult insects. The rate of protein synthesis increases by 80% in female adult *Schistocerca gregaria* within 4 h after injection of a corpora cardiaca extract, and returns to its original level within 24 h (Osborne, D. *et al.*, 1968). Conversely, fat body protein synthesis is reduced by 80% in female adult *S. gregaria* 8 days after neurosecretory cell cautery or allatocardiacectomy (Hill, L., 1965). Allatectomy, alone, decreases fat body protein synthesis to a level intermediate between neurosecretory deficient animals and untreated animals. This latter result indicates that the neurosecretory system acts directly on fat body protein synthesis and not indirectly via an allatotropic effect on JH secretion. Total protein in the fat body, hemolymph and ovary of adult female *Melanoplus sanguinipes* is 50% lower after neurosecretory cell cautery than after allatectomy, and vitellogenesis is prevented (Elliott, R. and Gillott, C., 1978). Ovarian proteins synthesized by the fat body are reduced in amount after either allatectomy or neurosecretory cell destruction but reappear after treatments with a JH analogue (Elliott, R. and Gillott, C., 1979). In addition, non-ovarian hemolymph proteins are absent after neurosecretory cell destruction. The fact that neurosecretory cell destruction reduces both ovarian and non-ovarian proteins suggests that the neuroendocrine system regulates two aspects of fat body protein synthesis (Elliott, R. and Gillott, C., 1979). First, a neuroendocrine, allatotropic hormone stimulates JH secretion which affects the formation of specific ovarian proteins (see R. Feyereisen, vol. 7). Second, a neuroendocrine regulation of general protein synthetic capacity in the fat body lowers the production of all fat body proteins.

A direct action on protein synthesis by neurohormones is indicated by the finding that fat body synthesis for five out of six hemolymph proteins, is increased by implants of brains or corpora cardiaca in decapitated *Leucophaea maderae* females (Scheurer, R., 1969). Of the proteins that respond to neurosecretions, four respond to brain implants and three respond to corpora cardiaca implants. These results suggest that, in addition to neuroendocrine effects on general protein synthesis, specific neurohormones may also regulate the formation of specific proteins.

Most studies on the neurohormonal regulation of fat body protein synthesis have been general and not concerned with regulation of specific proteins. Scheurer's studies demonstrate that greater definition is needed in future investigations, and relationships should be considered between specific neurohormones and the formation of specific proteins. Neuroendocrine controls of fat body protein synthesis are generally unclear and have been largely ignored up to this point with major emphasis focusing on ecdysteroids and JH. However, neuroendocrine regulations will assume greater importance in the future as individual neurohormones are isolated and characterized and defined studies undertaken on specific effects.

Ecdysteroids exert both general and specific effects on fat body protein synthesis. Treatment of 7–8-day-old larvae of *Calliphora erythrocephala* with 20-OH-ecdysone causes a 35% increase in fat body protein synthesis after 4 h (Neufeld, G. *et al.*, 1968; Thomson, J. *et al.*, 1971). All fat body proteins respond to the hormone and increase to the same degree with no specific effect being noted (Thomson, J. *et al.*, 1971). This generalized increase in protein synthesis agrees with observations that ecdysteroids stimulate rRNA formation in the fat body and enhance the overall protein synthetic capacity of the tissue (Natori, S., 1976; Scheller, K. and Karlson, P., 1977b).

Few studies have related ecdysteroids to the synthesis of any specific fat body proteins. Recently, however, the synthesis of five specific proteins (P_{1-5}) was identified for the fat body in response to high ecdysteroid titers in third-instar larvae of *Drosophila melanogaster* (Lepesant, J.-A. *et al.*, 1978). Of these proteins, P_1, P_3 and P_5 are newly translated from fat body mRNA in response to high ecdysteroid levels. Furthermore, cloned DNA probes show that 20-OH-ecdysone causes a 50-fold increase in the amount of poly(A)-RNA transcripts for P_1 in only the late third-instar larvae. P_2 is synthesized and secreted by the fat body when ecdysteroid titers are low, but P_2 is reabsorbed by the fat body and processed to form P_4 in late third-instar larvae when ecdysteroid titers increase. This indicates that P_2 formation occurs independent of ecdysteroids, but that P_2 processing to P_4 may be ecdysteroid-regulated. Lepesant, J.-A. *et al.* (1978) conclude that there are three effects by 20-OH-ecdysone: (1) gene specificity—induction of a single gene (P_1); (2) tissue specificity — only the fat body responds; and (3) stage specificity — the ecdysteroid effect on P_1 formation occurs only in the late third-instar larva and is not inducible in earlier larval instars.

Although the previous study demonstrates the formation of specific proteins by the fat body in response to ecdysteroids, it does not identify these proteins. Acid phosphatase is a specific protein that appears in response to 20-OH-ecdysone. Acid phosphatase occurs in adipocytes in association with the Golgi complex or in vacuoles, and is found in protein storage granules as pupariation nears in 4-day-old, third-instar larvae of *C. erythrocephala* (van Pelt-Verkuil, E. *et al.*, 1979). However, ligation blocks the formation of protein storage granules (van Pelt-Verkuil, E. *et al.*, 1979), and although treatment with 20-OH-ecdysone fails to stimulate the formation of granules in ligated larvae, acid phosphatase activity is nonetheless induced (van Pelt-Verkuil, E., 1979). Hence the appearance of acid phosphatase activity is a specific response to ecdysteroid and is related to, but independent of, protein granule formation.

3.3.2 LARVAL-SPECIFIC STORAGE PROTEINS

The fat body synthesizes and secretes large amounts of a specific protein during immature life in several insect species. It has been speculated that these proteins serve as storage reservoirs for amino acids which are recovered and used later for the construction of new adult tissues during metamorphosis (L. Levenbook, vol. 10). The synthesis of these proteins appears endocrine-regulated. Although storage proteins have been identified in only a few insect species, they may be generalized and widespread.

The first larval-specific storage protein identified as a fat body product was calliphorin. Munn, E. *et al.* (1967) found that squashes of *C. erythrocephala* pupae contain large numbers of particles that yield a single electrophoretic protein band. This abundant protein was termed calliphorin, and the fat body was confirmed as its source (Munn, E. *et al.*, 1969). Calliphorin is not found in eggs and first appears in 2-day-old larvae (Munn, E. *et al.*, 1971a). Calliphorin eventually constitutes 75% of the protein in the hemolymph of 6-day-old larvae, but it progressively disappears as the insect passes from the pupal stage into adult life. By day 3 of adult life. calliphorin constitutes only 10% of the total protein. Calliphorin is a single protein of molecular weight 5.28×10^5 daltons, composed of six subunits of 8.7×10^4 daltons each (Munn, E. *et al.*, 1971b). Calliphorin is high in tyrosine, phenylalanine and methionine, but it contains very little cysteine or cystine. Some carbohydrate is present (0.4–0.5%). Calliphorin is synthesized only by fat body tissue from 3- to 5-day-old *C. erythrocephala* larvae; however, the production of calliphorin by adipocyte mRNA in a cell-free, wheat-germ system confirms that the calliphorin mRNA transcript is present in adipocytes between days 3 and 7 of larval age (Sekeris, C. and Scheller, K., 1977). This shows that calliphorin mRNA remains intact for 2 days after calliphorin synthesis has stopped. This is unusual since, normally, protein synthesis stops because of mRNA degradation. Although RNA remains high in larval adipocytes on day 7 the ribosomes disappear, and it is the loss of ribosomal structures, not RNA, that accounts for the cessation in calliphorin production (Price, G., 1969).

Calliphorin accumulates rapidly in the hemolymph of *C. erythrocephala* larvae starting on day 3 and is reabsorbed by the fat body preparatory to pupariation when feeding ceases and wandering begins (Levenbook, L. and Bauer, A., 1980). Calliphorin is stored in protein storage bodies of adipocytes during pupariation (van Pelt-Verkuil, E., 1978; dePreister, W. and van der Molen, L., 1979; Levenbook, L. and Bauer, A., 1980) and disappears rapidly during late adult metamorphosis. Presumably it becomes a source of amino acids for adult tissue formation and is undetectable in 2-day-old adult flies (Levenbook, L. and Bauer, A., 1980).

Although calliphorin is the best-known of the larval storage proteins, others have been identified in immatures of several other insect species. Two proteins that cross-react with anti-calliphorin antibodies are synthesized by mRNAs isolated from the fat body of larval *D. melanogaster* (Sekeris, C. *et al.*, 1977). These *in vitro*-translated proteins comigrate electrophoretically with two proteins that are extracted with anti-calliphorin from *D. melanogaster* larvae. The putative, larval-storage proteins from *D. melanogaster* larvae are termed drosopholin. Manducin is also a storage protein that comprises 80% of the hemolymph proteins in late-instar larvae of *Manduca sexta* and is composed of six subunits with a total molecular weight of 5.4×10^5 daltons (Kramer, S. *et al.*, 1980). Manducin contains 20% tyrosine plus phenylalanine, 2% lipid and 3.5% carbohydrates. Finally, Kunkel, J. and Lawler, D. (1974) described a major protein found only in immature *Blattella germanica* cockroaches. A comparable larval-specific protein is also found in immature mantids and termites. The larval protein of *B. germanica* varies 10-fold during the course of an instar, and reaches its maximum just before each ecdysis. Like calliphorin, the larval-specific protein disappears within 2 days in adults and, presumably, serves an amino acid storage function for histogenesis.

Endocrine regulation for the synthesis of stage-specific storage proteins is indicated but unclear. Sekeris, C. and Scheller, K. (1977) suggested that JH may regulate mRNA translation for calliphorin based on the timing for calliphorin production, but their suggestion is speculative and not supported by empirical evidence. The studies of Pan, R. *et al.* (1979) demonstrated inhibition of calliphorin synthesis by ecdysteroids. This inhibition agrees with the *in vivo* endocrine situation that shows a rise in ecdysteroid titers and a cessation in storage protein synthesis due to ribosomal destruction at the time of last instar wandering (Price, G., 1969; Koolman, J., 1980). The increase in ecdysteroid titer also stimulates adipocytes to take up the storage proteins from the hemolymph at the time of larval–pupal transformation in *Bombyx mori* (Tojo, S. *et al.*, 1981). It can be speculated that since ecdysteroids suppress storage-protein synthesis, JH titers may exert antagonistic, stimulatory actions. However, in *B. mori*, JH suppresses storage protein synthesis, and the synthesis increases after allatectomy (Tojo, S. *et al.*, 1981). Therefore the endocrine regulation of storage protein synthesis appears complex and must await further investigation for clarification.

Identification of endocrine effects on storage protein synthesis may provide an approach to elucidating differential cellular responses by adipocytes to hormones that regulate metamorphosis. The large quantities of the storage proteins that are produced suggest a large amount of fat body RNA is mRNA transcript for these unique proteins. Since the mRNA exists independently of storage-protein synthesis (Sekeris, C. and Scheller, K., 1977), regulation of synthesis must occur at the level of translation rather than transcription. This system, with its single, defined protein made in large quantity, can be a useful model for examining endocrine regulations of fat body function in immature insects just as vitellogenin synthesis provides a model for studying the mode of action for JH on the fat body of adult insects.

3.3.3 Protein storage granules

The formation of protein storage granules at various times during development is a feature of fat body biochemistry that appears universal for those insect species studied. In general, proteins are synthesized by the fat body and released into the hemolymph during the course of larval growth. At the onset of pupation some of the hemolymph proteins are sequestered back into the fat body and stored in large granules until metamorphosis, at which time they disappear. Presumably, like the larval-specific storage proteins, the proteins in the granules are a source of reusable amino acids for histogenesis during metamorphosis. The formation and disappearance of the granules appears endocrine-dependent with ecdysone having the greatest influence.

Protein granules form from several sources in the adipocytes of *Calipodes ethlius* larvae. Protein granules composed of either protein, or protein plus RNA, form by autophagy of cytoplasmic organelles just before the onset of pupation and as a first step in cytoplasmic reorganization of the adipocyte (Locke, M. and Collins, J., 1965; Dean, R., 1978). A second type of protein granule forms by heterophagous uptake of hemolymph proteins into the adipocyte to form multivesicular bodies which fuse into granules (Locke, M. and Collins, J., 1968). The appearance of injected plant peroxidase in the protein granules, and the decline in circulatory proteins at the time of heterophagous protein granule formation, both argue for the sequestration of hemolymph proteins into the granules. Fat body protein synthesis declines during the formation of heterophagous protein granules, and the granules contain little newly made protein.

Ecdysone secretion by the prothoracic glands corresponds to the formation of both heterophagous and autophagous protein granules. Heterophagous protein granule formation consists of a period of ecdysteroid-independent protein sequestration from the hemolymph into multivesicular bodies of the adipocytes, and an ecdysteroid-dependent coalescence of the multivesicular bodies to form protein storage granules (Collins, J., 1969). Likewise, fat body isolated from *C. ethlius* larvae prior to ecdysone secretion will not form autophagous granules *in vitro* unless 20-OH-ecdysone is present (Dean, R., 1978); however, adipocytes isolated from larvae after ecdysone secretion will initiate autophagy in a hormone-free medium.

Ecdysone injected into ligated abdomens from *C. ethlius* larvae stimulates pupation and the formation of protein granules after the absorption of specific hemolymph proteins (Collins, J., 1969). Three types of antigenic proteins are found in hemolymph from *C. ethlius* larvae and are synthesized and released by the larval fat body *in vitro* (Collins, J., 1974, 1975). Protein types 1 and 2 are always present in larval fat bodies, but type 3 appears in the adipocytes only during the last 24 h before pupation when protein granules are forming (Collins, J., 1974). Thoracic ligation before ecdysone secretion blocks pupation, the appearance of type 3 protein in adipocytes and the formation of protein granules (Collins, J., 1974). Protein types 1 and 2 are absorbed by adipocytes in ligated larvae, but type 3 is excluded from larval adipocytes and remains in the hemolymph. Therefore, it appears that type 3 protein is sequestered and stored in protein granules specifically in response to ecdysteroids. The nature of the type 3 protein is not defined, although it can be speculated that it may be a larval-specific storage protein.

Other insect species also exhibit protein storage granule formation in their adipocytes. Calliphorin synthesis and storage in fat body protein granules was discussed in the previous section. In *Hyalophora cecropia*, protein storage granules are composed of two proteins with molecular weights of 4.8×10^5 and 5.3×10^5 daltons (Tojo, S. *et al.*, 1978). These proteins each have six subunits of 8.5×10^4 and 8.9×10^4 daltons, respectively. Both the molecular weights and their hexameric subunit structures are reminiscent of calliphorin and, like calliphorin, the constitutive protein is synthesized and secreted by the fat body of *H. cecropia* during early larval life and is reabsorbed and stored at the onset of pupation (Tojo, S. *et al.*, 1978).

Acid phosphatase is associated with protein granules, which suggests that autophagous protein granules are derived from lysosomes. Locke, M. and Collins, J. (1968) described the structural transformation of multivesicular bodies containing acid phosphatase activity into protein granules in adipocytes of *C. ethlius*. In the fat body of *Calliphora erythrocephala* the protein granule fraction contains high acid phosphatase activity (van Pelt-Verkuil, E. *et al.*, 1979). Autophagic vacuoles with acid phosphatase activity and heterophagous vacuoles of absorbed hemolymph proteins fuse in the fat body of *C. erythrocephala*, confirming a partly lysosomal character and mixed origin for these protein granules (van Pelt-Verkuil, E. *et al.*, 1979; Locke, M., 1980). As discussed previously, the appearance of acid phosphatase activity in the fat body of *C. erythrocephala* depends on exposure to 20-OH-ecdysone (van Pelt-Verkuil, E., 1979).

Protein granules form in the fat body of *Drosophila melanogaster* larvae at the onset of pupariation by sequestration of hemolymph proteins in response to 20-OH-ecdysone (Thomasson, W. and Mitchell, H., 1972). Transplantation of larval fat body into female flies also results in protein granule formation by the larval adipocytes (Butterworth, F. and Bodenstein, D., 1967). However, transplantation of larval fat body into male flies does not result in protein granule formation unless a larval ring gland (endocrine complex) is also implanted (Butterworth, F. *et al.*, 1967). Only the larval fat body forms protein granules in response to the ring gland; adult fat body is unresponsive. Furthermore, transplantation of ovaries into male flies along with larval fat body stimulates protein granule formation in the larval fat body (Butterworth, F. and Bodenstein, D., 1967). The ability of the ovaries to stimulate protein granule formation may be explained by the finding that ovaries of some Diptera secrete ecdysone

(Hagedorn, H. *et al.*, 1973), and the ovarian ecdysone stimulates the sequence of metamorphic events, including protein granule formation, in the transplanted larval adipocytes. This also explains the differences shown by larval fat body transplants in adult males and females.

Protein granules are also observed in adipocytes of adult, hemimetabolous insects. Generally, protein granules form in larvae immediately before pupation, and the proteins are presumably reused for histogenesis during metamorphosis. However, in adult insects the physiological role of protein granules is less certain. In adult male *Blaberus discoidalis* cockroaches protein granules begin forming in adipocytes on the day of adult emergence with a maximum content present in day old adults (Keeley, L., 1981). The granules disappear within 24–48 h thereafter. These granules clearly show engulfment of mitochondria and endoplasmic reticulum, indicating an autophagous origin and function. The protein granules apparently engulf and destroy existing organelles and reuse their structural components for cytoplasmic reformation, which suggests a post-emergence, imaginal metamorphosis that reorganizes the cytoplasm of immature adipocytes into mature, imaginal cells. In female *Diploptera punctata*, protein granules form and disappear in inverse synchrony with reproductive cycles (Stay, B. and Clark, J., 1971). Protein granules are abundant in adipocytes at adult emergence before vitellogenesis; disappear during vitellogenesis; then reform during oothecal gestation. In male *D. punctata* the initial granules disappear around day 4 of adult life, then reform and accumulate with no evident cyclicity. In the cases of both *B. discoidalis* and *D. punctata* the granules serve as reservoirs for proteins. In the first case the granules are apparently a protein source for reformation of imaginal cell structures in a manner comparable to their function in holometabolous insects. In the second case they act as a storage reservoir for proteins to be used for yolk formation.

3.3.4 Vitellogenic proteins

A major function of the adult fat body in females is the synthesis and release of vitellogenic proteins for yolk formation during oocyte maturation (Brookes, V., 1969; Pan, M. *et al.*, 1969; see J. Kunkel and J. Nordin, vol. 1). Vitellogenin synthesis accounts for 70–90% of the protein secreted by the mature fat body in female insects (Hartmann, R. *et al.*, 1972; Koeppe, J. and Ofengand, J., 1976), and vitellogenin comprises nearly 90% of the yolk proteins (Bell, W., 1970). Furthermore, vitellogenin synthesis is endocrine-regulated, usually by JH (J. Koeppe *et al.*, vol. 8). Particularly interesting is the fact that the fat body is not especially responsive to JH in immature insects, but it becomes a target tissue for the hormone in adults. The characteristics of endocrine-directed synthesis for large quantities of a uniform protein make the adult fat body useful as a model for studies on the cellular actions of JH. Vitellogenin synthesis has been reviewed several times recently (Wyatt, G. and Pan, M., 1978; Engelmann, F., 1979, 1980; Hagedorn, H. and Kunkel, J., 1979).

Immunology and electrophoresis demonstrate that vitellin and vitellogenin are identical with a basic molecular weight of about 5.5×10^5 daltons (Engelmann, F. *et al.*, 1976); Chen, T. *et al.*, 1978). Vitellin contains 12.3% mannose, 1.3% glucosamine, and 9.6% lipids as neutral lipid, phosphatidylcholine and phosphatidylethanolamine. The vitellogenin molecule consists of two immunologically distinct components with molecular weights of 2.65×10^5 and 2.5×10^5 daltons. The synthesis of these two subunits is directed by two distinct genes (Chen, T. *et al.*, 1978; Chen, T., 1980). Both vitellin and vitellogenin of *Locusta migratoria* consist of eight separable polypeptides falling into two molecular weight ranges of either 1.4×10^5 to 1.05×10^5 or 6.5×10^4 to 5.2×10^4 daltons (Chen, T. *et al.*, 1978). These two categories of molecular weights represent the products formed by adipocytes processing the initial polypeptide products (2.5×10^5 daltons) from the vitellogenin genes (Chen, T., 1980). Variabilities in protein processing account for the two size ranges in the resulting eight polypeptide subunits that comprise the final vitellogenin molecule.

Imaginal adipocytes increase their cytoplasmic capacity for protein synthesis in response to the endocrine milieu that promotes vitellogenesis. Feeding often acts as a stimulus to the endocrine system to initiate vitellogenic activity. In flies and mosquitoes, protein feeding activates the endocrine systems and ribosomes, mitochondria and rough endoplasmic reticulum accumulate in adipocytes, produce protein-filled secretion granules, then regress (Thomsen, E. and Thomsen, M., 1974; Behan, M. and Hagedorn, H., 1978; Tadkowski, T. and Jones, J., 1979). In *L. migratoria* the fat body stores nutrients for somatic growth during the first week of adult life (Couble, P. *et al.*, 1979). At about day 8, both rough endoplasmic reticulum and Golgi proliferate, and the tissue function changes from storage to protein synthesis and secretion. Not only are new proteins synthesized for vitellogenesis at this time, but proteins deposited in storage granules are mobilized by enzymes released from multivesicular bodies (Lauverjat, S., 1977). Allatectomy prevents the conversion from the storage to the synthetic phase in the locusts, whereas corpora allata implants or treatment with JH analogs stimulate protein production (Lauverjat, S., 1977; Couble, P. *et al.*, 1979). Protein granules are characteristic of immature adipocytes in newly molted female *Nauphoeta cinerea*, but they disappear 4 days after adult emergence when vitellogenin starts to appear (Wuest, J., 1978). The proteosynthetic organelles become apparent about day 3, with peripheral adipocytes more active for vitellogenesis than centrally located cells.

Fat bodies in those insect species that have brief adult lives and do not feed may produce vitellogenic metabolites independent of hormonal regulations. A classic example of this is the saturniid silkmoths which lack adult mouthparts and contain fully formed eggs at adult emergence. The fat body of *Antheraea polyphemus* produces a female-specific protein during metamorphosis that is sequestered by the ovaries (Blumenfeld, M. and Schneiderman, H., 1968). JH does not stimulate synthesis or release of the female-specific protein, but increases the hemolymph titers of the protein by inhibiting ovarian maturation and preventing its uptake into the ovaries. In *Hyalophora cecropia*, vitellogenin accumulates in the hemolymph twice during metamorphosis (Pan, M., 1971). It first appears in the pharate pupa 6–10 months before oocyte development; it next appears in the second week of adult metamorphosis during tissue reconstruction. It is interesting to note that ovaries implanted into male silkmoth pupae also develop mature eggs (Bhakthan, N. and Gilbert, L., 1972). It is not clear whether this is due to uptake of general cytoplasmic proteins and metabolites discharged from pupal adipocytes during metamorphosis, or whether the ovaries supply a specific stimulus to the male fat body. Allatectomy of larval instars or pupae of *H. cecropia* fails to alter vitellogenin synthesis in the pharate adult (Pan, M., 1977), and injections of JH into isolated pupal abdomens do not stimulate vitellogenin formation. At the time of writing, no endocrine regulations of vitellogenic protein synthesis have been identified for saturniid silkmoths.

Vitellogenic protein synthesis is hormonally regulated in most insect species. Wigglesworth, V. (1935) first indicated a role for the corpora allata in egg maturation of *Rhodnius prolixus*. Following this initial report it became generally accepted that JH secreted by the adult corpora allata was the hormone regulating fat body production of vitellogenic proteins (Engelmann, F., 1972; Hill, L., 1972). Furthermore, JH was focused on in adult insects because it was thought to be the only major hormone present. Ecdysteroids were believed to be absent in adult insects because the prothoracic glands deteriorate shortly after the adult molt. However, because of recent discoveries concerning ecdysone secretion from ovaries (Hagedorn, H. *et al.*, 1975) and the action of ecdysone on fat body vitellogenin synthesis (Fallon, A. *et al.*, 1974), the question of endocrine regulations of the fat body for ovarian maturation has been reopened amid controversy.

JH acts on the female fat body and ovaries to stimulate vitellogenesis. Vitellogenin appears in the hemolymph 40 h after treatment of female adult *Leucophaea maderae* with JH-active analogs (Brookes, V., 1969). This protein is not produced by fat bodies from males, nymphs of either sex or reproductively inactive females (Engelmann, F., 1969). Two JH-dependent vitellogenic proteins appear in the hemolymph of female adult *Periplaneta americana* at 4–5 days of adult age and accumulate in the hemolymph, after ovariectomy, to concentrations 20 times normal (Bell, W., 1969). Abdomens isolated from adult female *L. maderae* show an 18 h lag period before vitellogenin synthesis starts following injection of JH-1 (Koeppe, J. and Ofengand, J., 1976). The vitellogenin synthesis increases 50-fold by 72–96 h after JH treatment; then it starts to decline. JH is also necessary for ovarian uptake of the proteins, since injection of vitellogenin into allatectomized females does not, by itself, result in yolk formation (Bell, W., 1969).

JH appears to stimulate transcription. Fat body RNA doubles during the 8-day previtellogenic period in adult *Locusta migratoria* females and reaches its maximum on day 7 (Chen, T. *et al.*, 1979). Vitellogenin synthesis starts on day 7 and reaches its peak on day 13 when vitellogenin comprises 60% of the total proteins synthesized by the adipocytes. JH treatment of allatectomized *L. migratoria* females normally produces maximum vitellogenin synthesis 72 h later (Chen, T. *et al.*, 1979). However, a second application of JH within 10 days after the first renews vitellogenin synthesis after only a brief time lag due to the rapid appearance of polysomes that bear the vitellogenin transcript (Reid, P. and Chen, T., 1981). This is explained by the finding that total RNA, especially rRNA, increases in response to the first exposure of JH, whereas only a small amount of RNA is produced in response to the second exposure to JH (Chinzei, Y. *et al.*, 1980). The first exposure appears to stimulate transcription of vitellogenin genes along with genes for "bulk" RNA. The second exposure affects mainly vitellogenin gene transcription, while the majority of the ribosomal system is still intact and functional. Engelmann, F. (1971) demonstrated that actinomycin D blocks JH induction of vitellogenin synthesis, and existing vitellogenin mRNA transcripts persist for about 3 days. These findings suggest that the vitellogenin transcript is relatively short-lived, but ribosomes formed in response to the initial JH exposure remain available to resume protein synthesis in response to renewed JH induction of vitellogenin mRNA.

JH also increases endoplasmic reticulum formation in adipocytes. The rate of formation of endoplasmic reticulum increases by 3.5 to 5 times in JH-treated *L. maderae* females (Della-Cioppa, G. and Engelmann, F., 1980), and polysomes and heavy microsomes proliferate in reproductive females as compared to males or non-vitellogenic females (Engelmann, F., 1974). Endoplasmic reticulum formation increases the number of binding sites for polysomes during vitellogenin synthesis, and forms vesicles as part of the mechanism for vitellogenin secretion from the adipocyte (Della-Cioppa, G. and Engelmann, F., 1980). During vitellogenin synthesis the newly synthesized proteins are released from the polysomes into the lumen of the endoplasmic reticulum (Englemann, F. and Barajas, L., 1975) which is then physically transported to the adipocyte surface for secretion (Della-Cioppa, G. and Engelmann, F., 1980).

In mosquitoes, adipocyte RNA rises measurably within 2 h after a blood meal and is increased by 3-fold within 12 h (Hagedorn, H. *et al.*, 1973). Vitellogenin synthesis is detected within 3 or 4 h after blood feeding, and peaks at 28 h (Hagedorn, H. and Judson, C., 1972; Hagedorn, H. *et al.*, 1973). Ovarian ecdysone regulates the onset of vitellogenin synthesis after blood feeding in *Aedes aegypti* (Hagedorn, H. *et al.*, 1975). The ovaries secrete ecdysone as a prohormone which is converted by the fat body to 20-OH-ecdysone, the active hormone (see S. Smith or W. E. Bollenbacher and N. A. Granger, vol. 7). Fat body vitellogenin synthesis is responsive only to 20-OH-ecdysone, not to ovarian ecdysone *per se* (Fuchs, M. *et al.*, 1979). Fat bodies from newly emerged females were unresponsive to 20-OH-ecdysone, and developed responsiveness only after exposure to JH (Flanagan, T. and Hagedorn, H., 1977).

The cellular action of 20-OH-ecdysone is unclear (see L. Riddiford, vol. 8). Fong, W.-F. and Fuchs, M. (1976) found that α-amanitin and cordycepin, inhibitors of mRNA transcription and poly(A) mRNA formation respectively, do not prevent vitellogenin formation *in vivo*. However, actinomycin D, which inhibits transcription by blocking DNA, is inhibitory. In contrast, Kaczor, W. and Hagedorn, H. (1980) found that both α-amanitin and cordycepin inhibit vitellogenin production by the fat body *in vitro*. The results of these two studies are contradictory and remain to be resolved. However, it should be pointed out that Kaczor, W. and Hagedorn, H. (1980) reported that their mosquitoes would not feed after α-amanitin treatment, and although Fuchs, M. and Fong, W.-F. (1976) did not report trouble feeding their mosquitoes, they found that digestion was impaired. Hence the α-amanitin-treated mosquitoes studied by Fong, W.-F. and Fuchs, M. (1976) may have been abnormal if they were unable to digest their blood meal and receive the proper nutritional stimuli needed to trigger the sequence of *in vivo* endocrine events.

However, conflicting results have cast doubt on 20-OH-ecdysone as the principal regulatory hormone for vitellogenin synthesis in the fat body of mosquitoes. Physiological doses of 20-OH-ecdysone do not stimulate vitellogenin production in unfed *A. aegypti* females (Borovsky, D. and Van Handel, E., 1979), and doses of hormone (0.05–5 μg) that exceed physiological levels (275 pg, fed females) by hundreds or thousands of times stimulate less vitellogenin formation than blood feeding or the presence of ovaries in fed females (Hagedorn, H. *et al.*, 1975; Borovsky, D. and Van Handel, E., 1979). In contrast, JH (12.5 ng) at nearly physiological levels stimulates 70–80% ovarian development in blood-fed, decapitated *A. aegypti* females (Borovsky, D., 1981a). Finally, mature oocytes form in abdomens isolated from the autogenous mosquito, *Aedes atropalpus*, after treatments with 0.5 ng of JH-I followed by 4.8 ng of 20-OH-ecdysone (Kelly, T. *et al.*, 1981). Although only 50% of the oocytes develop the hormone doses are nearer to physiological levels. These results suggest that either hormone, alone, in high, non-physiological doses can stimulate the fat body to produce vitellogenic metabolites. However, the two hormones acting in concert in a properly timed sequence interact to increase the vitellogenic capacity beyond that of either hormone alone, and JH pretreatment sensitizes the fat body to 20-OH-ecdysone.

Finally, the ovaries may act as a "sink" to absorb vitellogenin and relieve a feedback inhibition of its further synthesis or release by the fat body. Engelmann, F. (1978) reported that the levels of circulating vitellogenin inhibit the release of further vitellogenin from the fat body of *Leucophaea maderae* but do not suppress its continued synthesis. However, Borovsky, D. (1981b) found that ovariectomy of vitellogenic female *A. aegytpi* stops further vitellogenin synthesis after vitellogenin levels increase in the hemolymph by 4.5 times. Implants of vitellogenic ovaries reactivate the vitellogenin production, although similar ovarian implants fail to stimulate protein synthesis in previtellogenic, blood-fed females. These results suggest that there may not be an ovarian factor that stimulates vitellogenin formation. Rather, the ovaries may relieve an end-product inhibition of vitellogenin synthesis or secretion by removing circulating vitellogenin.

Although feedback inhibition may affect vitellogenin production by the mosquito fat body, the wealth of evidence supports endocrine regulations. The data suggest a timed sequence of cooperative interactions between JH and 20-OH-ecdysone for optimal vitellogenin synthesis. JH is secreted before blood feeding in young adult female mosquitoes, and appears to prime the adipocytes for vitellogenin synthesis when exposed, later, to ovarian ecdysone. JH priming of the fat body may involve activation of the C_{20}-hydroxylation enzymes for converting ovarian-derived ecdysone to hormonally active 20-OH-ecdysone, or it may involve the formation of cytoplasmic receptor proteins to increase adipocyte sensitivity to 20-OH-ecdysone. However, the cellular effects by the two hormones that result in enhanced vitellogenin synthesis remain obscure and should prove to be a fruitful area of future investigation.

In *Drosophila melanogaster*, both JH and 20-OH-ecdysone stimulate the synthesis of three yolk proteins (Postlethwait, J. and Handler, A., 1979). Treatment with a JH analog affects the synthesis and secretion of the vitellogenins by both the fat body and ovaries, whereas 20-OH-ecdysone affects only the fat body (Jowett, T. and Postlethwait, J., 1980). This finding stresses the pre-eminence of 20-OH-ecdysone for stimulating fat body yolk protein synthesis in the Diptera.

Further support for the role of 20-OH-ecdysone in vitellogenin synthesis of Diptera is found with male *Sarcophaga bullata* (Huybrechts, R. and deLoof, A., 1977). JH-I does not induce vitellogenin formation by fat bodies from male *S. bullata*, but vitellogenin is produced by fat bodies of 20-OH-ecdysone-treated males 3 days after emergence, if the flies are first fed a high-protein (liver) diet. Ovaries implanted into male flies fail to induce vitellogenin formation or to become vitellogenic even after injections of 20-OH-ecdysone. However, both of the latter effects may fail because of the absence of JH. JH may be necessary for the "primer" effect on the fat body to convert ovarian-produced ecdysone into 20-OH-ecdysone. Furthermore, lack of yolk formation may result from inability of the oocytes to absorb vitellogenin in the absence of JH (Jowett, T. and Postlethwait, J., 1980).

The content of protein present in the fat body at adult emergence is a factor in autogeny. An autogenous strain of *Lucilia cuprina* was compared to wild-type individuals that were anautogenous (Williams, K. *et al.*, 1977). Protein stores in autogenous females are 20% larger at adult emergence and last longer than in anautogenous females. The increased protein stores are derived from the larval fat body which persists longer into adult life in autogenous flies than does the larval fat body of anautogenous flies.

An interesting and unique relationship exists in fleas concerning endocrine regulation of vitellogenin synthesis by the fat body. The fat body organization of the flea responds to the hormones that regulate reproductive cycles in the host rabbit (Rothschild, M. *et al.*, 1970). Exposure of the flea to estrogens, or luteinizing hormone, or feeding the flea on a pregnant or lactating doe or on newborn rabbits causes the fat body to organize and increase in size, and mating and reproduction ensue. Conversely, exposure to progesterones or feeding on nestling rabbits causes fat body deterioration and the suspension of reproduction. The fleas are sensitive to JH so that the vertebrate host hormones appear to interact with the endocrine system of the flea to regulate flea reproduction. In this way the fat body synchronizes flea reproduction with the availablity of host offspring for transference.

3.3.5 Diapause proteins

Patterns of fat body protein synthesis change during insect dormancy. Fat body protein synthesis declines to only 3% of its active larval level during diapause in saturniid pupae (Stevenson, E. and Wyatt, G., 1962). In female adults of *Leptinotarsa decemlineata* the fat body synthesizes in different ratios, at all times, both vitellogenic proteins and proteins related to diapause (Dortland, J., 1978; Dortland, J. and deKort, C., 1978). Under long-day conditions two vitellogenic proteins are synthesized most actively by the fat body; with short-day photoperiods three diapause proteins predominate (Dortland, J., 1978). Both types of proteins are present in either photoperiodic situation but the relative rate of synthesis for each type correlates to the photoperiod length. The vitellogenic proteins are secreted from the fat body, but the diapause proteins are stored (Dortland, J. and deKort, C., 1978).

JH influences the production of the vitellogenic proteins in *L. decemlineata* but not in the manner usual for most insects. JH stimulates by 32-fold the synthesis of vitellogenin-1 in fat bodies of dormant, short-day beetles with minimal synthetic activity (Dortland, J., 1979). However, after diapause termination, vitellogenin-1 increases by 160 times and vitellogenin-2 by 120 times, even in beetles that have undergone allatocardiacectomy (Dortland, J., 1979). Hence, although vitellogenin synthesis responds to JH when synthetic activity is depressed, JH is not needed to produce vitellogenin in diapause-terminating females. There is no vitellogenin synthesis response to 20-OH-ecdysone in diapausing females (Dortland, J., 1979). Furthermore, synthesis of the diapause proteins is not affected by JH in diapausing beetles (Dortland, J., 1979).

It might be noted that a JH dose of 50 μg per female was used to stimulate vitellogenin formation in diapausing *L. decemlineata* (Dortland, J., 1979). This dose exceeds normal JH levels and could be forcing a non-physiological response which may be similar to the effects of high 20-OH-ecdysone doses on vitellogenin synthesis by the mosquito fat body. At present, the endocrine regulations for the synthesis of the vitellogenic and diapause proteins in *L. decemlineata* are unclear, and further research on this topic may identify new endocrine relationships with fat body protein synthesis.

3.3.6 JH Carrier proteins and JH esterases

The fat body is the source for a low molecular weight, JH-specific carrier protein in *Manduca sexta* and a high molecular weight lipoprotein that also binds JH (Nowock, J. *et al.*, 1975; see W. Goodman and E. Chang, vol. 7). These proteins are synthesized in the fat body and secreted immediately. The low molecular weight carrier protein increases JH solubility in hemolymph and prevents degradation by non-specific esterases (Hammock, B. *et al.*, 1975; see B. Hammock, vol. 7). The putative carrier protein of *Locusta migratoria* is a lipoprotein with a molecular weight of 220,000 daltons (Emmerich, H. and Hartmann, R., 1973). In *M. sexta* the carrier protein is smaller with a molecular weight of 28,000–34,000 daltons (Kramer, K. *et al.*, 1976), and less than 1% of the high molecular weight lipoprotein is associated with

JH (Nowock, J. *et al.*, 1975). It may be that the high molecular weight JH-carrier protein reported for locusts corresponds with the non-specific lipoprotein of hornworms. In hornworms the large lipoprotein only acts as a JH-carrier protein in the unlikely event that the capacity of the smaller protein is exceeded by the quantity of available JH. There appears to be no apparent regulation of JH-carrier protein synthesis by the fat body and ample specific carrier protein exists in the hemolymph to bind more than the total amount of JH present at any given time (Whitmore, E. and Gilbert, L., 1972).

In addition to producing a carrier protein that protects JH, the fat body is also the source for the JH-specific esterase that degrades circulating JH (Wing, K. *et al.*, 1981; see B. Hammock, vol. 7). JH esterase reaches a major peak of activity on days 3 to 4 of the last larval instar in *M. sexta* when the JH titer is declining in preparation for metamorphosis (Weirich, G. *et al.*, 1973; Vince, R. and Gilbert, L., 1977). A second smaller peak of esterase activity occurs at the time of larval wandering.

The regulation of JH esterase by the fat body appears complex. Timed ligation studies suggest that JH esterase formation is neurohormonally controlled for the first peak of esterase activity in larvae of *Trichoplusia ni* and JH-controlled for the second peak of activity (Sparks, T. and Hammock, B., 1979). JH treatments suppress JH esterase formation in larvae of *Galleria mellonella* but induce the activity in pupae of both *G. mellonella* and *T. ni* (Reddy, G. *et al.*, 1979, Wing, K. *et al.*, 1981). However, further studies on *G. mellonella* suggest a brain neurohormone stimulates JH esterase production some time after 48 h of the last larval instar (McCaleb, D. and Kumaran, A., 1980). Jones, G. *et al.* (1981) suggest that both stimulatory and inhibitory neurohormones regulate JH esterase activity. The question of neurohormonal regulation of JH esterase is worthy of elucidation since it relates to regulations for metamorphosis and has a pragmatic potential for insect growth manipulation.

3.3.7 Protein kinases and cyclic nucleotides

Protein kinases phosphorylate proteins. Frequently, the protein substrate is an inactive enzyme that is activated by phosphorylation and regulates processes involved in metabolism or cellular growth and differentiation. The protein kinases respond to endocrine stimuli via the system of cyclic nucleotide second messengers (Sutherland, E., 1972; see W. Smith and W. Combest, vol. 8). In general, vertebrate protein kinases respond to cAMP. However, in the fat body of larval *Hyalophora cecropia* and *M. sexta* all protein kinases were reported to be activated by cGMP (Kuo, J. *et al.*, 1971). cGMP-dependent protein kinases appear to predominate in lepidopteran tissues, whereas cAMP-dependent protein kinases occur frequently in *Blaberus discoidalis* and *Drosophila melanogaster* (Kuo, J. *et al.*, 1971). Protein kinases of locusts are sensitive mainly to cGMP but to some degree to cAMP, and they require a high Mg^{2+} level for optimal activity (Pines, M. and Applebaum, S., 1978).

The relative levels of cAMP and cGMP may be significant to specific aspects of cell growth and differentiation, and it is interesting that the ratios of these two cyclic nucleotides change during metamorphosis. Both adenylate and guanylate cyclase are membrane-bound enzymes in adipocytes of *H. cecropia*, and they increase in activity at the larval–pupal transformation with little decrease in activity during pupal diapause (Filburn, C. and Wyatt, G., 1976). In *Bombyx mori*, adenylate cyclase activity is high at the start of the last larval instar; declines by 90% during pupation; then increases by two to three times above its minimal, pupal level just before adult emergence (Morishima, I., 1981). Guanylate cyclase changes inversely with adenylate cyclase in *B. mori* and parallels the ecdysteroid titers. Since most fat body protein kinases appear cGMP-dependent the data suggest that some aspects in the action of ecdysteroids on adipocytes may involve guanylate cyclase activation and second messenger effects. The reciprocity of adenylate and guanylate cyclase activities suggests that these enzymes may respond to hormones and control competing or antagonistic processes of development and metamorphosis by changing the relative ratios of the cyclic nucleotides.

3.4 Lipid metabolism

3.4.1 Synthesis

A large number of articles report on the lipid

composition of insects, but most of these are concerned with extracts of the whole body and are not specific for fat body (see Gilbert, L., 1967a). Since fat body is the major tissue depot for lipids, Kilby, B. (1963) states that studies on whole-body lipid composition also reflect fat body lipid composition without serious error. However, in keeping with the original focus of this review, only those studies dealing specifically with fat body lipids will be considered, with the emphasis on metabolism. The interested reader is referred to the comprehensive reviews of Gilbert, L. (1967a) and Downer, R. (1978) for details on the composition and general metabolism of insect lipids.

Triglycerides constitute the main storage form for fat body lipids. Palmitate is esterified to triglycerides rapidly in the fat bodies of larval and adult saturniid silkmoths but to diglycerides in the pupal fat body (Bhakthan, N. and Gilbert, L., 1970). Triglyceride formation exceeds diglyceride formation by 3-fold in larvae, but diglyceride synthesis exceeds triglyceride synthesis by 2-fold in pharate pupae. The most common fat body fatty acids are palmitic, palmitoleic, stearic, oleic, linoleic and linolenic (Stephen, W. and Gilbert, L., 1969). Only saturated and monoenoic fatty acids are synthesized from acetate by the fat body. Insects cannot synthesize polyenoic fatty acids, but monoenoic fatty acids form by direct desaturation of saturated precursors (Stephen, W. and Gilbert, L., 1969). Microsomal enzymes show desaturation activity in the fat body of *L. migratoria* with stearate as the principal substrate and NAD as the preferred cofactor (Tietz, A. and Stern, N., 1969).

The fat body synthesizes phospholipid. Phosphatidylcholine is the main fat body phospholipid in fifth-instar *M. sexta* larvae and comprises 41% of the lipid phosphorus in this species followed by phosphatidylethanolamine at 30.6% (Kulkarni, A. *et al.*, 1971). ATP and CTP both stimulate the formation of phosphatidylcholine in the fat body of larval *Phormia regina*, thus confirming the choline donor role for CTP in insect tissues (Shelley, R. and Hodgson, E., 1970, 1971a). Phosphatidylethanolamine predominates in *P. regina* fat body and is formed by an exchange reaction with choline from phosphatidylcholine (Taylor, J. and Hodgson, E., 1972). The data indicate two kinase enzymes for the initial phosphorylations of ethanolamine and choline in the fat body of *P. regina* larvae (Shelley, R. and Hodgson, E., 1971a, 1971b). One kinase is choline-specific; the other reacts with both choline and ethanolamine.

Lipogenesis from carbohydrate precursors occurs in the insect fat body. Radioactive label from [^{14}C]glucose appears in fatty acids as well as glycogen reserves in *H. cecropia* (Chino, H. and Gilbert, L., 1965a). The fat body of *Periplaneta americana* contains cytosolic ATP-citrate lyase (Storey, K. and Bailey, E., 1978) for the generation of acetyl CoA from carbohydrate-derived citrate that is released by mitochondria into the cytosol. The resulting acetyl CoA is then converted by acetyl CoA carboxylase to malonyl CoA by the addition of CO_2 from bicarbonate (Storey, K. and Bailey, E., 1978). In *Locusta migratoria*, lipogenesis from carbohydrate stores increases around day 6 of adult life during imaginal somatic growth (Walker, P. *et al.*, 1970). Both trehalose and glucose are converted to fatty acids via citrate and ATP-citrate lyase which increases on day 4 of adult life (Walker, P. and Bailey, E., 1970a,b).

Although lipogenesis from carbohydrates occurs readily, glyoxylate cycle enzymes are absent in insects (Storey, K. and Bailey, E., 1978) and there is no evidence for conversion of lipids to carbohydrates. In *H. cecropia* pupae there is no indication of fatty acid conversion to either glycogen or trehalose in adipocytes (Chino, H. and Gilbert, L., 1965a).

3.4.2 Mobilization

The fat body forms diglyceride and releases it as the major circulatory lipid (see H. Chino, vol. 10). Triglyceride content declines in the fat body of young adult *P. americana* due to hemolymph diglyceride formation from triglycerides composed mainly of saturated fatty acids (Nelson, D. *et al.*, 1967). Fat bodies from *Hyalophora cecropia*, *P. americana* and *Melanoplus differentialis* prelabeled *in vivo* with [^{14}C]palmitate (Chino, H. and Gilbert, L., 1965b) subsequently release the ^{14}C label in diglycerides when the isolated tissues are incubated in hemolymph. Diglyceride formation and release is energy-dependent and is blocked by inhibitors of energy production; free fatty acid release increases after energy inhibition. Fat body from chilled *H. cecropia* prelabeled with [^{14}C]glycerol and [^{3}H]palmitate release a higher ratio of ^{3}H/^{14}C than is stored in the fat body (Beenakkers, A. and Gilbert, L., 1968). The differences in fatty acid composition of released and stored glycerides suggest that fat body lipid release is a directed event for glycerides with a particular fatty acid composition. The glycerides for release are synthesized by an energy-dependent process specificially from a separate pool of stored lipids (triglycerides). Inhibition of energy production blocks specific glyceride formation but still permits the release of the hydrolyzed free fatty acids.

In general, triglycerides are the storage form for fat body lipids, while diglycerides are the mobile hemolymph lipids. Exceptions exist, however, that raise questions about the generality of this statement. Triglycerides are the primary form of lipid released by the fat body of *Galleria mellonella* larvae when incubated *in vitro* in dilute hemolymph (Wlodawer, P. and Lagwinska, E., 1967). *P. americana* fat body releases 85% free fatty acids when incubated in a medium containing bovine serum albumin (Cook, B. and Eddington, L., 1967). Both free fatty acids and triglycerides are released *in vitro* if the medium contains a minimum of 20% hemolymph, and triglyceride release is optimal in a medium of 50–100% hemolymph. Chang, F. and Friedman, S. (1971) surveyed the form of lipid released from the fat body in several insect species at various growth stages. Although the principal hemolymph lipid of *Manduca sexta* is diglyceride, *in vitro* fat body tissue from pupae or pharate adults releases mainly free fatty acids, and larval fat bodies fail to release lipids into any medium. Adult fat body releases either free fatty acids (*Leucophaea maderae*) or diglycerides (*Schistocera gregaria*), and pupal fat body from *H. cecropia* releases free fatty acids, diglycerides and triglycerides. These findings support the conclusion that diglycerides are the predominant form for lipids circulating in the hemolymph, but they suggest that lipids may be released from the fat body in several forms which are converted to diglycerides during circulation.

The fat body exhibits lipase activity for degradation of the storage triglycerides into circulatory lipids. Triglyceride hydrolysis in the fat body of *P. americana* produces diglycerides in the 1,2 configuration (Hoffman, A. and Downer, R., 1979). In the southern armyworm, *Prodenia eridania*, the adult fat body contains lipases for mono-, di- and triglycerides (Stevenson, E., 1972). Presumably the glyceride stores in adipocytes of adult *P. eridania* are hydrolyzed to free fatty acids which are released and circulated to the flight muscles for use as a flight fuel. Two major forms of lipases occur in the fat body of *L. migratoria* (Tietz, A. and Weintraub, H., 1978). An acidic lipase (optimum pH 5–6) degrades mono-, di-, and triglycerides and is present in both microsomal and soluble forms, and a monoglyceride lipase is present in either a soluble form with an optimum pH between 7 and 8.5 or a microsomal form with activities up to pH 9.

Circulating diglycerides are protein-bound and the lipoproteins are products of the fat body. The major lipid carrier is a globulin-like protein that combines with cholesterol, phospholipids and carotenes in addition to diglyceride and has a molecular weight of 700,000 daltons (Chino, H. *et al.*, 1967, 1969). This carrier protein binds 44% of the lipid (Chino, H. *et al.*, 1969). A second minor protein (500,000 daltons) lacks the carotene moiety and binds only 10% of the lipid. Lipoproteins comparable to those found in the hemolymph are produced by the fat body *in vitro* from fatty acid and amino acid precursors, and lipid-free extracts of the fat body can incorporate [^{14}C]palmitate into existing fat body apo-lipoproteins (Thomas, K., 1972). Puromycin blocks the *in vitro* incorporation of labeled amino acids into the lipoprotein, but it does not inhibit incorporation of the lipid moiety. This indicates *de novo* synthesis of the protein moiety by the fat body. Diglyceride conjugation to the apo-lipoprotein occurs within the adipocyte since diglyceride will not bind to free hemolymph proteins *in vitro* (Chino, H. *et al.*, 1969).

Ketone bodies are formed and released by the fat body as an energy source for the peripheral tissues in addition to glycerides and fatty acids. Both β-hydroxy-β-methyl glutaryl-CoA synthase and β-hydroxy-β-methyl glutaryl-CoA lyase have high activities in the fat body of *P. americana* (Shah, J. and Bailey, E., 1976). Acetoacetate is the major ketone in the hemolymph and the flight muscle, and hydroxybutyrate is the major fat body ketone (Hill, L. *et al.*, 1972). The ketone bodies are produced from the fat body and used as oxidative substrates by the fat body, flight muscles and testes. All three tissues use acetoacetate in preference to hydroxybutyrate. During starvation the levels of acetoacetate and hydroxybutyrate increase, not because of increased synthetic activity, but because of changes in the balance of the overall rates of production, use and excretion (Shah, J. and Bailey, E., 1976). Starvation reduces the oxidative capacity for ketone bodies in the flight muscles and fat body (Hill, L. *et al.*, 1972) and results in their accumulation.

3.4.3 Endocrine regulation

Lipids accumulate in the fat body of many insect species during early adult life but are released for vitellogenesis in response to JH. JH depresses adipocyte lipid synthesis but stimulates ovarian lipid synthesis (Gilbert, L., 1967b; Hill, L. and Izatt, M., 1974; Sroka, P. and Barth, R., 1976). Allatectomy results in fat body hypertrophy because triglyceride turnover decreases (Vroman, H. *et al.*, 1965). Furthermore, lipogenesis from carbohydrates may increase after allatectomy due to enhanced activities for ATP-citrate lyase and malic enzyme (Walker, P. and Bailey, E., 1971a,b). These changes in adipocyte metabolism are direct responses to the absence of JH following allatectomy since fat body hypertrophy does not occur after ovariectomy (Orr, C., 1964). Yolk precursors are still mobilized after ovariectomy, but the released precursors accumulate in the hemolymph and may suppress further precursor formation (Engelmann, F., 1978; Borovsky, D., 1981b). In addition, the increased osmotic pressure of the hemolymph due to the accumulated precursors may deter further feeding, especially in Diptera (Gelperin, A., 1966). Thus, although the precursors are not removed by the ovarian "sink", they do not accumulate in the fat body, and they suppress additional nutrient intake so that fat body hypertrophy does not result.

The release of lipid from the fat body for transfer to developing ovaries is accompanied by production of specific carrier proteins. Two proteins, designated A and B, are secreted by the fat body and bind to diglycerides in *Locusta migratoria* (Harry, P. *et al.*, 1979). The A-protein is a diglyceride-carrying lipoprotein present in the hemolymph throughout vitellogenesis. The B-protein is a female-specific lipoprotein that appears during vitellogenesis and is probably identical to vitellogenin. Protein A was not JH-regulated and did not remain in the oocyte after releasing its lipid.

Factors in the corpora cardiaca influence lipid oxidation and content in the fat body. The addition of corpora cardiaca extracts increases palmitate oxidation by 35% in the *in vitro* fat body from *L. maderae* (Wiens, A. and Gilbert, L., 1965). Carbohydrate and lipid stores accumulate in the fat body of adult *Pyrrhocoris apterus* following allatectomy but not after allatocardiacectomy (Janda, V. and Sláma, K., 1965). It is suggested that allatectomy disrupts reproduction-related metabolism in *P. apterus* and unused yolk precursors accumulate in the adipocytes. In contrast, the corpora cardiaca regulate "trophic" (digestive and basal) metabolism and allatocardiacectomy reduces digestion and the conversion of nutrients into metabolite stores.

An adipokinetic hormone (ADKH) from the corpora cardiaca stimulates the fat body to release diglycerides and increases the oxidation of lipids in preference to carbohydrates by the flight muscles in adult *Schistocerca gregaria* locusts (Mayer, R. and Candy, D., 1969; Robinson, N. and Goldsworthy, G., 1974). ADKH is a decapeptide and a product of the glandular lobe of the corpora cardiaca (Goldsworthy, G. *et al.*, 1972; Stone, J. *et al.*, 1976). Corpora cardiaca from *Tenebrio molitor* and *S. gregaria* are interchangeable for stimulating lipid mobilization in these two species. However, injection of *Periplaneta americana* gland extracts into *Locusta migratoria* causes lipid release, but fat body of *P. americana* does not release lipids in response to either its own corpora cardiaca or to glands from locusts (Goldsworthy, G. *et al.*, 1972). Instead, cockroach glands act on cockroach fat body in the reverse of the locust glands and cause lipid uptake and storage (Downer, R. and Steele, J., 1972).

It is particularly interesting to note that, although ADKH is one of two chemically defined insect neurohormones and much of its physiological significance is known, the means by which ADKH mobilizes fat body lipids is not clearly established. There is no measurable activation of fat body lipase in response to ADKH although Ca^{2+} and hemolymph proteins are needed for the lipid mobilization effect and cAMP appears involved (Spencer, I. and Candy, D., 1976). Fat body cAMP content increases within 5 min after corpora cardiaca extract administration and gives a linear dose-dependent response for glandular doses up to 0.01 gland pair (Gäde, G. and Holwerda, D., 1976). ADKH administration also results in an increase in fat body cAMP-activated protein kinase activity *in vitro* within 3.5 min (Pines, M. *et al.*, 1981). Incubation of fat body with cAMP *in vitro* activates lipase activity and increases diglyceride production by 2-fold. This sequence of *in vitro* events argues for the activation of fat body lipases in response to ADKH activation of the cAMP–second messenger system. However, it was not confirmed experimentally that ADKH did, in fact, stimulate lipase activity. Furthermore, only adult *L. migratoria* between 3 and 35 days old give a lipid mobilization response to ADKH even though fat body cAMP increases after ADKH treatment for all ages tested, and lipid is always available (Gäde, G. and Beenakkers, A., 1977; Mwangi, R. and Goldsworthy, G., 1977a).

The inability of locusts to respond at all ages to ADKH may result from the absence of a particular form of diglyceride-carrier protein. The usual diglyceride-carrier protein (A) is converted to a higher molecular weight protein (A+) after injections of ADKH-active corpora cardiaca extracts (Mwangi, R. and Goldsworthy, G., 1977b). The A protein, by itself, is unable to load diglyceride normally. As A+ forms, a lower molecular weight protein (C) disappears. Incubation of A+ in the presence of fat body gives rise to both A and C. Presumably, A combines with C in response to ADKH to form A+ for lipid transport. Young (< 3 days old) locust adults that are not responsive to ADKH, lack protein C and cannot form A+ after treatment with ADKH.

In conclusion, the ADKH response appears to consist of two components. Somehow ADKH stimulates the cAMP–second messenger system in adipocytes to cause lipid release and, although how this occurs is not demonstrated, lipase activation appears involved. At the same time two proteins, A and C, are needed in the hemolymph to form the carrier protein for the mobilized diglycerides. Presence of the C protein may not be ADKH-related but is a normal, timed, developmental event. Possibly the timed formation of these specific carrier proteins is regulated by other, presently unknown endocrine stimuli related to adult maturation.

3.5 Carbohydrate metabolism

3.5.1 General metabolsim

Like lipids, carbohydrates are important circulatory metabolites synthesized and released from the fat body for use as an energy source by peripheral tissues (see S. Friedman, vol. 10). Friedman, S. (1970, 1978); Chippendale, G. (1978); and Steele, J. (1981) have reviewed aspects of the role of the fat body in carbohydrate metabolism.

As found in other animals, the normal glycolytic pathways occur in insect adipocytes. Properties and activities are described for several major glycolytic enzymes from the fat body: hexokinase (Storey, K. and Bailey, E., 1977), phosphofructokinase (Walker, P. and Bailey, E., 1969; Storey, K. and Bailey, E., 1978), aldolase (Storey, K. and Bailey, E., 1978) and pyruvate kinase (Bailey, E. and Walker, P., 1969). Like the flight muscle, a glycerophosphate shuttle is active for the oxidation of cytoplasmic $NADH_2$ and lactate dehydrogenase activity is minimal (Storey, K. and Bailey, E., 1977). The TCA cycle and electron transport enzymes are described for adipocytes of *S. gregaria* (Hearfield, A. and Kilby, B., 1958; Keeley, L., 1971; Storey, K. and Bailey, E., 1978). In fact, there is little reason to doubt the presence in the fat body of all the common enzymes for glycolysis and the TCA cycle.

The pentose phosphate shunt may be a significant route for glucose degradation in the fat body. Phosphofructokinase has the least activity of any glycolytic enzyme in adipocytes of *P. americana* (Storey, K. and Bailey, E., 1977) and high activities by glucose-6-phosphate dehydrogenase and aldolase suggest the pentose phosphate pathway is a significant alternative for triose formation in adipocytes (Storey, K. and Bailey, E., 1977, 1978). Generally, however, the pentose phosphate pathway is considered minor for carbohydrate degradation in insects with less than 30% of the glucose being metabolized by this route (Chippendale, G., 1978). The pentose phosphate pathway is used mainly to produce $NADPH_2$ for fatty acid synthesis or for insecticide detoxification by the mixed-function oxidase pathway after sublethal poisoning.

In addition to the usual routes for carbohydrate degradation, the fat body contains enzymes for gluconeogenesis. Pyruvate carboxylase (mitochondrial), phosphoenolpyruvate carboxykinase (cytosolic), fructose-1,6-diphosphatase (cytosolic) and glucose-6-phosphatase (microsomal) are all key enzymes in the fat body of *P. americana* for the resynthesis of glucose from pyruvate (Storey, K. and Bailey, E., 1978).

Glycogen is the principal storage form for carbohydrate in the insect fat body and may comprise as much as 10–25% of the dry weight of the tissue (Wiens, A. and Gilbert, L., 1967b; Wimer, L., 1969). Glycogen is synthesized by glycogen synthase which is a bound enzyme stimulated by glucose-6-phosphate and Mg^{2+} (Murphy, T. and Wyatt, G., 1965).

Phosphorylase degrades fat body glycogen, and its activity is affected by several regulatory mechanisms. Inactive phosphorylase *b* is converted to active phosphorylase *a* by a kinase which is, itself, activated by ATP, Mg^{2+}, and Ca^{2+} but not by Pi (Yanagawa, H. and Horie, Y., 1978; Ashida, M. and Wyatt, G., 1979). As in the vertebrate liver, AMP increases the activity of fat body phosphorylase *b* (Stevenson, E. and Wyatt, G., 1964; Applebaum, S. and Schlesinger, H., 1973); however, the increase in activity of fat body phosphorylase *b* is not strictly AMP-dependent. Locust phosphorylase does not require AMP for activity, and all samples tested gave the same 75% increase in activity over a 200-fold range of AMP between 0.005 and 1 mM (Applebaum, S. and Schlesinger, H., 1973).

Furthermore, trauma to the fat body activates phosphorylase (Stevenson, E. and Wyatt, G., 1964). Standing *in vitro* increases the active enzyme level from 1% to 30% in tissue from diapausing silkmoth pupae. Since AMP does not increase in the isolated tissue (Stevenson, E. and Wyatt, G., 1964), the increase in phosphorylase activity is probably enzymically controlled by phosphorylase kinase. In support of kinase activation, both cAMP and phosphorylase *a* increase in the tissue after excision (Ziegler, R. *et al.*, 1979). Hence, trauma activation of phosphorylase may result from increased cAMP-dependent, phosphorylase kinase activity. Trauma-induced activation of phosphorylase results in trehalose synthesis (Ziegler, R. *et al.*, 1979).

Exposure to cold also activates phosphorylase but with the formation of glycerol rather than trehalose. Pupae held at 4° show an increase in fat body phosphorylase from a basal level of 8–10% active form up to 40% active form within 1–2 h (Ziegler, R. and Wyatt, G., 1975; Ziegler, R. *et al.*, 1979). This increased activity is maintained for 30 days along with glycerol accumulation in the hemolymph. The rate of glycerol production from glycogen by the fat body at 6° increases by 2–2.5 times above its rate at 25° (Ziegler, R. and Wyatt, G., 1975) and decreases when shifted back to 25°.

Cold activation of fat body phosphorylase occurs by a different mode of action than does trauma activation. Fat body held *in vitro* at 25° shows the trauma activation of phosphorylase (40%) that reverts to lower activity (7–10%) within an hour, whereas phosphorylase activity persists at high levels (80%) in fat bodies held *in vitro* at 4° (Ziegler, R. and Wyatt, G., 1975; Ziegler, R. *et al.*, 1979). Trauma activation correlates to increased levels of fat body cAMP and phosphorylase *a* (Ziegler, R. *et al.*, 1979); no increases are noted for cAMP or cGMP during cold activation of phosphorylase. Therefore, how cold activates fat body phosphorylase remains unresolved, but it does not appear cAMP-related.

Phosphorylated glucose intermediates regulate the synthesis and degradation of glycogen in insect adipocytes. Glucose-6-phosphate stimulates glycogen synthase in *Hyalophora cecropia* fat body (Murphy, T. and Wyatt, G., 1965) and inhibits phosphorylase in the locust fat body (Applebaum, S. and Schlesinger, H., 1973). These latter reports suggest that glucose-6-phosphate levels determine whether glucose is converted to either stored or mobile carbohydrate. If ample dietary glucose is available and trehalose levels are high, then accumulating levels of glucose-6-phosphate suppress further phosphorylase degradation of glycogen stores and stimulate glycogen synthase to store the accumulating glucose. Conversely, as supplies of glucose-6-phosphate are depleted by trehalose synthesis or glycolysis, phosphorylase is activated and new phosphorylated glucose intermediates are produced by glycogen degradation for either additional trehalose formation or energy production.

3.5.2 Trehalose synthesis and its regulation

Trehalose is the major circulatory carbohydrate of

most insects and is synthesized solely by the fat body. Trehalose was first identified as the principal hemolymph sugar in 10 species of insects representing five orders (Wyatt, G. and Kalf, G., 1957). In Lepidoptera, trehalose generally ranges from 200 to 1500 mg% in the hemolymph and constitutes 90% of the circulatory carbohydrates. The trehalose biosynthetic pathway was determined in the locust fat body and consists of the conjugation of glucose-6-phosphate and uridine diphosphoglucoside (UDPG) to form trehalose-6-phosphate which is subsequently dephosphorylated to trehalose (Candy, D. and Kilby, B., 1961).

The ability of the fat body to produce trehalose may differ between growth stages in some species. For example, glucose is the primary carbohydrate in the hemolymph of larval *Phormia regina* even though trehalose can be detected within larval adipocytes (Wimer, L., 1969). Trehalose is not detected in the hemolymph until pupariation.

Hemolymph trehalose levels are regulated by activation and feedback inhibition of key enzymes in the trehalose synthetic pathway. In *H. cecropia*, trehalose inhibits fat body trehalose-6-phosphate synthase by allosteric feedback inhibition to prevent the build-up of excessive trehalose levels (Murphy, T. and Wyatt, G., 1965). Trehalose-6-phosphate synthase is the key enzyme for trehalose synthesis and is activated by Mg^{2+} which lowers its K_m for the glucose-6-phosphate substrate. Trehalose competes with glucose-6-phosphate for an allosteric site on trehalose-6-phosphate synthase, and as trehalose concentrations approach 50 mM, trehalose strongly inhibits its synthetic enzyme (Murphy, T. and Wyatt, G., 1965). Just as trehalose inhibits its formation by inactivation of the key synthetic enzyme, so too, trehalose stimulates a phosphohydrolase in *P. regina* fat body to release Pi from glucose-6-phosphate (Friedman, S., 1968). Hence, increasing concentrations of trehalose stimulate the degradation of one of the main precursors for further trehalose biosynthesis.

Trehalose synthesis is also regulated by the relative affinities of trehalose-6-phosphate synthase and glycogen synthase for their common substrate: UDPG. In *H. cecropia* fat body, the K_m values for UDPG are 0.3 mM for trehalose-6-phosphate synthase and 1.6 mM for glycogen synthase (Murphy, T. and Wyatt, G., 1965). The lower K_m of trehalose-6-phosphate synthase for UDPG indicates that, given the competition for conversion of UDPG to either trehalose or glycogen, the trehalose pathway is most active with glycogen being formed only after trehalose levels have reached the concentrations needed for feedback inhibition of trehalose-6-phosphate synthase. The relative importance of the two competing pathways is confirmed *in vivo* in *Periplaneta americana* by injection of glucose (Spring, J. *et al.*, 1977). Injected glucose is converted to trehalose during the first 10–20 min; after 20 min, glycogen synthesis predominates. During the first 20 min circulating trehalose levels increase, but thereafter they remain constant. These results demonstrate that the production of circulating carbohydrate takes precedence over carbohydrate storage. Only after the needs for circulating carbohydrate are satisfied does excess carbohydrate convert to glycogen stores.

An additional regulation of trehalose synthesis occurs by differential permeability of the adipocyte plasma membrane. In *H. cecropia* larvae, trehalose penetrates readily into adipocytes; however, in pupae, trehalose is restricted from passing through the adipocyte plasma membrane (Jungreis, A. and Wyatt, G., 1972). This impermeability to trehalose is reversed in early pharate adults. Trehalose that is synthesized is retained in the pupal adipocytes and suppresses further trehalose synthesis by feedback inhibition. Therefore, low hemolymph trehalose levels are maintained in *H. cecropia* pupae. Furthermore, Mg^{2+} reverses the feedback inhibition by trehalose on trehalose-6-phosphate synthase (Murphy, T. and Wyatt, G., 1965). The loss of Mg^{2+} from adipocytes during the larval–pupal transition enhances the feedback inhibition, and blood trehalose levels correlate with Mg^{2+} levels in the adipocytes during pupal life (Jungreis, A. *et al.*, 1974).

3.5.3 Endocrine Regulation

Hemolymph trehalose levels are regulated by a corpora cardiaca neurohormone designated, recently, as trehalagon (Steele, J., 1980; see J. Steele, vol. 8). Injections of corpora cardiaca extracts elevate hemolymph trehalose at the expense of fat body glycogen in *Blaberus discoidalis* (Bowers, W. and Friedman, S., 1963) and *P. americana* (Steele, J.,

1961, 1963). The gland extracts increase fat body phosphorylase activity (Steele, J., 1963) by activation of adenyl cyclase to form cAMP (Gäde, G., 1977; Hanaoka, K. and Takahashi, S., 1977). Adenyl cyclase activity increases by 3-fold and doubles the cAMP content in the fat body within 5 min. These data suggest that trehalagon activates fat body phosphorylase via the cAMP second messenger system to degrade glycogen and produce the intermediates for trehalose synthesis.

Not all insects have a neuroendocrine factor that stimulates the production of trehalose by the fat body. For example, there is no evidence for a neuroendocrine factor affecting fat body phosphorylase in *H. cecropia* (Wiens, A. and Gilbert, L., 1967c; Jungreis, A., 1976), nor does fat body phosphorylase of *Hyalophora cecropia* respond to extracts of glands from *P. americana* that stimulate *P. americana* phosphorylase (Wiens, A. and Gilbert, L., 1967c). However, in *Manduca sexta*, another lepidopteran, corpora cardiaca extracts elevate hemolymph trehalose in the usual dose-dependent manner (Ziegler, R., 1979).

In cockroaches, corpora cardiaca extracts activate fat body phosphorylase at any time, and a maximum of activity is reached within 10 min after injection (Wiens, A. and Gilbert, L., 1967a). Injection of 0.025 corpora cardiaca increases hemolymph trehalose by 236% within 30 min in *P. americana* and maintains it at this level for 5 h, while the fat body glycogen content declines from 5.4 mg to 0.9 mg within 1 h (Hanaoka, K. and Takahashi, S., 1976).

In Diptera, responsiveness to corpora cardiaca extracts by fat body phosphorylase is a function of the nutritional state of the fly. Corpora cardiaca extracts increase hemolymph trehalose only in *Phormia regina* that are starved for 24 h (Friedman, S., 1967). This suggests that trehalose levels are normally maintained in flies by synthesis from nutrient-derived glucose. Only after starvation depletes the nutrient glucose does the fly rely upon trehalagon to insure the maintenance of hemolymph trehalose levels by the degradation of fat body glycogen reserves.

Finally, adult *Locusta migratoria* fail to show a clear trehalagon activation of phosphorylase during the first 6 days of adult life because glycogen stores are minimal (Goldsworthy, G., 1969). After 6 days of adult life, when somatic growth is finished, corpora cardiaca extracts elevate hemolymph trehalose levels because glycogen stores have accumulated.

These results demonstrate that phosphorylase activation by trehalagon is a function of the peculiar biology and physiology of a particular insect species or order. Insects that feed frequently, such as flies, have an ample supply of diet-derived glucose for use directly as trehalose precursors. Therefore, they require trehalagon only intermittantly as a back-up system to mobilize fat body glycogen stores in cases of starvation, or if the digestive source is too slow to meet the demands of rapid trehalose consumption. In contrast, the cockroach constantly uses the endocrine system to maintain trehalose levels by glycogen degradation. Finally, in locusts, sustained flight uses lipids for energy. Therefore, as discussed previously, carbohydrates are converted to lipids and lipid stores accumulate, hence glycogen stores are small and phosphorylase activity and its regulation are generally insignificant.

In most cases trehalose levels are high in insect hemolymph in comparison to glucose levels in mammalian blood. This is probably an adaptation by insects to the open circulatory system which distributes hemolymph metabolites to sites of consumption, in part by diffusion along concentration gradients. This is inefficient relative to the closed circulatory system and directed blood flow of vertebrates. Therefore a constant, high level of circulating carbohydrates is required in insects to compensate for circulatory inefficiency and to fulfill immediate emergency energy demands in peripheral tissues (especially in leg and flight muscles). Since trehalose is an inert disaccharide it provides two molar equivalents of glucose for use in metabolism relative to one molar equivalent of osmotic effect. The normally high concentrations of trehalose are sufficient to meet immediate needs. However, when depleted by consumption or starvation, trehalagon insures a rapid recovery by trehalose to its naturally high levels through the degradation of fat body glycogen reserves.

ADKH also activates fat body phosphorylase, but it does not appear to account for all of the hypertrehalosemic activity in corpora cardiaca extracts. Purified locust ADKH increases hemolymph trehalose in cockroaches, but only at doses 10 times greater than those needed for hyperlipemia in locusts (Jones, J. *et al.*, 1977), Corpora cardiaca extracts activate fat body phosphorylase in the locust, but treatments with synthetic ADKH indicate that all of the glandular activity is not accounted for by ADKH (Gäde, G., 1981). Furthermore, extracts of corpora cardiaca storage lobes also activate locust phosphorylase but have no lipemic effect. Likewise, corpora cardiaca from *M. sexta*, which lack an ADKH effect also stimulate phosphorylase activity in the locust fat body. In locusts the hypertrehalosemic effect and the adipokinetic effect remain connected throughout purification (Holwerda, D. *et al.*, 1977a). However, in *Periplaneta americana* both physiological responses occur but are associated with different chemical factors neither of which is identical to ADKH (Holwerda, D. *et al.*, 1977b). These findings suggest that ADKH may stimulate phosphorylase to degrade glycogen, but a distinct hypertrehalosemic hormone exists in addition, in some insect species, that maintains hemolymph carbohydrates by regulation of fat body phosphorylase.

An excitation-induced hypertrehalosemic response occurs in insects. Handling or other types of excitatory stresses cause elevation of hemolymph trehalose even in head-ligated *P. americana*, due to effects on the fat body by octopamine, a biogenic amine (Downer, R., 1979). Hemolymph trehalose levels double within 15 min after injection of octopamine. Octopamine (10^{-6} M) elevates fat body cAMP within 10 min *in vitro* (Gole, J. and Downer, R., 1979). Presumably, octopamine activates phosphorylase via cAMP for degradation of glycogen and formation of trehalose in the same manner as trehalagon.

A unique situation exists regarding amino acid metabolism in *L. decemlineata.* In this beetle the corpora cardiaca increase hemolymph glucose and decrease hemolymph alanine (Weeda, E., 1981). The alanine is used by the fat body along with fatty acid-derived acetyl CoA to form proline (Weeda, E. *et al.*, 1980). This is done by transaminating alanine with ketoglutarate to form pyruvate and glutamate. The glutamate is converted to proline and released into the hemolymph for transportation to the flight muscles where it serves as a flight fuel (Weeda, E. *et al.*, 1980). The pyruvate derived by deamination of alanine is converted by malic enzyme to malate in the presence of CO_2 and $NADPH_2$, and ultimately to citrate using fatty acid-derived acetyl CoA. The citrate is further converted to ketoglutarate and transaminated to glutamate for conversion to proline.

The proline-synthesizing ability of the fat body increases by 6-fold during the first 8 days of adult life in *Leptinotarsa decemlineata.* This corresponds to the time for development of flight capacity in the flight muscles. The ability to synthesize proline is sensitive to a corpora cardiaca extract which concomitantly decreases the hemolymph alanine (Weeda, E., 1981). Synthetic locust ADKH also exhibits this effect. Furthermore, synthetic ADKH and extracts of both beetle and locust corpora cardiaca stimulate glucose release from the beetle fat body, whereas cockroach gland extracts do not. These results suggest that, as in the locust, ADKH controls fat body mobilization of metabolites that are used for flight in the Colorado potato beetle.

3.6 Uric acid

Urocytes are cells, associated with the fat body, that store uric acid — the principal end product of nitrogen metabolism in many insects. Presumably uric acid is stored in the urocytes as a form of excretion. Urocytes appear to be degenerate cells lacking distinct cytoplasmic organelles and are located centrally in fat body lobes relative to the more peripheral adipocytes. In cockroach tissue the urocytes are associated with mycetocytes which may assist in remobilizing the stored nitrogen (see section on symbiotes).

Uric acid formation and degradation occur readily and with great variability in the fat body. Uric acid stores increase in the fat body after starvation, but the addition of xanthine, the immediate precursor for uric acid, to the diet of fed *P. americana* does not stimulate urate accumulation (Anderson, A. and Patton, R., 1955). In both *Bombyx mori* and *Aldrichna grahami* (blowfly), increased dietary protein stimulates both fat body xanthine dehydrogenase activity and urate formation (Ito, T. and Mukaiyama, F., 1964; Huynh, Q. *et al.*, 1979). However, dietary xanthine does not increase xanthine dehydrogenase activity in *A. grahami*; rather, additions of amino acids related to purine formation, or additions of adenine or hypoxanthine, increase xanthine dehydrogenase activity by up to 3 times (Huynh, Q. *et al.*, 1979). In *L. maderae*, [^{14}C]adenine is converted to [^{14}C]urate and stored in the fat body where it accumulates progressively in a protein-bound form with little turnover (Hopkins, T. and Lofgren, P., 1968). Only 2% of the initial ^{14}C label from [^{14}C]adenine is excreted in the feces after 12 days. This indicates that the urate stores are quite stable.

Metamorphosis affects the capacity for uric acid synthesis and storage in the fat body. Lafont, R. and Pennetier, J. (1975) find that uric acid stores in the pupae of *Pieris brassicae* butterflies are converted to allantoic acid just before imaginal eclosion, and it is allantoic acid that constitutes the major end-product for purine metabolism. At the transition from feeding to wandering in *Manduca sexta* larvae, there is also a transition from excretion of uric acid by the Malpighian tubules to storage in the fat body (Buckner, J. and Caldwell, J., 1980). The fat body stores uric acid during the first half of pupation in *B. mori*, but transfers it to the rectum for excretion in the meconium midway through pupation (Tojo, S., 1971). In *Hyalophora cecropia*, urate storage in the fat body at the larval–pupal transition is accompanied by storage of K^+ derived from the integument (Jungreis, A. and Tojo, S., 1973). The fat body uric acid probably results from uptake of potassium urate, although uric acid stores accumulate separately from potassium.

In general, urate storage may regulate osmolarity by sequestration of Na^+ and K^+ as stored urate salts in *P. americana* (Mullins, D. and Cochran, D., 1976). This lowers the need for the excretion of these cations along with concomitant water loss. By alternately sequestering or releasing Na^+ or K^+ from the fat body urate stores, hemolymph osmolarity can be regulated and stabilized (Mullins, D. and Cochran, D., 1974).

Uric acid stores appear responsive to the endocrine status of the insect. Bodenstein, D. (1953) reported that uric acid disappears from the fat body of *Periplaneta americana* after allatocardiacectomy. Corpora cardiaca implants reverse this effect, but corpora allata implants are without effect. These findings suggest neuroendocrine regulations for either protein–amino acid metabolism or uric acid metabolism. In *M. sexta*, uric acid deposits increase in the fat body during the last larval instar (Williams-Boyce, P. and Jungreis, A., 1980), but the capacity for urate synthesis decreases by 85% between the late feeding larva and the newly formed pupa. This decrease in urate synthetic ability can be prevented by thoracic ligation of the larva before the increase in ecdysteroid titer associated with the larval–pupal ecdysis. The decrease in urate synthesis occurs in ligated larval abdomens after injection of 20-OH-ecdysone. The presence of the hormone causes urate synthesis to decline as it would normally at the time for pupation. This effect may be a result of adipocyte metamorphosis rather than a direct effect by the hormone on the enzymes for uric acid synthesis.

3.7 Intracellular symbiotes and metabolism

Associated with the physiological role of urate stores are the symbiotes found in fat body mycetocytes and their role in metabolism. The best-studied fat body symbiotes are bacteria-like organisms of cockroaches referred to generally as bacteroids. Bacteroids in the fat bodies of all cockroach species are of one type identified as *Blattobacterium cuenoti*, and they possess features of both the Gram-positive and Gram-negative bacteria and of Rickettsia (Brooks, M., 1970). These organisms are situated in the mycetocytes that are located centrally in the fat body lobes and are surrounded by urocytes (see Fig. 2).

The spatial arrangement of the mycetocytes and urocytes suggests functional interactions which are confirmed by studies with aposymbiotic insects. Aposymbiotic nymphs and adult males of *Blattella germanica* contain significantly more fat body uric acid than do normal animals (Valovage, W. and Brooks, M., 1979). Newly emerged aposymbiotic adult females contain more fat body uric acid than normal females, but this situation reverses as normal adult females accumulate uric acid faster than aposymbiotic females. These results confirm that bacteroids affect uric acid stores, and indicate a causal relationship for the cellular arrangements within the fat body lobes.

Fat body symbiotes affect the growth and reproduction of their host. Female *B. germanica* exposed to high temperatures or fed antibiotics are incapable of transmitting their fat body symbiotes transovarially and produce aposymbiotic nymphs

(Brooks, M. and Richards, A., 1955). Aposymbiotic nymphs are unable to grow unless brewer's yeast is added to their diet; then they grow slowly and require 2 to 3 times longer to reach maturity. Adult aposymbiotic insects reproduce poorly in both sexes, and females produce abortive oothecae. Implants of normal fat body grow in aposymbiotic nymphs and improve the growth of the host insect (Brooks, M. and Richards, A., 1956); however, treatments of aposymbiotic nymphs with suspensions of free bacteroids show that the bacteroids do not invade mycetocytes or promote nymphal growth. Aposymbiosis also hinders reproduction in *Blaberus craniifer* (Garthe, W. and Elliott, M., 1971). Two out of three protein bands diminish in the hemolymph of aposymbiotic females. One of the bands is the female-specific protein, and although synthesis of the missing proteins recovers after 2 weeks, resulting oothecae are distorted, sterile and expelled early. Nymphs that emerge often die. Therefore, depletion of the bacteroids somehow reduces fat body protein synthesis contributing to reproduction.

Bacteroid symbiotes affect uric acid metabolism in nymphs. Adipocytes have low guanase activity for the conversion of guanine to xanthine for uric acid synthesis in nymphal *Leucophaea maderae* (Pierre, L., 1965). In addition, nymphal adipoctyes have low urate oxidase activity so that uric acid cannot be degraded to other end-products of nitrogen metabolism (Pierre, L., 1964). In contrast to the nymphs, both guanase and urate oxidase activities are high in adipocytes of adult *L. maderae* and in bacteroid symbiotes in the mycetocytes (Pierre, L., 1964, 1965). Nymphs may rely on the symbiote enzymes for urate metabolism since heavy deposits of urates are found in the fat body of aposymbiotic nymphs (Pierre, L., 1964). Symbiotes apparently also assist their host insect by recovering the nitrogen stored in the urates. Cultures of isolated bacteroids are reported to degrade uric acid through allantoin, allantoic acid, urea, glyoxylic acid and finally to pyruvate for entry into the TCA cycle (Donnellan, J. and Kilby, B., 1967). Thus, the urate nitrogen may be recovered back into the general metabolism. However, Brooks, M. (1970) questions the successful isolation of bacteroids and casts doubt on the results from studies reporting their use.

Uric acid is generally accepted as the main excretory product of metabolism in terrestrial insects. However, some doubt has been thrown on this doctrine by the finding that it is ammonia that is the most prevalent nitrogenous component in the excreta of *P. americana*, with no uric acid being present (Mullins, D. and Cochran, D., 1972, 1973). Consequently, the finding by Pierre, L. (1964) that low urate oxidase activity in adipocytes of nymphal *L. maderae* correlates to large fat body urate deposits lends credence to the concept that uric acid is not always a nitrogenous waste product. Rather, uric acid may be a nitrogen storage form that can be recovered by the symbiotes for reuse in growth processes of the host insect. In support of this, Mullins, D. and Cochran, D. (1975) demonstrate that uric acid is consumed more rapidly as a nitrogen source when *P. americana* is fed a diet with a negative nitrogen balance. Females fed a negative nitrogen diet mobilize urate stores faster than males, and a portion of the urate nitrogen is used for egg formation and is recovered in the ootheca. Since ammonia is the major excretory product of this insect, the urate nitrogen can be recovered and used in protein synthesis for growth and reproduction.

Alternative roles for bacteroids are unclear, although it is believed that they may contribute to the general metabolism of the fat body. Several *in vitro* studies claim to have elucidated the metabolic action of bacteroids (Pierre, L., 1962, 1964; Donnellan, J. and Kilby, B., 1967); however, Brooks, M. (1970) criticizes the techniques used to obtain the bacteroid cultures, and indicates that it is highly unlikely that the cultured organisms were derived from bacteroids. Therefore, studies with aposymbiotic insects may be useful for identifying biochemical capacities of bacteroids and their metabolic role *in vivo*.

Comparisons of normal and aposymbiotic insects suggest that isocitrate dehydrogenase, malate dehydrogenase, urate oxidase and guanine aminohydrolase are enzymes associated with symbiotes (Dubowsky, N. and Pierre, L., 1967; Tarver, R. and Pierre, L., 1967; Laudani, U. *et al.*, 1974). Also, succinoxidase and cytochrome *c* oxidase activities are greater in fat body of *Nauphoeta cinerea* when bacteroids are present as compared to aposymbiotic tissue (Laudani, U., *et al.*, 1974). This suggests that bacteroids have mitochondrial properties which may contribute to the total energy levels in the tissue. Some vitamins may be supplied by bacteroids since normal fat body homogenates synthesize ascorbic acid but not homogenates of aposymbiotic tissue (Pierre, L., 1962). Finally, evidence from aposymbiotic *Periplaneta americana* indicates that the micro-organisms are instrumental in ring cleavage of phenylalanine (Murdock, L. *et al.*, 1970). In contrast to the previous results, malic enzyme appears to be an adipocytic enzyme unrelated to the presence or absence of bacteroids (Dubowsky, N. and Pierre, L., 1966).

Although these studies suggest that bacteroids have a significant metabolic role, several cautions should be noted. It is quite likely that the bacteroids exhibit a variety of enzymic activities; however, in view of the broad metabolic functions of the adipocytes it is also likely that many of the enzymes ascribed primarily to symbiotes may also reside in adipocytes at levels greater than those determined. For example, if Brookes, M. (1970) is correct, then the micro-organisms in culture studies were not *Blattobacterium cuenoti*, and the results of culture studies are erroneous. This means that the only currently valid basis for assessing bacteroidal enzymes is by comparing enzyme activities in normal and aposymbiotic fat bodies. In this case it seems reasonable to question the side-effects on the adipocytic enzymes of the harsh procedures used to render the fat body aposymbiotic. These procedures may damage adipocyte enzymes and processes as well as destroying the bacteroids, thus making the affected process appear bacteroid-related.

4 CONCLUSIONS

The purpose of this review is to provide to the interested reader an indication of the breadth of the fat body's role in the normal biochemistry of the insect. Some areas of fat body biochemistry, for example detoxification mechanisms, were omitted intentionally since they will surely be covered in more depth by other speciality chapters in this series. Obviously, a focus on normal biochemistry was sufficient.

Although many studies on insect biochemistry involve the fat body, it is only recently that insect research has begun to take advantage of the unique features of the fat body and to design experiments using the fat body as a model based on these features. As an example are the studies on endocrine regulation of vitellogenin synthesis. The regulation of vitellogenin synthesis by JH promises to give the first insight into the mode of action of JH, and future studies on neuroendocrine regulations of metabolism will focus frequently on the fat body as a target tissue because of its central role in metabolism and homeostasis.

The fat body has advantages over vertebrate tissues as a model for studying general biochemical processes. The fat body is structurally simple, yet a metabolically complex tissue that shows the varieties of developmental and regulatory processes inherent to higher animals. Because the fat body is simple in organization it is particularly amenable to *in vitro* studies. For example, the presence of an open circulatory system in insects means that hemolymph percolates around and through the fat body lobes, and the adipocytes are arranged for maximum exposure to the circulatory fluid. Hence, suspension of fat body in a culture medium is similar in arrangement to the *in vivo* situation and permits exposure of the cells to the medium with a minimum of artifact. This is unlike vertebrate organs which require perfusion through the complex, closed circulatory system or exposure of the cells to the medium after destruction of the tissue by slicing, homogenization or cell dispersion.

Like vertebrate cells, adipocytes go through stages of growth and differentiation and are sensitive to hormonal regulations. Unlike vertebrates where phases of cell growth and differentiation are gradual and imperceptibly change from one stage to the next, the insect cells reach a stage of development, remain static, then synchronously shift to the next stage in response to endocrine stimuli as the insect molts. These shifts in growth and differentiation are easily observed externally and timed by the act of ecdysis. Hence, events of adipocyte differentiation can be precisely timed for experimentation. This easily observed synchrony is especially useful for studying endocrine regulations of developmental events *in vivo*.

The biochemical processes associated with the development, regulation and function of the fat body will be increasingly significant as they become better elucidated, not only to those interested in insects, but also to developmental and regulatory biologists. The fat body is an ideal model for research on general biochemical processes in higher eucaryotic animals.

ACKNOWLEDGEMENTS

The author wishes to thank Dr E. L. Thurston of the Texas A&M University Electron Microscopy Center for his assistance in the preparation of the micrographs, and Dr G. Bhaskaran and F. W. Plapp for their comments on the manuscript. Personal work cited by the author was supported by National Science Foundation grants PCM 74-03606 and PCM 81-03277 and National Institutes of Health grant AI-15190 and by the Texas Agricultural Experiment Station.

REFERENCES

ADAMS, T. S. and NELSON, D. R. (1969). Effect of corpus allatum and ovaries on amount of pupal and adult fat body in the housefly, *Musca domestica*. *J. Insect Physiol. 15*, 1729–1747.

AFZELIUS, B. A. (1970). Brown Adipose Tissue: Its Gross Anatomy, Histology, and Cytology. In *Brown Adipose Tissue*. Edited by O. Lindberg. Pages 1–31. American Elsevier, New York.

ANDERSON, A. D. and PATTON, R. L. (1955). *In vitro* studies of uric acid synthesis in insects. *J. Exp. Zool. 128*, 443–451.

APPLEBAUM, S. W. and SCHLESINGER, H. M. (1973). Regulation of locust fat body phosphorylase. *Biochem J. 135*, 37–41.

ASHIDA, M. and WYATT, G. R. (1979). Properties and activation of phosphorylase kinase from silkmoth fat body. *Insect Biochem. 9*, 403–409.

BAILEY, E. and WALKER, P. R. (1969). A comparison of the properties of the pyruvate kinases of the fat body and flight muscle of the adult male desert locust. *Biochem. J. 111*, 359–364.

BARTH, R. H., BUNYARD, P. P. and HAMILTON, T. H. (1964). RNA metabolism in pupae of the oak silkworm, *Antherea pernyi*: the effects of diapause, development and injury. *Proc. Nat. Acad. Sci. 52*, 1572–1580.

BAR-ZEV, A. and KAULENAS, M. S. (1975). The effect of β-ecdysone on *Gromphadorhina* adult female fat body at the transcriptional and translational levels. *Comp. Biochem. Physiol. (B) 51*, 355–361.

BEENAKKERS, A. M. TH. and GILBERT, L. I. (1968). The fatty acid composition of fat body and haemolymph lipids in *Hyalophora cecropia* and its relation to lipid release. *J. Insect Physiol. 14*, 481–494.

BEHAN, M. and HAGEDORN, H. H. (1978). Ultrastructural changes in the fat body of adult female *Aedes aegypti* in relationship to vitellogenin synthesis. *Cell. Tiss. Res. 186*, 499–506.

BELL, W. J. (1969). Dual role of juvenile hormone in the control of yolk formation in *Periplaneta americana*. *J. Insect Physiol. 15*, 1279–1290.

BELL, W. J. (1970). Demonstration and characterization of two vitellogenic blood proteins in *Periplaneta americana*: an immunochemical analysis. *J. Insect Physiol. 16*, 291–299.

BERRY, S. J., KRISHNAKUMARAN, A., OBERLANDER, H. and SCHNEIDERMAN, H. A. (1967). Effects of hormones and injury on RNA synthesis in Saturniid moths. *J. Insect Physiol. 13*, 1511–1537.

BHAKTHAN, N. M. G. and GILBERT, L. I. (1970). An autoradiographic and biochemical analysis of palmitate incorporation into fat body lipid. *J. Insect Physiol. 16*, 1783–1796.

BHAKTHAN, N. M. G. and GILBERT, L. I. (1972). Studies on the cytophysiology of the fat body of the American silkmoth. *Z. Zellforsch. 124*, 433–444.

BLUMENFELD, M. and SCHNEIDERMAN, H. A. (1968). Effect of juvenile hormone on the synthesis and accumulation of a sex-limited blood protein in the polyphemus silkmoth. *Biol. Bull. 135*, 466–475.

BODENSTEIN, D. (1953). Studies on the humoral mechanisms in growth and metamorphosis of the cockroach, *Periplaneta americana*. III. Humoral effects on metabolism. *J. Exp. Zool. 124*, 105–115.

BOROVSKY, D. (1981a). *In vivo* stimulation of vitellogenesis in *Aedes aegypti* with juvenile hormone, juvenile hormone analogue (ZR 515) and 20-hydroxyecdysone. *J. Insect Physiol. 27*, 371–378.

BOROVSKY, D. (1981b). Feedback regulation of vitellogenin synthesis in *Aedes aegypti* and *Aedes atropalpus*. *Insect Biochem. 11*, 207–213.

BOROVSKY, D. and VAN HANDEL, E. (1979). Does ovarian ecdysone stimulate mosquitoes to synthesize vitellogenin? *J. Insect Physiol. 25*, 861–865.

BOWERS, W. and FRIEDMAN, S. (1963). Mobilization of fat body glycogen by an extract of corpus cardiacum. *Nature 198*, 685.

BROOKES, V. J. (1969). The induction of yolk protein synthesis in the fat body of an insect, *Leucophaea maderae*, by an analog of the juvenile hormone. *Devel. Biol. 20*, 459–471.

BROOKS, M. A. (1970). Comments on the classification of intracellular symbiotes of cockroaches and a description of the species. *J. Invertebr. Pathol. 16*, 249–258.

BROOKS, M. A. and RICHARDS, A. G. (1955). Intracellular symbiosis in cockroaches. I. Production of aposymbiotic cockroaches. *Biol. Bull. 109*, 22–39.

BROOKS, M. A. and RICHARDS, A. G. (1956). Intracellular symbiosis in cockroaches. III. Re-infection of aposymbiotic cockroaches with symbiotes. *J. Exp. Zool. 132*, 447–465.

BUCKNER, J. S. and CALDWELL, J. M. (1980). Uric acid levels during last larval instar of *Manduca sexta*, an abrupt transition from excretion to storage in fat body. *J. Insect Physiol. 26*, 27–32.

BURKE, D. D., RUBIN, D. and WILLIS, J. H. (1976). Characterization of the DNA of *Hyalophora cecropia*. *J. Insect Physiol. 22*, 791–798.

BUTTERWORTH, F. M. (1972). Adipose tissue of *Drosophila melanogaster*. V. Genetic and experimental studies of an extrinsic influence on the rate of cell death in the larval fat body. *Devel. Biol. 28*, 311–325.

BUTTERWORTH, F. M. and BODENSTEIN, D. (1967). Adipose tissue of *Drosophila melanogaster*. II. The effect of the adult internal environment on growth, protein deposition, and histolysis in the larval fat body. *J. Exp. Zool. 164*, 251–266.

BUTTERWORTH, F. M., BODENSTEIN, D. and KING, R. C. (1967). Adipose tissue of *Drosophila melanogaster*. I. An experimental study of larval fat body. *J. Exp. Zool. 158*, 141–154.

BUYS, K. S. (1924). Adipose tissue in insects. *J. Morph. 38*, 485–527.

CANDY, D. J. and KILBY, B. A. (1961). The biosynthesis of trehalose in the locust fat body. *Biochem. J. 78*, 531–536.

CHANG, F. and FRIEDMAN, S. (1971). A developmental analysis of the uptake and release of lipids by the fat body of the tobacco hornworm, *Manduca sexta*. *Insect Biochem. 1*, 63–80.

CHEN, P. S. (1966). Amino acid and protein metabolism in insect development. *Adv. Insect Physiol. 3*, 53–132.

CHEN, T. T. (1980). Vitellogenin in locusts (*Locusta migratoria*): translation of vitellogenin mRNA in *Xenopus* oocytes and analysis of the polypeptide products. *Arch. Biochem. Biophys. 201*, 266–276.

CHEN, T. T., STRAHLENDORF, P. W. and WYATT, G. R. (1978). Vitellin and vitellogenin from locust (*Locusta migratoria*). Properties and post-translational modification in the fat body. *J. Biol. Chem. 253*, 5325–5331.

CHEN, T. T., COUBLE, P., ABU-HAKIMA, R. and WYATT, G. R. (1979). Juvenile hormone-controlled vitellogenin synthesis in *Locusta migratoria* fat body. Hormonal induction in vivo. *Devel. Biol. 69*, 59–72.

CHINO, H. and GILBERT, L. I. (1965a). Studies on the interconversion of carbohydrate and fatty acid in *Hyalophora cecropia*. *J. Insect Physiol. 11*, 287–295.

CHINO, H. and GILBERT, L. I. (1965b). Lipid release and transport in insects. *Biochim. Biophys. Acta 98*, 94–110.

CHINO, H., SUDO, A. and HARASHIMA, K. (1967). Isolation of diglyceride-bound lipoprotein from insect hemolymph. *Biochim. Biophys. Acta 144*, 177–179.

CHINO, H., MURAKAMI, S. and HARASHIMA, K. (1969). Diglyceride-carrying lipoproteins in insect haemolymph. Isolation, purification and properties. *Biochim. Biophys. Acta 176*, 1–26.

CHINZEI, Y. (1974). Biochemical properties of fat body DNA of the silkworm, *Bombyx mori. J. Insect Physiol. 20*, 2333–2346.

CHINZEI, Y. (1975). Biochemical evidence of DNA transport from the silk gland to the fat body of the silkworm, *Bombyx mori. J. Insect Physiol. 21*, 163–171.

CHINZEI, Y. and TOJO, S. (1972). Nucleic acid changes in the whole body and several organs of the silkworm, *Bombyx mori*, during metamorphosis. *J. Insect Physiol. 18*, 1683–1698.

CHINZEI, Y., POTTER, R. J., WHITE, B. N. and WYATT, G. R. (1980). Juvenile-hormone induced vitellogenin mRNA in the locust. *Canad. Fed. Biol. Sci. Proc. 23*, 128.

CHIPPENDALE, G. M. (1978). The Functions of Carbohydrates in Insect Life Processes. In *Biochemistry of Insects*. Edited by M. Rockstein. Pages 1–55. Academic Press, New York.

CLEMENTS, A. N. (1963). *The Physiology of Mosquitoes*. Pergamon Press, New York.

COLLINS, J. V. (1969). The hormonal control of fat body development in *Calpodes ethlius* (Lepidoptera, Hesperiidae). *J. Insect Physiol. 15*, 341–352.

COLLINS, J. V. (1974). Hormonal control of protein sequestration in the fat body of *Calpodes ethlius* Stoll. *Canad. J. Zool. 52*, 639–642.

COLLINS, J. V. (1975). Secretion and uptake of ^{14}C proteins by fat body of *Calpodes ethlius* Stoll. (Lepidoptera, Hesperiidae). *Differentiation 3*, 143–148.

CONGOTE, L. F., SEKERIS, C. E. and KARLSON, P. (1969). On the mechanism of hormone action. XIII. Stimulating effects of ecdysone, juvenile hormone, and ions on RNA synthesis in fat body cell nuclei from *Calliphora erythrocephala* isolated by a filtration technique. *Exp. Cell. Res. 56*, 338–346.

CONGOTE, L. F., SEKERIS, C. E. and KARLSON, P. (1970). On the mechanism of hormone action. XVIII. Alterations of the nature of RNA synthesized in isolated fat body cell nuclei as a result of ecdysone and juvenile hormone action. *Z. Naturforsch. 25b*, 279–284.

COOK, B. J. and EDDINGTON, L. C. (1967). The release of triglycerides and free fatty acids from the fat body of the cockroach, *Periplaneta americana. J. Insect Physiol. 13*, 1361–1372.

COUBLE, P., CHEN, T. T. and WYATT, G. R. (1979). Juvenile hormone-controlled vitellogenin synthesis in *Locusta migratoria* fat body: cytological development. *J. Insect Physiol. 25*, 327–337.

COUPLAND, R. E. (1957). Observations on the normal histology and histochemistry of the fat body of the locust (*Schistocera gregaria*). *J. Exp. Biol. 34*, 290–296.

DEAN, R. L. (1978). The induction of autophagy in isolated insect fat body by β-ecdysone. *J. Insect Physiol. 24*, 439–447.

DEAN, R. L., BOLLENBACHER, W. E., LOCKE, M., SMITH, S. L. and GILBERT, L. I. (1980). Haemolymph ecdysteroid levels and cellular events in the intermoult/moult sequence of *Calpodes ethlius. J. Insect Physiol. 26*, 267–280.

DELLA-CIOPPA, G. and ENGELMANN, F. (1980). Juvenile-hormone-stimulated proliferation of endoplasmic reticulum in fat body cells of a vitellogenic insect, *Leucophaea maderae* (Blattaria). *Biochem. Biophys. Res. Commun. 93*, 825–832.

DONNELLAN, J. F. and KILBY, B. A. (1967). Uric acid metabolism by symbiotic bacteria from the fat body of *Periplaneta americana. Comp. Biochem. Physiol. 22*, 235–252.

DORTLAND, J. F. (1978). Synthesis of vitellogenins and diapause proteins by the fat body of *Leptinotarsa*, as a function of photoperiod. *Physiol. Ent. 3*, 281–288.

DORTLAND, J. F. (1979). The hormonal control of vitellogenin synthesis in the fat body of the female Colorado potato beetle. *Gen. Comp. Endocr. 38*, 332–344.

DORTLAND, J. F. and DEKORT, C. A. D. (1978). Protein synthesis and storage in the fat body of the Colorado potato beetle, *Leptinotarsa decemlineata. Insect Biochem. 8*, 93–98.

DORTLAND, J. F. and ESCH, TH. H. (1979). A fine structural survey of the development of the adult fat body of *Leptinotarsa decemlineata. Cell Tiss. Res. 201*, 423–430.

DOWNER, R. G. H. (1978). Functional Role of Lipids in Insects. In *Biochemistry of Insects*. Edited by M. Rockstein. Pages 57–92. Academic Press, New York.

DOWNER, R. G. H. (1979). Induction of hypertrehalosemia by excitation in *Periplaneta americana. J. Insect Physiol. 25*, 59–63.

DOWNER, R. G. H. and STEELE, J. E. (1972). Hormonal stimulation of lipid transport in the American cockroach. *Periplaneta americana. Gen. Comp. Endocr. 19*, 259–265.

DUBOWSKY, N. and PIERRE, L. L. (1966). Malic enzyme activity in the fat body of the cockroach *Leucophaea maderae. Nature 210*, 1293–1294.

DUBOWSKY, N. and PIERRE, L. L. (1967). Activity of isocitric dehydrogenase in the fat bodies of the cockroach, *Leucophaea maderae. Nature. 213*, 209–210.

ELLIOTT, R. H. and GILLOTT, C. (1978). The neuroendocrine control of protein metabolism in the migratory grasshopper, *Melanoplus sanguinipes. J. Insect Physiol. 24*, 119–126.

ELLIOTT, R. H. and GILLOTT, C. (1979). An electrophoretic study of proteins of the ovary, fat body, and haemolymph in the migratory grasshopper, *Melanoplus sanguinipes. J. Insect Physiol. 25*, 405–410.

EMMERICH, H. and HARTMANN, R. (1973). A carrier lipoprotein for juvenile hormone in the haemolymph of *Locusta migratoria. J. Insect Physiol. 19*, 1663–1675.

ENGLEMANN, F. (1969). Female specific protein: biosynthesis controlled by corpus allatum in *Leucophaea maderae. Science 165*, 407–409.

ENGLEMANN, F. (1971). Juvenile-hormone-controlled synthesis of female-specific protein in the cockroach *Leucophaea maderae. Arch. Biochem. Biophys. 145*, 439–447.

ENGLEMANN, F. (1972). Juvenile hormone-induced RNA and specific protein synthesis in an adult insect. *Gen. Comp. Endocr. Suppl. 3*, 168–173.

ENGLEMANN, F. (1974). Polyribosomal and microsomal profiles of fat body homogenates from reproductively active and inactive females of the cockroach *Leucophaea maderae. Insect Biochem. 4*, 345–354.

ENGLEMANN, F. (1978). Synthesis of vitellogenin after long-term ovariectomy in a cockroach. *Insect Biochem. 8*, 149–154.

ENGLEMANN, F. (1979). Insect vitellogenin: identification, biosynthesis and role in vitellogenesis. *Adv. Insect Physiol. 14*, 49–108.

ENGLEMANN, F. (1980). Endocrine Control of Vitellogenin Synthesis. In *Insect Biology in the Future*. Edited by M. Locke and D. S. Smith. Pages 311–324. Academic Press, New York.

ENGLEMANN, F. and BARAJAS, L. (1975). Ribosome-membrane association in fat body tissues from reproductively active females of *Leucophaea maderae. Exp. Cell Res. 92*, 102–110.

ENGLEMANN, F. and MÜLLER, H. P. (1966). Fat body respiration as influenced by previously isolated corpora cardiaca. *Naturwissenschaften 53*, 388–389.

ENGLEMANN, F., FRIEDEL, T. and LADDUWAHETTY, M. (1976). The native vitellogenin of the cockroach *Leucophaea maderae. Insect Biochem. 6*, 211–220.

EVANS, J. J. T. (1967). Development and ultrastructure of the fat body cells and oenocytes of the Queensland fruit fly, *Dacus tryoni* (Frogg.). *Z. Zellforsch. Mik. Anat. 81*, 49–61.

FALLON, A. M., HAGEDORN, H. H., WYATT, G. R. and LAUFER, H. (1974). Activation of vitellogenin synthesis in the mosquito *Aedes aegypti* by ecdysone. *J. Insect Physiol. 20*, 1815–1823.

FILBURN, C. R. and WYATT, G. R. (1976). Adenylate and guanylate cyclases of cecropia silkmoth fat body. *J. Insect Physiol. 22*, 1635–1640.

FLANAGAN, T. R. and HAGEDORN, H. H. (1977). Vitellogenin synthesis in the mosquito: the role of juvenile hormone in the development of responsiveness to ecdysone. *Physiol. Ent. 2*, 173–178.

FONG, W.-F. and FUCHS, M. S. (1976). The differential effect of RNA synthesis inhibitors on ecdysterone induced ovarian development in mosquitoes. *J. Insect Physiol. 22*, 1493–1500.

FRAENKEL, G. and HSIAO, C. (1968). Morphological and endocrinological aspects of pupal diapause in a fleshfly, *Sarcophaga argyrostoma. J. Insect Physiol. 14*, 707–718.

FRIEDMAN, S. (1967). The control of trehalose synthesis in the blowfly, *Phormia regina* Meig. *J. Insect Physiol. 13*, 397–405.

FRIEDMAN, S. (1968). Trehalose regulation of glucose-6-phosphate hydrolysis in blowfly extracts. *Science. 159*, 110–111.

FRIEDMAN, S. (1970). Metabolism of carbohydrates in insects. *Chem. Zool. 5*, 167–197.

FRIEDMAN, S. (1978). Trehalose regulation, one aspect of metabolic homeostasis. *Ann. Rev. Ent. 23*, 389–407.

FUCHS, M. S. and FONG, W.-F. (1976). Inhibition of blood digestion by α-amanitin and actinomycin D and its effect on ovarian development in *Aedes aegypti*. *J. Insect Physiol. 22*, 465–471.

FUCHS, M. S., SCHLAEGER, D. A. and SHROYER, C. (1979). α-Ecdysone does not induce ovarian development in adult *Aedes aegypti*. *J. Exp. Zool. 207*, 153–159.

GÄDE, G. (1977). Effect of corpus cardiacum extract on cyclic AMP concentration in the fat body of *Periplaneta americana*. *Zool. Jb. Physiol. 81*, 245–249.

GÄDE, G. (1981). Activation of fat body glycogen phosphorylase in *Locusta migratoria* by corpus cardiacum extract and synthetic adipokinetic hormone. *J. Insect Physiol. 27*, 155–161.

GÄDE, G. and BEENAKKERS, A. M. TH. (1977). Adipokinetic hormone-induced lipid mobilization and cyclic AMP accumulation in the fat body of *Locusta migratoria* during development. *Gen. Comp. Endocr. 32*, 481–487.

GÄDE, G. and HOLWERDA, D. A. (1976). Involvement of adenosine 3′:5′-cyclic monophosphate in lipid mobilization in *Locusta migratoria*. *Insect Biochem. 6*, 535–540.

GARTHE, W. A. and ELLIOTT, M. W. (1971). Role of intracellular symbionts in the fat body of cockroaches: influence on hemolymph proteins. *Experientia 27*, 593.

GELPERIN, A. (1966). Control of crop emptying in the blowfly. *J. Insect Physiol. 12*, 331–345.

GILBERT, L. I. (1967a). Lipid metabolism and function in insects. *Adv. Insect Physiol. 4*, 69–211.

GILBERT, L. I. (1967b). Changes in lipid content during the reproductive cycle of *Leucophaea maderae* and effects of the juvenile hormone on lipid metabolism *in vitro*. *Comp. Biochem. Physiol. 21*, 237–257.

GOLDSWORTHY, G. J. (1969). Hyperglycaemic factors from the corpus cardiacum of *Locusta migratoria*. *J. Insect Physiol. 15*, 2131–2140.

GOLDWORTHY, G. J., MORDUE, W. and GUTHKELCH, J. (1972). Studies on insect adipokinetic hormones. *Gen. Comp. Endocr. 18*, 545–551.

GOLE, J. W. D. and DOWNER, R. G. H. (1979). Elevation of adenosine 3′,5′-monophosphate by octopamine in fat body of the American cockroach, *Periplaneta americana L. Comp. Biochem. Physiol.* (C). *64*, 223–226.

GRULA, M. A. and WEAVER, R. F. (1981). An improved method for isolation of *Heliothis zea* DNA-dependent RNA polymerase: separation and characterization of a form III RNA polymerase activity. *Insect Biochem. 11*, 149–154.

HAGEDORN, H. H. and JUDSON, C. L. (1972). Purification and site of synthesis of *Aedes aegypti* yolk protein. *J. Exp. Zool. 182*, 367–377.

HAGEDORN, H. H., FALLON, A. M. and LAUFER, H. (1973). Vitellogenin synthesis by the fat body of the mosquito *Aedes aegypti*: evidence for transcriptional control. *Devel. Biol. 31*, 285–294.

HAGEDORN, H. H., O'CONNOR, J. D., FUCHS, M. S., SAGE, B., SCHLAEGER, D. A. and BOHM, M. K. (1975). The ovary as a source of α-ecdysone in an adult mosquito. *Proc. Nat. Acad. Sci. 72*, 3255–3259.

HAGEDORN, H. and KUNKEL, J. G. (1979). Vitellogenin and vitellin in insects. *Ann. Rev. Ent. 24*, 475–505.

HAMMOCK, B., NOWOCK, J., GOODMAN, W., STAMOUDIS, V. and GILBERT, L. I. (1975). The influence of hemolymph-binding protein on juvenile hormone stability and distribution in *Manduca sexta* fat body and imaginal discs *in vitro*. *Mol. Cell. Endocr. 3*, 167–184.

HANAOKA, K. and TAKAHASHI, S. Y. (1976). Effect of a hyperglycemic factor on haemolymph trehalose and fat body carbohydrates in the American cockroach. *Insect Biochem. 6*, 621–625.

HANAOKA, K. and TAKAHASHI, S. Y. (1977). Adenylate cyclase system and the hyperglycemic factor in the cockroach, *Periplaneta americana*. *Insect Biochem. 7*, 95–99.

HANAOKA, K. and HAGEDORN, H. H. (1980). Brain Hormone Control of Ecdysone Secretion by the Ovary in a Mosquito. In *Progress in Ecdysone Research*. Edited by J. A. Hoffmann. Pages 467–480. American Elsevier, New York.

HARRY, P., PINES, M. and APPLEBAUM, S. W. (1979). Changes in the pattern of secretion of locust female diglyceride-carrying lipoprotein and vitellogenin by the fat body *in vitro*, during oocyte development. *Comp. Biochem. Physiol. (B) 63*, 287–293.

HARTMANN, R., WOLF, W. and LOHER, W. (1972). The influence of the endocrine system on reproduction behavior and development in grasshoppers. *Gen. Comp. Endocr. (Suppl.) 3*, 518–528.

HAYES, T. K. and KEELEY, L. L. (1981). Cytochromogenic factor: a newly-discovered neuroendocrine agent stimulating mitochondrial cytochrome synthesis in the insect fat body. *Gen. Comp. Endocr. 45*, 115–124.

HEARFIELD, A. H. and KILBY, B. A. (1958). Enzymes of the tricarboxylic acid cycle and cytochrome oxidase in the fat body of the desert locust. *Nature 181*, 546–547.

HILL, L. (1965). The incorporation of C^{14}-glycine into the proteins of the fat body of the desert locust during ovarian development. *J. Insect Physiol. 11*, 1605–1615.

HILL, L. (1972). Hormones and the control of metabolism in insects. *Gen. Comp. Endocr. Suppl. 3*, 174–183.

HILL, L. and IZATT, M. E. G. (1974). The relationships between corpora allata and fat body and haemolymph lipids in the adult female desert locust. *J. Insect Physiol. 20*, 2143–2156.

HILL, L., IZATT, M. E. G., HORNE J. A. and BAILEY, E. (1972). Factors affecting concentrations of acetoacetate and D-3-hydroxybutyrate in haemolymph and tissues of the adult desert locust. *J. Insect Physiol. 18*, 1265–1285.

HOFFMAN, A. G. D. and DOWNER, R. G. H. (1979). End product specificity of triacylglycerol lipases from intestine, fat body, muscle and haemolymph of the American cockroach, *Periplaneta americana L. Lipids 14*, 893–899.

HOLWERDA, D. A., VANDOORN, J. and BEENAKKERS, A. M. T. (1977a). Characterization of the adipokinetic and hyperglycaemic substances from the locust corpus cardiacum. *Insect Biochem. 7*, 151–157.

HOLWERDA, D. A., WEEDA, E. and VANDOORN, J. M. (1977b). Separation of the hyperglycemic and adipokinetic factors from the cockroach corpus cardiacum. *Insect Biochem. 7*, 477–481.

HOPKINS, T. L. and LOFGREN, P. A. (1968). Adenine metabolism and urate storage in the cockroach, *Leucophaea maderae*. *J. Insect Physiol. 14*, 1803–1814.

HUYBRECHTS, R. AND DELOOF, A. (1977). Induction of vitellogenin synthesis in male *Sarcophaga bullata* by ecdysterone. *J. Insect Physiol. 23*, 1359–1362.

HUYNH, Q. K., WADANO, A. and MIURA, K. (1979). Studies on nitrogen metabolism in insects: regulation mechanisms of xanthine dehydrogenase in the blowfly, *Aldrichina grahami*, (Diptera: Calliphoridae). *Insect Biochem. 9*, 287–292.

ISHIZAKI, H. (1965). Electron microscopic study of changes in the subcellular organization during metamorphosis of the fat-body cell of *Philosamia cynthia ricini* (Lepidoptera). *J. Insect Physiol. 11*, 845–855.

ITO, T. and MUKAIYAMA F. (1964). Relationship between protein content of diets and xanthine oxidase activity in the silkworm, *Bombyx mori* L. *J. Insect Physiol. 10*, 789–796.

JANDA, V. JR. and SLÁMA, K. (1965). Uber den Einfluss von Hormonen auf den Glykogen — Fett-und Stickstoffmetabolismus bei Adulten *Pyrrhocoris apterus* L. (Hemiptera) *Zool. Jb. (Physiol.) 71*, 345–358.

JONES, J., STONE, J. V. and MORDUE, W. (1977). The hyperglycaemic activity of locust adipokinetic hormone. *Physiol. Ent.* 2, 185–187.

JONES, G., WING, K. D., JONES, D. and HAMMOCK, B. D. (1981). The source and action of head factors regulating juvenile hormone esterase in larvae of the cabbage looper, *Trichoplusia ni*. *J. Insect Physiol. 27*, 85–91.

JOWETT, T. and POSTLETHWAIT, J. H. (1980). The regulation of yolk polypeptide synthesis in *Drosophila* ovaries and fat body by 20-hydroxyecdysone and a juvenile hormone analog. *Devel. Biol. 80*, 225–234.

JUNGREIS, A. M. (1976). Regulation of *Hyalophora cecropia* fat body hexokinase by hexose phosphates common to the pathway of glycolysis, glycogen and trehalose synthesis. *Comp. Biochem. Physiol, (B) 53*, 405–413.

JUNGREIS, A. M. and TOJO, S. (1973). Potassium and uric acid content in tissues of the silkmoth *Hyalophora cecropia*. *Amer. J. Physiol. 224*, 21–30.

JUNGREIS, A. M. and WYATT, G. R. (1972). Sugar release and penetration in insect fat body: relations to regulation of haemolymph trehalose in developing stages of *Hyalophora cecropia*. *Biol. Bull. 143*, 367–391.

JUNGREIS, A. M., JATLOW, P. and WYATT, G. R. (1974). Regulation of trehalose synthesis in the silkmoth *Hyalophora cecropia*: the role of magnesium in the fat body. *J. Exp. Zool. 187*, 41–45.

KACZOR, W. J. and HAGEDORN, H. H. (1980). The effects of α-amanitin and cordycepin on vitellogenin synthesis by mosquito fat body. *J. Exp. Zool. 214*, 229–233.

KEELEY, L. L. (1971). Endocrine effects on the biochemical properties of fat body mitochondria from the cockroach *Blaberus discoidalis*. *J. Insect Physiol. 17*, 1501–1515.

KEELEY, L. L. (1972). Biogenesis of mitochondria: neuroendocrine effects on the development of respiratory functions in fat body mitochondria of the cockroach *Blaberus discoidalis*. *Arch. Biochem. Biophys. 153*, 8–15.

KEELEY, L. L. (1977). Developmental and endocrine regulation of cytochrome levels in fat body mitochondria of the cockroach, *Blaberus discoidalis*. *Insect Biochem. 7*, 297–301.

KEELEY, L. L. (1978a). Endocrine regulation of fat body development and function. *Ann. Rev. Entomol.* 23, 329–352.

KEELEY, L. L. (1978b). Development and endocrine regulation of mitochondrial cytochrome biosynthesis in the insect fat body. δ-[^{14}C]Aminolevulinic acid incorporation. *Arch. Biochem. Biophys. 187*, 87–95.

KEELEY, L. L. (1981). Neuroendocrine Regulation of Mitochondrial Development and Function in the Insect Fat Body. In *Energy Metabolism and Its Regulation in Insects*. Edited by R. G. H. Downer. Pages 207–237. Plenum Press, New York.

KEELEY, L. L. and FRIEDMAN, S. (1967). Corpus cardiacum as a metabolic regulator in *Blaberus discoidalis* Serville (Blattidae). Long-term effects of cardiacectomy on whole body and tissue respiration and trophic metabolism. *Gen. Comp. Endocr. 8*, 129–134.

KELLY, T. J., FUCHS, M. S. and KANG, S.-H. (1981). Induction of ovarian development in autogenous *Aedes atropalpus* by juvenile hormone and 20-hydroxyecdysone. *Int. J. Invertebr. Reprod. 3*, 101–112.

KILBY, B. A. (1963). The biochemistry of the insect fat body. *Adv. Insect Physiol. 1*, 112–174.

KILBY, B. A. (1965). Intermediary Metabolism and the Insect Fat Body. In *Aspects of Insect Biochemistry*. Edited by T. W. Goodwin. Pages 39–48. Academic Press, New York.

KOEPPE, J. and OFENGAND, J. (1976). Juvenile hormone-induced biosynthesis of vitellogenin in *Leucophaea maderae*. *Arch. Biochem. Biophys. 173*, 100–107.

KOOLMAN, J. (1980). Ecdysteroids in the Blowfly, *Calliphora vicina*. In *Progress in Ecdysone Research*. Edited by J. A. Hoffmann. Pages 187–209. North-Holland, New York.

KRAMER, K. J., DUNN, P. E., PETERSON, R. C., SEBALLOS, H. L., SANBURG, L. L. and LAW, J. H. (1976). Purification and characterization of the carrier protein for juvenile hormone from the hemolymph of the tobacco hornworm *Manduca sexta* Johannson (Lepidoptera: Sphingidae). *J. Biol. Chem. 251*, 4979–4985.

KRAMER, S. J., MUNDALL, E. C. and LAW, J. H. (1980). Purification and properties of manducin, an amino acid storage protein of the haemolymph of larval and pupal *Manduca sexta*. *Insect Biochem. 10*, 279–288.

KRISHNAKUMARAN, A., BERRY, S. J., OBERLANDER, H. and SCHNEIDERMAN, H. A. (1967). Nucleic acid synthesis during insect development — II. Control of DNA synthesis in the cecropia silkworm and other saturniid moths. *J. Insect Physiol. 13*, 1–57.

KULKARNI, A. P., SMITH, E. and HODGSON, E. (1971). The phospholipids of *Manduca sexta* tissues and the incorporation, *in vivo*, of ethanolamine, choline, and inorganic phosphate. *Insect Biochem. 1*, 348–362.

KUNKEL, J. G. (1981). A Minimal Model of Metamorphosis: Fat Body Competence to Respond to Juvenile Hormone. In *Current Topics in Insect Endocrinology and Nutrition*. Edited by G. Bhaskaran, S. Friedman and J. G. Rodriguez. Pages 107–129. Plenum Press, New York.

KUNKEL, J. G. and LAWLER, D. M. (1974). Larval-specific serum protein in the order Dictyoptera-1. Immumologic characterization in larval *Blattella germanica* and cross-reaction throughout the order. *Comp. Biochem. Physiol. (B) 47*, 697–710.

KUO, J. F., WYATT, G. R. and GREENGARD, P. (1971). Cyclic nucleotide-dependent protein kinases. IX. Partial purification and some properties of guanosine-3′5′-monophosphate-dependent and adenosine-3′5′-monophosphate-dependent protein kinases from various tissues and species of Arthropoda. *J. Biol. Chem. 246*, 7159–7167.

LAFONT, R. and PENNETIER, J.-L. (1975). Uric acid metabolism during pupal-adult development of *Pieris brassicae*. *J. Insect. Physiol. 21*, 1323–1336.

LARSEN, W. J. (1970). Genesis of mitochondria in insect fat body. *J. Cell Biol. 47*, 373–383.

LARSEN, W. J. (1976). Cell remodeling in the fat body of an insect. *Tissue Cell, 8*, 73–90.

LAUDANI, U., FRIZZI, G. P., ROGGI, C. and MONTANI, A. (1974). The function of the endosymbiotic bacteria of Blattoidea. *Experientia 30*, 882–883.

LAUVERJAT, S. (1977). L'Evolution post-imaginale du tissu adipeux femelle de *Locusta migratoria* et son controle endocrine. *Gen. Comp. Endocr. 33*, 13–24.

LEPESANT, J.-A., KEJZLAROVA-LEPESANT, J. and GARAN, A. (1978). Ecdysone-inducible function of larval fat bodies in *Drosophila*. *Proc. Nat. Acad. Sci. 75*, 5570–5574.

LEVENBOOK, L. and BAUER, A. C. (1980). Calliphorin and soluble protein of haemolymph and tissues during larval growth and adult development of *Calliphora vicina*. *Insect Biochem. 10*, 693–701.

LINZEN, B. and WYATT, G. R. (1964). The nucleic acid content of tissues of cecropia silkmoth pupae. Relations to body size and development. *Biochem. Biophys. Acta. 87*, 188–198.

LOCKE, M. (1970). The molt/intermolt cycle in the epidermis and other tissues of an insect *Calpodes ethlius* (Lepidoptera, Hesperiidea). *Tissue Cell 2*, 197–223.

LOCKE, M. (1980). The Cell Biology of Fat Body Development. In *Insect Biology in the Future*. Edited by M. Locke and D. S. Smith. Pages 227–252. Academic Press, New York.

LOCKE, M. and COLLINS, J. V. (1965). The structure and formation of protein granules in the fat body of an insect. *J. Cell Biol. 26*, 857–884.

LOCKE, M. and COLLINS, J. V. (1968). Protein uptake into multivesicular bodies and storage granules in the fat body of an insect. *J. Cell Biol. 36*, 453–483.

DELOOF, A. and LAGASSE, A. (1970). Juvenile hormone and the ultrastructural properties of the fat body of the adult Colorado beetle, *Leptinotarsa decemlineata* Say. *Z. Zellforsch. 106*, 439–450.

LÜSCHER, M. (1968). Hormonal control of respiration and protein synthesis in the fat body of the cockroach *Nauphoeta cinerea* during oocyte growth. *J. Insect Physiol. 14*, 499–511.

LÜSCHER, M. and LEUTHOLD, R. (1965). Uber die hormonale Beeinflussung des respiratorischen Stoffwechsels bei der Schabe *Leucophaea maderae* (F.) *Rev. Suisse Zool. 72*, 618–623.

MANNIX, J. J. and KEELEY, L. L. (1980). Age and endocrine effects on fat body metabolite composition in adult male *Blaberus discoidalis* cockroaches. *J. Exp. Zool. 212*, 113–117.

MAYER, R. J. and CANDY, D. J. (1969). Control of haemolymph lipid concentration during locust flight: an adipokinetic hormone from the corpora cardiaca. *J. Insect Physiol. 15*, 611–620.

MCCALEB, D. C. and KUMARAN, A. K. (1980). Control of juvenile hormone esterase activity in *Galleria mellonella* larvae. *J. Insect Physiol. 26*, 171–177.

MORISHIMA, I. (1981). Adenylate and guanylate activities in fat body during development of the silkworm, *Bombyx mori*. *Insect Biochem. 11*, 713–716.

MÜLLER, H. P. and ENGELMANN, F. (1968). Studies on the endocrine control of metabolism in *Leucophaea maderae* (Blattaria) II. The effect of the corpora cardiaca on fat-body respiration. *Gen. Comp. Endocr. 11*, 43–50.

MULLINS, D. E. and COCHRAN, D. G. (1972). Nitrogen excretion in cockroaches: uric acid is not a major product. *Science. 177*, 699–701.

MULLINS, D. E. and COCHRAN, D. G. (1973). Nitrogenous excretory materials from the American cockroach. *J. Insect Physiol. 19*, 1007–1018.

MULLINS, D. E. and COCHRAN, D. G. (1974). Nitrogen metabolism in the American cockroach: an examination of whole body and fat body regulation of cations in response to nitrogen balance. *J. Exp. Biol. 61*, 557–570.

MULLINS, D. E. and COCHRAN, D. G. (1975). Nitrogen metabolism in the American cockroach-II. An examination of negative nitrogen balance with respect to mobilization of uric acid stores. *Comp. Biochem. Physiol.* (A) *50*, 501–510.

MULLINS, D. E. and COCHRAN, D. G. (1976). A comparative study of nitrogen excretion in twenty-three cockroach species. *Comp. Biochem. Physiol.* (A) *53*, 393–399.

MUNN, E. A., FEINSTEIN, A. and GREVILLE, G. D. (1967). A major protein constituent of pupae of the blowfly *Calliphora erythrocephala* (Diptera). *Biochem. J.* 102, 5p–6p.

MUNN, E. A., PRICE, G. M. and GREVILLE, G. D. (1969). The synthesis *in vitro* of the protein calliphorin by fat body from the larvae of the blowfly, *Calliphora erythrocephala. J. Insect Physiol. 15*, 1601–1605.

MUNN, E. A., and GREVILLE, G. D. (1971a). The soluble proteins of developing *Calliphora erythrocephala*, particularly calliphorin, and similar proteins in other insects. *J. Insect Physiol. 15*, 1935–1950.

MUNN, E. A., FEINSTEIN, A. and GREVILLE, G. D. (1971b). The isolation and properties of the protein calliphorin. *Biochem. J. 124*, 367–374.

MURDOCK, L. L., HOPKINS, T. L. and WIRTZ, R. A. (1970). Phenylalanine metabolism in cockroaches. *Periplaneta americana*: intracellular symbionts and aromatic ring cleavage. *Comp. Biochem. Physiol. 34*, 143–146.

MURPHY, T. A. and WYATT, G. R. (1965). The enzymes of glycogen and trehalose synthesis in silkmoth fat body. *J. Biol. Chem. 240*, 1500–1508.

MWANGI, R. W. and GOLDSWORTHY, G. J. (1977a). Age-related changes in the response to adipokinetic hormone in *Locusta migratoria. Physiol. Ent. 2*, 37–42.

MWANGI, R. W. and GOLDSWORTHY, G. J. (1977b). Diglyceride-transporting lipoproteins in *Locusta. J. Comp. Physiol. 114*, 177–190.

NATORI, S. (1976). Selective activation of RNA polymerase I in fat body nuclei of *Sarcophaga peregrina* larvae by β-ecdysone. *Devel. Biol. 50*, 395–401.

NELSON, D. R., TERRANOVA, A. C. and SUKKESTAD, D. R. (1967). Fatty acid composition of the glycerides and free fatty acid fraction of the fat body and haemolymph of the cockroach, *Periplaneta americana* (L). *Comp. Biochem. Physiol. 20*, 907–917.

NEUFELD, G. J., THOMPSON, J. A. and HORN, D. H. S. (1968). Short-term effects of crust-ecdysone (20-hydroxyecdysone) on protein and RNA synthesis in third instar larvae of *Calliphora. J. Insect Physiol. 14*, 789–804.

NOWOCK, J., GOODMAN, W., BOLLENBACHER, W. E. and GILBERT, L. I. (1975). Synthesis of juvenile hormone binding proteins by the fat body of *Manduca sexta. Gen. Comp. Endocr. 27*, 230–239.

OBERLANDER, H. (1976). Dissociation and Reaggregation of Fat Body Cells during Insect Metamorphosis. In *Invertebrate Tissue Culture: Applications in Medicine, Biology and Agriculture*. Edited by E. Kurstak and K. Maramorosch. Pages 241–246. Academic Press, New York.

ORR, C. W. M. (1964). The influence of nutritional and hormonal factors on the chemistry of the fat body, blood, and ovaries of the blowfly *Phormia regina* Meig. *J. Insect Physiol. 10*, 103–119.

OSBORNE, D. J., CARLISLE, D. B. and ELLIS, P. E. (1968). Protein synthesis in the fat body of the female desert locust, *Schistocerca gregaria* Forsk, in relation to maturation. *Gen. Comp. Endocr. 11*, 347–354.

PAN, M. L. (1971). The synthesis of vitellogenin in the cecropia silkworm. *J. Insect Physiol. 17*, 677–689.

PAN, M. L. (1977). Juvenile hormone and vitellogenin synthesis in the cecropia silkworm. *Biol Bull 153*, 336–344.

PAN, M. L., BELL, W. J. and TELFER, W. H. (1969). Vitellogenic blood protein synthesis by insect fat body. *Science, 165*, 393–394.

PAN, R., LEVENBOOK, L. and BAUER, A. C. (1979). Inhibitory effect of beta-ecdysone on protein by blowfly fat body *in vitro. Experientia, 35*, 1449–1451.

VAN PELT-VERKUIL, E. (1978). Increase in acid phosphatase activity in the fat body during larval and prepupal development in *Calliphora erythrocephala. J. Insect Physiol. 24*, 375–382.

VAN PELT-VERKUIL, E. (1979). Hormone mediated induction of acid phosphatase activity in *Calliphora erythrocephala* fat body prior to metamorphosis. *J. Insect Physiol. 25*, 965–973.

VAN PELT-VERKUIL, E., VAN RONGEN, E. and DEPRIESTER, W. (1979). Normal and experimentally induced lysosomal activity in the larval fat body of *Calliphora erythrocephala* Meigen. *Cell Tiss. Res. 203*, 443–455.

PIERRE, L. L. (1962). Synthesis of ascorbic acid by the normal fat-body of the cockroach, *Leucophaea maderae* (F.), and its symbionts. *Nature 193*, 904–905.

PIERRE, L. L. (1964). Uricase activity of isolated symbionts and the aposymbiotic fat body of a cockroach. *Nature 201*, 54–55.

PIERRE, L. L. (1965). Guanase activity of the symbionts and fat bodies of the cockroach, *Leucophaea maderae. Nature 208*, 666–667.

PINES, M. and APPLEBAUM, S. W. (1978). Cyclic nucleotide-dependent protein kinase activity of adult female locust fat body. *Insect Biochem. 8*, 183–187.

PINES, M., TIETZ, A., WEINTRAUB, H., APPLEBAUM, S. W. and JOSEFSSON, L. (1981). Hormonal activation of protein kinase and lipid mobilization in the locust fat body *in vitro. Gen. Comp. Endocr. 43*, 427–431.

POSTLETHWAIT, J. H. and HANDLER, A. M. (1979). The roles of juvenile hormone and 20-hydroxy-ecdysone during vitellogenesis in isolated abdomens of *Drosophila melanogaster. J. Insect Physiol. 25*, 455–460.

POSTLETHWAIT, J. H. and JONES, G. J. (1978). Endocrine control of larval fat body histolysis in normal and mutant *Drosophila melanogaster. J. Exp. Zool. 203*, 207–214.

PRICE, G. M. (1965). Nucleid acids in the larva of the blowfly *Calliphora erythrocephala. J. Insect Physiol. 11*, 869–878.

PRICE, G. M. (1969). Protein synthesis and nucleic acid metabolism in the fat body of the larva of the blowfly, *Calliphora erythrocephala. J. Insect Physiol. 15*, 931–944.

PRICE, G. M. (1973). Protein and nucleic acid metabolism in insect fat body. *Biol. Rev. 48*, 333–375.

DEPRIESTER, W. and VAN DER MOLEN, L. G. (1979). Premetamorphic changes in the ultrastructure of *Calliphora* fat cells. *Cell Tiss. Res. 198*, 79–93.

DEPRIESTER, W., VAN PELT-VERKUIL, E. and DELEEUW, G. (1979). Demonstration of acid phosphatase activity induced by 20-hydroxy-ecdysone in *Calliphora* fat body. *Cell Tiss. Res. 200*, 435–442.

PROTZEL, A., SRIDHARA, S. and LEVENBOOK, L. (1976). Ribosomal replacement and degradation during metamorphosis of the blowfly, *Calliphora vicina. Insect Biochem. 6*, 571–578.

REDDY, G., HWANG-HSU, K. and KUMARAN, A. K. (1979). Factors influencing juvenile hormone esterase activity in the wax moth, *Galleria mellonella. J. Insect Physiol. 25*, 65–71.

REID, P. C. and CHEN, T. T. (1981). Juvenile hormone-controlled vitellogenin synthesis in the fat body of the locust (*Locusta migratoria*): isolation and characterization of vitellogenin polysomes and their induction *in vivo. Insect Biochem. 11*, 297–305.

RIZKI, T. M. (1961). Intracellular localization of kynurenine in the fat body of *Drosophila. J. Biophys. Biochem. Cytol. 9*, 567–572.

RIZKI, T. M. (1964). Mutant genes regulating the inducibility of kynurenine synthesis. *J. Cell Biol. 21*, 203–211.

RIZKI, T. M. and RIZKI, R. M. (1962). Cytodifferentiation in the *rosy* mutant of *Drosophila melanogaster. J. Cell Biol. 12*, 149–157.

ROBINSON, N. L. and GOLDSWORTHY, G. J. (1974). The effects of locust adipokinetic hormone on flight muscle metabolism *in vivo* and *in vitro. J. Comp. Physiol. 89*, 369–377.

ROTHSCHILD, M., FORD, B. and HUGHES, M. (1970). Maturation of the male rabbit flea (*Spilopsyllus cuniculi*) and the oriental rat flea (*Xenopsylla cheopis*): some effects of mammalian hormones on development and impregnation. *Trans. Zool. Soc., Lond. 32*, 105–188.

SAHOTA, T. S. and MANSINGH, A. (1970). Cellular response to ecdysone: RNA and protein synthesis in larval tissues of oak silkworm, *Antheraea pernyi. J. Insect Physiol. 16*, 1649–1654.

SCHELLER, K. and KARLSON, P. (1977a). Synthesis of poly(A) containing RNA induced by ecdysterone in fat body cells of *Calliphora vicina. J. Insect Physiol. 23*, 435–440.

SCHELLER, K. and KARLSON, P. (1977b). Effects of ecdysteroids on RNA synthesis of fat body cells in *Calliphora vicina. J. Insect Physiol. 23*, 285–291.

SCHEURER, R. (1969). Endocrine control of protein synthesis during oocyte maturation in the cockroach, *Leucophaea maderae. J. Insect Physiol. 15*, 1411–1419.

SCHIN, K., LAUFER, H. and CARR, E. (1977). Cytochemical and electrophoretic studies of haemoglobin synthesis in the fat body of a midge, *Chironomus thummi. J. Insect Physiol. 23*, 1233–1242.

SEKERIS, C. E. and SCHELLER, K. (1977). Calliphorin, a major protein of the blowfly: correlation between the amount of protein, its biosynthesis and the titer of translatable calliphorin-mRNA during development. *Devel. Biol. 59*, 12–23.

SEKERIS, C. E., PERASSI, R., ARNEMANN, J., ULLRICH, A. and SCHELLER, K. (1977). Translation of mRNA from *Calliphora vicina* and *Drosophila melanogaster* larvae into calliphorin and calliphorin-like proteins of *Drosophila. Insect Biochem. 7*, 5–9.

SHAH, J. and BAILEY, E. (1976). Enzymes of ketogenesis in the fat body and the thoracic muscle of the adult cockroach. *Insect Biochem. 6*, 251–254.

SHELLEY, R. M. and HODGSON, E. (1970). Biosynthesis of phosphatidylcholine in the fat body of *Phormia regina* larvae. *J. Insect Physiol.* 16, 131–139.

SHELLEY, R. M. and HODGSON, E. (1971a). Choline kinase from the fat body of *Phormia regina* larvae. *J. Insect. Physiol. 17*, 545–558.

SHELLEY, R. M. and HODGSON, E. (1971b). Substrate specificity and inhibition of choline and ethanolamine kinases from the fat body of *Phormia regina* larvae. *Insect Biochem. 1*, 149–156.

SHIGEMATSU, H. (1958). Synthesis of blood proteins by the fat body in the silkworm *Bombyx mori*. *Nature 182*, 880–882.

SPARKS, T. C. and HAMMOCK, B. D. (1979). Induction and regulation of juvenile hormone esterases during the last larval instar of the cabbage looper, *Trichoplusia ni*. *J. Insect Physiol. 25*, 551–560.

SPENCER, I. M. and CANDY, D. J. (1976). Hormonal control of diacyl glycerol mobilization from fat body of the desert locust, *Schistocerca gregaria*. *Insect Biochem. 6*, 289–296.

SPRING, J. H., MATTHEWS, J. R. and DOWNER, R. G. H. (1977). Fate of glucose in haemolymph of the American cockroach, *Periplaneta americana*. *J. Insect Physiol. 23*, 525–529.

SRIDHARA, S. and LEVENBOOK, L. (1974). The contribution of the fat body to RNA and ribosomal changes during development of the blowfly *Calliphora erythrocephala* (Meig.) *Devel. Biol. 38*, 64–72.

SRIDHARA, S. and GILBERT, L. I. (1975). Alterations in DNA-dependent RNA polymerase activity during the development of the tobacco hornworm, *Manduca sexta*. *Devel. Biol. 45*, 7–20.

SRIDHARA, S. and GILBERT, L. I. (1978). Isolation and characterization of RNA polymerase B from the larval fat body of the tobacco hornworm, *Manduca sexta*. *Eur. J. Biochem. 90*, 161–169.

SROKA, P. and BARTH, R. H. (1976). Hormonal control of diglyceride metabolism during vitellogenesis in the cockroach, *Eublaberus posticus*. *J. Insect Physiol. 22*, 951–954.

STAY, B. and CLARK, J. K. (1971). Fluctuations of protein granules in the fat body of the viviparous cockroach, *Diploptera punctata*, during the reproductive cycle. *J. Insect Physiol. 17*, 1747–1762.

STEELE, J. E. (1961). Occurrence of a hyperglycemic factor in the corpus cardiacum of an insect. *Nature 192*, 680–681.

STEELE, J. E. (1963). The site of action of insect hyperglycemic hormone. *Gen. Comp. Endocr. 3*, 46–52.

STEELE, J. E. (1980). Hormonal Modulation of Carbohydrate and Lipid Metabolism in Fat Body. In *Insect Biology in the Future*. Edited by M. Locke and D. S. Smith. Pages 253–271. Academic Press, New York.

STEELE, J. E. (1981). The Role of Carbohydrate Metabolism in Physiological Function. In *Energy Metabolism in Insects*.. Edited by R. G. H. Downer. Pages 101–133. Plenum Press, New York.

STEPHEN, W. F., JR. and GILBERT, L. I. (1969). Fatty acid biosynthesis in the silkmoth, *Hyalophora cecropia*. J. Insect Physiol. 15, 1833–1854.

STEVENSON, E. (1972). Haemolymph lipids and fat body lipases of the southern armyworm moth. *J. Insect Physiol. 18*, 1751–1756.

STEVENSON, E. and WYATT, G. R. (1962). The metabolism of silkmoth tissues. 1. Incorporation of leucine into protein. *Arch. Biochem. Biophys. 99*, 65–71.

STEVENSON, E. and WYATT, G. R. (1964). Glycogen phosphorylase and its activation in silkmoth fat body. *Arch. Biochem. Biophys. 108*, 420–429.

STONE, J. V., MORDUE, W., BATLEY, K. E. and MORRIS, H. R. (1976). Structure of locust adipokinetic hormone, a neurohormone that regulates lipid utilization during flight. *Nature 263*, 207–211.

STOREY, K. B. and BAILEY, E. (1977). The intracellular distribution of enzymes of carbohydrate degradation in the fat body of the adult male cockroach. *Insect Biochem. 8*, 73–79.

STOREY, K. B. and BAILEY, E. (1978). Intracellular distribution of enzymes associated with lipogenesis and gluconeogenesis in fat body of the adult cockroach, *Periplaneta*. *Insect Biochem. 8*, 125–131.

SUTHERLAND, E. W. (1972). Studies on the mechanism of hormone action. *Science 177*, 401–408.

TADKOWSKI, T. M. and JONES, J. C. (1979). Changes in the fat body and oocysts during starvation and vitellogenesis in a mosquito, *Aedes aegypti* (L). *J. Morphol. 159*, 185–203.

TAKAHASHI, S. (1966). Studies on ribonucleic acid in the fat body of *Philosamia cynthia ricini* Donovan (Lepidoptera) during development. *J. Insect Physiol. 12*, 789–801.

TARVER, R. U. and PIERRE, L. L. (1967). Activity of malic dehydrogenase of the symbionts and fat bodies of a cockroach. *Nature 213*, 208–209.

TAYLOR, J. F. and HODGSON, E. (1972). Phospholipid ethanolamine biosynthesis in larval fat body of the blowfly, *Phormia regina*. *Insect Biochem. 2*, 243–248.

THOMAS, K. K. (1972). Studies on the synthesis of lipoproteins during larval-pupal development of *Hyalophora cecropia*. *Insect Biochem. 2*, 107–118.

THOMASSON, W. A. and MITCHELL, H. K. (1972). Hormonal control of protein granule accumulation in fat bodies of *Drosophila melanogaster* larvae. *J. Insect Physiol. 18*, 1885–1899.

THOMSEN, E. and THOMSEN, M. (1974). Fine structure of the fat body of the female *Calliphora erythrocepala* during the first egg maturation cycle. *Cell Tiss. Res. 152*, 193–217.

THOMSON, J. A., KINNEAR, J. F., MARTIN, M. D. and HORN, D. H. S. (1971). Effects of crustecdysone (20-hydroxyecdysone) on synthesis, release, and uptake of proteins by the larval fat body of *Calliphora*. *Life Sci. 10*, 203–211.

TIETZ, A. and STERN, N. (1969). Stearate desaturation by microsomes on the locust fat-body. *FEBS Lett. 2, 286–288.*

TIETZ, A. and WEINTRAUB, H. (1978). Hydrolysis of glycerides by lipases of the fat body of the locust *Locusta migratoria*. *Insect Biochem. 8*, 11–16.

TOJO, S. (1971). Uric acid production in relation to protein metabolism in the silkworm, *Bombyx mori*, during pupal-adult development. *Insect Biochem. 1*, 249–263.

TOJO, S., BETCHAKU, T., ZICCARDI, V. J. and WYATT, G. R. (1978). Fat body protein granules and storage proteins in the silkmoth, *Hyalophora cecropia*. *J. Cell Biol. 78*, 823–838.

TOJO, S., KIGUCHI, K. and KIMURA, S. (1981). Hormonal control of storage protein synthesis and uptake by the fat body in the silkworm, *Bombyx mori*. *J. Insect Physiol. 27*, 491–497.

TYSELL, B. and BUTTERWORTH, F. M. (1978). Different rate of protein granule formation in the larval fat body of *Drosophila melanogaster*. *J. Insect Physiol. 24*, 201–206.

VALOVAGE, W. D. and BROOKS, M. A. (1979). Uric acid quantities in the fat body of normal and aposymbiotic German cockroaches, *Blattella germanica*. *Ann. Ent. Soc. Amer. 72*, 687–689.

VINCE, R. K. and GILBERT, L. I. (1977). Juvenile hormone esterase activity in precisely timed last instar larvae and pharate pupae of *Manduca sexta*. *Insect Biochem. 7*, 115–120.

VROMAN, H. E., KAPLANIS, J. N. and ROBBINS, W. E. (1965). Effect of allatectomy on lipid biosynthesis and turnover in the female American cockroach, *Periplaneta americana* (L). *J. Insect Physiol. 11*, 897–904.

WALKER, P. R. and BAILEY, E. (1969). A comparison of the properties of the phosphofructokinase of the fat body and flight muscle of the adult male desert locust. *Biochem. J. 111*, 365–369.

WALKER, P. R. and BAILEY, E. (1970a). Metabolism of glucose, trehalose, citrate, acetate, and palmitate by the male desert locust during adult development. *J. Insect Physiol. 16*, 499–509.

WALKER, P. R. and BAILEY, E. (1970b). Changes in enzymes associated with lipogenesis during development of the adult male desert locust. *J. Insect Physiol. 16*, 679–690.

WALKER, P. R. and BAILEY, E. (1971a). Effect of allatectomy on the growth of the male desert locust during adult development. *J. Insect Physiol. 17*, 1125–1137.

WALKER, P. R. and BAILEY, E. (1971b). Effect of allatectomy on fat body lipogenic enzymes of the male desert locust during adult development. *J. Insect Physiol. 17*, 1359–1369.

WALKER, P. R., HILL, L. and BAILEY, E. (1970). Feeding activity, respiration, and lipid and carbohydrate content of the male desert locust during adult development. *J. Insect Physiol. 16*, 1001–1015.

WEEDA, E. (1981). Hormonal regulation of proline synthesis and glucose release in the fat body of the Colorado potato beetle, *Leptinotarsa decemlineata*. *J. Insect Physiol. 27*, 411–417.

WEEDA, E., KOOPMANSCHAP, A. B., DEKORT, C. A. D. and BEENAKKERS, A. M. TH. (1980). Proline synthesis in fat body of *Leptinotarsa decemlineata*. *Insect Biochem. 10*, 631–636.

WEIRICH, G., WREN, J. and SIDDALL, J. B. (1973). Developmental changes of the juvenile hormone esterase activity in haemolymph of the tobacco hornworm, *Manduca sexta*. *Insect Biochem. 3*, 397–407.

WEISSMANN, R. (1963). Untersuchungen über den larvelen und imaginalen Fett-korper der Imago von *Musca domestica*. *Mitt Schweiz. Ent. Ges. 35*, 185–210.

WEST, T. V., GRULA, M. A., WORMINGTON, W. M. and WEAVER, R. F. (1980). Isolation of three forms of DNA-dependent RNA polymerase from *Heliothis zea* fat body tissue. *Insect Biochem. 10*, 509–513.

WHITMORE, E. and GILBERT, L. I. (1972). Haemolymph lipoprotein transport of juvenile hormone. *J. Insect Physiol. 18*, 1153–1167.

WHITTEN, J. M. (1962). Breakdown and formation of connective tissue in the pupal stage of an insect. *Quart. J. Mic. Sci. 103*, 359–367.

WHITTEN, J. M. (1964). Haemocytes and the metamorphosing tissues in *Sarcophaga bullata, Drosophila melanogaster*, and other cyclorrhaphous Diptera Cyclorrhapha. *J. Insect Physiol. 10*, 447–469.

WIENS, A. W. and GILBERT, L. I. (1965). Regulation of cockroach fat body metabolism by the corpus cardiacum *in vitro*. *Science 150*, 614–616.

WIENS, A. W. and GILBERT, L. I. (1967a). Regulation of carbohydrate mobilization and utilization in *Leucophaea maderae*. *J. Insect Physiol. 13*, 779–794.

WIENS, A. W. and GILBERT, L. I. (1967b). Variations in the glycogen content of fat body, ovary, and embryo during the reproductive cycle of *Leucophaea maderae*. *J. Insect Physiol. 13*, 587–594.

WIENS, A. W. and GILBERT, L. I. (1967c). The phosphorylase system of the silkmoth *Hyalophora cecropia*. *Comp. Biochem. Physiol. 21*, 145–159.

WIGGLESWORTH, V. B. (1935). Function of the corpus allatum in insects. *Nature 136*, 338–339.

WILLIAMS, K. L., BARTON BROWNE, L. and VAN GERWIN, A. C. M. (1977). Ovarian development in autogenous and anautogenous *Lucilia cuprina* in relation to protein storage in the larval fat body. *J. Insect Physiol. 23*, 659–664.

WILLIAMS-BOYCE, P. K. and JUNGREIS, A. M. (1980). Changes in fat body urate synthesizing capacity during the larval-pupal transformation of the tobacco hornworm, *Manduca sexta*. *J. Insect Physiol. 26*, 783–789.

WIMER, L. T. (1969). A comparison of the carbohydrate composition of the hemolymph and fat body of *Phormia regina* during larval development. *Comp. Biochem. Physiol. 29*, 1055–1062.

WING, K. D., SPARKS, T. C., LOVELL, V. M., LEVINSON, S. O. and HAMMOCK, B. D. (1981). The distribution of juvenile hormone esterase and its interrelationship with other proteins influencing juvenile hormone metabolism in the cabbage looper, *Trichoplusia ni*. *Insect Biochem. 11*, 473–485.

WLODAWER, P. and LAGWINSKA, E. (1967). Uptake and release of lipids by the isolated fat body of the waxmoth larva. *J. Insect Physiol. 13*, 319–331.

WUEST, J. (1978). Histological and cytological studies on the fat body of the cockroach *Nauphoeta cinerea* during the first reproductive cycle. *Cell Tiss. Res. 188*, 481–490.

WYATT, G. R. and KALF, G. F. (1957). The chemistry of insect hemolymph. II. Trehalose and other carbohydrates. *J. Gen. Physiol. 40*, 833–847.

WYATT, G. R. and LINZEN, B. (1965). The metabolism of ribonucleic acid in cecropia silkmoth pupae in diapause, during development and after injury. *Biochim. Biophys. Acta. 103*, 588–600.

WYATT, G. R. and PAN, M. L. (1978). Insect plasma proteins. *Ann. Rev. Biochem. 47*, 779–817.

YANAGAWA, H. and HORIE, Y. (1978). Activating enzyme of phosphorylase b in the fat body of the silkworm, *Bombyx mori*. *Insect Biochem. 8*, 155–158.

ZIEGLER, R. (1979). Hyperglycaemic factor from the corpora cardiaca of *Manduca sexta* (L.) (Lepidoptera: Sphingidae). *Gen. Comp. Endocr. 39*, 350–357.

ZIEGLER, R. and WYATT, G. R. (1975). Phosphorylase and glycerol production activated by cold in diapausing silkmoth pupae. *Nature 254*, 622–623.

ZIEGLER, R., ASHIDA, M., FALLON, A. M., WIMER, L. T., WYATT, S. S. and WYATT, G. R. (1979). Regulation of glycogen phosphorylase in fat body of cecropia silkmoth pupae. *J. Comp. Physiol. B 131*, 321–332.

7 Connective Tissues

DOREEN E. ASHHURST

St George's Hospital Medical School, University of London, UK

1 EARLY RESEARCH ON INSECT CONNECTIVE TISSUES

The existence of connective tissue in insects was recognized over 100 years ago but only recently with the development of new investigative techniques has its widespread distribution and significance been appreciated. Indeed, until the 1950s there was a widely held assumption that no barrier exists between the haemolymph and the tissues.

The nervous system has been prominent in connective tissue studies from the earliest beginnings. The dorsal mass of connective tissue on the abdominal connectives of adult Lepidoptera was recognized as connective tissue by Burger, D. (1876) and Nusbaum, J. (1884). They identified it as a gelatinous structure containing cells into which the muscles of the ventral diaphragm are inserted. They also realized that it is continuous with the connective tissue around the rest of the nervous system. About this time, the sheath around the nervous system of *Oryctes nasicornis* was described by Michels, H. (1880) and later Schneider, K. (1902) named the connective tissue layer, the neural lamella, and the underlying glial cell layer, the perineurium.

Very little further work on connective tissues was reported over the next four decades. Lazarenko, T. (1925) recognized the presence of connective tissue binding the organs together in *Oryctes* larvae, but he considered that the connective tissue is produced by the breakdown of haemocytes. Similarly, Wermel, E. (1938) came to this conclusion in his work on silkmoths. In 1939, Scharrer, B. published a description of the neural lamella and perineurium of *Periplaneta americana* and suggested that the perineurial cells produce the neural lamella.

Renewed interest in the connective tissues in the 1950s arose from the realization that the ionic concentrations in the haemolymph would block insect nerves and, therefore, the possibility that a barrier exists between the haemolymph and axons was investigated. Thus Hoyle, G. (1952) described the neural lamella of *Locusta migratoria*, but at this time the composition of connective tissue matrix was still unknown. An early birefringence study by Richards, A. (1944) led him to suggest that collagen might be present in peripheral nerves of mosquitos, and Rudall, K. (1955) reported X-ray diffraction evidence for collagen in the nerve cords of mantids, but this latter author considered that the presence of large amounts of chitin in the exoskeleton precluded the presence of significant amounts of collagen in insects.

In an attempt to characterize the neural lamella and other connective tissues, Baccetti, B. (1955a,b, 1956a) performed histochemical studies which showed that neutral glycoproteins, and possibly some acid mucopolysaccharides, are present in the connective tissue matrices of the orthopteran, *Anacridium aegyptium*. His polarized light study (Baccetti, B., 1956b) suggested, however, that collagen might be present, as did a similar study by Richards, A. and Schneider, D. (1958) on cockroach and silkmoth nerve cords. This was confirmed soon after with the publication of electron micrographs of periodically banded fibrils in the connective tissue matrices of several insects of different orders (Gray, E., 1959; Hess, A., 1958; Smith, D., and Wigglesworth, V., 1959), and by the identification of hydroxyproline in hydrolysates of cockroach and locust nervous systems (Ashhurst, D., 1959, 1961a). Thus, the presence of a collagen-like protein in insects was established, but since the banding periodicities measured were very variable, there was much debate in the literature as to whether the fibrils are all similar.

From 1960 onwards there was a rapid increase in our understanding of vertebrate connective tissues, and also in the development of biochemical, histochemical and other investigative techniques. These techniques have been applied to insect connective tissues so that we can now say with confidence that insects possess connective tissue matrices of very similar composition to those of other animals, including the mammals. The discussion of the investigations which have led to this conclusion will form the major part of this chapter.

2 MORPHOLOGY OF THE CONNECTIVE TISSUE MATRICES

In the following discussion the connective tissue matrices found in insects will be divided into four main groups. Very few studies, primarily concerned with the characterization of the connective tissues, have been published and thus most information on their structure is concealed in papers on other topics. In this account of the different connective tissue matrices of insects, no attempt will be made to provide a comprehensive review of the occurrence of each type; instead a guide to their identification will be given. It should be noted that the type of connective tissue associated with an organ may vary in insects of different orders.

The classification of the connective tissue matrices will be based on that currently used for vertebrate connective tissue matrices. This will differ, particularly with respect to the basement membranes, from that used by Ashhurst, D. (1968).

2.1 Fibrous connective tissue matrices

Connective tissue matrices containing typically banded collagen fibrils are found in many locations. They frequently form a discrete layer lying on the outermost cells of an organ, which serves to separate the organ from the haemocoel. Layers of fibrous matrix are found around the whole nervous system (neural lamella) (Fig. 1), under the epidermis, around the fat body, reproductive system, and so on; descriptions are found in many papers (see, for example, Ashhurst, D., 1968; Gupta, B. and Berridge, M., 1966; Rinterknecht, E. and Lévi, P., 1966). Characteristically, cells are not normally found within these connective tissue matrices, which range in thickness from less than 1 μm to more than 10 μm.

Fibrous connective tissue matrix may be present within organs and between muscle fibres, though in rather small amounts (Clements, A. and May, T., 1974). There are few reports of large areas of connective tissue, but a layer which is about 80 μm thick in the mature adult is found around the ejaculatory duct of the male locust, *Locusta migratoria* (Fig. 2) (Ashhurst, D. and Costin, N., 1974; Martoja, R. and Bassot, J., 1965), and a similar, though somewhat thinner, layer occurs around the mesenteron of the adult cockroach, *Periplaneta americana*

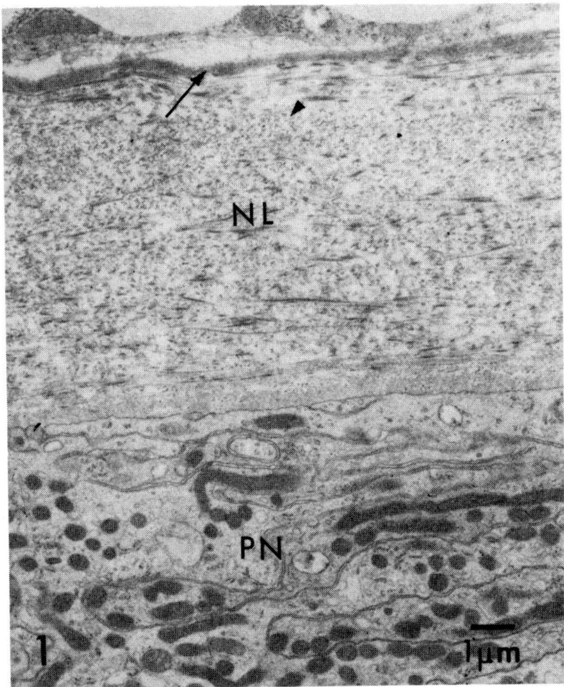

FIG. 1. Electron micrograph of the neural lamella (NL) and the perineurial cells (PN) of *Locusta migratoria*. The surface of neural lamella is covered by a thin layer of amorphous matrix (arrow). Irregularly shaped transverse sections of fibrils can be seen (arrowhead).

(François, J., 1978). The collagen fibrils in these layers are arranged in a random network and they contain cells, and also muscle fibres. The cells are morphologically indistinguishable from vertebrate fibroblasts (see section 4) (Ashhurst, D. and Costin, N., 1974; François, J., 1973, 1978).

It is usually assumed that connective tissue does not form tendon-like structures in insects. It was, however, revealed by Francois, J. (1968, 1971, 1972) that some endosternites in the head of Apterygota consist of parallel arrays of collagen fibrils (Fig. 3). The endosternites are attached to muscles or epidermal cells, as appropriate, by hemidesmosomes (Fig. 4). A similar, but much narrower band of connective tissue matrix joins two epidermal cells layers in the hypopharyngeal cavity of *Blaberus craniifer* (Moulins, M., 1968). Many more structures similar to these must await identification.

The collagen fibrils in the matrices mentioned so far are clearly banded. In most descriptions a figure for the periodicity of the banding pattern is given, and these vary between 55 and 70 nm. If, however, the patterns are examined carefully, it is apparent that they all conform to that of the locust and rat tail tendon fibrils in Fig. 5 and 22. Two bands, I and XI, are very obvious, but in sectioned fibrils bands III to X cannot be resolved clearly. It will be argued later (section 3.1) that the periodicity of this banding pattern in all fibrils is approximately 67nm; any variations from this figure measured on electron micrographs derive from shrinkage, etc., during the preparative procedures.

The diameter of the fibrils may vary according to the age of the tissue. Thus Ashhurst, D. and Costin, N. (1974) were able to follow the development of the collagen fibrils in the sexually mature adult locust ejaculatory duct from the 0-day adult in which they are thin (about 35 nm diameter), and rather indistinctly banded (Figs 7 and 8), to the very large, approximately 200 nm diameter, fibrils in the mature adult, which are irregularly shaped in cross-section (Fig. 6). It should be noted that fibrils, irregular in cross-section, can be seen in many published micrographs, and were noted in the locust neural lamella by Ashhurst, D. and Chapman, J. (1961), and in the ejaculatory duct by Martoja, R. and Bassot, J. (1965), but the latter authors considered them to be distinct from collagen. Examples of irregularly shaped fibrils occur in vertebrate

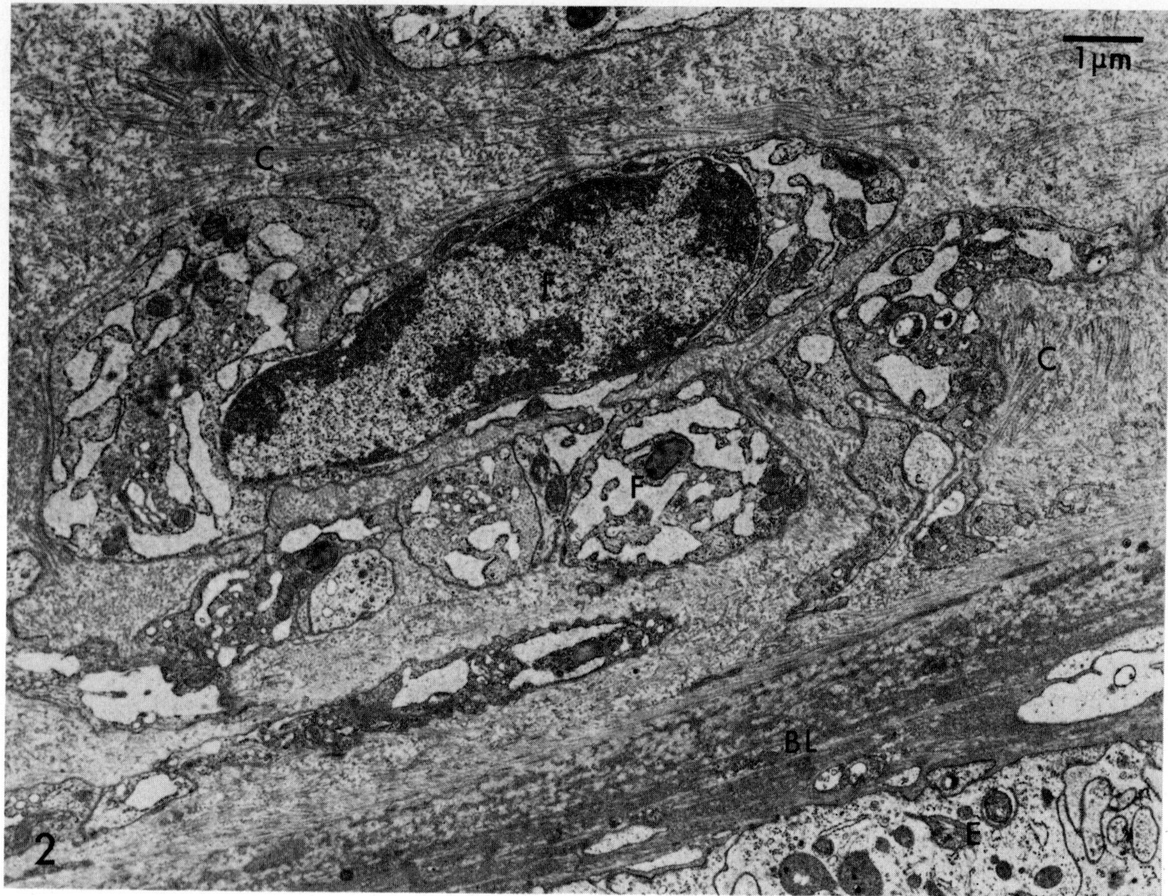

FIG. 2. Electron micrograph of the connective tissue around the ejaculatory duct of the 6-day adult male locust, *Locusta migratoria*. The matrix contains many collagen fibrils (C) and active fibroblasts (F). The basal layer (BL) lying on the epithelial cells (E) contains many closely packed collagen fibrils.

tissues, e.g. in the lungs of rats (Stephens, R. *et al.*, 1971).

The fibrous matrices of some orders, in particular the Coleoptera, Diptera, Lepidoptera and Protura, contain very thin, indistinctly banded fibrils (Ashhurst, D., 1968; Baccetti, B., 1961; de Biasi, S. and Pilotto, F., 1976; François, J., 1972; Locke, M. and Huie, P., 1972). Indications of banding are occasionally seen on the fibrils and figures ranging from 15 nm to 65 nm have been suggested for the periodicity. The fibrils are collagenous (see section 3.1), but in many tissues they are obscured by other matrix components which condense upon them, presumably during the preparative procedures (Fig. 10). The fibrils are typically found in layers such as the neural lamella (Sohal, R. *et al.*, 1972) and in the large mass of connective tissue found on the dorsal side of the abdominal connectives of adult Lepidoptera (Figs. 9 and 10) (Ashhurst, D., 1964; Ashhurst, D. and Costin, N., 1976; McLaughlin, B., 1974). The dorsal mass contains cells derived from glial cells, which have characteristics in common with typical vertebrate fibroblasts (see section 4).

It is pertinent to point out that many thin vertebrate collagen fibrils — for example those in the lens (type I collagen) and in cartilage (type II collagen) — are indistinctly banded. It appears that the number of heavy metal binding sites in a thin fibril is too small to permit sufficient stain to bind and visualize the bands clearly.

In some instances, layers of fibrous matrix may be bordered by layers of amorphous matrix. This occurs quite commonly in the Diptera; the neural lamella of *Musca domestica* has a 60 nm thick

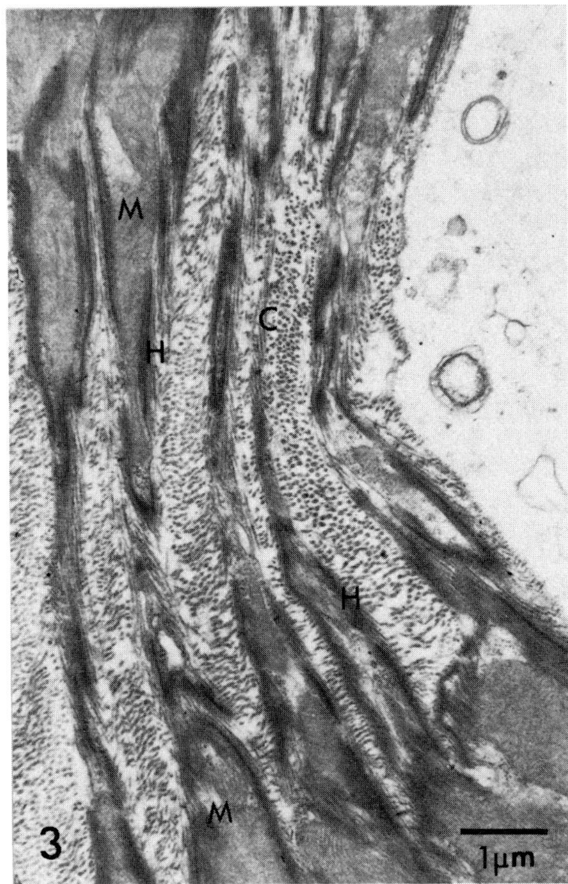

FIG. 3. Electron micrograph of the anterior maxillary endosternite of *Machilis burgundiae* showing collagen fibrils (C) forming a tendon-like structure between the muscle fibres (M). Hemidesmosomes (H) form the junctions between the connective tissue and muscle. (Micrograph by courtesy of Dr. J. François.)

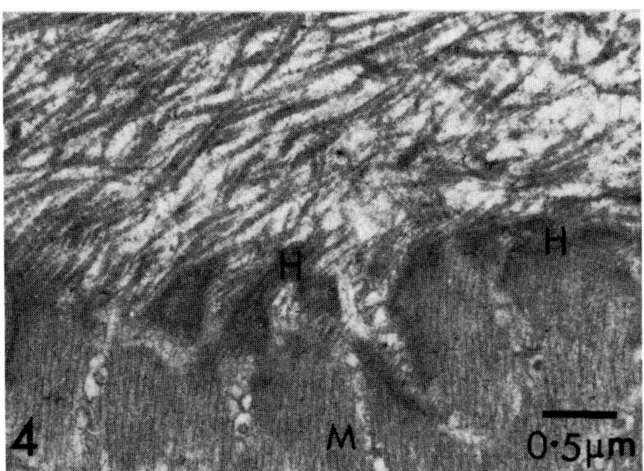

FIG. 4. A small area of an endosternite of *Thermobia domestica* with hemidesmosomes (H) anchoring the collagen fibrils to muscle fibres (M). (Micrograph by courtesy of Dr. J. François.)

amorphous layer peripheral to the fibrillar region (Sohal, R. *et al.*, 1972) and several regions of connective tissue in the rectum of *Calliphora erythrocephala* have alternating amorphous and fibrillar layers (Figs 11 and 12) (Gupta, B. and Berridge, M., 1966). A very thin amorphous layer may be found on the surface of the neural lamella of locusts and cockroaches (Fig. 1). The precise nature of this amorphous material is unknown. It may simply be a layer of non-collagenous proteins and glycosaminoglycans, or where it is a relatively greater component of the tissue, it could also contain a non-fibrous collagen.

The connective tissue around the ejaculatory duct of *Locusta* was classified as cartilage by Martoja, R. and Bassot, J. (1965) because in their preparations the cells appeared to be in lacunae. Subsequent work on this layer failed to substantiate this classification (Ashhurst, D. and Costin, N., 1974). There are no other suggested locations of cartilage in insects.

2.2 Basement membranes

Most vertebrate cells, with the exception of fibroblasts and blood cells, produce a basement membrane. This can be seen in light microscope preparations as a dense line under the basal membrane of an epithelial cell, or around a muscle cell; it is especially clear after the periodic acid-Schiff reaction since the basement membrane contains glycoproteins. In electron micrograph sections (Fig. 13), the basement membrane is seen to consist of two layers: the electron-lucent lamina rara (or lucida) which is about 50 nm thick, adjacent to the cell membrane, and the lamina densa, which appears amorphous, though some very fine filaments can often be discerned within it. The thickness of the lamina densa is variable, but it is not usually more than 100 nm thick. It contains type IV collagen (see section 3.1). Below the lamina densa there may be a dense meshwork of collagen fibrils, which has been called the reticular layer. This is produced by fibroblasts and not by the epithelial cells, and is not strictly part of the basement membrane. The term "basement membrane" is usually used to include the laminae rara and densa (see review by Heathcote, J. and Grant, M., 1981), but some workers restrict the term to the lamina densa (Kefalides, N. *et al.*, 1979).

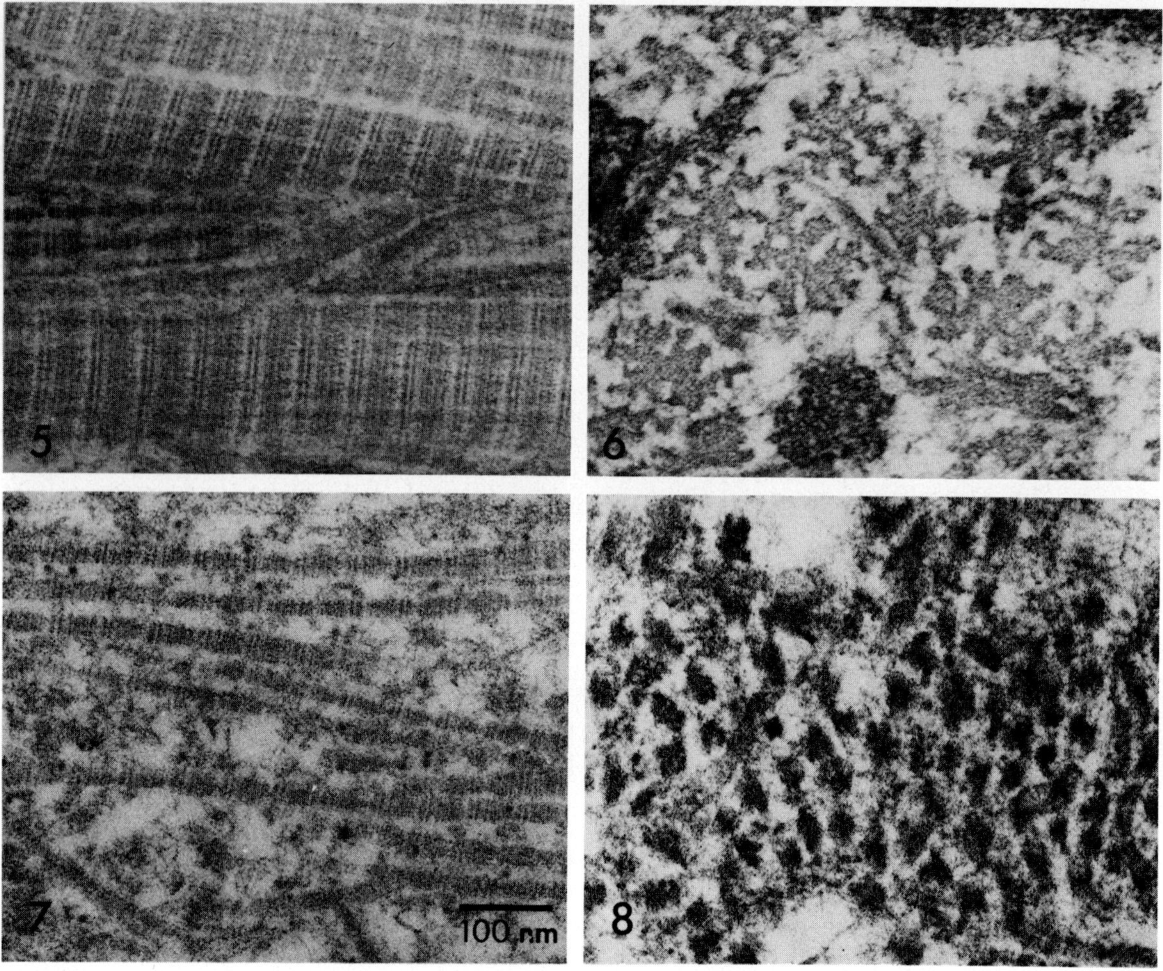

FIG. 5–8. Collagen fibrils in the connective tissue of the ejaculatory duct of the adult male locust. *Locusta migratoria*; all micrographs are at the same magnification. FIG. 5. Fibrils in a 26-day adult. The periodic banding pattern is clearly visible. FIG. 6. Transverse sections of the large fibrils in the 26-day adult. The cross-sections of the fibrils are very irregular. FIG. 7. Longitudinal sections of collagen fibrils in a 0-day adult. The fibrils are banded, but the pattern is not so clear as in the 26-day adult. FIG. 8. Transverse sections of the fibrils in a 0-day adult.

Basement membranes with this typical structure are found under many insect epithelia, around muscle fibres (Figs 14 and 15), and around fat body cells (Dutkowski, A., 1977); there is considerable variation in the thickness of the lamina densa. The basement membrane separates the muscle fibres from the haemolymph, but in other organs the basement membrane is adjacent to fibrous connective tissue. This situation is seen in the mesenteron of the cockroach (François, J., 1978).

A typical basement membrane may appear as a stage in the development of a connective tissue matrix. In *Schistocerca* embryos, for example, the nervous system is initially surrounded by a basement membrane (Fig. 26), but as more matrix is laid down the dense layer thickens and the lamina rara is obliterated (Ashhurst, D., 1965).

2.3 Thick non-fibrous layers of connective tissue matrix

There are many instances of layers of connective tissue matrix, which may be several micrometres thick, but which do not contain fibrils: they are also closely apposed to the adjacent cells. Thus, by virtue of their thickness and the absence of a lamina rara, they cannot be considered as basement membranes.

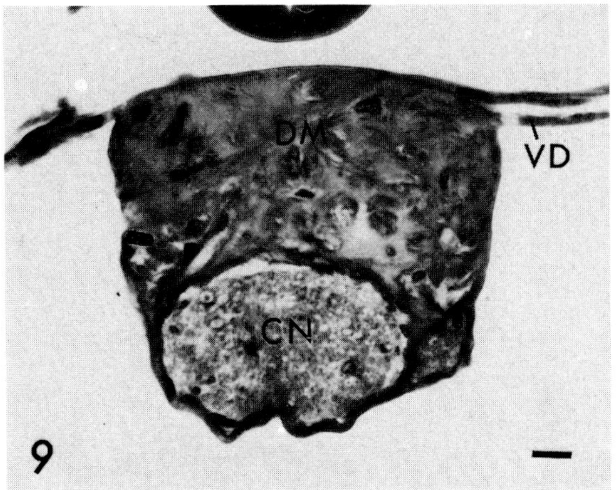

FIG. 9. A photomicrograph of the abdominal nerve cord of the adult waxmoth, *Galleria mellonella*. The dorsal mass (DM) of connective tissue lies on the fused connectives (CN). The muscle fibres of the ventral diaphragm (VD) are inserted into the dorsal mass. (Bar = 10 μm.)

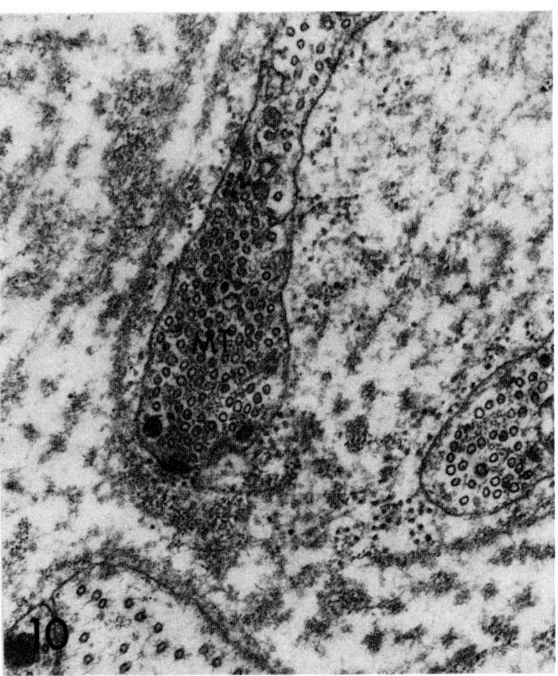

FIG. 10. An electron micrograph of a small area of the connective tissue mass of the 6-day pupal wax moth. The thin fibrils, covered by other matrix material, can been seen in both longitudinal and transverse section. The cell processes contain many microtubules (MT).

The occurrence and location of this type of matrix varies from insect to insect. In *Rhodnius prolixus* the epidermal cells lie on a matrix layer of this type (Fig. 16) (Wigglesworth, V., 1973). Several endocrine organs of the cockroach *Leucophaea maderae* are surrounded by sheaths of this type of matrix which also penetrates between the cells (Scharrer, B., 1963, 1964, 1971), and similar layers lie under the Malpighian tubules of the housefly, *Musca domestica* (Sohal, R., 1974). These are only a few examples to illustrate the wide distribution of thick, apparently non-fibrous layers of connective

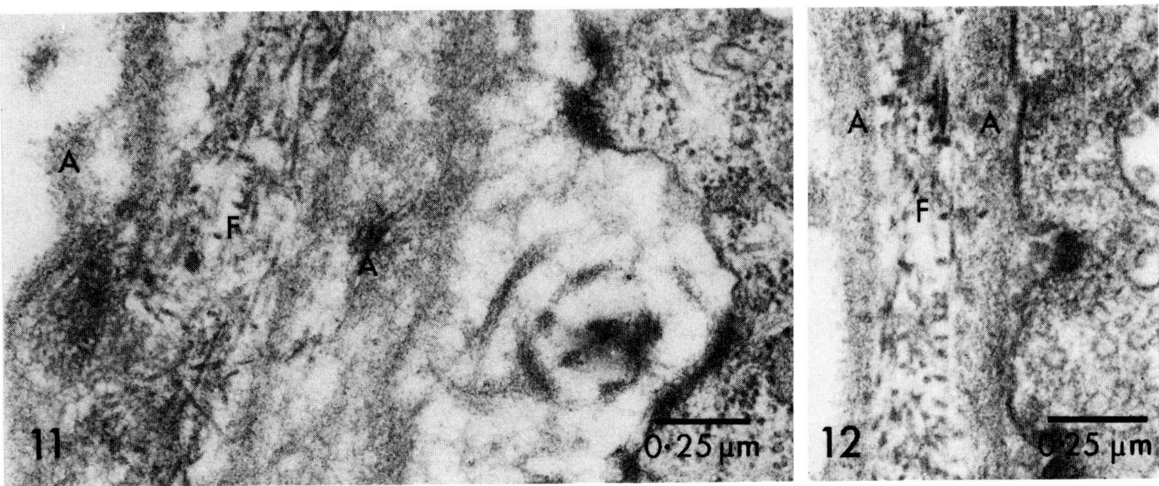

FIG. 11. Electron micrograph of a small area of a layer of connective tissue lying on a medullary cell of the rectum of *Calliphora erythrocephala*. Amorphous (A) and fibrous (F) layers alternate. (Micrographs by courtesy of Dr B. L. Gupta.)

FIG. 12. Electron micrograph of layer of connective tissue in the rectum of *Calliphora*. Again there are distinct amorphous (A) and fibrous (F) layers. (Micrograph by courtesy of Dr B. L. Gupta.)

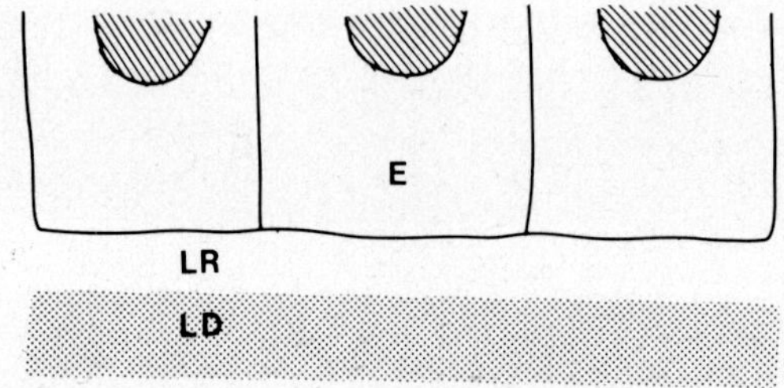

FIG. 13. A diagrammatic representation of a basement membrane. The lamina rara (LR) separates the lamina densa (LD) from the epithelial cells (E).

tissue matrix and it is noteworthy that typical fibrous matrices occur in all these insects, particularly in the neural lamella (Hess, A., 1958; Sohal, R. *et al.*, 1972; Smith, D. and Wigglesworth, V., 1959). Small areas between layers of cells are sometimes filled by this matrix. Recently, the so-called "basement membrane" of the compound eye of several insects has been examined (Odselius, R. and Elofsson, R., 1981). It consists of a layer of amorphous or finely fibrillar connective tissue matrix which, in different species, may vary from 1 μm to 40 μm in thickness and contains cell processes; the term basement membrane is clearly a misnomer in this instance.

Matrix, very similar to that described here, has been found forming connecting strands or ligaments between the heart, pericardial cells and other structures in Diptera and Lepidoptera (Crossley, A., 1972; Locke, M. and Huie, P., 1972); elastic fibres occur in these strands (Fig. 42) (see section 7). The composition of this connective tissue matrix is unknown. One presumes that a collagenous protein is present, but as yet this cannot be determined (see section 3.1). It is very difficult to detect the presence of glycosaminoglycans histochemically, as most of the layers are less than 1 μm thick. Thus many questions remain to be answered about this type of matrix.

2.4 Specialized connective tissue layers

In 1961 Bertram, D. and Bird, R. described a complex layered structure lying under the midgut epithelial cells of *Aedes aegypti*. Examination of longitudinal and tangential sections revealed that it consists of three or four layers, each of which is an array of cylindrical units with less electron-dense centres joined to form a network, which is expanded

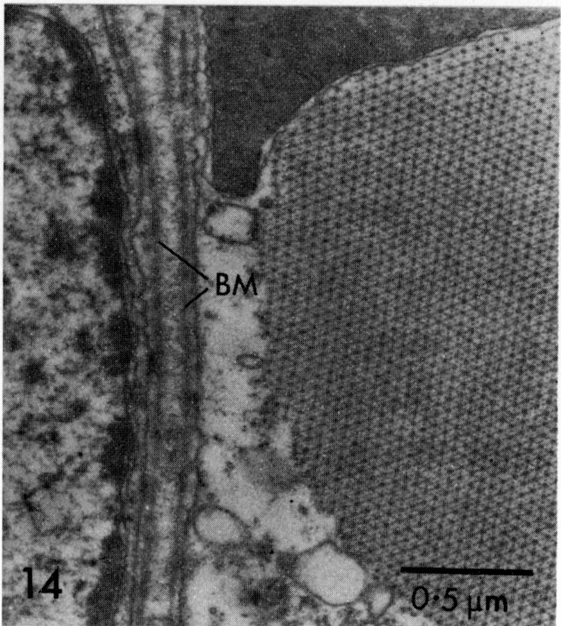

FIG. 14. Electron micrograph of the basement membranes (BM) of two adjacent muscle fibres of the giant water bug, *Lethocerus maximus*.

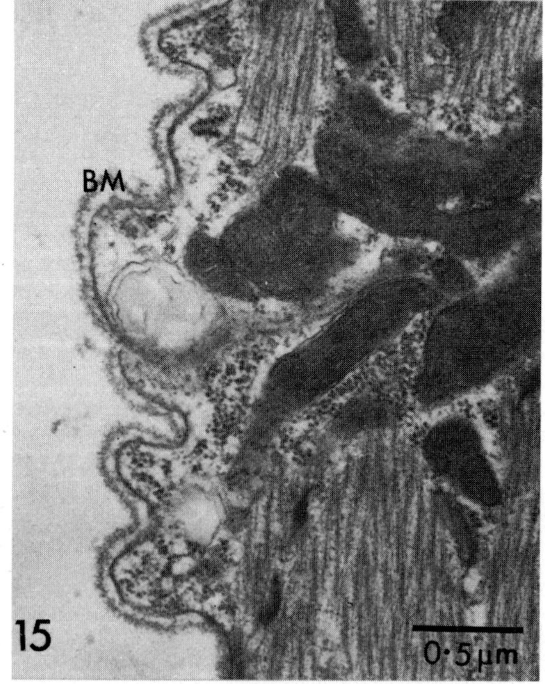

FIG. 15. Electron micrograph of the peripheral region of a muscle fibre of the ventral diaphragm of the 6-day pupa of the wax moth, *Galleria mellonella*, showing the basement membrane (BM).

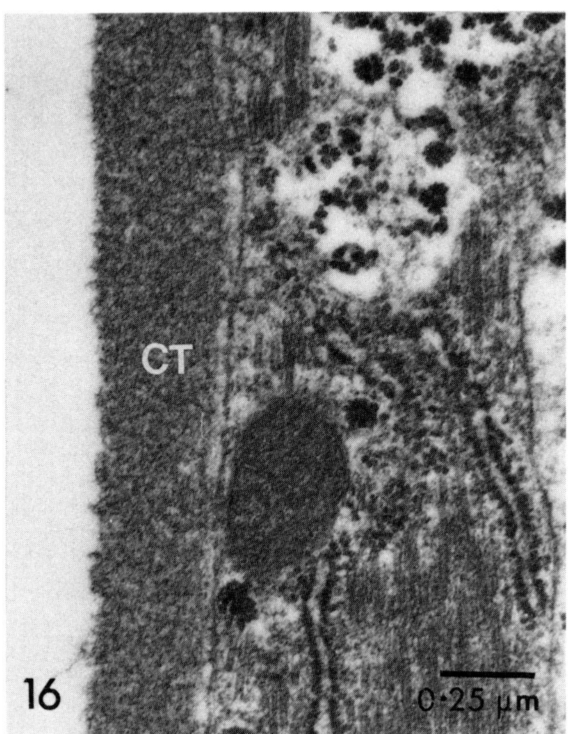

FIG. 16. Electron micrograph of the thick layer of connective tissue (CT) under the epidermal cells of a 0-day fifth-instar larva of *Rhodnius prolixus*.

and distorted after a blood-meal (Terzakis, J., 1967; Reinhardt, T. and Hecker, H., 1973). A similar structure has been seen in the midgut of fleas, but here the units are not joined together (Reinhardt, T. *et al.*, 1972; Richards, A. and Richards, P., 1968).

Grid-like layers were found under the midgut epithelium of six of the 13 species of Coleoptera examined by Holter, P. (1970). He found a single layer which appears in longitudinal sections as a ribbon-like structure transected by dark and light bands. In tangential sections, two types of units are distinguished; small electron-dense units are arranged around larger less dense units, and are held together by bridges. In *Oryctes* larvae the midgut epithelium lies on a three-layered structure (Fig. 17) (Bayon, C. and François, J., 1976; Hess, R. and Pinnock, T., 1975). Here the inner layer resembles that described by Holter, but the lighter units increase in size in the middle and again in the outer layer (Fig. 18).

A further type of modified layer is found in the midgut of the Heteroptera. In *Nepa cinerea* and *Ranatra linearis* (Gouranton, J., 1970), a typical lamina densa is separated from the cell membrane by an electron-lucent layer 35 nm thick and a second discontinuous layer, 90 nm thick, which consists of a series of polygonal plaques, separated by gaps of about 30 nm and bridged by thin filaments.

The chemical nature of these layers is unknown. The dense plaques and outer layer in *Ranatra* are digested by collagenase (Gouranton, J., 1970), but the layers in *Oryctes* are resistant (Bayon, C. and François, J., 1976). The layers in *Ranatra*, *Harpalis rufipes* and *Oryctes* are positive with the periodic acid-Schiff (PAS) test (Bayon, C. and François, J., 1976; Gouranton, J., 1970; Holter, P., 1970), but only in *Oryctes* are the layers positive with tests for glycosaminoglycans. Thus, while the positive PAS reaction suggests that these layers contain glycoprotein, their other constituents are yet to be determined.

The function of these networks is not understood; indeed Gouranton, J. (1970) considers it is impossible to predict. The only clue might come from the mosquitoes and fleas in which the midgut is very distended after a blood-meal; the network might permit the stretching necessary to accommodate the food.

Similar structures were described around the ovarioles of *Aedes aegypti* (Bertram, D. and Bird, R., 1961), but no more examples have been recorded so far in other insects; recently, modified layers have been found under the midgut epithelia of two Crustacea (Factor, J., 1981).

3 CONSTITUENTS OF CONNECTIVE TISSUE MATRICES

Connective tissue matrices are very diverse, since they range between the hard, dense, calcified matrix of bone and the soft, jelly-like vitreous of the eye. Nevertheless, they are composed of the same molecules: the fibrous proteins, collagen and elastin, glycosaminoglycans and many glycoproteins.

The collagens from animals of many phyla have been studied extensively (see reviews by Bailey, A. and Etherington, D., 1981; Miller, E., 1976; Prockop, D. *et al.*, 1979). At the present time, ten types of collagen molecule extracted from vertebrate tissues have been characterized, and yet more are under investigation. Some characteristics of these molecules are listed in Table 1. Those

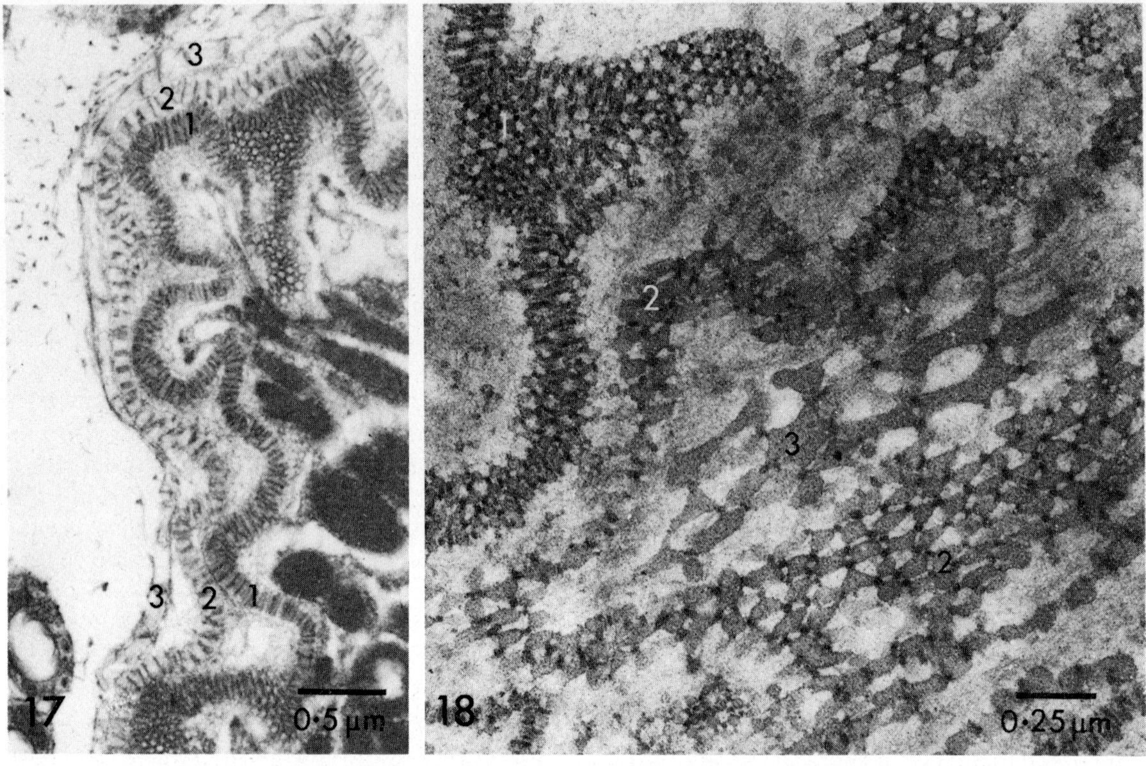

FIG. 17. Electron micrograph of the three specialized layers of connective tissue matrix (1, 2, 3) under the midgut epithelium of the larva of *Oryctes nasicornis*. (Micrograph by courtesy of Dr J. François.)

FIG. 18. An oblique section through the connective tissue matrix under the midgut epithelium of *Oryctes nasicornis* showing the *en face* views of the three layers (1, 2, 3). (Micrograph by courtesy of Dr J. François.)

invertebrate fibrous collagens investigated appear to be very similar to type I collagen, or type I trimer (Ashhurst, D. and Bailey, A., 1980; Kimura, S. and Matsuura, F., 1974; Minefra, S. *et al.*, 1975; Nordwig, A. *et al.*, 1970; Pikkarrainen, J. *et al.*, 1968). In addition, basement membrane collagens are present in invertebrates, but their similarity to mammalian type IV collagen is not fully established (Ashhurst, D. and Bailey, A., 1980; Hudson, B., 1978).

The collagen molecule, which is approximately 300 nm long and 1.5 nm in diameter, consists of three polypeptide chains, or α-chains. Each molecule has a central triple helical region, or domain. Each α-chain is coiled into a left-handed helix with about three amino acids per turn and the three α-chains of the molecule are then coiled around each other to form the right-handed superhelix; the short terminal regions are not triple-helical. The α-chains may be identical, or one may be dissimilar (Table 1).

Each α-chain consists of just over 1000 amino acid residues. In the helical domain every third amino acid is glycine, and hence glycine accounts for approximately one-third of the total amino acids of a collagen. Glycine is the smallest amino acid, and in the collagen triple helix glycine residues occupy the restricted space in the centre of the helix. The collagen molecule is also unusual since about 20% of the amino acids are proline, or hydroxyproline. These amino acids are important for the triple helical conformation: the imino acids (proline and hydroxyproline) confer a polyproline-type helix on the α-chain, while glycine stabilizes the triple helix by forming inter-chain hydrogen bonds. The hydroxylation of proline to form hydroxyproline occurs after the assembly of the α-chain, and its presence is essential for the stability of the triple helix at body temperatures. Hydroxyproline has been found in very few other proteins, notably vertebrate elastin, the C1q component of

Table 1: Some different types of vertebrate collagens

	α-chain composition	Fibrils	Typical locations in vertebrates
Type I	$\alpha 1(I)_2\alpha 2$	thick	skin, bone, tendon, arterial wall, uterus.
Type I trimer	$\alpha 1(I)_3$	thick	skin
Type II	$\alpha 1(II)_3$	thin	cartilage
Type III	$\alpha 1(III)_3$	thin	arterial wall, uterus, skin
Type IV	$\alpha 1(IV)_2\alpha 2(IV)_3$, $\alpha 1(IV)_3$ or $\alpha 2(IV)_3$	non-fibrous	basement membranes
Type V	$\alpha 1(V)_2\alpha 2(V)$	non-fibrous?	placenta, blood vessels

complement and the tail region of acetylcholinesterase. The remaining amino acids in the polypeptide chain are variable, but noteworthy among them is lysine. Some lysine residues are hydroxylated post-translationally to form hydroxylysine. Both lysine and hydroxylysine may be glycosylated and some residues are involved in the formation of the cross-links, dehydrodihydroxylysinonorleucine and dehydrohydroxylysinonorleucine, upon which the stability of the molecules and fibrils depends. The number of lysine and hydroxylysine residues is thus significant.

The molecules of the fibrous forms of collagen come together in a very precisely determined manner to form the characteristically banded fibrils. The repeating period (D) is approximately 67 nm; this figure has been determined by X-ray diffraction of native, hydrated, rat tail tendon collagen. Each molecule is 4.4 D in length. In the fibril the molecules are aligned in parallel but each is staggered by exactly 67 nm (D) with respect to its neighbours (Fig. 19). The gaps between the ends of the molecules are an intrinsic feature of this model, which is the modified quarter-stagger model of Hodge, A. and Petruska, J. (1963). The banding pattern seen in electron micrographs after both positive and negative staining is a reflection of the distribution of the polar, or charged, amino acids along the α-chains.

The vertebrate type IV basement membrane collagens do not form fibrils *in vivo*. The molecules are more diverse and are longer (approximately 400 nm) than those of the fibrous collagens (Heathcote, J. and Grant, M., 1981; Timple, R., *et al.*, 1981). It is suggested that the longer length may be due to the retention of the C-terminal extension peptide (see section 4.3) and that this may be responsible for the knob seen in electron micrographs

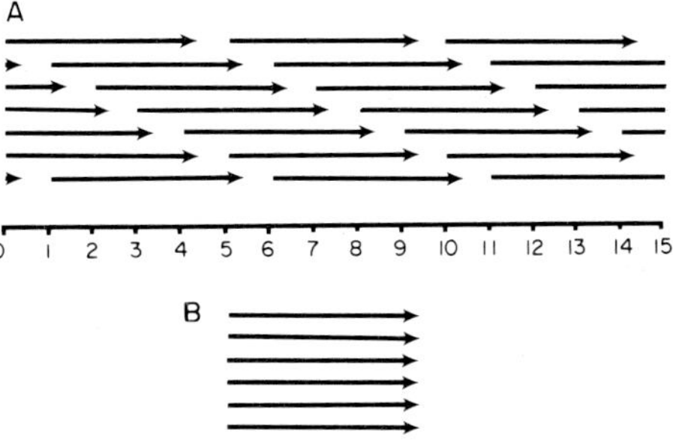

FIG. 19. **A**: Diagram to show the precise arrangement of the collagen molecules in a fibril according to the modified quarter-stagger model of Hodge, A. and Petruska, J. (1963); the gaps between the ends of the molecules are inherent in this model. Each spacing equals "D", the bending periodicity, and each molecule is 4.4 D in length. **B**: The arrangement of the collagen molecules in a segment-long-spacing crystallite.

at one end of the molecule (Fessler, J. *et al.*, 1982). The molecule consists of at least two dissimilar polypeptide chains, with molecular weights between 165,000 and 185,000. A model for the aggregation of type IV collagen molecules *in vivo* has been proposed recently (Timple, R., *et al.*, 1981). It is suggested that two molecules are joined at the C-terminal ends to form a dimer and that each N-terminal end of this dimer is linked to the N-terminal ends of three more dimers, so that a network of collagen molecules is produced.

The only invertebrate basement membrane collagen to be studied extensively to date is from the intestinal basement membrane of the roundworm, *Ascaris suum* (Hudson, B., 1978), and it was concluded that this invertebrate collagen is similar to vertebrate basement membrane collagens.

The glycosaminoglycans are a more heterogeneous group of substances. They are long-chain polymers composed of repeating, identical disaccharide units (Hardingham, T., 1981). Each disaccharide consists of an amino sugar, glucosamine or galactosamine, and a uronic acid moiety, either glucuronic or iduronic acid, or in some instances a sugar. The polymers have anionic charges along their length; these are the carboxyls of the uronic acid, and sulphates which may be attached to the amino sugar. The glycosaminoglycans include hyaluronate, chondroitin-4- and chondroitin-6-sulphates, dermatan sulphate, keratan sulphate, heparin and heparan sulphate; the repeating disaccharides are shown in Fig. 20. With the exception of hyaluronate, all the glycosaminoglycans have an obligatory association with a polypeptide, to form a proteoglycan. The glycosaminoglycan is attached to the polypeptide via a linkage region which consists of a trisaccharide unit (Gal-Gal-Xyl); the xylose residue is glycosidically linked to a serine residue. Typically, the polymers of chondroitin sulphate are much longer than those of keratan sulphate. The size and composition of the proteoglycans varies with the tissue (see Hardingham, T., 1981).

In cartilage the proteoglycans themselves are attached by a link protein to a single strand of hyaluronate, and the resulting aggregate, which in electron micrographs appears to have a feather-like form, may be several micrometres in both length and breadth. Heparan sulphate is associated with cell surfaces. Apart from one or two rare examples,

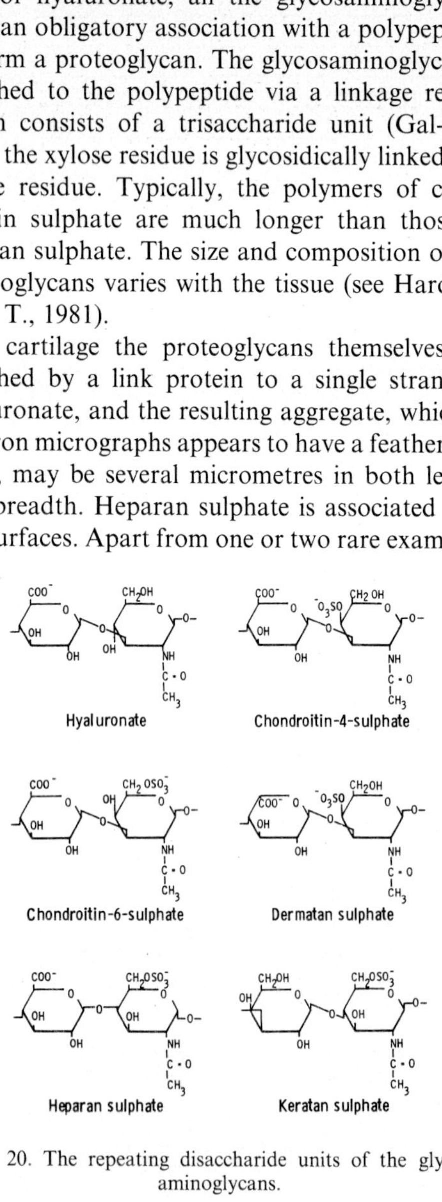

Hyaluronate

Chondroitin-4-sulphate

Chondroitin-6-sulphate

Dermatan sulphate

Heparan sulphate

Keratan sulphate

FIG. 20. The repeating disaccharide units of the glycosaminoglycans.

heparin is the only intracellular glycosaminoglycan and it occurs in the mast cell granules of vertebrates; it is released during trauma and is an anticoagulant of the blood. Hyaluronate may form long polymers more than 1 μm in length (Fessler, J. and Fessler, L., 1966) and its physical conformation depends on the ionic strength of the surrounding medium.

The physical properties of the glycosaminoglycans are related to their size and anionic charges. Because they form large polymers and aggregates and have a high charge density, they are highly expanded molecules which occupy a significant volume of the matrix. Thus it has been shown by Laurent and his co-workers (Laurent, T., 1977) that they form charged sieves which can impede large molecules, and repel other anionically charged ions and molecules which try to cross the matrix. The chondroitin sulphates are normally found together with fibrous collagens and may also affect the polymerization of collagen molecules to form fibrils. Another important property of glycosaminoglycans is their ability to bind water. The viscosity of hyaluronate solutions is determined by the amount of water bound, and its ability to form viscous solutions is important for its lubricating and shock-absorbing function in, for example, joints and the vitreous humour of the eye.

The glycoproteins of connective tissue are very heterogeneous, and have received far less attention than the collagens and glycosaminoglycans. Recently, two glycoproteins from vertebrate tissues, fibronectin and laminin, have been characterized. Fibronectin is found in a soluble form in the blood and plasma, but in an insoluble form on cell surfaces, and while both forms have been studied extensively, it was realized only recently that only one molecule exists (see review by Ruoslahti, E. *et al.*, 1981). The fibronectin molecule consists of two identical polypeptide chains, and while immunological evidence suggests that the soluble and insoluble forms are chemically very similar, some differences, probably due to variations in the glycosylation of the molecules, are evident on gel electrophoresis. It is suggested that fibronectin might form a bridge between the cells and matrix, since the molecule has binding sites for collagen, glycosaminoglycans, fibrinogen, actin and cell surfaces. Thus, it may be important in developmental and healing processes, but investigations along these lines have only just begun.

Laminin is a polypeptide of high molecular weight which consists of two subunits, one twice the size of the other. Carbohydrate accounts for about 15% of the weight of the molecule. It is a constituent of the lamina rara of the basement membrane (Foidart, J. *et al.*, 1980). Its occurrence, as demonstrated by immunofluorescence, in the early mouse embryo suggests that it may be important in morphogenetic movements (see Heathcote, J. and Grant, M., 1981).

3.1 Collagen molecules of insects

The identification of collagen in insects prior to the development of modern biochemical, biophysical and electron microscopical techniques was very difficult (see section 1). The first biochemical studies used crude paper chromatography from which the presence of hydroxyproline, and hence collagen, was established (Ashhurst, D., 1959, 1961a; Ashhurst, D. and Richards, A., 1964b). The next attempt to confirm the biochemical similarity of insect and other collagens was by Harper, E. *et al.*, (1967); their amino acid analyses of collagen from several cockroaches (Table 2) confirmed that a true collagenous protein is present.

In the last 15 years many new methods and parameters for the characterization of collagen molecules have been developed. Thus, the question which is asked about any collagen molecule is: is it similar to one of the well-established vertebrate collagen types, or is it a unique molecule? This question must be approached in a variety of ways. Firstly, it is possible to discover by electron microscopical examination of native fibrils and segment-long-spacing (SLS) crystallites whether the molecular lengths and distribution of polar amino acids along the α-chains are similar. Secondly, biochemical evidence from cyanogen bromide peptide patterns, amino acid analyses, salt precipitation and other techniques gives information which can be compared to that of other collagens.

The first problem encountered in attempting such an investigation of insect collagen is getting

Table 2: Amino acid composition of mammalian and insect collagen α-chains.

Amino acids	Amounts of amino acids (residues per 1000)					
	Human skin[1] α1(I) chains 2.5 M NaCl precipitate	Human skin[1] α1(I) chains 4.0 M NaCl precipitate	*L. migratoria*[2] 4.0 M NaCl precipitate	*P. americana*[3] 4.0 M NaCl precipitate	*P. americana*[4]	*L. migratoria*[2] basement membrane (calculated)
Hydroxyproline	93.0	98.0	109.3	107	41.2	90.3
Aspartic acid	46.3	45.4	46.7	50	66.4	83.0
Threonine	20.6	18.0	24.1	24	40.1	23.7
Serine	36.2	37.3	32.9	39	62.5	23.6
Glutamic acid	62.8	66.6	92.2	91	86.5	76.8
Proline	131.0	126.1	129.8	120	88.1	67.8
Glycine	320.1	326.2	324.8	320	235.2	300.7
Alanine	115.6	127.1	49.0	78	104.7	33.4
Valine	21.6	18.3	16.3	22	44.0	26.2
Methionine	7.3	7.6	n.d.	n.d.	3.8	n.d.
Isoleucine	9.6	9.5	18.5	16	25.3	42.8
Leucine	26.4	24.1	38.2	35	35.3	56.2
Tyrosine	2.3	1.9	1.6	n.d.	31.7	11.6
Phenylalanine	12.8	11.9	10.4	13	27.6	20.5
Hydroxylysine	4.6	9.6	24.1	16	26.4	64.6
Lysine	31.9	25.2	14.1	20	20.7	16.9
Histidine	3.9	3.6	4.6	4	18.2	0.17
Arginine	54.7	50.0	53.1	45	34.5	55.1

[1] From Uitto, J. (1979).
[2] From Ashhurst, D. and Bailey, A. (1980).
[3] From François, J. *et al.* (1980).
[4] From Harper, E. *et al.* (1967).

sufficient collagen from a restricted source. Ashhurst, D. and Costin, N. (1971a, 1974, 1976) had used the ejaculatory duct of adult male locusts (*Locusta migratoria*) for several investigations, and since the amount of connective tissue around the duct is relatively large, this source of collagen was used for extensive studies (Ashhurst, D. and Bailey, A., 1978, 1980).

Invertebrate collagens are notoriously insoluble; in order to reduce the cross-linking of the collagen, newly moulted adult locusts were fed on wheat seedlings sprayed with β-aminoproprionitrile, a lathrytic agent which prevents the formation of cross-links. After 6 days the ducts were removed, and acid-soluble collagen was extracted. By dialysing an acid extract against either ATP at acid pH, or phosphate buffer at pH 9.2, it is possible to form either SLS crystallites or reconstituted native fibrils, respectively (Fig. 19). These may then be compared with similarly prepared crystallites and fibrils made from vertebrate collagens. Micrographs of positively stained SLS crystallites and reconstitued fibrils made from locust collagen are seen in Figs 21 and 22 together with those made from rat tail tendon, type I collagen. It is immediately apparent that the banding patterns of both SLS crystallites, and both reconstituted fibrils, are identical. A similar investigation by François, J. *et al.* (1980), using pepsin-solubilized collagen from the mesenteric sheath of the cockroach (*Periplaneta americana*) yielded exactly the same result. Thus, the following conclusions can be drawn from the SLS crystallites: firstly, the locust and cockroach collagen molecules are the same length as the rat tail tendon molecules, and secondly, the distribution of polar amino acids along the α-chains is the same as that in mammalian type I molecules. The molecules form fibrils with a periodic banding pattern exactly similar to that of mammalian type I collagen, which means that when the molecules are assembled into fibrils, the molecular arrangement must be the same (see Fig. 19). Hence it follows that the periodicity of the insect fibrils must be identical to that of mammalian fibrils, that is 67 nm. Any variation from this figure measured on electron micrographs must be due to the preparative techniques*. The details of the banding pattern are not seen so readily in sectioned fibrils, but bands I and XI can usually be resolved, while bands III to X are less clear (Fig. 5).

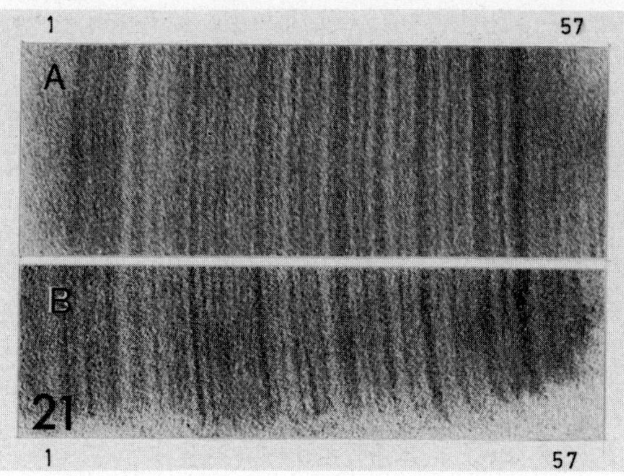

FIG. 21. Segment-long-spacing crystallites positively stained with phosphotungstic acid and uranyl acetate. **A**: Crystallite made from rat tail tendon collagen. **B**: Crystallite made from locust collagen. The numbering of the bands is that of Bruns, R. and Gross, J. (1973); 1 is at the N-terminal end of the molecules. (The magnification has been calculated and is approximately × 228,000.) (From Ashhurst, D. and Bailey, A., 1980.)

The amino acid analyses of several insect collagens and mammalian type I and type I trimer are given in Table 2. While amino acid analyses are somewhat variable, it is immediately apparent that the composition of insect collagens is similar to that of the mammalian collagens. The significant features are that in all, glycine accounts for approximately one-third of the total amino acids, hydroxyproline is present, and together with proline, makes up about 20% of the amino acids. Hydroxylysine occurs in amounts greater than in mammalian type I, but in type I trimer from human skin, the number of hydroxylysine residues is also raised (Uitto, J., 1979).

The locust and cockroach collagen molecules consist of one type of α-chain — that is, they are trimers. When the α-chains in the pepsin digest were

* All the preparative techniques for electron microscopy involve dehydration and other procedures which lead to shrinkage. Indeed it is this author's experience that not only is the shrinkage variable in identically prepared specimens, but it can be as great as 30%. Thus, most calibrations of high magnification electron micrographs of collagen are not calculated from the actual magnification used, but from parameters measured by other methods. Thus, the repeating periodicity for native, hydrated type I collagen is known to be 67 nm from X-ray diffraction measurements. The length of SLS crystallites is derived from the calculation of the actual distance between bands 6 and 22; this part of the molecule is the α1CNBr-8 peptide which is 79 nm long (Bruns, R. and Gross, J., 1973; Rauterberg, J. and Kühn, K., 1971).

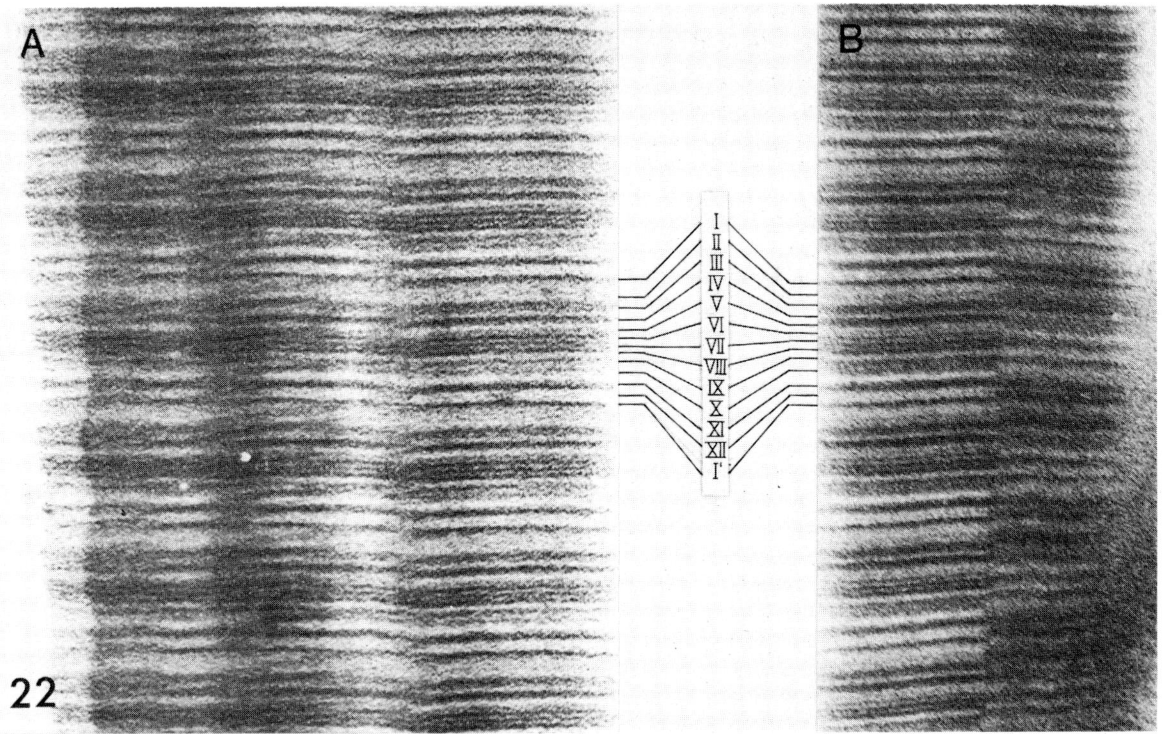

FIG. 22. Reconstituted collagen fibrils made from **A**, locust and **B**, rat tail tendon, collagens and positively stained with phosphotungstic acid and uranyl acetate. The notation of the banding is from Bruns, R. and Gross, J. (1974). (The calculated magnification is approximately × 254,000.) (From Ashhurst, D. and Bailey, A., 1980.)

isolated by salt fractionation, the locust material produced a minor precipitate at 2.4 M NaCl, and a major precipitate at 4.0 M NaCl. All the cockroach α-chains precipitated at 4.0 M NaCl. The precipitates were analysed by sodium dodecylsulphate/acrylamide gel electrophoresis; both the locust and cockroach 4.0 M NaCl precipitates produced a single band with the same mobility as the α1(I) chain (Fig. 23). This shows that only one type of α-chain is present and that it is similar to that of the mammalian type I α1 chain. The 2.4 M NaCl locust precipitate after mercaptoethanol treatment gave a faint band with a mobility greater than the major component. It is thought that this minor component is a basement membrane collagen, and it will be discussed later.

Another criterion on which collagens may be compared is the pattern of peptides produced after cleavage of the polypeptide chains at the methionine residues by cyanogen bromide. Neither the locust, nor the cockroach collagen has a pattern comparable to that of the α1(I) chain, but this indicates only that the position of the methionine residues differs. Sea anemone collagen, which has many properties in common with mammalian type I collagen, also has a different spectrum of CNBr peptides (Nowack, H. and Nordwig, A., 1974).

The molecular cross-link pattern of the locust collagen after borohydride reduction reveals that this collagen possesses the two reducible cross-links, dehydrodihydroxylysinonorleucine and dehydrohydroxylysinonorleucine; the cockroach collagen apparently possesses only the former. These cross-links are found in mammalian and other invertebrate collagens (see Ashhurst, D. and Bailey, A., 1980). They are formed by lysine and hydroxylysine residues, and thus the increased number of hydroxylysine residues in these and other invertebrate collagens is probably a significant factor in their greater insolubility when compared to vertebrate collagens since the dihydroxy cross-link

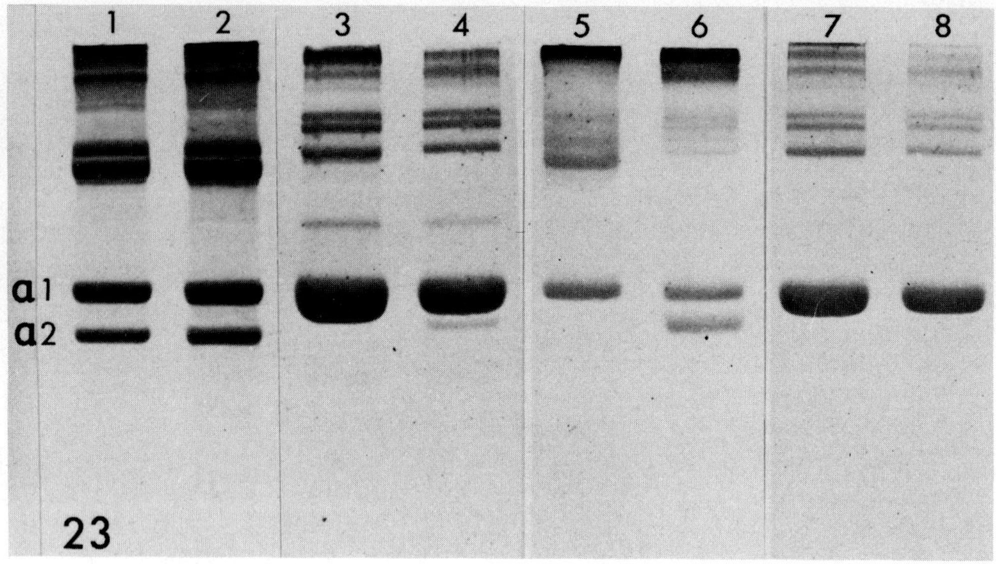

FIG. 23. Sodium dodecylsulphate/polyacrylamide-gel electrophoresis of the fractions obtained by precipitation of the pepsin digest of locust ejaculatory ducts with NaCl. Tracks 1 and 2: standard $\alpha 1$ and $\alpha 2$ chains from rat tail tendon collagen, without, and with, mercaptoethanol, respectively. Tracks 3 and 4: total pepsin digest of locust ducts, without, and with, mercaptoethanol, respectively; note the appearance of a minor component of higher mobility following mercaptoethanol treatment. Tracks 5 and 6: 2.4 M NaCl precipitate from pepsin digest of locust duct, without, and with, mercaptoethanol treatment; note the increased proportion of the mercaptoethanol-sensitive component. Tracks 7 and 8: 4.0 M NaCl precipitate from pepsin digest of locust duct, without and with, mercaptoethanol, respectively. (From Ashhurst, D. and Bailey, A., 1980.)

is more stable than the lysine derived monohydroxy cross-link.

François, J. and his co-workers (1980) also found that approximately 8.6% of the total dry weight of cockroach collagen is neutral sugar, which is consistent with the general finding that invertebrate collagens contain more hexoses than the vertebrate molecules.

On the basis of their data Ashhurst, D. and Bailey, A.(1980) concluded that the fibrous collagen molecule extracted from *Locusta* is very similar to that of vertebrate type I trimer collagen, differing only in the distribution of some amino acids. Despite many obvious points of similarity between the cockroach and type I trimer collagens, François, J. *et al.* (1980) considered, on the basis of the amino acid composition, that this collagen may be more similar to mammalian type II, i.e. the cartilage collagen. Their argument is based solely on the higher glutamic acid and hydroxylysine content of the cockroach collagen, and ignores their other significant evidence, which is firstly that the SLS crystallite banding pattern is the same as that of type I collagen, while type II has a completely different pattern (Stark, M. *et al.*, 1972), and secondly, that the cockroach collagen elutes on CM-cellulose chromatography later than the $\alpha 1$(I) chains, while type II α-chains elute with the type I α-chains. Thus, they base their conclusion on the relative amounts of two amino acids, and choose to ignore the weight of their other evidence. It would indeed seem peculiar if the cockroach collagen is so unique among invertebrate collagens.

The thin, indistinctly banded fibrils of the Lepidoptera and other orders have been an enigma. Some work currently in progress has shown that pepsin-solubilized collagen from the dorsal mass of the nerve cord of late pupae of *Manduca sexta* forms SLS crystallites similar to those of type I collagen and that $\alpha 1$-chains, with the same mobility as $\alpha 1$(I)-chains on sodium dodecylsulphate/acrylamide gels are precipitated by 2.4 M NaCl (unpublished observations). These results indicate that the moth collagen may be very similar to other insect collagens.

Several other invertebrate collagens, from

animals such as sea anemones, tapeworms, liver flukes, earthworms (body wall collagen), crabs, various molluscs, sea cucumbers and ascidians have been identified as trimers with α1-chains similar to those of type I collagen (Kimura, S. and Matsuura, F., 1974; Nordwig, A. *et al.*, 1970; Nowack, H. and Nordwig, A., 1974; Pikkarrainen, J. *et al.*, 1968). A few instances of both α1 and α2 chains in the collagens of crustaceans and molluscs have recently been documented (Kimura, S. and Matsuura, F., 1974; Pucci-Minafra, I. *et al.*, 1978). While the amount of data on each collagen varies, all that is available is in agreement with the conclusion that the fibrous collagens of the invertebrates have many properties in common with vertebrate type I collagen, and in some instances, more especially with type I trimer. Thus, it appears that a molecule similar to the type I trimer already existed in the coelenterates and that during the evolution of the higher invertebrates and vertebrates this molecule has been conserved with only minor alterations in some of the amino acids.

The basement membrane collagen of the locust is more elusive as it is present in such small amounts. It was detected on the gels after mercaptoethanol treatment, which breaks disulphide bonds, as a faint band with a mobility greater than the major component of the 2.4 M NaCl fractional precipitate. Its amino acid composition was calculated and is given in Table 2 (Ashhurst, D. and Bailey, A., 1980). The mobility of this component on dodecylsulphate/acrylamide gel electrophoresis (Fig. 23) is similar to that of α1(IV) chains and certain features of the amino acid analysis, for example, the high lysine/hydroxylysine ratio and the high contents of hydroxyproline and leucine, suggest that it is a basement membrane collagen. More recently a collagen has been isolated from locust leg muscle, which has similar properties to the basement membrane collagen from the ejaculatory duct (Fig. 24); work

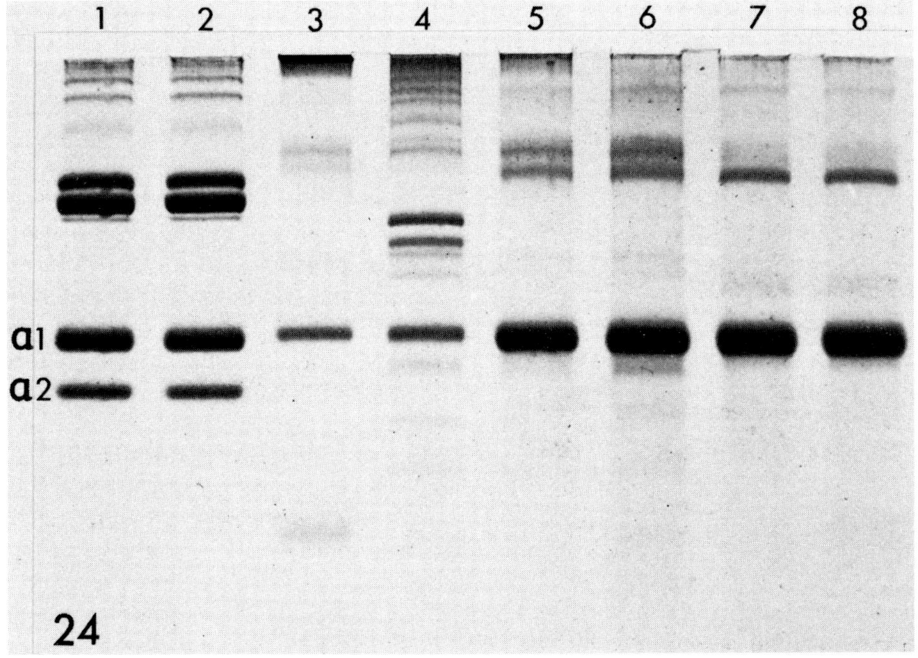

FIG. 24. Sodium dodecylsulphate/polyacrylamide-gel electrophoresis of the fractions obtained by precipitation of the pepsin digest of locust leg muscle with NaCl. Tracks 1 and 2: standard α1 and α2 chains from rat tail tendon collagen, without, and with, mercaptoethanol, respectively. Tracks 3 and 4: total pepsin digest of locust leg muscle, without, and with, mercaptoethanol, respectively. Tracks 5 and 6: total pepsin digest of locust ejaculatory ducts, without, and with, mercaptoethanol, respectively. Tracks 7 and 8: bovine type II collagen, without, and with, mercaptoethanol, respectively. The basement membrane collagen α-chains can be seen immediately below the position of the α1(I) chains on tracks 4 and 6.

is in progress to characterize this collagen further (Ashhurst, Bailey and Sims, unpublished observations).

A collagen from *Drosophila*, with morphological and biochemical properties similar to those of mammalian type IV collagen, has recently been characterized by Fessler, J. and his co-workers (Fessler, J. *et al.*, 1982; Fessler, J. *et al.*, 1984). The collagen has been extracted from both the media of cultured *Drosophila* KC cells and from *Drosophila* larvae. Electron micrographs of rotary shadowed isolated molecules show a molecule, approximately 465 nm long with a knob at one end. This appearance and size is similar to that of mouse type IV procollagen, and there is evidence that the knob is at the C-terminal end of the molecule (Bächinger, H. *et al*, 1982). Antibodies to this molecule are bound by the basement membranes of *Drosophila* larvae.

Another approach, used recently by Monson and her co-workers (Monson, J. *et al.*, 1982) was to screen a library of *Drosophila melanogaster* genomic DNA fragments with a chicken proα2(I) cDNA clone to isolate collagen-like genomic clones. The fragment isolated is composed of two large coding sequences which together specify a sequence of 469 amino acids. The polypeptide is composed almost entirely of the Gly-X-Y repeat characteristic of the helical portion of the molecule, but there are four minor interruptions in the sequence. Similar interruptions of the helical portion of the molecule occur in mouse basement membrane collagen, and hence it is suggested that this *Drosophila* collagen gene may encode part of a non-fibrous, basement membrane collagen molecule.

Thus evidence for the presence of insect basement membrane collagens, with properties similar to those of the mammalian type IV collagens, is gradually accumulating.

3.2 Glycosaminoglycans

Most information about the glycosaminoglycans of the connective tissue comes from histochemical studies. Baccetti, B. (1955a,b, 1956a) found that the neural lamella and the connective tissue matrices associated with the fat body, gut, Malpighian tubules, muscles and epidermis of the orthopteran *Anacridium aegyptium* are positive with the periodic acid-Schiff (PAS) test, but that only the matrices of the nervous system, fat body and gut stain metachromatically with toluidine blue. The connective tissue matrices of lice and the neural lamellae of the locust, *Locusta migratoria*, and cockroach, *Periplaneta americana*, appeared to differ in that they did not stain metachromatically with toluidine blue (Ashhurst, D., 1959, 1961a; Pipa, R. and Cook, E., 1958). This led to the conclusion that acidic carbohydrates were not universally present in insect connective tissue matrices. This finding seemed anomalous, since glycosaminoglycans are usually associated with fibrous connective tissues; collagen fibrils were identified in neural lamellae in 1958 (see section 1). A little later it was found that the connective tissue mass on the dorsal side of the abdominal nerve cord of the waxmoth, *Galleria mellonella*, is strongly metachromatic with toluidine blue, and that this metachromasia is hyaluronidase-labile (Ashhurst, D. and Richards, A., 1964b).

More critical histochemical tests for glycosaminoglycans were developed during the 1960s by Scott and Spicer and their co-workers (see Ashhurst, D., 1979b). These techniques enable the identification of the individual glycosaminoglycans and the rationale is presented in Table 3. Thus when this battery of tests was performed on the neural lamellae of several adult insects of different orders (*Carausius morosus*, *Galleria mellonella*, *Locusta migratoria*, *Periplaneta americana*) the results in Table 4 were obtained (Ashhurst, D. and Costin, N., 1971b,c). They suggest that all are similar in possessing chondroitin, and possibly dermatan, sulphates, but while those of the stick insect, locust and cockroach contain a highly sulphated hyaluronidase-resistant component, identified as keratan sulphate, the neural lamella and dorsal connective tissue mass of the waxmoth nervous system does not. The amount of keratan sulphate present usually increases as a matrix ages (Stockwell, R. and Scott, J., 1965) and as the connective tissue of the adult waxmoth develops over a few days and the adult lives only a short time, this ageing process may not occur. A recent study of the neural lamella during larval development in *Locusta migratoria* and *Periplaneta americana* showed that sulphated glycosaminoglycans increase during development and that keratan sulphate is present only in the last larval instar and adult (Ashhurst, D., 1984). The presence of glycosaminoglycans in the neural lamella of the waxmoth was recently confirmed by ultrastructural studies using ruthenium red (Dybowska, H. and Dutkowski, A., 1977, 1979).

The connective tissue matrices associated with other organs also give similar results with these histochemical tests. Thus, the connective tissues associated with the ejaculatory duct of the male locust (*Locusta migratoria*) (Table 4) and with the mesenteron of the cockroach (*Periplaneta*

Table 3: Histochemical reactions of glycosaminoglycans.

Test	Reactive groups	Reaction of GAGs
Alcian blue, pH 2.5	COO^- $SO_4{}^{2-}$	all GAGs, but some sulphate groups may be blocked by proteins
Alcian blue, pH 1.0	$SO_4{}^{2-}$	sulphated GAGs only
Alcian blue, pH 5.7 + 0.05 M $MgCl_2$	COO^- $SO_4{}^{2-}$	all GAGs
Alcian blue, pH 5.7 + 0.1 M $MgCl_2$	COO^- $SO_4{}^{2-}$	
Alcian blue, pH 5.7 + 0.2 M $MgCl_2$	$SO_4{}^{2-}$	sulphated GAGs
Alcian blue, pH 5.7 + 0.4 M $MgCl_2$	$SO_4{}^{2-}$	
Alcian blue, pH 5.7 + 0.6 M $MgCl_2$	$SO_4{}^{2-}$	highly sulphated GAGs only — heparan, keratan sulphate
Alcian blue, pH 5.7 + 0.8 M $MgCl_2$	$SO_4{}^{2-}$	
Alcian blue, pH 5.7 + 1.0 M $MgCl_2$	$SO_4{}^{2-}$	keratan sulphate *only*
High iron diamine	$SO_4{}^{2-}$	sulphated GAGs
Periodic acid-Schiff (standard test)	*vic*-glycol groups	negative
Testicular hyaluronidase	splits uronic acid	hyaluronate, chondroitin-4- and 6-sulphates labile
Neuraminidase	splits sialic acid	sialic acid residues labile

Key: GAG = Glycosaminoglycan

Table 4: Histochemical reactions of different fibrous connective tissues.

Histochemical tests	Adult neural lamellae				Dorsal mass of *Galleria* adult nerve cord[2]	Ejaculatory duct connective tissue of *Locusta*[3]	
	Periplaneta[1]	*Locusta*[1]	*Carausius*[2]	*Galleria*[2]		2-day adult	26-day adult
Alcian blue, pH 2.5	0–1+	1+	1–2+	1+	3–4+	1–3+	1–3+
Alcian blue, pH 1.0	0–3+	1–3+	2–4+	0–1+	0–3+	1–4+	2–4+
Alcian blue, pH 5.7 + 0.05 M $MgCl_2$	3+	4+	4+	4+	3–4+	4+	4+
Alcian blue, pH 5.7 + 0.1 M $MgCl_2$	3+	4+	4+	3+	3+	3+	4+
Alcian blue, pH 5.7 + 0.2 M $MgCl_2$	3+	3–4+	3–4+	2+	1–2+	1–4+	1–3+
Alcian blue, pH 5.7 + 0.4 M $MgCl_2$	1–2+	2–3+	3+	1+	0–±	1–3+	1–3+
Alcian blue, pH 5.7 + 0.6 M $MgCl_2$	1+	1–3+	1–2+	±	0	1+	1–2+
Alcian blue, pH 5.7 + 0.8 M $MgCl_2$	1+	1–2+	1–2+	0	0	1+	1–2+
Alcian blue, pH 5.7 + 1.0 M $MgCl_2$	±	1+	1–2+	0	0	0	1–2+
PAS (standard)	3+	3+	4+	4+	3–4+	1–4+	2–4+
PAS (after diastase)	3+	3+	4+	4+	3–4+	1–4+	1–4+
HID	2+	2+	1–3+	1–2+	2+	1–2+	1–3+
Hyaluronidase-AB, pH 2.5	±	0–1+	1+	0–1+	2+	+	±
Hyaluronidase control-AB, pH 2.5	±	0–1+	2+	1+	3–4+	2–3+	2–3+
Hyaluronidase-AB, pH 1.0	1+	1+	2–3+	±	0–1+	1+	
Hyaluronidase control-AB, pH 1.0	1+	2+	3–4+	1+	0–3+	1–2+	1–4+
Neuraminidase-AB, pH 2.5	±	0–1+	1–2+	1+	3–4+	1–3+	1–3+
Neuraminidase control-AB, pH 2.5	±	0–1+	1–2+	1+	3–4+	1–3+	1–3+

Key: AB = Alcian blue; HID = high iron diamine; PAS = periodic acid-Schiff; 1–4+ = increasing intensity of reaction or staining; ± = very weakly positive reaction or staining; 0 = negative, or no reaction or staining.

[1] Data from Ashhurst, D. and Costin, N., 1971b.

[2] Data from Ashhurst, D. and Costin, N., 1971c.

[3] Data from Ashhurst, D. and Costin, N., 1971a.

americana) contain increasing amounts of sulphated glycosaminoglycans as they develop through the larval instars to the adult, when chondroitin and keratan sulphates are present (Ashhurst, D. and Costin, N., 1971a; François, J., 1978).

Thus the anomaly of fibrous matrices seemingly devoid of chondroitin and keratan sulphates was resolved. The dye alcian blue is most commonly used to localize glycosaminoglycans; it is pertinent to reiterate here that alcian blue at pH 2.5, i.e. in 3% acetic acid, is often not bound by glycosaminoglycans due to interference by the polar groups on the proteins (Quintarelli, G. *et al*, 1964) although it is in these conditions that the dye is most commonly used. Thus a battery of tests is essential before any valid conclusions can be drawn (see Ashhurst, D., 1979b).

Histochemical tests locate the glycosaminoglycans in the connective tissues, but unfortunately they do not provide conclusive proof of the presence of a particular substance. This can only be done biochemically, but so far few attempts have been made. In early experiments, hyaluronate, which was presumed to be from the glial lacunar system, was found in extracts of cockroach ganglia (Ashhurst, D. and Patel, N., 1963) and Estes, Z. and Faust, R. (1964) concluded that only hyaluronate is present in the midgut of *Galleria mellonella* larvae. Sharief, F. *et al.* (1973) isolated hyaluronate, chondroitin sulphate and heparan sulphate from whole embryos and larvae of the fly, *Phormia regina*, and confirmed the presence of sulphate groups autoradiographically. They attempted to locate the glycosaminoglycans histochemically using alcian blue, pH 2.5 and colloidal iron, and found that the connective tissues associated with the gut, nervous system, fat body and salivary glands gave positive reactions. The positive reactions of other parts of these organs may be attributable to mucins, with either sialic acid carboxyls or sulphate groups, or in the case of the nuclei, to the phosphate groups on DNA. Alcian blue and other cationic dyes are bound by any anions in the tissues, not just those on the glycosaminoglycans.

A more thorough biochemical analysis of the glycosaminoglycans in the fly, *Calliphora erythrocephala*, during development from the larva to the adult fly was performed by Höglund, L. (1976a,b). The extracted glycosaminoglycans were fractionated and the amounts of hexosamine, hexuronic acids, neutral sugars, sialic acid, etc., in the eluted fractions were determined. The fractions were also subjected to electrophoresis and infrared spectroscopy and enzyme lability was investigated. Thus, hyaluronate, chondroitin, and chondroitin sulphate, and possibly heparan and keratan sulphates were identified in the adult fly (Höglund, L., 1976a). Hyaluronate, chondroitin, and heparan sulphates, together with a "larval acid mucopolysaccharide" ("larval AMPS") were found in the embryo from late cleavage onwards and in the larva (Höglund, L., 1976b). The "larval AMPS" is not sulphated, nor is it labile to testicular hyaluronidase. It contains galactosamine, glucosamine, hexuronic acid and galactose, and Höglund suggests that it may be a mixture of components.

No attempt was made to determine the tissue localization of the glycosaminoglycans in *Calliphora*. These biochemical studies do, however, confirm that glycosaminoglycans are present in insect tissues from an early stage in development, and that the sulphated glycosaminoglycans, chondroitin and keratan sulphates, more particularly the latter, occur in increasing quantities as the adult stage approaches.

The discussion of the glycosaminoglycans so far has been confined to those in the fibrous matrices. It is only recently that glycosaminoglycans have been detected in mammalian basement membranes (Bernfield, M. *et al.*, 1972; Kanwar, Y. and Farquhar, M., 1979), and similarly observations on insects are recent. François, J. (1978) reported that the basement membrane of the epithelial cells of the mesenteron of *Periplaneta* binds alcian blue under conditions which indicate the presence of sulphated glycosaminoglycans, and Dutkowski, A. (1977) found that the basement membrane of the fat body of *Galleria* binds ruthenium red, which again suggests the presence of glycosaminoglycans.

In her early study on cockroach ganglia, Ashhurst, D. (1961) found that the contents of the glial lacunar system stain metachromatically with toluidine blue, and this material was identified biochemically as hyaluronate (Ashhurst, D. and Patel, N., 1963). Further discussion on the glial lacunar system will be given in section 6.

There are suggestions in the literature that

glycosaminoglycans may be found in other locations. Marshall, A. (1968) characterized the mucus in the dwelling tubes of cercopoids and found evidence for glucosamine and glucuronic acid. Similarly, extracts of the peritrophic membrane of *Bombyx mori* contain glucosamine and glucuronic acid (Nisizawa, K. *et al.*, 1963). There are also many examples (see Mustafa, M. and Kamat, D., 1970) in which it is suggested on the basis of limited alcian blue staining that glycosaminoglycans are present in cells. It should, however, be borne in mind that glycosaminoglycans are rarely found within cells and that mucoproteins, i.e. mucus, may be acidic due to sialic acid or sulphate groups attached to the molecule.

3.3 Glycoproteins

The glycoproteins of vertebrate connective tissues have received little attention until recently and nothing is known of the biochemistry of the glycoproteins in insect connective tissue matrices. That glycoproteins are present is shown by the strongly positive reaction with the standard PAS test. The glycosaminoglycans do not give a positive reaction with the standard PAS test (Pearse, A., 1968), and so another group of carbohydrate-containing substances is responsible for the positive reaction.

That insect connective tissue matrices of different types are universally PAS-positive was observed in all the early histochemical studies (Ashhurst, D., 1959, 1961a; Baccetti, B. 1955a,b, 1956a; Bonhag, P. and Arnold, W. 1960; Pipa, R. and Cook, E., 1958; Wigglesworth, V., 1956) and this has been confirmed by all the subsequent studies (Ashhurst, D. and Costin, N., 1971a–c; François, J., 1978). Since collagen is glycosylated this may account for some of the PAS-positivity, but additional glycoproteins, as yet unidentifed, must be present. It is interesting, however, that fibronectin (LETS protein) could not be found among the cell surface proteins isolated from four lines of dipteran and lepidopteran cells in culture (Goldstein, N. and McIntosh, A., 1980). Laminin has, however, been isolated from *Drosophila*; the molecule has a similar appearance in electron micrographs to that of vertebrate laminin, and its sedimentation properties and amino acid analysis are also similar (Fessler, J. *et al.*, 1984). Further work on insect connective tissue glycoproteins must await the development of suitable experimental techniques.

4 FORMATION OF CONNECTIVE TISSUE MATRICES

The constituents of connective tissue matrices are synthesized and secreted by the cells on which they lie, or by fibroblasts within the matrix. Vertebrate fibroblasts produce collagen, glycosaminoglycans, elastin and the microfibrillar protein, and glycoproteins. A typical fibroblast is a spindle-shaped cell which has a large amount of rough endoplasmic reticulum with very dilated cisternae containing an electron-dense amorphous material. The cisternae may come very close to the cell membrane (Fig. 25) and lose the ribosomes on the cisternal membrane adjacent to the cell membrane. The Golgi complexes are rather small and usually consist of many small vesicles with no specific arrangement and only a few lamellae. There are no secretory granules in fibroblasts; the newly synthesized matrix molecules are not stored, but secreted straight into the matrix. A few lysosomes may be present. Insect fibroblasts (Figs 2, 25 and 31) share these features in common with vertebrate fibroblasts (Ashhurst, D. and Costin, N., 1974, 1976; François, J., 1973, 1978).

Some cells have other functions and produce connective tissue substances only at certain periods. Thus, all cells produce their own basement membrane (Hay, E. and Dodson, J., 1973) and smooth muscle cells produce the connective tissue which surrounds them (Ross, R. and Klebanoff, S., 1971). The situation is exactly the same in insects; the perineurial cells produce the neural lamella (Ashhurst, D., 1965), the epidermal cells produce the connective tissue matrix on which they lie (François, J., 1977), the underlying cells produce the tunica propria of the ovariole wall (Bonhag, P. and Arnold, W., 1961), and so on. Evidence for this derives from the presence of large amounts of dilated rough endoplasmic reticulum in the cytoplasm during periods of active matrix secretion.

The contribution of haemocytes, if any, to the connective tissues is most probably, as suggested by Whitten, J. (1962, 1964a), limited to the provision of nutrients to the cells responsible for their synthesis,

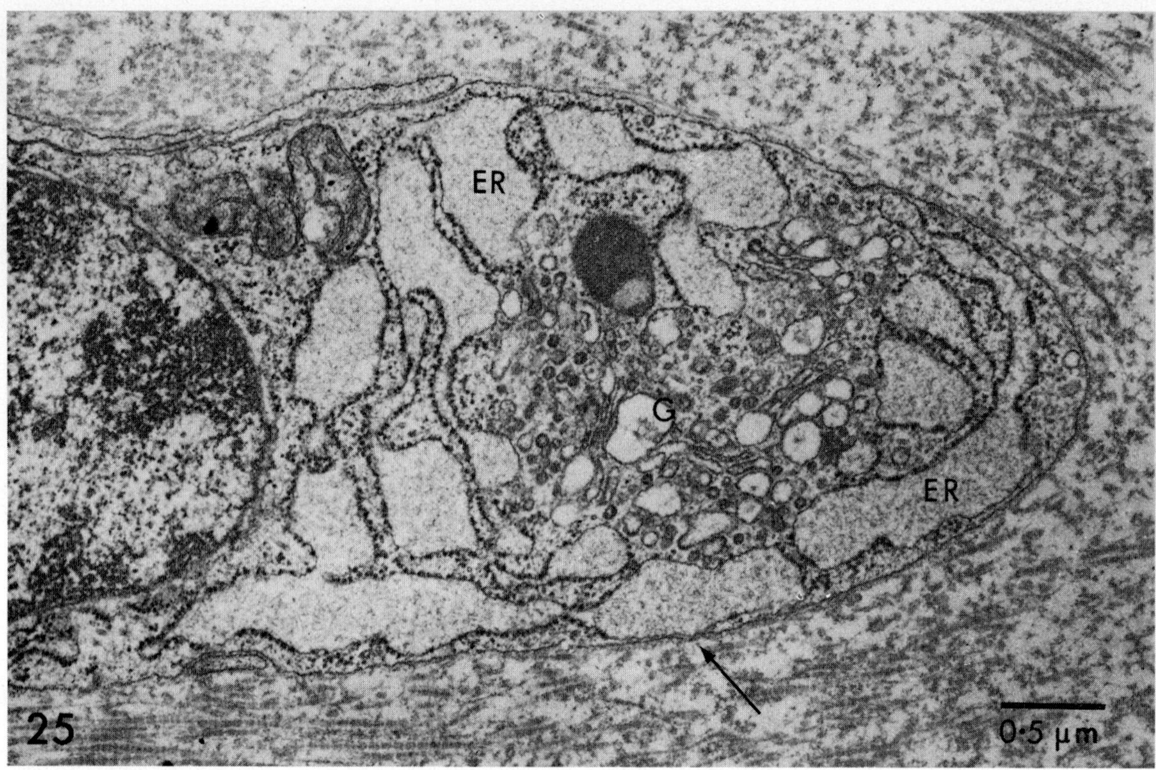

FIG. 25. An electron micrograph of part of the cytoplasm of a fibroblast in the connective tissue of the ejaculatory duct of a 6-day adult locust, *Locusta migratoria*. A Golgi complex (G) is surrounded by very dilated cisternae of rough endoplasmic reticulum (ER). Note that some of the cisternae come very close to the cell membrane (arrow) and have no ribosomes on this surface. (From Ashhurst, D. and Costin, N., 1974.)

though it has been suggested by several workers that haemocytes contribute directly to the insect connective tissue matrices (Beaulaton, J., 1968; Scharrer, B., 1972; Wigglesworth, V., 1956, 1973, 1979; Gupta, A. this volume). The evidence is based on the proximity of haemocytes to developing connective tissues and the PAS-positivity of the connective tissue matrices and granules within the haemocytes, or the presence of dilated rough endoplasmic reticulum in the blood cells. The evidence is at best circumstantial, and does not take into account our knowledge of the biochemistry and mode of formation of connective tissues in other animals. It has been evaluated fully elsewhere (Ashhurst, D., 1979*a*; Gupta, A. this volume) and will not be considered further here.

4.1 Embryonic development

Because they surround organs and bind the cells together, connective tissue matrices are usually formed late in development, although as will be discussed later (section 9), their importance in early stages of embryogenesis is becoming increasingly apparent.

Only one study of connective tissue formation in an embryo has been attempted, and this was confined to the neural lamella of the locust *Schistocerca gregaria* (Ashhurst, D., 1965). In conditions in which embryonic development lasts for 12 days, the neural lamella is first seen as a basement membrane in the 9-day embryo (Fig. 26). During the next 3 days more matrix is laid down, so that a typical, thin, fibrous neural lamella is present in the newly hatched nymph. During this period the perineurial cells enlarge and contain typical rough endoplasmic reticulum, but this is much reduced when the first-instar neural lamella is fully developed.

The neural lamella gradually thickens during the larval instars, but no-one has yet examined the

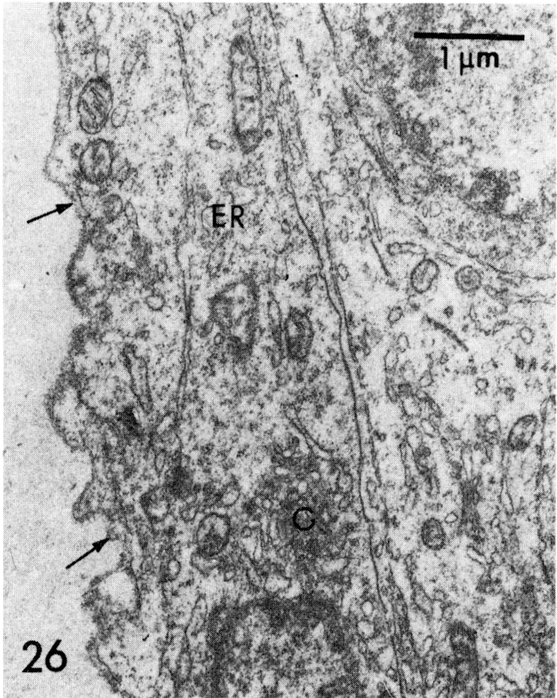

FIG. 26. Part of a perineurial cell of a 9-day embryo locust, *Schistocerca gregaria*. The cytoplasm contains some rough endoplasmic reticulum (ER) and a Golgi complex (G). The developing neural lamella can be seen (arrows); it appears like a basement membrane.

perineurial cells to see if there is renewed development of rough endoplasmic reticulum at, for example, the time of moulting when active matrix secretion might occur. The perineurial cells of first-instar larvae of *Galleria mellonella* contain a lot of rough endoplasmic reticulum, while it is reduced in the second and third instars (Osińska, H., 1981); unfortunately the larvae were not accurately aged within each instar and so it is impossible to conjecture whether the second and third instar cells might be in a quiescent intermoult period. There is evidence, however, that new endoplasmic reticulum might be formed in the perineurial cells of the tick, *Boophilus microplus*, after the adult moult (Binnington, K. and Lane, N., 1980).

4.2 Post-embryonic development

It is obvious from the preceding section that the connective tissues laid down in the embryo continue to develop in exopterygotes until the adult stage is reached, or in endopterygotes, until pupation.

The connective tissue around the mesenteron of the cockroach, *Periplaneta americana*, is present in the first instar as a layer 6–10 μm thick, in which active fibroblasts, muscle cells and nerves are embedded in a fibrous matrix (François, J., 1978). A basement membrane separates the epithelial cells from the connective tissue. In the succeeding instars, the layer gets thicker to reach the adult dimensions of 10–30 μm. The cellular content of the layer is similar throughout, and both active and inactive fibroblasts were observed in the adult. The initial stages of the development of this layer are in the embryo.

The connective tissue around the ejaculatory duct of the male locust, *Locusta migratoria*, in contrast, is formed during post-embryonic development (Ashhurst, D. and Costin, N., 1974). In the second larval instar the duct appears as a small tube surrounded by a few spindle-shaped cells (Fig. 27). By the fourth instar a layer of matrix is present along the basal membranes of the epithelial cells, and the spindle-shaped cells form a more compact layer (Fig. 28). Electron micrographs at this stage (Fig. 30) show that a layer of compact fibrous tissue is present which separates the differentiating fibroblasts from the epithelial cells. During the fifth instar, fibrous matrix begins to accumulate between the fibroblasts, but it is not until the first 6 days after the adult moult that most of the matrix is produced (Figs 2, 25 and 29), and the cells are fully active. In the sexually mature adult the cells are smaller with less rough endoplasmic reticulum; that is, they appear quiescent.

Connective tissues also develop during the pupal stage, since the removal of the larval connective tissue is necessary to permit the reorganization of the organs to their adult form. The nervous system is no exception. In the Lepidoptera the larval neural lamella is destroyed immediately after pupation (see section 8.1; Lockshin, R. vol. 2). While the perineurial cells are denuded, some migrate and form a cellular mass on the dorsal side of the abdominal connectives. As the neural lamella begins to reform, areas of matrix are produced between the cells of the dorsal mass. At this stage both the perineurial and dorsal mass cells contain large dilated cisternae of rough endoplasmic reticulum, but the Golgi complexes are very small (Fig. 31). Matrix production

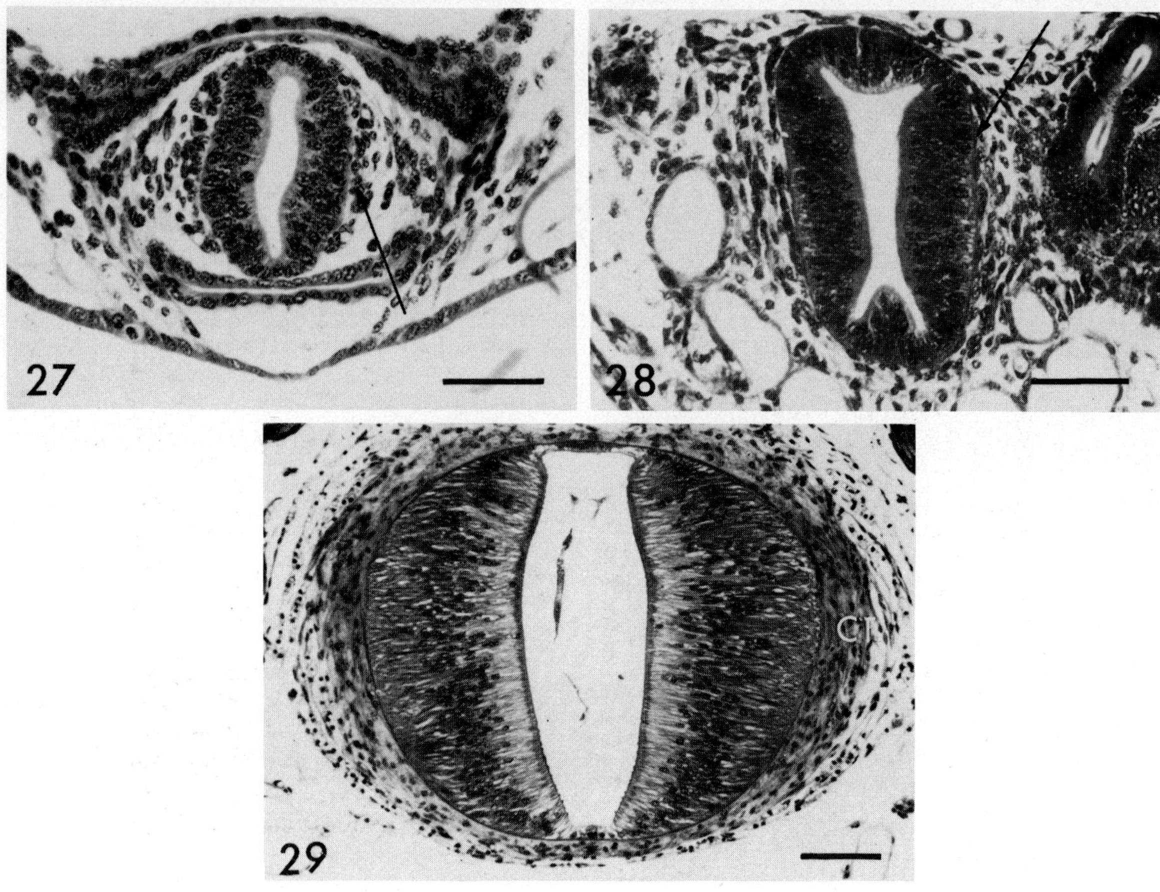

FIGS 27–29. Photomicrographs of the transverse sections of the ejaculatory duct of the male locust, *Locusta migratoria*. FIG 27 shows the duct in a second-instar larva. Cells have migrated to lie lateral to the duct (arrow); FIG 28 shows the duct in a fourth-instar larva. The duct is now surrounded by cells and the developing basal layer can be seen (arrow); FIG. 29 shows the 6-day adult duct, which is surrounded by a layer of fibrous connective tissue (CT). Bar on Figs. 27 and 28 = 50 μm; bar on Fig. 29 = 100 μm. (From Ashhurst, D. and Costin, N., 1974.)

ceases before the adult emerges (Fig. 9). The cells are now much reduced in size and contain few cytoplasmic organelles. Their shape is very irregular and adjacent cells are held together by numerous desmosomes with many associated microtubules (Ashhurst, D., 1964, 1970; McLaughlin, B., 1974). The matrix contains typical thin, indistinctly banded fibrils (Fig. 10).

4.3 Biosynthesis of collagen molecules

In common with other proteins the collagen polypeptide chains are formed on the ribosomes and then pass into the cisternae of the rough endoplasmic reticulum. At this stage the α-chains are extended at both the C- and N-terminals by about 150 residues. These terminal extensions are necessary for the proper alignment of the α-chains in the molecule. A number of post-translational modifications are made to the α-chains. These include the hydroxylation of proline and lysine, and the glycosylation of some lysine and hydroxylysine residues (Bailey, A. and Etherington, D., 1981; Prockop, D. *et al.*, 1979). The enzymes essential for these modifications are present in the cisternae of the rough endoplasmic reticulum. The collagen triple-helix is formed and the molecule is known as procollagen.

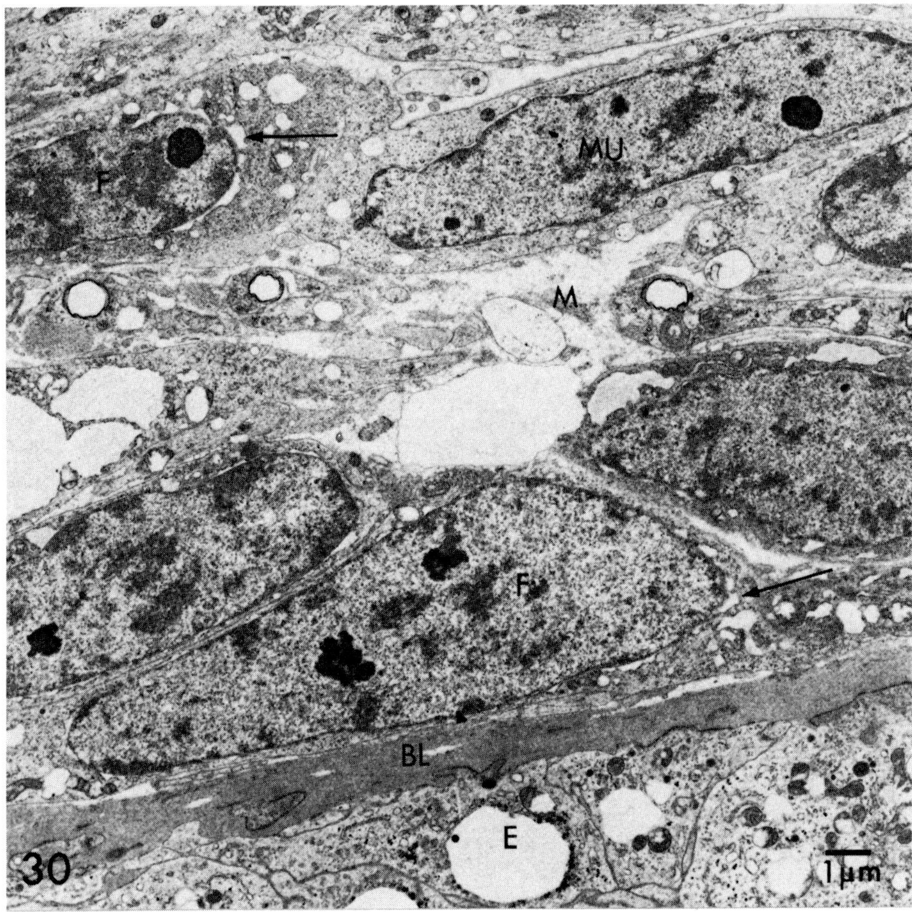

FIG. 30. Electron micrograph of part of the developing connective tissue around the ejaculatory duct of a fourth-instar larva of *Locusta migratoria*. The basal layer (BL) lies on the epithelial cells (E) and it contains a few fibrils. The cytoplasm of the differentiating fibroblasts (F) is restricted at this stage, but developing rough endoplasmic reticulum can be seen (arrows). Occasional developing muscle cells (MU) are also found around the duct. Small areas of matrix (M) occur between the fibroblasts. (From Ashhurst, D. and Costin, N., 1974.)

The procollagen is secreted from the cell into the extracellular matrix, where procollagen peptidases remove the C- and N-terminal extensions, and the cross-links are synthesized. Fibril formation then occurs.

During the period in which the work outlined was in progress, several investigations of the cellular pathway of collagen synthesis and secretion were made using electron autoradiographic techniques. A controversy arose as to the role of the Golgi complex in collagen synthesis, since in osteoblasts and odontoblasts there is evidence that the newly synthesized procollagen passes from the cisternae of the rough endoplasmic reticulum to the Golgi complex, and then it is secreted from the cell, whereas in chondroblasts and skin fibroblasts, it appears that some of the newly synthesized procollagen might pass straight to the extracellular matrix from the cisternae of the rough endoplasmic reticulum (see review by Ross, R., 1975).

The first attempt to determine the pathway in insect fibroblasts by electron autoradiography investigated the uptake and subsequent pathway of [^{3}H]proline in the fibroblasts of the ejaculatory duct of *Locusta migratoria* and of the dorsal mass of the nerve cord of *Galleria mellonella* (Ashhurst, D. and Costin, N., 1976); [^{3}H]proline is used as a tracer since proline is found in greater amounts in collagen

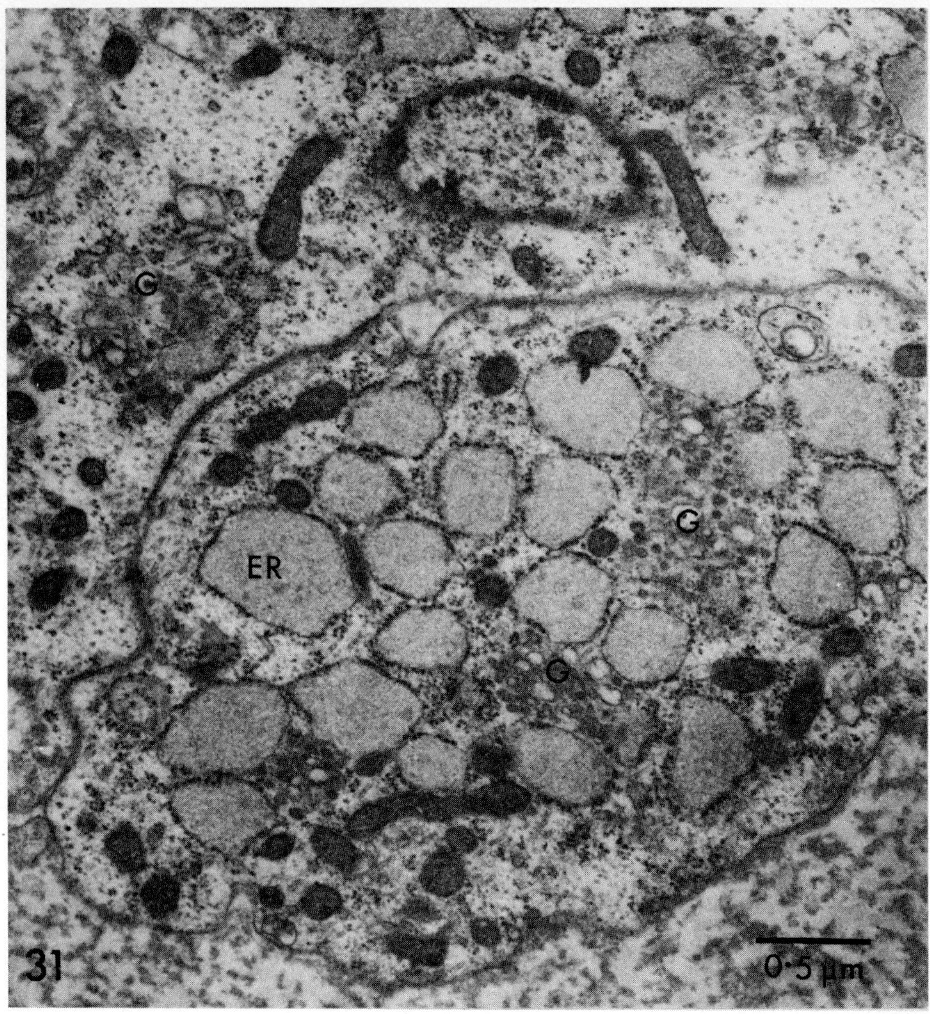

FIG. 31. Electron micrograph of parts of two fibroblast-like cells in the dorsal mass of the nerve cord of a 6-day pupa of *Galleria mellonella*. The cytoplasm contains large rounded cisternae of rough endoplasmic reticulum (ER) and small Golgi complexes (G).

than in other proteins. The resulting autoradiographs were analysed using the methods of Williams, M. (1973); these involve counting the silver grains over the cytoplasmic compartments and matrix and a stereological analysis of the relative areas occupied by each compartment on the micrographs used so that the grain counts may be expressed as the number of grains per unit area of each compartment. Both to determine grain locations and for the stereological analysis, circles, not points, are used; the diameter of the circle is determined from the data of Salpeter, M. *et al.* (1969) on the resolution of autoradiographs. For the *Locusta* and *Galleria* experiments, circles with a diameter of 4.2 mm (equal to 0.28 μm on the micrographs) were used; larger circles would have fallen over two compartments too frequently.

The results in the two species were similar; when the grain counts over the different compartments within the cells, and the extracellular matrix (Table 5) were expressed as the relative number of grains per unit area over each compartment, and plotted against time, the graphs shown in Figs 32 and 33 were obtained. If labelled molecules are passing from one compartment to another, through the cell, then the relative number of grains over compartment 1 should reach a maximum and start to decline, before the maximum number is reached in

Table 5: The data from an experiment on the uptake of [^3H]-proline by locust ejaculatory duct fibroblasts from which the graph in Fig. 32 was drawn

Compartment	Experimental times: 15 min			30 min			1 h			2 h			4 h		
	Area count	Grain count	Relative no. grains/unit area	Area count	Grain count	Relative no. grains/unit area	Area count	Grain count	Relative no. grains/unit area	Area count	Grain count	Relative no. grains/unit area	Area count	Grain count	Relative no. grains/unit area
Matrix	2759	18	0.14	2295	24	0.24	2320	48	0.41	3194	108	0.52	2838	124	0.77
Rough endoplasmic reticulum	211	53	5.37	215	40	4.27	134	21	3.12	167	41	3.78	96	19	3.50
Golgi complexes	123	16	2.78	92	9	2.25	71	14	3.92	77	5	1.00	74	7	1.67
Pooled counts for all other compartments*	4471	267	1.28	3554	195	1.26	2998	194	1.29	3741	312	1.28	2857	182	1.13
Total	7564	354	—	6156	268	—	5513	277	—	7179	466	—	5865	332	—

*For the complete data, see Ashhurst, D. and Costin, N., (1976).

compartment 2, and so on. Examination of the graphs in Figs 32 and 33 shows that this occurs if only the rough endoplasmic reticulum and matrix are considered, and that the figures for the relative number of grains per unit area over the Golgi complex are very erratic. This is due to the very low grain counts over the Golgi compartment and to their very small size (less than 0.5 μm diameter), which means that the Golgi complex compartment is very small (Table 5). It was therefore concluded by Ashhurst, D. and Costin, N. (1976) that most of the newly synthesized collagen in the locust and waxmoth fibroblasts passes directly from the cisternae of the rough endoplasmic reticulum to the matrix, and that only a minor portion passes through the Golgi complexes.

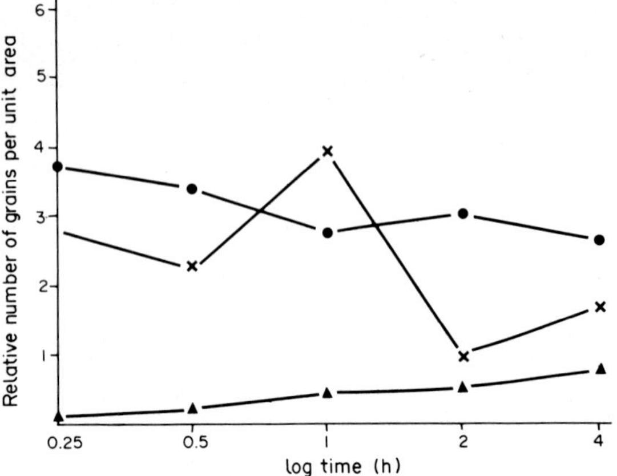

FIG. 32. A graph of the relative number of silver grains per unit area over the rough endoplasmic reticulum (●), Golgi complexes (×) and matrix (▲) in locust fibroblasts and matrix at different times after a pulse of [^{3}H]proline (see Table 5). (From Ashhurst, D. and Costin, N., 1976.)

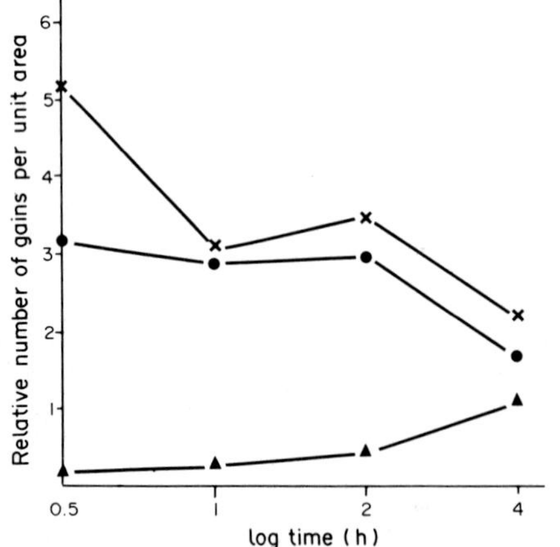

FIG. 33. A graph of the relative number of silver grains per unit area over the rough endoplasmic reticulum (●), Golgi complexes (×) and matrix (▲) in *Galleria* fibroblasts and matrix at different times after a pulse of [^{3}H]proline. (From Ashhurst, D. and Costin, N., 1976.)

A later investigation of the pathway in fibroblasts in *Thermobia domestica* (François, J., 1980) led to the suggestion that the Golgi complex is an obligatory part of the secretory pathway in these insect fibroblasts. The analysis was similar to that of Ashhurst, D. and Costin, N. (1976), but circles with a diameter of almost 0.5 μm were used, and this is as large as a *Thermobia* Golgi complex. It is thus difficult to see how so many circles in the area counts fell squarely over a Golgi complex (François, J., 1980, Table 1), and not over a junction between a Golgi complex and another compartment. Indeed, Ashhurst and Costin's experience was that even with smaller circles (0.28 μm diameter) very few fell over a Golgi complex only.

It is difficult to obtain unequivocal results with this type of study if one of the important cellular compartments is very small. The Golgi complex may be an obligatory step in the secretory pathway for at least part of the newly synthesized insect collagen. It is noteworthy that none of the enzymatic steps in the synthesis of collagen takes place in the Golgi complex (Prockop, D. *et al.*, 1979). It should be remembered that these cells also produce the glycosaminoglycans and glycoproteins of the matrix, and it is known that the sulphation of chondroitin sulphate occurs in the Golgi complex (Dziewiatkowski, D., 1962).

4.4 Vitamin C

That ascorbate is an essential co-factor for the hydroxylation of both proline and lysine by their respective hydroxylases is well established (Prockop, D. *et al.*, 1979). Thus a dietary intake of vitamin C is essential for animals, such as guinea-pigs and primates, which do not synthesize the vitamin.

No experiments have been performed to ascertain whether ascorbate is essential for collagen synthesis in insects, but it would be surprising if it is not. It is essential for the activity of prolyl hydroxylase in the shrimp, *Panaeus* sp. (Hunter, B. *et al.*, 1979). Many insects do synthesize ascorbate, but it appears that many phytophagous insects cannot, and require a dietary intake for normal growth and development to proceed. These insects include many Coleoptera, Homoptera, Lepidoptera and Orthoptera (Chippendale, G., 1975; Dadd, R., 1973; Kramer, K. *et al.*, 1978; Mittler, T. *et al.*, 1970). Other insects, including the cockroaches, synthesize vitamin C, or rather it is thought that it is synthesized by fat body symbiotes (Dadd, R., 1973). It is noteworthy that Day, M. (1949) detected ascorbate in the organs of several insects including *Blatta germanica* and *Locusta migratoria*.

5 COLLAGEN SILKS

The existence of a silk composed of collagen was first established from X-ray diffraction patterns of the cocoon silk of *Nematus ribesii* and later in the silks of other closely related species of Symphyta (Rudall, K. and Kenchington, W., 1971). Amino acid analyses of the silk collagens show that glycine accounts for about one-third of the total residues, and the hydroxylysine level is higher than in most other collagens (about 37 per 100 residues), but the proline content is lower and hydroxyproline is absent. The periodicity of the banding pattern is estimated as 55 nm, but the intraperiod pattern cannot be clearly seen in the published micrographs.

The silks of other related species are classified as polyglycine silks, since the silk of *Phymatocera aterrima* has 660 glycines per 1000 residues. These silks are clearly worthy of further attention.

6 GLIAL LACUNAR SYSTEM

The glial lacunar system was first described in ganglia of *Periplaneta americana* by Wigglesworth, V. (1960). He noticed the extensive system of extracellular spaces which in light microscope preparations (Figs 34 and 35) can be seen between the layers of the glial cells which surround the neuropile. He suggested that the lacunae might serve as a pool for the storage of nutrients. It was reported shortly after this initial description that the glial lacunar systems of *Periplaneta*, and of a number of other species of cockroaches — *Blaberus craniifer*, *Blatta orientalis*, *Blattella germanica* and *Diploptera punctata* — contain a metachromatic material, labile to hyaluronidase digestion (Ashhurst, D., 1961b; Pipa, R., 1961), which was identified biochemically as hyaluronate by Ashhurst, D. and Patel, N. (1963).

Since these initial findings in the cockroaches, it has been demonstrated that other insects, and also other arthropods, do possess a glial lacunar system (Martoja, R. and Cantacuzène, A., 1968), but that it is not so extensive as in cockroaches, where it occupies about 15–18% of the volume of the meso- and metathoracic ganglia in *Periplaneta* adults (Ashhurst, D., 1984).

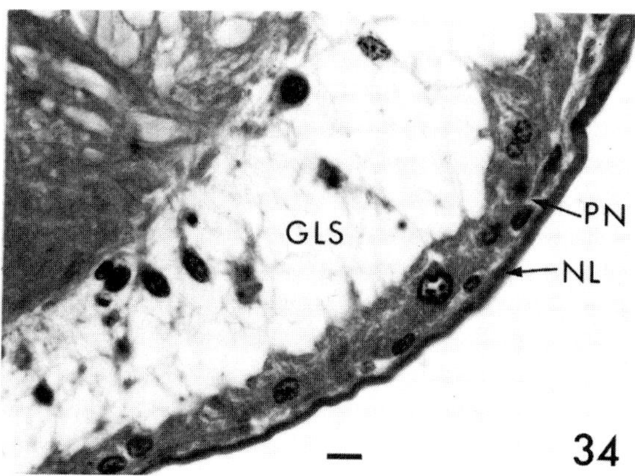

FIG. 34. Photomicrograph of part of a metathoracic ganglion of *Periplaneta americana*. The glial lacunar system (GLS) is very extensive and in this region of the ganglion lies immediately under the neural lamella (NL) and perineural cells (PN). (Bar = 10 μm.) (From Ashhurst, D. and Costin, N., 1971b.)

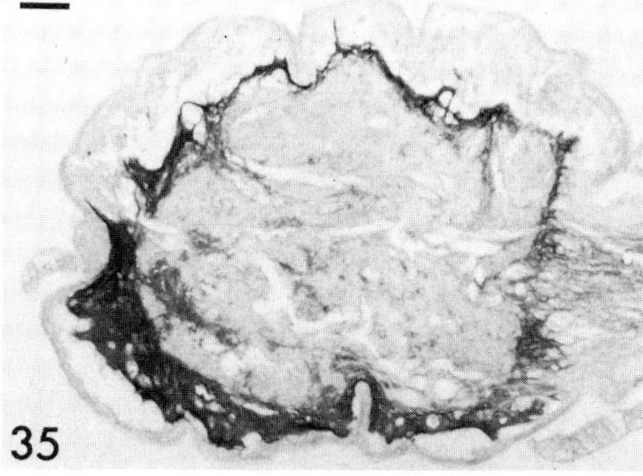

FIG. 35. A section of a metathoracic ganglion of *Periplaneta americana*. The glial lacunar system has bound alcian blue at pH 2.5 very strongly, the neural lamella, much less strongly. (Bar = 50 μm.) (From Ashhurst, D. and Costin, N., 1971b.)

The glial lacunar system is clearly demonstrated in preparations of ganglia stained with alcian blue at pH 2.5. In the locust, *Locusta migratoria* (Fig. 36), and the stick insect, *Carausius morosus* (Fig. 37), it is much less extensive than in *Periplaneta*, but histochemical studies of these three species using alcian blue and other histochemical techniques (Table 6) confirm that the glial lacunar system contains only hyaluronate (Ashhurst, D. and Costin, N., 1971b,c). Martoja, R. and Cantacuzène, A. (1966) thought that the glial lacunar system of *Locusta* is PAS-positive, and they suggested that it contains glycoproteins in addition to hyaluronate. Close examination of sections of locust ganglia on which both the PAS test and alcian blue staining at pH 2.5 have been performed indicates that the alcianophil areas are distinct from the PAS-positive areas, which are in fact the glial cells with much glycoprotein (not glycogen) in the cytoplasm (Ashhurst, D. and Costin, N., 1971b).

The glial lacunar system looks empty in electron micrographs apart from some fine amorphous material, or fibre-like concentrations of it (Fig. 38). The better preservation of the tissue enables the system to be traced into the neuropile where narrow extracellular channels ramify between the axons (Fig. 40); indications of these channels may actually be observed in alcian blue preparations (Fig. 39). It is noteworthy that although alcian blue staining fails to reveal a glial lacunar system in lepidopteran ganglia, narrow lacunae are present among the axons in all published electron micrographs (Lane, N., 1972).

The inclusion of the glial lacunar system among the connective tissue matrices is justified because the lacunae are extracellular and the substance within them is hyaluronate, a typical connective tissue molecule.

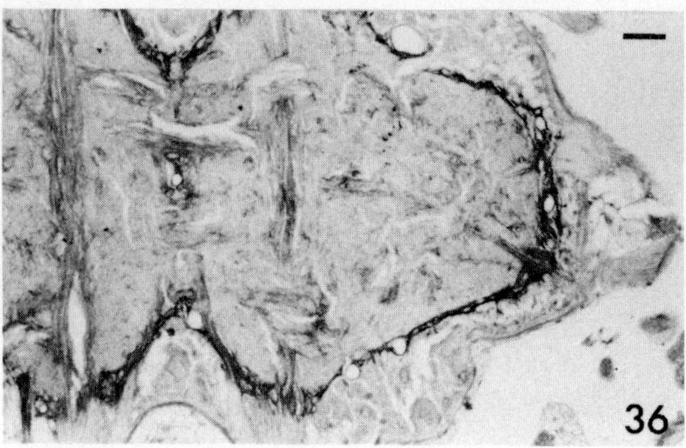

FIG. 36. A section of a metathoracic ganglion of *Locusta migratoria*. The glial lacunar system has bound alcian blue at pH 2.5 very strongly, but the neural lamella is much less strongly stained. (Bar = 50 μm.) (From Ashhurst, D. and Costin, N., 1971b.)

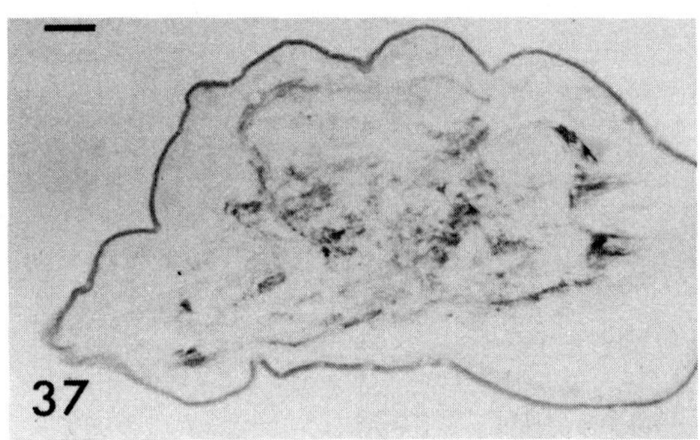

FIG. 37. A section of a thoracic ganglion of *Carausius morosus*. The glial lacunar system which has bound alcian blue at pH 2.5 is much less extensive than in Figs 35 and 36. The neural lamella, however, has bound more dye than those around the other ganglia. (Bar = 50 μm.) (From Ashhurst, D. and Costin, N., 1971c.)

Table 6: Histochemical reactions of the glial lacunar systems of the cockroach, locust and stick insect thoracic ganglia

	Glial lacunar system		
	Cock-roach[1]	Locust[1]	Stick-insect[2]
Alcian blue, pH 2.5	4+	3+	1–3+
Alcian blue, pH 1.0	0–1+	0	0
Alcian blue, pH 5.7 + 0.05 M $MgCl_2$	4+	4+	3+
Alcian blue, pH 5.7 + 0.1 M $MgCl_2$	4+	3+	2+
Alcian blue, pH 5.7 + 0.2 M $MgCl_2$	0	0	0
Alcian blue, pH 5.7 + 0.4 M $MgCl_2$	0	0	0
Alcian blue, pH 5.7 + 0.6 M $MgCl_2$	0	0	0
Alcian blue, pH 5.7 + 0.8 M $MgCl_2$	0	0	0
Alcian blue, pH 5.7 + 1.0 M $MgCl_2$	0	0	0
PAS	0	0	0
HID	0	0	0
Hyaluronidase-AB pH 2.5	1–2+	2+	0–1+
Hyaluronidase control-AB pH 2.5	4+	3+	1–3+
Neuraminidase-AB pH 2.5	4+	3+	1–3+
Neuraminidase control-AB pH 2.5	4+	3+	1–3+

Key: [1] Data from Ashhurst, D. and Costin, N., 1971b.
[2] Data from Ashhurst, D. and Costin, N., 1971c.
AB = Alcian blue; HID = high iron diamine; PAS = periodic acid-Schiff; 1–4+ = increasing intensity of reaction or staining; ± = very weakly positive reaction or staining; 0 = negative, or no reaction or staining.

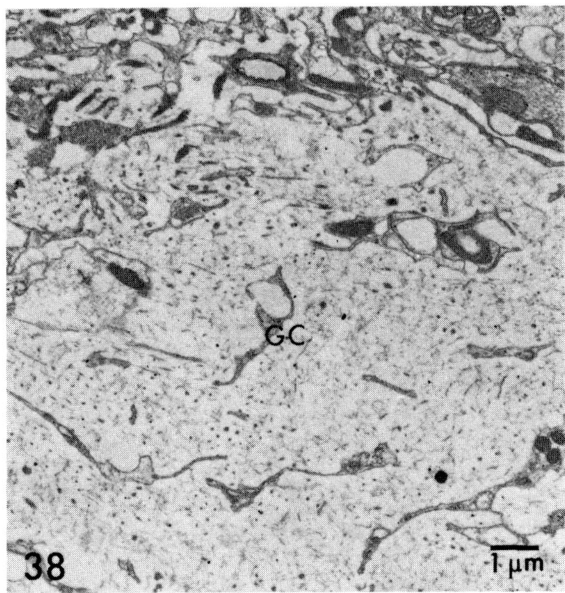

FIG. 38. An electron micrograph of a small area of the glial lacunar system of *Periplaneta americana*. The hyaluronate in the lacunae has precipitated to give a fine granular appearance. Some fine glial processes (GC) ramify through the lacunae.

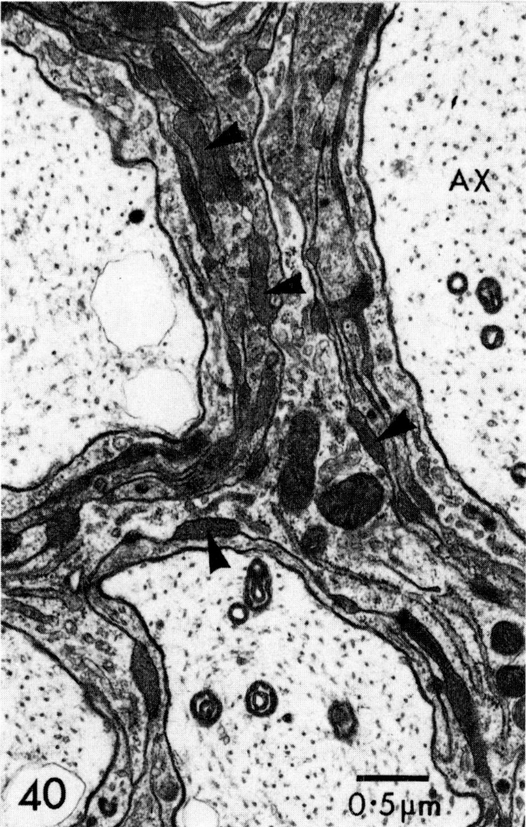

FIG. 40. Electron micrograph of a small area of the neuropile in a thoracic ganglion of *Locusta migratoria* to show the extracellular spaces (arrow-heads), filled with an amorphous material, which ramify between the glial cell process which separate the large axons (AX).

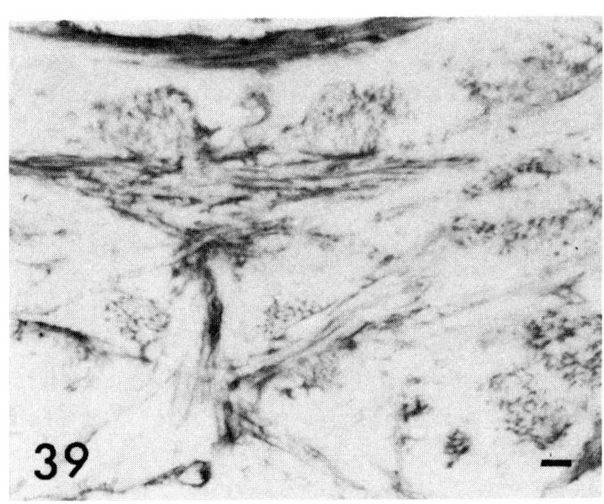

FIG. 39. Photomicrograph of a small area of the neuropile of a thoracic ganglion of *Locusta migratoria* after staining with alcian blue pH 2.5. The dye has been bound by hyaluronate in the narrow channels which permeate the neuropile. (Bar = 10 μm.) (From Ashhurst, D. and Costin, N., 1971b.)

6.1 Development of the glial lacunar system

It was noted by Martoja, R. and Cantacuzène, A. (1966) that the glial lacunar system can be identified from the end of the first larval instar in *Locusta* and *Gryllus*. A recent examination of developing meso- and metathoracic ganglia of *Locusta* and *Periplaneta* has shown that the glial lacunar system is present in *Periplaneta* at hatching, but not until the second instar in *Locusta* (Ashhurst, D. 1984). The variation with the results mentioned above may be because the first instar locusts used by Ashhurst were not at the end of the instar. From the earliest stages the histochemical reactions of the material in the lacunae are the same as those in the adult. That is, hyaluronate is present in the glial lacunar system throughout the life of the insect.

The volume of the glial lacunar system of the meso- and metathoracic ganglia of *Periplaneta* increased steadily from about 1% in the newly hatched larva to between 15 and 18% of the total volume of the thoracic ganglia in the adult (Ashhurst, D., 1984).

6.2 Function of the glial lacunar system

The mechanism of nerve conduction in insects poses many questions since the ionic concentrations in the haemolymph would affect the conduction of potentials adversely if a selective barrier, located in the glial cells, did not separate the haemolymph and axons. The ionic composition of the fluid bathing the axons is controlled by ion pumps in the glial cells. The extracellular clefts around each axon are very narrow, and it is suggested that the hyaluronate in the glial lacunar system acts as an extracellular cation reservoir, particularly for sodium ions (for a full discussion of this topic see review by Treherne, J. and Schofield, P., 1981 and Treherne, J. vol. 5). Varying concentrations of potassium, calcium, magnesium and phosphate ions have been recorded in the glial lacunar system of *Locusta migratoria* (Martoja, R., 1973). Thus, the glial lacunar system with its content of hyaluronate appears to be essential for the proper functioning of the insect nervous system.

7 ELASTIC FIBRES

A second type of connective tissue fibre was described by Baccetti, B. and Bigliardi, E. (1969) in the wall of the dorsal vessel of the orthopteran *Aiolopus strepens*. Since the fibres appeared to be labile to elastase digestion, they were identified as elastic fibres. Later, Locke, M. and Huie, P. (1972) found similar fibre bundles in the neural lamella and pericardial connective tissue matrix of *Calpodes ethlius* larvae (Figs 41 and 42), and Crossley, A. (1972) saw them in "connecting ligaments" between the pericardial cells of *Calliphora erythrocephala*. They also occur in the connective tissue matrix around the mesenteron of *Periplaneta americana* (François, J., 1978).

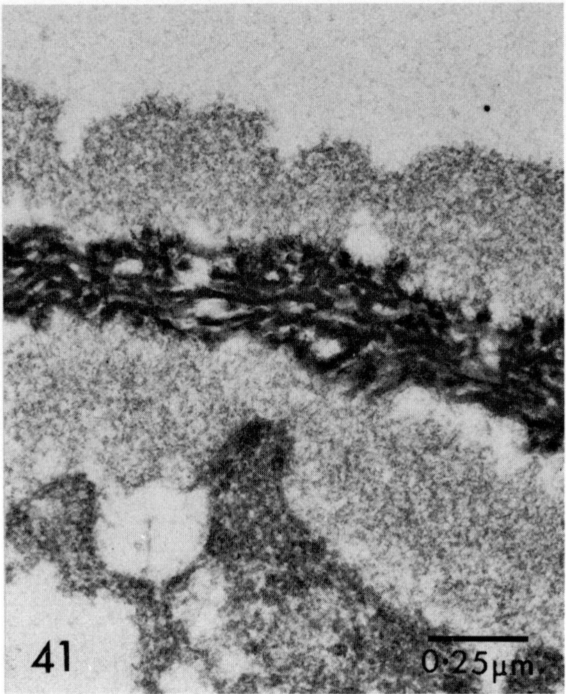

FIG. 41. Electron micrograph of part of the pericardial sheath of a larva of *Calpodes ethlius*, showing an "elastic" fibre bundle stained with phosphotungstic acid. (From Locke, M. and Huie, P., 1972.)

The fibre bundles are not easily distinguished in sections of material prepared for electron microscopy by the usual procedures. Locke, M. and Huie, P. (1972, 1975) found that they are very obvious when a peroxidase reaction is performed on the tissue, when sections of tissue fixed in glutaraldehyde only are stained with phosphotungstic acid, or when tannic acid is added to the glutaraldehyde fixative. Using these methods they were able to distinguish two types of fibril in the fibre bundles: thick fibrils, about 40 nm in diameter, are separated by a 70 nm space in which microfibrils, about 6 nm in diameter, are found.

The identification of these fibre bundles as a type of elastic fibre is based entirely on the known elasticity of the tissues. The presence of these fibre bundles in the neural lamella of the abdominal nerve cord of the adult female locust confirmed this hypothesis (Locke, M. and Huie, P., 1975) because the abdomen stretches to about three times its normal length during oviposition. It appears doubtful, however, whether these fibrils should be termed

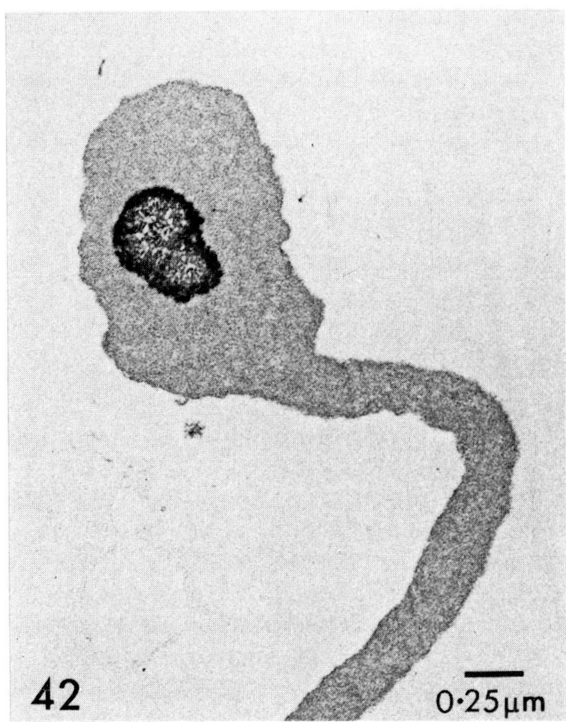

FIG. 42. Electron micrograph of a pericardial strand of a larva of *Calpodes ethlius*. An "elastic" fibre bundle can be seen in transverse section, stained with phosphotungstic acid. (From Locke, M. and Huie, P., 1972.)

elastic fibres, as this nomenclature implies that they are homologous to vertebrate elastic fibres.

Vertebrate elastic fibres have an amorphous, electron-lucent core, composed of the protein elastin, which is surrounded by microfilaments. Despite an extensive study of invertebrates from different phyla, Sage, H. and Gray, W. (1979, 1980) were unable to find either the protein elastin, or fibres which stain with the standard methods used for vertebrate elastic fibres. Thus, they conclude that elastin is found exclusively in vertebrates. The fibres with elastic properties in the invertebrates are not, therefore, homologous to vertebrate elastic fibres. Thus the composition of the fibres described in insects, and of those of other invertebrates identified as "elastic fibres" by their affinity for the dye spirit blue (Elder, H., 1973), has yet to be established. The lability of the fibres of *Aiolopus* to elastase digestion, mentioned above, is probably of little significance, since commerical samples of connective tissue enzymes all contain large amounts of non-specific proteolytic enzymes.

8 CONNECTIVE TISSUE ENZYMES

There is, as yet, no information about the degradative enzymes in the connective tissues; some must be present, as the connective tissues have to be modified during growth and development. There are, however, a few reports of connective tissue enzymes which are produced to destroy the connective tissues of other animals; these enzymes must be very commonly produced by insects.

A collagenase has been isolated from first-instar larvae of *Hypoderma lineatum*. The larvae of *Hypoderma* are endoparasites of cattle, and after penetrating the skin they migrate through the connective tissues, aided by the collagenase produced by the salivary glands. The enzyme has been characterized (Lecroisey, A. *et al.*, 1979). It cleaves the helical region of the collagen molecule and has a molecular weight of 24,000. The amino acid composition suggests that it is similar to the serine proteases, as is the collagenase found in the fiddler crab, *Uca pugilator* (Eisen, A. *et al.*, 1973). It differs from the vertebrate collagenases, which are metalloenzymes, since it is not inhibited by EDTA. It is suggested by Lecroisey, A. *et al.* that the insect and crab enzymes which function as digestive enzymes rather than tissue enzymes facilitating morphogenesis, may be serine proteases, related to trypsin.

Hyaluronidases are also produced by insects. An extract of the testis of *Drosophila melanogaster* was found to reduce the viscosity of hyaluronate (Mutchmor, J. and Richards, A., 1956), though its enzymatic nature was not conclusively confirmed. Hyaluronidases have also been isolated from the testis and mucus accessory glands of the male bee, *Apis mellifera*, and the hornet, *Vespa orientalis* (Allalouf, D. *et al.*, 1974, 1975). The enzymes from both sources are similar; they have a pH optimum of about 4.5, they degrade hyaluronate and chondroitin-6-sulphate, and in contrast to mammalian testicular hyaluronidase, they are not inactivated by a temperature of 56° for 3 h. It is suggested that they serve to remove mucus from the female reproductive tract. More recently, the hyaluronidase in bee venom with a molecular weight of 60,000 was purified (Kemeny, D. *et al.*, 1981). Since the purpose of the research was to obtain pure samples of the enzyme for work on the human allergens in bee venom, it was not characterized.

8.1 Breakdown of connective tissue

While some modification of the connective tissues must occur during growth and development in all insects, the most extensive breakdown of connective tissue matrices occurs during the metamorphosis of the endopterygote insects. The disintegration and removal of the neural lamella around the nervous system of Lepidoptera at the beginning of the pupal stage is well documented (Ali, F., 1973; Ashhurst, D. and Richards, A., 1964a; McLaughlin, B., 1974; Nordlander, R. and Edwards, J., 1969; Pipa, R. and Woolever, P., 1965), and it appears to be an essential prerequisite for the shortening and reorganization of the larval nerve cord to the adult form. Adipohaemocytes (granular haemocytes) are in close proximity to the neural lamella at this time, and while their role in phagocytosing the remnants of the connective tissue is establised (Pipa, R. and Woolever, P., 1965; Gupta, A. this volume), there is as yet no evidence that they release lytic enzymes. Acid phosphatase is present in the granules of the adipohaemocytes of *Galleria* (Ashhurst, unpublished observation) and thus if the granules contain this lysosomal enzyme, it is very probable that they also contain other hydrolytic enzymes, which could be released at the onset of metamorphosis. A similar process occurs in the Diptera during metamorphosis; Whitten, J. (1962, 1964a) observed the lysis of the connective tissues in dissections of *Sarcophaga* pupae and the removal of the fragments by granular haemocytes.

The hormonal control of the changes in the nerve cord of *Galleria mellonella* was investigated by Pipa, R. (1967). He found by transplanting nerve cords from pupae at known ages into larvae, pupae and adults, that the process of axon shortening and connective tissue removal is initiated during a short critical period in the early pupa. It has been demonstrated in a series of experiments on *Galleria* in which the brain and associated glands were removed from pupae of different ages (Pipa, R., 1969) that the brain, and to a lesser extent the corpora cardiaca and corpora allata, have a role in the control of axon shortening; the evidence suggests that the brain may act by producing a prothoracotropic hormone. In brainless pupae, shortening, and the associated connective tissue removal and later re-formation, are induced by injection of synthetic α-ecdysone (Pipa, R., 1969).

9 FUNCTIONS OF CONNECTIVE TISSUES

Connective tissues are supportive. They bind cells together, hold organs in position and in many animals form the main skeletal tissue. It is only relatively recently that their wider significance has been appreciated, as information about the properties of the glycosaminoglycans has accummulated. The glycosaminoglycan–protein complexes, or proteoglycans, are very large and have a high density of anionic charges along the length of the glycosaminoglycan chains. As a result the molecules are very highly expanded *in vivo* and occupy a significant proportion of the volume of the matrix. They form an anionically charged meshwork which appears to act as a sieve affecting the movement of ions and molecules within the matrix (Comper, W. and Laurent, T. 1978; Laurent, T. 1977); anionically charged ions and molecules may be repelled, cationically charged ions may be hindered while crossing the matrix, or excluded from it, by virtue of their size and shape. Chondroitin sulphates also appear to effect the formation of collagen fibrils (Mathews, M., 1975).

The mechanical strength of a connective tissue matrix will depend on the density and orientation of the collagen fibrils within it, and those with non-fibrous forms of collagen will be less strong. The presence of proteoglycans in matrices means that they form a potentially selective barrier, and in all animals, exchange between the blood and tissues occurs across a layer of connective tissue. It has been emphasized throughout this chapter that all insect organs are covered by a connective tissue matrix which effectively separates the cells from the haemolymph. The role of the neural lamella has been extensively investigated, and while Eldefrawi, M. and O'Brien, R. (1966, 1967) found evidence for a mechanism discriminating against large size and polarity, which slows the passage of molecules through the neural lamella, the actual barrier to the free passage of ions and molecules is located in the perineurial and glial cells (see Lane, N., vol. 5). It appears, however, that connective tissue might form the barrier which protects muscle fibres from adverse concentrations of glutamate in the haemolymph (Clements, A. and May, T., 1974). The ion-binding properties of hyaluronate have been implicated in the formation of a cation reservoir in the glial lacunar system (see Treherne, J. vol. 5).

It was suggested by Whitten, J. (1964b) that the thin connective tissue strands observed in many insects might form channels for the passage of hormones and other substances between organs. This idea has not been substantiated by other workers.

More recently it has become apparent that connective tissue matrices are essential for morphogenetic movements and cell differentiation. Basement membranes are found in vertebrate embryos prior to gastrulation (Hay, E. 1973) and sulphated glycosaminoglycans must be present in the basement membrane of developing salivary glands for normal acini to be formed (Bernfield, M. *et al.*, 1972). In a study of limb bud development, Toole, B. (1972) found that the presumptive chondroblasts move into position in a matrix of hyaluronate, but this must be removed by hyaluronidase before differentiation and the secretion of the cartilage matrix can occur. These are some examples of early work in this area, which has expanded rapidly to confirm the intimate association of the connective tissues in developmental processes. While such studies on insects have not commenced, it is significant that Höglund, L. (1976b) in his biochemical study found glycosaminoglycans in *Calliphora* embryos from late cleavage onwards.

10 CONCLUSIONS

Research on insect connective tissues over the last decade has shown that the chemistry of the connective tissue matrices of insects is essentially similar to that of other animals. Perhaps the most significant finding is that the fibrous collagen molecule is of the same type as that of many other invertebrates and vertebrates, namely the type I trimer. The next step is to characterize the glycosaminoglycans and glycoproteins in more detail biochemically, since histochemical identification has inherent limitations.

The greatest problem to be overcome in any biochemical work is the small amount of material available from any known site, since the characterization of an extracted substance should include knowledge of its precise location in the insect. The newly developed immunocytochemical techniques offer a means by which small quantities of substances can be located within a tissue, and it is obvious that these could give us a great deal of information about the components of insect connective tissue matrices. Their widespread use will, however, depend upon the cross-reactivity of antibodies made to collagen and the other macromolecules from vertebrate tissues with the insect antigens, since it is not easy to extract and purify sufficient antigen from insects to immunize a rabbit, or other mammal. While monoclonal antibodies have been made to some components of the *Drosophila* matrix (see sections 3.1 and 3.3), it is not yet known if these will cross-react with the molecules of other species of Diptera, or those of other insect orders. To produce antibodies to the connective tissue molecules of many different species of insect would be laborious indeed. Specific, cross-reacting antibodies made to the collagens, proteoglycans and glycoproteins of avian and mammalian species will offer the most economical way to locate these macromolecules in the various insect tissues. These new techniques will provide many new possibilities for the development of insect connective tissue research. Throughout this chapter I have tried to stress the similarities between the connective tissues of insects and other animals. While insects do have some peculiar connective tissue matrices, such as the layers found under some midgut epithelia (section 2.4), in general, insect connective tissues are similar in their morphology, biochemistry and development to the connective tissues of other animals.

ACKNOWLEDGEMENTS

I should like to thank the editors and publishers of the following journals for their permission to use published micrographs and graphs in this chapter: the *European Journal of Biochemistry*, the *Histochemical Journal*, the *Journal of Cell Science* and *Tissue and Cell*.

REFERENCES

Ali, F. A. (1973). Post-embryonic changes in the central nervous system and perilemma of *Pieris brassicae* (L), (Lepidoptera: Pieridae). *Trans. Roy. Ent. Soc. Lond. 40*, 463–498.

Allalouf, D., Ber, A. and Ishay, J. (1974). Hyaluronidase in the mucus accessory glands of the drone (*Apis mellifera*). *Experientia 30*, 835–836.

Allalouf, D., Ber, A. and Ishay, J. (1975). Properties of testicular hyaluronidase of the honey bee and oriental hornet: comparison with insect venom and mammalian hyaluronidase. *Comp. Biochem. Physiol. 50B*, 331–337.

Ashhurst, D. E. (1959). The connective tissue sheath of the locust nervous system; a histochemical study. *Quart. J. Mic. Sci. 100*, 401–412.

Ashhurst, D. E. (1961a). A histochemical study of the connective tissue sheath of the nervous system of *Periplaneta americana*. *Quart. J. Mic. Sci. 102*, 455–461.

Ashhurst, D. E. (1961b). An acid mucopolysaccharide in cockroach ganglia. *Nature 191*, 1224–1225.

Ashhurst, D. E. (1964). Fibrillogenesis in the wax-moth, *Galleria mellonella*. *Quart. J. Mic. Sci. 105*, 391–403.

Ashhurst, D. E. (1965). The connective tissue sheath of the locust nervous system: its development in the embryo. *Quart. J. Mic. Sci. 106*, 61–73.

Ashhurst, D. E. (1968). The connective tissues of insects. *Ann. Rev. Ent. 13*, 45–74.

Ashhurst, D. E. (1970). An insect desmosome. *J. Cell. Biol. 46*, 421–425.

Ashhurst, D. E. (1979a). Haemocytes and connective tissue: a critical assessment. In *Insect hemocytes: Development, Form, Functions and Techniques*. Edited by A. P. Gupta. Pages 319–330. Cambridge University Press, Cambridge.

Ashhurst, D. E. (1979b). Histochemical methods for hemocytes. In *Insect Hemocytes: Development, Form, Functions and Techniques*. Edited by A. P. Gupta. Pages 581–599. Cambridge University Press, Cambridge.

Ashhurst, D. E. (1984). The glycosaminoglycans of the thoracic ganglia of the nymphal stages of the cockroach, *Periplaneta americana*, and the locust, *Locusta migratoria*: a histochemical study. *J. Insect. Physiol. 30*, 803–810.

Ashhurst, D. E. and Bailey, A. J. (1978). Insect collagen. *Coll. Int. CNRS 287*, 79.

Ashhurst, D. E. and Bailey, A. J. (1980). Insect collagen, morphological and biochemical characterization. *Eur. J. Biochem. 103*, 75–83.

Ashhurst, D. E. and Chapman, J. A. (1961). The connective tissue sheath of the nervous system of *Locusta migratoria*: an electron microscope study. *Quart. J. Mic. Sci. 102*, 463–467.

Ashhurst, D. E. and Costin, N. M. (1971a). Insect mucosubstances. I. The mucosubstances of developing connective tissue in the locust, *Locusta migratoria*. *Histochem. J. 3*, 279–295.

Ashhurst, D. E. and Costin, N. M. (1971b). Insect mucosubstances. II. The mucosubstances of the central nervous system. *Histochem. J. 3*, 297–310.

Ashhurst, D. E. and Costin, N. M. (1971c). Insect mucosubstances. III. Some mucosubstances of the nervous system of the wax-moth (*Galleria mellonella*) and the stick insect (*Carausius morosus*). *Histochem. J. 3*, 379–387.

Ashhurst, D. E. and Costin, N. M. (1974). The development of a collagenous tissue in the locust, *Locusta migratoria*. *Tissue Cell 6*, 279–300.

Ashhurst, D. E. and Costin, N. M. (1976). The secretion of collegen by insects. Uptake of [^{3}H]-proline by collagen-synthesizing cells in *Locusta migratoria* and *Galleria mellonella*. *J. Cell Sci. 20*, 377–403.

Ashhurst, D. E. and Patel, N. G. (1963). Hyaluronic acid in cockroach ganglia. *Ann. Ent. Soc. Amer. 56*, 182–184.

Ashhurst, D. E. and Richards, A. G. (1964a). A study of the changes occurring in the connective tissue associated with the central nervous system during the pupal stage of the wax-moth, *Galleria mellonella*. L. *J. Morphol. 114*, 225–236.

Ashhurst, D. E. and Richards, A. G. (1964b). The histochemistry of the connective tissue associated with the central nervous system of the pupa of a moth, *Galleria mellonella*, L. *J. Morphol. 114*, 237–246.

Baccetti, B. (1955a). Richerche sulla fine struttura del perilemma nel sistema nervosa degli insetti, *Redia 40*, 197–212.

Baccetti, B. (1955b). Sulla presenza a struttura di una tunica involgente i corpi grassi degli insetti. *Redia 40*, 269–278.

Baccetti, B. (1956a). Richerche preliminare sui connettivi e sulle membrane basali degli insetti. *Redia 41*, 75–104.

Baccetti, B. (1956b). Lo stroma di rostegna di organi degli insetti esaminato a luce polarizzata. *Redia 41*, 259–276.

Baccetti, B. (1961). Indagini comparative sulla ultrastruttura della fibrilla collagene nei diversi ordini degli insetti. *Redia 46*, 1–7.

Baccetti, B. and Bigliardi, E. (1969). Studies on the fine structure of the dorsal vessel of Arthropods. I. The "heart" of an orthopteran. *Z. Zellforsch. 99*, 13–24.

Bächinger, H. P., Doege, K.-J., Petschek, J. P., Fessler, L. I. and Fessler, J. H. (1982). Structural implications from an electron microscopic comparison of procollagen V with procollagen I, pC-collagen I, procollagen IV, and a *Drosophila* procollagen. *J. Biol. Chem. 257*, 14590–14592.

Bailey, A. J. and Etherington, D. J. (1981). Metabolism of collagen and elastin. *Comp. Biochem. 19B*, 299–460.

Bayon, C. and François, J. (1976). Ultrastructure de la lame basale du mesenteron chez la larve d'*Oryctes nasicornis* L. (Coleoptera: Scarabaeidae). *Int. J. Insect Morph. Embryol. 5*, 205–217.

Beaulaton, J. (1968). Étude ultrastructurale et cytochemique des glandes prothoracique de vers à soie aux quatrième et cinquièmes âges larvaires. I La tunica propria et ses relations avec les fibres conjonctives et les hémocytes. *J. Ultrastruct. Res. 23*, 474–498.

Bernfield, M. R., Banerjee, S. D. and Cohn, R. H. (1972). Dependence of salivary epithelial morphology and branching morphogenesis upon acid mucopolysaccharide–protein (proteoglycan) at the epithelial surface. *J. Cell Biol. 52*, 674–689.

Bertram, D. S. and Bird, R. G. (1961). Studies on mosquito-borne viruses in their vectors. I. The normal fine structure of the midgut epithelium of the adult female *Aedes aegypti* (L) and the functional significance of its modification following a blood meal. *Trans. Roy. Soc. Trop. Med. Hyg. 55*, 404–423.

Binnington, K. C. and Lane, N. J. (1980). Perineurial and glial cells in the tick *Boophilus microplus* (Acarina: Ixodidae): freeze-fracture and tracer studies. *J. Neurocytol. 9*, 343–362.

Bonhag, P. F. and Arnold, W. J. (1961). Histology, histochemistry and tracheation of the ovariole sheaths in the american cockroach, *Periplaneta americana* (L). *J. Morph. 108*, 107–130.

Bruns, R. R. and Gross, J. (1973). Band pattern of the segment-long-spacing form of collagen. Its use in the analysis of primary structure. *Biochemistry 12*, 808–815.

Bruns, R. R. and Gross, J. (1974). High resolution analysis of the modified quarter-stagger model of the collagen fibril. *Biopolymers 13*, 931–941.

Burger, D. (1876). Ueber das sogenenannte Bauchgefaess der Lepidoptera, nebst einigen Beobachtungen ueber das sympathische Nervensystem dieser Insectenordnung. *Neiderländ Arch. Zool. 3*, 97–125.

Chippendale, G. M. (1975). Ascorbic acid: an essential nutrient for a plant feeding insect, *Diatraea grandiosella*. *J. Nutr. 105*, 499–507.

Clements, A. N. and May, T. E. (1974). Studies on locust neuromuscular physiology in relation to glutamic acid. *J. Exp. Biol. 60*, 673–705.

Comper, W. D. and Laurent, T. C. (1978). Physiological function of connective tissue polysaccharides. *Physiol. Rev. 8*, 255–315.

Crossley, A. C. (1972). The ultrastructure and function of pericardial cells and other nephrocytes in an insect, *Calliphora erythrocephala*. *Tissue Cell. 4*, 529–560.

Dadd, R. H. (1973). Insect nutrition: current developments and metabolic implications. *Ann. Rev. Ent. 18*, 381–420.

Day, M. F. (1949). The distribution of ascorbic acid in the tissues of insects. *Aust. J. Sci. Res. B. 2*, 19–31.

De Biasi, S. and Pilotto, F. (1976). Ultrastructural study of collagenous structures in some Diptera. *J. Submic. Cytol. 8*, 337–345.

Dutkowski, A. B. (1977). The ultrastructure and ultracytochemistry of the basement membrane of the *Galleria mellonella* fat body. *Cell Tissue Res. 176*, 417–429.

DYBOWSKA, H. E. and DUTKOWSKI, A. B. (1977). Ruthenium red staining of the neural lamella of the brain of *Galleria mellonella*. *Cell Tissue Res.* *176*, 275–284.

DYBOWSKA, H. E. and DUTKOWSKI, A. B. (1979). Developmental changes in fine structure and some histochemical properties of the neural lamella of *Galleria mellonella* (L) brain. *J. Submic. Cytol.* *11*, 25–37.

DZIEWIATKOWSKI, D. D. (1962). Intracellular synthesis of chondroitin sulfate. *J. Cell Biol.* *13*, 359–364.

EISEN, A. Z., HENDERSON, K. O., JEFFREY, J. J. and BRADSHAW, R. A. (1973). A collagenolytic protease from the hepatopancreas of the fiddler crab *Uca pugilator*. Purification and properties. *Biochemistry* *12*, 1814–1822.

ELDEFRAWI, M. E. and O'BRIEN, R. D. (1966). Permeability of the abdominal nerve cord of the american cockroach to fatty acids. *J. Insect Physiol.* *12*, 1133–1142.

ELDEFRAWI, M. E. and O'BRIEN, R. D. (1967). Permeability of the abdominal nerve cord of the american cockroach *Periplaneta americana (L.)* to quaternary ammonium salts. *J. Exp. Biol.* *46*, 1–12.

ELDER, H. Y. (1973). Distribution and functions of elastic fibers in the invertebrates. *Biol. Bull.* *144*, 43–63.

ESTES, Z. E. and FAUST, R. M. (1964). Studies on the mucopolysaccharides of the greater wax moth, *Galleria mellonella* (Linnaeus). *Comp. Biochem. Physiol.* *13*, 443–452.

FACTOR, J. R. (1981). Unusually complex basement membranes in the midgut of two decapod crustaceans, the stone crab (*Menippe mercenaria*) and the lobster (*Homarus americanus*). *Anat. Rec.* *200*, 253–258.

FESSLER, J. H. and FESSLER, L. I. (1966). Electron microscopic visualization of the polysaccharide hyaluronic acid. *Proc. Nat. Acad. Sci. USA* *56*, 141–147.

FESSLER, J. H., BÄCHINGER, H. P., LUNDSTRUM, G., and FESSLER, L. I. (1982). Biosynthesis of some procollagens. In *New Trends in Basement Membrane Research.* Edited by K. Kuehn, H. Schoene and R. Timpl. Pages 145–153. Raven Press, New York.

FESSLER, J. H., LUNSTRUM, G., DUNCAN, K. G., CAMPBELL, A. G., STERNE, R. BACHINGER, H. P. and FESSLER, L. I. (1984). In *The Role of Extracellular Matrix in Development.* Edited by R. L. Trelstad, pages 207–219. Alan R. Liss, New York.

FOIDART, J. M., BERE, E. W., YAAR, M., RENNARD, S. I., GULLINO, M., MARTIN, G. R. and KATZ, S. I. (1980). Distribution and immunoelectron microscopic localization of laminin, a non-collagenous basement membrane glycoprotein. *Lab. Invest.* *42*, 336–342.

FRANÇOIS, J. (1968). Nature conjonctive du "tentorium" des Diploures (Insectes, Apterygotes). Étude ultrastructurale. *C.R. Acad. Sci. Paris* *267*, 1976–1978.

FRANÇOIS, J. (1971). L'endosquelette céphalique des insectes aptérygotes; étude anatomique, histochimique et ultrastructurale. I. Collemboles et Diploures. *Arch. Anat. Mic. Morph. Exp.* *60*, 389–406.

FRANÇOIS, J. (1972). L'endosquelette céphaliques des insectes aptérygotes. II Protoures et Thysanoures. *Arch. Anat. Mic. Morph. Exp.* *61*, 279–300.

FRANÇOIS, J. (1973). Sur la presence de fibroblastes caracteristiques chez le Thysanoure *Thermobia domestica. C.R. Acad. Sci. Paris* *277*, 2505–2507.

FRANÇOIS, J. (1977). Development of collagenous endoskeletal structures in the firebrat, *Thermobia domestica* (Packard) (Thysanura: Lepismatidae). *Int. J. Insect Morph. Embryol.* *6*, 161–170.

FRANÇOIS, J. (1978). The ultrastructure and histochemistry of the mesenteric connective tissue of the cockroach *Periplaneta americana* L. (Insecta, Dictyoptera). *Cell Tiss. Res.* *189*, 91–107.

FRANÇOIS, J. (1980). Secretion of collagen by insects: autoradiographic study of L-proline ^{3}H-5 incorporation by the firebrat *Thermobia domestica. J. Insect Physiol.* *26*, 125–133.

FRANÇOIS, J., HERBAGE, D. and JUNQUA, S. (1980). Cockroach collagen: isolation, biochemical and biophysical characterization. *Eur. J. Biochem.* *112*, 389–396.

GOLDSTEIN, N. I. and MCINTOSH, A. H. (1980). The proteins and glycoproteins of insect cells in culture. *Insect Biochem.* *10*, 419–427.

GOURANTON, J. (1970). Étude d'une lame basale presentant une structure d'un type nouveau, *J. Mic.* *9*, 1029–1040.

GRAY, E. G. (1959). Electron microscopy of collagen-like connective tissue fibrils of an insect. *Proc. Roy. Soc. B.* *150*, 233–239.

GUPTA, B. L. and BERRIDGE, M. J. (1966). Fine structural organization of the rectum in the blowfly, *Calliphora erythrocephala* (Meig.) with special reference to connective tissue, tracheae and neurosecretory innervation in the rectal papillae. *J. Morph.* *120*, 23–82.

HARDINGHAM, T. E. (1981). Proteoglycans; their structure, interactions and molecular organization in cartilage. *Trans. Biochem. Soc.* *9*, 489–497.

HARPER, E., SEIFTER, S. and SCHARRER, B. (1967). Electron microscopic and biochemical characterization of collagen in blattarian insects, *J. Cell Biol.* *33*, 385–394.

HAY, E. D. (1973). Origin and role of collagen in the embryo. *Amer. Zool.* *13*, 1085–1101.

HAY, E. D. and DODSON, J. W. (1973). Secretion of collagen by corneal epithelium. I. Morphology of the collagenous products produced by isolated epithelia grown on frozen-killed lens. *J. Cell Biol.* *57*, 190–213.

HEATHCOTE, J. G. and GRANT, M. E. (1981). The molecular organization of basement membranes. *Int. Rev. Connective Tissue Res.* *9*, 191–264.

HESS, A. (1958). The fine structure of nerve cells and fibres, neuroglia, and sheaths of the ganglion chain in the cockroach, *Periplaneta americana. J. Biophys. Biochem. Cytol.* *4*, 731–742.

HESS, R. T. and PINNOCK, D. E. (1975). The ultrastructure of a complex basal lamina in the midgut of larvae of *Oryctes rhinoceros* L. *Z. Morph. Tiere.* *80*, 277–285.

HODGE, A. J. and PETRUSKA, J. A. (1963). Recent studies with the electron microscope on ordered aggregates of the tropocollagen macromolecule. In *"Aspects of protein structure".* Edited by G. N. Ramachandran. Pages 289–300. Academic Press, New York and London.

HÖGLUND, L. (1976a). Changes in acid mucopolysaccharides during the development of the blowfly. *Calliphora erythrocephala. J. Insect Physiol.* *22*, 917–923.

HÖGLUND, L. (1976b). The comparative biochemistry of invertebrate mucopolysaccharides V Insecta (*Calliphora erythrocephala*). *Comp. Biochem. Physiol.* *53B*, 9–14.

HOLTER, P. (1970). Regular grid-like substructures in the midgut epithelial basement membrane of some Coleoptera. *Z. Zellforsch.* *110*, 373–385.

HOYLE, G. (1952). High blood potassium in insects in relation to nerve conduction. *Nature* *169*, 281–282.

HUDSON, B. G. (1978). Chemistry of *Ascaris suum* intestinal basement membrane. In *Biology and Chemistry of Basement membranes.* Edited by N. A. Kefalides. Pages 253–263. Academic Press, New York.

HUNTER, B., MAGARELLI, P. C., LIGHTNER, D. V. and COLVIN, L. B. (1979). Ascorbic acid-dependent collagen formation in penaeid shrimps. *Comp. Biochem. Physiol.* *64B*, 381–385.

KANWAR, Y. S. and FARQUHAR, M. G. (1979). Presence of heparan sulfate in the glomerular basement membrane. *Proc. Natl. Acad. Sci. USA* *76*, 1303–1307.

KEFALIDES, N. A., ALPER, R. and CLARK, C. C. (1979). Biochemistry and metabolism of basement membranes. *Int. Rev. Cytol.* *16*, 167–228.

KEMENY, D. M., BANKS, B. E. C., LAWRENCE, A. J., PEARCE, F. L. and VERNON, C. A. (1981). The purification of hyaluronidase from the venom of the honey bee (*Apis mellifera*). *Biochem. Int.* *2*, 145–152.

KIMURA, S. and MATSUURA, F. (1974). The chain compositions of several invertebrate collagens. *J. Biochem. (Tokyo)* *75*, 1231–1240.

KRAMER, K. J., HENDRICKS, L. H., LIANG, Y. T. and SEIB, P. A. (1978). Effect of ascorbic acid and related compounds on the tobacco hornworm, *Manduca sexta* Johannson (Lepidoptera; Sphingidae). *J. Agric. Food Chem.* *26*, 874–878.

LANE, N. J. (1972). Fine structure of a lepidopteran nervous system and its accessibility to peroxidase and lanthanum. *Z. Zellforsch.* *131*, 205–222.

LAURENT, T. C. (1977). Interaction between proteins and glycosaminoglycans. *Fed. Proc.* *36*, 24–27.

LAZARENKO, T. (1925). Beiträge zur vergleichenden Histologie des Blutes und des Bindesgewebes. II. Die morphologische Bedeutung der Blut- und Bindegewebe elemente der Insekten. *Z. Mik. Anat. Forsch.* *3*, 409–499.

LECROISEY, A., BOULARD, C. and KEIL, B. (1979). Chemical and enzymatic characterization of the collagenase from the insect *Hypoderma lineatum. Eur. J. Biochem.* *101*, 385–393.

LOCKE, M. and HUIE, P. (1972). The fiber components of insect connective tissue. *Tissue Cell* *4*, 601–612.

LOCKE, M. and HUIE, P. (1975). Staining of the elastic fibers in insect connective tissue after tannic acid/gluteraldehyde fixation. *Tissue Cell 7*, 211–216.

MARSHALL, A. T. (1968). The chemical nature of Malpighian tubule mucofibrils in cercopoid dwelling tubes. *J. Insect Physiol. 14*, 1435–1444.

MARTOJA, R. (1973). Localisation, par microsonde électronique, des ions K. Ca, Mg et PO_4 dans les ganglions nerveux des Orthoptères. *J. Insect Physiol. 19*, 1–18.

MARTOJA, R. and BASSOT, J. M. (1965). Existence d'un tissue conjonctif de type cartilagineux chez certain insectes orthoptères. *C.R. Acad. Sci. Paris 261*, 2954–2957.

MARTOJA, R. and CANTACUZÈNE, A. M. (1966). Donées histologiques sur les mucoprotéines du système glial des Orthoptéroïdes et des Blattoptéroïdes. *C.R. Acad. Sci. Paris 263*, 152–155.

MARTOJA, R. and CANTACUZÈNE, A. M. (1968). Sur les mucopolysaccharides acides des centres nerveux des Arthropodes. *C.R. Acad. Sci. Paris 267*, 1607–1610.

MATHEWS, M. B. (1975). *Connective tissue. Macromolecular structure and Evolution*. Springer Verlag, Berlin.

MCLAUGHLIN, B. J. (1974). Fine structural changes in a Lepidopteran nervous system during metamorphosis. *J. Cell Sci. 14* 369–387.

MICHELS, H. (1880). Beschreibung des Nervensystems von *Oryctes nasicornis* im Larven-, Puppen- und Koferzustande. *Z. Wiss. Zool. 34*, 641–702.

MILLER, E. J. (1976). Biochemical characteristics and biological significance of the genetically-distant collagens. *Molec. Cell Biochem. 13*, 165–192.

MINEFRA, S., PUCCI-MINAFRA, I., CASANO, C. and GIANGUZZA, F. (1975). Chromatographic characterization of soluble collagen in sea urchin embryos (*Paracentrotus lividus*). *Boll. Zool. 42*, 205–208.

MITTLER, T. E., TSITSIPIS, J. A. and KLEINJAN, J. E. (1970). Utilization of dehydroascorbic acid and some related compounds by the aphid *Myzus persicae* feeding on an improved diet. *J. Insect Physiol. 16*, 2315–2326.

MONSON, J. M., NATZLE, J., FRIEDMAN, J. and MCCARTHY, B. J. (1982). Expression and novel structure of a collagen gene in *Drosophila*. *Proc. Natl. Acad. Sci. USA, 79*, 1761–1765.

MOULINS, M. (1968). Étude ultrastrucurale d'une formation de soutien epidermo-conjonctive inédite chez les Insectes. *Z. Zellforsch. 91*, 112–134.

MUSTAFA, M. and KAMAT, D. N. (1970). Mucopolysaccaride histochemistry of *Musca domestica*. I. A report on the occurrence of a new type of KOH-labile alcianophilia. *Histochemie 21*, 54–63.

MUTCHMOR, J. A. and RICHARDS, A. G. (1956). The degradation of hyaluronic acid by male gonads of *Drosophila*. *Canad. J. Zool. 34*, 200–205.

NISIZAWA, K., YAMAGUCHI, T., HANDA, N., MAEDA, M. and YAMAZAKI, H. (1963). Chemical nature of a uronic acid-containing polysaccharide in the peritrophic membrane of the silkworm. *J. Biochem. 54*, 419–426.

NORDLANDER, R. H. and EDWARDS, J. S. (1969). Postembryonic brain development in the monarch butterfly, *Danaus plexippus plexippus*, L. I. Cellular events during brain morphogenesis. *Wilhelm Roux' Archiv. 162*, 197–217.

NORDWIG, A., ROGALL, E. and HAYDUK, U. (1970). The isolation and characterization of collagen from three invertebrate tissues. In *Chemistry and Molecular Biology of the Intercellular Matrix*. Edited by A. E. Balazs. Volume I, pages 27–41. Academic Press, London.

NOWACK, H. and NORDWIG, A. (1974). Sea anemone collagen. Isolation and characterization of the cyanogen-bromide peptides. *Eur. J. Biochem. 45*, 333–342.

NUSBAUM, J. (1884). Bau, Entwicklung und morphologische Bedeutung der Leydig'schen Chorda der Lepidopteran. *Zool. Anz. 7*, 17–21.

ODSELIUS, R. and ELOFSSON, R. (1981). The basement membrane of the insect and crustacean compound eye: definition, fine structure, and comparative morphology. *Cell. Tiss. Res. 216*, 205–214.

OSIŃSKA, H. E. (1981). Ultrastructural study of the postembryonic development of the neural lamella of *Galleria mellonella* (L.) (Lepidoptera). *Cell Tiss. Res. 217*, 425–433.

PEARSE, A. G. E. (1968). *Histochemistry, Theoretical and Applied*. 3rd edition. Churchill, London.

PIKKARRAINEN, J., RANTANEN, J., VASTAMÄKI, M., LAMPIAHO, K., KARI, A. and KULONEN, E. (1968). On collagens of invertebrates with special reference to *Mytilus edulis*. *Eur. J. Biochem. 4*, 555–560.

PIPA, R. L. (1961). Studies on the hexapod nervous system. III. Histology and histochemistry of cockroach neuroglia. *J. Comp. Neurol. 116*, 15–26.

PIPA, R. L. (1967). Insect neurometamorphosis. III. Nerve cord shortening in a moth, *Galleria mellonella* (L) may be accomplished by humoral potentiation of neuroglial motility. *J. Exp. Zool. 164*, 47–60.

PIPA, R. L. (1969). Insect neurometamorphosis. IV Effects of the brain and synthetic α-ecdysone upon interganglionic connective shortening in *Galleria mellonella* (L). (Lepidoptera). *J. Exp. Zool. 170*, 181–192.

PIPA, R. L. and COOK, E. F. (1958). The structure and histochemistry of the connective tissue of the sucking lice. *J. Morph. 103*, 353–377.

PIPA, R. L. and WOOLEVER, P. S. (1965). Insect neurometamorphosis. II. The fine structure of perineurial connective tissue, adipohaemocytes, and the shortening ventral nerve cord of a moth, *Galleria mellonella* (L). *Z. Zellforsch. 68*, 80–101.

PROCKOP, D. J., KIVIRIKKO, K. I., TUDERMAN, L. and GUZMAN, N. A. (1979). The biosynthesis of collagen and its disorders. *N. Engl. J. Med. 301*, 13–23.

PUCCI-MINAFRA, I., GALANTE, R. and MINEFRA, S. (1978). Identification of collagen in the Aristotle's lantern of *Paracentrotus lividus*. *J. Submic. Cytol. 10*, 53–63.

QUINTARELLI, G., SCOTT, J. E. and DELLOVO, M. C. (1964). The chemical and histochemical properties of Alcian Blue. III. Chemical blocking and unblocking. *Histochemie 4*, 99–112.

RAUTERBERG, J. and KÜHN, K. (1971). Acid soluble calf skin collagen. Characterization of the peptides obtained by cyanogen bromide cleavage of its α1-chain. *Eur. J. Biochem. 19*, 398–407.

REINHARDT, T. C. and HECKER, H. (1973). Structure and function of the basal lamina and of the cell junctions in the midgut epithelium (stomach) of female *Aedes aegypti* L. (Insecta, Diptera). *Acta Tropica 30*, 213–236.

REINHARDT, T. C., SCHULZ, U., HECKER, H. and FREYVOGEL, T. A. (1972). Zur Ultrastrucktur des Mitteldarmepithels bei Floken (Insecta, Siphonaptera). *Rev. Suisse Zool. 79*, 1130–1137.

RICHARDS, A. G. (1944). The structure of the living insect nerves and nerve sheaths as deduced from the optical properties. *J NY Ent. Soc. 52*, 285–310.

RICHARDS, A. G. and RICHARDS, P. A. (1968). Flea *Ctenophthalmus*: heterogeneous, hexagonally organized layer in the midgut. *Science 160*, 423–424.

RICHARDS, A. G. and SCHNEIDER, D. (1958). Über den komplexen Bau der Membranen des Bindegwebes von Insekten. *Z. Naturforsch. 13b*, 680–687.

RINTERKNECHT, E. and LÉVI, P. (1966). Étude au microscope electronique du cycle cuticulaire au cours du 4ème stade larvaires chez *Locusta migratoria*. *Z. Zellforsch. 72*, 390–407.

ROSS, R. (1975). Connective tissue cells, cell proliferation and synthesis of extracellular matrix — a review. *Phil. Trans. Roy. Soc. Lond. B. 271*, 247–259.

ROSS, R. and KLEBANOFF, S. J. (1971). The smooth muscle cell. I, In vivo synthesis of connective tissue proteins. *J. Cell Biol. 50*, 159–171.

RUDALL, K. M. (1955). The distribution of collagen and chitin. *Symp. Soc. Exp. Biol. 9*, 49–71.

RUDALL, K. M. and KENCHINGTON, W. (1971). Arthropod silks; the problem of fibrous proteins in animal tissues. *Ann. Rev. Ent. 16*, 73–96.

RUOSLAHTI, E., ENGVALL, E. and HAYMAN, E. G. (1981). Fibronectin: current concepts of its structure and functions. *Coll. Res. 1*, 95–128.

SAGE, H. and GRAY, W. R. (1979). Studies on the evolution of elastin, I. Phylogenetic distribution. *Comp. Biochem. Physiol. 64B*, 313–327.

SAGE, H. and GRAY, W. R. (1980). Studies on the evolution of elastin. II. Histology. *Comp. Biochem. Physiol. 66B*, 13–22.

SALPETER, M. M., BACHMANN, L. and SALPETER, E. E. (1969). Resolution in electron microscope radioautography. *J. Cell Biol. 41*, 1–20.

SCHARRER, B. C. J. (1939). The differentiation between neuroglia and connective tissue sheath in the cockroach (*Periplaneta americana*). *J. Comp. Neurol. 70*, 77–88.

SCHARRER, B. (1963). Neurosecretion. XIII. The ultrastructure of the corpus cardiacum of the insect *Leucophaea maderae*. *Z. Zellforsch. 60*, 761–796.

SCHARRER, B. (1964). The fine structure of blattarian prothoracic glands. *Z. Zellforsch. 64*, 301–326.

SCHARRER, B. (1971). Histophysiological studies on the corpus allatum of *Leucophaea maderae*. V. Ultrastructure of sites of origin and release of a distinctive cellular product. *Z. Zellforsch. 120*, 1–16.

SCHARRER, B. (1972). Cytophysiological features of hemocytes in cockroaches. *Z. Zellforsch. 129*, 301–319.

SCHNEIDER, K. C. (1902). *Lehrbuch der Vergleichenden Histologie der Tiere*. G. Fischer, Jena.

SHARIEF, F. S., PERDUE, J. M. and DOBROGOSZ, W. J. (1973). Biochemical, histochemical and autoradiographic evidence for the existence and formation of acid mucopolysaccharides in larvae of *Phormia regina*. *Insect Biochem. 3*, 243–262.

SMITH, D. S. and WIGGLESWORTH, V. B. (1959). Collagen in the perilemma of insect nerve. *Nature 183*, 127–128.

SOHAL, R. S. (1974). Fine structure of the malpighian tubules in the housefly, *Musca domestica*. *Tissue Cell 6*, 719–728.

SOHAL, R. S., SHARMA, S. P. and COUCH, E. F. (1972). Fine structure of the neural sheath, glia and neurons in the brain of the housefly, *Musca domestica*. *Z. Zellforsch. 135*, 449–459.

STARK, M., MILLER, E. J. and KÜHN, K. (1972). Comparative electron microscope studies on the collagens extracted from cartilage, bone and skin. *Eur. J. Biochem. 27*, 192–196.

STEPHENS, R. J., FREEMAN, G. and EVANS, M. J. (1971). Ultrastructural changes in connective tissue of rats exposed to NO_2. *Arch. Intern. Med. 127*, 873–883.

STOCKWELL, R. A. and SCOTT, J. E. (1965). Observations on the acid glycosaminoglycan (mucopolysaccharide) content of the matrix of aging cartilage. *Ann. Rheum. Dis. 24*, 341–349.

TERZAKIS, J. A. (1967). Substructure in an epithelial basal lamina (basement membrane). *J. Cell Biol. 35*, 273–278.

TIMPLE, R., WIEDEMANN, H., VAN DELDEN, V., FURTHMAYR, H. and KÜHN, K. (1981). Network model for the organization of type IV collagen molecules in basement membranes. *Eur. J. Biochem. 120*, 203–211.

TOOLE, B. P. (1972). Hyaluronate turnover during chondrogenesis in the developing chick limb and axial skeleton. *Develop. Biol. 29*, 321–329.

TREHERNE, J. E. and SCHOFIELD, P. K. (1981). Mechanisms of ionic homeostasis in the central nervous system of an insect. *J. Exp. Biol. 95*, 61–74.

UITTO, J. (1979). Collagen polymorphism: isolation and partial characterization of $\alpha1(1)$-trimer molecules from normal human skin. *Arch. Biochem. Biophys. 192*, 371–379.

WERMEL, E. M. (1938). Über die Struktur des Bindesgewebes der Seidenspinnerraupen. *Bull. Biol. Méd. Exp.* URSS. *5*, 10–13.

WHITTEN, J. M. (1962). Breakdown and formation of connective tissue in the pupal stage of an insect. *Quart. J. Micr. Sci. 103*, 359–367.

WHITTEN, E. M. (1964a). Haemocytes and the metamorphosing tissues in *Sarcophaga bullata, Drosopholia melanogaster*, and other cyclorrhaphous Diptera. *J. Insect Physiol. 10*, 447–469.

WHITTEN, J. M. (1964b). Connective tissue membranes and their apparent role in transporting neurosecretory and other secretory products in insects. *Gen. Comp. Endocrinol. 4*, 176–192.

WIGGLESWORTH, V. B. (1956). The haemocytes and connective tissue formation in an insect, *Rhodnius prolixus* (Hemiptera). *Quart. J. Mic. Sci. 97*, 89–98.

WIGGLESWORTH, V. B. (1960). The nutrition of the central nervous system in the cockroach, *Periplaneta americana* L. *J. Exp. Biol. 37*, 500–512.

WIGGLESWORTH, V. B. (1973). Haemocytes and basement membrane formation in *Rhodnius*. *J. Insect Physiol. 19*, 831–844.

WIGGLESWORTH, V. B. (1979). Secretory activities of plasmatocytes and oenocytes during the moulting cycle in an insect (*Rhodnius*). *Tissue Cell 11*, 69–78.

WILLIAMS, M. A. (1973). Electron microscopic autoradiography: its application to protein biosynthesis. In *Techniques in Protein Biosynthesis*. Edited by P. N. Campbell and J. R. Sargent. Vol. 3, pages 125–190. Academic Press, London and New York.

8 Structure and Physiology of the Circulatory System

T. A. MILLER

University of California, Riverside, California, USA

CIP VOL 3-T

ORDERS OF INSECTS USED IN PREPARATION OF THIS CHAPTER, AFTER IMMS, A. (1970)

Subclass I. Apterygota

Order

Thysanura	Protura
Diplura	Collembola

Subclass II. Pterygota

Exopterygota	*Endopterygota*
Order	*Order*
Ephemeroptera	Neuroptera
Odonata	Mecoptera
Plecoptera	Lepidoptera
Grylloblattodea	Trichoptera
Orthoptera	Diptera
Phasmida	Siphonaptera
Dermaptera	Hymenoptera
Embioptera	Coleoptera
Dictyoptera	Strepsiptera
Isoptera	
Zoraptera	
Psocoptera	
Mallophaga	
Siphunculata	
Hemiptera	
Thysanoptera	

1 DEDICATION

Professor Jack Colvard Jones has maintained an interest in the circulatory system of insects for many years. Besides his own work on the physiology of the cockroach and mosquito, Professor Jones has provided several exhaustive reviews on the circulatory system of insects and related subjects. The three most prominent are the extensive 1954 paper on the mosquito, the 1964 first and 1974 second editions of Rockstein's *Physiology of Insecta* and the valuable book on *The Circulatory System of Insects*, which took several years to complete and was published by Charles Thomas in 1977. His scholarship is characterized by an extremely thorough and painstaking study of the literature beginning with the 17th century and extending until recent times, and an exhaustive treatment of the subject in review.

All of these reviews taken together represent an extremely useful background of knowledge for anyone interested in the circulatory system of insects, and they represent the starting and reference point for the present chapter.

Professor Jones retired from the Entomology Department at the University of Maryland in 1981, after a long and productive career. It is a special pleasure for me to dedicate this article to Professor Jones in appreciation for his mammoth efforts concerning this subject over the years, and for how much easier this has made it for everyone else.

Along with Professor Jones' retirement, the very sad news of Professor Manfred Gersch's death came with the end of 1981. Professor Gersch retired in 1974 as Professor and head of the Institute of Zoology of the Friedrich-Schiller University in Jena, a post he held since 1953.

His work led the present field of neuroendocrine regulation in animals. He was among the first to show neurohormonal control mechanisms in insects. Because of this it is particularly unfortunate that Dr Gersch will not be on hand to see his students and colleagues identify the structure of neurohormone D, the cardioaccelerator isolated from cockroaches.

Professor Gersch was a member of the Academy of Sciences of the DDR, the Saxon Academy of Sciences and the "Deutsche Akademie der Naturforscher Leopoldine" in Halle. He was awarded the National Prize of the German Democratic Republic in 1958, and until his death, he was editor of *Zoologische Jahrbücher* and *Zoologische Anzeiger*.

2 INTRODUCTION

The poorly developed circulatory system of insects was considered by Raabe, M. (1978) to be consistent with their having an extensive neurohemal system. She pointed out that the latter is duplicated in a paired way in each segment and responds locally to sensory input.

An elaborate tracheal system in arthropods is said to be correlated with a "simple" circulatory system (see pp.142–3 of Mill, P., 1972; P. Mill, this volume). The circulatory system in insects is almost entirely of the open category where hemolymph is confined to the body cavity or hemocoele. If a completely closed cardiovascular system is considered complex in the sense of being highly organized, then the circulatory system of insects would be considered simple. Since a number of accessory pulsatile organs and septa are present to help direct the flow of hemolymph, when taken together with their controlling innervation, these organs represent a rather extensive circulatory system. Functionally, the circulatory system plays a vital role in temperature regulation in some adult insects by use of a variety of devices including heat exchangers, and bidirectional flow of hemolymph in the dorsal vessel. Thus the term "simple" circulatory system is to be used advisedly because by itself the modifying term simple is misleading at best.

The dorsal vessel and the pericardial sinus in particular are important because of the almost exclusive presence here of the pericardial cells which function as ductless glands which sieve the hemolymph of certain substances and convert them to products for further refinement or excretion elsewhere (A. Crossley, this volume). The pericardial sinus can have an elaborate substructure with its own microcirculation under presumed neural control. In many Orthoptera, perfusion from the heart to the perivisceral sinus is facilitated directly by excurrent ostia as in the locust or indirectly via segmental vessels as in the American cockroach, *Periplaneta americana* (McIndoo, N., 1939; Nutting, W., 1951).

3 DORSAL VESSEL

Jones, J. (1977) refers to the entire dorsal pulsatile organ as the dorsal vessel. He distinguishes the abdominal dorsal vessel as the heart. Despite the presence of ostia and chambers in the thoracic portion of the dorsal vessel, the portions of the dorsal vessel in the thorax and head are normally referred to as the thoracic aorta or cephalic aorta.

Imms, A. (1970) refers to the heart as that part of the dorsal vessel containing ostial valves, and the aorta as a simple tube without specialization. Imms, A. (1970) suggests the most primitive condition as being three thoracic heart segments and nine abdominal heart segments. This can perhaps be seen best in Orthoptera (Figs 1, *Blaberus trapezoideus* and 2, *Grylloblatta campodeiformis*) (Nutting, W., 1951) and the thoracic chambers are clear in Collembola (Frish, K., 1978) where alary muscles and ostia are present in all of the segments behind the head (Fig. 3). The term heart, then, is ambiguous unless specified for each insect because the heart would extend anteriorly to the prothorax in some insects such as Collembola (Frish, K., 1978)

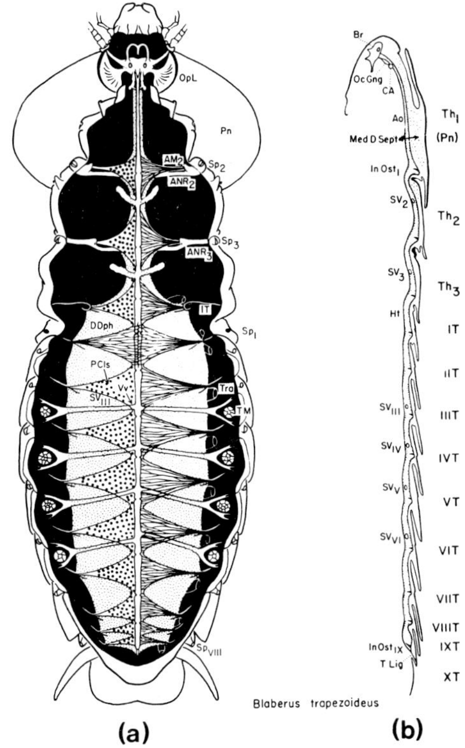

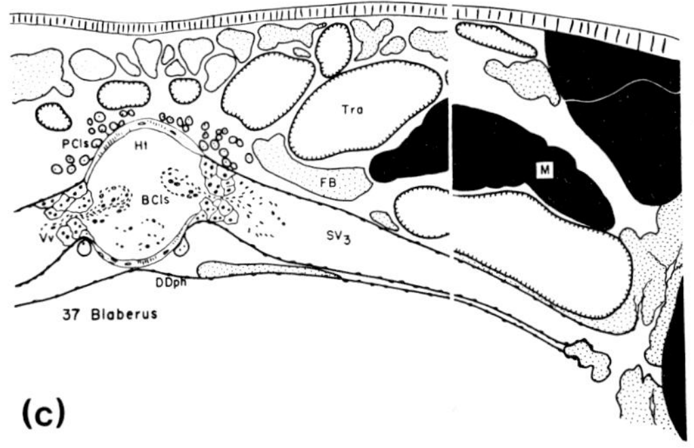

FIG. 1. Three views of an adult female cockroach, *Blaberus trapezoideus*. A ventral dissection (**a**) showing the dorsal vessel, then the dorsal vessel in profile (**b**) and finally a cross-section (**c**) through the fourth abdominal chambers. (From Nutting, W., 1951.)

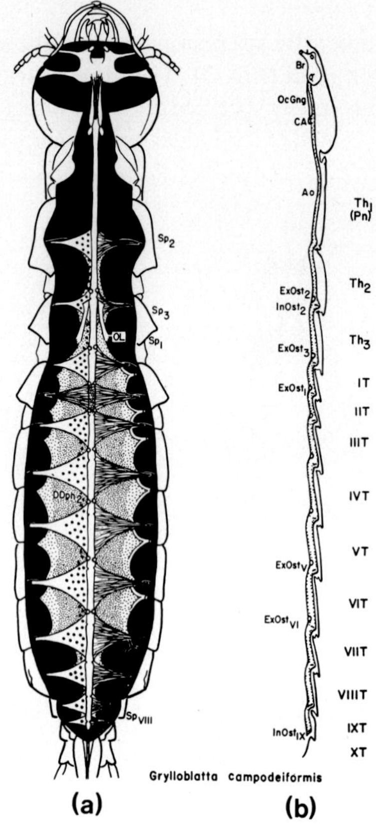

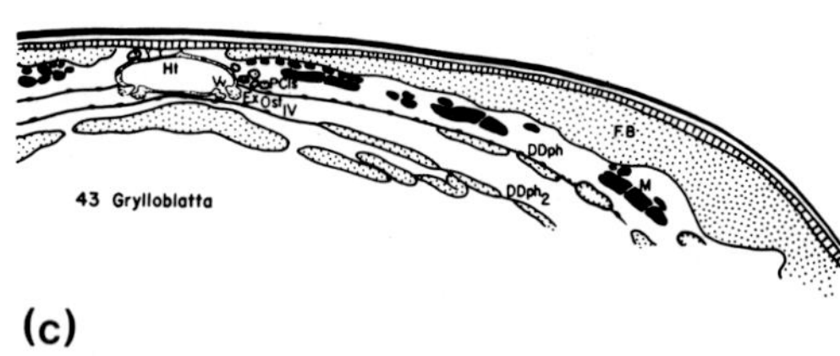

FIG. 2. Three views of the dorsal vessel of *Grylloblatta campodeiformis*, adult female (21 mm long overall). Six ventral dissection (**a**) and profile (**b**) of the dorsal vessel are shown plus a cross-section (**c**) of the dorsal aspect of the fourth abdominal segment showing excurrent ostia.

and not past the rear of the abdomen in others such as Hemiptera (Hinks, C., 1966) (Fig. 4).

The aorta, or anterior dorsal vessel, may end in a frontal sac (Bayer, R., 1968; Jones, J., 1977) or open simply, or the anterior aorta may be modified to divide into vessels supplying the antenna (cf. section 13). As the aorta passes near the retrocerebral complex, it may receive neurosecretory nerve endings directly from the medium neurosecretory neurons in the brain or via the corpora cardiaca (section 3.3).

Despite elegant early descriptions of dorsal vessels in insects, Gerould, J. (1938) refers to the Snodgrass, R. (1935) text on insect morphology as being the first comprehensive description of the insect heart. The insect dorsal vessel is generally a simple structure, usually one cell thick with myocardial cells oriented in a uniform opposing position in bilateral symmetry. Reports of the dorsal vessel of Lepidoptera describe a spiral-oriented myocardium (Gerould, J., 1938 for *Bombyx mori*) (Sanger, J. and McCann, F., 1968a for *Hyalophora cecropia*) with somewhat less order. Longitudinal muscle bands sometimes accompany the circular muscle bands of the dorsal vessel (Jones, J., 1954; Hinks, C., 1966) and muscle layers of the aorta (Unnithan, G. *et al.*, 1971).

Nutting, W. (1951) drew phylogenetic relationships between insect orders based on structures of the dorsal vessel. He found mantids and cockroaches to be closely related, based on the presence of lateral blood (segmental) vessels in both groups. He also judged *Grylloblatta* to be related to the saltatorial Orthoptera by virtue of the presence of eight pairs of excurrent ostia. The incurrent and excurrent ostial system of *Grylloblatta* was suggestive of the basic ancestral pattern from which all orthopteroid variations may have derived (Fig. 2).

In the larvae of the honeybee, *Apis mellifera*, the dorsal vessel is located just beneath the dorsal cuticle in the thorax and abdomen; however, in adult bees the heart has become modified (Fig. 5). Wille, A. (1958) characterized two main types of dorsal vessel: one with a straight portion in the thorax running just dorsal to the gut and the other more or less arching dorsally into the space between the dorsolongitudinal flight muscles which occupy the middle portion of the thorax. This arching was most developed in bees of the genus *Xylocopa*. Wille, A. (1958) also found varying degrees of coiling of the dorsal vessel in advanced bees in the petiole area.

Frish, K. (1978) noted the appearance of ostia in the dorsal vessel of the thorax was a primitive condition with evolution tending to reduce the thoracic dorsal vessel to a simple unmodified tube. The dorsal vessel of the collembolan *Anurida maritima* with six ostial pairs represents a very simple arrangement

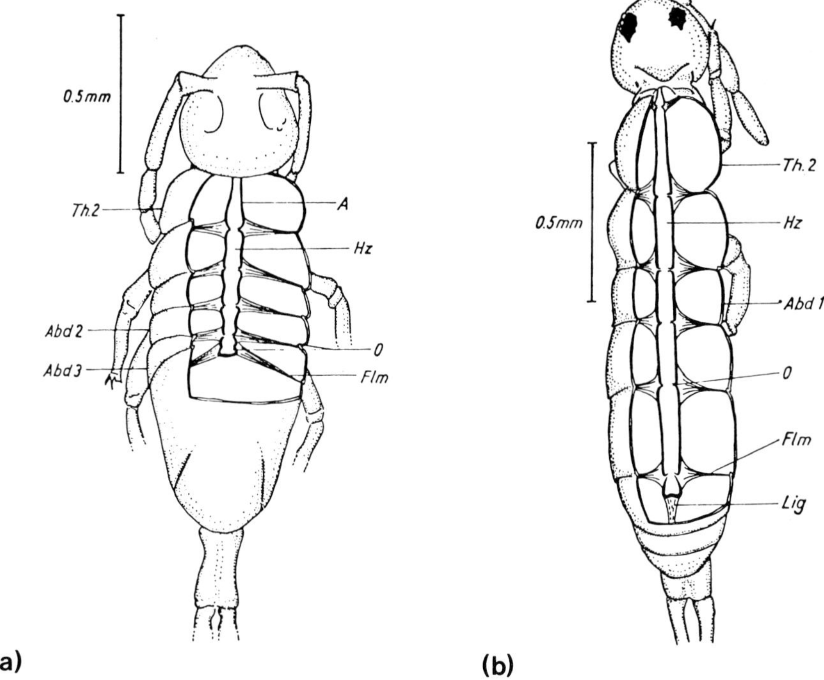

FIG. 3. **a:** A schematic drawing of the heart of a collembolan, *Anurida maritima*, from the dorsal aspect. Ostial valves coincide with the alary muscle arrays. **b:** A schematic drawing of the heart of *Octaletes neptuni*, dorsal aspect. Note the modified position forward in the body. (After Frish, K., 1978.)

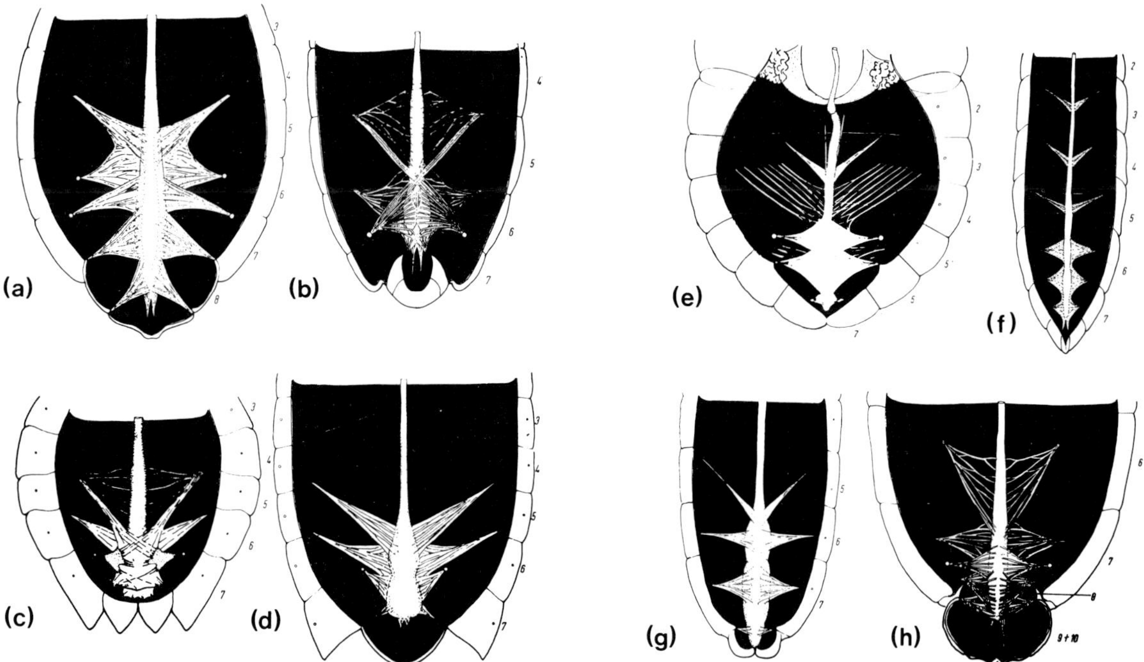

FIG. 4. Dorsal views of 8 Heteroptera. *Dolichonabis limbatus* (**a**), *Dysdercus fasciatus* (**b**), *Enoplops scapha* (**c**), *Elasmostethus interstincus* (**d**), *Eurygaster intergriceps* (**e**), *Leptopterna dolobrata* (**f**), *Gerris lacustris* (**g**), and *Phonoctonus nigrofasciatus* (**h**). In most cases the dorsal vessel has retained only a few alary mucles in the abdomen. (From Hinks, C., 1966.)

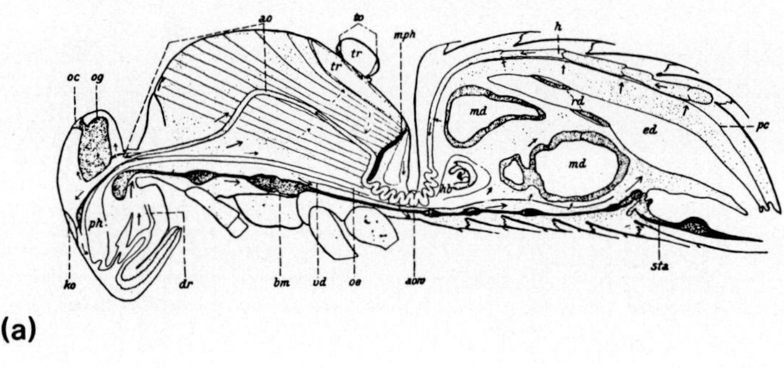

(a)

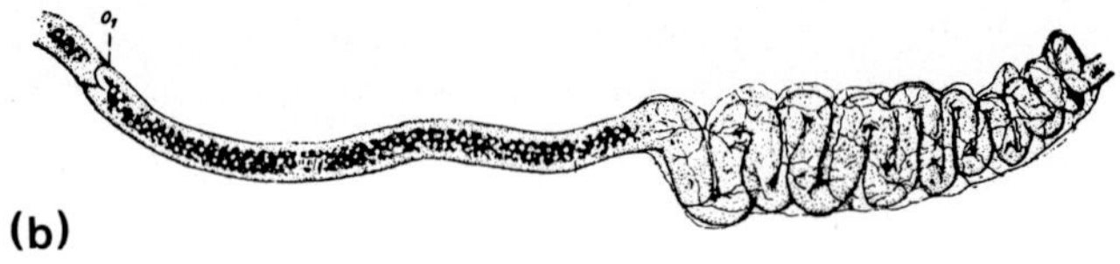

(b)

FIG. 5. **(a)** A longitudinal section of the honeybee, *Apis mellifera*, showing the position of the dorsal vessel. mph, Phragma; ao, aorta; tr, trachea; md, midgut; sta, sting apparatus; h, dorsal vessel; bm, thoracic ganglion; oe, oesophagus. Note in particular the convoluted portion of the dorsal vessel between the thorax and abdomen (aorv) which is shown in some detail in **(b)**. The coiled portion of the dorsal vessel is covered with a tracheated sheath and an ostia is indicated on the forward part of the aorta. (From Freudenstein, K., 1928.)

(Fig. 3a). *Actaletes neptuni* is similar except that the caudal end of the heart lies far forward of the terminal end of the abdomen (Fig. 3b), and the dorsal vessel contains five pairs of ostia.

Comparing Collembola in Apterygota with the pterygote insects, Frish listed up to 11 pairs of ostia in Archaeognatha and nine in Diplura with 12 pairs of ostia in Blattodea, eight pairs in *Dytiscus*, seven to eight in Lepidoptera and four in *Musca*. Thus, reduced numbers of ostial valves occur amongst all groups of both advanced and primitive insects. Wille, A. (1958) noted the failure of some authors to find all of the thoracic ostia in some bees because these ostia are difficult to identify. Frish, K. (1978) counts four ostia in the abdomen of adult *Apis* while Wille, A. (1958) pictures five ostia in workers of *Apis mellifera*. One must be cautious in accepting numbers of ostia as reported.

Wille's study concluded that the straight portion of the dorsal vessel in the thorax was the primitive condition, and that the series of coils or loops were present only among specialized bees. He also interpreted an evolutionary trend toward reduction of the posterior section of the dorsal vessel and its ostia. Moreover, in the more specialized bees of the family, Apidae, the two ostia commonly occurring in the thorax are less evident than in more primitive bees.

The Meliponini, or stingless bees, show enlargement of the terminal chambers of the dorsal vessel. Wille notes that the two main groups of stingless bees, *Melipona* and *Trigonia*, can be distinguished on the basis of their dorsal vessels.

Baccetti, B. and Bigliardi, E. (1969) reported two layers of a connective tissue cover the external surface of the dorsal vessel of the grasshopper, *Aiolopus strepens*. They found that the elements of this matrix included materials that stained positively for polysaccharides in light microscopy and for mucopolysaccharides in electron microscopy. The latter procedure included extraction by pancreatic elastase. Collagen fibrils were isolated from the same tissue (D. Ashhurst, this volume). This external sheath gives off processes that connect it to similar tissues investing pericardial cells and fat body, the nervous system and surrounding muscles.

A connective tissue sheath is also said to line the lumen of the dorsal vessel in *Aiolopus strepens*, along with hemocytes. My own experience with the ultrastructure of the dorsal vessel of *Periplaneta americana* confirms the presence of an occasional hemocyte tightly opposed to the surface of the lumen of the dorsal vessel and such a condition is reportedly common for pupal *Sarcophaga bullata* when the larval heart is being transformed into an adult heart (Sedlak, B. and Whitten, J., 1976). However, besides a uniformly thin acellular basal lamina, immediately external to the myocardium, the lumen appears to be devoid of any regular cellular layer.

G. Meyer's (1958) report on the aorta of *Apis mellifera* was one of the first ultrastructure studies of an insect myocardium. Intercalated discs were reported in myocardium of the cockroach (Edwards, G. and Challice, C., 1960), house-fly (Sohal, 1971, see Jones, J., 1977) and moth (Sanger, J. and McCann, F., 1968a). The myofibrils of the various pumps and diaphragms in the circulatory system of insects vary somewhat in ultrastructural appearance. The muscles of all pulsatile organs are striated. No exception to this has been found, even in the segmental vessel valve muscle (Miller, T., 1975a).

There is a relationship between the rate of contraction and the length of A-band of the various muscles in circulatory pumps and diaphragms (Table 1). The myocardial tissues examined have all been reported with A-bands of near 2.0 μm (Fig. 6). These tend to be relatively uniform myofibrils with Z lines fairly closely adjacent, whereas the slower-contracting alary muscles tend to have somewhat longer A-bands and fewer mitochondria than the myocardium, and have a less well-defined order in the way the Z lines are lined up in adjacent myofilament arrays (Fig. 7). Also, the sarcoplasmic reticulum is more abundant in myocardium compared to alary or diaphragm muscle. Thus, the ultrastructure of the slower-contracting diaphragm muscles appears to conform to tonic muscle while the myocardium appears in comparison similar to phasic muscle.

3.1 Ostial valves

Gerould, J. (1938) described the structure and function of ostial valves in insects. The ostial valves in most insects are simple openings that: (1) permit hemolymph to flow into the dorsal vessel and

Table 1: Length of A-band of various pulsatile organs and diaphragms in insects. Taken partly from Miller, T. et al. *(1979)*

Muscles and species	A-band (μm)	Source
Ventral diaphragm		
Locusta migratoria	3-5[a]	Dierichs, R., 1972
Apis mellifera	3.7	Morison, G., 1928
Hyperneural muscle		
Periplaneta americana	2.3	Miller, T. and Adams, M. 1974
Alary muscle		
Locusta migratoria	6.4	Miller, T. *et al.* 1979
Periplaneta americana	3-3.5	Adams, M. *et al.* 1973
Hyalophora cecropia	5.5	Sanger, J. and McCann, F., 1968b
Apis mellifera	8.0	Morison, G., 1928
Myocardium		
Locusta migratoria	2.0	Benedeczky, unpublished
Periplaneta americana	2.0	Miller, T., 1975a
Hyalophora cecropia	1.8	Sanger, T. and McCann, F. 1968a
Aiolopus strepens	2.0	Baccetti, B. and Bigliardi, E. 1969
Apis mellifera	2.0	Morison, G. 1928
Aorta		
Sympetrum danae	3.0	Jensen, H. 1976
Ampulla (Antennal APO)	3.0 (short type)	
Periplaneta americana	5.0 (long type)	(Beattie, unpublished, cf. section 13.2)

[a] Uncontracted sarcomere length • 5 μm (Dierichs, R. 1972). A sarcomere of 5 μm has an A-band of 3.8 μm.

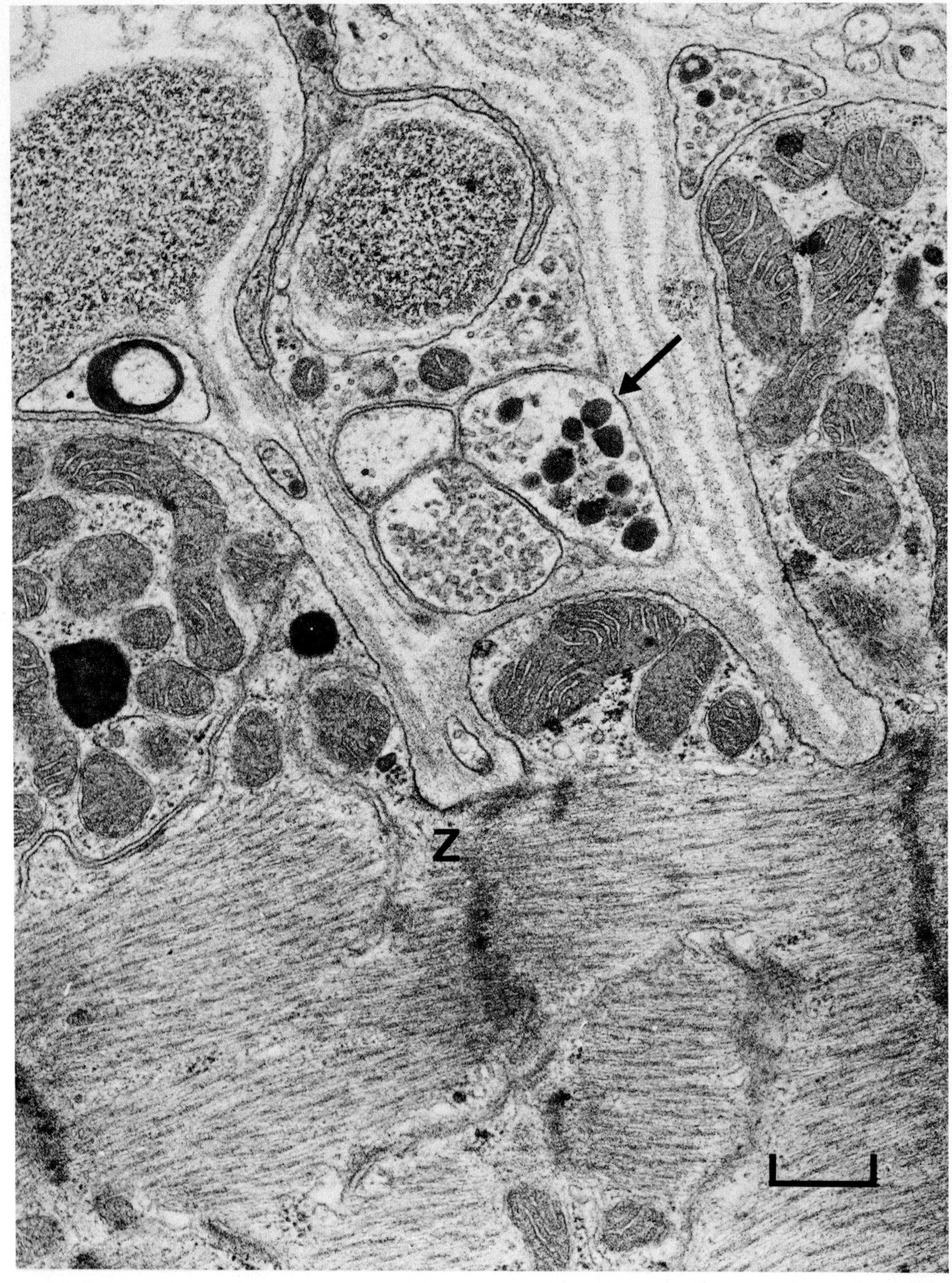

FIG. 6. The external surface of the myocardium *Periplaneta americana* showing folding at the level of each Z-line. The muscle is fully contracted with sarcomeres of near 2 μm and a trio of axon profiles, one with electron-dense granules (arrow), approaches one of the folds between adjacent Z-lines. The section was taken slightly askew of the longitudinal.

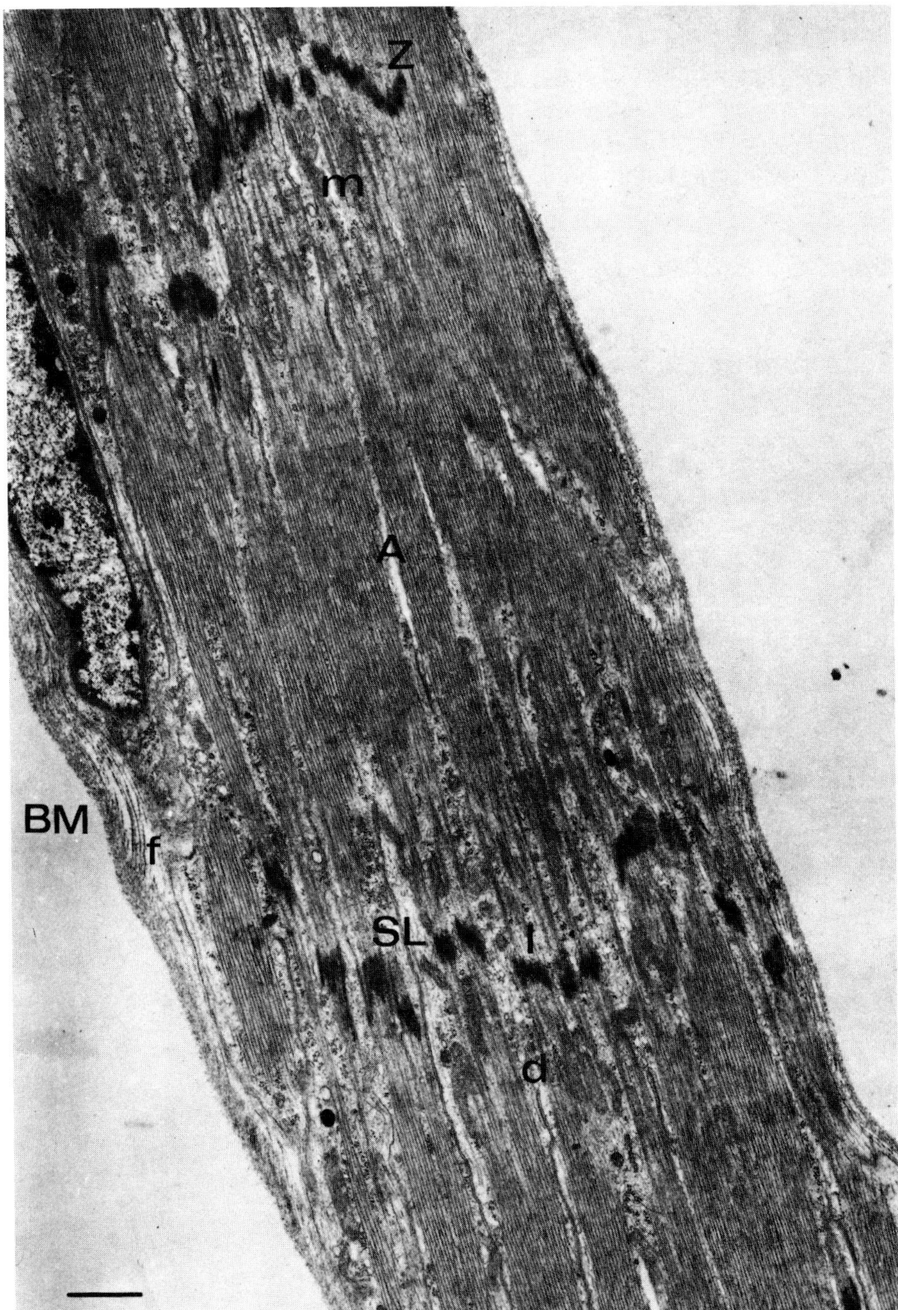

FIG. 7. Longitudinal section of alary muscle fiber slightly askew of the filaments. Only a few mitochondria (m) are seen near the Z-lines. A-bands (A), I-bands (I) and Z-lines (Z) can be distinguished along with dyads (d) and collagen-like fibrils (f) in the adjacent basement membrane (BM). Sarcolemma (SL) is present as invaginated sheets. Calibration mark: 1 μm. (From Miller, T. *et al.*, 1979.)

prevent backflow (incurrent ostia); or (2) allow free flow of hemolymph out of the dorsal vessel usually without valves (excurrent ostia). Excurrent ostia in Orthoptera are simply holes in the heart which allow flow of hemolymph from the dorsal vessel directly into the perivisceral sinus (Nutting, W.,

1951). Curiously, those orders of aquatic insects whose nymphal or larval respiratory system is modified by some form of gill-like apparatus to take oxygen from the water may also possess ostia that appear to perform a true catch-valve function to prevent backflow of hemolymph in the dorsal vessel of the anterior abdomen (Gerould, J., 1938).

There are nine pairs of ostia in the abdominal part of the dorsal vessel of *Bombyx mori*, one pair in the mesothorax and one pair in the metathoracic part of the dorsal vessel (Gerould, J., 1938). Three thoracic ostia and nine abdominal ostia appear to be the maximum number, with a tendency to lose thoracic ostia in the pro- and then mesothorax. Extreme loss of ostia occurs in some Hemiptera (Hinks, C., 1966). The naming of the "heart" based on the presence of ostial valves defining "chambers" (actually these are pseudochambers because they are rarely closed internally) is therefore ambiguous and different in many orders. The most accurate term is dorsal vessel.

The correlation between the presence of ostia with alary muscles suggests that the ostia regulate the flow of hemolymph in conjunction with the dorsal diaphragm. Exceptions to this occur in the thoracic dorsal vessel.

The excurrent ostia are often particularly difficult to see unless one uses dye in the perfusing saline. The excurrent ostia in *Locusta migratoria* are simple non-contractile openings though which hemolymph flows passively. Rhythmic cardiac contractions cause no excess movement of the tissue around the excurrent ostia because no flaps of tissues are present here (personal observations). Indeed one wonders at the efficiency of such a pulsating tube. Perhaps the gut tissue pressing dorsally against the dorsal diaphragm from the perivisceral sinus acts as a valve to cover the excurrent ostial openings in the intact locust.

3.2 Segmental vessels (see also the section on innervation)

Five pairs of segmental vessels were reported in the second to sixth segments of the dorsal vessel in the abdomen of *Periplaneta americana* in large male or female nymphs (McIndoo, N., 1939). However, segmental vessels in adults of both sexes are reportedly also present in the meso- and metathorax (McIndoo, N., 1939). The segmental vessels in the thorax are small and evidently difficult to find.

I have corresponded a few times with both Bruce Johnson and Terry Beattie since 1968. Johnson developed the theme of delivery of neurosecretory material via axons to organs (see Miller, T., 1975b for more references) and Beattie, T. (1976) described the ultrastructure of the segmental vessels and antennal pulsatile organs of the American cockroach in his doctoral thesis. The segmental vessel work was never formally reported. As Beattie is no longer actively pursuing these subjects, I have decided to paraphrase freely from our 1972 correspondence because his results are not readily available.

The segmental blood vessel walls of *Periplaneta americana*, as shown in Fig. 8, are composed of fibroblasts and longitudinal strands of connective tissue (bundles of collagen fibers). The walls of the vessel also support numerous neurosecretory axons (three types), but no motor/sensory axons could be found (Fig. 9). The NS axons terminate before the lateral extremities of the vessels and appear to consititute an extension of the neurohemal organ associated with the lateral cardiac nerves.

The valves of the excurrent ostia at the mouth of the segmental vessels are composed of a peculiar

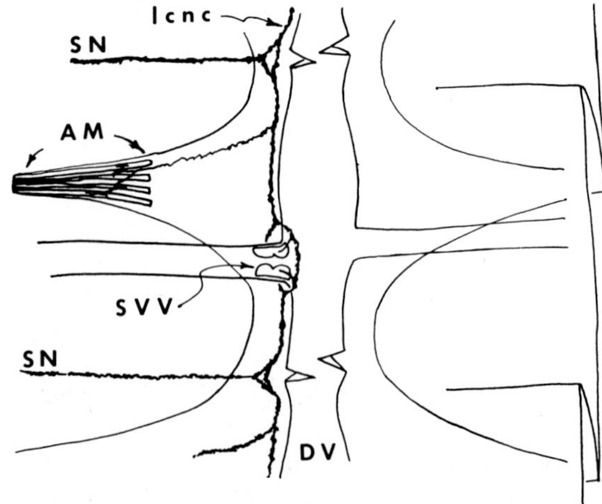

FIG. 8. A drawing of the dorsal vessel (DV) of cockroach, *Periplaneta americana*, showing the location of the segmental vessel valve (SVV) in the segmental vessel of the fourth abdominal heart chamber. SN, segmental nerve; AM, alary muscle; lcnc, lateral cardiac nerve cord. Ostial valves are shown, but not labelled.

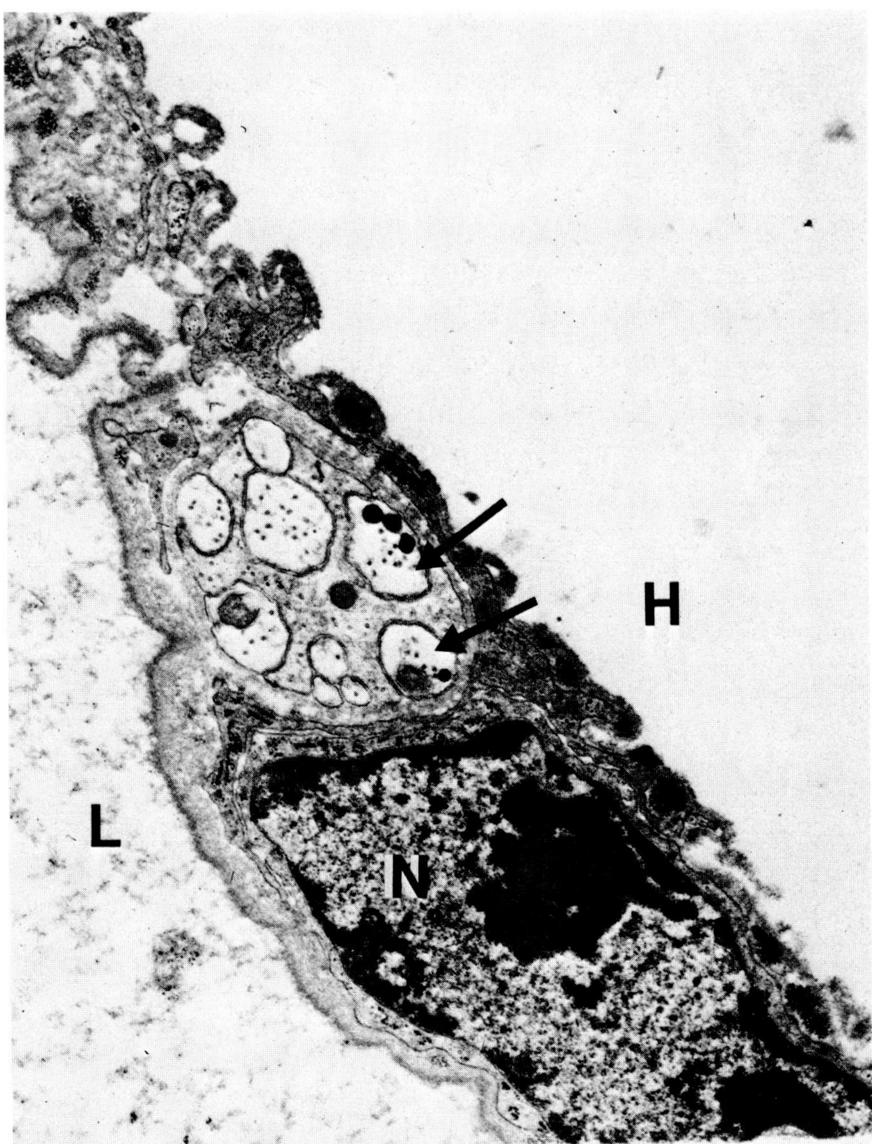

FIG. 9. Ultrastructure of a portion of the segmental vessel shown in Fig. 8. Note the neurosecretory axon profiles (arrow). These axons ramify over the vessel "wall" and form release sites. Note the acellular nature of the segmental vessel wall which is apparently composed of basal lamina secreted by the cellular tissue whose nucleus is shown (N). L, lumen of the vessel; H, hemocoel.

type of muscle cell (Miller, T., 1975a) which is almost spherical and appears to be multinucleate (Fig. 10). Each has only a small number of myofibrils and the major portion of the cytoplasm is filled with granular material (glycogen?). The myofibrils tend to be oriented at all angles (Fig. 11) and only occasionally is there any regular arrangement as in normal muscle cells. The myofibrils are usually aggregated on one side of the cell. This side has the plasmalemma greatly infolded and interdigitates with adjacent muscle cells. The plasmalemma of the remaining part of the cell is relatively smooth. In these interdigitated regions numerous granule-containing axons are to be found along with tracheoles and small bundles of connective tissue. The granule-containing axons synapse with the muscle cells (Miller, T., 1975a). The synapses are either on the surface of the muscle cell with a glial

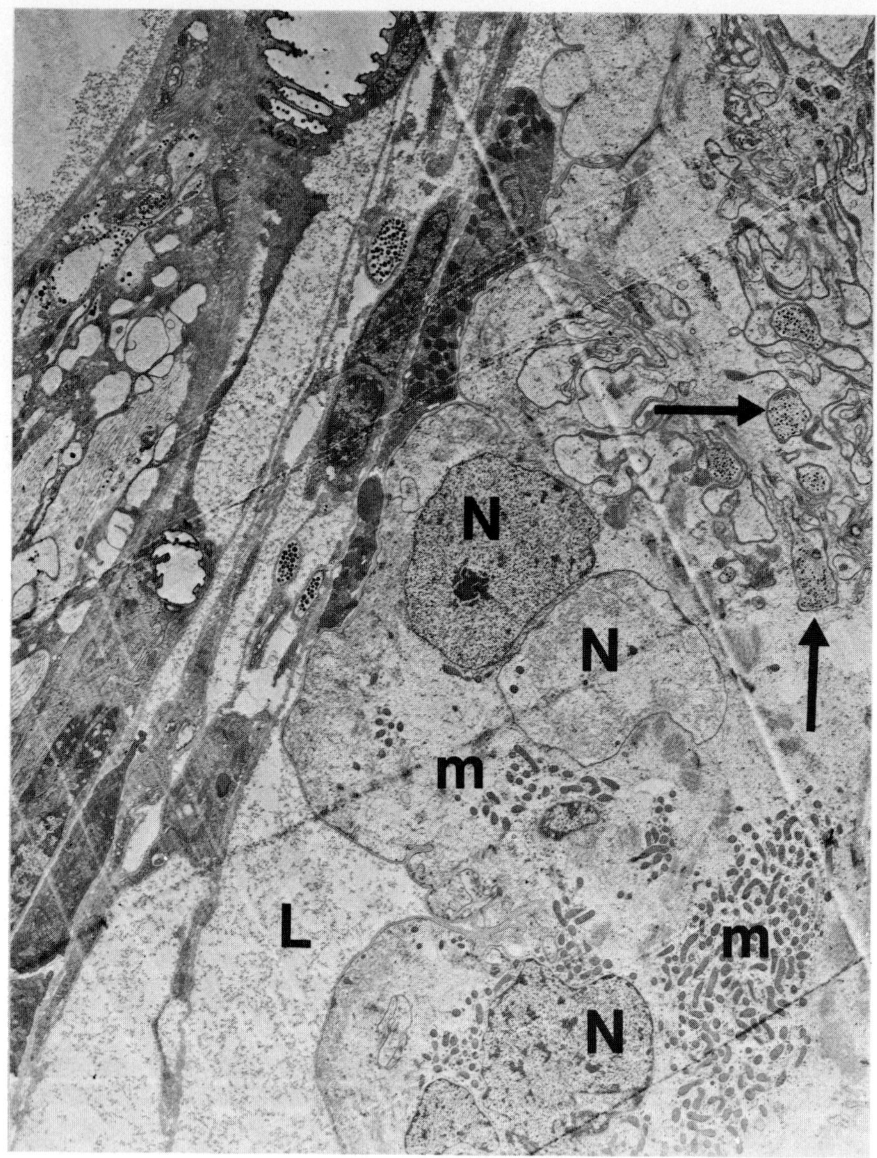

FIG. 10. Segmental vessel valve muscle located at the opening of a segmental vessel into the dorsal vessel of *Periplaneta americana*. Note the round shape of the cell and the axon profiles found deeply invaginated into the middle of the cell (arrows). The cell contains numerous nuclei (N), and groups of mitochondria (m). L, lumen of segmental vessel.

layer surrounding the non-synapsing part of the axon, or the axon penetrates the muscle cell and there is no associated glial tissue (Fig. 10). The granules are *ca.* 100–150 nm diameter with a mean of *ca.* 120 nm. The synaptic vesicles are variable in shape (spherical to ellipsoidal). The synaptic cleft is 30–35 nm across. In some clefts a layer of dense material is present on the postsynaptic membrane. This layer is *ca.* 10 nm thick and has fine striations running across it at about 15 μm intervals. The neuromuscular synapes on the heart wall have similar structures.

The perikarya of these axons is not known but *may* be situated in the lateral cardiac nerves. The neurons when present tend to be associated with the segmental vessels. Also, McIndoo, N. (1939) and Alexandrowicz, J. (1926), as well as some methylene blue preps (Beattie, 1972, personal

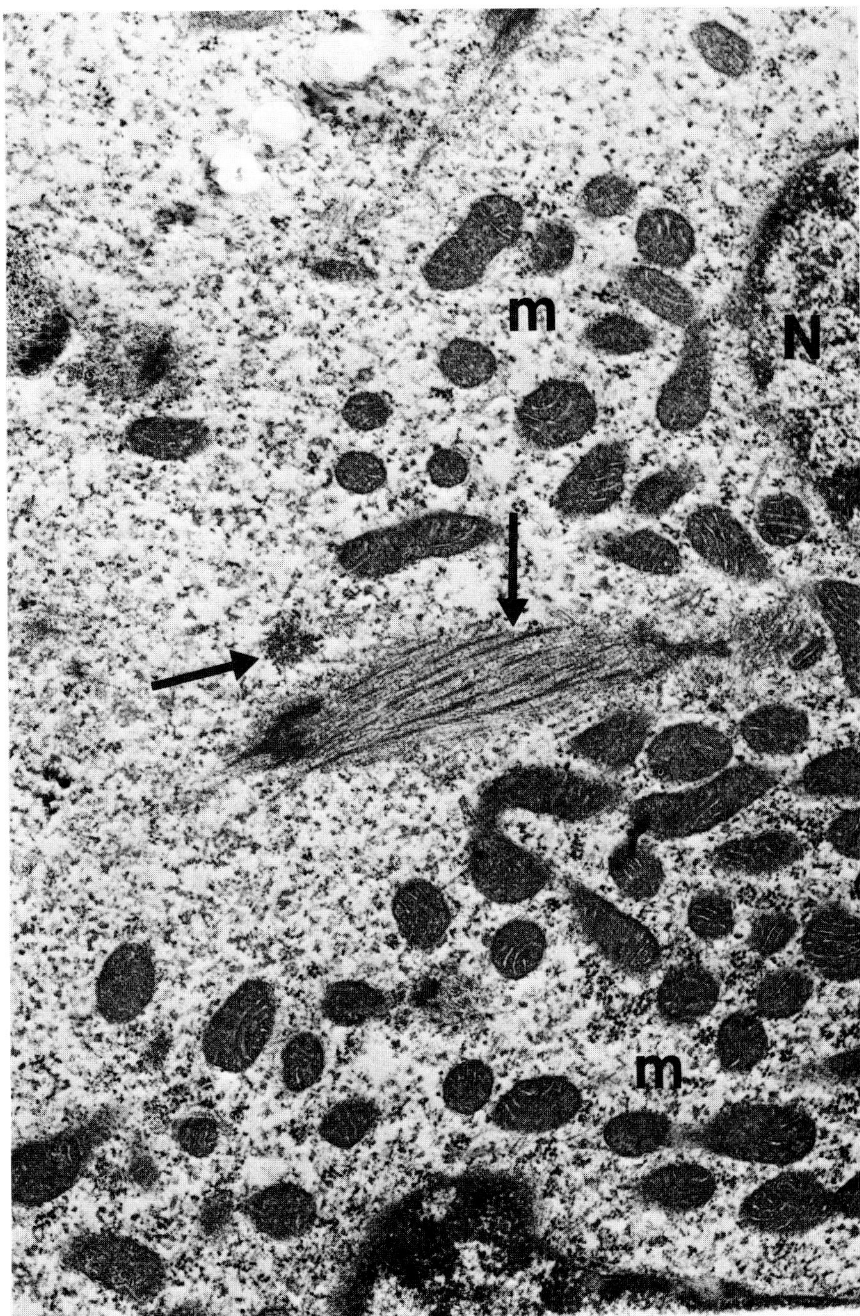

FIG. 11. A close-up of the segmental vessel valve shown in Figs. 8 and 10. Note the mitochondria (m), nuclei (N), and that the myofilaments are oriented perpendicular to one another (arrows).

communication), have shown processes running from cell bodies to the base of the segmental vessels. These processes may be the source of the valve muscle innervation.

The action of the segmental vessel valve muscle can sometimes be appreciated in semi-isolated nerve preparations. The contraction of the excurrent ostia valves is 180° out of phase with the heart (and incurrent ostia) contraction. When the valves are open, heartbeat contractions of the dorsal vessel are accompanied by a puff of fluid, seen as a ballooning increase in the diameter of the segmental vessel itself. Since the vessels are transparent, the motion described is seen as a movement by the bounding tissues, usually fat body. When the valves close, the fluid propulsion and puffing in the segmental vessels is no longer seen during heartbeat (Miller, personal observations).

It would appear as if a separate control is required for the excurrent valves. The excurrent valves do not operate between all heart beats, but tend to operate in cycles, so that the valves open for about six heartbeats, partially open on the next beat, and then remain closed for the next six successive heartbeats. The cycle begins again with the valve partially opening for one beat and then fully open for the next six, and so on. The ratio of opening to closings is not fixed for different valves in the same animal and may be 5 : 8. However, for a particular valve the ratio is fairly constant and may vary by one for one cycle, i.e. 6 : 6, then 6 : 7 and then 6 : 6. This consistency of the cycle of particular valves would seem to indicate a local control, but whether or not this is a property of semi-isolated preparations is difficult to ascertain. The cuticle is not sufficiently transparent to see the valves.

A few nerve cell bodies were found, and these appeared to be similar to "ordinary" neurons and only contained a few dense granules of the same size as those in axons innervating the valves. The neurons which were near the segmental vessels may have given rise to the valve axons, but there is no proof of this and it needs further investigation.

The incurrent ostia, which close in phase with the heart, are composed of muscle similar to the heart wall. Two points of difference were found. Firstly, the overall electron-density of the incurrent ostia muscle is greater than that of the heart muscle, and secondly, there appears to be more mitochondria packed into the cytoplasmic bulges (the bulges are in register with the sarcomeres) and the bulges have an angular profile, whereas the bulges of the heart wall are rounded. No trace of innervation of the incurrent ostia was found. However, the junction between the incurrent ostia muscle cells and the heart muscle cells is specialized. There is a well-developed desmosome between the two types of muscle cell. It is similar to those in *Hyalophora cecropia* heart (Sanger, J. and McCann, F., 1968a), except that there do not appear to be any septate desmosomes or any other membrane specialization which may indicate electrical coupling between adjacent cells.

This type of junction was also found between the heart muscle and muscle cells of the excurrent valve. In this case at least it would appear to be a mechanical coupling only, since these two types of muscle cell contract out of phase. Also, the valve cells appear to bond onto connective tissue strands by a similar structure (but only 1/2 is present, of course). This type of muscle/connective tissue junction was also found in the antennal accessory heart complex.

The major portion of neurosecretory axons found in the cardiac nerves is derived from segmental innervation. The number of axons coming from the retrocerebral complex is quite small. Most work has been concentrated on the abdominal portion of the heart. The pathway of the segmental nerves to the heart appears to involve the median nerve.

The neurosecretory axons appear to come via the median/transverse nerve of the preceding segment. neurosecretory axons coming from the ganglion of the same segment appear to turn down a link nerve and do not proceed to the dorsal part of the segment. The link nerve appears to be a neurohemal organ with intrinsic neurosecretory cells, rather similar to that found in *Carausius morosus* by Finlayson, L. and Osborne, M. (1968). The nature of the multipolar cardiac neuron described by McIndoo, N. (1939) is unknown, except for its gross morphology. The anatomy of the nervous system seems to be best described by Shankland, D. (1965) being more exact than most other workers quoted by Guthrie, D. and Tindall, A. (1968). Whether or not the motor innervation of the heart follows a similar pathway has not been determined.

3.3 Aorta — neurohemal organ

The wall of the aorta just behind the brain in *Oncopeltus fasciatus* contains nerve endings of at least three types of neurosecretory axon (Unnithan, G. *et al.*, 1971). These axons are thought to arise from cell bodies in the pars intercerebralis region of the brain where four classes of cells were recognized on the basis of the shape and size of neurosecretory granules.

There was no evidence that the intrinsic cells of the corpora cardiaca sent axons to the aorta. Type II and III neurosecretory axons were found making apparent synaptic contact with the muscle cells of the aorta (Unnithan, G. *et al.*, 1971). Although the possibility does exist that these axons may arise from elsewhere, the study strongly suggests direct innervation of the aorta muscles from the brain. If we term these apparent motor endings as "neurosecretory" or units, we may consider them to be similar to those found innervating the dorsal vessel and alary muscles (see Miller, T., 1979a).

Axons arise from the median neurosecretory cells of the brain of *Calliphora vicina*, leave the brain (as NCC-I of some authors), penetrate the corpora cardiaca and pass on to terminate on the wall of the aorta (Normann, T., 1975) (Fig. 12). Some axons in this cardiac "recurrent" nerve (see Normann, T., 1975) continue past the aorta to supply the corpora allata. Awasthi, V. (1975) suggested that in the earwig, *Labedura riparia*, the median neurosecretory cells in the pars intercerebralis give rise to nerve bundle NCC-I which travels to and forms a neurohemal release site with the cephalic aorta of the dorsal vessel. He also noted that the lateral neurosecretory cells in the brain gave rise to nerve bundle NCC-II which formed a neurohemal release site at the corpora cardiaca (see Orchard, I. and Loughton, B. vol. 7).

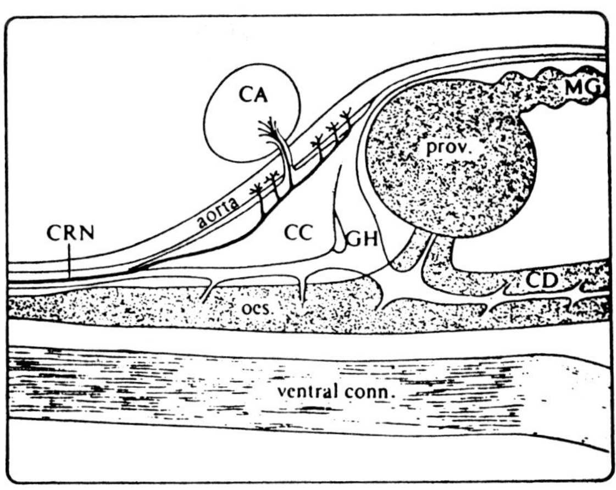

FIG. 12. Drawing of the retrocerebral complex of blowfly. CRN, cardiac recurrent nerve, supplies the aorta and the corpora allata. CC, corpora cardiaca; GH, hypocerebral ganglion; prov., proventriculus; MG, midgut; CD, crop; oes, esophagus. (From Normann, T., 1975.)

In the bark bug, *Halys dentatus* F., a pentatomid, the nervi corpori cardiaci I (NCC-I) runs past the corpora cardiaca to innervate the aortal wall directly. The aorta does not store material. Neurosecretory material is stored in and released from the axons associated with the neurohemal structure which includes the aorta (Srivastava, R., 1970). The association of NCC-I with the aorta is also reported in the beetle, *Aulacophora foevicollis* Lucas (Srivastava, R., 1970).

The prominent nature of the neurohemal structure associated with the anterior aorta in some adult insects may be related to neuroendocrine control of trehalose levels (and carbohydrate metabolism) for control of energetics for flight or locomotion in general (Normann, T., 1975). Low levels of hemolymph trehalose appear to be inversely related to the activity of the corpora allata and the median neurosecretory nerve cells in the pars intercerebralis of the brain. Larvae have low trehalose but very active corpora allata during development, and adults have high hemolymph trehalose but little or no activity in corpora allata except for a suspected role in egg development (Engelmann, F., 1970).

Aside from the specialization at the aorta, the crickets, *Melanogryllus desertus* and *Acheta domestica*, have a unique neurohemal area on the ventral median surface of the brain (Geldiay, S. and Karaçali, S., 1980). Nervi corpori cardiaci I (NCC-I) leave the pars intercerebralis and pass through the neurohemal area (Fig. 13). There is some suggestion that this neurohemal organ is a primitive condition (Geldiay, S. and Karaçali, S., 1980).

3.4 Histolysis of myocardium

Histolysis of the larval heart during metamorphosis has been reported, and it is claimed that modification occurs but not complete histolysis (Jones, J., 1977). In *Sarcophaga bullata*, the flesh-fly, while most larval tissue is replaced from imaginal discs with adult tissue, the dorsal vessel appears to be different, perhaps inferentially at least, because a heartbeat can always be recorded in the pupae. Sedlak, B. and Whitten, J. (1976) reported a reorganization of the myocardium in pupal flesh-flies with a dramatic decrease in numbers of mitochondria occurring in 2-day-old pupae along with reduction in

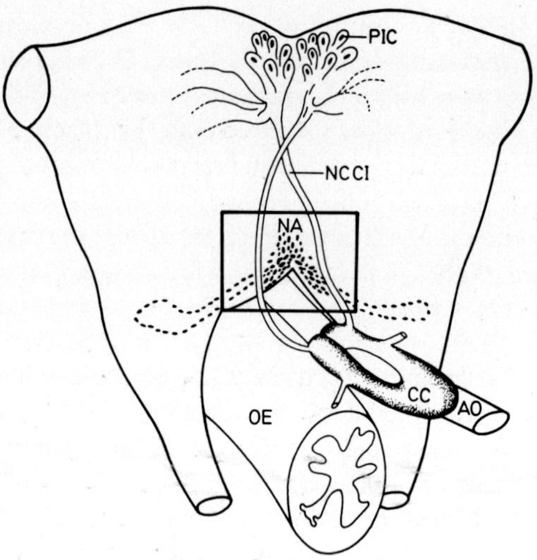

FIG. 13. Neurohemal organ (NA) associated with the nervi corpori cardiaci (NCC-I) axons in the brain near the retrocerebral complex of *Melanogryllus desertus*. Posterior view of the brain. The dotted lines show the extent of the neurohemal area. (From Geldiay, S. and Karaçali, S., 1980.)

myofilaments. Hemocytes lined the heart lumen most conspicuously on days 5 and 6. By day 10 myofibrils became well-organized and by day 11 the mitochondria again became as numerous as in the active larval. Jensen, P. (1973) has provided a fate map of the changes occurring in each cell of the anterior heart of *Calliphora vicina* during metamorphosis.

The pupal heartbeat is characterized by periods of activity interspersed with periods of quiescence which can last several hours (Miller, T., 1979a). In some cases heartbeat reversal occurs in pupae, but according to Jones, J. (1954) not often. My own impression from published work is that the characteristic heartbeat reversals in endopterygote insects are begun in the pupal stages and merely continue into the adult stage. Heartbeat reversal is the rule in *Musca domestica* pupae (Fig. 14a) and *Trichoplusia ni* pupae (Fig. 14b), and these stages are marked by intermittent activity of the heart.

3.5 Recording heartbeat

Electrical and optical methods were originally developed for recording the electrical activity or movements of the insect dorsal vessel. Methods have been described for recording heartbeats directly from the eyepieces of dissecting microscopes with photocells or other light-sensitive devices, and these require the subject to be immobilized (Tachibana, K. and Nagashima, C., 1957).

The two most important methods for recording insect heartbeat have proven to be impedance conversion, as pioneered by Bernd Heinrich, and contact thermography, developed by Lutz Wasserthal. These and the optical methods for recording insect heartbeat are detailed elsewhere (Miller, T., 1979b).

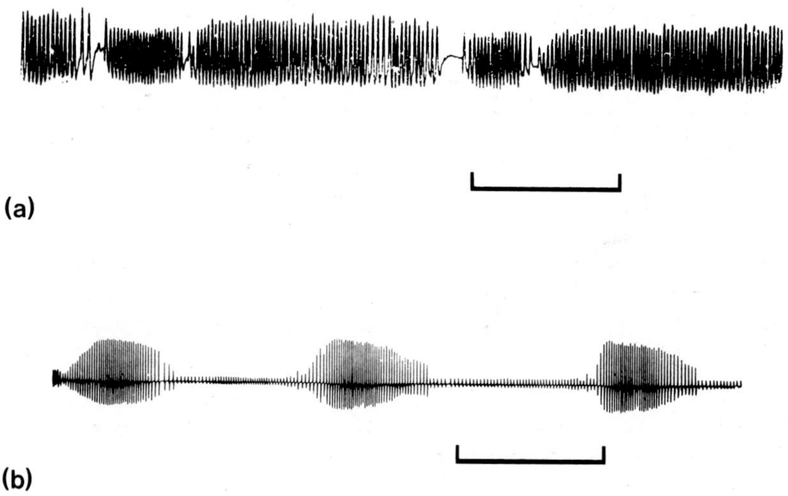

FIG. 14. Heartbeat pattern of pupae of the house-fly, *Musca domestica* (**a**) and cabbage looper moth, *Trichoplusia ni* (**b**). Note the periodic cessation of heartbeat in both cases, and a recurrent pattern which is more evident in the looper (**b**). The time mark is 20 s. Recordings were made by impedance conversion.

The impedance conversion method has proven especially valuable for a rapid, simple recording of movements and can be adapted to record any movement of tissues, even those vanishingly small. The recordings obtained depend greatly on placement of the radiofrequency probes used for electrodes and can give an accurate measure of the relative position of the heart. The electrodes may be waxed through the cuticle of the insect or placed near a tissue that is submerged in saline in semi-isolated preparations. Since the probes do not actually contact the tissue being monitored, there is no force involved, only a change in impedance that is dictated by the geometry of the probes and the tissue. The device can be calibrated in ohms by replacing input probes with a potentiometer. Probably the most important trait of impedance conversion is a lack of interference from external sources, and this make it extremely versatile.

Lhotsky, S. *et al.* (1975) reported different amplitudes and waveform changes between successive single cycles in very irregular heartbeats of the locust heart as recorded optically from a microscope eyepiece. It is possible that even very small sideways movements of the dorsal vessel could explain the recordings presented in the 1975 paper. If so, then the heartbeat in *Schistocerca gregaria* would be almost identical with the heartbeat of *Locusta migratoria migratorioides* (Miller, T. and S.-Rózsa, K., 1981).

The suggestions by Lhotsky, S. *et al.* (1975) that heartbeat is somehow correlated with, or coupled to, respiration must be viewed with caution since the respiratory movements may have impressed themselves on the optical heartbeat recordings independently of any heartbeat changes.

For the microscope eyepiece recording method to accurately reflect movements of the dorsal vessel, the segment viewed must be rigidly held. Even then it is possible for respiratory rhythms to alter the position of the dorsal vessel with respect to the cuticle, and thus introduce artifacts. Because of these difficulties, the invasive impedance or non-invasive thermographic recording methods may be superior to optical recording methods, although these, too, are subject to interference from struggling or other movements in the whole insect. Certainly impedance recordings of the heartbeat from the abdomen of the cockroach are obscured by large abdominal movements.

3.6 Cardiac pacemaker automatism

The term "myogenic" means originating in the muscle. When used to describe cardiac automatism it describes those animal hearts whose contractions are an inherent property of the heart muscle. Myogenic hearts contract rhythmically in the absence of any or all nervous impact.

The neurogenic heart, on the other hand, contracts in response to a volley of excitatory nervous impulses originating in a cardiac ganglion center, and the cardiac ganglia are normally intimately associated with the heart itself. The most studied examples of the neurogenic heart have been those of decapod crustaceans or chelicerata such as the horseshoe crab, *Limulus polyphemus* (Anderson, M. and Cooke, I., 1971; Lang, F., 1971; Abbott, B. *et al.,* 1969). Recent reports suggest that other chelicerata, e.g. spiders and scorpions, are also neurogenic (Sherman, R. and Pax, R., 1969; Bursey, C. and Sherman, R., 1970; Zwicky, K., 1968).

The classification of the automatism of insects hearts encountered difficulty almost from the beginning. Because of the extensive cardiac nervous system described by Alexandrowicz, J. (1926), and because of drug responses, the cockroach heart, *Periplaneta americana*, was considered to be neurogenic (Prosser, L. and Brown, F., 1961; Wigglesworth, V., 1965). Regardless of this, careful work had shown an underlying myogenicity in the hearts of *Anax junius* and *Belostoma flumineum* (Maloeuf, N., 1935), and in *Periplaneta* itself (Richards, A., 1963a). In retrospect, the American cockroach was probably the worst insect heart to study for automatism. Because of its complex cardiac nervous system and great sensitivity to mechanical and chemical influence, the behavior of the heart of *Periplaneta americana*, when studied in the semi-isolated preparation, could and actually did convince a number of workers that it behaved as a classical neurogenic heart.

McIndoo, N. (1945) relied largely on the descriptions of Maloeuf, N. (1935) in his interpretations on automatism. McIndoo also quoted B. Coon's (1944) finding that pyrethrum and nicotine nerve poisons permit the heart of *P. americana* to beat long after appendages are paralyzed — indicating that "... the nervous control of heart pulsations is secondary to [its] myogenic action". Thus,

McIndoo interpreted the previous work as evidence of myogenicity in 1945.

For some reason, the early evidence for myogenicity was ignored by the Needham, A. (1950) article on automatism, and by Krijgsman, B. and Krijgsman-Berger, N. (1951). Perhaps somewhere in the heat of the work the distinction between cockroaches and crayfish, between insects and crustacea, was forgotten. The lesson to be learned here, of course, is that one can never ignore the original workers, who in this case (along with Richards, A., 1963a) had the correct interpretation. Needham, A. (1950) referred to the heart of adult insects as "usually neurogenic" with Prosser and Davenport as authorities. We know now that these preconceptions were based largely on classifications of the hearts of a crustacean. However, even Needham, A. (1950) pointed out that the normal simultaneous contractions throughout the neurogenic heart of *Limulus polyphemus* became sequential (peristaltic) contractions upon removal of the cardiac ganglion (a note which seems to have been overlooked).

On the way to rediscovering cockroach myogenicity, Itzak Parnas, in 1968, suggested that if the cockroach heart were indeed myogenic, perfusion of the semi-isolated preparation with tetrodotoxin would block all nervous activity and leave muscle activity unaffected. Ned Smith (1969) showed that this was indeed the case. He perfused 10^{-6} g ml^{-1} tetrodotoxin on the semi-isolated heart of *Periplaneta americana,* leading to an unaltered heartbeat while blocking all nervous activity in the cardiac neurons.

The incorrect concept of a neurogenic cockroach heart was so strongly entrenched that, even when presented with the overwhelming evidence of myogenicity, some were of the opinion that the neurogenic classification needed to be modified somewhat and thus "myogenic innervated" was proposed to describe the cockroach heart (McCann, F., 1970). Others suggested that perhaps one was actually stretching the cockroach heart or subjecting it to abnormally low calcium concentrations to produce "myogenicity" (see comments by Prosser, on pp.209–210 in Miller, T., 1973).

The main reason for the latter suggestion was that abnormal stretching or chemical treatment of the deganglionated (i.e. cardiac ganglion removed) heart of *Limulus polyphemus,* could be shown to develop a rhythmic contraction. Very recently, Hoshi, T. and Watson, W. (1981) reported that treatment of the deganglionated heart of *Limulus polyphemus* with 10^{-6} M proctolin produced rhythmic but non-synchronous contractions of the myocardium. The contractions resembled those in the normal heartbeat and persisted in sodium-free saline or in the presence of tetrodotoxin, which was sufficient to block all nervous activity. Carlson, A. and Meek, A. (1908) had found rhythmic contractions in the heart of *Limulus polyphemus* at a stage in embryonic development before the appearance of the cardiac neurons or nervous innervation of the heart. Thus, it was claimed that any heart has inherent properties of myogenicity.

The idea that the heart muscle of all animals is myogenic finds support even among skeletal muscles of insects. At various times following section of nerve bundles supplying femoral muscles in locust or cockroach, a rhythmic contraction can occur (Case, J., 1956; Beránek, R. and Novotný, I., 1958; Usherwood, P., 1963).

A tonic bundle of muscles in the extensor tibia muscle of locusts *Schistocerca gregaria* or *Locusta migratoria* normally do develop rhythmic contractions without neurotomy. This rhythmic contraction has been used as a sensitive bioassay for proctolin or octopamine (O'Shea, M. and Adams, M., 1981), and although the role of the spontaneous rhythmicity in the locust extensor tibia muscle is unknown, it is thought to assist circulation in the leg (Evans, P. and O'Shea, M., 1978; Evans, P., 1980).

Still another case of rhythmically active muscle is known in Diptera. The body-wall muscles of prepupal larvae of cyclorrhaphan Diptera sometimes begin rhythmic contractions upon dissection (Irving, S. and Miller, T., 1980a). After perfusion with saline the spontaneous activity gradually decreases, but treatment of quiescent preparations with octopamine or proctolin at less than 10^{-8} M induces spontaneous membrane depolarizations very similar to those recorded from insect hearts. Piek, T. and Mantel, P. (1977) some time ago showed that proctolin could induce spontaneous contractions in the tonic bundle of fibers in the extensor tibia muscle of the locust.

Despite the propensity of various animal muscles to develop heart-like myogenic activity, the suggestions that insect hearts are not strictly myogenic are ill-advised, because the overwhelming evidence points to a basic myogenic activity in the intact insect. In fact, there is no evidence to support the possibility that rhythmicity somehow resides in a cardiac ganglionic center which drives the myocardium by regular bursts of motor activity. No such cardiac ganglionic center has been demonstrated despite the fact that many orthopteroid insects possess paired lateral cardiac nerve cords. In fact, the paired cardiac nerve cords in *Periplaneta americana* function independently of one another (Miller, T., 1968; Miller, T. and Usherwood, P., 1971).

The use of the term "myogenic-innervated" rather than "myogenic" may be an attempt to somehow acknowledge or account for the extensive innervation of the hearts of some insects. It is becoming more obvious that many insect hearts are indeed highly regulated to beat at various rates, amplitudes, or divert the flow of hemolymph in any of a number of directions, or to beat intermittently and even stop for seconds or hours. But behind this regulation the heart itself develops the basic rhythmic contraction, i.e. it is myogenic. At present, any other classification or modification of this simple description of automatism seems pedantic.

Electrogenesis of the spontaneously beating insect dorsal vessel always shows a pacemaker potential characteristic of myogenic hearts (Fig. 15). During contractions of the cockroach heart the cardiac neurons in the lateral cardiac nerve cord fired during the peak of the relaxation (diastasis) and were inhibited from firing during contraction (Fig. 16).

Perhaps the key condition or pattern of heartbeat automatism in hexapods can best be appreciated from an examination of some holometabolous insects. In the pupal stages of some insects, regardless of whether the pupae are active such as the mosquito, the heartbeat is intermittent and highly patterned (Fig. 14b). In particular, the heartbeat can cease completely for varying periods, then become active, and this alteration between cessation and activity appears not to be under ordinary excitatory or inhibitory nervous control, at least in the housefly pupae, because separation of the heart from the central nervous system does not change the pattern of heartbeat (Miller, T., 1979a). This operation presumably eliminates all regulatory innervation since the whole of the ventral ganglia are condensed in the complex thoracic ganglion mass, and neither

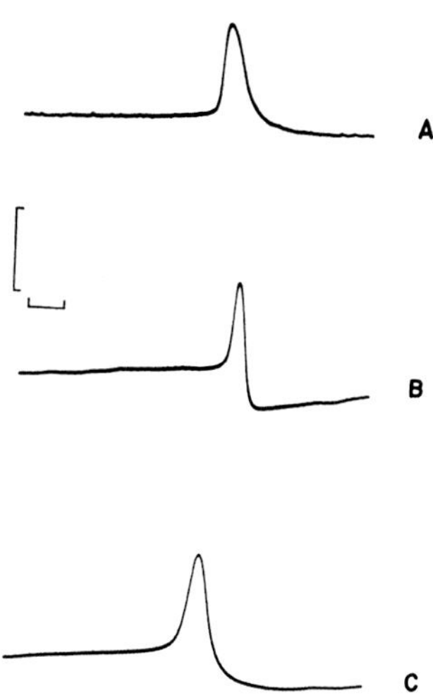

FIG. 15. Spontaneous depolarization of the myocardium of the cockroach, *Periplaneta americana* (**A**), the locust, *Schistocerca gregaria* (**B**) and the stick insect, *Carausius morosus* (**C**). A generator or prepotential phase occurs in all three cases. In **A**, the cockroach heart was treated with strontium ion and miniature postsynaptic potentials can be seen on the base line. Horizontal calibration: **A**, 50 ms; **B**, 100 ms; **C**, 200 ms. Vertical: 20 mV. (From Miller, T., 1971.)

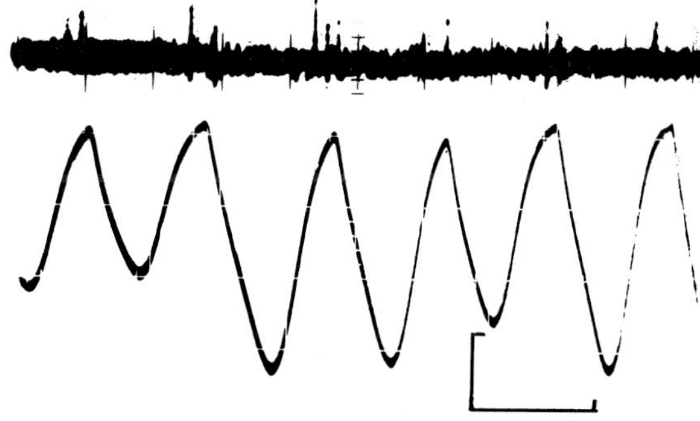

FIG. 16. Simultaneous recording of nervous activity in one lateral cardiac nerve cord of the cockroach, *Periplaneta americana*, recorded by suction electrode (upper trace), and movement of the dorsal vessel (bottom trace) recorded by capacitance actograph. Rephotographing lost some detail in the nerve recording, but nervous impulses are clearly evident and occur only during diastole. Contraction or systole is down. Calibration: horizontal, 1 s; vertical, 10 μV.

lateral cardiac nerve cords nor cardiac neurons have ever been reported for the house-fly (Miller, T., 1979a).

3.6.1 Semi-isolated Heart Preparations

Perfusion of the semi-isolated heart preparation with saline solutions often leads to an immediate arrest or drastic alteration in spontaneous activity with recovery of heartbeat following sooner or later. The results reported by Maloeuf, N. (1935) in perfusing the semi-isolated heart preparation of larval *Anax junius* Drury with Pringle's saline are fairly typical.

An illustrative example of the difference between saline solutions and hemolymph comes from an observation on the myogenic skeletal muscle movements of skeletal muscles of grasshopper (Hoyle, G., 1974). Once the myogenic rhythm is induced, it is immediately blocked upon perfusion of the leg with ordinary saline solution. Presumably this saline medium washes out the hemolymph which supported the contractions (perhaps minute amounts of proctolin were washed out).

The brief cessation normally seen in semi-isolated heartbeats upon addition of saline solution suggests an interruption in the equilibrium between the hemolymph which is washed away, and the myocardial cells that are spontaneously active. Careful attention to the behavior of the heartbeat at the moment saline is added to a semi-isolated heart preparation can sometimes indicate the suitability of a particular combination of ions without resorting to rather extensive measurements of specific ion activities.

3.6.2 Electrogenesis

When salines too high in potassium ions are perfused onto the freshly dissected cockroach heart, the initial reaction is increased heartbeat rate developing quickly into sustained contraction (systolic arrest) due to depolarization. The greater the K^+ concentration above hemolymph levels, the stronger and more prolonged the contraction. If saline K^+ levels below those in the hemolymph are perfused on the freshly dissected heart, the initial reaction is often relaxation and decreased heartbeat rate due to hyperpolarization. For the American cockroach a dependence of the resting membrane potential on external potassium concentration has been shown (Fig. 17), and McCann, F. (1963) reported a hyperpolarization of *Hyalophora cecropia* myocardium in low K concentrations.

Similarly, the cockroach heart is sensitive to divalent ion concentrations. Basal heartbeat rates are increased in low divalent ion and decreased when divalent ion concentrations are above average levels, and in these responses Ca^{2+} is replaceable by Mg^{2+} or Sr^{2+} (Miller, T. and Metcalf, R., 1968). Two millimolar $MnCl_2$ added to the saline caused a progressive decrease in the rate of rise of the evoked action potential in moth myocardium, plus a concomitant decline in the maximum amplitude of the spike-like depolarization. Treatment with manganese ion, or alteration of external calcium concentrations, did not alter resting membrane potential appreciably (McCann, F., 1971). Omission of sodium ion from the saline solution, or treatment with tetrodotoxin, has no effect on either the moth heartbeat or the electrical responses of the myocardium. A progressive decrease in calcium ion concentration in the saline also causes a decrease in rate of rise of the electrically-evoked muscle

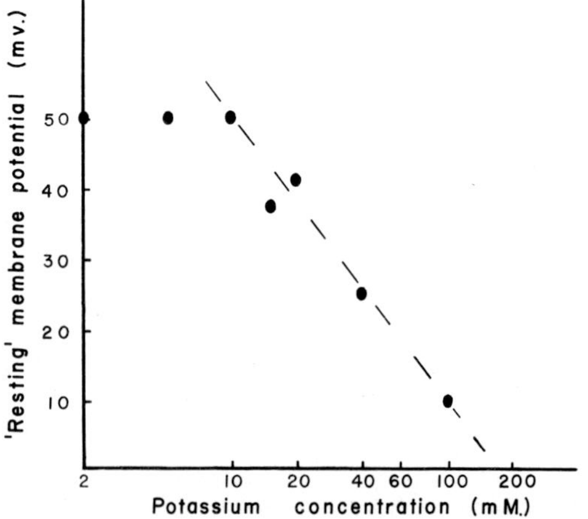

Fig. 17. Plot of equilibrium values of external K concentration $[K]_o$, *vs.* the membrane "resting" potential (E_m) of the heart of *Periplaneta americana*. Perfusion with High $[K]_o$ caused an amount of contracture which was directly correlated with the increased amount of potassium. The dotted line indicates the position predicted by the Nernst equation for the following values: $(E_m) = 39.5 \log [K]_o - 39.5 \log [K]_i$, where $[K]_i$ is internal potassium concentration. If the system represents a true K-electrode, then $[K]_i = 170\,\text{mM}$ for this example. (Modified from Miller, T., 1971.)

potential until zero potential amplitude was recorded in the presence of zero calcium (McCann, F., 1971).

The straightforward responses of cockroach heart to K^+ and Ca^{2+} indicate that potassium permeability of the myocardium dictates "resting" electrogenesis very similar to insect skeletal muscle (Usherwood, P., 1969), and Ca^{2+} reportedly dictates depolarizing electrogenesis (McCann, F., 1971; S.-Rózsa, K. and V.-Szöke, I., 1972) very similar to the case in insect skeletal muscles (Yamamoto, D. and Washio, H., 1980, 1981; Yamamoto, D. *et al.*, 1981).

Beyond these simple conclusions, research workers have not concentrated their attention upon the electrogenesis in insect myocardium apart from the continuing efforts of Frances McCann. Her choice of adult *H. cecropia* myocardium for these experiments is good from the standpoint of a lack of cardiac neurons which might complicate such studies. Extensive intercalated discs which are present in *H. cecropia* myocardium might produce complicated current fluxes between cells and the currents in turn might be dependent upon ion concentrations in the bathing medium, but a similar cautionary note would hold true for any insect myocardium.

When McCann began her pioneering on the heart function in *H. cecropia* adults, *Telea polyphemus* and *Samia cynthia*, she emphasized what at that time was the oddity of an apparent lack of use of sodium ion in the electrogenesis of the myocardium, a situation quite distinct from that of the vertebrate heart (McCann, F., 1963, 1965). Today, of course, it is suspected that calcium spikes are common features of many insect muscles including heart, and the insect heart is insensitive to the removal or replacement of sodium ions, or to perfusion with drugs such as procaine (Maloeuf, N., 1935) or tetrodotoxin (Smith, N., 1969).

One feature of electrogenesis that has not been examined is the possibility of a metabolically driven cellular potential. Wareham, S. *et al.* (1974) originally described an 18 mV resting membrane potential that was supported by the presence of bicarbonate ion in coxal muscle of the cockroach, *Periplaneta americana*. The remainder of the membrane potential at rest was dependent upon K^+ ion differences across the muscle membrane in the normal manner.

We used bicarbonate ion in the saline of the house-fly maggot, *Musca domestica*, for studying neuromuscular transmission, and found a substantial contribution to the reversal potential of slow postsynaptic potentials by bicarbonate ions while changes in the bicarbonate ion had no effect on the fast postsynaptic potential (Irving, S. and Miller, T., 1980b). The maggot hemolymph contains about 10 mM bicarbonate ion.

The possible contribution of bicarbonate ion to the cardiac membrane potentials has never been investigated; however, McCann, F. (1967) showed that the resting potential of the myocardium of adult *H. cecropia*, the silk moth, could be reduced by 35–40% of its original value by poisoning with metabolic inhibitors such as dinitrophenol. This is similar to the original observations of Wareham, S. *et al.* (1974) and may suggest a metabolic component of the action potential of insect myocardium. This question needs further investigation.

3.6.3 Saline solutions

The best reason for using a given saline in the semi-isolated heart preparations of various insect still seems to be: it works; i.e., one uses the saline that maintains the most apparently normal heartbeat for the longest period of time (see Jones, J., 1977 for tables of salines reported for various insect heart preparations). So far, "normal" physiology has been studied in this manner and the results appear straightforward.

Saline solutions for insect hearts almost never employ large concentrations of amino acids, even though these are said to comprise the majority of the osmotic species in the hemolymph (Wigglesworth, V., 1967; D. Mullins, this volume). Despite the K^+ and Ca^{2+} requirements outlined above, insect hearts are reported to remain functioning apparently normally for up to an hour when bathed in saline solutions of very simple composition, for example, isotonic $MgCl_2$, $CaCl_2$ or NaCl salts or distilled water (Jones, J., 1977). One wonders about the amount of mixing in these cases, because although the reasons for these responses have not been studied carefully, very few workers have actually removed the dorsal vessel from the cuticle and studied it in complete isolation to ensure maximum mixing of added saline with the myocardium.

Barsa, M. (1954) reported that the semi-isolated

heart of the grasshopper, *Chortophaga viridifasciata*, beat normally when superfused with distilled water, but although the pulsations persisted for an average of 45 min, the heartbeat became progressively slower and more faint. This response suggests that the insect heart can withstand a considerable insult to its ionic environment, but that activity cannot be maintained for extended periods. Indeed, the criterion of pulsations maintained for 24 h has been used as the test of a good saline (Miller, T., 1975a; Jones, J., 1977).

3.6.4 Peristalsis and heartbeat reversal

There is one main type of insect heartbeat: a peristaltic wave of contraction in the dorsal vessel. The peristaltic wave of contraction is usually anterograde, i.e., traveling anteriorly. Because each cell of the myocardium has the potential ability to support rhythmic contractions, centers of dominance have been identified which appear to recruit most or all of the remainder of the myocardium at the beginning of each cycle of contraction.

Probably the most characteristic property of the insect heartbeat, certainly one of the most studied, is that of reversal of heartbeat where anterograde peristaltic contractions occur for a period, then reverse or retrograde peristalsis occurs; this pattern may continue through most or all of a stage of development. It may be particularly common in all stages except larval of holometabolous insects (Normann, T., 1972; Jones, J., 1954; Tenney, S., 1953; Gerould, J., 1938; Campan, R., 1972). Yokoyama, T. (1932) reported heartbeat reversal in embryos of *Bombyx mori*, and Davis, C. (1961) reported heartbeat reversal in a gyrinid prolarvae.

The dorsal vessel of the adult blowfly, *Calliphora vomitoria* beats at a high rate with anterograde peristalsis of up to 376 beats min^{-1}. Periodically the heartbeat pauses, then changes to retrograde peristalsis at the lower rate of about 175 beats min^{-1}. When the abdomen and thorax were separated, the thoracic dorsal vessel beat with a persistent retrograde peristalsis of 167 beats min^{-1} and the abdominal dorsal vessel persisted with anterograde peristalsis at 335 beats min^{-1} (Campan, R., 1972). This strongly suggests anterior and posterior pacemaker sites.

Campan, R. (1972) called the reversed heartbeat "reciprocal inhibition" where first one center is dominant then another. It is not possible to tell what role the nervous system might have in heartbeat reversal. Since the heartbeat rhythm did persist after removal of the central nervous system, it does appear to be basically autonomic. Tenney, S. (1953) described anterior and then posterior pacemakers exerting periodic dominance one upon the other, exactly as Campan.

Campan, R. (1972) concluded that since changes in incident light caused changes in the average heartbeat rates of blowfly and cricket, that this was evidence for a nervous influence. Isolating the heart from the thoracic ganglion centers, or cutting the ventral nerve cord at the neck, eliminated light-induced changes in the character of the heartbeat, further suggesting nervous influences upon the heartbeat.

In a comparative study of the larvae of *Tenebrio molitor*, *Aedes, Anopheles* and *Culex* mosquitoes, *Musca domestica*, *Drosophila melanogaster*, *Sarcophaga bullata*, *Chironomus* and *Sciara coprophila*, the heartbeat was anterograde without reversals (Jones, J., 1954). The heartbeat of pupae, on the other hand, beat predominantly forward, but beats were punctuated by periodic reversals in heartbeat direction to retrograde. Reversal of heartbeat occurred in all intact adults, and persisted even in semi-isolated heart preparations (Jones, J., 1954).

In *Mamestra brassicae* the larvae and early pharate adult stages are characterized by heartbeat with unidirectional anterograde pulsations (Queinnec, Y. and Campan, R., 1972). In old pharate adults heartbeat reversals began, and occurred through the remainder of the pharate stage. The appearance of reversals in the pupal stage was thought to signify restructuring of the myocardium as described by Gerould, J. (1938). Reverse peristalsis of the dorsal vessel is dominant in *Bombyx mori* at the time of shedding of the larval skin in the prepupae.

A slow rhythmic anterograde heartbeat chiefly irrigates the dorsal sinus and the wing bases. Campan, R. (1972) thought this might have something to do with preparation for flight. In *Musca domestica,* heartbeat changes are not obviously correlated with flight (Miller, T., 1979a), but this interesting idea needs further examination.

Tachibana, K. (1975) reported a presystolic notch during retrograde pulsations of the heart of

Drendrolinius spectabilis, but during anterograde heartbeat no notch was recorded. The presystolic notch phenomenon, which is a slight relaxation immediately preceding systolic contraction, has been reported many times (see Jones, J., 1977). Although a few explanations have been offered the notch may be a simple property of the myogenic heartbeat where tension is lost briefly against the pull of elastic tissues in the dorsal diaphragm. The notch is very unlikely to be caused by an active pull of the alary muscles. Presystolic notch may indeed remain a speculative peculiarity until the forces on the myocardium are studied. The measurements are now within reach with present technology (refer to force transducer described in Vero, M. and Miller, T., 1979).

A reversed heartbeat is rare in undisturbed larvae of holometabolous insects; however, the heartbeat of silkworm larvae, *Bombyx mori,* was induced to reverse peristalsis by warming the anterior dorsal vessel and cooling the posterior part of the dorsal vessel (Yokoyama, T., 1932). Thus, although the posterior regions of the dorsal vessel in larvae often have the dominant pacemaker, reverse peristalsis can be demonstrated.

Although heartbeat reversal and anterograde peristalsis are common, many insects exhibit a heartbeat in the dorsal vessel that is a synchronized (non-peristaltic) contraction. Chief among these insects are those of the orthopteroid complex, including cockroach, cricket and grasshopper. Curiously, these insects are also notable in having lateral cardiac nerve cords and a complement of intrinsic cardiac neurons.

Despite the common occurrence of a synchronized heartbeat in the Orthoptera, the underlying basic heartbeat pattern can still be demonstrated. The dorsal vessel of the grasshopper, *Melanoplus differentialis*, when viewed from above or from below after removal of the ventrum and viscera, contracted simultaneously, or by anterograde peristalsis or even retrograde peristalsis (Jahn, T. *et al.,* 1937). Under conditions of stress, such as during eclosion, the cockroach heartbeat becomes asynchronous.

Roussel, M. (1973) reported a "crowded" cardiac rhythm in *Locusta migratoria* where in the first few days of adult life the solitary form has a higher heartbeat rate than the gregarious phase. The differences, however, were about 10 beats min^{-1} (i.e., 70 compared to 80 beats min^{-1}). The gregarious form was thought to respond actively with a reduction in heartbeat rate, since crowding always gave rise to a decreased heartbeat, but imposing solitary conditions on the gregarious locust had no effect on heartbeat rate.

4 CARDIAC INNERVATION

Although there are a few thorough studies on the innervation of the organs of circulation in insects one must still be very concerned about completeness. Similarly there are few studies that demonstrate the effects of nerve stimulation on organs, and virtually none on the central nervous pathways that are responsible for the control of organs involved in circulation.

4.1 Segemental innervation

The dorsal vessels in the majority of insect orders lack lateral cardiac nerve cords. They are conspicuously rare among the advanced orders of holometabola. In the latter, the dorsal vessel appears to be innervated by ordinary segmental nerves or from the median nerves of the ventral nerve cord or both. These cases have not been examined in detail, and need to be further explored, especially the precise origin of innervating axons.

Anderson, M. and Finlayson, L. (1978) described the innervation of the dorsal vessel of the tsetse fly, *Glossina morsitans* (Fig. 18). From the pattern of innervation they reported it is difficult to tell whether these are ordinary segmental nerves or part of the median nervous system. They do show a neuron cell body (labelled *e*) near the junction of the nerve with the dorsal vessel (Fig. 18b) and this bears a remarkable resemblance to similar structures in *Carausius morosus* (Opoczyńska-Sembratowa, Z., 1936). The ultrastructure of the *j* branch (Fig. 18b and c) leading to the alary muscle shows axon profiles with electron-dense granules; in fact, *j* is a neurosecretory neuron somata also (Anderson, 1982, personal communication). It is not known whether *j* innervates the alary muscle.

The innervation of the house-fly, *Musca domestica,* shows a similar pattern except the branch to the alary muscles is not so clear. In *Calliphora vicina*

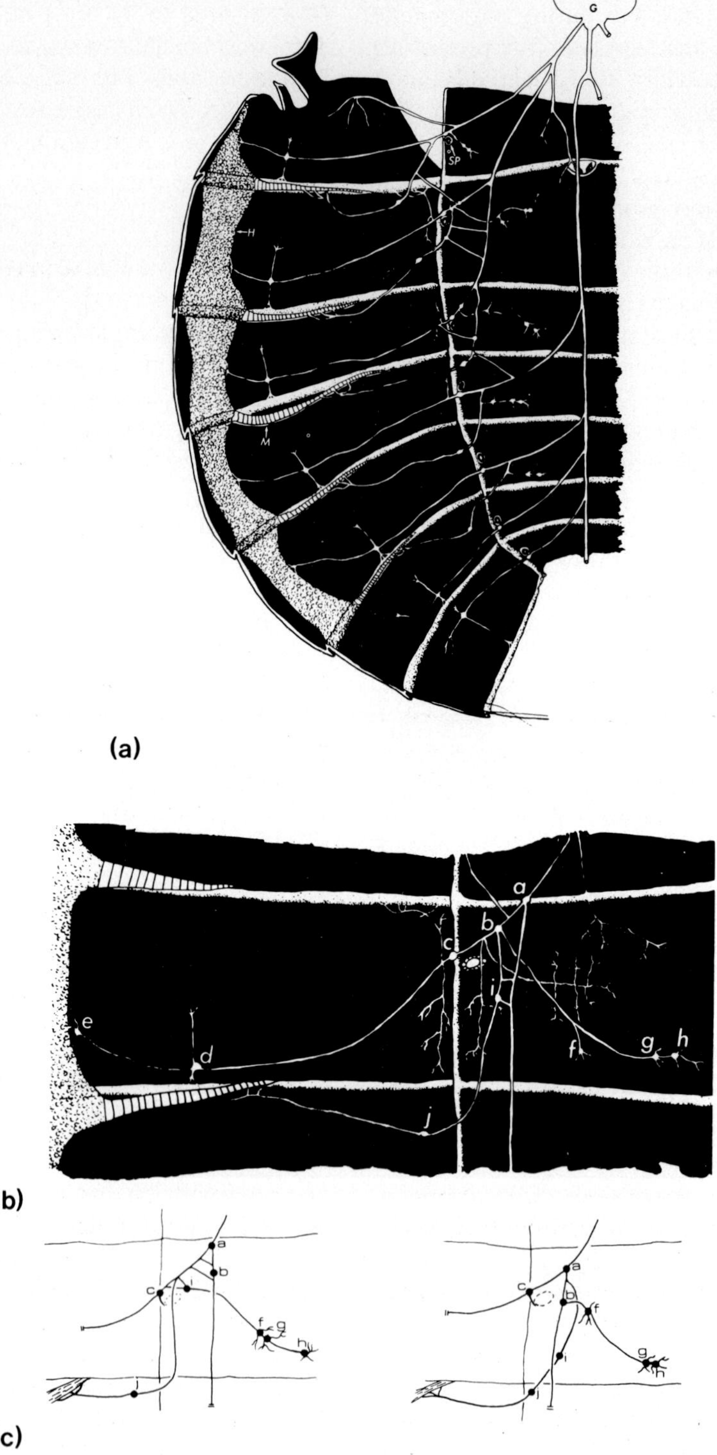

FIG. 18. **(a)** Innervation of the dorsal vessel of adult female tsetse fly, *Glossina morsitans*, in the abdomen. G, thoracic ganglion mass; SP, spiracle; H, dorsal vessel; M, alary muscle. **(b)** Close-up of segmental innervation of the second abdominal segment of abdomen. The letters identify peripheral neuron somata with variations of the innervation pattern shown on the two additional drawings underneath **(c)**. (After Anderson, M. and Finlayson, L., 1978.)

the neuromuscular junction of the dorsal vessel shows axons with electron-dense granules, suggesting neurosecretory units (Normann, T., 1972); however, Normann (1981, personal communication) thinks there are both excitatory and inhibitory units innervating the dorsal vessel because the normally rapid acceleration and deceleration of frequency that the intact heart exhibits reverts to a steady median heartbeat upon cutting of segmental innervation (cf. Fig. 21).

Jones, J. (1954) could not find segmental nerves supplying the abdominal dorsal vessel of adult *Anopheles quadrimaculatus* despite examination of whole mounts, fresh tissue, or serial sections. He did find nerves supplying the alary muscles, but could not confirm whether or not these axons sent branches on to innervate the dorsal vessel. The latter case does occur among some other insects (Ke, O., 1932; Rehm, E., 1939).

Despite being unable to find innervation of the dorsal vessel of *Anopheles quadrimaculatus,* Jones, J. (1954) reported that wiggling or sudden swimming movements were correlated with elevated heartbeat rates which then reverted quickly to a normal rate for resting larvae. A similar description was reported for *Thermobia* (Edwards, G. and Nutting, W., 1950), and struggling in *Periplaneta americana* results in a period of elevated heartbeat rate (Richards, A., 1963a). All of these reports suggest nervous cardioregulation, but such a conclusion is incompatible with the failure of Jones, J. (1954) and others to find innervation of the dorsal vessel of anopheline mosquitoes. The control of heartbeat in mosquitoes certainly deserves a detailed investigation. One needs to shock centrally and record from the dorsal vessel in some manner.

Sanger, J. and McCann, F. (1968a,b) reported a complete absence of nerves associated with the dorsal vessel or alary muscles of *Hyalophora cecropia.* In contrast to this Senff, R. (1971) mentioned obtaining an increase in heartbeat rate in response to nerve stimulation of *H. cecropia,* and Lockshin, R. and Williams, C. (1965) reported branches passing from the median nerves to the "dorsal heart" and to the "alary muscles" of *Antheraea pernyi, Antheraea polyphemus, H. cecropia, Philosamia cynthia* and *Rothschildia orizaba* silkmoths. Also, *Bombyx mori* has a well-developed cardiac nervous system (Kuwana, Z., 1932; McIndoo, N., 1945), and an extensive innervation of the heart and alary muscles was reported (Wasserthal, L. and Wasserthal, W., 1977; Wasserthal, W. and Wasserthal, L., 1980).

From the Lockshin, R. and Williams, C. (1965) article, details of innervation were not given, so it is difficult to determine what they meant by: nerves supply the "dorsal heart". They do mention nerves to the alary muscles. To account for the failure of Sanger, J. and McCann, F. (1968b) to find innervation of the alary muscles, one might assume they simply neglected to section near the lateral insertion. If one examines the alary muscles of *P. americana, Locusta migratoria* or the beetle, *Cyclocephala pasadenae* (Adams, M. *et al.,* 1973; Miller, T., 1975a; Miller, T. *et al.,* 1979) in the ultrastructure, only sectioning near the cuticular insertion is likely to yield axon profiles, and the innervation there is sparse indeed. Here again the entire question of innervation and control of the dorsal vessel in adult or larval *Hyalophora cecropia* needs re-examination. The best approach would be to follow up the Senff, R. (1971) note on response to nerve stimulation since over 100 years of microscopic observations have failed to clear up this point.

The relationship of the moths, *Hyalophora, Philosamia, Antheraea, Rothschildia* and *Bombyx,* where only the latter has extensive cardiac innervation, bears some resemblance to the interrelationships in the innervation of nematoceran Diptera as described by Yaguzhinskaya, L. (1954). The alary muscles of the tipulid, *Pachyrrhina cornicina,* receives innervation from the ventral nerve cord via a segmental or median nerve (Fig. 19). The axons ramify over the alary muscle array and extend to give what are termed motor terminals on the dorsal vessel.

In contrast, the innervation of the heart of *Anopheles maculipennis* is far more sparse, with branches traceable to the cuticular insertion of the alary muscles only (Fig. 20). Jones, J. (1954) reported the same conclusion in his observation on adult anopheline mosquitoes.

The two orders, Lepidoptera and Diptera, thus contain examples of possible extreme reductions in the innervation of the dorsal vessel. If indeed the dorsal vessel is not innervated, then we need to examine what regulates the dorsal vessel in both *Hyalophora* and *Anopheles.*

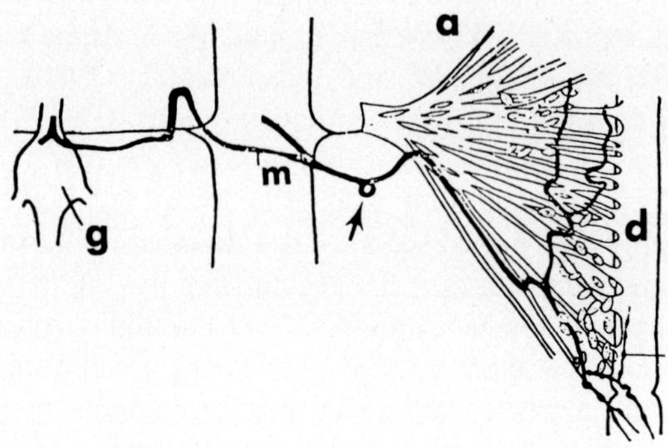

FIG. 19. Drawing of innervations of the dorsal vessel in a typical abdominal segment of tipulid fly, *Pachyrrhina cornicina.* The ventral ganglion (g) and a branch of the median nerve (m) are shown. A nerve cell body is shown (arrow) near where segmental innervation joins the alary muscle array (a), and nerve branches ramify over the alary muscle and the dorsal vessel (d). (After Yaguzhinskaya, L., 1954.)

Nervous or other regulation of the *in vivo* heartbeat of insects has only been demonstrated in a few cases. Heinrich, B. (1970b, 1971) showed a close relationship between heartbeat and temperature regulation in Hymenoptera and Lepidoptera. The most recent demonstration of regulation of insect heartbeat was that of Wasserthal, L. (1980, 1981) who described increased heartbeat coinciding with wing expansion in newly eclosed moths, *Papilio machaon* L. Prior to the two demonstrations of heartbeat control, only the article of Steiner, G. (1932) described responses to nerve stimulation, in this case of the heartbeat of *P. americana*.

As described above, segmental nerve stimulation in *P. americana* leads to transitory responses. The heartbeat increases briefly, but the responses are short-lived and habituate rapidly. The same occurs when the lateral cardiac nerve cord of *P. americana* is stimulated (Miller, T., 1969). The nervous basis for heartbeat control in insects is thus completely undescribed as yet. Heartbeat "accelerator" nerves have not been found. Indeed, in the locust heart, *Locusta migratoria,* stimulation of the segmental nerve usually causes no response in the heartbeat at all (Miller, T. and S.-Rózsa, K., 1981).

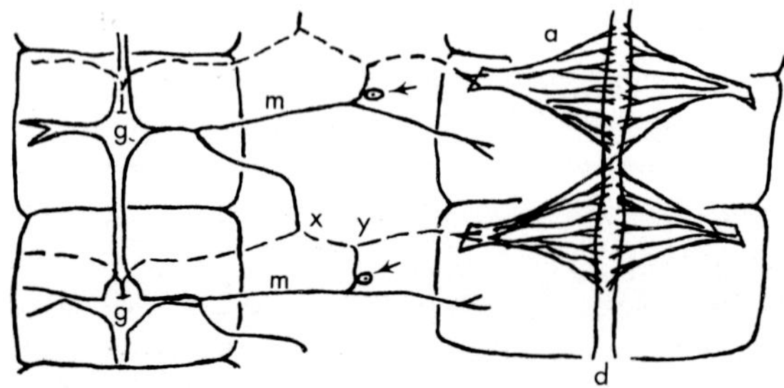

FIG. 20. Innervation of the alary muscle of the mosquito, *Anopheles maculipennis*. The median nerves (dashed lines) leave the ventral ganglia (g), and eventually supply the alary muscles (a). At two points (marked *x* and *y*) the segmental nerve bundles join branches from the segmental nerve (m). A peripheral neuron somata is present in the second link nerve (arrow) making the actual origin of the alary muscle nerves uncertain. (Redrawn after Yaguzhinskaya, L., 1954.)

It is possible that each insect dorsal vessel is regulated in a unique way, as suggested by their normal heartbeat. For example, *Manduca sexta* (Heinrich, B., 1971) and the bumblebee, *Bombus vosnesenskii* (Heinrich, B., 1976) appear to respond to the overheating that occurs with flying by increasing their circulatory hemolymph flow to cool their bodies; the adult house-fly, *Musca domestica*, on the other hand, has a widely varying heartbeat rate and amplitude during flying or walking, with a steady heartbeat rate and amplitude being abnormal (Miller, T., 1979a). The locust, *Locusta migratoria*, and cockroach, *P. americana,* appear to have a well-regulated constant heartbeat with simultaneous contractions; and the nymphalid, *Papilio machaon* (Wasserthal, L., 1980) has a reversing heartbeat with obvious peristalsis.

The heartbeat in nymph or adult cockroach, *Periplaneta americana,* is variable in frequency. It can respond by doubling in rate when the whole insect is disturbed, but a constant steady heartbeat is maintained in the quiescent insect (Richards, A., 1963a). When the heart is separated from the central nervous system, the variability of the heartbeat decreases. The heartbeat of the house-fly, on the other hand, is extremely variable at rest, and the variability is eliminated when the dorsal vessel is separated from the central nervous system (Fig. 21). These two examples, cockroach and house-fly, suggest that cardioregulation by the central nervous system occurs. Examples of thermoregulation in Lepidoptera, Hymenoptera and Odonata (see. B. Heinrich review, 1974b) also suggest some nervous control of heartbeat.

Although many insect hearts are greatly affected by the activity of the insect, the locust seems to be an exception, because the heart rate remains extraordinarily constant under a variety of conditions (Roussel, M., 1973) despite the fact that central

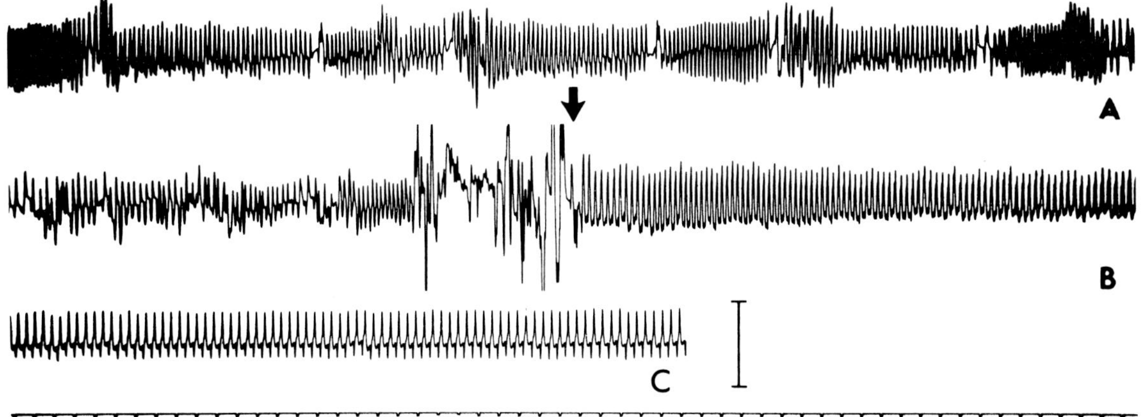

FIG. 21. Adult house-fly, *Musca domestica*, abdominal heartbeat before **A**, at **B**, and **C** after severing the abdomen from the thorax. In **B** the heartbeat rate becomes regular at around 300 beats min^{-1}, then gradually declines over several minutes. Cardioregulatory neurons are thought to innervate the heart in the abdomen from the thoracic ganglion mass. (From Miller, T., 1979a.)

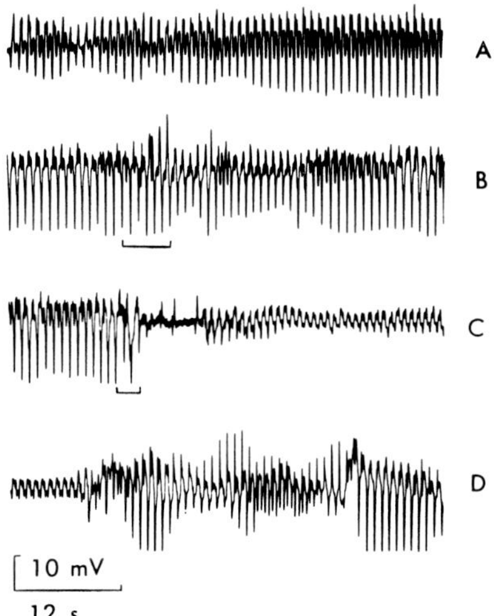

FIG. 22. AC impedance recording of the heartbeat in the third abdominal segment of intact and restrained adult female *Locusta migratoria* before (**A**), during (**B** and **C**) and after (**D**) dorsal dissection of the thorax to expose the metathoracic ganglion. The bar indicates in **B** when the ventral diaphragm was torn to expose the ganglion, and in **C** when a dissection needle was scratched across the dorsal surface of the ganglion. (From Miller, T. and S.-Rózsa, K., 1981.)

nervous input appears to constantly alter the tension in the dorsal diaphragm (Miller, T. and S.-Rózsa, K., 1981) (Fig. 22).

The supplying of neurosecretory material via axons to various organs has been described in many insects (Miller, T., 1975b), including delivery of material to the heart (Raabe, M. *et al.,* 1974). The delivery of octopamine to the tonically rhythmic part of the extensor tibia muscle of the jumping leg in the locust has been implied (Evans, P. and O'Shea, M., 1978). O'Shea, M. and Adams, M. (1981) have also implied the delivery of proctolin from lateral white neurons in the ventral ganglia of cricket and cockroach to the vicinity of the abdominal dorsal vessel. Proctolin is thought to be manufactured in nerve cells of the terminal abdominal ganglion of *Periplaneta americana* and delivered to the rectal longitudinal muscle (Penzlin, H. *et al.,* 1981). The latter scheme is exactly as suggested by Brian Brown who, with Alvin Starratt, isolated and identified proctolin as a possible neurotransmitter at the rectal longitudinal muscle of *P. americana* (Brown, B. and Starratt, A., 1975). One should hasten to add that the actual delivery of these materials has not been demonstrated. When they are present in the somata and the length of the axon, one assumes they are transported and released at the ends upon stimulation. The release has not been shown.

Rosenberg, J. and Seifert, G. (1978) reported that the median cardiac nerve of the heart of Onychophora (*peripatoides leucharti*) ran into the lumen of the heart. This anatomical arrangement seems more comparable to cardiac innervation in Arachnida, Crustacea and *Limulus polyphemus* than to those insects which have paired lateral cardiac nerve cords. On the other hand, the synapses present on the myocardium of Onychophora are of the ordinary or of the neurosecretory category, and these are the same as have been reported for the cockroach heart, *Periplaneta americana* (Miller, T. and Thomson, W., 1968).

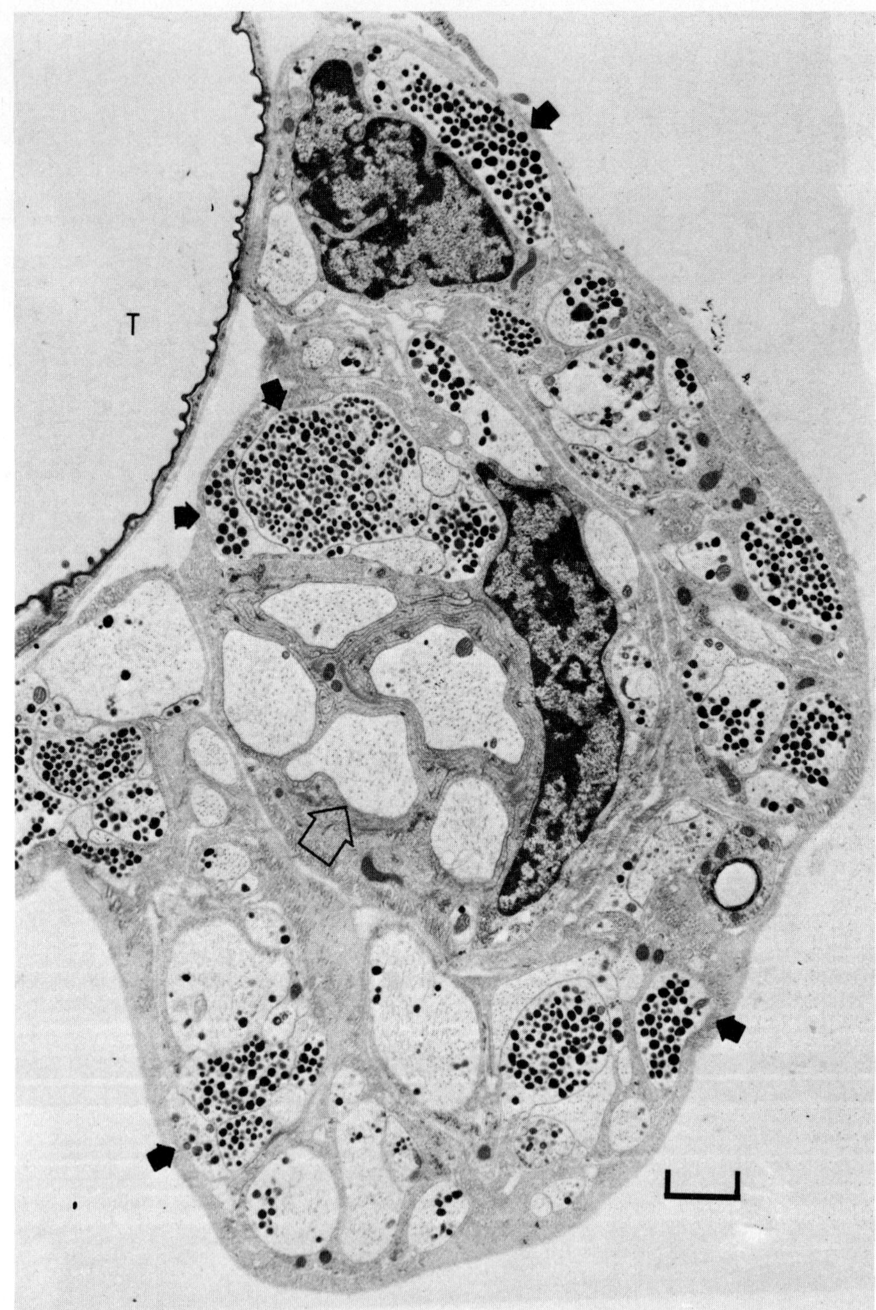

FIG. 23. Ultrastructure of the fourth segmental nerve in the abdomen of *Periplaneta americana* just before its juncture with the lateral cardiac nerve cord. Note the five ordinary axons of about 1 μm diameter with glial covering near the center of the nerve (clear arrow) and numerous neurosecretory axon profiles containing electron-dense granules elsewhere (filled arrows). T, a very large trachea. Calibration: 1 μm.

To appreciate properly the innervation of the insect heart, one must be aware of the presence and role of neurosecretory neurons as they occur in the peripheral nervous system (Miller, T., 1975b). The staining and histochemical demonstration of neurosecretory axons has always been fraught with difficulty and inconsistency.

The term "neurosecretory" neurons or axons

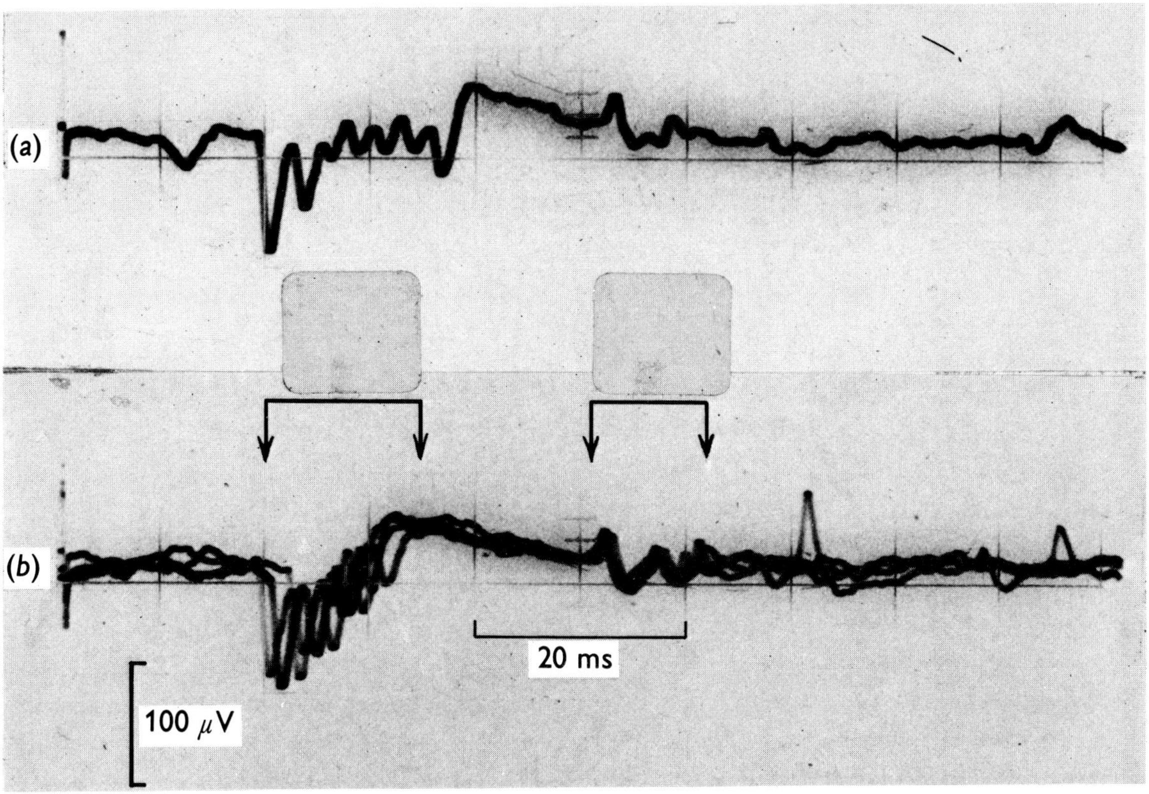

(a)

(b)

20 ms

100 μV

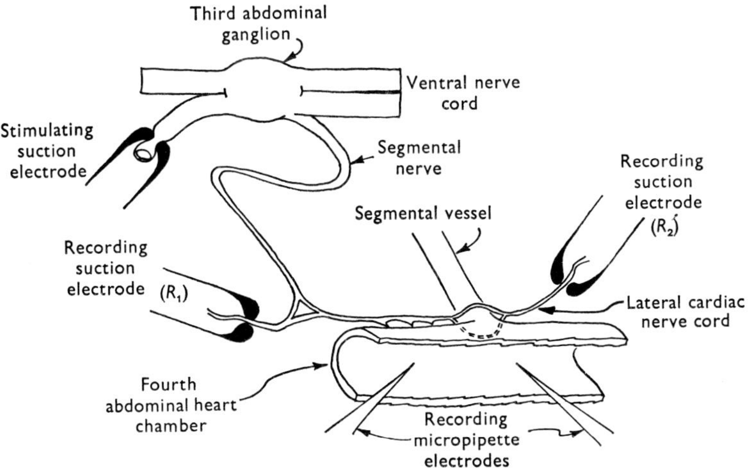

FIG. 24. Stimulation of the ventral connective of the third abdominal ganglion (inset) and recording nerve activity in the lateral cardiac nerve cord (R_1 on the inset). **a:** Single response; **b:** repeated responses reveal first large-amplitude potentials and later a group of small-amplitude potentials indicated by the double-arrowed bars. The large potentials are thought to be conducted over ordinary axons (cf. Fig. 23) and the small potentials over neurosecretory potentials. (After Miller, T. and Usherwood, P., 1971.)

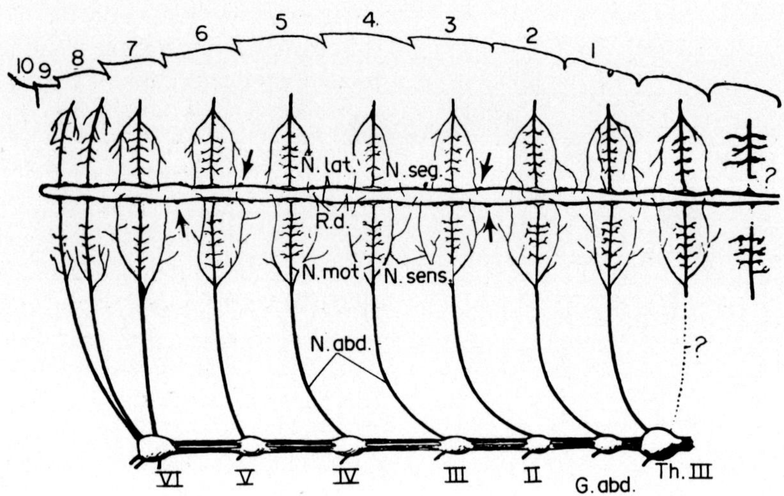

FIG. 25. Innervation of the cockroach, *Periplaneta orientalis*, as reported in the classic work of Alexandrowicz, J. (1926). Segmental innervation of the abdominal dorsal vessel from the ventral nerve cord. Roman numerals are numbers assigned to existing ganglia. Arabic numerals are bodily segment numbers (and the numbers of abdominal chambers). Individual neuron cell bodies (arrows) are drawn in the lateral cardiac nerve cords (N lat). The segmental nerve is labelled motor (N mot) or sensory (N sens). Alexandrowicz also showed a dorsal nerve for the heart (Rd) whose function is obscure.

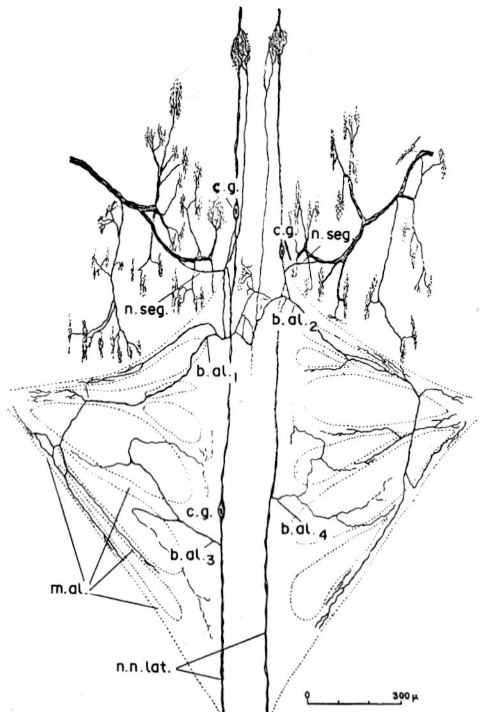

FIG. 26. Innervation of one chamber of the dorsal vessel of *Carausius morosus* in the abdomen (anterior is at the top). Lateral cardiac nerve cords (n.n. lat.) parallel to the dorsal vessel. Segmental nerves (n.seg.) joint to the lateral cardiac nerves and apparently supply the alary muscles (m.al.). Ramifications around ostial valves are seen near the top. C.g— cardiac neurons. (From Opoczyńska-Sembratowa. Z., (1936.)

applies to those which contain large electron-dense granules whose diameter is either greater than 1000 Å (Type A of Knowles, F., 1967), or less than 1000 Å (Type B of Knowles, F., 1967), or to granules with less density, but whose large size easily distinguishes them from synaptic vesicles of 450 Å diameter (B. Loughton and I. Orchard, vol. 7). Neurosecretory axons also conspicuously lack a glial coat which is so characteristic of "ordinary" axons. Further, neurosecretory axons are often found in the basal lamina of axon bundles or other tissues adjacent to the hemolymph (Fig. 23). Nervous impulses propagated in neurosecretory axons are comparatively slow, and the axons reportedly appear varicose and blue under the dissecting microscope.

Because only a few substances have been identified (proctolin and adipokinetic hormone, Starratt, A., 1979) which are associated with neurosecretory axons, it is premature to speculate on what the granules are, whether carrier protein or the actual putative peptides themselves.

4.2 Lateral cardiac nerve cords

The lateral cardiac nerve cords were described in a classic study by Alexandrowicz, J. (1926) who reported 40 nerve cell bodies in the lateral cardiac

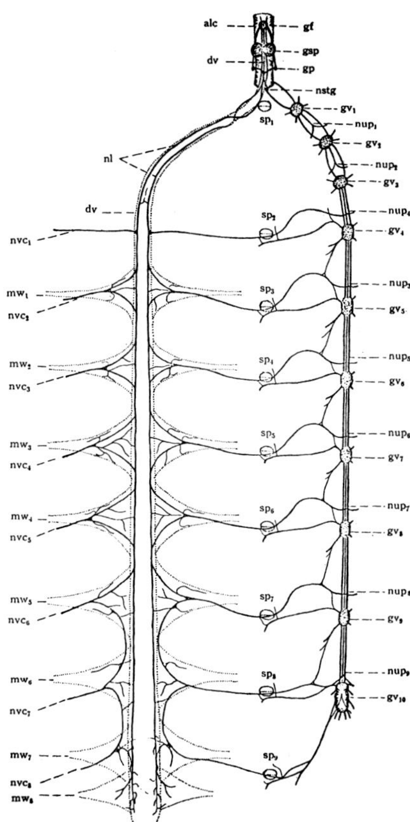

FIG. 27. Innervation of the dorsal vessel of the larvae of the silkworm, *Bombyx mori*, showing the entire dorsal vessel (dv) (left) and the ventral nerve cord (right) associated with the thorax and abdomen. Note that both the median and segmental nerves are associated with the cardiac innervation, and that nerve branches diverge from the segmental nerves to innervate (apparently) the alary muscles. SP, spiracles numbered 1–9; gv, ventral ganglia numbered 1–10; mvc, segmental/median nerves numbered 1–8; dv, dorsal vessel; nl, lateral cardiac nerve; mw indicates alary muscles. (From Kuwana, Z., 1932.)

nerve cords in the cockroach, *Periplaneta orientalis*. About the same number of neurons were reported in the lateral cardiac nerve cords in the stick insect, *Carausius morosus* (Opoczyńska-Sembratowa, Z., 1936). Alexandrowicz, J. (1926) was accurate in his model of motor innervation in that nerve impulses travel to the adjacent myocardium and to distant chambers along common axons, but he did not know that the neurons are stretch-sensitive and that the impulses travel in both directions from the somata, anteriorly and posteriorly in the lateral cardiac nerve cords (Miller, T. and Usherwood, P., 1971).

In *Periplaneta americana*, McIndoo, N. (1945) reported probably not more than 32 neuron somata in both the left and right lateral cardiac nerve cords. In all cases the location of the somata is relatively indefinite; except when segmental vessels are present, the somata do tend to be located near the confluence of the segmental vessel with the myocardium (McIndoo, N., 1939).

The lateral cardiac nerve cords are not present in all insects. They are reported for *Aeshna* (Zawarzin, A., 1911); *Periplaneta*, *Blaberus* and other cockroaches (Alexandrowicz, J., 1926; see also Fig. 25; McIndoo, N., 1945); *Bombyx mori*, the silkworm (Kuwana, Z., 1932; see Fig. 27); the stick insect, *Carausius morosus* (Opoczyńska-Sembratowa, Z., 1936; see Fig. 26); the grasshopper and locust, *Melanoplus*, *Schistocerca* (both *nitens* and *gregaria*), *Locusta migratoria* (Tyrer, M., 1971; Roussel, M., 1972; see Fig. 28); crickets (Davenport, D., 1949); and for the adult stage only of some

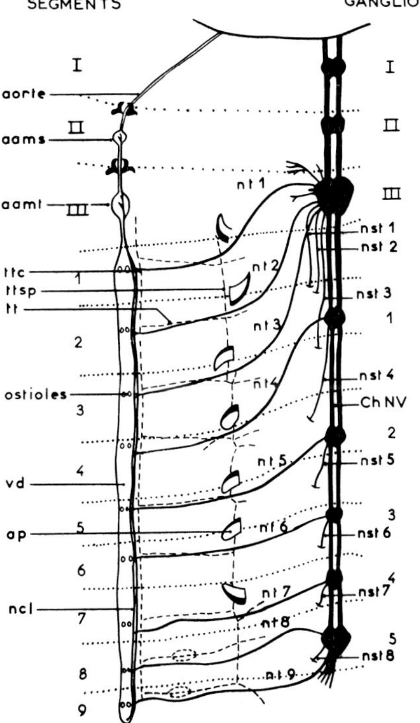

FIG. 28. Cardiac innervation of *Locusta migratoria* adult from the thoracic (I–III) and abdominal (1–5) ventral ganglia. Dashed lines draw major trachea, dotted lines show body divisions (numbered 1–9 in abdomen and I–III in thorax). nt, tergal nerve; ncl, lateral cardiac nerve cord; nst, sternal nerve; ChNV, ventral nerve cord; ap, apodeme; tt, various tracheal trunks and branches; Vd, dorsal vessel; aa, ampullae; aorte, aorta. The median nerves are omitted. (From Roussel, M., 1972.)

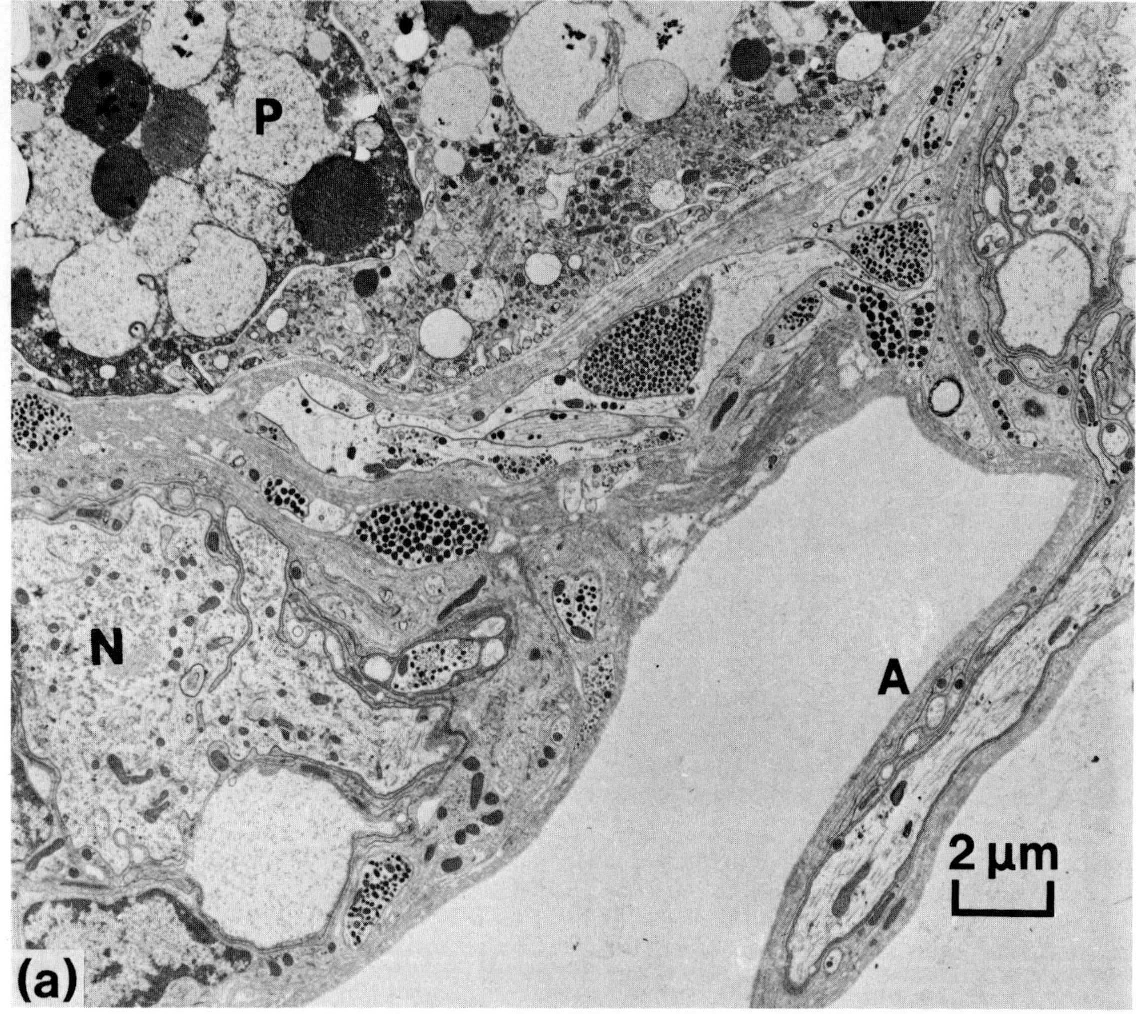

FIG. 29. **(a)** Ultrastructure of a cardiac neuron (N) in the lateral cardiac nerve cord of *Periplaneta americana*. Note the conspicuous absence of neurosecretory granules with axon (A). P, pericardial cell. Neurosecretory axon profiles are nearby in the lateral cardiac nerve cord.

Coleoptera (Toshio Narahashi, personal communcation).

When present, the lateral cardiac nerve cords receive innervation from the segmental nerves and from their own intrinsic cardiac neurons. The intrinsic cardiac neurons are thought to be of two types. The first type is the "ordinary" cardiac neuron with conspicuous Golgi bodies, but no evidence of neurosecretory granules (Fig. 29). The stick insect, *Carausius*, reportedly has larger cardiac neurons in the rear and smaller cardiac neuron somata in the anterior part of the lateral cardiac nerve cord (Opoczyńska-Sembratowa, Z., 1936).

The branch of the segmental nerve that joins the lateral cardiac nerve cord in *Carausius morosus* was devoid of any neurosecretory inclusions (Fifield, S. and Finlayson, L., 1978). Furthermore, the neurosecretory material in the lateral cardiac nerve cords of *C. morosus* are reportedly different from material in any other part of the segment. Fifield, S. and Finlayson, L. (1978) considered the neurosecretory material in the lateral cardiac nerve cords of *C. morosus* to be synthesized locally, possibly by neurosecretory neurons in the lateral cardiac nerve cords as described by Opoczyńska-Sembratowa, Z. (1936).

The segmental nerve complement in *Carausius*,

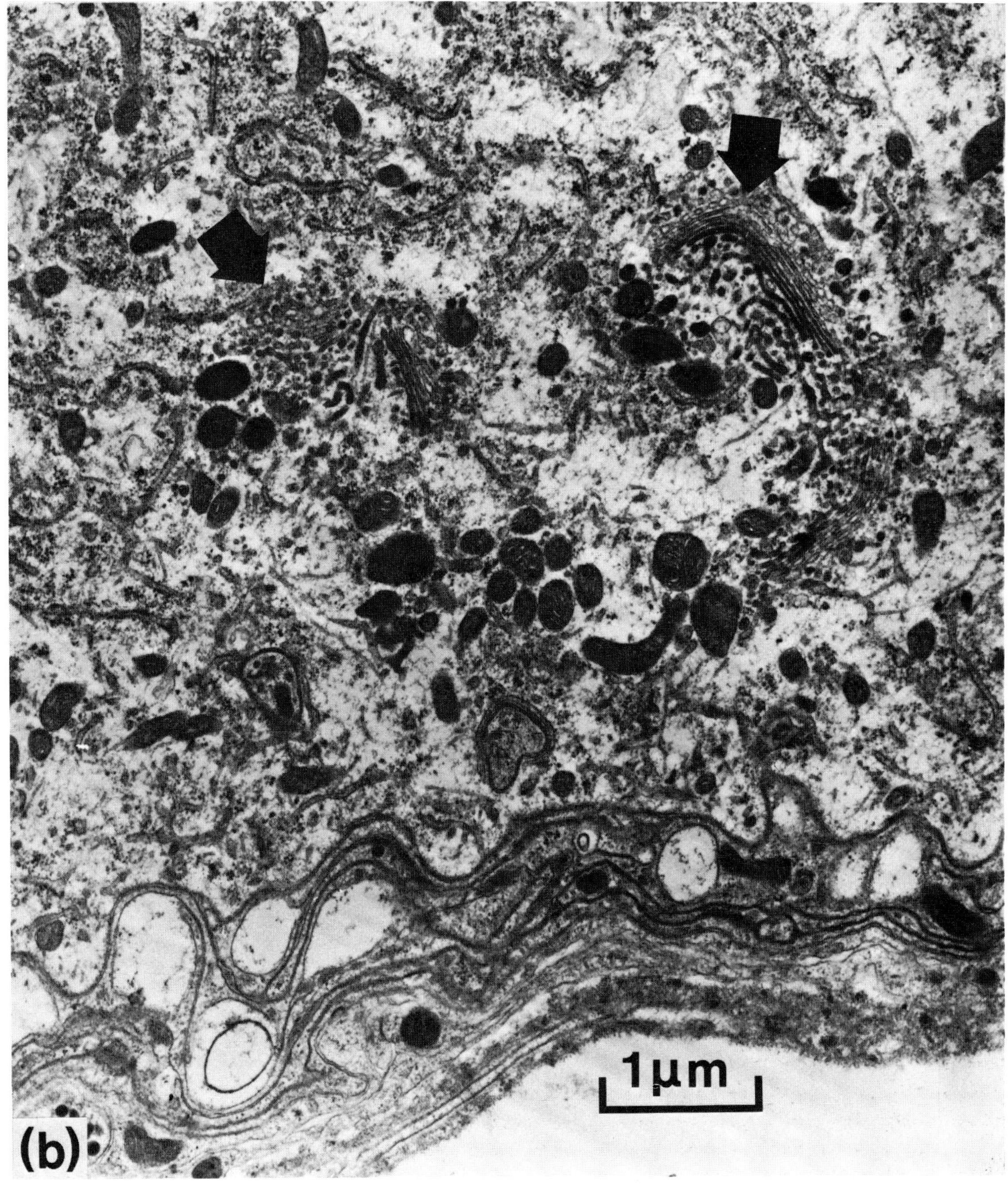

FIG. 29. **(b)** A close-up of the somata shows conspicuous Gogli apparati (arrow). **a:** calibration 2 µm; **b:** 1 µm.

then, is unlike that of grasshopper and locust (Tyrer, M., 1971) and cockroach (Fig. 23), both of which show neurosecretory axons and ordinary axons joining the lateral cardiac nerve cords. In cockroachs and grasshoppers the lateral cardiac nerve cord itself contains large amounts of neurosecretory axon profiles. Johnson, B. (1966) considered these to be neurohemal organs, and the segmental vessels of the cockroach contain numerous neurosecretory axon terminals (see section 3.2 on segmental vessels) which appear to be specialized release sites directed at lateral hemolymph circulation.

Although the origin of all of the axons descending via the segmental nerves to the cockroach heart is not known, at least one central unit innervates the myocardium directly, because central stimulation will produce a slowly propagating potential which puts an excitatory postsynaptic potential on the myocardium (Miller, T. and Usherwood, P., 1971). The paired lateral "white" neurons in the ventral ganglia of the cockroach and cricket send axons via the segmental nerve to the dorsal vessel, but the final destination of these axons is not known (Mike E. Adams, 1981, personal communication). The lateral white neuron is probably one of the neurosecretory axons in the segmental nerve (Fig. 23).

When one records from the lateral cardiac nerve of *Periplaneta americana* in the semi-isolated (abdominal heart) preparation of the dorsal vessel, two types of spontaneous or evoked nervous activity are measured (Fig. 24), a large-amplitude extracellular potential and a small-amplitude potential. The large extracellular potential is thought to originate from intrinsic cardiac neurons of the type shown in Fig. 29 and conducted along the ordinary axons. These units innervate the myocardium, and extracellular recordings of their spontaneous activity are seen to be correlated with excitatory postsynaptic potentials (EPSPs) in the myocardium.

The EPSPs in adjacent chambers of the dorsal vessel are correlated as long as the lateral cardiac nerve cord is intact. When the lateral cord is cut, EPSPs are independent in adjacent chambers. Although the myocardium may briefly be driven to contract more quickly by stimulating the segmental nerve (Fig. 30), the greatest response occurs as soon as the stimuli begin, and quickly habituates with continued stimulation. It has not been possible to "drive" the heartbeat at an elevated rate in a sustained manner by prolonged stimulation of either the segmental nerve or the lateral cardiac nerve cord (Miller, T. and Usherwood, P., 1971).

Central stimulation of the ventral nerve cord is sometimes followed by a recruitment of the cardiac neurons in the cockroach lateral nerve cord (Fig. 31a and b). It was found that such stimulation could recruit neurons in these cords uniformly up to a stimulation rate of $5\,s^{-1}$ (Fig. 31c). Above this, they failed to follow the stimulus one for one, but did continue firing at their maximum rate. When the stimulus was stopped, the cardiac neurons abruptly became relatively silent for a time before again beginning to spike spontaneously at their basal rate. These results suggest a connection between the central nervous system and the cardiac neurons, and may help account for the large numbers of axons found in the segmental nerve (Fig. 29). Unfortunately, the nature of the neural connections between the cardiac neurons and central units is not understood. We can only suspect synaptic input. Possibly the cardiac neurons are innervated by the ordinary (non-neurosecretory) axon profiles seen in the ultrastructure.

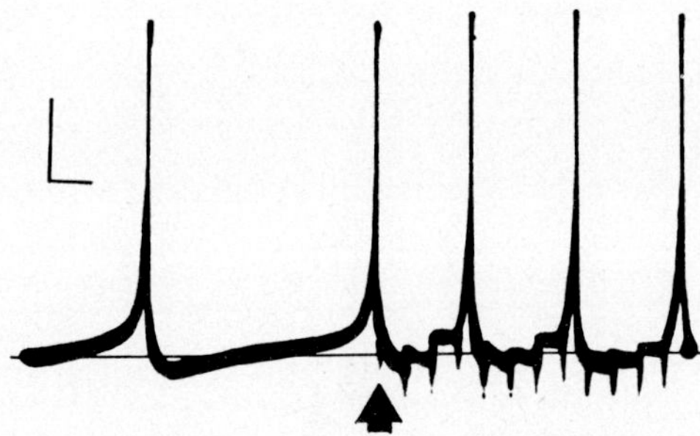

FIG. 30. Intracellular recording of the myocardium during stimulation of the segmental nerve. Continuous stimulation at 5 Hz begins at the arrow. The generator potential is abruptly shortened by depolarization resulting from postsynaptic potentials. The synaptic input is accommodated quickly, and the heartbeat gradually decreases from an initial peak during constant stimulation. Calibration marks: 10 mV, and 500 ms. (From Miller, T., 1973.)

4.2.1 CARDIAC NEURONS

The point has been made above that the dorsal vessels of *P. americana* and *L. migratoria* contract synchronously, and that this is different from dorsal vessels exhibiting peristaltic waves of contraction. It is perhaps no accident that synchronous contractions are correlated with the presence of lateral cardiac nerve cords and cardiac neurons.

Unfortunately it is only for *P. americana* that the behavior of the cardiac neurons has been described in any detail (Miller, T., 1968). During the cardiac cycle the cardiac neurons begin to fire spontaneously as soon as the myocardium begins to relax in diastole. The firing increases in rate as relaxation continues, then as soon as contraction starts in early systole, the cardiac neurons immediately stop firing (cf. Fig. 17), thus unequivocally showing that the heartbeat is not induced by nerve impulses. The

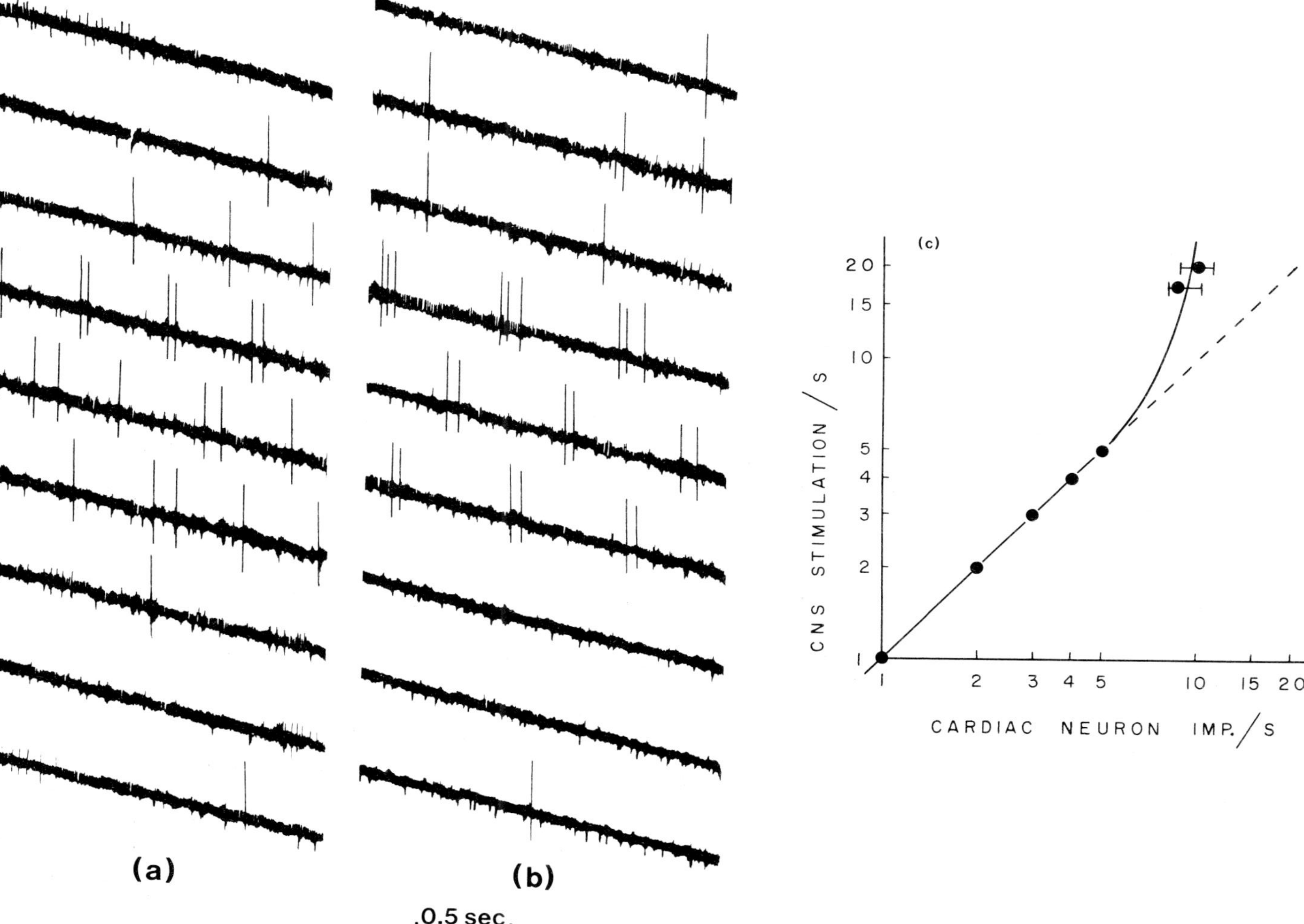

FIG. 31. Entrainment of intrinsic cardiac neurons by central nervous stimulation. The left anterior connective of the third abdominal ganglion was stimulated at 2 Hz at just above threshold for responses (**a**), and at 1.4 Hz at slightly higher intensity (**b**). The first three traces of these continuous records are control activity. Stimulation occurred during the middle three traces, and the final three traces are controls. The cardiac neurons responded with a pattern of two (**a**) and three (**b**) nervous discharges per stimulation. Accommodation or habituation is clearly shown during the second sweep of stimulation when the pattern alternates 1 and 2 in (**a**) and is composed of only two nervous impulses in (**b**). The final three traces following stimulation are conspicuous in their relative lack of activity, again showing some habituation to the central nervous stimulation. (Reproduced with permission from *J. Exp. Biol. 54*, 329 (1971). (**c**) A plot of the stimulation frequency versus cardiac neuron impulse frequency as measured from experiments, including that shown in (**a**) and (**b**).)

cardiac neurons react to stretch whenever the dorsal vessel is artificially expanded (Miller, T., 1968; Smith, N., 1969). The overall activity of the cardiac neurons, then, is probably dictated by input from other cardiac neurons and from the central nervous system and the state of stretch of the myocardium. The heartbeat with cardiac neurons can be irregular if the firing of the cardiac neurons is uncoordinated, but becomes synchronized when the cardiac neurons fire in coordination with the cardiac cycle. With the cardiac neurons removed or paralyzed with tetrodotoxin, the cockroach heartbeat becomes much smoother, similar to dorsal vessels lacking lateral cardiac nerve cords. Regardless of whether the myocardium is innervated from intrinsic cardiac neurons or not, the primitive condition seems to be a vessel supplied with an abundant bilateral neurosecretory innervation (Miller, T., 1975b), whether or not cardiac neurons are present.

The aortic diverticula of adult dragonfly, *Sympetrum danae*, is innervated by axons whose terminals contain electron-dense granules of 750–1800 Å diameter, as well as synaptic vesicles (Jensen, H., 1976). Normann, T. (1972) reported electron-dense granules in axons innervating the myocardium of *Calliphora vicina*, axons containing electron-dense granules supply the cockroach, *P. americana*, dorsal vessel from the ventral nerve cord (Miller, T., 1975a,b), and the segmental nerve in cockroach supplies the lateral cardiac nerve cord with a variety of axons, both neurosecretory and ordinary (non-neurosecretory) (cf. Fig. 23). Tyrer, M. (1971) described both neurosecretory and ordinary axons in the segmental nerve joining the lateral cardiac nerve cord of *Melanoplus differentialis*. Except for the total number of axons the segmental nerves supplying the lateral cardiac nerve cords in *M. differentialis* and *P. americana* are similar.

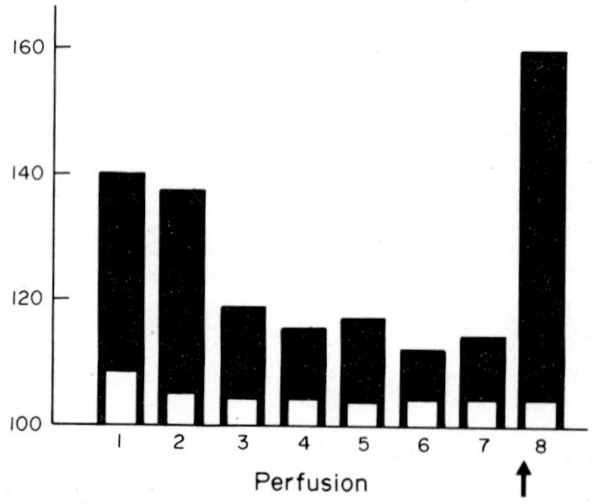

FIG. 32. Percentage increase in basal heartbeat rate of the semi-isolated cockroach heart preparation after treatment with perfusates from the isolated head preparation. Saline solution was perfused over brain–retrocerebral complex tissues before (1–7) or after (8) stimulation of the brain with 15 V, 0.3 ms pulses at 1 Hz for 15 min (black histogram). Perfusates collected under identical conditions, but with the corpora cardiaca removed, are plotted as white histograms. (After Kater, S., 1968.)

Besides the obvious lack of electron-dense granules, ordinary axons as seen with the electron microscope tend to have a more obvious glial wrapping and neurosecretory axons tend to appear nearer the bounding stromal matrix at the edge of the bundle (cf. Fig. 23). It is thought that whereas the neurosecretory units are only thinly separated from contact with the hemolymph, ordinary axons are well insulated from it.

5 NEUROHORMONAL CONTROL OF HEARTBEAT

Cardioactive factors have been demonstrated for a number of years in insects (Sternburg, J., 1963; Brown, B., 1965; Davey, K., 1964; Richter, K., 1971). Many studies on these factors have focused on the American cockroach, *Periplaneta americana*, and cardioaccelerator activity has been demonstrated in extracts from most parts of the nervous system (Ralph, C., 1962), including the lateral cardiac nerve cords (Johnson, B. and Bowers, C., 1963). Prominent amongst the nervous structures studied for possible cardioacceleratory activity has been the corporus cardiacum (Evans, J., 1962; Davey, K., 1963, 1964) because of its known properties as a neurohemal organ.

Although Unger, H. (1956) was the first to demonstrate neurohormonal control of the insect heart, perhaps the demonstration of neurohormonal control of cockroach heartbeat that came closest to the *in vivo* situation was that reported by S. Kater in 1968. In this short article, Kater described mounting the head and retrocerebral complex of *Periplaneta americana* in such a manner that he could drip saline solution over the corpora cardiaca and collect the simple perfusate in a small container (termed the isolated head preparation). He electrically stimulated the brain with the retrocerebral complex intact or with various specific nervous connections between the complex and the brain severed. The perfusates were bioassayed on a semi-isolated heart preparation according to the procedures of C. L. Ralph (1962), except that a mechanical transducer was used to record heartbeat. The heart preparation was stabilized for 2 h before bioassays began, and a basal heartbeat rate was established as the mean rate over this time.

After stimulating electrodes were positioned on the brain in the isolated head preparation, perfusates were collected over intervals of 15 min each without stimulation, and bioassayed on the heart preparation (Fig. 32, 1–7, black histograms). Then the brain was stimulated with 15V, 0.3 ms duration pulses at 1 Hz during the 15 min collection of perfusate. When bioassayed, this "stimulated" perfusate caused an increase in the heartbeat rate of about 60% of the control rate. Note the initial high percentage increase in heartbeat rate of near 40% caused by unstimulated control perfusates (Fig. 32, 1–2), and the subsequent moderate increases of 10–15% produced by later perfusates in a series (Fig. 32, 3–7).

When the corpora cardiaca were removed from the preparation, stimulation failed to produce an active perfusate. Specific nervous connections between the brain and retrocerebral complex were severed, only cutting NCC-I led consistently to reduced responses. Thus, it was suggested that a neural pathway between the brain and the corpora cardiaca mediated the release of a cardioacceleratory substance.

Kater's work was reported in the days before it was clear that presence of calcium ion was necessary for release, and thus his experiments did not include a perfusage without calcium as a control; nor was a perfusate reported employing high potassium to depolarize the tissues and evoke release. These are now established procedures (Normann, T., 1974). However, Kater's experiments were certainly a clear demonstration of the release of a cardioacceleratory substance and, more importantly, they did establish that such a material could be released into the hemolymph rather than being obtained solely from tissue extracts, and thus the plausibility of neurohormonal control of heartbeat was established.

Kater, S. (1968) reported further on the details of his bioassay response and the neurohormonal substance. He found the cardioactive material in the perfusate to be heat-stable, and labile to 1 h incubation with pronase, trypsin or chymotrypsin. The substance was unstable: it lost 80% of its bioassay activity upon standing, and freezing did not retard this loss. Proctolin is a myotropic peptide isolated from *Periplaneta americana* (Brown, B. and Starratt, A., 1975) with potent cardioacceleratory properties (Miller, T., 1979a). Since proctolin was heat-stable but insensitive to trypsin and chymotrypsin, it was thought to be different from the substance Kater studied.

The substance studied by Kater increased frequency as well as amplitude in the semi-isolated heart bioassay. The response started within 5 s of application, but did not reach a peak until 6 min later. Significantly, in the presence of the active substance, the cardiac frequency maintained an elevated rate of about 85% of the maximal peak increase for 5–8 h. Thus, the bioassay response was sustained rather than being transient, but the effect could be washed out at any time within 5 m, upon which the basal heartbeat rate was re-established.

Thus, from Kater's work we are left with the impression that a material may be released from the corpora cardiaca in response to electrical stimulation, carried in the hemolymph to the vicinity of the dorsal vessel, resulting in an increased basal heartbeat rate. The increased heartbeat rate would last as long as the unstable substance persisted in the hemolymph. We have little information on the actual use in the intact *Periplaneta americana* of this neurohormonal system to increase its basal heartbeat rate; however, Davey, K. (1962, 1963) and others have reported increased heartbeat rates in *P. americana* coinciding with various events such as gross mechanical stimulation (Davey, K., 1963) and notably, upon feeding (Davey, K., 1962). Remarkably, an increased heartbeat rate observed in intact *P. americana* upon feeding did not occur if the corpora cardiaca were first removed (Davey, K., 1962).

The feeding experiments by Davey were conducted with restrained *P. americana*, a procedure not recommended because of a significant mortality in controls. However, intact unrestrained *P. americana* are known to show an increased rate of heartbeat in response to disturbance (Richards, A., 1963a). Disturbance leads to an immediate increase in heartbeat rate which is then sustained for several minutes, and Davey, K. (1962) reported an immediate increase in heartbeat rate upon acceptance of food. This is similar to S. Kater's (1968) bioassay data which reports a 5 s delay in bioassay response. However, the dorsal vessel is not downstream from the retrocerebral complex. The dorsal vessel in *P. americana* is, in fact, the furthest tissue away from the corpora cardiaca in terms of circulation unless one assumes a reversal in the direction of hemolymph flow (known to occur in Lepidoptera and other holometabolous insects, but a condition not commonly reported in *P. americana*).

A description by Davey, K. (1964) indicated that when cardioacceleratory factors are released from the corpora cardiaca they interact with the pericardial cells. The latter in turn were supposed to produce a secondary substance which stimulated the heart. This somewhat unwieldy scheme involving pericardial cells as intermediary was never fully accepted, and at present the pericardial cells are not generally held to play a role in cardioregulation (Hertel, W., 1971; Normann, T., 1972).

The presence of cardioacceleratory substances in various tissues of the cockroach has been reported in many studies, and seems beyond doubt. One must question, however, the existence of a neurohormonal control mechanism based in the corpora cardiaca, when the lateral cardiac nerve cord in *P. americana* and other cockroaches and other insects, is itself an impressive neurohemal organ containing cardioacceleratory factors (Johnson, B. and Bowers, C., 1963) and it or segmental nerves, or the perisympathetic organ system (Raabe, M. *et al.*, 1974), are far better situated to perfuse the dorsal vessel downstream. The identity of the carioacceleratory factor in the lateral cardiac nerve cord of *Periplaneta americana* has not been revealed as yet. It is not certain whether neurohormone D is present there (Klaus Richter, personal communication, 1981). However, certain paired neurons (termed lateral white, "LW", cells), which occur in the ventral ganglia of *P. americana* and the cricket, *Gryllus bimaculatus*, send axons to the abdominal body wall musculature and to the cardiac nervous system (O'Shea, M. and Adams, M., 1981). A combination of immunohistochemistry, microbioassay of the contents of the lateral white neuron somata, and high-pressure liquid chromatography have shown the presence of proctolin in the lateral white neurons (O'Shea, M. *et al.*, 1981; Bishop, C. *et al.*, 1981). Thus, if LW axons occur in the lateral cardiac nerves, then proctolin may be present in the lateral cardiac nerve cords.

For various reasons the control of heartbeat by corpora cardiaca in locust, *Locusta migratoria*, does not seem a viable scheme because diluted extract has only a small effect on the heartbeat rate (Mordue, W. and Goldsworthy, G., 1969). Goldsworthy, G. and Mordue, W. (1974) concluded that although the semi-isolated heart preparation of the locust provided a highly sensitive bioassay for different peptide neurohormones, such assays do not necessarily reflect what occurs *in vivo*. Normann, T. (1972) reached a similar conclusion in studies on cardioregulation in blowfly, *Calliphora vicina*.

Neurohormone D is said to be similar to factor C of Mordue, W. and Goldsworthy, G. (1969). However, the latter authors stressed the possible lack of physiological role for these peptide factors on the heartbeat, even though corpora cardiaca extracts were fairly potent in increasing the rate of heartbeat. Mordue (personal communication, 1974; and Goldsworthy, G. and Mordue, W., 1974) considered the actions of corpora cardiaca extract on semi-isolated heartbeat of *Locusta migratoria* to be highly useful as bioassays, but he cautioned against making too much of a possible cardioaccelerating neurohormone since injection of the extracts into the whole adult locust leads to little change in heartbeat rate. Brown, B. (1965) issued a similar caution for peptides on cockroach heart. Further caution may be necessary because a change in heartbeat rate may not occur, and changes in heartbeat amplitude or movements of the dorsal diaphragm may have a significant effect on regulation of circulation in the locust (see Miller, T. and S.-Rózsa, K., 1981).

Perhaps one of the longest-standing investigations into cardioacceleratory substances is that of the late Manfred Gersch and his associates in Jena, East Germany. Upon stimulation of NCC-I connecting the brain and corpora cardiaca, neurohormone D and a nerve-stimulating factor are released (Gersch, M. *et al.*, 1970). The Kater, S. (1968) substance was reported to be heat-stable, sensitive to pronase and chymotrypsin. Neurohormone D has a molecular weight of between 750 and 1000, is not resistant to heat and acids, and is inactivated by pronase and chymotrypsin. It does not migrate in an electric field and shows eight amino acid residues

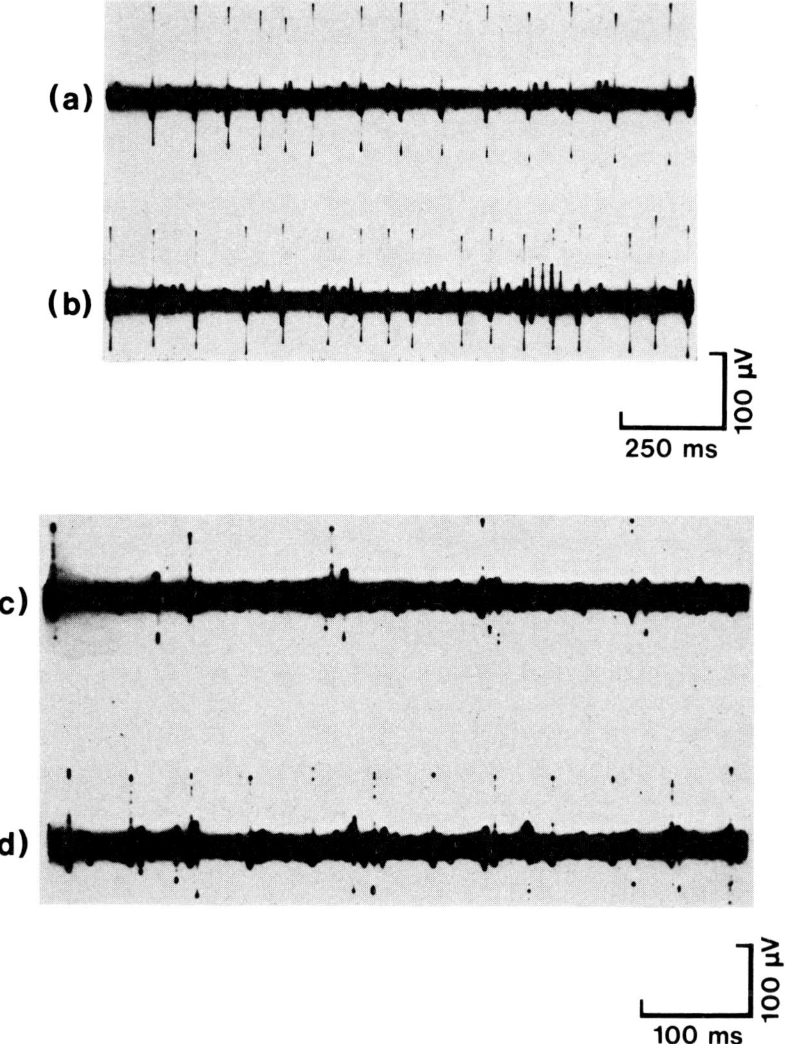

FIG. 33. Response of cardiac neurons to the actions of neurohormone D (**a,b**) and acetylcholine (**c,d**) when perfused on the lateral cardiac nerve cords of cockroach, *Periplaneta americana*. Note the large spikes, presumably from activity of intrinsic cardiac (motor) neurons, are unaffected, but the smaller potential activity is increased somewhat possibly. The figures were provided by Klaus Richter, 1981, from the work of the Jena group on neurohormone D. Acetylcholine increases the firing rate of intrinsic cardiac neurons in the lateral cardiac nerve cord of the cockroach, *Periplaneta americana*. **c;** Control; **d;** after application of acetylcholine. (Richter, 1981, unpublished).

upon hydrolysis, dansylation and microchromatography of the relative values: Asp 2, Glu 1, Pro 1, Phe 1, Ser 1, Val 1. (Richter, K., 1981; Baumann, E. and Gersch, M., 1982). Thus, only the heat instability of neurohormone D distinguishes it from the cardioacceleratory factor described by Kater, S. (1968). Neurohormone D was isolated from the hemolymph of cockroaches that had been stimulated electrically. Both neurohormone D and corpora cardiaca extracts caused an immediate increase in the rate of heartbeat in denervated heart preparations, suggesting that the site of action is the myocardium (Hertel, W., 1971).

When extracts of corpora cardiaca known to contain neurohormone D are placed on the spontaneously active cardiac neurons in the lateral cardiac nerve cord, the activity of the cardiac neurons does not change appreciably (Fig. 33a,b). Note, however, that "neurosecretory" activity may be increased somewhat. In contrast to the lack of action of neurohormone D on the activity of cardiac neurons, acetylcholine is known to have a potent stimulatory action on the spontaneous firing rate of the cardiac neurons (Miller, T., 1968). Recordings from the lateral cardiac nerve cord of *P. americana* demonstrate this very consistent response to acetylcholine (Fig. 33c,d).

When the equivalent of 0.1 to 0.2 pairs of corpora cardiaca were injected into the intact cockroach, mild reactions included increased heartbeat rate of 40–60% above the control rate, followed by a gradual decline to the original rate (Hertel, W., 1971). The entire response lasted 5–8 min even when a severe reaction was recorded resulting in systolic arrest for up to 5 min. Curiously, this time period of 5 min corresponds to the time for material to travel around the cockroach body (Coon, B., 1944). Thus, neurohormone D appears to be a myotropic substance.

Injection of various neurohormonal materials into the intact cockroach requires the insect to be held. Usually, this causes a slightly elevated heartbeat rate from about 85 beats min^{-1} to around 94 beats min^{-1} at 25°. Upon injection of corpora cardiaca extracts, or of neurohormone D, or injection of proctolin into the dorsal midline of the abdomen near the dorsal vessel, the heartbeat rate immediately increases (Fig. 34).

An immediate increase in heartbeat is followed in all cases by a gradual return to the basal heartbeat rate. Note the potency of a small amount of corpora cardiaca extract (Fig. 34, filled circles). Richter, (1981, personal communication) and his colleagues feel that the elevated heartbeat rate seen in stressed cockroaches (Davey, K., 1964) is due to an increased titre of neurohormone D in the hemolymph (Gersch, M. *et al.*, 1981).

Neurohormone D also reportedly stimulates the antennal hearts (accessory pulsatile organ of antennae) of *Periplaneta americana* in a way comparable with its action on the dorsal vessel (Richter, 1981, personal communication). On the other hand, neurohormone D is active on the proctodaeum and hyperneural muscles of *P. americana*, but in both cases it is less active than the extracts of corpora cardiaca. Thus, there may be some differences between actions on strictly pulsatile organs and other

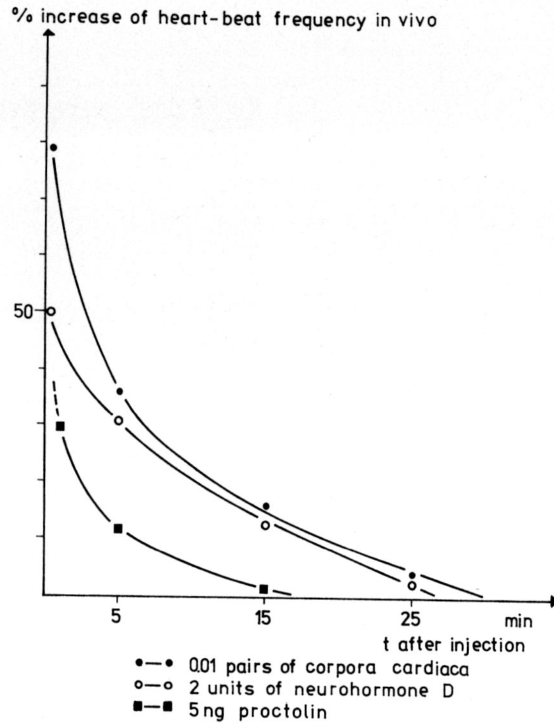

FIG. 34. Percentage increase in heartbeat rate counted from intact *Periplaneta americana* upon injection of the extract of 0.01 pairs of corpora cardiaca (filled circles), two units of neurohormone D (open circles) or 5 ng proctolin (filled squares). (After Gersch, M. *et al.*, 1981; courtesy of Klaus Richter, 1981.)

visceral muscles. The hyperneural muscle in cockroach appears to be electrically inexcitable, and yet is neurally driven, rather than being a spontaneously active myogenic organ (Miller, T. and James, J., 1976; Miller, T. and Adams, M., 1974).

5.1 Neurohormonal control of eclosion heartbeat

Wasserthal, L. (1975a,b), using his technique of recording heartbeat by contact thermography, recorded the peculiar heartbeat reversals of several Lepidoptera. One such recorded heartbeat pattern showed that for about $2\frac{1}{2}$ h around the time of emergence of the adult from the pupal case and expanding of the wings, the heartbeat pattern changed completely.

Before ecdysis to the adult stage in *Caligo brasiliensis*, a nymphalid butterfly, the heartbeat rate increased while maintaining anterograde peristalsis (Fig. 35). During the period of altered

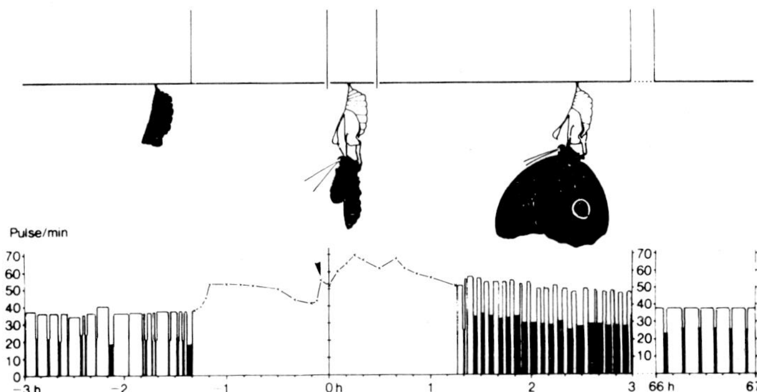

FIG. 35. Heartbeat frequency of the nymphalid *Caligo brasiliensis* before (left), during (middle) and after (right) adult emergence. Hours before emergence are given negative numbers on the abscissa, and positive after emergence. The arrow shows the point at which the exuvia was split at the beginning of emergence near hour 0. Note the heart reversal (anterograde peristalsis shown as white bars, retrograde peristalsis shown as black bars) abruptly stopped 1 hr and 20 min prior to emergence and was replaced by an elevated heartbeat rate with anterograde peristalsis. About 1 h 15 min after beginning emergence, when the wings were fully expanded, the heartbeat reverted back to periodic reversal (From Wasserthal, L., 1980.)

heartbeat the internal pressure increased and the pupal cuticle was split. The imago emerged and expanded its wings fully. Afterwards the heartbeat reverted back to its ordinary condition of alternating period of forward peristalsis (anterograde) and periods of reverse peristalsis in the posterior direction. Wasserthal, L. (1976) noted that during retrograde peristalsis hemolymph is pulled from the wings, and that only during anterograde peristalsis is hemolymph supplied to the wing veins.

Similar changes in the heartbeat have been reported also at eclosion of the tobacco hornworm, *Manduca sexta*, by Tublitz, N. and Truman, J. (1981). They extracted a substance from the ventral nerve cord complex, and in particular from the transverse nerves of the median nervous system of pharate adults, which, when bioassayed on a semi-isolated heart preparation, caused an abrupt increase in heartbeat rate. The substance was pronase-sensitive and heat-insensitive with an apparent molecular weight of less than 1000 daltons, and had properties dissimilar to proctolin.

When hemolymph from eclosing *M. sexta* adults was collected and bioassayed on the *M. sexta* heart preparation, a cardioactive substance was shown to be present. Collection and bioassay of hemolymph either before or after the 75 min eclosion period showed no cardioactive properties. Simultaneous with the appearance of a cardioacceleratory factor in the hemolymph during the eclosion period, the amount of cardioacceleratory material extractable from the ventral nerve cord decreased by some 80%.

This peculiar "eclosion heartbeat" demonstrable in Lepidoptera represents the first plausible *in vivo* case of neurohormonal control of heartbeat in insects. Many cases of cardioacceleration have been shown before, particularly in the cockroach *Periplaneta americana*, but none of the reports before those of Wasserthal, and Tublitz and Truman, demonstrate cardioacceleration clearly connected with a natural event except in an inferential way.

It is worth noting too that this eclosion heartbeat is a discrete event which lasts about 75 min in *Manduca sexta* and just over 2 h in the nymphalid *Caligo brasiliensis*. Thus, neurohormonal control of heartbeat in this case is a special event coinciding with wing expansion or eclosion. It would be of interest to examine similar events in other holometabolous insects. It is possible that an eclosion-timed change in the heartbeat may be a general phenomenon.

6 INTRINSIC MYOGENICITY IN SKELETAL MUSCLES

Small intrinsic rhythmic movements of the grasshopper or locust metathoracic tibia have been noted for some years (see Evans, P., (1980), page 380, for background). When one cuts the metathoracic (jumping) leg from an adult *Schistocera gregaria* and mounts the femur in Plasticine so that the tibia is oriented "up", either soon, or after a pause of up to several minutes, the tibia can be seen to rhythmically move back and forth, usually with a very low amplitude. Sometimes the movement is barely perceptible and sometimes it cannot be seen.

Other cases of rhythmic contractions in skeletal muscles have been documented for muscles whose motor nerve supply had been cut some days previously (Case, J., 1956; Beránek, R. and Novotný, I., 1958; Usherwood, P., 1963). Although equally curious, the case of neurotomy-induced rhythmic contractions may be a condition that develops as the muscle responds to motor nerve section and thus may be different from the intrinsic rhythm described above.

The intrinsic rhythm can develop almost immediately upon severing the metathoracic leg of *Schistocerca gregaria*, *Schistocerca nitens*, and *Locusta migratoria*. The only cautionary note is that the leg must be cut as close to the thorax as possible. Of the three Orthoptera mentioned above, I have been shown the leg rhythms in *S. gregaria* and can report personally that *S. nitens* and *L. migratoria* adult males or females exhibit these isolated leg movements. Thus, the intrinsic leg rhythm, or myogenic rhythm, in the distal aspect of the metathoracic extensor tibia muscle appears to be common in locusts.

As he has done on several previous occasions, Graham Hoyle did the pioneering work on this new preparation. In this case, it was in collaboration with Mick O'Shea (Hoyle, G. and O'Shea, M., 1974), and subsequently his other colleagues. Somehow, the intrinsic rhythm of the metathoracic extensor tibia was discovered to be inhibited by stimulation of one of the dorsal unpaired median (DUM) neurons in the metathoracic ganglion.

The so-called DUM neurons are dorsal, unpaired and project ventrally to a single branch from which paired axons leave the ganglion and ramify at the proximal end of the left and right extensor tibia muscles. Discrete "release sites" were found at the muscle, but not neuromuscular junctions (Hoyle, G. *et al.*, 1980).

The lack of discrete innervation by synaptic contact goes along with the peculiar function of the DUM neurons which, when stimulated, do not alter responses to ordinary motor innervation of the fast or slow variety except for a subtle modulation of slow responses (Evans, P. and O'Shea, M., 1977). Instead, stimulation of the DUM cell inhibits the spontaneous rhythm of the toxic bundle of fibers at the proximal end of the extensor tibia muscle. Although there is a suggestion that the intrinsic rhythm plays a role in circulation, there is no evidence for this and such rhythms have not been investigated in other insects.

A series of paired neurons that stain with neural red (lateral white or LW cells) have been discovered in each of the ventral ganglia of the cricket, locust and cockroach. Axons are sent to the periphery, to the lateral cardiac nerve cords in the case of the cockroach, *Periplaneta americana* (O'Shea, M. and Adams, M., 1981). DUM cells are also reported for each of the ventral ganglia in these same orthoptera. To date, O'Shea and Adams (December 1982, personal communication) have not discovered what role these LW neurons have, nor where the DUM project to from the ventral ganglia; however, their function may be as subtle as that of the original DUM neurons in the metathoracic ganglion of locust.

Of related interest is the recent work on light organ innervation and control (Christensen, T. and Carlson, A., 1981). The two most posterior abdominal ganglia of the male firefly, *Photuris versicolor*, contain clusters of three or four large DUM neuron somata. The neurons send bilateral processes out of the ganglia to ramify over the dorsal surface of the lantern tissue. Although synapses in the lantern organ are of the neurosecretory type, and reportedly identical to those described in the exterior tibia muscle (Hoyle, G. *et al.*, 1980), a complete description of innervation and control is still ahead (Christensen, T. and Carlson, A., 1981). In particular, Hollingworth, R. and Murdock, L. (1980) showed octopamine-like effects of chlordimeform insecticides on firefly light organ of *Photinus* species.

7 BIOGENIC AMINES

The DUM neurons responsible for spontaneous tibia movements in the locust are reported to contain 0.1 pmol of octopamine (Evans, P., 1980). Octopamine mimics stimulation of the DUM neuron when perfused directly on the proximal extensor muscle (Evans, P., 1980). Octopamine also

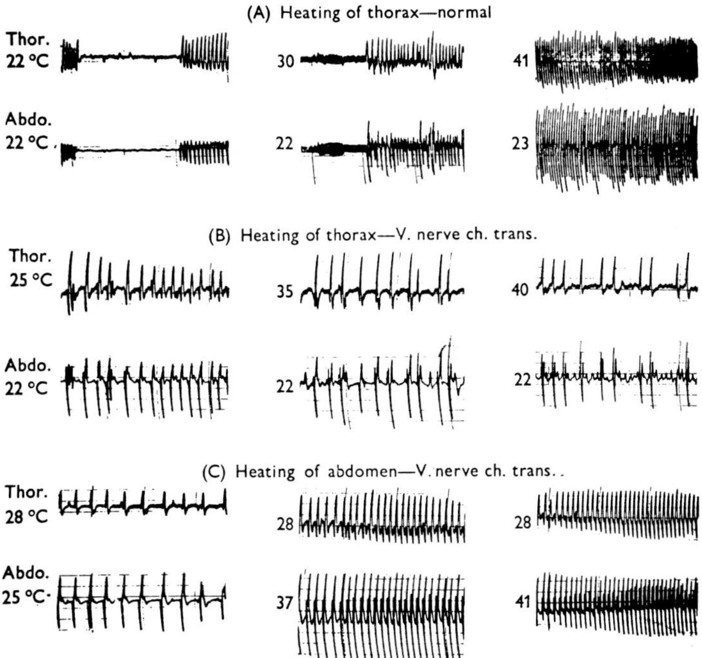

FIG. 36. Heartbeat of the sphinx moth, *Manduca sexta*, impedance recorded simultaneously from the thorax (upper traces) and from the second abdominal segment (lower traces). All of the recordings are derived from the same moth. The amplification and position of the electrodes were not changed throughout **A**, **B** and **C**. T_A was 22°. Thoracic temperature and T_{Ab} are indicated at the side of each record spanning 32 s. **A:** Thoracic heating of the normal moth; **B:** heating of the thorax after the ventral nerve cord was transected at the second abdominal segment; **C:** heating of the abdomen after the ventral nerve cord was transected. (From Heinrich, B., 1971.)

produces light in the lantern of *Photuris versicolor* mimicking stimulation of the DUM neuron there (Christensen, T. and Carlson, A., 1981). Thus octopamine is a putative neurotransmitter or neuromodulator of the DUM neurons in two cases. Octopamine is also active on the cockroach heart in increasing the rate of heartbeat. Gersch, M. *et al.* (1974) reported green and yellow fluorescence in varicosities of the lateral cardiac nerve cords of the cockroach, *Blaberus craniifer* Burm. This was interpreted as evidence of the presence of dopamine and 5-hydroxytryptamine (5-HT). The chief evidence for this comes from the Falck-Hillarp technique (Klemm, H., 1980). The catecholamine and 5-HT reaction products were also found in the frontal ganglion, recurrent nerve, hypocerebral ganglion and the corpora cardiaca.

Lafon-Cazal, M. and Arluison, W. (1976) injected tritiated 5-HT into *Locusta migratoria* and found that the tritium disintegrations in autoradiographs were localized in association with electron-dense granules of 100 nm diameter. When labelled dopamine was injected, the autoradiographic activity was associated more with granules of 200 nm diameter than 100 nm granules. The heart of the locust is said to contain large quantities of 5-HT and dopamine (Hiripi, L. and S.-Rózsa, K., 1973). In bioassay, 5-HT is more active than dopamine on locust heart preparations (S.-Rózsa, K. and V.-Szöke, I., 1972).

8 THERMOREGULATION AND CIRCULATION

In the adult sphinx moth, *Manduca sexta*, when the thorax was heated externally by a beam of light for several minutes the freqency, amplitude and duration of the heartbeat in both the abdomen and thorax increased, when the temperature of the thorax reached 40–43° (Heinrich, B., 1970b). When the temperature of the thorax rose above 40° the temperature of the abdomen near the dorsal vessel stayed below 35° and the frequency of heartbeat in the thorax and abdomen remained the same with anterograde peristalsis even when the difference in temperature was made to differ by 20°. Transsection of the ventral nerve cord immediately reduced the frequency of pulsations of the dorsal vessel and the frequency and amplitude of the abdominal heart no longer responded to thoracic heating (Fig. 36).

One may infer from the nerve transection experiment that overheating of the thorax causes an increase in heartbeat rate and amplitude via cardioregulatory units in the ventral nerve cord, and that this in turn caused an increased flow of hemolymph through the aorta loop in the thorax. Increased circulation was thought to pull heat from the thoracic musculature back into the abdomen assisted by pulsations of the ventral diaphragm. Thus, Heinrich, B. (1970b) considered the aortic loop in the thorax of *M. sexta* adults to act as a cooling coil during thermoregulation, and he thought the abdomen, being poorly insulated, could act as a heat radiator.

Given that the heat produced was independent of the abdominal temperature, this cardioregulatory mechanism was thought to explain why adult *M. sexta* could maintain a thoracic temperature at 42° ± 1° during free flight at ambient temperatures of from 17° to 32°. Moths with the dorsal vessel tied off overheated and stopped flying at air temperatures of 23°, and were unable to control thoracic temperature (Heinrich, B., 1970a).

These early experiments from Heinrich's dissertation work made two major contributions. First they provided the vastly improved impedance conversion recording method and second they clearly albeit indirectly, demonstrated cardioregulation with the ventral nerve cord as the principal regulatory pathway. Steiner, G. (1932) had earlier shown an increased heartbeat rate in cockroach, *Periplaneta americana*, by shocking the ventral nerve cord, and many authors had reported increased heartbeat in the American cockroach (e.g. Richards, A., 1963a) in response to prodding. Despite these and many other observations, including those describing neurohormonal control mechanisms (Gersch, M., 1973), Heinrich's work clearly points to a control of frequency, amplitude and duration of heartbeat, and to the key role played by the dorsal vessel in a specific function, thermoregulation.

The work by Heinrich also incidentally suggests specific experiments to study control of the dorsal vessel. By using the restrained *M. sexta* preparation (Heinrich, B., 1970b) one might be able to gain access to and describe the units in the central nervous system responsible for cardioregulation.

Thermoregulation is a common feature in advanced Hymenoptera. The swarm (Heinrich, B., 1981) and brood (Heinrich, B., 1974a) temperatures are controlled in bees and bumblebees. Thermoregulation also occurs in katydid males, *Euconocephalus nasutus*, in muscles used for stridulation, and thermoregulation has been reported in Lepidoptera, Hymenoptera, Coleoptera and other Orthoptera besides katydids (Heinrich, B., 1974b). Similar temperature increases due to endothermy in Diptera are termed modest (Heinrich, B., 1974b, Heinrich, B. and Pantle, C., 1975). The thermoregulation in Lepidoptera was described above for *Manduca sexta*, in which the looping course of the aorta in the thorax was considered to function as a cooling coil. The bumblebee, *Bombus vosnesenskii*, seems to have taken advantage of circulatory anatomy for a slightly different purpose. The dorsal vessel in the bumblebee usually pumps hemolymph in an anterior direction, while the ventral diaphragm directs blood posteriorly (Fig. 37). The dorsal vessel is in contact with the ventral diaphragm in the petiole region between thorax and abdomen (Heinrich, B., 1976), after which the dorsal vessel forms a loop between the flight muscles in the thorax before entering the head capsule.

The peculiar anatomy of the dorsal vessel in the petiole of the bumblebee is said to constitute a countercurrent heat exchanger. At low thoracic temperatures the ventral diaphragm beats at various rates and the dorsal vessel beats at relatively higher frequencies, but at low amplitudes. When the thoracic temperature exceeded 43°, both dorsal vessel and ventral diaphragm pulsated at identical rates and high amplitudes, and the abdomen was vigorously pumped at the same frequency (Heinrich, B., 1976) (Fig. 38).

Since the bumblebee could efficiently transfer heat to the abdomen during overheating, and since this ability was lost when the abdominal dorsal vessel was made inoperative, circulation was thought to play a key role in thermoregulation. Furthermore, the retention of heat in the thorax by the countercurrent heat exchangers was thought to

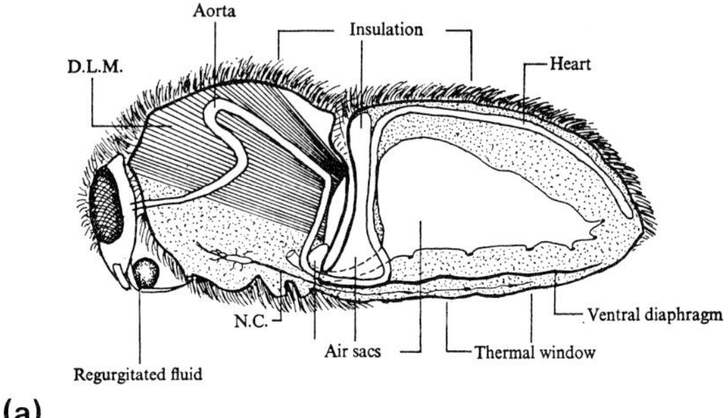

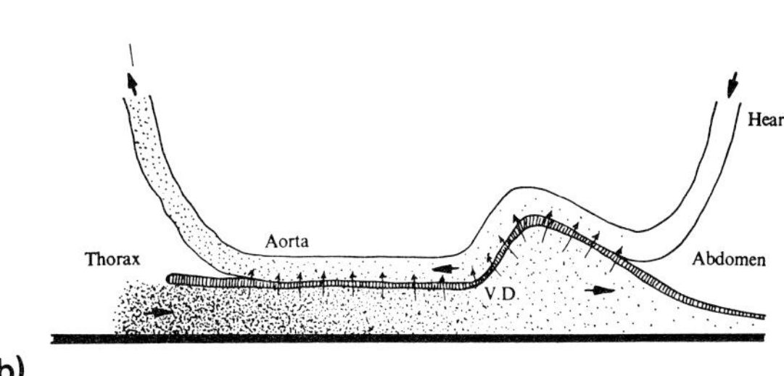

FIG. 37. Drawing of the bumblebee (a), showing details of the aorta in the petiole regions (b) and hypothesized countercurrent heart flow. (From Heinrich, B., 1976.)

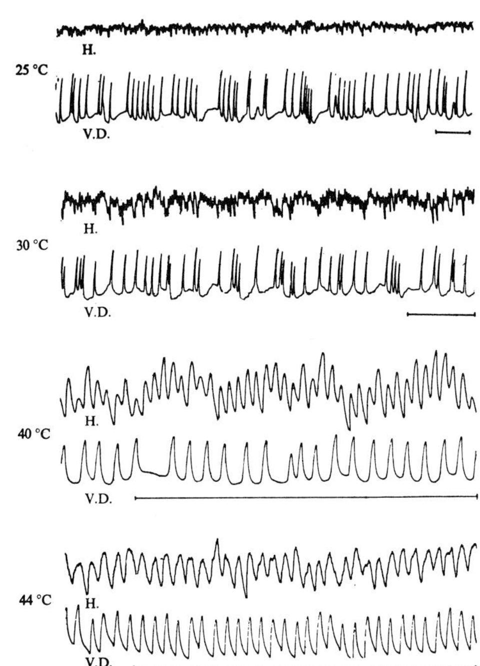

FIG. 38. Concurrent activity of the heart (H.) and the ventral diaphragm (V.D.) at four different thoracic temperatures (indicated at left) during heating of the thorax in a bee. Five-second intervals are indicated by horizontal lines. The electrodes and amplification were not changed throughout the experiment. (From Heinrich, B., 1976.)

be circumvented by pulsing hemolymph rapidly through the dorsal vessel anteriorly past the petiole rather than continuously, not allowing the heat exchanger to operate efficiently (Heinrich, B., 1976). The anterior portion of the ventral diaphragm may provide a switching mechanism, alternately allowing air and hemolymph to pass anteriorly into the thorax, then allowing hemolymph to pulse posteriorly into the abdomen.

Thermoregulation in the adult dragonflies *Anax junius* and *Aeshna multicolor* appears to be accomplished exclusively by control of hemolymph circulation. Ligation of the dorsal vessel anywhere in the abdomen abolished heat transfer from the thorax into the abdomen (Heinrich, B. and Casey, T., 1978). By way of contrast, the adult dragonfly *Libellula saturata* appeared to thermoregulate by postural adjustment and showed little capacity to transfer heat from thorax to abdomen (Heinrich, B. and Casey, T., 1978).

It should be noted that *Libellula saturata* do not hunt for prey by flying; rather, they alight and wait for prey to pass by before attacking. The *Anax* and *Aeshna* dragonflies continually fly to encounter prey.

9 CONTROL OF HEMOLYMPH PRESSURE

Bayer, R. (1968) measured body pressures at various extra- and intracardiac spaces in *Locusta migratoria*. The lowest pressure in the hemocoel was found in the dorsal ampullae of the metascutellem concerned with hemolymph circulation in the wings, $-15.2\,cmH_2O$. The maximum pressure, $+11.3\,cmH_2O$, was measured in the ventral thorax.

The pressure within the dorsal vessel was measured in the thorax (aorta), and in the anterior and posterior ends of the abdomen. The mean aorta pressure was $+3.2\,cmH_2O$ in diastole and $+8.6\,cmH_2O$ in systole during a heartbeat frequency of 79.5 beats min^{-1}. The diastolic pressure in the anterior abdominal heart was near zero and the systolic pressure was $+9.4\,cmH_2O$, while the dorsal vessel at the rostral end of the abdomen fluctuated near zero with pressure amplitudes of $8.5\,mmH_2O$ (Bayer, R., 1968).

Air intake is sometimes important in assisting expansion of newly emerged imagos. If the proboscis of the newly emerged blowfly is ligated, preventing air intake, expansion is prevented (Houlihan, D. and Breckenridge, L., 1981). To expand the wings in newly eclosed adult *Bombyx mori*, air is swallowed to assist the increase in internal pressure. However, in this case if air swallowing is prevented, the wings expand, but take longer to do so (Moreau, R., 1974). Wing expansion is prevented in *B. mori* if the pro- and mesothoracic ganglia are separated by severing the paired connectives, but not if only one is cut (Moreau, R., 1974). When the meso- and metathoracic ganglia were separated by cutting their common connective, the anterior wings expanded, but the hind wings did not.

The hemolymph pressure of insects is known to rise at various times. Most normally these times are associated with ecdysis or oviposition, and wing expansion certainly is accompanied by an increase in hemocoele pressure (Sláma, K., 1976). If any generalizations can be made, they are that larval (endoptergyote) insects maintain positive hemocoele pressures, while adult insects tend to have zero or negative pressure once the tanning process is complete (Sláma, K., 1976). Exposure of larvae of endopterygota to diethylether vapors causes a drop of pressure to zero, suggesting that the maintenance of pressure results from neuromuscular tonus of the body wall musculature (Sláma, K., 1976). Larvae and adults of exopterygote insects show more variable pressures with peak pulses associated with ventilatory movements and negative pressures coinciding with starvation or other living conditions (Sláma, K., 1976).

Improved methods of recording pressure over extended periods have made it possible to subdivide ecdysis into specific events associated with abdominal contractions, taking air into the alimentary canal, etc. Exuvial rupture commonly corresponds with high hemocoele pressures of 400–600 mmH_2O (Sláma, K., 1976).

Very high internal pressures accompanying puparium formation in some Diptera (300 mmHg in *Sarcophaga*; Zdárek, J. *et al.*, 1979) may result from neuroendocrine control of muscular contractions of the body wall muscles (see Irving, S. and Miller, T., 1980b; Zdárek, J., 1980). The control of hemocoele pressure appears to be achieved largely by the mesothoracic ganglion via the ventral ganglia to abdominal intersegmental muscles in pupal *Tenebrio molitor* (Sláma, K. *et al.*, 1979). To a lesser extent the diuresis process may affect pressure by changing the hemolymph volume, albeit slowly. In adult female *Pyrrhocoris apterus* the hemolymph pressure increased during drinking, and decreased upon formation of a liquid drop of urine in the rectal reservoir (Sláma, K., 1976).

10 RESPIRATION AND CIRCULATION

Gerould, J. (1938) pointed out that one could not adequately describe circulation in insects without considering the respiratory system. He used the example of aquatic insects with their peculiar adaptations in the structure of the dorsal vessel. Indeed, in the immature stages of aquatic insects, one sees slightly more development of arterial structures and ostial valves (Gerould, J., 1938) which more closely resemble circulatory structures in other arthropods (Clarke, K., 1973).

Lhotsky, S. *et al.* (1975) inferred a connection between circulation and respiration in *Schistocerca gregaria* from their intact heartbeat recordings. Abdominal pumping is accomplished by the intersegmental muscles of the abdomen, whose main role is ventilation. There is some suggestion now that abdominal pumping plays an active role in the circulation of hemolymph in adult Lepidoptera (Wasserthal, L., 1980) and possibly in Hymenoptera (Heinrich, B., 1976). However, Heinrich (1981, personal communication) pointed out that when dorsal vessel, ventral diaphragm and abdominal ventilatory movements are occurring simultaneously, it can be difficult to determine what exactly is moving the hemolymph.

Very recently, Wasserthal, L. (1981) suggested that expiration in the anterior body of the giant silkmoth, *Attacus atlas* L., is coordinated with a tidal flow of hemolymph. Expiration in the thorax occurs when hemolymph accumulates anteriorly during anterograde pulsations of the dorsal vessel. Active removal of hemolymph from the anterior body coincides with inspiration there, and with a retrograde peristalsis of the dorsal vessel and a flow of hemolymph into the abdomen in the perineural sinus. A septal valve at the abdomen–thorax junction is said to compartmentalize the body during these tidal hemolymph movements, and accumulation of hemolymph in the abdomen accompanies expiration there along with bouts of abdominal pulsations.

11 ALARY MUSCLES

The primitive number of alary muscle pairs in the orthopteroid complex is 12. All forms studied by Nutting, W. (1951) had 10 pairs in the abdomen (Figs. 1 and 2). He found that non-flying Tettigonioidea had 12 pairs of alary muscles with strong fliers lacking alary muscles in the thorax. Thus a reduction of thoracic alary muscles in favour of longitudinal flight muscles in strong fliers in the orthopteroid complex is suggested. Consequently, Nutting placed little value on numbers of alary muscles as indicators of group relationships in view of these correlations with flight muscles.

In Lepidoptera and some Hymenoptera, the presence of flight muscles is accompanied by modification of the aorta in the meso- and metathorax to include well-developed accessory pulsatile organs (APO's). The APOs are generally associated with the wing base structure and are normally confluent with the anal veins (Wasserthal, L., 1980; Hessel, J., 1966, 1969). Neither winged queen ants nor alate workers have a modified aorta (Wheeler, W., 1910). Instead, the aorta extends into the head traversing straight through the thorax in much the same manner as the house-fly, *Musca domestica*.

Nutting, W. (1951) noted that the alary "muscles" were probably largely composed of apparent connective tissue with elastic properties rather than muscle cells. The biggest difference among members of the orthopteroid complex in appearance of alary muscles is the broad origin of the muscle along the tergum in Acridoidea rather than a point origin as exists in all other Orthoptera examined except *Tridactylus*. In most if not all Orthoptera whose alary muscles have a restricted point of origin, the alary muscle fibers extend medially only part of the way to the dorsal vessel. Anastomosing connective tissues then ramify over the ventral midline of the dorsal vessel, or diverge from the plane of the dorsal diaphragm to attach along the sides of the dorsal vessel. The alary muscle fibers in Acridoidea not only originate over a broad area on the lateral tergite, but the fibers extend all of the way to the ventral midline of the dorsal vessel (Nutting, W., 1951).

Cruz-Landim, C. (1978) noted a close relationship between pericardial cells and the alary muscles of the bee. A similar close relationship exists between the dorsal diaphragm of *Locusta migratoria* and pericardial cells, but whether this does indeed suggest a functional relationship (mentioned as a possibility by Cruz-Landim, C., 1978) needs further study. As noted above, the pericardial cells are normally restricted to, and therefore associated with, the pericardial sinus whether or not they are associated with the dorsal diaphragm. The latter may be coincidental.

The role of alary muscles in heartbeat has been interpreted differently in several studies. Maloeuf, N. (1935) reported a general collapse of the dorsal vessel when the alary muscles were detached from their cuticular connection. He also found that rhythmic contractions could be supported by the collapsed dorsal vessel. Snodgrass, R. (1935) related that while it was often assumed that muscles of the dorsal diaphragm actively assist the diastole (relaxation) phase of contraction of the dorsal vessel, in some cases the dorsal vessel supports heartbeats without the alary muscles connected. Also, the alary muscles are able to contract spontaneously with a rhythm of a much longer period than that maintained by the dorsal vessel itself (de Wilde, J., 1947; Miller, T. and Metcalf, R., 1968; Miller, T. and S.-Rózsa, K., 1981).

Contractions of alary muscles are claimed to have been recorded during heartbeat in a restrained adult *Schistocerca gregaria*. Optical recording of heartbeat in *S. gregaria* showed a rather complex cardiac cycle (Lhotsky, S. *et al.*, 1975). In particular, the presystolic notch (a slight relaxation immediately preceding contraction and described by Yeager, J., 1939) is variously considered to be due to relaxation of the myocardium (Yeager, J., 1939), dilation from influx of hemolymph from another part of the dorsal vessel (Jones, J., 1977, see p.44) or due to alary muscle contraction (Lhotsky, S. *et al.*, 1975). The latter case appears to be a movement artifact in the optical record rather than a true alary muscle contraction.

Duwez, Y. (1938) found that cutting all eight pairs of alary muscles slowed the heartbeat of *Dytiscus marginalis*, but did not stop the heart. The electrical properties of the heart were similar before and after ablation of the alary muscle suspension. Distension of the heart modified the frequency of heartbeat, but was independent of automatism.

The heartbeat rate is unaffected when the alary muscles of *Anopheles quadrimaculatus* were severed (Jones, J., 1954). Severing of the alary muscles caused the dorsal vessel in the abdomen to assume a new position and to gyrate in a whip-like manner with each heartbeat.

De Wilde, J. (1947) reviewed the controversy of the role of the alary muscles in heartbeat. He found that the alary muscle contraction was of a generally slow period, and the heartbeat relatively fast, so that both muscles were not coordinated and the diastole of the heartbeat was a relaxation assisted by elastic forces. This view is supported by the writer.

The insect dorsal vessel can be sensitive to position. De Wilde, J. (1947) noted a stretch-sensitivity of the heart of some insects. Duwez, Y. (1938) referred to alterations in heartbeat of *Dytiscus marginalis* caused by removing the main ventral suspensions of the dorsal diaphragm with its musculature, the alary muscles. My own observations on *Periplaneta americana* confirm that changes in position brought about by changes in suspensory tissues can modify the resultant heartbeat. However, at least for *P. americana*, we now know the cardiac neurons to be sensitive to the stretch of the myocardium (Miller, T., 1968; Smith, N., 1969) and any change in the suspension of the dorsal vessel brought about by cutting alary muscles or any other part of the dorsal diaphragm will alter the firing pattern of the cardiac neurons, thus modifying the overall heartbeat.

The heartbeat may recover from the initial sectioning of the dorsal diaphragm with time. For example, *P. americana* recovers from unilateral section of one side of the dorsal diaphragm to give a flattened dorsal vessel suspended between cuticular attachments along the dorsal midline and the remaining dorsal diaphragm insertion laterally. The resulting heartbeat is similar in rate and amplitude to the intact heartbeat, except for being planar (personal observations). The advantage of the flattened heartbeat preparation is a greatly facilitated access to one lateral cardiac nerve cord along its entire length which lies exposed against the unoperated side of the dorsum in the dorsal-side-down semi-isolated heart preparation.

Yeager, J. (1938, 1939) reported a lack of response to stimulation of the alary muscles of the cockroach, *Periplaneta americana*. He developed a mechanical–optical method of recording heartbeat and alary muscle movements. The Yeager method was adopted by de Wilde in his early investigation of the alary muscles of adult American cockroach, *Periplaneta americana*, adult water beetles, *Dytiscus marginalis* and *Hydrophilus piceus* and the last stage larvae of the geometrid moth, *Smerinthus ocellata*; the sphingid, *Sphinx ligustri*; one of the Notodontidae, *Phalera bucephala*; *Bombyx mori*, the saturniid silk moth; *Cossus cossus*; the goat moth, a Cossidae; and adults of *Smerinthus ocellata*; *Lymantria dispar*, the gypsy moth; and *B. mori*.

De Wilde's technique (Fig. 39) for recording the alary muscle movements also recorded heartbeat. Contractions of the heart of *Cossus cossus* larvae alternated with contractions of the alary muscle. The alary muscle displaced the recording needle a total of about three-quarters of displacement caused by the heart; although de Wilde wrote that it was a strength measure, the method of recording argues that he measured movement. This amount of movement and its high rate, nevertheless, is remarkable, but is dependent upon the condition of the semi-isolated heart. With a heart in poor condition its amplitude of contraction should be abnormally low.

The adult diving beetle, *Hydrophilus piceus*, on the other hand, showed a heartbeat rate of 80 beats $\min^{-1}$ with slow alary muscle contractions every 9 s (6.7 contractions per minute), so that the heartbeat recording was modulated by the slow contractions of alary muscle. Thus here are two cases of spontaneous alary muscle recording, one slow (*H. piceus*) and the other relatively fast (*C. cossus*).

De Wilde, J. (1947) attempted to shock the alary muscles of the insects he studied, but was mostly successful with *C. cossus* larvae. His attempts to shock the alary muscles of *D. marginalis* and of *P. americana* were unsuccessful. In fact, in these two insects, the diving beetle and American cockroach, the alary muscles failed to show spontaneous activity. De Wilde suggested his experimental

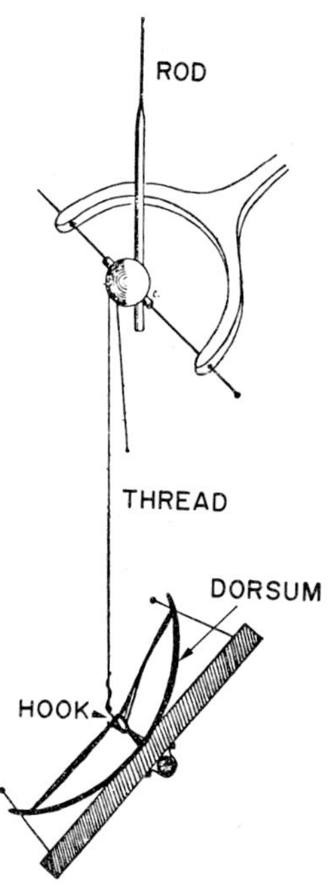

FIG. 39. The mechanical device used by de Wilde for studying alary muscle contraction (after de Wilde, J., 1947), a method adopted from Yeager, J. (1938). The hook was attached to the heart, and the pointed rod interrupted a focused beam of light to throw a shadow on moving film. The dorsum-down heart preparation was tipped so that the thread was pulled down and left by the alary muscle or down and to the right by heart contractions.

conditions caused this reticence to contraction. James Moss (1982, personal observations) can produce spontaneous alary muscle rhythms in *P. americana* simply by using a saline with pH below 7.0. Salines with a pH value above 7.0 were never seen to support rhythmic contractions.

It should be noted that J. de Wilde's (1947) remarkable study on the alary muscle, which is a classic work because of the number of different insects reported, was all conducted using a very simple saline solution. He kept the heart tissues moist with a few drops of "Ringer's", 0.65% NaCl (about 111 mM), and reported activity lasting from 6 to 24 h. If one pays too close attention to more modern papers, one would surely never use this saline, because we now suspect that calcium is necessary for depolarizing electrogenesis. However, the literature has many references to the use of simple sodium chloride salts for heart preparations, and evidently the myocardium is capable of functioning by making the appropriate adjustments to the ionic medium.

The alary muscles of the worker bee, *Melipona quadrifasciata anthidioides* Lep., are innervated by two types of axon — one containing synaptic vesicles and the second containing electron-dense granules (Cruz-Landim, C., 1978). These were termed motor and hormonal innervation, respectively. Electron-dense granules were also reported in the nerve terminals of alary muscle of the cockroach, *P. americana*, (Adams, M. *et al.*, 1973); the beetle, *Cyclocephala pasadenae* (Miller, T., 1975a); and in *Locusta migratoria* (Miller, T. *et al.*, 1979) (Fig. 40). It is difficult to say whether these nerve endings in the beetle, cockroach, locust and bee alary muscles are basically the same neuromuscular junction, using the same neurotransmitter, but in the ultrastructure, at least, they are similar. Cruz-Landim, C. (1978) observed a second type of axon characterized by an absence of granules. Without electrophysiological evidence of dual-innervation, this ending may have been identical to the first granular type. Cruz-Landim may simply have sectioned at a portion of the axon which was devoid of electron-dense granules. Granules are not necessarily plentiful in all parts of an axon (Adams, M. *et al.*, 1973).

Spontaneous contractions of the alary muscles of *P. americana* do not alter the normal heartbeat but change the position of the dorsal vessel. The rhythm of spontaneous contractions of alary muscles in the heart of *P. americana* is always considerably slower than that of the heartbeat. This large difference in rhythm constitutes the most convincing evidence that the alary muscles do not develop a force coincident with diastole of the myocardium, and that the myocardium functions independently of the tension in the dorsal diaphragm.

A similar heartbeat/alary muscle contraction pattern has been recorded in *Locusta migratoria migratorioides* where the alary muscles in the dorsal diaphragm also generates a contractile rhythm that is slow compared to the heartbeat (Fig. 41a). Bayer, R. (1968) reported the dorsal diaphragm in the anterior abdomen of adult *Locusta migratoria migratorioides* to move about 2.7 times every minute or one movement for 25 heartbeats of the dorsal vessel. Walling, L. (1908) observed the same thing.

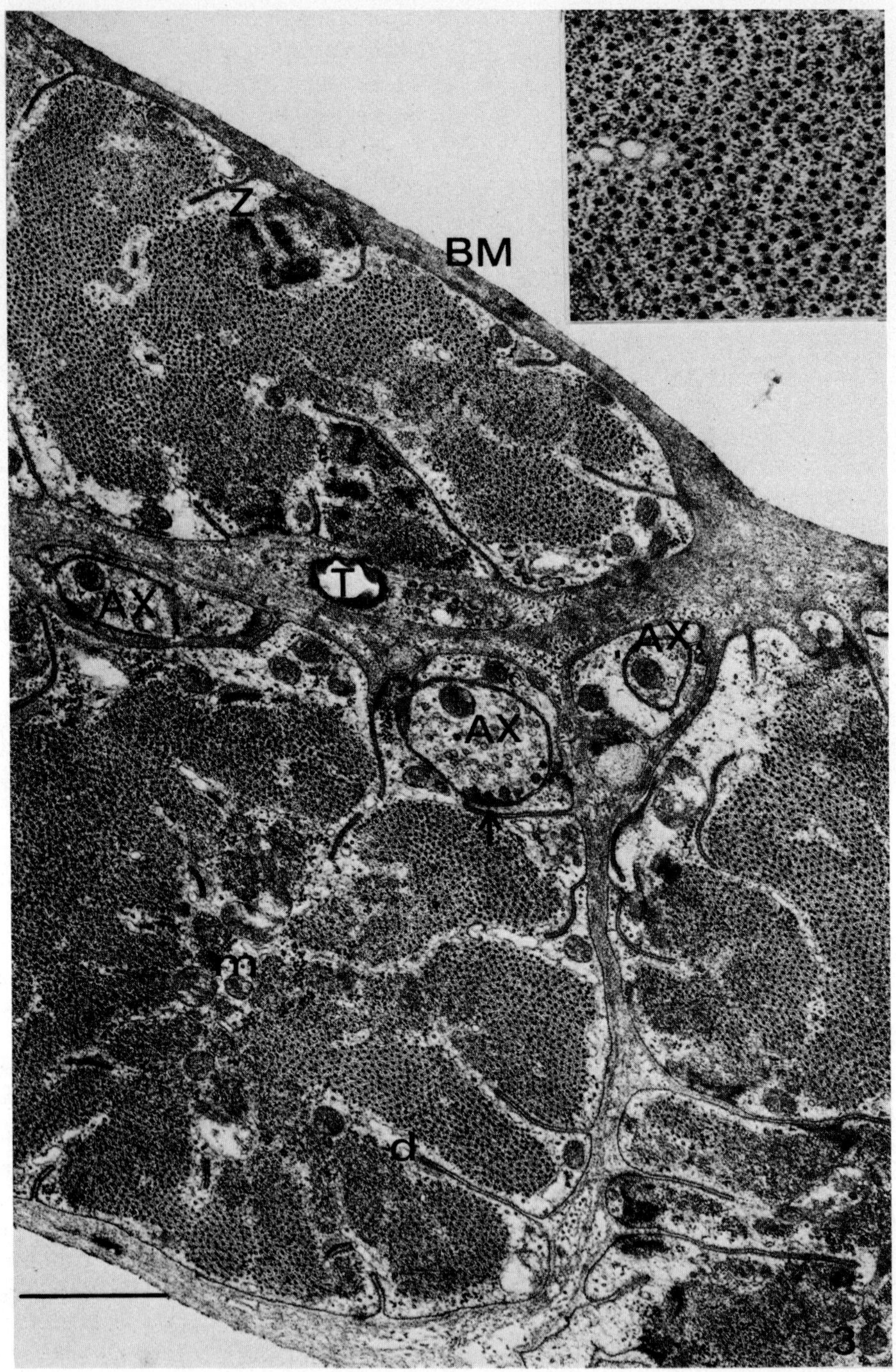

Fig. 40. Survey cross-section of alary muscle fiber. Two muscles are separated by a tracheoblast (T); three axons (AX) ramify near the muscles making synaptic contact (arrow). Z-lines (Z), basal lamina (BM); dyads (d); mitochondria (m). The insert shows arrangement of thick and thin filaments. Note the diaphragm is two cells thick here. This muscle was fixed during contraction from continuous shocks at 13 Hz for 105 s. Calibration mark: 1 μm. (From Miller, T. *et al.* (1979.)

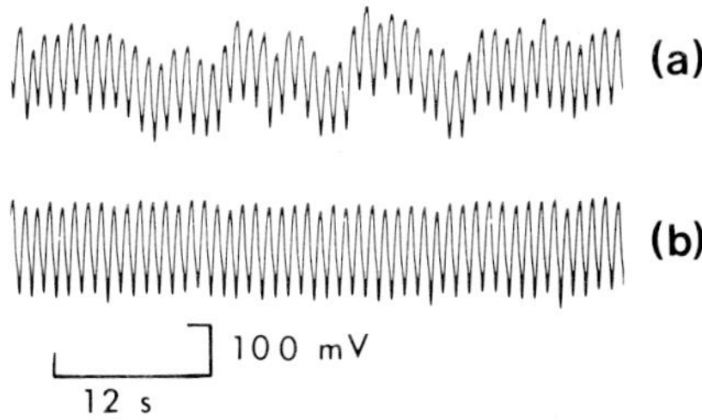

FIG. 41. D.C. impedance recording of the heart movement in the fifth abdominal segment of an adult immediately after dissection before (**a**) and after (**b**) removal of the ventral nerve cord. Note the irregular amplitude in (**a**) indicating impulses from the central nervous system descending the intact fourth dorsal nerve, and marked regularity of heartbeat (**b**) after removal of centrally originated motor nerve activity to the alary muscles. Systolic contraction is up, and diastole is down. (From Miller, T. and S.-Rózsa, K., 1981.)

Severing the segmental nerves and those separating the alary muscles from their connections to the ventral nerve cord of *L. migratoria* caused the slow contractile rhythm to be irrevocably lost (Fig. 41b). The locust alary muscle rhythm, then, appears to be neurally driven and dependent on motor connections with the ventral nerve cord. In fact, the alary muscle contractions are normally neurally evoked (Miller, T. and S.-Rózsa, K., 1981).

The spontaneous alary muscle contractions of the cockroach, *P. americana*, referred to above, were recorded in the semi-isolated heart preparation and thus exclude the ventral nerve cord. It is possible that the cardiac neurons have something to do with the alary muscle contractions in the cockroach, but such connections have never been described. I have not been able to neurally evoke contractions in cockroach alary muscles by stimulating the segmental nerve, but this has not been done in a systematic way. I noted, however, that neurally evoked contractions in locust alary muscles worked far better in young adult males and females (within 2 weeks of imaginal molt) and worked poorly or not at all in older preparations (Miller, T. and S.-Rózsa, K., 1981). The reason for this age-effect is still unexplained.

12 VENTRAL DIAPHRAGM

Accessory pulsatile organs and fibromuscular septae are largely structures associated with post-larval insect stages (Jones, J., 1954). In the holometabolous insects the mosquito is probably typical. No ventral diaphragm is evident in the larval stage, but in the pupal stage a well-developed ventral diaphragm forms and in the adult the ventral diaphragm makes a boundary between, and defines, a perineural sinus and a perivisceral sinus, especially in the abdomen.

Lockshin, R. and Williams, C. (1965) report ventral "alary" muscles that appear in the newly emerged silkmoth imago and attach directly onto the ventral ganglia. Gerould, J. (1938) earlier reported the same for *Bombyx mori*. Ventral diaphragms are absent in immature stages of Lepidoptera.

The authoritative work on the ventral diaphragm in insects has to be that of A. Glenn Richards (1963b) whose 30-page treatise on the subject is a most useful starting point because of the sheer numbers of insects included. Richards examined all orders of Insecta except Protura (104 genera from 81 families) to develop his generalizations.

Characteristic of the ventral diaphragm is its intimate and extensive association with the ventral nerve cord in insects. Although the ventral diaphragm is often present in the form of a continuous layer which divides the abdomen into separate perineural and perivisceral sinuses, it can be reduced to single or a few muscles oriented transversely (as in Trichoptera) or longitudinally (as in some Blattaria). The overall rule appears to be that the ventral diaphragm, or its vestigial remnants, is present only if a ventral nerve cord is also present in the abdomen, and even when present it may not cover the entire cord. The reverse is not true. Some insects with a ventral nerve cord in the abdomen lack a ventral diaphragm or muscles associated with the nerve cord (e.g., some Orthoptera, e.g. *Blaberus giganteus*; Isoptera; Trichoptera). In fact, the diaphragm does not always accompany the entire ventral nerve cord. Thus, while the ventral diaphragm is typical of adult mosquitoes which have ganglia in the abdomen, its absence is characteristic of adult Cyclorrhapa Diptera which lack a ventral nerve cord in the abdomen (Andersen, M. and Finlayson, L., 1978). A ventral diaphragm is present in the embryo of grasshopper or honeybee (Richards, A., 1963b). A ventral diaphragm is found in the thorax of some but not all Acrididae and some larval Hymenoptera, but is usually absent in insect larvae and in the thorax of most insects. In

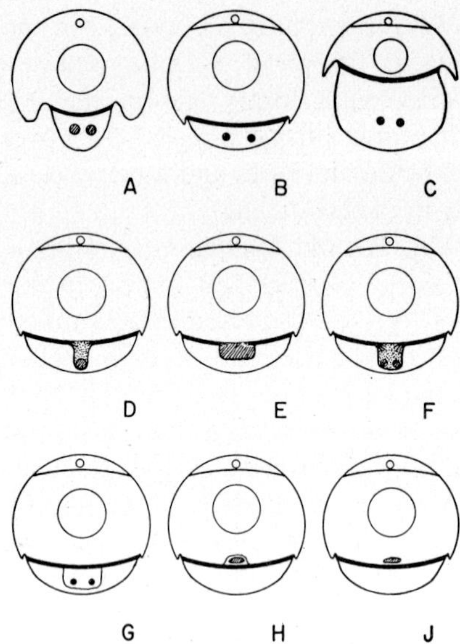

FIG. 42. Diagrammatic cross-sections of various insects with the ventral diaphragm shown by the heavy line. **A:** The usual situation at the junction of thorax and abdomen; **B:** the usual situation of a ventral diaphragm in the abdomen (Acrididae, Odonata, Neuroptera, Mecoptera, Hymenoptera); **C:** the unusual situation with greatly enlarged perineural space, as in Ichneumonid wasps; **D:** the situation at interganglionic connectives in the abdomen of Lepidoptera; **E:** the situation at ganglia in most Lepidoptera (*Philosamia* is exceptional in having the connective tissue mass over both ganglia and connectives); **F:** the situation in moths with double interganglionic connectives in the abdomen (Saturniidae, Bombycidae, Sphingidae); **G:** the situation in *Grylloblatta*, where the nerve cord is buried in a lobe of fat body which hangs from the ventral diaphragm; **H:** the situation in the fly, *Proctacanthella*, in the metathorax and first abdominal segment; **J:** the situation in the fly, *Proctacanthella*, posterior to the first abdominal ganglion. (From Richards, A., 1963b.)

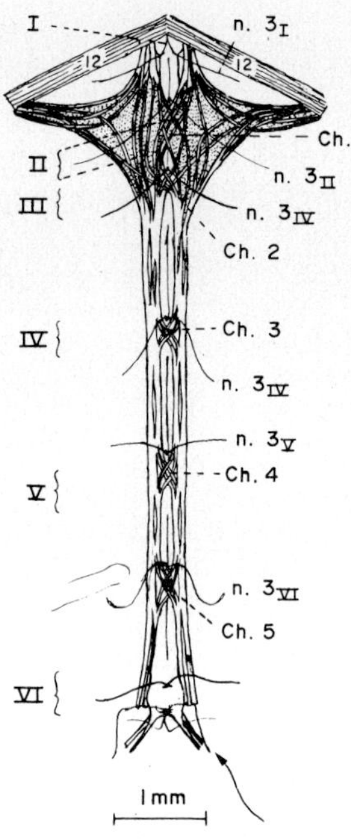

FIG. 43. Dorsal aspect of the hyperneural muscle covering the ventral nerve cord in the abdomen of *Periplaneta americana*. Ch. 1, etc. are numbered chiasmas or crossing-over points of the muscle fibers; G, ventral ganglia; n, median nerves numbered. (After Shankland, D., 1965.)

all but two cases (Fig. 42 H,J) the ventral diaphragm lies dorsal to or above the ventral nerve cord (Fig. 42 A–C), and in all cases the diaphragm is separate from the nerve cord except in Lepidoptera (Fig. 42 D–F).

Richards, A. (1963b) eliminated Blattaria (Dictyoptera) from his lists of insects containing a ventral diaphragm because the diaphragm was considered present only if it defined a perineural sinus. In *Periplaneta americana* (Fig. 43) and *Leucophaea maderae* (Fig. 44) (Engelmann, F., 1963), although a peculiar longitudinal muscle is present, termed the hyperneural muscle in *P. americana* (Shankland, D., 1965), the muscular lattice does form a more-or-less solid diaphragm structure near its connection with the thorax (Shankland, D., 1965) (Fig. 43), and thus should be considered a true ventral diaphragm, even if it is present only in the first abdominal segment.

On the basis of cursory examination of vitally stained fresh preparations, the following species have a hyperneural muscle similar to that in *Periplaneta americana*: *Periplaneta australasiae* (Fabricius), *Periplaneta brunnea* Burmeister and *Periplaneta fuliginosa* (Serville) (Arthur Appel, 1981, personal observations).

12.1 Function of the ventral diaphragm

Where the ventral diaphragm actively contracts in a peristaltic manner, (Jones, J., 1954 and Yaguzhinskaya, L., 1954, in *Anopheles*; Freudenstein, K., 1928 and Snodgrass, R., 1935, in Hymenoptera; Gerould, J., 1938, in *Bombyx mori*; Richards, A., 1963b in *Galleria mellonella*), circulation is evidently aided. This certainly appears to be true in the cases where circulation appears to be involved in thermoregulation in some Diptera, Odonata,

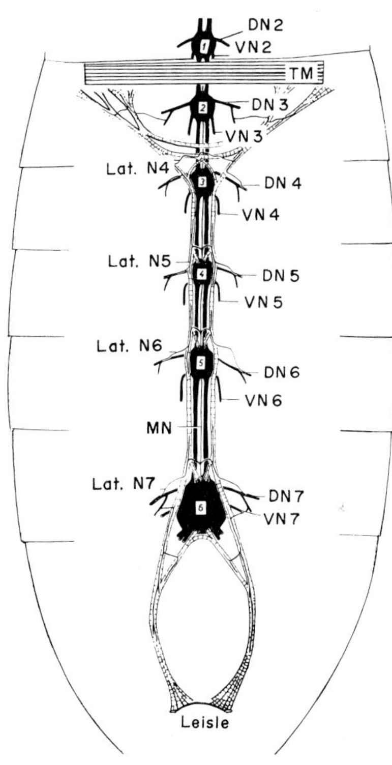

FIG. 44. Dorsal aspect of the ventral abdomen of cockroach, *Leucophaea maderae*. The ventral abdominal ganglia are numbered 1–6 and the vestigial hyperneural muscle is shown attached to the ganglia. Innervation of the hyperneural muscle from the median nerve (lat. N 5, etc.) in each segment is evident. (From Engelmann, F., 1963.)

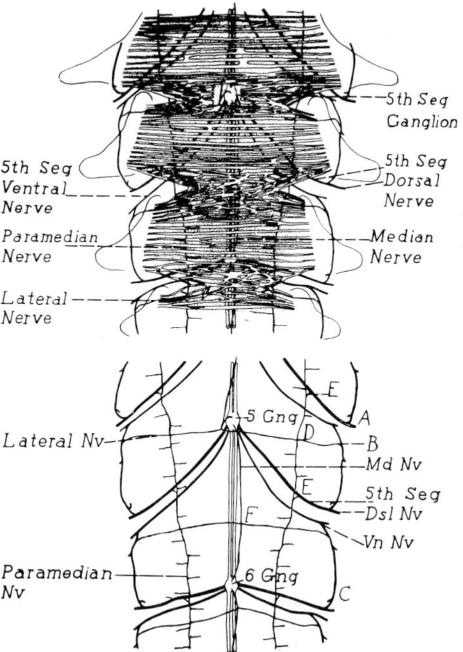

FIG. 45. The structural arrangement (upper figure) and innervation pattern (lower figure) of the Acridoidea, *Dissosteira carolina*, in the fourth, fifth and sixth abdominal segments, viewed dorsally. 5 Gng, fifth abdominal ganglion, etc.; Nv, ventral nerve; Dsl Hv, dorsal nerve; Md Nv, median nerve. The diaphragm immediately over the fifth abdominal ganglion (in the upper figure) deliberately left clear to show detail underneath, and in the lower figure the diaphragm is missing to show innervation. The letters A, B, C, D and E indicate connections between nerve bundles, and F indicates the branches of the median nerve. (Taken from Schmitt, J., 1954.)

Hymenoptera and Lepidoptera (Heinrich, B., 1974b, 1976; Heinrich, B. and Casey, T., 1978; Heinrich, B. and Pantle, C., 1975; Wasserthal, L., 1975c).

Another function of the diaphragm seems to be the active movement of the ventral nerve cord itself. In some Lepidoptera the diaphragm and nerve cord form a distinct structure which has been termed the "chord of Leydig" (Gerould, J., 1938), and during contractions the nerve cord is actively whipped back and forth. If the ventral diaphragm is so directly concerned with the presence of the ventral nerve cord, one might postulate that the nerve cord is the sole reason for a diaphragm being present; the diaphragm, therefore, serves the nerve cord either to slosh it around actively or to direct hemolymph rapidly over the nervous tissue for a constant flushing. In a few cases (Dermaptera, Richards, A., 1963; the lubber grasshopper, *Romalea microptera*, Jones, 1982, unpublished), the ventral nerve cord is said to be completely covered by fat body which impedes the access of the hemolymph (Richards, A., 1963b).

The innervation and electrical properties of the muscles of the ventral diaphragm have been investigated on only a few occasions. The hyperneural muscles of *Periplaneta americana* were shown to resemble electrically inexcitable muscle. The muscles showed non-facilitation, summating postsynaptic potential responses to motor nerve stimulation (Miller, T. and Adams, M., 1974). The voltage–current curve from hyperneural muscle is similar to that of high-resistance fibers in the superior rectal longitudinal muscles of *P. americana* (Belton, P. and Brown, B., 1969).

The ventral diaphragm in the Acridoidea has

been examined. The structure and innervation of the ventral diaphragm in the Carolina locust, *Dissosteira carolina*, was described (Schmitt, J., 1954) and is typical (Fig. 45). Paired dorsal and ventral nerve trunks leave each ventral ganglion (Fig. 45, Dsl Nv and Vn Nv). The dorsal nerve supplies the paramedian nerves and is attached by a large nerve trunk to the ventral nerve in the next anterior segment. The dorsal nerve then continues as a large obvious trunk, which in *Locusta migratoria* floats freely in the perineural sinus, as the main segmental nerve supplying the dorsal aspect of the segment. The dorsal nerve also ultimately supplies the lateral cardiac nerve cord and the alary muscle, although the latter has been overlooked (cf. Miller, T. and S.-Rózsa, K., 1981).

Schmitt, J. (1954) traced homologies in segmental innervation in various Orthoptera and there appears to be considerable consistency in these patterns, perhaps extending beyond the Orthoptera (M. E. Anderson, 1982, personal communication).

Paired branches leave the median nerve (at F, Fig. 45) in each segment and course laterally as the lateral nerve (Fig. 45; Lateral Nv.). Connections exist between the lateral nerve and the paramedian nerve where both cross (Fig. 45 d) and also between the lateral nerve and the "link" nerve (Fig. 45 b) connecting the dorsal nerve with the next anterior ventral nerve (Fig. 45 a,c). The link nerve occurs in *Carausius morosus* (Fifield, S. and Finlayson, L., 1978) and a similar connection occurs in blowfly larvae (Finlayson, L. and Osborne, M., 1968).

The ventral diaphragm of the grasshopper, *Romalea microptera*, is well innervated by the median nerve in the abdomen (Jones, J., 1981). The motor connections to the ventral diaphragm muscles in the locust, *Locusta migratoria*, leave the ventral ganglia via the dorsal nerves, branch to form the paramedian nerve in such a way that the muscles in the ventral diaphragm are innervated from motor neurons in the same and the next anterior segment (Fig. 46). This innervation pattern in *Locusta migratoria* was confirmed both by electrophysiological measurements and by axonal iontophoresis using cobalt or nickel ions (Peters, M., 1977) and agrees with the pattern found by Guthrie, D. (1962) in *Schistocerca gregaria*. The nerve endings on the ventral diaphragm are localized laterally as shown by Figs. 45 and 46, and although this is

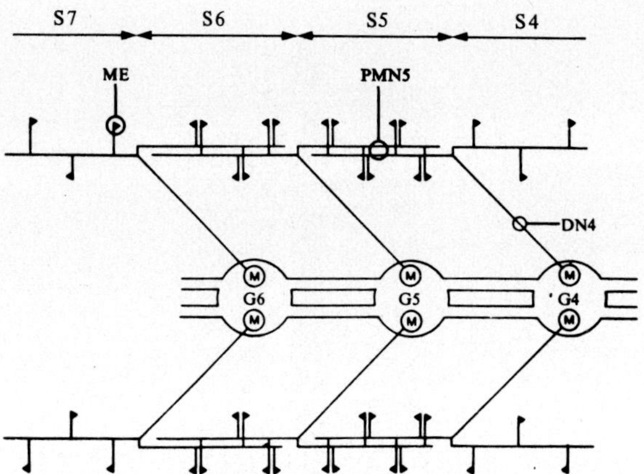

FIG. 46. Diagram of motor innervation of the ventral diaphragm of *Locusta migratoria*. Paired motor neuron cell bodies (M) are present in the ganglia (G) and send axons out to the dorsal nerve (DN4) in the appropriate segment (S4). The axons branch anteriorly and posteriorly to form the paramedian nerve (PMN) and terminate (ME) on the muscle of the same and the next posterior abdominal segment. (From Peters, M., 1977.)

unusual compared to skeletal muscles, it is very similar to nerve terminals on alary muscles (muscles of the dorsal diaphragm) in *Periplaneta americana* and *Locusta migratoria* which are also located on the lateral margins of the dorsal diaphragms.

When the dorsal and ventral nerves were cut in *Schistocerca gregaria*, a retrograde peristaltic rhythm of 8 to 12 waves per minute was doubled to 20–25 min (Guthrie, D., 1962). Mechanical disturbances of the ventral diaphragm caused an immediate and sustained contractile response with or without segmental nerves intact, and although a sensory cell was found on the ventral diaphragm near each paramedian nerve, its role in response to mechanical disturbance is unclear (Guthrie, D., 1962). The ventral diaphragm of *Locusta migratoria* shows an electrically excited component (Peters, M., 1977), and this is compatible with the claim that the *Schistocera gregaria* diaphragm can contract myogenically (Guthrie, D., 1962). Excitability properties and histological and electrical connections between muscle fibers described by both Dierichs, R. (1972) and Peters, M. (1977) tend to suggest that the resultant rhythm and responses of the ventral diaphragm are rather complex and cannot be explained simply as a play between demonstrated excitatory and hypothesized inhibitory nerve fibers, mentioned by Guthrie, D. (1962) as one possibility.

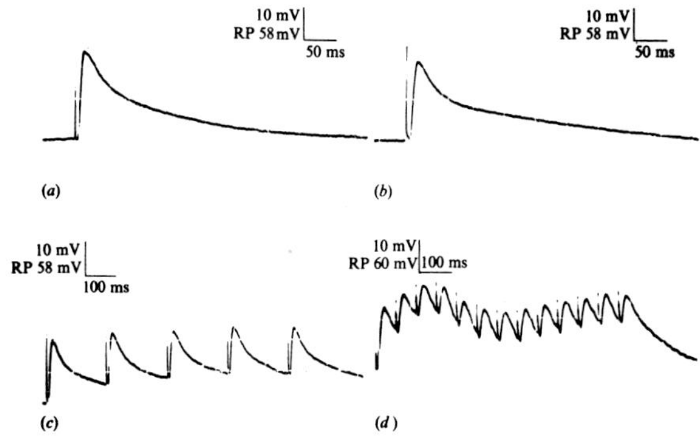

FIG. 47. Excitatory postsynaptic potentials (EPSPs) recorded from a ventral diaphragm muscle in segment 5 upon stimulation of either DN5 **(a)** or DN4 **(b)** (refer to the diagram in Fig. 46, Peters, M., 1977, for location of the nerves). Stimulation of DN4 at 4 Hz shows no facilitation **(c)** and stimulation at 15 Hz shows summation **(d)**. (From Peters, M., 1977.)

Peters, M. (1977) found excitatory innervation in the paramedian nerves (Fig. 47 a,b) with clear summation produced by multiple shocks (Fig. 47 c,d). Under semi-isolated conditions the electrical activity in a given paramedian nerve is about a 1s burst of a dozen or more spikes repeated at the rate of contraction of the diaphragm. The individual nervous impulses in pairs of paramedian nerves are coordinated but not superimposable, indicating that they are separate nerves. The activities of nerve bursts in adjacent segments show a delay in the posterior segment, suggesting peristaltic coordination between burst drivers in the ventral ganglion. There also exists a correlation between activity in respiratory muscles and contractions of the ventral diaphragm, suggesting further central coordination.

It is clear that future work on the circulatory system must be pursued along the lines suggested by Figs 46 and 47. Once motor neurons are described, as those by Peters, M. (1977), it is necessary to determine the central nervous connections between sensory input (sensory nerve described by Guthrie, D., 1962) and interconnections between motor neurons to the ventral diaphragm and other motor or regulatory neurons such as those to pulsatile organs, dorsal vessel and ventilatory muscles. Only when this picture begins to unfold will we have an idea how the various pumps are controlled. L. Wasserthal's (1980) work, at least, suggests that a considerable amount of coordination exists between the many pumps of the circulatory system.

13 ACCESSORY PULSATILE ORGANS (APOs)

As a rule, hemolymph is circulated to appendages unidirectionally by various tubes, septa, valves and pumps. The pumps are collectively termed accessory pulsatile organs (APOs). Four such organs were thought to occur in each wing of *Musca domestica* in the efferent veins. The anterior edge of the wing is the efferent pathway with the posterior part of the wing containing return pathways. Zeichner, B. (1980) did not find any structures within the four "wing heart" regions which could cause contractions (no muscle strand, no connective tissue, no ligaments; nothing at all).

He concluded from experiments that the pulsations within the wing pulsatile regions "are a passive response to a pressure decrease within the wing veins by the scutellar [pulsatile] organ" (J. C. Jones, 1982, personal communication).

Those parts of the wings deprived of hemolymph become brittle and break away. Circulation in the wings of the cockroach was studied by Clare, S. and Tauber, O. (1940) by cutting veins in the wings and observing the changes in circulation.

Wasserthal, L. (1982) has now proposed a major new idea. A tidal flow of hemolymph is said to occur in the wing veins of the adult moth, *Attacus atlas*, in which the efflux of hemolymph from wing veins corresponds with the influx of air into wing veins. Tracheoles expand in diameter to take up the volume of the veins occupied by the effluxing hemolymph, and the air is pulsed from trachea in the body. This proposal is a radical departure from accepted view of circulation in the wings.

During development in the pupae of *Bombyx mori* the metathoracic part of the aorta sinks ventrally along with the tergal infolding in front of and behind the metathorax. The infoldings secrete the phragmata of the adult, which provide insertion plates for the large dorsolongitudinal flight muscles. Metathoracic pulsatile organs appear to develop independently from the aorta and maintain a more rapid pulsatile rate than the dorsal vessel (Gerould, J., 1938).

13.1 Leg circulation

The circulation of hemolymph in the leg of the heteropteran, *Triatoma phyllosoma pallidipennis* (Stal.) is assisted by a septum and a pulsatile organ as shown diagrammatically in Fig. 48 (Kaufman, W. and Davey, K., 1971). The septum is described as dividing the femur and tibia into dorsal and ventral "sinuses". The hemolymph is said to enter the femur via the ventral sinus, and is pushed peripherally by a pulsatile muscle located just below the femur–tibia join in the tibia. A valvular membrane (not shown in Fig. 48) is located in the dorsal sinus of the tibia which is thought to direct the return flow of hemolymph into the femur.

The pulsatile muscle in the tibia is not innervated by branches from the main crural nerve which lies in the ventral sinus and divides in the tibia into the posterior tibia nerve and median tibial nerve, both of which continue to the tarsus and both of which receive presumably sensory branches from the epidermis.

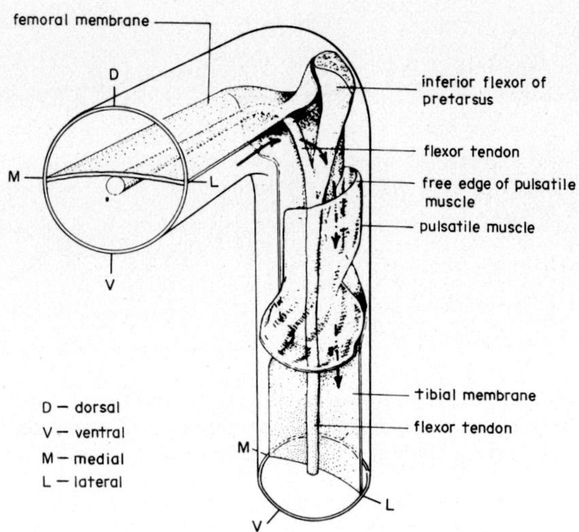

FIG. 48. Schematic drawing of the femur–tibia area of the right leg of *Triatoma phyllosoma pallidipennis* (Stal.) showing the septum and pulsatile organ which direct (arrows show direction of flow) hemolymph along the ventral (V) part of the leg and back to the body on the dorsal side of the septum. (From Kaufman, W. and Davey, K., 1971.)

Three accumulations of neurons are present in the femur and tibia whose functions are unknown. One of these groups is composed of two or more neurons and is located in the ventral sinus of the tibia just at the pulsatile organ. These neurons are termed the pulsatile ganglion based on their position only because their function is not known; they may possibly supply axons to the pulsatile organ, although no nerve endings were seen (Kaufman, W. and Davey, K., 1971).

Amputation of the leg, or severing the crural nerve, causes the pulsatile organ to cease beating. Resumption of pulsatile activity thereafter does occur intermittently but at a lower pulse rate. Pulsations also appear to cease during times when the main skeletal muscles of the leg are used. Delivering gross shocks to the thoracic ganglion mass either induced a quiescent organ to pulse, or increased or decreased the pulse rate of an active organ, depending on the frequency of stimulation. The rate of pulsations of some APOs were increased by applications of dopamine or serotonin at less than 10 nM concentrations, but responses were inconsistent.

From work on *Triatoma* and earlier work on aquatic Hemiptera, the tibial pulsatile organ and septum appear to be common. Moreover, the two sinuses in the leg are thought to direct the flow of hemolymph by a muscular compression and aspiration of the adjacent sinuses near the proximal end of the tibia.

13.2 Antennal circulation

Antennal pulsatile organs (APOs) have been described in insects since the last century. The organs responsible for circulation of the hemolymph in the antennal appendages bear a remarkable similarity of structure in the various orders of insects from cockroach (Pawlowa, M., 1895) and Hemiptera (Pinet, J., 1964) to Diptera (Clements, A., 1956; Dudel, H., 1977, 1978a,b) and Coleoptera (Pass, G., 1980). In all cases described, an ampulla gathers hemolymph from the blood space at the anterior end of the aorta. A muscular pump either surrounds the ampulla (Pinet, J., 1964) as in *Rhodnius prolixus*, or muscles are either attached to the ampulla as is usual in Diptera (Dudel, H., 1977, 1978a,b) or lie flat over the ampulla sac as in cockchafer beetle, *Melolontha melolontha*, (Pass, G., 1980).

In most cases the innervation of the ampulla muscle, and therefore the control of the pulsations of the APO, have not been described; neither has the pharmacology of these pumps been investigated to anywhere near the extent done by Kaufman, W. and Davey, K. (1971) on the leg APO of a *Triatoma* bug. However, Pass, G. (1980) found a peripheral ganglion on the antennal nerve of the cockchafer beetle that appeared to be associated with the antennal APO, and Beattie, T. (1976) found an extensive ramification of neurosecretory nerve endings on the ampulla of the cockroach *Periplaneta americana* which led him to postulate this as the site of a neurohemal organ.

The following is quoted freely from personal communications with T. M. Beattie dating from 1972 (published partly in Beattie, T., 1976) and is intended to be so recognized.

> The structure of the accessory heart associated with the antennae of the cockroach, *Periplaneta americana*, indicates that it is more than just a pump. Pawlowa's original description of the anatomy appears to be accurate except for the innervation of the system.
>
> *Anterior aorta*: the aorta is open ventrally and forms an inverted "V". It has two types of neurosecretory axon (presumably derived from the corpora cardiaca). The muscle cells of the aorta are attached to those of the ampulla muscle.
>
> The neurosecretory axons reach the ampulla muscle and *perhaps* continue to the ampulla where they terminate in large Herring bodies (these terminal swellings are up to 10 μm across).
>
> *Ampulla muscle*: the muscle, which dilates the ampulla, contains two types of muscle cell: a long sarcomere type (A band = *ca.* 5 μm) and a short sarcomere type (A band = *ca.* 3 μm). Innervation is by granule-containing axons. The origin of these nerves is not known for certain. They may be derived either from the retrocerebral complex via the anterior aorta or from the nervus connectivus which passes through the ampulla muscle on its course from the frontal ganglion to the protocerebrum, according to Willey, R. (1961).
>
> *Ampulla wall*: the ampulla wall has well developed connective tissue layers, one of which is much thicker than the other two. This connective tissue is probably responsible for contraction of the ampulla after dilation induced by contraction of the ampulla muscle. There is no muscle in the ampulla wall.
>
> *Convoluted portion of antennal vessel*: the walls of this portion of the vessel are thickened and are composed of a double layer of cells. The inner layer of cells is highly polarized in its structure, having a basal nucleus and apical projections. Much of the cytoplasm between the nucleus and the apical projections is filled with mitochondria. Also, the lateral cell boundaries have numerous junctional specializations (desmosomes, septate desmosomes, tight or gap junctions). The outer layer of cells is less specialized but does appear to create an extracellular compartment between it and the inner layer of cells. This portion of the vessel has many morphological features in common with other epithelia involved in ion and water transport. The distal portion of the antennal vessel is not specialized and its wall is only one cell thick. Also, its diameter is much less than the convoluted portion. The accessory heart system may be involved in regulating the ionic environment of the antennae interior. Support for this is given by a piece of work done by Pichon, Y. (1970), who showed that the ionic concentraion of Na^+ and K^+ in the dorsal vessel (heart) is significantly different from that in the antennae. This speculative suggestion of Beattie's needs further examination

Antennal pulsatile organs are located at the base of each antennae of the cockchafer beetle, *Melolontha melolontha* (Pass, G., 1980). Each organ consists of an ampullae or sac without contractile properties and a long blood vessel, also non-contractile, that supplies the antenna (Fig. 49). Motive force is supplied by a compressor muscle (Fig. 49b) that operates against a band of elastic tissue. Hemolymph gains entry by means of a valvular ostium (Fig. 49 arrow). Pass, G. (1980) also described a new nerve mass (Fig. 49 a; G) that supplies the APO.

Clements, A. (1956) found the antennal APO to be present in species representatives from most families of Nematocera Diptera except three species, one from each of the families: Simuliidae, Bibionidae (*Dilophus febrilis*), and Thaumaleidae which were examined and lacked the antennal APO. However, Dudel, H. (1977) found a large median ampulla in *Dilophus febrilis* and figures this in his compendium on the antennae of Diptera (Fig. 50a).

Table 2: Antennal pulsatile organs (APOs) in 17 species of Diptera. The APO is either a single ampulla (S), single but with lobes (L) or paired (P) structures. The major groups of Diptera are identified as Orthorrhapha (N, Nematocera or B, Brachycera) or Cyclorrhapha (C). The subdivisions of C are C_1, Ashiza; C_2, Schizophora; C_3, Pupipara. Schizophora are identified as Acalyptratae (C_{2a}) or Calyptratae (C_{2b}). (From Dudel, H. 1977, 1978a.)

Ampullae	Group	Family	Species
S	N	Bibionidae	*Dilophus febrilis*
S	B	Stratiomyidae	*Nemotelus pantherinus*
P	B	Rhagionidae	*Rhago strigosa*
L	B	Tabanidae	*Tabanus* sp.
P	B	Bombyllidae	*Bombylius venosus*
P	B	Bombyllidae	*Hemipenthus morio*
S	B	Empididae	*Empis tesselata*
S	B	Dolichopodidae	*Dolichopus analus*
S	C_1	Pipunculidae	*Pipunculus* sp.
S	C_1	Syrphidae	*Eristalis pertinax*
S	C_{2a}	Micropezidae	*Micropeza corrigiolata*
P	C_{2a}	Diopsidae	*Diasemopsis fasciata*
S	C_{2a}	Sciomyzidae	*Limnia fumigata*
L	C_{2a}	Carnidae	*Carnus hemopterus*
P	C_3	Hippoboscidae	*Melophagus ovinus*
S	C_{2b}	Tachinidae	*Echinomgia fera*
S	C_{2b}	Calliphoridae	*Calliphora erythrocephala* (= *C. vicina*)

Thus, one is inclined to consider seriously the last sentence in Clements' paper which mentions the likelihood of finding accessory pulsatile organs supplying the antennae of most insects.

The extensive survey done by Dudel, H. (1977, 1978a,b) on the internal and external morphology of the antenna of Diptera shows the extraordinarily uniform structure of the antennal pulsatile organ. Although the organ either sends a separate blood vessel to each antenna from a single ampulla (12 of 17 species examined), or each antenna is supplied by a separate ampulla (5 of 17 species examined) (see Table 2). Note from Table 2 that the pairing of ampullae arises in diverse groups of Diptera: the sheep ked, *Melophagus ovinus* (Fig. 50b); Diopsidae, *Diasemopsis fasciata* (Fig. 50c); three Brachycera, *Rhagio strigosa* (Fig. 50d); and two beeflies, *Bombylius venosus* (Fig. 50) and *Hemipenthus morio*.

The fact that the ampullar structures in Tabanidae and Carnidae (Fig. 50f) are somewhat bilobed suggests an intermediate condition. The distribution and non-uniform occurrence of paired ampullae also suggests that the appearance of the ampullae depends simply on the physical separation of the bases of the antennae themselves, rather than on the ampullae. The ampullae can, however, occupy a middle position and supply the separate antenna with long blood vessels.

14 SUMMARY

Just as Manfred Gersch described it over 20 years ago, neurohormonal control of heartbeat appears finally to have been conclusively demonstrated in *Manduca sexta* during wing expansion in the newly emerged adult by Tublitz and Truman. Similar events during eclosion by adult nymphalid butterflies have also been demonstrated by Wasserthal.

Wasserthal has made a great contribution in describing long-term insect heartbeat activity by the non-invasive method of contact thermography. Another method, impedance conversion, first introduced by Heinrich, has revolutionized the recording of movements of insect tissues, both *in vivo* and *in vitro*. Heinrich's work has also implied a very strong nervous regulation of the heartbeat principally to control the dorsal vessel in the abdomen.

Both Wasserthal and Heinrich have revealed new aspects to our knowledge of the circulatory system; Wasserthal in his descriptions of tidal hemolymph movement and normal circulatory activity and Heinrich concerning the role of circulation in thermoregulation.

Sláma has measured the internal pressure of insects during various life cycles. The control of hemolymph pressure, described in section 9, and the implication of a vital role played by the

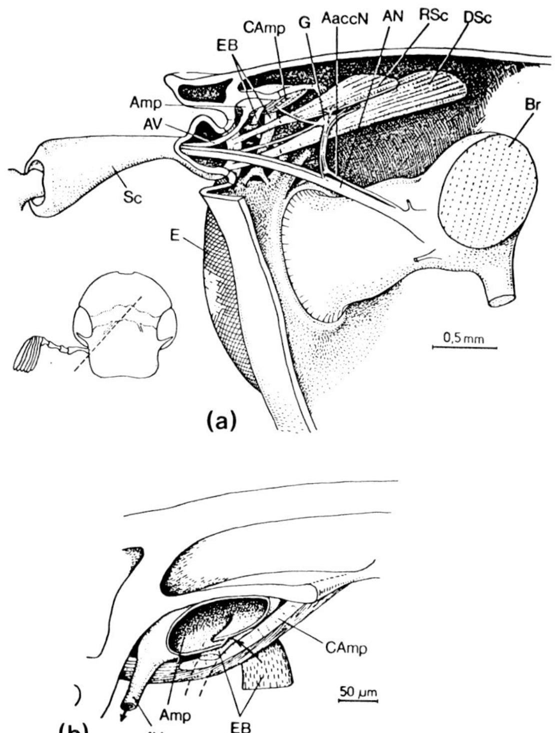

FIG. 49. Antennal pulsatile organ of cockchafter beetle, *Melolontha melolontha*. **a:** Three-quarter view of the head with the dorsal side down (inset shows the line of section); **b:** close-up of the ampulla of the APO. E, ege; CAmp, circular muscle of the ampulla; EB, elastic band of tissue which acts as a restoring force to expand the ampulla; AV, antennal blood vessel; Br, brain; Ac, scapes; G, ganglion supplying the ampulla; RSc, DSc, muscle of the scape; AN, antennal nerve; AaccN, accessory nerve connecting G and the brain. (From Pass, G., 1980.)

intersegmental muscles of the ventilatory system in the abdomen suggests a connection between circulation and ventilation (respiration). The contributions of Wasserthal, Heinrich and Sláma taken together give us fresh glimpses of circulation in insects as it relates to development and bodily functions in general. These landmark works suggest a rather impressive coordination between all of the fibromuscular septae, accessory pumps, diaphragms and ventilatory musculature.

Very fittingly, Peters has shown us the first glimpse of central neurons responsible for controlling circulatory pumps in his elegant work on the innervation and control of the ventral diaphragm of *Locusta migratoria*. This latter, of course, is one of the most challenging future research directions, i.e., the eventual complete identification of all motor units controlling the muscular pumps and diaphragms in insects.

One might predict that new principles of physiology await discovery in the coming work on innervation and control of pumps. There is very little doubt that neurosecretory innervation, so prevalent here, uses entirely new types of neurotransmitter molecules; perhaps peptides, perhaps amines, perhaps even wholly new means of controlling pumps. Normann has already suggested that an inhibition and excitation occurs in control of the blowfly heartbeat. This possible inhibitory innervation is certainly a first in visceral muscles in insects, and cries out for closer examination. In fact, not one neurotransmitter has been proposed for any of the circulatory pumps in the adult insect with the possible exception of proctolin.

Technical advances in recording small forces or small movements from muscular pumps, or in recording blood pressure, have removed all hurdles from these studies. It is no longer possible to view even minute circulatory organs such as the APOs as inaccessible.

Beattie has suggested that the antennal pulsatile organ in cockroach releases neurohormones that are carried to the antenna lumen and perhaps bring about changes in the sensory neurons there. This original idea obviously requires further examination.

The intriguing idea of Raabe that the open circulatory system in insects gives rise naturally to a proliferation of a neurosecretory system based on neurohormonal release has been gradually taking on greater significance as the peripheral neurosecretory system is described by Beattie (mentioned above), Johnson, Finlayson, Raabe and her colleagues, the colleagues of Manfred Gersch in Jena, and many others.

To put in perspective what discoveries in this field might mean, one merely can consider what the discovery of proctolin by Brian Brown did. Within 5 years proctolin has been found to have potent myotropic actions in a number of invertebrates, and through the imaginative efforts of O'Shea and his colleagues, and Penzlin and his colleagues, immunohistochemistry has begun to pinpoint suspected proctolinergic neurons in the insect central nervous system.

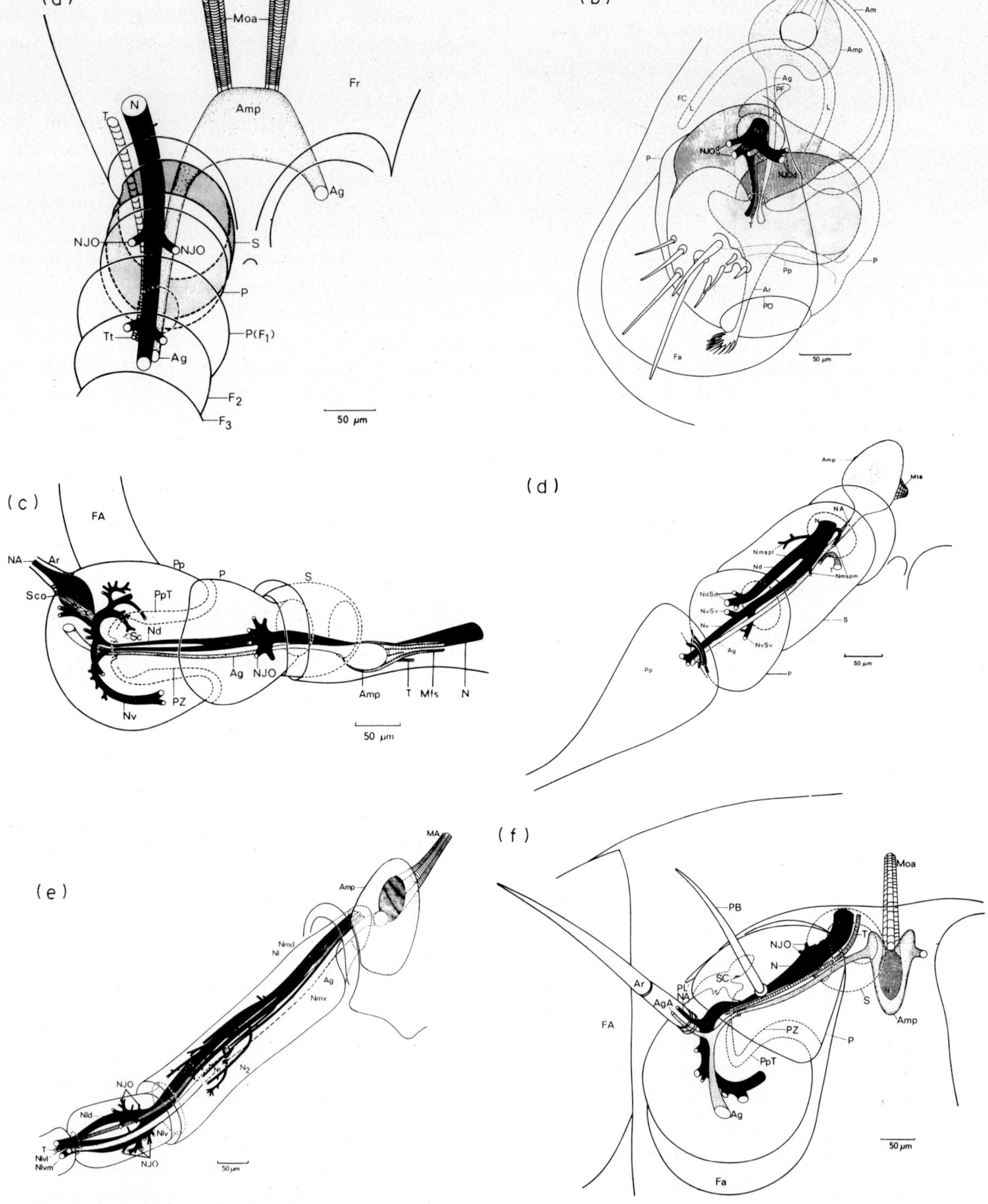

FIG. 50. Accessory pulsatile organ (APO) of the antenna of *Dilophus febrilis*; **a:** family Bibeonidae of the Diptera; **b:** APO of the antenna of the sheep ked, *Melophagus ovinus* (from Dudel, H., 1978b); **c:** APO of the antenna of *Diasemopsis fasciata*, (Diopsidae: Diptera) (from Dudel, H., 1978a); **d:** APO of the antenna of *Rhagio strigosa* (from Dudel, H., 1977); **e:** APO of the antennae of *Bombylius venosus* (from Dudel, H., 1977); **f:** APO of the antennae of *Carnus hemopterus* (Carnidae: Diptera), (from Dudel, H., 1978a). Ag, blood vessel; Ap, apodeme; Amp, ampulla; Ar, arista; F refers to flagellum; Fr, frons; M refers to muscles, and Moa or Maa would be muscles of, or associated with, the ampullae; N refers to nerve or nerve branches, for example, NJO is the branch supplying the Johnston's organ; P, pedicellus; T refers to trachea; S, scape; SC, sensillum campaniform. (From Dudel, H., 1977.)

Anyone even remotely familiar with the proctolin work cannot help but be impressed by the impact this discovery had and continues to have. The availability of neurohormone D, once it is identified, is sure to cause an equally impressive impact, as indeed would the identification of any number of the neurosecretory materials associated with the circulatory system. A great deal remains to be done.

ACKNOWLEDGEMENTS

The author recognizes collaborations and assistance from Janos Salanki, Katalin Salanki-Rózsa, Klaus Richter, Wieland Hertel, Professor Penzlin, Nathan Tublitz, Lutz Wasserthal, Bernd Heinrich, Alvin Starratt, Professor Mordue, Randall House, Wellcome Research Laboratories in England, Fred Lang, J.-P. Roussel, Tom Normann, Moray Anderson, Professor Finlayson, Stan Kater, Malte Peters, Frances McCann, Joe Sanger, Mike Adams, Bill Thomson, Istvan Benedeczky, Terry Beattie, and a special note of gratitude to Professor Jack Jones for an extremely helpful critical reading.

REFERENCES

ABBOTT, B. C., LANG, F. and PARNAS, T. (1969). Physiological properties of the heart and cardiac ganglion of *Limulus polyphemus*. *Comp. Biochem. Physiol.* *28*, 149–158.

ADAMS, M. E., MILLER, T. and THOMSON, W. W. (1973). Fine structure of the alary muscles of the American cockroach. *J. Insect Physiol.* *19*, 2199–2207.

ALEXANDROWICZ, J. S. (1926). The innervation of the heart of the cockroach (*Periplaneta orientalis*). *J. Comp. Neurol.* *41*, 291–310.

ANDERSON, M. and COOKE, I. (1971). Neural activation of the heart of the lobster *Homarus americanus*. *J. Exp. Biol.* *55*, 449–468.

ANDERSON, M. E. and FINLAYSON, L. H. (1978). Topography and electrical activity of peripheral neurons in the abdomen of the tsetse fly (*Glossina*) in relation to abdominal distension. *Physiol. Ent.* *3*, 157–167.

AWASTHI, V. B. (1975). Neurosecretory system of the earwig *Labedura riparia*, and the role of the aorta as a neurochemical organ. *J. Insect Physiol.* *21*, 1713–1719.

AWASTHI, V. B. (1976). Contribution to the neurohaemal organs in pterygote insects. *Z. Mikrosk.-Anat. Forsch., Leipzig* *90*, 48–78.

BACCETTI, B. and BIGLIARDI, E. (1969). Studies on the fine structure of the dorsal vessel of arthropods. I. The heart of an Orthoperan. *Z. Zellforsch.* *99*, 13–24.

BARSA, M. C. (1954). The behavior of isolated hearts of the grasshopper, *Chortophaga viridifasciata*, and the moth, *Samia walkeri*, in solutions with different concentrations of sodium, potassium, calcium, and magnesium. *J. Gen. Physiol.* *38*, 79–92.

BAUMANN, E. and GERSCH, M. (1982). Purification and identification of neurohormone D, a heart accelerating peptide from the corpora cardiaca of the cockroach *Periplaneta americana*. *Insect Biochem.* *12*, 7–14.

BAYER, R. (1968). Untersuchungen am Kreislaufsystem der Wanderkeuschrecke (*Locusta migratoria migratorioides* R. et F. Orthopteroidea) mit besonderer Berücksichtigung des Blutdruckes. *Z. Vergl. Physiol.* *58*, 76–136.

BEATTIE, T. M. (1976). Autolysis in axon terminals of a new neurohaemal organ in the cockroach *Periplaneta americana*. *Tissue Cell* *8*, 305–310.

BELTON, P. and BROWN, B. E. (1969). The electrical activity of cockroach visceral muscle fibers. *Comp. Biochem. Physiol.* *28*, 853–863.

BERÁNEK, R. and NOVOTNÝ, I. (1958). Spontaneous electrical activity in the denervated muscles of the cockroach *Periplaneta americana*. *Nature* *182*, 957–958.

BISHOP, C. A., O'SHEA, M. and MILLER, R. J. (1981). Neuropeptide proctolin (H–Arg–Tyr–Leu–Pro–Thr–OH): Immunological detection and neuronal localization in insect central nervous system. *Proc. Nat. Acad. Sci.* Submitted.

BROWN, B. E. (1965). Pharmacologically active consitutents of the cockroach corpus cardiacum: resolution and some characteristics. *Gen. Comp. Endocr.* *5*, 387–401.

BROWN, B. E. and STARRATT, A. N. (1975). Isolation of proctolin: a myotropic peptide from *Periplaneta americana*. *J. Insect Physiol.* *21*, 1879–1881.

BURSEY, C. R. and SHERMAN, R. G. (1970). Spider cardiac physiology. I. Structure and function of the cardiac ganglion. *Comp. Gen. Pharmacol.* *1*, 160–170.

CAMPAN, R. (1972). Light-induced heart-beat disturbances. Comparative study in *Calliphora vomitoria* (Diptera) and *Nemobius sylvestris* (Orthophoptera). *Monitore Zool. Ital.* *6*, 269–289.

CANARD, M. and QUEINNEC, Y. (1971). Modifications du rythme cardiaque au cours de la diapause de *Chrysopa perla* (L.). *C.R. Acad. Sci. Paris*, Ser. D, *273*, 1960–1963.

CARLSON, A. J. and MEEK, A. (1908). On the mechanisms of the embryonic heart rhythm in *Limulus*. *Amer. J. Physiol.* *21*, 1–10.

CASE, J. F. (1956). Spontaneous activity in denervated insect muscle. *Science* *124*, 1079–1080.

CHRISTENSEN, T. A. and CARLSON, A. D. (1981). Symmetrically organized dorsal unparied median (DUM) neurons and flash control in the male firefly, *Photuris versicolor*. *J. Exp. Biol.* *93*, 133–147.

CLARE, S. and TAUBER, O. E. (1940). Circulation of hemolymph in the wings of the cockroach, *Blattella germanica*. I. In normal wings. *Iowa State Coll. J. Sci.* *14*, 107–127.

CLARKE, K. U. (1973). The Biology of the Arthropoda. London. Edward Arnold. 270 pages.

CLEMENTS, A. N. (1956). The antennal pulsatile organs of mosquitoes and other Diptera. *Quart. J. Mic. Sci.* *97*, 429–435.

COON, B. F. (1944). Effects of paralytic insecticides on heart pulsations and blood circulation in the American cockroach as determined with a fluorescein indicator. *J. Econ. Ent.* *37*, 785–789.

CRUZ-LANDIM, C. (1978). Fine structure of some nerve endings and neuromuscular junctions in *Melipona quadrifasciata anthidioides* Lep. (Apidae, Meliponinae). *Rev. Brasil. Biol.* *38*, 99–107.

DAVENPORT, D. (1949). Studies in the pharmacology of the heart of the orthopteron, *Stenopelmatus*. *Physiol. Zool.* *22*, 35–44.

DAVEY, K. G. (1962). The release by feeding of a pharmacologically active factor from the corpus cardiacum of *Periplaneta americana*. *J. Insect Physiol.* *8*, 205–208.

DAVEY, K. G. (1963). The release by enforced activity of the cardiac accelerator from the corpus cardiacum of *Periplaneta americana*. *J. Insect Physiol.* *9*, 375–381.

DAVEY, K. G. (1964). The control of visceral muscles in insects. *Adv. Insect Physiol.* *2*, 219–245.

DAVID, J. and ROUGIER, M. (1972). Physiology of the explanted dorsal vessel of insects. In *Invertebrate Tissue Culture*. Edited by C. Vago. Vol. 2, pages 211–244. Academic Press, New York.

DAVIS, C. C. (1961). Periodic reversal of heart beat in the prolarva of a gyrinid. *J. Insect Physiol.* *7*, 1–4.

DIERICHS, R. (1972). Elektronenmikroskopische Unterschungen des ventralen Diaphragmas von *Locusta migratoria* und der langsamen Kontraktionswelle nach Fixation durch Gefriersubstitution. *Z. Zellforsch.* *126*, 402–420.

DUDEL, H. (1977). Vergleichende funktions = anatomische Untersuchungen über die Antennen der Dipteran. I. Bibiomorpha, Homoeodactyla, Asilomorpha. *Zool. Jb. Anat.* *98*, 203–308.

DUDEL, H. (1978a). Vergleichende funktions = anatomische Untersuchungen über die Antennen der Dipteran. II. Cyclorrapha (Aschiza and Schizophora-Acalyptratae). *Zool. Jb Anat. 99*, 224–298.

DUDEL, H. (1978b). Vergleichende funktions = anatomische Untersuchungen über die Antennen der Dipteren. III. Calyptratae (I.O. Cyclorrapha). *Zool. Jb Anat. 99*, 301–370.

DUWEZ, Y. (1938). L'automatisme cardiaque chez le Dytique. *Arch. Int. Physiol. 46*, 389–403.

EDWARDS, G. A. and CHALLICE, C. E. (1960). The ultrastructure of the heart of the cockroach, *Blatella germanica. Ann. Ent. Soc. Amer. 53*, 369–383.

EDWARDS, G. A. and NUTTING, W. L. (1950). The influence of temperature upon the respiration and heart activity of *Thermobia* and *Grylloblatta. Psyche 12*, 33–44.

ENGELMANN, F. (1963). Die Innervation der Genital- und Postgenitalsegmente bei Weibchen der Schabe *Leucophaea maderae. Zool. Jb. Anat. 81*, 1–16.

ENGELMANN, F. (1970). *The Physiology of Insect Reproduction*. Pergamon Press, Oxford.

EVANS, J. J. T. (1962). Insect neurosecretory material separated by differential centrifugation. *Science 136*, 314–315.

EVANS, P. D. (1980). Biogenic amines in the insect nervous system. *Adv. Insect Physiol. 15*, 317–473.

EVANS, P. D. and O'SHEA, M. (1977). The identification of an octopaminergic neuron which modulates neuromuscular transmission in the locust. *Nature 270*, 257–259.

EVANS, P. D. and O'SHEA, M. (1978). The identification of an octopaminergic neuron and the modulation of a myogenic rhythm in the locust. *J. Exp. Biol. 73*, 235–260.

FIFIELD, S. M. and FINLAYSON, L. H. (1978). Peripheral neurons and peripheral neurosecretion in the stick insect, *Carausius morosus. Proc. Roy. Soc. B 200*, 63–85.

FINLAYSON, L. H. and OSBORNE, M. P. (1968). Peripheral neurosecretory cells in the stick insect (*Carausius morosus*) and the blowfly larva (*Phormia terrae-novae*). *J. Insect Physiol. 14*, 1793–1801.

FREUDENSTEIN, K. (1928). Das Herz and Circulationssystem der Honigbiene (*Apis mellifera*) L. *Z. Wiss. Zool. 132*, 404–475.

FRISH, K. (1978). Das Herz der Collembola (Insecta) — Ein Beitrag zur Anatomie der Collembola. *Zool. Anz.*, Jena *201*, 177–198.

GELDIAY, S. and KARACALI, S. (1980). The neurosecretory system of the adult *Melanogryllus desertus* Pall. (Orthoptera, Gryllidae). *Cell Tiss. Res. 211*, 235–240.

GEROULD, J. H. (1938). Structure and action of the heart of *Bombyx mori* and other insects. *Acta Zool. 19*, 297–352.

GERSCH, M. (1973). General aspects and problems in comparative endocrinology. *Gen. Comp. Endocrinol. 20*, 425–435.

GERSCH, M., HENTSCHEL, E. and UDE, J. (1974). Aminerge substanzen in lateralen Herznerven und im stomatogastrischen Nervensystem der Schabe *Blaberus craniifer* Burm. *Zool. Jb. Physiol. 78*, 1–15.

GERSCH, M., RICHTER, K., BÖHM, G-A and STÜRZEBECKER, J. (1970). Selektive Ausschuttung von Neurohormonen nach elektrischer Reizung der Corpora Cardiaca von *Periplaneta americana in vitro. J. Insect Physiol. 16*, 1991–2013.

GERSCH, M., BAUMANN, E., BIRKENBEIL, H., PASS, G. and PENZLIN, H. (1981). Die Wirkungsweise des Herzaktiven Peptide Neurohormon D von *Periplaneta americana*. (In press.)

GOLDSWORTHY, G. J. and MORDUE, W. (1974). Neurosecretory hormones in insects. *J. Endocrinol. 60*, 529–558.

GUTHRIE, D. M. (1962). Control of the ventral diaphragm in an insect. *Nature 196*, 1010–1012.

GUTHRIE, D. M. and TINDALL, A. R. (1968). *The Biology of the Cockroach*. Edward Arnold, London.

HEINRICH, B. (1970a). A thoracic temperature stabilization by blood circulation in a free-flying moth. *Science 168*, 580–582.

HEINRICH, B. (1970b). Nervous control of the heart during thoracic temperature regulation in a sphinx moth. *Science 169*, 606–607.

HEINRICH, B. (1971). Temperature regulation in the sphinx moth, *Manduca sexta. J. Exp. Biol. 54*, 141–166.

HEINRICH, B. (1974a). Thermoregulation in bumblebees. *J. Comp. Physiol. 88*, 129–140.

HEINRICH, B. (1974b). Thermoregulation in endothermic insects. *Science 185*, 747–756.

HEINRICH, B. (1976). Mechanisms of beat exchange between thorax and abdomen in bumblebees. *J. Exp. Biol. 64*, 561–585.

HEINRICH, B. (1981). Energitics of honeybee swarm thermoregulation. *Science 212*, 565–566.

HEINRICH, B. and CASEY, T. M. (1978). Heat transfer in dragonflies: 'fliers' and 'perchers'. *J. Exp. Biol. 74*, 17–36.

HEINRICH, B. and PANTLE, C. (1975). Thermoregulation in small flies (*Syrphus* sp.); basking and shivering. *J. Exp. Biol. 62*, 599–610.

HERTEL, W. (1971). Untersuchungen zur neurohormonalen Steuring des Herzens der Amerikanischen Schabe *Periplaneta americana* (L.). *Zool. Jb. Physiol. 76*, 152–184.

HERTEL, W. (1975). Untersuchungen zur Beeinflussung des Herzschlages von *Periplaneta americana* L. *Zool. Jb. Physiol. 79*, 70–76.

HESSEL, J. H. (1966). A preliminary comparative anatomical study of the mesothoracic aorta of the Lepidoptera. *Ann. Ent. Soc. Amer. 59*, 1217–1227.

HESSEL, J. H. (1969). The comparative morphology of the dorsal vessel and accessory structures of the Lepidoptera and its phylogenetic implications. *Ann. Ent. Soc. Amer. 62*, 353–370.

HINKS, C. F. (1966). The dorsal vessel and associated structures in some Heteroptera. *Trans. Roy. Ent. Soc. 118*, 375–392.

HIRIPI, L. and S.-RÓZSA, K. (1973). Fluorimetric determination of 5HT and catecholamines in the central nervous system and heart of *Locusta migratoria migratorioides. J. Insect Physiol. 19*, 1481–1485.

HOLLINGWORTH, R. M. and MURDOCK, L. L. (1980). Formamidine pesticides: Octopamine-like actions in a firefly. *Science 208*, 74–76.

HOSHI, T. and WATSON, W. (1981). Proctolin induces myogenicity in the deganglionated *Limulus* heart. Abstracts, *Soc. neurosci. 7*, 254.

HOULIHAN, D. F. and BRECKENRIDGE, L. (1981). Stretch-induced growth of blowfly muscle. *J. Insect Physiol. 27*, 521–526.

HOYLE, G. (1974). A function for neurons (DUM) neurosecretory on skeletal muscle of insects. *J. Exp. Zool. 189*, 401–406.

HOYLE, G. and O'SHEA, M. (1974). Intrinsic contractions in insect skeletal muscle. *J. Exp. Zool. 189*, 407–412.

HOYLE, G., COLQUHOUN, W. and WILLIAMS, M. (1980). Fine structure of an octoaminergic neuron and its terminals. *J. Neurobiol 11*, 103–126.

IMMS, A. D. (1970). *A General Textbook of Entomology*, 9th edition. Revised by O. W. Richards and R. G. Davies. Methuen, London.

IRVING, S. N. and MILLER, T. A. (1980a). Ionic differences in "fast" and "slow" neuromuscular transmission in body wall muscles of *Musca domestica* larvae. *J. Comp. Physiol. 135*, 291–298.

IRVING, S. N. and MILLER, T. A. (1980b). Octopamine and proctolin mimic spontaneous membrane depolarizations in *Lucilia* larvae. *Experientia 36*, 566–567.

JAHN, T. L., CRESCITELLI, F. and TAYLOR, A. B. (1937). The electrocardiograme of the grasshopper (*Melanoplus differentialis*). *J. Cell Comp. Physiol. 10*, 439–460.

JENSEN, H. (1976). Ultrastructure of the aortic diverticula of the adult dragonfly *Sympetrum danae* (Odonata: Anisoptera). *Cell Tiss. Res. 168*, 177–191.

JENSEN, P. V. (1973). Structure and metamorphosis of the larval heart of *Calliphora erythrocephala. Det. Kong. Denske Viden. Sel. Biol. Skr. 20*, 1–19

JOHNSON, B. (1966). Fine structure of the lateral cardiac nerves of the cockroach *Periplaneta americana* (L.). *J. Insect Physiol. 12*, 645–653.

JOHNSON, B. and BOWERS, C. (1963). Transport of neurohormones from the corpora cardiaca in insects. *Science 141*, 264–266.

JONES, J. C. (1954). The heart and associated tissues of *Anopheles quadrimaculatus* Say (Diptera: Culicidae). *J. Morphol. 94*, 71–123.

JONES, J. C. (1977). *The Circulatory System of Insects*. C. C. Thomas, Springfield.

JONES, J. C. (1981). *The Anatomy of the Grasshopper*. Charles C. Thomas, Springfield.

KATER, S. B. (1968). Carioaccelerator release in *Periplaneta americana* (L.). *Science 160*, 765–767.

KAUFMAN, W. R. and DAVEY, K. G. (1971). The pulsatile organ in the tibia of *Triatoma phyllosomia pallidipennis. Canad. Ent. 103*, 487–496.

KE, O. (1932). Contributions to the anatomy and physiology of the dorsal vessel of the silkworm. *Bull. Sci. Terk. Univ. 5*, 1–16. (In Japanese with a summary in English.)

KLEMM, H. (1980). Histochemical demonstration of biogenic monoamines (Falck–Hillarp method) in the insect nervous system. In *Neuroanatomical Techniques*. Edited by N. J. Strausfeld and T. A. Miller. Pages 52–73. Springer-Verlag, New York.

KNOWLES, F. G. W. (1967). Neuronal properties of neurosecretary cells. In *Neurosecretion*. IV. International Symposium on Neurosecretion. Edited by F. Strtinsky. Pages 8–19. Springer-Verlag, Berlin.

KRIJGSMAN, B. J. and KRIJGSMAN-BERGER, N. E. (1951). Physiological investigations into the heart function of arthropods. The heart of *Periplaneta americana*. *Bull. Ent. Res.* *42*, 143–155.

KUWANA, Z. (1932). Morphological studies of the nervous system of the silkworm, *Bombyx mori* Linne. I. The innervation of the dorsal vessel. *Japan Imp. Seric. Expt. Sta. Bul.* *8*, 109–120. (In Japanese with resumé in English.)

LAFON-CAZAL, M. and ARLUISON, W. (1976). Localization of monoamines in the corpora cardiaca and the hypocerebral ganglion of locusts. *Cell Tiss. Res.* *172*, 517–527.

LANG, F. (1971). Intracellular studies on pacemaker and follower neurons in the cardiac ganglion of *Limulus*. *J. Exp. Biol.* *54*, 815–826.

LHOTSKY, S., PHANEUF, C., LANGLOIS, J. M., GAGNE, S. and POUSSART, D. (1975). Optocardiography: *in vivo* measurements of the insect cardiac activity with a new optical method. *J. Insect Physiol.* *21*, 237–248.

LOCKSHIN, R. A. and WILLIAMS, C. M. (1965). Programmed cell death — III. Neural control of the breakdown of the inter-segmental muscles of silkmoths. *J. Insect Physiol.* *11*, 601–610.

LYSENKO, O. and SLÁMA, K. (1979). Monitoring the course of bacterial infections using hemolymph pressure pulse records. *Proc. Int. Coll. Invert. Pathol.* Pages 119–120. Prague.

MALOEUF, N. S. R. (1935). the myogenic automatism of the contraction of the heart of insects. *Ann. Ent. Soc. Amer.* *28*, 322–337.

MCCANN, F. V. (1963). Electrophysiology of an insect heart. *J. Gen. Physiol.* *46*, 803–821.

MCCANN, F. V. (1965). Unique properties of the moth myocardium. *Ann. New York Acad. Sci.* *127*, 84–99.

MCCANN, F. V. (1967). The effect of metabolic inhibitors on the moth heart. *Comp. Biochem. Physiol.* *20*, 399–409.

MCCANN, F. V. (1970). Physiology of insect hearts. *Ann. Rev. Entomol.* *15*, 173–200.

MCCANN, F. V. (1971). Calcium action potentials in insect myocardial fibers. *Comp. Biochem. Physiol.* *40A*, 353–357.

MCINDOO, N. E. (1939). Segmental blood vessels of the American cockroach (*Periplaneta americana* (L.). *J. Morphol.* *65*, 323–351.

MCINDOO, N. E. (1945). Innervation of insect hearts. *J. Comp. Neurol.* *83*, 141–155.

MEYER, G. F. (1958). Der fienere ban der aorta im thorax der honigbiene. *Z. Zelforsch.* *48*, 635–638.

MILL, P. J. (1972). *Respiration in the Invertebrates*. Macmillan, London.

MILLER, T. (1968). Response of cockroach cardiac neurons to cholinergic compounds. *J. Insect Physiol.* *14*, 1713–1717.

MILLER, T. (1969). Initiation of activity in the cockroach heart. *Experientia Suppl.* *15*, 206–218.

MILLER, T. (1973). Regulation of the heartbeat of the American cockroach. In *Neurobiology of the Invertebrates*. Edited by J. Salanki. Pages 195–212. Akademiai Kiddo, Budapest.

MILLER, T. (1975a). Insect visceral muscle. In *Insect Muscle*. Edited by P. N. R. Usherwood. Pages 545–606. Academic Press, London (Invited chapter.)

MILLER, T. (1975b). Neurosecretion and the control of visceral organs in insects. *Ann. Rev. Entomol.* *20*, 133–149. (Review article.)

MILLER, T. A. (1979a). Nervous *versus* neurohormonal control of insect heartbeat. *Am. Zool.* *19*, 77–86.

MILLER, T. A. (1979b). *Insect Neurophysiological Techniques*. Springer, New York.

MILLER, T. and ADAMS, M. E. (1974). Ultrastructure and electrical properties of the hyperneural muscle of *Periplaneta americana*. *J. Insect Physiol.* *20*, 1925–1936.

MILLER, T. and JAMES, J. (1976). Chemical sensitivity of the hyperneural nerve–muscle preparation of the American cockroach. *J. Insect Physiol.* *22*, 981—988.

MILLER, T. A. and METCALF, R. L. (1968). Site of action of pharmacologically active compounds on the heart of *Periplaneta americana* L. *J. Insect Physiol.* *14*, 383–394.

MILLER, T. and S.-RÓZSA, K. (1981). Control of the alary muscles of locust dorsal diaphragm. *Physiol. Eng.* *6*, 51–59.

MILLER, T. and THOMSON, W. W. (1968). Ultrastructure of cockroach cardiac innervation. *J. Insect Physiol.* *14*, 1099–1104.

MILLER, T. and USHERWOOD, P. N. R. (1971). Studies of cardioregulation in the cockroach *Periplaneta americana*. *J. Exp. Biol.* *54*, 329–348.

MILLER, T. A., BENEDECZKY, I. and S.-RÓZSA, K. (1979). Ultrastructure of the muscles of the dorsal diaphragm in *Locusta migratoria*. *Cell Tiss. Res.* 203, 93–105.

MORDUE, W. and GOLDSWORTHY, G. J. (1969). The physiological effects of corpus cardiacum extracts in locusts. *Gen. Comp. Endocr.* *12*, 360–369.

MOREAU, R. (1974). Variations de la pression interne au cours de l'emérgence et de l'expansion des ailes chez *Bombyx mori* et *Pieris brassicae*. *J. Insect Physiol.* *20*, 1475–1480.

MORISON, G. D. (1928). The muscles of the adult honeybee (*Apis mellifera* L.). *Quart J. Mic. Sci.* *71*, 563–651.

NATALIZI, G. M. and FRONTALI, N. (1966). Purification of insect hyperglycaemic and heart accelerating hormones. *J. Insect Physiol.* *12*, 1279–1287.

NEEDHAM, A. E. (1950). The neurogenic heart and ether anesthesia. *Nature* *166*, 9–11.

NORMANN, T. C. (1972). Heart activity and its control in the adult blowfly, *Calliphora erythrocephala*. *J. Insect Physiol.* *18*, 1793–1810.

NORMANN, T. C. (1974). Calcium-dependence of neurosecretion by exocytosis. *J. Exp. Biol.* *61*, 401–409.

NORMANN, T. C. (1975). Neurosecretory cells in the insect brain and production of hypoglycaemic hormone. *Nature* *254*, 259–261.

NUTTING, W. L. (1951). A comparative anatomical study of the heart and accessory structures of the orthopteroid insects. *J. Morph.* *89*, 501–598.

OPOCZYŃSKA-SEMBRATOWA, Z. (1936). Recherches sur l'anatomie et L'innervation du coeur de Carausius morosus. *Bull. Int. Akad. Cracovie* *B5*, 411–436.

O'SHEA, M. and ADAMS, M. E. (1981). Pentapeptide (proctolin: Arg-Tyr-Leu-Pro-Thr) associated with an identified neuron. *Science* *213*, 567–569.

O'SHEA, M., ADAMS, M. E. and BISHOP, C. A. (1981). Identification of proctolin-containing neurons: a combination of immunohistochemistry, HPLC, intracellular dye marking and electrophysiology. *Fed. Proc.* (in press).

PASS, G. (1980). The anatomy and ultrastructure of the antennae circulatory organs in the cockchafer beetle *Melolontha melolontha* L. (Coleoptera, Scarabaeidae). *Zoomorphology* *96*, 77–89.

PAWLOWA, M. (1895). Über ampullenartige Blut circulation sorgane im Kopfe verschiedener Orthopteren. *Zool. Ang.* *18*, 7–13.

PENZLIN, H., AGRICOLA, H., ECKERT, M. and KUSCH, T. (1981). Distribution of proctolin in the sixth abdominal ganglion of *Periplaneta americana* L. and the effect of proctolin on the ileum of mammals. Advances in Physiological Sciences, Vol. 22: *Neurotransmitters in Invertebrates*. Edited by K. S.-Rózsa. Pages 525–535. Pergamon Press, Oxford.

PETERS, M. (1977). Innervation of the ventral diaphragm of the locust (*Locusta migratoria*). *J. Exp. Biol.* *69*, 23–32.

PICHON, Y. (1970). Ionic content of haemolymph in the cockroach, *Periplaneta americana*. *J. Exp. Biol.* *53*, 195–209.

PIEK, T. and MANTEL, P. (1977). Myogenic contractions in locust muscle induced by proctolin and by wasp, *Philanthus triangulum*, venom. *J. Insect Physiol.* *23*, 321–325.

PINET, J. M. (1964). Les coeurs accessoires antennaires de *Rhodnius prolixus* Stil (Heteroptera, Reduviidae). *Bull. Soc. Zool. Fr.* *89*, 443–449.

PROSSER, L. C. and BROWN, F. A. Jr. (1961). *Comparative Animal Physiology*, Little, Brown, New York.

QUEINNEC, Y. and CAMPAN, R. (1972). Heartbeat frequency variations in the moth, *Mamestra brassicae*, during ontogeny. *J. Insect Physiol.* *18*, 1739–1744.

QUEINNEC, Y. and CAMPAN, R. (1975). Influence de la maturité sexuelle sur l'activité et la réactivité cardiaques de *Calliphora vomitoria* (Diptère, Calliphoridae). *J. Physiol. Paris* *70*, 457–466.

RAABE, M. (1978). Perisympathetic organs in insects: their physiological meaning. In *Neurosecretion and Neuroendocrine Activity, Evolution Structure and Function*. Edited by W. Bargmann, A. Oksche, A. Polenov and B. Scharrer. Page 390. Springer-Verlag. Berlin.

RAABE, M., BAUDRY, N., GRILLOT, J. P. and PROVANSAL, A. (1974). The perisympathetic organs of insects. In *Neurosecretion — The Final Neuroendocrine Pathway*. VIth International Symposium on Neurosecretion. Pages 59–71. Springer-Verlag, Berlin.

RALPH, C. L. (1962). Heart accelerators and decelerators in the nervous system of *Periplaneta americana* (L.). *J. Insect Physiol. 8*, 431–439.

REHM, E. (1939). Die innervation der innern organe von *Apis mellifica*. Zugleich ein Beitrag zur Frage des sog. Sympathetischen Nervensystems der Insekten. *Z. Morph. Ökol. Tiere 36*, 89–122.

RICHARDS, A. G. (1963a). The effect of temperature on the heartbeat frequency in the cockroach, *Periplaneta americana*. *J. Insect Physiol. 9*, 597–606.

RICHARDS, A. G. (1963b). The ventral diaphragm of insects. *J. Morphol. 113*, 17–47.

RICHTER, K. (1971). Zur wirkung von neurohormon D auf die membranpotentiale von muskelzellen des myokards von *Blaberus craniifer* Burm. (Insecta: Blatteria), ain Beitrag zur Frage der regulation des insektenherzens. *Zool. Jb. Physiol. 76*, 51–63.

RICHTER, K. (1981). On the role of peptides in synaptic transmission. *Adv. Physiol. Sci. 22*, 537–555.

ROSENBERG, J. and SEIFERT, G. (1978). Feinstruktur der Innervierung des Dorsalgefässes von *Peripatooides leuckarti* (Saenger, 1869) (Onychopora, Peripatosidae). *Zool. Anz. Jena. 201*, 21–30.

ROUSSEL, M. P. (1972). Innervation du corur et régulation nerveuse du rythme carioque chez *Locusto migratoria* L. *Arch. Zool. Exp. Gen. 113*, 265–294.

ROUSSEL, M. P. (1973). Induction de la difference phasaire du rythme cardiaque chez *Locusta migratoria migratorioides* R. et F. *Nat. Canad. 100*, 11–18.

SANGER, J. W. and MCCANN, F. V. (1968a). Ultrastructure of the myocardium of the moth, *Hyalophora cecropia*. *J. Insect Physiol. 14*, 1105–1111.

SANGER, J. W. and MCCANN, F. V. (1968b). Ultrastructure of moth alary muscles and their attachment to the heart wall. *J. Insect Physiol. 14*, 1539–1544.

SCHMITT, J. B. (1954). The nervous system of the pregenital abdominal segments of some Orthoptera. *Ann. Ent. Soc. Amer. 47*, 677–682.

SEDLAK, B. J. and WHITTEN, J. (1976). Changes in heart ultrastructure during development of the flesh fly, *Sarcophaga bullata*. *Devel. Biol. 54*, 308–313.

SENFF, R. E. (1971). Cardiac acceleration in response to stimulation of the segmental nerves in the adult cockroach, *Periplaneta americana*. *Comp. Biochem. Physiol. 40A*, 1009–1013.

SHANKLAND, D. L. (1965). Nerves and muscles of the pregenital abdominal segments of the American cockroach, *Periplaneta americana* (L.). *J. Morph. 117*, 353–386.

SHERMAN, R. G. and PAX, R. A. (1969). Electrical activity in single muscle cells of a spider heart. *Comp. Biochem. Physiol. 28*, 487–489.

SLÁMA, K. (1976). Insect haemolymph pressure and its determination. *Acta Ent. Bohemslov. 73*, 65–75.

SLÁMA, K., BAUDRY-PARTIAOGLOU, N. and PROVANSAL-BAUDEZ, A. (1979). Control of extracardiac haemolymph pressure pulses in *Tenebrio molitor*. *J. Insect Physiol. 25*, 825–831.

SMITH, N. A. (1969). Observations on the neural rhythmicity in the cockroach. *Experientia, Suppl. 15*, 200–205.

SNODGRASS, R. E. (1935). *Principles of Insect Morphology*. McGraw-Hill New York.

SRIVASTAVA, R. C. (1970). Morphology of the neurosecretory system and the retrocerebral endocrine glands of adult *Halys dentatus* (Heteroptera: Pentatomidae). *Ann. Ent. Soc. Am. 63*, 1372–1376.

S.-RÓZSA, K. and V.-SZÖKE, I. (1972). The effect of bioactive substances in the heart muscle cell membranes of *Locusta migratoria migratorioides* R.F. *Acta Physiol. Acad. Sci. Hung. 41*, 27–36.

STARRATT, A. N. (1979). Proctolin, an insect neuropeptide. *Trends neurosci. 2*, 15–17.

STEINER, G. (1932). Die Automatie und die zentrale Beeinflussung des Herzens von *Periplaneta americana*. *Z. Vergl. Physiol. 16*, 290–304.

STERNBURG, J. (1963). Autointoxication and some stress phenomena. *Ann. Rev. Ent. 8*, 19–38.

TACHIBANA, K. (1975). Normal and reverse pulsation of the dorsal vessel in insects. *Jap. J. Appl. Ent. Zool. 19*, 139–143.

TACHIBANA, K. and NAGASHIMA, C. (1957). Studies on the heartbeat of insect. *Jap. J. Appl. Ent. Zool. 1*, 155–163.

TENNY, S. M. (1953). Observations on the physiology of the lepidopteran heart with special reference to reversal of the beat. *Physiol. Comp. Oecol. 3*, 286–306.

TUBLITZ, N. and TRUMAN, J. E. (1981). An insect cardioactive peptide modulates heart rate during development. Abstracts, *Soc. for Neurosci. 7*, 253.

TYRER, M. N. (1971). Innervation of the abdominal intersegmental muscles in the grasshopper. *J. Exp. Biol. 55*, 305–314.

UNGER, H. (1956). Neurohormonale Steureing der Herztätigkut bei Insekten. *Naturwissenschaften. 43*, 66–67.

UNNITHAN, G. C., BERN, H. A. and NAYAR, K. K. (1971). Ultrastructural analysis of the neuroendocrine apparatus of *Oncopeltus fasciatus* (Heteroptera). *Acta Zool. 52*, 117–143.

USHERWOOD, P. N. R. (1963). Response of insect muscle to denervation — II. Changes in neuromuscular transmission. *J. Insect Physiol. 9*, 811–825.

USHERWOOD, P. N. R. (1969). Electrochemistry of insect muscle. *Adv. Insect Physiol. 6*, 205–278.

VERO, M. and MILLER, T. (1979). Sensitive tension and force transducer. *Med. Biol. Eng. Computg. 17*, 662–666.

WALLING, L. V. (1908). The anatomy of the acrididaean heart and its histological structure. *Kansas Univ. Sci. Bull. 4*, 359–367.

WAREHAM, S. C., DUNCAN, C. J. and BOWLER, K. (1974). The resting potential of cockroach muscle membrane. *Comp. Biochem. Physiol. 48A*, 765–797.

WASHIO, H. (1972). Calcium inward currents in insect muscle fibers. *Canad. J. Physiol. Pharmac. 50*, 1114–1116.

WASSERTHAL, L. T. (1975a). Herschlag-Umkehr bei Insekten und die Entwicklung der imaginalen Herzrhythmik. *Verh. Dtsch. Zool. Ges. 67*, 95–99.

WASSERTHAL, L. T. (1975b). Periodische Herschlag-Umkehr bei Insekten. *Umschau 75*, 93–94.

WASSERTHAL, L. T. (1975c). The rôle of butterfly wings in regulation of body temperature. *J. Insect Physiol. 21*, 1921–1930.

WASSERTHAL, L. T. (1976). Heartbeat reversal and its coordination with accessory pulsatile organs and abdominal movements in Lepidoptera. *Experientia 32*, 577–579.

WASSERTHAL, L. T. (1980). Oscillating haemolymph 'circulation' in the butterfly, *Papilio machaon* L. revealed by contact, thermography and photocell measurements. *J. Comp. Physiol. 139*, 145–163.

WASSERTHAL, L. T. (1981). Oscillating haemolymph 'circulation' and discontinuous trachael ventilation in the giant silk moth *Attacus atlas* L. *J. Comp. Physiol. 145*, 1–15.

WASSERTHAL, L. T. (1982). Antagonism between haemolymph transport and tracheal ventilation in an insect wing (*Attacus atlas* L.). *J. Comp. Physiol. 147*, 27–40.

WASSERTHAL, L. T. and WASSERTHAL, W. (1977). Innervation of heart and alary muscles in *Sphinx ligustri* L. (Lepidoptera). *Cell Tiss. Res. 184*, 467–486.

WASSERTHAL, W. and WASSERTHAL, L. T. (1980). Multinucleate neurons with neurohaemal and synapsing axons at the heart and alary muscles of the butterfly *Caligo beltrao* Illiger (Lepidoptera). *Cell Tiss. Res. 212*, 351–362.

WHEELER, W. M. (1910). *Ants*. Columbia University Press, New York.

WIGGLESWORTH, V. B. (1965). *Principles of Insect Physiology*. 6th edition. Methuen, New York.

DE WILDE, J. (1947). Contributions to the physiology of the heart of insects, with special reference to the alary muscles. *Arch. Neerl. Physiol. 28*, 530–542.

WILLE, A. (1958). A comparative study of the dorsal vessel of bees. *Ann. Ent. Soc. Am. 51*, 538–546.

WILLEY, R. B. (1961). The morphology of the stomodeal nervous system in *Periplaneta americana* (L.) and other Blattaria. *J. Morph. 108*, 219–261.

YAGUZHINSKAYA, L. V. (1954). New data on the physiology and anatomy of the heart in Diptera. *Biull. Mosk. Obsch. Isp. Prir. Otdel. Biol. 59*, 44–50.

YAMAMOTO, D. and WASHIO, H. (1980). Ionic selectivity of the calcium channels in insect larval muscle fibres. *J. Exp. Biol. 85*, 333–375.

YAMAMOTO, D. and WASHIO, H. (1981). Voltage clamp studies on insect skeletal muscle. II. The outward currents. *J. Exp. Biol. 92*, 13–22.

YAMAMOTO, D., FUKAMI, J.-I. and WASHIO, H. (1981). Voltage clamp studies on insect skeletal muscle. I. The inward current. *J. Exp. Biol. 92*, 1–12.

YEAGER, J. F. (1938). Mechanographic method of recording insect cardiac activity with reference to the effect of nicotine on isolated heart preparations from *Periplaneta orientalis*. *J. Agric. Res. 46*, 267–276.

YEAGER, J. F. (1939). Electrical stimulation of isolated heart preparations from *Periplaneta americana*. *J. Agr. Res. 59*, 121–137.

YOKOYAMA, T. (1932). Studies on the heart-beat in the silkworm *Bombyx mori* L., with special reference to the cause of its reversal. *Bull. Imp. Seric. Expt. Sta. 8*, 43–102. (In Japanese with English summary.)

ZDÁREK, J. (1980). Neurohormonal factors involved in the control of pupariation. In *Neurohormonal Techniques in Insects*. Edited by T. A. Miller. Pages 154–178. Springer, New York.

ZDÁREK, J., SLAMA, K. and FRAENKEL, G. (1979). Changes in internal pressure during puparium formation in flies. *J. Exp. Zool. 207*, 187–196.

ZAWARZIN, A. (1911). Hisologische Studien uber Insekten. I. Das Herz der Aeschnalarven. *Z. Wiss. Zool. 97*, 481–510.

ZEICHNER, B. C. (1980). A study of the wing pulsatile regions of *Musca domestica*. M.S. thesis, University of Maryland, College Park, MD.

ZWICKY, K. T. (1968). Innervation and pharmacology of the heart of *Urodacus*, a scorpion. *Comp. Biochem Physiol. 24*, 799–808.

9 Chemistry and Physiology of the Hemolymph

D. E. MULLINS

Virginia Polytechnic Institute and State University, Blacksburg, Virginia, USA

1 INTRODUCTION

Insect hemolymph is best described as the extracellular circulating fluid that fills the body cavity or hemocoel. It is physically isolated from direct contact with body tissues by a thin, permeable membrane which lines the hemocoel (D. Ashhurst, this volume). Hemocytes, which are cells contained in hemolymph (A. Gupta, this volume), and pericardial cells (A. Crossley, this volume), which are associated with the dorsal heart, are exceptions because they are in direct contact with this fluid. Organs contained in the hemocoel, particularly the aggregations of fat body cells (R. Dean, *et al.*, this volume), are in intimate contact with hemolymph.

Studies on insect hemolymph have revealed that major differences exist between it and the body fluids of organisms from other animal phyla. Some of these include high levels of organic molecules with lower levels of inorganic molecules, the presence of trehalose as the major carbohydrate, a relatively high organic phosphate content, and a wide variety of enzyme activities. In addition, since the insect tracheal system design is such that efficient gas exchange is provided to tissues without involving the circulatory system in any major way (P. Mill, this volume), there is a lack of oxygen or carbon dioxide carriers in the hemolymph. However, the hemolymph of certain chironomid larvae does contain hemoglobin.

Comparison of insect hemolymph composition with their phylogenetic, ontogenetic and physiological states reveals that there are differences. Furthermore, the specific composition of hemolymph may vary depending on such things as diet, temperature, disease, etc. Figure 1 illustrates some of the factors which may influence the specific hemolymph composition at certain points in an insect's life. It can be surmised from this figure that the specific composition will reflect the interaction of several factors.

The hemolymph has traditionally been viewed as a homeostatic fluid. With this view, it was implied that the regulation of hemolymph constituents is static or passive, rather than dynamic or active. The concept that hemolymph homeostasis is dynamic, and that turnover times for certain materials may be rapid, has been advanced only recently (Friedman, S., 1978; S. Friedman, vol. 10; Jungreis, A., 1980).

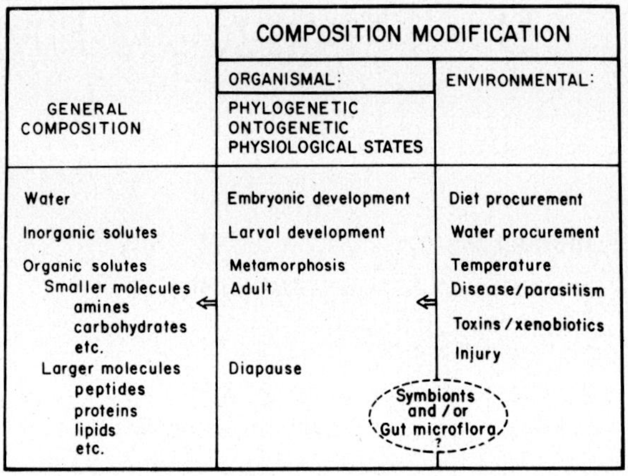

GENERAL COMPOSITION	COMPOSITION MODIFICATION	
	ORGANISMAL: PHYLOGENETIC, ONTOGENETIC, PHYSIOLOGICAL STATES	ENVIRONMENTAL:
Water	Embryonic development	Diet procurement
Inorganic solutes	Larval development	Water procurement
Organic solutes	Metamorphosis	Temperature
Smaller molecules: amines, carbohydrates, etc. ⇐	Adult ⇐	Disease/parasitism
		Toxins/xenobiotics
		Injury
Larger molecules: peptides, proteins, lipids, etc.	Diapause	Symbionts and/or Gut microflora ?

FIG. 1. Interrelationships of hemolymph with factors which may modify its composition.

The static view of hemolymph homeostasis is quite apparent in the older literature. This is understandable because in many studies the methods necessitated measurement of constituent levels from a sample obtained from one insect or pooled from a group of insects. Data based on single samples or sampling times may not provide information on the fluctuation of constituent levels, or their flow through metabolic pools in the insect. However, with developments and improvements in micromethodology, and the employment of radiotracer techniques, studies designed to examine the flux of hemolymph constituents have become feasible. There are several previous reviews on subjects pertaining to the chemistry and/or physiology of insect hemolymph which provide good reference material. Some of these are: Wyatt, G. (1961); Sutcliffe, D. (1963); Florkin, M. and Jeuniaux, C. (1964, 1974); Stobbart, R. and Shaw, J. (1964, 1974); Berridge, M. (1970); Jeuniaux, C. (1971); Crossley, A. (1975); and Wyatt, G. and Pan, M. (1978).

In addition, there are numerous chapters in volumes of this series which contain aspects inclusive of the involvement of hemolymph. As one examines the many functions and activities in which hemolymph is involved, it becomes quite clear that the topic of hemolymph chemistry and physiology is broad. Because of the broad nature of the subject material relating to the functions and composition of hemolymph, it is intended that this chapter will provide a general overview of this topic. Use will be made of reference material contained in previously published reviews as well as referral to related topics occurring in this series. The reader seeking detailed information not provided in this chapter should refer to these sources.

2 METHODS USED IN HEMOLYMPH STUDIES

2.1 Sample collection

Numerous approaches and methods have been used to study the composition and physiology of insect hemolymph. The results reported in some studies undoubtedly were influenced to some extent by the sample collection techniques which were employed. Therefore, those seeking specific information on hemolymph contained in the literature may find it worthwhile to examine closely the sampling and analytical techniques which were used. Some of the general problems associated with sample collection, approaches used in sample collection, and analytical methods will be discussed in this section.

Problems associated with hemolymph sample collections include: (1) obtaining sufficient quantities for reliable analysis, (2) preventing rapid coagulation, (3) preventing melanization or darkening of the sample, and (4) accounting for the presence and chemical contribution provided by hemocytes in whole serum. The size of the sample volume which may be obtained from an insect depends on such things as insect size and stage. In some cases it has been necessary to pool samples collected from individuals, not only to obtain adequate volumes but also to reduce sample variability (Brown, S. and Mazzone, H., 1977). Hemolymph of many, but not all insects, coagulates rapidly upon contact with air. Often it is instantaneous (Gregoire, C., 1974). The process may be effectively slowed by cooling the insects and the collection apparatus (Boctor, I. and Salem, S., 1973), by collection into ethylenediamine tetra-acetate (EDTA) (Wheeler, R., 1963), by sample collection from a stage (pupal) in which rapid clotting does not occur (Chino, H. and Gilbert, L., 1971), or by avoiding air contact during the collection process using capillary tubing and oil (Weidler, D. and Sieck, G., 1977). Hemolymph may darken when it comes into contact with air due to the action of phenoloxidases contained in whole serum. Several methods may serve to slow the melanization process. These include collection from chilled insects in a cold environment (Boctor, I. and Salem, S., 1973) or the addition of phenoloxidase inhibitors such as phenylthiourea (Wyatt, G. and Pan, M., 1978) or glutathione (Chino, H. and Gilbert, L., 1971). The presence of hemocytes in whole serum may complicate the determination of the chemical composition of hemolymph. Some reports indicate that hemocytes may release materials into the plasma upon lysis such as K^+ (Brady, J., 1967), aspartate and glutamate (Evans, P. and Crossley, A., 1974), and trehalase (Katagiri, C., 1977). Many studies have been done on cell-free preparations. The contribution of the proportionally small number of hemocytes to the composition of whole serum has been viewed as relatively insignificant (Wyatt, G., 1961). However, reviews on hemolymph (including this one) do not generally distinguish between information obtained from whole hemolymph versus plasma. In studies concerned with the composition of plasma, hemocytes should be removed by centrifugation (Wyatt, G. and Pan, M., 1978).

Sample collection may be divided into two categories, based on volume size and the purpose for which it is obtained. Small sample volumes (nl to μl range) may be sufficient for determination of the composition; whereas, the larger samples (μl to ml range) may be required for purification and characterization of specific components. Collection of small samples generally involves collection from various sites such as severed antennae or appendages (Leader, J. and Bedford, J., 1978; Brady, J., 1967), or abdominal heart punctures with capillary tubes (Dallmann, S. and Herman, W., 1978). In some cases, special efforts to prevent melanization have been performed by using tubes precoated with phenylthiourea (Cheung, P. *et al.*, 1978), and changes in osmotic concentration by collecting hemolymph from insects held under oil (Downing, N., 1978). The volume of the sample contained in a capillary may be determined on a linear or gravimetric basis. Hemocytes and tissue debris may be removed after centrifugation by breaking off the plugged portion of the tube (Weidler, D. and Sieck, G., 1977), providing cell-free plasma for chemical analysis. Collection of larger hemolymph volumes may involve puncturing the insect body, or severing an appendage and removing the hemolymph by expressing it (Gringorten, J. and Friend, W., 1979) or by low-speed centrifugation of the insect (Sternburg, J. and Corrigan, J., 1959). Care should be taken to prevent contamination of the fluid by gut contents when centrifugation is used. The mouth and anus may be sealed with paraffin (Sternburg, J. and Corrigan, J., 1959) or the head may be supported in an absorbent tissue (Chino, H. and Downer, R., 1979) during the collection procedure. Flushing the hemocoel by injecting EDTA-saline has been used to obtain larger volumes of hemolymph (Chino, H. and Downer, R., 1979). Hemocytes and cellular debris may then be removed by centrifugation before proceeding to the isolation and purification of hemolymph components. In general, it appears that sample collection using lowered temperatures, minimal exposure to air, rapid collection and removal of cellular components allow for improved analytical determinations.

2.2 Determination of volume

Calculation of total hemolymph solute content requires determination of both solute concentration and volume. Several different methods have been employed in estimating hemolymph volume. Some of these are: cell dilution, exsanguination and dye or radiolabelled ^{14}C-dilution. Cell dilution determinations are unreliable (Jones, J., 1977). Exsanguination methods are based on insect weight determinations made before and after hemolymph removal. This method generally results in an underestimate of volume because it is impossible to remove all of the hemolymph adhering to the tissues. In many cases, particularly in dehydration or inanition, this is the only method available because sufficient hemolymph for other kinds of determinations (dye, [^{14}C]inulin) cannot be obtained (Nicolson, S. *et al.*, 1974). Techniques based on dilution involve injections of a known amount of material into the hemocoel, and, after sufficient time has elapsed for it to become thoroughly mixed, a known volume of hemolymph is removed. Comparison of the initial injected amount with the sample concentration provides a dilution factor from which the total volume can be obtained. The use of dyes may give variable results and tends to provide overestimates of hemolymph volume (Jones, J., 1977; Levenbook, L., 1979). There may be several reasons for this. Depending upon the material used, it may be subject to removal by excretion, metabolism or adsorption to tissues. The isotope dilution technique using [^{14}C]inulin appears to be more reliable. Studies have shown that inulin metabolism and excretion in insects is very low or negligible, even over extended time intervals (Wharton, D. *et al.*, 1965a), but gradually it may become concentrated in the fat body and gonads (Loughton, B. and Tobe, S., 1969).

2.3 Determination of osmolality

Hemolymph osmotic pressure, expressed in terms of osmolality, is a measure of the total number of moles of solute per litre of hemolymph. The osmolal concentration is determined by the colligative properties of hemolymph which depend upon, or vary as a function of, the number of solute molecules contained in it, and not upon the nature of these molecules. One of the earliest micromethods developed was the measurement of the colligative properties of solutions using freezing point depression, or by measurement of fluid vapour pressure (Little, C., 1977). There have been other methods used in determinations of hemolymph osmolality such as evaporation against a standard salt solution in capillaries containing mammalian blood cells (osmometers) and by cooling of a thermocouple junction (Buck, J., 1953). Development of freezing point methods has provided the capability of determinations on 0.1–1.0 nl volumes, and has made it the general method of choice in hemolymph studies. The standard methodology was developed by Ramsay, J. and Brown, R. (1955) and has been so successful that several instruments based on this technique are commercially available. The mode of operation of these instruments is more rapid but can produce results of doubtful validity if solutions of different protein content and surface tensions are used (Little, C., 1977).

2.4 Determination of chemical composition

Advances in analytical technology have led to improved methods for identification and quantitation of inorganic and organic compounds in small sample volumes. Determination of cations such as Na^+,

K^+, Ca^{2+} and Mg^{2+} in samples (nl to μl range; with sensitivities for some in the picomole range) is possible using normal flame photometry, helium glow photometry, atomic absorption (flame and non-flame) as well as microprobe analysis (Little, C., 1977; Staddon, B. and Everton, I., 1980). Measurement of anions contained in small sample volumes has not yet developed to the extent that an overall method equivalent to the photometric methods used for multiple cation determinations exists (Little, C., 1977). As a consequence, determination of Cl^-, HCO_3^-, HPO_4^- must be done independently and a variety of methods have been employed in studies on hemolymph. The development and use of micro-electrodes has provided for pH determination in 10 μl hemolymph volumes (Mack, S. and Vanderberg, J., 1978) and continuous (24–96 h) *in vivo* measurement of hemolymph pH, Na^+ and K^+ (Lettau, J. *et al.*, 1977).

Determinations of organic compounds contained in hemolymph has been achieved by a variety of methods. The presence of many previously unsuspected components was revealed in earlier studies using chromatographic techniques (Wyatt, G., 1961). Microanalytical methods available for non-protein materials such as amino acids, carbohydrates, and lipids now include a broad spectrum of chromatographic techniques including paper, thin-layer, gas–liquid, high pressure liquid and column chromatography. Derivatization of certain materials has improved their separation and identification in low concentrations. For example, a microbiochemical procedure involving condensation of radiolabelled dansyl chloride with tissue extracts/fluids containing free amino or phenolic hydroxyl groups and separation on 3×3 cm polyamide layers may allow picomole quantities of a wide range of amino acids to be analyzed in less than 1 mg (1 μl) of tissue (Osborne, N. and Neuhoff, V., 1974). Development of electrophoretic and immunochemical methods has allowed for improved plasma protein separation, identifications, and quantitation of radiolabel incorporation (Wyatt, G. and Pan, L., 1978; Engelmann, F., 1979).

2.5 Determination of pressure and flow patterns

Because hemolymph is not confined to a closed vascular system, its relative pressure may be regarded as equivalent to the pressure of the main part of the hemocoel. This internal pressure corresponds closely to the pressure of the whole body (Sláma, K., 1976). Changes in pressure which may occur as a result of certain physiological processes (i.e. ecdysis, oviposition, wing expansion) are influenced primarily by interaction between somatic muscular tones and internal body volumes (hemolymph and alimentary canal). Despite the apparent importance of hemolymph pressure in the life processes of insects, there is much to be learned about this area of insect physiology. Measurement of hemolymph pressure was initially achieved by direct manometric techniques and later by employing the use of electronic transducer manometers (Davey, K. and Treherne, J., 1964). These methods provided useful information on pressure changes resulting from certain insect activities. The recent development of an apparatus using miniature tensometers provides for recording changes in hemolymph pressure (ranging from -50 to over 400 mm pH_2O) over prolonged time periods (Sláma, K., 1976).

Distribution patterns of hemolymph solutes have been the subject of numerous studies. However, little information exists on the time required for hemolymph to complete a cycle through the hemocoel (Jones, J., 1977). Techniques employed to determine the rate of solute mixing and circulation in the hemocoel have relied upon injection of dyes (i.e. fluorescein, indigo carmine), colloids (ink), and radiolabelled materials. The migration of these materials through the hemocoel was determined by direct visual observation, or by analysis of hemolymph samples obtained from different body regions at various intervals. Problems encountered with this approach may include such things as loss of materials due to excretion (dyes), sequestration (colloids) or stress caused by entrainment for direct observation or removal of samples. A new technique has recently been developed employing the use of externally attached microthermistors which monitor heating and cooling trends effected by hemolymph flow patterns (Wasserthal, L., 1980). This approach appears to be quite useful in clarifying some of the changes in hemolymph flow patterns coordinated with tracheal ventilation, periodic heartbeat reversal and different modes of abdominal movement (Wasserthal, L., 1981).

3 HEMOLYMPH CIRCULATION

In order for hemolymph to perform many of its functions there is a requirement for it to circulate throughout the insect hemocoel. This may be achieved by body movements and circulatory pumps. Each species possesses systems adapted for its specific needs; hence, there are many variations on the patterns and structures which may be involved. A comprehensive coverage of this topic will be found in Chapter 8 of this volume (T. Miller). In the embryo, the circulatory system arises from mesoderm. Embryonic hemolymph is derived from blastocoelic fluid arising from the coelomic sacs as they are emptied (Jones, J., 1977). The volume will vary depending on such things as the species, stage, diet, age, and reproductive status of insects. Jones, J. (1977) has tabulated information on hemolymph volumes found to occur in numerous species. He reported that the amount of hemolymph which can be removed from an insect may range from less than 1 μl (various species) to 20 ml (large queen termite). The volume may vary considerably within a larval instar. For example, in third-instar *Calliphora vicina* larvae 3 and 7 days post-oviposition, it varies from 4.7 to 26.3 μl, respectively (Levenbook, L., 1979). It should be noted that even though the hemolymph volume may increase considerably, it is maintained at relatively constant levels with respect to body size and weight. In the *Calliphora* larvae referred to above, the corresponding weights for the 3- and 7-day-old individuals were 12 and 82 mg. The hemolymph volume represents about 39% of the larval body weight (3-day) and 32% of the larval body weight in the 7-day-old individuals. The volume increase was 5.6-fold, while the corresponding body weight increased 6.8-fold.

Most insects possess a dorsal vessel or heart in each stage of development which is considered to be the primary circulatory pump (Jones, J., 1964). It may be divided into three anatomical zones, and at least three general vessel types have been described (Jones, J., 1977). The dorsal vessel usually extends from the abdomen through the thorax to the head. The abdominal portion is designated as the heart, which is the only region generally containing segmented swellings with openings (ostia). The thoracic and cephalic portion of the dorsal vessel is designated as the aorta. A brief description of dorsal vessel embryonic development and information on anatomical characteristics of this structure in the Insecta are available (Jones, J., 1977). The design of the dorsal vessel is such that it functions to pump hemolymph from the posterior to the anterior region of the insect. Hemolymph enters the vessel in the abdominal region via the ostia, which act as valves that close as the vessel contracts. Rhythmic contractions of the vessel initiate the general directional flow anteriorly. Upon reaching the cephalic region, hemolymph flows posteriorly and laterally towards the insect abdomen. The return flow pattern is determined by the internal morphology of the hemocoel which features apodemes, diaphragms, and internal organs. The physiology of the dorsal vessel has not been clearly established. The pacemaker mechanism was originally considered to be neurogenic, but evidence supporting myogenic or hormonal control has indicated there is much yet to be learned on the *in vivo* function of these possible influences (Miller, T., 1974; T. Miller, this volume). Also, it is known that some insects may discontinue pumping hemolymph anteriorly and reverse dorsal vessel peristalsis which results in the posterior flow of hemolymph (Jones, J., 1974, 1977; McCann, F., 1970; Miller, T., 1974). Evidence for the periodic dorsal vessel pumping reversal has been regarded with reserve. However, recent studies by Wasserthal, L. (1980, 1981) have provided some convincing evidence indicating that in certain resting Lepidoptera, hemolymph direction reversal is an activity coordinated with other processes. Using contact thermography, the effects of dorsal vessel pumping in basking *Papilio machaon* can be demonstrated (Fig. 2). Since the thoracic temperature in *P. machaon* is 1–1.5° above abdominal temperatures, placement of externally applied thermistors near the aorta and caudal heart allowed for continuous monitoring of hemolymph directional flow. It can be seen in this figure that periodic heart reversal caused marked effects in the hemolymph flow pattern as indicated by the cyclical temperature changes (Wasserthal, L., 1980). Closer study of dorsal vessel pumping has revealed that reversal is coordinated with abdominal movements (expansion) and both of these activities appear, in turn, to be coordinated with tracheal ventilation (Wasserthal, L., 1980, 1981). These observations suggest that, in at least some species, the control of

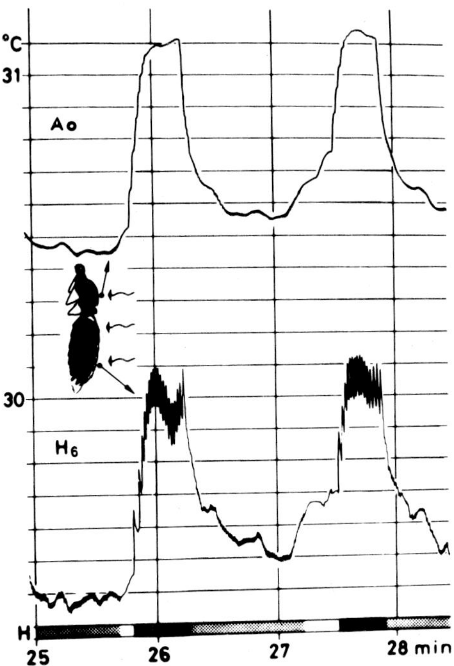

FIG. 2. Demonstration of periodic heartbeat reversals, utilizing externally applied temperature (°C) probes. Measurements were made in a basking butterfly (*Papilio machaon*) female, 4 days old. The reverse pulsations are observed as warming pulses at the cooler caudal heart (H_6) caused by inflow of relatively warmer hemolymph from the thorax. A_0 = thermistor site upon the aorta; ambient temperature 21°. Shading of the bar at the bottom indicates; *gray* = period of forward beating; *white* = pause; *black* = period of reverse beating of the heart (H). (From Wasserthal, L., 1980.)

hemolymph circulation and its functional relationship with other physiological processes (respiration) are more complex than originally thought.

A group of cells, the pericardial cells, are associated with the dorsal heart and aorta, and contain various granules, vacuoles and pigments. These cells ingest and store various dyes and foreign proteins from the hemolymph (Wigglesworth, V., 1970; A. Crossley, this volume). As a result, they are involved in modification of hemolymph composition.

Hemolymph circulation may also be assisted by accessory pulsatile pumps located near or within various insect appendages. According to Jones, J. (1977), these organs may consist of: small ovoid ampulla, small fan-shaped muscles, membrane-bound tubes formed from connective tissue sheaths, thin pulsating filaments, or a combination of several of these types. Antennal and leg pulsatile organs occur in many species, and their function is thought to facilitate circulation in these structures. An additional type of accessory pulsatile pump which certain insects may possess are the thoracic pulsatile organs. These appear to be specialized exoskeletal muscles which contract rhythmically (Jones, J., 1977). In *P. machaon* they are paired lying dorsolaterally in the winged segments and are not connected to the aorta (Wasserthal, L., 1980). These organs display similar activity periods and pauses, but are not closely synchronized among themselves or with the forward pulsations of the heart.

Heart pumping rates for a large number of insects has been summarized by Jones, J. (1974). There are several factors which affect these rates in insects. The heart rate tends to decrease with each successive nymphal or larval instar, becoming slower during the pupal stage. In the adult, the rate may be less than, greater than, or the same as, that in the last instar (Jones, J., 1974). Increases in somatic and visceral muscular and feeding activity and temperature result in more rapid pumping rates, whereas decreased activity and fasting result in lowered rates. Light and anoxia, as well as factors of endocrine origin, may also affect heart rates (Jones, J., 1974).

The summary of hemolymph circulation presented above supports the generalization that the system in its various forms is capable of providing for the demands which are made upon it. It is assumed that circulation is "efficient", as it is represented in each particular insect form. There is little direct information available on this aspect of hemolymph physiology. Studies designed to determine the length of time required for hemolymph to make a complete circuit within the hemocoel have been made. These rates depend on the fluid volume pumped by, and the respective rates of, the pumping organs, the activity of the insect, the total hemolymph volume as well as the physiological state of the insect. Jones, J. (1977) has pointed out that most of the information available on complete circulation of hemolymph generally deals with insects of unknown age and physiological status. Injection of dyes or radiolabelled materials indicates that they may be translocated from major body regions within 1 min, reaching more peripheral points of the hemocoel such as the mesothoracic pulvilli with times exceeding 1 min. Longer periods of time ranging from minutes to hours are required for an injected marker to become evenly distributed in the hemolymph (Jones, J., 1977).

Evaluation of circulation efficiency with respect to a particular physiological event, such as hemolymph cooling during flight or removal of toxic or undesirable compounds from hemolymph, is also complex, because it involves the activities of other tissues and organ systems. These will be discussed later in appropriate sections of this chapter.

4 FUNCTIONS OF HEMOLYMPH

Hemolymph performs several functions in insects. It serves as the transport milieu for the exchange of essential materials between cells, tissues and organs, and may serve as a storage reserve for certain of these materials. It is the medium in which hemocytes are allowed access to sites in the hemocoel for performance of their many activities. The physical properties of hemolymph allow for its use either as a hydraulic or heat-exchanging fluid. Both of these actions are essential to survival under certain various physiological states. Of major importance is the function of hemolymph as the controlled internal environment which is regulated with respect to ionic and chemical composition. This regulation results from the coordinated interaction of the insects' cellular and organ components such as hemocytes, alimentary canal, salivary glands, fat body, etc. The various functions of insect hemolymph will now be discussed.

4.1 Transport milieu

4.1.1 Hemocytes

Hemolymph provides for hemocyte circulation in the hemocoel. Because of the important relationships hemocytes share with hemolymph physiology and composition, some general comments on their nature and functions will be provided here and in other sections of this chapter. Comprehensive reviews (text) which have recently been published (Gupta, A., 1979b; A. Gupta, this volume) provide detailed information on cellular elements in hemolymph.

Insect hemocytes comprise several types of cells which, being part of the circulatory system, are of mesodermal origin (Arnold, J., 1974; Mori, H., 1979). A diversity of different types may be observed in the hemolymph at any given time. A classification scheme for hemocytes which are most frequently found includes seven types (Jones, J., 1977; Gupta, A., 1979a). They are: prohemocytes, plasmatocytes, granulocytes, coagulocytes (or cystocytes), adipohemocytes, spherulocytes (or spherule cells) and oenocytoids. In addition, there are some types which may be found infrequently, or not at all. These are podocytes, vermiform cells and granulocytophagous cells (Jones, J., 1977). Crossley, A. (1979) has provided a hypothetical differentiation pathway of mesodermal cells (inclusive of hemocytes) which is based on synthetic activities. He pointed out that biochemical and ultrastructural classification of hemocytes and their distinction from other mesodermal cells can be quite complex. This is because protein synthetic capabilities may remain constant or change over time, making distinct biochemical or ultrastructural characterization difficult. Hemocyte circulation in the hemolymph is in a dynamic state. They may cease to circulate and aggregate, becoming hemocytopoietic tissue where mitosis may occur, or they may become sessile hemocytes (hemocytic reservoirs) where mitosis does not occur (Jones, J., 1977).

There are a variety of functions which have been attributed to hemocytes. Four basic functions include phagocytosis of small particles, encapsulation of large foreign materials (H. Boman and P. Götz, this volume), hemolymph coagulation, and storage and distribution of nutritive materials (Arnold, J., 1974). In addition, there are a variety of activities associated with these basic functions in which hemocytes may become involved. Most of these activities involve chemical metabolism or exchanges. Phagocytosis, encapsulation and coagulation will be discussed in section 4.5. The role of hemocytes in storage, metabolism and distribution of nutritive materials, and their influence on hemolymph homeostasis, may be considerable. For example, in *Periplaneta americana* a significant proportion of whole hemolymph K^+ may be contained within hemocytes (Brady, J., 1967). Although amino acids contained in the hemocyte fraction of third-instar *Calliphora vicina* hemolymph accounted for only 6% of the total free amino acid content, a high percentage of glutamate (62%) and aspartate (69%) appeared to be sequestered in hemocytes (Evans, P. and Crossley, A., 1974). These findings, along with the increased hemocyte glutamate levels observed after saline–glutamate injections, were one of the first indications that insect hemocytes may participate in hemolymph amino acid homeostasis (see section 5.3.2.(a)).

4.1.2 Dissolved Matter: Metabolites, Nutrients and Excretory Materials

Hemolymph solutes include inorganic and organic materials which have originated from endogenous and exogenous sources. Circulation of these materials ensures access and transport to appropriate sites of metabolism, storage or excretion. Hemolymph can be viewed as the primary recipient of nutrient materials from exogenous sources via the gut, and the contributor of endogenous excretory materials via the Malpighian tubule–rectum complex. Nutrient procurement may involve mechanical and chemical processing that are subjected to controlled uptake. Associated with procurement is the probability that unavoidable uptake of materials which are of little nutritive value, or are potentially toxic, may occur. Duffey, S. (1980) has pointed out that organisms are adapted to sequester essential chemicals from the variety encountered in their diets. In providing for an efficient means for acquisition of essential nutrients, the processes involved (physical and biochemical) may not be absolutely specific. As a result, materials entering the hemolymph from the gut may be of variable usefulness and are subjected to the selective and regulatory processes within the insect. Some of these processes may involve rapid biochemical conversion of specific materials entering the hemolymph. For example, when radiolabelled [^{14}C]L-glutamate was slowly infused into either the gut or the hemocoel of locusts, the major radioactive product was glutamine, and the elevation of hemolymph L-glutamine appeared to be suppressed (Murdock, L. and Koidl, B., 1972). Contrarily, incorporation or rapid conversion of other amino acids, such as D-glutamate, L-alanine and glycine in hemolymph, apparently was not suppressed.

Nutrients which have entered the hemocoel may be transported to various tissues where they may be utilized or stored as a metabolic reserve. Figure 3 provides an example showing the metabolic interrelationships of some common metabolites as they might be utilized in a flying insect. The interrelationships indicated are achieved by hemolymph acting as a recipient and transit medium for these materials.

Included in the transit role attributed to hemolymph is its major involvement in excretory and some secretory processes. Excretion refers to those processes by which substances that interfere with normal animal functions, either by their presence or toxic action, are removed from the metabolic pool (Maddrell, S., 1971). Secretion refers to those processes by which substances are permanently removed from the metabolic pool to achieve some useful purpose outside the pool (Maddrell, S., 1971; Cochran, D., 1975). The Malpighian tubule–rectum complex is the principal excretory system which is involved in voiding excess water, organic and inorganic molecules contained in the hemolymph. The functional design of this system is such that what essentially represents a primary filtrate of hemolymph is secreted by the distal portion of the Malpighian tubules. It is believed that active transport of K^+ accompanied by anions (Cl^-), is primarily responsible for producing fluid (containing solutes) flow across the Malpighian tubule into its lumen. The ultrafiltrate, containing smaller solute molecules, travels proximally through the Malpighian tubule lumen to and through the hindgut to the rectum. As it traverses this route, water and various solutes may be selectively reabsorbed into the hemolymph. More detailed coverage of these processes may be found in reviews by Maddrell, S. (1971, 1977); Cochran, D. (1975) and chapters in volume 4.

Several points should be made with regard to the functional relationship of the excretory system with hemolymph. The first of these relates specifically to functional design. Maddrell, S. (1981) has recently discussed this topic. The secretion-driven filtration process producing the ultrafiltrate prevents removal of large molecules contained in the hemolymph, while allowing large quantities of useful but smaller

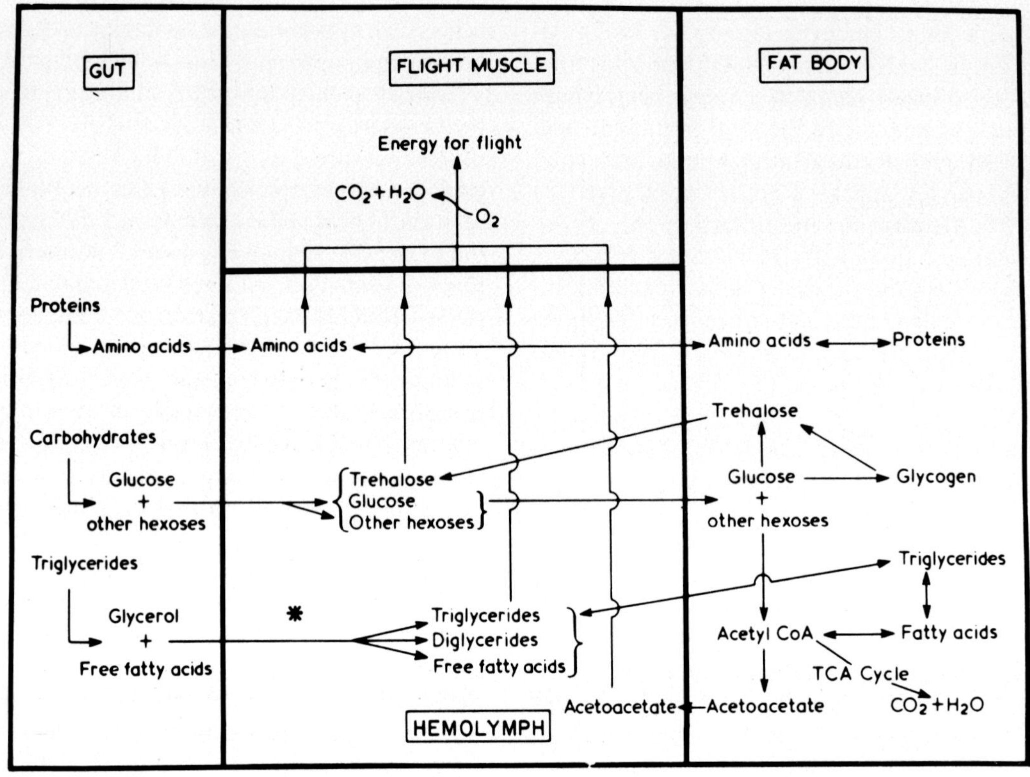

FIG. 3. Metabolic interrelationships between various insect tissue compartments. (Adapted from Bailey, E., 1975.)
* Recent findings have indicated that lipids are transported in the hemolymph primarily as diacylglycerol lipoproteins (DGLP) in several insect species (see section 5.3.2(e)).

molecules to move into the interior of the excretory system where they must be reabsorbed in order to be conserved by the insect (Fig. 4). This process may appear inefficient. However, if these activities provide an effective means by which toxic materials can be removed from hemolymph (not being selectively reabsorbed along the hindgut–rectal epithelium, resulting in elimination from the insect), then the energetics involved might be viewed as efficient. Excretion of certain toxic substances may be slow using this process. However, Malpighian tubules also possess specific active transport mechanisms that can rapidly eliminate a wide variety of substances contained in hemolymph which insects repeatedly need to excrete (Maddrell, S., 1977, 1981). These mechanisms may provide an effective means by which extremely toxic compounds may be eliminated. A second point, which should be made, also relates to the functional design of the excretory system and the important role of hemolymph. Emphasis should be placed on the fact that recycling of water and ions by the Malpighian tubule–rectum complex and returning them to the hemolymph, not only provides a means of removing wastes, but is also intimately associated with osmoregulation. Hence, hemolymph is an integral part of the excretory system. Osmoregulation will be considered in section 4.2.1.

Associated with excretion is the possibility that some useful materials may be lost. However, such unavoidable losses are usually very small. In *Rhodnius prolixus*, only trace amounts of amino acids are lost in the urine during rapid diureses after a blood meal even through the hemolymph is rich in amino acids (Maddrell, S. and Gardiner, B., 1980). The efficiency of absorption of nutrients into hemolymph and subsequent excretion of excess materials may be quite high. For example, ingestion of a blood meal by the tsetse fly *Glossina austeni* results in rapid excretion of water and salts contained in the blood meal. In 30 min, a volume of fluid equivalent to 80% of the unfed weight passes

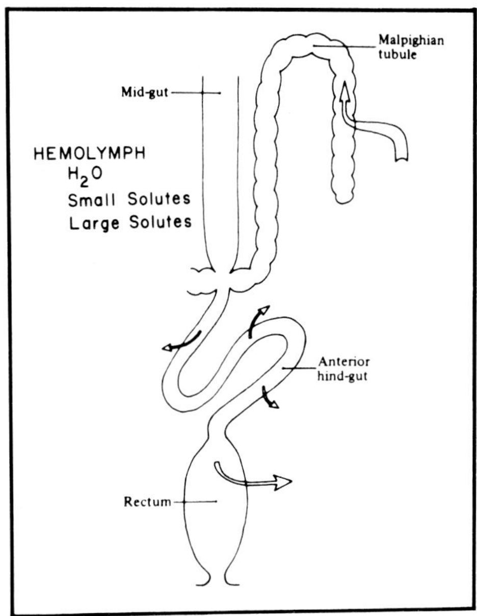

FIG. 4. Schematic representation of the generalized insect excretory system and the participation of hemolymph in its functions. Fluid movements in the system are featured. Fluid is transported into the Malpighian tubules where many hemolymph solutes can diffuse across the walls into the lumen. Much of the water may then be recovered (recycling) in the anterior hindgut and rectum together with other useful substances. Since the hemolymph provides and receives these materials during the "recycling" processes, it is intimately associated with the excretory system. (Adapted from Maddrell, S., 1981.)

through the hemolymph (Gee, J., 1975). *Trichoplusia ni* (seventh-instar larvae) fed [^{32}P]-labelled diet for 1 min and then transferred to a non-labelled diet, showed that 70 s after feeding was initiated, radiolabel circulated in the hemolymph and excretion through Malpighian tubules was detectable 1 min later (Ru, N. and Kloft, W., 1976). In addition, most of the radiolabel was eliminated within 2 h after feeding ceased.

Insect-derived materials destined for secretion may also be transported from their synthesis/storage sites to points of release or final deposition. They include parental investment materials (vitellins, brood sac milk, certain accessory gland materials, etc.), pigments and other such materials (Mullins, D. and Cochran, D., 1983). Some of these will be discussed in appropriate sections of this review.

4.1.3 Regulatory factors and specialized carriers

Insect development, tissue maintenance, homeostasis and reproduction are under hormonal control. Regulatory factors may include: glandotropic neurohormones (prothoracicotropic and allatotropic hormones), morphogenetic neurohormones (bursicon, pupariation diapause hormones), myotropic neurohormones (factors affecting the kinetics of various organs), metabolic neurohormones (adipokinetic, insulin-like, diuretic hormones) and ethotopic neurohormones (eclosion hormone, pupariation factors) (Fraenkel, G., 1980; see volumes 7 and 8). The importance of hormonal regulation of a variety of metabolic and physiological events is reflected by its appearance in several chapters in this series. In addition, several other recent reviews reporting on advances in insect endocrinology are available (Steele, J., 1976; Gee, J., 1977; Bodnaryk, R., 1978; Miller, T., 1980; Stone, J. and Mordue, W., 1980).

Specific nervous tissue cells have the ability to synthesize and release neurohormones into hemolymph and produce an effect on target tissues or organs. Hence, the role of these active factors, which synchronize physiological events such as development and have influences on anabolic and catabolic metabolism, is predicated on their distribution by hemolymph. The neurohormones are synthesized and packaged into neurosecretory granules in the cell bodies and may then be transported down the axons to storage and release sites termed neurohaemal areas. Examples of these materials are diuretic hormones which influence Malpighian tubule secretion rates. They are synthesized in some insects (*Dysterus, Schistocerca* and *Carausius*) by brain neurosecretory cells and are stored/released into hemolymph from the corpus cardiacum (neurohaemal organ). Other species synthesize a diuretic hormone in neurosecretory cells fused within the metathoracic ganglionic mass and release it from neurohaemal areas on abdominal nerves (*Rhodnius*). In *Glossina,* a similar hormone from the thoracic ganglion is released from neurosecretory axon endings in the abdomen (Gee, J., 1977). The release of these diuretic hormones into the hemolymph in turn affects the insects' regulatory activities on hemolymph through acceleration of primary filtrate secretion by the Malpighian tubules. Other examples of hormone releases which may affect the composition of hemolymph after their release are the adipokinetic hormone and hyperglycaemic factors which are found in the corpora cardiaca. Adipokinetic hormone is a small neuropeptide which stimulates the release of fat body diglycerides into hemolymph for transport to, and utilization by, *Schistocerca* flight muscle during flight (Stone, J. and Mordue, W., 1980). Hyperglycaemic factor(s) obtained from *Periplaneta americana* corpus cardiaca extracts cause elevations in their hemolymph trehalose levels, but the actual physiological role of these factor(s) is unknown (Stone, J. and Mordue, W., 1980). In addition to the hormone transport role, hemolymph may also be involved in influencing the steady-state concentrations of these factors. Hemolymph carboxylesterases which degrade juvenile hormone or its analogs have been reported (see DeKort, C. and Granger, N., 1981). Similarly, there is evidence that insect hemolymph may contain enzymes capable of destroying diuretic hormone which would assist in reducing the hormone level when elevated levels are no longer needed (Gee, J., 1977).

Transport of insoluble materials in hemolymph may be achieved by the use of carrier proteins. Insect hemolymph contains several lipoprotein components, one of which (diacylglycerol-carrying lipoproteins) has been shown to have the capability of receiving fat body diglycerides for transport to various sites (see Wyatt, G. and Pan, M., 1978; Chino, vol. 10). In addition, binding of insecticides to macromolecules in *Periplaneta* has been demonstrated (Boyer, A., 1975; Skalsky, H. and Guthrie, F., 1977).

4.2 Storage reservoir

A major function attributed to hemolymph is that of a storage reservoir for many materials essential for a variety of insect processes. The materials which may be stored reflect the metabolic or physiological state of the insect. In this regard it may merely serve as a generalized storage compartment during the more or less routine processes, or as a major reservoir during certain phases of molting or reproduction.

4.2.1 Metabolic pools

Insect body water content and water concentration must be maintained within certain limits for maintenance of physiological activity (Arlian, L. and Veselica, M., 1979). Water is the most concentrated molecule, as well as the most common one in actively metabolizing tissues, and has many functions. In serving as the matrix for metabolic and physiological activities it is responsible for: solvation; the three-dimensional configuration of macromolecules; bioelectrical currents associated with ion diffusion; nutrient, waste and metabolic intermediate movement by diffusion or bulk flow; lubrication of reproductive, digestive and other tracts; and other functions such as temperature control through evaporative cooling (Wharton, G. and Arlian, L., 1972). Mellanby, K. (1939) first suggested that a major water storage compartment in insects is the hemolymph. For it to efficiently function as a reserve, Edney, E. (1977) has pointed out that it should be possible for water to move alone from the tissues to the hemolymph and in the opposite direction in responding to tissue requirements. Movement of water without corresponding solute movements would result in large changes in hemolymph volume and osmotic pressure, to which the tissues must be tolerant. Alternatively, if excess solutes remaining in the hemolymph were excreted or sequestered in some manner, large changes in osmotic pressure would not occur. Many studies have revealed that large changes in hemolymph volume may occur when insects are exposed to various conditions (Edney, E., 1977). Further, there is a considerable amount of data accumulating that indicate that terrestrial insects are capable of strongly regulating hemolymph osmolality even with large fluctuations in hemolymph volume (Berridge, M., 1970; Coutchie, P. and Crowe, J., 1979; Machin, J., 1981). For example, Wall, B. (1970) found that during dehydration and subsequent rehydration of adult male *Periplaneta americana,* although the hemolymph volume changed significantly (a loss of 61% hemolymph water during dehydration and a regain to 109% of the original hemolymph water during rehydration), the osmolality remained relatively constant (normal: 400 mOsm; dehydrated: 447 mOsm; rehydrated: 372 mOsm). During this cycle, solutes were removed from the hemolymph (but not excreted) during dehydration and reappeared upon rehydration. The sequestration/release site(s) of the inorganic solutes

(K^+, Na^+) is/are not known, but fat body urates may be involved (Hyatt, A. and Marshall, A., 1977; Tucker, L., 1977). Tenebrionid beetle larvae *Onymacris* can also survive prolonged dehydration as well as rapid water uptake without exchanging solutes with their external environment (Machin, J., 1981). It appears that tissue water is partially regulated at the expense of hemolymph water, probably as a result of mobilization of osmotically active solutes in hydrated animals and by their storage and inactivation in dehydrated individuals. Coutchie, P. and Crowe, J. (1979) have provided information on all the major osmotically active hemolymph solutes at different levels of hydration for *Onymacris*. Figure 5 compares hemolymph levels of these solutes with the various beetle hydration states. It can be seen that all materials are regulated, and there is no single solute responsible for regulating water distribution between the hemolymph and the tissues. Although amino acids, chloride, sodium and trehalose appear to be equally responsible, little direct evidence is available regarding how the solutes (inorganic) are stored/released during dehydration/rehydration (Machin, J., 1981). The observed balance of amino acid and trehalose solutes might be due to the establishment of an equilibrium between larger polymers of these materials. Amino acids may be polymerized to, or released from, proteins or polypeptides (Djajakusumah, T. and Miles, P., 1966; Collett, J., 1976a,b). Trehalose might be converted into glycogen or metabolized or synthesized by various biochemical pathways (Coutchie, J. and Crowe, P., 1979).

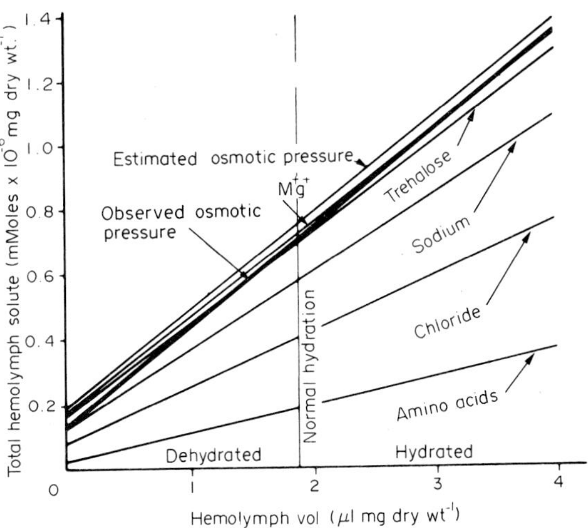

FIG. 5. Demonstration of the effects of hemolymph volume changes on the major hemolymph solutes in a tenebrionid beetle *(Onymacris)* during dehydration and rehydration. Note that the total hemolymph solutes do not remain constant, but change with hemolymph volume. The osmotic pressure is strongly regulated. Observed and estimated osmotic pressures are expressed as mOsm × 10^{-6} (mg dry wt)$^{-1}$. (From Machin, J., 1981.)

Rapid increases in insect metabolic demands illustrate the function of hemolymph as a metabolic reserve. For example, the dynamics of the hemolymph trehalose pool during *Locusta migratoria* flight has been investigated using [^{14}C]trehalose pulse labelling (Van Der Horst, D. *et al.*, 1978). They found (Fig. 6) that preflight trehalose fell from 25 to 11 mg ml^{-1} hemolymph in 20 min of flight, and was maintained at the latter level for 100 min. Contrarily, the specific activity of [^{14}C]trehalose remained constant during the first 30 min of flight, after which it decreased despite the constancy in hemolymph trehalose concentration. It appears that initially the source of metabolized trehalose was the hemolymph (not body carbohydrate), because this would have resulted in a decline in hemolymph pool-specific activity. However, once the constant flight level concentration was established (rate of utilization equalling mobilization), the specific activity diminished logarithmically with time. This indicated that as [^{14}C]trehalose was removed from hemolymph it was replenished from non-labelled (body) carbohydrates (Van Der Horst, D. *et al.*, 1978).

4.2.2 MOLTING

Insect growth and development is achieved, in part, by molting, resulting in periodic shedding of the cuticle. It is complex and includes various behavioral, physiological and biochemical activities which have been the subject of numerous studies (Jungreis, A., 1979; Reynolds, S., 1980). Although the physiology of molting is not completely understood, hemolymph obviously plays an important role in the process. This involvement includes serving as a source for certain materials used in degrading and reabsorbing cuticular components (endocuticle) which are conserved during the process, and as a reserve for materials destined for incorporation into the new cuticle (S. Anderson, K. Kramer, *et al.*, this

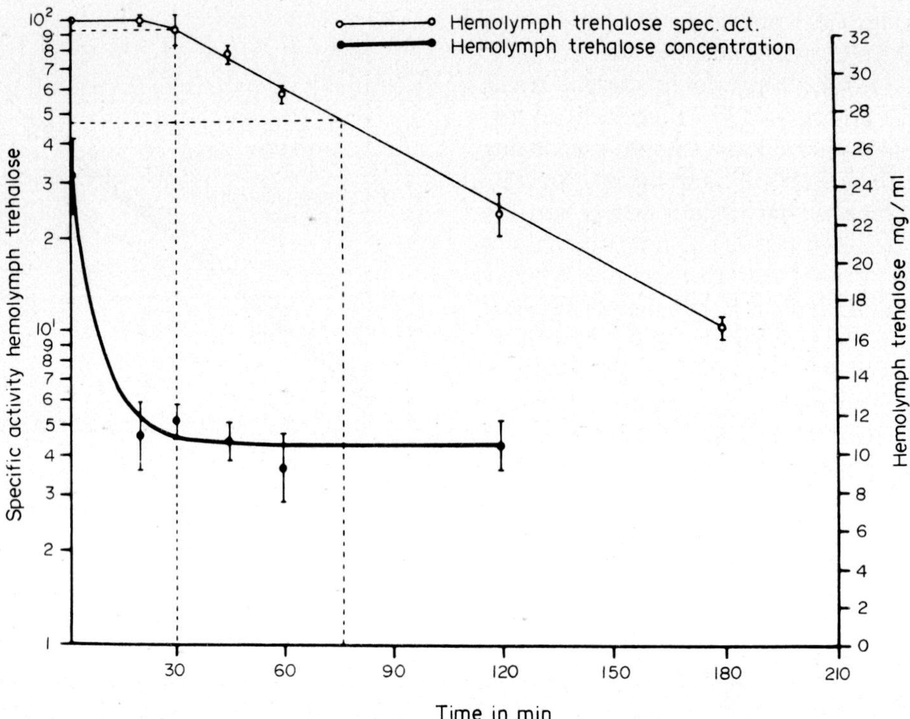

FIG. 6. Demonstration of the dynamics of the trehalose pool during flight of the locust (*Locusta migratoria*). Changes in the specific activity of injected [^{14}C]trehalose (plotted semi-logarithmically against time) are compared with the hemolymph trehalose pool. The preflight concentration of 25.1 mg trehalose ml^{-1} hemolymph falls to 11.4 mg trehalose ml^{-1} after 20 min of flight. This level is maintained throughout sustained flight (10.6 mg trehalose ml^{-1} after 120 min). However, the changes in specific activity of the hemolymph trehalose pool using [^{14}C]trehalose pulse-labelling, are essentially opposite to the change in the trehalose concentration. During the first 30 min of flight the specific activity remains virtually constant, despite the sharp decline in hemolymph trehalose. After this period the specific activity decreases in spite of the constancy of the trehalose concentration. Based on the apparent isotope dilution, initial flight activity utilizes trehalose from the hemolymph pool, and, in sustained flight, the trehalose pool level is maintained by an influx from other sources. (From Van Der Horst, D. *et al.*, 1978.)

volume). In addition, increases in hemolymph volume may assist in escape from the exuvia and inflating the new cuticle which will be discussed in section 4.3.

The molting process is under endocrine control (Richards, G., 1981; volumes 7 and 8). The decision to molt, and specific instructions to the epidermis for synthesis of specific enzymes and proteins associated with the formation of larval, pupal or adult cuticle, are mediated by hormones. The hormone levels associated with producing cuticular features specific to a particular insect stage have been quantified. However, details of the regulatory mechanisms involving two hormones (juvenile hormone and 20-hydroxy ecdysone) need clarification (Jungreis, A., 1979). During the molting process, the epidermal cells separate from the old cuticle (apolysis) forming a space (exuvial space) between it and the new epicuticle (cuticulin and homogenous layer) (Locke, M., 1974; M. Locke, vol. 2). The exuvial space is occupied by molting gel and in some insects by an ecdysial membrane (Zacharuk, R., 1976; K. Kramer, *et al.*, this volume). There is some disagreement among various workers regarding the origin of the exuvial space, but it appears that it is formed following secretion of ecdysial droplets (derived from fusion of Golgi secretory vesicles into the ecdysial membrane) and the cuticulin layer (Jungreis, A., 1979; M. Locke, vol. 2). The molting gel gives rise to molting fluid. In *Hyalophora cecropia* (silkmoths) the volume of fluid increases in the exuvial space one order of magnitude at the expense of the hemolymph (Jungreis, A., 1974). According to Wigglesworth, V. (1972) molting fluid arises mainly by exudation from the epidermal cells, and dermal gland involvement at earlier stages in molting is uncertain. During the larval–pupal ecdysis in *Hyalophora cecropia* and *Manduca sexta*, K^+ HCO_3^- and cuticle chitinase are released into the exuvial space (Jungreis, A., 1979). It is thought that the K^+ HCO_3^- serves to activate proenzymes and indirectly provide substrates for preactivated enzymes present in the molting gel. Comparison of *H. cecropia* molting fluid and hemolymph obtained during the larval–pupal transformation has indicated some major differences (Jungreis, A., 1974). For example, osmotic pressure was 70% greater in the molting fluid. This led Jungreis to propose that the epidermis actively transports K^+ into the exuvial space accompanied by passive bulk flow of H_2O, resulting in increased molting fluid volume.

The major function of the formed molting fluid is to degrade the procuticle which occurs to varying degrees in different insects. It is generally thought that only the endocuticle is degraded, and the layers remaining after digestion are lost at ecdysis (Wigglesworth, V., 1972; Zacharuk, R., 1976). The solubilized products of the cuticular digestion, and almost all of the molting fluid, are reabsorbed by the insect prior to ecdysis. Reabsorption probably occurs through the body surface in most insects (Zacharuk, R., 1976). It may occur via pinocytotic uptake through the apical membranes of the epidermal cells which process them (digestion in multivesicular bodies and lysosomes) followed by release into the cytosol (Jungreis, A., 1979). These products may then be sequestered in the epidermal cells or transported (via the hemolymph) to the fat body for storage. Some of these materials (i.e. sugars) may be reused in synthesis of new cuticle (Wigglesworth, V., 1972). Other cuticular materials, such as pigments, may be transferred to the hemolymph and finally sequestered in the midgut lumen (Jungreis, A., 1974, 1979). There are reports that molting fluid may also be swallowed by some insects (Wigglesworth, V., 1972; Zacharuk, R., 1976).

It should be noted that during the molting period parts of the tracheal system are also filled with fluid, which is assumed to be similar in composition to molting fluid (Wigglesworth, V., 1972). The tracheal system is filled with this fluid prior to hatching from the egg, and in new trachea formed at each molt. It is removed at hatching but the mechanism by which it is absorbed from the trachea is obscure (Reynolds, S., 1980).

Hemolymph volume is large at ecdysis but is reduced after the molt (Reynolds, S., 1980). In *Schistocerca gregaria* it appears that the hemolymph volume increase occurring prior to ecdysis is derived from cellular and gut water, since the total percentage of dry weight remains constant (Lee, R., 1961). In other insects (*Periplaneta* and *Calpodes*) the excretory system is oriented towards water conservation (antidiuretic) activities during the pre-ecdysial period (Reynolds, S., 1980). The hemolymph volume decreased from 240 μl to about 173 μl (28% decline) in *Periplaneta* during a 7 h interval following ecdysis (Mills, R. and Whitehead, D., 1970), but in *Schistocerca* the volume remained high for 24 h, followed by a decline 36–48 h post ecdysis (Lee, R., 1961). The post-ecdysial hemolymph volume reduction appears to be due primarily to excretion via the Malpighian tubule–rectal complex, but other routes of fluid loss may be involved, such as increased transpiration rates or labial gland secretions (Reynolds, S., 1980).

Secretion of new cuticle deposited under the old cuticle occurs after apolysis. The portion of the cuticle deposited prior to ecdysis usually can be distinguished (as exocuticle) from that laid down after shedding of the exuvia (endocuticle) (Reynolds, S., 1980). Materials used for producing the new cuticular layers are derived from several sources. Some of the materials absorbed from the molting fluid may be reused for new cuticle synthesis. For example, labelled sugars contained in larval *Hyalophora cecropia* appear to be specifically incorporated into pupal cuticle (Bade, K. and Wyatt, G., 1962). Changes in hemolymph proteins have been observed during the molting process (Steinhauer, A. and Stephen, W., 1959; Fox, F. and Mills, R., 1969; Duhamel, R. and Kunkel, J., 1978). There is evidence that hemolymph proteins contribute to cuticle formation (Hill, L. and Goldsworthy, G., 1968; Tobe, S. and Loughton, B., 1969; Fox, F. *et al.*, 1972). However, the relationship between hemolymph and cuticular proteins has not been clarified (Ruh, M. and Willis, J., 1974). Koeppe, J. and Gilbert, L. (1973) extracted immunologically similar *Manduca sexta* hemolymph carrier proteins (functioning as a dopamine or dopamine–metabolite carrier from hemolymph to cuticle) from the cuticle. They suggested that these proteins may pass through the epidermal cells unaltered. Further,

Geiger, J. *et al.* (1977) have reported possible involvement of hemocytes in cuticular protein synthesis. They were able to obtain labelled *Periplaneta americana* hemolymph proteins, purify, concentrate and inject them into freshly molted individuals. After 1 h, radiolabel was detected in serum, hemocyte and cuticular proteins. This information suggests that similar proteins may exchange or pass rapidly among these insect components, but there is no evidence that synthesis of these proteins actually occurs in the hemocytes themselves (Crossley, A., 1979).

Concentration levels of other materials such as tyrosine and phenylalanine which may be utilized in the cuticular tanning process also fluctuate during the molting cycle. In *Leucophaea maderae* and *Periplaneta americana,* tyrosine concentrations are low during the inter-ecdysial period, increase rapidly at apolysis but decline as tanning of the new cuticle proceeds (Wirtz, R. and Hopkins, T., 1977). These whole-insect changes are mirrored by corresponding cyclical patterns in the integument, hemolymph and fat body. It appears that tyrosine and/or phenylalanine are sequestered in an as yet unelucidated storage form during the pre-ecdysial (larval feeding) period. Tyrosine is sequestered in the form of β-alanyl-tyrosine (sarcophagine) which accumulates throughout larval growth and becomes the major non-protein, ninhydrin-positive substance in *Sarcophaga bullata* larval hemolymph. It decreases rapidly during the hardening and darkening (sclerotization) of the puparia (Bodnaryk, R., 1978). Similarly, in *Musca domestica* γ-glutamyl-phenylalanine is rapidly and extensively consumed during the pupal sclerotization process, its concentration reduced from 275 moles per insect (white puparium) to virtually undetectable levels after 12 h (darkened puparium) (Bodnaryk, R., 1974, 1978). There are other reports of sequestered aromatic compounds contained in hemolymph which may be important precursors of sclerotizing agents. Examples of these are: tyrosine-*O*-phosphate (*Drosophila* larvae), dopamine-3-*O*-sulfate (*Periplaneta americana*), "celerin" (proposed identity L-tyrosyl-*O*-acetyldopamine from *Celerio euphorbiae*) (Bodnaryk, R., 1978), and β-glucosyl-*O*-tyrosine (*Drosophila buckii*) (Chen, P. *et al.*, 1978; P. Chen, vol. 10; S. Andersen, this volume).

4.2.3 Developmental Changes

Development of holometabolous insects includes transformation from larval to adult form. The process involves destruction of most larval tissues and organs (histolysis) and construction of adult structures and organs (histogenesis). During metamorphosis, profound biochemical changes occur and here the hemolymph plays an important role. The concentration of certain pools (e.g. amino acids) may remain relatively constant, but there is an important class of hemolymph proteins ("storage") which do undergo significant changes (Chen, P., 1966; Mansingh, A. and Baqaya, V., 1971; L. Levenbook, vol. 10).

Protein storage in the hemolymph and fat body appears to be a common phenomenon during growth and development. Generally, a high rate of protein synthesis (fat body) occurs during the larval feeding stage. The mature larvae contain high protein levels in both the hemolymph and fat body. As protein synthesis declines during the pharate pupal stage, protein distribution changes; the specific hemolymph proteins are sequestered in the fat body (Chippendale, G., 1973b; Thomson, J., 1975; R. Dean, *et al.*, this volume). In *Calliphora* at pupariation more than 80% of the total nitrogen is protein nitrogen, and almost 60% of the total soluble protein present in the late larval stage of *Calliphora erythrocephala* is a specific protein, calliphorin (Kinnear, J. and Thomson, J., 1975). Kinnear and Thomson have found that during brown blowfly (*Calliphora stygia*) development there are four major hemolymph proteins which are immunologically, electrophoretically and structurally distinct. During metamorphosis the total plasma protein levels change quite significantly as follows: late feeding larvae, 200 mg ml^{-1} (hemolymph volume, 53 μl); pupariation, 108 mg ml^{-1} (40 μl); newly emerged adult, 29 mg ml^{-1} (20 μl).

In addition to calliphorin, specific storage proteins have been identified in *Bombyx mori* (Tojo, S. *et al.*, 1980) and *Manduca sexta* (manducin) (Kramer, K. *et al.*, 1980). Their function is unknown, but because of their relative abundance in larval and early pupal developmental stages and virtual absence in adults, it is thought that they have an important function. Experiments using [^{14}C]calliphorin injected into late-instar *Calliphora* larvae resulted in $[^{14}C]O_2$ release during adult development, and adult proteins were labelled (Kinnear, J. and Thomson, J., 1975). It is surmized (at least for *Calliphora*) that storage proteins synthesized during the feeding stage of last-instar larvae are utilized as an energy and amino acid source for protein synthesis during metamorphosis (Thomson, J., 1975; Wyatt, G. and Pan, M., 1978).

It should be pointed out that even though hemolymph storage protein is best documented during metamorphosis of holometabolous insects, it is possible that a similar phenomenon may occur in other developmental stages of other insects (Wyatt, G. and Pan, M., 1978).

4.2.4 Parental investment

Parental investment includes any activity which may increase survivability of the offspring (Thornhill, R., 1976; Alexander, R. and Borgia, G., 1979). Investment of hemolymph-borne or stored materials is important in insect reproduction since males and females may use this means to contribute to the material reserves of the young. In females, the major investment includes egg-yolk materials derived primarily from the hemolymph during oogenesis (R. King and S. Büning, J. Kunkel and J. Nordin, vol. 1). Also, certain viviparous females may transfer large quantities of nutrients to the developing embryos contained in their brood sacs (Ingram, M. *et al.*, 1977; Tobe, S., 1978; K. Davey, vol. 1). In addition to these provisions for embryo development, female accessory gland secretions may function to glue eggs together, form oothecal coverings or cocoons, and dissolve the walls of the spermatophore in the genital tract (Adiyodi, K. and Adiyodi, R., 1974). In males, parental investment may include such things as nourishment from glands, spermatophores, and mating (genital) plugs (Leopold, R., 1976; Thornhill, R., 1976). Recently Friedel, T. and Gillott, C. (1977) found that during copulation radiolabelled *Melanoplus sanguinipes* male accessory reproductive gland protein is transferred to the female. Most of the material is removed from the spermatheca and enters the hemolymph; some of it may be accumulated unchanged in the ovaries. In many cases details on the precise role of hemolymph in transport and storage of parental investment materials or their precursors is lacking, but the involvement of hemolymph is implied by their presence in it. Most of the information available regarding this subject is on oogenesis. Therefore, some discussion of the relation of hemolymph to oogenesis is included here.

There are two major types of oogenesis in insects, distinguished primarily by the cellular mechanisms that are used in synthesis of the reserves of RNA-containing organelles and other components of the ooplasm. In panoistic ovaries, the oocyte nucleus provides all of the RNA contained in the mature egg. Contrarily, meroistic ovaries (composed of various subcategories) contain a cluster of nurse cells or trophocytes which are derived from germinal tissue but do not form the mature egg. They are responsible for providing most of the RNA, and other substances also, to the developing oocyte and mature egg (Mahowald, A., 1972). The formation and subsequent incorporation of yolk into developing oocytes (vitellogenesis) has been a major research area for many years. Yolk materials are deposited in the ooplasm as globules or platelets, as proteins or lipids. It is also known that yolk-precursor substances may be provided by trophocytes, follicle cells, the oocyte or obtained directly from hemolymph (De Wilde, J. and De Loof, A., 1973).

A majority of studies on vitellogenesis have dealt with proteinaceous yolk formation and its incorporation into the yolk. Vitellogenins are the predominant female-specific yolk protein precursors that are produced extraovarially and which become the major yolk proteins, the vitellins (Wyatt, G. and Pan, M., 1978; Engelmann, F., 1979; Hagedorn, H. and Kunkel, J., 1979; J. Kunkel and J. Nordin, vol. 1). Hemolymph is the most important source of vitellogenin, which is synthesized and released by fat body tissue, and its uptake may occur in all types of ovaries (Mahowald, A., 1972). Fat body vitellogenin synthesis in some species may be induced by juvenile hormone released under appropriate physiological or environmental conditions (Wyatt, G. and Pan, M., 1978; J. Koeppe, *et al.*, vol. 8). Large amounts of vitellogenin may be produced during vitellogenesis. For example, during a 15-day period *Leucophaea maderae* may synthesize as much as 72 mg and deposit it in oocytes as yolk, representing an average rate of $0.21\,mg\,h^{-1}$ (Brookes, V., 1976). During the early stages of vitellogenesis in this insect, vitellogenin accounts for most of the protein synthesized by the fat body at that time. The titre of vitellogenin in hemolymph may fluctuate considerably, and is a function of the age of the insect, its rate and timing of production by the fat body and subsequent rate and timing of uptake by the ovaries. In *Hyalophora cecropia* it appears at the larval–pupal molt, attaining a high level through pupal diapause, followed by a rapid decline beginning with yolk deposition in the growing oocytes (see Engelmann, F., 1979). Vitellogenin appears in *Nauphoeta cinerea* hemolymph 2 days prior to the onset of its incorporation into oocytes (Buhlmann, G., 1976). The hemolymph of diapausing milkweed bugs, *Oncopeltus faciatus*, contains a vitellogenin that is modified in some manner after diapause and is subsequently sequestered by the oocyte where it forms the major yolk protein (Kelley, T. and Telfer, W., 1977).

The vitellogenins make up the largest amount (60–90%) of the yolk protein, but other proteins may be incorporated into the yolk (Engelmann, F., 1979; Hagedorn, H. and Kunkel, J., 1979). Vitellogenin uptake may occur quite rapidly, entering the oocyte by micropinocytosis, a process which appears to be quite selective. Using radiolabelled vitellogenins from *Hyalophora cecropia* and *Blattella germanica*, Kunkel, J. and Pan, M. (1976) found that *B. germanica* oocytes (*in vivo*) rapidly sequestered their own vitellogenins, but sequestered *H. cecropia* vitellogenins at a low rate (comparable to that of a non-vitellogenin protein). Similarly, *H. cecropia* oocytes sequestered their own vitellogenins rapidly *in vitro,* but those obtained from *B. germanica* were sequestered at a negligible rate. Vitellogenins may be sequestered in the oocyte at levels 20–100 times those found in hemolymph. However, due to the mechanisms involved, all proteins contained in hemolymph may be taken up (Hagedorn, H. and Kunkel, J., 1979).

Uptake of other materials into the yolk is possible. Recently it has been reported that significant amounts of radiolabelled inulin injected into the hemocoel of *Oncopeltus fasciatus* females was found in the ovaries. Approximately 15% of the injected label was sequestered in the eggs during the first ovarian cycle (Aldrich, J. *et al.*, 1981). This material does not occur in insects, but its incorporation into oocytes under experimental conditions may serve as an indicator of hemolymph-borne materials normally taken up by ovaries. The mechanism of inulin incorporation into oocytes is not known. It may be due to the selective process, or simply due to passive incorporation during vitellogenin pinocytosis (Anderson, L. and Telfer, W., 1970). The precise mechanisms of carbohydrate and lipid incorporation into yolk are also not clear (Engelmann, F., 1970; Mahowald, A., 1972; De Wilde, J. and De Loof, A., 1973). Studies of their incorporation in relation to their source, transport and storage in hemolymph would be useful.

Oocyte reabsorption, or oosorption, is characterized by the cessation of vitellogenesis. It may be caused by a variety of factors such as nutritional deficiencies, virginity or absence of males, social pressure, ovipositional blocks, seasonal factors, parasitism, etc. (Bell, W. and Bohm, M., 1975). Generally, the processes of oosorption involve the progressive decline in oocyte volume to a point where protein and lipid yolk spheres disappear (De Wilde, J. and De Loof, A., 1973; Bell, W. and Bohm, M., 1975). It is probable that the products released by oosorption find their way into the hemolymph. Starvation in *Periplaneta americana* has been correlated with increases in hemolymph vitellogenin which apparently are not due to synthesis, but to vitellogenins derived from oocyte reabsorption (Bell, W., 1971).

4.3 Hydrostatic pressure

The open design of the insect body cavity is such that contraction of a portion of it results in increased hemolymph hydrostatic pressure which can be transmitted throughout the hemocoel. Changes in hemolymph pressure play important roles at various points in the lives of many insects. The most pronounced of these occur during molting, egg and pupal escape and wing inflation. Other activities which may involve pressure changes are: extension of genitalia and the proboscis, shedding of wings and other appendages, reflexive bleeding, and assisting in appendage hemolymph circulation (Jones, J., 1977; Provansal, A. *et al.*, 1977; T. Miller, this volume). Development of hydrostatic pressure is directly related to hemolymph volume, gut volume (containing swallowed air or food) and the general body musculature tonus. Specific hemolymph

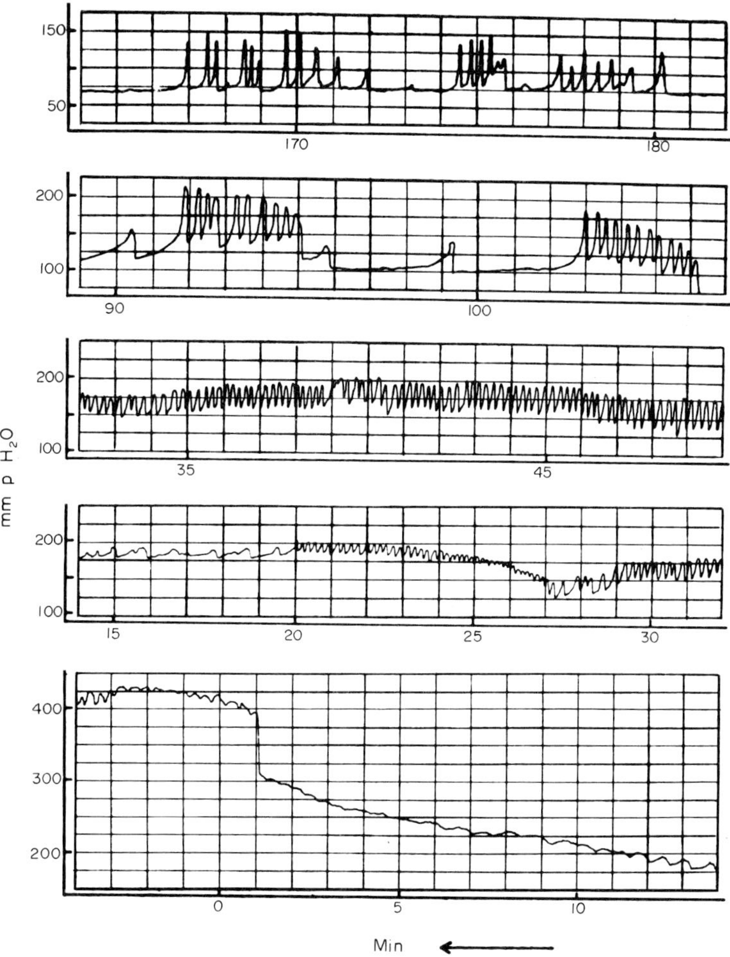

FIG. 7. Demonstration of hemolymph pressure changes during larval–pupal ecdysis of *Dermestes vulpinus*. The time is measured backwards from the moment of exuvial rupture. (From Sláma, K., 1976.)

pressure pulses, as well as more prolonged increases in pressure, may be generated by contraction of abdominal muscles. The pressures may vary considerably, from negative values (−50 to −100 mmH_2O) to fairly high positive ones (400 to 600 mmH_2O) depending upon the hydration and activity state of the insect (Sláma, K., 1976).

4.3.1 Molting, Egg and Pupal Escape

During molting, hemolymph pressure used to assist in bursting the old cuticle may be involved in two ways. These are:

(1) a general application of pressure resulting from an increase in the pharate insect volume (swallowing air or water); and

(2) local application of pressure, which is considered to be the most important (Reynolds, S., 1980).

Increases in hemolymph pressure due to air swallowing may range from 20 to 95 mmHg (270 to 1286 mmH_2O), but development of high hemolymph pressures may not be directly correlated with swallowing air or fluids (Reynolds, S., 1980). Figure 7 shows the changes in hemolymph pressure during larval–pupal ecdysis of *Dermestes vulpinus*. It can be seen that over the 3 h interval to the point of actual exuvial rupture, a series of peaks are apparent, which are associated with a continuous increase in the basic pressure level. At rupture, the pressure reaches 400 mmH_2O (Sláma, K., 1976).

Insects may utilize highly localized pressure applied anteriorly to burst open their egg shells. In *Aedes aegypti* the pharynx is responsible for pumping hemolymph into the head, causing it to swell. The presence of a small hatching spine located on the dorsal head surface in conjunction with the increased hydrostatic pressure applied to the chorion results in eggshell fracture and allows the mosquito larvae to escape (Jones, J., 1977). In *Schistocerca gregaria* it appears that splitting of the egg occurs by a concentration of pressure anteriorly derived from general peristalsis. This stretches the eggshell above the thorax and causes it to split transversely above the thorax (Reynolds, S., 1980).

Jones, J. (1977) has described the complex sequence of events associated with the emergence of an adult fly. During this period hemolymph pressures may range from 60 to 120 mmHg (812 to 1625 mmH_2O) (Cottrell, C., 1962). Hemolymph pressure changes during metamorphosis of *Tenebrio molitor* have been studied recently (Provansal, A. *et al.*, 1977). The hemolymph pressure pulses range from 50 to 250 mmH_2O during the 2.5 h interval prior to ecdysis. As the point of pupal rupture approaches, the pressure increases, as well as the amplitude of the contractions (similar to those illustrated in Fig. 7). At pupal rupture, the pressure reaches 400 to 600 mmH_2O and the old cuticle splits along the middle pronotal and mesonotal suture. After pupal escape the pressure fluctuates, dropping temporarily. Later a series of new hemolymph pressure pulses appear (above 600 mmH_2O) which are associated with expansion of soft appendages of the newly emerged adult. The periodic pulsations of hemolymph pressure in *T. molitor* are independent of the heartbeat and are attributed to abdominal pumping. Control of the abdominal pump responsible for the extracardiac pressure pulsations is independent of the brain, and the nerve impulses, which directly control pumping activities, arise from the mesothoracic ganglion (Sláma, K. *et al.*, 1979).

4.3.2 Wing inflation

The involvement of hemolymph in expansion of wings after adult ecdysis has been known for some time, dating back to the 1600s (Jones, J., 1977). Inflation of wings with hemolymph may occur immediately after exuvial escape (locusts) or after some delay (blowflies and hornworms) (Reynolds, S., 1980). In the hornworm, *Manduca sexta*, wing-spreading behavior has two components. The first involves movements of the thorax which shifts wing position. The second component includes wing expansion by hemolymph. Pressure is generated by abdominal tonic contraction and air-swallowing appears to be of little importance during the process (Truman, J. and Endo, P., 1974). Hemolymph pressures generated during wing expansion have been measured. In *Bombyx mori* and *Pieris brassicae* they have been found to reach about 51 and 31 mmHg (691 and 420 mmH_2O), respectively (Moreau, R., 1974). In *Calliphora erythrocephala* and *Sarcophaga barbata*, maximum pressure values of 60 and 95 mmHg (812 and 1290 mmH_2O), have been observed (Cottrell, C., 1962). In *Schistocerca gregaria* the process of wing expansion occurs when the gut is fully inflated with air, providing for increased pressure. Additional pressure is applied with tonic compression controlled by an expansional motor program (Hughes, T., 1980). It has been observed that wings removed from ecdysing adult locusts (and other insects) will expand on their own (Reynolds, S., 1980). Therefore, the significance of hemolymph pressure in post-ecdysial wing expansion is not entirely clear.

Certain Embioptera use hemolymph pressure to stiffen their wings prior to flight initiation which is not associated with ecdysis (Jones, J., 1977). This group of insects has undergone a special modification of their wing structures during their adaptation to inhabit silken galleries. In contrast to other winged insects such as ants and termites, which inhabit earthen and wooden galleries and which lose their wings after nuptial flights, Embioptera reduced the disadvantage of having rigid wings by evolving greater wing flexibility. Although in some species the wings have been reduced to setae, others have well-developed, sclerotized hemolymph sinuses in certain wing veins which allow for them to become rigid. During flight, hemolymph pressure increases as the wings are expanded and the turgid sinuses stiffen the wings. When the wings are returned to the resting position over the dorsum, the sinuses flatten and the wings become flexible (Ross, E., 1970).

4.4 Thermoregulation and frost protection

As a group, insects have been successful in inhabiting areas where extreme environmental temperatures occur. In order to live under these conditions they may possess the ability to make rapid behavioral and physiological adjustments or synchronize their life cycles with long-term seasonal changes. Two general areas included here are thermoregulation and frost protection, because they share an involvement with hemolymph in two rather distinct ways. The relationship between hemolymph and thermoregulation involves changes in hemolymph circulation patterns during certain activities. On the other hand, the relationship between hemolymph and frost protection are derived from changes in hemolymph composition. (M. May, vol. 4.)

4.4.1 Thermoregulation

Heinrich, B. (1981a) has recently pointed out that temperature regulation involves the assumption that an animal maintains its body temperature below some maximum set point or above some minimum set point (or both), for at least a portion of its activity period, even though environmental thermal conditions may vary. Ectotherms acquire heat from the environment by radiation, convection or conduction which determines their body temperature. Endotherms acquire heat from their own metabolism for establishment of body temperatures above ambient temperatures. Both of these types of thermoregulators differ from poikilotherms whose body temperatures are variable and dependent on ambient temperatures (Heinrich, B., 1981a). Traditionally, insects have been viewed as ideal poikilotherms. However, recent work has revealed that thermoregulation does occur in insects, some using both endothermic and ectothermic processes. In addition, there is reason to believe that representatives of nearly every major insect order may be found capable of thermoregulation (May, M., 1979). There are several benefits which may be derived from the ability to thermoregulate which have been discussed by Heinrich, B. (1981b). These include such things as broadening thermal niches, allowing for necessary reproductive activities, more rapid larval development, enhanced food-gathering capabilities and predator escape, and combating disease. Of these, the broadening of an insect's thermal activity niche is the most important from the standpoint of hemolymph involvement during flight activities. In most insects flight muscle is the exclusive metabolic heat source of thermoregulatory significance (May, M., 1979). However, recently it has been reported that elephant beetles (*Megasoma elephas*) increase their metabolism when ambient temperature is lowered. This metabolic response is apparently not associated with locomotion or other detectable activity and body temperature oscillates with oxygen consumption (Morgan, K. and Bartholomew, G., 1982).

During flight, large, well-insulated moths and bees are able to maintain thoracic temperature independent of ambient temperatures over a wide range (Kammer, A. and Heinrich, B., 1978). Further, their metabolic rate is independent of ambient temperatures over the range in which their thoracic temperatures can be stabilized. The maximum efficiency for flight muscle is determined by the enzyme systems which are involved; and therefore the requirement for thermoregulation of the thoracic flight musculature. During flight, heat production is high because of the metabolic rate required for this activity. Heat released in the thorax may represent 80–90% of the energy produced by the flight musculature (Kammer, A. and Heinrich, B., 1978). Stabilization of thoracic temperature during continuous flight is achieved by controlling heat loss rates and not heat production rates. Maintenance of relatively constant thoracic temperatures during flight is dependent on optimizing heat loss rates at high ambient temperatures or heat retention rates at low ambient temperatures. Large insects (e.g. moths, bumblebees, beetles) that regulate their thoracic temperature at high ambient temperatures utilize hemolymph circulation to transfer excess heat from the thorax to the abdomen where the heat is readily lost to the environment. Contrarily, when ambient temperatures are low, heat transfer to the abdomen via the hemolymph is reduced (Kammer, A., 1981). For example, the bumblebee (*Bombus vosnesenskii*) has a thoracic–abdominal structural design such that heat generated in the thorax may be efficiently retained at low ambient temperatures or released at high ambient temperatures (Fig. 8a). A narrow

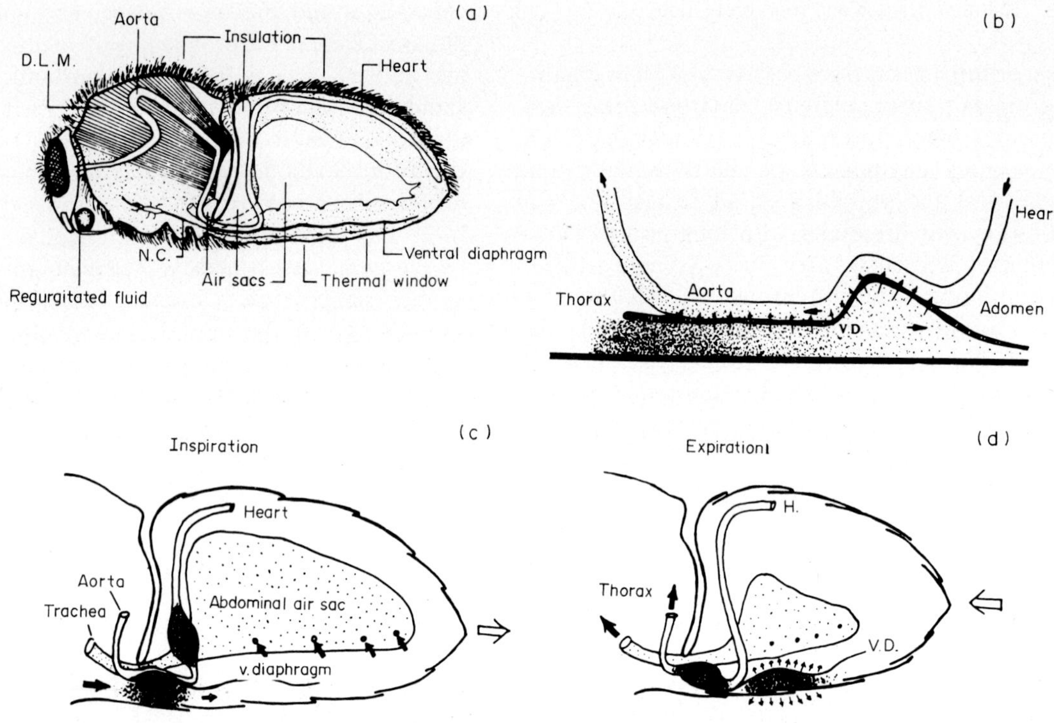

FIG. 8a–c. Probable involvement of hemolymph in bumblebee (*Bombus vosnesenskii*) thermoregulation. **a:** Anatomy of *Bombus vosnesenskii* relevant to thermoregulation. The insect is covered (thorax and dorsal portion of the abdomen) with insulation. Note the ventral abdominal "thermal window" (absence of insulation); note also, the droplet of fluid on the proboscis which may be used in evaporative cooling (see text). D.L.M. = dorsal longitudinal muscle. N.C. = nerve cord. **b:** Schematic drawing of the anatomical arrangement of the aorta and ventral diaphragm in the petiole (junction between the thorax and abdomen), and the probable countercurrent heat exchange which would return heat (represented by dots) to the thorax when hemolymph is flowing simultaneously in both directions. V.D. = ventral diaphragm. **c,d:** Schematic representation of probable air and hemolymph flow cycle through the petiole during inspiration (**c**) and expiration (**d**) at high thoracic temperatures, showing the elimination of the countercurrent heat exchanger. In part, due to the synchronized pulsation of the ventral diaphragm (V.D.) that acts as a two-way valve, hemolymph (heavy stippling) passes as a pulse through the petiole from the thorax to the abdomen during inspiration. During expiration, hemolymph pulses in the opposite direction as the V.D. valve at least partially closes the fluid passage out of the thorax, while at the same time enlarging both the fluid and air passages into it. ⇨ = abdominal movements; ➡ movement of hemolymph/air; → direction of heat flow. (From Heinrich, B., 1976.)

passage contained within the petiole connecting the thorax with the abdomen is constructed in a manner such that countercurrent exchange in hemolymph flowing between these regions may allow for heat retention in the thorax (Fig. 8a,b). However, it appears that the countercurrent exchanges can be circumvented when a bumblebee becomes overheated (high ambient and thoracic temperatures) by shunting heated hemolymph to the abdomen (Heinrich, B., 1976). Pumping activities of the heart, ventral diaphragm and abdomen are involved in directing this change in hemolymph flow (Fig. 8c,d). At low thoracic temperatures the heart and the ventral diaphragm beat at variable rates and amplitude, and are asynchronous with one another. When the thorax temperature exceeds 43° both the heart and ventral diaphragm beating rates increase in frequency and amplitude, becoming more or less synchronous. In addition, heated bumblebees pump their abdomens vigorously at the same frequency as the heart and ventral diaphragm pulsations.

Ectothermal regulation in insects is achieved primarily by body orientation or postural control of solar energy uptake, microhabitat selection and alteration of diel activity patterns (May, M., 1979). Positive orientation (basking) occurs in many insects when body or air temperatures are suboptimal by positioning the longitudinal body axis perpendicular to solar radiation. It has been suggested that hemolymph flowing through the wings of basking butterflies could transfer heat absorbed in the wings to the thorax, facilitating thoracic heating (Clench, H., 1966). Although it would appear to be an attractive hypothesis, supportive experimental evidence is lacking and alternative explanations of wing involvement in basking butterflies have been proposed (Wasserthal, L., 1975; May, M., 1979; Casey, T., 1981).

The role of evaporative cooling as a means of thermoregulation has been examined in insects. Generally, the water reserves (hemolymph) of most insects are not of sufficient size to allow evaporative cooling as normal means for thermoregulation (May, M., 1979). In situations where high body temperatures are reached, transpiration also increases, providing for some cooling. However, evaporative cooling by transpiration may be viewed as a survival mechanism allowing for insect survival under conditions of acute heat exposure, and not as a regulatory mechanism. In these situations the hemolymph would most likely be a major source of water. There are reports of insects which extrude fluid droplets from various body parts (e.g. bumblebees — mouthparts; aphids — anus; sawfly larvae — anus) which may contribute to body heat reduction in active insects with elevated body temperatures (May, M., 1979). Under these conditions evaporative cooling would be favorable if the insect has ready access to a water supply or is capable of storing large water reserves.

4.4.2 Frost protection

In order for insects to survive low temperatures encountered in cold regions of the world, they usually must be capable of making certain physiological adjustments. Low temperatures may affect insects in several ways, but a critical factor important in frost protection is whether freezing, or how freezing, of an insect occurs. It appears that insects have developed two general strategies which allow them to achieve variable degrees of frost protection. These are supercooling (avoidance of freezing) and frost resistance (tolerance to freezing) (Asahina, E., 1969; Ohyama, Y. and Asahina, E., 1972). Hemolymph may contain cryoprotectants which apparently play important roles in preventing cold injury by providing for either one, or perhaps both, of these types of frost protection.

Supercooling results from the unavailability of nucleation sites, from which ice crystals may form. When supercooled insects are exposed to ice crystals or appropriate nucleation sites, they will freeze instantly (Salt, R., 1961). Most insects supercool to some extent, but unless they are protected by substantial amounts of antifreeze materials, they are not able to supercool to temperatures near $-30°$ (Heinrich, B., 1981b). There are several materials which may be accumulated in hemolymph which may allow for supercooling as well as lowering the freezing point of hemolymph below this level. Following the discovery that immature diapausing insects may contain glycerol (Wyatt, G. and Kalf, G., 1958), subsequent work has suggested that this polyhydric alcohol (and other polyols such as sorbitol, mannitol and threitol) was responsible for increasing supercooling in insects (cold-hardening) (Miller, L. and Smith, J., 1975; S. Friedman, vol. 10). For example, in *Bracon cephi* glycerol accumulations (as great as 5 molal) produced hemolymph supercooling points as low as $-47.2°$ with a melting point depression of $-17.5°$ (Salt, R., 1959). Exposure of *Galleria mellonella* pupae to cold temperatures will induce an increase in hemolymph glycerol content (from 64 to 135 mg%) which is associated with an increase in osmotic pressure from 300 to 542 mOsm (Marek, M., 1979). Because glycerol and sorbitol production is usually associated with diapause it is likely that their production is under hormonal control (Tsumuki, H. and Kanehisa, K., 1979; Steele, J., 1981; O. Yamashita and K. Hasegawa, vol. 1). It has been found that both of these polyols are synthesized at the expense of glycogen during diapause or periods of exposure to low temperatures, but they are reconverted into glycogen when diapause or low-temperature exposure is terminated (Wyatt, G., 1967; Steele, J., 1981; J. Steele, vol. 10).

In addition to the polyols, some insects may contain protein antifreeze agents which function in a manner similar to those occurring in polar fish (Heinrich, B., 1981b). One such protein has been isolated and partially characterized from *Tenebrio molitor* larvae (Patterson, J. and Duman, J., 1979). The hemolymph of nine other insect species (eight Coleoptera and one Orthoptera) which overwinter under the bark of dead trees, also have been shown to contain antifreeze proteins (thermal-hysteresis proteins) (Duman, J., 1979). According to Heinrich, B. (1981b) most of the antifreeze materials found to occur in insects not only lower the freezing point (non-colligatively), but also the supercooling point. However, the mechanisms responsible for supercooling are not well understood.

Insects exposed to extremely low temperatures cannot avoid becoming frozen, even in the presence of antifreeze materials and with the capability for a high degree of supercooling. Some insects do in fact survive ice formation in their bodies. The processes involved in the avoidance of extensive tissue damage by freezing and thawing is also not well understood. It is thought that death by freezing is caused by several factors such as protein denaturation and disruption of cellular components by ice crystals (Heinrich, B., 1981b). It appears that elevated polyol levels (glycerol) and the presence of nucleation agents may allow for the freezing of insect tissue in a non-detrimental manner. Thus, in addition to its function as an antifreeze–supercooling agent, glycerol appears to confer protection to tissue when it freezes, perhaps by preventing protein denaturation. Freezing-tolerant insects appear to freeze at comparatively high temperatures (Miller, L., 1969) even with relatively high levels of polyols in their body fluids (Lee, R. *et al.*, 1981). This observation has led several groups of workers to examine insect fluids for the presence of cryoprotectant nucleation agents. Zachariassen, K. and Hammel, H. (1976) found that hemolymph obtained from freeze-tolerant beetles (*Eleodes blanchardi*) had supercooling points ranging from −5.5 to −7.0°, whereas hemolymph obtained from freeze-sensitive beetles *Iphthimus laevissimus* froze at temperatures ranging from −12 to −20°. Nucleating factors obtained from *E. blanchardi* hemolymph have been found effective in freezing solutions containing organic solutes at their melting point depression levels, reducing supercooling effects (Lee, R. *et al.*, 1981). The effects of some nucleating agents are partially, but not completely, destroyed by heating (Zachariassen, K. and Hammel, H., 1976; Somme, L., 1978), and it appears that some nucleating agents may be proteins (Duman, J., 1980). In overwintering *Dendroides canadensis* larvae, concentrations of the cryoprotectants, glycerol and sorbitol, ice-nucleating agents and thermal-hysteresis proteins (antifreeze proteins) all peak in the winter. The hemolymph ice-nucleating agents limit undercooling to approximately 1° below the hemolymph freezing point (Duman, J., 1980). It appears that the function of the nucleating agents found in hemolymph is to ensure the formation of extracellular, rather than lethal, intracellular, ice crystals.

Obviously the processes involved in frost tolerance are quite complex, and much more study is needed before we have a clear understanding of this phenomenon. Information gathered on this topic so far seems to indicate that frost tolerance is related to changes which occur between the intra- and extracellular compartments. It appears that the extracellular compartment always freezes before the intracellular compartments, resulting in less cellular component damage (Heinrich, B., 1981b).

4.5 Protection and defense

Insect hemolymph may contain certain components which provide for protection and defense from various hazards which they may encounter. These hazards fall into three general categories: physical injury, entry of foreign bodies and compounds (Boman, H. and Götz, P. this volume), and protection from predation. Physical injury may result from damage to the integument caused by fortuitous mechanical wounding or from the activities of predators or parasites. Disease organisms, foreign compounds and parasites which gain access to the hemocoel may be exposed to a series of responses mediated by components contained in, or activities associated with, hemolymph. Finally, insects may utilize hemolymph specifically as an agent to ward off attack from a would-be predator. This may simply involve the release of hemolymph to the exterior, confusing and impairing the predator (reflexive bleeding), or it may involve sequestration of natural products in hemolymph which are distasteful or toxic to their enemies.

4.5.1 Physical injury

Damage to the integument of many insects results in a defensive response involving hemolymph coagulation, which serves to seal off the wound. As a result, hemolymph loss is reduced, bacterial access to the hemocoel is impaired and wound repair is facilitated (Gregoire, C., 1970; Ratcliffe, N. and Rowley, A., 1979). In most cases hemocytes play an active role in this process. Numerous reviews on hemocytes discuss coagulation along with the numerous roles which have been ascribed to them. Some of these are: Jones, J. (1964, 1977); Gregoire, C. (1970, 1974); Arnold, J. (1974); Whitcomb, R. *et al.* (1974); Crossley, A. (1975, 1979); Ratcliffe, N. and Rowley, A. (1979); Shapiro, M. (1979) and A. Gupta in this volume. In this section the injury response exclusive of reactions to invasion of foreign elements and materials (section 4.5.2), will be summarized.

Studies on hemolymph coagulation have revealed there are two distinct processes involved. These are: cell agglutination (coagulation) and plasma coagulation. The contribution of both of these processes during clotting may vary depending on the insect species. Hemocyte coagulation (coagulocytes) involvement in clot formation may follow one of four general patterns (I, II, III and IV) (Gregoire, C., 1974). Coagulocytes corresponding to pattern I form islands of coagulation after the release of cytoplasmic material. Plasma surrounding these islands of coagulation (having a granular appearance resulting from these circular areas of greater density) becomes organized into networks of granular fibrils. In pattern II coagulation, coagulocytes extrude long thread-like pseudopodial processes which anchor to, and may entrap, particulate hemolymph materials. As the resulting cytoplasmic meshwork develops, plasma reacts within the network forming elastic, contractile, and transparent veils that do not display distinct islands as in pattern I. Pattern III coagulation includes reactions of both patterns I and II. Coagulocytes produce both cytoplasmic meshworks and circular areas of greater density. Pattern IV hemocyte activity does not include the formation of pseudopodia or the release of cytoplasmic material, and no detectable change occurs in the vicinity of these cells upon exposure to a foreign surface (Gregoire, C., 1974; Crossley, A., 1975). Although the general patterns of hemolymph coagulation have been described, and have been the topic of many studies, details of the processes involved and the composition of hemolymph clots have not been determined (Wyatt, G. and Pan, M., 1978). Generally, it appears that in many insect species coagulation is a continuous process which begins with rapid changes taking place in coagulocytes that are sensitive to contact with foreign surfaces. Following these initial changes, the plasma gels, form a coagulum of variable size and viscosity (Gregoire C., 1974). In addition, coagulocytes may not be the only reactive cells that change during hemolymph coagulation. There is evidence that in some species other hemocytes (granulocytes and spherulocytes) display functional and structural relationships with coagulocytes (Crossley, A., 1979).

It appears that some insects have a fibrinogen-like plasma clotting protein (coagulogen), which is involved in the polymerization process that produces the clot (Brehelin, M., 1979; Barwig, B. and Bohn, B., 1980). In *Locusta migratoria*, plasma coagulation has been characterized as being a high-molecular-weight lipo-glyco-proteic complex (Brehelin, M., 1979).

Hemolymph coagulation as an initial step in wound healing has been observed in various tissues (Gregoire, C., 1974). In addition to hemocyte participation in coagulum formation, they may increase in numbers, phagocytize cellular debris and participate in the formation of new tissue (Arnold, J., 1974; Crossley, C., 1979; Shapiro, M., 1979). Details on the events involved in wound repair are available in only a few insect species (Ratcliffe, N. and Rowley, A., 1979). It appears that plasmatocytes are the major cell types which are involved, but spherulocytes and granulocytes may also participate in the repair processes (Arnold, J., 1974; Rowley, A. and Ratcliffe, N., 1978; Ratcliffe, N. and Rowley, A., 1979).

It has been suggested that hemocytes involved in the wound healing process respond to a wound or injury factor produced by damaged hemocyte or epidermal cells (Bohn, H., 1975; Ratcliffe, N. and Rowley, A., 1979; Shapiro, M., 1979). Cherbas, L. (1973) has reported on an injury factor (haemokinin), purified from plasma and epidermal tissue. She found that this factor increased the motility and adhesiveness of hemocytes in saturniid pupae. The possibility that factors released at the site of wounding mediate in wound repair is quite likely. In addition to action at or near the site of injury, wound or injury factor(s) could affect other tissues located elsewhere in the insect. For example, Stevenson, E. and Wyatt, G. (1962) found the protein synthetic activity of fat body and midgut tissues increased after injury to the integument of diapausing *Hyalophora cecropia* pupae. Neither the source of this stimulus, nor the nature and significance of the proteins produced, are known (Wyatt, G., 1980). However, Crossley, A. (1979), has discussed some of the reports on appearance of injury proteins in hemolymph and the possible roles they may perform.

4.5.2 Disease organisms, parasites and foreign materials

Organisms or foreign material gaining entry to the hemocoel come into contact with defense mechanisms associated with the hemolymph. It is generally believed that the hemolymph defense mechanisms rely heavily on both cellular and humoral activities, but their relative relationship with one another has yet to be determined (Whitcomb, R. *et al.*, 1974; Ratcliffe, N. and Rowley, A., 1979). Insect diseases are produced by pathogenic bacteria, fungi, rickettsiae, protozoa and nematodes. Insect parasites may also invade the hemocoel in an attempt to establish a living relationship within the host (Ratcliffe, N. and Rowley, A., 1979; Shapiro, M., 1979). Cellular defense mechanisms include phagocytosis, encapsulation and nodule formation (see Boman, H. and Götz, P. this volume). Phagocytosis occurs when specific hemocytes (predominantly plasmatocytes) encounter smaller particulate material, such as bacteria, fungi, protozoa and India ink. Phagocytosis may be preceded by rapid coagulation of foreign particles introduced into the hemocoel (within 10 s of injection of latex beads and ink particles into *Calpodes ethlius* larvae) (Neuwirth, M. and Lai-Fook, J., 1977).

Encapsulation occurs when foreign materials gain entry to the hemocoel which are too large for hemocyte phagocytosis. Materials which may be encapsulated include such things as nylon fibers, nematodes, parasitoids, large protozoa, etc., and a variety of hemocyte types have been reported as participants in this type of reaction (Ratcliffe, N. and Rowley, A., 1979). Formation of nodules involves a defense reaction against large amounts of small foreign material entering the hemocoel, which overwhelms the normal phagocytic response. They were considered to be the product of both phagocytic and encapsulation processes (Whitcomb, R. *et al.*, 1974). However, recent work, reviewed by Ratcliffe, N. and Rowley, A. (1979), indicates that nodule formation occurs in two stages: rapid coagulation (granulocytes and/or coagulocytes) and specific attachment of large numbers of plasmatocytes which form the outer sheath.

Although insects have been found to lack an antibody system functionally equivalent to that known to occur in higher vertebrates (Whitcomb, R. *et al.*, 1974), they apparently possess humoral agents that provide protective functions (Wyatt, G. and Pan, M., 1978). Hemagglutinating activity of insect hemolymph against vertebrate erythrocytes has been observed in several species of insects (Scott, M., 1972; Amirante, G. *et al.*, 1976; Amirante, G. and Mazzalai, F., 1978; Komano, H. *et al.*, 1980; Rowley, A. and Ratcliffe, N., 1980). Although the physiological role of hemagglutinins is not known, they appear to be associated with defense mechanisms (recognition of foreignness, enhancement of phagocytosis and encapsulation) (Hapner, K. and Jermyn, M., 1981). Komano, H. *et al.* (1980) have purified a lectin obtained from *Sarcophaga peregrina* larvae hemolymph after integumental injury, which agglutinates sheep red blood cells. These workers suggest that the lectin hemagglutinating activity may assist in phagocytosis of foreign materials that might gain entry to the hemocoel during wounding, or may play a role in repairing injured tissues. Induction of humoral immunity to soluble proteins (honeybee and snake venom) injected into an adult male *Periplaneta americana* has been reported (Karp, R. and Rheins, L., 1980; Rheins, L. *et al.*, 1980). The responses they observed peaked after 2 weeks and apparently were specific to the immunizing toxin, not to a heterologous toxin. In addition, passive transfer studies indicated that protection could be transferred to non-immunized (virgin) cockroaches with cell-free hemolymph containing the hemolymph immune factor (Rheins, L. *et al.*, 1980).

Humoral lysozyme activity may also play an important role as a defense mechanism against certain bacteria. Lysozymes are enzymes which are capable of lysing the cell walls of certain bacteria (primarily Gram-positive). They are distributed widely among many organisms including insects (Whitcomb, R.

et al., 1974), and have been partially characterized (Powning, R. and Davidson, W., 1973, 1976). Hemolymph lysozyme levels have been shown to rise in response to injury or introduction of various foreign materials into the hemocoel (Anderson, R. and Cook, M., 1979). Because of its normal presence and the observed inducible nature of its production, it may function to some extent as an antimicrobial agent. For example, after active immunization of *Galleria mellonella* larvae, a rise and fall in protective immunity against *Pseudomonas aeruginosa* and hemolymph lysozyme action results; indicating that lysozyme could serve as a form of an acquired immune response against bacterial infections (Jarosz, J., 1979). In addition to lysozyme and hemagglutinins, the presence of peroxidases and phenoloxidases in hemolymph or at an injury site has led certain workers to suggest that they also might be involved in providing some protection against foreign agents (Whitcomb, R. *et al.*, 1974; Sroka, P. and Vinson, S., 1978; Crossley, A., 1979). However, the role of these and other humoral factors, as well as their relationship to the cellular factors which are involved, is not clear.

4.5.3 Hemolymph as a defensive agent

Hemolymph can be utilized as a defensive agent by some insects in two general ways, both of which deter the action of predators. The first method is restricted to hemolymph composition or the sequestration of deterrent materials. The second method primarily involves the release of hemolymph from the insect, which may also contain deterrent materials. There are a variety of compounds that may be sequestered in hemolymph (and other body components) which render an insect distasteful or toxic to a predator. Although many insects which utilize this type of protection have warning coloration, a disadvantage resides in the likelihood that several may be consumed or severely injured before a predator learns to avoid them (Mellanby, K., 1939). They may be capable of synthesizing these compounds *de novo* or may obtain them from their food sources (Duffey, S., 1980; Blum, M., 1981).

The second method involving hemolymph release from various body parts to deter or foil an attack is referred to as reflexive bleeding or autohemorrhage. It has been recognized as a defense mechanism since Cuenot, L. (1896) and Hollande, A. (1911) made their observations on various insects which exhibited this activity. Some insects may utilize hemolymph which does not contain any apparent chemical (toxic) repellancy as an effective physical deterrent, but still may be lethal to predators (Blum, M., 1981). For example, when attacked by imported fire ants, larval cucumber beetles (*Diabrotica unidecimpunctata howardi* and *Diabrotica balteata*), autohemorrhage, releasing hemolymph from the head–prothorax and the abdomen while moving their bodies in a characteristic twisting and spinning motion (Wallace, J. and Blum, M., 1971). As a result, ants attacking the larvae become entangled, and often permanently affixed (resulting in their death). The loss of hemolymph due to reflexive bleeding may be considerable, without any noticeable ill effects (Happ, G. and Eisner, T., 1961; Benfield, E., 1974). In cucumber beetles, up to 13% of the body weight may be lost in this manner (Wallace, J. and Blum, M., 1971). However, some insects may retract exuded hemolymph after removal of the autohemorrhagic stimulus, suffering little fluid loss (Mellanby, K., 1939). The deterent effectiveness of hemolymph exuded during autohemorrhage may be enhanced by incorporation of distasteful or toxic natural products (Blum, M. and Sannasi, A., 1974; Blum, M., 1981) or by mixing it with secretions from defense glands (Jones, J., 1977).

5 HEMOLYMPH COMPOSITION

The general relationships affecting hemolymph composition in an insect are summarized in Fig. 1. It is important to note that virtually any material to which an insect is exposed may find its way into hemolymph. This includes exogenous materials derived from feeding activities, or the uptake of materials through the exoskeleton or trachea. It is likely that materials derived from symbiont activities or hindgut microflora may also contribute to this mixture. In this section an attempt will be made to summarize the hemolymph composition derived from a diverse and vast literature base. The variety of insects (stages), analytical techniques, components studied, and reporting methods which exist in the literature, as well as space limitations, allow for only a brief summary of what has been reported.

5.1 General content and physical properties

5.1.1 HEMOLYMPH VOLUME AND WATER CONTENT

As indicated in section 3, insect hemolymph volumes vary from 1 to 20,000 μl and it is a major water storage compartment (section 4.2.1). In normally hydrated adult *Locusta*, hemolymph contains about 20% of the total body water (Loveridge, J., 1975), and in last-instar *Heliothis* larvae it represents about 33% of the body weight (Burton, R. *et al.*, 1972). Hemolymph volume may fluctuate in response to many activities or conditions. Some of these include: drinking and feeding, (Edney, E., 1977; Bernays, E. and Chapman, R., 1974), flight (Gringorten, J. and Friend, W., 1979), starvation (Wharton, D. *et al.*, 1965a), molting (Nicolson, S., 1980a; section 4.2.2), desiccation (Nicolson, S. *et al.*, 1974), insecticide exposure (Samaranayaka, M., 1977) and irradiation (Wharton, D. *et al.*, 1965b).

5.1.2 SPECIFIC GRAVITY AND OSMOTIC PRESSURE

Hemolymph solute concentrations vary considerably among insects. Specific gravity values range between 1.012 and 1.070 (Buck, J., 1953; Terra, W. *et al.*, 1974; Dahlman, D., 1974; Mack, S. and Vanderberg, J., 1978). Osmolality values which have been reported range between 215 and 593 mOsm, corresponding to freezing point depressions of 0.40–1.10° (Buck, J., 1953; Joshua, H. *et al.*, 1973). Extreme examples of increased osmotic pressures have been noted in insects adapted for frost protection (section 4.4.2). Insects appear to be quite capable of strongly regulating osmotic pressure even under circumstances where water volume may fluctuate over a wide range (Nicolson, S., 1980b; section 4.2.2). In addition, ontogenetic differences in hemolymph osmolality have been observed (Joshua, H. *et al.*, 1973; Cohen, D. and Patana, R., 1982).

5.1.3 PH AND CO_2

Reports on the hydrogen ion concentration in insect hemolymph show values ranging between 6.0 and 8.2 (Buck, J., 1953; Bedford, J. and Leader, J., 1975), but which are generally restricted to a narrower range between 6.4 and 6.8 (Wyatt, G., 1961; Joshua, H. *et al.*, 1973; Burton, R. *et al.*, 1972). The buffering capacity is minimal in the physiological range (Buck, J., 1953; Wyatt, G., 1961). Although hemolymph pH remains fairly constant under resting or normal physiological conditions, activities such as flight and stress (bacterial toxin) may result in changes (Downer, R., 1981; Nishiitsutsuji-Uwo, J. and Endo, Y., 1981).

Since hemolymph does not play a significant role in CO_2 transport, the CO_2 concentration levels are restricted to that amount dissolved in hemolymph (physical) and in equilibrium with bicarbonates (chemical) (Wyatt, G., 1961). Buck, J. (1953) has tabulated a series of CO_2 levels which occur in insect hemolymph. More recent reports on individual species are available (Terra, W. *et al.*, 1974; Fischl, J. *et al.*, 1975; Bedford, J. and Leader, J., 1975; Staddon, B. and Everton, I., 1980; see Table 1).

5.2 Inorganic solutes

A considerable amount of information is available on the inorganic components of insect hemolymph (Buck, J., 1953; Sutcliffe, D., 1962, 1963; Florkin, M. and Jeuniaux, C., 1964, 1974; Jeuniaux, C., 1971). The major emphasis of the two most recent of these reviews was primarily on the inorganic hemolymph constituents of numerous species and a discussion of phylogenetic relationships. As a result, they serve as an excellent resource on this topic. Table 1 includes some more recent information on inorganic hemolymph components of insects representing several orders. They tend to follow the same general phylogenetic composition patterns as those reported in the earlier reviews. Some generalizations have been made regarding the relative inorganic constituents and their importance as osmolar effectors in the various groups of Insecta (Florkin, M. and Jeuniaux, C., 1974). The participation of inorganic ions as osmoeffectors tends to decrease with the more advanced phylogenetic level of the insect. The apterygotes have a hemolymph composition which is quite similar to that found in other arthropods, with Na^+ and Cl^- being most important. The sum of Na^+, K^+, Ca^{2+} and Mg^{2+}

Table 1: Concentration of inorganic ions and osmotic pressure occurring in hemolymph of some insects

Ion/Molecule[1]	Order and species					
	Odonata: *U. carovei*[2]	Orthoptera: *H. maori*[3]	Dermaptera: *A. littorea*[4]	Heteroptera: *O. faciatus*[5]	Diptera: *S. venustrum*[6]	Lepidoptera: *S. exigua*[7]
Sodium	152.2	95.0	193.0	21.6	109.6	24
Calcium	7.6	8.0	14.6	7.6	36.9	6
Potassium	4.6	5.5	5.4	6.2	26.1	53
Magnesium	7.8	19.0	11.4	36.4	7.4	66
Iron	—	—	—	—	<0.001	—
Copper	—	—	—	—	<0.001	—
Lead	—	—	—	—	<0.001	—
Manganese	—	—	—	—	<0.001	—
Chloride	121.0	65.6	116.9	33.0	25.4	48
Phosphate	4.7	—	3.5	15.0	50.9	38
Bicarbonate	27.6	34	—	4.5	—	—
Dissolved CO_2	28.0	35.6	—	—	—	—
Osmotic pressure	348	480	449	383	256	358

[1] Ion concentrations expressed as mEq l^{-1}; CO_2 as mM; osmotic pressure as mOsm.
[2] *Uropetala carovei* (Odonata), field-collected last-instar dragonfly nymphs (Bedford, J. and Leader, J. 1975).
[3] *Hemideina maori* (Orthoptera), field-collected adult wetas (Leader, J. and Bedford, J. 1978).
[4] *Anisolabis littorea* (Dermaptera), field-collected ?-stage earwigs (Leader, J. and Bedford, J. 1972).
[5] *Oncopeltus faciatus* (Heteroptera), lab-reared adult milkweed bugs (Staddon, B. and Everton, I. 1980).
[6] *Simulium venustrum* (Diptera), field-collected larval blackflies (Gordon, R. and Bailey, C. 1976).
[7] *Spodoptera exigua* (Lepidoptera), artificial diet-reared fourth-instar beet armyworms (Cohen, A. and Patana, R. 1982).

contributes almost half of the osmotic pressure in the more primitive pterygotes (e.g. Table 1: Odonata, Orthoptera, Dermaptera, Heteroptera). In this group, Na^+ and Cl^- are the major osmoeffectors, with the other cations playing a minor role. However, *Oncopeltus fasciatus* represents an exception due to its high Mg^{2+} content (Table 1, 36.4 mEq l^{-1}; Florkin, M. and Jeuniaux, C., 1974). This high value may be a characteristic of phytophagous Heteroptera (Staddon, B. and Everton, I., 1980). In the lower endopterygote orders (Megaloptera, Neuroptera, Mecoptera and Diptera), the sum of the cations contribute nearly half of the osmotic pressure. Sodium is the principal osmoeffector and Cl^- plays a minor role (being partially replaced by amino acids and other small organic molecules). There are exceptions, as can be seen by the values reported for *Simulium venustrum* (Diptera; Table 1). Although Na^+ represents the major cation, K^+ and Ca^{2+} levels are atypical of Diptera (Gordon, R. and Bailey, C., 1976). In the higher endopterygote orders (Lepidoptera, Hymenoptera and Coleoptera), the organic molecules become much more important as osmoeffectors. The levels of Na^+ are comparatively much lower, but Mg^{2+} and K^+ are comparatively higher than those reported for other groups of insects (Lepidoptera, Table 1; Florkin, M. and Jeuniaux, C., 1974). It is possible that the higher levels of Mg^{2+} and K^+ found in these groups may be attributed to evolution, along with the angiosperms. This assumes that with progressive adaptation to plant diets (high K^+ and Mg^{2+}), regulation of the "steady-state" hemolymph cationic species developed into the specialized pattern of low Na^+, high K^+ and Mg^{2+} which is characteristic (but not exclusively adhered to) of these groups (Wyatt, G., 1961; Florkin, M. and Jeuniaux, C., 1974).

Florkin, M. and Jeuniaux, C. (1974), and other workers, have pointed out that even though there is a considerable amount of information available on ion content in various insects, there is still a need for more specific information on the ontogenetic and physiological differences which may occur within a single species. In some studies where comprehensive analyses (stages, physiological state and measurements on a variety of components) have been done, significant differences have been found to occur. This type of information is of considerable importance in developing tissue culture media, and insect salines, etc. (Chen, A. and Friedman, S., 1975; Schin, K. and Moore, R., 1977; Firling, C. and

Table 2: Concentration of glucose and trehalose in hemolymph of some insects

Order	Species	Stage	glucose (μmol ml^{-1})	trehalose (μmol ml^{-1})	Reference
Thysanura	*Lepisma saccharina*	?	0.24	0.26	Bedford, J. 1977
Ephemeroptera	*Coloburiscus humeralis*	?	2.03	0.63	Bedford, J. 1977
Odonata	*Uropetala carovei*	?	0.47	2.09	Bedford, J. 1977
	Procordulia smithii	?	0.78	1.44	Bedford, J. 1977
Dictyoptera	*Periplaneta americana*	adult	0.56	—	Matthews, J. *et al.* 1976
	Periplaneta americana	adult	—	7.0	Kramer, K. *et al.* 1978
	Periplaneta americana	adult	—	46.1	Matthews, J. and Downer, R. 1973
	Periplaneta americana	?	3.08	18.87	Bedford, J. 1977
Isoptera	*Stolotermes ruficeps*	?	3.04	1.12	Bedford, J. 1977
Plecoptera	*Stenoperla prasina*	?	1.82	1.53	Bedford, J. 1977
Orthoptera	*Clitarchus hookeri*	?	2.49	4.89	Bedford, J. 1977
	Locusta migratoria	adult	12.1	77.9	Jutsum, A. and Goldsworthy, G. 1976
Dermaptera	*Anisolabis littorea*	?	trace	trace	Bedford, J. 1977
Heteroptera	*Oncopeltus faciatus*	nymph	—	6.0	Kramer, K. *et al.* 1978
Lepidoptera	*Manduca sexta*	larvae	—	46.0	Kramer, K. *et al.* 1978
	Manduca sexta	larvae	—	30.1	Jungreis, A. 1980
	Manduca sexta	larvae	5.1	71.8	Dahlman, D. 1973
	Manduca quinquemaculata	larvae	—	25.0	Kramer, K. *et al.* 1978
	Sphinx chersis	larvae	—	133.0	Kramer, K. *et al.* 1978
Coleoptera	*Tenebrio obscurus*	larvae	—	8.0	Kramer, K. *et al.* 1978

Kobilka, B., 1979). Information on fluctuations of insect hemolymph composition in response to various conditions or activities is available. Studies on these effects include: ontogenetic effects (Jungreis, A. *et al.*, 1973; Schin, K. and Moore, R., 1977; Woodring, J. *et al.*, 1977; Cohen, A. and Patana, R., 1982; Shimizu, I., 1982), diet (Pichon, Y. and Boistel, J., 1963; Jungreis, A. *et al.*, 1973; Shimizu, I., 1982), drinking-saline (Heit, M. *et al.*, 1973; Bradley, T. and Phillips, J., 1977), dehydration (Machin, J., 1981), distribution patterns in various hemocoel areas (Pichon, Y., 1970), temperature stress (Ahearn, G., 1970; Cohen, A. and Patana, R., 1982), xenobiotics (Racioppi, J. and Dahlman, D., 1980) and disease (Steinkraus, K. *et al.*, 1973).

5.3 Organic constituents

5.3.1 Carbohydrates

Carbohydrates are important in insects, since they function as a major energy source and are utilized as a major exoskeleton component. There are many reviews available on the biochemistry of these materials, and recently certain aspects of their physiological chemistry have been discussed (Steele, J., 1981; J. Steele, vol. 8; S. Friedman, vol. 10; D. Candy, vol. 10). Wyatt, G. (1967) has provided an excellent summary of the carbohydrate composition of insect hemolymph. In addition, Florkin, M. and Jeuniaux, C. (1974) have reviewed this topic.

When compared with vertebrates, the total carbohydrate content of insect hemolymph is considered to be high. Levels generally are greater than 0.5% but may reach 8.1% (Wyatt, G., 1967). The most characteristic carbohydrate found in insect hemolymph is trehalose (α-D-glucopyranosyl-α-D-glucopyranoside), a non-reducing disaccharide. However, in certain insects it may be absent or present in very low amounts (Evans, D. and Dethier, V., 1957; Wimer, L., 1969; Bedford, J., 1977). In honeybees it may represent as much as 95% of the total carbohydrates during postembryonic development (Tsao, W. and Shuel, R., 1973). It may increase in concentration during larval development (Boctor, I., 1974). Values for trehalose range from 0 (non-feeding *Celerio euporibae* larvae) to 65 mg ml^{-1} (189 mM) in *Anthophora* larvae (Wyatt, G., 1967). Glucose is present in the hemolymph of many insects, but generally at lower levels; from 0 to 20 mg ml^{-1}

(111 mM) (Wyatt, G., 1967). Table 2 includes some more recent information of glucose and trehalose concentrations, and reflects the variability which may occur among and within species. Glycogen may be found in hemolymph but in very small amounts (Wyatt, G., 1967; Chippendale, G., 1973a). Other sugars reported in insect hemolymph include arabinose, cellobiose, fructose, fucose, galactose, maltose, mannose, ribose and sucrose (Wyatt, G., 1967; Florkin, M. and Jeuniaux, C., 1974; Phillipe, M. *et al.*, 1976). In many cases, sugars found in the hemolymph reflect those which are contained in their diet (Hansen, O., 1964; Maurizio, A., 1965). In addition to these sugars there are other carbohydrates or derivatives which have been reported to occur in hemolymph. These include: hexosamines (*N*-acetylgalactosamine and *N*-acetylglucosamine) in *Blaberus craniifer* (Philippe, M. *et al.*, 1976) and other insects (Florkin, M. and Jeuniaux, C., 1974), β-glucosyl-*O*-tyrosine in silkworm pupae, and *Leucania separata* and *Drosophila busckii* larvae (Chen, P. *et al.*, 1978; Isobe, M. *et al.*, 1981), glucuronic acid and inositol in *Anopheles stephensi* (Mack, S. *et al.*, 1979b), and *scyllo*-inositol in *Schistocerca gregaria* (Candy, D., 1967).

Much of the biochemical work on carbohydrates has centered around flight muscle metabolism (see section 4.1.2; Fig. 3). The fat body is the major site of trehalose synthesis (Jungreis, A. and Wyatt, G., 1972; Friedman, S., 1978). The fat body can remove glucose from hemolymph quite rapidly. For example, radiolabelled glucose injected into *Periplaneta americana* is rapidly removed (50%, 20 min post-injection) and converted to trehalose and finally stored as glycogen in the fat body (Spring, J. *et al.*, 1977). These glycogen stores are mobilized from the fat body under various physiological conditions (Matthews, J. and Downer, R., 1974). Hemolymph carbohydrate levels are under hormonal control (Friedman, S., 1978). Extracts of corpora cardiacum contain hyperglycemic factor(s) (Steele, J., 1963; Loughton, B. and Orchard, I., 1981; Weeda, E., 1981). *Manduca sexta* may possess two carbohydrate-regulating peptides, a glucagon-like and an antagonistic insulin-like peptide which can be obtained from the corpus cardiacum–corpus allatum complex (Candy, D., 1981; K. Kramer, vol. 7). The glucagon-like peptide promotes the release of trehalose into the hemolymph (hyperglycemia) whereas the insulin-like peptide appears to decrease hemolymph trehalose levels (hypoglycemia). Hemolymph carbohydrate levels have been shown to fluctuate under various conditions. Daily fluctuations in hemolymph sugar levels have been reported in *Periplaneta americana* (Hilliard, S. and Butz, A., 1969). During flight certain insects draw upon the carbohydrate pool as an initial energy source; as a result the trehalose levels decline (Bailey, E., 1975; Kammer, A. and Heinrich, B., 1978; Jutsum, A. and Goldsworthy, G., 1976; Van Der Horst, D. *et al.*, 1980; Figs 3 and 6). Starvation in *Manduca sexta*, *Locusta migratoria* and *Oxya japonica* produces significant reduction of hemolymph carbohydrate (Dahlman, D., 1973; Mwangi, R. and Goldsworthy, G., 1977; Lim, S. and Lee, S., 1981). Parasitization of some insects may cause decreases (Schmidt, S. and Platzner, E., 1980; Dahlman, D., 1975) or increases in hemolymph carbohydrate (Dahlman, D. and Vinson, S., 1976). Hyperglycemia can be produced by various forms of anesthesia, saline injection, or handling (without anesthesia) (Matthews, J. and Downer, R., 1973; Hanaoka, K. and Takahashi, S., 1976) and induced by various amines, most notably octopamine (Downer, R., 1979).

5.3.2 Nitrogenous constituents

(a) *Amino acids.* A characteristic feature of insect hemolymph is its high amino acid content (P. Chen, vol. 10). At one time it was proposed that this property be considered as a taxonomic character for the Insecta (Florkin, M., 1959). The literature regarding hemolymph amino acids is quite vast. However, several sources are available which provide tabulations of the findings (Duchateau, G. and Florkin, M., 1958; Florkin, M., 1959; Florkin, M. and Jeuniaux, C., 1974; Mansingh, A. and Baqaya, V., 1971; Schoffeniels, E. and Gilles, R., 1970; Jeuniaux, C., 1971). More recent information has been reported (see Table 3; and for other species such as: *Uropetala carovei*: Bedford, J. and Leader, J., 1975; *Rhodnius prolixus*: Barrett, F., 1974; Barrett, F. and Friend, W., 1975; *Calpodes ethlius*: Barrett, F. and Lai-Fook, J., 1976; *Chironomus tentans*: Firling, C., 1977; *Simulium venustrum*: Gordon, R. and Bailey, C., 1976; etc.). It should be pointed out that as improved methods and analytical techniques have become available, some of the older information may be subject to modification. For example, information found in Table 3 compares the plasma *vs.* hemocyte content of hemolymph obtained from two species and which was subjected to very rigid collection and analytical procedures. It can be seen that had whole, or partially separated (containing damaged hemocytes), hemolymph been analyzed, glutamate would have been most likely detected (and reported) as a plasma component. Another problem which occurs when attempting to draw generalizations from the literature is the lack of complete information regarding the insects examined (physiological state, stage, etc.) (Florkin, M. and Jeuniaux, C., 1974). In some studies where this was considered, variation in hemolymph content has been found to occur. Although the information available on insect amino acid content is complex and diverse, some generalizations can be made. Aminoacidemia is more pronounced in endopterygotes than for exopterygotes (Florkin, M. and Jeuniaux, C., 1974). In exopterygotes the concentrations of various amino acids differ from about 5 to 70 mg 100 ml^{-1}; glutamic acid + glutamine and glycine being somewhat more concentrated. In endopterygotes, aspartate, asparagine, phenylalanine, leucine and isoleucine are usually present at low levels, with glutamic acid + glutamine and proline, being the most significant components (Mansingh, A. and Baqaya, V., 1971). Hemolymph is usually found to contain all of the amino acids that are commonly present in proteins (Table 3). However, a variety of other amino acids may be found. These include: derivatives of tryptophan and tyrosine (e.g. kynurenine, 3-OH-kynurenine, dihydroxyphenylalanine) (Mansingh, A. and Baqaya, V., 1971), methylcysteine, homoarginine and hydroxyproline (Florkin, M. and Jeuniaux, C., 1974). The isometric configuration of the amino acids found in insects is usually the L form. However, D-alanine and D-serine have been reported in hemolymph (Mansingh, A. and Baqaya, V., 1971). It appears that *Oncopeltus faciatus* is able to synthesize D-alanine *de novo* (Ayers, G. *et al.*, 1974).

Changes in amino acid patterns are influenced by several factors which include stage of development, diet, insect activities, disease and exposure to toxins. Developmental changes in amino acid concentrations may be associated with molting (Florkin, M. and Jeuniaux, C., 1974), metamorphosis (Wirtz, R. and Hopkins, T., 1974) and diapause (Boctor, I., 1981). In developing *Melanoplus sanguinipes*, various hemolymph amino acid concentrations fluctuate from the egg to adult stage with large increases in asparagine, proline, alanine, glycine, glutamine and valine occurring in the third-instar nymph (Roberts, R. and Smith, H., 1971a). Free hemolymph amino acids increase from 982 (second instar) to 1190 μM (sixth instar) in developing *Spodoptera littoralis*, with a decline in glutamine and tyrosine, but, an increase in histidine, citrulline and gamma-aminobutyric acid (GABA) (Boctor, I. and Salem, S., 1973). Total hemolymph amino acids decline in aging *Schistocerca gregaria* females and males (Kulkarni, A. and Mehrota, K., 1970). The effects of diet on hemolymph levels have been demonstrated. Differences in the amino acid spectra from laboratory-fed and field-collected adult *Melanoplus* have been reported (Roberts, R. and Smith, H., 1971b). Starvation results in a slight increase in hemolymph amino acids in *Philosamia cynthia* (Bosquet, G., 1977), but in *Spodoptera exigua* they decrease (Cohen, A. and Patana, R., 1982). Disease and parasitism may result in reduced amino acid levels in some species (Mack, S. *et al.*, 1979a; Wang, D. and Moeller, F., 1970) but increases in others (Rutherford, T. and Webster, J., 1978). Exposure of insects to certain insecticides results in significant reduction of some amino acids, including depletion of hemolymph proline (Corrigan, J. and Kearns, G., 1963; Kulkarni, A. and Mehrota, K., 1973).

Our knowledge about the relationship of amino acid metabolism to the various insect metabolic pools is still limited. However, progress is being made. Several workers have concerned themselves with studies on the regulation of the hemolymph amino acid pools. Collett, J. (1976b) has found that hemolymph amino acid pools in *Calliphora* are regulated, even under stressful conditions. Studies designed to determine the half-lives of various amino acids injected into the hemolymph have revealed that they each appear to have characteristic retention-time patterns (Feir, D. *et al.*, 1973; Jungreis, A., 1980). The role of proline in flight metabolism is of considerable interest because it is

Table 3: Concentrations of free amino acids contained in whole and separated hemolymph obtained from several insect species in μmol l^{-1}.

Amino acid	P. americana[1]		C. erythrocephala[2]		H. armigera[3]
	Plasma[4]	*Hemocytes*	*Plasma*	*Hemocytes*	*Whole hemolymph*
Phosphoserine	—	—	*10.0*	*5.9*	*NR*
Taurine	*t*	—	*155.0*	*19.2*	*NR*
Phosphoethanolamine	—	—	*t*	—	*NR*
Aspartate	*1.3*	*0.9*	*3.2*	*22.7*	*7.5*
Threonine	*4.9*	*t*	*301.0*	*59.9*	*25.9*
Serine	*11.5*	*1.1*	*209.0*	*51.5*	*22.0*
Asparagine	*0.6*	*t*	*520.0*	*95.9*	*128.2*
Glutamate	*t*	*0.9*	—	*82.7*	*3.6*
Glutamine	*85.0*	*8.7*	*1790.0*	*440.0*	*63.1*
Proline	*58.1*	*11.7*	*670.0*	*128.0*	*17.4*
Glycine	*197.0*	*13.5*	*300.0*	*29.4*	*22.1*
Alanine	*13.9*	*2.9*	*2360.0*	*401.0*	*13.8*
Valine	*26.9*	*0.9*	*366.0*	*43.7*	*55.4*
Methionine	*2.1*	—	*40.3*	*8.4*	*16.5*
Cystathionine	—	—	*t*	*t*	*NR*
Isoleucine	*3.8*	*0.7*	*90.1*	*16.0*	*NR*
Leucine	*8.8*	*0.9*	*947.0*	*16.0*	*34.3*
Tyrosine	*32.7*	*2.6*	*466.0*	*78.7*	*7.4*
Phenylalanine	*5.7*	*0.7*	*204.0*	*21.3*	*15.8*
β-alanine	—	—	*2.0*	—	*NR*
Ornithine	—	—	*8.6*	*t*	*34.6*
γ-aminobutyric acid	—	—	*t*	—	*NR*
Lysine	*29.6*	*2.0*	*129.0*	*24.5*	*28.7*
Histidine	*36.2*	*3.4*	*104.0*	*t*	*28.8*
Arginine	*18.2*	*3.5*	*70.0*	*40.6*	*5.4*
3-methyl histidine	*NR*	*NR*	*3.0*	—	*NR*
α-aminobutyric acid	*NR*	*NR*	*7.8*	—	*NR*
Citrulline	*NR*	*NR*	*NR*	*NR*	*5.4*
Cystine	*NR*	*NR*	*NR*	*NR*	*62.8*

[1] *Adult male* Periplaneta americana *(Irving, S.* et al. *1979).*
[2] *Third-instar* Calliphora erythrocephala *larvae (Irving, S.* et al. *1979).*
[3] *Pharate adult (18-week-old)* Heliothis armigera *(Boctor, I. 1981).*
[4] *NR = not reported; t = trace amounts.*

a major energy source in some insects (i.e. *Glossina*, *Leptinotarsa* and *Stomoxys*). During sustained flight their hemolymph proline levels are closely regulated (Bursell, E., 1981). In addition, there is a need for more information on hemolymph levels of both GABA and glutamate. This is important since they may be involved in neurotransmission, or injury response (GABA) (Irving, S. *et al.*, 1979; Jungreis, A. *et al.*, 1980).

(b) *Amines and ammonia.* Various amines and ammonia may be found in the hemolymph. Several biogenic amines such as dopamine, 5-hydroxytryptamine and octopamine occur in insect nervous tissue (Evans, P., 1980; P. Evans, vol. 11) and hemolymph (Ishay, J. *et al.*, 1974; Muszynska-Pytel, M. and Cymborowski, B., 1978; Goosey, M. and Candy, D., 1980), but their functional role is unclear. Several roles have been proposed for some of these amines, including involvement as neurotransmitters, neuromodulators and neurohormones (Steele, J., 1976; Evans, P., 1980; Orchard, I., 1982; Loughton, B. and Orchard, I., vol. 7). In addition to these amines, a neuroactive factor(s) (autoneurotoxin), tentatively characterized as an aromatic amine, has been found in the hemolymph of stressed *Periplaneta americana* (Hawkins, W. and Sternburg, J., 1964). Severe forms of stress: chemical (Shambaugh, G., 1969; Flattum, R. *et al.*, 1973), mechanical stimulation (Cook, B. and Holt, G., 1974), or mild electrical stimulation (Beament, J., 1958), result in the release of this substance. Accumulation in the hemolymph causes paralysis and may lead to death (Sternburg, J., 1963). The chemical identity of the active factor(s) is not known; however, it appears to be distinct from all commonly known neuropharmacologically active agents (Cook, B., 1967).

Since ammonia may be an excretory product in both aquatic and terrestrial insects, it is most likely a common hemolymph constituent. Ammonia is generally found in insect excreta when it is examined closely (Cochran, D., 1975; D. G. Cochran, vol. 4). It shares an important role in transamination reactions (glutamate dehydrogenase) involving α-ketoglutarate and glutamate, which are both principal participants in nitrogen metabolism (Corrigan, J., 1970). Because it is a degradation product of many nitrogenous compounds, it is difficult to obtain reliable values representative of those which actually may occur in hemolymph. As a result, values may vary considerably. Some values which have been reported include: *Sialis lutaria*, 0.36 mM (Staddon, B., 1955); *Chironomus tentans*, 1.35 mM (Firling, C., 1977); *Phormia regina*, 11 μmol g^{-1} fresh weight hemolymph (Levenbook, L., 1966).

(c) *Purines and pteridines.* Purines and pteridines may be found in the hemolymph because many of them are byproducts of intermediary metabolism which may be destined to become excretory products, or utilized in pigmentation (Corrigan, J., 1970; Cochran, D., 1975; H. Kaysen, vol. 10). Uric acid is generally regarded as a major nitrogenous end-product which is produced in the fat body and other tissues from various nitrogenous intermediates. In order for this product to be eliminated from the insect, or stored in various compartments, it must be transported by the hemolymph. The concentration levels are generally low because urates are quite insoluble. The hemolymph concentrations may be influenced by feeding (Barrett, F. and Friend, W., 1966; Ramakrishnan, N. and Pawar, V., 1975), photoperiodicity (Hilliard, S. and Butz, A., 1969), development (Buckner, J. and Caldwell, J., 1980) and parasitization (Condon, W. and Gordon, R., 1977). In addition to uric acid, uric acid riboside has been found (equimolar concentrations) in ligated *Galleria mellonella* larvae (Krzyzanowska, M. and Niemierko, W., 1979, 1980). These workers have proposed that this riboside may be involved in facilitating (transport) uric acid excretion. Hemolymph pteridines may fluctuate during developmental stages of an insect. For example, Harmsen, R. (1966) has found that both leucopterin and isoxanthopterin are stored in the fat body during the first 5 days after pupation. Prior to adult emergence, these pteridines appear in the hemolymph and are transported to the wings for storage.

(d) *Peptides.* Insects utilize peptides in a diverse array of physiological and metabolic processes. Some of these include cuticle sclerotization, amino acid transport, insecticide detoxication, neurotransmission and metabolic regulation (Bodnaryk, R., 1978; vols 8 and 11, this series). Although they play major roles in insect physiological processes and their functions imply transit (presence) in the hemolymph, information on their presence/concentration is largely lacking. Those peptides which have a neurotransmitter or hormonal function are present in low concentrations and are detected by bioassay. However, dipeptides like sarcophagine (β-alanyl-tyrosine), which accumulate during larval growth of *Sarcophaga*, may become the major non-protein ninhydrin-positive material in their hemolymph. This dipeptide, and other similar materials, appear to represent specific storage forms of tyrosine or phenylalanine destined for use during sclerotization of their puparia (section 4.2.2).

Polypeptides appear to play an important role as storage reserves (metabolic pools) or as transport forms for amino acids. Collett, J. (1976a,b) has found that adult *Calliphora* contain a large heterogeneous pool of small peptides whose composition and rapid turnover appear to be subject to a group of peptidases (whose activities are regulated). The amino acid pool is also regulated in this insect, and it appears that hemolymph peptides serve as an effective metabolic reserve for essential amino acids. Tobe, S. (1978) has found that pregnant female *Glossina austeni* appear to preferentially sequester tyrosine, phenylalanine, aspartate and lysine into their hemolymph, and that these peptides are transferred to the developing larva (midgut) at an exceptionally high rate.

The role of peptides as regulatory factors has been recently reviewed (Stone, J. and Mordue, W., 1980; Candy, D., 1981; section 4.1.3; vols 7 and 8). This has been a very active area of research, and despite formidable technical difficulties progress is being made. Peptides which elevate the blood trehalose level (hyperglycemic hormones) have not

Table 4: Some types of hemolymph proteins and activities or factors which may affect their presence in some insects.

Category	Species	Observation	Reference
Storage	*Calliphora vicina*	major soluble protein in third instar and pharate adults	Levenbook, L. and Bauer, A. 1980
	Galleria mellonella	identification and purification of 4 high mol. wt proteins	Miller, S. and Silhacek, D. 1982
Vitellogenins	*Periplaneta americana*	structure, synthesis and processing	Sams, G. *et al.* 1980
	Locusta migratoria	purification and properties of vitellogenin and vitellin	Chinzei, Y. *et al.* 1981
Transport	*Periplaneta americana*	purification, characterization and function of DGLP[1]	Chino, H. *et al.* 1981b
	Locusta migratoria	purification and properties of DGLP[1]	Gellissen, G. and Emmerich, H. 1980
	Locusta migratoria	interconversions of DGLP[1]	Van Der Horst, D. *et al.* 1981
	Leptinotarsa decemlineata	juvenile hormone carrier lipoproteins	Kramer, K. and De Kort, C. 1978
Binding/ Transport?	*Locusta migratoria*	specific binding of ^{3}H-ecdysterone to high mol wt. protein	Feyereisen, R. 1977
	Periplaneta americana	some ion binding to hemolymph macromolecules	Weidler, D. and Sieck, G. 1977
	Periplaneta americana	binding of serotonin to serum proteins and hemocytes	Kirksey, C. *et al.* 1974
	Periplaneta americana	affinity of pesticides to hemolymph macromolecules	Skalsky, H. and Guthrie, F. 1977
	Periplaneta americana	DDT-lipoprotein complex — possible transport?	Winter, C. *et al.* 1975
Chromoproteins	*Rhynchosciara americana*	isolation of a violet carotenoprotein	Terra, W. *et al.* 1981
Hemoglobin	*Chironomus thummi*	major fourth instar-specific hemoglobin purified to homogeneity	Trewitt, P. and Bergtrom, G. 1981
Thermal hysteresis	*Dendroides canadensis*	probable protein which is involved in cryoprotection	Duman, J. 1980
Inhibitors	*Bombyx mori*	chymotrypsin inhibitors	Sasaki, T. 1978
Coagulogen	*Locusta migratoria*	functional equivalent of fibrinogen?	Brehelin, M. 1979
	Leucophaea maderae	immunochemical analysis of plasma and hemocyte coagulogens	Bohn, H. *et al.* 1981
Enzymes	*Apis mellifica*	hemolymph-α-glucosidase kinetics	Bounias, M. 1980
	Nauphoeta cinerea	soluble amine oxidases increased by stress	Shambaugh, G. *et al.* 1974
	Manduca sexta	preparation of homogeneous juvenile hormone-specific esterase	Coudron, J. *et al.* 1981
	Periplaneta americana	triacylglycerol lipases	Hoffman, A. and Downer, R. 1979
Development	*Hyalophora cecropia*	protein changes during development	Patel, N. 1971
	Formica rufa	protein changes during metamorphosis	Schmidt, G. and Schwankl, W. 1975
Diet	*Manduca sexta*	differences: natural/synthetic diet	Dahlman, D. 1974
	Glossina morsitans	absorb undigested albumins, globulin fractions	Nogge, G. and Giannetti, M. 1979
Parasitization	*Pieris rapae*	*Apanteles glomeralus*: change in band patterns	Smilowitz, Z. and Smith, C. 1977
Infection	3 spp. Lepidoptera	*Beauveria bassiana*: change in band patterns	Gardner, W. *et al.* 1979
	Spodoptera mauritia	nuclear polyhedrosis virus: change in band patterns	Takei, G. and Tamashiro, M. 1975

[1] DGLP diacylglycerol-carrying lipoprotein — a "Lipophorin" (Chino, H. *et al.* 1981a).

yet been purified to characterization and their physiological roles have not been established. However, adipokinetic hormone, which is involved in diacylglycerol release and transport in *Locusta,* has been purified and synthesized (Candy, D., 1981). As was indicated earlier (section 5.3.1), *Manduca sexta* hemolymph apparently contains both glucagon- and insulin-like peptides (Kramer, K. *et al.*, 1980). Recently, the insulin-like peptides have been purified and appear to possess properties similar to those of vertebrate insulin (Kramer, K. *et al.*, 1982; K. Kramer, vol. 7).

(e) *Proteins.* Proteins are among the most complex of all known chemical compounds and also the most characteristic of living organisms (P. Chen, vol. 10). Insect hemolymph protein chemistry has been a very active area of study in the past 20 years. As a result, a considerable amount of information is available. Wyatt, G. and Pan, M. (1978) have provided an excellent comprehensive review on plasma proteins which includes a systematic summary of literature on this topic. In addition, several treatments dealing with insect proteins may be found in vol. 10 of this series. The range of insect proteins which are present in hemolymph is variable, depending on the resolving power of the analytical techniques employed as well as the specific insect (physiological state, stage, etc.) examined. Table 4 contains a few types of proteins and activities which may affect hemolymph protein content and concentration. For more details see Wyatt, G. and Pan, M. (1978).

Major changes in protein content occur during development (Chen, P., 1966; L'Helias, C., 1970; Agosin, M., 1978; Wyatt, G., 1980). Hemolymph protein levels generally increase during each instar, but decline during molting. The levels are low in early instars (1–2 g 100 ml^{-1}), but increase considerably to 6–10 g 100 ml^{-1} (Lepidoptera) or 20 g 100 ml^{-1} (Diptera) (Wyatt, G. and Pan, M., 1978). In many insects examined this rise in concentration may be attributed to a few distinct proteins (storage proteins and development: Table 4; section 4.2.3; L. Levenbook, vol. 10) Protein levels may decline prior to larval–pupal ecdysis, rise in the pupae, followed by another decline during adult development (Wyatt, G. and Pan, M., 1978). Major changes in hemolymph protein content occur during the female reproductive cycle. This involves the production, storage and transport of vitellogenins destined for the ovaries and subsequent sequestration by developing oocytes (vitellogenins: Table 4; section 4.2.4; J. Kunkel and J. Nordin, vol. 1).

Hemolymph contains several lipoprotein components which appear to play major role(s) in the transport of various materials. Diacylglycerol-carrying lipoproteins (DGLP) have been detected and characterized in several insects (Table 4; H. Chino, vol. 10). In *Periplaneta,* DGLP appears to function in diacylglycerol and cholesterol transport, and has been implicated in the transport of other lipophilic materials (Chino, H., 1981). Because of their apparent involvement in insect physiological and biochemical processes, a generic name, lipophorin, has been suggested for this class of lipoproteins (Chino, H. *et al.*, 1981a). Juvenile hormone is transported by carrier proteins (Table 4; section 4.1.3; W. Goodman and E. Chang, vol. 7). Binding of other hemolymph-borne materials has been indicated, but in most cases more information is needed before their functional significance can be ascertained.

Hemolymph may contain a variety of enzymes (Table 4; Wyatt, G., 1961; Florkin, M. and Jeuniaux, C., 1974; Jeuniaux, C., 1971; Agosin, M., 1978; Wyatt, G. and Pan, M., 1978). However, in most cases the enzyme characteristics were determined on whole blood and the precise localization (the plasma *vs.* cellular compartment) is not known. Other proteins which are known to occur in some insects include those which are apparently involved in clotting and defense (sections 4.5.1 and 4.5.2), frost protection (section 4.4.2), hemoglobins, chromoproteins, and phenoloxidases (Table 4, Wyatt, G. and Pan, M., 1978; Crossley, A., 1979; Wyatt, G., 1980; S. Andersen, this volume; C. Sekeris and E. Fragoulis, vol. 8).

Environmental factors which may affect hemolymph protein components include such things as diet, parasites, and disease (Table 4). These as well as other changes in protein composition are undoubtedly under endocrine control, but the precise regulatory mechanisms involved remain to be elucidated.

5.3.3 Lipids

Lipids are generally defined as compounds which are poorly soluble in water, but soluble in organic solvents (R. Downer, vol. 10). This classification includes a wide variety of compounds which may have quite divergent physiological or biochemical functions. Because of their low solubility, lipophilic materials either are present in hemolymph at very low concentrations or must be attached to a carrier material. The total lipid content of hemolymph has been found to range between 1.5 and 5.5% (Jeuniaux, C., 1971; Florkin, M. and Jeuniaux, C., 1974). Hemolymph lipid levels may fluctuate under a variety of conditions. These include muscular activity (Beenakkers, A. *et al.*, 1981; A. Beenakkers, vol. 10), development (Downer, R. and Matthews, J., 1976), starvation (Jutsum, A. *et al.*, 1975), and disease (Bennett, G. and Shotwell, O., 1972; Bennett, G. *et al.*, 1972). The processes of lipid absorption are not well understood. Insects do not appear to utilize emulsifiers during digestion, and chylomicral transport of lipids does not occur (Wyatt, G. and Pan, M., 1978; Turunen, S., 1979; see vol. 4). Instead, it appears that triglycerides undergo partial hydrolysis in the gut and are transported in the hemolymph as diacylglycerols complexed to lipoproteins (Chino, H. *et al.*, 1981b). Transport of diacylglycerols from midgut to fat body (where they are stored as triglycerides) and fat body to flight muscle has been demonstrated (Beenakkers, A. *et al.*, 1981). In addition, it appears that diacylglycerol-lipoproteins (DGLP) are involved in hemolymph transport and sequestration (deposition) of lipids in developing cockroach oocytes (Sroka, P. and Barth, R., 1976) and the *Glossina* uterine glands (Langley, P. *et al.*, 1981). The mobilization of triglycerides from fat body and release to hemolymph is under endocrine control (adipokinetic hormone, section 5.3.2(d); J. Steele, vol. 8; W. Mordue, vol. 7).

Cholesterol appears to be transported in hemolymph plasma, predominantly in a non-esterified form (Downer, R. and Chino, H., 1979). It appears that in addition to diglycerides and cholesterol, DGLP may serve to transport other lipophilic materials (Chino, H., 1981). There are reports in which some degree of binding of lipophilic materials to hemolymph proteins has been found (section 10 5.3.2(e)).

Some of the lipophilic compounds which have been found in hemolymph include juvenile hormones (binding to lipoproteins, section 5.3.2(e)) and phospholipids (De Kort, C. and Granger, N., 1981; Florkin, M. and Jeuniaux, C., 1974). Various lipophilic hydrocarbons have been isolated from hemolymph. *Popillia japonica* larvae may contain as many as 21 saturated hydrocarbons (Bennett, G. *et al.*, 1972). In cockroaches *cis;cis*-6,9 heptacosadiene, pentacosane, 3-methylpentacosane, heptacosane, nonacosane and C_{41}-C_{43} hydrocarbons have been found (Florkin, M. and Jeuniaux, C., 1974).

5.3.4 Organic acids, organic phosphates, polyols and miscellaneous compounds

It has been known for a long time that insect hemolymph contains large amounts of organic acids (Wyatt, G., 1961). The major organic acids found are those metabolites associated with the tricarboxylic acid cycle. These include citrate, α-ketoglutarate, succinate, fumarate, malate and oxaloacetate (Jeuniaux, C., 1971). Citrate is generally found as a constituent at relatively high concentrations (e.g. 0.186 mM in mature *Rhynchosciara americana* larvae, Terra, W. *et al.*, 1974), but may reach levels as high as 20.5 mM in *Prodenia eridania* larvae (see Florkin, M. and Jeuniaux, C., 1974). Other organic acids have been reported to occur in hemolymph. These include glyoxylic, acetoacetic acid, lactate, pyruvate and ascorbate (Jeuniaux, C., 1971; Fischl, J. *et al.*, 1975; Kramer, K. *et al.*, 1981). Carbon dioxide anesthesia elevates lactate and pyruvate levels in *Ephestia cautella* pupae (Navarro, S. and Friedlander, A., 1975).

The relatively high level of organic phosphates found in insect hemolymph has been noted as being exceptional among animals (Wyatt, G., 1961). It is present primarily in the form of acid–soluble organic phosphates (e.g. 26–44 mM total acid-soluble phosphate in *Hyalophora cecropia* — see Jeuniaux, C., 1971). The more common organic phosphates found include α-glycerolphosphate, phosphorylethanolamine, glycerophosphoethanolamine, phosphorylcholine, sorbitol-6-phosphate and glucose-6-phosphate (Jeuniaux, C., 1971). Phosposerine also has been recently reported in a number of insects (blackflies, Gordon, R. and Bailey, C., 1976; *Lymantria dispar* and *Neodiprion sertifer*, Brown, S. *et al.*, 1977; *Anopheles stephensi*, Mack, S. *et al.*, 1979a and several other Lepidoptera, Mitsuhashi, J., 1978). The concentrations of hemolymph organic phosphates are variable, changing considerably under conditions of metamorphosis, diapause, anoxia and low-temperature exposure (Florkin, M. and Jeuniaux, C., 1974).

Polyols such as glycerol, mannitol and threitol are found in hemolymph. These compounds are important cryoprotectant materials which become quite concentrated in cold-adapted insects (section 4.2.2). In addition, glycerol is an important metabolite involved in lipid transport and metabolism. Its concentration may increase under conditions such as flight, where lipid metabolism is elevated. Candy, D. *et al.* (1976) have found that during flight, hemolymph glycerol in male *Schistocerca gregaria* increases 10-fold (about 0.34–3.4 μmol ml^{-1}).

Some materials found in the hemolymph may be the result of absorption of compounds produced by gut microflora. Recently Cheung, P. *et al.* (1982) found a diamine, putrescine, in the *Heliothis zea* larvae which apparently is produced by gut microbes. It has been shown that *Schistocerca* rectal tissues can actively absorb amino acids (Balshin, M. and Phillips, J., 1971) and acetate (Baumeister, T. *et al.*, 1981). Bracke, J. and Markovetz, A. (1980) have shown that bacterial products (e.g. acetic acid, proprionic and butyric acid) are transported through the colon wall in *Periplaneta*. With these and other similar observations, it is quite likely that future studies will reveal a significant role for gut microbes in insects, their metabolic products appearing in the hemolymph.

6 CONCLUSIONS

Insect hemolymph differs from the body fluids of organisms from other animal phyla. Characteristically, hemolymph contains high levels of organic phosphates and amino acids, lower levels of inorganic molecules, and trehalose is usually present as the major carbohydrate. Because insects utilize a tracheal respiratory system, a close relationship between the respiratory and circulatory systems is not required. As a result there is no need for respiratory pigments; however, hemoglobin does occur in hemolymph of some chironomid larvae. In addition, it appears that lipid transport is achieved by complexing lipophilic materials with lipoprotein carriers; chylomicral transport apparently does not occur in insects.

Insect hemolymph has many functions which include roles in transport of nutrients and hormonal factors, and the storage of various materials. It plays an important role in excretion, defense, molting and metamorphosis, and in some cases, thermoregulation. In addition, hemolymph provides a medium in which hemocytes are allowed access to the tissues and organs contained in the hemocoel for the performance of their numerous activities.

The functional design of the circulatory system is such that it provides versatility, allowing for it to perform the many roles which have been ascribed to it. Based on flow pattern and rate studies, hemolymph circulation has been characterized as being sluggish, and, with this, a degree of inefficiency is implied. However, it is apparent that the circulatory system is quite effective, allowing insects to meet the biochemical and physiological demands placed upon it under most circumstances (i.e. flight). Therefore, it can be assumed that circulation is efficient, as it is represented in particular insect forms. In addition, hemolymph can be viewed as a dynamic fluid which shares an intimate metabolic relationship with tissues and organs with which it is associated. This relationship is well illustrated by studies which have indicated that many of the metabolic pools are strongly regulated.

Hemolymph is an integral part of the insect and can be viewed as an extension of the intracellular compartment. Because of this, future studies designed to clarify such things as the metabolic relationships between hemocytes and hemolymph plasma, the sources of plasma enzymes, half-lives of metabolites and hormonal factors, would be particularly useful. Although progress has been made on the characterization of certain plasma proteins, much more investigation is needed. This includes further characterization and clarification of the roles of storage and carrier proteins as well as the mechanisms which control their steady-state, or turn-over rates.

ACKNOWLEDGEMENTS

Appreciation is expressed to Dr D. G. Cochran for the interest, encouragement and discussion he provided in the preparation of this chapter. Recognition and thanks are extended to Drs Cochran, J. L. Eaton and R. D. Fell for critical review of the final manuscript. Thanks are also due to Mrs K. J. Mullins who assisted in the preparation of the figures, and Mrs Bea Martin who typed the manuscript.

REFERENCES

ADIYODI, K. G. and ADIYODI, R. G. (1974). Comparative physiology of reproduction in arthropods. In *Advances in Comparative Physiology and Biochemistry*. Edited by O. Lowenstein. Vol. 5, pages 36–107. Academic Press, New York.

AGOSIN, M. (1978). Functional role of proteins. In *Biochemistry of Insects*. Edited by M. Rockstein. Pages 93–144. Academic Press, New York.

AHEARN, G. A. (1970). Changes in hemolymph properties accompanying heat death in the desert tenebrionid beetle, *Centrioptera muricata*. *Comp. Biochem. Physiol. 33*, 845–857.

ALDRICH, J. R., SODERLUND, D. M., BOWERS, W. S. and FELDLAUFER, M. F. (1981). Ovarian sequestration of (^{14}C)-inulin by an insect: an index of vitellogenesis. *J. Insect Physiol. 27*, 379–382.

ALEXANDER, R. D. and BORGIA, G. (1979). On the origin and basis of the male–female phenomenon. In *Sexual Selection and Reproductive Competition*. Edited by M. S. Blum and N. A. Blum. Pages 417–440. Academic Press, New York.

AMIRANTE, G. A. and MAZZALAI, F. G. (1978). Synthesis and localization of hemoagglutinins in hemocytes of the cockroach *Leucophaea maderae* L. *Devel. Comp. Immunol. 2*, 735–740.

AMIRANTE, G. A., DEBERNARDI, F. L. and MAGRETTI, P. C. (1976). Immunochemical studies on heteroagglutinins in the haemolymph of cockroach *Leucophaea maderae* L. (Insecta: Dictyoptera). *Boll. Zool. 43*, 63–67.

ANDERSON, L. M. and TELFER, W. H. (1970). Extracellular concentrating of proteins in the cecropia moth follicle. *J. Cell. Physiol. 76*, 37–54.

ANDERSON, R. S. and COOK, M. L. (1979). Induction of lysozyme-like activity in the hemolymph and hemocytes of an insect, *Spodoptera eridania*. *J. Invert. Pathol. 33*, 197–203.

ARLIAN, L. G. and VESELICA, M. M. (1979). Water balance in insects and mites. *Comp. Biochem. Physiol. 64*, 191–200.

ARNOLD, J. W. (1974). Hemocytes of insects. In *The Physiology of Insecta*. Edited by M. Rockstein. Vol. 5, pages 201–254. Academic Press, New York.

ASAHINA, E. (1969). Frost resistance in insects. *Adv. In. Physiol. 6*, 1–49.

AYERS, G. S., BROWN, T. M. and MONROE, R. E. (1974). Studies on D-alanine in the Heteroptera. *J. Ent. 49*, 1–5.

BADE, K. L. and WYATT, G. R. (1962). Metabolic conversions during pupation of cecropia silkworm. I. Deposition and utilization of nutrient reserves. *Biochem J. 83*, 470–478.

BAILEY, E. (1975). Biochemistry of insect flight. Part 2. Fuel supply. In *Insect Biochemistry and Functions*. Edited by D. J. Candy and B. A. Kilby. Pages 89–176. Chapman & Hall, London.

BALSHIN, M. and PHILLIPS, J. E. (1971). Active absorption of amino acids in the rectum of the desert locust, *Schistocerca gregaria*. *Nature, 233*, 53–55.

BARRETT, F. M. (1974). Changes in the concentration of the free amino acids in the hemolymph of *Rhodnius prolixus* during the 5th instar. *Comp. Biochem. Physiol. 48B*, 241–250.

BARRETT, F. M. and FRIEND, W. G. (1966). Studies on the uric acid concentration in the hemolymph of fifth-instar larvae of *Rhodnius prolixus* (Stal) during growth and metamorphosis. *J. Insect Physiol. 12*, 1–7.

BARRETT, F. M. and FRIEND, W. G. (1975). Differences in the concentration of free amino acids in the haemolymph of adult male and female *Rhodnius prolixus*. *Comp. Biochem. Physiol. 52B*, 427–431.

BARRETT, F. M. and LAI-FOOK, J. (1976). Free amino acids in the haemolymph of *Calpodes ethlius* during the fifth instar. *Comp. Biochem. Physiol. 53B*, 205–210.

BARWIG, B. and BOHN, H. (1980). Evidence for presence of two clotting proteins in insects. *Naturwissenschaften 67*, 47–48.

BAUMEISTER, T., MEREDITH, J., JULIEN, W. and PHILLIPS, J. (1981). Acetate transport by locust rectum *in vitro*. *J. Insect Physiol. 27*, 195–201.

BEAMENT, J. W. L. (1958). A paralyzing agent in the blood of cockroaches. *J. Insect Physiol. 2*, 199–214.

BEDFORD, J. J. (1977). The carbohydrate levels of insect hemolymph. *Comp. Biochem. Physiol. 57A*, 83–86.

BEDFORD, J. J. and LEADER, J. P. (1975). Hemolymph composition in the larva of the dragonfly *Uropetala carovei*. *J. Ent. 50A*, 1–7.

BEENAKKERS, A. M. T., VAN DER HORST, D. J. and VAN MERREWIJK, W. J. A. (1981). Role of Lipids in Energy Metabolism. In *Energy Metabolism in Insects*. Edited by R. G. H. Downer. Pages 53–100. Plenum Press, New York.

BELL, W. J. (1971). Starvation-induced oocyte resorption and yolk protein salvage in *Periplaneta americana*. *J. Insect Physiol. 17*, 1099–1111.

BELL, W. J. and BOHM, M. K. (1975). Oosorption in insects. *Biol. Rev. 50*, 373–396.

BENFIELD, E. F. (1974). Autohemorrhage in two stoneflies (Plecoptera) and its effectiveness as a defense mechanism. *Ann. Ent. Soc. Amer. 67*, 739–742.

BENNETT, G. A. and SHOTWELL, O. L. (1972). Haemolymph lipids of healthy and diseased Japanese beetle larvae. *J. Insect Physiol. 18*, 53–62.

BENNETT, G. A., KLEIMAN, R. and SHOTWELL, O. L. (1972). Hydrocarbons in hemolymph from healthy and diseased Japanese beetle larvae. *J. Insect Physiol. 18*, 1343–1350.

BERNAYS, E. A. and CHAPMAN, R. F. (1974). Changes in haemolymph osmotic pressure in *Locusta migratoria* larvae in relation to feeding. *J. Ent. 48A*, 149–155.

BERRIDGE, M. J. (1970). Osmoregulation in terrestrial arthropods. In *Chemical Zoology*. Edited by M. Florkin and B. T. Scheer. Vol. 5, pt. A, pages 287–319. Academic Press, New York.

BLUM, M. S. (1981). *Chemical Defenses of Arthropods*. Academic Press, New York.

BLUM, M. S. and SANNASI, A. (1974). Reflex bleeding in lampyrid, *Photinus pyralis*: Defensive function. *J. Insect Physiol. 20*, 451–460.

BOCTOR, I. Z. (1974). Carbohydrates in the hemolymph of the cotton leafworm *Spodoptera littoralis* larvae (Lepidoptera: Noctuidae). *Comp. Biochem. Physiol. 49B*, 79–82.

BOCTOR, I. Z. (1981). Changes in the free amino acids of the hemolymph of diapause and nondiapause pupae of the cotton bollworm *Heliothis armigera* (Lepidoptera: Noctuidae). *Experentia 37*, 125–126.

BOCTOR, I. Z. and SALEM, S. I. (1973). Free amino acids of the hemolymph of the cotton leafworm. *Spodoptera littoralis* larvae (Lepidoptera, Noctuidae). *Comp. Biochem. Physiol. 45B*, 785–790.

BODNARYK, R. P. (1974). Kinetic aspects of the breakdown of gamma-L-glutamyl-L-phenylalanine during sclerotization of the puparium of *Musca domestica*. *Insect Biochem. 4*, 439–454.

BODNARYK, R. P. (1978). Structure and function of insect peptides. In *Advances in Insect Physiology*. Edited by J. E. Treherne, M. J. Berridge and V. B. Wigglesworth. Vol. 13, pages 69–132. Academic Press, London.

BOHN, H. (1975). Growth-promoting effect of haemocytes on insect epidermis *in vitro*. *J. Insect Physiol. 21*, 1283–1293.

BOHN, H., BARWIG, B. and BOHN, B. (1981). Immunochemical analysis of hemolymph clotting in the insect *Leucophaea maderae*. *J. Comp. Physiol. 143B*, 169–184.

BOSQUET, G. (1977). Hemolymph modifications during starvation in *Philosamia cynthia* Walkeri. II. Amino acids and peptides. *Comp. Biochem. Physiol. 58A*, 377–382.

BOUNIAS, M. (1980). *In vivo* control of hemolymph alpha-glucosidase kinetics by glycemic components according to the developmental stages of worker bees, *Apis mellifica*. *Comp. Biochem. Physiol. 66B*, 335–342.

BOYER, A. C. (1975). Sorption of tetrachlorvinphos insecticide (Gardona) to the hemolymph of *Periplaneta americana*. *Pest. Biochem. Physiol. 5*, 135–141.

BRACKE, J. W. and MARKOVETZ, A. J. (1980). Transport of bacterial end products from the colon of *Periplaneta americana*. *J. Insect. Physiol. 26*, 85–89.

BRADLEY, T. J. and PHILLIPS, J. E. (1977). The effect of external salinity on drinking rate and rectal secretion in the larvae of the saline-water mosquito. *Aedes taeniorhychus*. *J. Exp. Biol. 66*, 97–110.

BRADY, J. (1967). The relationship between blood and blood-cell density in insects. *J. Exp. Biol. 47*, 313–326.

BREHELIN, M. (1979). Hemolymph coagulation in *Locusta migratoria*: evidence for a functional equivalent of fibrinogen. *Comp. Biochem. Physiol. 62B*, 329–334.

BROOKES, V. J. (1976). Protein synthesis in the fat body of *Leucophaea maderae* during vitellogenesis. *J. Insect Physiol. 22*, 1649–1657.

BROWN, S. E. and MAZZONE, H. M. (1977). Electrophoretic studies on proteins in the egg and hemolymph of the gypsy moth with reference to isoenzymes. *J. N.Y. Ent. Soc. 85*, 26–35.

BROWN, S. E., PATTON, R. L., ZERILLO, R. T., DOUGLAS, S. M., BREILLATT, J. P. and MAZZONE, H. M. (1977). *J. N.Y. Ent. Soc. 85*, 36–42.

BUCK, J. B. (1953). Physical properties and chemical composition of insect blood. In *Insect Physiology*. Edited by K. D. Roeder. Pages 147–190. John Wiley & Sons, New York.

BUCKNER, J. S. and CALDWELL, J. M. (1980). Uric acid levels during the last larval instar of *Manduca sexta*, an abrupt transition from excretion to storage in fat body. *J. Insect Physiol. 26*, 27–32.

BUHLMANN, G. (1976). Haemolymph vitellogenin, juvenile hormone, and oocyte growth in the adult cockroach *Nauphoeta cinerea* during the first preovipositional period. *J. Insect Physiol. 22,* 1101–1110.

BURSELL, E. (1981). The role of proline in energy metabolism. In *Energy Metabolism in Insects*. Edited by R. G. H. Downer. Pages 135–154. Plenum Press, New York.

BURTON, R. L., HOPPER, D. G., SAUER, J. R. and FRICK, J. H. (1972). Properties of the hemolymph of the corn earworm. *Comp. Biochem. Physiol. 42B,* 713–716.

CANDY, D. J. (1967). Occurrence and metabolism of *scyllo*-inositol in the locust. *Biochem. J. 103,* 666–671.

CANDY, D. J. (1981). Hormonal regulation of substrate transport and metabolism. In *Energy Metabolism in Insects*. Edited by R. G. H. Downer. Pages 19–52. Plenum Press, New York.

CANDY, D. J., HALL, L. J. and SPENCER, I. M. (1976). The metabolism of glycerol in the locust, *Schistocerca gregaria* during flight. *J. Insect Physiol. 22,* 583–587.

CASEY, T. M. (1981). Behavioral mechanisms of thermoregulation. In *Insect Thermoregulation*. Edited by B. Heinrich. Pages 79–114. John Wiley & Sons, New York.

CHEN, A. C. and FRIEDMAN, S. (1975). An isotonic saline for the adult blowfly *Phormia regina* and its application to perfusion experiments. *J. Insect Physiol. 21,* 529–536.

CHEN, P. S. (1966). Amino acid and protein metabolism in insect development. In *Advances in Insect Physiology*. Edited by J. W. L. Beament, J. E. Treherne and V. B. Wigglesworth. Vol. 3, pages 53–132. Academic Press, London.

CHEN, P. S., MITCHELL, H. K. and NEUWEG, M. (1978). Tyrosine glucoside in *Drosophila buskii. Insect Biochem. 8,* 279–286.

CHERBAS, L. (1973). The induction of an injury reaction in cultured haemocytes from saturniid pupae. *J. Insect Physiol. 19,* 2011–2023.

CHEUNG, P. Y. K., GRULA, E. A. and BURTON, R. L. (1978). Hemolymph responses in *Heliothis zea* to inoculation with *Bacillus thuringiensis* Kurstaki or *Micrococcus lysodeikticus. J. Invert. Pathol.* 31, 148–156.

CHEUNG, P. Y. K., GRULA, E. A., SATAYMURTHY, N. and BERLIN, K. D. (1982). Presence of free putrescine in the haemolymph of corn earworm *(Heliothis zea)* larvae. *Insect Biochem. 12,* 41–48.

CHINO, H. (1981). Lipid transport by hemolymph lipoprotein. In *Energy Metabolism in Insects*. Edited by R. G. H. Downer. Pages 155–167. Plenum Press, New York.

CHINO, H. and DOWNER, R. G. H. (1979). The role of diacylglycerol in absorption of dietary glycerides in the American cockroach *Periplaneta americana* L. *Insect Biochem. 9,* 379–382.

CHINO, H. and GILBERT, L. I. (1971). The uptake and transport of cholesterol by haemolymph lipoproteins. *Insect Biochem. 1,* 337–347.

CHINO, H., DOWNER, R. G. H., WYATT, G. R. and GILBERT, L. I. (1981a). Lipophorins, a major class of lipoproteins of insect haemolymph. *Insect Biochem. 11,* 491.

CHINO, H., KATASE, H., DOWNER, R. G. H. and TAKAHASHI, K. (1981b). Diacylglycerol-carrying lipoprotein of hemolymph of the American cockroach: Purification, characterization and function. *J. Lipid Res. 22,* 7–15.

CHINZEI, Y., CHINO, H. and WYATT, G. R. (1981). Purification and properties of vitellogenin and vitellin from *Locusta migratoria. Insect Biochem. 11,* 1–7.

CHIPPENDALE, G. M. (1973a). Diapause of the southwestern cornborer, *Diatraea grandiosella*: Utilization of fat body and haemolymph reserves. *Ent. Exp. Appl. 16,* 395–406.

CHIPPENDALE, G. M. (1973b). Metabolic reserves of larvae and pupae in the angoumois grain moth *Sitotroga cerealella. Insect Biochem. 3,* 1–10.

CLENCH, H. K. (1966). Behavioral thermoregulation in butterflies. *Ecology 47,* 1021–1034.

COCHRAN, D. G. (1975). Excretion in insects. In *Insect Biochemistry and Function*. Edited by D. J. Candy and B. A. Kilby. Pages 177–281. Chapman & Hall, London.

COHEN, A. C. and PATANA, R. (1982). Ontogenetic and stress-related changes in hemolymph chemistry of beet army worms. *Comp. Biochem. Physiol. 71A,* 193–198.

COLLETT, J. I. (1976a). Peptidase-mediated storage of amino acids in small peptides. *Insect Biochem. 6,* 179–185.

COLLETT, J. I. (1976b). Some features of the regulation of free amino acids in adult *Calliphora erythrocephala. J. Insect Physiol. 22,* 1395–1403.

CONDON, W. J. and GORDON, R. (1977). Effects of the mermithid nematode, *Mermis nigrescens* on the levels of hemolymph and fecal uric acid in its host; the migratory locust. *Locusta migratoria. Canad. J. Zool. 55,* 690–692.

COOK, B. J. (1967). An investigation of factor S, a neuromuscular excitatory substance from insects and crustacea. *Biol. Bull. 133,* 526–538.

COOK, B. J. and HOLT, G. G. (1974). Neurophysiological changes associated with paralysis arising from body stress in the cockroach *Periplaneta americana. J. Insect Physiol. 20,* 21–40.

CORRIGAN, J. J. (1970). Nitrogen metabolism in insects. In *Comparative Biochemistry of Nitrogen Metabolism*. Edited by J. W. Campbell. Vol. 1, pages 387–488. Academic Press, Inc., London.

CORRIGAN, J. J. and KEARNS, G. W. (1963). Amino acid metabolism in DDT-poisoned American cockroaches. *J. Insect Physiol. 9,* 1–12.

COTTRELL, C. B. (1962). The imaginal ecdysis of blowflies. Observations on the hydrostatic mechanisms involved in digging and expansion. *J. Exp. Biol. 39,* 431–448.

COUDRON, J. A., DUNN, P. E., SEBALLOS, H. L., WHAREN, R. E., SANBURG, L. L. and LAW, J. H. (1981). Preparation of homogeneous juvenile hormone specific esterase from the haemolymph of the tobacco hornworm *Manduca sexta. Insect Biochem. 11,* 453–461.

COUTCHIE, P. A. and CROWE, J. H. (1979). Transport of water vapor by tenebrionid beetles. II. Regulation of the osmolality and composition of the hemolymph. *Physiol. Zool. 52,* 88–100.

CROSSLEY, A. C. (1975). Cytophysiology of Insect Blood. In *Advances in Insect Physiology*. Edited by J. E. Treherne, M. J. Berridge and V. B. Wigglesworth. Vol. 11, pages 117–221. Academic Press, New York.

CROSSLEY, A. C. (1979). Biochemical and ultrastructural aspects of synthesis, storage and secretion in hemocytes. In *Insect Hemocytes*. Edited by A. P. Gupta. Pages 423–473. Cambridge University Press, Cambridge.

CUENOT, L. (1896). La Saignee reflexe et moyens de defense de quelques insectes. *Arch. Zool. Exp. Gen. 24,* 655–680.

DAHLMAN, D. L. (1973). Starvation of the tobacco hornworm *Manduca sexta*. I. Changes in hemolymph characteristics of 5th stage larvae. *Ann. Ent. Soc. Am. 66,* 1023–1029.

DAHLMAN, D. L. (1974). Hemolymph characteristics of developing adult tobacco hornworms reared as larvae on tobacco leaf or synthetic diet. *Comp. Biochem. Physiol. 49A,* 369–375.

DAHLMAN, D. L. (1975). Trehalose and glucose levels in hemolymph of diet-reared tobacco leaf-reared and parasitized tobacco hornworm larvae. *Comp. Biochem. Physiol. 50A,* 165–167.

DAHLMAN, D. L. and VINSON, S. B. (1976). Trehalose levels in the hemolymph of *Heliothis virescens* parasitized by *Campoletis sonorensis. Ann. Ent. Soc. Am. 69,* 523–524.

DALLMANN, S. H. and HERMAN, W. S. (1978). Hormonal regulation of hemolymph lipid concentration in the monarch butterfly, *Danaus plexipus. Gen. Comp. Endocr. 36,* 142–150.

DAVEY, K. G. and TREHERNE, J. E. (1964). Studies on crop function in the cockroach *(Periplaneta americana* L.). III. Pressure changes during feeding and crop emptying. *J. Exp. Biol. 41,* 513–524.

DE KORT, C. A. D. and GRANGER, N. A. (1981). Regulation of the juvenile hormone titer. *Ann. Rev. Ent. 26,* 1–28.

DE WILDE, J. and DE LOOF, A. (1973). Reproduction. In *Physiology of Insecta*. Edited by M. Rockstein. Vol. 1, pages 11–95. Academic Press, New York.

DJAJAKUSUMAH, T. and MILES, P. W. (1966). Changes in the relative amounts of soluble protein and amino acid in the hemolymph of the locust *Chortoicetes terminifera* Walker, (Orthoptera: Acrididae) in relation to dehydration and subsequent hydration. *Austr. J. Biol. Sci. 19,* 1081–1094.

DOWNER, R. G. H. (1979). Induction of hypertrehalosemia by excitation in *Periplaneta americana. J. Insect Physiol. 25,* 59–63.

DOWNER, R. G. H. (1981). Physiological and Environmental Considerations in Insect Bioenergetics. In *Energy Metabolism in Insects*. Edited by R. G. H. Downer. Pages 1–17. Plenum Press, New York.

DOWNER, R. G. H. and CHINO, H. (1979). Cholesterol and cholesterol ester in haemolymph of the American cockroach, *Periplaneta americana* L. *Can. J. Zool. 57,* 1333–1336.

Downer, R. G. H. and Matthews, J. R. (1976). Patterns of lipid distribution and utilization in insects. *Amer. Zool. 16,* 733–745.

Downing, N. (1978). Measurements of the osmotic concentrations of stylet sap, hemolymph and honeydew from an aphid under osmotic stress. *J. Exp. Biol. 77,* 247–250.

Duchateau, G. and Florkin, M. (1958). A survey of aminoacidemia with special reference to the high concentration of free amino acids in insect hemolymph. *Arch. Int. Physiol. 66,* 573–591.

Duffey, S. S. (1980). Sequestration of plant natural products by insects. *Ann. Rev. Ent. 25,* 447–477.

Duhamel, R. C. and Kunkel, J. G. (1978). A molting rhythm for serum proteins of the cockroach *Blatta orientalis. Comp. Biochem. Physiol. 60B,* 333.

Duman, J. G. (1979). Thermal-hysteresis factors in overwintering insects. *J. Insect Physiol. 25,* 805–810.

Duman, J. G. (1980). Factors involved in overwintering survival of the freeze tolerant beetle *Dendroides canadensis. J. Comp. Physiol. 136,* 53–59.

Edney, E. B. (1977). *Water Balance in Land Arthropods.* Springer-Verlag, Berlin.

Engelmann, F. (1970). Gonadal development. In *The Physiology of Insect Reproduction.* Edited by G. A. Kerkut. Vol. 44, pages 41–56. Pergamon Press, Oxford.

Engelmann, F. (1979). Insect vitellogenin: identification biosynthesis and role in vitellogenesis. In *Advances in Insect Physiology.* Edited by J. E. Treherne, M. J. Berridge and V. B. Wigglesworth. Vol. 14, pages 49–108. Academic Press, London.

Evans, D. R. and Dethier, V. G. (1957). The regulation of taste thresholds for sugars in the blowfly. *J. Insect Physiol. 1,* 3–17.

Evans, P. D. (1980). Biogenic amines in the insect nervous system. In *Advances in Insect Physiology.* Edited by M. J. Berridge, J. E. Treherne and V. B. Wigglesworth. Vol. 15, pages 317–473. Academic Press, London.

Evans, P. D. and Crossley, A. C. (1974). Free amino acids in the haemocytes and plasma of the larvae of *Calliphora vicina. J. Exp. Biol. 61,* 463–472.

Feir, D., McClain, E. and Lupo, V. (1973). Retention of amino acids in the hemolymph of the large milkweed bug. *Ann. Ent. Soc. Amer. 66,* 229–231.

Feyereisen, R. (1977). A specific binding protein for the molting hormone ecdysterone in locust hemolymph. *Experientia 33,* 1111–1113.

Firling, C. E. (1977). Amino acid and protein changes in the hemolymph of developing 4th instar *Chironomus tentans. J. Insect Physiol. 23,* 17–22.

Firling, C. E. and Kobilka, B. K. (1979). A medium for the maintenance of *Chironomus tentans* salivary glands *in vitro. J. Insect Physiol. 25,* 93–103.

Fischl, J., Ishay, J. and Talmor, N. (1975). Monosaccharidase activity and pyruvate, lactate and carbon dioxide content of the *Vespa orientalis* hemolymph. *Comp. Biochem. Physiol. 50B,* 71–74.

Flattum, R. F., Watkinson, I. A. and Crowder, I. A. (1973). The effect of insect "autoneurotoxin" on *Periplaneta americana* (L.) and *Schistocerca gregaria* (Forskal) Malpighian tubules. *Pest. Biochem. Physiol. 3,* 237–242.

Florkin, M. (1959). The free amino acids of insect hemolymph. In *Biochemistry of Insects.* Edited by L. Levenbook. Pages 63–77. Pergamon Press, London.

Florkin, M. and Jeuniaux, C. (1964). Hemolymph Composition. In *The Physiology of Insecta.* Edited by M. Rockstein. Vol. 3, pages 110–148. Academic Press, New York.

Florkin, M. and Jeuniaux, C. (1974). Hemolymph composition. In *The Physiology of Insecta.* Edited by M. Rockstein. Vol. 5, pages 255–307. Academic Press, New York.

Fox, F. R. and Mills, R. R. (1969). Changes in haemolymph and cuticle proteins during the moulting process in the American cockroach. *Comp. Biochem. Physiol. 29,* 1187–1195.

Fox, F. R., Seed, J. R. and Mills, R. R. (1972). Cuticle sclerotization by the American cockroach: immunological evidence for the incorporation of blood proteins into cuticle. *J. Insect Physiol. 18,* 2065–2070.

Fraenkel, G. S. (1980). Foreword and overview. In *Neurohormonal Techniques in Insects.* Edited by T. A. Miller. Pages ix–xv. Springer-Verlag, New York.

Friedel, T. and Gillott, C. (1977). Contribution of male-produced proteins to vitellogenesis in *Melanoplus sanguinipes. J. Insect Physiol. 23,* 145–151.

Friedman, S. (1978). Trehalose regulation, one aspect of metabolic homeostasis. *Ann. Rev. Ent. 23,* 389–407.

Gardner, W. A., Sutton, R. M. and Noblet, R. (1979). Effects of infection by *Beauveria bassiana* on hemolymph proteins of noctuid larvae. *Ann. Ent. Soc. Am. 72,* 224–228.

Gee, J. D. (1975). Diuresis in the tsetse fly *Glossina austeni. J. Exp. Biol. 63,* 381–390.

Gee, J. D. (1977). The hormonal control of excretion. In *Transport of Ions and Water in Animals.* Edited by B. L. Gupta, R. B. Moreton, J. L. Oschmann and B. J. Wall. Pages 265–281. Academic Press, London.

Geiger, J. G., Krolak, J. M. and Mills, R. R. (1977). Possible involvement of cockroach haemocytes in the storage and synthesis of cuticle proteins. *J. Insect Physiol. 23,* 227–230.

Gellissen, G. and Emmerich, H. (1980). Purification and properties of a diglycerol binding lipoproteins, Lp I of the hemolymph of adult male *Locusta migratoria. J. Comp. Physiol. 136B,* 1–9.

Goosey, M. W. and Candy, D. J. (1980). The D-octopamine content of the hemolymph of the locust, *Schistocerca americana gregaria* and its elevation during flight. *Insect Biochem. 10,* 393–397.

Gordon, R. and Bailey, C. H. (1976). Free amino acids, ions and osmotic pressure of the hemolymph of 3 species of black flies. *Canad. J. Zool. 54,* 399–404.

Gregoire, C. (1970). Haemolymph Coagulation in Arthropods. In *The Haemostatic Mechanism in Man and Other Animals.* Edited by R. G. Macfarlane. *Symp. Zool. Soc. Lond.* No. 27. Academic Press, New York.

Gregoire, C. (1974). Hemolymph coagulation. In *The Physiology of Insecta.* Edited by M. Rockstein. Vol. 5, pages 309–360. Academic Press, New York.

Gringorten, J. L. and Friend, W. G. (1979). Hemolymph volume changes in *Rhodnius prolixus* during flight. *J. Exp. Biol. 83,* 325–333.

Gupta, A. P. (1979a). Hemocyte types: their structures, synonomies, inter-relationships, and taxonomic significance. In *Insect Hemocytes.* Edited by A. P. Gupta. Pages 86–127. Cambridge University Press, Cambridge.

Gupta, A. P. (Editor) (1979b). *Insect Hemocytes.* Edited by A. P. Gupta. Cambridge University Press, Cambridge.

Hagedorn, H. H. and Kunkel, J. G. (1979). Vitellogenin and vitellin in insects. *Ann. Rev. Ent. 24,* 475–505.

Hanaoka, K. and Takahashi, S. Y. (1976). Effect of a hyperglycemia factor on haemolymph trehalose and fat body carbohydrates in the American cockroach. *Insect Biochem. 6,* 621–625.

Hansen, O. (1964). Effect of diet on the amount and composition of locust blood carbohydrates. *Biochem. J. 92,* 333–337.

Hapner, K. D. and Jermyn, M. A. (1981). Haemagglutinin activity in the haemolymph of *Teleogryllus commodus* (Walker). *Insect Biochem. 11,* 287–295.

Happ, G. M. and Eisner, T. (1961). Hemorrhage in a coccinellid beetle and its repellent effect in ants. *Science 134,* 329–331.

Harmsen, R. (1966). A quantitative study of the pteridines in *Pieris brassicae.* I. During post-embryonic development. *J. Insect Physiol. 12,* 9–22.

Hawkins, W. B. and Sternburg, J. (1964). Some chemical characteristics of a DDT-induced neuroactive substance from cockroaches and crayfish. *J. Econ. Ent. 57,* 241–247.

Heinrich, B. (1976). Heat exchange in relation to blood flow between thorax and abdomen in bumblebees. *J. Exp. Biol. 64,* 561–585.

Heinrich, B. (1981a). Definitions and thermoregulatory taxonomy. In *Insect Thermoregulation.* Edited by B. Heinrich. Pages 4–17. John Wiley & Sons, New York.

Heinrich, B. (1981b). Ecological and evolutionary perspectives. In *Insect Thermoregulation.* Edited by B. Heinrich. Pages 235–302. John Wiley & Sons, New York.

Heit, M., Sauer, J. R. and Mills, R. R. (1973). The effects of high concentrations of sodium in the drinking medium of the American cockroach, *Periplaneta americana* (L.). *Comp. Biochem. Physiol. 45A,* 363–370.

Hill, L. and Goldsworthy, G. J. (1968). Growth, feeding activity, and the utilization of reserves in larvae of locusta. *J. Insect Physiol. 14,* 1085–1098.

HILLIARD, S. D. and BUTZ, A. (1969). Daily fluctuations in the concentrations of total sugar and uric acid in the hemolymph of *Periplaneta americana*. *Ann. Ent. Soc. Amer. 62*, 71–74.

HOFFMAN, A. G. D. and DOWNER, R. G. H. (1979). End product specificity of triacylglycerol lipases from intestine, fat body, muscle and hemolymph of the American cockroach, *Periplaneta americana* (L.). *Lipids, 14*, 893–899.

HOLLANDE, A. C. (1911). L'autohemorrhee' ou le rejet du sang chez les insectes (toxicologie du sang). *Arch. d'Anat. Microsc. 13*, 171–318.

HUGHES, T. D. (1980). The imaginal ecdysis of the desert locust, *Schistocerca gregaria*. III. Motor activity underlying the expansional and post-expansional behavior. *Physiol. Ent. 5*, 141–152.

HYATT, A. D. and MARSHALL, A. T. (1977). Sequestration of haemolymph sodium and potassium by fat body in the water-stressed cockroach, *Periplaneta americana*. *J. Insect Physiol. 23*, 1437–1441.

INGRAM, M. J., STAY, B. and CAIN, G. D. (1977). Composition of milk from the viviparous cockroach *Diploptera punctata*. *Insect Biochem. 7*, 257–267.

IRVING, S. N., WILSON, R. G. and OSBORNE, M. P. (1979). Studies on L-glutamate in insect haemolymph. III. Amino acid analysis of the haemolymph of various arthropods. *Physiol. Ent. 4*, 231–240.

ISHAY, J., ABRAHAM, Z., GRUNFIELD, Y. and GITTER, S. (1974). Catecholamines in social wasps. *Comp. Biochem. Physiol. 48A*, 369–373.

ISOBE, M., KONDO, N., IMAI, K., YAMASHITA, O. and GOTO, T. (1981). Glucosyltyrosine in silkworm hemolymph as a transient metabolite of insects. *Agr. Biol. Chem. 45*, 687–692.

JAROSZ, J. (1979). Simultaneous induction of protective immunity and selective synthesis of hemolymph lysozyme protein in larvae of *Galleria mellonella*. *Biol. Zbl. 98*, 459–471.

JEUNIAUX, C. (1971). Hemolymph–Arthropoda. In *Chemical Zoology*. Edited by H. Florkin and B. T. Sheer. Vol. 6, pages 63–118. Academic Press, New York.

JONES, J. C. (1964). The circulatory system of insects. In *The Physiology of Insecta*. Edited by M. Rockstein. Vol. 3, pages 1–107. Academic Press, New York.

JONES, J. C. (1974). Factors affecting heart rates in insects. In *The Physiology of Insecta*. Edited by M. Rockstein. Vol. 5, pages 119–167. Academic Press, New York.

JONES, J. C. (1977). *The Circulatory System of Insects*. Charles C. Thomas Publishing Co., Springfield, Ill.

JOSHUA, H., FISCHL, J., HENIG, E., ISHAY, J. and GITTER, S. (1973). Cytological, biochemical and bacteriological data of the hemolymph of *Vespa orientalis*. *Comp. Biochem. Physiol. 45B*, 167–175.

JUNGREIS, A. M. (1974). Physiology and composition of moulting fluid and midgut lumenal contents in the silkmoth, *Hyalophora cecropia*. *J. Comp. Physiol. 88*, 113–127.

JUNGREIS, A. M. (1979). Physiology of moulting in insects. In *Advances in Insect Physiology*. Edited by J. E. Treherne, M. J. Berridge and V. B. Wigglesworth. Vol. 14, pages 109–183. Academic Press, London.

JUNGREIS, A. M. (1980). Hemolymph as a dynamic tissue. In *Insect Biology in the Future*. Edited by M. Locke and D. S. Smith. Pages 273–294. Academic Press, New York.

JUNGREIS, A. M. and WYATT, G. R. (1972). Sugar release and penetration in insect fat body relations to regulation of haemolymph trehalose in developing stages of *Hyalophora cecropia*. *Biol. Bull. 143*, 367–391.

JUNGREIS, A. M., JATLOW, P., WYATT, G. R. (1973). Inorganic ion composition of haemolymph of the cecropia silkmoth changes with diet and ontogeny. *J. Insect Physiol. 19*, 225–233.

JUNGREIS, A. M., MCCALEB, D. C. and KUMARAN, A. K. (1980). Hemolymph gamma-amino butyric acid titers in the greater wax moth *Galleria mellonella* changes during normal or super-numerary larval development and following injury. *Comp. Biochem. Physiol. 67C*, 167–172.

JUTSUM, A. R. and GOLDSWORTHY, G. J. (1976). Fuels for flight in *Locusta*. *J. Insect Physiol. 22*, 243–249.

JUTSUM, A. R., AGARWAL, H. C. and GOLDSWORTHY, G. J. (1975). Starvation and hemolymph lipids in *Locusta migratoria migratoriodes*. *Acridia 4*, 47–56.

KAMMER, A. E. (1981). Physiological mechanisms of thermoregulation. In *Insect Thermoregulation*. Edited by B. Heinrich. Pages 115–158. John Wiley & Sons, New York.

KAMMER, A. E. and HEINRICH, B. (1978). Insect flight metabolism. In *Advances in Insect Physiology*. Edited by J. E. Treherne, M. J. Berridge and V. B. Wigglesworth. Vol. 13, pages 133–228. Academic Press, London.

KARP, R. D. and RHEINS, L. A. (1980). Induction of specific humoral immunity to soluble proteins in the American cockroach (*Periplaneta americana*). II. Nature of the secondary response. *Devl. Comp. Immunol. 4*, 629–639.

KATAGIRI, C. (1977). Localization of trehalose in the haemolymph of the American cockroach *Periplaneta americana*. *Insect Biochem. 7*, 351–353.

KELLEY, T. J. and TELFER, W. H. (1977). Antigenic and electrophoretic variants of vitellogenin in *Oncopeltus* blood and their control by juvenile hormone. *Devel. Biol. 61*, 58–69.

KINNEAR, J. F. and THOMSON, J. A. (1975). Nature, origin and fate of major hemolymph proteins in *Calliphora*. *Insect Biochem. 5*, 531–552.

KIRKSEY, C. D., MILLS, R. R. and KIMBROUGH, T. D. (1974). Binding of 5-hydroxytryptamine (serotonin) to serum proteins and haemocytes by the American cockroach. *Insect Biochem. 4*, 17–22.

KOEPPE, J. K. and GILBERT, L. I. (1973). Immunochemical evidence for the transport of hemolymph protein into the cuticle of *Manduca sexta*. *J. Insect Physiol. 19*, 615–624.

KOMANO, H., MIZUNO, D. and NATORI, S. (1980). Purification of lectin induced in the hemolymph of *Sarchophaga peregrina* larval injury. *J. Biol. Chem. 255*, 2919–2924.

KRAMER, K. J., SPIERS, R. D. and CHILDS, C. N. (1978). A method for separation of trehalose from insect hemolymph. *Anal. Biochem. 86*, 692–696.

KRAMER, K. J., TAGER, H. S. and CHILDS, C. N. (1980). Insulin-like and glucagon-like peptides in insect (*Manduca sexta*) hemolymph. *Insect Biochem. 10*, 179–182.

KRAMER, K. J., CHILDS, C. N., SPIERS, R. D. and JACOBS, R. M. (1982). Purification of insulin-like peptides from insect hemolymph and royal jelly. *Insect Biochem. 12*, 91–98.

KRAMER, K. J., SPIERS, R. D., LOOKHART, G., SEIB, P. A. and LIANG, Y. T. (1981). Sequestration of ascorbic acid by the larval labial gland and hemolymph of the tobacco hornworm *Manduca sexta* (Lepidoptera: Sphingidae). *Insect Biochem. 11*, 93–96.

KRAMER, S. J. and DE KORT, C. A. D. (1978). Juvenile hormone carrier lipoproteins in the hemolymph of the Colorado potato beetle, *Leptinotarsa decemlineata*. *Insect Biochem. 8*, 87–92.

KRAMER, S. J., MUNDALL, E. C. and LAW, J. H. (1980). Purification and properties of manducin, an amino acid storage protein of the hemolymph of larval and pupal *Manduca sexta*. *Insect Biochem. 10*, 279–288.

KRZYZANOWSKA, M. and NIEMIERKO, W. (1979). Accumulation, translocation and excretion of uric acid and uric acid riboside by ligated larvae of *Galleria mellonella*. *Insect Biochem. 9*, 11–18.

KRZYZANOWSKA, M. and NIEMIERKO, W. (1980). Purine and uric acid riboside in the ligated larvae of *Galleria mellonella* L. *Insect Biochem. 10*, 323–330.

KULKARNI, A. P. and MEHROTA, K. N. (1970). Amino acid nitrogen and proteins in the hemolymph of adult desert locusts *Schistocerca gregaria*. *J. Insect Physiol. 16*, 2181–2199.

KULKARNI, A. P. and MEHROTA, K. N. (1973). Effects of dieldrin and sumathion on the amino acid nitrogen and proteins in the hemolymph of the desert locust, *Schistocerca gregaria*. *Pest. Biochem. Physiol. 3*, 420–434.

KUNKEL, J. G. and PAN, M. L. (1976). Selectivity of yolk protein uptake: Comparison of vitellogenins in two insects. *J. Insect Physiol. 22*, 809–818.

LANGLEY, P. A., BURSELL, E., KABAYO, J., PIMLEY, R. W., TREWERN, M. A. and MARSHALL, J. (1981). Hemolymph lipid transport from fat body to uterine gland in pregnant females of *Glossina morsitans morsitans*. *Insect Biochem. 11*, 225–232.

LEADER, J. P. and BEDFORD, J. J. (1972). The composition of the haemolymph of the New Zealand earwig *Anisolabis littorea*. *J. Insect Physiol. 18*, 2229–2235.

LEADER, J. P. and BEDFORD, J. J. (1978). Hemolymph composition of 2 species of New Zealand Weta, *Hemideina* (Orthoptera Stenopelmatidae). *Comp. Biochem. Physiol. 61A*, 173–176.

LEE, R. E. JR., ZACHARIASSEN, K. E. and BAUST, J. G. (1981). Effects of cryoprotectants on the activity of hemolymph nucleating agents in physical solutions. *Cryobiology 18*, 511–514.

LEE, R. M. (1961). The variation of blood volume with age in the desert locust (*Schistocerca gregaria* Forsk.). *J. Insect Physiol. 6,* 36–51.

LEOPOLD, R. A. (1976). The role of male accessory glands in insect reproduction. *Ann. Rev. Ent. 21,* 199–221.

LETTAU, J., FOSTER, W. A., HARKER, J. E. and TREHERNE, J. E. (1977). Diel changes in potassium activity in the haemolymph of the cockroach, *Leucophaea maderae*. *J. Exp. Biol. 71,* 171–186.

LEVENBOOK, L. (1966). Hemolymph amino acids and peptide during larval growth of the blowfly *Phormia regina*. *Comp. Biochem. Physiol. 18,* 341–351.

LEVENBOOK, L. (1979). Hemolymph volume during growth of *Calliphora vicina* larvae. *Ann. Ent. Soc. Am. 72,* 454–455.

LEVENBOOK, L. and BAUER, A. C. (1980). Calliphorin and soluble protein of hemolymph and tissues during larval growth and adult development of *Calliphora vicinia*. *Insect Biochem. 10,* 693–701.

L'HELIAS, C. (1970). Chemical aspects of growth and development in insects. In *Chemical Zoology*. Edited by M. Florkin and B. T. Scheer. Vol. 5A, pages 343–400. Academic Press, London.

LIM, S. J. and LEE, S. S. (1981). The effect of starvation on hemolymph metabolites, fat body and ovarian development in *Oxya japonica* (Acrididae: Orthoptera). *J. Insect Physiol. 27,* 93–96.

LITTLE, C. (1977). Microsample analysis. In *Transport of Ion and Water in Animals*. Edited by B. L. Gupta, R. M. Moreton, J. L. Oschman and B. J. Wall. Pages 15–28. Academic Press, London.

LOCKE, M. (1974). The structure and function of the integument in insects. In *The Physiology of Insecta*. Edited by M. Rockstein. Vol. 6, pages 123–213. Academic Press, New York.

LOUGHTON, B. G. and ORCHARD, I. (1981). The nature of the hyperglycaemic factor from the glandular lobe of the corpus cardiacum of *Locusta migratoria*. *J. Insect Physiol. 27,* 383–385.

LOUGHTON, B. G. and TOBE, S. S. (1969). Blood volume in the African migratory locust. *Can. J. Zool, 47,* 1333–1336.

LOVERIDGE, J. P. (1975). Studies on the water relations of adult locusts. III. The water balance of non-flying locusts. *Zool. Afr. 10,* 1–28.

MACHIN, J. (1981). Water compartmentalization in insects. *J. Exp. Zool. 215,* 327–333.

MACK, S. R. and VANDERBERG, J. P. (1978). Hemolymph of *Anopheles stephensi* from non-infected and *Plasmodium berghei*-infected mosquitoes, I. Collection procedure and physical characteristics. *J. Parasitol. 64,* 918–923.

MACK, S. R., SAMUELS, S. and VANDERBERG, J. P. (1979a). Hemolymph of *Anopheles stephensi* from uninfected and *Plasmodium berghei*-infected mosquitoes. II. Free amino acids. *J. Parasitol. 65,* 130–136.

MACK, S. R., SAMUELS, S. and VANDERBERG, J. P. (1979b). Hemolymph of *Anopheles stephensi* from non-infected and *Plasmodium berghei*-infected mosquitoes. III. Carbohydrates. *J. Parasitol. 65,* 217–221.

MADDRELL, S. H. P. (1971). The mechanisms of insect excretory systems. In *Advances in Insect Physiology*. Edited by J. W. L. Beament, J. E. Treherne and V. B. Wigglesworth. Vol. 8, pages 199–331. Academic Press, New York.

MADDRELL, S. H. P. (1977). Insect Malpighian tubules. In *Transport of Ions and Water in Animals*. Edited by B. L. Gupta, R. B. Moreton, J. L. Oschman and B. J. Wall. Pages 541–569. Academic Press, London.

MADDRELL, S. H. P. (1981). The functional design of the insect excretory system. *J. Exp. Biol. 90,* 1–15.

MADDRELL, S. H. P. and GARDINER, B. O. C. (1980). The retention of amino acids in the hemolymph during diuresis in *Rhodnius*. *J. Exp. Biol. 87,* 315–329.

MAHOWALD, A. P. (1972). Oogenesis. In *Developmental Systems: Insects*. Edited by S. J. Counce and C. H. Waddington. Vol. 1, pages 1–47. Academic Press, London.

MANSINGH, A. and BAQAYA, V. (1971). Amino acids in insects: changes during ontogeny and various physiological and pathological conditions. *Labdev. J. Sci. Tech. 9-B,* 158–182.

MAREK, M. (1979). Influence of cooling and glycerol on metabolism of proteins and esterase isoenzymes in hemolymph of pupae *Galleria mellonella* (L.). *Comp. Biochem. Physiol. 63A,* 489–492.

MATTHEWS, J. R. and DOWNER, R. G. H. (1973). Hyperglycaemia induced by anaesthesia in the American cockroach *Periplaneta americana*. *Canad. J. Zool. 51,* 395–397.

MATTHEWS, J. R. and DOWNER, R. G. H. (1974). Origin of trehalose in stress-induced hyperglycaemia in the American cockroach, *Periplaneta americana*. *Canad. J. Zool. 52,* 1005–1010.

MATTHEWS, J. R., DOWNER, R. G. H. and MORRISON, P. E. (1976). Estimation of glucose in haemolymph of the American cockroach, *Periplaneta americana*. *Comp. Biochem. Physiol. 53A,* 165–168.

MAURIZIO, A. (1965). Untersuchungen uber das Zuckerbild Hamolymphe der Honigbiene (*Apis mellifica* L.). I. Das Zuckerbild des Blutes Erwachsener Bienen. *J. Insect Physiol. 11,* 745–763.

MAY, M. L. (1979). Insect thermoregulation. *Ann. Rev. Ent. 24,* 313–349.

MCCANN, F. W. (1970). Physiology of insect hearts. *Ann. Rev. Ent. 15,* 173–200.

MELLANBY, K. (1939). The functions of insect blood. *Biol. Rev. 14,* 243–260.

MILLER, L. K. (1969). Freezing tolerance in an adult insect. *Science 166,* 105–106.

MILLER, L. K. and SMITH, J. S. (1975). Production of threitol and sorbitol by an adult insect: association with freezing tolerance. *Nature 258,* 519–520.

MILLER, S. G. and SILHACEK, D. L. (1982). Identification and purification of storage proteins in tissues of the greater wax moth. *Galleria mellonella* (L.). *Insect Biochem. 12,* 277–292.

MILLER, T. A. (1974). Electrophysiology of the insect heart. In *The Physiology of Insecta*. Edited by M. Rockstein. Vol. 5, pages 169–200. Academic Press, New York.

MILLER, T. A. (Editor) (1980). *Neurohormonal Techniques in Insects*. Springer-Verlag, New York.

MILLS, R. R. and WHITEHEAD, D. L. (1970). Hormonal control of tanning in the American cockroach: changes in blood cell permeability during ecdysis. *J. Insect Physiol. 16,* 331–340.

MITSUHASHI, J. (1978). Free amino acids in the hemolymph of some lepidopterous insects. *Appl. Ent. Zool. 13,* 318–320.

MOREAU, R. (1974). Variations de la pression interne au cours de l'emergence et de l'expansion des ailes chez *Bombyx mori* et *Pieris brassicae*. *J. Insect Physiol. 20,* 1475–1480.

MORGAN, K. R. and BARTHOLOMEW, G. A. (1982). Homeothermic response to reduced ambient temperature in a scarab beetle. *Science 216,* 1409–1410.

MORI, H. (1979). Embryonic hemocytes: origin and development. In *Insect Hemocytes*. Edited by A. P. Gupta. Pages 3–27. Cambridge University Press, Cambridge.

MULLINS, D. E. and COCHRAN, D. G. (1983). Nitrogen metabolism. In *Insect Endocrinology*. Edited by H. Laufer and R. G. H. Downer. Part VI chapter 5 pages 451–464. Alan R. Liss, Inc., New York.

MURDOCK, L. and KOIDL, B. (1972). Blood metabolites after intestinal absorption of amino acids in locusts. *J. Exp. Biol. 56,* 795–808.

MUSZYNSKA-PYTEL, M. and CYMBOROWSKI, B. (1978). The role of serotonin in regulation of the circadian rhythms of locomotor activity in the cricket *Acheta domesticus*. I. Circadian variations in serotonin concentration in the brain and hemolymph. *Comp. Biochem. Physiol. 59C,* 13–15.

MWANGI, R. W. and GOLDSWORTHY, G. J. (1977). Interrelations between hemolymph lipid and carbohydrate during starvation in *Locusta*. *J. Insect Physiol. 23,* 1275–1280.

NAVARRO, S. and FRIEDLANDER, A. (1975). The effect of carbon dioxide anesthesia on the lactate and pyruvate levels in the hemolymph of *Ephestia cautella* (WLK.) pupae. *Comp. Biochem. Physiol. 50B,* 187–189.

NEUWIRTH, M. and LAI-FOOK, J. (1977). Cytochemistry of the very early stages of *in situ* coagulation in *Calpodes ethlius* (Lepidoptera). *Canad. J. Zool. 55,*1767–1772.

NICOLSON, S. W. (1980a). Diuresis and its hormonal control in butterflies. *J. Insect Physiol. 26,* 841–846.

NICOLSON, S. W. (1980b). Water balance and osmoregulation in *Onymacris plana,* a tenebrionid beetle from the Namib Desert. *J. Insect Physiol. 26,* 315–320.

NICOLSON, S., HORESFIELD, P. M., GARDINER, B. O. C. and MADDRELL, S. H. P. (1974). Effects of starvation and dehydration on osmotic and ionic balance in *Carausius morosus*. *J. Insect Physiol. 20,* 2061–2069.

NISHIITSUTSUJI-UWO, J. and ENDO, Y. (1981). Mode of action of *Bacillus thuringiensis* delta-endotoxin, changes in hemolymph pH and ions of *Pieris lymantria* and *Ephestia*. *Appl. Ent. Zool. 16,* 225–230.

NOGGE, G. and GIANNETTI, M. (1979). Midgut absorption of undigested albumin and other proteins by tsetse *Glossina m morsitans* (Diptera: Glossinidae). *J. Med. Ent. 16*, 263.

OHYAMA, Y. and ASAHINA, E. (1972). Frost resistance in adult insects. *J. Insect Physiol. 18*, 267–282.

ORCHARD, I. (1982). Octopamine in insects: neurotransmitter, neurohormone and neuromodulator. *Can. J. Zool. 60*, 659–669.

OSBORNE, N. N. and NEUHOFF, V. (1974). Amino acid and serotonin content in the nervous system, muscle and blood of the cockroach *Periplaneta americana. Brain Res. 80*, 251–264.

PATEL, N. G. (1971). Protein synthesis during insect development. *Insect Biochem. 1*, 391–427.

PATTERSON, J. L. and DUMAN, J. G. (1979). Composition of a protein antifreeze from larvae of the beetle *Tenebrio molitor. J. Exp. Zool. 210*, 361–367.

PHILIPPE, M., FOURNET, B. and SCHREVEL, J. (1976). Conjugated glucides of *Blaberus craniifer* hemolymph, variations during imaginal ecdysis. *J. Insect Physiol. 22*, 1405–1407.

PICHON, Y. (1970). Ionic content of haemolymph in the cockroach *Periplaneta americana. J. Exp. Biol. 53*, 195–209.

PICHON, Y. and BOISTEL, J. (1963). Modifications of the ionic content of haemolymph and of the activity of the *Periplaneta americana* in relation to diet. *J. Insect Physiol. 9*, 887–891.

POWNING, R. F. and DAVIDSON, W. J. (1973). Studies in insect bacteriolytic enzymes. Part 1. Lysozyme in haemolymph of *Galleria mellonella* and *Bombyx mori. Comp. Biochem. Physiol. 45B*, 669–681.

POWNING, R. F. and DAVIDSON, W. J. (1976). Studies on insect bacteriolytic enzymes. Part 2. Some physical and enzymatic properties of lysozyme from hemolymph of *Galleria mellonella. Comp. Biochem. Physiol. 55B*, 221–228.

PROVANSAL, A., BAUDRY-PARTIAOGLOU, N. and SLAMA, K. (1977). Hemolymph pressure pulses in the metamorphosis of *Tenebrio molitor. Acta Ent. Bohem. 74*, 362–374.

RACIOPPI, J. V. and DAHLMAN, D. L. (1980). Effects of L-canavanine on *Manduca sexta* (Sphingidae: Lepidoptera) larval hemolymph solutes. *Comp. Biochem. Physiol. 67C*, 35–39.

RAMAKRISHNAN, N. and PAWAR, V. M. (1975). Effect of diets on the hemolymph protein on the larvae of *Spodoptera litura. Ind. J. Ent. 37*, 172–178.

RAMSAY, J. A. and BROWN, R. H. J. (1955). Simplified apparatus and procedure for freezing-point determinations upon small volumes of fluid. *J. Sci. Instrum. 32*, 372–375.

RATCLIFFE, N. A. and ROWLEY, A. F. (1979). Role of hemocytes in defense against biological agents. In *Insect Hemocytes.* Edited by A. P. Gupta. Pages 331–414. Cambridge University Press, Cambridge.

REYNOLDS, S. E. (1980). Integration of behavior and physiology in ecdysis. In *Advances in Insect Physiology*. Edited by M. J. Berridge, J. E. Treherne and V. B. Wigglesworth. Vol. 15, pages 475–595. Academic Press, London.

RHEINS, L. A., KARP, R. D. and BUTZ, A. (1980). Induction of specific humoral immunity to soluble proteins in the American cockroach, *Periplaneta americana.* I. Nature of the primary response. *Devel. Comp. Immunol. 4*, 447–458.

RICHARDS, G. (1981). Insect hormones in development. *Biol. Rev. 56*, 501–549.

ROBERTS, R. B. and SMITH, H. W. (1971a). Free amino acids in the eggs and hemolymph of the grasshopper *Melanoplus sanguinipes* (Orthoptera: Acrididae). I. Identification and quantitative changes throughout development. *Ann. Ent. Soc. Am. 64*, 693–702.

ROBERTS, R. B. and SMITH, H. W. (1971b). Free amino acids in the eggs and hemolymph of the grasshopper *Melanoplus sanguinipes* (Orthoptera: Acrididae). 2. Changes in composition as a result of diet and population and artificial stress. *Ann. Ent. Soc. Am. 64*, 703–708.

ROSS, E. S. (1970). Biosystematics of the Embioptera. *Ann. Rev. Ent. 15*, 157–172.

ROWLEY, A. F. and RATCLIFFE, N. A. (1978). A histological study of wound healing and hemocyte function in the wax-moth *Galleria mellonella. J. Morphol. 157*, 181–200.

ROWLEY, A. F. and RATCLIFFE, N. A. (1980). Insect erythrocyte agglutinins. *In vitro* opsonization experiments with *Clitumnus extradentatus* and *Periplaneta americana* haemocytes. *Immunol. 40*, 483–492.

RU, N. and KLOFT, W. J. (1976). Fast absorption and elimination of ^{32}P-labelled food in larvae of the cabbage cooper *Tricoplusia ni* (Lepidoptera, noctuidae). *Ent. Germ. 2*, 242–248.

RUH, M. F. and WILLIS, J. H. (1974). Synthesis of blood and cuticular proteins in late pharate adults of the cecropia silkmoth. *J. Insect Physiol. 20*, 1277–1285.

RUTHERFORD, T. A. and WEBSTER, J. M. (1978). Some effects of *Mermis nigrescens* on the hemolymph of *Schistocerca gregaria. Canad. J. Zool. 56*, 339–347.

SALT, R. W. (1959). Role of glycerol in the cold-hardening of *Bracon cephi* (Grahan). *Can. J. Zool. 37*, 59–69.

SALT, R. W. (1961). Principles of insect cold hardiness. *Ann. Rev. Ent. 6*, 55–74.

SAMARANAYAKA, M. (1977). The effect of insecticides on osmotic and ionic balance in the locust *Schistocerca gregaria. Pest. Biochem. Physiol. 7*, 510–516.

SAMS, G. R., BELL, W. J. and WEAVER, R. F. (1980). Vitellogenin: Its structure, synthesis and processing in the cockroach, *Periplaneta americana. Biochem. Biophys. Acta. 609*, 121–135.

SASAKI, T. (1978). Chymotrypsin inhibitors from hemolymph of the silkworm *Bombyx mori. J. Biochem. 84*, 267–276.

SCHIN, K. and MOORE, R. D. (1977). Cation concentrations in the hemolymph of the fly, *Chironomus thummi* during development. *J. Insect Physiol. 23*, 723–729.

SCHMIDT, G. H. and SCHWANKL, W. (1975). Changes in hemolymph proteins during the metamorphosis of both sexes and castes of polygynous *Formica rufa* (Insecta: Hymenoptera). *Comp. Biochem. Physiol. 52B*, 365–380.

SCHMIDT, S. P. and PLATZNER, E. G. (1980). Changes in body tissues and hemolymph composition of *Culex pipiens* in response to infection by *Romanomermis culicivorax. J. Invert. Pathol. 36*, 240–254.

SCHOFFENIELS, E. and GILLES, R. (1970). Nitrogenous constituents and nitrogen metabolism in arthropods. In *Chemical Zoology*. Edited by M. Florkin and B. T. Scheer. Vol. 5A, pages 199–227.

SCOTT, M. T. (1972). Partial characterization of the hemagglutinating activity in hemolymph of the American cockroach, *Periplaneta americana. J. Invert. Pathol. 19*, 66–71.

SHAMBAUGH, G. (1969). Toxicity and effects on motor co-ordination of some neurotropic drugs on the cockroach *Nauphoeta cinerea. Ann. Ent. Soc. Amer. 62*, 370–375.

SHAMBAUGH, G. F., ZELLER, L. H. and BYLICA, S. M. (1974). Soluble amine oxidases in cockroach blood. *Insect Biochem. 4*, 185–196.

SHAPIRO, M. (1979). Changes in Hemocyte Populations. In *Insect Hemocytes.* Edited by A. P. Gupta. Pages 475–523. Cambridge University Press, Cambridge.

SHIMIZU, I. (1982). Variation of cation concentrations in the hemolymph of the silkworm *Bombyx mori* with diet and larval-pupal development. *Comp. Biochem. Physiol. 71A*, 445–447.

SKALSKY, H. and GUTHRIE, F. E. (1977). Affinities of parathion, DDT, dieldrin and carbaryl for macromolecules in the blood of the rat and the American cockroach. *Pest. Biochem. Physiol. 7*, 284–296.

SLÁMA, K. (1976). Insect hemolymph pressure and its determination. *Acta Ent. Bohem. 73*, 65–75.

SLÁMA, K., BAUDRY-PARTIAOGLOU, N. and PROVANSAL-BAUDEZ, A. (1979). Control of extracardiac hemolymph pressure pulses in *Tenebrio molitor. J. Insect Physiol. 25*, 825–832.

SMILOWITZ, Z. and SMITH, C. L. (1977). Hemolymph proteins of developing *Pieris rapae* larvae parasitized by *Apanteles glomeratus. Ann. Ent. Soc. Am. 70*, 447–454.

SOMME, L. (1978). Nucleating agents in the haemolymph of third instar larvae of *Eurosta solidagensa* (Fitch). (Dipt. Tephritidae). *Norw. J. Ent. 25*, 187–188.

SPRING, J. H., MATTHEWS, J. R. and DOWNER, R. G. H. (1977). Fate of glucose in haemolymph of the American cockroach *Periplaneta americana. J. Insect Physiol. 23*, 525–529.

SROKA, P. and BARTH, R. H. (1975). Lipid metabolism during oocyte growth in the ovoviviparous cockroaches *Eublaberus posticus* and *Nauphoeta cinerea. Insect Biochem. 5*, 637–645.

SROKA, P. and VINSON, S. B. (1978). Phenol oxidase activity in the hemolymph of parasitized and unparasitized *Heliothis virescens. Insect Biochem. 8*, 399–402.

STADDON, B. W. (1955). The excretion and storage of ammonia by the aquatic larvae of *Sialis lutaria* (Neuroptera). *J. Exp. Biol. 32*, 84–94.

STADDON, B. W. and EVERTON, I. J. (1980). Hemolymph of the milkweed bug *Oncopeltus faciatus* (Heteroptera, Lygaeidae). Inorganic constituents and amino acids. *Comp. Biochem. Physiol. 65A*, 371–374.

STEELE, J. E. (1963). The site of action of insect hyperglycemic hormone. *Gen. Comp. Endocr. 3*, 46–52.

STEELE, J. E. (1976). Hormonal control of metabolism in insects. In *Advances in Insect Physiology*. Edited by J. E. Treherne, M. J. Berridge and V. B. Wigglesworth. Vol. 12, pages 239–323. Academic Press, London.

STEELE, J. E. (1981). The role of carbohydrate metabolism in physiological function. In *Energy Metabolism in Insects*. Edited by R. G. H. Downer. Pages 101–133. Plenum Press, New York.

STEINHAUER, A. L. and STEPHEN, W. P. (1959). Changes in blood proteins during the development of the American cockroach *Periplaneta americana*. *Ann. Ent. Soc. Amer. 52*, 733–738.

STEINKRAUS, K. H., FIELD, C. C., KOCHANSKY, M. C., KAEGEBEIN, M. E. and TASHIRO, H. (1973). Cations in hemolymph and alimentary tract tissues of healthy and milky diseased European chafer *Amphimallon majalis* larvae. *Appl. Microbiol. 26*, 72–77.

STERNBURG, J. (1963). Autointoxication and some stress phenomena. *Ann. Rev. Ent. 8*, 19–38.

STERNBURG, J. and CORRIGAN, J. (1959). Rapid collection of insect blood. *J. Econ. Ent. 52*, 538–539.

STEVENSON, E. and WYATT, G. R. (1962). The metabolism of silkmoth tissues. I. Incorporation of leucine into protein. *Arch. Biochem. Biophys. 99*, 65–71.

STOBBART, R. H. and SHAW, J. (1964). Salt and water balance: excretion. In *The Physiology of Insecta*. Edited by M. Rockstein. Vol. 3, pages 189–258. Academic Press, New York.

STOBBART, R. H. and SHAW, J. (1974). Salt and water balance. In *The Physiology of Insecta*. Edited by M. Rockstein. Vol. 5, pages 361–446. Academic Press, New York.

STONE, J. V. and MORDUE, W. (1980). Isolation of insect neuropeptides. *Insect Biochem. 10*, 229–239.

SUTCLIFFE, D. W. (1962). The composition of haemolymph in aquatic insects. *J. Exp. Biol. 39*, 325–343.

SUTCLIFFE, D. W. (1963). The chemical composition of haemolymph in insects and some other arthropods in relation to their phylogeny. *Comp. Biochem. Physiol. 9*, 121–135.

TAKEI, G. H. and TAMASHIRO, M. (1975). Changes observed in hemolymph of the lawn armyworm *Spodoptera mauritia acronyctoides* during growth development and exposures to a nuclear polyhedrosis virus. *J. Invert. Pathol. 26*, 147–158.

TERRA, W. R., DE BIANCHI, A. G. and LARA, F. J. S. (1974). Physical properties and chemical composition of the hemolymph of *Rhychosciara americana* (Diptera) larvae. *Comp. Biochem. Physiol. 47B*, 117–129.

TERRA, W. R., FERREIRA, C., DE BIANCHI, A. G. and ZINNER, K. (1981). A violet carotenoprotein-containing echinenone isolated from the hemolymph of the fly *Rhynchosciara americana*. *Comp. Biochem. Physiol. 68B*, 89–93.

THOMSON, J. A. (1975). Major patterns of gene activity during development in holometabolus insects. In *Advances in Insect Physiology*. Edited by J. E. Treherne, M. J. Berridge and V. B. Wigglesworth. Vol. 11, pages 321–398. Academic Press, London.

THORNHILL, R. (1976). Sexual selection and paternal investment in insects. *Amer. Nat. 110*, 153–163.

TOBE, S. S. (1978). Changes in free amino acids and peptides in the hemolymph of *Glossina austeni* during the reproductive cycle. *Experientia 34*, 1462–1463.

TOBE, S. S. and LOUGHTON, B. (1969). An autoradiographic study of haemolymph protein uptake by the tissues of the fifth instar locust. *J. Insect Physiol. 15*, 1331–1346.

TOJO, S., NAGATA, M. and KOBAYASHI, M. (1980). Storage proteins in the silkworm *Bombyx mori*. *Insect Biochem. 10*, 289–303.

TREWITT, P. M. and BERGTROM, G. (1981). Immunological characterization of the haemoglobins of *Chironomus thummi* (Diptera). *Insect Biochem. 11*, 634–644.

TRUMAN, J. W. and ENDO, P. T. (1974). Physiology of insect ecdysis: neural and hormonal factors involved in wing-spreading behavior of moths. *J. Exp. Biol. 61*, 47–55.

TSAO, W. and SHUEL, R. W. (1973). Studies in the mode of action of royal jelly in honey bee development. Part 9. The carbohydrates and lipids in the hemolymph and the fat body of developing larvae. *Canad. J. Zool. 51*, 1139–1148.

TSUMUKI, H. and KANEHISA, K. (1979). Glycerol concentrations in hemolymph of hibernating larvae of the rice stern borer *Chilo suppressalis*: Effects of ligation and cold tolerance. *Appl. Ent. Zool. 14*, 497–499.

TUCKER, L. E. (1977). Regulation of ions in the haemolymph of the cockroach *Periplaneta americana* during dehydration and rehydration. *J. Exp. Biol. 71*, 95–110.

TURUNEN, S. (1979). Digestion and absorption of lipids in insects. *Comp. Biochem. Physiol. 63A*, 455–460.

VAN DER HORST, D. J., HOUBEN, N. M. D. and BEENAKKERS, A. M. T. (1980). Dynamics of energy substrates in the hemolymph of *Locusta migratoria* during flight. *J. Insect Physiol. 26*, 441–448.

VAN DER HORST, D. J., VAN DOORN, J. M. and BEENAKKERS, A. M. T. (1978). Dynamics in the hemolymph trehalose pool during flight of the locust *Locusta migratoria*. *Insect Biochem. 8*, 413–416.

VAN DER HORST, D. J., VAN DOORN, J. M., DEKEIJZER, A. N. and BEENAKKERS, A. M. T. (1981). Interconversions of diacylglycerol-transporting lipoproteins in the haemolymph of *Locusta migratoria*. *Insect Biochem. 11*, 717–723.

WALL, B. J. (1970). Effects of dehydration and rehydration on *Periplaneta americana*. *J. Insect Physiol. 16*, 1027–1042.

WALLACE, J. B. and BLUM, M. S. (1971). Reflex bleeding: a highly refined defensive mechanism in *Diabrotica* larvae (Coleoptera: Chrysomelidae). *Ann. Ent. Soc. Amer. 64*, 1021–1024.

WANG, D. I. and MOELLER, F. E. (1970). Comparison of the free amino acid composition in the hemolymph of healthy and *Nosema*-infected female honey bees. *J. Invert. Pathol. 15*, 202–206.

WASSERTHAL, L. T. (1975). The role of butterfly wings in regulation of body temperature. *J. Insect Physiol. 21*, 1921–1930.

WASSERTHAL, L. T. (1980). Oscillating hemolymph circulation in the butterfly *Papilio machaon* revealed by contact thermography and photocell measurements. *J. Comp. Physiol. 139B*, 145–164.

WASSERTHAL, L. T. (1981). Oscillating haemolymph "circulation" and discontinuous tracheal ventillation in the giant silkmoth *Attacus atlas* (L.). *J. Comp. Physiol. 145*, 1–15.

WEEDA, E. (1981). Hormonal regulation of proline synthesis and glucose release in the fat body of the Colorado potato beetle *Leptinotarsa decemilineata*. *J. Insect Physiol. 27*, 411–417.

WEIDLER, D. J. and SIECK, G. C. (1977). A study of ion binding in the hemolymph of *Periplaneta americana*. *Comp. Biochem. Physiol. 56A*, 11–14.

WHARTON, D. R. A., WHARTON, M. L. and LOLA, J. (1965a). Blood volume and water content of the male American cockroach, *Periplaneta americana* L.: methods and the influence of age and starvation. *J. Insect Physiol. 11*, 391–404.

WHARTON, D. R. A., WHARTON, M. L. and LOLA, J. (1965b). Weight and blood volume changes induced by irradiation of the American cockroach. *Radiat. Res. 25*, 514–525.

WHARTON, G. W. and ARLIAN, L. G. (1972). Utilization of Water by Terrestrial Mites and Insects. In *Insect and Mite Nutrition*. Edited by J. C. Rodriquez. Pages 154–165. Elsevier–North Holland, Amsterdam.

WHEELER, R. E. (1963). Studies on the total haemocyte count and haemolymph volume in *Periplaneta americana* L. with special reference to the last moulting cycle. *J. Insect Physiol. 9*, 223–235.

WHITCOMB, R. F., SHAPIRO, M. and GRANADOS, R. R. (1974). Insect defense mechanisms against microorganisms and parasitoids. In *The Physiology of Insecta*. Edited by M. Rockstein. Vol. 5, pages 447–536. Academic Press, New York.

WIGGLESWORTH, V. B. (1970). The pericardial cells of insects: analogue of the reticulo-endothelial system. *J. Reticuloendo. Soc. 7*, 208–216.

WIGGLESWORTH, V. B. (1972). *The Principles of Insect Physiology*. 7th edition. Chapman & Hall, London.

WIMER, L. T. (1969). A comparison of the carbohydrate composition of the hemolymph and fat body of *Phormia regina* during larval development. *Comp. Biochem. Physiol. 29*, 1055–1062.

WINTER, C. E., GIANNOTTI, O. and HOLZHACKER, E. L. (1975). DDT–lipoprotein complex in the American cockroach hemolymph: A possible way of insecticide transport. *Pest. Biochem. Physiol. 5*, 155–162.

WIRTZ, R. A. and HOPKINS, T. L. (1974). Tyrosine and phenylalanine concentrations in haemolymph and tissues of the American cockroach *Periplaneta americana* during metamorphosis. *J. Insect Physiol. 20,* 1143–1154.

WIRTZ, R. A. and HOPKINS, T. L. (1977). Tyrosine and phenylalanine concentrations in cockroaches *Leucophaea maderae* (F.) and *Periplaneta americana* (L.) in relation to cuticle formation and ecdysis. *Comp. Biochem. Physiol. 56A,* 263–266.

WOODRING, J. P., CLIFFORD, C. W., ROE, R. M. and MERCIER, R. R. (1977). Relation of blood composition to age in the larval female house cricket, *Acheta domesticus. J. Insect Physiol. 23,* 559–567.

WYATT, G. R. (1961). The biochemistry of insect haemolymph. *Ann. Rev. Ent. 6,* 75–102.

WYATT, G. R. (1967). Biochemistry of sugars and polysaccarides in insects. In *Advances in Insect Physiology*. Edited by J. W. L. Beament, J. E. Treherne and V. B. Wigglesworth. Vol. 4, pages 287–360. Academic Press, London.

WYATT, G. R. (1980). The fat body as a protein factory. In *Insect Biology in the Future*. Edited by M. Locke and D. S. Smith. Pages 201–225. Academic Press, New York.

WYATT, G. R. and KALF, G. F. (1958). Organic compounds in insect hemolymph. *Proc. 10th Intern. Cong. Entomol. Montreal.* (1956). *2,* 333.

WYATT, G. R. and PAN, M. L. (1978). Insect plasma proteins. *Ann. Rev. Biochem. 47,* 779–817.

ZACHARIASSEN, K. E. and HAMMEL, H. T. (1976). Nucleating agents in the hemolymph of insects tolerant to freezing. *Nature 262,* 285–287.

ZACHARUK, R. Y. (1976). Structural changes of cuticle during moulting. In *The Integument*. Edited by H. R. Hepburn. Pages 299–321. Elsevier, Amsterdam.

10 Cellular Elements in the Hemolymph

A. P. GUPTA

Rutgers University–Cook College, New Brunswick, New Jersey, USA

1 HISTORICAL INTRODUCTION

Insect hemocytes have been the subject of investigation for nearly 150 years. Among arthropods, hemocytes have been most extensively studied in insects, followed by those of crustaceans, arachnids, and myriapods (Gupta, A., 1979a,e). Hemocytes of a few onychophorans have also been described (Gupta, A., 1979a; Ravindranath, M., 1981). From such nondescript references as "round and spindle-shaped cells" by Leydig, L. (1859) and "amoeboid cells" by Magretti, P. (1881), and Cuénot, L. (1896) to the presently recognized, morphologically distinct types, such as prohemocytes, plasmatocytes, granulocytes, spherulocytes, oenocytoids, coagulocytes, and adipohemocytes (Gupta, A., 1979b), insect hemocyte classification has come a long way indeed!

Hemocyte classifications both in insects and other arthropods have been variously based on morphology, functions, and staining or histochemical reactions of hemocytes. Thus, it is not unusual to find the same hemocyte type or its various forms being referred to by different names in various arthropods, by different authors — a situation that has inevitably resulted in a confusing mass of terminology. Consequently, it becomes very difficult to compare hemocytes of one species with those of others. This has particularly hindered any phylogenetic consideration of the evolution of hemocyte types in various arthropod groups and the Onychophora. Clearly, there is a need for a uniform hemocyte classification for insects as well as other arthropod groups. The insect hemocyte classification that is generally used has evolved over a period of more than half a century. According to Millara, P. (1947), Cuénot, L. (1896) was the first to classify insect hemocytes into four categories, and was later followed in this attempt by Hollande, A. (1909, 1911) and others. Wigglesworth, V. (1939) summarized most of the earlier classifications and presented a classification that was widely accepted. He modified it later (Wigglesworth, V., 1959). On the American side, J. Yeager's (1945) work stimulated considerable interest in the study of hemocytes. Jones, J. (1962) revised and greatly improved Yeager's classification. For the most recent arthropod and insect hemocyte classifications refer to Gupta, A. (1979a-e). A summary of the seven main hemocyte types in various insect orders is

Table 1: Summary of hemocyte types in various orders, based on published and unpublished information, and personal observations

Species	Prohemo-cyte PR	Plasmato-cyte PL	Granulo-cyte GR	Spherulo-cyte SP	Adipohemo-cyte AD	Coagulo-cyte CO	Oeno-cytoid OE
Collembola	—	—	GR	—	—	—	—
Thysanura	—	PL	GR	SP	—	CO	—
Emphemeroptera	PR	PL	GR	—	—	—	OE
Odonata	—	PL	GR	—	—	—	—
Orthoptera (= Cursoria)	PR	PL	GR	(SP)	—	—	—
Dermaptera	PR	PL	GR	SP	—	CO	—
Blattaria	PR	PL	GR	SP	AD	CO	OE
Mantodea	PR	PL	GR	SP	—	—	—
Plecoptera	PR	PL	GR	SP	—	CO	—
Hemiptera	PR	PL	GR	—	AD	—	OE
Phthiraptera*	PR	PL	GR	—	—	—	OE
Hymenoptera	PR	PL	GR	SP	AD	CO†	OE
Coleoptera	PR	PL	GR	SP	AD	CO	OE
Megaloptera	PR	(PL)	GR	—	—	CO	OE
Neuroptera	PR	PL	(GR)	SP	AD	—	OE
Trichoptera	PR	PL	GR	—	—	—	—
Lepidoptera	PR	PL	GR	SP	AD	CO	OE
Diptera	PR	PL	GR	SP	AD	CO	OE

Terms in parentheses were not used by the original authors, but have been adopted by me after scrutiny of the original micrographs and figures (from Gupta, A., 1979).
*From Saxena, A. and Agarwal, G. (1979).
†From Cruz Landim, C. (1971).

presented in Table 1. Terms in parentheses were not used by the original authors, but have been adopted by me after scrutinizing original descriptions and/or figures. Hemocytes categorized as amoebocytes and/or phagocytes by the original authors have been assigned mostly to the category of plasmatocyte (PL), although they could be included in granulocyte (GR) or spherulocyte (SP), and/or adipohemocyte (AD), inasmuch as these latter three forms also are supposedly phagocytic in certain insects.

2 MORPHOLOGY OF HEMOCYTES

2.1 Hemocyte terminologies

Anyone venturing into the study of insect hemocytes knows only too well how discouraging it is in the beginning to have to contend with a bewildering array of terminologies that exists in the literature. The tendency, and perhaps the irresistible urge, on the part of some workers to introduce new terms for various variant forms of the morphologically distinct types of hemocytes is mainly responsible for the plethora of terms. I believe that descriptions of "new" hemocytes, based on superficial dissimilarities from the seven morphologically distinct types of hemocytes, should be avoided. Various synonymies of the presently recognized seven types of hemocytes are presented in Table 2, for in order to discuss hemocytes and their physiological significance in various insects, it is necessary to homologize terminologies used by different authors on the bases of descriptions, observed functions, line drawings, and micrographs of hemocytes used by those authors.

2.2 Main types of hemocytes

There is disagreement among insect hematologists as to the number of hemocyte types in various insects. From one or a few to as many as nine or more types have been described, particularly by light microscopy. Ultrastructurally, however, only seven types have so far been identified in various insects: prohemocyte (PR), plasmatocyte (PL), granulocyte (GR), spherulocyte (SP), adipohemocyte (AD), oenocytoid (OE), and coagulocyte (CO). Of these seven, AD has been reported only by Devauchelle, G.

Table 2: Various synonyms for the seven types of hemocytes

Synonyms	Synonym of	Authors
Adipocyte	AD	Wigglesworth, V., 1965, 1979a
Adipohemocyte	GR	Shrivastava, S. and Richards, A., 1965; Lea, M. and Gilbert, L., 1966
Adipohemocyte	SP	Harpaz, F., *et al.* 1969
Adipoleucocyte	AD	Hollande, A., 1911
Adipo-spherule	AD	Kollmann, M., 1908
Amoebocyte	PL	many authors
Amoebocyte	GR	many authors
Amoebocyte with granular cytoplasm	GR	Cuénot, L., 1896
"Cellule sphéruleuses" or "cellules à sphérules"	SP	Paillot, A., 1919; Paillot, A. and Noel, R., 1928
Coagulocyte	OE	Hoffmann, J. and Stoekel, M., 1968
Crescent cell	OE	Gupta, A., 1983
Crystal cell	OE	Rizki, M., 1953, 1962; Nappi, A., 1970
Crystalloid cell	OE	Selman, J., 1961
Cystocyte (= CO)	GR	Devauchelle, G., 1971
Cystocyte	CO	Yeager, J., 1945; and other authors
Dark, hyaline hemocyte	OE	Selman, J., 1961
Eleocyte	SP	Rowley, A. and Ratcliffe, N., 1981
Eruptive cell	SP	Yeager, J., 1945
Formative cell	PR	Müller, K., 1925
Giant fusiform cell	PL	Tuzet, O. and Manier, F., 1959
Hyaline cell	GR	several authors
Hyaline cell	SP	Whitten, J., 1963
"Jeune globule"	PR	Bruntz, L., 1908
"Jeune leucocyte"	PR	Millara, P., 1947
Lamellocyte	PL	Rizki, M., 1962; and other authors
Later stages of spherule	AD	Whitten, J., 1963
Leucocyte	PL	Kollmann, M., 1908; Metalnikov, S., 1908
Lymphocyte	PL	many authors
Macronucleocyte	PR	Paillot, A., 1919
Micronucleocyte	PL	Paillot, A., 1919
Nematocyte	PL	Yeager, J., 1945; Rizki, M., 1953; Arnold, J., 1982
Non-granular spindle cell	OE	Wigglesworth, V., 1933, 1955
Non-phagocytic giant hemocyte	OE	Wigglesworth, V., 1933, 1955
Oenocyte	OE	Metalnikov, S. and Gaschen, H., 1922; Müller, K., 1925; Tateiwa, J., 1928; Metalnikov, S. and Chorine, V., 1929; Bogojavlensky, K., 1932; Cameron, G., 1934
Oenocyte-like cell	OE	Yeager, J., 1945
Oenocytoid	SP	Dennell, R., 1947
Phagocyte	PL	many authors
Phagocyte	GR	many authors
Plasmatocyte-like cell	PR	Jones, J., 1959
Podocyte	PL	Yeager, J., 1945; Devauchelle. G., 1971; and other authors
Prohemocytoid	PR	Jones, J. and Tauber, O., 1954
Proleucocyte	PR	Hollande, A., 1911; and other authors
Proleucocytoid	PR	Yeager, J., 1945
"Pycnoleucocyte"	GR	Wille, H. and Vecchi, M., 1966
Radiate cell	PL	Lutz, K., 1895
Rhegmatocyte	SP	Hrdy, I., 1957
Smooth-contour chromophil cell	PR	Yeager, J., 1945
Spherocyte	SP	Bogojavlensky, K., 1932
Spheroidocyte	AD	Yeager, J., 1945; Arnold, J., 1952; Rizki, M., 1953; Jones, J., 1959
Spherulocyte	GR	François, J., 1975a
Star-shaped amoebocyte	PL	Graber, V., 1871
Thombocytoids	CO	Zachary, D. and Hoffmann, J., 1973
Vermiform cell	PL	Yeager, J., 1945; Lea, M. and Gilbert, L., 1966; Devauchelle, G., 1971
Young granulocyte	PR	François, J., 1974
Young plasmatocyte	PR	Gupta, A. and Sutherland, D., 1966; Gupta, A., 1969

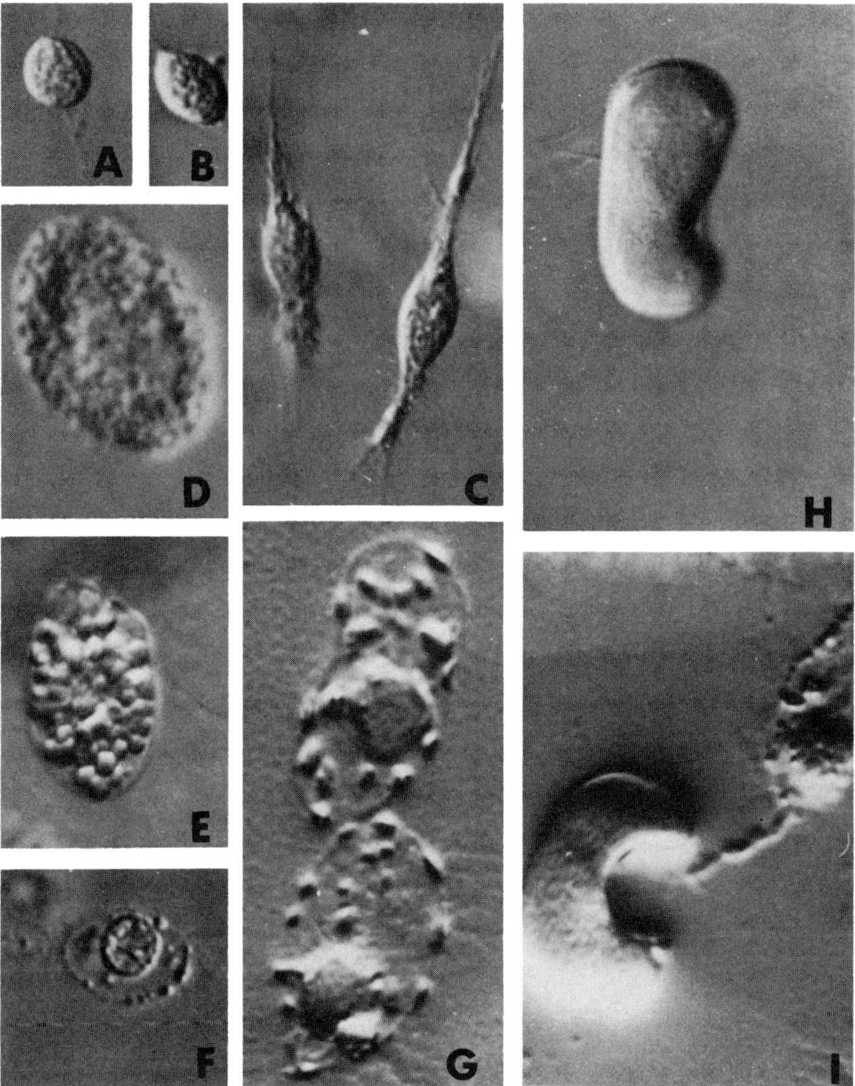

FIG. 1. Various hemocyte types in *Gromphadorhina portentosa*. **A:** Prohemocyte, (×4333). **B:** Prohemocyte, (×5633). **C:** Plasmatocytes, (×6500). **D:** Granulocyte, (×5687). **E:** Spherulocyte, (×5687). **F:** Coagulocyte (×3250). **G:** Coagulocytes, (×6314). **H:** Enucleate oenocytoid, (×5525). **I:** Oenocytoid with extruded nucleus, (×7150). (Nomarski phase-contrast; originals).

(1971) and CO by Goffinet, G. and Grégoire, Ch. (1975) and Ratcliffe, N. and Price, C. (1974). Podocyte (PO) and vermicyte (VE) have not been recognized as distinct types in electron microscopic studies so far, primarily because ultrastructurally they appear similar to PLs (Devauchelle, G., 1971). A general description of various hemocyte types, based on both light and electron microscopic studies, follows.

2.2.1 Prohemocyte (PR)

The term that has survived to date with little or no change since its adoption by Hollande, A. (1911) is proleucocyte. Prohemocytes are small, round, oval or elliptical cells with variable sizes (6–10 μm wide and 6–14 μm long). The plasma membrane is generally smooth (Figs 1A, B, 2A) but may show vesiculation (Fig. 4A). The nucleus is large, centrally

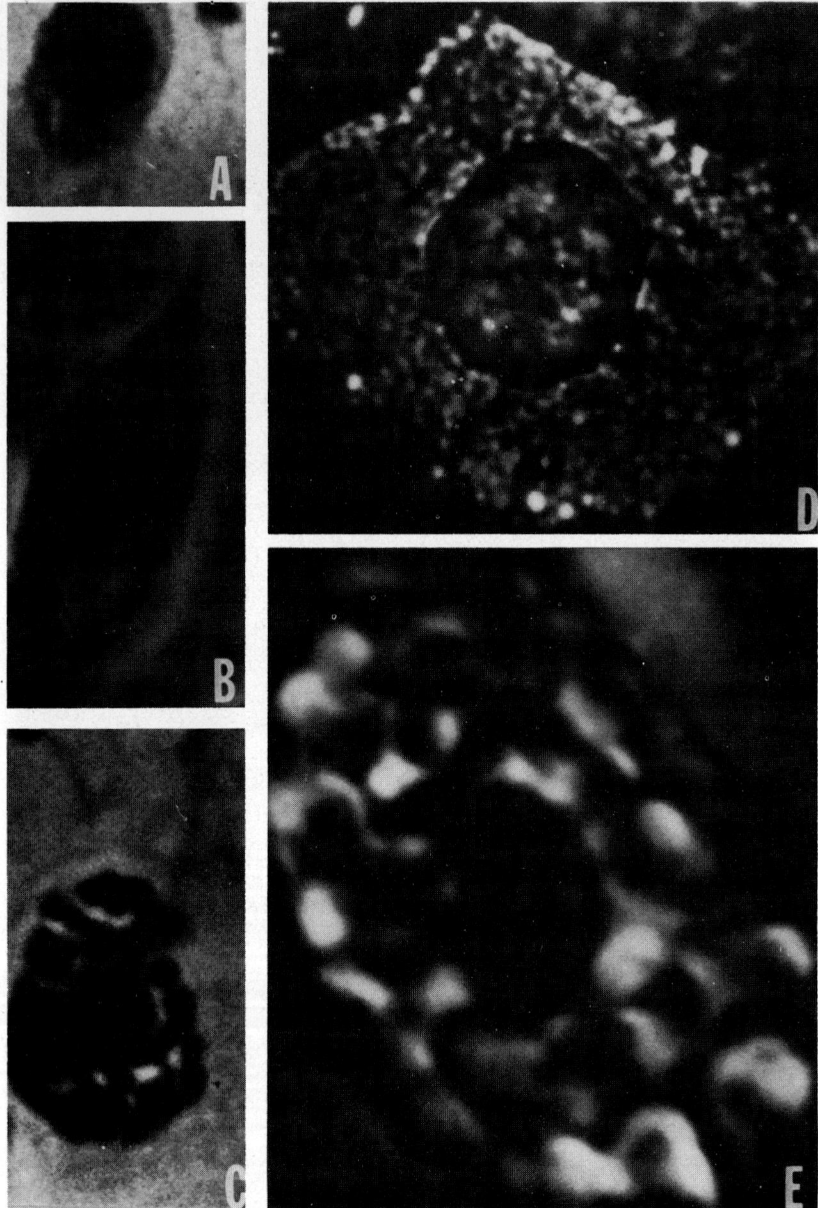

FIG. 2. **A:** Prohemocyte of *Periplaneta americana* (*ca.* ×8000). **B:** Plasmatocyte of *Periplaneta americana* (*ca.* ×10,000). **C:** Spherulocyte of *Nauphoeta cinerea* (*ca.* ×8500). **D:** Granulocyte of *Locusta migratoria*. **E:** Spherulocyte of *Nauphoeta cinerea* (*ca.* ×25,000). (From Gupta, A., 1979c.)

located, and almost filling the cell; nuclear size is variable (3.6–12 μm) in various insects; more than one nuclei, and nucleoli may be present. A thin or dense, homogeneous and intensely basophilic layer of cytoplasm surrounds the nucleus, the nucleo-cytoplasmic ratio being 0.5–1.9 or more. The cytoplasm may contain granules, droplets or vacuoles (Fig. 4A).

The laminar nature of the plasma and nuclear membranes may not be evident. The cytoplasm generally contains a low concentration of endoplasmic reticulum (ER), mitochondria, and Golgi

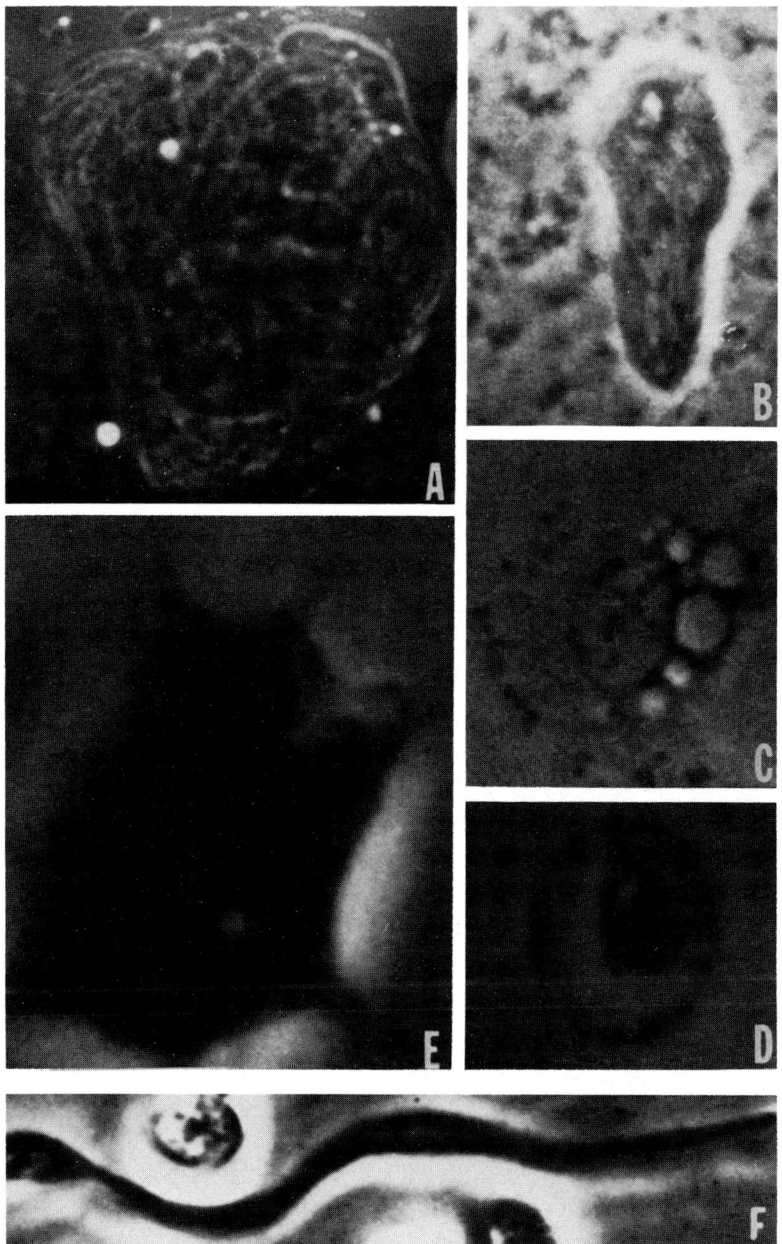

FIG. 3. **A:** Oenocytoid of *Locusta migratoria*, showing cytoplasmic filaments. **B:** Oenocytoid of *Periplaneta americana* (*ca.* × 6800). **C:** Adipohemocyte of *Periplaneta americana* nymph (*ca.* × 7500). **D:** Coagulocyte of *Epicauta cinerea* (*ca.* × 6800). **E:** Podocyte of *Periplaneta americana* nymph (*ca.* × 7225). **F:** Vermicyte of *Periplaneta americana* (*ca.* × 1190). (From Gupta, A., 1979c.)

bodies. However, free ribosomes, rough endoplasmic reticulum (RER), and even mitochondria may be numerous. Centrioles — indicating the mitotic nature of PRs — and microtubules have been observed in the cytoplasm.

PRs are generally found in groups, and appear indistinguishable from young or small PLs. They may be numerous, rare or absent, depending on the development and physiological state of the insect at the time of observation. PRs are seldom seen *in vivo*

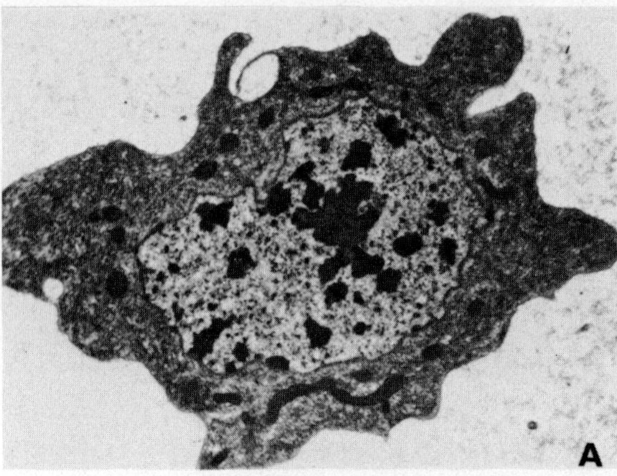

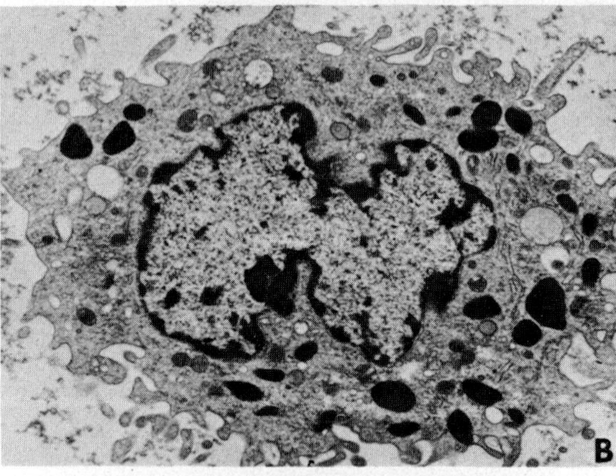

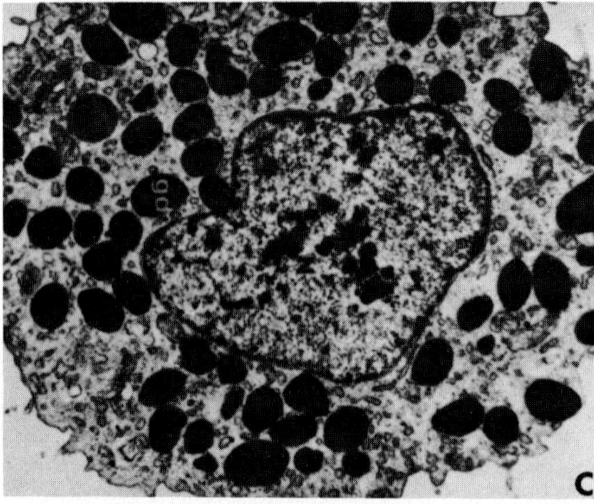

FIG. 4. **A:** Prohemocyte of *Pectinophora gossypiella* (*ca.* ×8760). **B:** Plasmatocyte of *Carausius morosus*, showing micropapillae and lobate nucleus (×6570). **C:** Granulocyte of *Melolontha melolontha* (×6300). (From Gupta, A., 1979c.)

and do not show transitional stages as other hemocyte types do.

2.2.2 Plasmatocyte (PL)

These are small to large polymorphic cells with variable sizes (3.3–5 μm wide and 3.3–40 μm long). The plasma membrane may have micropapillae, filopodia or other irregular processes, as well as pinocytotic or vesicular invaginations (Figs 1C, 2B, 4B). The nucleus may be round or elongate, and is generally centrally located. It may be lobate (Fig. 4B), vary in size (3–9 μm wide and 4–10 μm long) in various insects, and appear punctate. Scattered chromatin masses may be present along with the nucleolus (Fig. 4B). Occasionally, binucleate PL may be found.

The laminar nature of the plasma and nuclear membranes may or may not be visible. The cytoplasm is generally abundant, and may be granular or agranular, is basophilic and rich in organelles. Generally, there is well-developed and extensive RER, which may form greatly distended cisternae or a vacuolar system. Golgi bodies (= dictyosomes = golgiosomes or internal reticular apparatus) and lysosomes (membrane-bound, electron-dense bodies, 0.1–1.30 μm in size) may be numerous; lysosomes can be identified by the presence in them of the reaction products of the hydrolytic marker enzymes, acid phosphatase and thiamine pyrophosphatase (Scharrer, B., 1972) and are often associated with the RER or the vacuolar system. The Golgi bodies produce the electron-dense granules (generally 0.5 μm in diameter) that one observes in the PLs. Microsomes and cisternae of the ER (= "ergastoplasme" of French authors) may be present. Free ribosomes (polysomes or polyribosomes) or those attached to microsomes or RER may be present; intracytoplasmic microtubules are present, sometimes arranged in bundles.

PLs are generally abundant, and in some insects may be indistinguishable from PRs and GRs. Several types (most often the transitional forms) of PLs have been described on the basis of their sizes and shapes.

2.2.3 Granulocyte (GR)

These are small to large, spherical or oval cells (Figs 1D, 2D, 4C, 5A) with variable sizes (10–45 μm long and 4–32 μm wide, rarely larger). The plasma

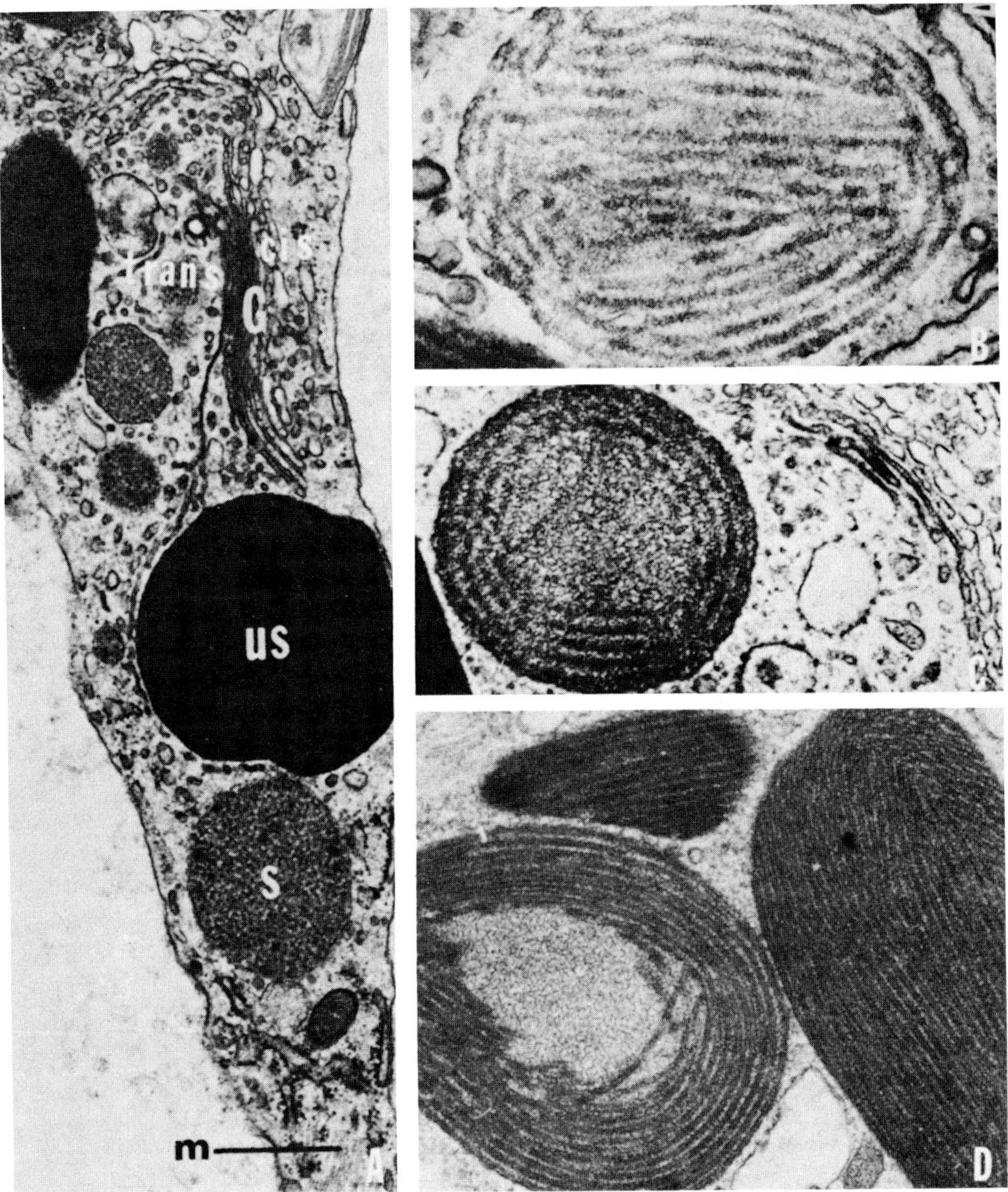

FIG. 5. **A:** Portion of granulocyte of *Leucophaea maderae*, showing derivation of structured (= premelanosome-like) granule(s) from Golgi (G), structureless or unstructured (= melanin-like) granules (US), and intracytoplasmic microtubules (m). Note *cis* and *trans* faces of Golgi. (× 50,000). **B:** Structured granule from granulocyte of *Leucophaea maderae*, showing internal microtubules, (× 40,000). **C:** An earlier stage of development of internal microtubules (× 50,000). **D:** Section of a structured granule showing concentric arrangement of internal microtubules, (× 38,000).

membrane may or may not have micropapillae, filopodia or other irregular processes. The nucleus may be relatively small (as compared with that in the PL), round or elongate, and is generally centrally located. Nuclear size is variable (2–8 μm long and 2–7 μm wide).

The laminar nature of the plasma and nuclear membranes may not be visible. The cytoplasm is characteristically granular (Figs 1D, 2D, 4C). Several types of membrane-bound granules have

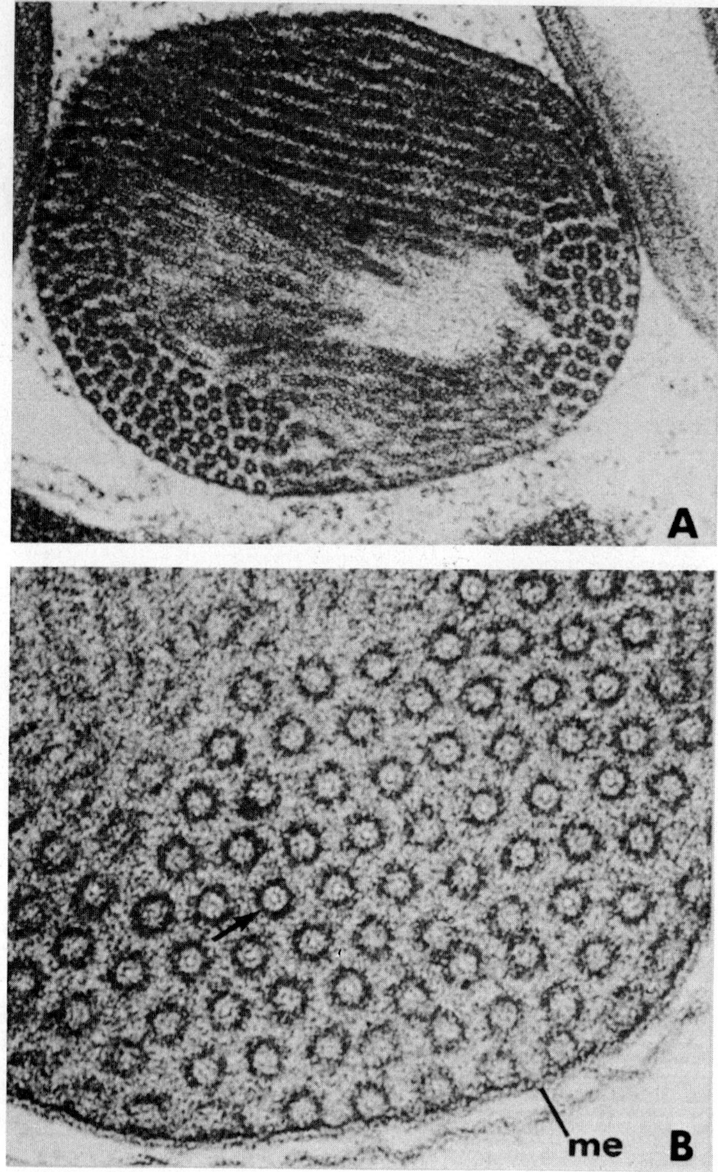

Fig. 6. **A:** Section of structured granule of a granulocyte of *Leucophaea maderae*, showing arrangement of microtubules about 25 nm in diameter, (× 63,000). **B:** Highly magnified view of microtubules of structured granules. Note micro-microtubules (5 nm in diameter (arrow) within microtubules and limiting membrane (me) of granule, (× 240,000)). (From Gupta, A., 1979c.)

been described in the GRs of various insects (Figs 5A, B, C, D, 6A, B). Recently, Goffinet, G. and Grégoire, Ch. (1975) summarized and synonymized various types of granules into three categories, based on their observations in *Carausius morosus*. The following summary and synonymy of granules are based on these and other works:

(1) structureless, electron-dense granules (= unstructured inclusions (type 1) of Baerwald, R. and Boush, G., 1970; melanosome-like granules of Hagopian, M., 1971; opaque body of Moran, D., 1971; type 2 bodies of Scharrer, B., 1972; and electron-dense granules of Raina, A., 1976, and others);

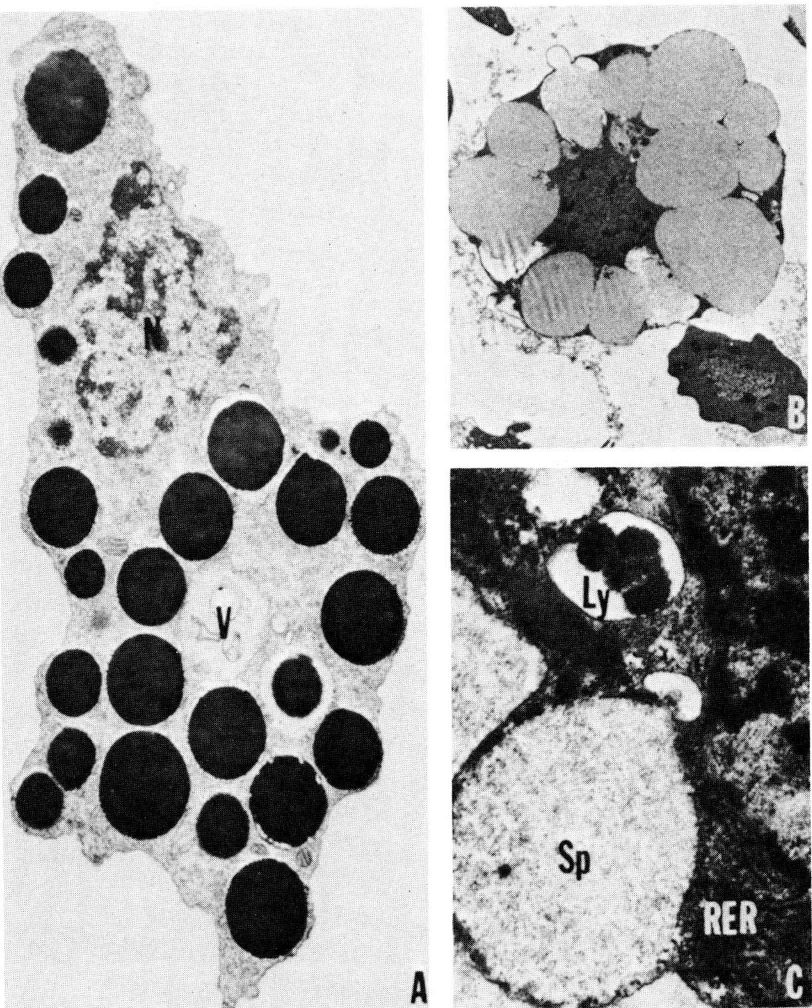

FIG. 7. **A:** Spherulocyte of *Melolontha melolontha*, showing eccentric nucleus (N) and numerous spherules and a vacuole (V), (×9450). **B:** Spherulocyte of *Bombyx mori* (*ca.* ×7000). **C:** Portion of spherulocyte of *Pectinophora gossypiella*, showing rough endoplasmic reticulum (RER), spherule (Sp) with granular contents, and lysosome (Ly). (×21,000). (From Gupta, A., 1979c.)

(2) structureless, thinly granular bodies [= type 3 of Scharrer, B., 1972; electron-lucent granules of Raina, A., 1976]; and

(3) structured granules (= "globules" or "granules multibullaires" of Beaulaton, J., 1968; "grain denses structures" in the AD of Devauchelle, G., 1971 and Landureau, J. and Grellet, P., 1975b; "corpus fibrillaires" of Hoffmann, J. *et al.*, 1968b, 1970; cylinder inclusions (type 2), regular-packed inclusions (type 3) and inclusions with band-like units (type 4) of Baerwald, R. and Boush, G., 1970; "Granula mit tubularer Binnenstruktur" of Stang-Voss, C., 1970; premelanosome-like granules of Hagopian, M., 1971; tubule-containing bodies or TCB of Moran, D., 1971; type 1 of Scharrer, B., 1972; and granules with a microtubular structure of Ratcliffe, N. and Price, C., 1974).

The length or the diameter of the structureless granules varies from 0.15 to 3 μm or more in various insects, while that of the structured granules varies from 0.5 to 2 μm. The shape of the granules may be spherical, ovoid, elongate or irregularly polygonal (Figs 4C, 5A, 7B, C, D). The diameter of the microtubules within the structured granules varies from 15 to 80 nm in various insects. Internally, the microtubules may show micro-microtubules about 5 nm in diameter (Hagopian, M., 1971) (Fig. 6B). Akai, H. and Sato, S. (1973) have also described "subunits of fibrils" in their so-called secretory vesicles. The number of microtubules/granule may vary from nine to eighty. From the accounts provided by Hagopian, M. (1971), Scharrer, B. (1972), Akai, H. and Sato, S. (1973), and François, J. (1975a), it appears that the granules are derived from the Golgi bodies (Fig. 5A), the microtubules developing during the later stages of morphogenesis. It is conceivable that the structureless, electron-dense granules represent the final stage of development of these granules in which the structured nature becomes obliterated. Supposedly, the granules are eventually released into the hemolymph. Histochemically, most of the granules contain sulfated, periodate-reactive sialomucin and other glycoproteins or neutral mucopolysaccharides (Costin, N., 1975; François, J., 1974, 1975a). Occasionally, lipid droplets may be present, especially in older GRs.

In addition to the structureless and structured granules, the cytoplasm is rich in free ribosomes (polysomes), Golgi bodies, both ER and RER, and lysosomes. Mitochondria are generally few in number. Marginal bundles of intracytoplasmic microtubules are also present (Fig. 5A).

The GRs of hemimetabolous insects are generally larger and with more prominent granules than those in holometabolous insects (Arnold, J., 1979) and Akai, H. (1969) even found two types of GRs in the same insect (*Philosamia cynthia ricini*).

2.2.4 Spherulocyte (SP)

These are ovoid or round cells (Figs 1E, 2E, 7A, B) with variable sizes (9–25 μm long and 5–10 μm wide), and usually larger than GRs. The plasma membrane may or may not have micropapillae, filopodia or other irregular processes. The nucleus is generally small (5–9 μm long and 2.5 μm wide), central or eccentric, rich in chromatin bodies and generally obscured by the membrane-bound, electron-dense, intracytoplasmic spherules, which are characteristic of these cells.

The number of spherules may vary from few to many, and the diameter from 1.5 to 5 μm. The spherules contain granular, fine-textured, filamentous or flocculent material (Raina, A., 1976). The granules within the spherules may vary from 15 to 17 nm in diameter (Akai, H. and Sato, S., 1973). In addition to the spherules, the cytoplasm contains polyribosomes (Fig. 8C), Golgi bodies (moderately to well-developed) (Fig. 8A), membrane-bound vacuoles (= lysosomes) (Fig. 7C), numerous, randomly distributed microtubules, elongated mitochondria, and RER (Figs 7C, 8A, C). Devauchelle, G. (1971) has also described a more or less loose network of fibrils in the cytoplasm (Fig. 8B). SPs release the material in their spherules into the hemolymph by exocytosis.

Histochemically, the spherules have been reported to contain neutral or acid mucopolysaccharide and glycomucoproteins by several authors (see Gupta, A., 1979c). Much earlier, Hollande, A. (1909) reported that the spherules contain "lipochrome" (a kind of carotenoid lipid). The presence of tyrosinase has been reported by Dennell, R. (1947), Jones, J. (1956), and Rizki, M. and Rizki, R. (1959). Recently, Costin, N. (1975) reported the presence of non-sulfated sialomucin, in addition to glycoproteins and neutral mucopolysaccharides.

2.2.5 Adipohemocyte (AD)

These are small to large, spherical or oval cells (Fig. 3C) with variable sizes (7–45 μm in diameter). The plasma membrane may or may not have micropapillae, filopodia, or other irregular processes. The nucleus is relatively small (as compared with that in PL or SP), round or slightly elongate, and is centrally or eccentrically located. Nuclear size is variable (4–10 μm in diameter). It may appear to be concave, biconvex, punctate, or lobate.

The laminar nature of the plasma and nuclear membranes may not be visible. The cytoplasm contains characteristic, small to very large refringent fat droplets (0.5–15 μm in diameter), and other non-lipid granules (0.5–9 μm in diameter), and vacuoles, which, according to Arnold, J. (1974), become filled with lipids under certain conditions. In addition, the cytoplasm contains well-developed Golgi bodies, mitochondria, and polyribosomes.

Histochemically, ADs are reported to contain PAS-positive substance in granules (Ashhurst, D.

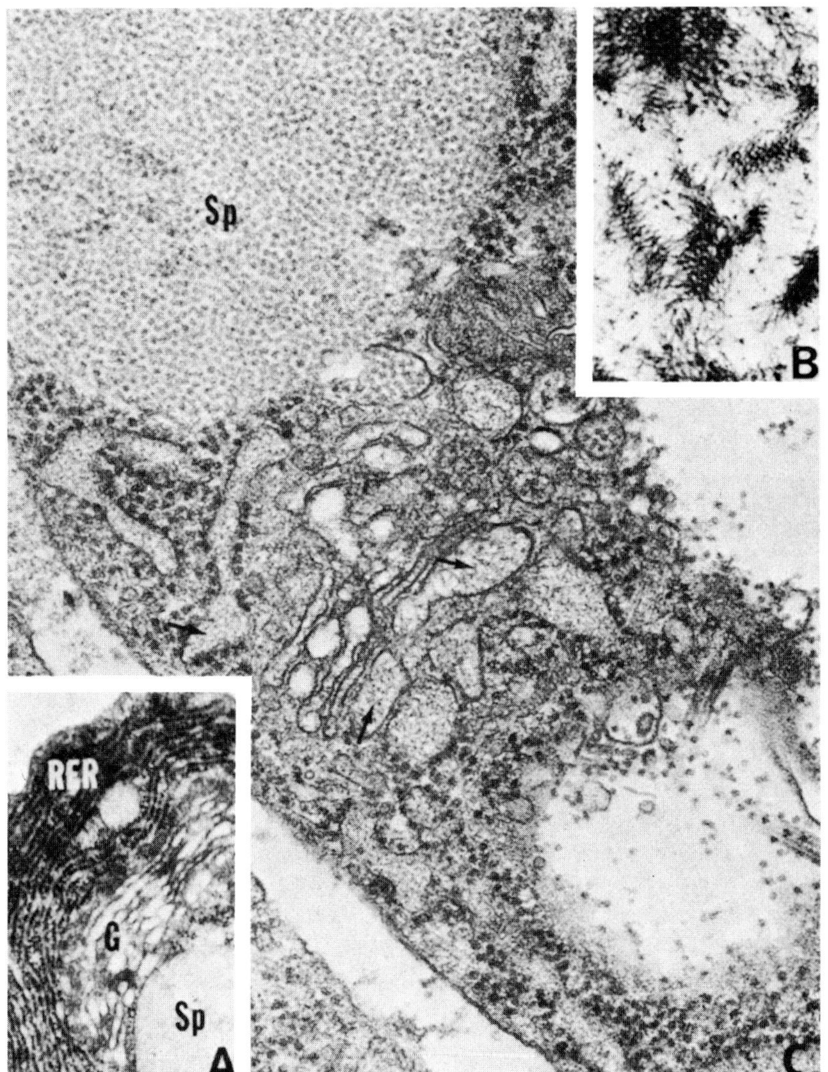

FIG. 8. **A:** Portion of spherulocyte of *Pectinophora gossypiella* showing rough endoplasmic reticulum (RER) and Golgi (G) involved in formation of spherule (Sp) (*ca.* × 15,750). **B:** Portion of spherulocyte of *Melolontha melolontha*, showing loose network of intracytoplasmic fibrils. **C:** Portion of spherulocyte of *Bombyx mori*, showing spherule (Sp) with fine granules, rough endoplasmic reticulum containing fibrous material in its cisternae (arrows), and ribosomes (× 50,400). (From Gupta, A., 1979c.)

and Richards, A., 1964; Lea, M. and Gilbert, L., 1966). Costin, N. (1975) did not recognize ADs as a type in her study.

2.2.6 Oenocytoid (OE)

These are small to large, thick, oval, crescent-shaped, spherical or elongate cells (Figs 1H, I, 3A,B, 10A,B) with widely variable sizes (16–54 μm or more) and shapes. The plasma membrane is generally without micropapillae, filopodia or other irregular processes. The nucleus is generally small, round or elongate, and generally eccentrically located (Figs 3B, 10A). Nuclear size may vary (3–15 μm). Occasionally, two nuclei may be present.

The laminar nature of the plasma and nuclear membranes may not be visible. The cytoplasm is generally thick and homogeneous, and has several kinds of plate-, rod-, or needle-like inclusions. According to Costin, N. (1975), the OE is

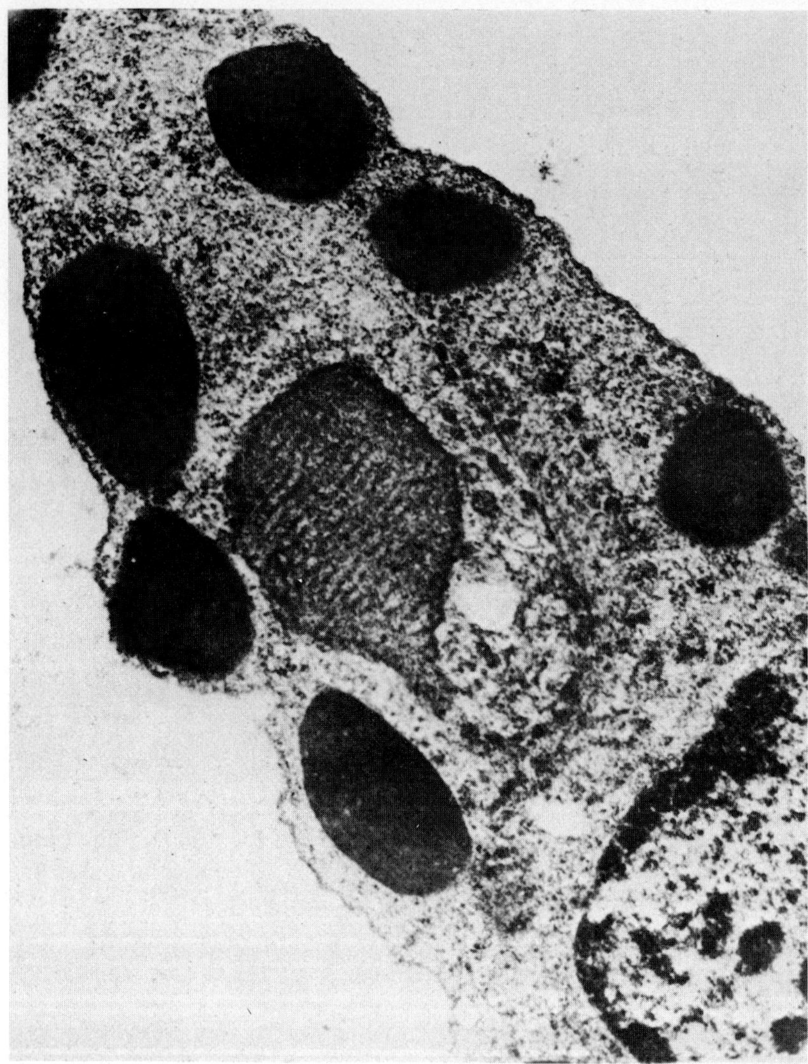

FIG. 9. Supposed adipohemocyte of *Melolontha melolontha*. Note, however, that it looks like a granulocyte, (×41,500). (From Gupta, A., 1979c.)

distinguished by an elaborate system of filaments that fills the cytoplasm (Figs 3A, 10C), and is visible under a phase-contrast microscope. Histochemically, the filaments resemble the cytoplasm, and hence are not visible in stained preparations. Hoffmann, J. (1966) and Hoffmann, J. *et al.* (1968b) have also reported such filaments. In addition to the above intracytoplasmic inclusions and filaments, a few electron-dense spherules may be present in the cell periphery (Devauchelle, G., 1971). With the exception of polyribosomes and the abundant, large mitochondria, which are conspicuous, other organelles, such as ER and Golgi, are poorly developed. Supposedly, lysosomes are absent.

Histochemically, OEs are reported to contain tyrosinase (Dennell, R., 1947), protein (Akai, H. and Sato, S., 1973), and PAS-positive-only granules, indicating the presence of glycoproteins or neutral mucopolysaccharides, and sulfated, periodate-reactive sialomucin (Costin, N., 1975).

One peculiarity of OEs seems to be their highly labile nature. They are particularly fragile *in vitro*, and lyse quickly, ejecting material into the hemolymph. They are non-phagocytic.

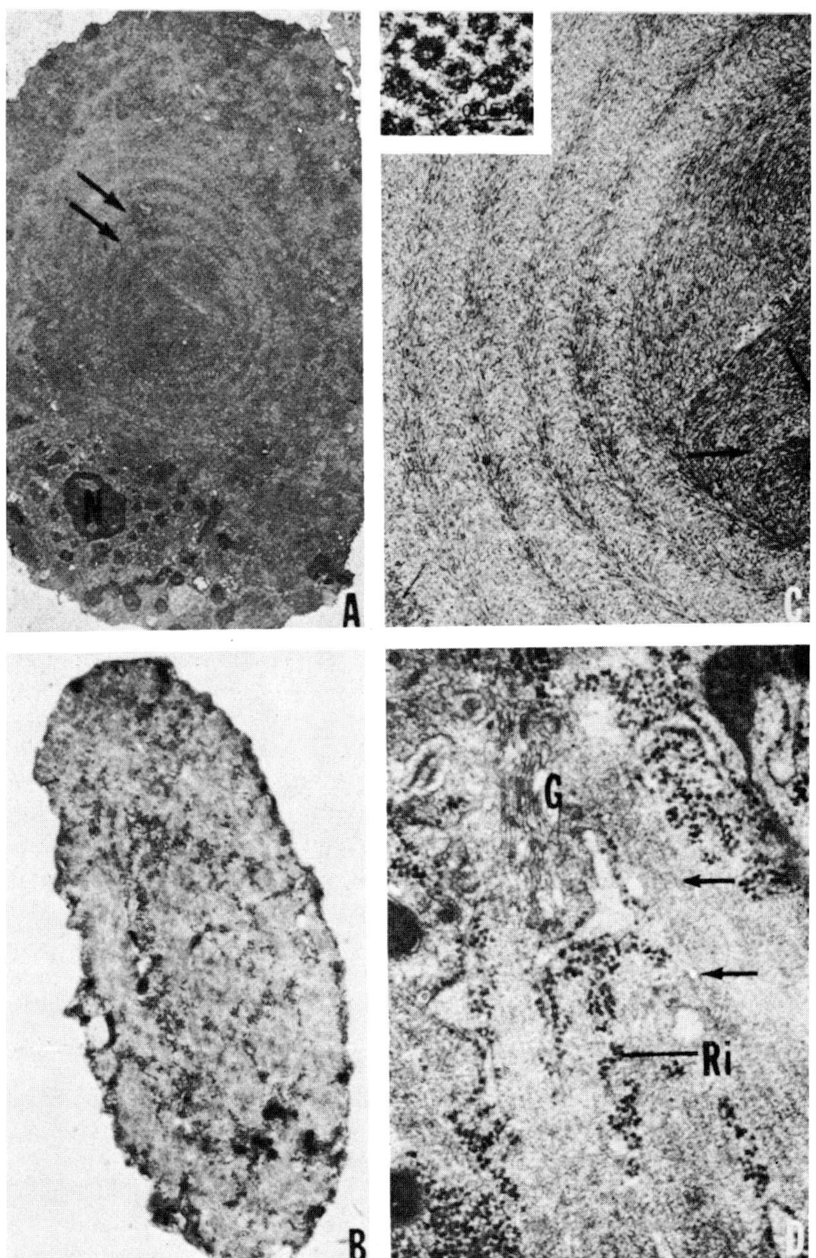

FIG. 10. **A:** Oenocytoid of *Bombyx mori*, showing concentric arrangement of intracytoplasmic fibrils (arrows) and eccentric nucleus (N) (× 3000). **B:** Oenocytoid of *Pectinophora gossypiella*. (× 6600). **C:** Portion of oenocytoid of *Bombyx mori*, showing highly magnified view of concentric rings of intracytoplasmic fibrils. Note also unoriented fibrils (arrows), (*ca.* × 30,000). Inset (*ca.* × 225,000) shows fibrils in cross-section. **D:** Portion of oenocytoid of *Pectinophora gossypiella*, showing longitudinally arranged intracytoplasmic microtubules (arrows), Golgi (G), and ribosomes (Ri), (× 35,475). (From Gupta, A., 1979c.)

2.2.7 Coagulocyte (CO)

These are generally small to large (3–30 μm long), spherical, hyaline, fragile, and unstable cells, combining the features of GRs and OEs (Arnold, J., 1974). The plasma membrane is generally without any micropapillae, filopodia or other irregular processes. The nucleus is relatively small (5–11 μm

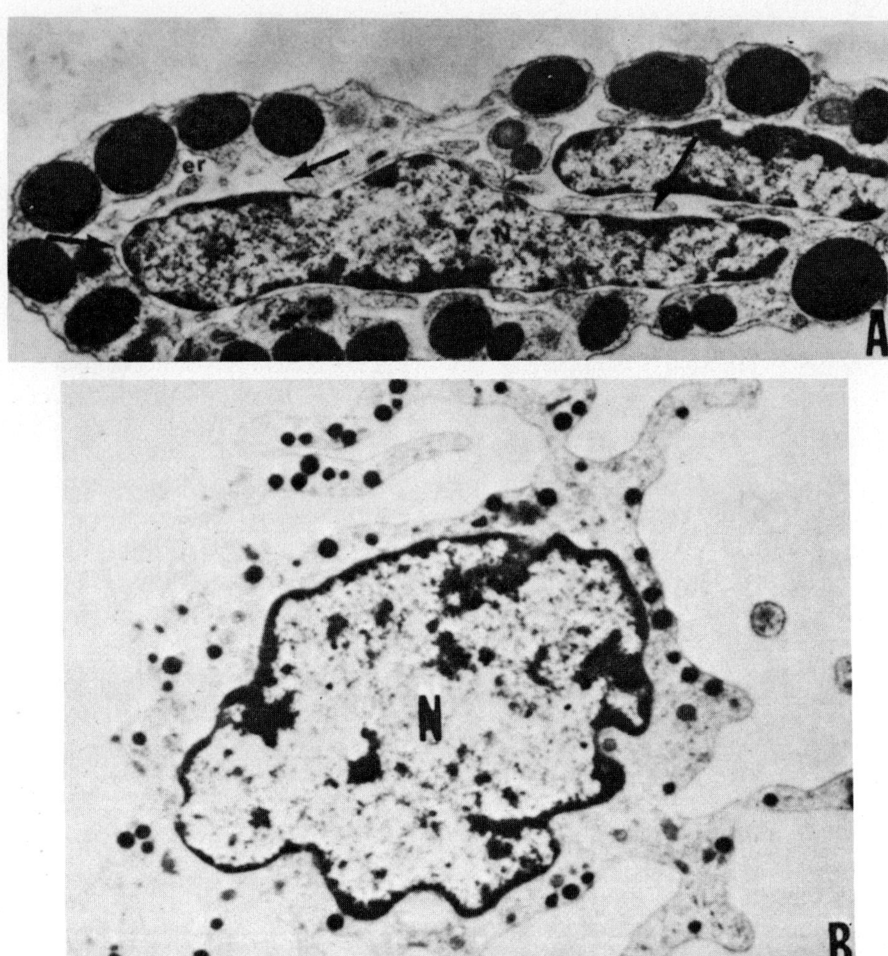

FIG. 11. **A:** Coagulocyte, showing perinuclear cisternae (arrows) and those of endoplasmic reticulum (ER). Note presence of electron-dense (structureless) granules such as are found in granulocytes, (×20,000). **B:** Podocyte, showing pseudopodia. Note resemblance to prohemocyte on young plasmatocyte; N = nucleus, (×9500). (From Gupta, A., 1979c.)

long), generally eccentric, oval, sharply outlined, and under phase-contrast may appear cartwheel-like owing to the arrangement of the chromatin in that fashion (Figs 1F,G, 2D). According to Goffinet, G. and Grégoire, Ch. (1975), there is a pronounced perinuclear cisterna (Fig. 11A), which supposedly distinguishes these cells from other types.

The laminar nature of the plasma and nuclear membranes may not be visible. The plasma membrane may show microruptures. The cytoplasm is hyaline and rich in polyribosomes, but has few mitochondria, and moderately developed ER. In addition, the cytoplasm has some spherical or elongate granular inclusions, about 1 μm in diameter (Fig. 11A). François, J. (1975a) has described four types of such granules in *Thermobia domestica*: (1) electron-dense, homogeneous granules, generally resembling those in GRs; (2) moderately electron-dense, homogeneous granules; (3) heterogeneous granules with a central or lateral dense zone, the remaining portion being homogeneously granular, and (4) structured granules, with internal microtubules (15 nm in diameter), arranged in a parallel fashion and 40 nm apart. Goffinet, G. and Grégoire, Ch. (1975) also described structured granules in the COs of *Carausius morosus*. It is obvious that there is a very close resemblance between GRs and COs.

Histochemically, COs are clearly distinguishable

from PLs and GRs according to the periodic acid-Schiff (PAS) test (Costin, N., 1975). According to her, "compared with the cytoplasm of other types of blood cells, that of coagulocyte has much reduced basophilia." It is very weakly PAS-positive.

2.3 Other hemocyte types

2.3.1 Podocyte (PO)

These hemocytes have not been recognized as a separate category in any ultrastructural study, and are not ordinarily observed in the hemocyte samples under the light microscope. They should be regarded as a variant form of PL. Arnold, J. (1979) commented that they might be due to fixation artifacts. According to Arnold, J. (1974), they have been correctly identified only in *Prodenia eridania* (Yeager, J., 1945; Jones, J., 1959). However, I have observed them in *Periplaneta americana* nymph (Gupta, A., 1969). More often than not, radiate PLs with pseudopodia are mistaken for POs.

These hemocytes are very large (Fig. 3E), extremely flattened, PL-like cells with several cytoplasmic extensions (Fig. 11B). The nucleus is generally large and centrally located, and may appear punctate.

They are derived from PLs. Gupta, A. and Sutherland, D. (1966) have suggested their transformation from PLs. This seems to be supported by M. Rizki's (1962) observation that in *Drosophila melanogaster* when POs increase in differential counts, PLs decrease. M. Rizki's (1953) POs appear to be PLs. Whitten, J. (1963) questioned the concept of POs, and Devauchelle, G. (1971) considered them variant forms of PLs.

2.3.2 Vermicyte (VE)

This form is generally called vermiform cell and should not be regarded as a separate category. As the name suggests, these are extremely elongated cells with slightly granular or agranular cytoplasm. The nucleus may be located centrally or eccentrically (Fig. 3F).

Their origin is unknown. However, it is conceivable that they are derived from PLs, as has been suggested by Gupta, A. and Sutherland, D. (1966). Lea, M. and Gilbert, L. (1966) considered them a variant form of PLs. According to Arnold, J. (1974), "they seem to occur mainly just prior to pupation, but never in large numbers", and may be caused by faulty fixation (Arnold, J., 1979).

2.3.3 Additional miscellaneous hemocyte types

In addition to the nine hemocyte types, several authors have, from time to time, reported hemocytes, many or all of which have not been generally accepted — for example: haemocytoblast of Bogojavlensky, K. (1932); "leucoblast" of Arvy, L. and Gabe, M. (1946) and Arvy, L. (1954); proleucocytoid and prohaemocytoid of Yeager, J. (1945) and Jones, J. (1950) respectively. Yeager also introduced the term nematocyte. Rizki, M. (1962) in his works used the terms lamellocyte and crystal cell. The latter was also adopted by Whitten (1963). According to Arnold, J. (1974), the crystal cell and lamellocytes are considered to be the variants of OEs and PLs respectively. Gupta, A. (1969) also suggested that the crystal cell is probably OE. Terms such as seleniform cells (Poyarkoff, E., 1910); miocytes (Tillyard, R., 1917); splanchnocytes (Muttkowski, R., 1924); "teratocytes" (Hollande, A., 1920); "pycnonucleocyte" (Morganthaler, P., 1953; Wille, H. and Vecchi, M., 1966); nucleocyte, and rhegmatocyte (Yeager, J., 1945; Hrdy, I., 1957) are rarely encountered in the literature; and most likely several of these cells are not even hemocytes. Jones, J. (1965) introduced the term granulocytophagous cell in his work on *Rhodnius prolixus*, and Ritter, H. (1965) and Scharrer, B. (1965a) "anucleate crescent body" and crescent cell respectively in the cockroach, *Gromphadorhina portentosa*. Zachary, D. and Hoffmann, J. (1973) described the hemocyte "thrombocytoid" that takes part in encapsulation in *Calliphora erythrocephala* (Zachary, D. *et al.*, 1975).

2.4 Controversies about, and interrelationships of, various hemocyte types

Despite the fact that most of the seven types of hemocytes are morphologically distinct, controversies about their identity and the manner of differentiation or origin persist. For discussion refer to Gupta, A. (1979c).

2.5 Identification key for hemocyte types in hanging-drop preparations

The identification key presented in Table 3 is for the novice who is studying insect, or for that matter any other arthropod, hemocytes for the first time. It is often very frustrating for a beginner to try to identify various types of hemocytes in a hemolymph sample or film under light microscope. This key should be helpful in guiding the beginner to become acquainted with the seven main types of hemocytes described in this chapter. Because not all hemocyte types are readily observed in any one species, at all developmental stages, and under all physiological conditions, it is important to examine hemolymph samples from different species at various developmental stages and under different physiological conditions. The method of study is no less important. The key is based on hemocyte observations under phase-contrast and Nomarski phase-contrast microscopes in hanging-drop preparations of fixed or unfixed hemolymph from various insects (Gupta, A. and Sutherland, D., 1966, 1967; Gupta, A., 1968, 1969).

2.5.1 Selected species for examination

Although the key can be used to identify hemocytes from any species, I suggest the following insects be used in the beginning: adult *Blaberus* spp. and *Gromphadorhina portentosa* for typical PRs, PLs, GRs, and SPs; larvae of *Galleria mellonella* and *Porthetria dispar* for typical PRs, PLs, OEs, SPs, and ADs, particularly in larvae about to molt, which is when GRs accumulate lipids and appear as ADs; and larval *Tenebrio molitor* for ADs, particularly after the larvae have been chilled at 5° for 20–24 h. For the unusual OEs (the so-called crescent cell of other authors), take samples from *G. portentosa*.

2.5.2 Procedure for hanging-drop preparation

(1) Take a square coverslip and put a tiny drop of saline-versene (NaCl, 0.9 g; KCl, 0.942 g; $CaCl_2$, 0.082 g; $NaHCO_3$, 0.002 g; distilled water, 100 ml + 2% versene).
(2) Cut the tip of the antenna or leg (or proleg) and let a drop of hemolymph flow into the saline-versene drop.
(3) Carefully turn the coverslip upside down and place it over the depression or cavity slide. Seal the sides of the coverslip with petroleum jelly.
(4) Examine the hanging-drop preparation under the phase-contrast or Nomarski phase-contrast microscope.
(5) You will notice that the hemocytes are more evenly distributed near the periphery than at the center of the hemolymph drop. It will therefore be easier to focus them sharply near the periphery.
(6) Make a fresh preparation for examination every 8–10 min because hemocytes begin to deteriorate after bleeding.
(7) For longer-lasting preparations, you may try hemolymph from an insect that has been heat-fixed in water at 60° for about 5 min.

3 ORIGIN OF HEMOCYTES

3.1 Embryonic origin

Mori, H. (1979) has discussed the embryonic origin and development of hemocytes in insects, and this section is largely based on his review and a few other works. Dohrn, A. (1876) was supposedly the first to describe embryonic blood cells in *Bombyx*, and Heymons, R. (1895) provided a detailed description of embryonic hemocytes in Dermaptera and Orthoptera. Other accounts of embryonic hemocytes are those of Roonwal, M. (1937) and Johannsen, O. and Butt, F. (1941).

At one time or another, embryonic hemocytes have been thought to be derived from yolk cells, serosal cells, cells at the junction of somatic and visceral musculature, wall of the heart, median mesoderm cells (i.e. cells at the median part of the inner layer), and suboesophageal body. Of these six described sites of origin, the median mesoderm cells are considered to be the most likely by the majority of authors and form the basis of the so-called median mesoderm theory, first proposed by Korotneff, A. (1885); the other two theories mentioned by Mori, H. (1979) are the coelomic sac theory and the suboesophageal body theory. Most authors believe that embryonic hemocytes are mesodermal in origin and are derived from the median part of the inner layer (Mori, H., 1979). That embryomic hemocytes are derived from the coelomic sacs has been suggested by Patten, W. (1884) and Roonwal, M. (1937), but has not been clearly demonstrated. Toyama, K. (1902) and Wada, S. (1955a,b) suggested that embryonic hemocytes are derived from a cluster of cells in the tritocerebral segments or beneath the stomodaeum (the so-called suboesophageal body) in *Bombyx mori*, but Okada, M. (1960) disputed this theory and suggested their mesodermal origin.

Mori, H. (1979) seems to be the only author who has cytologically characterized the embryonic hemocytes in *Gerris*. He found PRs, PLs, and GRs. According to him, PRs are not numerous but found throughout embryogenesis at various places in the embryo, PLs were larger or smaller in size and are probably derived from the first-generation PRs. Found at first (at 9 h) in the epineural sinus, they were observed in various parts of the embryo at later stages. The GRs are the largest of the three embryonic hemocyte types and appeared at 24 h in the epineural sinus; they are probably derived from the PRs.

Sohi, S. (1968, 1979) reported that it is possible to grow embryonic hemocytes in culture.

3.2 Postembryonic origin

3.2.1 Mitosis

Mitosis is frequently mentioned as a means of postembryonic multiplication of hemocytes. However, the normally low mitotic index (MI) (most often less than 1%) in circulating hemocytes casts doubt on the potential of mitosis for providing or maintaining large numbers of circulating hemocytes. That mitosis probably plays an important role in maintaining abundant hemocytes is suggested by reports that under certain stress conditions (e.g. injury), MI is very high. For example, Feir, D. and McClain, E. (1968) found that in *Oncopeltus fasciatus*, bleeding, injury, and the injection of adrenaline in the fifth instar nymph increased the MI during a 48–72 h period. Wigglesworth, V. (1937) found frequent mitoses in the circulating hemocytes in wounded *Rhodnius prolixus*. Similar observations of increased mitotic activity, following injury, have been made by Davis, R. and Schneiderman, H. (1960); Harvey, W. and Williams, C. (1961); Lea, M. and Gilbert, L. (1961); Bowers, B. and Williams, C. (1964); Crossley, A. (1964); Clark, R. and Harvey, W. (1964); Shrivastava, S. and Richards, A. (1965); Wittig, G. (1966); Feir, D. and McClain, E. (1968); and Shapiro, M. (1979a). Similar responses have been observed during phagocytosis, encapsulation, and nodule formation (see sections 7.1.1, 7.1.2 and 7.1.3). Shapiro, M. (1979a), however, has concluded that it is unlikely that mitosis alone could account for the increase in the number of circulating hemocytes.

Sohi, S. (1979) commented that because hemocytes can grow and multiply *in vitro* for an indefinite period in the absence of any hemopoietic organs, they can maintain their numbers *in vivo* through mitosis without the involvement of hemopoietic organs.

Based on the data on the duration of mitosis (20–30 min from metaphase to cytokinesis) provided by Clark, R. and Harvey, W. (1964) and Lea, M. and Gilbert, L. (1966), and assuming that a complete mitosis would require approximately 1 h to complete, Arnold, J. (1979) suggested the one could theoretically estimate the peak productivity of hemocyte by the formula: $(MI \times HIC \times 24)/1000$, where MI = mitotic index (i.e. cells in division/1000 cells) and HIC = total number of hemocytes in circulation, derived by multiplying THC (total number of hemocytes per mm^3) and blood volume. Knowing the duration (number of days) of each instar, one could further estimate the maximum hemocyte production or the maximum number of cells of any one type by multiplying DHC (differential hemocyte counts) and HIC during any instar. However, he pointed out several pitfalls in this procedure. Feir, D. and McClain, E. (1968) did not find any consistent diurnal periodicity in the MI in *O. fasciatus*.

Mitosis has been most frequently reported in the PRs (see Feir, D., 1979 and Arnold, J., 1979) and this is not surprising in view of their nature as stem cells that give rise to certain other hemocyte types. Jones, J. and Liu, D. (1968) erroneously thought that mitoses are exclusive to PRs. At one time or another PLs, GRs, SPs, COs, ADs, and OEs have also been observed in mitosis (see section 2.2).

Table 3: Identification key for insect hemocyte types

1.	Nucleus compact, large in relation to cell size, centrally located, almost filling the cell; very thin, peripheral layer of homogeneous cytoplasm around the nucleus; cells may be round, oval, or elliptical, but always small (compared with other cells in sample) (Figs. 1A,B, 2A)	*Prohemocyte* (*PR*)
1′.	Nucleus not compact, generally small in relation to cell size, not nearly filling the cell	2
2(1′).	Nucleus with chromatin arranged in cartwheel-like fashion, generally eccentric, oval, and sharply outlined; cytoplasm hyaline, generally scant, may contain some spherical or elongate granular inclusions; cell sometimes with cystlike blebs in process of exocytosis (Figs. 1F,G, 3D) (Beware! COs may be confused with OEs and with GRs in some insects)	*Coagulocyte* (*CO*)
2′.	Nuclear chromatin not arranged in cartwheel-like fashion, nucleus eccentric or central; cytoplasm not hyaline, abundant, homogeneous, and without any plate-, rod-, or needle-like inclusions and filaments	3
3(2′).	Cytoplasm generally agranular or slightly granular; nucleus round or elongate and central, and may or may not appear punctate; cells polymorphic and variable in size in various insects (Figs. 1C, 2B)	*Plasmatocyte* (*PL*)
3′.	Cytoplasm generally agranular, thick, and homogeneous with or without several kinds of plate-, rod-, or needle-like inclusions and filaments; "vacuoles" may or may not be present; nucleus generally small, round or elongate, and generally eccentric; cells variable in size and shape, generally lyse quickly *in vitro* ejecting material into the hemolymph (Figs. 1H,I, 3A,B)	*Oenocytoid* (*OE*)
4(3′).	Cytoplasm distinctly granular	4′
4′.	Cytoplasm prominently and characteristically granular; granules may or may not be numerous; nucleus comparatively small (compared with that in plasmatocyte) and compact, round or elongate, and generally central (Figs. 1D, 2D) (Note that GRs in lower orders are generally larger than in higher orders.)	*Granulocyte* (*GR*)
5(4′).	Granules in cytoplasm considerably enlarged and appear as distinct spherules or droplets	5′
5′.	Spherules non-refringent, generally obscuring the nucleus, number of spherules varying from few to many; nucleus rather small, central or eccentric; cells ovoid or round with variable sizes, usually larger than granulocytes, and may be observed releasing material from spherules into hemolymph by exocytosis (Figs. 1E, 2C,E)	*Spherulocyte* (*SP*)
6(5′).	Spherules or droplets refringent owing to presence of lipid; nucleus relatively small (compared with that in plasmatocytes or spherulocytes), round or slightly elongate, central or eccentric, and may or may not appear concave, biconvex, punctate, or lobate; cytoplasm may contain other non-lipid granules (Fig. 3C)	*Adipohemocyte* (*AD*)

3.2.2 Hemopoietic organs

In many insects, as in higher animals, hemocytes are produced in specialized hemopoietic tissues or organs ("hemocytopoietic" would be more precise because, as far as we know, these organs in insects produce only cells and not the plasma as well, as would be implied by such words as hemopoietic, hematopoietic and hematogenetic (Dorland, A., 1974) that are used in higher animals). Cuénot, L. (1896) was perhaps the first to observe these organs as "phagocytic organ", the most recent reviews on hemopoietic organs being those of Jones, J. (1970) and Hoffmann, J. *et al.* (1979).

O'Connor, G. and Feir, D. (1968) suggested four criteria for characterizing a hemopoietic tissue: (1) demonstration of higher levels of DNA synthesis and/or mitotic activity, (2) a more consistent number of cells, (3) a definite location, and (4) the presence of a membrane or some clear organ morphology. Jones, J. (1970) has also suggested more or less similar criteria, adding that in a true hemopoietic organ, one should demonstrate specific correlations between the numbers of cells being produced in and released from these organs and statistically valid increases in the number of circulating cells. Some of the recent descriptions of such organs in insects are those by Hoffmann, J. *et al.* (1969); Klein, M. and Coppel, H. (1969); Akai, H. and Sato, S. (1971); François, J. (1975a); Arnold, J. and Hinks, C. (1976); Hinks, C. and Arnold, J. (1977); and Hoffmann, J. *et al.* (1979).

Ultrastructurally, the hemopoietic organs in *Gryllus bimaculatus* and *Locusta migratoria* consist of an external basement membrane or material; a peripheral cellular layer, whose cells show characteristic features of fibroblasts; a cortical zone composed of reticular cells, some of which take part in phagocytosis, while others divide and eventually produce various types of hemocytes; and a medullary zone of differentiated hemocytes (Hoffmann, J. *et al.*, 1979). The organ in *Melolontha melolontha* does not show any specific organization, according to these authors. Akai, H. and Sato, S. (1971) reported openings in the acellular sheath of the hemopoietic organs in *Bombyx mori* through which the hemocytes pass to the hemolymph. The outer sheath is absent in the hemopoietic cell clusters in *Calliphora erythrocephala* and no cellular organization exists (Hoffmann, J. *et al.*, 1979), but reticular cells are present. The organ in *Thermobia domestica* is an irregular accumulation of cells and consists of star-shaped reticular cells, fibroblasts and islets of hemocytes (François, J., 1975a).

There is disagreement as to the types of hemocytes that are differentiated in the hemopoietic organs in various insects. Apparently all hemocytes are not produced in these organs. Akai, H. and Sato, S. (1971) reported that during larval–pupal molt well-differentiated PLs, GRs, SPs, and OEs are present; during intermolt periods these organs contain only electron-opaque cysts. In *G. bimaculatus*, the organ produces PLs, GRs, and COs in isogenic islets (Hoffmann, J. *et al.*, 1979). Arnold, J. (1979) suggested that the GRs and SPs are not produced in the hemopoietic organs of the noctuid, *Euxoa declarata*, but instead probably originate as separate lines in the embryo. Furthermore, according to him, the hemopoietic organ in this insect differentiates two cell lines: (1) germinal cells → PRs → PLs, (2) germinal cells → OEs. Thus, including the two separate cell lines of GRs and SPs, there are four postembryonic cell lines in *E. declarata*. In the PR–PL line the PRs divide and produce PLs both within the hemopoietic organs and in the hemolymph, and the PLs do not transform into any other type; instead they mature and degenerate. The OEs behave similarly, and so do the self-cloning GRs and SPs. Arnold, J. (1979) concluded that "true hemocytes are isogenic clones", and that the "fundamental basis for hemocyte classification and the acceptance of a particular form of cell as a type lies in the cell's origin". Unfortunately this concept, if applied to the known types of hemocytes of cells, would not account for all the hemocyte types that are presently recognized. Finally, Mitsuhashi, J. (1972) cultured hemopoietic organs of *Papilio xuthus* and thought that a few cells produced by these organs looked like GRs and SPs. He did not report any other type.

3.3 Differentiation of various hemocyte types

I discussed in section 3.1 the fact that most authors believe that in the embryo, hemocytes differentiate from the median mesodermal cells. Mori, H. (1979) identified PRs, PLs, and GRs in *Gerris* and suggested that both PLs and GRs differentiated from the PRs during various stages of embryogenesis. Arnold, J. (1979) concluded that GRs and SPs in the larvae of *Euxoa declarata* are embryonic cell lines, each probably derived from the same stem cell. Unfortunately no one has attempted to establish the differentiation pathways of hemocytes in insect embryos. By contrast, there are many studies, both in immature stages and adults, that deal with the differentiation of various types from one another (see sections 2.3 and 3.2.2 for discussion). There is no agreement among various authors as to how the various types of hemocytes originate, and it appears that probably several differentiation pathways exist in various insects. Arnold, J. (1979) neatly categorized the prevailing opinions into two theories: (1) the single-cell theory, and (2) the multiple-cell theory.

3.3.1 Single-cell theory

Because of the presence of transitional stages of some hemocyte types that possess certain common structural features, and of the existence of only PRs and PLs (supposedly the PRs transforming only to PLs) in tissue culture (Mitsuhashi, J. 1966: Sohi, S., 1971) and in the hemopoietic organs (Arnold, J., 1979), many authors have either suggested, or it can be inferred from their works, that the various hemocyte types are merely stages (with separate functions) of a single germinal type (cell line), generally the PR (Smith, H., 1938; Ogel, S., 1955; Selman, J., 1961; Gupta, A. and Sutherland, D., 1966; Shrivastava, S. and Richards, A., 1965; Devauchelle, G., 1971; Moran, D., 1971; Scharrer, B., 1972; Lai-Fook, J., 1973; Olson, K. and Carlson, S., 1974; Ratcliffe, N. and Price, C., 1974; Crossley, A., 1975; Landureau, J. and Grellet, P., 1975a; Beaulaton, J. and Monpeyssin, M., 1977; Wigglesworth, V., 1979a). By and large, the conditions and the causes of the transformation from one stage (hemocyte type) to another are speculative.

3.3.2 Multiple-cell theory

Because of the existence of structurally different hemocyte types, whose immature nature and old stages are present at the same time in the hemolymphs, and the presence, in hemopoietic organs and cell cultures, of more than one cell line, many authors believe, or their works suggest, that each of the hemocyte types represents a separate immutable cell line that differentiates from a single germinal stem, generally the PR (Arnold, J. and Salkeld, E., 1967; Wittig, G., 1968; Harpaz, F. *et al.*, 1969; Akai, H. and Sato, S., 1973; Zachary, D. and Hoffmann, J., 1973; François, J., 1974; Arnold, J. and Hinks, C., 1976; Raina, A., 1976; and Beaulaton, J. and Monpeyssin, M., 1976, 1977). As in the single-cell theory, the conditions and causes of differentiation of many different cell lines from a single germinal stem cell are unknown.

3.3.3 Hormonal effects on hemocyte differentiation

Several authors have suggested that juvenile hormone (JH) and ecdysone (see vols 7 and 8) affect hemograms (total and differential hemocyte counts; THC, DHC). None of the works, however, has clearly demonstrated the effects of these hormones on the differentiation and multiplication of hemocytes. Jones, J. (1967) reported that hormones partly regulate the cyclic changes in hemocyte types and number. Hoffmann, J. and Joly, L. (1969) and Hoffmann, J. (1970) found that the corpora allata (CA) influence both the differentiation and production of hemocytes in *Locusta migratoria.* In a preliminary study of the effects of synthetic JH-I (0.33 μl injected) on the THC and DHC in the treated (during sixth-instar nymphs), newly emerged, female, adult *Blattella germanica*, we found that these roaches had considerably lower (50%) THC than the untreated controls, both after 1-day and 7-day post-emergence periods. Differential counts of PRs, PLs, GR–SP complex, ADs, and COs were also considerably lower, except the GR–SP complex that showed marked (61%) increase after 1-day post-emergence, but decreased to a level lower than in the controls after 7-days post-emergence. These results are difficult to interpret at this time, but we reported elsewhere (Ramaswamy, S. and Gupta, A., 1981) that JH treatment of the sixth-instar (last) nymph leads to retention of certain nymphal characteristics (e.g. antennal and labial palp sensilla) in the treated female *B. germanica* that behave as nymphs and do not mate. Thus, it is tempting to suggest that the overall lower THC and DHC in the treated females also reflect a nymphal characteristic.

Because of the conflicting reports (Patton, R. and Flint, R., 1959; Jones, J., 1967) about the THC prior to, during, and after ecdysis, it is difficult to correlate either the increase or the decrease in the THC to the ecdysteroid titer during these periods. For example, Nittono, Y. (1960) found that in the fourth larval and pupal stages of *Bombyx mori*, the THC peaked prior to molting; Webley, D. (1951), on the other hand, reported a decrease just prior to ecdysis, a rise soon after ecdysis, and a subsequent decline. Bahadur, J. and Pathak, J. (1970) made similar observations in *Halys dentata*, but noted an increase in mid-instar. Hoffmann, J. and Joly, L. (1969) and Hoffmann, J. (1970) reported that prothoracic glands stimulate both the differentiation and production of hemocytes during the last nymphal instar of *Locusta migratoria migratorioides.* Judy, K. (1969) reported morphogenetic activities in the larval and pupal hemocytes of *Manduca sexta in vitro* under the influence of 20-hydroxyecdysone, and suggested that a similar response may occur *in vivo* to an elevated titer of this hormone.

Any direct role of any of the two morphogenetic hormones on the differentiation and multiplication would have to await more experimental work specifically designed to test the effects of these hormones.

4 EVOLUTION OF HEMOCYTE TYPES

4.1 The plesiomorphic hemocyte and its evolution into other types

I have reported elsewhere (Gupta, A., 1979a) that the granulocyte (GR) is the plesiomorphic hemocyte, and it is the only hemocyte type that has been reported in all major arthropod groups, including all studied insect orders, and the Onychophora. The following account of the presence of GR in Insecta is based on J. Arnold's (1974) reinterpretation of various works, my own (Gupta, A., 1969) survey of the hemocyte literature in many insects, and recent transmission electron microscopic (TEM) studies of hemocytes. As far as is known, only Bruntz, L. (1908); Millara, P. (1947); Barra, J. (1969); and François, J. (1974, 1975a) have worked on the hemocytes of the Apterygota; and on the basis of these works, both Collembola (it is controversial whether they should be included in Apterygota) and Thysanura possess GR. All higher orders of insects possess GRs (see Table 1). Some of the recent TEM studies (Hoffmann, J. *et al.*, 1968b; Baerwald, R. and Boush, G., 1970; Hagopian, M., 1971; Moran, D., 1971; Scharrer, B., 1972; Ratcliffe, N. and Price, C., 1974; Goffinet, G. and Grégoire, Ch., 1975; Beaulaton, J. and Monpeyssin, M., 1976; Brehélin, M. *et al.*, 1976; Ratcliffe, N. *et al.*, 1976b; Rowley, A. and Ratcliffe, N., 1976; Schmit, A. and Ratcliffe, N., 1977) have reported (or can be interpreted to show) GRs in various insects. The most highly specialized neuropteroid orders also possess GRs.

Since it has been reported by several authors that one hemocyte type can and does differentiate into another type, it is conceivable that during evolution the plesiomorphic GR differentiated into other hemocyte types. It can be postulated also that the GR originated from the so-called prohemocyte (PR) or stem cell and passed through the plasmatocyte (PL) stage before becoming a distinct GR type. In taxa in which only GRs have been observed (e.g. Xiphosura), the PR and PL are merely evanescent stages, and have not achieved distinctness as types. In taxa that are reported to possess other types besides PR, PL, and GR, the last perhaps further differentiated into SP, AD, CO, and OE, although not necessarily in that order. The post-GR differentiation is generally accompanied by distinct PRs and PLs. Furthermore, in the more highly evolved taxa any of the types may be suppressed. The main differentiation pathways as postulated above may be represented as follows:

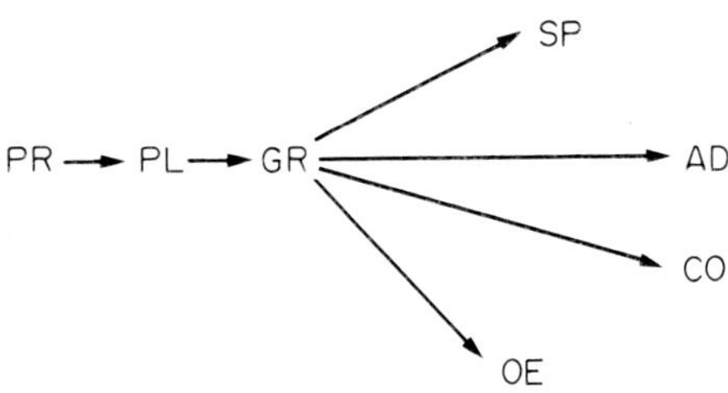

5 HEMOCYTE TYPES IN INSECTS AND OTHER ARTHROPODS

5.1 Hemocyte types in insects

Several years ago (Gupta, A., 1979c), I reported that the hemocytes in several orders of insects have not been studied. For example, as of that year, among Apterygota, only Thysanura had been studied. Among the orthopteroid groups, Isoptera and Embioptera awaited studies. In the hemipteroid complex no account of hemocytes was available in Zoraptera, Corrodentia, and Thysanoptera, and finally in the neuropteroid groups hemocytes were yet to be studied in Raphidioidea, Mecoptera, and Siphonaptera. It seems that the situation has changed very little since then, for hemocytes in most of the above groups are still awaiting study.

In terms of the number of species studied in various orders, Lepidoptera, Hymenoptera, Coleoptera, and Diptera appear to be the most extensively studied groups. In addition to the Heteroptera, Homoptera, and Odonata, of which only a few species have so far been studied, Dermaptera, Plecoptera, Trichoptera, and Thysanoptera are the most poorly studied groups. Hemocytes of several insect orders are unknown.

With the exception of the GR (and possibly also PLs) all other types of hemocytes are not present in all insect orders (see Table 1). According to Arnold, J. (1974), all hemocyte types have been reported only in *Prodenia eridania* (Yeager, J., 1945; Jones, J., 1959). Most insects seem to possess PR, PL, and GR.

5.2 Hemocyte types in other arthropods

For a detailed review of hemocyte types one may refer to reviews by Gupta, A. (1979a,e); Bauchau, A.

Table 4: Summary of hemocyte types in various groups of arthropods and Onychophora

Groups	Prohemocyte PR	Plasmatocyte PL	Granulocyte GR	Spherulocyte SP	Adipohemocyte AD	Coagulocyte CO	Oenocytoid OE	Others
Aquatic								
Chelicerata (Xiphosura)	—	—	GR	—	—	—	—	cyanoblast cyanocyte
Crustacea	PR	PL	GR	SP	AD	CO	OE	cyanocyte
Terrestrial								
Chelicerata	PR	PL	GR	SP	AD	CO	OE	(cyanocyte)
Myriapoda	PR	PL	GR	SP	AD	CO	OE	crystal cell
Insecta	PR	PL	GR	SP	AD	CO	OE	for numerous other terms see section 2.3.3
Onychophora	PR	PL	GR	SP	—	—	OE	—

Cells that could not be homologized have been included in the category of "others" (from Gupta, A., 1979a)

(1981); and Ravindranath, M. (1981). The same seven types of hemocytes that are present in insects are also present in various arthropod groups (Table 4), not all types being present in all groups. Aquatic Chelicerata (Xiphosura) possess only GR and a hemocyanin-producing hemocyte, the cyanocyte (CYN). Among Crustacea the more highly evolved groups, such as Decapoda and some Isopoda, have all seven types, but the primitive branchiopods (*Artemia* and *Daphnia*) possess only PL and GR. Among terrestrial Chelicerata, Scorpionida, Arnaeida, and Acarina have developed among them all seven types as well as the CYN. Myriapoda also have the seven types. Onychophora, however, have only five types: PR, PL, GR, SP, and OE.

5.3 Taxonomic significance of hemocyte types

The prospect of using variations in hemocyte types in various insect orders for phylogenetic considerations is severely limited owing to: (1) lack of uniform terminology, and hence the difficulty of establishing comparisons, and (2) paucity of comprehensive studies of hemocyte types in large numbers of species within various orders to enable one to draw meaningful conclusions on the basis of important variations. However, hemocyte types and their numerical variations in many insect orders are known. Does the number of hemocyte types in various orders have any phylogenetic significance? Arnold, J. (1972a,b, 1976, 1982) and Arnold, J. and Hinks, C. (1975) have used hemocytes in insect taxonomy, and several authors have suggested (see Gupta, A., 1979a,c,e) that some phylogenetic relationship among various arthropod groups can be demonstrated on the basis of the occurrence of the GR and other hemocyte types. For detailed discussion see Gupta, A. (1979c).

6 HEMOGRAMS

Information on the hemocyte population within an insect is essential for many types of physiological studies. The hemocyte picture of an insect at a particular time, or from one time to another, is called the hemogram, and consists of the measurements of the total hemocyte count (THC), differential hemocyte count (DHC), the absolute hemocyte count (AHC), and the blood (hemolymph) volume (BV).

6.1 Total hemocyte count (THC)

The THC is determined in a standard amount of blood, usually 1 mm^3. According to Shapiro, M. (1979b), Tauber, O. and Yeager, J. (1934) were the first to study the THC in several insects and later (1935) in several orders. Over the years their method, based on the method used for mammalian blood counts, has been used by numerous authors (see Shapiro, M., 1979b for listing). The method generally consists of drawing hemolymph into a Thoma white blood cell pipette, diluted (1 : 100) (Gupta, A. and Sutherland, D., 1968) in physiological saline, and counted in a standard hemocytometer according to the following formula:

$$\frac{\text{Hemocytes in five 1 mm squares} \times \text{dilution} \times \text{depth of chamber}}{\text{Number of 1 mm squares counted (i.e. five)}}$$

The depth factor is usually 10, but different dilution factors have been used by various authors, ranging from 20 (hemolymph drawn to 0.5 mark) to 150 or higher (Shapiro, M., 1979b).

6.2 Differential hemocyte count (DHC)

The DHC (i.e. the relative numbers of different types of hemocytes) may be determined in hemolymph-smear preparations, histological sections, and hanging-drop preparations (see section 2.5.2), but the latter method is recommended because there is little or no distortion in the morphology of the cells in this preparation. The DHC is made per fixed numbers of hemocytes (say 200) counted.

6.3 Absolute hemocyte count (AHC)

The AHC represents an estimate of both the THC and DHC in relation to the BV, and is estimated by multiplying the THC by the BV.

6.4 Hemolymph (blood) volume (BV)

Because the BV has a direct bearing on the THC, it is essential to determine the BV, which is usually done by dye-dilution methods (Shapiro, M., 1979b). Using essentially J. Yeager and S. Munson's (1950) and M. Lea's (1961) (cited by Shapiro, M., 1979b) modified methods, Shapiro, M. (1979b) calculated the BV in *Galleria* by the following formula:

$$V = \frac{d\,(c' - c'')\,1}{c''}$$

where V = BV in microliters; d = volume of amaranth dye (1% aqueous solution) injected in microliters; c' = original concentration (percentage); c'' = concentration of the dye after circulation (percentage). To obtain the BV, the V value is divided by the body weight of the insect.

It must be pointed out that great variability in the hemograms within the same insect or among insects exists, and therefore great caution must be exercised both in their determinations and interpretations. The variability may be due to the differences in techniques used, imprecision in the use of these techniques, inherent variability within the insects, physiological state of the insect, its age, sex, and external factors, such as disease, starvation, wounds, etc. For a discussion of the effects of these factors on the hemogram, one may refer to Arnold, J. (1974) and Shapiro, M. (1979a,b,c).

7 FUNCTIONS OF HEMOCYTES

7.1 Hemocytic immunity

Hemocytic immunity (see chapter 12, this volume) in insects is accomplished by phagocytosis, encapsulation, nodule formation, immunologic factors, coagulation, and poison detoxification mechanisms; processes that are triggered in nature as defense reactions against foreign biological agents (immunogens) and toxic substances. Artificially injected inanimate objects also induce some of these reactions. Immunity that kills the biological agent is called antiblastic immunity.

7.1.1 Phagocytosis

Hemocytic phagocytosis is the endocytic process of ingestion of foreign micro-organisms, such as viruses, bacteria, fungi, and protozoa, by certain types of hemocytes. The process is also called athrocytosis or pinocytosis. PLs and GRs are the main hemocyte types involved in phagocytosis, although OEs, SPs, and ADs have also been occasionally reported as phagocytes. As stated earlier (section 4), in arthropods as a whole, GR is the most primitive or plesiomorphic hemocyte type and performs phagocytosis. Furthermore, in arthropods, in which PLs, SPs, and ADs are present as distinct types, the phagocytic function may have

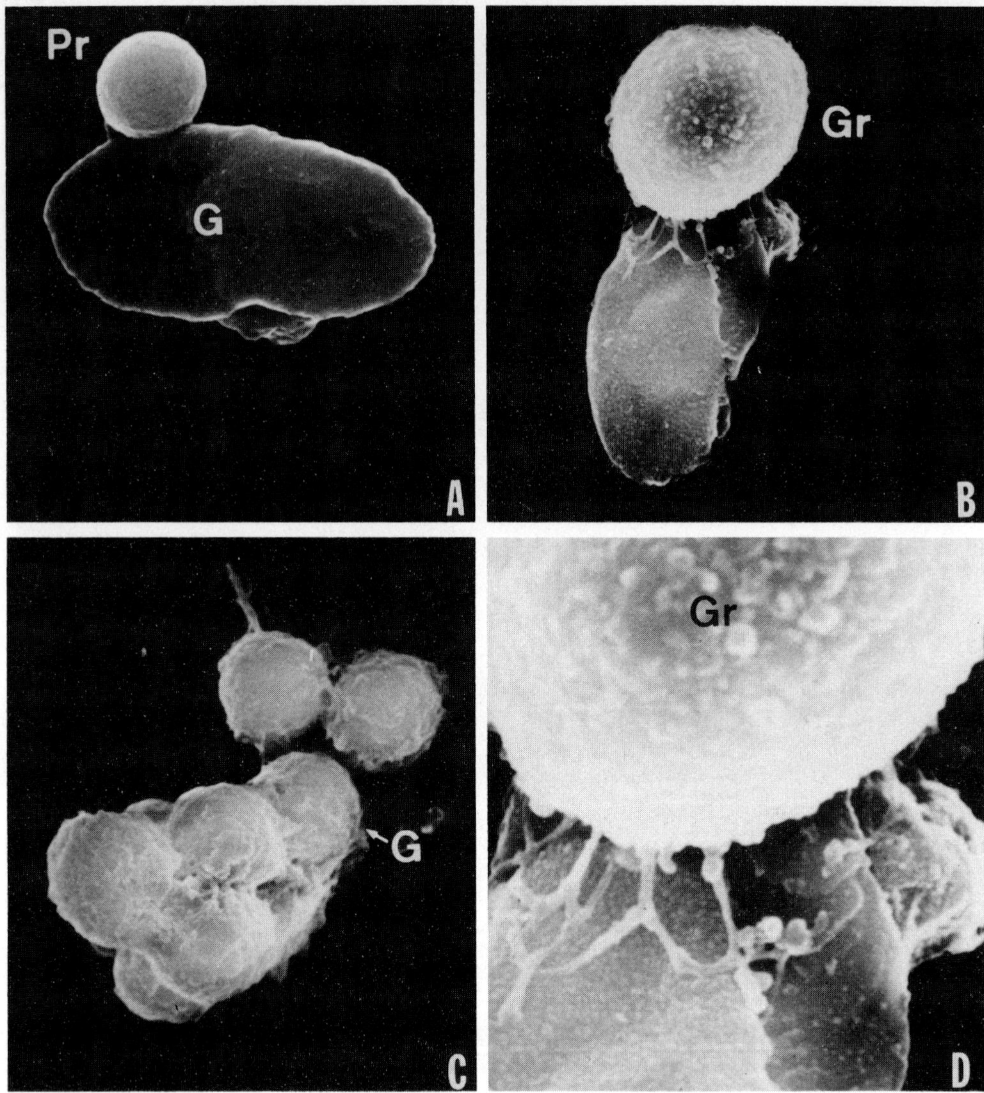

FIG. 12. **A:** Attachment of goose erythrocyte (G) to prohemocyte (Pr) of *Bombyx mori* (× 4000). **B:** Attachment of goose erythrocytes to granular cell (Gr) (× 4400). **C:** Several granulocytes surrounding an erythrocyte (G) (× 4000). **D:** A higher magnification showing details of attachment (× 8000). (Courtesy of Dr H. Wago.)

shifted from GRs to other types. It is also likely that some of these were incorrectly identified.

According to Ratcliffe, N. and Rowley, A. (1979), E. Metchnikoff's works on *Daphnia* (1884) and other arthropods (1892) were the first studies of phagocytosis; since then the phenomenon has been observed by many workers. There are several reviews on phagocytosis (Jones, J., 1962; Salt, G., 1970; Arnold, J., 1974; Whitcomb, R. *et al.*, 1974; Bang, F., 1975; and Ratcliffe, N. and Rowley, A., 1979), and one may refer to these for various details. Although most of the early work, and the majority of the recent work on phagocytosis, was done *in vivo* (injection method) recent *in vitro* techniques, such as cell cultures and monolayer hemocyte preparations, have provided the details of the *modus operandi* of phagocytosis.

Phagocytosis is accomplished in three stages: (1) recognition of the foreign body (immunogen) by the hemocyte, (2) its ingestion, and (3) final disposal or

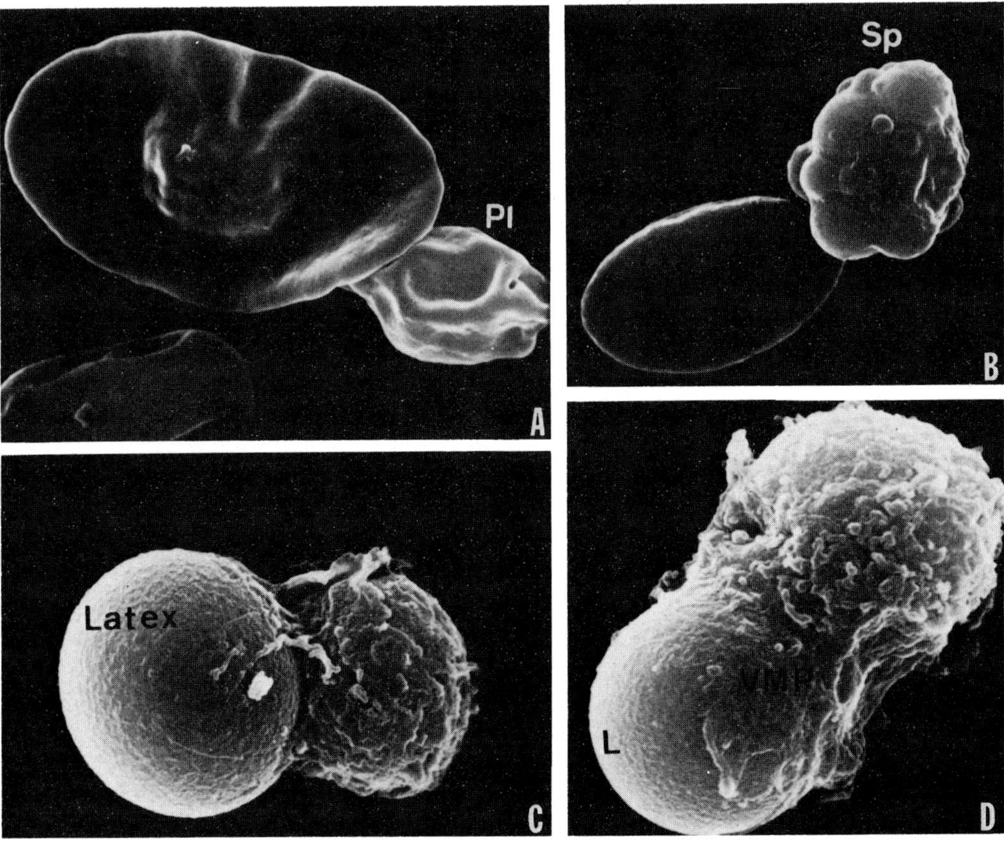

FIG. 13. **A:** Attachment of goose erythrocyte to the plasmatocyte (Pl) of *Bombyx mori* (× 9800). **B:** Same with spherulocyte (Sp), (× 5300). **C:** Attachment of 5.7 μm latex particle to granular cell (× 6700). **D:** Internalization of the latex particle by the granular cell; L = latex; VMP = veil-like membrane process (× 6700). (Courtesy of Dr H. Wago.)

clearance from the body (Bang, F., 1975; Ratcliffe, N. and Rowley, A., 1979).

(1) *Recognition of the foreign body by the hemocyte.* Perhaps the most important measure of the competence of an immunologic system is its ability to distinguish between "self" and "non-self" tissue. This competence is most highly developed in mammals and is due to immunoglobulins (serum proteins) that act as serum recognition factors (opsonins or antibodies). The presence of these serum immunoglobulins is necessary in the so-called "immune" phagocytosis, in contrast to "non-immune" phagocytosis that occurs in the absence of serum recognition factors. Insects possess hemolymph-dependent and hemolymph-independent recognition factors, the latter being partly attributed to hemocyte surface receptors (Ratcliffe, N. and Rowley, A., 1979; Wago, H., 1981). Wago demonstrated that in the last-instar larva of *Bombyx mori*, the capacity of the PLs and GRs to adhere to sheep erythrocytes was increased in the presence of hemolymph, although hemolymph-independent binding also occurred *in vitro* (Figs 12 and 13).

Although insects can distinguish between "self" and "non-self" tissue, they do not possess vertebrate immunoglobulins (γ-globulins). The hemolymph recognition factors in insects are called hemagglutinins and are secreted by certain hemocyte types (see section 7.1.4). In addition to the hemagglutinins, several types of chemical surface receptors have been reported in insects, primarily on the basis of the affinity of hemocytes for vertebrate erythrocytes. For example, Scott, M. (1971b) reported that in *Periplaneta americana* the attachment of the hemocytes to mammalian erythrocytes is due to trypsin-labile surface receptors. Anderson, R. (1976a,b) suggested the presence of

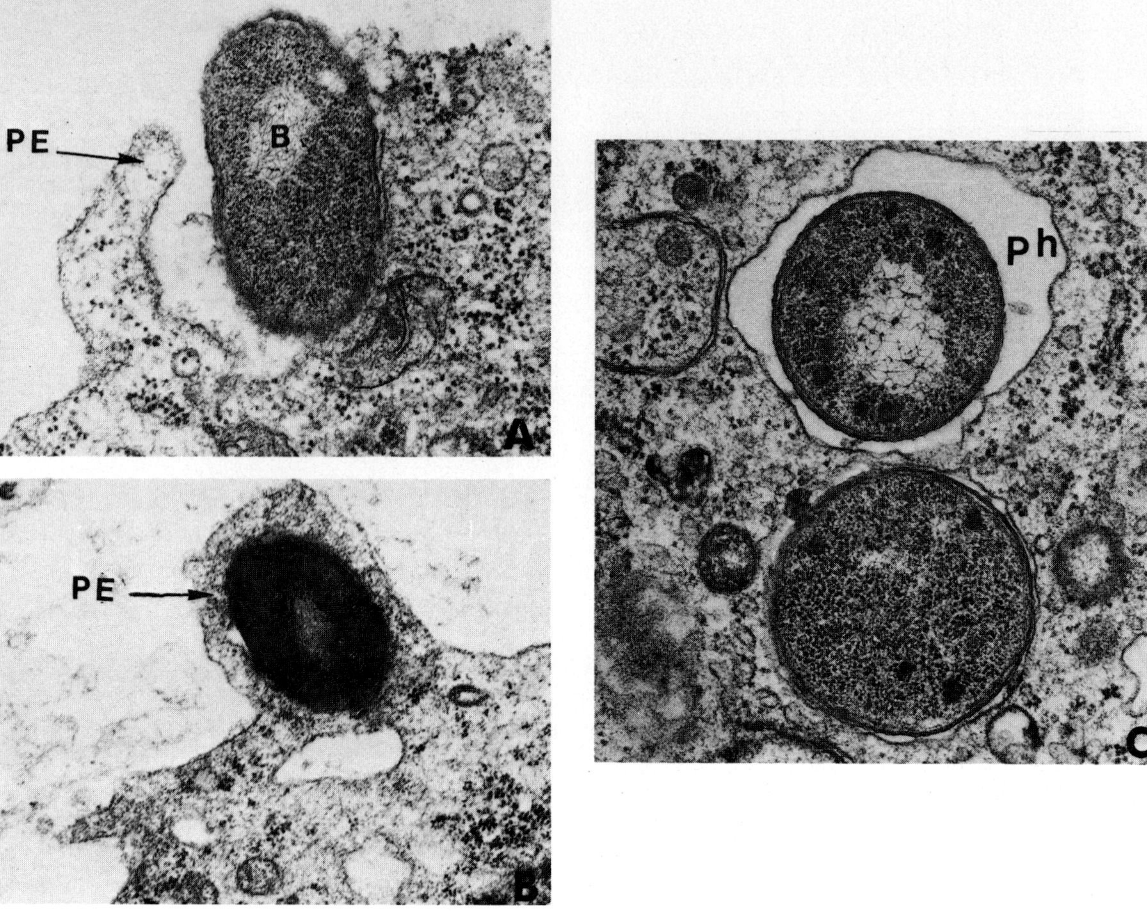

FIG. 14. **A:** *Galleria mellonella* plasmatocyte ingesting *Escherichia coli* (B) (×40,000). **B:** Later stage of **A** (×32,000). **C:** Intracellular *Escherichia coli*. (B) in peripheral cytoplasm of plasmatocyte of *Calliphora erythrocephala*. One bacterium is surrounded by a large phagosome. Other phagosome is much smaller (×38,000). PE = protoplasmic extension; Ph = phagosome. (Courtesy of Dr. N. A. Ratcliffe.)

non-specific surface receptors on the hemocytes of *Spodoptera eridania* and *Estigmene acrea*. Ratcliffe, N. and Rowley, A. (1979) reported that in *Pieris brassicae*, the SPs (= specialized GRs in this species) have a layer of "sticky" acid mucopolysaccharide that is probably responsible for binding sheep erythrocytes, *Escherichia coli*, and *Staphylococcus aureus* to these hemocytes. It should be noted, however, that Wago, H. and Ichikawa, Y. (1979) reported that in *B. mori* larva, the initial reaction of the GRs to injected goose erythrocytes was due to random contact and not due to any chemotaxis.

(2) *Ingestion of foreign body by the hemocyte.* Once the foreign body is recognized and contact with it established, ingestion commences. According to Salt, G. (1970) and Ratcliffe, N. and Rowley, A. (1979), ingestion is carried out either by (a) forming pinocytotic or coated vesicles to engulf particles, such as viruses (Leutenegger, R., 1967); (b) encircling the particles by pseudopodia (Werner, R. and Jones, J., 1968; Kislev, N. *et al.*, 1969) (Fig. 14A,B); or (c) spreading the plasma membrane around the particle (Grimstone, A. *et al.*, 1967). The ultrastructural details of the internalization processes, however, are poorly known in insects. Pastan, I. and Willingham, M. (1981) most recently described the details of receptor-mediated endocytosis in cultured vertebrate fibroblasts and suggested that similar processes occur in other types of

cells. According to them, the fibroblasts have a wide variety of mobile, specific surface receptors. After a ligand (L) (e.g. virus) binds with the receptor (R), the L–R complex moves to a bristle-coated pit (150 nm in diameter) embedded in the plasma membrane, and then enters the cell in an uncoated vesicle, the receptosome (250 nm in diameter), which then shuttles it to the interior of the cell. The bristles of the coated pit are composed of the protein, clathrin (Pearse, B., 1976). Pastan, I. and Willingham, M. (1981) chose the term receptosome to emphasize that the L–R complex appears in this organelle and not in the lysosome. Materials that do not bind to specific receptors enter the cell via another type of organelle, the pinsome ($\sim$0.08–2 μm in diameter), which fuses with the lysosome. The receptosome eventually transfers its contents to the lysosome via the Golgi–Golgi endoplasmic reticulum. Although Pastan and Willingham did not describe the mechanism of this transfer, it is probably akin to the transport pathway suggested by Rothman, J. (1981) on the basis of the structure of the Golgi apparatus. The Golgi typically consists of a stack of flattened, plate-like, membrane-bound cisternae (about 1 μm in diameter) (comparable to a distillation tower) from which many small vesicles (about 5–10 nm in diameter) bud off or fuse (or both), especially at the rims. The stack of cisternae is arranged in two compartments, the *cis* and the *trans* (Fig. 5A); the former, consisting of most of the cisternae, is nearer the nucleus and the endoplasmic reticulum (ER); the latter, consisting in many cases of only the last one or two cisternae of the stack, is on the opposite face of the *cis* portion and acts as the receiver of the distilled product. The protein molecules synthesized by the ER enter the stack at its *cis* face and exit at the *trans* face. Thus, the newly formed secretion granules (vesicles) are always associated with the *trans* face, which ensures that only the most refined fraction is secreted for distribution in the cell. The lysosomes are formed and their enzymes transferred to them in the region of the Golgi endoplasmic reticulum.

In insects, coated pits, filled with yolk protein, were described by Roth, T. and Porter, K. (1964) in the plasma membrane of *Aedes aegypti* oocytes, indicating that the pits were involved in "selective adsorption" from the extracellular space. Crossley, A. (1975) described flask-shaped (120–165 nm in diameter) coated vesicles in the plasma membrane of the hemocytes that approach cells undergoing phagocytosis in *Calliphora erythrocephala.* (Fig. 15A). Coated vesicles have also been found in insect pericardial cells (Bowers, B., 1964; Crossley, A., 1972), because of the involvement of these cells in endocytosis. Crossley experimentally demonstrated that these coated vesicles in the pericardial cells seem to be involved in selective uptake of some molecules, and rejection of others.

Pastan, I. and Willingham, M. (1981) speculated that substances that enter the cell by coated pits and receptosomes might include, among others, hormones and growth factors, and according to them, the action of many peptide hormones occurs on the cell surface.

According to Ratcliffe, N. and Rowley, A. (1979), insect hemocytes absorb soluble, proteinaceous, foreign material by coated vesicles and incorporate particulate matters, such as viruses, bacteria and some tracer substances by "true phagocytosis". In this case the particulate matters (e.g. *E. coli*), upon ingestion by PLs, in *Galleria mellonella* become enclosed in a membrane-bound phagosome (Fig. 14C) which later fuses with the lysosome (the so-called multivesicular body, MVB). It appears that in this case the ingested particulate matter in the phagosome was not channeled through the Golgi complex.

(3) *Final disposal or clearance of the foreign body.* Once transferred to the lysosome, the micro-organism is broken down by the mucolytic and other enzymes and antimicrobial agents. The objects generally phagocytized are bacteria, protozoa, fungi, erythrocytes, yeast cells, and autolysing tissue.

Effectiveness of phagocytosis. The effectiveness of phagocytosis depends, among other factors, on the phagocytic index (the number of phagocytizing cells), availability of circulating hemocytes, nature of the micro-organism, and the frequency of attack. It has been reported that when the number of phagocytizing cells is high, ingestion, and hence removal of the micro-organism from the hemolymph, is hastened (Wheeler, R., 1962; Wittig, G., 1966; Leutenegger, R., 1967). The availability of circulating hemocytes is important in the effectiveness of phagocytosis for it is known that insects become more susceptible to infection if the activity of the phagocytizing cells is blocked (Bettini, S. *et al.*, 1951; Stairs, G., 1964). However, Hoffmann, J. *et al.* (1979) reported that in *Locusta migratoria*, circulating hemocytes do not play a major role in taking up bacteria. Wago, H. and Ichikawa, Y. (1979) reported that the rate of phagocytosis of injected goose erythrocytes in *Bombyx mori* larvae was highest in the fifth larval instar. This was probably due to the increased number of circulating hemocytes (the GRs) in this instar. Unfortunately, little or no precise information is available regarding the phagocytic rate at any given time in insects. Most recently, however, Steinkamp, J. *et al.* (1982), using a new technique, measured the percentage of phagocytizing cells in rats, and suggested that this technique could be used in many cell systems.

Apparently, not all micro-organisms invading insects are equally affected by the phagocytizing hemocytes. Phagocytosis appeared to be ineffective against the bacteria, *Bacillus thuringiensis* and *Bacillus cereus*, microsporidia, *Nosema* sp., some viruses, and certain fungi, *Sorosporella uvella*, *Coelomomyces* sp. (see Arnold, J., 1974). By

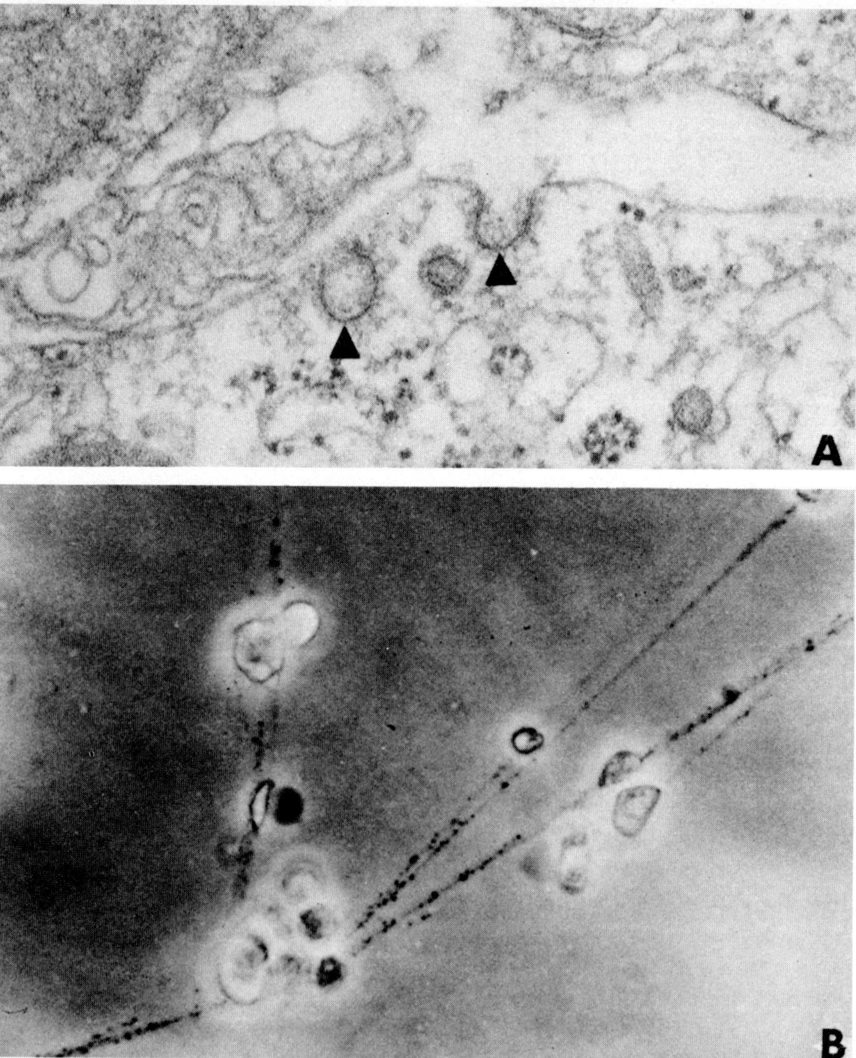

FIG. 15. **A:** Coated vesicles (arrow-heads) on the plasma membrane of a phagocytic hemocyte of *Calliphora erythrocephala*. These vesicles detach and pass into the cell. Note amorphous material inside the vesicles (hemolymph aspect of the vesicles) and coating "bristles" (cytoplasmic aspect of hemocyte) (arrow-heads). (×88,000). **B:** Coagulation pattern II in *Epicauta cinerea*, showing long, straight, thread-like cytoplasmic strands (×8700).

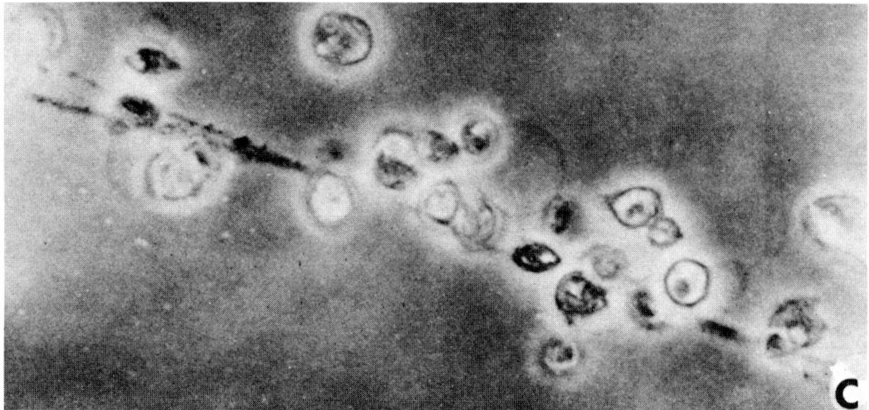

Fig 15 (Continued): **C:** Same showing "veil", (×8700). **A** (courtesy of Dr A. C. Crossley): **B,C** (from Gupta, A., 1969).

contrast, single hemocytes can ingest a number of particles (Harshbarger, J. and Heimpel, A., 1968).

The role of the so-called phagocytic organs or tissues in the effectiveness of phagocytosis (and perhaps also in encapsulation) deserves mention. To the extent that these organs in many insects are also hemopoietic in nature and thus produce hemocytes (e.g. PLs, GRs, and OEs in *L. migratoria*) (Hoffmann, J. *et al.*, 1968a, 1979), they surely play a role in phagocytosis, a fact that was first observed by Cuénot, L. (1895, 1896, 1897). He reported (1896) that the phagocytic tissue of the orthopteroid insects takes up bacteria (for further discussion on hemopoietic organs see section 3.2.2). According to Hoffmann, J. *et al.* (1979), it is the reticular cells of the hemopoietic tissue in *L. migratoria* that engulf the bacteria, not the circulating hemocytes. Furthermore, these reticular cells can be induced to produce large amounts of proteinaceous antibacterial substance by injecting immunizing doses of pathogens. These authors have suggested the possibility that, as in vertebrates, some of the reticular cells may be differentiated into immunocompetent cells, and that it is these cells that synthesize the antibacterial factor, as evidenced by marked dilations of the cisternae of the rough endoplasmic reticulum and the presence of crystalloid inclusions in these cells.

7.1.2 Encapsulation

Hemocytic encapsulation is a mechanism that isolates the foreign micro-organism, if it is too large to be phagocytized. Encapsulation is evoked by both biological and non-biological (inert) materials. As in phagocytosis, PLs and GRs are the primary hemocyte types involved, although SPs, COs, and OEs, or the so-called thrombocytoids in Diptera (Zachary, D. *et al.*, 1975) have also been reported taking part in encapsulation. There are several reviews on encapsulation (Shapiro, M., 1969; Salt, G., 1970; Nappi, A., 1974, 1975; Poinar, G., 1974; Ratcliffe, N. and Rowley, A., 1979) and one may refer to them for various details.

Encapsulation is accomplished in two stages: (1) recognition of the foreign body by the hemocyte, (2) capsule formation and melanization.

(1) *Recognition of the foreign body by the hemocyte.* Encapsulation is initiated by the GRs and/or COs, which must perform two functions: (a) recognize the foreign body and establish contact with it; (b) lyse and release recognition factor(s) in the hemolymph to attract PLs that would form the capsule. I have stated in section 7.1.1 that recognition of the foreign body is accomplished by both the surface receptors on the plasma membrane of the homocyte, in this case the GRs and/or COs, and the hemagglutinins that are secreted by the PLs and GRs and are present in the hemolymph. According to Ratcliffe, N. and Rowley, A. (1979), the attraction of the PLs to the foreign body to initiate capsule formation occurs due to chemotactic stimuli or factors released by the lysing GRs and/or COs. Discharge of material from GRs/COs or other hemocytes during the initial phase of encapsulation apparently occurs (Poinar, G. *et al.*, 1968; Crossley,

A., 1975; Rizki, M. and Rizki, R., 1976; Ratcliffe, N. and Gagen, S., 1976, 1977; Schmit, A. and Ratcliffe, N., 1977). Ratcliffe, N. and Rowley, A. (1979) stated that in *Galleria mellonella* during encapsulation "sticky" acid mucopolysaccharide-like substance(s) "appears to be discharged onto the foreign surfaces by GRs, ... and, independently of or in combination with material(s) from the foreign bodies, apparently specifically attracts PLs to the alien surface." Earlier, Brewer, F. and Vinson, S. (1971) suggested that tyrosine-containing proteins or polyphenols may be involved in opsonization of alien surfaces. Parish, C. (1977) hypothesized that 5-glycosyltransferases, secreted by hemocytes, form subunits of "recognition factors" and polymerize, after being secreted into the hemolymph, into hexamers that in turn react with the foreign body. The processes of recognition and attraction of the PLs to the foreign body are summarized below:

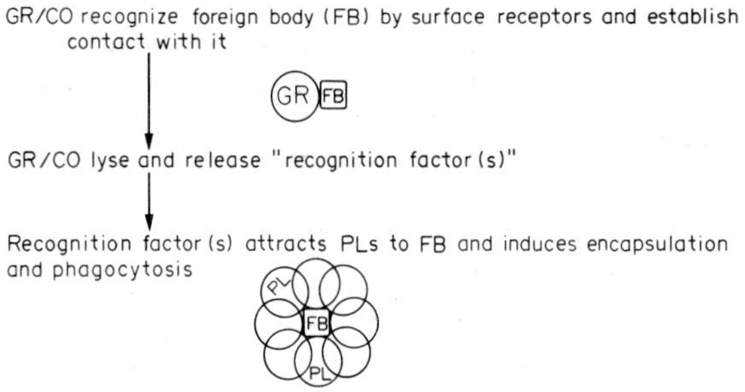

GR/CO recognize foreign body (FB) by surface receptors and establish contact with it

GR FB

↓

GR/CO lyse and release "recognition factor (s)"

↓

Recognition factor (s) attracts PLs to FB and induces encapsulation and phagocytosis

PL FB PL

(2) *Capsule formation and melanization.* Cellular capsules vary in terms of the structural changes in the hemocytes taking part in capsule formation and the number of cell layers (regions) in the completed capsule. After the release of the recognition factor(s) from the lysing GRs/COs, PLs (and/or GRs) are attracted to the foreign body and form a layer (inner) around it. In many insects this layer shows somewhat flattened and necrotic PLs. As more and more PLs adhere to the developing capsule, they undergo more intense flattening, but lesser necrosis, and form the so-called middle layer of the capsule. Apparently, both the flattening and necrosis of the PLs cease after the middle layer has attained a certain thickness, until finally the outermost layer consists of only loosely attached, normal PLs. These foregoing processes of capsule formation have been observed in *Ephestia kühniella* (Grimstone, A. *et al.*, 1967) and *Galleria mellonella* (Schmit, A. and Ratcliffe, N., 1977), and probably occur in other insects. The degree of flattening and their layered arrangements vary in various insects (*Carausius morosus, Carausius extradentatus, Blaberus craniifer, Tenebrio molitor*) (Ratcliffe, N. and Rowley, A., 1979). In *Periplaneta americana*, the capsule formed in response to *Moniliformis dubius* (Acanthocephala) is almost non-cellular (Mercer, E. and Nicholas, W., 1967), although Misko (1972), cited by Ratcliffe, N. and Rowley, A., (1979) found a different kind of capsule in this insect in response to the nematode *Caenorhabditis briggsae*, indicating that the type of capsule formed might depend on the foreign body.

Apart from the flattening of the PLs, several other changes have been observed in them. For example, PLs surrounding the foreign body may (Misko, 1972; cited by Crossley, A., 1979) or may not (Poinar, G. *et al.*, 1968; Misko, 1972) form a syncytium. The formation of membrane-bound, electron-dense vesicles in the PLs of the inner layer have been reported in many insects and seem to contain precursors of melanin, which causes the melanization in the capsules of many insects (Misko, 1972 as cited by Crossley, A., 1979). The PLs of the inner layer of the *G. mellonella* capsule contain cytolysosomes and a dense intercellular substance, which in *E. kühniella* is composed of acid mucopolysaccharide fused with a protein secreted by the PLs (Reik, 1968, cited by Ratcliffe, N. and Rowley, A., 1979). The PLs of the middle and the outer layers contain numerous microtubules, mitochondria, and ribosomes (Brehélin, M. *et al.*, 1975; François, J., 1974). According to Baerwald, R. (1979), these microtubules may be associated with the desmosomes. In addition to the intracellular structural changes in the PLs, several kinds of intercellular cell junctions, such as desmosomes (local "weld" spots without channels for intercellular ionic communication) or gap junctions (with cell-to-cell communicating channels that allow molecules of low molecular weight (300–500) to pass freely) have been observed (Baerwald, R., 1979). The structural changes in the PLs, after they contact the foreign body, apparently do not occur in all insects, for in the larvae of *Drosophila*, parasitized by the wasps, *Pseudeucoila bochei* and *Pseudeucoila mellipis*, structural and behavioral changes occurred in the free hemocytes prior to capsule formation (Walker, I., 1959; Nappi, A. and Streams, F., 1969; Nappi, A., 1973).

According to Baerwald, R. (1979), interspecific variation in the cellular make-up of capsules is to be expected owing to the variation in the free, circulating hemocytes in various species. Generally, three layers have been reported in the capsules of several insects: *E. kühniella* (Grimstone, A. *et al.*, 1967); *Galleria mellonella* (Schmit, A. and Ratcliffe, N., 1977); beetle larvae *Diabrotica* sp. (Poinar, G. *et al.*, 1968: four layers if one counts these authors' innermost non-cellular melanin layer). In contrast, no layers are present in the capsule in the phasmids, *Carausius morosus* and *C. extradentatus* (Ratcliffe, N. and Rowley, A., 1979).

Melanization does not necessarily occur in all cellular capsules. Those around inanimate objects are always without melanization and so are some around animate objects. More often than not, however, melanization occurs during encapsulation, and may even be essential for capsule formation in certain cases (Nappi, A., 1973). Melanization always occurs in the humoral capsule (discussed later in this section) (Ratcliffe, N. and Rowley, A., 1979). The question of whether melanization originates in the animate foreign body or in the host's hemocytes is still debated, although it has been documented in many cases that the hemocytes provide both the phenolic substrates and the phenol-oxidizing enzymes, the GRs and/or PLs (Ferron, P., 1978) supposedly providing the phenols, and the OEs supplying the enzymes (Nappi, A. and Stoffolano, J., 1971, 1972; Crossley, A., 1979). It should be stated, however, that the hemocytic origin of both the phenols and the phenoloxidases is controversial and GRs, COs, and OEs have been reported by one author or another to provide both (Neuwirth, M., 1973; Schmit, A. *et al.*, 1977). Mazzone, H. (1976) reported that cultured gypsy moth cells produced melanin for long periods. For more detailed discussion of melanization in capsules the reader is referred to Lipke, H. (1975); Nappi, A. (1975); Crossley, A. (1979); and Ratcliffe, N. and Rowley, A. (1979).

In addition to cellular capsules, capsules of non-hemocytic origin (humoral capsules) have been reported (Schell, S., 1952; Burns, W., 1961; Couturier, A., 1963). These humoral capsules are formed in insects (e.g. Diptera) that have few circulating cells, and thus apparently without the participation of hemocytes. According to Ratcliffe, N. and Rowley, A. (1979), humoral capsules are formed only against living parasites, such as nematodes and fungi, and not against inanimate objects.

The factors that affect encapsulation are the nature of the foreign body (animate or inanimate), availability of free hemocytes, age, sex, and nutritional state of the host (Ratcliffe, N. and Rowley, A., 1979).

(a) *Avoidance strategies of invading organisms against encapsulation* Although encapsulation provides the insect with defense against many invading micro-organisms, this defense is ineffective against certain parasitoids that successfully live within insects. Obviously, these parasitoids have evolved certain strategies that inactivate the process of encapsulation. The simplest way to discuss this will be to divide the process of encapsulation into five stages and point out which of these stages are successfully avoided by parasitoids by means of certain adaptations. This is summarized as follows:

GRs/COs: Hemocytes involved in encapsulation.
Stage I: Recognition of foreignness (the parasitoid) by the GRs/COs.
Stage II: Attraction of GRs/COs to the parasitoid and lysis upon contact.
Stage III: Release of recognition factor by the lysing GRs/COs.
Stage IV: Recognition factor chemotactically attracts PLs to parasitoid–GR/CO complex.
Stage V: Encapsulation commences.

The egg of *Nemeretis canescens* avoids being perceived as foreign by its host, *Ephestia kühniella* by means of a coating of calyx particles (Salt, G., 1965; Rotheram, S., 1967) that suppress the completion of Stage I. Bedwin (1970) (cited by Ratcliffe, N. and Rowley, A., 1979) showed that the high ornithine content of these particles, with its chelating properties, may be responsible for suppressing the attachment of GRs/COs to the eggs. *Cardiochiles nigriceps* and *Campoletis sonorensis* also employ calyx particles to suppress encapsulation by injecting them into the host at the time of oviposition (Vinson, S. and Scott, J., 1974; Norton, W. and Vinson, S., 1976). By avoiding Stage I, these parasitoids manage to suppress the remaining stages of encapsulation. Kitano, H. (1969a,b) reported that the eggs of the braconid, *Apanteles glomeralus*, oviposited in the larvae of *Pieris rapae crucivora*, cause reduced adherence of PLs, thus affecting stage IV. Finally, the teratocytes (giant cells) derived from the trophamnions of the egg (Salt, G., 1970), by secreting "encapsulation-inhibiting factor" (Kitano, H., 1969a,b) probably affect Stage V, and thus ward off encapsulation. The various adaptations described in the foregoing may be equally effective against nodule formation.

(b) *Effectiveness of encapsulation* The effectiveness of encapsulation seems to depend on the taxonomic position and the developmental stage of the host, and availability of circulating hemocytes at the time of attack. While Lepidoptera seem to be capable of encapsulating any parasite that has not developed avoidance strategies (Arnold, J., 1974), Hemiptera do not have efficient encapsulating capabilities (Day, M. and Bennetts, M., 1953; Beard, R., 1942). Apparently, the effectiveness of encapsulation differs during development stages (Puttler, B., 1967), among species (Salt, G., 1957; Streams, F., 1968), in populations of different localities (Muldrew, J., 1953; Puttler, B., 1967), and depends on the availability of hemocytes (Puttler, B., 1967).

7.1.3 Nodule formation:

Nodule formation is different from phagocytosis but in several respects similar to encapsulation, and occurs in response to both animate and inanimate particulate material that cannot normally be phagocytized. Both encapsulation and nodule formation may occur within the same insect in response to the same parasite (Ratcliffe, N. and Rowley, A., 1979), and it may be difficult to distinguish between the two defense reactions (Salt, G., 1970). Nodules or "granuloma" can be described as multicellular sheaths formed variously by PLs, GRs, or COs (Vey, A. *et al.*, 1968; Gagen, S. and Ratcliffe, N., 1976; Ratcliffe, N. and Gagen, S., 1976, 1977).

Typically, a nodule is formed around a melanized core of coagulum that is produced by the degranulating GRs/COs and that entraps the particulate material and the GRs. As in encapsulation, PLs are soon attracted to this core and form a sheath, indicating a chemotactic response (Ratcliffe, N. and Rowley, A., 1979) to the recognition factors released by the lysing GRs/COs. According to these authors the nodules in *Galleria mellonella* and *Pieris brassicae*, in response to certain bacteria, contained the same three distinct layers, as in encapsulation, the inner layer being composed of flattened PLs, the middle layer of more flattened PLs, and the outer layer of normal PLs. Melanization of the nodules seems to occur in the same way as in encapsulation.

7.1.4 Secretion of immunologic factors

I have stated earlier that mammals possess serum protein (γ-globulins), called immunoglobulins (Igs) (antibodies or opsonins) that act as humoral recognition factors in removing the foreign antigen. Although insects do not possess Igs they do produce humoral recognition factors, called hemagglutinins, that are secreted by certain types of hemocytes.

(a) *Hemagglutinins and anti-bacterial and anti-viral factors.* Since annelids have evolved the capacity to distinguish between "self", "closely related non-self" and distantly related "non-self" (Cooper, E., 1969a,b), it is not surprising that their descendants, the arthropods, also possess this capacity. This is accomplished by chemical surface receptors on the plasma membrane of certain types of hemocytes and humoral recognition factors, called hemagglutinins. Perhaps the most fully characterized arthropod hemagglutinin is that of the primitive arthropod, *Limulus polyphemus* (Tyson, C. and Jenkin, C., 1973, 1974; Paterson, W. and Stewart, J., 1974; Tripp, M., 1975). It agglutinates vertebrate erythrocytes and has similar electrophoretic mobility to γ- and β-globulins (Cohen, E. *et al.*, 1965). Tyler, A. and Metz, C. (1945) demonstrated that the serum of the lobster, *Panulirus interruptus* has a protein-agglutinating material. Such substances have been demonstrated in several other crustaceans (Tyler, A. and Scheer, C., 1946; Sindermann, C. and Mairs, D., 1959; Cushing, J. *et al.*, 1963; McKay, D. and Jenkin, C., 1970; Miller, V. *et al.*, 1972).

The presence of hemagglutinins in insects has been demonstrated by several authors (Bernheimer, A., 1952; Briggs, J., 1958; Feir, D. and Walz, M., 1964; Gilliam, M. and Jeter, W., 1970; Scott, M.,

1971a,b, 1972; Anderson, R. *et al.*, 1972, 1973; Ratcliffe, N. *et al.*, 1976a; Rowley, A., 1977). Some of the most recent reviews on hemagglutinins in insects and other arthropods and invertebrates are those of Cohen, E. (1974); Cooper, E. (1974); Lackie, A. (1980); Yeaton, R. (1981); Pistole, T. (1982); and Ratner, R. and Vinson, S. (1983). However, I believe the most interesting works are those of Amirante, G. (1976); Valvassori, R. and Amirante, G. (1976), and Amirante, G. and Mazzalai, F. (1978), who demonstrated for the first time that in the cockroach, *Leucophaea maderae*, the hemagglutinins are present in two types of hemocytes: in the vacuoles and on the plasma membrane of GRs and in the cytoplasm of SPs. Amirante, G. (1976) also showed that in this cockroach the hemagglutinins are in the β-globulin range. It should be mentioned here that Marek, M. (1970b) reported an "immunoglobulin" in *G. mellonella* that is induced by cooling, and is within the "range of vertebrate γ-globulin." Nelstrop, A. *et al.* (1970) concluded that although proteins with electrophoretic mobility comparable to vertebrate γ-globulin may be detected in insects (e.g. *Locusta*), they should not be characterized as immunoglobulins just on the basis of their electrophoretic mobility in the γ-globulin range. Considering the lower phylogenetic position of arthropods, one would not expect their hemagglutinins to be comparable to the γ-globulins of the most highly evolved vertebrates! It is not surprising, therefore, that attempts (Scott, M., 1971a; Anderson, R. *et al.*, 1973) to demonstrate properties of hemagglutinins comparable to those of vertebrate immunoglobulins have failed! At any rate, one may draw several conclusions from Amirante's and his associates' works.

(1) Hemolymph hemagglutinins in insects, and probably in other arthropods also, seem to be of hemocytic origin. Note that mammalian immunoglobulins also are produced by plasma cells (β-lymphocytes) in the blood.
(2) Because they have been demonstrated to be present on the plasma membrane of GRs, they most likely play a major role as surface receptors in foreign body recognition, in addition to their opsonin role as humoral recognition factors. This is supported by the fact that GRs are the first to be attracted to the foreign body during phagocytosis, encapsulation, and nodule formation.
(3) The hemolymph-independent attachment of vertebrate erythrocytes to hemocytes (Ratcliffe, N. and Rowley, A., 1979; Wago, H., 1981) is most likely due to these hemagglutinins and that insects, like vertebrates, do possess serum-independent-like surface recognition factors on certain hemocytes.

It is also interesting to note here that arthropods possess at least some components of the vertebrate complement system, necessary for the operation of the many immunologic reactions, such as antigen–antibody reaction and agglutination of cells. Day, N. *et al.* (1970) demonstrated the presence of complement component 3 proactivator in *Limulus*, and apparently a similar factor is present in *Blaberus* (Anderson, R. *et al.*, 1972) and *Galleria* (Aston, W. *et al.*, 1976). Since *Limulus* has only one type of hemocyte, the GR (the so-called amoebocytes of other authors) (Gupta, A., 1979a), it is reasonable to suggest that this complement factor is probably secreted by the GRs in this animal, and probably in other arthropods. Since immunoglobulins activate the complement system, it is likely that the hemagglutinin in *Limulus* activates the 3 proactivator in this arthropod. Note, however, that the hemocytic origin of both the hemagglutinins and the 3 proactivator in *Limulus* has not yet been confirmed.

It should be noted here that the immunity provided by the hemagglutinins of arthropods may be comparable to the non-specific immunity in man provided by the microphages (polymorphonuclear leucocytes), also called granulocytes.

Various antibacterial and antiviral toxins have been reported in insects (see Chadwick, J., 1975), but the direct role of hemocytes in the secretion of these toxins is unexplored. Hoffmann, J. *et al.* (1979) speculated that some of the reticular cells of the hemopoietic organs in *Locusta migratoria* perhaps differentiate into immunocompetent cells and synthesize the antibacterial factor.

(b) *Lysozymes* Lysozymes are mucolytic enzymes located in the lysosomes that are acid phosphatase-positive organelles in hemocytes, pericardial cells

(Crossley, A., 1979), and ecdysial (prothoracic) glands. Lysozymes, along with hemagglutinins and other complement-like substances in the hemolymph, play an important role in insect immunity. According to Houk, E. and Griffiths, G. (1980), hemolymph lysozyme may be important in eliminating any symbiotes that escape from the mycetocytes. Many studies have demonstrated that during phagocytosis, the lysosome fuses with the phagosome that contains the ingested micro-organism (Ratcliffe, N. and Rowley, A., 1979). According to Crossley, A. (1975), lysozyme is readily identifiable in the hemolymph of injured *Galleria* and rises from 25–500 μg ml^{-1} hemolymph to 9000 μg ml^{-1} in insects stressed by injection of Gram-positive bacteria. It is also found in *Calliphora erythrocephala* (Crossley, A., 1972) and *Locusta* (Hoffmann, J. *et al.*, 1977).

7.1.5 Coagulation

Hemolymph coagulation in insects, and indeed in other arthropods (Gupta, A., 1979a), is caused by:

(1) hemocyte agglutination (*Calliphora*: Åkesson, B., 1953; Crossley, A., 1975);
(2) plasma gelation (*Locusta*: Brehélin, M., 1972; and *Periplaneta*: Franke, H., 1960a,b,; Yeager, J. and Knight, H., 1933); or
(3) a combination of both.

The hemocyte type most often implicated in causing coagulation is the coagulocyte (CO), first named by Grégoire, Ch. and Florkin, M. in 1950, and often called cystocyte by others. It must be noted, however, that GRs, SPs, PLs, and the so-called thrombocytoids of Diptera have been reported to cause coagulation in various arthropods, including insects. Lai-Fook, J. (1970) found that in *Rhodnius prolixus*, the PLs resemble COs, GRs, and young OEs ultrastructurally. Coagulation patterns, on the basis of the reactions of and structural changes in the COs during coagulation, have been extensively studied by Grégoire, Ch. (1971) in various arthropods, and most recently by Grégoire, Ch. and Goffinet, G. (1979) in insects. The latter authors have recognized four patterns in insects. In the so-called pattern I, the COs, without lysing, discharge small amounts of cytoplasmic materials that later form circular islets around them, owing to plasma gelation. In pattern II the COs extrude long, straight, thread-like, granular, cytoplasmic strands that eventually form a network in which other types of hemocytes become agglutinated (Fig. 15B,C). This network is probably formed by lipoprotein (Siakotos, A., 1960a,b) and shows "veils", pattern III is actually a combination of patterns I and II, both patterns occurring either separately in the same coagulating sample or superimposed on each other. According to Grégoire, Ch. and Goffinet, G. (1979), starved or diseased specimens show only Pattern II reaction. In pattern IV, according to these authors, the COs do not apparently undergo any changes. This pattern is absent in blister beetles (Gupta, A., 1969). It is likely that the changes in the labile hemocytes involved in coagulation may be the effect rather than the cause of coagulation (Gupta, A. and Sutherland, D., 1966).

The most controversial and unresolved question about hemolymph coagulation in insects concerns the nature and the origin of the clottable protein and those of the coagulation factor. Do the hemocytes involved in coagulation produce both the ingredients of coagulation? The origin and nature of the clottable protein remains debatable. Dumont, N. *et al.* (1966) studied the clotting process in *Limulus* and concluded that the plasma in this animal is not capable of coagulation and that the granules of the GRs (= amoebocytes) provide the material that actively participates in the formation of the clot (see Gupta, A., 1979a). It should be noted that Howell, W. (1885); Loeb, L. (1904, 1910); and Maluf, N. (1939) suggested much earlier that the material that actually forms the clot is contained within the hemocytes and not in the plasma. This was recently supported by Murer, E. *et al.* (1975) who claimed that the granules of the GRs (= amoebocytes) in *Limulus polyphemus* contain all the factors, including the clottable protein. Moran, D. (1971) reported that the tubule-containing bodies (= the structured granules, Gupta, A., 1979a,d) in *Blaberus* disappear as clotting progresses, and the presence of many structured granules correlates with high coagulability. The nature of the clottable protein remains unresolved. Solum, N. (1970) has characterized the clottable protein in *Limulus*, and according to Crossley, A. (1975), this protein in several crustaceans is a

PAS-positive muco- or glycoprotein. Brehélin, M. (1977) also reported that, in *Locusta*, the coagulum is formed from a glycoprotein. PAS-positive contents have been observed in several insects (Wigglesworth, V., 1956b; Ashhurst, D. and Richards, A., 1964; Scharrer, B., 1972). The most interesting finding is that of Durliat, M. and Vranckx, R. (1976) who reported three soluble proteins in the hemocyte of the crusteacean, *Astacus leptodactylus*, one of which they referred to as fibrinogen. This is the first report of the presence of this protein, which is considered to be the main component of the clot in vertebrates, in an arthropod, and probably it is only a matter of time before it is reported also in the GRs/COs of insects.

There seems to be general agreement that the factor causing coagulation originates in the hemocytes (GRs, COs). It should be stated at the outset that substances such as prothrombin, thrombin, or thromboplastin have not yet been found in arthropods, although the possibility of their occurrence has been suggested (Cuénot, L., 1881; Heilbrunn, L., 1961). Gilliam, M. and Shimanuki, H. (1970) have even claimed that the clotting system of honeybee and man are similar. However, according to Crossley, A. (1975), the coagulation mechanisms of crustaceans and vertebrates involve different chemistry. Although the role of crustacean GR (or its differentiated form, the CO) in coagulation has been either directly reported, or has been interpreted to be so, by several authors (see Gupta, A., 1979a for listing), the actual coagulation factor has not been identified. It is of interest to note that Stang-Voss, C. (1971) recognized a separate category of clotting cells in *Astacus* (although these cells look like PLs or early stages of GRs) and suggested that these cells store hemagglutinin. Stutman, L. and Dolliver, M. (1968) reported that the crustacean hemocyte seems to be more potent than vertebrate cells in clotting, and for a given number of cells, crustacean hemolymph coagulates two to twenty times faster than human blood. Finally, Ghidalia, W. *et al.* (1977) reported that in the decapod crustacean, *Macropipus puber*, the coagulation process depends not only on a non-specific, thermolabile cellular factor but also on several plasmatic factors.

The role of calcium in hemolymph coagulation is debatable. Apparently, in some insects blood can coagulate in the absence of Ca^{2+} (Muttkowski, R., 1924; Beard, R., 1950), although Grégoire, Ch. (1953) found that sequestering Ca^{2+} ions by potassium oxalate (0.2%) prevented coagulation in *Periplaneta* and *Gryllotalpa*, and Marchalonis, J. and Edelman, G. (1968) reported that Ca^{2+} potentiates the agglutinating activity of the hemagglutinin.

Inhibition of coagulation. Several kinds of metabolic inhibitors and anticoagulants are known to inhibit coagulation *in vivo* or *in vitro*. Sensitivity of coagulation mechanisms to metabolic inhibitors (e.g. cyanide and azide) and to antithrombins (e.g. heparin and hirudin) have been reported (Shull, W. *et al.*, 1932; Beard, R., 1950), but Crossley, A. (1975) did not observe inhibition of the hemocytic agglutination in *Calliphora* by certain concentrations of cyanide (10^{-3} M), azide (10^{-2} M), thiourea (5×10^{-3} M), cycloheximide (10^{-4} M), EDTA (ethylenediamine tetra-acetic acid) (5×10^{-3} M), or saturated aqueous cytochalasin B, when injected into the hemolymph 2 or 30 min before bleeding the insect for clotting assay. Acetic acid vapor (Shull, W. and Rice, P., 1933), X-irradiation (Grégoire, Ch. 1955), sodium hydrosulphite, 0.1 M ascorbic acid (Beard, R., 1950), and veronal buffer (Bowen, T. and Kilby, B., 1953) are known to inhibit coagulation.

7.2 Detoxification of poisons

Virtually nothing is known about the direct role of hemocytes in the detoxification of poisons (Jones, J., 1957; Arnold, J., 1974; Feir, D., 1979; Zlotkin, vol. 10). Most likely, hemocytes indirectly play a role in detoxification mechanisms via esterases. Patton, R. (1961) suggested that non-specific esterases in hemocytes might be involved in detoxifying parathion. Whitten, J. (1968) also suggested an active role of hemocytes in detoxification of contact insecticides that penetrated the tarsal cuticle in the fly, *Sarcophaga bullata*. Yeager, J. *et al.* (1942) have also suggested that hemocytes detoxify poisons. Exposure to sublethal doses of the insecticide chlordane caused increases in the total hemocyte count (THC) and differential counts (DHC) of PLs, GRs, COs, and SPs in *Periplaneta americana* (Gupta, A. and Sutherland, D., 1967). Yeager, J. and Munson, S. (1942) demonstrated some pathological effects on PLs and COs in *Prodenia eridania* larvae after feeding them turnip leaf–cornstarch "sandwiches" with sodium fluoride and mercuric chloride.

7.3 Hemocytes as carriers of growth hormones

I have commented elsewhere (see Preface in Gupta, A., 1979d) that much of the physiological significance of hemocytes remains to be understood, particularly their role in the insect endocrine system. The role of hemocytes as carriers of hormones to the target tissues or organs is largely conjectural, although hemocytes may have assumed this function with the evolution of the circulatory system, for the first time, in the phylum Nemertenea (ribbon worms) (Gupta, A., 1983). Wigglesworth, V. (1956b) was probably the first to suspect that the PLs might be the carriers of the brain hormone (prothoracicotropic hormone; Bollenbacher and Granger, vol. 7) to the prothoracic glands in *Rhodnius prolixus*. Wigglesworth based his suggestion on the presence of large numbers of PLs around the corpus cardiacum (CC) during the early stages of molting and their staining reactions comparable to the neurosecretory (NS) material in the pars intercerebralis. Hodgson, E. and Geldiay, S. (1959) suggested that the hemocytes found in the brain of hyperactive cockroaches may be involved in transporting NS material. Scharrer, B. (1965b) reported the presence of hemocytes in all parts of the prothoracic glands of the nymphs of *Leucophaea maderae*, and suggested that these hemocytes may be involved in the transport of hormones and metabolites to and from the endocrine cells of the prothoracic gland. Dogra, G. (1970) proposed that the hemocytes in the mole cricket, *Gryllotalpa africana* Beauvois, transport NS material to the target organs from the CC. He noted an increase in the number of hemocytes containing stainable NS material during the maturation of the ovaries, and fewer such hemocytes prior to oogenesis and during the "postovulatory" period. As far as I know, only Takeda, N. (1977) has clearly demonstrated the role of hemocytes in the transport of NS material from the brain to the prothoracic glands in the moth, *Monema flavescens*. Wigglesworth, V. (1979a) suggested that the PLs in *Rhodnius* might be the major site for conversion of ecdysone to 20-hydroxyecdysone (Smith and Bollenbacher, vol. 7) and might act as carriers of the latter hormone to the epidermis during molting. No clear proof exists, however, as to whether or how the 20-hydroxyecdysone is transported from its source(s) to the epidermal cells. Following molting, within hours, the cuticle hardens and develops the characteristic color pattern of the insect. Sclerotization occurs as a result of the interaction of quinones with the soft proteins of the cuticle. It is most likely that the OEs transport tyrosine (one of the precursors of quinones) and the phenoloxidases, which mediate the interaction of quinones with the cuticle protein, and probably also bursicon (the hormone that governs the sclerotization process; Reynolds, vol. 3) to the new cuticle. Experimentally, however, these transports remain to be confirmed.

7.4 Injury or wound repair

Some authors still question the direct role of hemocytes in injury repair, although it has been confirmed in several insects. On the basis of his work on *Rhodnius*, Wigglesworth, V. (1937, 1972, 1979a) believes that hemocytes are not directly involved in wound repair. The repair, according to him, is done by the epidermal cells; these cells around the wound divide mitotically, migrate, and send out cytoplasmic processes to seal off the wound. However, GRs are known to produce a "conditioning factor" that stimulates the migration of epidermal cells (Bohn, H., 1977). The PLs in this case merely provide a protective sheet below the migrating epidermal cells. However, Lai-Fook, J. (1968, 1970) in an ultrastructural study of *Rhodnius*, showed that PLs and GRs were directly involved in wound repair. Gupta, A. (1970) observed SPs at the wound site in the blister beetle, *Epicauta cinerea*, but these cells were perhaps engaged in phagocytosis of the disintegrating wound tissue. It is known that PLs form membrane tissue over wounds *in vivo* and *in vitro* (Clark, R. and Harvey, W., 1964; Walters, D., 1970). This epitheloid character of hemocytes in injury has also been noted by Dawe, C. *et al.* (1967). The direct role of PLs and GRs in wound repair has also been reported by Ratcliffe, N. *et al.* (1976b) and Rowley, A. and Ratcliffe, N. (1978) in *Galleria mellonella*. These authors also noticed the flattening of PLs as occurs in encapsulation and nodule formation. The attraction of hemocytes (PLs) to the wound site has been attributed to an injury factor, produced by damaged epidermal cells or hemocytes (GRs/PLs) (Harvey, W. and Williams, C., 1961; Bohn, H., 1974), and called "haemokinin" by Cherbas, L. (1973). The evidence of the direct participation of hemocytes also comes from their role in (1) causing localized coagulation to seal off the wound, (2) synthesis of injury protein; and (3) increase in their mitotic activity as a result of the wound. Ratcliffe, N. and Rowley, A. (1979) reported that in *G. mellonella*, GRs caused hemolymph coagulation within 10 min of injury. Day, M. (1952) and Grégoire, Ch. (1970) also recognized the role of coagulation in sealing off the wound. Taylor, R. (1969) suggested that the phenol-oxidizing system in the hemocyte plays a major role in sealing wounds. Crossley, A. (1975) reported that in *Calliphora*, hemocytes containing phenol-oxidizing enzymes accumulated at wound sites. Telfer, W. and Williams, C. (1955, 1960) reported an increase in the incorporation of [^{14}C]glycine into the hemolymph plasma proteins in diapausing *Hyalophora* pupae due to prior injury in the integument. Berry, S. *et al.* (1964) also suggested that injury stimulated protein synthesis. Coles, G. (1965) reported an increase in non fat-body protein in *Rhodnius* following injury. Marek, M. (1968) identified an injury protein, termed X_2 (an esterase apoenzyme), and a "cooling protein" in *Galleria* (see also the discussion on lysozyme in section 7.1.4).

Many authors have reported an increased mitotic activity in circulating hemocytes following injury (see section 3.2.1 on mitosis). Although many of these authors reported a general increase in the total hemocyte counts (THC), the differential counts of various hemocyte types in various insects following injury is highly variable, making it difficult to generalize which types, if any, specifically increase in number. It is reasonable to assume, however, that in most insects it is the GRs and PLs that increase.

7.5 Formation of connective tissue and basement membrane

All internal organs are invested with a connective tissue sheath, including the basement membrane of the integument (although it is usually a thicker layer containing collagen fibrils; Ashhurst, personal communication) and the neural lamella of the central nervous system (see Ashhurst, this volume). According to Ashhurst, D. (1979), the term "connective tissue" includes all material present in intercellular spaces. The constituents of the connective tissue matrices are collagenous proteins, glycoproteins, and acid mucopolysaccharides (= glycosaminoglycans) (Ashhurst, D., 1979). Locke, M. and Huie, P. (1972) distinguished four connective tissue components in *Calpodes*: matrix, collagen, fibers less than 6 nm diameter, and 40 nm fibers. Contrary to the opinions of other authors, Ashhurst, D. (1979) believes that connective tissue in insects is formed by either the glial cells of the perineurium or the fibroblasts. Cells comparable to fibroblasts, which secrete connective tissue matrices in vertebrates, have been reported in insects (Ashhurst, D. and Costin, N., 1974; François, J., 1974; Hoffmann, J. *et al.*, 1979), and like the PLs, have well-developed rough endoplasmic reticulum (François, J., 1975a).

For details of the formation of the connective tissue and basement membrane the reader is referred to Ashhurst's chapter in this volume. In this section I merely wish to point out the difference of opinion between some of those who have suggested that hemocytes (generally PLs, but also ADs) are involved in the formation of the connective tissue (Lazarenko, T., 1925; Wermel, E., 1938; Wigglesworth, V., 1956a,b, 1972; Pipa, R. and Cook, E., 1958; Whitten, J., 1964, 1968; Beaulaton, J., 1968; Scharrer, B., 1972) and Ashhurst, D. (1979) and others who believe that they are not. Ashhurst argues that because the insect connective tissues are similar to those of vertebrates, the hemocytes involved in the formation of this tissue must show: (1) well-developed, dilated, rough endoplasmic reticulum (RER); (2) active Golgi complexes; and (3) absence of secretory granules, criteria she considers typical of cells that secrete collagen and glycosaminoglycans in vertebrates. According to her, these criteria have not been indisputably demonstrated in hemocytes by the proponents of the role of hemocytes in the formation of these tissues. She also points out that PAS-positive hemocytes, invoked by other authors to suggest the role of hemocytes in the formation of these tissues, are not only not specific, but also do not prove that such PAS-positive hemocytes are contributing to the formation of connective tissue. Ashhurst is correct in pressing for more convincing experimental, histochemical, and ultrastructural evidence to demonstrate the role of hemocytes in connective tissue secretion. However, it must be stated that although more meticulously performed histochemical and experimental data may eventually establish the role of hemocytes in these processes, the hemocytes secreting connective tissue may not necessarily possess the characteristic ultrastructural features common in the connective-tissue-secreting cells in vertebrates, for it is known that the steroid-secreting cells in insects (e.g. prothoracic gland) do not possess the same ultrastructural characteristics as those in vertebrates (Doane, W., 1973).

Because basement membrane is a connective tissue, it should not be separated from the latter (Ashhurst, personal communication; this volume). A true basement membrane (a thin amorphous layer 50–80 nm thick) is not generally found in all insects; it is present in some (e.g. around muscle fibers in belostomatids). In vertebrates the basement membrane is secreted by cells next to which it is found (Hay, E. and Dodson, J., 1973). Thus, by analogy the epidermal cells of the integument and epithelial cells of other organs that have a basement membrane might secrete these membranes in insects, not the hemocytes. However, Wigglesworth, V. (1979b) reported that PLs contribute to the formation of the basement membrane in *Rhodnius prolixus*, and Percy, J. (1978) concluded that layer 2 of the basement membrane of the sex pheromone gland of *Trichoplusia ni* is formed by the GRs.

7.6 Synthesis and storage of nutrients

7.6.1 Proteins, glyco- and lipoproteins, and amino acids

Most studies of protein synthesis in insects relate to hemolymph proteins, and provide little or no information about the direct role of hemocytes, if any, in the synthesis of proteins. Chippendale, G. (1970) concluded that in *Diatraea*, hemocytes are of minor importance in the synthesis of bulk hemolymph proteins. Wigglesworth, V. (1979a), on the other hand, considers hemocytes to be active sites for the synthesis of proteins. This is supported by several studies. Bühlmann, G. (1974) reported that in the cockroach, *Nauphoeta cinerea*, hemocytes take up labeled amino acids and synthesize vitellogenins and other proteins. Geiger, *et al.* (1977) (cited by Crossley, A., 1979) also showed that, in *Periplaneta*, hemocytes take up injected [^{14}C]leucine (μl) from the hemolymph and synthesize cuticle proteins. Goltzené, F. and Hoffmann, J. (1974) showed that X-irradiation of the hemopoietic tissue (and hence destruction of hemocytes) affected both the growth of oocytes and the synthesis of hemolymph proteins.

Certainly the role of hemocytes in the production of injury proteins that include several esterases, lysozymes, and possibly globulins (Crossley, A., 1979) is documented. Experiments by Berry, S. *et al.* (1964) on *Hyalophora cecropia* pupae indicated that the hemocytes of injured pupae synthesize and release injury proteins into the hemolymph. Coles, G. (1965) suggested that the non-fat-body injury protein in *Rhodnius* was synthesized by hemocytes. Marek, M. (1969a,b) assumed that the "cooling protein" in *Galleria mellonella* pupae (without their cocoons) that were cooled for 3–7 days at 4°, was synthesized by hemocytes. This cooling protein, termed "immunoglobulin" by Marek, M. (1969b), is an esterase apoenzyme and its electrophoretic mobility and molecular weight are within the range of variation of vertebrate γ-globulin. It is reported that the complement component 3 proactivator found in the hemolymph of *Limulus* by Day, N. *et al.* (1970) and hemagglutinins in *Blaberus craniifer* (Anderson, R. *et al.*, 1972) and other insects are proteins synthesized by hemocytes. Similarly, the lysozymes in *Galleria* and *Calliphora erythrocephala* (Crossley, A., 1979) are produced by hemocytes.

It appears that hemocytes produce a certain macromolecular protein factor (MF) that either triggers spermiogenesis or is essential for it, for Landureau, J. and Szöllösi, A. (1974) have shown that *Blattella germanica* hemocytes in culture liberate MF that initiates spermiogenesis in isolated cysts of saturniid moths in diapause.

The role of hemocytes in the synthesis and secretion of glyco- and lipoproteins, which form matrices of hemolymph coagulum nodules and connective tissue, has been reported by several authors. Brehélin, M. (1972) found that in the fifth-instar *Locusta* nymphs the 3-fold increase in the rate of coagulation was correlated with the increase in the number of GRs and COs, and he later (Brehélin, M., 1977) reported that the coagulum in this insect is formed from a glycoprotein (MW 500,000–1,000,000) and that the GRs produce a thermolabile coagulation factor. Histochemical tests have revealed the presence of glycoprotein in certain hemocytes: SPs in *Bombyx mori* (Akai, H. and Sato, S., 1973); GRs and COs in *Thermobia domestica* (François, J., 1974, 1975a); PLs, COs, SPs, and OEs in *Locusta migratoria* (Costin, N., 1975); and GRs in *Antheraea pernyi* (Beaulaton, J. and Monpeyssin, M., 1976).

Wigglesworth, V. (1979a) has suggested that possibly the OEs are involved in the transfer of lipoproteins from the oenocytes (non-hemocytic ectodermal cells) of the abdominal wall to the integument throughout the body, because the OEs are at the peak of their secretory activity when the oenocytes are liberating their contents at the time of epicuticle formation. OEs may also be involved in phenol metabolism preliminary to sclerotization of the cuticle.

Characteristically, insect hemolymph is rich in amino acids, which play an important role as osmotic effectors. However, few studies are available on the role of hemocytes in the synthesis and/or sequestering of these amino acids. Evans, P. and Crossley, A. (1974) reported that in *Calliphora*, over 60% of the amino acids glutamate and aspartate are present in the hemocytes. When they injected glutamate its concentration within the hemocytes rapidly increased, indicating that they sequestered the amino acid from the hemolymph. Irving, S. *et al.* (1976) similarly reported the presence of glutamate in the hemocytes of *Lucilia.* The ability of hemocytes to sequester glutamate from the hemolymph assumes great significance when one considers the role of glutamate as a transmitter at neuromuscular junctions (Kerkut, G. *et al.*, 1965; Usherwood, P. and Cull-Candy, S., 1975). Because injection of L-glutamate into blowfly larvae causes reversible paralysis (Evans, P. and Crossley, A., 1974; Irving, S. *et al.* 1976), hemocytes may play a role in regulating the amount of glutamate, either by sequestering it from, or releasing it into, the hemolymph.

The presence of tyrosine in OEs has been reported by many authors (Dennell, R., 1947; Crossley, A., 1975; Gupta, A., 1979c; Wigglesworth, V., 1979a) and in SPs by Srdić, Ž. and Gloor, W. (1979).

7.6.2 Carbohydrates

The most characteristic carbohydrate in insects is trehalose, but there is no information on the presence of this disaccharide in hemocytes. The same is true of glucose. According to Friedman, S. (1978), the major source of synthesis of trehalose in insects is the fat body (Friedman, vol. 10). Glycogen, on the other hand, has been shown to be present in hemocytes. Babers, F. (1941) concluded that in *Prodenia*, most of the glycogen is present in the hemocytes. Ashhurst, D. and Richards, A. (1964) also reported the presence of glycogen in *Galleria* hemocytes. Although Costin did not find glycogen in the hemocytes of the fifth-instar nymph of *Locusta*, Brehélin, M. *et al.* (1975) demonstrated its presence in the GRs by electron microscopy. According to Crossley, A. (1975), the β form of glycogen has a distinctive ultrastructure and thus is easier to detect under the electron microscope. Ultrastructurally, glycogen has been reported in the hemocytes on *Rhodnius* (Lai-Fook, J., 1968); *Calliphora* (Crossley, A., 1968); *Blaberus* (Moran, D., 1971); and *Antheraea* (Beaulaton, J. and Monpeyssin, M., 1976).

Several workers have demonstrated the presence of mucopolysaccharides in several hemocyte types. The SPs of several species of Dictyoptera (Gupta, A. and Sutherland, D., 1967) and *Bombyx* (Akai, H. and Sato, S., 1973), GRs of *Galleria* (Ashhurst, D. and Richards, A., 1964; Neuwirth, M., 1973), and GRs and SPs of *Thermobia* (François, J., 1974, 1975a). According to Hoffmann, J. (1967), in *Locusta* all hemocyte types, except the PRs, contain mucopolysaccharides. Later, Costin, N. (1975) confirmed the presence of these compounds in all hemocyte types in this insect. Gupta, A. and Sutherland, D. (1967) discussed the significance of the presence of mucopolysaccharides in the SPs of Dictyoptera. The two common components of acid mucopolysaccharides are β-D-glucosamine and β-D-glucuronic acid (Pearse, A., 1961). It is reasonable to suppose that the acid mucopolysaccharides undergo hydrolysis to these two components. Passonneau, J. and Williams, C. (1953) reported the presence of glucosamine in the molting fluid of the *Hyalophora cecropia* silkworm. The glucosamine could be used in glycolysis, for it is known that in *Locusta* some hexokinases can phosphorylate

D-glucosamine (Gilmour, D., 1961), or it could enter into the formation of new chitin molecules. Hackman, R. (1954) and Anderson, S. (1979) reported small amounts of D-glucosamine in enzymatic hydrolyzates of chitin (Kramer *et al.*, this volume). This discovery is interesting in view of the fact that the number of SPs, which contain acid mucopolysaccharides, increases gradually shortly after molting. According to Gilmour, D. (1961), glucuronic acid has not been identified in insects, although mucoid substances have been reported (Day, M., 1949).

7.6.3 Lipids

It is generally recognized that ADs contain lipid droplets. Zeller, H. (1938) reported that hemocytes released lipid droplets in the developing wing of the pupa of *Anagasta.* Munson, S. and Yeager, J. (1944) reported that in *Prodenia eridania*, as glycogen disappears, lipid droplets appear. Larval hemocytes of *Ephestia kühniella* (Arnold, J., 1952) and *Galleria mellonella* (Ashhurst, D. and Richards, A., 1964) accumulate lipid before pupation. Gupta, A. and Sutherland, D. (1966) found that cooling *Tenebrio molitor* larvae to 5° for 20–24 h showed a preponderance of ADs, other types of hemocytes being rare. I have stated elsewhere (Gupta, A., 1979b) that in *Galleria* and *Porthetria dispar*, just before pupation, the GRs accumulate lipids and appear as ADs. Raina, A. (1976) also reported a gradual transformation of GRs into ADs in the larvae of *Pectinophora gossypiella.* Gupta, A. and Sutherland, D. (1966) also reported the transformation of GRs into ADs in several insects. Under the electron microscope, lipid droplets have been demonstrated by Grimstone, A. *et al.* (1967) and Smith, D. (1968) in *Ephestia* and by Crossley, A. (1968, 1975) and Zachary, D. and Hoffmann, J. (1973) in *Calliphora erythrocephala.*

7.7 Metabolism of proteins, phenols, amino acids, carbohydrates, and lipids

7.7.1 Proteins, phenols, amino acids

The presence of several types of enzymes in hemocytes suggests their role in protein, peptide or amino acid metabolism (Chen, vol. 10). Collett, J. (1976) suggested that in *Calliphora*, certain hemocytes produce peptidases that mediate the conversion of peptides into amino acids. Crossley, A. (1979) suggested that perhaps a peptidase produces the neurotransmitter glutamate by hydrolysis of inactive peptides in the hemolymph.

Tryosinases (phenol oxidases), essential for the conversion of tyrosine to dopa (dihydroxylphenylamine) and dopamine, have been reported to be present in the hemocytes (GRs, OEs, COs, SPs) of many insects (Dennell, R., 1947; Rizki, M., 1957a,b; Vercauteren, R. and Aerts, F., 1958; Rizki, M. and Rizki, R., 1959; Decleir, W. *et al.*, 1960; Kawase, S., 1960; Crossley, A., 1964, 1975; Evans, J., 1967a,b; Mills, R. *et al.*, 1968; Hoffmann, J., *et al.*, 1970; Preston, J. and Taylor, R., 1970; Neuwirth, M., 1973; Maier, W., 1972; Hughes, L. and Price, G., 1975a,b, 1976; Andreadis, T. and Hall, D., 1976; Mazzone, H., 1976; Schmit, A. *et al.*, 1977). The conversion of tyrosine to quinone involves first the activation of an inert proenzyme (prophenoloxidase) by a protein (peptide) activator, which in *Antheraea pernyi* comes from the hemocytes (Evans, J., 1967b). According to Rizki, M. and Rizki, R. (1959), both the substrate and the enzyme are present in the OEs (crystal cells); the enzyme, according to Crossley, A. (1979), may be an inactive prophenoloxidase. Other hemocytic enzymes involved in phenol metabolism have been reported from *Calliphora* (mono- and *o*-diphenoloxidase) (Crossley, A., 1964, 1975), *Periplaneta* (monoamine oxidase that is involved in the conversion of tyramine to protocatechuic acid) (Whitehead, D., 1970), and *Glossina* (decarboxylase that is active on tyrosine and possibly dopa) (Whitehead, D., 1973).

Hemocytes, under the influence of bursicon, are known to take up labeled tyrosine from the hemolymph (Mills, R. and Whitehead, D., 1969), which is then converted to quinone precursors (Post, L., 1972). *Sarcophaga* hemocytes are known to synthesize dopa and dopamine *in vitro* (Whitehead, D., 1970) from tyrosine by an *o*-diphenoloxidase. Direct transformation of tyrosine to tyramine by tyrosine decarboxylase was reported by Whitehead, D. (1969) in the hemocytes of *Periplaneta.* According to Mills, R. and Whitehead, D. (1969), dopamine may be further converted to

N-acetyldopamine by the hemocytes before transfer to the cuticle, and Wigglesworth, V. (1979a) believes that OEs might be involved in the metabolism of phenols for sclerotin formation. Bernier, I. *et al.* (1973) and Landureau, J. and Grellet, P. (1975b) reported the secretion of chitinase by cultured cells of *Periplaneta americana*.

Messner, B. (1976) demonstrated histochemically the presence of the oxidases of pyridine nucleotides NADH and NADPH in the hemocytes of insects; and Anderson, R. (1974) also reported peroxidase in 2% of *Blaberus craniifer* hemocytes. The presence of peroxidase in hemocytes is interesting in view of the fact that Klebanoff, S. (1967) described a peroxidase-mediated antimicrobial system in vertebrate leucocytes and so did Homan-Müller, J. *et al.* (1974) in the phagocytizing human microphages (granulocytes).

7.7.2 Carbohydrates

Although the presence of trehalose in hemocytes has not been demonstrated to date, trehalase (an α-glucosidase) is apparently present in the hemocytes as well as in the hemolymph, and brings about the hydrolysis of the hemolymph trehalose. Friedman, S. (1960) showed that in *Phormia regina*, the activity of hemolymph trehalase is increased upon dilution of the hemolymph with water. Later, Matthews, J. *et al.* (1975) demonstrated the same increase in activity in *Periplaneta* upon dilution with buffer, and interpreted the increased activity due to the lysis (and release) of hemocytes. These authors also reported that two-thirds of the trehalase activity was contributed by the hemocytes.

Phosphatase has been found in the lysosomes of several insects (see section 7.1.4 on lysozymes) and it is likely that it takes part also in the breakdown of glycogen to glucose-1-phosphate. According to Crossley, A. (1979), the role of hemocytes in sugar homeostasis is unknown.

7.7.3 Lipids

Although ADs contain lipid droplets, their role in lipid metabolism is unknown.

8 SUMMARY

There is disagreement among insect hematologists about the number of hemocyte types in various insects. From one or few to as many as nine or more types have been described, particularly by light microscopy. Ultrastructurally, however, only seven types have so far been identified in various insects: prohemocyte (PR), plasmatocyte (PL), granulocyte (GR), spherulocyte (SP), adipohemocyte (AD), oenocytoid (OE), and coagulocyte (CO). Of these seven, AD and CO have been described by only a few authors. All seven types have been described under various other names. It is suggested that of the seven types of hemocytes, the GR is the most primitive type and evolved into other hemocyte types.

Hemocytes perform many functions. PLs and GRs actively participate in phagocytosis, encapsulation, and nodule formation. These hemocytes possess some kind of surface receptors that play a role in foreign body recognition. Although insects lack vertebrate immunoglobulins (γ-globulins) that act as humoral recognition factors in removing foreign antigen, they do produce humoral recognition factors, called hemagglutinins that are secreted by certain hemocytes. In addition to hemagglutinins, lysozymes, located in the lysosomes of the hemocytes, play an important role in insect immunity. Hemolymph coagulation is brought about by hemocyte agglutination, plasma gelation or a combination of both. COs and GRs supposedly contain the factor that causes coagulation. Virtually nothing is known about the direct role of hemocytes in the detoxification of poisons. Hemocytes probably carry hormones to target organs but no work has been reported on juvenile hormone. PLs and GRs participate in injury repair by causing localized coagulation to seal off the wound and synthesizing injury protein. The role of hemocytes in the synthesis and formation of connective tissues and basement membrane continues to be controversial. Hemocytes synthesize proteins, glyco- and lipoproteins, and amino acids. Whether they synthesize trehalose and glucose is unknown. Several kinds of mucopolysaccharides are present in various hemocytes, and lipids have been reported in the ADs. Hemocytes metabolize peptides, phenols, and probably also trehalose.

ACKNOWLEDGEMENTS

I thank Drs A. C. Crossley, N. A. Ratcliffe, and H. Wago for providing me with some photomicrographs. I appreciate the assistance of Christine Wheeler and L. K. Hazarika in the preparation of the illustrations, and of Evelyn Weinmann and Joan Gross in typing and word processing of the manuscript. New Jersey Agricultural Experiment Station Publication No. F–08112–04–82 supported by State Funds and by US Hatch Act Funds.

REFERENCES

Akai, H. (1969). Ultrastructure of haemocytes observed on the fat-body cells in *Philosamia* during metamorphosis. *Jap. J. Appl. Zool. 13*, 17–21.

Akai, H. and Sato, S. (1971). An ultrastructural study of the haemopoietic organs of the silkworm, *Bombyx mori. J. Insect Physiol. 17*, 1665–1676.

Akai, H. and Sato, S. (1973). Ultrastructure of the larval hemocytes of the silkworm *Bombyx mori* L. (Lepidoptera: Bombycidae). *Int. J. Insect Morph. Embryol. 2*, 107–231.

Åkesson, B. (1953). Observations on the hemocytes during the metamorphosis of *Calliphora erythrocephala. Ark. Zool. 6*, 203–211.

Amirante, G. A. (1976). Production of heteroagglutinins in haemocytes of *Leucophaea maderae L. Experientia 32(4)*, 526–528.

Amirante, G. A. and Mazzalai, F. G. (1978). Synthesis and localization of hemagglutinins in haemocytes of the cockroach *Leucophaea maderae L. Devel. Comp. Immunol. 2(4)*, 735–740.

Anderson, R. S. (1974). Metabolism of insect hemocytes during phagocytosis. In *Contemporary Topics in Immunobiology*. Edited by E. L. Cooper. Vol. 4, pages 47–54. Plenum Press, New York.

Anderson, R. S. (1976a). Expression of receptors by insect macrophages. In *Phylogeny of Thymus and Bone Marrow-Bursa Cells*. Edited by R. K. Wright and E. L. Cooper. Pages 27–34. Elsevier-North Holland, Amsterdam.

Anderson, R. S. (1976b). Macrophage function in insects. In *Proc. 1st Int. Colloq. Invertebr. Pathol.* Edited by T. A. Angus, P. Faulkner and A. Rosenfield. Pages 215–219. Queens University, Kingston, Ontario.

Anderson, R. S., Day, N. K. B. and Good, R. A. (1972). Specific hemagglutinin and a modulator of complement in cockroach hemolymph. *Infect. Immun. 5*, 55–59.

Anderson, R. S., Holmes, B. and Good, R. A. (1973). *In vitro* bactericidal capacity of *Blaberus cranifer* hemocytes. *J. Invert. Pathol. 22(1)*, 127–135.

Anderson, S. O. (1979). Biochemistry of insect cuticle. *Ann. Rev. Ent. 24*, 29–61.

Andreadis, T. G. and Hall, D. W. (1976). *Neoaplectana carpocapsae*: Encapsulation in *Aedes aegypti* and changes in host haemocytes and hemolymph proteins. *Exp. Parasitol. 39*, 252–261.

Arnold, J. W. (1952). The haemocytes of the Mediterranean flour moth, *Ephestia kühniella Zell.* (Lepidoptera: Pyralidae). *Canad. J. Zool. 30*, 352–364.

Arnold, J. W. (1972a). Haemocytology in insect biosystematics: The Prospect. *Canad. Ent. 104*, 655–659.

Arnold, J. W. (1972b). A comparative study of the haemocytes (blood cells) of cockroaches (Insecta: Dictyoptera: Blattaria), with a view of their significance in taxonomy. *Canad. Ent. 104*, 309–348.

Arnold, J. W. (1974). The hemocytes of insects. In *The Physiology of Insecta*. Edited by M. Rockstein. Vol. 5, 2nd edition, pages 201–254. Academic Press, New York.

Arnold, J. W. (1976). Biosystematics of the genus *Euxoa* (Lepidoptera: Noctuidae). VIII. The hemocytological position of *E. rockburnei* in the "declarata" group. *Canad. Ent. 108*, 1387–1390.

Arnold, J. W. (1979). Controversies about hemocyte types in insects. In *Insect Hemocytes*. Edited by A. P. Gupta. Pages 231–258. Cambridge University Press, Cambridge.

Arnold, J. W. (1982). Larval hemocytes in Noctuidae (Insecta: Lepidoptera). *Int. J. Insect Morph. Embryol.* 1982 (In press).

Arnold, J. W. and Hinks, C. F. (1975). Biosystematics of the genus *Euxoa* (Lepidoptera: Noctuidae) III. Hemocytological distinctions between two closely related species, *E. campestris* and *E. declarata. Canad. Ent. 107*, 1095–1100.

Arnold, J. W. and Hinks, C. F. (1976). Haemopoiesis in Lepidoptera I. The multiplication of circulating haemocytes. *Canad. J. Zool. 54(6)*, 1003–1012.

Arnold, J. W. and Salkeld, E. H. (1967). Morphology of the haemocytes of the giant cockroach, *Blaberus giganteus*, with histochemical tests. *Canad. Ent. 99*, 1138–1145.

Arvy, L. (1954). Données sur la leucopoïèse chez *Musca domestica L. Proc. Roy. Ent. Soc. Lond. A29*, 39–41.

Arvy, L. and Gabe, M. (1946). Identification des diastrases sanguines chez quelques insectes. *C.R. Soc. Biol. (Paris) 140*, 757–758.

Ashhurst, D. E. (1979). Hemocytes and connective tissue: a critical assessment. In *Insect Hemocytes*. Edited by A. P. Gupta. Pages 319–330. Cambridge University Press, Cambridge.

Ashhurst, D. E. and Costin, N. M. (1974). The development of a collagenous connective tissue in the locust, *Locusta migratoria. Tissue Cell 6*, 279–300.

Ashhurst, D. E. and Richards, A. G. (1964). Some histochemical observations on the blood cells of the wax-moth *Galleria mellonella L. J. Morph. 114*, 247–253.

Aston, W. P., Chadwick, J. S. and Henderson, M. J. (1976). Effect of cobra venom factor on the *in vivo* immune response in *Galleria mellonella* to *Pseudomonas aeruginosa. J. Invert. Pathol. 27(2)*, 171–176.

Babers, F. H. (1941). Glycogen in *Prodenia eridania*, with special reference to the ingestion of glucose. *J. Agric. Res. 62*, 509–530.

Baerwald, R. J. (1979). Fine structure of hemocyte membranes and intercellular junctions formed during hemocyte encapsulation. In *Insect Hemocytes*. Edited by A. P. Gupta. Pages 155–188. Cambridge University Press, Cambridge.

Baerwald, R. J. and Boush, G. M. (1970). Fine structure of hemocytes of *Periplaneta americana* (Orthoptera: Blattidae), with particular reference to marginal bundles. *J. Ultrastruct. Res. 31*, 151–161.

Bahadur, J. and Pathak, J. P. N. (1970). Changes in the total haemocyte counts of the bug *Halys dentata*, under certain specific conditions. *J. Insect Physiol. 17*, 329–334.

Bang, F. B. (1975). Phagocytosis in invertebrates. In *Invertebrate Immunity*. Edited by K. Maramorosch and R. E. Shope. Pages 137–151. Academic Press, New York.

Barra, J. A. (1969). Tégument des Collemboles: Présence d'hémocytes à granules dans le liquide exuvial au cours de la mue (Insectes, Collemboles). *C.R. Acad. Sci. Paris 269D*, 902–903.

Bauchau, A. G. (1981). Crustaceans. In *Invertebrate Blood Cells*. Edited by N. A. Ratcliffe and A. F. Rowley. Vol. 2, pages 385–420. Academic Press, New York.

Beard, R. L. (1942). On the formation of the tracheal funnel in *Anasa tristis De G.* induced by the parasite *Trichopoda pennipes. Fabr. Ann. Ent. Soc. Amer. 35*, 68–72.

Beard, R. L. (1950). Experimental observations on coagulation of insect hemolymph. *Physiol. Zool. 23*, 47–57.

Beaulaton, J. (1968). La tunica propria et ses relations avec les fibres conjonctives et les hémocytes. *J. Ultrastruct. Res. 23*, 474–498.

Beaulaton, J. and Monpeyssin, M. (1976). Ultrastructure and cytochemistry of hemocytes of *Antheraea pernyi Guer* (Lepidoptera, Attacidae) during the fifth larval stage. I. Prohaemocytes, plasmatocytes and granulocytes. *J. Ultrastruct. Res. 55*, 143–156.

Beaulaton, J. and Monpeyssin, M. (1977). Ultrastructure et cytochimie des hémocytes d'*Antheraea pernyi Guer.* (Lepidoptera: Attacidae). II. Cellules à sphérules et oenocytoïdes. *Biol. Cell. 28(1)*, 13–18.

Bernheimer, A. W. (1952). Hemagglutinins in caterpillar bloods. *Science (Wash., D.C.) 115*, 150–151.

BERNIER, I., LANDUREAU, J. C., GRELLET, P. and JOLLES, P. (1973). Characterization of chitinases from hemolymph and cell cultures of the cockroach *Periplaneta americana*. *Comp. Biochem. Physiol. 47B*, 41–44.

BERRY, S. J., KRISHNAKUMARAN, A. and SCHNEIDERMAN, H. A. (1964). Control of synthesis of RNA and protein in diapausing and injured *Cecropia* pupae. *Science (Wash., D.C.) 146*, 928–940.

BETTINI, S., SARKARIA, D. S. and PATTON, R. L. (1951). Observations on the fate of vertebrate erythrocytes and hemoglobin injected into the blood of the American cockroach (*Periplaneta americana*). *Science (Wash., D.C.) 113*, 9–10.

BOGOJAVLENSKY, K. S. (1932). The formed elements of the blood of insects. *Archiv. Russ. Anat. Hist. Embryol. 11*, 361–386. (In Russian).

BOHN, H. (1974). Growth promoting effect of haemocytes on insect epidermis *in vitro*. *J. Insect Physiol. 21*, 1283–1293.

BOHN, H. (1977). Enzymatic and immunological characterization of the conditioning factor for epidermal outgrowth in the cockroach *Leucophaea maderae*. *J. Insect Physiol. 23*, 1063–1073.

BOWEN, T. J. and KILBY, B. A. (1953). Electrophoresis of locust haemolymph. *Arch. Int. Physiol. 61*, 413–416.

BOWERS, B. (1964). Coated vesicles in the pericardial cells of the aphid (*Myzus persicae* Sulz.). *Protoplasma 59*, 351–367.

BOWERS, B. and WILLIAMS, C. M. (1964). Physiology of insect diapause, XIII. DNA synthesis during the metamorphosis of the *Cecropia* silkworm. *Biol. Bull. (Woods Hole) 126*, 205–219.

BREHÉLIN, M. M. (1972). Étude du mechanisme de la coagulation de l'hémolymph d'un acridien, *Locusta migratoria migratorioides. (R and F.)*. *Acrida 1*, 167–175.

BREHÉLIN, M. (1977). Sur la coagulation de l'hémolymph chez *Locusta migratoria*. *Ann. Parasitol. Hum. Comp. 52*, 98–99.

BREHÉLIN, M., ZACHARY, D. and HOFFMANN, J. A. (1976). Fonctions des granulocytes typiques dans la cicatrisation chez l'orthoptère *Locusta migratoria L. J. Mic. Biol. Cell. 25(2)*, 133–136.

BREHÉLIN, M., HOFFMANN, J. A., MATZ, G. and PORTE, A. (1975). Encapsulation of implanted foreign bodies by hemocytes in *Locusta migratoria* and *Melolontha melolontha*. *Cell Tissue Res. 160*, 283–289.

BREWER, F. D. and VINSON, S. B. (1971). Chemicals affecting the encapsulation of foreign material in an insect. *J. Invert. Pathol. 18*, 287–289.

BRIGGS, J. D. (1958). Humoral immunity in lepidopterous larvae. *J. Exp. Zool. 138*, 155–188.

BRUNTZ, L. (1908). Nouvelles recherches sur l'excrétion et la phagocytose chez les Thysanoures. *Archiv. Zool. Exp. Gén. 38*, 471–488.

BÜHLMANN, G. (1974). Vitellogenin in adulten Weibchen der Schabe *Nauphoeta cinerea*: Immunologische Untersuchungen über Herkunft und Einbau. *Rev. Suisse Zool. 81*, 642–647.

BURNS, W. C. (1961). Penetration and development of *Allassogonoporus vespertilionis* and *Acanthatrium oregonense* (Trematoda: Lecithodendriidae) cercariae in caddisfly larvae. *J. Parasitol. 47*, 927–938.

CAMERON, G. R. (1934). Inflammation in the caterpillars of Lepidoptera. *J. Pathol. Bacteriol. 38*, 441–446.

CHADWICK, J. S. (1975). Hemolymph changes with infection or induced immunity in insects and ticks. In *Invertebrate Immunity*. Edited by K. Maramorosch and R. E. Shope. Pages 241–271. Academic Press, New York.

CHERBAS, L. (1973). The induction of an injury reaction in cultured haemocytes from saturniid pupae. *J. Insect Physiol. 19*, 2011–2033.

CHIPPENDALE, G. M. (1970). Metamorphic changes in haemolymph and midgut proteins of the southwestern corn borer *Diatraea grandiosella*. *J. Insect Physiol. 16*, 1909–1920.

CLARK, R. M. and HARVEY, W. R. (1964). Cellular membrane formation by plasmatocytes of diapausing cecropia pupae. *J. Insect Physiol. 11*, 161–175.

COHEN, E., ROWE, A. W. and WISSLER, F. C. (1965). Heteragglutinins of the horseshoe crab *Limulus polyphemus*. *Life Sci. 4*, 2009–2016.

COHEN, E. (1974). Biomedical perspectives of agglutinins of invertebrates and plant origin. *Ann. N.Y. Acad. Sci. 234*, 1–412.

COLES, G. C. (1965). The haemolymph and moulting in *Rhodnius prolixus* *Stäl. J. Insect Physiol. 11*, 1317–1323.

COLLETT, J. I. (1976). Small peptides, a life-long store of amino acid in adult *Drosophila* and *Calliphora*. *J. Insect Physiol. 22*, 1433–1440.

COOPER, E. L. (1969a). Chronic allograft rejection in *Lumbricus terrestris*. *J. Exp. Zool. 171*, 69–74.

COOPER, E. L. (1969b). Allograft rejection in *Eisenia foetida*. *Transplantation 8*, 220–223.

COOPER, E. L. (editor) (1974). *Contemporary Topics in Immune Biology*, vol. 4. *Invertebrate Immunology*. Plenum Press, New York.

COSTIN, N. M. (1975). Histochemical observation of the haemocytes of *Locusta migratoria*. *Histochem. J. 7*, 21–43.

COUTURIER, A. (1963). Recherches sur des Mermithidae, Nematodes parasites du Hanneton commun (*Melolontha melolontha L.* Coleop. Scarab.) *Ann. Epiphyt. 14*, 203–267.

CROSSLEY, A. C. S. (1964). An experimental analysis of the origins and physiology of haemocytes in the blue bottle blowfly *Calliphora erythrocephala (Meig.)*. *J. Exp. Zool. 157*, 375–397.

CROSSLEY, A. C. (1968). The fine-structure and mechanism of breakdown of larval intersegmental muscles in the blowfly *Calliphora erythrocephala*. *J. Insect Physiol. 14*, 1389–1407.

CROSSLEY, A. C. (1972). The ultrastructure and function of pericardial cells and other nephrocytes in an insect: *Calliphora erythrocephala*. *Tissue Cell 4*, 529–560.

CROSSLEY, A. C. S. (1975). The cytophysiology of insect blood. *Adv. Insect Physiol. 11*, 117–222.

CROSSLEY, A. C. (1979). Biochemical and ultrastructural aspects of synthesis, storage, and secretion in hemocytes. In *Insect Hemocytes*. Edited by A. P. Gupta. Pages 423–473. Cambridge University Press, Cambridge.

CRUZ LANDIM, C. da (1971). Hemocitos de rainha de *Apis mellifera adersonii* (Hymenoptera-Apidae). Estudo ao microsciopio optico e eletronico. *Ann. Congr. Latinoamer. Ent. 14(2)*, 238–245.

CUÉNOT, L. (1881). Études sur le sang et les glandes lymphatiques dans le série animale. *Arch. Zool. Exp. Gén. 9*, 365–475, 592–670.

CUÉNOT, L. (1895). Études physiologiques sur les crustacés decapodes. *Arch. Biol. Liège 13*, 245–303.

CUÉNOT, L. (1896). Études physiologiques sur les Orthoptères. *Arch. Biol. 14*, 293–341.

CUÉNOT, L. (1897). Les globules sanguins et les organes lymphoïdes des Invertebrès. (Revue critique et nouvelles recherches). *Arch. Anat. Mic. 1*, 153–192.

CUSHING, J. E., CALAPRICE, N. L. and TRUMP, G. (1963). Blood group reactive substances in some marine invertebrates. *Biol. Bull. (Woods Hole) 125*, 69–80.

DAVIS, R. P. and SCHNEIDERMAN, H. A. (1960). An autoradiographic study of wound healing in diapausing silkworm pupae. *Anat. Rec. 137*, 348.

DAWE, C. J., MORGAN, W. D. and SLATICK, M. S. (1967). Cellular response of a cockroach *Leucophaea maderae* to transplants of cell culture lines of vertebrates. *Fed. Proc. 26*, 1698–1706.

DAY, M. F. (1949). The occurrence of mucoid substance in insects. *Aust. J. Sci. Res. (B) 2*, 421–427.

DAY, M. F. (1952). Wound healing in the gut of the cockroach *Periplaneta americana*. *Aust. J. Sci. Res. (B) 5*, 282–289.

DAY, M. F. and BENNETTS, M. J. (1953). Healing of gut wounds in the mosquito *Aedes aegypti* (L.) and the leafhopper *Orosius argentatus (Ev.)*. *Aust. J. Biol. Sci. 6*, 580–585.

DAY, N. K. B., GEWURZ, H., JOHANNSEN, R., FINSTAD, J. and GOOD, R. A. (1970). Complement and complement-like activity in lower vertebrates and invertebrates. *J. Exp. Med. 132*, 941–950.

DECLEIR, W., AERTS, F. and VERCAUTEREN, R. E. (1960). The localization of polyphenoloxidase in hemocytes. *11th Int. Congr. Ent. 3*, 176–179.

DENNELL, R. (1947). A study of the insect cuticle: The formation of the puparium of *Sarcophaga falculata Pand.* (Diptera). *Proc. Roy. Soc. B. 134*, 79–110.

DEVAUCHELLE, G. (1971). Étude ultrastructure des hémocytes du Coléoptère *Melolontha melolontha (L.)*. *J. Ultrastruct. Res. 34*, 492–516.

DOANE, W. W. (1973). Role of hormones in insect development. In *Developmental Systems: Insects*. Edited by S. J. Counce and C. H. Waddington. Vol. 2, pages 291–497. Academic Press, New York.

DOGRA, G. S. (1970). Functional significance of the neurosecretory material in the haemocytes of the adult female *Gryllotalpa africana Beauvois* (Orthoptera: Gryllatalpidae). *Anat. Anz. 126*, 355–362.

DOHRN, A. (1876). Notizen zur Kenntnis der Insektenentwicklung. *Z. Wiss. Zool. 26*, 112–138.

DORLAND, A. W. N. (1974). *The American Illustrated Medical Dictionary*. Saunders, Philadelphia.

DUMONT, N. J., ANDERSON, E. and WINNER, G. (1966). Some cytologic characteristics of the hemocytes of *Limulus* during clotting. *J. Morph. 119*, 181–208.

DURLIAT, M. and VRANCKX, R. (1976). Analyses of the hemocyte proteins from *Astacus leptodactylus. C.R. Acad. Sci. Paris 282(24)D*, 2215–2218.

EVANS, J. J. T. (1967a). The activation of prophenoloxidase during melanization of the pupal blood of the Chinese oak silkmoth *Antheraea pernyi. J. Insect Physiol. 13*, 1699–1711.

EVANS, J. J. T. (1967b). Natural substrates of the phenoloxidase in the pupal blood of the silkmoths, *Antheraea pernyi* and *A. eucalypti. J. Insect Physiol. 14*, 277–291.

EVANS, P. D. and CROSSLEY, A. C. (1974). Free amino acids in the haemocytes and plasma of the larva of *Calliphora vicina. J. Exp. Biol. 61*, 463–472.

FEIR, D. (1979). Cellular and humoral responses to toxic substances. In *Insect Hemocytes*. Edited by A. P. Gupta. Pages 415–421. Cambridge University Press, Cambridge.

FEIR, D. and MCCLAIN, E. (1968). Induced changes in the mitotic activity of hemocytes of the large milkweed bug, *Oncopeltus fasciatus. Ann. Ent. Soc. Amer. 61*, 416–421.

FEIR, D. and WALZ, M. A. (1964). An agglutinating factor in insect hemolymph. *Ann. Ent. Soc. Amer. 57*, 388.

FERRON, P. (1978). Biological control of insect pests by entomogenous fungi. *Ann. Rev. Ent. 23*, 409–442.

FRANÇOIS, J. (1974). Etude ultrastructurale des hémocytes du Thysanoure *Thermobia domestica* (Insecte, Aptérygote). *Pedobiologia 14*, 157–162.

FRANÇOIS, J. (1975a). Hemocytes and the hematopoietic organ of *Thermobia domestica* (Thysanura: Lepismatidae). *Int. J. Insect Morph. Embryol. 4*, 477–494.

FRANÇOIS, J. (1974). L'encapsulation hemocytaire experimentale chez le lépisme *Thermobia domestica. J. Insect Physiol. 21*, 1535–1546.

FRANKE, H. (1960a). Licht- und elektronenmikroskopische Untersuchungen über die Blutgerinnung bei *Periplaneta orientalis. Zool. Jahrb. Abt. Allg. Zool. Physiol. Tiere 68*, 499–518.

FRANKE, H. (1960b). Licht- und elektronenmikroskopische Untersuchungen über die Blutgerinnung bei *Periplaneta orientalis. Zool. Jahrb. Abt. Allg. Zool. Physiol. Tiere 69*, 131–132.

FRIEDMAN, S. (1960). The purification and properties of trehalase isolated from *Phormia regina Meig. Archiv. Biochem. Biophys. 87*, 252–258.

FRIEDMAN, S. (1978). Trehalose regulation, one aspect of metabolic homeostasis. *Ann. Rev. Ent. 23*, 389–407.

GAGEN, S. J. and RATCLIFFE, N. A. (1976). Studies on the *in vivo* cellular reactions and fate of injected bacteria in *Galleria mellonella* and *Pieris brassicae* larvae. *J. Invert. Pathol. 28(1)*, 17–22.

GHIDALIA, W., VENDRELY, R., COIRAULT, Y., MONTMORY, C. and PROUWARTELLE, O. (1977). Hemolymph coagulation in *Macropipus puber (L.)* Crustace Decapoda. Respective roles of hemolymph and plasma. *C.R. Acad. Sci. Paris 284(1)D*, 69–72.

GILLIAM, M. and JETER, W. S. (1970). Synthesis of agglutinating substances in adult honeybees against *Bacillus* larvae. *J. Invert. Pathol. 16*, 69–70.

GILLIAM, M. and SHIMANUKI, H. (1970). Coagulation of hemolymph of the larval honeybee (*Apis mellifera L.*) *Experientia 26*, 908–910.

GILMOUR, D. (1961). *Biochemistry of Insects*. Academic Press, New York.

GOFFINET, G. and GRÉGOIRE, CH. (1975). Coagulocyte alterations in clotting hemolymph of *Carausius morosus L. Arch. Int Physiol. Biochem. 83(4)*, 707–722.

GOLTZENÉ, F. and HOFFMANN, J. A. (1974). Control of hemolymph protein synthesis and oocyte maturation by the corpora allata in female adults of *Locusta migratoria* (Orthoptera): Role of the blood-forming tissue. *Gen. Comp. Endocr. 22*, 489–498.

GRABER, V. (1871). Ueber die Blutkorperchen der Insekten. *Sitzb. Akad. Math-Nat. Wiss. Wien 64*, 9–44.

GRÉGOIRE, CH. (1953). Blood coagulation in Arthropods. III. Reactions of insect haemolymph to coagulation inhibitors of *Biol. Bull. (Woods Hole) 104*, 372–393.

GRÉGOIRE, CH. (1955). Coagulation de l'hémolymph chez les insectes irradiés par les rayons X. *Archiv. Int. Physiol. Biochem. 68*, 246–248.

GRÉGOIRE, CH. (1970). Haemolymph coagulations in arthropods. In *The Haemostatic Mechanism in Man and Other Animals*. Edited by R. G. Macfarlane. Pages 45–74. Symp. Zool. Soc. Lond, No. 27. Academic Press, New York.

GRÉGOIRE, CH. (1971). Haemolymph coagulation in arthropods. In *Chemical Zoology*. Edited by M. Florkin. Vol. 6, pages 145–186. Academic Press, New York.

GRÉGOIRE, CH. and FLORKIN, M. (1950). Blood coagulation in arthropods. I. The coagulation of insect blood, as studied with the phase contrast microscope. *Physiol. Comp. Oecol. 2(2)*, 126–139.

GRÉGOIRE, CH. and GOFFINET, G. (1979). Controversies about the coagulocyte. In *Insect Hemocytes*. Edited by A. P. Gupta. Pages 189–229. Cambridge University Press, Cambridge.

GRIMSTONE, A. V., ROTHERAM, S. and SALT, G. (1967). An electron-microscope study of capsule formation by insect blood cells. *J. Cell Sci. 2*, 281–292.

GUPTA, A. P. (1968). Hemocytes of *Scutigerella immaculata* and the ancestry of Insecta. *Ann. Ent. Soc. Amer. 61(4)*, 1028–1029.

GUPTA, A. P. (1969). Studies of the blood of Meloidae (Coleoptera). I. The haemocytes of *Epicauta cinerea (Forster)*, and a synonympy of haemocyte terminologies. *Cytologia 34(2)*, 300–344.

GUPTA, A. P. (1970). Midgut lesions in *Epicauta cinerea* (Coleoptera: Meloidae). *Ann. Ent. Soc. Amer. 63*, 1786–1788.

GUPTA, A. P. (1979a). Arthropod hemocytes and phylogeny. In *Arthropod Phylogeny*. Edited by A. P. Gupta. Pages 669–735. Van Nostrand Reinhold, New York.

GUPTA, A. P. (1979b). Identification key for hemocyte types in hanging-drop preparations. In *Insect Hemocytes*. Edited by A. P. Gupta. Pages 527–529. Cambridge University Press, Cambridge.

GUPTA, A. P. (1979c). Hemocyte types: their structures, synonymies, interrelationships, and toxonomic significance. In *Insect Hemocytes*. Edited by A. P. Gupta. Pages 85–127. Cambridge University Press, Cambridge.

GUPTA, A. P. (editor). (1979d). Hemocyte types: their structures, synonymies, interrelationships and toxonomic significance. *Insect Hemocytes*, pages 85–127. Cambridge University Press, Cambridge.

GUPTA, A. P. (1979e). Origin and affinities of Myriapoda. In *Myriapod Biology*. Edited by M. Camatini. Pages 373–390. Academic Press, New York.

GUPTA, A. P. (1983). Neurohemal and neurohemal-endocrine organs and their evolution in arthropods. In *Neurohemal Organs of Arthropods*. Edited by A. P. Gupta. Pages 17–50. Charles C. Thomas, Publishers, Springfield, Illinois.

GUPTA, A. P. and SUTHERLAND, D. J. (1965). Observations on the spherule cells in some Blattaria (Orthoptera). *Bull. Ent. Soc. Amer. 11*, 161.

GUPTA, A. P. and SUTHERLAND, D. J. (1966). *In vitro* transformations of the insect plasmatocyte in some insects. *J. Insect Physiol. 12*, 1369–1375.

GUPTA, A. P. and SUTHERLAND, D. J. (1967). Phase contrast and histochemical studies of spherule cells in cockroaches (dictyoptera). *Ann. Ent. Soc. Amer. 60(3)*, 557–565.

GUPTA, A. P. and SUTHERLAND, D. J. (1968). Effects of sublethal doses of chlordane on the hemocytes and midgut epithelium of *Periplaneta americana. Ann. Ent. Soc. Amer. 61(4)*, 910–918.

HACKMAN, R. H. (1954). Studies on chitin. I. Enzymatic degradation of chitin and chitin esters. *Aust. J. Biol. Sci. 7*, 168–178.

HAGOPIAN, M. (1971). Unique structures in the insect granular hemocytes. *J. Ultrastruct. Res. 36* 646–658.

HARPAZ, F., KISLEV, N. and ZELCER, A. (1969). Electron-microscopic studies on hemocytes of the Egyptian cottonworm, *Spodoptera littoralis* (Boisduval) infected with a nuclear-polyhedrosis virus, as compared to noninfected hemocytes. I. Noninfected hemocytes. *J. Invert. Pathol. 14*, 175–185.

HARSHBARGER, J. C. and HEIMPEL, A. M. (1968). Effects of zymosan on phagocytosis in larvae of the greater wax moth, *Galleria mellonella, J. Invert. Pathol. 10*, 176–179.

HARVEY, W. R. and WILLIAMS, C. M. (1961). The injury metabolism of the cecropia silkworm. I. Biological amplification of the effects of localized injury. *J. Insect Physiol. 7*, 81–99.

HAY, E. D. and DODSON, J. W. (1973). Secretion of collagen by corneal epithelium. I. Morphology of the collagenous products produced by isolated epithelia grown on frozen-killed lens. *J. Cell Biol. 57*, 190–213.

HEILBRUNN, L. V. (1961). The evolution of the haemostatic mechanism. In *Functions of the Blood*. Edited by R. G. Macfarlane and A. H. T. Robb-Smith. Pages 283–301. Academic Press, New York.

HEYMONS, R. (1895). *Die Embryonalentwicklung von Dermateren und Orthopteven unter besonderer Berucksichtigung der Keimblatterbildung*. G. Fisher, Jena.

HINKS, C. F. and ARNOLD, J. W. (1977). Haemopoiesis in Lepidoptera. II. The role of the haemopoietic organs. *Canad. J. Zool. 55*, 1740–1755.

HODGSON, E. S. and GELDIAY, S. (1959). Experimentally induced release of neurosecretory materials from roach corpora cardiaca. *Biol. Bull (Woods Hole) 117*, 275–283.

HOFFMANN, D., BREHÉLIN, M. and HOFFMANN, J. A. (1977). Premiers résultats sur les réactions de défense antibactériennes de larves et d'imagos de *Locusta migratoria. Ann. Parasitol. Hum. Comp. 52*, 87–88.

HOFFMANN, J. A. (1966). Étude des oenocytoïdes chez *Locusta migratoria* (Orthoptera). *J. Mic. (Paris) 5*, 269–272.

HOFFMANN, J. A. (1967). Étude des hémocytes de *Locusta migratoria L.* (Orthoptera). *Archiv. Zool. Exp. Gén. 108*, 251–291.

HOFFMANN, J. A. (1970). Régulations endocrines de la production et de la différenciation des hémocytes chez un insecte Orthoptère: *Locusta migratoria migratorioides. Gen. Comp. Endocr. 15*, 198–219.

HOFFMANN, J. A. and JOLY, L. (1969). Action du système endocrine et de l'ovaire sur l'hémogramme de *Locusta migratoria L.* Rôle des copora allata. Action de l'ovaire. *C.R. Acad. Sci. Paris 268D*, 1218–1220.

HOFFMANN, J. A. and STOEKEL, M. E. (1968). Sur le modification ultrastructurales des coagulocytes au cours de la coagulation de l' hémolymphe chez un insecte Orthopteroïde: *Locusta migratoria. C.R. Séances Sci. Biol. Strasbourg 162*, 2257–2259.

HOFFMANN, J. A., PORTE, A. and JOLY, P. (1968a). Présence d'un tissue hématopoïétique au niveau du diaphragme dorsal de *Locusta migratoria* (Orthoptère). *C.R. Séances Acad. Sci. Paris 266D*, 1882–1883.

HOFFMANN, J. A., PORTE, A. and JOLY, P. (1969). L'hématopoïèse chez les insectes Orthoptères. *C.R. Séances Soc. Biol. 163*, 2701–2703.

HOFFMANN, J. A., PORTE, A. and JOLY, P. (1970). On the localization of phenoloxidase activity in coagulocytes of *Locusta migratoria L.* (Orthoptera). *C.R. Acad. Sci. Paris 270D*, 629–631.

HOFFMANN, J. A., STOEKEL, M. E., PORTE, A. and JOLY, P. (1968b). Ultrastructure des hémocytes de *Locusta migratoria* (Orthoptère). *C.R. Hebd. Séances Acad. Sci. Paris 266D*, 503–505.

HOFFMANN, J. A., ZACHARY, D., HOFFMANN, D. and BREHÉLIN, M. (1979). Postembryonic development and differentiation: hemopoietic tissues and their functions in some insects. In *Insect Hemocytes*. Edited by A. P. Gupta. Pages 29–66. Cambridge University Press, Cambridge.

HOLLANDE, A. C. (1909). Contribution á l'étude du sang des Coléoptères. *Arch. Zool., Exp. Gén (Ser 5) 21*, 271–294.

HOLLANDE, A. C. (1911). Étude histologiques comparée du sang des insectes à hémorrhée et des insectes sans hémorrhée. *Arch. Zool. Exp. Gén (Ser 5) 6*, 283–323.

HOLLANDE, A. C. (1920). Oenocytoïdes et teratocytes du sang des chenilles. *C.R. Acad. Sci. Paris 170*, 1341–1344.

HOMAN-MÜLLER, J. W. T., WEENING, R. S. and ROOS, D. (1974). Production of hydrogen peroxide by phagocytizing human granulocytes. *J. Lab. Clin. Med. 85*, 198–207.

HOUK, E. J. and GRIFFITHS, G. W. (1980). Intracellular symbiotes of the Homoptera. *Ann. Rev. Ent. 25*, 161–187.

HOWELL, W. H. (1885). Observations upon the chemical composition and coagulation of the blood of *Limulus polyphemus, Callinectes hastatus*, and *Cucumaria* sp. *Johns Hopkins Univ. Cir. 5*, 4–5.

HRDY, I. (1957). Comparison of preparation and staining techniques of insect blood cells. *Acta Soc. Ent. Czech. 54*, 305–311.

HUGHES, L. and PRICE, G. M. (1975a). Tyrosinase activity in the larva of the fleshfly *Sarcophaga barbarta. Trans. Biochem. Soc. 3*, 72–75.

HUGHES, L. and PRICE, G. M. (1975b). The inhibition of catecholase activity by cuticle extracts in larvae of the fleshfly *Sarcophaga barbata. Trans. Biochem. Soc. 3*, 549–551.

HUGHES, L. and PRICE, G. M. (1976). Hemolymphal activation of protyrosinase and site of synthesis of hemolymph protyrosinase in larvae of the fleshfly *Sarcophaga barbarta. J. Insect Physiol. 22*, 1005–1011.

IRVING, S. N., OSBORNE, M. P. and WILSON, R. G. (1976). Virtual absence of L-glutamate from the haemoplasm of arthropod blood. *Nature (Lond.) 263*, 431–433.

JOHANNSEN, O. A. and BUTT, F. H. (1941). *Embryology of Insects and Myriapods*. McGraw-Hill, New York.

JONES, J. C. (1950). The normal hemocyte picture of the yellow mealworm *Tenebrio molitor Linnaeus. Iowa State Coll. J. Sci. 24*, 356–361.

JONES, J. C. (1956). The hemocytes of *Sarcophaga bullata Parker. J. Morph. 99(2)*, 233–257.

JONES, J. C. (1957). DDT and the hemocyte picture of the mealworm, *Tenebrio molitor L. J. Cell. Comp. Physiol. 50*, 423–428.

JONES, J. C., (1959). A phase contrast study of the blood cells in *Prodenia* larvae (Order Lepidoptera). *Quart. J. Mic. Sci. 100(10)*, 17–23.

JONES, J. C. (1962). Current concepts concerning insect hemocytes. *Amer. Zool. 2*, 209–246.

JONES, J. C. (1965). The hemocytes of *Rhodnius prolixus (Stäl). Biol. Bull. (Woods Hole) 129*, 282–294.

JONES, J. C. (1967). Normal differential count of haemocytes in relation to ecdysis and feeding in *Rhodnius. J. Insect Physiol. 13*, 1133–1141.

JONES, J. C. (1970). Hemocytopoiesis in insects. In *Regulation of Hematopoiesis*. Edited by A. S. Gordon. Pages 7–65. Appleton, New York.

JONES, J. C. and LIU, D. P. (1968). A quantitative study of mitotic divisions of haemocytes of *Galleria mellonella* larvae. *J. Insect Physiol. 14*, 1055–1061.

JONES, J. C. and TAUBER, O. E. (1954). Abnormal hemocytes in mealworms (*Tenebrio molitor L.*). *Ann. Ent. Soc. Amer. 47(3)*, 428–444.

JUDY, K. J. (1969). Cellular respone to ecdysterone *in vitro. Science (Wash., D.C.) 165*, 1374–1375.

KAWASE, S. (1960). Tyrosinase in the silkworm during pupation period. *J. Insect Physiol. 5*, 335–340.

KERKUT, G. A., SHAPIRO, A. and WALKER, R. G. (1965). The effect of acetylcholine, glutamic acid and GABA on contraction of the perfused cockroach leg. *Comp. Biochem. Physiol. 16*, 37–48.

KISLEV, N., HARPAZ, F. and ZELCER, A. (1969). Electron-microscopic studies on hemocytes of the Egyptian cottonworm, *Spodoptera littoralis* (Boisduval) infected with a nuclear-polyhedrosis virus, as compared to nonifected hemocytes. II. Non-infected hemocytes. *J. Invert. Pathol. 14*, 245–257.

KITANO, H. (1969a). Defensive ability of *Apanteles glomeratus L.*, (Hymenoptera: Braconidae) to hemocytic reaction of *Pieris rapae crucivora Boisduval* (Lepidoptera: Pieridae). *Appl. Ent. Zool. 4(1)*, 51–55.

KITANO, H. (1969b). Experimental studies on the parasitism of *Apanteles glomeratus L.* with special reference to its encapsulation-inhibiting capacity. *Bull. Tokyo Gekugei Univ. 21(4)*, 95–156.

KLEBANOFF, S. J. (1967). A peroxidase-mediated antimicrobial system in leucocytes. *J. Clin. Invest. 46*, 1078.

KLEIN, M. G. and COPPEL, H. C. (1969). Hemocytopoietic organs in larvae of the introduced pine sawfly, *Diprion similis. Ann. Ent. Soc. Amer. 62(6)*, 1259–1261.

KOLLMANN, M. (1908). Recherches sur les leucocytes et le tissue lymphoïde des invertébres. *Ann. Sci. Nat. Zool. 9*, 1–238.

KOROTNEFF, A. (1885). Die Embryologie der *Gryllotalpa. Z. Wiss. Zool. 41*, 507–604.

LACKIE, A. M. (1980). Invertebrate Immunity. *Parasitology 80*, 393–412.

LAI-FOOK, J. (1968). The fine-structure of wound repair in an insect *Rhodnius prolixus. J. Morph. 124*, 37–78.

LAI-FOOK, J. (1970). Haemocytes in the repair of wounds in an insect *Rhodnius prolixus. J. Morph. 130*, 297–314.

LAI-FOOK, J. (1973). The structure of the haemocytes of *Calpodes ethlius* (Lepidoptera). *J. Morph. 139*, 79–104.

LANDUREAU, J. C. and GRELLET, P. (1975a). Obtentions de lignées permanentes d'hémocytes de Blatte: Caractéristiques physiologiques et ultrastructurales. *J. Insect Physiol. 21*, 137–151.

LANDUREAU, J. C. and GRELLET, P. (1975b). New permanent cell lines from cockroach hemocytes: Physiological and ultrastructural characteristics. *J. Insect Physiol. 21*, 137–152.

LANDUREAU, J. C. and SZÖLLÖSI, A. (1974). Démonstration, par la méthode de culture *in vitro*, du rôle des hémocytes dans la spematogénèse d'un insecte. *C.R. Acad. Sci. Paris 278D*, 3359–3362.

LAZARENKO, T. (1925). Beitrage zur vergleichenden Histologie des Blutes und des Bindegewebes. *Z. Mikr. Anat. Forsch. 3*, 409–499.

LEA, M. S. and GILBERT, L. I. (1961). Cell division in diapausing silkworm pupae. *Amer. Zool. 1*, 368–369.

LEA, M. S. and GILBERT, L. I. (1966). The hemocytes of *Hyalophora cecropia* (Lepidoptera). *J. Morph. 118(2)*, 197–216.

LEUTENEGGER, R. (1967). Early events of *Sericesthis* iridescent virus infection in hemocytes of *Galleria mellonella (L.). Virology 32*, 109–116.

LEYDIG, L. (1859). Zur Anatomie der Insekten. *Arch. Anat. Physiol. Med. 1859*, 33–89.

LIPKE, H. (1975). Melanin in host-parasite interaction. In *Invertebrate Immunity*. Edited by K. Maramorosch and R. E. Shope. Pages 327–336. Academic Press, New York.

LOCKE, M. and HUIE, P. (1972). The fiber components of insect connective tissue. *Tissue Cell 4*, 601–612.

LOEB, L. (1904). On the spontaneous agglutination of blood cells of arthropods. *Univ. Penn. Med. Bull. 16*, 441–443.

LOEB, L. (1910). Über die Blutgerinnung bei Wirbellosen. *Biochem. Z. 24*, 478–495.

LUTZ, K. G. (1895). Das Bluten der Coccinelliden. *Zool. Anz. 18*, 244–245.

MAGRETTI, P. (1881). Del prodotto di secrezione in alcuni meloidi esame microscopico. *Boll. Sci. 1*, 23–27.

MAIER, W. A. (1972). Die Phenoloxidase von *Chironomus thummi* und ihre Beeinflussung durch Parasitare Mermithiden. *J. Insect Physiol. 19*, 85–95.

MALUF, N. S. R. (1939). The blood of arthropods. *Quart. Rev. Biol. 14*, 149–191.

MARCHALONIS, J. J. and EDELMAN, G. M. (1968). Isolation and characterization of a hemagglutinin from *Limulus polyphemus*. *J. Mol. Biol. 32*, 453–465.

MAREK, M. (1968). *In vitro* synthesis of injury protein and companion protein in organs of prepupa of *Galleria mellonella L. Comp. Biochem. Physiol. 29*, 1231–1237.

MAREK, M. (1969a). Effect of actinomycin D on synthesis of cooling protein in haemolymph of pupae of *Galleria mellonella L. Comp. Biochem. Physiol. 34*, 221–229.

MAREK, M. (1969b). Effects of stress of cooling on the synthesis of immunoglobulins in haemolymph of *Galleria mellonella L. Comp. Biochem. Physiol. 35*, 615–622.

MATTHEWS, J. R., DOWNER, R. G. H. and MORRISON, P. E. (1975). α-Glucosidase activity in haemolymph of the American cockroach *Periplaneta americana*. *J. Insect Physiol. 22*, 157–163.

MAZZONE, H. M. (1976). Influence of polyphenol oxidase on hemocyte cultures of the gypsy moth. In *Invertebrate Tissue Culture. Applications in Medicine, Biology and Agriculture*. Edited by E. Kurstak and K. Maramorosch. Pages 275–278. Academic Press, New York.

MCKAY, D. and JENKIN, C. R. (1970). Immunity in the invertebrates: The role of serum factors in phagocytosis of erythrocytes by haemocytes of the freshwater crayfish (*Parachaeraps bicarinatus*). *Aust. J. Exp. Biol. Med. Sci. 48*, 139–150.

MERCER, E. H. and NICHOLAS, W. L. (1967). The ultrastructure of the capsule of the larval stages of *Moniliformis dubius* (Acanthocephala) in the cockroach *Periplaneta americana*. *Parasitology 57*, 169–174.

MESSNER, B. (1976). NADH oxidase NADPH oxidase activity in the blood cells of hemimetabolous insects. *Acta Histochem. 56*, 261–269.

METALNIKOV, S. (1908). Recherches expérimentales sur les Chenilles de *Galleria mellonella*. *Archiv. Zool. Exp. (Ser. 4) 8*, 489–588.

METANLIKOV, S. and CHORINE, V. (1929). On the natural and acquired immunity of *Pyrausta nubilalis Hb*. *Int. Corn Borer Invest. (Chicago). 2*, 22–38.

METALNIKOV, S. and GASCHEN, H. (1922). Immunité cellulaire el humoral chez la chenille. *Ann. Inst. Pasteur 36*, 233–252.

METCHNIKOFF, E. (1884). Über eine Sprosspilzkrankheit der Daphnien: Beitrag zur lehre über den Kampf der Phagocyten gegen Krankheitserreger. *Verchow's Arch. Pathol. Anat. Physiol. 96*, 177–195.

METCHNIKOFF, E. (1892). *Leçons sur la Pathologie Comparée de l'Inflammation*. Masson et cie, Paris.

MILLARA, P. (1947). Contribution a l'étude cytologique et physiologique des leucocytes d'Insectes. *Bull. Biol. Fr. Belg. 81*, 129–153.

MILLER, V. H., BALLBACK, R. S., PAULEY, G. B. and KRASSNER, S. M. (1972). A preliminary physiochemical characterization of an agglutinin found in the hemolymph of the crayfish *Procambarus clarkii*. *J. Invert. Pathol. 19*, 83–93.

MILLS, R. R. and WHITEHEAD, D. L. (1969). Humoral control of tanning in the American cockroach: Changes in blood cell permeability during ecdysis. *J. Insect Physiol. 16*, 331–340.

MILLS, R. R., ANDROUNY, S. and FOX, F. R. (1968). Correlation of phenoloxidase activity with ecdysis and tanning hormone release in the American cockroach. *J. Insect Physiol. 14*, 603–611.

MITSUHASHI, J. (1966). Tissue culture of the rice borer, *Chilo suppressalis Walker* (Lepidoptera: Pyralidae). II. Morphology and *in vitro* cultivation of hemocytes. *Appl. Ent. Zool. 1*, 5–20.

MITSUHASHI, J. (1972). Primary culture of the haemocytopoietic tissue of *Papilio xuthus Linne*. *Appl. Ent. Zool. 7(1)*, 39–41.

MORAN, D. T. (1971). The fine structure of cockroach blood cells. *Tissue Cell 3*, 413–422.

MORGANTHALER, P. W. (1953). Blutuntersuchungen bei Bienen. *Mitt. Schweiz. Ent. Ges. 26*, 247–257.

MORI, H. (1979). Embryonic hemocytes: origin and development. In *Insect Hemocytes*. Edited by A. P. Gupta. Pages 3–27. Cambridge University Press, Cambridge.

MULDREW, J. A. (1953). The natural immunity of the larch sawfly (*Pristiphora erichsonii (HTg)* to the introduced parasite *Mesoleuis tenthredinis Morley*, in Manitoba and Saskatchewan. *Canad. J. Zool. 31*, 313–332.

MÜLLER, K. (1925). Über die Korpuskularen Elemente der Blutflussigkeit bei der erwaschsenen Honigbiene (*Apis mellifica*). *Erlanger Jahr. Blenenkunde 3*, 5–27.

MUNSON, S. C. and YEAGER, J. F. (1944). Fat inclusion in the blood cells of the Southern armyworm *Prodenia eridania*. *Ann. Ent. Soc. Amer. 37*, 396–400.

MURER, E. H., LEVIN, J. and HOLME, R. (1975). Isolation and studies of the granules of the amoebocytes of *Limulus polyphemus*, the horseshoe crab. *J. Cell Physiol. 86*, 533–542.

MUTTKOWSKI, R. A. (1924). Studies on the blood of insects. III. The coagulation and clotting of insect blood. *Bull. Brooklyn Ent. Soc. 19*, 128–44.

NAPPI, A. J. (1970). Hemocytes of larvae of *Drosophila euronotus* (Diptera: Drosophilidae). *Ann. Ent. Soc. Amer. 63*, 1217–1224.

NAPPI, A. J. (1973). The role of melanization in the immune reaction of larvae of *Drosophila algonquin* against *Pseudeucoila bochei*. *Parasitology 66*, 23–32.

NAPPI, A. J. (1974). Insect hemocytes and the problems of host recognition of foreignness. In *Contemporary Topics in Immunobiology*. Edited by E. L. Cooper. Vol. 4, pages 207–224. Plenum Press, New York.

NAPPI, A. J. (1975). Parasite encapsulation in insects. In *Invertebrate Immunity*. Edited by K. Maramorosch and R. E. Shope. Pages 293–326. Academic Press, New York.

NAPPI, A. J. and STOFFOLANO, J. G. Jr (1971). *Heterotylenchus autumnalis*. Hemocytic reactions and capsule formation in the host, *Musca domestica*. *Exp. Parasitol. 29*, 116–125.

NAPPI, A. J. and STOFFOLANO, J. G. Jr. (1972). Haemocytic changes associated with the immune reaction of nematode-infected larvae of *Orthellia caesarion*. *Parasitology 65*, 295–302.

NAPPI, A. J. and STREAMS, F. A. (1969). Hemocytic reactions of *Drosophila melanogaster* to the parasites *Pseudeucoila mellipes* and *P. bochei*. *J. Insect Physiol. 15*, 1551–1566.

NELSTROP, A. E., TAYLOR, G., and COLLARD, P. (1970). Comparative studies of serum and haemolymph proteins. *Comp. Biochem. Physiol. 35*, 191–196.

NEUWIRTH, M. (1973). The structure of the hemocytes of *Galleria mellonella* (Lepidoptera). *J. Morph. 139(1)*, 105–124.

NITTONO, Y. (1960). Studies on the blood cells in the silkworm, *Bombyx mori L. Bull. Seric. Exp. Stn. Tokyo 16*, 171–266. (In Japanese, English summary).

NORTON, W. N. and VINSON, S. B. (1976). Egg membrane synthesis of a hymenopteran parasitoid. *Amer. Zool. 15*, 828.

O'CONNOR, G. M. Jr. and FEIR, D. (1968). Application of quantitative criteria for haemopoietic activity to insect haemocytes. *J. Insect Physiol. 14 (12)*, 1779–1784.

OGEL, S. (1955). A contribution to the study of the blood cells in Orthoptera. *Comm. Fac. Sci. Univ. Ankara (Istanbul) 4C (2)*, 15–41.

OKADA, M. (1960). Embryonic development of the rice stem-borer, *Chilo suppressalis*. *Sci. Rep. Tokyo Kyoiku Daigaku Sect. B. 9*, 243–296.

OLSON, K. and CARLSON, S. D. (1974). Surface fine-structure of hemocytes of *Periplaneta americana*: A scanning electron microscope study. *Ann. Ent. Soc. Amer. 67(1)*, 61–65.

PAILLOT, A. (1919). Le karyocinetose, nouvelle réaction d'immunité naturelle observé chez les chenilles de macrolépidoptères. *C.R. Acad. Sci. Paris. 169*, 396–398.

PAILLOT, A. and NOEL, R. (1928). Recherches histophysiologiques sur les cellules péricardiales et les éléments du sang des larves d'insectes (*Bombyx mori* et *Pieris brassicae*). *Bull. Hist. Appl. Physiol. Pathol.* *5*, 105–128.

PARISH, C. R. (1977). Simple model for self- non-self discrimination in invertebrates. *Nature (Lond.)* *267*, 711–713.

PASSONNEAU, J. V. and WILLIAMS, C. M. (1953). The moulting fluid of the *Cecropia* silkworm. *J. Exp. Biol.* *30*, 545–560.

PASTAN, I. H. and WILLINGHAM, M. C. (1981). Journey to the center of the cell: role of receptosome. *Science (Wash., D.C.)* *214*, 504–509.

PATERSON, W. D. and STEWART, J. E. (1974). *In vitro* phagocytosis by hemocytes of the American lobster (*Homarus americanus*). *J. Fish. Res. Bd. Canada* *31*, 1051–1056.

PATTEN, W. (1884). The development of phryganids, with a preliminary note on the development of *Blatta germanica*. *Quart. J. Mic. Sci.* *24*, 549–602.

PATTON, R. L. (1961). The detoxication function of insect hemocytes. *Ann. Ent. Soc. Amer.* *54*, 696–698.

PATTON, R. L. and FLINT, R. A. (1959). The variation in the bloodcell count of *Periplaneta americana (L.)* during a molt. *Ann. Ent. Soc. Amer.* *52*, 240–242.

PEARSE, A. G. E. (1961). *Histochemistry: Theoretical and Applied.* 2nd edition. Little, Brown & Co., Boston.

PEARSE, B. M. F. (1976). Clathrin: A unique protein associated with intracellular transfer of membrane by coated vesicles. *Proc. Nat. Acad. Sci. U.S.A.* *73*, 1255–1259.

PERCY, J. (1978). Haemocytes associated with basement membrane of sex pheromone gland of *Trichoplusia ni* (Lepidoptera: Noctuidae). Ultrastructural observations. *Canad. J. Zool.* *56 (2)*, 238–245.

PIPA, R. L. and COOK, E. F. (1958). The structure and histochemistry of the connective tissue of the sucking lice. *J. Morph.* *103*, 353–385.

PISTOLE, T. G. (1982). *Limulus* lectins: Analogues of vertebrate immunoglobulins. In *Physiology and Biology of Horseshoe Crabs: Studies on Normal and Environmentally Stressed Animals.* Pages 283–288. Alan R. Riss, New York.

POINAR, G. O. Jr. (1974). Insect immunity to parasite nematodes. In *Contemporary Topics in Immunobiology*. Edited by E. L. Cooper. Vol. 4, pages 167–178. Plenum Press, New York.

POINAR, G. O. Jr., LEUTENEGGER, R. and GOTZ, P. (1968). Ultrastructure of the formation of a melanotic capsule in *Diabrotica* (Coleoptera) in response to a parasitic nematode (Mermithidae). *J. Ultrastruct. Res.* *25*, 293–306.

POST, L. C. (1972). Bursicon: Its effect on tyrosine permeation into insect haemocytes. *Biochem. Biophys. Acta* *290*, 424–428.

POYARKOFF, E. (1910). Recherches histologiques sur la métamorphose d'un Coléoptère (La Galeruque d l'Orme). *Arch. Anat. Mic.* *12*, 333–474.

PRESTON, J. W. and TAYLOR, R. L. (1970). Observations on the phenoloxidase system in the haemolymph of the cockroach *Leucophaea maderae*. *J. Insect Physiol.* *16*, 1729–1744.

PRICE, C. D. and RATCLIFFE, N. A. (1974). A reappraisal of insect haemocyte classification by the examination of blood from fifteen insect orders. *Z. Zellforsch.* *147*, 537–549.

PUTTLER, B. (1967). Interrelationship of *Hypera postica* (Coleoptera: Curculionidae) and *Bathyplectes curculionis* (Hymenoptera: Ichneumonidae) in the Eastern United States with particular reference to the encapsulation of the parasite eggs by the weevil larvae. *Ann. Ent. Soc. Amer.* *60*, 1031–1038.

RAINA, A. K. (1976). Ultrastructure of the larval haemocytes of the pink bollworm *Pectinophora gossypiella* (Lepidoptera: Gelechiidae). *Int. J. Insect Morphol. Embryol.* *5*, 187–195.

RAMASWAMY, S. B. and GUPTA, A. P. (1981). Effects of juvenile hormone on sense organs involved in mating behaviour of *Blattella germanica (L.)* (Dictyoptera: Blattellidae). *J. Insect Physiol.* *27(9)*, 601–608.

RATCLIFFE, N. A. and GAGEN, S. J. (1976). Cellular defense reactions of insect hemocytes *in vivo*: Nodule formation and development in *Galleria mellonella* and *Pieris brassicae* larvae. *J. Invert. Pathol.* *28(3)*, 373–382.

RATCLIFFE, N. A. and GAGEN, S. J. (1977). Studies on the *in vivo* cellular reactions of insects: An ultrastructural analysis of nodule formation in *Galleria mellonella*. *Tissue Cell* *9(1)*, 73–85.

RATCLIFFE, N. A. and PRICE, C. D. (1974). Correlation of light and electron microscope hemocyte structure in the Dictyoptera. *J. Morph.* *144*, 485–497.

RATCLIFFE, N. A. and ROWLEY, A. F. (1979). Role of hemocytes in defense against biological agents. In *Insect Hemocytes*. Edited by A. P. Gupta. Pages 331–414. Cambridge University Press, Cambridge.

RATCLIFFE, N. A., GAGEN, S. J., ROWLEY, A. F. and SCHMIT, A. R. (1976a). Studies on insect cellular defense mechanisms and aspects of the recognition of foreigness. In *Proc. 1st Int. Colloq. Invertebr. Pathol.* Edited by T. A. Angus, P. Faulkner and A. Rosenfield. Pages 210–214. Queens University, Kingston, Ontario.

RATCLIFFE, N. A., GAGEN, S. J., ROWLEY, A. F. and SCHMIT, A. R. (1976b). The role of granular hemocytes in the cellular defense reactions of the wax moth, *Galleria mellonella*. In *Proc. 6th Europ. Congr. Electron Micr.* Edited by Y. Ben-Shaul. Pages 295–297. Tal International, Israel.

RATNER, R. and VINSON, S. B. (1983). Phagocytosis and encapsulation: Cellular immune response in Arthropoda. *Amer. Zool.* *23*, 185–194.

RAVINDRANATH, M. H. (1981). Onychophorons and myriapods. In *Invertebrate Blood Cells*. Edited by N. A. Ratcliffe and A. F. Rowley. Vol. 2, pages 327–354. Academic Press, New York.

RITTER, H. Jr. (1965). Blood of a cockroach: Unusual cellular behavior. *Science (Wash., D.C.)* *147*, 518–519.

RIZKI, M. T. M. (1953). The larval blood cells of *Drosophila willistoni*. *J. Exp. Zool.* *123*, 397–411.

RIZKI, M. T. M. (1957a). Alterations in the haemocyte population of *Drosophila melanogaster*. *J. Morph.* *100*, 437–458.

RIZKI, M. T. M. (1957b). Tumor formation in relation to metamorphosis in *Drosophila melanogaster*. *J. Morph.* *100*, 459–472.

RIZKI, M. T. M. (1962). Experimental analysis of hemocyte morphology in insects. *Amer. Zool.* *2*, 247–256.

RIZKI, M. T. M. and RIZKI, R. M. (1959). Functional significance of crystal cells in the larva of *Drosophila melanogaster*. *J. Biophys. Biochem. Cytol.* *5*, 235–240.

RIZKI, M. T. M. and RIZKI, R. M. (1976). Cell interactions in hereditary melanotic tumor formation in *Drosophila*. In *Proc. 1st. Int. Colloq. Invertebr. Pathol.* Edited by T. A. Angus, F. Faulkner and A. Rosenfield. Pages 137–141. Queens University, Kingston, Ontario.

ROONWAL, M. L. (1937). Studies on the embryology of the African migratory locust, *Locusta migratoria migratorioides*. II. Organogeny. *Phil. Trans. Roy. Soc. Lond. B. Biol. Sci.* *227*, 175–244.

ROTH, T. F. and PORTER, K. R. (1964). Yolk protein uptake in the oocyte of the mosquito *Aedes aegypti*. *J. Cell Biol.* *20*, 313–332.

ROTHERAM, S. (1967). Immune surface of eggs of a parasitic insect. *Nature (Lond.)* *214*, 700.

ROTHMAN, J. E. (1981). The Golgi apparatus: two organelles in tandem. *Science (Wash., D.C.)* *213*, 1212–1219.

ROWLEY, A. F. (1977). The role of the haemocytes of *Clitumnus extradentatus* in haemolymph coagulation. *Cell Tissue Res.* *182(4)*, 513–524.

ROWLEY, A. F. and RATCLIFFE, N. A. (1976). The granular cells of *Galleria mellonella* during clotting and phagocytic reactions *in vitro*. *Tissue Cell* *8*, 437–446.

ROWLEY, A. F. and RATCLIFFE, N. A. (1978). A histochemical study of wound healing and hemocyte function in the wax-moth *Galleria mellonella*. *J. Morph.* *157*, 181–200.

ROWLEY, A. F. and RATCLIFFE, N. A. (1981). Insects. In *Invertebrate Blood Cells*. Edited by N. A. Ratcliffe and A. F. Rowley. Vol. 2, pages 421–488. Academic Press, New York.

SALT, G. (1957). Experimental studies in insect parasitism. X. The reactions of some endopterygote insects to alien parasite. *Proc. Roy. Soc. Lond. B* *147*, 167–184.

SALT, G. (1965). Experimental studies in insect parasitism. XIII. The haemocytic reaction of a caterpillar to eggs of its habitual parasite. *Proc. Roy. Soc. B.* *162*, 303–318.

SALT, G. (1970). *The Cellular Defense Rections of Insects.* Cambridge University Press, Cambridge.

SAXENA, A. K. and AGARWAL, G. P. (1979). Free haemocytes of *Lipeurus lawrensis tropicalis Peters*, an ischnoceran Mallophaga (sens. lat. Phthiraptera) infesting poultry birds. *Indian J. Ent.* *41 (3)*, 231–236.

SCHARRER, B. (1965a). The fine structure of an unusual hemocyte in the insect *Gromphadorhina portentosa*. *Life Sci.* *4*, 1741–1744.

SCHARRER, B. (1965b). Hemocytes within prothoracic glands of insects. *Amer. Zool.* *5(2)*, 235–236.

SCHARRER, B. (1972). Cytophsiological features of haemocytes in cockroaches. *Z. Zellforsch. 129*, 301–319.

SCHELL, S. C. (1952). Tissue reactions of *Blattella germanica L.* to the developing larva of *Physaloptera hispida Schell*, 1950 (Nematoda: Spiruroidea). *Trans. Amer. Mic. Soc. 71*, 293–302.

SCHMIT, A. R. and RATCLIFFE, N. A. (1977). The encapsulation of foreign tissue implants in *Galleria mellonella* larvae. *J. Insect Physiol. 23*, 175–184.

SCHMIT, A. R., ROWLEY, A. F. and RATCLIFFE, N. A. (1976). The role of *Galleria mellonella* hemocytes in melanin formation. *J. Invert. Pathol. 29*, 232–234.

SCOTT, M. T. (1971a). Recognition of foreignness in invertebrates. II. *In vitro* studies of cockroach phagocytic haemocytes. *Immunology 21*, 817–827.

SCOTT, M. T. (1971b). A naturally occurring hemagglutinin in the hemolymph of the American cockroach. *Arch. Zool. Exp. Gen. 112*, 73–80.

SCOTT, M. T. (1972). Partial characterization of the hemagglutinating activity in hemolymph of the American cockroach (*Periplaneta americana*). *J. Invert. Pathol. 19*, 66–71.

SELMAN, J. (1961). The fate of the blood cells during the life history of *Sialis lutaria L. J. Insect Physiol. 8*, 209–214.

SHAPIRO, M. (1969). Immunity of insect hosts to insect parasites. In *Immunity to Parasitic Animals*. Edited by G. J. Jackson, R. Herman and I. Singer. Vol. 1, pages 211–228. Appleton, New York.

SHAPIRO, M. (1979a). Changes in hemocyte population. In *Insect Hemocytes*. Edited by A. P. Gupta. Pages 475–523. Cambridge University Press, Cambridge.

SHAPIRO, M. (1979b). Techniques for total and differential hemocyte counts and blood volume and mitotic index determinations. In *Insect Hemocytes*. Edited by A. P. Gupta. Pages 539–548. Cambridge University Press, Cambridge.

SHAPIRO, M. (1979c). Hemocyte techniques: advantages and disadvantages. In *Insect Hemocytes*. Edited by A. P. Gupta. Pages 549–561. Cambridge University Press, Cambridge.

SHRIVASTAVA, S. C. and RICHARDS, A. G. (1965). An autoradiographic study of the relation between hemocytes and connective tissue in the wax moth, *Galleria mellonella L. Biol. Bull. (Woods Hole) 128*, 337–345.

SHULL, W. E. and RICE, P. L. (1933). A method for temporary inhibition of blood coagulation in insects. *J. Econ. Ent. 26*, 1083–1089.

SHULL, W. E., RILEY, M. K. and RICHARDSON, C. H. (1932). Some effects of certain toxic gases on the blood of the cockroach *Periplaneta orientalis (L.)*. *J. Econ. Ent. 25*, 1070–1072.

SIAKOTOS, A. N. (1960a). The conjugated plasma proteins of the American cockroach. *J. Gen. Physiol. 43*, 999–1013.

SIAKOTOS, A. N. (1960b). The conjugated plasma proteins of the American cockroach. II. Changes during molting and clotting processes. *J. Gen. Physiol. 43*, 1015–1030.

SINDERMANN, C. J. and MAIRS, D. F. (1959). A major blood group system in Atlantic sea herring. *Copeia 3*, 228–232.

SMITH, D. S. (1968). *Insect Cells: Their Structure and Function*. Oliver & Boyd, Edinburgh.

SMITH, H. W. (1938). The blood of the cockroach, *Periplaneta americana* L. *Cell structure and degeneration, and cell counts. Studies of contact insecticides XIII. Tech. Bull. No. 71*, New Hampshire Agric. Exp. Stn., Durham.

SOHI, S. S. (1968). *In vitro* cultivation of *Choristoneura fumiferana* (Clemens) (Lepidoptera: Tortricidae). *Canad. J. Zool. 46(1)*, 11–13.

SOHI, S. S. (1971). *In vitro* cultivation of hemocytes of *Malacosoma disstria* Hübner (Lepidoptera: Lasiocampidae). *Canad. J. Zool. 49*, 1355–1358.

SOHI, S. S. (1979). Hemocyte cultures and insect hemocytology. In *Insect Hemocytes*. Edited by A. P. Gupta. Pages 259–277. Cambridge University Press, Cambridge.

SOLUM, N. O. (1970). Coagulation in *Limulus* — some properties of *Limulus polyphemus* blood cells. *Symp. Zool. Soc. Lond. 27*, 207–216.

SRDIĆ, Ž. and GLOOR, W. (1979). Spherule cells in *Drosophila* species. *Experientia 35*, 1226–1249.

STAIRS, G. R. (1964). Changes in the susceptibility of *Galleria mellonella (Linnaeus)* larvae to nuclear-polyhedrosis virus following blockage of phagocytes with India ink. *J. Insect Pathol. 6*, 373–376.

STANG-VOSS, C. (1970). Zur Ultrastrucktur der Blutzellen wirbelloser Tiere. I. Über die Hämocyten der Larve des Mehlkafers *Tenebrio molitor. Z. Zellforsch. 103*, 589–605.

STANG-VOSS, C. (1971). Zur Ultrastruktur der Blutzellen wirbelloser Tiere. V. Über die Hämocyten von *Astacus astacus* (Crustacea). *Z. Zellforsch. 122*, 68–75.

STEINKAMP, J. A., WILSON, J. S., SAUNDERS, G. C. and STEWART, C. C. (1982). Phagocytosis: Flow cytometric quantitation with fluorescent microspheres. *Science (Wash., D.C.) 215*, 64–66.

STREAMS, F. A. (1968). Defense reactions of *Drosophila* species (Diptera: Drosophilidae) to the parasite *Pseudeucoila bochei* (Hymenoptera: Cynipidae). *Ann. Ent. Soc. Amer. 61*, 158–164.

STUTMAN, L. J. and DOLLIVER, M. (1968). Mechanism of coagulation in *Gecarcinus lateralis. Amer. Zool. 8*, 481–489.

TAKEDA, N. (1977). Brain hormone carrier haemocytes in the moth *Monema flavescens. J. Insect Physiol. 23*, 1245–1254.

TATEIWA, J. (1928). Le formule leucocytaire du sang des chenilles normales et immunisées de *Galleria mellonella. Ann. Inst. Pasteur 42*, 791–806.

TAUBER, O. E. and YEAGER, J. F. (1934). On the total blood (hemolymph) cell count of the field cricket *Gryllus assimilis pennsylvanicus Burm. Iowa State Coll. J. Sci. 9*, 13–24.

TAUBER, O. E. and YEAGER, J. F. (1935). On the total hemolymph (blood) cell counts of insect. I. Orthoptera, Odonata, Hemiptera and Homoptera. *Ann. Ent. Soc. Amer. 28*, 229–240.

TAYLOR, R. L. (1969). A suggested role for the polyphenol-phenoloxidase system in invertebrate immunity. *J. Invert. Pathol. 14*, 427–428.

TELFER, W. H. and WILLIAMS, C. M. (1955). Incorporation of radioactive glycine into the blood proteins of *Cecropia* silkworm. *Anat. Rec. 122*, 441–442.

TELFER, W. H. and WILLIAMS, C. M. (1960). The effect of diapause, development and injury on the incorporation of radioactive glycine into the blood proteins of the *Cecropia* silkworm. *J. Insect Physiol. 5*, 61–72.

TILLYARD, R. J. (1917). *The Biology of Dragonflies (Odonata or Paraneuroptera*. Cambridge University Press, Cambridge.

TOYAMA, K. (1902). Contribution to the study of silkworm. I. On the embryology of the silkworm. *Bull. Coll. Agric. Tokyo Imp. Univ. 5*, 73–118.

TRIPP, M. R. (1975). Humoral factors and molluscan immunity. In *Invertebrate Immunity*. Edited by K. Maramorosch and R. E. Shope. Pages 201–223. Academic Press, New York.

TUZET, O. and MANIER, F. F. (1959). Recherches sur les hémocytes fusiforme géants (= "vermiform cells" de Yeager) de hémolymphe d'insectes homolmetaboles et heterometaboles. *Ann. Sci. Nat. Zool. Biol. Anim. 12(1)*, 81–89.

TYLER, A. and METZ, C. B. (1945). Natural heteroagglutinins in the serum of the spiny lobster, *Panulirus interruptus*. I. Taxonomic range of activity, electrophoretic and immunizing properties. *J. Exp. Zool. 100*, 387–406.

TYLER, A. and SCHEER, C. B. (1946). Natural heteroagglutinins in the serum of the spiny lobster, *Panulirus interruptus*. II. Chemical and antigenic relations to blood proteins. *Biol. Bull. (Woods Hole) 89*, 193–200.

TYSON, C. J. and JENKIN, C. R. (1973). The importance of opsonic factors in the removal of bacteria from the circulation of the crayfish (*Parachaeraps bicarinatus*). *Aust. J. Exp. Biol. Med. Sci. 51*, 609–615.

TYSON, C. J. and JENKIN, C. R. (1974). Phagocytosis of bacteria *in vitro* by (*Parachaeraps bicarinus*). *Aust. J. Exp. Biol. Med. Sci. 52*, 341–348.

USHERWOOD, P. N. R. and CULL-CANDY, S. G. (1975). Pharmacology of somatic nerve-muscle synapses. In *Insect Muscle*. Edited by P. N. R. Usherwood. Pages 201–280. Academic Press, New York.

VALVASSORI, R. and AMIRANTE, G. A. (1976). Caracteristiques immunochimiques et ultrastructurales des hémcoytes de *Leucophaea maderae L.* (Insecta Blattoidea). *Monit. Zool. Ital. (N.S.) 10*, 403–412.

VERCAUTEREN, R. E. and AERTS, F. (1958). On the cytochemistry of the hemocytes of *Galleria mellonella* with special reference to polyphenoloxidase. *Enzymologia 20*, 167–172.

VEY, A., QUIOT, J. M. and VAGO, C. (1968). Formation *in vitro* de réaction d'immunité cellulaire chez les insectes. In *Proc. 2nd Int. Colloq. Invertebr. Tissue Culture*. Edited by C. Barigozzi. Pages 254–263. Instituto Lombardo di Scienze e Lettere, Milan.

Vinson, S. B. and Scott, J. R. (1974). Parsitoid egg shell changes in a suitable and unsuitable host. *J. Ultrastruct. Res. 47*, 1–15.

Wada, S. (1955a). Zur Kenntnis der Keimblatterherkunft der subosophgealkorpers am Embryo des Seindenraupe, *Bombyx mori L. J. Seric. Sci. Jap. 24(2)*, 114–117. (In Japanese, German Summary).

Wada, S. (1955b). Zur Frage des Subosophagealkorpers als hamopoietische Gewebes am Embryo des Seidenraupe, *Bombyx mori L. J. Seric. Sci. Jap. 24 (516)*, 311–313. (In Japanese; German summary.)

Wago, H. (1981). The role of hemolymph in the initial cellular attachment to foreign cells by the hemocytes of the silkworm, *Bombyx mori. Devel. Comp. Immunol. 5*, 217–227.

Wago, H. and Ichikawa, Y. (1979). Changes in the phagocytic rate during the larval development and manner of hemocyte reactions to foreign cells in *Bombyx mori. Appl. Ent. Zool. 14(4)*, 397–403.

Walker, I. (1959). Die Abwehrreaktion des Wirtes *Drosophila melanogaster* gegen die Zoophage Cynipidae *Pseudeucoila bochei Weld. Rev. Suisse Zool. 68*, 569–632.

Walters, D. R. (1970). Hemocytes of saturniid silkworms: Their behavior *in vitro* and *in vivo* in response to diapause, development and injury. *J. Exp. Zool. 174*, 441–450.

Webley, D. P. (1951). Blood cell counts in the African migratory locust (*Locusta migratoria migratorioides Recihe and Fairmaire*). *Proc. Roy. Ent. Soc. Lond. A26*, 25–37.

Wermel, E. M. (1938). Die Explanation des Blutes von Seidenspinnerraupen. *Bull. Biol. Med. Exp. USSR 5*, 6–9.

Werner, R. A. and Jones, J. C. (1968). Phagocytic haemocytes in unfixed *Galleria mellonella* larvae. *J. Insect Physiol. 15*, 425–437.

Wheeler, R. E. (1962). Studies on the total haemocyte count and haemolymph volume in *Periplaneta americana (L.)* with special reference to the last moulting cycle. *J. Insect Physiol. 9*, 223–235.

Whitcomb, R. F., Shapiro, M. and Granados, R. R. (1974). Insect defense mechanisms against microorganisms and parasitoids. In *The Physiology of Insecta*. Edited by M. Rockstein. Vol. 5, 2nd edition, pages 447–536. Academic Press, New York.

Whitehead, D. L. (1969). New evidence for the control mechanism of sclerotization in insects. *Nature (Lond.) 224*, 721–723.

Whitehead, D. L. (1970). The role of haemocytes in the biosynthesis of protocatechuate in the cockroach colleterial system. *Biochem. J. 119*, 65–66.

Whitehead, D. L. (1973). The events preceding formation of the puparium in *Glossina* larvae. *Trans. Roy. Soc. Trop. Med. Hyg. 67*, 300–301.

Whitten, J. M. (1963). Haemocytes and the metamorphosing tissues in *Sarcophaga bullata, Drosophila melanogaster*, and other cyclorrhaphous Diptera. *J. Insect Physiol. 10*, 447–470.

Whitten, J. M. (1968). Haemocyte activity in relation to epidermal cell growth, cuticle secretion and cell death in a metamorphosing cyclorrhaphan pupa. *J. Insect Physiol. 15*, 763–778.

Wielowiejsky, H. (1886). Ueber das Blutgewebe der Insekten. *Z. Wiss. Zool. 43*, 512–536.

Wigglesworth, V. B. (1933). The physiology of the cuticle and of ecdysis in *Rhodnius prolixus* (Triatomidae, Hemiptera), with special reference to the function of the oenocytes and of the dermal glands. *Quart. J. Mic. Sci. 76*, 269–319.

Wigglesworth, V. B. (1937). Wound healing in an insect, *Rhodnius prolixus* (Hemiptera). *J. Exp. Biol. 14*, 364–381.

Wigglesworth, V. B. (1939). *The Principles of Insect Physiology*. 1st edition. Methuen, London.

Wigglesworth, V. B. (1955). The role of the haemocytes in the growth and moulting of an insect, *Rhodnius prolixus* (Hemiptera). *J. Exp. Biol. 32*, 649–663.

Wigglesworth, V. B. (1956a). The function of the amoebocytes during moulting in *Rhodnius. Ann. Sci. Nat. Zool. (Ser. 11) 18*, 139–144.

Wigglesworth, V. B. (1956b). The haemocytes and connective tissue formation in an insect, *Rhodnius prolixus* (Hemiptera). *Quart. J. Mic. Sci. 97*, 89–98.

Wigglesworth, V. B. (1959). Insect blood cells. *Ann. Rev. Ent. 4*, 1–16.

Wigglesworth, V. B. (1965). *The Principles of Insect Physiology*, 6th edn. Methuen, London.

Wigglesworth, V. B. (1972). Haemocytes and basement membrane formation in *Rhodnius. J. Insect Physiol. 19*, 831–844.

Wigglesworth, V. B. (1979a). Hemocytes and growth in insects. In *Insect Hemocytes*. Edited by A. P. Gupta. Pages 303–318. Cambridge University Press, Cambridge.

Wigglesworth, V. B. (1979b). Secretory activities of plasmatocytes and oenocytoids during the moulting cycle in an insect (*Rhodnius*) *Tissue Cell 11(1)*, 69–78.

Wille, H. and Vecchi, M. A. (1966). Étude sur l'hémolymphe de l'abeille (*Apis mellifica L.*). I. Les frottis de sang de l'abeille adelle adulte d'éte. *Mitt. Schweiz. Ent. Ges. 34*, 69–97.

Wittig, G. (1966). Phagocytosis by blood cells in healthy and diseased caterpillars. II. A consideration of the method of making hemocyte counts. *J. Invert. Pathol. 8*, 461–477.

Wittig, G. (1968). Electron microscopic characterization of insect hemocytes. *Proc. 26th Annu. Meeting EMSA. 1968*, 68–69.

Yeager, J. F. (1945). The blood picture of the southern armyworm (*Prodenia eridania*). *J. Agric. Res. 71*, 1–40.

Yeager, J. F. and Knight, H. H. (1933). Microscopic observation on blood coagulation in several different species of insects. *Ann. Ent. Soc. Amer. 26*, 591–602.

Yeager, J. F. and Munson, S. C. (1942). Changes induced in the blood cells of the southern armyworm (*Prodenia eridania*) by the administration of poisons. *J. Agric. Res. 64(6)*, 307–332.

Yeager, J. F. and Munson, S. C. (1950). Blood volume of the roach *Periplaneta americana* determined by several methods. *Arthropoda 1*, 255–265.

Yeager, J. F., McGovran, E. R., Munson, S. C. and Mayer, E. L. (1942). Effects of blocking hemocytes with Chinese ink and staining nephrocytes with trypan blue upon resistance of the cockroach *Periplaneta americana (L.)* to sodium arsenite and nicotine. *Ann. Ent. Soc. Amer. 35(1)*, 23–40.

Yeaton, R. W. (1981). Invertebrate lectins: I. Occurrence. *Dev. Comp. Immunol. 5*, 391–402.

Zachary, D. and Hoffmann, J. A. (1973). The haemocytes of *Calliphora erythrocephala (Meig.)* (Diptera). *Z. Zellforsch. 141*, 55–73.

Zachary, D., Brehélin, M. and Hoffmann, J. A. (1975). Role of "thrombocytoids" in capsule formation in the diptertan *Calliphora erythrocephala. Cell Tissue Res. 162*, 343–348.

Zeller, H. (1938). Blut und Feltkorper im Flugel der Mehlmotte *Ephestia kühniella. Z. Morphol. Oekol. Tiere 34*, 663–738.

11 Insect Immunity

PETER GÖTZ

Free University of Berlin, FRG

and

HANS G. BOMAN

University of Stockholm, Sweden

1 THE USE OF THE TERM "IMMUNITY"

The use of the expression immunity for the reactions of insects against infection is justified for several reasons:

(1) It has already been widely used mainly in the English literature. For this reason the term "insect immunity" has been adopted as key words in the computerized documentation of scientific papers.
(2) The alternative expression "defense reactions of insects" has a double meaning since it also refers to reactions by which insects defend themselves against other insects and against predators (e.g. the sting of aculeate Hymenoptera or the secretion of chemicals by certain beetles and bugs).
(3) The use of the expression insect immunity is in agreement with the historical use as well as with the definition of "immunity" given in recent encyclopaedias as "ability of an organism to resist diseases".

2 CELLULAR IMMUNITY

2.1 Insect hemocytes involved in cellular defense reactions

The first problem encountered when discussing cellular immunity in insects is the fact that authors investigating cellular defense reactions have used different classifications of hemocytes (A. Gupta, this volume). This makes it necessary first to define the hemocyte types to which we will refer in the following sections. The classification of hemocytes is obviously more difficult in insects than in vertebrates. Reasons for this are the diversity of insects (more than 800,000 species described) and the fact that insect hemocytes apparently are not as definitely differentiated as vertebrate blood cells. Clearly, distinct types of hemocytes do exist in insects but intermediate stages with characteristics of more than one type can often be found. In addition, and partly as a consequence of these difficulties, insufficient criteria have been used to classify hemocytes in insects. Shape and size, nuclear–cytoplasmic relations, lack or presence of cytoplasmic extensions and lack or presence of inclusions according to light or electron microscopic observations have frequently been the only criteria for classifying hemocytes in insects. Further ultrastructural details and clearly defined biological functions have either not been used or were not considered with desirable accuracy.

No wonder that a large variety of insect hemocyte types has accumulated in the literature. Fortunately, in the last few years several excellent reviews and comparative studies of hemocytes have appeared which significantly improved the situation in insect hematology (Jones, J., 1970, 1977; Gupta, A., 1979, this volume; Ratcliffe, N. and Rowley, A., 1979a; Rowley, A. and Ratcliffe, N., 1981). All of these authors tried to reduce the number of insect hemocyte types, establishing five or six basic types occurring in every insect and three or four additional forms, which are found only in some insect species.

This analytical effort should be continued intensively. Comparative investigations with consideration of ultrastructural and functional aspects are urgently needed. The danger of oversimplification has to be avoided as well as taxonomic minutia. At the present status of knowledge it is most important to elucidate the main qualities and functions of the basically different types of hemocytes.

The following description of the hemocytes, involved in cellular immunity, follows the outlines given by Jones, J. (1977), Gupta, A. (1979), and Rowley, A. and Ratcliffe, N. (1981).

Plasmatocytes are medium-sized (10–15 μm), round, ovoid, spindle-like or irregularly formed cells. They may possess few to many extensions (pseudopodia) of various sizes and shapes. Filiform pseudopodia can be many times as long as the main cellular body. The cytoplasm usually contains only a few inclusions. Plasmatocytes are a rather stable type of cell; under *in vitro* conditions they show little tendency to disintegrate. However, on contact with substrates (e.g. with the surface of a glass slide) plasmatocytes readily flatten and spread out; such cells have also been referred to as lammellocytes (Rizki, M., 1953; Nappi, A., 1970).

Under the electron microscope plasmatocytes present a well-developed RER (rough endoplasmic reticulum) and Golgi vesicles, many mitochrondria and free ribosomes. The scarce membrane-bound inclusions possess characteristic features of phagocytosis such as acid phosphatase activity (Crossley, A., 1975). Some vacuoles contain debris of intracellular digestion (phagosomes).

Vermiform cells (Jones, J., 1977), podocytes and thrombocytoids (Zachary, D. and Hoffmann, J., 1973) should be included among the plasmatocytes as specialized forms of this hemocyte type. Plasmatocytes represent 30–60% of the circulating hemocytes (in certain cases up to 90%; Schmit, A. *et al.*, 1977).

Plasmatocytes are considered to act as phagocytes, and to participate in cellular encapsulation and nodule formation forming layers of extremely flattened cells around the melanizing core.

Granular cells are rather large (8–20 μm), round to ovoid cells, packed with granules. *In vitro* the highly unstable cells rapidly degranulate, leaving clear vacuoles in their cytoplasm. As a result of degranulation the formerly opaque granular cells become hyaline. In contrast to plasmatocytes the granular cells do not spread on surfaces, and only seldom form cytoplasmic extensions. The electron microscope reveals membrane-bound granules which are either less electron-dense with a microtubular substructure or completely electron-dense. The microtubular substructure seems to disappear during maturation of the granules, and reappears during degranulation (Rowley, A. and Ratcliffe, N., 1981). The cytoplasm of granular cells is rich in ribosomes, mitochondria, vacuoles and vesicles. The RER near Golgi complexes is often swollen, indicating intense synthetic activity of the cell. Granular cells are apparently easy to distinguish from plasmatocytes in Lepidoptera, but not in other insect orders. According to some authors, adipohemocytes might be mature forms of granular cells.

In a number of insects investigated, 30–60% of the total hemocytes are granular cells. In insect defense reactions granular cells have been found responsible as initiators of nodule formation. They may also control the beginning of cellular encapsulation, forming the innermost hemocyte layer near the surface of foreign objects.

Cystocytes (also called coagulocytes) are more or less round cells, 8–15 μm in diameter, containing many inclusions with variable substructure. *In vitro*, these cells are highly fragile and quickly degranulate. Following degranulation, cystocytes become characteristically surrounded with flocculent material. Compared to granular cells the cystocytes are more fragile and react faster.

In ultra-thin sections cystocytes and granular cells are difficult to distinguish. A typical feature of cystocytes might be the swelling of the perinuclear space, which is only common in this hemocyte type.

Cystocytes have not been observed in certain insect orders such as Diptera, Hymenoptera or Lepidoptera. In other groups cystocytes represent 30–60% of the total hemocyte population.

Cystocytes are regarded as being responsible for hemolymph coagulation. They have also been reported to be capable of phagocytosis, cellular encapsulation and nodule formation (Rowley, A. and Ratcliffe, N., 1981).

Oenocytoids are extremely large (12–30 μm), round to oval cells. They remain very stable *in vitro* and rarely change their shape or extend pseudopodia. The usual cell organelles are poorly developed in oenocytoids. Mitochondria and RER occur rarely and Golgi complexes are poorly differentiated. The cytoplasm, however, is rich in free ribosomes. Bundles of microtubules have been found in oenocytoids of *Galleria mellonella*. The so-called "crystal cells", known from *Drosophila* (Rizki, M. and Rizki, R., 1959) might belong to this cell type. Crystal cells contain phenoloxidase and obviously contribute to melanization. The function of other oenocytoids which represent 1–2% of the total hemocyte number, is unknown. The remaining hemocyte types, prohemocytes and spherule cells, are not directly involved in defense reactions, as far as we know.

2.2 Phagocytosis

Phagocytosis is one of the basic abilities of the animal cell. It is a typical feature of every protozoan and of the nutritional cells in low metazoa. With increasing differentiation of the intestinal system, phagocytosis no longer serves for nutrition, but rather to eliminate debris and foreign materials from the body cavity, either by transportation or by intracellular digestion. Thus, phagocytosis becomes confined to the defense system of metazoan organisms and is restricted to its specialized cells.

Successful phagocytosis comprises a series of cellular activities, which will be discussed first.

2.2.1 Recognition

During phagocytosis hemocytes behave selectively with respect to the objects around them. Certain particles cause the reaction of phagocytes, others do not. There is no general concept which could explain how phagocytic activity is induced. Alteration in the surroundings of a foreign object (e.g. by uptake or release of substances), but also electrostatic or other chemical binding forces on the foreign surface, are conceivable signals. Recognition from a distance requires the perception of differences around the foreign object. Positive chemotaxis toward a foreign particle can only occur if a gradient of alterations can be recognized by the hemocyte.

2.2.2 Chemotaxis

Within the hemocoel of an insect chemotaxis might not be considered as being neccessary. Turbulences, narrow passages and filtering phagocytic tissues offer ample opportunity for random contact between circulating or fixed hemocytes and invading foreign particles. Existence of chemotaxis could be assumed by mistake, if thread-like extensions of hemocytes, which can easily be overlooked, attach to the foreign body. The movement of such cells toward the foreign object should not be interpreted as chemotaxis. This term is exclusively restricted to active movement along a chemical gradient.

A number of authors investigating phagocytosis vainly tried to find evidence for chemotactic attraction caused by foreign materials. In the few cases where directional movement of hemocytes has been observed, the moving cell reacted either to aggregates of granulocytes and bacteria (Rowley, A. and Ratcliffe, N., 1981), or to conidia of *Aspergillus niger* involved in cellular encapsulation (Vey, A. and Vago, C., 1971). In both cases the signals to which the migrating hemocytes reacted could originate from other hemocytes already in contact with the foreign material and not from the foreign object itself.

2.2.3 Attachment

Attachment must be based on physicochemical binding leading to adhesion of the foreign particles to the hemocytes. Adhesion in turn could influence the cellular membrane and cause further reactions of the phagocyte. Mechanisms of high specificity would be realized, if receptor molecules of foreign objects bind with adequate binding sites on the hemocyte surface. Attachment could also be mediated by humoral multivalent substances which bind with surface molecules of both the foreign object and the phagocyte.

Little information exists concerning recognition and attachment in insect phagocytosis. More knowledge about phagocytosis is available from mammalian leucocytes and also from molluscan hemocytes. To study phagocytosis in molluscs, erythrocytes have so far been the major object used. Of course, exposure to red blood cells from man, sheep or guinea pig is not a naturally occuring situation for invertebrate phagocytes, but red blood cells are easy to collect, are not pathogenic, are well-defined and are much easier to observe under the light microscope than viruses or bacteria. Their use to elucidate cellular and humoral mechanisms of phagocytosis is therefore legitimate. However, before general conclusions about immune mechanisms can be drawn, further experiments are necessary to prove that results from such a model system can be extrapolated to biological situations.

From the investigations using red blood cells it was concluded that molluscan phagocytes are equipped with membrane-bound recognition factors, "which enable them to discriminate between self and nonself" (Sminia, T. and van der Knaap, W., 1981). This extensive interpretation needs experimental justification. It has to be shown that the mechanisms found with erythrocytes are also effective against pathogens and parasites. For the investigations mentioned, monolayers of hemocytes adhering to glass surfaces were prepared. The phytohemagglutinin concanavalin A (Con-A), combined with fluorescein or peroxidase, was used to label the polysaccharide determinants on the membrane of the phagocyte. It was found that the receptors for Con-A disappeared after treatment of the phagocytes with proteolytic enzymes (in *Lymnaea stagnalis*). Simultaneously, the phagocytic activity of the treated phagocytes also disappeared. Incubation of Con-A with certain sugars inhibited the binding of the Con-A molecules to the phagocytic membrane. From these experiments it was concluded that the Con-A binding sites are anchored in the surface membrane of *Lymnaea* phagocytes by proteins, and that the discriminating binding structure is an oligosaccharide composed of those sugar molecules which had an inhibitory effect on Con-A binding (Sminia, T. *et al.*, 1981).

The distribution of the Con-A determinants on the surface of the cell membrane was investigated by Yoshino, T. *et al.* (1979). These authors demonstrated that the Con-A receptors in active cells (i.e. at 20°, but not at 4° and not after treatment with metabolic inhibitors such as sodium azide) change their position on the cell surface. These phenomena, which are called patching and capping, are believed to be the result of lateral movement and clustering of the receptor molecules in the hemocyte membrane. Subsequent to capping, internalization of labeled elements was observed in ultra-thin sections of hemocytes. The events described could represent mechanisms responsible for attachment and ingestion of particles. Capping and internalization of receptors is also known from mammalian lymphocytes (see Yoshino, T. *et al.*, 1979). It is important to note that receptor molecules with identical specificities have been found on erythrocytes and on bacteria. Thus, the hemocyte monolayer could also provide a suitable assay for studying phagocytosis of bacterial mutants with known surface alterations. However, one has to realize that a hemocyte monolayer represents a selection from the total hemocyte population, consisting only of those hemocytes which have attached to the glass surface.

2.2.4 Efficiency of phagocytosis

Investigation of phagocytosis began a hundred years ago when Metchnikoff, E. (1884) injected yeast cells (*Monospora* sp.) into the phyllopod *Daphnia magna* and observed the clearance of the hemolymph resulting from the phagocytic activity of the hemocytes. Since then, many materials have been injected into different crustaceans and insects to study the efficiency of phagocytosis and to investigate its mechanisms. Insect hemocytes have been found to ingest all kinds of micro-organisms such as viruses, bacteria, protozoa and fungi and many other particulate materials including India ink, different dyes, various organic and inorganic powders, polystyrene and polysaccharide beads and also erythrocytes of different vertebrates.

Efficiency of phagocytosis has been evaluated either as clearance efficiency or increased mortality after overloading the phagocytes with non-pathogenic particles, causing a so-called phagocytic blockade. Leutenegger, R. (1967) calculated that 2.5×10^8 injected virus particles have been eliminated within 3 h from the hemolymph of *Galleria mellonella*. Stairs, G. (1964) injected India ink into larvae of *G. mellonella* and found the hemocytes packed with engulfed ink particles. If waxmoth larvae in this condition were injected again with an otherwise harmless dose of virus particles, lethal virus infections developed.

Improved methods to investigate the effect of phagocytosis have been developed in recent years, based on the progress in cell and tissue culture methodology. Anderson, R. *et al.* (1973) incubated hemocyte cultures of the cockroach *Blaberus craniifer* with bacteria. By plating samples of the bacteria–hemocyte suspension after different times of incubation, the rate of surviving bacteria could be determined. The effect of phagocytosis was found to differ, depending on the bacterial strains tested. Ratcliffe, N. and Rowley, A. (1979b) introduced different bacteria into hemocyte cultures of *G. mellonella.* Only bacteria of one strain (*Sarcina lutea*) were found to be killed after phagocytosis. Ingested bacteria of other strains (e.g. *Escherichia coli, Klebsiella aerogenes, Staphylococcus aureus* 502 A) multiplied within the phagocytic hemocytes and were released into the culture medium. In such hemocytes only limited lysosomal activity (measured as acid phosphatase activity) could be demonstrated. In other insects, e.g. *Calliphora erythrocephala, in vitro* phagocytosis was more efficient, and *E. coli* was rapidly ingested and killed. Active lysosomes had accumulated around the phagosomes within 15 min after ingestion of the bacteria.

It is surprising that in some insects phagocytosis should be such an inefficient defense reaction against bacteria, but the efficiency of an immune reaction cannot be estimated from *in vitro* experiments alone. Phagocytosis with cultured hemocytes does not account for the contribution coming from fixed phagocytes ("phagocytic tissues") and also neglects the additional effects which result from cooperation of phagocytosis with other cellular and humoral reactions occurring within the intact hemocoel.

2.2.5 Ingestion and digestion

Ingestion of particles by insect hemocytes corresponds to the events known from vertebrate cells. After attachment of the particle to the surface of the phagocyte, cytoplasmic pseudopodia extend from the cell periphery and surround the particle. As they fuse, they enclose the particle in a membrane-bound vacuole which moves to the interior of the cell. If lysosomes fuse with the phagocytic vacuole and release their digestive enzymes into it, digestion of its contents may begin. This results in swelling and subsequent disintegration of the engulfed organisms. Indigestible remnants may finally be released from the phagocytic cell by exocytosis.

2.3 Cellular encapsulation

Cellular encapsulation is a reaction of hemocytes to living or non-living foreign objects, which leads to a multicellular envelope (= capsule) surrounding this object (Figs 1 and 2). It was found to be effective against all kinds of parasites such as trematodes, cestodes, nematodes, parasitic insects and fungi. Cellular encapsulation is also provoked by a variety of implants (Fig. 3) such as eggs and tissues or inanimate materials, e.g. glass rods, cotton, nylon threads, latex particles etc. (Salt, G., 1960, 1963; Brehelin, M. *et al.*, 1975). Independent of the insect species and of the type of foreign material involved, the resulting capsules are very similar. In general their fine structure corresponds to the description given in the first electron microscopical investigation of cellular encapsulation (Grimstone, S. *et al.*, 1967).

The multicellular envelope, enclosing pieces of Araldite in larvae of *Ephestia kühniella*, consisted of three different zones. The innermost layers, about 10 cells thick, were formed of tightly packed hemocytes, which were not visibly flattened but showed pronounced signs of necrotic and autolytic alterations. The middle region of the cellular envelope was composed of about 20–40 layers of extremely flattened hemocytes and the outer region consisted of about 10 layers of cells which had hardly changed their normal appearance. Cellular elements which are typical for tight cell contacts, such as microtubules and desmosomes, were predominant in the middle region of the capsule. The intercellular space between the flattened cells was filled with electron-dense material. The only hemocytes discovered in the cellular envelope were plasmatocytes. Melanization, which is typical for capsules around living organisms, did not visibly occur in the case of Araldite encapsulation in *E. kühniella*.

The formation of a melanotic capsule was then studied in the beetle *Diabrotica undecempunctata* in response to a mermithid nematode (Poinar, G. *et al.*, 1968). The hemocytes found in contact with the nematode 2 h after penetration were classified as plasmatocytes. At the sites of contact the hemocyte membrane had disappeared and the cytoplasm stuck to the cuticle of the nematode, thereby following every irregularity of the cuticle's surface. During the following hours the cells attached to the cuticle underwent further autolysis. Adjacent to the surface of the nematode electron-dense material appeared. Under the light microscope this area was

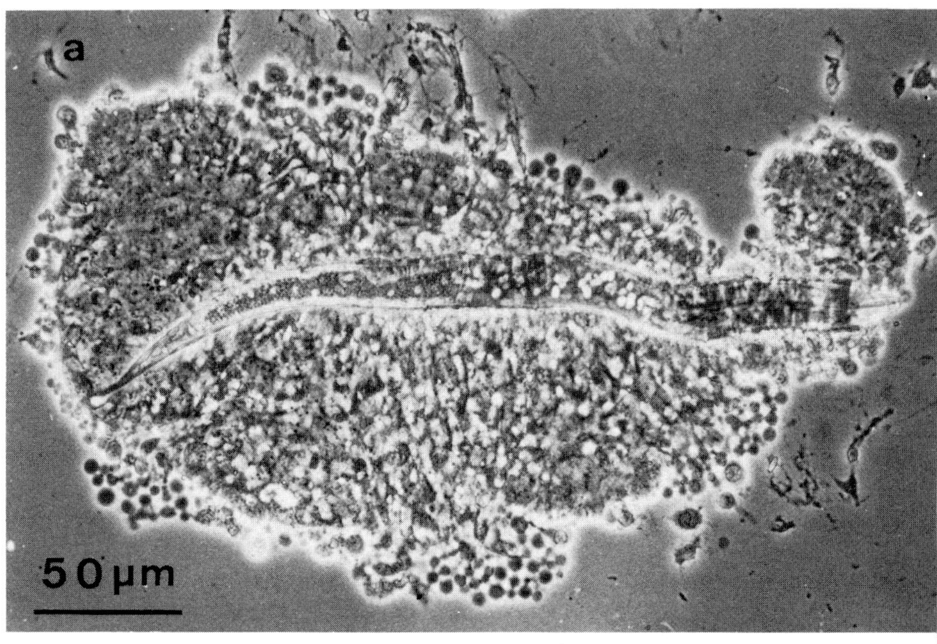

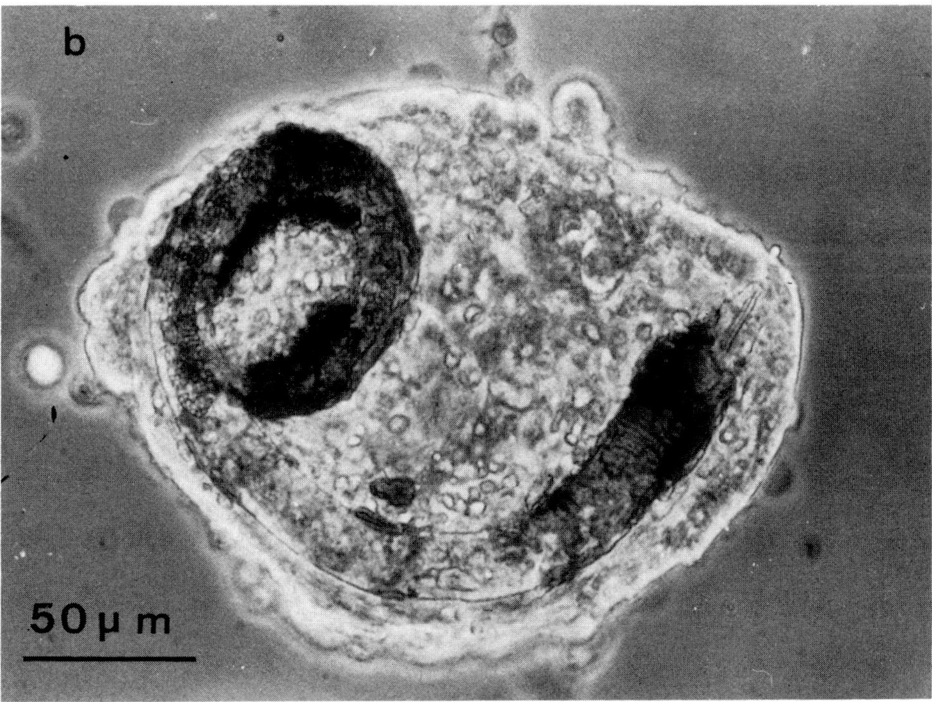

FIG. 1. Cellular encapsulation of nematodes (*Filipjevimermis leipsandra*) in the hemocoel of the cucumber beetle (*Diabrotica undecempunctata*). **a**: 12–24 h after invasion: attaching hemocytes have formed a multicellular envelope around the parasite; **b**: 48 h after invasion: the parasitic nematode has died within the compact hemocytic capsule. Note melanization near the surface of the nematode.

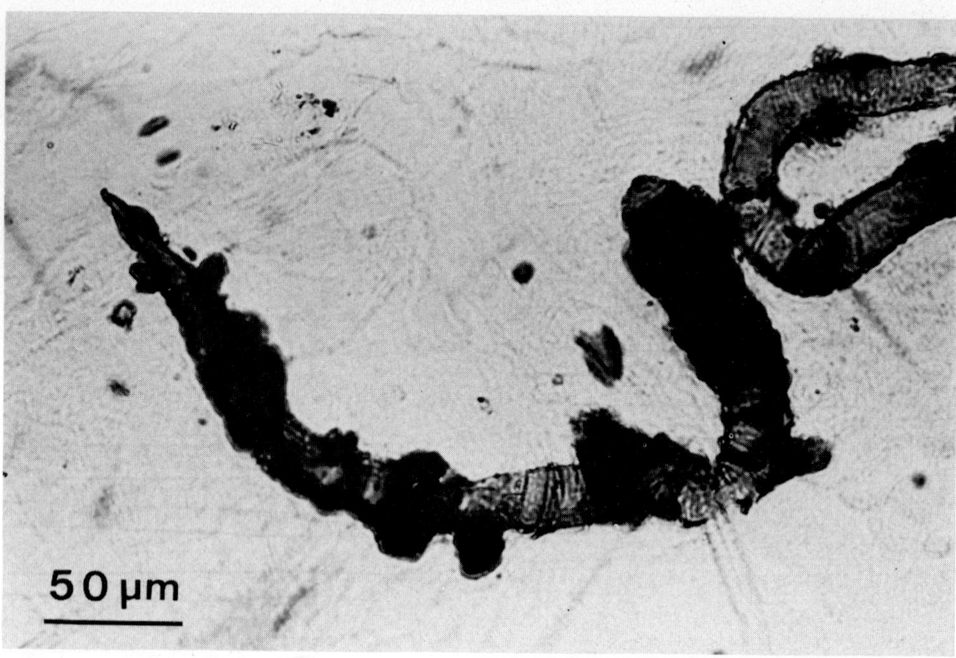

FIG. 2. Cellular encapsulation of *Filipjevimermis leipsandra* in *Diabrotica undecempunctata*. The carcasses of killed nematodes covered with crusts of completely melanized hemocytes remain in the insect hemocoel without causing further reactions.

brown to black in color and the material was therefore considered to be melanin. This assumption received further support by the positive results of different histochemical tests for melanin. The melanin partly originated from electron-dense inclusions appearing within the cytoplasm of the attached cells and partly from direct transformation of the cytoplasm which stayed in contact with the foreign surface. The melanin layer completely sealed the surface of the nematode, forming a negative of the surface structure of the cuticle. Next to the disintegrated and partly melanized inner cells followed a middle layer of flattened cells and finally an outer layer with hemocytes of more or less normal appearance. The melanization process continued for several days and occasionally included the whole cellular envelope which then changed into a solid black crust (Fig. 2).

Many further investigations concerning cellular encapsulation in insects have been performed since the classical studies of Salt, G. (1956, 1963), Grimstone, S. and Salt, G. (1967) and Poinar, G. *et al.* (1968). If one tries to survey and summarize the present status of knowledge about cellular encapsulation, the following points have to be discussed.

2.3.1 EARLY EVENTS IN CELLULAR ENCAPSULATION

Cellular encapsulation occurs much faster than first investigators thought. It is true that the conspicuous melanized capsule around a foreign body can only be encountered 1 day or more after invasion or implantation, but the very first events of cellular encapsulation take place within minutes (Fig. 4). Attachment of hemocytes to foreign objects was observed within 1 min, a complete layer of cells after 5 min and an advanced multicellular envelope within 1 h (Burghause, F., 1980; Burghause, F. in prep.). On average, the typical three-layered cellular envelope is formed within 2–24 h.

2.3.2 HEMOCYTES INVOLVED

The type of cells responsible for cellular encapsulation of foreign objects apparently varies between different insect species. Most often granular cells have been found to form the capsules (e.g. during encapsulations in *Ephestia, Galleria, Melolontha, Locusta* and *Periplaneta*), but in some insects other cell types such as plasmatocytes, thrombocytoids, lamellocytes or oenocytoids were reported to

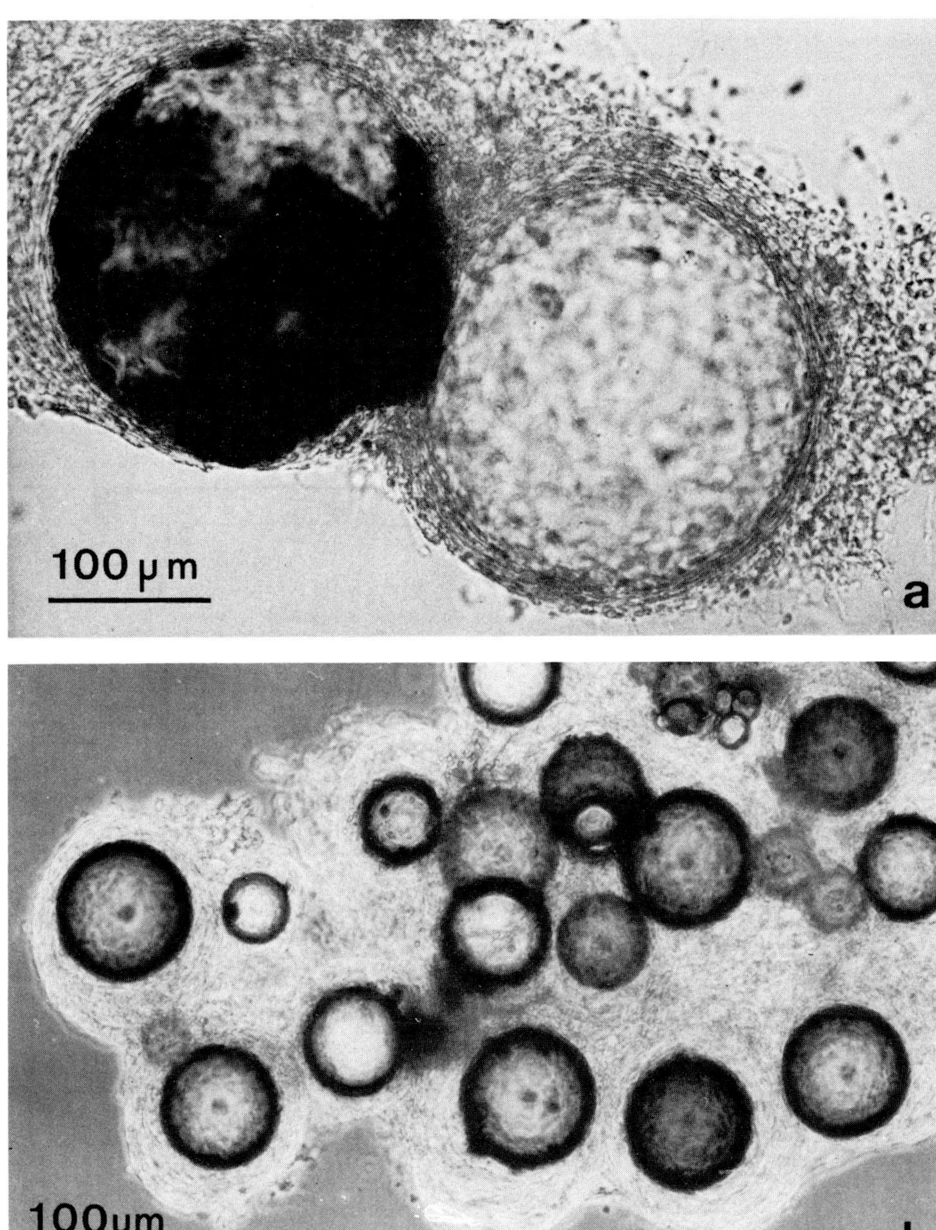

FIG. 3. Cellular encapsulation of inanimate materials injected into the hemocoel of larvae of *Galleria mellonella* (courtesy of W. Lingg). **a**: Cellular encapsulation and partial melanization of two beads of DEAE cellulose (Sephadex A); **b**: heavy cellular encapsulation and melanization of beads of polystyrol (Dowex 50W × 4).

participate in cellular encapsulation (Baerwald, R., 1979). This inconsistency may be an expression of an insufficient classification of insect hemocytes which gives different names to cells with identical function. It could also mean, however, that the mechanisms of cellular encapsulations are not identical in different groups of insects. Careful analysis of the cell types which react with foreign objects during encapsulation is needed. We also ought to know if, in a given insect species, the same

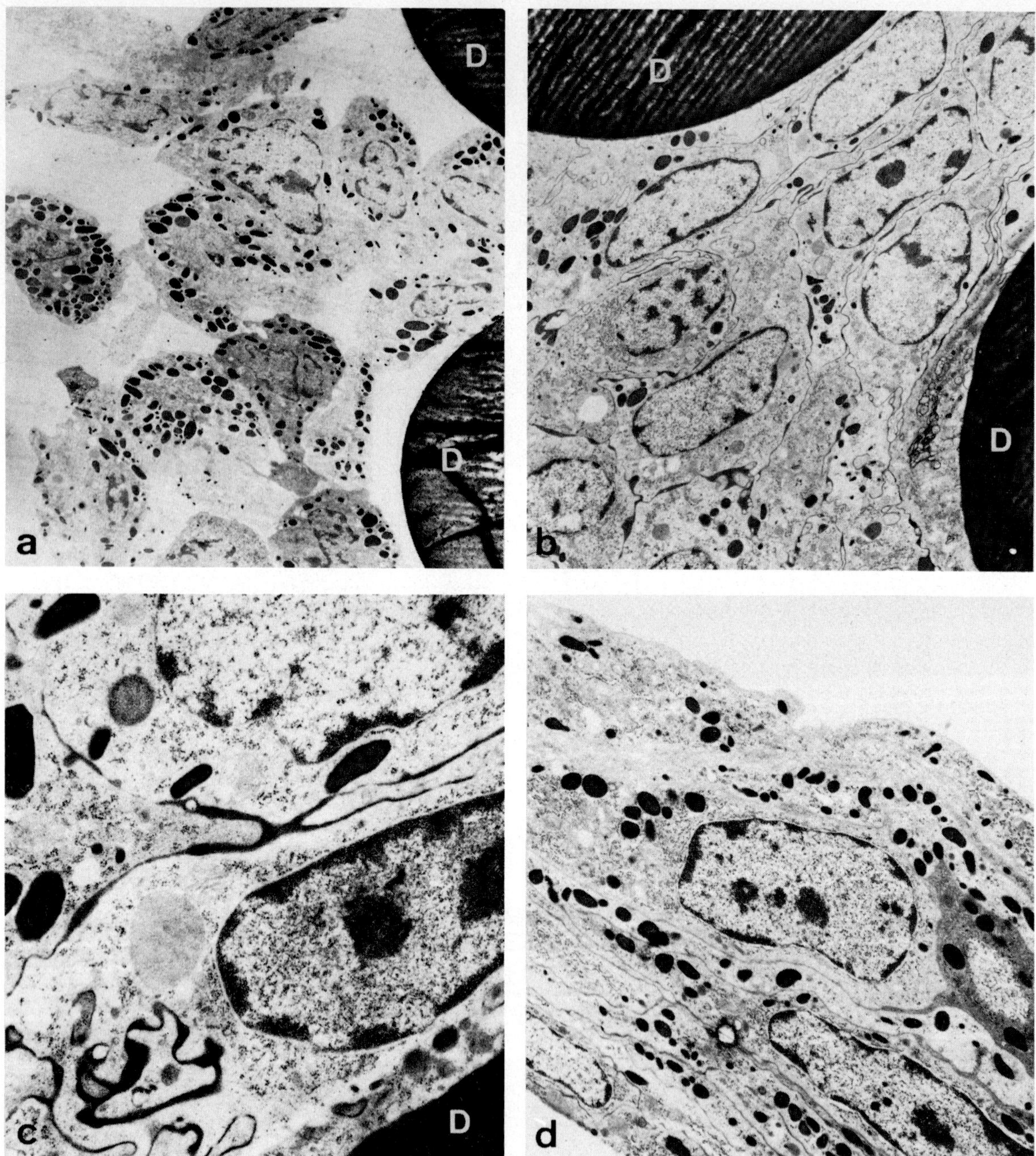

FIG. 4. Ultrastructure of capsule formation in *Galleria mellonella* at different times after injection of foreign materials (D = Dowex 50W × 4) (courtesy of Dr F. Burghause). **a**: After 5 min: accumulation and first attachment of hemocytes. (× 1800); **b**,**c**: 30 min: compact package of the hemocytes. Note: the number of granules per hemocyte has decreased: electron-dense material appears in the intercellular space. (**b** = × 2700, **c** = × 32000); **d**: 1 h after injection: flattening of the hemocyte in progress (× 5600).

type of hemocytes reacts against different foreign objects.

Some investigators found more than one hemocyte type involved in the formation of a cellular capsule. In *Galleria mellonella*, during encapsulation of an implanted pieces of nerve cord from *Schistocerca gregaria*, the first cells adhering were cystocytes and granular cells (Schmit, A. and Ratcliffe, N., 1977). The attached cells underwent rapid degranulation and cell lysis. About 30 min after implantation of the foreign tissue, plasmatocytes attached at those sites where lysis of the granular cells and cystocytes had occurred before.

Biphasic encapsulation was also observed in *Clitumnus extradentatus* in response to implanted fragments of Araldite resin (Schmit, A. and Ratcliffe, N., 1978). Within 5 min cystocytes and granular cells aggregated and lysed on the implant surface. Subsequently, plasmatocytes and granular cells accumulated around the implant. The final result of this encapsulation differed from the capsules described in most other insect species. It was composed of an inner melanized region of flattened cells and an outer region of degenerating unflattened hemocytes.

Nappi, A. (1973), in a light microscopical investigation of parasitized *Musca domestica* and *Orthellia caesarion*, observed activation of fixed hemocytes which entered the circulation upon invasion of the nematode parasite. The first cells aggregating were oenocytoids which fused to a pigmented layer adhering to the cuticle of the parasite. This initiated subsequent aggregation and fusion of other hemocytes.

Disintegration of those hemocytes that attached first to the foreign surface has been reported by many investigators (e.g. Poinar, G. *et al.*, 1968), but a biphasic encapsulation by different cell types, such as Schmit, A. and Ratcliffe, N. (1977, 1978) and Nappi, A. (1973) observed, was not noted in other investigations. It remains unclear whether these details have been missed or did not occur. A thorough comparative investigation should prove or disprove the hypothesis that cellular encapsulation is initiated by a granular type of hemocyte and continued by plasmatocytes forming the multicellular envelope.

2.3.3 Mechanism of capsule formation

Regardless of whether or not the cellular capsule is formed by different types of hemocytes, only the cells of the first layer come in direct contact with the foreign surface. It is important to note that all further hemocytes are exposed to different conditions than the cells which first attach. The signals initiating their attachment could originate from the hemocytes already adhering. Since the accumulation of hemocytes does not continue indefinitely, the triggering stimuli must decrease with increasing thickness of the cellular envelope. The condensation of the accumulated cell layers could contribute to a reduction of the aggregation stimuli, which rise either from the foreign object or from the disintegrating innermost cells.

The decreasing tendency of the outer cells to alter their normal shape is in agreement with the concept of a stimulus gradient from the center to the periphery of the capsule. During capsule formation some of the outer aggregated hemocytes leave the cellular envelope and return to the circulating hemolymph. The cell number (and the volume) of a capsule has its peak 12–24 h after introduction of the foreign object, and decreases during the following days.

The solidification of the cellular envelope, starting from a loose accumulation of cells and resulting in the firm package of extremely flattened cells, represents an active process. Mechanisms of cell movement and cell junction must be involved. The ultrastructural elements for such activities, e.g. microtubules, intercellular gap junctions and desmosomes, have clearly been demonstrated. Baerwald, R. (1979) therefore recommends encapsulation experiments with compounds known to stop cell movement and cell division (e.g. by destruction of microtubules). Hemocyte cultures might also offer an opportunity to study how monodispersed cells can be induced to attach and to form cell junctions which normally would not occur in free hemocytes.

2.3.4 Melanization

Melanization is not a regular feature of cellular encapsulation. Only living organisms, tissues (xenografts) and certain inanimate materials provoke both encapsulation and melanization. The majority of inanimate materials tested were encapsulated without, or with little formation of,

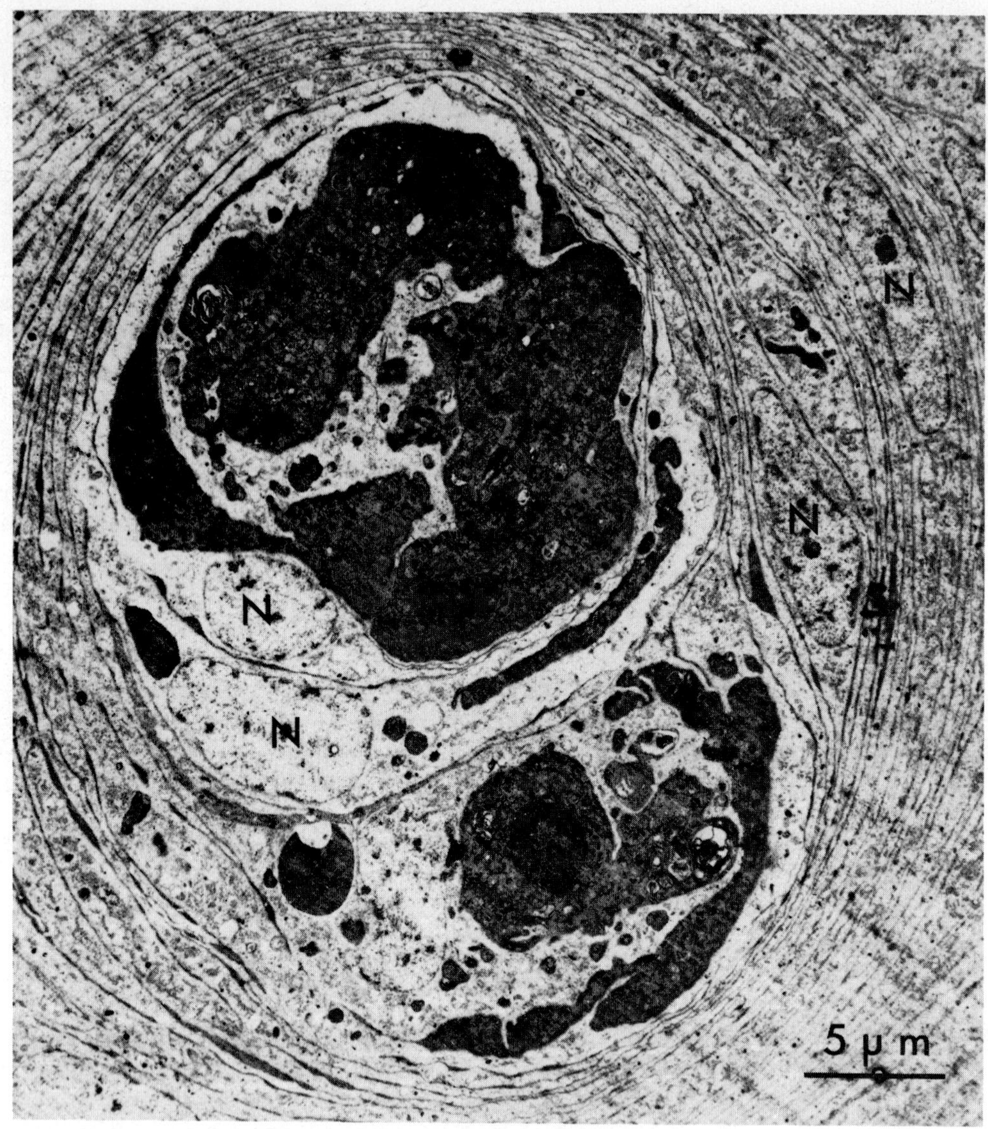

FIG. 5. Nodule formation after injection of fungal spores (*Mucor hiemalis*) into larvae of *Galleria mellonella*. The core of such nodules consists of aggregated fungal spores and disintegrated hemocytes which become surrounded by concentric layers of extremely flattened hemocytes. N = hemocyte nucleus.

melanin. The degree of melanization also varies with the insect species involved (Burghause, in preparation). The phenomenon of melanization with its many implications will be discussed in section 2.6.

2.4 Nodule formation

Nodule formation occurs after high doses of particulate material such as bacteria, fungal spores, protozoa or suspensions of living and non-living particles have been injected into the hemocoel of an insect (Fig. 5). Nodules have also been found in field-collected insects (Ratcliffe, N. and Rowley, A., 1979a).

The most detailed analysis of nodule formation was given by Gagen, S. and Ratcliffe, N. (1976) and Ratcliffe, N. and Gagen, S. (1977) working with

Galleria mellonella injected with *Bacillus cereus*. On random contact with the bacteria, granular cells discharged a flocculent material surrounding bacteria and granular hemocytes. Within 30 min the aggregates of discharged material, entrapped bacteria and partly disintegrating granular cells compacted and began to melanize, mainly in the region around the bacteria. Simultaneously the entire aggregates acted like foreign bodies, provoking their cellular encapsulation by further cells (plasmatocytes in the case of *G. mellonella*). Within 12–24 h a typical cellular capsule was formed consisting of hardly changed hemocytes in the outer region, extremely flattened cells in the middle region and less flattened cells with sites of melanization near the core. Only some of the hemocytes contained phagocytosed bacteria.

On the basis of these results the earlier interpretation of nodule formation as cellular encapsulation by hemocytes which had disintegrated after heavy phagocytosis, (Götz, P., 1973), can no longer be considered valid. As Rowley, A. and Ratcliffe, N. (1981) pointed out, nodule formation is initiated by degranulation of granular cells reacting to the bacteria present. Most bacteria within the nodule were not phagocytosed but directly entrapped within the discharged material.

These results from *Galleria* were principally affirmed with three other insect species (*Clitumnus extradentatus*, *Tenebrio molitor* and *Schistocerca gregaria*). Two types of hemocytes, namely granular cells and cystocytes, were found discharging their inclusions in response to injected bacteria. The plasmatocytes, subsequently forming the multicellular envelope, were involved in little, or in *C. extradentatus* heavy, phagocytic activity (Rowley, A. and Ratcliffe, N., 1981).

Further detailed studies are necessary to reveal whether this pattern of nodule formation generally occurs in insects. If so, nodule formation would offer an opportunity to compare the hemocyte types of different insects and to identify those with identical functions. Furthermore, the question of whether nodule formation always follows the same pattern, independent of the type of particulate material involved could also be answered.

2.5 Humoral encapsulation

Humoral encapsulation is an unusual type of defense reaction to foreign material which so far has only been observed in certain Diptera. The reaction commands special interest because it may represent a cell-free model system for biochemical events which also occur during cellular encapsulation. Furthermore, humoral encapsulation is fast (complete capsules are deposited within minutes), it is highly efficient, and it occurs *in vivo* as well as *in vitro* and therefore can easily be followed under the microscope.

Cell-free encapsulation was first reported by Bronskill, J. (1962) and Esslinger, J. (1962) in mosquito larvae infected with parasitic nematodes. The authors noted direct deposition of minute particles of pigmented material that they called "melanin". This deposition resulted in a continuous layer that completely enclosed the nematode. Following melanin deposition, hemocytes accumulated and the innermost of these attached cells also melanized. On cross-sections through definite capsules 4–5 h after invasion, a two-layered envelope was detected, consisting of an inner layer of melanin clumps and an outer layer of host hemocytes, containing melanin particles in their cytoplasm.

Electron microscopic investigation of humoral encapsulation in *Chironomus* larvae confirmed the non-cellular origin of the capsule material (Götz, P., 1969). Electron-dense material was found deposited directly from the hemolymph on the surface of the nematode (a mermithid). Debris of disintegrating hemocytes occurred only outside the capsule. The contact between the capsule material and the foreign surface was so close that the inner side of the capsule forms a negative of the surface structures of the nematode's cuticle. It appears obvious that such a tight seal must interrupt the exchange of nutrients and oxygen between the enclosed organism and its surroundings. A correspondence between humoral encapsulation and melanization as occurring during cellular encapsulation was postulated (Götz, P., 1969).

Poinar, G. and Leutenegger, R. (1971) described a biphasic formation of the melanized sheath around the nematode *Neoaplectana carpocapsae* in larvae of *Culex pipiens*. At the beginning of

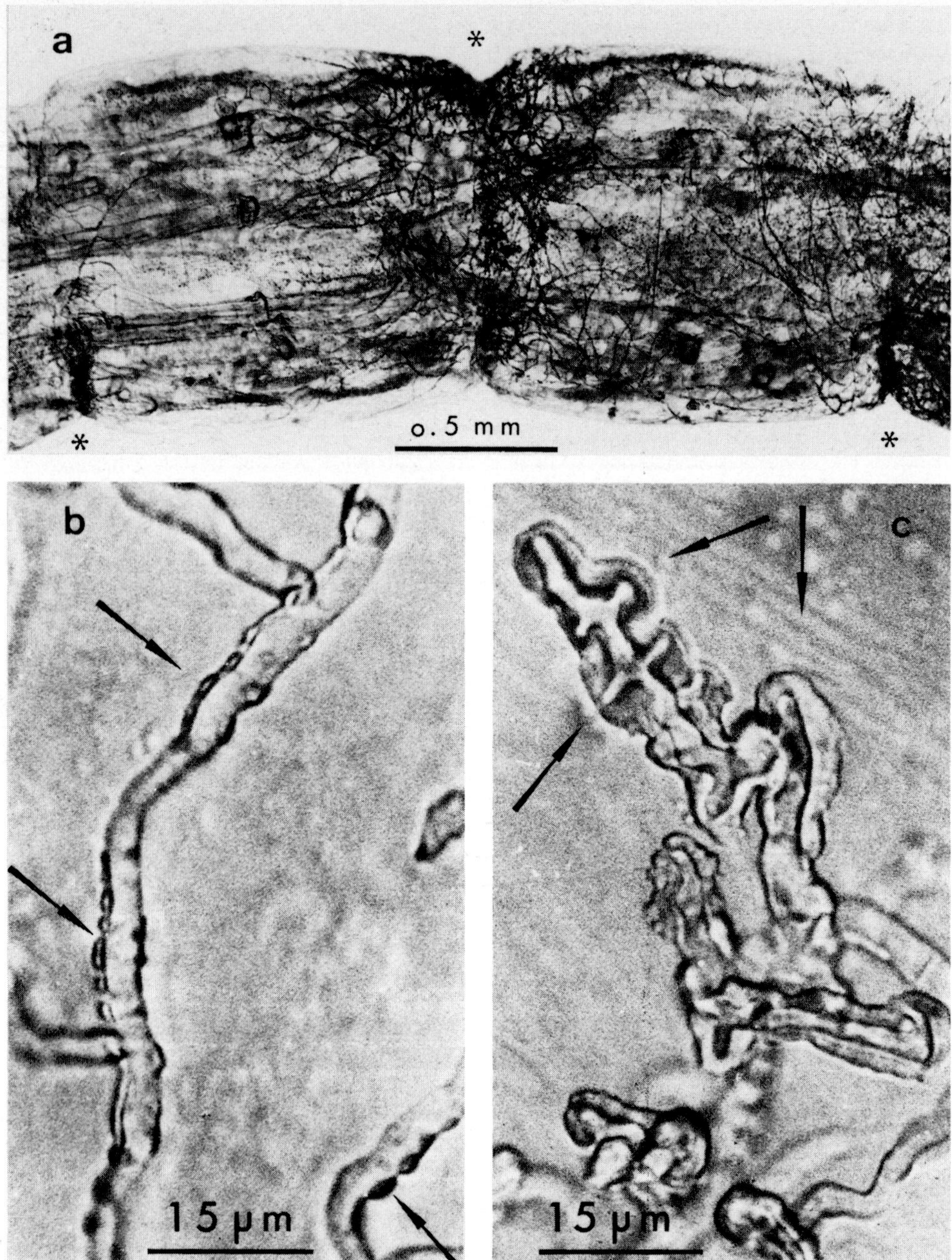

FIG. 6. Insufficient humoral encapsulation of the fungus *Beauveria bassiana* invading larvae of *Chironomus riparius*. Two days after exposing a *Chironomus* larva to fungal spores the germinating hyphae have penetrated the cuticle, mainly in the intersegmental areas (stars). Humoral encapsulation (arrows), occurring in the hemocoel (**b**) as well as within the cuticle (**c**), is not capable of stopping the development of this fast-growing fungus.

encapsulation (25 min after invasion of the nematodes) a homogeneous deposit with low electron-density was found surrounding the parasites. After 1 h, electron-dense granules had appeared within this homogeneous layer, preferentially in the region adjacent to the nematode's surface. The definite capsule was encountered 5–10 h after invasion and it was

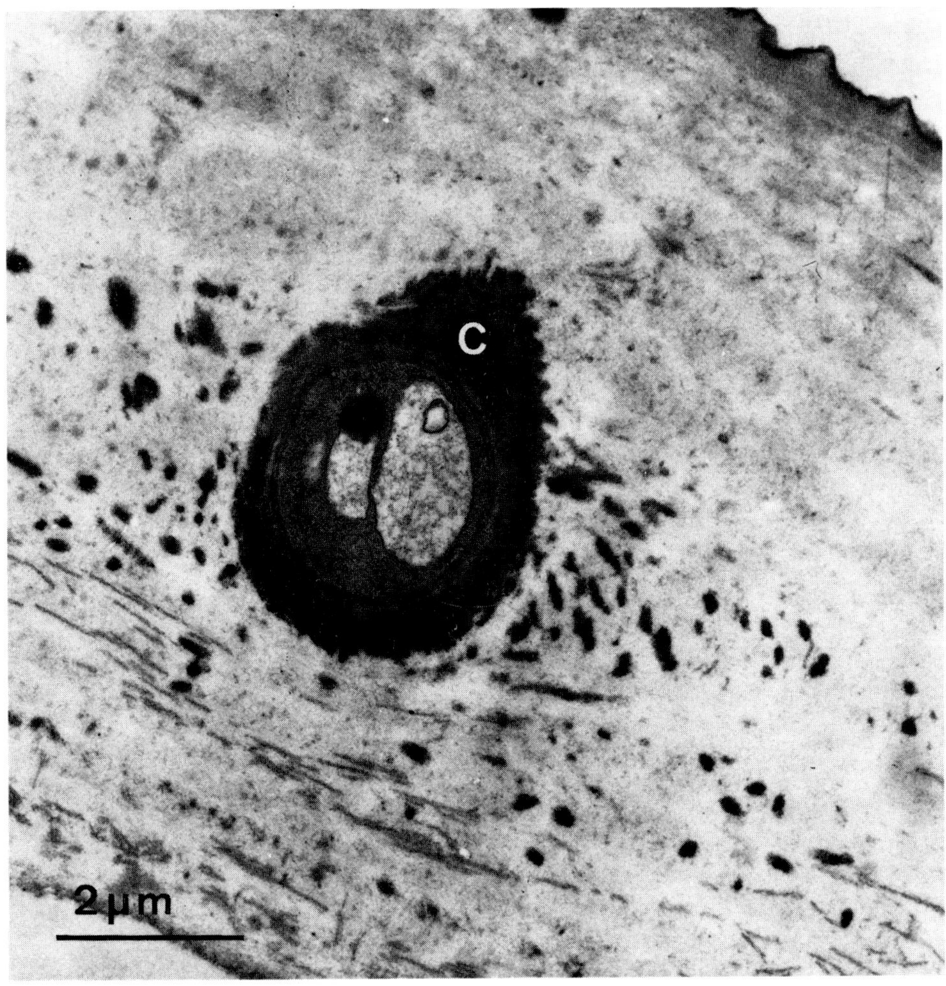

FIG. 7. Ultrastructure of humoral encapsulation of *Beauveria bassiana* within the cuticle of a *Chironomus riparius* larva. Cross-section through hypha surrounded with melanin deposit. C = capsule material.

composed of an inner region with electron-dense material, a non-melanized, or slightly melanized, middle layer and an outer layer that contained tracheoles and cellular debris.

2.5.1 EFFICIENCY OF HUMORAL ENCAPSULATION:

Incubation of living nematodes in drops of isolated hemolymph impressively illustrates the efficiency of humoral encapsulation. At the beginning the nematodes swim around freely, but as soon as the deposition of capsule material starts, i.e. after 2–3 min, the nematodes become glutinous and stick to the glass surface, first with their ends only, then with their whole body. Eventually the nematodes are entrapped in a solidifying capsule which completely stops their ability to move. Sometimes a capsule cracks and a nematode manages to escape from it, but since deposition of material starts again on the newly exposed surface, the nematode might become encapsulated a second or a third time (Götz, P., 1976).

As many as 45 nematode larvae have been found encapsulated within one *Chironomous* larva. Nevertheless, some nematodes occasionally escape from encapsulation and develop into mature parasites. Intensity and success of encapsulation vary with the host and parasite species involved, with the age of the larvae at the time of infection, and obviously this also depends on individual

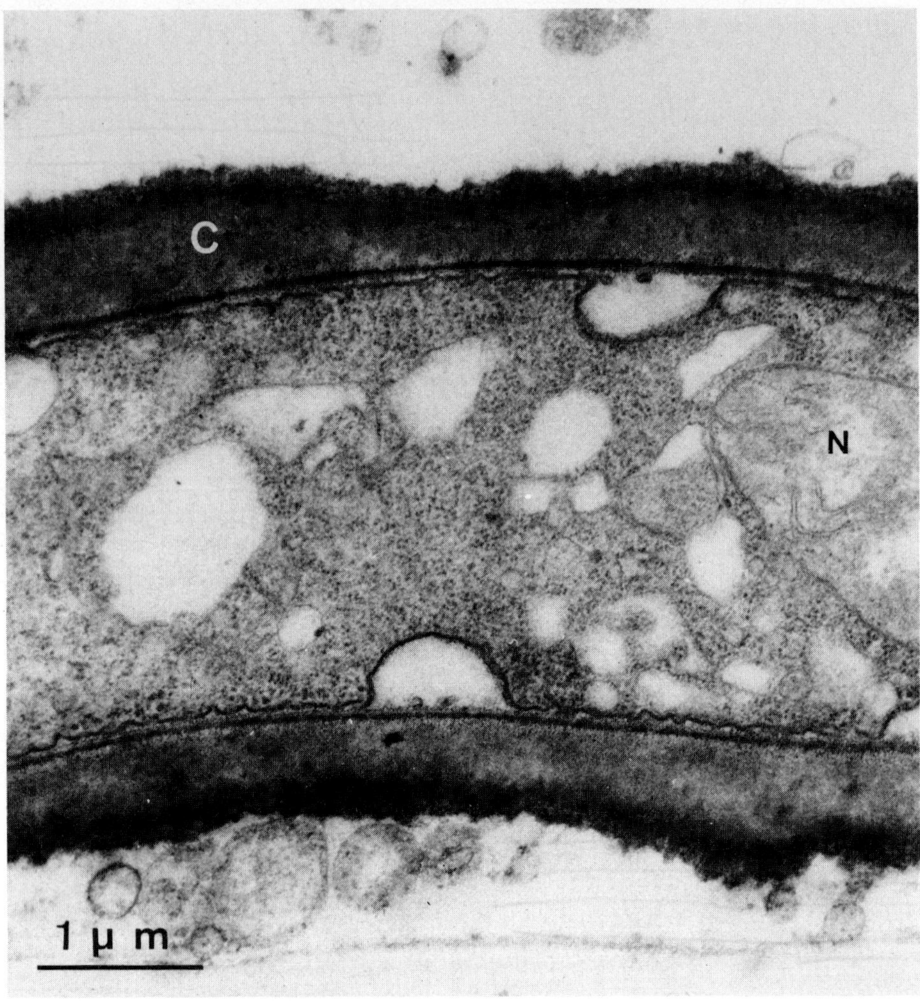

FIG. 8. Ultrastructure of humoral encapsulation of *Beauveria bassiana* in the hemocoel of a *Chironomus riparius* larva. Longitudinal section of encapsulated hypha. Note decomposition of cell content. C = capsule material; N = nucleus.

differences between the larvae (Götz, P., 1964).

Humoral encapsulation also proved to be very efficient against fungi and bacteria. Conidiospores of insect pathogenic fungi (*Metarrhizium anisopliae*, *Aspergillus niger*, *Beauveria bassiana*) injected into the hemocoel of *Chironomus* larvae were encapsulated within minutes. Only *B. bassiana*, the fastest-growing of the fungi tested, competed well with the speed of capsule formation and often escaped complete encapsulation (Figs 6–8). The hyphae of *B. bassiana*, which are able to actively penetrate the cuticle, provoked heavy melanization within the cuticle as well as in the hemocoel (Götz, P. and Vey, A., 1974).

In a recent study *Chironomus* larvae were tested with 28 bacterial strains, some of them being highly pathogenic for insects (e.g. *Serratia marcescens*, *Pseudomonas aeruginosa*, *Xenorhabdus nematophilus*, *Bacillus thuringiensis*, *Bacillus popilliae*). *Chironomus* larvae were found to be very insensitive to all of them (Fig. 9). More than 10^5 cells had to be injected to significantly increase the mortality of *Chironomus* larvae. However, injection of less than 10^2 cells of the same bacterial strains was sufficient to induce a lethal bacteriaemia in control animals, larvae of *Galleria mellonella* (Götz, P. *et al.*, in preparation).

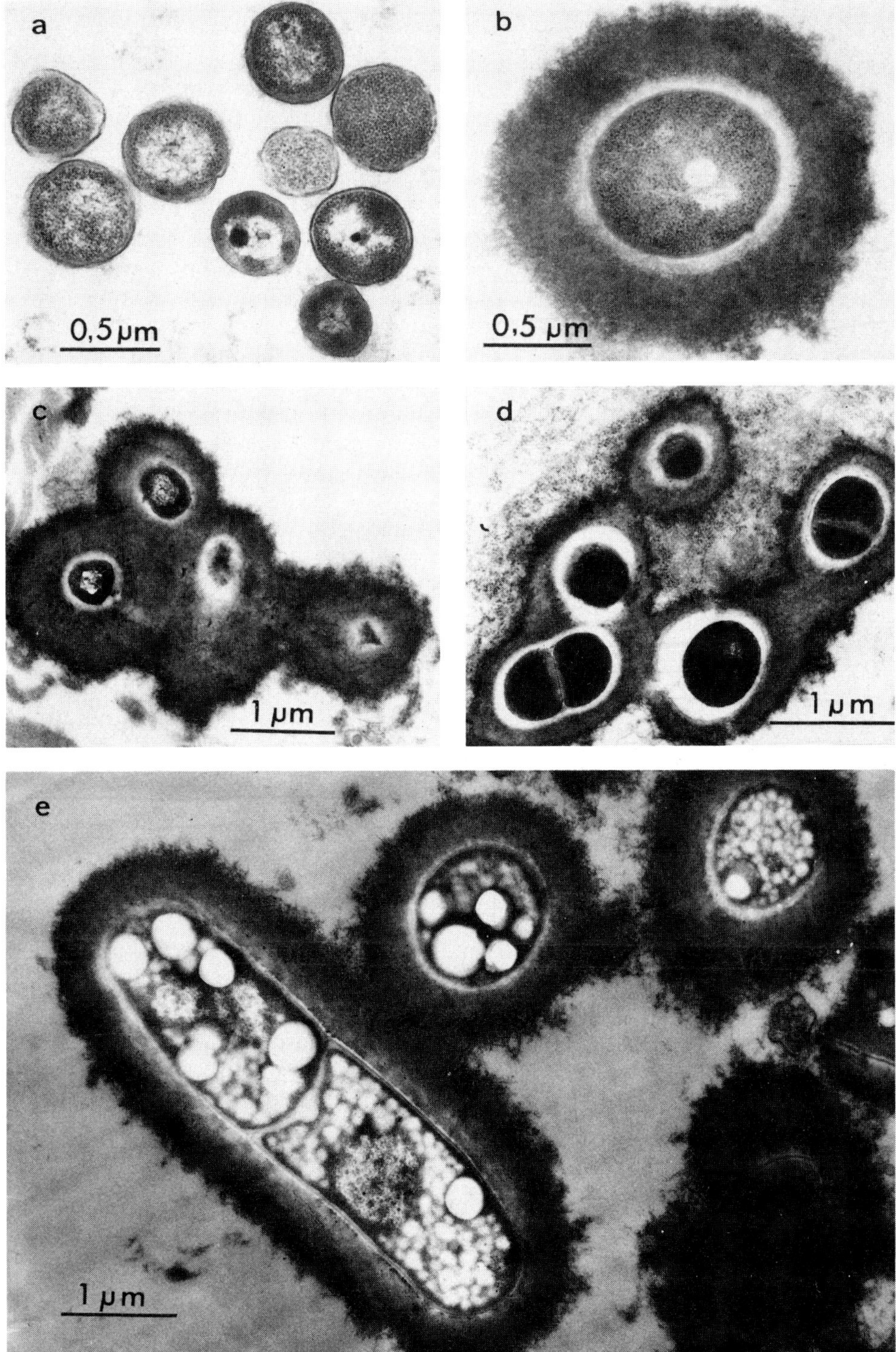

FIG. 9. Humoral encapsulation of different bacterial strains injected into the hemocoel of larvae of *Chironomus riparius*: *Bacillus thuringiensis* 1 min (**a**), 10 min (**b**), and 16 h (**e**) after injection, *Pseudomonas aeruginosa* after 1 h (**c**) and *Staphylococcus aureus* after 24 h (**d**). **a,b**: Capsule formation occurs within 10 min; **c–e**: during the following hours the capsule material becomes more electron-dense and the enclosed bacteria decompose.

Characteristically, the majority of pathogens that can be encountered in free-living *Chironomus* larvae are intracellular parasites such as viruses, rickettsiae and microsporidia. Extracellular parasites such as bacteria, fungi and protozoa are not common in these insects. Only nematodes seem to be large enough to withstand humoral encapsulation and occasionally play a significant role in the control of *Chironomus* populations (Wülker, W., 1961).

2.5.2 Humoral encapsulation occurs only in insects with small numbers of hemocytes

Götz, P. *et al.* (1977) when investigating insects from 12 different orders, found a good correlation between the number of hemocytes and the type of defense reaction. Humoral encapsulation was encountered in certain dipteran genera (*Culex, Anopheles, Chironomus, Chaoborus, Psychoda, Eristalomya,* etc.) along with small numbers of circulating cells (500–5100 mm^{-3}). The hemocyte counts of dipteran species with cellular encapsulation, however, ranged from 2300 to 49,000 mm^{-3}.

2.5.3 Encapsulation of inanimate materials

Occurrence and intensity of humoral encapsulation depends on the type of material introduced into the hemolymph of *Chironomus* larvae. Heavy encapsulation is elicited by all kinds of living organisms and tissues, including injured *Chironomus* tissues. Only a few inanimate materials were found to provoke heavy encapsulation; these were anionic and neutral polydextrane (Sephadex C 50 and G 50), silk, agar gel, cellulose and polyacrylamide gel. Weak provocators are, for example, cationic polydextrane (Sephadex A 50), methacrylate, epoxide resin and keratin (hairs and feathers). No visible encapsulation occurred when glass, activated carbon, chalk, iron powder, paraffin oil, poly vinylchloride, polyethylene, polystyrol, etc. were incubated in *Chironomus* hemolymph (Wilke, U., 1979).

2.5.4 Chemical nature of capsule material

The material resulting from humoral encapsulation has always been called melanin, but originally this designation was only based on the brown or black color of the substance. Chemically, melanins are difficult to identify. They cannot be degraded to defined compounds and are completely destroyed when rigid methods are applied such as treatment with warm alkali.

A number of tests known to be characteristic for melanins were therefore performed with capsules resulting from humoral encapsulation in *Chironomus* hemolymph (Vey, A. and Götz, P., 1975). Some of the properties identified were: resistance of the capsule material to organic and inorganic solvents, resistance to strong acids, solubility in strong bases, bleaching by oxidizing agents. A selection of enzymes specific for different proteins, lipids and polysaccharides was also applied to the capsules, but none of these was able to digest the capsule. Deposition of capsule material was inhibited when glutathione was added to the hemolymph together with foreign materials which normally provoke humoral encapsulation. Addition of phenylthiourea (PTU) only prevented the coloration and hardening, but not the deposition of capsule material. From these and further histochemical tests the authors concluded that the capsule represents a protein–polyphenol complex which is produced under the control of phenoloxidases (Vey, A. and Götz, P., 1975). Further discussion of the phenoloxidase system follows.

2.6 Phenoloxidases and their possible role in cellular and humoral encapsulation

The possibility that phenoloxidases might be involved in insect defense reactions has been considered by several authors. Taylor, R. (1969) pointed out that the phenol–phenoloxidase system could be capable of "killing micro-organisms and parasites, isolating foreign bodies and sealing and repairing wounds".

In insects, phenoloxidases are responsible for tanning and the related processes known as sclerotization (Richards, A., 1978). These enzymes oxidize tyrosine derivates, e.g. dopamines, to quinones which rapidly react with each other and with secondary and terminal amines of proteins. The result of such cross-linking and polymerization are very stable protein–polyphenol complexes which cannot be dissolved without complete destruction (see S. Andersen, this volume).

This makes it very difficult to analyze the melanization process, but it is also the basis for the extraordinary mechanical qualities of its end products, which are perfectly suited to harden cuticular structures (see S. Andersen, this volume) and might be able to entrap and seal foreign organisms as well.

Depending on the substrate specificity of a given phenoloxidase, and on the proteins which happen to be present when the quinones are formed, the end products resulting from phenoloxidase activity differ considerably. Since the variety of enzymes, substrates and proteins within the cuticle is limited, cuticular quinone sclerotization is a relatively well-defined reaction. However, if phenoloxidases become activated in the hemolymph, e.g. in the presence of disintegrating hemocytes, the variety of reacting proteins is immense. Thus, one can observe the whole content of a living cell undergoing melanization.

The products of quinone sclerotization normally are brown to red-brown in color. Decarboxylation of dopaquinone or deacetylation of *N*-acetyldopamine leads to exclusive polymerization of the indole-quinone formed, resulting in pure, black-colored melanin.

Hemolymph phenoloxidase has been studied in insects repeatedly where it is found in hemocytes and serum (Richards, A., 1978). In *Calliphora erythrocephala*, hemolymph phenoloxidase occurs as an inactive proenzyme that can be activated by an enzyme which is present in the cuticle. Three different prophenoloxidases have been distinguished in the hemolymph of *Drosophila*. Two of them oxidize L-dopa (3,4-dihydroxyphenylalanine) but not tyrosine. The third oxidizes both tyrosine and dopa. Some of the prophenoloxidases found are not activated by the cuticle enzyme. The presence of different phenoloxidases in insects suggest that these enzymes have still other functions besides the hardening and tanning of the new cuticle after molting.

A number of preliminary experiments have been performed to test the role of phenoloxidase in insect immunity. Salt, G. (1956) injected crystals of PTU, a known inhibitor of phenoloxidase, into *Carausius morosus*. More live parasites were found in stick insects injected with PTU than in the control animals. The injection of PTU prevented melanization but not cellular encapsulation. Brewer, F. and Vinson, S. (1971) used different inhibitors of phenoloxidase, which they injected into larvae of *Heliothis zea* parasitized with *Cardiochiles nigriceps*. In these experiments both hemocytic encapsulation and melanization were reduced. Nappi, A. (1973) observed more surviving *Pseudocoila* parasites when *Drosophila algonquin* larvae were kept on a diet containing PTU. Unfortunately it could not be determined from these experiments if the reduced efficiency of the defense against parasites resulted directly from inhibition of melanin synthesis or from a general impairment of the host because of an inadequate diet. Control animals on the PTU diet showed higher mortality than PTU-free controls.

Phenoloxidases and their relation to defense reactions have been extensively studied in crustaceans. The results obtained so far furnish interesting aspects and parallels for corresponding investigations in insects; they are therefore presented here in some detail. Spores and hyphae of the crayfish plaque fungus *Aphanomyces astaci* are heavily encapsulated in crayfish hemolymph *in vivo* and *in vitro* (Unestam, T. and Nylund, J.-E., 1972). Hemocytes attach to the fungal surface upon chance contact. Following degranulation of labile granular cells, a light-refracting layer appears on the surface of the fungus, which melanizes heavily within a few hours. Under the electron microscope the light-refracting layer shows a fibrillar fine structure. With progressing melanization the deposit becomes more and more electron-dense. The material attached to the fungal cell wall has polyphenoloxidase activity.

When crayfish (*Astacus astacus*) hemolymph is centrifuged the phenoloxidase activity can be found in the serum. Such serum also produced light-refracting deposits on fungal surfaces, but melanization only occurs when substrate, e.g. dopa or fresh hemolymph is subsequently added. The authors concluded that the crayfish hemolymph prophenoloxidase is stored in the hemocyte granules and is discharged during sticking or clotting of the cells. The release of substrate from hemocytes occurs much slower; serum preparations therefore contain no substrate. The prophenoloxidase has to attach to a surface before its activation is possible (Unestam, T. and Nylhén, L., 1974).

Hemolymph prophenoloxidase prepared from homogenized crayfish hemocytes not only attaches to fungal surfaces but, after approximately 1.5 h at 22°, serum prophenoloxidase spontaneously precipitates and attaches to a variety of materials such as glass, Araldite or polystyrene. Trypsin prevents the attachment, indicating that it is mediated by a proteinaceous compound. Calcium ions are necessary for successful attachment (Söderhäll, K. *et al.*, 1979).

The attached prophenoloxidase still needs a specific activator. Purified fungal cell walls and even aqueous extracts from cell walls and from zymosan (a cell wall preparation of *Saccharomyces cerevisae*) are excellent activators. The active compounds in these extracts were determined to be β-1,3-glucans, which are common in the glycoproteins of many fungal cell walls.

Heavy melanization on and around hyphae of *A. astacus* is also elicited when the fungus actively penetrates the soft cuticle of the crayfish. Evidence was collected that the cuticle prophenoloxidase also attaches to fungal surfaces where it becomes activated by the glycoproteins of the fungal cell wall.

2.7 Resistance to cellular defense reactions

Phagocytosis, nodule formation, cellular and humoral encapsulation are capable of suppressing or killing all kinds of insect pathogens and parasites. One should expect that these mechanisms were sufficient to rule out every possible invader, but in reality there is a notable number of organisms which exist at the expense of insects. They are successful because they have developed sophisticated methods to inhibit or avoid the immune reactions of their insect hosts.

The pathogens or parasites existing today are the result of continuous selection. During evolution those parasites which were either too effective or not effective enough have been exterminated. The ineffective parasites could not produce enough offspring to survive, and the too-effective ones reduced the survival chances of their hosts and thereby risked their own existence.

Such mechanisms of optimization were also effective during the evolution of insect immunity. Development of highly efficient defense mechanisms necessarily includes increasing adaptation to the specific properties of some particular parasites. Efficiency against special parasites, however, implicates increasing susceptibility to other parasites with different qualities for which the specialized defense system is not prepared. Even if such different parasites do not yet exist they may develop any time; mutation and selection always offer plenty of opportunities for their evolution.

As a result of such evolutionary mechanisms, hosts and parasites coexist in a labile equilibrium. A survey of different forms of resistance to insect immunity illustrates the dynamics in the relationship between hosts and parasites and also contributes to a better understanding of the immune system involved.

2.7.1 Resistance to phagocytosis

Insect hemocytes have been found infected with viruses, bacteria, microsporidia and fungi. It has been assumed that growth and multiplication of these pathogens within the hemocytes followed unsuccessful phagocytosis. Obviously the pathogens were able to resist intracellular digestion, which normally occurs after ingestion by phagocytic hemocytes. In the case of bacterial infections, reduced lysosomal activity of infected cells (Ratcliffe, N. and Rowley, A., 1979a) as well as release of proteases and lipases by the bacteria have been reported (Ratcliffe, N. and Rowley, A., 1979b; Lysenko, O., 1976).

2.7.2 Resistance to cellular encapsulation

In his excellent review Salt, G. (1968) has listed a number of different possibilities of how insect parasitoids can reduce the danger of being encapsulated. Rapid growth and development of parasites is one method to reduce the efficiency of cellular encapsulation. An impressive example is provided by the ichneumonid *Phaeogenes nigridens*, which completes each of its first three larval stages within 1 day.

Dilution of the insect hemocytes among a surplus of targets represents another means of weakening the insects' immunity. In a number of parasitic Hymenoptera the developing embryos are surrounded by a cellular membrane which probably has a trophic function. When the embryo hatches from the egg, the cellular membrane breaks down and the individual cells, usually several hundred, become distributed to all parts of the host's body cavity. Since these cells continue to grow and may reach enormous size, they have been called teratocytes. It has to be assumed that the development of giant teratocytes acts on the host mainly through attrition, but the possibility that the large number of foreign surfaces weakens the insect's ability to encapsulate the parasitic larva has to be considered also.

Resistance due to a protecting coat was encountered in the wasp *Nemeritis canescens* and other ichneumonids. The usual hosts of *N. canescens* are larvae of the flour-moth *Ephestia kuehniella*. The hemocytes of these larvae vigorously encapsulate parasitoids other than *Nemeritis* and all types of foreign bodies implanted in their hemocoel, but they do not react to *Nemeritis* eggs and the larvae which hatch from these eggs. However, immediate encapsulation of implanted *Nemeritis* eggs occurred after experimental alteration of the egg surface, e.g. by mechanical abrasion or treatment with fat solvents. A gland at the base of the ichneumonid oviduct was found to excrete a protective fluid which coated the eggs while passing through the calyx (Salt, G., 1968). The mechanisms behind this resistance to cellular encapsulation have been the object of repeated investigations during the last 15 years (e.g. Führer, E., 1973; Kitano, H. and Nakatsuji, N., 1978). These efforts led to many different discoveries, including the finding that the protection is due to a virus which is introduced into the host by the female parasitoid during oviposition. The virus infects the fat body of the host and interferes with the production of immune factors (Vinson, S. *et al.*, 1979; Stoltz, D. and Vinson, S., 1979; Edson, K. *et al.*, 1981). A conclusive judgment of the phenomena described is not yet possible.

Since only hemocytes are capable of encapsulation, temporary occupation of an organ affords protection from this immune reaction. Nematodes of the genus *Filipjevimermis* make use of this opportunity by migration to a ganglion of the nervous system as soon as they have penetrated the insect's body (Strelkov, A., 1964; Götz, P. and Poinar, G., 1968). There the parasites grow very rapidly within a few days (Fig. 10). When they finally leave the ganglion the parasites can develop undisturbed in the hemocoel of their insect host (e.g. larvae of the cucumber beetle *Diabrotica undecempunctata*). All parasite larvae which do not succeed in escaping into a ganglion as soon as they enter the host are killed by encapsulation (Figs 1 and 2). We do not know why the parasites are not encapsulated when they return into the hemocoel a few days later. Unfortunately, tests have not been performed to determine whether the encapsulation capacity of the hosts decreases during the period that the parasites are in the nervous system. It is also conceivable that the parasite's cuticle changes in the interim (e.g. by assimilating substances peculiar to the host) and therefore no longer stimulates the host's immune system.

A similar behavior was reported by Strickland, E. (1923) for larvae of the tachinid fly *Gonia capitata*, a parasite of caterpillars of *Porosagrotis orthogonia*. After invading the hemocoel via the gut the parasitic larvae penetrate a lobe of the brain where they remain for a few days. Many larvae are encapsulated before they can reach the brain, but when the surviving parasites return to the hemocoel some days later, they are seldom found to provoke hemocyte reactions. More tachinid species were found meanwhile to spend the first days after invasion within the nervous system or the salivary gland of their host. Other parasitic Hymenoptera lay their eggs with great precision directly inside host organs.

There is a large diversity of further avoidance mechanisms which have been observed, especially with insect parasitoids. The possibility that parasites might also directly influence and damage hemocytes and thus inhibit successful encapsulation has also been discussed, but since such interference is difficult to demonstrate no author has succeeded in proving it.

3 HUMORAL IMMUNITY

Humoral immunity in insects has been studied for more than two decades and reviews have been published by Chadwick, J. (1975); Chadwick, J. and Aston, P. (1979); Boman, H. (1981); and Boman, H. and Hultmark, D. (1981). Some of the humoral immune factors considered are normally present either in the hemolymph or in hemocytes. In the latter case they are released without *de novo*

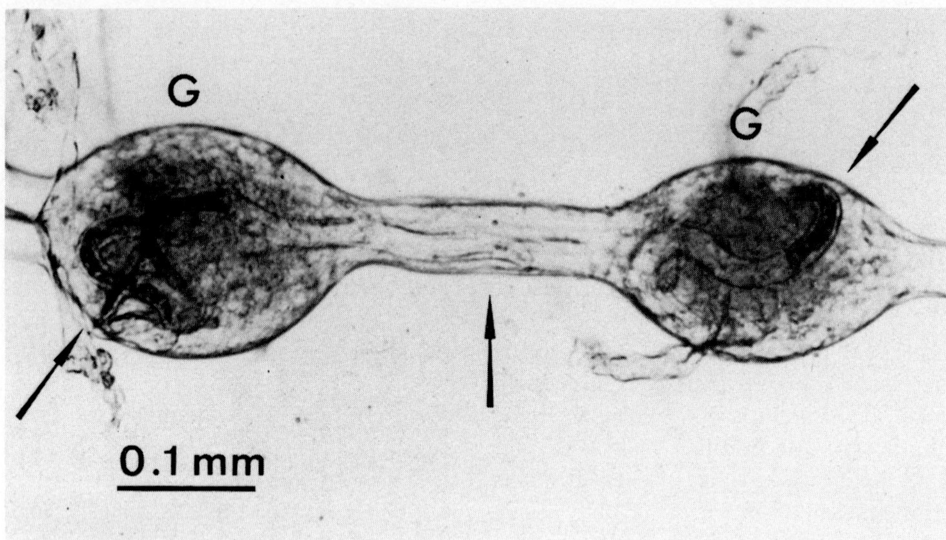

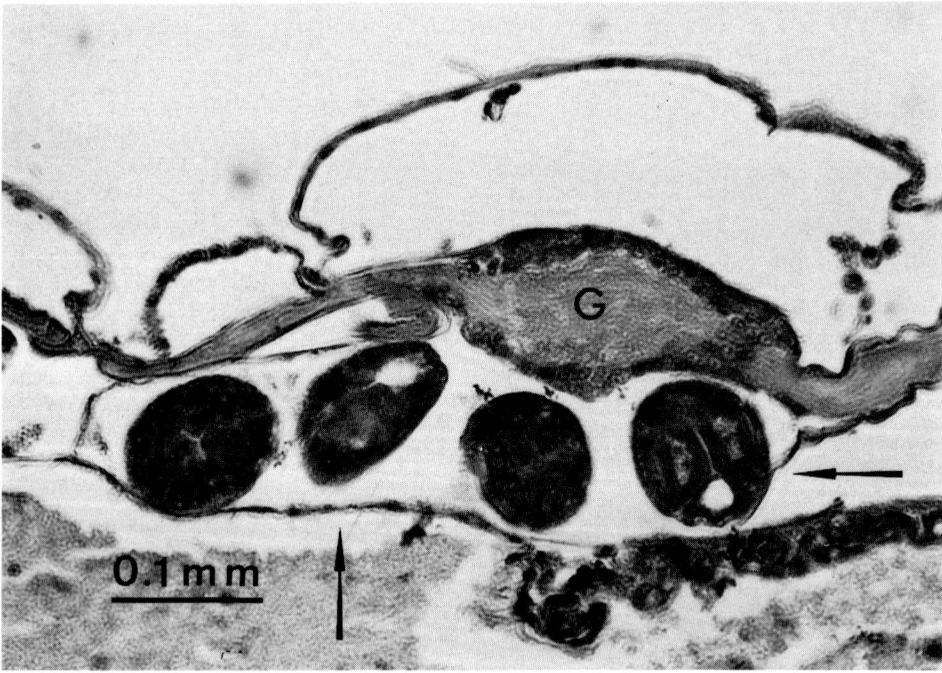

FIG. 10. Avoidance of hemocytic reaction by temporary occupation of a host organ. Top: infective larvae of the nematode *Filipjevimermis leipsandra* (arrows) within the nerve cord of a larva of the cucumber beetle *Diabrotica undecempunctata*. Bottom: three days after invasion: the fast-growing nematode lies curled up (cross-sections marked by arrows) inside the enlarged neural sheath of the ganglion which protects the parasite against hemocyte attack. G = ganglion.

synthesis of proteins. In the following text we will first treat factors existing prior to an infection either as the final proteins or as precursor proteins. Later, we discuss factors proven to be inducible, that is those which require *de novo* synthesis of RNA and proteins.

3.1 Factors normally present in insect hemolymph

The two main types of factors which are normally present in insect hemolymph are phenoloxidase and lectins. Inducibility has been reported for lectins but not for phenoloxidase.

3.1.1 Prophenoloxidase and its activation

Phenoloxidase has already been discussed in connection with cellular and humoral encapsulation. The enzyme is widely distributed in plants and in many arthropods, and the literature is extensive (see review by Nappi, A., 1975).

Phenoloxidase is a highly reactive enzyme and in many organisms it is often stored in the form of prophenoloxidase. Ashida, M. and Dohke, K. (1980) have shown that prophenoloxidase from the silkworm *Bombyx mori* has a molecular weight of 80 K daltons and that it probably exists as a dimer. It is activated by a highly specific proteolytic enzyme which removes a 5 K dalton polypeptide from each subunit. It has been reported that many different classes of substances, including carbohydrates, organic solvents, detergents and proteolytic enzymes, can activate phenoloxidase. It seems likely that the natural control function is mediated by proteolytic cleavage of prophenoloxidase. The enzyme responsible for this activation is in turn carefully controlled; some reports indicate that it can be either activated or inhibited by different carbohydrates.

Purified phenoloxidase from the house-fly *Musca domestica* was found to have a very high molecular weight (around 340 K daltons) (Yamaura, I. *et al.*, 1980). The latter authors emphasize that phenoloxidase has a strong tendency to stick to surfaces and also to polymerize. Some of the melanin precursors have been found to be fungistatic (Söderhäll, K. and Ajaxon, R., 1982); however, the overall view must be that the role of phenoloxidase as an antibiotic substance is not very well established.

3.1.2 Lectins

Proteins which agglutinate mammalian erythrocytes were earlier called hemagglutinins but the term lectins is now generally accepted. Such proteins have long been known in many plants and animals, and they were demonstrated in 46 different species of lepidopterous larvae 30 years ago (Bernheimer, A., 1952). The mechanism behind the agglutination of red cells has in many cases been shown to be a highly specific multivalent capacity to bind to certain sugar residues on the erythrocyte membranes. However, the natural function of most of the lectins has remained a puzzle.

Only recently an insect lectin from the flesh-fly *Sarcophaga* was purified to homogeneity (Komano, H. *et al.*, 1980, 1981). The protein was found to be inducible by an injury during the larval stage but it was synthesized constitutively in the pupa. The molecule was composed of six subunits and showed a specificity for galactose. Further studies suggest that the assembly of the molecule starts from precursor(s) of the subunit(s) which are cleaved by a proteolytic enzyme (Komano, H. *et al.*, 1981).

Castro and Hammarström (personal communication) recently purified the lectin of *Hyalophora cecropia* discovered by Bernheimer, A. (1952). The molecule is composed of four subunits and, like the *Sarcophaga* lectin, it is specific for galactose. The *H. cecropia* lectin was found to be more abundant in growing larvae than in diapausing pupae. Both of the lectins were easy to purify by affinity chromatography and the specificities were determined by inhibition studies using different sugars.

It is reasonable to assume that the biological function of the lectins is to agglutinate invading micro-organisms which carry the respective sugar residues on their cell surfaces. Such clumps of foreign cells could then be easily susceptible either to phagocytosis or to encapsulation followed by melanization. However, only in two cases have insect lectins been shown to act against micro-organisms. Seaman, G. and Robert, N. (1968) reported an immobilization of *Tetrahymena pyriformis* by the American cockroach, a reaction presumably caused by lectin(s). Pereira, M. *et al.* (1981) recently demonstrated that the assassin bug, which is the vector for *Trypanosoma cruzi*, contained two lectins which agglutinated trypanosomes.

3.2 Factors inducible by an infection

It is an old observation that insects given an injection of heat-killed pathogenic bacteria, after a short lag period acquire an immunity against the viable bacterium (Briggs, J., 1958; Stephens, J., 1959). These pioneering experiments, as well as several subsequent studies, were to a large extent influenced by concepts of vertebrate immunology which we now know to be invalid. An additional difficulty in elucidating the molecular mechanisms of insect immunity was the fact that many early experiments were performed on insects unsuited for biochemical work.

3.2.1 Research strategies

Disregarding lysozyme, no real progress was made before diapausing pupae of the giant silkmoth *H. cecropia* were introduced as a model system. Their advantage is that they can be immunized without an immediate diapause break (Boman, H. *et al.*, 1974; Faye, I. *et al.*, 1975). More precisely, this means that it is possible to turn on the genes responsible for immunity and to label the RNA and proteins made without having much biosynthetic background from the rest of the total genome of the animal. An additional improvement introduced at the same time was the immunization with live non-pathogenic bacteria. The reasons for using live bacteria are the following:

(1) under natural conditions insects are infected by live bacteria; dead bacteria cannot be expected to be a real threat for an insect;
(2) a live vaccine with a suitable antibiotic marker can be monitored simultaneously with the challenging bacterial strain carrying a different specific marker — such an experiment could be used directly to test the specificity of the immune reaction;
(3) natural infections occur with rather low numbers of bacteria (10^2–10^4 cells) and such low levels of infecting agents can only be detected by viable count assays;
(4) there is a strong argument against using heat-killed bacteria, because their surface proteins are denaturated.

Consequently, experiments with dead bacteria could give false information concerning the specificity of the immune reactions. If dead bacteria are to be used for an immunological experiment the optimal method of killing is irradiation with ultraviolet light, a procedure known to cross-link DNA and to leave all surface carbohydrates and proteins unaffected.

3.2.2 Early events in the induction of immunity

As separation and assay methods improved it became possible to show that an infection of a pupa produced two early responses: a clearing of the bacteria from hemolymph and the onset of RNA synthesis (Faye, I., 1978; Abu-Hakima, R. and Faye, I., 1981; Boman, H. *et al.*, 1981). In *H. cecropia* pupae the time required for the synthesis of RNA was only 5 h; after this period, the expression of immunity could no longer be inhibited by actinomycin D.

The next step in the development of immunity is protein synthesis which gives rise to increasing antibacterial activity in the hemolymph. Isotope incorporation into proteins can be demonstrated as early as 4 h after immunization. The amount of newly synthesized protein will increase for a time, depending on the species. In the two giant silkmoths *H. cecropia* and *Antheraea pernyi*, the peak time was 7–8 days, while in a third species, *Samia cynthia*, it was about half that time. The activity then gradually declined and disappeared after a time about equal to that required for reaching the maximum. The overall pattern, and the profile of newly synthetized proteins, can be demonstrated by SDS-polyacrylamide gel electrophoresis. This method was used to define the newly synthesized proteins which were designated P1–P9. Figure 11 illustrates that there are two major labeled components (immune proteins P4 and P5) which were also the first two components to be purified.

Immune protein P4 is a single polypeptide with a size of 48 K daltons (Rasmuson, T. and Boman, H., 1979). It is synthesized during an infection but it is also present both in normal hemolymph and in pupal tissue. Nothing is known about its function.

Immune protein P5 was originally isolated as one protein composed of four subunits with an equal size of about 24 K daltons (Pye, A. and Boman, H., 1977). Recent work (Hultmark *et al.*, unpublished) has shown P5 to be a family of proteins. They can all be separated by isoelectric focusing and for each component antibacterial activity can be demonstrated by overlaying the gel with viable indicator bacteria. It is possible that the proteins recently isolated correspond to the subunits earlier described.

The largest part of the antibacterial activity against a variety of bacteria can be accounted for by the P9 group of proteins called cecropins and discussed below. P7 has been identified as lysozyme

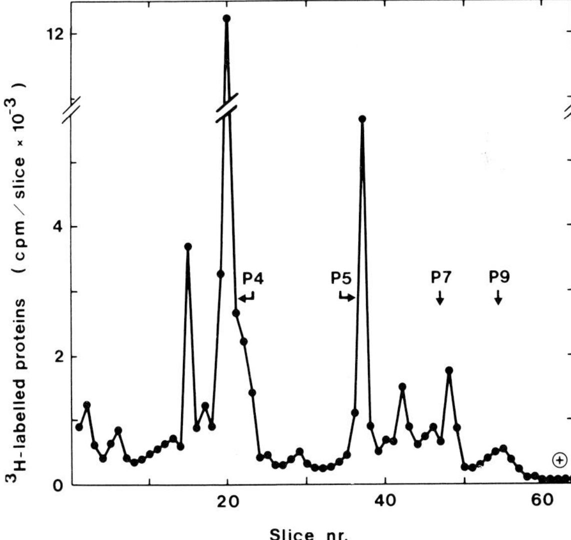

FIG. 11. SDS polyacrylamide gel (gradient from 7.5 to 15% concentration) electrophoresis of [^{3}H]-labeled immune proteins from *Hyalophora cecropia*. The proteins were originally designated P1–P8 depending on their mobility in a 7.5% gel (Faye, I. *et al.*, 1975). The pupae were vaccinated with about 10^6 viable cells of *Enterobacter cloacae* and simultaneously given a mixture of generally [^{3}H]-labeled lysine, arginine and leucine. With basic amino acids most of the cpm are found in P4, with [^{35}S]methionine or [^{3}H]leucine P4 and P5 are about equally labeled. With [^{35}S]methionine one can hardly detect the P9 peak because only P9B contains one Met (cf. Fig. 13). Immune proteins P4, P5, P7 (the lysozyme) and four forms of P9 (cecropins A, B, C and D) have been purified.

(Hultmark, D. *et al.*, 1980) and it will also be discussed subsequently.

3.2.3 Cecropins

The cecropins constitute a family of basic proteins with molecular weights around 4000 K daltons and with a broad antibacterial activity. These properties make it easy to demonstrate cecropins in polyacrylamide gel electrophoresis at pH4. Following separation the gel is overlaid by a suitable indicator bacterium suspended in soft agar (Hultmark, D. *et al.*, 1980). After an overnight incubation the cecropins and the P5 immune proteins are visualized as spots without bacterial growth. Figure 12 shows that the *H. cecropia* pupae contain three major cecropins (A, B, and D) which are easy to separate in this manner. In the Chinese oak silkmoth *Antheraea pernyi*, cecropin D was generally the major antibacterial product while A and B were present in lower concentrations (Qu, X.-M. *et al.*, 1982). It has been possible to purify and sequence cecropin A, B and D from *Hyalophora cecropia* and cecropin B and D from *A. pernyi*. The sequences given in Fig. 13 show that these five cecropins are so homologous that they must have evolved by gene duplication of a common ancestral gene. All cecropins contain a stretch of 10 hydrophobic amino acids (residues 22–32) and they all have blocked C-terminal residues. In case of cecropin A and B, the N-terminal part is strongly basic while in cecropin D it is moderately basic. Between cecropin A and B there are five principal differences affecting molecular charge, all located in the central and C-terminal portions of the molecules (residues 13, 17, 31, 33 and 37). In case of the cecropins A and D six differences occur in the central and N-terminal part. They concern five alterations in charge (residues 1, 3, 7, 14 and 21) and one which inhibits alpha helix formation (residue 4). These differences between the primary structures of cecropin A, B and D indicate that they are produced by separate genes (Steiner, H. *et al.*, 1981; Hultmark, D. *et al.*, in preparation).

The general design of the cecropin molecule is similar to that of the bee venom toxin melittin; only the polarity is reversed because in melittin the C-terminus is basic and the N-terminus is hydrophobic. Cecropin A was shown to lyse *Escherichia coli* but not Chang liver cells while melittin lysed both (Steiner, H. *et al.*, 1981). Cecropins therefore seem to be specific for procaryotic membranes while melittin can destroy both procaryotic and eucaryotic cells. It could be assumed that part of the structure of the cecropins carries information for the avoidance of self-destruction.

Calculation and model building indicate that cecropin A in a lipophilic solvent forms a nearly perfect amphipathic alpha helix, i.e. a cylindrical molecule with all charged groups on one longitudinal side and all hydrophobic side-chains on the opposite side. Since both hemolymph and membranes are lipophilic this secondary structure could be of importance for the membrane-disrupting activity of the cecropins (Steiner, H., 1982; Merrifield, R. *et al.*, 1982).

The antibacterial activity of the cecropins has been recorded in different ways. Three of these depend on the use of buffer-suspended bacteria and

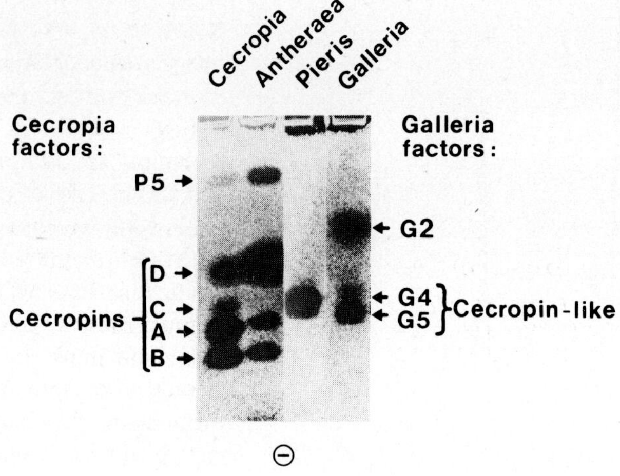

FIG. 12. Polyacrylamide gel electrophoresis under native conditions at pH4 of antibacterial factors in hemolymph from four different insects, Cecropia (*Hyalophora cecropia*) lane 1; Antheraea (*Antheraea pernyi*) lane 2; Pieris (*Pieris napi*) lane 3 and Galleria (*Galleria mellonella*) lane 4. After the separation, the gel was overlaid with a suspension of *Escherichia coli* D31 in soft agar as described by Hultmark, D. *et al.*, (1980). Antibacterial activity is indicated by the dark spots without bacterial growth. Cecropins A, B, C and D and immune protein P5 from *H. cecropia* are indicated by arrows. The *H. cecropia* factors are described by Hultmark, D. *et al.* (1982) and the *Galleria* factors G4 and G5 by Hoffmann, D. *et al.* (1981).

		1 5 10
A.p.	Cecropin B	H_2N-Lys-Trp-Lys-Ile-Phe-Lys-Lys-Ile-Glu-Lys-Val-Gly-Arg-
H.c.	Cecropin B	H_2N-Lys-Trp-Lys-Val-Phe-Lys-Lys-Ile-Glu-Lys-Met-Gly-Arg-
H.c.	Cecropin A	H_2N-Lys-Trp-Lys-Leu-Phe-Lys-Lys-Ile-Glu-Lys-Val-Gly-Gln-
H.c.	Cecropin D	H_2N-Trp-Asn-Pro-Phe-Lys-Glu-Leu-Glu-Lys-Val-Gly-Gln-
A.p.	Cecropin D	H_2N-Trp-Asn-Pro-Phe-Lys-Glu-Leu-Glu-Arg-Ala-Gly-Gln-
		15 20 25
A.p.	Cecropin B	Asn-Ile-Arg-Asn-Gly-Ile-Ile-Lys-Ala-Gly-Pro-Ala-Val-Ala-
H.c.	Cecropin B	Asn-Ile-Arg-Asn-Gly-Ile-Val-Lys-Ala-Gly-Pro-Ala-Ile-Ala-
H.c.	Cecropin A	Asn-Ile-Arg-Asp-Gly-Ile-Ile-Lys-Ala-Gly-Pro-Ala-Val-Ala-
H.c.	Cecropin D	Arg-Val-Arg-Asp-Ala-Val-Ile-Ser-Ala-Gly-Pro-Ala-Val-Ala-
A.p.	Cecropin D	Arg-Val-Arg-Asp-Ala-Ile-Ile-Ser-Ala-Gly-Pro-Ala-Val-Ala-
		30 35
A.p.	Cecorpin B	Val-Leu-Gly-Glu-Ala-Unknown C-terminal
H.c.	Cecropin B	Val-Leu-Gly-Glu-Ala-Lys-Ala-Ile-Leu-Ser-CO-R
H.c.	Cecropin A	Val-Val-Gly-Gln-Ala-Thr-Gln-Ile-Ala-Lys-CO-R
H.c.	Cecropin D	Thr-Val-Ala-Gln-Ala-Thr-Ala-Leu-Ala-Lys-CO-R
A.p.	Cecoprin D	Thr-Val-Ala-Gln-Ala-Thr-Ala-Leu-Ala-Lys-CO-R

FIG. 13. Amino acid sequences of cecropins B and D from *Antheraea pernyi* (A.p.) and cecropins A, B and D from *Hyalophora cecropia* (H.c.). H.c. cecropin A was used as a reference in the comparison of the other sequences. Identical amino acids are boxed by solid lines. Underlined residues show differences between the two B forms but identities between A.p. cecropin B and H.c. cecropin A.

Table 1: Antibacterial spectra of the major cecropins A, B and D

Species	Strain	Parent	Lethal concentration (μM) of cecropin		
			B	A	D
Escherichia coli	D21	D11	0.32	0.19	0.43
	D31	D21	0.32	0.21	0.40
Serratia marcescens	Db11	Db10	4.5	4.2	14
	Db1108	Db11	2.4	2.5	12
	Db1121	Db11	2.2	4.4	12
Strain 122[a]			2.9	3.2	19
Pseudomonas aeruginosa	OT97		1.9	4.8	100[b]
Xenorhabdus nematophilus	Xn21		1.6	1.4	19[b]
Bacillus megaterium	Bm11		0.44	0.70	41[b]
Bacillus subtilis	Bs11		18[b]	61[b]	95[b]
Bacillus thuringiensis	Bt11		133[b]	80[b]	95[b]
Streptococcus fecalis	AD-4		7.3	15	95[b]
	DS16		24[b]	74[b]	88[b]
Micrococcus luteus	Ml11		1.3[b]	4.6[b]	21[b]

[a] Isolated from a dead *Hyalophora cecropia* larva in this laboratory and tentatively identified as a *Serratia*.
[b] Value estimated from zones smaller than the critical diameter.
Lethal concentrations of the cecropins were determined for different bacteria with the inhibition zone assay as described by Hultmark, D. *et al.* (in preparation).

the action of the cecropins is monitored either as a decrease in viable counts (Boman, H. *et al.*, 1974), as a decrease in light absorption (Hultmark, D. *et al.*, 1980), or as an increase in ultraviolet-absorbing substances released from the cells (Steiner, H. *et al.*, 1981). The first method records bacterial killing while the other two record cell lysis. It is finally possible to determine the activity of cecropins on live growing bacteria either by measuring inhibition zones on thin agar plates (Hoffmann, D. *et al.*, 1981) or by measuring the minimum inhibitory concentration (MIC) with two-fold dilution steps of the cecropins. For routine measurements the inhibition zone assay may be the method of choice.

Table 1 illustrates the antibacterial activity against eight different bacteria, selected to represent both the normal bacterial flora and established insect pathogens. It is clear that cecropins can act on both Gram-positive and Gram-negative bacteria, although in general the latter group was more sensitive. Cecropin A and B were clearly more active than cecropin D against *Pseudomonas aeruginosa* and *Xenorhabdus nematophilus*, two Gram-negative insect pathogens. The cecropins have not yet been tested in different combinations nor for synergistic effects with the P5 proteins or lysozyme. It is therefore too early to discuss the precise biological function(s) of the cecropins.

Certain bacteria such as *Micrococcus luteus* and *Bacillus megaterium* are highly sensitive to lysozyme. In order to demonstrate cecropin action against these organisms possible contamination by lysozyme must be excluded. Synthetic cecropins would offer the best method to avoid lysozyme involvement. First results with this approach were obtained with a synthetic polypeptide with a structure identical to residues 1–33 of cecropin A (Merrifield, R. *et al.*, 1982). When tested against *E. coli* D31 the synthetic compound was about half as active as natural cecropin A. The inhibition zone assay with the synthetic material showed no activity against *B. megaterium*. However, on acid electrophoresis (when much more material was used) activity against *B. megaterium* could clearly be demonstrated at the position of the synthetic cecropin A (1–33). These results represent the first satisfactory demonstration that cecropins act on a lysozyme sensitive Gram-positive bacterium. In addition the experiments also indicate the importance of the intact C-terminal part of the molecule for its activity against *B. megaterium* (Merrifield, R. *et al.*, 1982).

3.2.4 Lysozyme

Using *Galleria mellonella* as a test organism it was earlier claimed that lysozyme was the main antibacterial immune factor in hemolymph (Mohrig, W. and Messner, B., 1968a) as well as in the gut (Mohrig, W. and Messner, B., 1968b). Lysozyme from *G. mellonella* and *Bombyx mori* was the first immune protein from an insect to be isolated in pure form (Powning, R. and Davidson, W., 1973, 1976). This purification work was never followed up by any experiments on the biological role of the enzyme, and different opinions concerning the importance of lysozyme have been expressed (see reviews by Chadwick, J., 1975; Boman, H., 1981).

Table 1 shows that cecropin A and B acted on *Micrococcus luteus*, the classical assay organism for lysozyme. Care will therefore have to be exercised in the interpretation of antibacterial activity obtained with insect hemolymph and this organism. Recent reports on the protective effect of induced lysozyme are difficult to evaluate because assays for cecropins were not available (Jarosz, J., 1979; Jarosz, J. and Spiewak, N., 1979; Ourth, D. and Smalley, D., 1980).

The *H. cecropia* lysozyme was isolated by Hultmark, D. *et al.* (1980). As could be expected it did not act on four different Gram-negative bacteria tested (Boman, H. and Steiner, H., 1981). It seems that the main target of the cecropins are bacterial membranes, while lysozyme acts on the murein skeletons. Thus, both cecropins and lysozymes are needed for a complete and effective destruction of bacterial cell walls.

It is fairly easy to purify insect lysozymes and they are therefore suitable for evolutionary studies on different insect species. Initial results showed insect lysozyme to be of the hen type (Jollès, J. *et al.*, 1979).

3.2.5 Further humoral reactions possibly related to immunity

The literature contains a number of further reports on phenomena which could be related to immunity. The mechanisms behind the reactions described are not clarified, and often their effects on microorganisms or toxins are not convincing enough for a role in immunity.

Bacillus thuringiensis is known to be very resistant to insect immunity; neither cecropins nor lysozyme act on this bacterium. However, *B. thuringiensis* induces a weak immune response of unknown nature when injected into *Locusta migratoria* (Hoffmann, D., 1980). This may be one of two unidentified antibacterial activities against *B. thuringiensis* which previously could be distinguished using selective inhibitors (Rasmuson, T. and Boman, H., 1977). Peroxidase reactions are known to have antibacterial effects in vertebrates and have been reported to also occur in insects (Messner, B., 1981). Using heat-inactivated venoms as vaccines, an inducible protective effect against bee and snake venom was found in the cockroach *Periplaneta americana* (Karp, R. and Rheins, L., 1980; Rheins, L. *et al.*, 1980). A clear specificity was demonstrated in this work but the interpretation would have been much easier if the experiments had been performed with a single pure toxin like melittin (which constitutes about 50% of the bee venom).

3.2.6 Injury reactions

It is likely that in nature insects are often injured in one way or other, and that such injuries may be followed by infections. It is therefore reasonable to expect related reactions to injuries and to infections.

Insect surgery has been an important tool in physiological research (Schneiderman, H., 1967). Sham operations showed that an injury triggers the onset of many biosynthetic reactions in insects (Telfer, W. and Williams, C., 1960; Cherbas, L., 1973). Even an injury as small as that caused by a fine injection needle is enough for significant stimulation of protein synthesis (Faye, I. *et al.*, 1975). Thus, the immune response to injected bacteria must always be compared to a sham injection with saline.

The relationships between the humoral responses to injuries and to infections in insects are not yet understood but the current state of knowledge can be summarized as follows:

(1) On the RNA level there is no qualitative difference between the injury and the immune response that can be detected with the methods yet available (Boman, H. *et al.*, 1981).
(2) On the protein level there are qualitative similarities but quantitative differences in the pattern of newly synthetized proteins (Faye, I. *et al.*, 1975; Boman, H. and Steiner, H., 1981).
(3) On the level of antibacterial activity there is a significant difference. The immune response is a potent antibacterial activity, chiefly due to cecropins, while the injury response induces either no or only weak antibacterial activity.

The overall conclusion from these findings would be that the first part of immune and injury responses may be identical. The differences seen may be due mainly to tuning effects at the level of translational or post-translational events.

3.3 Resistance to humoral immune reactions

All higher forms of life need immune systems to protect themselves against invading microorganisms or other parasites. Since this is a universal phenomenon one can expect strong selection pressures for mechanisms to escape immune reactions. Different groups of protozoa have developed sophisticated means to avoid the mammalian immune recognition (see review by Bloom, B., 1979). Two main types of avoidance mechanisms have been found to counteract insect immunity and they were originally referred to as "passive and active resistance".

Passive resistance is said to occur if a bacterium is resistant against the immune system despite the fact that the organism carries a genetic potential for sensitivity. This was first detected in a strain of *Serratia marcescens* pathogenic for *Drosophila*. This organism was found to be highly resistant to *Hyalophora cecropia* immunity. It was possible to isolate a number of spontaneous phage-resistant mutants which turned out to be pleotropic because they had become sensitive to cecropins and simultaneously lost most of their virulence in *Drosophila* (Flyg, C. *et al.*, 1980). Thus, since the mutants were spontaneous, the parental strain must have had targets for the cecropins, and these must have been passively protected — presumably by some sort of penetration barrier.

Active resistance is due to destruction of immunity by an active process; so far, the only mechanism found has been proteolytic digestion of the cecropins and the immune proteins P5. The first such "immune inhibitor" was isolated from *B. thuringiensis* and referred to as InA (Edlund, T. *et al.*, 1976; Sidén, I. *et al.*, 1979). However, the survival value of InA for the organism is not yet fully understood. A very complex situation was found in the entomophilic nematode *Neoaplectana carpocapsae*, which is associated with a symbiotic bacterium. The bacterium, *Xenorhabdus nematophilus*, is highly sensitive to the cecropins A and B (see Table 1), but the nematode is able to help its symbiont by producing an immune inhibitor that quickly destroys the cecropins and the immune proteins P5 (Götz, P. *et al.*, 1981).

4 GENERAL DISCUSSION OF INSECT IMMUNITY

4.1 The conceptual problem

Simplistically speaking, one can state that scientists who have worked with insect immunity belonged to two different groups, characterized by very different attitudes toward invertebrate immunity.

The scientists of the first group have borrowed their experimental methods and terminology from vertebrate immunity. They may have transplanted tissues or injected red blood cells, heat-killed bacteria or unfractionated and denatured venoms. As a rule these workers do not acknowledge that these types of experiments are totally unphysiological. When qualified mammalian immunologists investigate immune reactions against red blood cells, they always include competition experiments with well-defined carbohydrates or polypeptides. From such data, conclusions can be obtained concerning the cellular sites which are recognized by the host immune system. However, when these types of experiments are conducted on invertebrates there is not much to conclude without a subsequent molecular analysis. The use of terms from mammalian immunology may then be misleading because it suggests an understanding which does not exist.

The second group takes as their departure point the assumption that there are fundamental differences between the immune system of mammals and those of invertebrates. The investigators of the second group try to design experiments in such a way that they resemble the natural infections which their experimental animals are likely to encounter. They inject controlled doses of live bacteria. The crucial aim of these workers is to explain their results in terms of a possible survival value for the host. They realize that an injection is an artificial operation which in nature most likely corresponds to an injury. Consequently, it is always important to compare the immune response obtained with controls in which an equal injury is created.

It is regrettable that there is not much understanding between scientists belonging to the two categories described above. Disregarding semantic problems, it should be possible to use many of the experimental methods of mammalian immunology if only the results are interpreted in terms of biochemical or biological functions of the insect.

4.2 Insect immunity *versus* vertebrate immunity

Insects are relatively well protected against invasion of foreign organisms. No epithelia are directly exposed to invaders since the cuticle completely covers the outside as well as the foregut, hindgut and the tracheal tubes. Even the intestinal epithelium of the midgut in most insects is guarded by the peritrophic membrane, a structure which encloses the gut contents.

Vertebrates, especially birds and mammals, offer much easier access for invading organisms which can easily penetrate the mucous epithelia of the intestinal system, the respiratory system, the excretory system and the genital organs. In fact, these tracts are the main entrance for pathogens, as clearly demonstrated by examples from human and veterinary medicine.

4.2.1 Insect immunity is different from vertebrate immunity

It is well established today that insects lack lymphocytes, the principal elements in vertebrate immunity (de Pasquier, L., 1976). Insects therefore have no capacity to produce immunoglobulins, which mediate the high specificity of vertebrate immune reactions. Reports that insects contain complement-like systems are also most likely based on analogies of limited importance.

Although we are far from a complete knowledge of the immune mechanisms existing in insects, some principal features are clearly visible. Insect immunity is based on:

(1) The structural and biochemical properties of the insect cuticle and intestine, which act as barriers against many possible invaders.
(2) The high capacity of insect hemocytes to phagocytose, entrap and encapsulate invading organisms. Nodule formation and encapsulation combined with melanization are highly efficient reactions which as far as we know have no parallel in other animals besides arthropods.
(3) Humoral reactions of the insect hemolymph in which enzymes and antibacterial proteins act against foreign organisms. Some of these substances are always present (e.g. phenoloxidase, lysozyme); others are furnished by *de novo* synthesis upon stimulation (e.g. cecropins).

The specificity of the mammalian immune system is very high; in fact so high that in certain situations it may create biological disadvantages (e.g. by causing allergic and autoaggressive diseases). Insect immunity, on the other hand, does not show the same degree of specificity. In insects there are no immune reactions against implanted tissues of the same organism or of related species (isografts and allografts) and only limited reactions against xenografts (Carton, Y., 1976; Lackie, A., 1979, 1981; Thomas, I. and Ratcliffe, N., 1982).

A large variety of organisms and substrates are suited to provoke the reactions of insect hemocytes or enzyme systems present in the cuticle and the hemolymph. It seems likely that relatively simple and common surface properties of foreign organisms act as the signal provoking an immune reaction (e.g. β-glucans of fungal cell walls). However, it is also conceivable that the so-called "foreignness" consists of the lack of inhibiting properties with which the cells and tissues within the hemocoel of an insect are equipped.

4.2.2 Insect immunity is as efficient as it has to be

Pathogens and parasites play about the same role for insects as they do for other animal groups, including mammals. It is therefore misleading to classify insect immunity as underdeveloped and of minor quality as compared to the immune reactions of higher animals. From an evolutionary point of view pathogens and parasites represent one possibility of life which necessarily evolves wherever organisms exist. As result of a coevolution, parasites and hosts coexist in an equilibrium, their numerical ratios oscillating around a mean value. Parasites, pathogens (and predators) represent important factors for the regulation of the population density of every group of organisms.

5 SUMMARY

Insect immunity was defined as the insect's "ability to resist diseases". Although insect immunity lacks the high specificity, inducibility and memory capacity of the vertebrate immune system, it is able to efficiently control invading micro-organisms and parasites. Insects are sheltered by strong natural barriers such as the outer cuticle and the cuticular intima of large parts of the insect gut. Powerful cellular immune reactions of insects lead to nodule formation and encapsulation, reactions which derive their efficiency from their combination of cellular ensheathment and melanin formation. Among the humoral factors, a newly discovered class of inducible antibacterial immune proteins, the cecropins, contributes substantially to the antibacterial capacity of insects.

Many questions remain to be resolved in insect immunity. An efficient system of cooperating hemocytes, well-controlled enzymes and inducible gene activities await elucidation. The answers which could result from successful research in insect immunity are of general importance in the biological sciences; they would contribute to a better understanding of host–parasite relationships; they are also related to applied science by helping to develop agents of high pathogenicity for the biological control of insect pests.

REFERENCES

Abu-Hakima, R. and Faye, I. (1981). An ultrastructural and autoradiographic study of the immune response in *Hyalophora cecropia* pupae. *Cell Tiss. Res. 217*, 311–320.

Anderson, R. S., Holmes, B. and Good, R. A. (1973). In vitro bactericidal capacity of *Blaberus craniifer* hemocytes. *J. Invertebr. Pathol. 22*, 127–135.

Ashida, M. and Dohke, K. (1980). Activation of pro-phenoloxidase by the activating enzyme of the silkworm *Bombyx mori. Insect Biochem. 10*, 37–47.

Baerwald, R. J. (1979). Fine structure of hemocyte membranes and intercellular junctions formed during hemocyte encapsulation. In *Insect Hemocytes*. Edited by A. P. Gupta. Pages 155–188. Cambridge University Press, Cambridge.

Bernheimer, A. W. (1952). Hemagglutinins in caterpillar bloods. *Science 115*, 150–151.

Bloom, B. R. (1979). Games parasites play: how parasites evade immune surveillance. *Nature 279*, 21–26.

Boman, H. G. (1981). Insect responses to microbial infection. In *Microbial Control of Insects, Mites and Plant Diseases*. Edited by D. Burges. Pages 769–784. Academic Press, New York.

Boman, H. G. and Hultmark, D. (1981). Cell-free immunity in insects. *Trends Biochem. Sci. 6*, 306–309.

Boman, H. G. and Steiner, H. (1981). Humoral immunity in Cecropia pupae. In *Current Topics in Microbiology and Immunology*. Edited by W. Henle *et al.* Vol. 94/95, pages 75–91. Springer, Berlin.

Boman, H. G., Nilsson-Faye, I. and Rasmuson, T. (1974). Why is insect immunity interesting? In *Energy, Biosynthesis and Regulation in Molecular Biology*. Edited by D. Richter. Pages 103–114. Walter de Gruyter, Berlin.

Boman, H. G., Boman, I. A. and Pigon, A. (1981). Immune and injury responses in *Cecropia* pupae. RNA isolation and comparison of protein synthesis *in vivo* and *in vitro*. *Insect Biochem. 11*, 33–42.

Brennan, B. M. and Cheng, T. C. (1975). Resistance of *Moniliformis dubius* to the defense reactions of the American cockroach, *Periplaneta americana*. *J. Invert. Pathol. 26*, 65–73.

Brewer, F. D. and Vinson, S. B. (1971). Chemicals affecting the encapsulation of foreign material in an insect. *J. Invertebr. Pathol. 18*, 287–289.

Briggs, J. D. (1958). Humoral immunity in lepidopterous larvae. *J. Exp. Zool. 138*, 155–188.

Bronskill, J. F. (1962). Encapsulation of rhabditoid nematodes in mosquitoes. *Canad. J. Zool. 40*, 1269–1275.

Burghause, F. (1980). The cellular encapsulation of foreign material in Gryllus. In *Verh. Dtsch. Zool. Ges. 1980*. Edited by W. Rathmayer. Page 329. Fischer, Stuttgart.

Burghause, F. (in preparation). Cellular encapsulation of various foreign bodies by four insect species.

Carton, Y. (1976). Isogenic, allergenic and xenogenic transplants in an insect species. *Transplantation 21*, 17–22.

Chadwick, J. S. (1975). Hemolymph changes with infection or induced immunity in insects and ticks. In *Invertebrate Immunity*. Edited by K. Maramorosch and R. E. Shope. Pages 241–271. Academic Press, New York.

Chadwick, J. S. and Aston, P. (1979). An overview of insect immunity. In *Animal Models of Comparative and Developmental Aspects of Immunity and Disease*. Edited by M. E. Gershwin and E. L. Cooper. Pages 1–14. Pergamon Press, New York.

Cherbas, L. (1973). The induction of an injury reaction in cultured hemocytes from Saturniid pupae. *J. Insect Physiol. 19*, 2011–2023.

Crossley, A. C. S. (1975). The cytophysiology of insect blood. *Adv. Insect Physiol. 11*, 117–222.

Du Pasquier, L. (1976). Phylogenesis of the vertebrate immune system. In *The Immune System*. Edited by F. Melchers and K. Rajewsky. Pages 101–115. Springer, Berlin.

Edlund, T., Sidén, I. and Boman, H. G. (1976). Evidence for two immune inhibitors from *Bacillus thuringiensis* interfering with the humoral defence system of Saturniid pupae. *Infect. Immun. 14*, 934–941.

Edson, K. M., Vinson, S. B., Stoltz, D. B. and Summers, M. D. (1981). Virus in a parasitoid wasp. Suppression of the cellular immune response in the parasitoids host. *Science 211*, 582–583.

Esslinger, J. H. Y (1962). Behaviour of microfilariae of *Bruggia pahangi* in *Anopheles quadrimaculatus*. *Amer. J. Trop. Med. Hyg. 1*, 749–758.

Faye, I. (1978). Insect immunity IV: Early fate of bacteria injected in Saturniid pupae. *J. Invertebr. Pathol. 31*, 19–26.

Faye, I., Pye, A., Rasmuson, T., Boman, H. G. and Boman, I. A. (1975). Insect Immunity II. Simultaneous induction of antibacterial activity and selective synthesis of some hemolymph proteins in diapausing pupae of *Hyalophora cecropia* and *Samia cynthia*. *Infect. Immun. 12*, 1426–1438.

Flyg, C., Kenne, K. and Boman, H. G. (1980). Insect pathogenic properties of *Serratia marcescens*: phage-resistant mutants with a decreased resistance to Cecropia immunity and a decreased virulence to *Drosophila*. *J. Gen. Microbiol. 120*, 173–181.

Führer, E. (1973). Sekretion von Mucopolysacchariden im weiblichen Geschlechtsapparat von Pimpla turionella L. (Hym., Ichneumonidae). *Z. Parasitenkd. 41*, 207–214.

Gagen, S. J. and Ratcliffe, N. A. (1976). Studies on the in vivo cellular reactions and fate of injected bacteria in *Galleria mellonella* and *Pieris brassicae* larvae. *J. Invertebr. Pathol. 28*, 17–24.

Götz, P. (1964). Der Einfluβ unterschiedlicher Befallsbedingungen auf die mermithogene Intersexualität von *Chironomus* (Dipt.). *Z. Parasitenkd. 24*, 484–545.

Götz, P. (1969). Die Einkapselung von Parasiten in der Hämolymphe von Chironomus-Larven (Diptera). *Zool. Anz., Suppl. 33, Verh. Zool. Ges.*, 610–617.

Götz, P. (1973). Immunreaktionen bei Insekten. *Naturwiss. Rundschau 26*, 367–375.

Götz, P. (1976). Parasit-Wirt-Beziehungen zwischen dem Nematoden *Hydromermis contorta* und der Zuckmücke *Chironomus thummi*. *Publ. Wiss. Filmen 9*, 67–87.

Götz, P. and Poinar, Jr., G. O., (1968). Entwicklung einer Mermithide (Nematode) im Nervensystem von Insekten. *Z. Parasitenkd. 31*, 10.

Götz, P. and Vey, A. (1974). Humoral encapsulation in Diptera (Insecta): defence reaction of *Chironomus* larvae against fungi. *Parasitology 68*, 193–205.

Götz, P., Roettgen, J. and Lingg, W. (1977). Encapsulement humoral en tant que réaction de défense chez les Diptères. *Ann. Parasitol. Hum. Comp. 52*, 95–97.

Götz, P., Boman, A. and Boman, H. G. (1981). Interactions between insect immunity and an insect pathogenic nematode with symbiotic bacteria. *Proc. R. Soc. Lond. B 212*, 333–350.

Götz, P., Enderlein, G. and Roettgen, J. (In preparation). Humoral encapsulation of bacteria in *Chironomus* larvae (Diptera).

GRIMSTONE, S. R., ROTHERAM, S. and SALT, G. (1967). An electron-microscope study of capsule formation by insect blood cells. *J. Cell Sci. 2*, 281–292.

GUPTA, A. P. (1979). Hemocyte types: their structures, synonymies, interrelationships and taxonomic significance. In *Insect Hemocytes*. Edited by A. P. Gupta. Pages 85–128. Cambridge University Press, Cambridge.

HOFFMAN, D. (1980). Induction of antibacterial activity in the blood of the migratory locust *Locusta migratoria* L. *J. Insect Physiol. 26*, 539–549.

HOFFMAN, D., HULTMARK, D. and BOMAN, H. G. (1981). Insect Immunity: *Galleria mellonella* and other Lepidoptera have Cecropia-P9-like factors active against Gram negative bacteria. *Insect Biochem. 11*, 537–548.

HULTMARK, D., STEINER, H., RASMUSON, T. and BOMAN, H. G. (1980). Insect immunity. Purification and properties of three inducible bactericidal proteins from hemolymph of immunized pupae of *Hyalophora cecropia*. *Eur. J. Biochem. 106*, 7–16.

HULTMARK, D., ENGSTRÖM, H., BENNICH, H., KAPUR, R. and BOMAN, H. G. (In preparation). Insect immunity. Isolation and structure of cecropin D and four minor antibacterial components from *Cecropia* pupae.

JAROSZ, J. (1979). Simultaneous induction of protective immunity and selective synthesis of hemolymph lysozyme protein in larvae of *Galleria mellonella*. *Biol. Zbl. 98*, 459–471.

JAROSZ, J. and SPIEWAK, N. (1979). Comparative levels of lysozyme activity in larvae and pupae of *Galleria mellonella* after particulate and soluble materials injection. *Cytobios 26*, 203–219.

JOLLÈS, J., SCHOENTGEN, F., CROIZIER, G., CROIZIER, L. and JOLLÈS, P. (1979). Insect lysozymes from three species of Lepidoptera: Their structural relatedness to the C (chicken) type lysozyme. *J. Mol. Evol. 14*, 267–271.

JONES, J. C. (1970). Hemocytopoiesis in insects. In *Regulation of Hematopoiesis*. Edited by A. S. Gordon. Vol. 1, pages 7–65. Appleton, New York.

JONES, J. C. (1977). *The Circulatory System of Insects*. Thomas, Springfield, Illinois.

KARP, R. D. and RHEINS, L. A. (1980). Induction of specific humoral immunity to soluble proteins in the American cockroach (*Periplaneta americana*). II. Nature of the secondary response. *Dev. Comp. Immunol. 4*, 629–639.

KITANO, H. and NAKATSUJI, N. (1978). Resistance of *Apanteles* eggs to the haemocytic encapsulation by their habitual host, *Pieris*. *J. Insect Physiol. 24*, 261–271.

KOMANO, H., MIZUNO, D. and NATORI, S. (1980). Purification of lectin induced in the hemolymph of *Sarcophaga peregrina* larvae on injury. *J. Biol. Chem. 255*, 2919–2924.

KOMANO, H., MIZUNO, D. and NATORI, S. (1981). A possible mechanism of induction of insect lectins. *J. Biol. Chem. 256*, 7087–7089.

LACKIE, A. M. (1979). Cellular recognition of foreign-ness in two insect species, the American cockroach and the desert locust. *Immunology 36*, 909–914.

LACKIE, A. M. (1981). Immune recognition in insects. *Dev. Comp. Immunol. 5*, 191–204.

LEUTENEGGER, R. (1967). Early events of *Sericesthis* iridescent virus infection in hemocytes of *Galleria mellonella* (L). *Virology 32*, 109–116.

LYSENKO, O. (1976). Chitinase of *Serratia marcescens* and its toxicity to insects. *J. Invertebr. Pathol. 27*, 385–386.

MERRIFIELD, R. B., VIZIOLI, L. D. and BOMAN, H. G. (1982). Synthesis of the antibacterial peptide cecropin A (1–33). *Biochemistry 21*, 5020–5031.

MESSNER, B.. (1981). Eine Bakterizidie in den Homogenaten von *Locusta migratoria* und *Galleria mellonella* (Insecta). *Zool. Jb. Physiol. 85*, 164–172.

METCHNIKOFF, E. (1884). Über eine Sprosspilzerkrankung der Daphnien. Beitrag zur Lehre über den Kampf der Phagocyten gegen Krankheitserreger. *Virchows Arch. 96*, 177–195.

MOHRIG, W. and MESSNER, B. (1968a). Immunreaktionen bei Insekten. I. Lysozym als grundlegender antibacterieller Faktor im humoralen Abwehrmechanismus der Insekten. *Biol. Zbl. 87*, 439–470.

MOHRIG, W. and MESSNER, B. (1968b). Immunreaktionen bei Insekten. II. Lysozym als antimikrobielles Agens im Darmtrakt von Insekten. *Biol. Zbl. 87*, 705–718.

NAPPI, A. J. (1970) Defense reactions of *Drosophila euronotus* larvae against the hymenopterous parasite *Pseudocoila bochei*. *J. Invertebr. Pathol. 16*, 408–418.

NAPPI, A. J. (1973). Hemocytic changes associated with the melanization of some insect parasites. *Exp. Parasitol. 33*, 285–302.

NAPPI, A. J. (1975). Parasite encapsulation in insects. In *Invertebrate Immunity*. Edited by K. Maramorosch and R. E. Shope. Pages 293–326. Academic Press, New York.

OURTH, D. D. and SMALLEY, D. L. (1980). Phagocytic and humoral immunity of the adult cotton boll weevil, *Anthonomus grandis* (Coleoptera: Curculionidae), to *Serratia marcescens*. *J. Invertebr. Pathol. 36* 104–112.

PEREIRA, M. E. A., ANDRADE, A. F. B. and RIBEIRO, J. M. C. (1981). Lectins of distinct specificity in *Rhodnius prolixus* interact selectively with *Trypanosoma cruzi*. *Science 211*, 597–599.

POINAR JR., G. O. and LEUTENEGGER, R. (1971). Ultrastructural investigation of the melanization process in *Culex pipiens* (Culicidae) in response to a nematode. *J. Ultrastruct. Res. 36*, 149–158.

POINAR JR., G. O., LEUTENEGGER, R. and GÖTZ, P. (1968). Ultrastructure of the formation of a melanotic capsule in *Diabrotica* (Coleoptera) in response to a parasitic nematode (Mermithidae). *J. Ultrastruct. Res. 25*, 293–306.

POWNING, R. F. and DAVIDSON, W. J. (1973). Studies on insect bacteriolytic enzymes. I. Lysozyme in haemolymph of *Galleria mellonella* and *Bombyx mori*. *Comp. Biochem. Physiol. 45B*, 669–681.

POWNING, R. F. and DAVIDSON, W. J. (1976). Studies on insect bacteriolytic enzymes. II. Some physical and enzymatic properties of lysozyme from haemolymph of *Galleria mellonella*. *Comp. Biochem. Physiol. 55B*, 221–228.

PYE, A. E. and BOMAN, H. G. (1977). Insect Immunity. III. Purification and partial characterization of immune protein P5 from hemolymph of *Hyalophora cecropria* pupae. *Infect. Immun.* 17, 408–414.

QU, X.-M., STEINER, H., ENGSTRÖM, A., BENNICH, H. and BOMAN, H. G. (1982). Insect immunity. Isolation and structure of cecropins B and D from pupae of the Chinese oak silk moth, *Antheraea pernyi*. *Eur. J. Biochem. 127*, 219–224.

RASMUSON, T. and BOMAN, H. G. (1977). The assay and the specificity problem in insect immunity. In *Developmental Immunobiology*. Edited by J. B. Solomon and J. D. Horton. Pages 83–90. Elsevier, Amsterdam.

RASMUSON, T. and BOMAN, H. G. (1979). Insect immunity. V. Purification and some properties of immune protein P4 and haemolymph of *Hyalophora cecropria* pupae. *Insect Biochem. 9*, 259–264.

RATCLIFFE, N. A. and GAGEN, S. J. (1977). Studies on the *in vivo* cellular reactions of insects: an ultrastructural analysis of nodule formation in *Galleria mellonella*. *Tissue Cell 9*, 73–85.

RATCLIFFE, N. A. and ROWLEY, A. F. (1979a). Role of hemocyte in defence against biological agents. In *Insect Hemocytes*. Edited by A. P. Gupta. Pages 331–414. Cambridge University Press, Cambridge.

RATCLIFFE, N. A. and ROWLEY, A. F. (1979b). Structure and function of blood cells of insects. *Dev. Comp. Immunol. 3*, 189–221.

RHEINS, L. A., KARP, R. D. and BUTZ, A. (1980). Induction of specific humoral immunity to soluble proteins in the American cockroach (Periplaneta americana). I. Nature of the primary response. *Dev. Comp. Immunol. 4*, 447–458.

RICHARDS, A. G. (1978). The chemistry of insect cuticle. In *Biochemistry of Insects*. Edited by M. Rockstein. Pages 205–232. Academic Press, New York.

RIZKI, M. T. M. (1953). The larval blood cells of *Drosophila willistoni*. *J. Exp. Zool. 123*, 397–411.

RIZKI, M. T. M. and RIZKI, R. M. (1959). Functional significance of the crystal cells in the larva of *Drosophila melanogaster*. *J. Biophys. Biochem. Cytol. 5*, 235–240.

ROWLEY, A. F. and RATCLIFFE, N. A. (1981). Insects. In *Invertebrate Blood Cells*. Edited by N. A. Ratcliffe and A. F. Rowley. Pages 421–488. Academic Press, London.

SALT, G. (1956). Experimental studies in insect parasitism. IX. The reaction of a stick insect to an alien parasite. *Proc. R. Soc. Lond. B 146*, 93–108.

SALT, G. (1960). Experimental studies in insect parasitism. XI. The haemocytic reaction of a caterpillar under varied conditions. *Proc. R. Soc. Lond. B 151*, 446–467.

SALT, G. (1963). The defense reactions of insects to metazoan parasites. *Parasitology 53*, 527–642.

SALT, G. (1968). The resistance of insect parasitoids to the defense reactions of their hosts. *Biol. Rev. 43*, 200–232.

SCHMIT, A. R. and RATCLIFFE, N. A. (1977). The encapsulation of foreign tissue implants in *Galleria mellonella* larvae. *J. Insect Physiol. 23*, 175–184.

SCHMIT, A. R. and RATCLIFFE, N. A. (1978). The encapsulation of Araldite implants and recognition of foreignness in *Clitumnus extradentatus*. *J. Insect Physiol. 24*, 511–521.

SCHMIT, A. R., ROWLEY, A. F. and RATCLIFFE, N. A. (1977). The role of *Galleria mellonella* hemocytes in melanin formation. *J. Invertebr. Pathol. 29*, 232–234.

SCHNEIDERMANN, H. A. (1967). Insect surgery. In *Methods In Developmental Biology*. Edited by F. H. Wilt and N. K. Wessels. Pages 753–766. Thomas Y. Crowell, New York.

SEAMAN, G. R. and ROBERT, N. L. (1968). Immunological response of male cockroaches to injection of *Tetrahymena pyriformis*. *Science 161*, 1359–1361.

SIDÉN, I., DALHAMMAR, G., TELANDER, B., BOMAN, H. G. and SOMERVILLE, H. (1979). Virulence factors in *Bacillus thuringiensis*: Purification and properties of a protein inhibitor of immunity of insects. *J. Gen. Microbiol. 114*, 45–52.

SMINIA, T. and VAN DER KNAAP, W. (1981). The internal defense system of the freshwater snail *Lymnaea stagnalis*. *Dev. Comp. Immunol. 5*, Sup. 1, 87–97.

SMINIA, T., WINSEMIUS, A. A. and VAN DER KNAAP, W. P. W. (1981). Recognition of foreigness by blood cells of the fresh water snail *Lymnaea stagnalis*, with special reference to the role and structure of the cell coat. *J. Invertebr. Pathol. 38*, 175–183.

SÖDERHÄLL, K. and AJAXON, R. (1982). Effect of quinones and melanin on mycelial growth of *Aphanomyces* spp. and extracellular protease of *Aphanomyces astaci*, a parasite on crayfish. *J. Invertebr. Pathol. 39*, 105–109.

SÖDERHÄLL, K., HÄLL, L., UNESTAM, T. and NYLHÉN, L. (1979). Attachment of phenoloxidase to fungal cell walls in arthropod immunity. *J. Invertebr. Pathol. 34*, 285–294.

STAIRS, G. R. (1964). Changes in the susceptibility of *Galleria mellonella* (Linnaeus) larvae to nuclear-polyhedrosis virus following blockage of the phagocytes with India ink. *J. Insect Pathol. 6*, 373–376.

STEINER, H. (1982). Secondary structure of the cecropins: Antibacterial peptides from the moth *Hyalophora cecropia*. *FEBS Lett. 137*, 283–287.

STEINER, H., HULTMARK, D., ENGSTRÖM, A., BENNICH, H. and BOMAN, H. G. (1981). Sequence and specificity of two antibacterial proteins involved in insect immunity. *Nature 292*, 246–248.

STEPHENS, J. M. (1959). Immune response of some insects to some bacterial antigens. *Canad. J. Microbiol. 5*, 203–228.

STOLTZ, D. B. and VINSON, S. B. (1979). Penetration into caterpillar cells of virus-like particles injected during oviposition by parasitoid ichneumonid wasps. *Canad. J. Microbiol. 25*, 207–216.

STRELKOV, A. (1964). The biology of Filipjevimermis singularis sp. nov. to be found in the Rybinsk reservoir. *Vestn. Leningr. Univ. Ser. Biol. 19*, 55–69.

STRICKLAND, E. H. (1923). Biological notes on parasites of prairie cutworms. *Bull. Dep. Agric. Dom. Canada 26*, 1–40.

TAYLOR, R. L. (1969). A suggested role for the polyphenol-phenoloxidase system in invertebrate immunity. *J. Invertebr. Pathol. 14*, 427–428.

TELFER, W. H. and WILLIAMS, C. M. (1960). The effect of diapause, development and injury on the incorporation of radioactive glycine into the blood proteins of the Cecropia silkworm. *J. Insect Physiol. 5*, 61–72.

THOMAS, I. G. and RATCLIFFE, N. A. (1982). Integumental grafting and immunorecognition in insects. *Dev. Comp. Immunol. 6*, 643–654.

UNESTAM, T. and NYLUND, J.-E. (1972). Blood reactions *in vitro* in crayfish against a fungal parasite, *Aphanomyces astaci*. *J. Invertebr. Pathol. 19*, 94–106.

UNESTAM, T. and NYLHÉN, L. (1974). Cellular and noncellular recognition of and reactions to fungi in crayfish. In *Contemporary Topics in Immunology*. Edited by E. L. Cooper. Vol. 4, pages 189–206. Plenum Press, New York.

UNESTAM, T. and SÖDERHÄLL, K. (1977). Soluble fragments from fungal cell walls elicit defense reactions in crayfish. *Nature (Lond.) 267*, 45–46.

VEY, A. and GÖTZ, P. (1975). Humoral encapsulation in Diptera (Insecta): Comparative studies in vitro. *Parasitology 70*, 77–86.

VEY, A. and VAGO, C. (1971). Reactions anticryptogamique de type granulome chez les insectes. *Ann. Inst. Pasteur 121*, 527–532.

VINSON, S. B., EDSON, K. M. and STOLTZ, D. B. (1979). Effect of a virus associated with the reproductive system of the parasitoid wasp, *Campoletis sonorensis*, on host weight gain. *J. Invertebr. Pathol. 34*, 133–137.

WILKE, U. (1979). Humolale Infektabwehr bei Chironomuslarven. In *Verh. Dtsch. Zool. Ges.* 1979. Edited by W. Rathmayer. Page 315. Fischer, Stuttgart.

WÜLKER, W. (1961). Untersuchungen über die Intersexualität der Chironomiden (Dipt.) nach *Paramermis*-Infektion. *Arch. Hydrobiol., Suppl. 25*, 127–181.

YAMAURA, I., YONEKURA, M., KATSURA, Y., ISHIGURO, M. and FUNATSU, M. (1980). Purification and some physicochemical properties of phenoloxidase from the larvae of housefly. *Agric. Biol. Chem. 44*, 55–59.

YOSHINO, T. P., RENWRANTZ, L. and CHENG, T. C. (1979). Binding and redistribution of surface membrane receptors for concanavalin A on oyster hemocytes. *J. Exp. Zool. 207*, 439–449.

ZACHARY, D. and HOFFMANN, J. A. (1973). The haemocytes of *Calliphora erythrocephala* (Meig.) (Diptera). *Z. Zellforsch. mikrosk. Anat. 141*, 55–74.

ADDENDUM

The following are important references published after completion of this article.

HULTMARK, D., ENGSTRÖM, A., ANDERSSON, K., STEINER, H., BENNICH, H. and BOMAN, H. G. (1983). Insect immunity. Attacins, a family of antibacterial proteins from *Hyalophora cecropia*. *The EMBO Journal 2*, 571–576.

LEE, J.-Y., EDLUND, T., NY, T., FAYE, I. and BOMAN, H. G. (1983). Insect immunity. Isolation of cDNA clones corresponding to attacins and immune protein P4 from *Hyalophora cecropia*. *The EMBO Journal 2*, 577–583.

OKADA, M. and NATORI, S. (1983). Purification and characterization of an antibacterial protein from hemolymph of *Sarcophaga peregrina* (flesh fly) larva. *Biochem. J. 211*, 727–734.

RENWRANTZ, L., SCHÄNCKE, W., HARM, H. ERL, H., LIEBSCH, H. and GERCKEN, J. (1981). Discriminative ability and function of the immunobiological recognition system of the snail *Helix pomatia*. *J. Comp. Physiol. 141*, 477–488.

SINDEN, I. and BOMAN, H. G. (1983). *Escherichia coli* mutants with an altered sensitivity to cecropin D. *J. Bacteriology 154*, 170–176.

SÖDERHÄLL, K. and SMITH, V. J. (1983). Separation of the haemocyte populations of *Carcinus meanas* and other marine decapods, and prophenoloxidase distribution. *Developm. Comp. Immunol. 7*, 229–239.

12 Nephrocytes and Pericardial Cells

A. C. CROSSLEY

University of Sydney, New South Wales, Australia

1 DEFINITION, RECOGNITION AND DISTRIBUTION OF NEPHROCYTES AND PERICARDIAL CELLS

1.1 Definition

Insect nephrocytes contribute to the regulation of haemolymph composition by the removal of some molecules and the introduction of others. This responsibility is shared with other cellular systems, especially haemocytes (A. Gupta, this volume), fat body (R. Dean *et al.*, this volume) and excretory organs (T. Bradley, vol. 4), and the aim of this chapter is to focus on the special structural and functional properties of nephrocytes, and assess their contribution to the physiology of the whole insect.

Nephrocytes are cells ensheathed in basement membrane, either individually or in groups, and suspended in the haemocoel by connective tissue strands (Fig. 1). Each cell is a podocyte (section 2.2), with its plasma membrane folded to form invaginations, or "foot processes", and is engaged in selectively sequestering non-particulate colloids from the haemolymph by endocytosis, and releasing other molecules by exocytosis. The term "nephrocyte" was introduced by Kowalevsky, A. (1892) and reviewed by de Ribaucourt, H. (1901). It is derived from the Greek roots *nephros* = kidney, and *kýtos* = container. Many other terms have been used to describe members of the nephrocyte family (section 1.3), which includes pericardial cells, but the general functional term "nephrocyte" is retained here.

Nephrocytes can be distinguished from sessile haemocytes by their enclosure in a basement membrane (haemocytes are always naked), and their selective rejection of particulate material in endocytosis. (Nephrocytes are pinocytotic rather than phagocytic, section 4.3). Thus, haemocytes and nephrocytes form complementary parts of a scavenging and recycling system applied to the haemolymph. In these roles they are in some ways analogous to the vertebrate reticuloendothelial system (section 4.2). Nephrocytes also participate in the synthesis of haemolymph solutes, a function they share with many other tissues, notably the fat body and haemocytes (section 5), but nephrocytes can be distinguished both *in vivo* and in the electron microscope.

1.2 Recognition

In vivo nephrocytes can readily be recognized by injection of non-particulate colloidal dyes. If the dye is filtered through a 0.45 μm pore size membrane filter to remove particles, and if the dye loading and toxicity is low, only nephrocytes are coloured. For the recognition of nephrocytes under the dissection microscope a simple procedure is to use Trypan blue filtered through a 0.45 μm type HA Millipore filter to remove particulate material. Very heavy loading of haemolymph with dye will induce staining of many organs, such as Malpighian tubules, salivary glands and fat body. But if the dye concentration is kept below about 0.01% dye in haemolymph, and if particulate dye is excluded by filtration, only nephrocytes become coloured. Orders of insects which have been explored by injection of vital dyes for the mapping of nephrocytes are indicated in Table 1 (INJ), but further work is needed, particularly in Coleoptera and Hymenoptera.

In the electron microscope nephrocytes can be recognized by their "podocyte" configuration. Deeply indented channels are guarded by filter slits, and spawn coated vesicles into the cytoplasm. In the cytoplasmic cortex lie 65 nm diameter tubular elements and distinctive α-vacuoles, whilst the medulla is replete with storage vacuoles which differ from those of the fat body, and in fact from all other cell types (Figs 2 and 3) (section 2).

The biochemistry of nephrocytes is in its infancy, but holds promise of techniques for recognition of

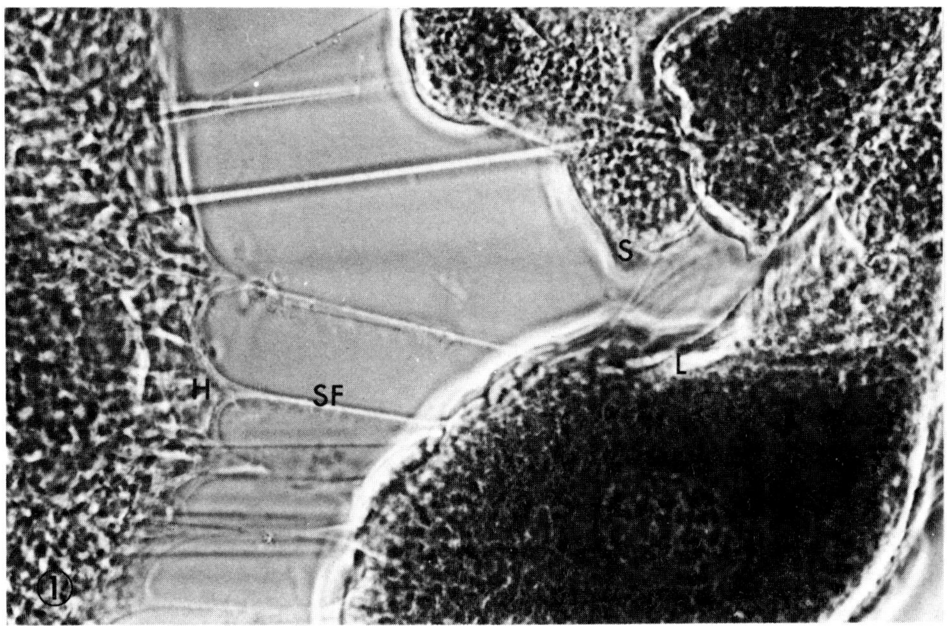

FIG. 1. Light micrograph of dorsal pericardial nephrocyte in the larva of the blowfly *Calliphora vicina*, showing attachment to the heart (H) by suspensory fibrils (SF). Large posterior (L) and small anterior (S) nephrocytes are shown. (×530.)

nephrocytes in the future. At present the most useful biochemical marker for nephrocytes is lysozyme (section 5.4).

1.3 Distribution of nephrocytes in insects

Nephrocytes have been described under a variety of names in most groups of insects (Table 1) and are probably present in all insects. An early description refers to cells which form a garland around the oesophagus at its junction with the proventriculus ("Guirlandenzellen", Weismann, A., 1865), and before the end of the 19th century these "garland cells" had been demonstrated to ingest dyes and salts of silver and of iron injected into the haemolymph, functioning as "accumulation kidneys" (Kowalevsky, A., 1886, 1889). The introduction of ammonia carmine vital staining by Kowalevsky led to the detection of a family of pinocytotic cells which he called "nephrocytes" (Kowalevsky, A., 1892). In a series of papers Bruntz, L. (1904, 1908a,b) described pericardial, ventral and disseminated nephrocytes in a variety of insects, crustaceans and *Peripatus*.

The dorsal pericardial nephrocytes were first noticed by Leydig, F. (1857) as mesodermal cells arranged along the heart and aorta, and have often subsequently been referred to simply as "pericardial cells". Other terms which refer to the same dorsal nephrocytes are "diaphragm cells" and "paracardial cells" (section 3.1, Table 1).

The ventral nephrocytes lie suspended beneath the oesophagus, and have been called "garland cells", "wreath cells", and "circum-" or "suboesophageal cells" (Table 1). In some insects there are nephrocytes scattered amongst fat body cells which are termed "disseminated nephrocytes". The "organ rameux" of Lepidoptera is made up of disseminated nephrocytes (Table 1). Nephrocytes do not enter the circulation, and are not usually included in haemolymph samples, but they frequently contaminate samples of fat body, especially in Lepidoptera.

In the primitive apterygote insect *Petrobius* nephrocytes are present in the pericardial sinus from the second thoracic segment to the eighth abdominal segment. In some insects nephrocytes are present in the head (e.g. *Perla marginata*, Schwermer, W., 1914; and *Myzus persicae*, Bowers, B., 1964). In the holometabolous dipteran insect *Calliphora erythrocephala* the ventricle of the larval heart in abdominal segments 5 to 8 is provided with 12–14 pairs of

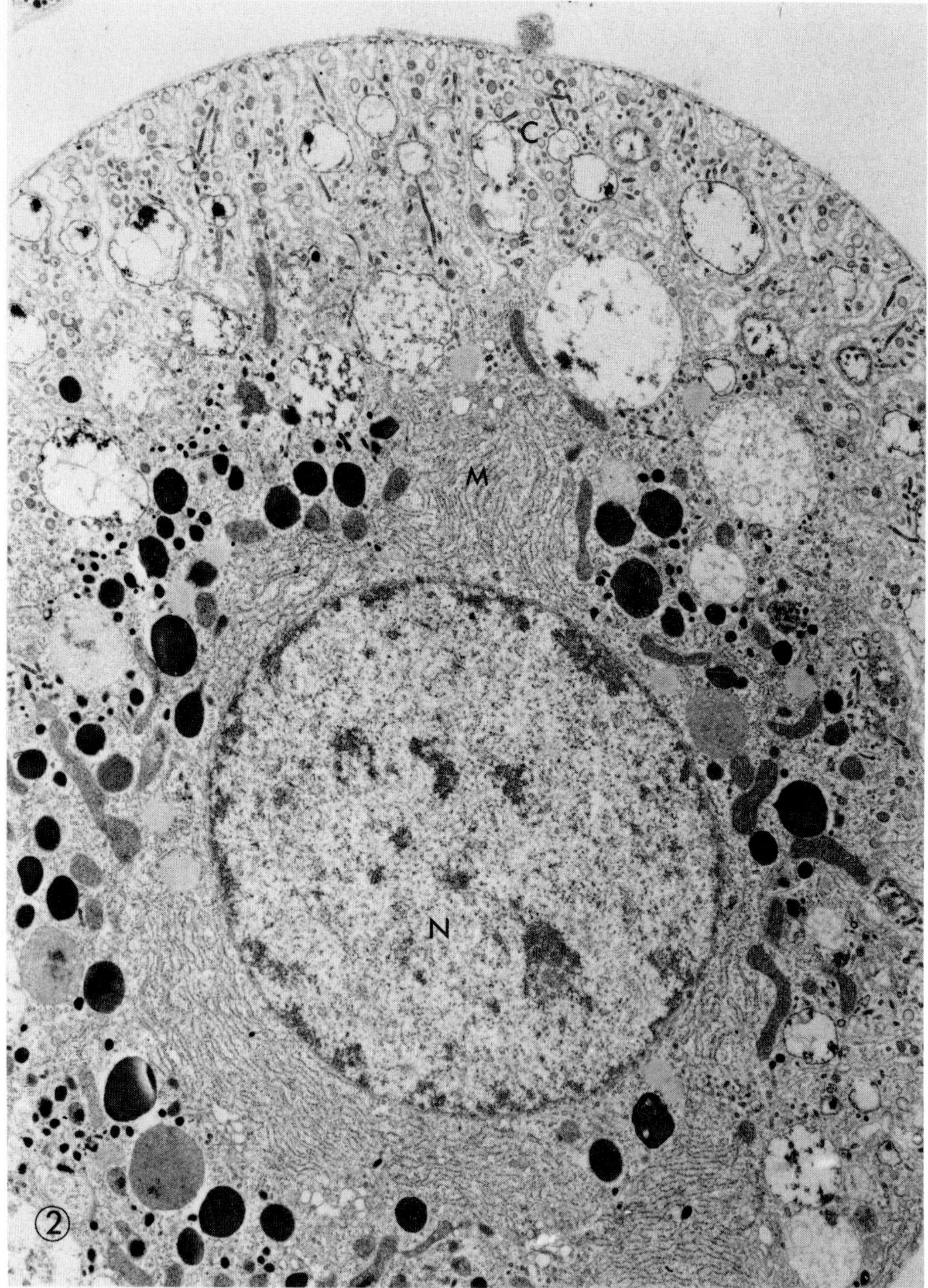

FIG. 2. Survey electron micrograph of a dorsal pericardial nephrocyte in the adult *Calliphora*. Informal zonation of organelles around the nucleus is shown. The cortex (C) of invaginated channels and vesicles grades into the medulla (M) replete with endoplasmic reticulum and vacuoles. (× 12,000.)

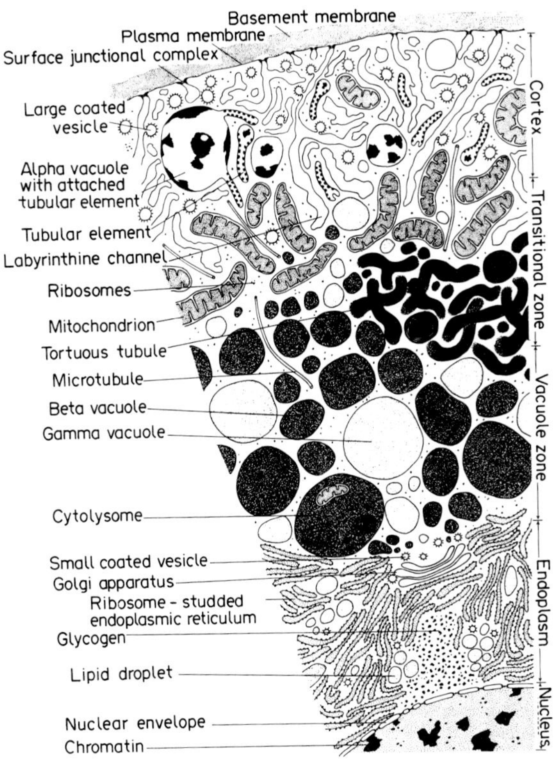

FIG. 3. Diagram of organelle zonation in a nephrocyte.

large (140–200 μm diameter) pericardial nephrocytes, which come to contain polytene chromosomes. The anterior small heart in abdominal segments 2 to 5 is provided with many hundreds of small nephrocytes which do not become polytene. These small dorsal nephrocytes extend to the insertion of the most anterior alary muscle at the border of abdominal segments 2 and 3, but do not extend over the aorta (Crossley, A., 1972a, 1983; Jensen, P., 1973). Small ventral nephrocytes occur in thoracic segments 2 and 3 in *Calliphora* and have polytene chromosomes (Thomson, J., 1975).

Embryological evidence suggests that nephrocytes derive from the bilateral coelomosac mesoderm present in each metameric segment (section 3.1). In larval and adult insects nephrocytes are usually to be found in sites swept by a heavy flow of haemolymph, such as the pericardium or haemolymph sinuses.

Table 1 summarizes the most recent or complete works on the distribution and anatomy of nephrocytes for each Order of insects. There do not appear to be descriptions of nephrocytes for the Orders: Ephemeroptera, Isoptera, Mantodea, Zoraptera, Grylloblattodea, Plecoptera, Phasmatodea, Embioptera, Psocoptera, Thysanoptera, Megaloptera, Neuroptera, Strepsiptera, Siphonaptera, Trichoptera, or surprisingly for the Hymenoptera. Table 1 also indicates the number of nuclei present in each nephrocyte, and whether tracer dye injection or ultrastructural studies have been included in the report.

2 FUNCTIONAL ANALYSIS OF NEPHROCYTE ULTRASTRUCTURE

2.1 The basement membrane

Nephrocytes are anchored in the haemocoel by connective tissue strands, and each cell or small group of cells is ensheathed in a basement membrane 100 nm or more thick. In *Bombyx mori* the basement membrane on nephrocytes reaches a thickness of 800 nm, and after freeze-fracture appears as a thick outer granular layer overlying a homogenous layer applied to the plasma membrane (Lavanseau, L. *et al.*, 1981). The outer granular layer of the basement membrane extends over the elastic suspensory fibrils which anchor the nephrocytes in position, but the suspensory fibrils have additional core components described by Hoffmann, J. (1967) and Crossley, A. (1972a) (Fig. 1). It is not clear whether the basement membrane is secreted by nephrocytes or receives a contribution from haemocytes. The basement membrane imposes a significant barrier to molecules passing from the haemolymph to the nephrocyte plasma membrane (sections 4.3 and 7.2.1).

2.2 Plasma membrane

The plasma membrane of nephrocytes in all insects is extensively invaginated to form a network of "labyrinthine channels" (Figs 3 and 4) and this feature is thus very helpful in the ultrastructural recognition of nephrocytes (Table 2). Although the plasma membrane is folded, the external contour of the cell remains rather smooth, because junctional complexes link the faces of plasma membrane at each invagination (Fig. 5). The junctional complexes at the cell surface, together with the overlying

Table 1: Summary of distribution and anatomy of nephrocytes in insects.

Taxonomic group	Species	Reference	Locations	Nuclei	Tracer	Anatomy
Cl. Collembola	*Tomocerus minor, Lepidocyrtus curvicollis*	Humbert, W., 1975	Ceph. Abd.	4–7	—	E.M.
O. Thysanura	*Lepisma, Machilis*	Bruntz, L., 1908a,b	P/C, V	< 4	INJ	—
	Petrobius maritimus	Gabe, M., *et al.* 1973	V, Dissem.	1–2	INJ	—
O. Odonata	Larvae	Wigglesworth, V., 1970	Dissem	—	—	—
Blattoid-orthopteroid orders	*Blattella germanica*	Edwards, G. and Challice, C.1960	Diaphragm.	1	—	E.M.
	Melanoplus differentialis	Kessel, R., 1961a,b, 1962	P/C, Oesoph.	2–6	INJ	E.M.
	Periplaneta americana	Davey, K., 1961a,b,c, 1962	P/C	2–5	INJ	—
	Locusta migratoria	Hoffmann, J. and Levi, C., 1965 Hoffmann, J., 1967	P/C	> 2	INJ	E.M.
O. Demaptera	*Forficula*	Cuénot, L., 1896	P/C	2	INJ	—
O. Mallophaga	*Pediculus humanus*	Nuttall, G., and Keilin, D., 1921	P/C, V, Dissem, Oesoph.	2–3	INJ	—
O. Hemiptera	*Cimex lectulanus*	Puri, I., 1924	P/C, Dissem, Oesoph.	1	INJ	—
	Rhodnius prolixus	Wigglesworth, V., 1943	P/C		INJ	—
	Myzus persicae	Bowers, B., 1964	P/C, V, Ceph.	1–2	—	E.M.
O. Coleoptera	*Bathysciinae*	Deleurance, S. and Charpin, P., 1971	Oesoph.	1	—	E.M.
O. Mecoptera	*Panorpa communis*	Schwinck, I., 1951, 1952	P/C Oesoph.	2	—	—
O. Lepidoptera	*Vanessa*	Hollande, A., 1922	P/C	1–5	INJ	—
	Galleria	Cameron, G., 1934	P/C	2–4	INJ	—
	Hyalophora cecropia	Sanger, J., and McCann, F., 1968	P/C	3–7	INJ	E.M.
	Bombyx mori	Lavanseau, L., *et al.*, 1981	Organe Rameux	1	INJ	E.M.
O. Diptera	*Musca vomitoria*	Weismann, A., 1865	V	2	—	—
	Lonchaea chorea	Keilin, D., 1917, 1924	P/C, V	1	INJ	—
	Culex	Pal, R., 1944	P/C, V	2–4	INJ	—
	Drosophila melanogaster	Mills, R. and King, R., 1965	P/C	1	—	E.M.
		Aggarwal, S. and King, R., 1966	V	1–2	—	E.M.
	Calliphora vicina	Crossley, A., 1972a	P/C, V	1	INJ	E.M.

Abbreviations:
Locations: Abd., abdominal; Ceph., cephalic; Dissem., disseminated; Oesoph., peri, circum, and suboesophageal; P/C, pericardial; V, ventral
Tracer: INJ, indicates that endocytosis has been explored by injection of tracer.
Anatomy: E.M., indicates that electron micrographs are included in the report.

basement membrane, form a gateway to an extracellular compartment of labyrinthine channels extending into the cell (Figs 2–4). The cortical region of the cell thus consists of elongated stalks or "pedicels" of cytoplasm separated by extracellular labyrinthine channels. This cellular architecture is found in many cells engaged in ultrafiltration, and is termed "podocyte form". The pedicels of cortical cytoplasm are elongated, and the entrances to labyrinthine channels take the form of narrow slits, reminiscent of the "filter slits" of the podocytes of the vertebrate glomerulus (section 7.2). Indeed the basement membrane and junctional complexes of nephrocytes represent extracellular barriers that must filter the fluid entering labyrinthine channels (section 4.3) and may act much like the filter slits of the glomerulus.

The walls of labyrinthine channels facing the extracellular lumen consist of plasma membrane lined with tufts of filiform material about 35 nm long, arranged side by side (Sanger, J. and McCann, F., 1968). This glycocalyx forms the outermost interface between the cell and its environment, and has in most cells unusual properties with regard to charge distribution and water- and ion-binding capacity (Berridge, M. and Oschman, J., 1972). The plasma membrane lining the labyrinthine channels is the site of very active pinocytosis, as evidenced by early electron micrographs (Kessel, R., 1962; Bowers, B., 1964) and later tracer studies (section 4.3).

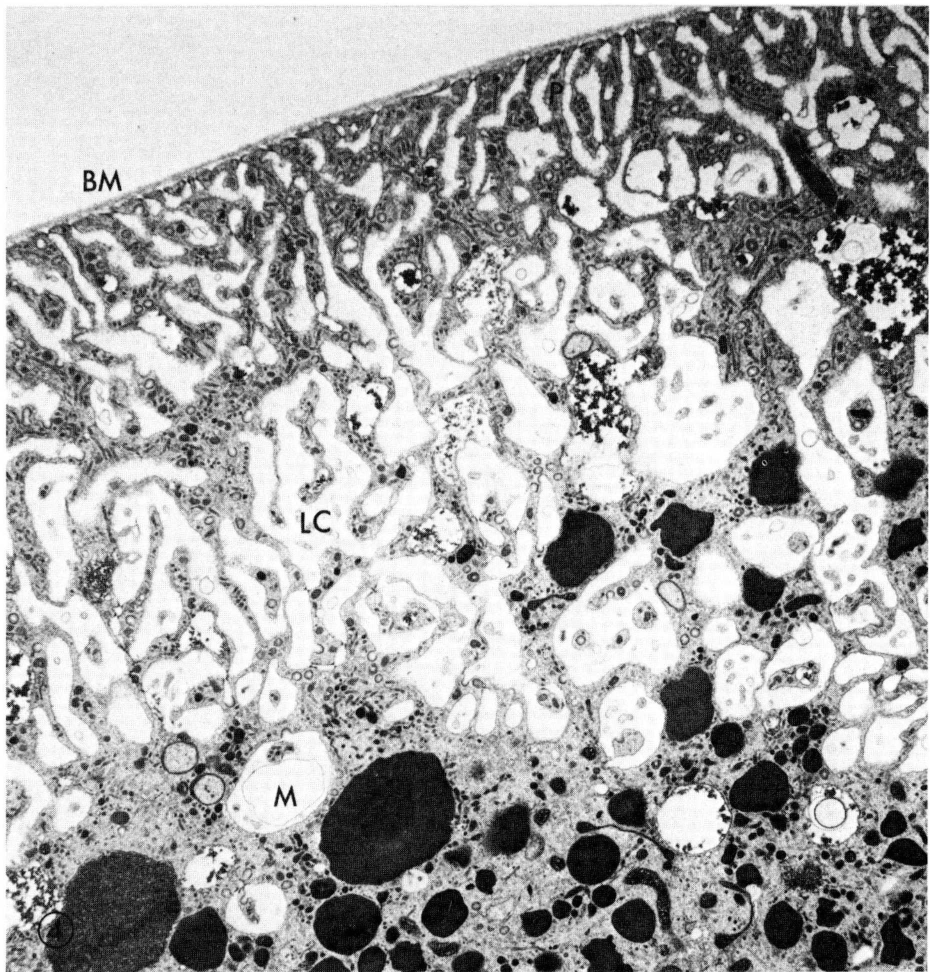

FIG. 4. Invagination of the plasma membrane of a nephrocyte (*Calliphora* larva) to form labyrinthine channels (LC). Junctional complexes beneath the basement membrane (BM) hold the plasma membrane in an even contour, linking adjacent pedicels (P) of cytoplasm. Numerous vacuoles occupy the cytoplasm of the medulla (M). (× 12,000.)

2.3 Surface junctional complexes

Junctional complexes linking opposite faces of plasma membrane along lines of invagination were first described in *Myzus persicae* by Bowers, B. (1964) who called them "desmosomes". As in typical desmosomes the cytoplasmic face of the plasma membrane is lined with a strip of electron-dense material, but the structures linking opposite membranes take the form of perforated diaphragms and differ from the usual cementing material of desmosomes. The gap between membranes at the nephrocyte junctional complex ranges from 25 nm in *Myzus* (Bowers, B., 1964) to 40 nm in *Tomocerus minor* (Humbert, W., 1975). The gap is bridged by one or more diaphragms, or laminae, of cementing material. In *Myzus* there are commonly two laminae 15 nm apart, bearing a polygonal pattern, but up to six laminae have been observed (Bowers, B., 1964). In *Calliphora* and *Tomocerus* there are usually two laminae, each about 8 nm thick (Crossley, A., 1972a; Humbert, W., 1975) (Figs 4 and 5).

Patterning appears in favourable profiles of all nephrocyte surface junctional complexes, but it is diffuse and further ultrastructural analysis is needed. In *Calliphora* pericardial nephrocytes the patterning (Fig. 6) arises from rows of regularly spaced pegs of relatively electron-dense material extending from each opposed membrane. The centre-to-centre spacing of the pegs is about 28 nm, and

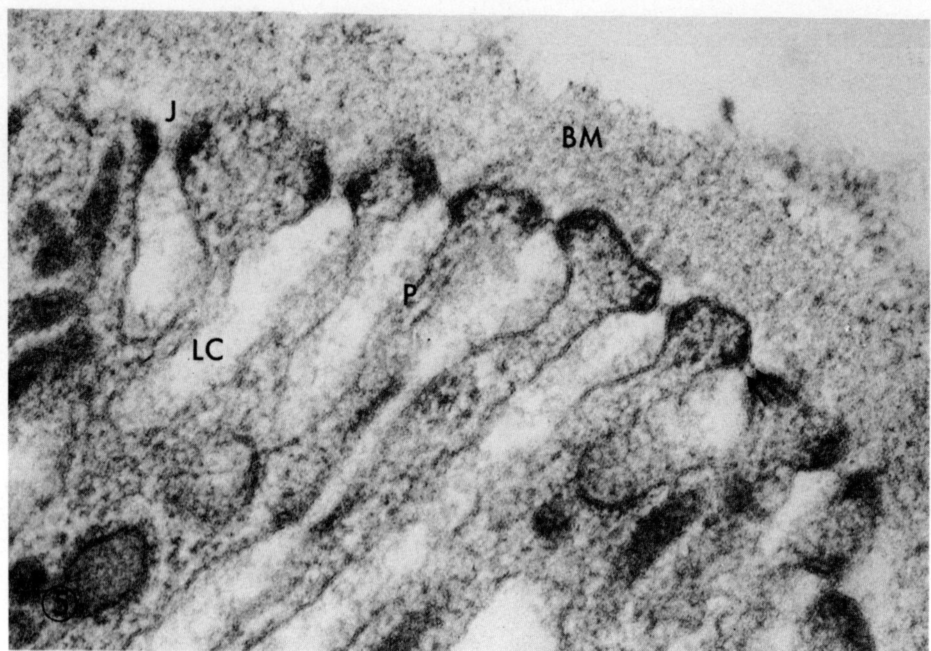

FIG. 5. Junctional complexes (J) forming "filter slits' guard the entrances to labyrinthine channels (LC) between pedicels (P). Note the thick basement membrane (BM) which acts as a pre-filter of material entering the filter slits which are occluded by one or two laminae (arrows). (*Calliphora* larva; × 100,000.)

they consist of stalks about 8 nm wide with swollen tips. Opposed stalks are staggered, and meet or overlap slightly at the midline to form a serrated electron-dense line, which may incorporate extra cementing material. As indicated in the diagram (Fig. 7) the overall impression is of a zipper with spaces between adjacent teeth. Since the spaces are electron-lucent, and about 13 × 20 nm, they are assumed to be "pores", permitting access to the extracellular spaces of labyrinthine channels.

Nephrocyte junctional complexes differ from both smooth and pleated septate junctions and scalariform junctions of other types of insect cell (see review of Noirot-Timothée, C. and Noirot, C., 1980). In the insect goblet cell, where septate junctions have been analysed by lanthanum tracer and freeze-etch techniques, the intermembrane septa are resolved as a dense array of 3 × 10 nm rods (Flower, N. and Filshie, B., 1976), and are thus smaller than in nephrocyte junctional complexes, and have different periodicity. In insect goblet cells the pedicels appear to function as valves, regulating the flux of materials into the extracellular cavity enclosed by the cell, by means of configurational changes mediated by microfilaments (Anderson, E. and Harvey, W., 1966; Flower, N. and Filshie, B., 1976). An analogous situation could exist in nephrocytes, which also contain microfilaments in association with intracellular junctional complexes (Fig. 6) and a lanthanum freeze-etch study of nephrocytes could provide useful clues in the puzzle of nephrocyte fluid dynamics (section 7.3).

In other phyla comparable surface junctional complexes occur in the vertebrate glomerulus (section 7.2) and in the podocytes and pore cells of gastropod molluscs. In all these cell types a meshwork of rods connects pedicels of cytoplasm to form slit diaphragms. In molluscs the "pores" formed are 9 × 11 nm in the podocyte and 20 × 22 nm in the pore cell, and as in insect nephrocytes, the opposing rods alternate like the teeth of a zipper (Boer, H. and Sminia, T., 1976). Functional analogies between mollusc podocytes and insect nephrocytes are discussed in section 6.2. The function of surface junctional complexes is explored in section 4.3.

2.4 Inner junctional complexes

The inner reaches of labyrinthine channels in the nephrocytes of some insects are bridged by typical

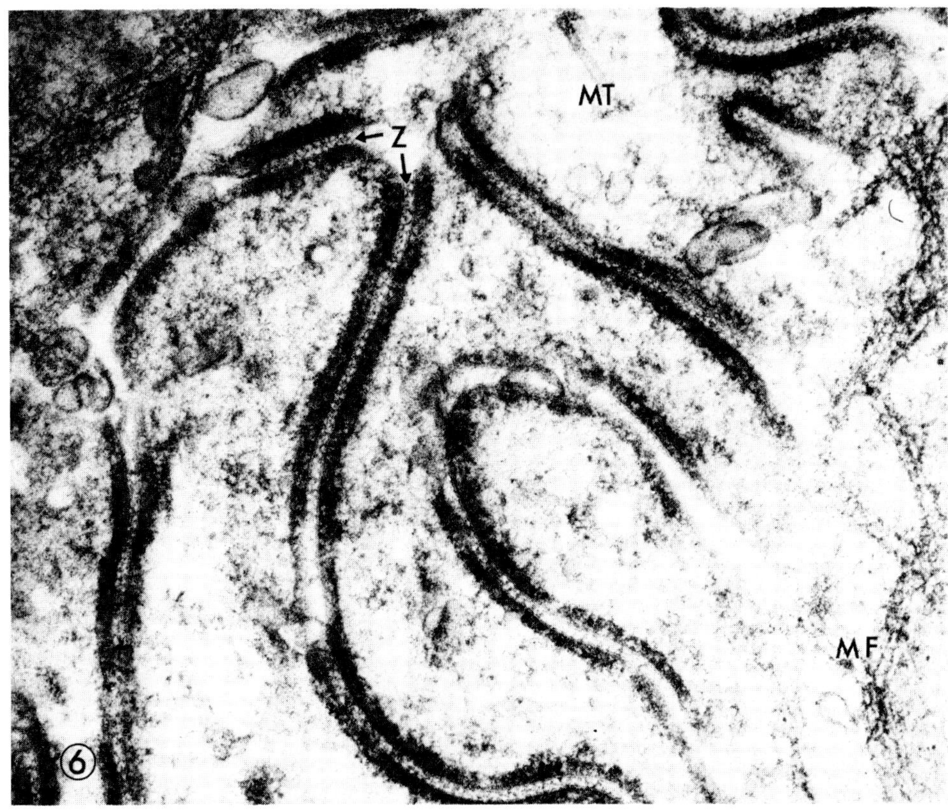

FIG. 6. Tangential section of the nephrocyte shown in Fig. 5, indicating the zipper-like (Z) polygonal pattern present on laminae of surface junctional complexes. Microtubules (MT) and microfilaments (MF) are present in the sub-membrane region. (×92,000.)

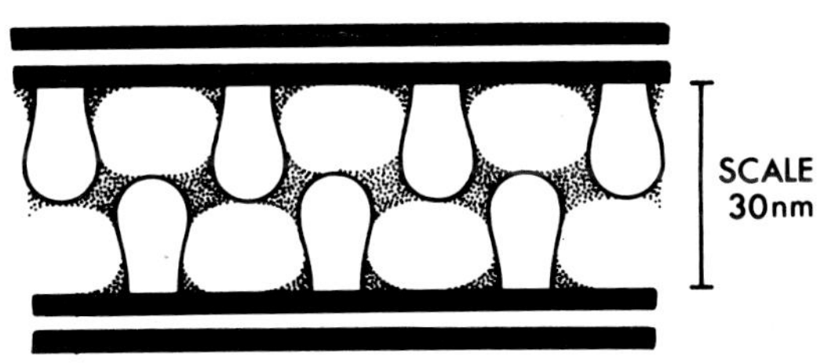

FIG. 7. Diagram to show author's interpretation of "pore" structure in the surface junctional complex of nephrocyte filter slits.

septate desmosomes (e.g. *Locusta migratoria*, Hoffmann, J., 1967; *Hyalophora cecropia*, Sanger, J. and McCann, F., 1968). Septate desmosomes are usually involved in intercellular junctions (Noirot-Timothée, C. and Noirot, C., 1980), but intracellular septate junctions have been described in a cell of the insect nervous system (Stuart, A. and Satir, P., 1968). We should, however, be alert to the possibility that some supposedly multinucleate nephrocytes may in fact be multicellular groups, interwoven and linked by septate junctions.

2.5 Labyrinthine channels and coated vesicles

The labyrinthine channels of nephrocytes are an extension of the extracellular milieu and, as the site of micropinocytosis and probably exocytosis, can be considered as a "transfer zone" where the surface area of the cell is amplified (Lavanseau, L. *et al.*, 1981).

Dynamic micropinocytosis is evidenced by profiles of pits indented in the plasma membrane, and of small vesicles adjacent to it (Figs 4 and 8). As in coated vesicles of vertebrate cells, the glycocalyx extends inside the coated pits or vesicles (Roth, T. and Porter, K., 1964). Micropinocytotic vesicles have been recorded in all ultrastructural studies of insect nephrocytes, beginning with that of Kessel, R.

Table 2: Chronology of ultrastructural studies of insect nephrocytes

Year	Authors	Species	Order	Cell types	Cortex	Micropino-cytotic vesicles	Tubular elements	Vacuole types	α	β	γ	Nucleus no.
1962	Kessel, R.	*Melanoplus differentialis*	Orthoptera	P	I	+	–	2	α*	β*	γ*	2
1964	Bowers, B.	*Myzus persicae*	Hemiptera	P	I	C	+	2	+	+	+	1–2
1965	Mills, R. and King, R.	*Drosophila melanogaster*	Diptera	P	I	+	–	3	α* (=B_1)	β* (=A)	+	1
1966	Aggarwal, S. and King, R.	*Drosophila melanogaster*	Diptera	V	I	C	–	2	+	+	–	1–2
1967	Hoffmann, J.	*Locusta migratoria*	Orthoptera	P	I	C	60 nm	3	α*	–	–	2
1968	Sanger, J. and McCann, F.	*Hyalophora cecropia*	Lepidoptera	P	I	C	60–75 nm		α*	–	+	3–7
1972	Crossley, A.	*Calliphora vicina*	Diptera	P	I	C	60–75 nm	3	α	β	γ	1†
1973	Gabe, M. Cassier, P. and Fain-Maurel, M.	*Petrobius maritimus*	Apterygota	P	I	C	60 nm		+	+	+	1
1975	Humbert, W.	*Tomocerus minor*	Collembola	C+A	I	C	+	3	α	β	α	1
1981	Lavanseau, L. Lahargue, J. Surleve-Bazeille, J.	*Bombyx mori*	Lepidoptera	D	I	C	+	3	α	β	γ	1

\+ = Appears in micrographs but not described.
– = Not visible in micrographs.
† Two nuclei are present in ventral nephrocytes.
α, β, γ, Terminology introduced by Crossley, A., (1972); earlier matching descriptions are indicated*.

(1962) for *Melanoplus differentialis*. The micrographs of Bowers, B. (1964) were the first to resolve, and draw attention to, a coating on the cytoplasmic face of pits in the plasma membrane and of vesicles in the nephrocyte cortex. Later studies indicate that most or all the cortical vesicles of nephrocytes are derived from coated pits (Fig. 9, Table 2).

In *Calliphora erythrocephala* pericardial nephrocytes (Figs 3 and 7) values for the diameter of spherical vesicles at the unit membrane range from 50 to 165 nm in profiles judged to pass close to the vesicular "equator", by virtue of near-transverse sectioning of the membrane. Most coated vesicles in the cortex are between 120 and 160 nm in diameter, but smaller coated vesicles 50–80 nm in diameter are encountered commonly in the vicinity of Golgi centres and occasionally at the cell cortex (Crossley, A., 1972a) (Fig. 11).

In *Bombyx mori* disseminated nephrocytes the coated vesicles are 130–170 nm in diameter, and freeze-etching techniques reveal a coating of globules 27 nm thick and 25 nm apart in a polygonal pattern on the cytoplasmic face of the vesicular membrane (Lavanseau, L., *et al.*, 1981).

The smaller coated vesicles may belong to a separate population with a distinct function. There is ample evidence that larger coated vesicles are involved in adsorptive endocytosis (section 4.4), but the smaller coated vesicles, which are numerous at Golgi centres and at the cell surface, may be involved in the transfer of newly synthesized proteins within the cell. Two size classes of coated vesicle with a similar functional demarcation were reported in the rat vas deferens (Friend, D. and Farquhar, M., 1967) (section 5.2).

2.6 Tubular elements

Thick-walled tubular elements abound in the cortical cytoplasm of most nephrocytes (Table 2). Tubular elements are 60–75 nm in diameter and may approach a micron in length. A single unit membrane encloses a wall of dense material about 20 nm thick, which includes a repeating globular subunit (Fig. 10), but the core of the tube is

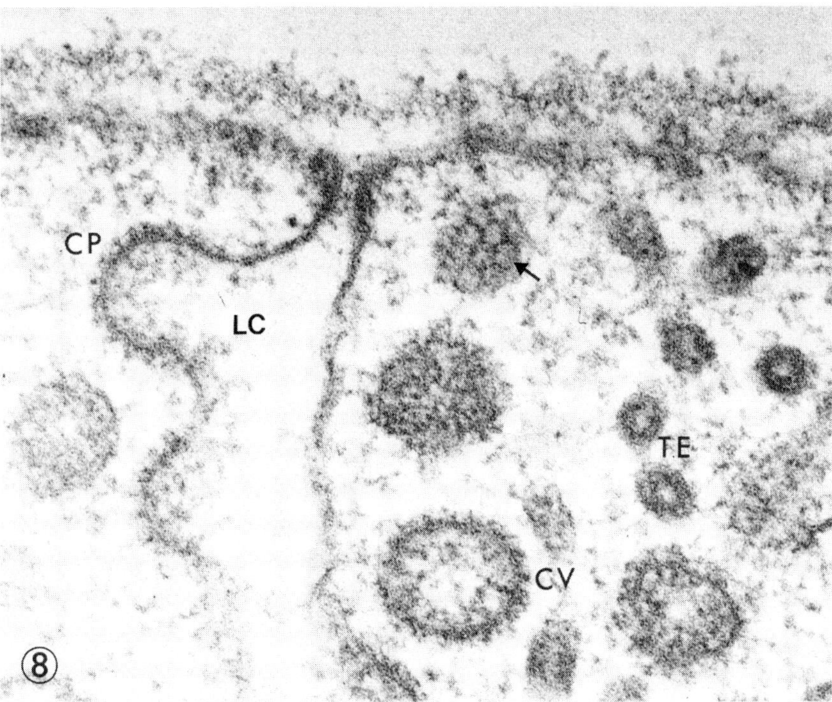

FIG. 8. Electron micrograph of coated pits (CP) and vesicles (CV) forming on labyrinthine channels (LC) of nephrocytes. The hexagonal clathrin coating on vesicles is seen in tangential sections (arrow). Numerous tubular elements (TE) are cut in transverse section. (*Calliphora* larva; × 120,000.)

electron-lucent (Bowers, B., 1964). The wall subunits include relatively electron-dense granules about 11 nm in diameter, spaced about 17 nm apart, which are embedded in a less dense matrix. Sections of material cut at various angles to the axis of the tubular element indicate that the granules may be arranged in a multiple-start helix (Crossley, A., 1972a; Gabe, M., *et al.*, 1973; Humbert, W., 1975).

Tubular elements are of indeterminate length (values of 0.4–1.3 μm are given in the literature) but do not appear to branch. They are often expanded at one end to form a flask-shaped cavity or ampoule (Figs 9 and 10) (Hoffmann, J., 1967; Crossley, A., 1972a; Humbert, W., 1975). Some tubular elements are connected directly to α-vacuoles (section 2.10), and indeed the flask-shaped ampoule may expand into a vacuole, since both are partially lined with tubular element granules (Fig. 10) (Sanger, J. and McCann, F., 1968; Crossley, A., 1972a; Lavanseau, L. *et al.*, 1981).

Tubular elements about 65 nm in diameter are usually numerous in nephrocytes (Table 2) and may occur in all nephrocytes. However, they are rarely reported in other cell types, and their distribution is of particular interest in analysis of cell function. Thick-walled 65 nm diameter tubular elements occur in insect oocytes (Anderson, E., 1969) and in the proximal tubules of vertebrate kidneys (Tisher, C. *et al.*, 1969), both cell types implicated in endocytosis. Functional hypotheses are considered in section 7.3.

2.7 Zonation of cytoplasmic organelles

An informal radial zonation of organelles occurs in the cytoplasm, and is striking in large spherical nephrocytes. The cortical "transfer zone" rich in coated vesicles, tubular elements and α-vacuoles is separated from the endoplasmic reticulum of the perikaryon by a "transitional zone". The transitional zone contains numerous mitochondria, vacuoles, and Golgi centres, as indicated diagrammatically in Fig. 3. Zonation of organelles presumably arises from the logistics of membrane and protein synthesis, and intracellular sorting dynamics (sections 4 and 5).

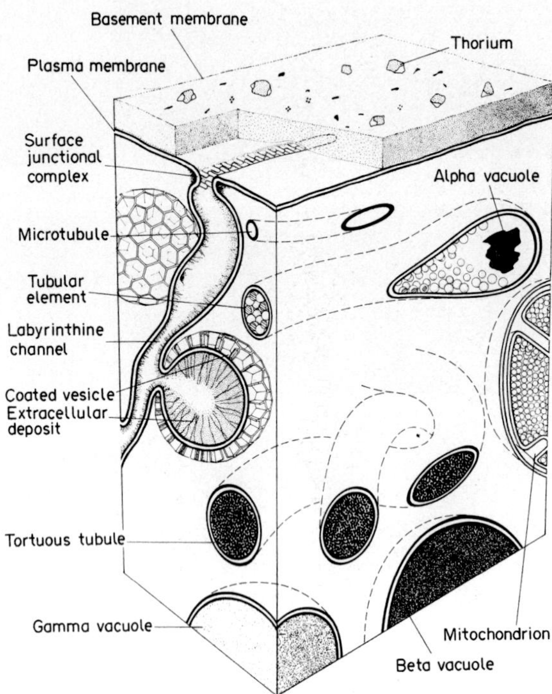

FIG. 9. Diagram showing the structure of nephrocyte cortical organelles and α, β and γ vacuoles. Thorium tracer and other particulate substances are excluded from the labyrinthine channels by the basement membrane. Labyrinthine channels are lined with a glycocalyx, and are the site of active coated vesicle micropinocytosis.

2.8 Microtubules and microfilaments

Microtubules 24 nm in diameter appear in most profiles of nephrocyte cytoplasm, and they extend from the cortex to the perinuclear region. No obvious axial or circumferential orientation has been reported, and there has been no attempt to probe function.

Microfilaments 6–8 nm in diameter are present in the cortex of nephrocytes but their nature and function in nephrocytes remains unexplored. Attempts at labelling with heavy meromyosin would be useful in view of a possible function in regulation of filter slit aperture (section 2.3).

2.9 Organelles of metabolism and synthesis

As first noted by Kessel, R. (1962) almost all nephrocytes examined in the electron microscope (Table 2) contain abundant rough endoplasmic reticulum (RER). In *Calliphora* large pericardial nephrocytes, about one-fifth of the cytoplasm is RER in the form of flattened and tubular cisternae, mostly located in the perikaryon (medulla) (Fig. 2). Only in Collembola is RER sparse (Humbert, W., 1975). Membrane and protein synthesis is a dominant nephrocyte activity (section 5).

Golgi centres are numerous, but relatively small, averaging 1 μm in length (Kessel, R., 1962). Coated vesicles of small size (*ca.* 50 nm diameter) are associated with many Golgi centres, especially those implicated in the formation of primary lysosomes (Fig. 11). The term GERL (Golgi–endoplasmic reticulum–lysosome) of cytochemists (e.g. Novikoff, A., 1976) may be appropriate here.

Some nephrocytes contain knots of anastomosing electron-dense tortuous tubules, averaging 90 nm in diameter. Some tubules are expanded and contain myeloid figures. These tortuous tubules appear to be a form of smooth endoplasmic reticulum, but nothing is known of their function, (Crossley, A., 1972a; Fig. 1.).

Mitochondria are numerous in the transitional zone, particularly around the inner extremities of the labyrinthine channels, but they do not associate directly with plasma membrane, as in cells involved in bulk fluid transport. Mitochondria are on average 0.23×1.7 μm in size (in *Calliphora* Fig. 2), but long pleomorphic forms up to 5 μm are occasionally observed.

2.10 Vacuoles: α, β and γ

Insect nephrocytes contain at least three morphologically distinct types of vacuole. Crossley, A. (1972a) suggested the use of the Greek letters α, β and γ to designate defined vacuole types, and as this terminology has been adopted in most subsequent descriptions is retained here (Table 2) (Fig. 13).

Alpha vacuoles are 0.1 to 3.0 μm in diameter, are enclosed within a single unit membrane, and are confined to the transfer zone at the cell cortex. Each vacuole contains condensed electron-dense deposits in an electron-lucent matrix (Figs 2, 4 and 9), and the proportion of dense to lucent content remains approximately constant as the vacuoles change in size. The electron-dense component is often seen as a single eccentric mass. Tubular elements are

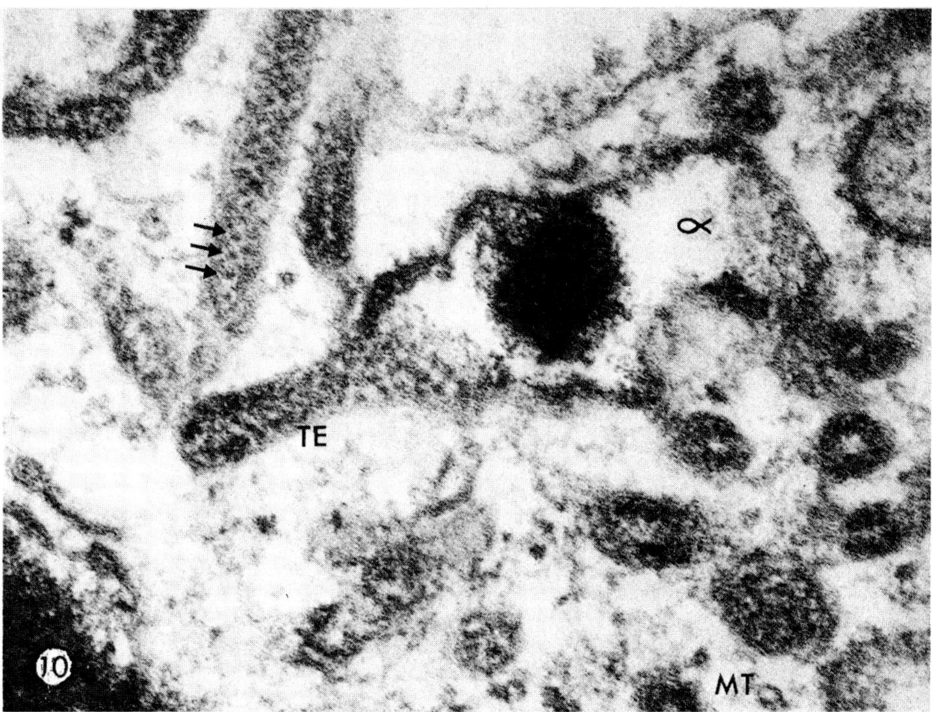

FIG. 10. Tubular elements (TE) have thick walls built up from granular subunits (arrows). Tubular elements open into α-vacuoles (α). MT = microtubule. (*Calliphora* larva; × 170,000.)

frequently sectioned to reveal continuity with α-vacuoles (Crossley, A., 1972a) and the flask-shaped extremities of tubular elements (section 2.6) may expand to form α-vacuoles (Fig. 10) (section 2.6). α-Vacuoles form part of the reception system for molecules sequestered in endocytosis (section 4.3).

β-Vacuoles range from 0.1 to 2.0 μm in greatest dimension, are enclosed within a single unit membrane, and are conspicuously electron-dense (Figs 11 and 13). Beta vacuoles are homogeneous and non-granular when small, but larger profiles may contain fragments of membrane or organelle, indicating that they may be cytolysomes. This interpretation is strengthened by the electron-histochemical localization of acid phosphatase activity in β-vacuoles (Crossley, A., 1972a).

Gamma vacuoles range from 0.5 to 3.0 μm in greatest dimension, are limited by a single unit membrane, and are filled with finely granular material of only moderate electron density (Fig. 13). The material enclosed in these vacuoles is similar in fine structure to the contents of RER, and is probably protein. The complement of γ-vacuoles varies with moulting cycles, and a functional interpretation is considered in connection with protein synthesis (section 5).

2.11 Stored lipid and carbohydrate

Lipid is much less abundant in nephrocytes than in fat body cells, but lipid droplets are scattered amongst the protein vacuoles of the transition zone, as evidenced by low affinity for electron stains, and a sharp but non-membranous phase boundary (Fig. 11). Lipid is reported in *Calliphora, Bombyx* and *Tomocerus* (refs in Table 2).

Clusters of polygonal granules 25–33 nm in greatest dimension occur in the perinuclear zone. This material is interpreted as glycogen on evidence from ultrastructure and light microscope histochemistry. Glycogen is usually obvious in nephrocytes from *Calliphora* and *Bombyx*, is occasionally seen in Collembola but is reportedly absent in *Melanoplus differentialis* and *Panorpa communis* (Schwinck, I., 1951; Kessel, R., 1962; Crossley, A., 1972a; Humbert, W., 1975; Lavanseau, L. *et al.*, 1981) (see also section 3.3).

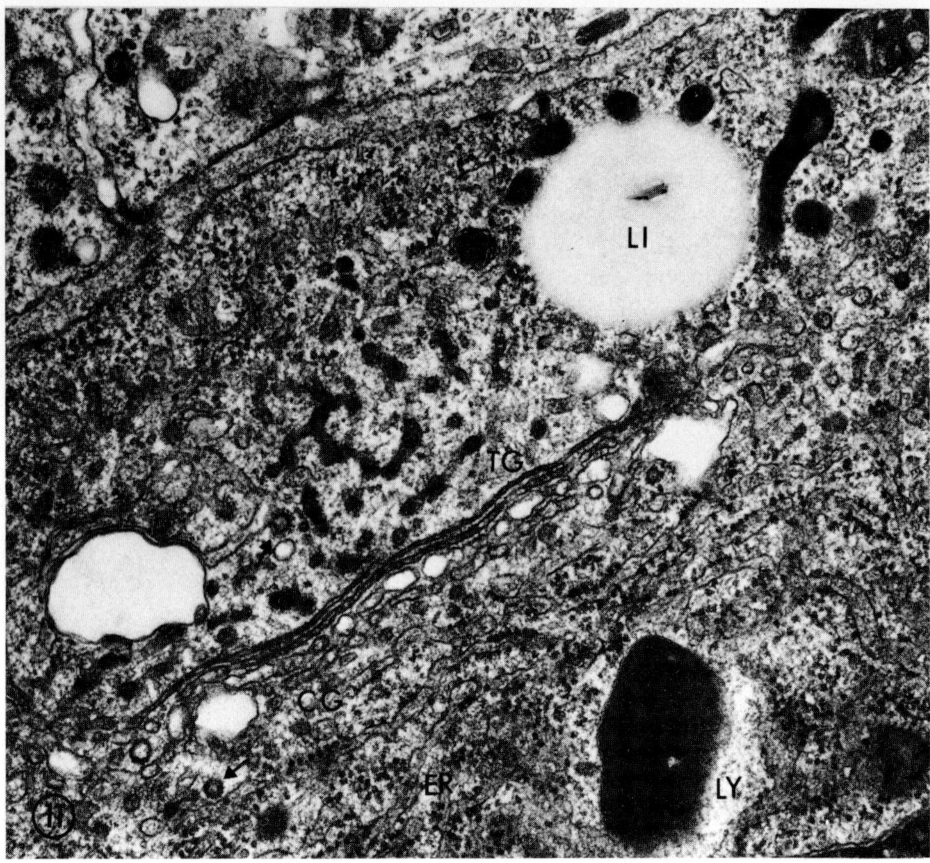

Fig. 11. Golgi complex processing the electron-dense product that forms the content of beta vacuoles (lysosomes—LY). Endoplasmic reticulum (ER) is associated with the *cis*-Golgi (CG). Clouds of electron-dense vesicles form from the *trans*-Golgi (TG) (= GERL). Small coated vesicles are present on both faces of the Golgi (arrows). LI = Lipid droplets. (*Calliphora* pupa; × 53,000.)

2.12 Nuclei and nucleoli

The number of nuclei reported in each nephrocyte varies from one to seven (Tables 1 and 2), but some multicellular clusters may be mistaken for syncytia as discussed in section 2.4. The origin of the multinucleate condition in nephrocytes has led to disputes in the literature, as for muscle syncytia (Crossley, A., 1965). Cell fusion, mitosis and amitosis have all been proposed (see section 3.4). Convincing multinuclearity is revealed in some light and electron micrographs. In *Calliphora stygia* the pericardial nephrocytes are uninucleate with multiple nucleoli, whereas the ventral nephrocytes are binucleate with single nucleoli but both types have polytene chromosomes (Thomson, J. and Gunson, M., 1970). The nucleus is often lobed (e.g. Collembola, Humbert, W., 1975) and may partially enclose cytoplasmic zones (Sanger, J. and McCann, F., 1968). Although many micrographs of nephrocyte nuclei have been published there has not been much effort to analyse ultrastructure (Fig. 2).

3 DEVELOPMENT AND METAMORPHOSIS OF NEPHROCYTES

3.1 Embryology

The embryology of ventral and dorsal nephrocytes is described in detail for *Calandra oryzae* (Coleoptera), and early controversies on nephrocyte origins are discussed by Tiegs, O. and Murray, F. (1938). The ventral nephrocytes of *Calandra* are termed the "sub-oesophageal body" and arise from mesoderm just anterior to the mandible. In the early embryo the suboesophageal body resembles a coelomic sac. These cells associate at first with the yolk, then with midgut anlagen, then form a strand of 9–12 nephrocytes in the larva, surviving into the adult. The ultrastructure of the sub-oesophageal body in *Bathysciinae* (Coleoptera) is described by Deleurance, S. and Charpin, P. (1971).

The dorsal nephrocytes of *Calandra* are termed "paracardial nephrocytes" and arise from mesoderm forming the dorsolateral walls of the two rows of coelomic sacs which meet in the midline above the gut. These nephrocytes appear on the fourth day of embryonic development, and remain segmented during differentiation (Tiegs, O. and Murray, F., 1938).

The dorsal pericardial nephrocytes are derived from somatic mesoderm associated with heart in *Forficula* (Dermaptera) (Hollande, A., 1922), and in *Dacus tryoni* (Diptera), where the anterior dorsal nephrocytes are described as "lymph glands", and the posterior dorsal nephrocytes as "pericardial cells" (Anderson, D., 1963).

3.2 Larval development

In *Vanessa* (Lepidoptera) cell division by mitosis occurs in nephrocytes in the egg and in the first larval instar, but multinucleate cells appear in later instars. The smaller nephrocytes remain uninucleate, whilst the larger become multinucleate syncytia following karyokinesis without cell division (Hollande, A., 1922).

In *Dacus* (Diptera) both anterior and posterior pericardial nephrocytes increase in size slowly during the first two larval instars, and then grow rapidly early in the third instar, but there is no cell proliferation in either group (Anderson, D., 1963). In *Calliphora erythrocephala* (Diptera), however, the small anterior pericardial nephrocytes multiply rapidly from early in the third instar to the time of puparium formation. At the same time the posterior pericardial nephrocytes enlarge without cell division, but become polytene (Jensen, P., 1973). In *Calliphora stygia* the ventral nephrocytes are both binucleate and polytene (Thomson, J. and Gunson, M., 1970) We can conclude that the problems of control of increased cytoplasmic activity and growth meet with different solutions in the nuclei of different nephrocytes, and may be under feedback control relating to both cytoplasmic volume and activity.

Schwinck, I. (1951) attributes an increase in the number of pericardial nephrocytes in larval *Panorpa communis* to the settlement of new cells from the haemolymph, but confusion with haemocytes and myoblasts is a possibility (section 3.4).

3.3 Metamorphosis

In Diptera the large, polytene, posterior pericardial nephrocytes are histolysed at metamorphosis (Keilin, D., 1924). These nephrocytes are attacked by haemocytes, commencing with the posterior pair, at the same time as the heart ventricle is attacked (Jensen, P., 1973). However, even in *Calliphora* some nephrocytes survive metamorphosis intact (Fig. 2), and dye sequestered in the larva remains visible until the adult dies of old age. It is possible that some tracer dye is passaged through the haemolymph plasma after disintegration of haemocytes in late pupae, only to be sequestered again by nephrocytes in the pharate adult.

In Orders other than Diptera many nephrocytes survive metamorphosis, and even in *Galleria* where many nephrocytes are destroyed by haemocytic phagocytosis, some survive, and Hollande, A. (1922) suggests that excessive experimental dye loading may trigger phagocytosis of defunct nephrocytes.

Nephrocytes change in structure in relation to moulting cycles, and these changes are intense at metamorphosis (Schwinck, I., 1951). The ultrastructural changes in nephrocytes at metamorphosis have not been studied in detail for any insect, but there is some information for *Calliphora* (Diptera) (Crossley, A., 1972a). The cortex of nephrocytes reorganizes during the 24 h following pupariation, before pupation proper. Organelles associated with endocytosis, such as coated vesicles, tubular elements and α-vacuoles are reduced in number, and the labyrinthine channels become tubular (diameter 75–100 nm) and packed closely together. Gamma vacuoles, presumed to contain stored protein, are dissipated by exocytosis at the same time. The greater part of the ribosome-studded endoplasmic reticulum disappears shortly after puparium formation, but since many smooth-membraned cisternae remain, the initial event may be detachment of ribosomes. During the pupal period numerous cytolysomes remain, and formation of primary lysosomes in association with Golgi (? GERL) centres continues (Fig. 11). Since endocytosis is reduced, it is presumed that these cytolysomes are concerned with autophagy, and this is supported by the appearance of numerous isolation vacuoles. Glycogen accumulates in the cytoplasmic matrix during the pupal period.

3.4 Formation of nephrocytes of the adult

Adult nephrocytes are, in general, derived from larval cells, sometimes with modification at metamorphosis. A striking modification in *Pieris* is the appearance alongside large (15 μm diameter) larval nuclei of small (6 μm diameter) pupal nuclei (Bruntz, L., 1904). The origin of these nuclei has generated a controversy which is discussed at length by Hollande, A. (1922). In *Calliphora* muscle syncytia the small nuclei are derived from imaginal discs by cell migration and fusion, as evidenced by electron micrographs (Crossley, A., 1965, 1972b,c). Jensen, P. (1973) comments that "myoblasts" penetrate between pericardial nephrocytes during pupation in *Calliphora*, and form the pericardial septum. It is possible that some of these migratory histoblasts are not committed to muscle differentiation, and may provide imaginal nuclei for nephrocytes. Many small nuclei are observed in association with nephrocytes in pupae of *Calandra*, and although interpreted as haemocytes (Murray, F. and Tiegs, O., 1935), these, too, may be imaginal histoblasts. The evidence for *Vanessa* (Hollande, A., 1922) is also consistent with fusion of cells derived from the haemolymph with existing nephrocytes. However, in *Galleria* the large larval nuclei disappear during metamorphosis and are replaced by numerous small adult nuclei. Absence of mitotic figures led Hollande, A. (1922) to postulate subdivision of larval nuclei by amitosis, but as in *Vanessa* cell fusion seems a more probable explanation even though the origin of the imaginal histoblasts remains to be explored.

4 ENDOCYTOSIS BY NEPHROCYTES

4.1 Early observations

Kowalevsky, A. (1886) discovered that nephrocytes in larvae of Diptera which had been fed on artificially coloured food became stained. Cochineal powder and silver salts found their way from the gut into the cytoplasm of nephrocytes. Injection of ammonia carmine or litmus into the haemolymph led to coloration of nephrocytes, whose cytoplasm gave an acid reaction (Kowalevsky, A., 1892). Dye injection experiments were carried out for many species (Table 1) and careful work with the light microscope showed that dye was first adsorbed to peripheral membranes, then into the cortex, and finally stored in the perinuclear region. Ammonia carmine was particularly useful because the acidity of nephrocytes caused release of insoluble free carmine (Hollande, A., 1922).

Nephrocytes often become naturally coloured as the result of uptake of materials derived from the food. For example the nephrocytes in *Lonchaea chorea* become coloured dark brown when the larvae feed on decomposing mangel-wurzels (Keilin, D., 1924). The nephrocytes of blood-sucking insects become coloured by haemoglobin leaked into the haemolymph. In *Rhodnius prolixus* haemoglobin is converted to biliverdin in pericardial nephrocytes, colouring them blue-green (Wigglesworth, V., 1943). In *Melanoplus differentialis* the pericardial nephrocytes are pigmented, with pink, red, purple or blue hue in a single insect. Older grasshoppers have red-brown nephrocytes, and the intensity of colour increases with age (Kessel, R., 1962).

Cuénot, L. (1896) pointed out that pericardial nephrocytes must subsequently release ingested materials since they do not increase in size.

4.2 Nephrocytes as a component of an insect "reticuloendothelial system"

The analogy between insect nephrocytes and the vertebrate reticuloendothelial system was drawn in the 1930s when it was realized that nephrocytes selectively sequestered small electronegative colloids (Poll, M., 1934; Grassé, P. and Lesperon, L., 1935; Lesperon, L., 1937). A wide range of proteins which have lost their native configuration are taken up by the cells of the vertebrate reticuloendothelial system (Thorbecke, G. *et al.*, 1960). In articulating the analogy Wigglesworth, V. (1970) emphasized that nephrocytes do not take up larger particles such as indian ink, bacteria, or erythrocytes, and thus have only a part of the scavenging propensities of the vertebrate reticuloendothelial system.

An important component of the vertebrate system is the family of Kupffer cells, which make up 15% of liver cells in mammals, and function as fixed macrophages in liver sinusoids and elsewhere. They are phagocytic towards most circulating foreign particles including living micro-organisms, thorotrast, and HRP (Fahimi, H., 1970).

The ultrastructure of Kupffer cells provides an example of adaptation to endocytosis of particles together with solutes. Invaginations of the plasma membrane form vermiform tubules of strikingly uniform diameter (100 nm), woven into a labyrinth, with multiple openings to the extracellular space but without an overlying basement membrane. The vermiform tubules ("micropinocytosis vermiformis") of Kupffer cells are associated with coated vesicles, and it is proposed that the tubules sweep proteins toward the interior of the cell by membrane flow, whilst coated vesicles mediate transport across the membrane (Orci, L. *et al.*, 1967; Matter, A. *et al.*, 1968).

An analogy can be drawn between the vermiform tubules of Kupffer cells and the labyrinthine channels of nephrocytes, although the latter are generally of less uniform but greater diameter (100–350 nm) (except in pupae: see section 3.3) and are always guarded by basement membrane and junctional complexes.

Kupffer cells selectively bind and internalize a wide size range of particles bearing glycoproteins terminating in *N*-acetylglucosamine, or mannose residues (Hubbard, A. *et al.*, 1979). In insects the uptake of particles is a function of phagocytic haemocytes. The demarcation between the cells of the insect nephrocyte–haemocyte system is discussed in section 4.3, below.

4.3 Selective sequestration

Nephrocytes are selective in their sequestration of materials from the haemolymph. Lesperon, L. (1937) found absorption of substances by nephrocytes in Lepidoptera to be a function of particle size and charge, electronegative colloids being preferentially absorbed. She found that acid dyes were usually eliminated by the Malpighian tubules, but basic dyes such as trypan blue were taken up by pericardial nephrocytes and oenocytes, although the distribution depended to some extent on dose.

In *Locusta migratoria* latex beads 0.9 μm in diameter are never taken up by pericardial nephrocytes, but are accepted by haemocytes. Iron saccharate fails to enter either labyrinthine channels or coated vesicles, but nevertheless appears in peripheral (α) vacuoles (Hoffmann, J., 1967).

Pericardial nephrocytes of *Calliphora* reject colloidal thorium dioxide stabilized with dextran, a *ca.* 15 nm electronegative colloid, which is taken up by haemocytes. Lanthanium hydroxide, a smaller (*ca.* 5–10 nm) similarly charged colloid passes both basement membrane and junctional complexes *in vivo*, and is admitted to the labyrinthine channels. Neither colloid is adsorbed at nephrocyte-coated vesicles, although thorium dioxide is accepted by haemocyte-coated vesicles. However, nephrocyte-coated vesicles sequester two proteins, HRP (MW 40,000) and ferritin (*ca.* 10–11 nm), and these pass rapidly to α-vacuoles. HRP is later detected in β-vacuoles (cytolysomes) where its activity is slowly lost. No tracer molecules are detected in *Calliphora* tubular elements, nor do tubular elements contain acid phosphatase. Ultrafiltration in *Calliphora* is thus a multistep process with a primary filter related to dimension and charge, and a secondary filter involving molecular recognition.

Phagocytosis of large colloids and whole cells is a haemocyte role, (A. Gupta, this volume), and there is a size demarcation for the uptake of similarly charged particles between haemocytes and nephrocytes (Crossley, A., 1972a, 1975; Brehélin, M. and Hoffmann, J., 1980). Nevertheless there is an ambiguous size zone where colloids are sequestered by both systems. An example of this is the uptake of iron saccharate (ferric oxide, saccharated) at all concentrations by both nephrocytes and haemocytes in *Rhodnius* and *Locusta* (Wigglesworth, V., 1970; Hoffmann, J., 1967; Brehélin, M. and Hoffmann, J., 1980).

Protein endocytosis occurs at the surface of most insect cells, including fat body epidermis and

Malpighian tubules (Locke, M. and Collins, J., 1967). Indeed there appears to be a link between the storage excretion role of nephrocytes and the dissipative excretion role of Malpighian tubules (section 7.1).

The exclusion of particles of large size or electropositive charge is evidently a function of the basement membrane–junctional complex (see section 7.2), but the problem remains of how nephrocytes sequester some molecules whilst rejecting others of similar size and charge. Since the haemolymph is a complex reservoir of nutrients in transit, including yolk and cuticle precursors, molecular recognition is a significant problem for all cells, but particularly for nephrocytes which sequester denatured or exogenous molecules. This recognition process occurs at the surface of coated pits lining the labyrinthine channels of nephrocytes, and is discussed in the next section.

4.4 The mechanism of endocytosis by nephrocytes

Endocytosis is regulated quantal uptake of exogenous molecules from the cell's environment, via plasma-membrane-derived vesicles and vacuoles. Adsorptive uptake is both a selective and a concentrating device, whereby the cell takes in large amounts of specific solute, without ingesting a correspondingly large volume of solution (Silverstein, S. *et al.*, 1977). Micropinocytotic vesicles have been recorded in all ultrastructural studies of nephrocytes (section 2.5) and recent microscopy has revealed that these vesicles are "coated" on the cytoplasmic face with a polygonal array of protein subunits. Coated vesicles were first described in insect oocytes (Roth, T. and Porter, K., 1962, 1964) and in insect nephrocytes (Bowers, B., 1964), but have subsequently been recorded in many vertebrate cells. Coated vesicle structure is strongly conserved in evolution, and structural and functional studies on vertebrate coated vesicles appear to be relevant to nephrocytes, and vice-versa. Nephrocyte zonation provides a useful spatial reference for studying coated vesicle function.

Coated vesicles were isolated from pig brain and partially purified by Pearse, B. (1975), who described the principal constituent of the cell coat: a protein of MW 180,000 which she termed clathrin. Clathrin is a non-glycosylated protein whose amino-acid sequence is highly conserved between different species (Pearse, B. and Bretscher, M., 1981). Other smaller polypeptides are isolated with clathrin and remain after removal of membrane fragments by permeation chromatography and of ATP-ase activity by gel electrophoresis (Pfeffer, S. and Kelly, R., 1981; Rubenstein, J. *et al.*, 1981). Crowther, R. and Pearse, B. (1981) have developed a model for the packing of clathrin molecules into hexagons and pentagons covering coated pits and vesicles. The principal assembly unit is a symmetrical trimer with three hinged legs, termed a "triskelion". The model involves a variable joint at the vertex of the molecule and a hinged joint in each leg, allowing packing into pentagons and hexagons on the surface of vesicles of different sizes.

In fibroblasts it is calculated that each cell takes up the equivalent of its entire surface area as pits every 50 min or so, and it has been suggested that the internalized membrane returns to the surface at the leading edge, thus contributing to locomotion. Clathrin presumably acts both as a framework for budding, and also to concentrate specific receptors on the pit whilst excluding non-receptor proteins (Bretscher, M., 1981). In non-motile nephrocytes it is not known where the membrane is returned, but the scale of endocytosis suggests that recycling must be rapid. One possibility is that tubular elements form discontinuous links to the plasma membrane, and return fluid, lipid, and membrane components to the cell surface (see section 7.3).

Receptor-mediated endocytosis along the coated vesicle pathway can be stimulated by ligand challenges, leading to an increase in the fraction of the plasma membrane occupied by coated pits. Receptors appear to act as molecular filters, selecting certain macromolecules and excluding others. Movement of ligand to intracellular destinations is dependent on the actin-based microfilament system of the cell (Salisbury, J. *et al.*, 1981). These concepts, which are suggested by nephrocyte structure and function but not yet fully explored experimentally, are summarized in Fig. 12.

4.5 Endocytosis in nephrocytes and in oocytes compared

The involvement of coated vesicles in the selective uptake of haemolymph protein was first observed in developing mosquito oocytes. During yolk

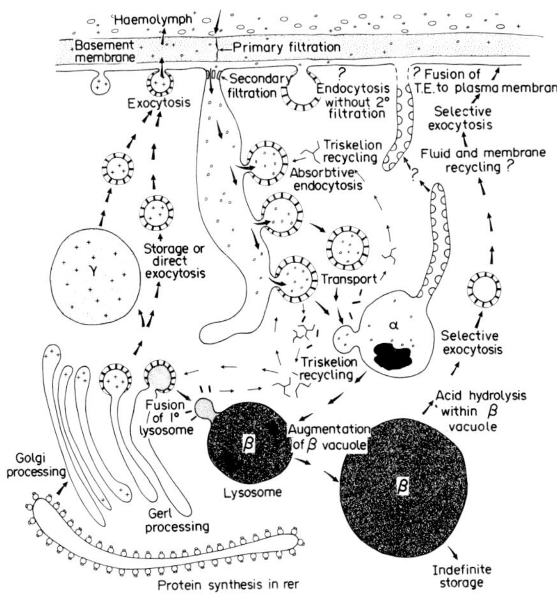

FIG. 12. Diagram showing hypothetical transport routes within a nephrocyte.

deposition the number of coated pits rose 15-fold, to reach 300,000 per oocyte (Roth, T. and Porter, K., 1964). Later in the same year in a study of vitellogenesis in *Gryllus*, Favard-Séréno, C. (1964) observed that electron-dense yolk precursor taken up by coated pits was amassed by vacuoles (average diameter 75 μm) which were connected to 30 nm diameter tubules (termed "réseau de canicules") assumed to be smooth endoplasmic reticulum. In *Hyalophora cecropia* oocytes the coated vesicles are 130–170 nm in diameter, and tubular elements 42–63 nm in diameter are contiguous with yolk vacuoles. It is again suggested that the tubules are part of the endoplasmic reticulum, and that they contribute a yolk component synthesized in the oocyte (Stay, B., 1965). Oocytes in *Periplaneta americana* form coated vesicles 85–125 nm diameter in previtellogenic oocytes, but vesicles up to 220 nm in diameter form during vitellogenesis. Tubular elements are also present, but no dimensions are given, and they are considered to be part of the endoplasmic reticulum (Anderson, E., 1969).

Although the tubules described in oocytes are less numerous than the tubular elements of nephrocytes, they are similar in size and structure, and are associated in some unknown way with the coated vesicle pathway. Tubular element function may be similar in both cells, and not directly related to the endoplasmic reticulum.

Presumably nephrocytes selectively reject yolk proteins, since they do not accumulate yolk. However, most foreign proteins introduced into the haemocoel can be recovered from yolk in small amounts (Telfer, W., 1960), so the selectivity is not absolute.

5 SECRETORY ACTIVITY OF NEPHROCYTES

5.1 Protein for export: histochemistry

Histochemical methods for protein applied to nephrocytes give positive indications of dynamic storage and release rhythms in numerous species, but very little effort has been made to identify specific proteins. In *Panorpa communis* (Mecoptera) pericardial and perioesophageal nephrocyte protein secretion rhythms, monitored by light microscope histochemistry, paralleled epidermal cell activity, and ceased during diapause and pupation. Both starvation and interruption of cephalic secretion by ligation halted synthesis of protein granules, and after 24 h pre-existing secretion was exhausted, but starvation and humoral effects were not distinguished in these experiments (Schwinck, I., 1951, 1952). There is a build-up of protein inclusions and subsequent degranulation in nephrocytes of Trichoptera, Lepidoptera, Coleoptera, and Diptera at the time of metamorphosis (Wigglesworth, V., 1970; Crossley, A., 1972a).

5.2 Protein for export: ultrastructure

All insect nephrocytes have a well-developed ribosome-studded endoplasmic reticulum, numerous Golgi centres, and in most it is possible to distinguish protein storage vacuoles (type γ) from lysosomes (type β) on a basis of ultrastructure and of enzyme histochemistry (Fig. 13). In *Calliphora erythrocephala* circumstantial evidence suggests that the granular material common to both cisternae of the RER and to γ-vacuoles is protein for export. Gamma vacuoles are acid phosphatase negative, and disappear shortly after puparium formation when their contents are presumably released into the haemolymph (Crossley, A., 1972a).

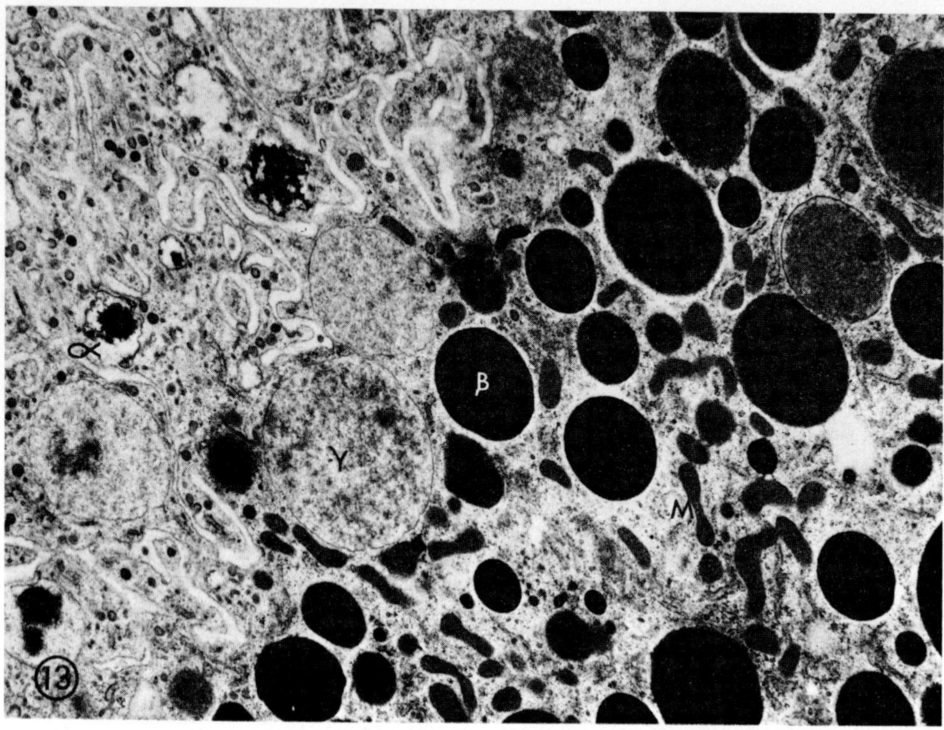

FIG. 13. Vacuolar region of nephrocyte from a feeding *Calliphora* larva showing the distinctive ultrastructure of alpha (α), beta (β) and gamma (γ) vacuoles. M = Mitochondrion. ($\times$ 9600.)

In their study of protein absorption in the rat vas deferens, Friend, D. and Farquhar, M. (1967) showed that large (> 100 nm) coated vesicles serve as heterophagosomes to transport absorbed protein to lysosomes, whilst some small (< 75 nm) coated vesicles transport newly synthesized protein from the Golgi complex towards vacuoles and the cell surface. The small coated vesicles never transport exogenous peroxidase tracer in endocytosis. More recently it has been demonstrated that a class of small coated vesicles (40–60 nm) transport newly synthesized casein to the cell surface in rat mammary epithelium, which is not involved in endocytosis (Franke, W., *et al.*, 1976). In insect oocytes it also appears to be the small coated vesicles (42–63 nm) which contribute internally synthesized protein to yolk spheres, whereas larger vesicles (130–170 nm) are involved in endocytosis (Stay, B., 1965). In insect nephrocytes small (50–80 nm) coated vesicles form at the Golgi apparatus and appear to be part of the outward transport route for newly synthesized proteins, whilst larger (120–160 nm) coated vesicles are involved in endocytosis, as evidenced by uptake of tracer. Many of the class of small coated vesicles form in the GERL region (Fig. 11) and are probably involved in sorting and transport of lysosomal enzymes, but numerous small coated vesicles also accumulate at the plasma membrane and may be involved in exocytosis of secretory proteins (Crossley, A., 1972a). These concepts are summarized diagrammatically in Fig. 12.

5.3 Protein for export: biochemistry

Many haemolymph polypeptides are made in the fat body, including vitellogenins, lipid transport proteins, hormone binding proteins and storage proteins (Wyatt, G. and Pan, M., 1978; L. Keeley, this volume; also see vol. 10), but others such as lysozyme are made in nephrocytes (Crossley, A., 1972a), and the demarcation between these cells has received little attention. A particular complication in the study of protein synthesis in Lepidoptera is the presence of disseminated nephrocytes amongst fat body cells, but the relative contribution of each

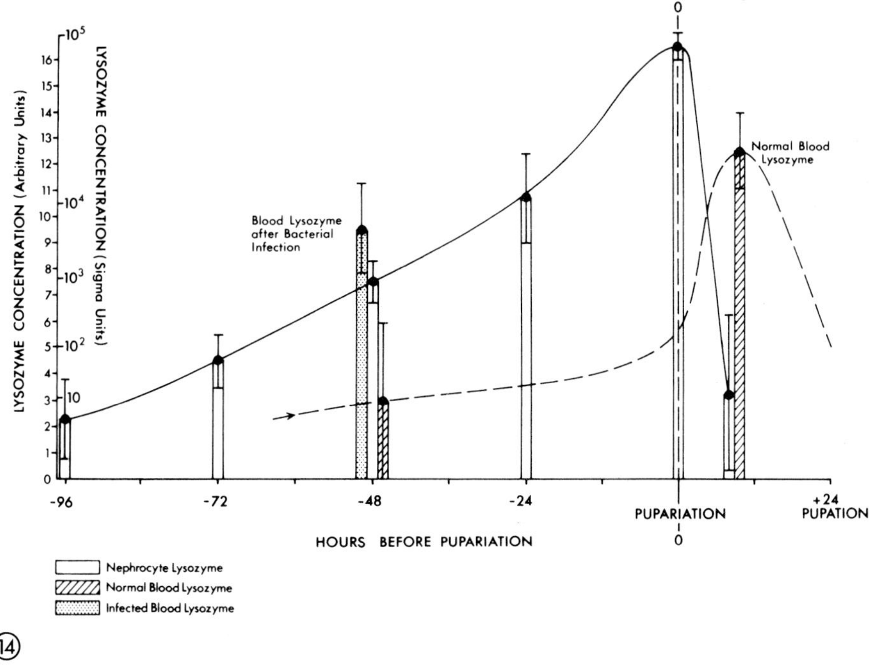

FIG. 14. Changing lysozyme concentration in dorsal nephrocytes and in blood from *Calliphora vicina*. Data obtained from measurement of the area cleared on standardized plates of *Micrococcus* walls in agar (see Crossley, A., 1972a). One Sigma unit is defined as that amount of enzyme which will cause a ΔOD_{450} of 0.001 in a *Micrococcus lysodeikticus* suspension in 1 min at pH 6.24 in 2.6 ml reaction mix and lightpath of 1 cm. Bacteria used to infect *Calliphora* were live *E. coli*.

cell type has not been accounted for in biochemical studies, and most workers appear to be unaware of the presence of nephrocytes.

5.4 Production of lysozyme and other bacteriocides

In Diptera such as *Calliphora* all dorsal and ventral nephrocytes produce, store and release the enzyme lysozyme. This has been shown by incubation of cell extracts with a muramic acid-containing test system. The lysozyme titre within nephrocytes rises gradually during the third larval instar and reaches a peak 3.5 h after puparium formation, then rapidly declines. At 10 h after puparium formation little activity remains in the nephrocytes but the titre in the haemolymph reaches a peak (Fig. 14). The falling titre of lysozyme coincides with the degranulation of nephrocytes, but it is not known which type of granule is involved. Injection of living bacteria (*Bacillus cereus* 10^9 cells l^{-1}) causes a significant increase in lysozyme titre within the nephrocytes but not in the haemolymph. Since nephrocytes are non-phagocytic it is unlikely that lysozyme can contact its substrate in the form of a bacterial cell within the nephrocyte. But it is possible that nephrocytes may take up fragments of bacterial cell wall released by phagocytic haemocytes. In any case the release of lysozyme by nephrocytes at pupariation appears to be a generalized bacteriostatic response linked in some way to the detection of bacterial molecules (Crossley, A., 1972a).

Lysozyme is present in insect haemolymph at a concentration of 25–500 gamma units ml^{-1}, but the titre rises to 9000 gamma units ml^{-1} within 24 h of injection of bacteria (Mohrig, W. and Messner, B., 1968). As pointed out by Chadwick, J. (1970) lysozyme is only one element in insect immunity. Recently a lectin which is released at pupariation has been isolated from *Sarcophaga peregrina* (Komono, H. *et al.*, 1980). The origin of this protein is unknown, but its release coincides with degranulation in dipteran nephrocytes. In silkmoth pupae at least eight haemolymph polypeptides are induced by immunization with bacteria or viruses and their origins are largely unknown (Faye, I. *et al.*, 1975; see Götz, P. and Boman, H., this volume).

Since nephrocytes can be maintained for several days *in vitro* (unpublished observations), analyses of protein synthesis *in vitro* should be possible and informative. Methods recently developed for culture of insect fat body *in vitro* (e.g. Philippe, C., 1982) should also be suitable for nephrocytes. Until more biochemical studies are undertaken, our knowledge of nephrocyte synthetic capabilities will remain largely speculative.

5.5 Nephrocyte-heart pharmacology

Davey, K. (1962) discovered that *Periplaneta americana* pericardial nephrocytes exposed to extracts of corpus cardiacum changed in size, developed a branching nucleus, and elaborated numerous "secretory granules". Extraction of activated pericardial nephrocytes yielded a substance with pharmacological activity on the insect heart. This substance was initially interpreted as a serotonin-like indolalkylamine, but later analysis failed to support this interpretation, and a possible alternative role for nephrocytes involving amino acid metabolism was suggested (Davey, K., 1961a,b,c, 1964). No account was taken in these experiments of endocytosis of test extracts, which could account for both the morphological changes in nephrocytes, and the release of pharmacologically active factors. It would be useful to have a comparison of nephrocyte response to extracts of corpus cardiacum and to non-specific protein, such as albumen. Extracts containing nephrocytes give a negative assay for indolalkylamines (Colhoun, E., 1963) and further work is needed to resolve the role of nephrocytes in heart pharmacology. It would be interesting to explore the sensitivity of nephrocytes to biogenic amines, particularly since nephrocyte activities appear to be under humoral rather than direct nervous control. As precedents one can recall the adenylate cyclase activated by octopamine in *Periplaneta* fat body which has a glycogenolytic effect, or the dopamine-mediated salivary gland secretion in *Manduca sexta* (Bodnaryk, R., 1982).

5.6 Nephrocytes and synthesis of haemolymph respiratory pigments

Nephrocyte-like podocytes are involved in the synthesis of haemolymph pigments in other invertebrates. In pulmonate molluscs, for example, the pore cells synthesize the respiratory pigment haemoglobin, which appears to be released directly from the endoplasmic reticulum without transit through the Golgi system (Skelding, J. and Newell, P., 1975). In other molluscs haemoglobin and haemocyanin are synthesized in pore cells (Sminia, T. *et al.*, 1972). It is not known whether insect nephrocytes are involved in synthesis of haemoglobin. In *Chironomus thummi* fat body cells are involved in haemoglobin synthesis, but the role of nephrocytes and haemocytes has not been explored (Bergtrom, G. *et al.*, 1976).

6 THE PHYLOGENETIC SIGNIFICANCE OF NEPHROCYTES: THE EVOLUTION OF INSECT EXCRETORY SYSTEMS

6.1 The ancestry of insects

Evidence from comparative morphology and embryology supports the view that the onychophoran–miriapod–hexapod assemblage is a unitary phylogenetic group, the Uniramia, derived from a common lobopod ancestry. The Chelicerata and Crustacea are distinct taxa in this polyphyletic assessment (Manton, S., 1977). However, the pattern of spiral cleavage development which includes 4d mesoderm and a subsequent larva of the trochophore type links many phyla including molluscs, annelids, pogonophorans and uniramians with the platyhelminthes (Anderson, D., 1982). In the following sections the evolution of insect excretory systems is considered against this polyphyletic background.

6.2 Primitive ultrafilter systems in invertebrates and the evolution of the podocyte

The podocyte cell type, in which interdigitating feet of peripheral cytoplasm (the pedicels) support thin diaphragms and a basement membrane to form a filter slit, can be traced back to the earliest ultrafilter systems of invertebrates, and on to the vertebrate glomerulus (section 7.2). The insect nephrocyte is an elaboration of the podocyte.

Flatworms have excretory and osmoregulatory protonephridia, where the ciliated flame bulb forms a pre-podocytic ultrafilter (Berridge, M. and Oschman, J., 1972). Annelids have podocyte-like cells in the perioesophageal plexus, which supports thin membranes between the haemolymph and the coelom: a primitive filter slit. Flow across this filter slit is induced by the ciliated nephrostome, which acts as a pump (Koechlin, N., 1966).

Molluscs have elaborate podocytes of three types. In podocytes of the auricle of the prosobranch *Viviparus viviparus*, zip-fastener like structures between the pedicels leave 9×11 nm spaces to form filter slits strongly reminiscent of insect podocytes (Boer, H. *et al.*, 1973). A second podocyte type in gastropods such as *Lymnaea* is the pore cell of connective tissue, which functions in endocytosis and in production of blood pigments (haemocyanin or haemoglobin). In pore cells the filter-slit aperture appears to be 20×22 nm, which accords with the measured maximum size for filtered colloids in endocytosis (Boer, H. and Sminia, T., 1976). Finally, bivalve molluscs such as *Mytilus edulis* have pericardial cells of podocyte form, which are active in coated vesicle endocytosis of molecules such as HRP. These podocytes have no distinct slit-diaphragm and differ in many details from insect nephrocytes (Moore, M. *et al.*, 1980).

Chelicerates have coxal glands developed from coelomic end-sacs in association with ectodermal exit ducts (Anderson, D., 1973). In mites (Acari) the coxal gland consists of a sacculus composed of podocytes involved in ultrafiltration of haemolymph, and a resorption region, the labyrinth (Groepler, W., 1969, Woodring, J., 1973).

In Crustacea the end-sac epithelium of antennal and maxillary excretory organs is built of podocytes involved in ultrafiltration. In *Artemia salina* maxillary glands the podocytes also have a well-developed coated vesicle endocytosis pathway (Tyson, G., 1968). In *Uca mordax* antennal glands the podocytes also resorb sodium from the ultrafiltrate (Schmidt-Nielsen, B. *et al.*, 1968). The cells of the coelomosac of the crayfish kidney are podocytes which are also involved in endocytosis (Kümmel, G., 1964; Peterson, D. and Loizzi, R., 1974).

6.3 Podocytes in the Uniramia

All the podocytes considered in section 6.2 form part of ultrafilter systems in which coelomic fluid or haemolymph is passed through a resorption duct to the external environment. In the Uniramia we can trace the development of a new integrated role for the podocyte as a self-contained ultrafiltration–resorption–secretion system. In this integrated role the podocyte monitors and regulates the composition of haemolymph in conjunction with a separate renal organ.

In onychophorans such as *Peripatoides leucharti*, podocytes occur both in the sacculus of the nephridium and in the perivisceral and pericardial sinuses. Both types of podocyte contain numerous coated vesicles and 60 nm diameter tubular elements (Seifert, G. and Rosenberg, J., 1976, 1977). The perivisceral and pericardial cells strongly resemble insect nephrocytes, and indeed they were originally termed nephrocytes by Bruntz, L. (1903). In *Peripatus* the nephrocytes are pinocytotic (but not phagocytic) and also appear to synthesize protein for export. However *Peripatus acacioi* nephrocytes associate in groups of one to four cells enclosed within the same basement membrane, and circulate in the haemolymph forming the "pericardial globules" (Campiglia, S. and Lavallard, R., 1975).

Excretion in Onychophora is thus primarily by excretion through a coelomosac connected to an external duct, with a secondary role played by nephrocytes. Both forms of excretion are accomplished by podocytes, but nephrocytes collect ultrafiltrate into subcompartments of the haemocoel enclosed within the nephrocytes themselves: labyrinthine channels, which are not connected to an extracellular duct. Podocyte structure is retained, with amplification of organelles associated with endocytosis, but the nephrocyte is no longer tied to a segmental duct, and can wander through the haemocoel. Taken together with embryological evidence (section 3.1), the situation in Onychophora provides further support for a theory in which nephrocytes arise from segmental excretory organs.

Chilopods such as *Lithobius* have nephrocytes forming the "ecdysial glands", so named because ultrastructural changes presage moulting. These nephrocytes have coated vesicles, tubular elements and lysosomes, resembling insect nephrocytes in their adaptation to both synthesis and endocytosis. Diplopods such as *Orthomorpha* have nephrocytes in the pericardial and perivisceral sinuses of all except the last two metameres of the trunk. These cells have well-developed coated vesicle endocytosis, tubular elements with a helicoidal inner thickening, and closely resemble insect nephrocytes. In *Polyxenus* (Diplopoda) the cells of the collar gland surrounding the foregut are also podocytes, and undergo ultrastructural changes at moulting, as do insect nephrocytes in the same location (Seifert, G. and Rosenberg, J., 1974). Thus centipedes and millipedes each possess well-developed nephrocytes analogous to those seen in insects.

6.4 The evolution of insect excretory systems

The Collembola lack Malpighian tubules, but possess paired coelomoduct ultrafiltration kidneys, the labial nephridia. Collembola also have cephalic and abdominal ductless nephridia. The labial nephridia have sacculi of podocytes, complete with coated vesicles and tubular elements, but evidently are only secondarily involved in endocytosis, since the filtered haemolymph passes into a resorptive segment with a striking resemblance to the vertebrate nephron (Humbert, W., 1974, 1975; Verhoef, H. *et al.*, 1979).

In apterygote insects such as the Thysanura, the segmental excretory organs (labial kidneys) retain an important role in excretion, but a few short Malpighian tubules are also present (Gabe, M. *et al.*, 1973). Thysanura also possess numerous nephrocytes, and these have been implicated in the processing of excretory products subsequently released by labial kidneys (Bruntz, L., 1908b). *Petrobius maritimus* (Thysanura) has pericardial nephrocytes in the dorsal sinus from the second thoracic segment to the eighth abdominal segment, providing circumstantial evidence for origins in a metameric segmental organ derived from a coelomoduct. Although these nephrocytes are smaller (15 μm diameter) than most nephrocytes of pterygote insects, their podocyte configuration is typical, and micropinocytotic vesicles are present together with tubular elements (Gabe, M. *et al.*, 1973).

Within the class Insecta, only the apterygote forms have ultrafiltration–resorption kidneys like the vertebrate nephron (e.g. Diplura, Francois, J., 1972; Thysanura, Haupt, J., 1969). Pterygote insects have secretion–resorption kidneys: the Malpighian tubule–rectum system (Wall, B. and Oschman, J., 1975). The evidence summarized above (sections 6.1–6.3) suggests that the podocyte which evolved at first as an excretory ultrafilter, loses connection with any excretory duct but retains functional adaptation for endocytosis and secretion as it becomes a nephrocyte. This transformation begins with the Onychophora and can be followed through the Uniramia. Distinctly different transformations occur in the Chelicerates and Crustacea, but there are parallel changes in the Mollusca.

7 THE FUNCTIONAL RELATIONSHIP OF NEPHROCYTES TO OTHER EXCRETORY SYSTEMS

7.1 Nephrocyte–Malpighian tubule relationships

Malpighian tubules in most insects that have been investigated rapidly absorb and secrete compounds of molecular weight below 400 daltons. Larger molecules penetrate more slowly, and inulin (MW 5200) is found in secreted fluid at concentrations of 1–5% of that in the bathing fluid (Maddrell, S. and Gardiner, B., 1974). Some Malpighian tubules are impermeable to proteins. In *Calpodes*, injected horseradish peroxidase (MW 44,000) enters the fat body but is excluded from Malpighian tubules by the basement membrane (Locke, M. and Collins, J., 1967, 1968), but in the dragonfly *Libellula* the same tracer penetrates the basement membrane and is apparently taken up by Malpighian tubule cells (Kessel, R., 1970). However Malpighian tubules do not show ultrastructural specialization for protein transport, and there are no coated vesicles or tubular elements associated with basal infolds. Although small metabolites such as amino acids and sugars are passed into the urine and selectively resorbed by the rectum, protein is not processed in this way. Malpighian tubules actively concentrate acid dyes, which are probably handled quite separately from fluid, and also accumulate concentrations of phosphate and metals as mineralized deposits in their cytoplasm (Ramsay, J., 1958; Maddrell, S., 1971, 1977).

There are examples of cooperation between Malpighian tubules and nephrocytes in the excretion of small colloids. Dyes that are only slowly excreted by Malpighian tubules are first taken up by pericardial nephrocytes, then slowly released as the Malpighian tubules lower the concentration in the haemolymph (Palm, N., 1952). Malpighian tubules also excrete biliverdin released from pericardial nephrocytes following catabolism of haemoglobin in *Rhodnius prolixus* (Wigglesworth, V., 1943).

7.2 Nephrocytes as renal analogues in ultrafiltration and endocytosis

Insect nephrocytes appear to combine the renal functions performed by the glomerulus (ultrafiltration) and by the proximal tubule (endocytosis) of the vertebrate kidney, as evidenced by striking ultrastructural parallels between nephron and nephrocyte.

The organization of the insect nephrocyte cortex into foot processes (pedicels) is reminiscent of the podocytes of the glomerulus in the vertebrate kidney. However the filtration slits in insect nephrocytes are formed by intracellular junctional complexes, not intercellular junctional complexes as in the vertebrate glomerulus. In the glomerulus, as in insect nephrocytes, the narrow (30–80 nm) slits between adjacent pedicels of cytoplasm are bridged by a diaphragm (Pease, D., 1955; Yamada, E., 1955). This diaphragm has a highly ordered isoporous substructure, consisting of a zipper-like pattern of cross-bridges. The cross-bridges leave two rows of rectangular pores *ca.* 4×14 nm in cross-section, which appear in both chemically fixed and freeze-etched preparations for the electron microscope (Rodewald, R. and Karnovsky, M., 1974; Karnovsky, M. and Ryan, G., 1975).

The glomerular capillary wall is a filtration barrier which permits the passage of molecules up to a maximum effective radius of 3.6 nm (Renkin, E. and Robinson, R., 1974). Molecules smaller than 12,000 daltons pass freely whilst those larger than 65,000 daltons are mostly retained. Tracer studies indicate that a component of the basement membrane (the lamina densa) rather than the slit membrane itself is the main barrier to proteins. Electrical charge as well as size contribute to the filtration properties of the glomerulus, since cationized ferritin accumulates in the glomerular basement membrane more than does anionic ferritin, intrinsic negative charges are apparently present in the basement membrane. The filtration slit diaphragm constitutes a further barrier, with a major function in determining hydraulic conductivity (Laliberté, F. *et al.*, 1978; Kanwar, Y. and Farquhar, M., 1979; Rennke, H. *et al.*, 1975; Shea, S. and Morrison, A., 1975). Removal of glycosaminoglycans from the glomerular basement membrane by enzyme digestion leads to a dramatic increase in protein permeation, and a molecular filter composed largely of heparin sulphate is the key filtration element, rich in anionic sites (Kanwar, Y. *et al.*, 1980).

The ultrastructure of the permeability barriers of the nephrocytes of insects, available tracer studies, and analyses of insect connective tissues (D. Ashhurst, this volume) demonstrating the presence of glycosaminoglycans, all provide evidence for the similarity of the two systems. Further tracer studies and digestion experiments with glycosaminoglycan-degrading enzymes applied to nephrocytes are needed to extend the analogy (Ashhurst, D. and Costin, N., 1971; Crossley, A., 1972a).

A second analogy to the vertebrate kidney relates to the epithelial cells of the proximal convolution of the nephron. These cells are involved in reabsorption of proteins from the glomerular filtrate by endocytosis. Protein absorption droplets appear in proximal tubule cells when the reabsorptive and digestive capacity of the cell is overtaxed. Miller, F. (1960) using haemoglobin as a tracer in mouse kidney demonstrated a functional continuity between the apical cell surface, the dense apical tubules, and the apical vacuoles. Haemoglobin breakdown was indicated by the appearance of ferritin in apical vacuoles, where it remained for several days. Colloidal tracers such as trypan blue and thorium dioxide are also taken up by proximal tubule cells in the rat kidney, where they first appear in small apical vesicles and later in lysosomes (Trump, B., 1961). Homologous proteins (represented by [^{125}I]-labelled albumen) are also sequestered by coated vesicles and transferred to lysosomes (Maunsbach, A., 1966a).

Tubular elements (65 nm diameter) of the type seen in insect nephrocytes (section 2.6) have not been much investigated in the vertebrate nephron, but they certainly exist in the proximal tubule cells of the Rhesus monkey, where they are termed "dense tubules" (Tisher, C. *et al.*, 1969; Fig. 11), and are in direct connection with apical vacuoles. Ferritin injected into rat proximal tubule enters the epithelial cells at coated pits, but is not detected in 65–70 nm "dense tubules" (Maunsbach, A., 1966b). Tubular elements are more numerous in insect nephrocytes, and this may be the material of choice to elucidate their function.

7.3 Fluid dynamics: a problem and a hypothesis

In the vertebrate glomerulus fluid is forced through basement membrane and filter slit by the hydrostatic pressure of the blood. In the insect nephrocyte a rather similar ultrafiltration structure is present without an extracellular secretory duct, and hence without a hydrostatic pressure differential. In Malpighian tubules of insects fluid is believed to be drawn through a basement membrane towards the closed end of an extracellular channel by a standing osmotic gradient (osmotic filtration) (Berridge, M. and Oschman, J., 1969). It is likely that standing osmotic gradients form in labyrinthine channels of the nephrocyte cortex, provided that molecules are prevented from accumulating within channels, as they are believed to be, by coated vesicles. However, two problems remain. How is excess fluid (plasma) removed from the nephrocyte following solute endocytosis? And further, how is membrane returned to the cell surface to compensate for intake at endocytosis (section 4.4.2.)?

Do coated vesicle molecular components shuttle back and forth between plasma membrane and α-vacuoles, or is the tubular element (60 nm diameter) the key to the recycling problem? Tubular elements certainly connect for long periods to α-vacuoles, and could draw off both fluid and membrane. However, very few profiles suggest continuity of tubular elements and plasma membrane so this must be a transient event if it occurs, as suggested in Fig. 12. Further experiments with tracer molecules are needed, and it would be particularly useful to know how clathrin is distributed in nephrocytes. Many of the possible membrane compartment pathways are indicated in Fig. 12.

REFERENCES

Aggarwal, S. K. and King, R. C. (1966). The ultrastructure of the wreath cells of *Drosophila melanogaster* larvae. *Protoplasma 63*, 343–352.

Anderson, D. T. (1963). The larval development of *Dacus tryoni* (Frogg.) (Diptera: Trypetidae). 1. Larval instars, imaginal discs and haemocytes. *Austral. J. Zool. 11*, 202–218.

Anderson, D. T. (1973). *Embryology and Phylogeny in Annelids and Arthropods*. Pergamon Press, Oxford.

Anderson, D. T. (1982). Origins and relationships among the animal Phyla. *Proc. Linn. Soc. N.S.W. 106*, 151–166.

Anderson, E. (1969). Oogenesis in the cockroach *Periplaneta americana*, with special reference to the specialization of the fate of coated vesicles. *J. Microsc. 8*, 721–738.

Anderson, E. and Harvey, W. R. (1966). Active transport by the Cecropia midgut, II. Fine structure of the midgut epithelium. *J. Cell Biol. 31*, 107–134.

Ashhurst, D. E. and Costin, N. M. (1971). Insect mucosubstances. I. The mucosubstances of developing connective tissue in the locust, *Locusta migratoria*. *Histochem. J. 3*, 279–295.

Bergtrom, G., Laufer, H. and Rogers, R. (1976). Fat-body: a site of hemoglobin synthesis in *Chironomus thummi* (Diptera). *J. Cell Biol. 69*, 264–274.

Berridge, M. J. and Oschman, J. L. (1969). A structural basis for fluid secretion by Malpighian tubules. *Tissue Cell 1*, 247–272.

Berridge, M. J. and Oschman, J. L. (1972). *Transporting Epithelia*. Academic Press, New York.

Bodnaryk, R. P. (1982). Biogenic amine-sensitive adenylate cyclases in insects. *Insect Biochem. 12*, 1–6.

Boer, H. H., Algera, N. H. and Lommerse, A. W. (1973). Ultrastructure of possible sites of ultrafiltration in some gastropods, with particular reference to the auricle of the freshwater prosobranch *Viviparus viviparus* L. *Z. Zellforsch. 143*, 329–341.

Boer, H. H. and Sminia, T. (1976). The sieve structure of slit diaphragms of podocytes and pore cells of gastropod molluscs. *Cell. Tiss. Res. 170*, 221–229.

Bowers, B. (1964). Coated vesicles in the pericardial cells of the aphid *Myzus persicae* Sulz. *Protoplasma 59*, 351–367.

Brehélin, M. and Hoffmann, J. A. (1980). Phagocytosis of inert particles in *Locusta migratoria* and *Galleria mellonella*. Study of ultrastructure and clearance. *J. Insect Physiol. 26*, 103–111.

Bretscher, M. S. (1981). Surface uptake by fibroblasts and its consequences. *Cold Spring Harbor. Symp. Q. Biol. 46*, 707–712.

Bruntz, L. (1903) Excrétion et phagocytose chez des Onychophores. *C. R. Acad. Sci. 136*, 1148–1150.

Bruntz, L. (1904). Contribution a l'étude de l'excrétion chez les Arthropodes. *Arch. Biol. 20*, 217–422.

Bruntz, L. (1908a). Nouvelles recherches sur l'excrétion et la phagocytose chez les thysanoures. *Arch. Zool. Exp. et Gen. Ser. IV 8*, 471–488.

Bruntz, L. (1908b). Les reins labiaux et les glandes céphaliques des Thysanoures. *Arch. Zool. Exp. et Gen. Ser. IV 39*, 195–238.

Cameron, G. R. (1934). Inflammation in the caterpillars of Lepidoptera. *J. Path. Bact. 38*, 441–466.

Campiglia, S. and Lavallard, R. (1975). Contribution à l'hématologie de *Peripatus acacioi* Marcus et Marcus (Onycophora) II. Structure et ultrastructure des globules péricardiaux. *Ann. Sci. Nat., Zool., Paris 17*, 93–120.

Chadwick, J. S. (1970). Relation of lysozyme concentration to acquired immunity against *Pseudomonas aeruginosa* in *Galleria mellonella*. *J. Invert. Path. 15*, 455–456.

Colhoun, E. H. (1963). Synthesis of 5-hydroxytryptamine in the American cockroach. *Experientia 19*, 9–10.

CROSSLEY, A. C. (1965). Transformations in the abdominal muscles of the blue blow-fly *Calliphora erythrocephala* (Meig.), during metamorphosis. *J. Embryol. Exp. Morph. 14*, 89–110.

CROSSLEY, A. C. (1972a). The ultrastructure and function of pericardial cells and other nephrocytes in an insect *Calliphora erythrocephala*. *Tissue Cell 4*, 529–560.

CROSSLEY, A. C. (1972b). Ultrastructural changes during transition of larval to adult intersegmental muscle at metamorphosis in the blowfly *Calliphora erythrocephala*. I. Dedifferentiation and myoblast fusion. *J. Embryol. Exp. Morph. 27*, 43–74.

CROSSLEY, A. C. (1972c). Ultrastructural changes during transition of larval to adult intersegmental muscle at metamorphosis in the blowfly Calliphora erythrocephala. II. The formation of adult muscle. *J. Embryol. Exp. Morph. 27*, 75–101.

CROSSLEY, A. C. (1983). Haemolymph colloid control: the role of nephrocytes. *Devel. Comp. Immunol. Suppl. 3*, 673–678.

CROWTHER, R. A. and PEARSE, B. M. F. (1981). Assembly and packing of clathrin into coats. *J. Cell Biol. 91*, 790–797.

CUÉNOT, L. (1896). Etudes physiologiques sur les Orthopteres. *Arch. Biol. 14*, 293–341.

DAVEY, K. G. (1961a). Substances controlling the rate of the heart of *Periplaneta*. *Nature 192*, 284.

DAVEY, K. G. (1961b). The release by feeding of a pharmacologically active factor in the corpus cardiacum of *Periplaneta americana*. *J. Insect Physiol. 8*, 205–208.

DAVEY, K. G. (1961c). The mode of action of the heart accelerating factor from the corpus cardiacum of insects. *Gen. Comp. Endocrinol. 1*, 24–29.

DAVEY, K. G. (1962). Changes in the pericardial cells of *Periplaneta americana* induced by exposure to homogenates of the corpus cardiacum. *Quart. J. Micr. Sci. 103*, 349–357.

DAVEY, K. G. (1964). The control of visceral muscles in insects. *Adv. Insect Physiol. 2*, 219–245.

DELEURANCE, S. and CHARPIN, P. (1971). Sur le corpes sous-oesophagien des Coléoptères Bathysciinae. Description et infrastructure. *C.R. Acad. Sci. Paris 272*, 1109–1112.

EDWARDS, G. A. and CHALLICE, C. E. (1960). The ultrastructure of the heart of the cockroach *Blatella germanica*. *Ann. Ent. Soc. Amer. 53*, 369–383.

FAHIMI, H. D. (1970). The fine structural localization of endogenous and exogenous peroxidase activity in Kupffer cells of rat liver. *J. Cell Biol. 47*, 247–61.

FAVARD-SÉRÉNO, C. (1964). Phénomène de pinocytose au cours de la vitellogenèse protéique chez le grillon (Orthoptere). *J. Microscopie 3*, 323–338.

FAYE, I., PYE, A., RASMUSON, T., BOMAN, H. G. and BOMAN, I. A. (1975). Insect immunity. II Simultaneous induction of antibacterial activity and selective synthesis of some haemolymph proteins in diapausing pupae of *Hyalophora* and *Samia*. *Infect. Immun. 12*, 1426–1438.

FLOWER, N. E. and FILSHIE, B. K. (1976). Goblet cell membrane differentiations in the midgut of a lepidopteran larva. *J. Cell Sci. 20*, 357–375.

FRANÇOIS, J. (1972). Ultrastructure du rein labial cephalique de *Campodea chardardi*. (Diplura, Insecta). *Z. Zellforsch. 127*, 34–49.

FRANKE, W. W., LUDER, M. R., KARTENBECK, J., ZERBAN, H. and KEENAN, T. W. (1976). Involvement of vesicle coat material in casein secretion and surface regeneration. *J. Cell Biol. 69*, 173–195.

FRIEND, D. and FARQUHAR, M. (1967). Functions of coated vesicles during protein adsorption in the rat vas deferens. *J. Cell Biol. 35*, 357–376.

GABE, M., CASSIER, P. and FAIN-MAUREL, M. A. (1973). Données morphologiques sur les organes excréteurs abdominaux de *Petrobius maritimus* Leach (Insecte—aptérygote). *Archiv. Anat. Micr. Morphol. Exp. 62*, 101–143.

GRASSÉ, P. P. and LESPERON, L. (1935). Accumulation des colorants acides chez le ver à soie par des tissus differents selon la voie d'accès. *C.R. Acad. Sci. Paris. 201*, 618–620.

GROEPLER, W. (1969). Feinstruktur der coxalorgane bei der Gattung *Ornithodorus* (Acari — Argasidae). *Z. Wiss. Zool. 178*, 235–275.

HAUPT, J. (1969). Zur feinstruktur der Labialniere des silberfischschens *Lepisma saccharina*. *Zool. Beitr. N.F. 15*, 139–170.

HOFFMANN, J. A. (1967). Contribution a l'étude de la cellule péricardiale de *Locusta migratoria* (Orthoptera). *Bull. Biol. Fr. Belg. 1*, 3–12.

HOFFMANN, J. A. and LEVI, C. (1965). Étude au microscope électronique du vaisseau dorsal de *Locusta migratoria*. *C.R. Acad. Sci. Paris 260*, 6988–6990.

HOLLANDE, A. C. (1922). La cellule péricardiale des insectes. Cytologie, histochemie, rôle physiologique. *Archiv. Anat. Mic. 18*, 85–307.

HUBBARD, A. L., WILSON, G., ASHWELL, G. and STUKENBROK, H. (1979). An electron microscope autoradiographic study of the carbohydrate recognition systems in rat liver. *J. Cell Biol. 83*, 47–64.

HUMBERT, W. (1974). Localization structure et genèse des concrétions minérales dans le mésentéron des Collemboles Tomoceridae (Insecta, Collembola). *Z. Morph. Tiere 78*, 93–109.

HUMBERT, W. (1975). Ultrastructure des nephrocytes cephaliques et abdominaux chez *Tomocerus minor* (Lubbock) et *Lepidocyrtus curvicollis* (Bourlet) (Collemboles). *Int. J. Insect Morph. Embryol. 4*, 307–318.

JENSEN, P. V. (1973). Structure and metamorphosis of the larval heart of *Calliphora erythrocephala*. *Biol. Skr. K. Danske Vidensk Selskab. 20*, 1–19.

KANWAR, Y. S. and FARQUHAR, M. G. (1979). Anionic sites in the glomerular basement membrane. *In vivo* and *in vitro* localization to the Laminae Rarae by cationic probes. *J. Cell Biol. 81*, 137–153.

KANWAR, Y. S., LINKER, A. and FARQUHAR, M. (1980). Increased permeability of the glomerular basement membrane to ferritin after removal of glycosaminoglycans (Heparan sulfate) by enzyme digestion. *J. Cell Biol. 86*, 688–693.

KARNOVSKY, M. J. and RYAN, G. B. (1975). Substructure of the glomerular slit diaphragm in freeze-fractured normal rat kidney. *J. Cell Biol. 65*, 233–236.

KEILIN, D. (1917). Recherche sur les anthomyides à larves carnivores. *Parasitology 9*, 325–450.

KEILIN, D. (1924). On the nephrocytes of the larvae and pupae of *Lonchaea chorea* (Diptera — Acalypterae). *Ann. Mag. Nat. Hist. 13*, 219–223.

KESSEL, R. G. (1961a). Electron microscope observations on the submicroscopic vesicular component of the suboesophageal body and pericardial cells of the grasshopper *Melanoplus differentialis* Thomas. *Exp. Cell Res. 22*, 108–119.

KESSEL, R. G. (1961b). Cytological studies on the suboesophageal body cells and pericardial cells in embryos of the grasshopper *Melanoplus differentialis* Thomas. *J. Morph. 109*, 289–321.

KESSEL, R. G. (1962). Light and electron microscope studies on the pericardial cells of nymphal and adult grasshoppers, *Melanoplus differentialis*. *J. Morph. 110*, 79–103.

KESSEL, R. G. (1970). The permeability of dragonfly Malpighian tubule cells to protein using HRP as tracer. *J. Cell Biol. 47*, 299–303.

KOECHLIN, N. (1966). Ultrastructures du plexus sanguin périoesophagien: ses relations avec la néphridie de *Sabella pavonina* (Savigny). *C.R. Acad. Sci. Paris 262D*, 1266–1269.

KOMONO, H., MIZUNO, D. and NATORI, S. (1980). Purification of lectin induced in the haemolymph of *Sarcophaga peregrina* larvae on injury. *J. Biol. Chem. 255*, 2919–2924.

KOWALEVSKY, A. (1886). Zum Verhalten des Rückengefässes und des girlanden-förmigen Zellenstrangs der Musciden während der Metamorphose. *Biol. Centralbl. 6*, 74–79.

KOWALEVSKY, A. (1889). Ein Beitrag zur Kenntnis der Excretions-organe. *Biol. Centralbl. 9*, 33–47.

KOWALEVSKY, A. (1892). Sur les organes excréteurs chez les Arthropodes terrestres. *Congr. Intern. Zool. (2) Moscow 1*, 187–234.

KÜMMEL, G. (1964). Das Cölom — sackchen der Antennaldrüse von *Cambarus affinis* Say. (Decapoda, Crustacea). Eine electronmikroskopische Untersuchung mit einer Diskussion über die Function. *Zoologische Beitrage. 10*, 227–252.

LALIBERTÉ, F., SAPIN, C., BELAIR, M.-F., DRUET, P. and BARIETY, J. (1978). The localization of the filtration barrier in normal rat glomeruli by ultrastructural immunoperoxidase techniques. *Biol. Cell. 31*, 15–26.

LAVANSEAU, L., LAHARGUE, J. and SURLEVE-BAZEILLE, J. E. (1981). Ultrastructure of the 'Organe Rameux' of the silkworm *Bombyx mori* L. (Lepidoptera: Bombycidae). *Int. J. Insect Morphol. Embryol. 10*, 235–245.

LESPERON, L. (1937). Recherches cytologiques et experimentales sur la sécrétion de la soie. *Archiv. Zool. Exptl. Gén. 79*, 1–156.

LEYDIG, F. (1857). *Lehrbuch der Histologie*. G. Grote, Hamm.

LOCKE, M. and COLLINS, J. V. (1967). Protein uptake in multivesicular bodies in the molt intermolt cycle of an insect. *Science N.Y. 155*, 467–469.

LOCKE, M. and COLLINS, J. V. (1968). Protein uptake into multivesicular bodies and storage granules in the fat-body of an insect. *J. Cell Biol. 36*, 453–483.

MADDRELL, S. H. P. (1971). Mechanisms of insect excretory systems. *Adv. Insect Physiol. 8*, 199–331.

MADDRELL, S. H. P. (1977). Insect Malpighian tubules. In *Transport of Ions and Water in Animals*. Edited by B. L. Gupta, R. B. Moreton, J. L. Oschmann and B. J. Wall. Pages 541–569. Academic Press, New York.

MADDRELL, S. H. P. and GARDINER, B. O. C. (1974). The passive permeability of insect Malpighian tubules to organic solutes. *J. Exp. Biol. 60*, 641–52.

MANTON, S. M. (1977). *The Arthropoda. Habits, functional morphology and evolution.* Oxford University Press, Oxford.

MATTER, A., ORCI, L., FORSSMANN, W. G. and ROUILLER, C. H. (1968). The stereological analysis of the fine-structure of the "micropinocytosis vermiformis" in Kupffer cells of the rat. *J. Ultrastruct. Res. 23*, 272–279.

MAUNSBACH, A. B. (1966a). Absorbtion of I^{125} labelled homologous albumin by rat kidney proximal tubule cells. *J. Ultrastruct. Res. 15*, 197–241.

MAUNSBACH, A. B. (1966b). Absorbtion of ferritin by rat kidney proximal tubule cells. Electron microscopic observations on the initial uptake phase in cells of microperfused single proximal tubules. *J. Ultrastruct. Res. 16*, 1–12.

MILLER, F. (1960). Haemoglobin absorbtion by the cells of the proximal convoluted tubule in mouse kidney. *J. Biophys. Biochem. Cytol. 8*, 689–718.

MILLS, R. P. and KING, R. (1965). The pericardial cells of *Drosophila melanogaster*. *Quart. J. Mic. Sci. 106*, 261–268.

MOHRIG, W. and MESSNER, B. (1968). Immunereaktionen bei Insekten. *Biol. Zentralbl. 87*, 439–470.

MOORE, M. N., BUBEL, A. and LOWE, D. M. (1980). Cytology and cytochemistry of the pericardial gland cells of *Mytilus edulis* and their lysosomal responses to injected horseradish peroxidase and anthracene. *J. Mar. Biol. Assoc. U.K. 60*, 135–150.

MURRAY, F. V. and TIEGS, O. W. (1935). The metamorphosis of *Calandra oryzae*. *Quart. J. Mic. Sci. 77*, 405–495.

NOIROT-TIMOTHÉE, C. and NOIROT, C. (1980). Septate and scalariform junctions in arthropods. *Int. Rev. Cytol. 63*, 97–139.

NOVIKOFF, A. B. (1976). The endoplasmic reticulum: A cytochemist's view (A Review). *Proc. Natl. Acad. Sci. USA 73*, 2781–2787.

NUTTALL, G. H. F. and KEILIN, D. (1921). On the nephrocytes of *Pediculus humanus*. *Parasitology 13*, 184–92.

ORCI, L., PICTET, R. and ROUILLER, C. H. (1967). Image ultrastructurale de pinocytose dans la cellule de Kupffer du foie de rat. *J. Mic. 6*, 413–418.

PAL, R. (1944). Nephrocytes in some *Culicidae: Diptera*. *Ind. J. Ent. 6*, 143–148.

PALM, N. B. (1952). Storage and excretion of vital dyes in insects with special regard to trypan blue. *Arkiv für Zoologi 3*, Ser. 2, 195–272.

PEARSE, B. M. F. (1975). Coated vesicles from pig brain: purification and biochemical characterization. *J. Mol. Biol. 97*, 93–98.

PEARSE, B. M. F. and BRETSCHER, M. (1981). Membrane recycling by coated vesicles. *Ann. Rev. Biochem. 50*, 85–102.

PEASE, D. C. (1955). Fine structures of the kidney seen by electron microscopy. *J. Histochem. Cytochem. 3*, 295–308.

PETERSON, D. R. and LOIZZI, R. F. (1974). Ultrastructure of the crayfish kidney. Coelomosac, labyrinth, nephridial canal. *J. Morph. 142*, 241–264.

PFEFFER, S. R. and KELLY, R. B. (1981). Identification of minor components of coated vesicles. Use of permeation chromatography. *J. Cell Biol. 91*, 385–391.

PHILIPPE, C. (1982). Culture of fat-body of *Periplaneta americana*: tissue development and establishment of cell lines. *J. Insect Physiol. 28*, 257–265.

POLL, M. (1934). Recherches histophysiologiques sur les tubes de Malpighi due *Tenebrio molitor*. L. *Rec. Inst. Zool. Torley-Rousseau 5*, 73–126.

PURI, I. M. (1924). Studies on the anatomy of *Cimex lectularius*. *Parasitology 16*, 84–97.

RAMSAY, J. A. (1958). Excretion by the Malpighian tubules of the stick insect *Dixippus morosus* (Orthoptera, Phasmidae): amino acids, sugars, and urea. *J. Exp. Biol. 35*, 871–91.

RENKIN, E. M. and ROBINSON, R. R. (1974). Glomerular filtration. *N. Engl. J. Med. 290*, 785–792.

RENNKE, H. K., COTRAN, R. S. and VENKATACHALAM, M. A. (1975). Role of molecular charge in glomerular permeability. Tracer studies with cationized ferritins. *J. Cell Biol. 67*, 638–646.

RIBAUCOURT, H. DE (1901). Les néphrocytes. *C.R. Soc. Biol. T. 53*, 43.

RODEWALD, R. and KARNOVSKY, M. J. (1974). Porous substructure of the glomerular slit diaphragm in the rat and the mouse. *J. Cell Biol. 60*, 423–433.

ROTH, T. F. and PORTER, K. R. (1962). Specialized sites on the cell surface for protein uptake. *5th Int. Congr. E.M. LL-4*. Academic Press, New York.

ROTH, T. F. and PORTER, K. R. (1964). Yolk protein uptake in the oocyte of the mosquito *Aedes aegypti*. *J. Cell Biol. 20*, 313–332.

RUBENSTEIN, J. L. R., FINE, R. E., LUSKEY, B. D. and ROTHMAN, J. E. (1981). Purification of coated vesicles by agarose gel electrophoresis. *J. Cell Biol. 89*, 357–361.

SALISBURY, J. L., CONDEELIS, J. S., MAIHLE, N. J. and SATIR, P. (1981). Receptor mediated endocytosis by clathrin-coated vesicles: evidence for a dynamic pathway. *Cold Spring Harbor Symp. 46*, 733–742.

SANGER, J. W. and MCCANN, F. V. (1968). Fine structure of pericardial cells of the moth *Hyalophora cecropia* and their rôle in protein uptake. *J. Insect Physiol. 14*, 1839–1845.

SCHMIDT-NEILSEN, B., GERTZ, K. H. and DAVIS, L. E. (1968). Excretion and ultrastructure of the antennal gland of the fiddler crab *Uca mordax*. *J. Morph. 125*, 473–496.

SCHWERMER, W. (1914). Beitrage zur Biologie und Anatomie von *Perla marginata*. *Zool. Jahrb, Anat. Ontog. 37*, 267–312.

SCHWINCK, I. (1951). Veränderungen der Epidermis, der Perikardialzellen und der Corpora Allata in der Larven-Entwicklung von *Panorpa communis* L. unter Normalen and Experimentellen Bedingungen. *Wilhelm Roux Arch. Entw. Mech. 145*, 62–108.

SCHWINCK, I. (1952). Zur Function der Pericardialzellen weitere experimentelle Untersuchungen an der Larvae von *Panorpa communis*. *Naturwissenschaften 39*, 160.

SEIFERT, G. and ROSENBERG, J. (1974). Elektronenmikroscopische intersuchungen der Haütungsdrüsen (Lymphstränge) von *Lithobius forficatus* L. (Chilopoda). *Z. Morphol. Okol. Tiere. 78*, 263–279.

SEIFERT, G. and ROSENBERG, J. (1976). Feinstruktur des 'sacculus' der Nephridien von *Peripatoides leuckarti* (Onychophora: Peripatopsidae). *Ent. Germ. (Stuttgart) 3*, 202–211.

SEIFERT, G. and ROSENBERG, J. (1977). Die Ultrastructure de Nephrozyten von *Peripatoides leuckarti* (Onychophora: Peripatopsidae). *Zoomorphologie 86*, 169–181.

SHEA, S. M. and MORRISON, A. B. (1975). A stereological study of the glomerular filter in the rat. *J. Cell Biol. 67*, 436–443.

SILVERSTEIN, S. C., STEINMANN, R. M. and COHN, Z. A. (1977). Endocytosis. *Ann. Rev. Biochem. 46*, 669–722.

SKELDING, J. M. and NEWELL, P. F. (1975). On the functions of the pore cells in connective tissue of terrestrial pulmonate molluscs. *Cell. Tiss. Res. 156*, 381–390.

SMINIA, T., BOER, H. H. and NIEMANTSVERDRIET, A. (1972). Haemoglobin-producing cells in freshwater snails. *Z. Zellforsch. 135*, 563–568.

STAY, B. (1965). Protein uptake in the oocytes of the cecropia moth. *J. Cell Biol. 26*, 49–62.

STUART, A. M. and SATIR, P. (1968). Morphological and functional aspects of an insect epidermal gland. *J. Cell Biol. 36*, 527–549.

TELFER, W. H. (1960). The rate of entry and localization of blood proteins in the oocytes of Saturniid moths. *Biol. Bull. 118*, 338–351.

THOMSON, J. A. (1975). Major patterns of gene activity during development in holometabolous insects. *Adv. Insect Physiol. 11*, 321–398.

THOMSON, J. A. and GUNSON, M. M. (1970). Development changes in the major inclusion bodies of polytene nuclei from larval tissues of the blowfly *Calliphora stygia*. *Chromosoma (Berl.) 30*, 193–201.

THORBECKE, G. J., MAURER, P. H. and BENACERRAF, B. (1960). The affinity of the reticulo endothelial system for various modified serum proteins. *Brit. J. Exp. Pathol. 41*, 190–197.

Tiegs, O. W. and Murray, F. V. (1938). The embryonic development of *Calandra oryzae*. *Quart. J. Mic. Sci. 80*, 159–284.

Tisher, C. C., Rosen, S. and Osborne, G. B. (1969). Ultrastructure of the proximal tubule of the Rhesus monkey kidney. *Amer. J. Pathol. 56*, 469–517.

Trump, B. (1961). An electron microscope study of uptake transport, and storage of colloidal materials by the cells of the vertebrate nephron. *J. Ultrastruct. Res. 5*, 291–310.

Tyson, G. E. (1968). The fine-structure of the maxillary gland of the brine shrimp *Artemia salina*: the end-sac. *Z. Zellforsch. 86*, 129–138.

Verhoef, H. A., Bosman, C., Bierenbroodspot, A. and Boer, H. H. (1979). Ultrastructure and function of the labial nephridia and the rectum of *Orchesella cincta* L. Collembola. *Cell Tiss. Res. 198*, 237–246.

Wall, B. J. and Oschman, J. L. (1975). Structure and function of the rectum in insects. In *Fortschritte der Zoologie*. Edited by A. Wessing. Pages 193–222. Gustav Fischer Verlag, Jena.

Weismann, A. (1865). Die nachembryonale Entwicklung der Musciden nach Beobachtungen an *Musca vomitoria* and *Sarcophaga carnaria*. *Z. Wiss. Zool. 14*, 187–336.

Wigglesworth, V. B. (1943). The fate of haemoglobin in *Rhodnius prolixus* (Hemiptera) and other blood-sucking insects. *Proc. Roy. Soc. B. 131*, 313–339.

Wigglesworth, V. B. (1970). The pericardial cells of insects: analogue of the reticuloendothelial system. *J. Reticuloendothelial Soc. 7*, 208–216.

Woodring, J. P. (1973). Comparative morphology, functions and homologies of the coxal glands in oribitid mites. (Arachnida-Acari). *J. Morphol. 139*, 407–430.

Wyatt, G. R. and Pan, M. L. (1978). Insect plasma proteins. *Annu. Rev. Biochem. 47*, 779–817.

Yamada, E. (1955). The fine structure of the renal glomerulus of the mouse. *J. Biophys. Biochem. Cytol. 1*, 551–566.

13 Structure and Physiology of the Respiratory System

PETER J. MILL

University of Leeds, UK

1 INTRODUCTION

During the last 15 years a number of reviews of various aspects of insect respiration have appeared; some fairly broad-based, others very specialized (e.g. Hinton, H., 1968a, 1969; Guthrie, D. and Tindall, A., 1968; Chapman, R., 1969; Bursell, E., 1970; Steen, J., 1971; Mill, P., 1972, 1974, 1977; Whitten, J., 1972; Miller, P., 1974, 1980, 1981a,b), and it is useful to draw together the various aspects of the structure and physiology of the insect respiratory system.

The respiratory system of insects consists of a series of tubes called tracheae, which convey oxygen to the tissues. The larger tracheae are normally lined with epicuticle, which is shed at each ecdysis. In the early larval stages of some insects the tubes contain liquid, but in the vast majority of cases they are filled with air. In terrestrial and many aquatic insects the tracheae open to the exterior via spiracles (open system), but in some aquatic larvae there are no spiracles (closed system) and the tracheae divide peripherally to form a plexus, either over the general body surface, so allowing cutaneous respiration, or within specialized gills (tracheal gills).

In those insects with functional spiracles various devices are utilized to conserve water, while at the same time allowing for an interchange of respiratory gases. A number of aquatic insects with an open tracheal system carry gas gills with them; these may be temporary or permanent. Gas gills are also found in the chorion of the eggshells of many insects. All of these aspects are discussed in the following pages, together with details as to how the ventilatory system works and how it is controlled. A few insects also possess haemoglobin and the functions of this are dealt with.

The respiratory rate is directly related to the metabolic requirements of the animal and, although metabolism is dealt with in detail in volume 10, various aspects are included in this chapter for completeness. The total metabolism (M) is expressed in terms of oxygen consumption/unit time and, assuming that body weight (W) is directly related to volume

$$M = kW^b$$

where k and b are constants. If the total metabolism is linearly related to the surface area (S) of the animal then

$$M = kW^{2/3}$$

Expressed in terms of metabolic rate (R), which is defined as oxygen consumption/unit weight/unit time, then

$$R = kW^{-1/3}$$

The relationship expressed by the last two equations is known as the surface law.

However, if total metabolism is linearly related to weight then

$$M = kW \quad \text{and} \quad R = k$$

(Keister, M. and Buck, J., 1964; Mill, P., 1972, 1974). However, insects are not homogeneous and it is often difficult to draw conclusions from the experimental data, and Edwards, R. (1958) has pointed out the confusion which this has led to; also the relationship may change during the course of development. For example, there is a good correlation between oxygen consumption and weight in early instars of *Locusta migratoria* but not in later ones, where oxygen consumption remains almost constant while weight increases (Clarke, K., 1957). Nevertheless it is often useful to refer to respiratory rate, especially when comparing individuals of the same species under different conditions, or of different species under identical conditions.

2 TRACHEAL SYSTEMS

2.1 General organization and structure

The tracheae are basically organized on a segmental plan with the spiracles, when present, occurring either on or between segments. Primitively there were probably 12 pairs of spiracles. However, 10 pairs is the most seen post-embryonically: two pairs on the thorax and eight pairs on the abdomen, and their origins have been discussed by Matsuda, R. (1975). Insects which have the maximum number of functional spiracles are called Holopneustic; insects with less than ten functional pairs (Hemipneustic) can be divided into those with no functional spiracles (Apneustic) and those in which one or more pairs have disappeared, as distinct from those which have become non-functional (Hypopneustic). Many spiracles lie on a special plate called a peritreme. Typically each spiracle leads into an atrium, although spiracle 1 in many species (e.g. odonates, orthopterans, dictyopterans and some coleopterans) has two openings leading to separate atria (Miller, P., 1960b; Poonawalla, Z., 1966). A trachea arises from the base of each atrium (in a few species directly from the spiracular opening) and divides into dorsal, visceral and ventral branches, which further subdivide to supply the various tissues and organs.

In many apterygotes the tracheal system in each half-segment is a discrete entity, but in most insects interconnections form between adjacent segments, thereby producing longitudinal trunks; there are typically paired dorsal, lateral and ventral longitudinal trunks. The dorsal and ventral trunks are united segmentally with those of the opposite side by transverse commissures; hence the whole system can receive air from a single spiracle.

The opening of the trachea may be several millimetres in diameter in the largest insects, but their inner ends have a diameter of only 1–2 μm. Depending on its size each trachea is surrounded by one or two epithelial cells. The tracheal lumen is lined with an epicuticular intima (which may have a wax monolayer on its surface). The lining is smooth when the tracheae are first formed; it expands equally in all directions, resulting in an increase in diameter but buckling in the longitudinal axis to produce a thickened spiral or rings, called the taenidium, which presumably serves to keep the tracheae distended, and which in large-diameter tracheae may be strengthened by the addition of taenidial bars. The ridges in the tracheoles are formed in a similar manner. Underlying the outer layer there is often a layer containing protein in which the micelles are axially oriented, except in the taenidia where they are arranged tangentially. Finally there is an endocuticular layer (Locke, M., 1957, 1958a; Amos, W. and Miller, P., 1965; Wigglesworth, V., 1970).

Often some of the tracheae are dilated to increase the reservoir of air; for example the dorsal branches connecting the dorsal and lateral longitudinal trunks in the cockroach abdomen (Miall, L. and Denny, A., 1886; Baudet, J. and Sellier, E., 1975) (Fig. 1). Within the cockroaches (suborder Blattaria or Blattodea) the basic adult features of the tracheal system are present in the first instar but the system becomes more complex (i.e. an increase in the number of tracheal branches, dilated tracheae and tracheoles) with increase in size; the complexity also increases through the family sequence blattids–blatellids–blaberids, but is not related to flying ability (Baudet, J. and Sellier, E., 1975).

Air sacs occur in the tracheae of many larger insects, particularly flying insects, and reach their peak of development in the Diptera and Hymenoptera. However, they are lacking in cockroaches (Facheu, M. and Sellier, E., 1971). They are dilated regions in which the taenidium is reduced or absent (Miller, P., 1974; Kerry, C., 1982; Kerry, C. and Mill, P., in preparation) and Tonapi, G. (1957) has shown a good correlation between their development and the size and activity of various hymenopteran species. Their function is to increase the volume of tidal air which can be changed during each ventilatory cycle and to reduce the diffusion path to the tissues.

In locusts and dragonflies each flight muscle receives a primary tracheal trunk with an associated air sac. In the centro-radial and latero-radial types the primary trachea is respectively intramuscular and superficial, and ends in an air sac. Both give rise to secondary tracheae which branch radially from the primary trachea. In the latero-linear type the air sac is juxtaposed to the muscle surface and the secondary tracheae lie parallel to each other. In the locust all three types occur (Fig. 2). In large insects many of the flight muscles have a tracheal shunt between the primary supply and the terminal air sac, which facilitates diffusion when the animal is at rest (Weis-Fogh, T., 1964a). The tracheal openings can generally be closed, either by a spiracular valve (e.g. Odonata and Orthoptera) or by a mechanism at the junction of the atrium and the trachea (section 2.3.1). This ability to close the spiracles is of particular importance both for limiting water loss in dry conditions and for providing the basis for the unidirectional air flow which occurs in the main tracheal trunks of larger insects.

In some species, e.g. *Rhodnius prolixus*, the parallel-sided tracheae branch in such a way that the summed cross-sectional area of the diffusion path remains constant (Locke, M., 1958b), but in others may be reduced from the outside inwards, e.g. silkworm larvae and dragonflies (Nunome, J., 1944; Weis-Fogh, T., 1964b). In *R. prolixus*,

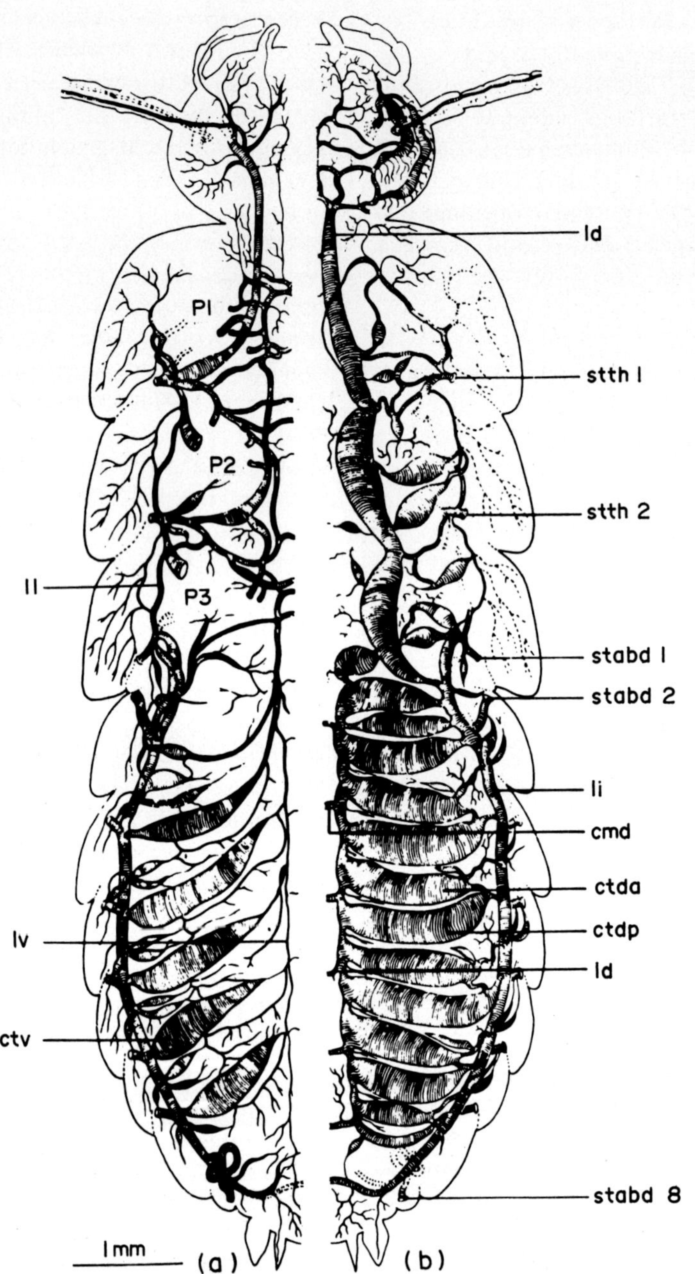

FIG. 1. The morphology of the tracheal system of a first instar *Gromphadorhina portentosa*. (**a**) Ventral view; (**b**) dorsal view. cmd, dorsal commissure; **cdta**, **cdtp**, anterior and posterior dorsal transverse connectives; **ctv**, ventral transverse connective; **ld**, **ll** and **lv**, dorsal, lateral and ventral longitudinal trunks; P1, P2, P3, positions of legs; **stabd**, abdominal spiracle; **stth**, thoracic spiracle. (From Baudet, J. and Sellier, E., 1975.)

increase in length of the tracheae occurs mainly from stretching the tracheae in the preceding instar (Locke, M., 1958b). The proportion of the body occupied by the tracheal system varies both in different insects and in different stages of development. For example, in *Locusta* its volume may decrease from 42% to 3.8% of body volume during the course of a stadium and this is caused by the

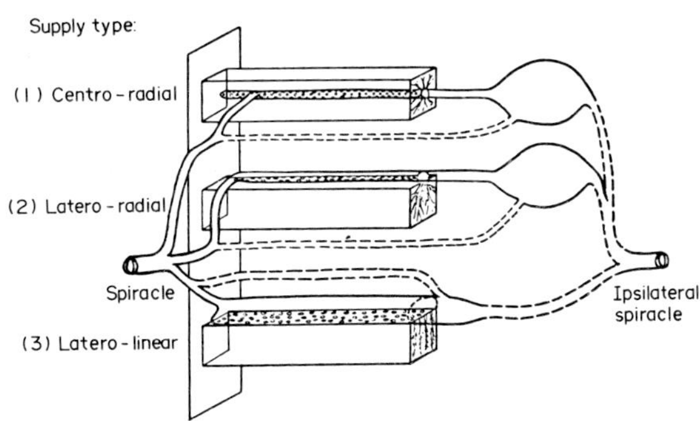

FIG. 2. The three types of tracheal supply to the wing muscles of the locust. The primary supply is drawn horizontal; the secondary supply is indicated on the vertical cut ends of the muscles. (From Weis-Fogh, T., 1964a.)

gradual occlusion of the abdominal air sacs as the body tissues grow during the stadium. At ecdysis the tracheal volume increases again as the overall body size increases and the pressure on the air sacs is hence reduced (Clarke, K., 1958). The higher the metabolic rate of an organ, the greater is its tracheal supply, and flight muscles (Weis-Fogh, T., 1964b) and the expiratory dorsoventral muscles of dragonfly larvae (Mill, P. and Lowe, D., 1971) are particularly well endowed. There is also a rich tracheal supply to the ganglia of locusts (Burrows, M., 1980), crickets (Longley, A. and Edwards, J., 1979) and dragonfly larvae.

Estimates of the total volume of the tracheal system have recently been made by measuring the rate of wash-out of the inert gas, argon, in the cockroaches *Byrsotria fumigata* and *Gromphadorhina brunneri* and in the pupa of the lepidopteran *Hyalophora cecropia*. The values obtained were $82.0 \pm 7.0\,\mu l\,g^{-1}$ (adult male *Byrsotria*), $127\,\mu l\,g^{-1}$ (larval *Byrsotria*), $83\,\mu l\,g^{-1}$ (adult *Gromphadorhina*) and $48\,\mu l\,g^{-1}$ (range 39–$59\,\mu l\,g^{-1}$) (pupal *Hyalophora*) (Bridges and Scheid, in Miller, P., 1981a; Bridges, Kestler and Scheid, in Miller, P., 1981a). These values contrast with the much higher figure of $265\,\mu l\,g^{-1}$ for a fifth-instar *Locusta migratoria* (Clarke, K., 1958). Weis-Fogh, T. (1964a) recorded a total figure of about 500–950 $\mu l\,g^{-1}$ for a mature locust of average size.

Tracheoles arise from the interior end of each tracheal branch. They are intracellular and taper to a diameter of 0.1–0.5 μm and extend for up to several hundred μm. Although the tracheoles are lined with taenidia, there is probably no wax layer (Beament, J., 1964) and no underlying protein epicuticle, and the taenidia are not strengthened as in the tracheae (Whitten, J., 1968; Wigglesworth, V., 1970; Miller, P., 1974). Furthermore the tracheolar linings are not normally shed at ecdysis (Wigglesworth, V., 1959; Whitten, J., 1968, 1972).

The developing tracheal system becomes air-filled because its initial fluid contents are actively absorbed by the walls of the tracheoles; the slight negative pressure which this causes is thought to liberate gas from the fluid (Wigglesworth, V., 1953a). In small insects diffusion of oxygen meets the metabolic requirements of the tissues. Diffusion of oxygen is several orders of magnitude greater in air than in water and only the tracheolar extremities of the system are filled with liquid (Wigglesworth, V., 1930, 1932). When a tissue is active the released metabolites raise the osmotic pressure within the tissue and this effects the withdrawal of liquid from the tracheoles and so the air penetrates closer to their tips, thereby allowing faster diffusion of oxygen to the tissues (Wigglesworth, V., 1953a). In larger insects active pumping movements of the thorax and/or abdomen ventilate the peripheral regions of the system and so the diffusion pathway to the tissues is reduced; this reduction is enhanced by the presence of air sacs. Indeed it has been shown that the length of the diffusion pathway is sufficiently short for diffusion to be able to supply adequate oxygen to even the most active of tissues provided that the primary tracheal supply is strongly ventilated (Krogh, A., 1920a,b; Weis-Fogh, T., 1964b).

In flying locusts ventilation of the thorax is directly associated with the wing movements (autoventilation) and this is discussed in section 3.3.4. However, the thoracic tracheal system in locusts and dragonflies is partially isolated during flight and can provide the flight muscles with all of their requirements. Thoracic pumping also provides most of the requirements during flight in large moths and beetles, but hymenopterans and dipterans rely on abdominal pumping (Fraenkel, G., 1932a; Krogh, A. and Zeuthen, E., 1941; Krogh, A., 1948; Burton, A., 1962; Miller, P., 1962; Weis-Fogh, T., 1964a, 1967).

The intracellular tracheoles generally lie juxtaposed to the cells of the tissues which they are supplying, and the supply is particularly rich in muscles such as the expiratory dorsoventral muscles of dragonfly larvae, which contract rhythmically for long periods and hence presumably have a fairly high metabolic demand (Mill, P. and Lowe, D., 1971), and the thoracic spiracle closer muscles of *Periplaneta americana* where Miller, P. (1981a) has suggested that it may be related to the responsiveness of these muscles to tracheal gases.

Within the tissues the maximum distance for diffusion is in the order of 10 μm and hence the maximum spacing of tracheoles is about 20 μm. It follows that the diameter of the muscle fibres cannot exceed about 20 μm unless the tracheoles indent the surface (Weis-Fogh, T., 1964b). Indeed, in dragonflies the fibres of the tubular flight muscles have a diameter less than this (Smith, D., 1966, 1968) and the mitochondria are probably all less than 10 μm from the nearest tracheole. However, in many cases where the tissue has a particularly high metabolic demand, e.g. in the flight muscles of dipterans and hymenopterans, the tracheoles actually indent the surface membrane of the muscle fibres and lie within 2–3 μm of each other (Smith, D., 1961, 1963), but are still surrounded by their own tracheoblast. The basement membranes of the tracheoblast and muscle fibre sometimes fuse (e.g. Richards, A. and Korda, F., 1950; Edwards, G. *et al.*, 1958; Brosemer, R. *et al.*, 1963; Smith, D., 1968). In locust neuropile the maximum distance from a tracheole is about 8–9 μm (Burrows, M., 1980).

Although carbon dioxide is released via the spiracles it diffuses through the tissues more rapidly than oxygen and, depending on the thickness of the cuticle, some of the loss may be transcuticular (Schneidermann, H. and Williams, C., 1955). Thus 2.5% is lost via this route in *Phormia* larvae (Buck, J. and Keister, M., 1956) and 10% in *Calliphora* larvae (Fraenkel, G. and Herford, G., 1938).

2.2 Closed tracheal systems

The larvae of many aquatic insects have a closed tracheal system, i.e. one in which the spiracles are non-functional or absent. Hence, oxygen enters the tracheal system either over the general body surface or via specialized tracheal gills.

2.2.1 Cutaneous respiration

Some aquatic larvae with a closed tracheal system (e.g. *Chironomus* larvae) rely on oxygen diffusing in over the general body surface and have a well-developed tracheal plexus lying beneath a rather thin cuticle (Fox, H., 1920). However, the layer of water adjacent to the body wall contributes to the diffusion barrier and this "boundary layer" (ΔX) (Krogh, A., 1919) decreases in thickness with increase in the flow of water over the body. Nevertheless, even in flowing water the drag effect of the body surface produces a relatively stagnant layer close to the body (Fig. 3a) and the water velocity increases asymptotically with distance from the surface (Paganelli, C. *et al.*, 1967). Rahn, H. and Paganelli, C. (1968) defined the boundary layer as "that thickness of perfectly still water which has the same diffusion resistance as the actual boundary layer", while Paganelli, C. *et al.* (1967) stated that

$$\Delta X = \frac{A\,\alpha_O\,D_O\,\Delta P_O}{J_O}$$

where A is the membrane area (cm^2), α_O, solubility of oxygen in water (cm^3 cm^{-3} atm^{-1}), D_O, diffusion coefficient of oxygen in water (cm^2 s^{-1}), ΔP_O, oxygen pressure difference between the environment (outside the boundary layer) and the inside, and J_O, observed inward diffusion of oxygen (cm^3 s^{-1}). From a plot of computed values of ΔX against water flow they deduced that the effective boundary layer in still water is about 300 μm, and that this decreases asymptotically to about 10 μm with increase in the rate of flow (Fig. 3b).

Some insect larvae reduce their boundary layer by producing oscillations of the body, e.g. the larvae of some species of *Tanytarsus* (Diptera) which live in tubes in fast-flowing water (Keilin, D., 1944) and *Tubifex* which live in slow-flowing water. Additionally *Tubifex* exposes more of its body above the surface of the mud as the pO_2 decreases.

In some species the tracheal supply to the body wall is restricted to specific zones. Thus the larva of *Atrichopogon trifasciatus* (Diptera), which lives on partially submerged stones, has eight such respiratory zones which can be pulled downwards by muscular action and virtually enclosed if the larva is in danger of becoming desiccated.

One might expect that the metabolic rate of animals relying on cutaneous respiration would be linked to surface area (section 1); however,

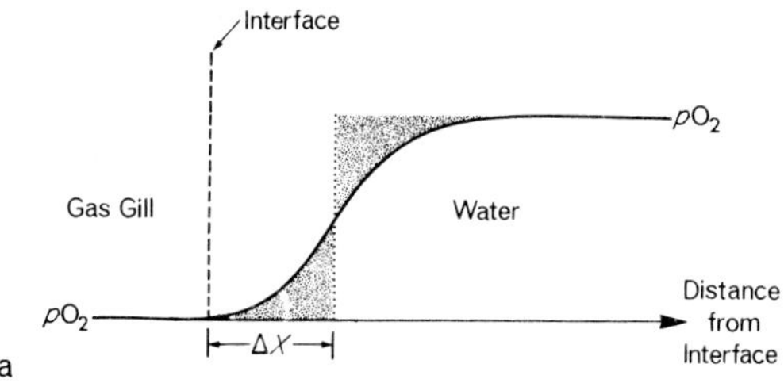

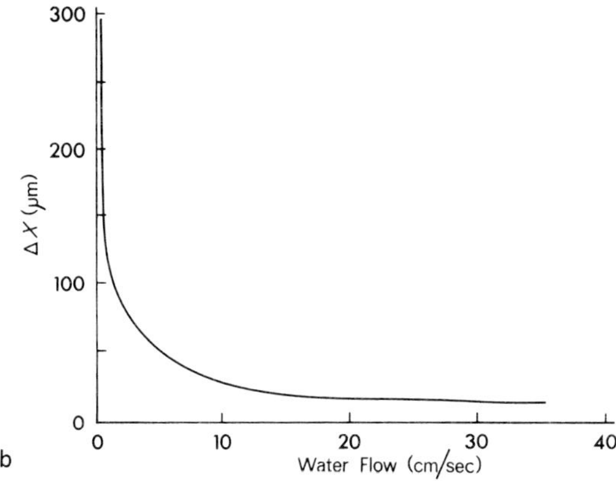

FIG. 3. (**a**) Diagrammatic representation of the pO_2 gradient next to the surface of a gas gill. ΔX is the effective thickness of the boundary layer. (The two shaded areas are equal.) (**b**) Computed relationship between the effective thickness of the boundary layer and the current velocity. (From Mill, P., 1974; after Paganelli, C. *et al.*, 1967.)

Edwards, R. (1958) has shown that fourth-instar larvae of *Chironomus riparius* have a metabolic rate which is proportional neither to surface area nor to body weight.

2.2.2 TRACHEAL GILLS — MORPHOLOGY AND OCCURRENCE

In many aquatic insects the superficial tracheal plexi are confined to gills and these tracheal gills are, with few exceptions, found only in larvae (Imms, A., 1957). In some cases aquatic insect larvae can live quite successfully after their gills have been removed, and indeed their removal may have little effect on oxygen consumption (Morgan, A. and O'Neil, H., 1931; cf. Wigglesworth, V., 1953b; MacNeill, N., 1960). However, tracheal gills are richly supplied with tracheal capillaries, have a very thin cuticle, and do not contain any structures which have a high metabolic rate. Hence they presumably take up more oxygen from the environment than they need for their own requirements and are thus of respiratory importance (Koch, H., 1936). Furthermore, they are often used to produce a current of water, and this ventilatory current is of importance in reducing the thickness of the boundary layer. Maybe in some instances gills are only essential under conditions of stress.

2.2.2.1 *Lateral abdominal gills.* In ephemeropteran larvae the gills are borne on the abdomen and their surface area is inversely proportional to the environmental pO_2 (Dodds, G. and Hisaw, F., 1924). There is much variation throughout the order. In many species they comprise seven pairs of simple lamellae which project laterally from the abdomen. In some (e.g. *Baetis*) each gill consists of a single lamella, but in others (e.g. *Cloëon*) some are double (Fig. 4a, b; Plate 1). Possibly the greater surface area is required in *Cloëon* because it lives in ponds, in contrast to *Baetis* which lives among weeds in fairly fast-flowing water.

A number of species (such as *Ecdyonurus*) which live among rocks in fast-flowing water have a dorsoventrally flattened body, and in these some of the lamellae have a basal tuft of branchial filaments (Fig. 4e). In *Leptophlebia* and *Habrophlebia* the gills are bifid with simple and multifurcate branching respectively (Fig. 4c,d). In the burrowing *Ephemera* and *Caenis* there are six pairs of gills. In the former they are held over the dorsal surface of the abdomen and consist of biramous lamellae fringed with long filaments (Fig. 4f). In *Caenis* the second pair are very large and quadrangular and cover the posterior four pairs, producing a branchial chamber which is guarded by fringes of setae to keep out particles of sand and mud. In *Ephemerella* and *Prosopistoma* there are only five pairs of gills and, in the latter, a branchial chamber is formed by the wing buds and the thoracic and abdominal terga.

Megalopteran larvae also possess lateral abdominal gills, which are filamentous and again variable in number. *Corydalis*, which lives under stones in fast-flowing water, has eight pairs of unjointed or imperfectly jointed gills, while *Sialis*, which lives in still or slow-flowing water, has seven pairs of five-segmented gills together with a similar terminal filament on the ninth abdominal segment (Plate 1b). In *Corydalis* the first seven abdominal segments also bear ventral, spongy tufts of accessory tracheal gills.

2.2.2.2 *Caudal gills.* The terminal filament of larval *Sialis* is mentioned in the previous section; otherwise

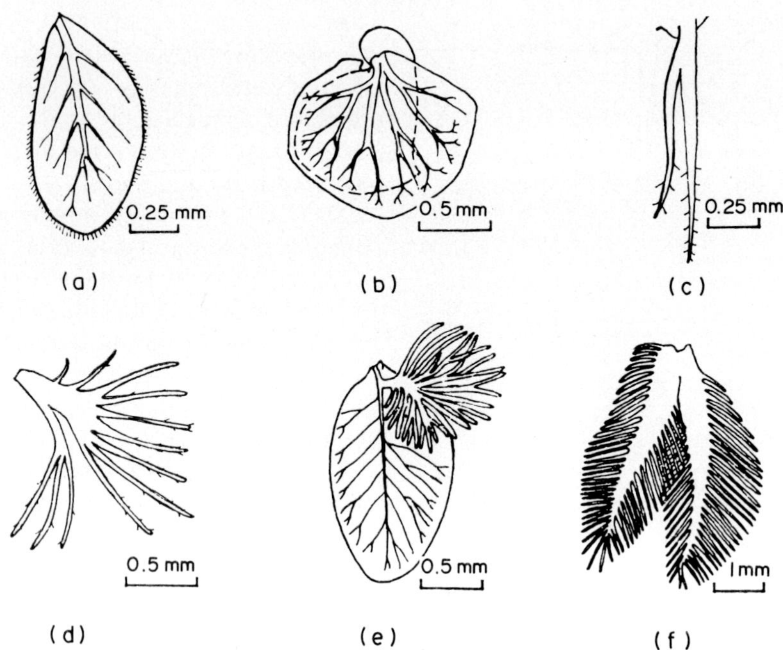

FIG. 4. Lateral abdominal gills of mayflies. All are of the 4th gill, except for that of *Leptophlebia cincta* which is of the 3rd gill. (**a**) *Baetis rhodani*; (**b**) *Cloëon simile*; (**c**) *Leptophlebia cincta*; (**d**) *Habrophlebia fusca*; (**e**) *Ecdyonurus fluminum*; and (**f**) *Ephemera vulgata*. (From Mill, P., 1972; after Eaton, A. (1883–8).)

caudal gills seem to be a prerogative of zygopteran larvae, although some zygopterans (Epallagidae and Polythoridae) do possess lateral abdominal gills as well. Typically there are three caudal gills: one median and two lateral. In young larvae they are ferula (MacNeill, N., 1960) and the lateral ones retain this shape in older individuals belonging to the Calopteryginae. The medial gills of Calopteryginae, and all the gills of other groups, develop to become lamellate or saccoid (Plate 2) (Tillyard, R., 1917). In many species the gills each have a proximal and a distal zone and the latter may increase in length relative to the proximal zone (duplex gills). However, this "protrusive growth" does not occur in all species (simplex gills). In duplex gills the cuticle of the distal zone is thinner than that of the proximal zone and bears relatively fewer spines.

2.2.2.3 *Rectal gills.* Anisopteran larvae are unique in that their tracheal gills lie in the modified anterior region of the rectum, the branchial chamber. Basically there are six longitudinal (main) folds, supported on either side by cross-folds. Primitively all of the folds bear gills (simplex system), but in most cases the main folds are either non-functional or absent (duplex system). The shape and attachment of the gills are variable and serve as a basis for their classification. Thus, in the simplex system the free edge of each gill may consist of an undulated membrane (simplex-undulate) (Fig. 5a) (e.g. Cordulegastrinae) or is divided into long filaments (simplex-papillate) (e.g. most Gomphinae). Those with non-functional main folds include the Brachytronini, which have broad-based gills with overlapping lamellae (duplex-implicate) (Fig. 5b), and the Aeshnini, which have basally constricted gills. The latter may be lamellate (duplex-foliate) as in *Aeshna* (Fig. 5c) or the distal portion of each gill may be dilated and bear papillae (duplex-papillate) as in *Anax* (Fig. 5d). In the Libellulidae the main folds have been lost and the gills are broad-based lamellae attached to the cross-folds. In the Synthemini there are 12 gills in each longitudinal row and each gill has a large basal pad (duplex-archilamellate) (Fig. 5e), but in other members of the family the gills are more numerous and their basal pads are small (duplex-neolamellate) (Fig. 5f) (Tillyard, R., 1917).

2.2.2.4 *Other gills.* Many other insect larvae possess filamentous gills. In plecopterans tufts of gills may

PLATE 1. *Top*: Lateral abdominal gills of *Cloëon*. *Middle*: Lateral and caudal segmented abdominal gills of *Sialis*. *Bottom*: Respiratory siphon and tracheal gills of *Culex*. (Photographs by P. J. Evennett.)

PLATE 2. Caudal gills of damsel fly larvae. *Top left*: Simplex gills of *Pyrrhosoma nymphula*. *Top right*: Duplex gills of *Coenagrion puella*. *Bottom left and right*: Duplex gills of *Enallagma cyathigerum*. In all three cases note the tracheal trunks entering the gills. (Photographs by P. J. Evennett.)

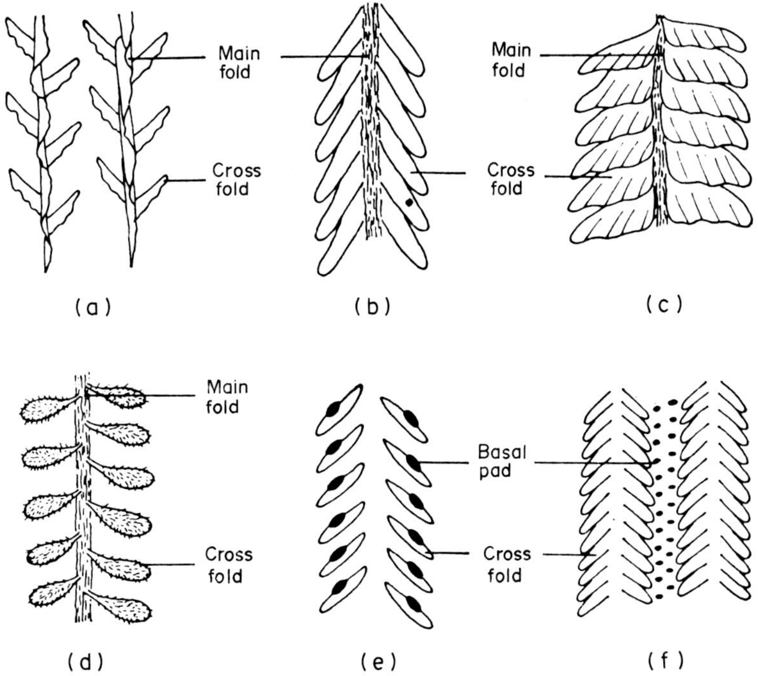

FIG. 5. Rectal gills found in the branchial chamber of anisopteran dragonfly larvae. (**a**) Simplex, undulate (*Austrogomphus*); (**b**) duplex, implicate (*Austroaeshna*); (**c**) duplex, foliate (*Aeshna*); (**d**) duplex, papillofoliate (*Anax*); (**e**) duplex, archilamellate (*Synthemis*); (**f**) duplex, neolamellate (*Diplacodes*). (From Mill, P., 1972; after Tillyard, R., 1917.)

occur on the head, thorax, abdomen or the bases of the legs, while in some megalopterans (e.g. *Corydalis* and *Neuromus*) they are found on the ventral surface of the abdomen (Imms, A., 1957). Filamentous gills also occur in many trichopterans, notably those with portable cases, where they are found in segmental groups on various regions of the abdomen, and there may be as many as 60 gills (e.g. *Macronema zebratum*) (Morgan, A. and O'Neil, H., 1931). In case-bearing trichopterans the number of gill filaments increases with body size. However, in larvae without cases (e.g. limnephilids) the number of gills is inversely related to environmental pO_2, the changes occurring when the animal moults (Dodds, G. and Hisaw, F., 1924; Wichard, W., 1974a,b). Similarly, in *Molanna angustata*, while the size of the gills is related to body size, the variable number found in different populations is in part determined by the environment (Wichard, W., 1977). Some dipterans (e.g. *Phalacrocerca*) and lepidopterans (e.g. *Nymphula*) possess gills on most body segments and *Nymphula maculalis* has over 400 gills (Buckler, W., 1875; Miall, L., 1895; Welch, P., 1916). The ephemeropterans *Jolia* and *Oligoneuria* bear gills on their heads, while the coleopterans *Peltodytes* and *Hygrobia* have, respectively, jointed gills on the dorsal surface of the thorax and abdomen and ventral gills near the base of each leg and on the first three abdominal segments (Imms, A., 1957).

2.2.3 TRACHEAL GILLS — STRUCTURE

In all tracheal gills that have been studied so far the tracheoles are situated within an epithelium lying just beneath a very thin layer of cuticle (Sadones, J., 1896; Tillyard, R., 1916). Each tracheole is surrounded by the thin cytoplasmic sheath of its tracheoblast and lies in an extracellular tube formed by an invagination of the basal plasma membrane (Fig. 6). However, there are variations in their general arrangement within the gills (Wichard, W., 1979a). In the filamentous type of gill the tracheoles are arranged almost parallel to each other and to the long axis of the filament. In trichopterans they are of approximately equal diameter and are more or less equidistantly spaced (Wichard, W., 1973, 1977), whereas in plecopterans and ephemeropterans they are of variable diameter with interspaces generally less than the tracheolar diameters (Wichard, W. and

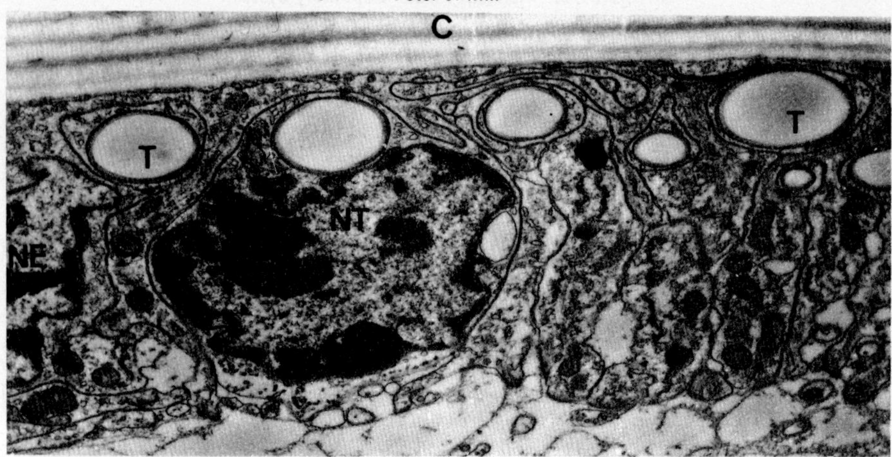

Fig. 6. Electron micrograph of a cross-section through a caudal gill of the larva of the damselfly (*Epallage fatime*). c, Cuticle; NE, nucleus of epithelial cell; NT, nucleus of a tracheoblast; T, tracheole. (From Wichard, W., 1979b.)

Komnick, H., 1974a; Wichard, W., 1979a). Both types are thus very efficient, with little oxygen diffusing between the tracheoles. In the lamellar gills of zygopterans (caudal gills) and ephemeropterans (lateral gills) the tracheoles branch and ramify almost radially throughout the gill. The interspaces decrease towards the periphery of the gill and hence the effective respiratory surface is less than the surface area of the gill (Wichard, W., 1979a,b). However, in the lamellar gills of anisopterans (rectal gills) each tracheole forms a complete loop within the gill (Sadones, J., 1896; Tillyard, R., 1917). They run almost parallel to each other, and are so densely packed that their interspaces are comparatively small; hence they are similar to the filamentous gills in terms of efficiency (Greven, H. and Rudolph, R., 1973; Wichard, W. and Komnick, H., 1974b; Wichard, W., 1979a). In addition to their respiratory function, the caudal gills of ephemeropterans have numerous chloride cells in their epithelium which are concerned with osmoregulation (Wichard, W. and Komnick, H., 1971). In the rectal gills of anisopterans the chloride cells are concentrated into one small and one large pad on the base of each gill (Wichard, W. and Komnick, H., 1974c; Komnick, H., 1977, 1978).

2.2.4 Tracheal gills — physiology

The physical aspects of tracheal gill ventilation are described in section 3, while here their role in oxygen uptake is considered. In the aquatic environment the oxygen tension (pO_2) can be extremely variable. Some species, conformers, have a metabolic rate which decreases as the oxygen tension decreases below that at which it is in equilibrium with air ($7.9\,cm^3\,l^{-1}$ water at 10°). Other species, regulators, are able to compensate, at least over a range of pO_2 below air saturation, and hence maintain their metabolic rate. The pO_2 below which regulators can no longer maintain this rate is called the critical level. (In increasingly supersaturated water a level of pO_2 is ultimately reached beyond which conformers can maintain a steady metabolic rate.)

2.2.4.1 *Lateral abdominal gills*

(a) Burrowing larvae. Current evidence indicates that the gills of burrowing ephemeropterans such as *Ephemera* and *Hexagenia* contribute significantly to oxygen uptake both by direct uptake via the gills and by reducing the thickness of the boundary layer by increasing the flow of water over the body. The evidence for the former stems from comparisons of the metabolic rate of normal larvae with those from which the gills have been removed. Thus Morgan, A. and Grierson, M. (1932) showed that 43% of oxygen uptake is via the gills in *Hexagania recurvata*, while Wingfield, C. (1939) obtained a value of 73% for *Ephemera vulgata*. While these values may reflect real differences between the two species, it could reflect the different experimental procedures used. Thus Wingfield used gill-less larvae 3–4 h after the gills had been removed, whereas Morgan and Grierson in most cases waited 14–16 weeks before taking any measurements. However, the current produced

by gill-beating will also reduce the boundary layer (section 2.2.1), and it has been shown that the frequency of gill-beating in *Ephemera* is inversely proportional to the environmental pO_2 (Babák, E. and Foustka, O., 1907; Wingfield, C., 1939; Eriksen, C., 1968). Thus, although Morgan, A. and Grierson, M. (1932) and Wingfield, C. (1939) used fairly high oxygen concentrations (5–9 cm^3 l^{-1} and 6.5–8 cm^3 l^{-1} respectively), Wingfield noted that the gills were still beating slowly. (Morgan and Grierson did not give any information on this.) More recently Eriksen, C. (1963b) has compared the oxygen uptake of anaesthetized normal and gill-less larvae of *Ephemera simulans* and *Hexagenia limbata* and found that about 45% and 47% respectively of the total oxygen uptake is via the gills at a pO_2 of 6 cm^3 l^{-1}.

E. simulans occurs in coarser sediments than *H. limbata* (Eriksen, C., 1968) and both of these species (Eriksen, C., 1963a), as well as *Ephemera danica* (Wautier, J. and Pattée, E., 1955), have a lower metabolic rate when they are provided with a substrate, and this reaches a minimum when the substrate particle size is similar to that of the animal's natural environment. Furthermore, *E. simulans*, when given a choice, shows a marked preference for substrate particles of the same size at which its metabolic rate is lowest (Eriksen, C., 1963a). The presence of a substrate also affects the ability of burrowing larvae to regulate their metabolism. Hence, although the earlier work of Fox, H. *et al.* (1937) indicated that *E. vulgata* is a respiratory conformer, they did not use a substrate. Eriksen, C. (1963b) has since shown that, whereas *E. simulans* and *H. limbata* also behave as respiratory conformers in the absence of a substrate, they become respiratory regulators with a critical level at about 1 cm^3 oxygen l^{-1} when provided with optimal substrates (Fig. 7).

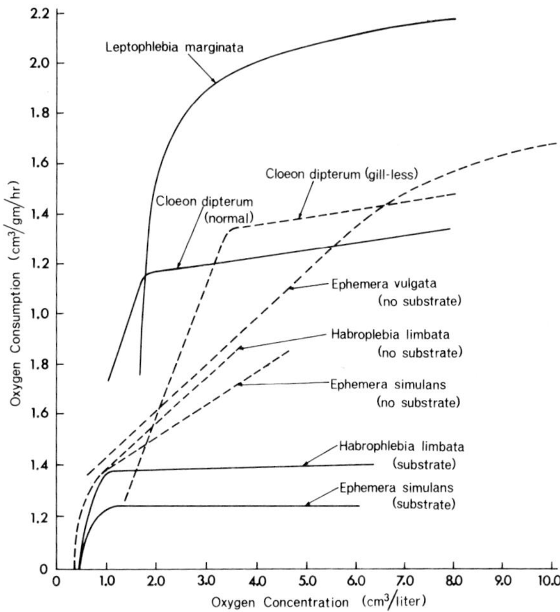

FIG. 7. The relationship between oxygen consumption and environmental oxygen concentration for various ephemeropteran larvae. (From Mill, P., 1974; after Fox, H. *et al.*, 1937 (*Leptophlebia marginata* and *Ephemera vulgata* at 10°); Wingfield, C., 1937 (*Cloëon dipterum* at 10°); and Eriksen, C., 1963b (*Ephemera simulans* and *Hexagenia limbata* at 13°).)

(b) Larvae living in ponds and lakes. In *Cloëon* the gills beat rhythmically at low oxygen concentrations; intermittently at higher concentrations. These larvae and those of *Leptophlebia* are respiratory regulators with a critical level at about 1.5 and 2.5 cm^3 oxygen l^{-1} respectively at 10° (Fox, H. *et al.*, 1937; Wingfield, C., 1937). Gill-less larvae of *Cloëon* are also able to regulate their metabolism but have a higher critical level (about 3 cm^3 oxygen l^{-1} at 10°) than normal larvae, although above this level their metabolic rate is similar to that of normal larvae. However, if the water is stirred, gill-less larvae have the same metabolic rate as normal larvae at all oxygen concentrations. This indicates that the gills are only important at oxygen concentrations below about 3 cm^3 l^{-1} and that their primary function is to reduce the boundary layer (Wingfield, C., 1937, 1939). Considering the structure of the gills (section 2.2.3) the latter conclusion is surprising.

(c) Larvae living in flowing water. Little information is available on the larvae of species such as *Baetis* and *Rhythrogena*. As might be expected they are less resistant to low environmental pO_2 than pond-dwelling species and *Baetis* can only tolerate oxygen concentrations down to 4 cm^3 l^{-1}. The gills of *Baetis* do not beat and the larvae are basically conformers, although there is evidence that large larvae can regulate their metabolism above an oxygen concentration of about 5 cm^3 l^{-1}. Above this concentration normal and gill-less larvae have similar metabolic rates (Fox, H. *et al.*, 1935, 1937; Wingfield, C., 1939).

2.2.4.2 *Caudal gills.* The available evidence indicates that, while oxygen uptake does occur over the general body surface, the caudal gills of zygopterans are of respiratory importance, particularly at low environmental pO_2 (e.g. Pennak, R. and McColl, C., 1944; Zahner, R., 1959).

(a) Larvae living in ponds and lakes. In *Coenagrion* larvae figures for oxygen uptake via the gills range from 32 to 60% of total uptake (Koch, H., 1934; Harnisch, O., 1958). Larvae of *Pyrrhosoma nymphula* have a metabolic rate which increases with both body size and temperature. They are effectively respiratory conformers, but do show some ability to regulate their metabolism above an oxygen concentration of 50% saturation (Lawton, J., 1971). Various behavioural adaptations occur at low oxygen concentration. Thus some species make ventilatory movements (Lieftinck, M., 1940; Gardner, A., 1956; Corbet, P., 1962), others (e.g. *Lestes viridis*) lie just beneath the surface of the water (Robert, P.-A., 1958).

(b) Larvae living in streams and rivers. When the environmental oxygen concentration is decreased larvae of *Calopteryx virgo* and *Calopteryx splendens* spread their gills and wing sheaths and move their bodies from side to side. At still lower concentrations they lie at the surface of the water, expose their anterior ends or even leave the water (Zahner, R., 1959).

2.2.4.3 *Rectal gills.* Larval anisopterans have a fairly thick cuticle and it is unlikely that any significant oxygen uptake occurs over the body wall. Hence the respiratory role of the rectal gills has not been questioned although there is little direct experimental evidence. MacNeill, N. (1960) has argued that the body wall is an important site of oxygen uptake in zygopterans because of the vulnerability of their external caudal gills, but ceases to be of importance in anisopterans because the gills are protected.

Anisopteran larvae irrigate their rectal gills by ventilatory movements which produce a tidal flow of water in and out of the branchial chamber via the anus (section 3.2). The ventilatory frequency increases with both decrease in oxygen concentration and increase in temperature (Matula, J., 1911).

Metabolic rate increases with increase in oxygen concentration (*Anax imperator*—Klekowski, R. and Kamler, E., 1968) and with increase in temperature (*Anax junius* — Petitpren, M. and Knight, A., 1970). At temperatures between 13° and 20° Petitpren, M. and Knight, A. (1970) showed that metabolic rate decreases with increase in body weight, but that at higher temperatures (27–34°) is weight-independent (Fig. 8). Consequently carbon dioxide output increases with temperature, although the larvae eventually acclimatize, and decreases with size (or age) (*Aeshna umbrosa* — Sayle, M., 1928). Anisopteran larvae also adopt behavioural strategies other than increasing their ventilatory frequency to compensate for low oxygen concentrations. Thus *Aeshna grandis* comes to the surface and pumps air in and out of the branchial chamber or, towards the end of the last instar, exposes the thoracic spiracles (Wallengren, H., 1914b). The latter also occurs prior to emergence (Lucas, W., 1900; Calvert, P., 1929; Byers, C., 1930), when there is both an increase in water temperature and in metabolic rate, which presumably increase the oxygen requirements of the larvae (Lutz, P. and Jenner, C., 1960; Corbet, P., 1962).

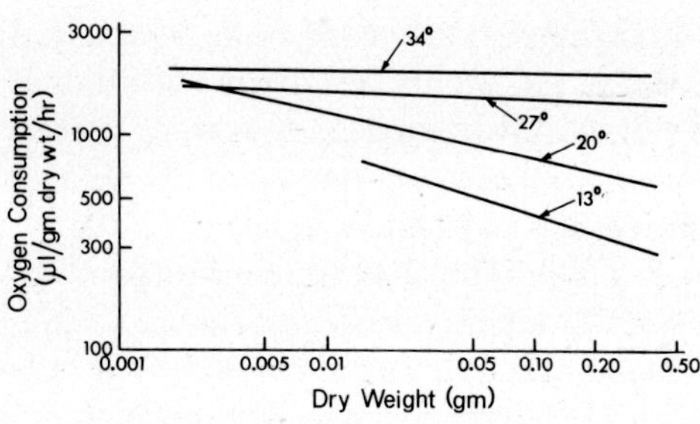

FIG. 8. The relationship between oxygen consumption and dry weight in larvae of *Anax junius* at different temperatures (double log transform). The slopes are −0.32 at 13° ($r = -0.5995$) and −0.24 at 20° ($r = -0.7259$). At the two higher temperatures the slopes do not differ significantly from zero ($r = -0.1245$ at 27° and −0.0504 at 34°). (From Mill, P., 1974; after Petitpren, M. and Knight, A., 1970.)

2.2.4.4 *Other gills.* The larvae of some species of trichopterans apparently rely exclusively on cutaneous respiration (Sleight, C., 1913) and Morgan, A. and O'Neil, H. (1931) demonstrated that oxygen consumption in larvae of *Macronema zebratum* was virtually unaffected by removal of all the gills. However, a respiratory role for the tracheal gills of various larvae was suggested by Réaumur, R. (1734–42), Dufour, L. (1847) and Houghton, W. (1868). Most of the experimental work has been carried out on species which live in flowing water.

Many trichopteran larvae use rhythmic undulations of the abdomen to produce a ventilatory current through their case, and species such as *Phryganea grandis*, *Limnephilus flavicornis* and *Polycentropus flavomaculatus*, which live in slow-flowing water, show an increase in the frequency of the undulations with decrease in environmental pO_2 (van Dam, L., 1937; Fox, H. and Sidney, J., 1953; Philipson, G., 1954). In *P. grandis* the undulations only occur intermittently in air-saturated water but with decreasing pO_2, or increasing temperature, the pauses become shorter and then disappear (van Dam, L., 1937).

In species such as *Hydropsyche angustipennis* and *Hydropsyche instabilis*, (which do not have a case and catch their food in a net), *Halesus digitatus*, *Pycnopsyche guttifer* and *Pycnopsyche lepida*, all of which live in fast-flowing water, the frequency of the undulations increases with decreases in current velocity, presumably to compensate for the latter (Fig. 9a) (Philipson, G., 1954; Ambühl, H., 1959; Feldmeth, C., 1970a). The larvae of *P. guttifer* and *P. lepida* live near the edge of streams where the average water velocity is about 5 cm s^{-1}, although larvae of the latter species move into more rapidly flowing water, where the average velocity is about 14.5 cm s^{-1}, during their last two instars (Cummins, K., 1964; Feldmeth, C., 1970a). The metabolic rate of anaesthetized larvae of both these species remains fairly constant at current velocities above about 2 cm s^{-1}, but decreases below this, presumably because of the increase in thickness of the boundary layer. However, in normal larvae both the metabolic rate and locomotor activity are maximal in the current velocity range in which each species is found, and this is a result of acclimation rather than genetic differences (Fig. 9b) (Feldmeth, C., 1970a,b).

In these fast-flowing water species the relationship between undulating frequency and environmental pO_2 is less clear. Thus, if larvae of *H. instabilis* are kept in fast-stirred water the undulations are inhibited, but commence and increase in frequency with decrease in pO_2 below about 4 ml oxygen l^{-1}. However, if the water is still (or slowly stirred) the frequency shows a marked decline (Philipson, G., 1954).

The trichopteran *Philanisus plebeius* is unusual in that its larvae live in the littoral and sublittoral zones. Both the undulatory frequency and the rate of oxygen uptake are maximal at 25° in normal

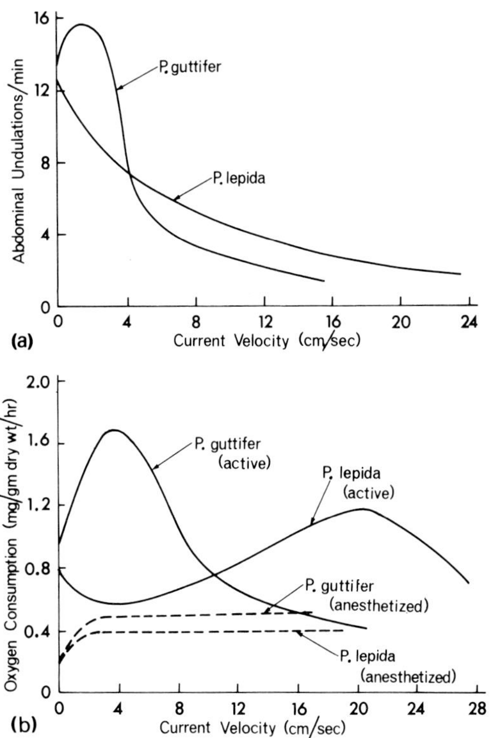

FIG. 9. (**a**) The relationship between ventilation frequency and current velocity in *Pycnopsyche guttifer* and *Pycnopsyche lepida*. (**b**) The relationship between oxygen consumption and current velocity in normal and anaesthetized larvae of *Pycnopsyche guttifer* and *Pycnopsyche lepida*. (From Mill, P., 1974; after Feldmeth, C., 1970a.)

seawater. Exposure to low environmental pO_2 causes an initial increase in ventilatory frequency but this adapts almost completely in 5–10 min. The adapted frequency increases slightly with decreases in pO_2 down to about 50% saturation and then decreases rapidly, with complete inhibition at low pO_2. This is presumably of survival value in animals which are often exposed (Leader, J., 1970).

Environmental pCO_2 over the range 0–29% saturation has no effect on the undulatory frequency of *P. grandis* (van Dam, L., 1937). Nevertheless Morgan, A. and O'Neil, H. (1931) have shown that the gills of *M. zebratum* are important sites for the removal of carbon dioxide.

The larvae of the china-mark moths *Nymphula stratiotalis* and *N. maculalis*, which live in shelters made of leaves, ventilate their shelters using rapid undulations of the anterior end of the body for short periods at regular intervals (Buckler, W., 1875;

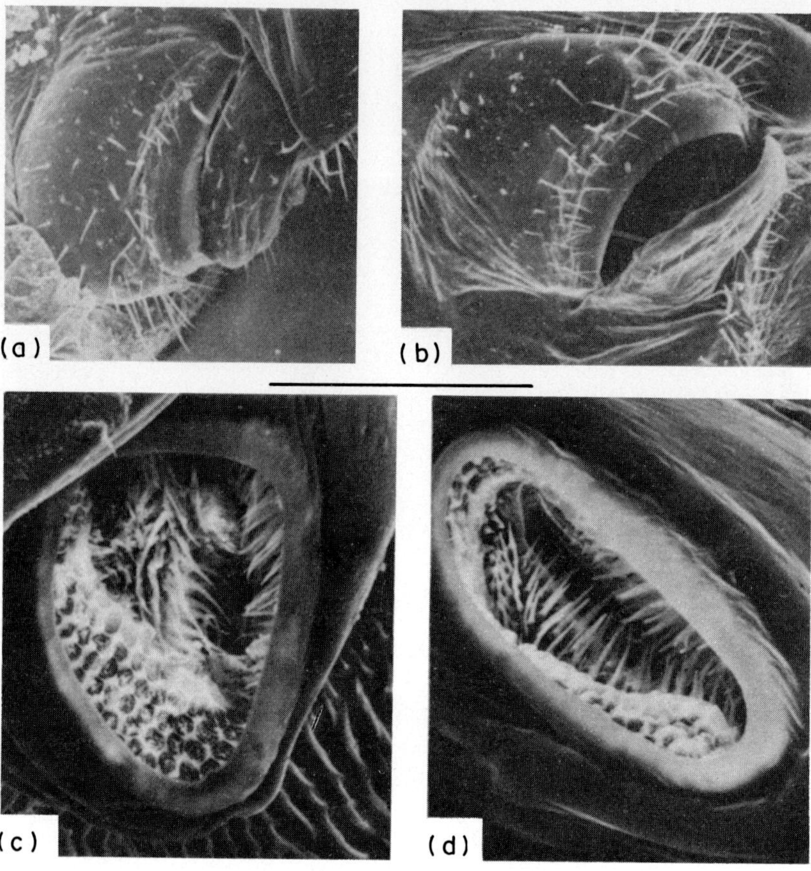

FIG. 10. Scanning electron micrographs of external views of the spiracles of *Periplaneta americana*. (**a**) Spiracle 2 with the valves closed. (**b**) Spiracle 2 with the valves open. (**c**) Spiracle 10 showing the honeycomb structure and bristles in the outer atrium. (**d**) Spiracle 7 showing the valves below the outer atrium. Scale bar: (**a**) and (**b**), 500 μm; (**c**) and (**d**), 125 μm. (From Miller, P., 1981a.)

Welch, P. and Sehon, G., 1928). In *Nymphula maculalis* the frequency of the periods of undulation decreases when a current of water is passed through the shelter, or part of the shelter is removed, and increases with decrease in pO_2 or increase in temperature, the undulations becoming continuous below a pO_2 of $1.0\,cm^3\,l^{-1}$ or above about 30.5°. The undulations cease below 4.5° (Welch, P. and Sehon, G., 1928).

2.3 Open tracheal systems

The vast majority of insects have an open tracheal system and various modifications and adaptations occur to prevent excessive water loss in terrestrial animals, particularly those living in a dry environment, and to prevent water from entering the tracheal system in aquatic species.

2.3.1 Spiracular Structure

In most large insects the spiracle comprises one or two atrial valves (or lips) which open into the atrium, and the closing mechanism operates by pulling the valves together, or one towards the other, either at their outer edges as in spiracles 1 and 2 of *Periplaneta americana* (Fig. 10a,b) or subterminally and so constricting the atrium, which can then be viewed as having outer and inner regions, as in spiracles 3–10 of *Periplaneta* (Fig. 10c,d) (Miller, P., 1981a).

The spiracle is always closed by muscular action and opened by cuticular elasticity, often aided by muscular action. For example the second spiracles in *Schistocerca*, spiracles 1–3 of *Periplaneta* and all of the spiracles in *Aeshna* have only closer muscles, whereas in *Schistocerca* each spiracle except the

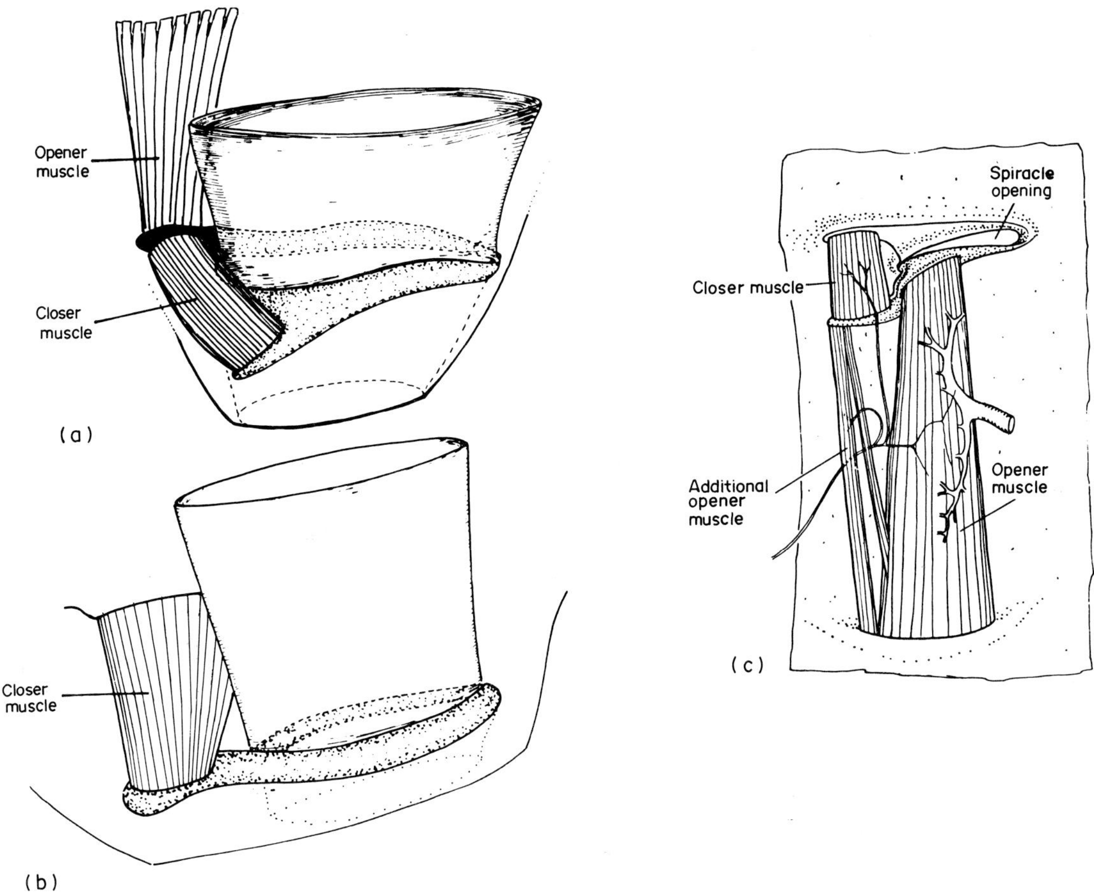

FIG. 11. Diagrams of spiracles. (**a**) Spiracle 7 of *Periplaneta americana* showing the opener (o) and closer (c) muscles. (**b**) Spiracle 3 of *Periplaneta americana*, which only has a closer muscle (c). (**c**) Spiracle 7 of *Blaberus giganteus* showing the large, divided opener muscle and the small closer muscle. (**a** and **b** from Miller, P., 1981a; **c** from Miller, P., 1974.)

second, and spiracles 4–10 of *Periplaneta* have both a closer muscle and an opener muscle. The muscles often operate directly on the valves but in spiracles 4–10 of *Periplaneta* they are attached to a sclerotized bar which is an extension of the anterior valve (Fig. 11) (Miller, P., 1974, 1981a). There are exceptions to this general rule. Thus in the abdominal spiracles of *Blaberus* which, like *Periplaneta*, have both closer and opener muscles, it is closing which is aided by cuticular elasticity, and in some blaberid cockroaches the opener muscle is divided into two parts which may represent a functional separation into tonic and phasic parts (Miller, P., 1973, 1974; Kaars, C., 1979; Nelson, M., 1979). In some insects the first spiracle is more complex. In *Periplaneta* it has a tripartite atrium from which five major tracheal trunks arise and in *Sphodromantis* there are two orifices with a separate closer muscle for the part of the valve covering each one (Miller, P., 1971a). Some insects possess an internal lever which closes the tracheae where they join the atrium. For example, in *Hyalophora* the closer muscle activates a lever which is restored to its resting position by an elastic ligament (Beckel, W., 1958). In some insects there are three or four muscles which control the valves (Kaars, C., 1979).

In many insects the spiracles are covered by interlocking hairs or bristles; others have cuticular projections and bristles in the atria; while some have a "sieve plate" overlying their spiracles (e.g. Hamilton, M., 1931; Hassan, A., 1944; Keilin, D., 1944; Imms, A., 1957). The situation is fairly simple in adult dragonflies, with each spiracle in anisopterans being covered with hairs (although these are absent in zygopterans) (Paulpandian, A., 1964). The spiracles of *Periplaneta americana* all contain a honeycomb-like structure in their atria (outer atria in spiracles 3–10). This is a hydrofuge structure, which hence may serve to resist water entry into the tracheal system, and comprises a network of elevated ribs of cuticle bordering small areas within which there are a number of small cuticular projections. In addition the outer atria of spiracles 3–10 contain bristles (Fig. 10) (Miller, P., 1981a). The sieve plate consists of a cuticular sheet containing pores varying between about 0.2 μm and about 4 μm in diameter which is supported by cuticular trabeculae (Nunome, J., 1944; Hinton, H., 1967a). Richter, P. (1969a,b) has suggested that one of the evolutionary trends within the Scarabeoidea (Coleoptera) has involved changes in the filtering system from simple atrial spines (as are also found in the Cerambycidae and Dytiscidae) to the development of a solid wall with embedded trabeculae surrounding a slit-like subatrial opening (abdominal spiracles) and the development of a porous or solid filter together with a reduction in the size of the spiracular opening in the mesothoracic spiracles. (The metathoracic spiracles are simple in structure.)

The above structures may serve to prevent dust from entering the tracheal system (Chapman, R., 1969) and sieve plates may also help reduce water loss by, for example, restricting bulk air movements (caused by body movements) through the spiracles (Miller, P., 1974). However, this could be a problem when oxygen demand is high and indeed, in those coleopterans which have spiracular filters, the spiracles which supply the flight muscles, and which therefore need any restrictions on air flow to be minimal, either lack a filter or have only a coarse one (Miller, P., 1974). Diffusion, however, is not impeded by sieve plates since this depends on the total pore perimeter rather than on pore area (Keister, M. and Buck, J., 1961). Miller, P. (1974) has suggested other possible functions, e.g. to prevent the entry of water or parasites.

The spiracles may also be protected by other parts of the body. Thus in many coleopterans and some hemipterans the elytra or wings protect the abdominal spiracles, access to which may be only via an opening at the tip of the abdomen (Scudder, G., 1963; Ahearn, G., 1970). Indeed, in the desert tenebrionid *Eleodes armata* only 2% of the water loss is via the spiracles at 25°, although this increases to 20% at 40° (Ahearn, G., 1970).

2.3.2 Spiracular adaptations of aquatic insects

The body wall is virtually impervious to water in aquatic insects which have an open tracheal system (Krogh, A., 1943; Blomquist, G. and Dillwith, J., this volume). Many insects need to make frequent visits to the surface to replenish the oxygen in their tracheal system, but others have evolved gas gills which have allowed them to become partly or completely independent of surface breathing, and a few species are able to utilize air spaces in plants.

In all cases some mechanism is required which will allow air to pass through the spiracles, while at the same time preventing the entry of water. For a long time it has been known that some dipteran larvae produce an oily secretion from peristigmatic glands and so create a dry "hydrofuge" region around each spiracle (e.g. Leydig, F., 1859; Dolley, W. and Farris, E., 1929; Keilin, D. *et al.*, 1935; Phillips, M., 1939). In other insects such as *Notonecta* and the larva of the dipteran *Hedriodiscus truquii* each spiracle is surrounded by a group of hydrofuge hairs, so that while the animal is submerged the hairs curve over the spiracle and so prevent entry of water, but when they touch the surface film they spread out because of the effect of surface tension (Fig. 12) (e.g. Brocher, F., 1909, 1913; Wigglesworth, V., 1953b; Stockner, J., 1971).

Various mechanical devices are also employed. Thus in a number of dipteran larvae belonging to the Tipulidae and Anisopodidae the spiracles open onto a post-abdominal disc which is withdrawn when the animal submerges, so that the surrounding fleshy perispiracular lobes close over it (Keilin, D., 1944). In many insect larvae the functional spiracles lie at the end of a respiratory siphon, which in most cases is at the posterior end of the abdomen (e.g. in the dipteran *culex*) (Plate 1), but may be

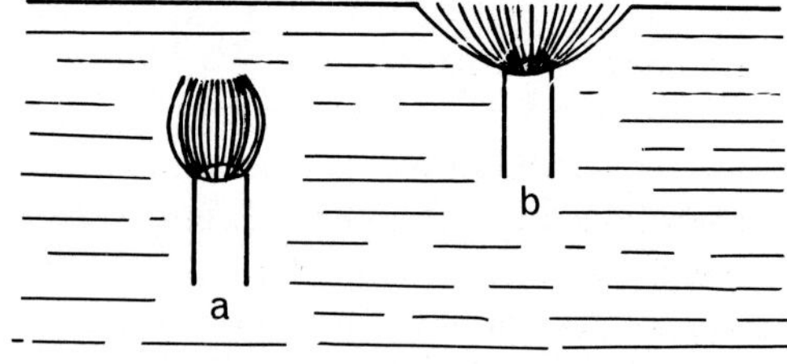

Fig. 12. Diagram to show the arrangement of hydrofuge hairs around a spiracle (**a**) under water and (**b**) at the surface. (From Mill, P., 1974; after Wigglesworth, V., 1953b.)

Table 1: Density of hair pile (hairs mm^{-2}) of gas gill

Temporary gas gills		
Type III		
Coleoptera		
Dryopoidea	6×10^2–8×10^3	Thorpe, W. and Crisp, D. (1949)
Hydrophilidae	1×10^2–1×10^5	Thorpe, W. and Crisp, D. (1949)
Permanent gas gills		
Type II		
Coleoptera		
Elmis maugei	3×10^4–1.6×10^5	Thorpe, W. and Crisp, D. (1949)
Hemiptera		
Hydrocyrius columbiae	1×10^4–2×10^4	Miller, P. (1961)
Type I		
Coleoptera		
Eubrychius (= Phytobius) velatus	1.8×10^6–2.0×10^6	Thorpe, W. and Crisp, D. (1949)
Hemiptera		
Aphelocheirus aestivalis	4.3×10^6	Hinton, H. (1976)

thoracic (e.g. in the dipterans *Notiphila* and *Mansonia*) or be formed by the antennae (e.g. in the coleopteran *Hydrophilus*). In culicid larvae the spiracles are closed over as described above (Keilin, D., 1944). In the larvae of the dipterans *Eristalis* and *Ptychoptera*, which live on the bottom, the siphon is very extensible (Réaumur, R., 1734–42) and is extended by the contraction of its circular muscles (Miall, L., 1895).

In the larvae of *Donacia simplex* (Coleoptera) and the larvae of the dipterans *Chrysogaster*, *Notiphila*, *Mansonia* and *Taeniorhynchus richiardii*, the siphon is pointed and is able to penetrate the gas spaces of plant roots (e.g. Edwards, F., 1919; Keilin, D., 1944; Varley, G., 1937; Houlihan, D., 1969). Furthermore, in *D. simplex* the larva bites a hole in a root while it is constructing a cocoon in which to pupate, and it is possible that the pupa and adult continue to receive oxygen from the plant, even after scar tissue forms over the hole (Böving, A., 1910; Ege, R., 1915b; Houlihan, D., 1970). Other insects, such as the coleopteran *Elmis* and the pupating larva of the lepidopteran *Hydrocampa*, bite into plant air spaces (e.g. Brocher, F., 1912b).

2.3.3 Temporary (compressible) gas gills

A number of aquatic insects carry a bubble of air with them when they are submerged, and it has long been considered that these bubbles are capable of extracting oxygen from the surrounding water (e.g. Comstock, J., 1887; Babák, E., 1912). In many instances the bubble slowly declines in volume and then has to be reformed by a visit to the surface. Such compressible gas gills therefore only serve as a temporary air store. They are often held by a dense pile of body hairs and/or under the elytra.

Insects which have a dense hair pile include coleopterans belonging to the Dryopoidea and Hydrophilidae (Table 1) as well as many hemipterans (Nepidae, Corixidae, Notonectidae, Pleidae, Veliidae, Naucoridae, Belostomatidae) and some trichopterans (see Brocher, F., 1912a; Ege, R., 1915a; Thorpe, W., 1950). The bubble may be so large that, in the case of *Dryops* and some hydrophilids, it covers all but the extremities of the animal. In some hydrophilids, in the hemipterans *Gerris* and *Velia*, in some trichopterans, and in the larvae of the lepidopterans *Palustra* and *Diacrisia* there are two sets of hairs (long and short) (Myers, J., 1935; Thorpe, W. and Crisp, D., 1949; Thorpe, W., 1950). The bubble held by the long hairs decreases in volume fairly rapidly, but that held by the short hairs is more permanent. Furthermore, after the bubble supported by the longer hairs has been used up, the long hairs may help to prevent the short hairs from becoming wet (Thorpe, W. and Crisp, D., 1949; Thorpe, W., 1950).

When the animal is submerged it starts to withdraw oxygen from its gas gill into its spiracles, and so the oxygen tension in the gill starts to decrease. When this falls below that in the surrounding water the difference in oxygen tension (ΔP_O) will cause oxygen to diffuse from the water into the gill, thereby augmenting the supply. However, the reduction of oxygen tension in the gill produces a difference in nitrogen tension (ΔP_N) so that nitrogen diffuses out of the gill. Furthermore, the decrease in total gill pressure caused by the removal of oxygen will cause the gill to shrink in volume since it must remain at hydrostatic pressure (Rahn, H. and Paganelli, C., 1968).

Since the gill initially contains air, the primary oxygen supply (V_{iO}) is about 0.2 of the volume of the store Ege, R. (1915a), and Rahn, H. and Paganelli, C. (1968) pointed out that the lifetime of the gill based solely on this stored oxygen (T_s) is given by:

$$T_s = \frac{V_{iO}}{q}$$

(where V_{iO} is the initial volume of oxygen and q is its rate of consumption by the animal).

However, there will be an additional period of time during which oxygen is being taken up from the water (T_w) and this period is given by:

$$T_w = \frac{V_{iN}}{q}$$

(where V_{iN} is the initial volume of nitrogen). Hence the total gill lifetime (T) is given by:

$$T = T_s + T_w$$

If nitrogen were to diffuse out of the gill at the same rate that oxygen diffuses in, then the volume of oxygen taken up would be equal to the initial volume of nitrogen (V_{iN}), which is about 0.8 of the total volume. This is approximately four times the initial volume of oxygen and hence would increase the gill lifetime by about four times. However, oxygen diffuses in faster than nitrogen diffuses out, i.e. the invasion coefficient of oxygen is greater than that of nitrogen. Hence T_w will be increased by the ratio between the invasion coefficients of oxygen and nitrogen and so the equation becomes

$$T_w = \frac{V_{iN}}{q} \cdot \frac{i_O}{i_N}$$

(where i_O and i_N are the invasion coefficients of oxygen and nitrogen respectively). Ege, R. (1915a) determined values for the invasion coefficients and obtained a value of 3.2 for their ratio. The amount of oxygen taken up from the water (V_{wO}) is given by

$$V_{wO} = V_{iN} \cdot \frac{i_O}{i_N} \times \text{gill volume}$$

(where V_{iN} is the proportion of the total initial gill volume). Hence:

$$\begin{aligned} V_{wO} &= 0.8 \times 3.2 \times \text{gill volume} \\ &= 2.56 \times \text{gill volume} \end{aligned}$$

The total volume of oxygen utilized (V_O) is given by

$$\begin{aligned} V_O &= (V_{iO} + V_{wO}) \times \text{gill volume} \\ &= (0.2 + 2.56) \times \text{gill volume} \\ &= 2.76 \times \text{gill volume} \end{aligned}$$

The ratio V_O/V_{iO} is called the "gill factor" (and is also given by the ratio between total gill lifetime (T) and the gill lifetime due to the stored oxygen alone (T_s), i.e. T/T_s (Rahn, H. and Paganelli, C., 1968)). Thus:

$$\begin{aligned} \text{Gill factor} &= \frac{V_O}{V_{iO}} \\ &= \frac{2.76 \times \text{gill volume}}{0.2 \times \text{gill volume}} = 13.8 \end{aligned}$$

Alternatively, this may be restated in terms of the contribution which the stored oxygen makes to the total amount utilized, i.e. $(0.2/2.76) \times 100 = 7\%$. Rahn, H. and Paganelli, C. (1968) estimated a value of less than 5%. Yet another way of expressing the data is in terms of the extent to which the gill lifetime is extended in comparison with that due to the stored oxygen alone, i.e. V_{wO}/V_{iO} or T_w/T_s. Thus:

$$\frac{V_{wO}}{V_{iO}} = \frac{2.56 \times \text{gill volume}}{0.2 \times \text{gill volume}} = 12.8$$

Hence the gill can extract 12.8 times as much oxygen from the water as it possesses initially in its store (see Ege, H., 1915a).

Assuming that the stored oxygen is negligible compared to the total amount used, Rahn, H. and Paganelli, C. (1968) derived the following equations from a model of a temporary gas gill

$$\Delta P_O = \frac{\Delta X}{i_O} \cdot \frac{V_{iO}}{A}$$

$$T = \frac{\Delta X}{i_N A} \cdot \frac{V_{iN}}{\Delta P_N}$$

(where ΔX is the height of the gill and A is the gill area), and hence:

$$T = \frac{V_{iN}}{q} \cdot \frac{i_O}{i_N}$$

They determined i_O and i_N from the product of the solubility and the diffusion coefficients of oxygen and nitrogen respectively. The values which they used were 4.07×10^{-5} and 2.04×10^{-5} mmHg^{-1} for the solubilities (Hodgman, C., 1958) and 2.28×10^{-5} and 2.08×10^{-5} cm^2 s^{-1} for the diffusion coefficients (Gertz, V. and Loeschcke, H., 1954) of oxygen and nitrogen, respectively, in water at 20°. This gives a ratio of 2.19 for i_O/i_N.

Since $T_s = V_{iO}/q$ and $T = (V_{iN}/q)(i_O/i_N)$ and the gill factor $= T/T_s$

$$\text{Gill factor} = \frac{V_{iN}}{V_{iO}} \cdot \frac{i_O}{i_N}$$

$$= \frac{0.8}{0.2} \times 2.19 = 8.76$$

Ege, R. (1915a) used a higher value for the invasion coefficient (i.e. 3.2) which would give a gill factor, using the above equation, of 12.8, which is slightly lower than the value (13.8) which he obtained based on volumes (see p. 534). However, the approximation used by Rahn, H. and Paganelli, C. (1968) to obtain T (i.e. $T = T_w$ to within 5%) does lead to a slight underestimation of T, and hence of the gill factor.

Just below the surface the gill is at approximately atmospheric pressure and, if the water is saturated with air, ΔP_N is equal and opposite to ΔP_O. The rate of decrease in volume in these circumstances is given by Crisp, D. (1964) as

$$\frac{dv}{dt} = Ai_O(\Delta P_O) - q + Ai_N(\Delta P_N)$$

Gill lifetime is shortened both by an increase in oxygen consumption and by increase in depth, but for different reasons. Assuming that reduction in volume is achieved by decreasing the height of the gill while maintaining a constant area, increase in oxygen consumption will increase ΔP_O and so cause a decrease in the steady state level of oxygen tension in the gill. Thus ΔP_N increases, the diffusion rate of nitrogen increases and hence the gill lifetime is shortened.

However, increase in depth has very little effect on the steady-state level of pO_2. Rather, increase in depth increases ΔP_N, causing an increase in the rate at which nitrogen diffuses out of the gill; this is augmented by a transient increase in gill pO_2 above water pO_2 (caused by the dive) and so some oxygen is lost from the gill (Rahn, H. and Paganelli, C., 1968). This is perhaps best explained by using an O_2–N_2 diagram. Thus Fig. 13a shows the relationship between pN_2 and pO_2. The 45° diagonals indicate the proportions of nitrogen and oxygen at the stated pressures. The dashed line has a slope of about 4 and indicates the proportions of nitrogen and oxygen in air. In Fig. 13b (which is an enlargement of part of Fig. 13a) point A indicates the partial pressures of nitrogen and oxygen in air at 1 atm and in air-saturated water at all depths. If the animal stays immediately below the surface its respiratory requirements will cause a reduction in gill pO_2. Since the system is equilibrated at 1 atm there will be a corresponding increase in gill pN_2 to keep the overall gill pressure constant at 1 atm. This will continue along the line AB until a steady state is reached at B. B is further to the left the higher the metabolic rate (and the smaller the area of the gill), but at all rates $\Delta P_N = \Delta P_O$. If the animal dives to a depth of 1 m after replenishing its gill at the surface, the gill pressure will be about 1.1 atm and the point A′ indicates the initial partial pressures of nitrogen and oxygen. As before the uptake of oxygen by the animal from the gill will cause a decrease in pO_2 and an increase in pN_2, this time along the line A′B′ until a steady state is reached. B′ will lie vertically above B if the metabolic requirements are the same in both cases. The above-mentioned transient loss of oxygen caused by the dive occurs because the increase in overall pressure in the gill increases the pO_2, whereas the water pO_2 is equilibrated at 1 atm (i.e. A′ is to the right of A). Hence oxygen will be lost to the water between points A′ and X (Rahn, H. and Paganelli, C., 1968). These gas gills can be destroyed by excess pressures in the region of 0.5 atm, which corresponds to a depth of about 5 m.

Ege, R. (1915a) determined values of gill pO_2 and pCO_2 at the start of a dive and at various intervals during the dive for a variety of dytiscids, notonectids and corixids, and these data are presented in Table 2. The maximum initial pCO_2 recorded was 2.8% and pCO_2 remained low during a dive. Values of the average difference in oxygen tension $(\Delta P_O)_{av}$ for *Hydrophilus* are given in Table 3.

The gas gill of the elmid beetle *Potamodytes tuberosus* consists of a large air bubble which decreases in size at low current velocities, but serves

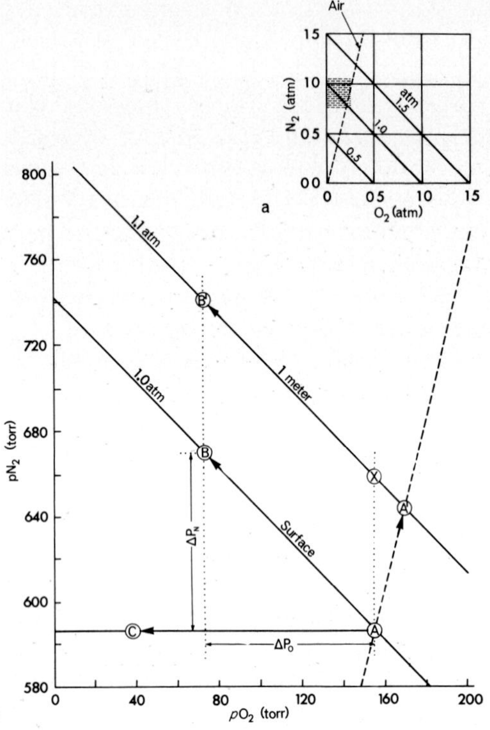

FIG. 13. O_2–N_2 diagrams. (**a**) The relationship between nitrogen and oxygen at pressures up to a total of 1.5 atm. The dashed line indicates their relative proportions in air. (**b**) An enlargement of the shaded area in (**a**). A indicates the relative proportions of nitrogen and oxygen in air at 1.0 atm. In a compressible gas gill their proportions change during a dive in the direction of B, provided the animal remains just below the water surface (i.e. at a pressure of 1.0 atm). If the animal dives to 1 m (a pressure of 1.1 atm), the initial proportions are given by A′ and they change during the dive in the direction of B′. To the right of X the gas gill loses oxygen to the water. In an incompressible gas gill the change in the proportions of the two gases is in the direction of C at all depths. (From Mill, P., 1974; after Rahn, H. and Paganelli, C., 1968.)

as a permanent gas gill in the animal's normal environmental situation of fast-flowing air-saturated water (Stride, G., 1955). This can be explained in

*Table 2: The oxygen tension (pO_2) in gas gills during a dive**

Air	pO_2 20.36%	
At start of dive	pO_2 16.0–19.6%	
Species	**Time after diving (min)**	**pO_2 (%)**
Dytiscidae		
Colymbetes fuscus	1	15.4
Acilius sp.	2	11.9
Colymbetes sp.	5	9.2
Dytiscus marginalis	5	3.0
Ilybius sp.	20	0.7
Notonectidae		
Notonecta sp.		
Dorsal space	1	11.0
	2	5.3
Canals (ventral surface)	1	15.4
	2	3.9
Corixidae		
Corixa geoffroyi		
Passive	10	9.4
	15	5.8
Active	5	6.5

*Data from Ege, R. (1915a)

terms of the Bernouilli effect: $p + \frac{1}{2}\rho V^2 = \text{constant}$ (where p = pressure, V = velocity, and ρ is a gravitational constant). Thus, as the water is deflected over the animal's body and hence increases in velocity there is an accompanying decrease in pressure. The water will thus become supersaturated and give up oxygen and nitrogen to the gas gill. Indeed, the size of the bubble increases with current velocity (Stride, G., 1958).

2.3.4 PERMANENT (INCOMPRESSIBLE) GAS GILLS

Permanent gas gills are called plastrons (Thorpe, W. and Crisp, D., 1947a; cf. Brocher, F., 1912a,b) and are held in place either by hairs or by cuticular modifications. In the case of the hair plastron, the hairs are more dense and regular than they are in

Table 3: Average differences in oxygen tension $(\Delta P_O)_{av}$ (atm)

Temporary gas gills			
Hydrophilus	0.12–0.17	inactive	Thorpe, W. and Crisp, D. (1949)
Permanent gas gills			
Hair plastrons			
Macroplea (= *Haemonia*)	0.108	inactive	Thorpe, W. and Crisp D. (1949)
Aphelocheirus	0.0055	inactive	Thorpe, W. and Crisp, D. (1949)
Spiracular gill plastrons			
Simulium ornatum	0.08–0.16	active adult	Hinton, H. (1968)
	0.03–0.06	active young adult	Hinton, H. (1968)

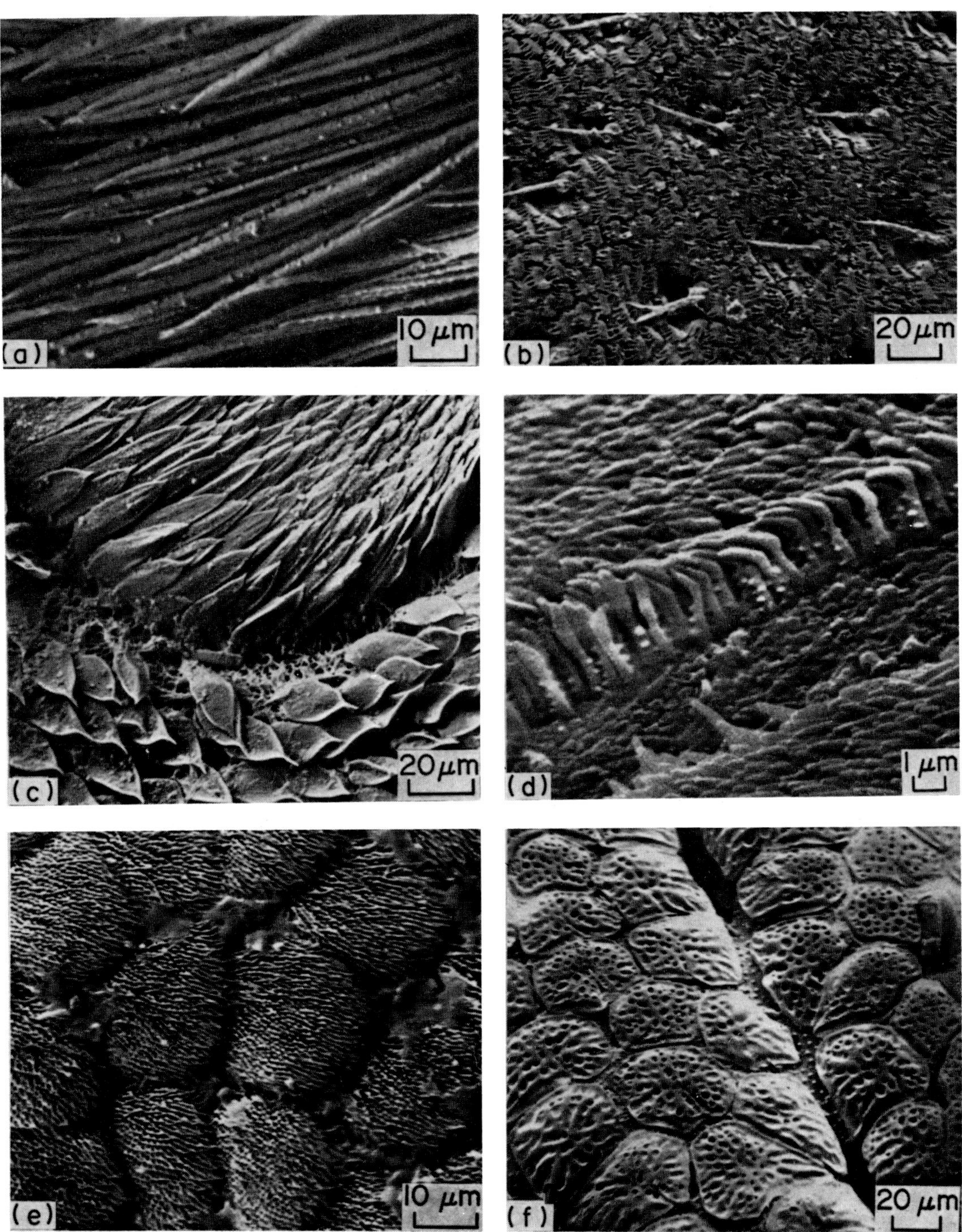

FIG. 14. Scanning electron micrographs of plastrons of (**a**) *Limnius volkmari*, propleura, (**b**) *Stegoelmis* sp., side of metasternum, (**c**) *Cryphocricos mexicanus*, organ of first abdominal segment, (**d**) *Aphelocheirus aestivalis*, first abdominal tergite, (**e**) *Eubrychius velatus*, first abdominal sternite and (**f**) *Neochitina eichhorinae*, elytra. (From Hinton, H., 1976.)

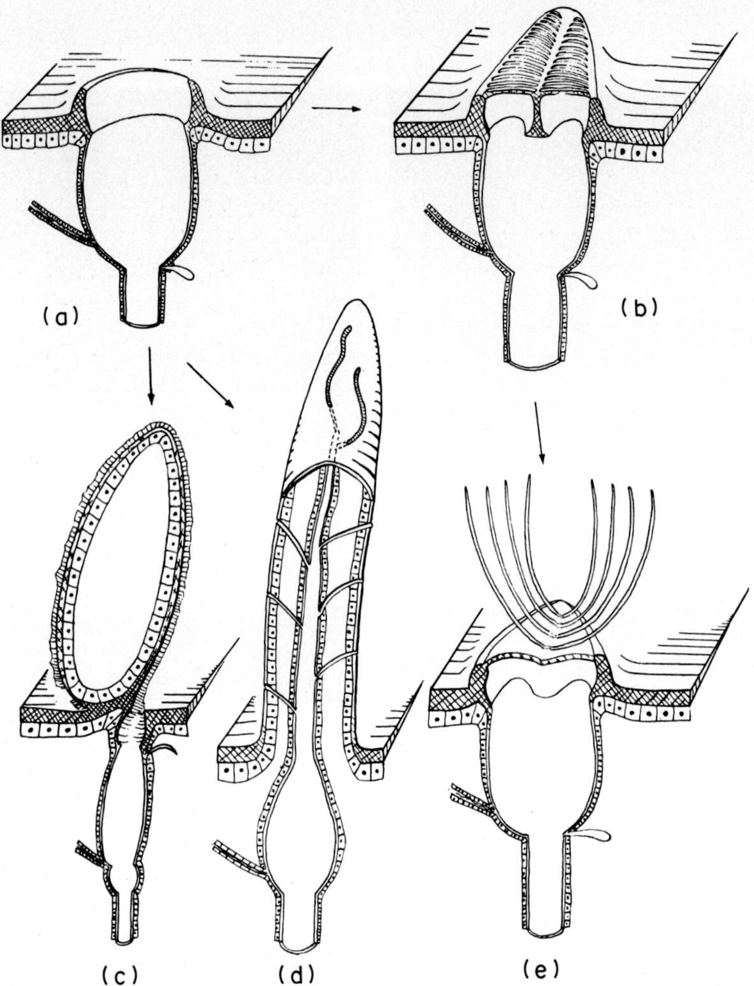

FIG. 15. Diagrams to show the origin of different types of spiracular gills. The simulid (**c**) and tipulid (**d**) types of spiracular gill could have originated from a spiracle as in (**a**). A possible stage (**e**) in the development of the type of spiracular gill found in *Psephenoides*, which could have originated from a spiracle as in (**b**). (From Hinton, H., 1966b.)

temporary gas gills (Table 1). They have been divided into two groups (I and II) by Thorpe, W. and Crisp, D. (1949) and Thorpe, W. (1950), depending on the density of the hairs (Table 1).

The aquatic elmids *Elmis maugei* and *Riolus capreus* belong to group II. The plastron in these species covers most of the lateral and lateroventral surfaces of the animal and the spiracles open into a lateral groove on either side of the body which communicates with the plastron and the subelytral space. There are two layers of hairs, a short dense layer (microplastron) and a less dense layer of longer hairs (macroplastron). As in hydrophilids (which only have temporary gas gills) the air film held by the longer hairs is a temporary store, but the microplastron is normally sufficient on its own (Thorpe, W., 1950). In *Elmis maugei* and *Limnius volkmari* the hairs are lanceolate (Fig. 14a), but in some elmids (e.g. *Stegoelmis*) the short hairs are scale-like (plastron scales) and bear projections (Fig. 14b) (Hinton, H., 1976). The belostomatid water-bug *Hydrocyrius columbiae* also has hairs of two different lengths. The longer ones end in flattened blades and lie over the short hairs. Although the latter are no more dense than in the temporary gas gills of some hydrophilids, they are regular and are bent over at their tips. Again, the plastron communicates with a subelytral bubble. Nevertheless, the gas gill of this animal is somewhat intermediate in function and can be replenished via a posterior retractile siphon (Miller, P., 1961). The beetle *Haemonia mutica* (Donaciinae) also has hairs which are bent over at their tips (Thorpe, W. and Crisp, D., 1949), while in the naucorid hemipterans belonging to the genus *Cryphocricos* there are longer, leaf-like hairs together with shorter, much finer hairs (Fig. 14c) (Parsons, M. and Hewson, R., 1974; Hinton, H., 1976).

The animals belonging to group I have the most efficient hair plastrons; these include some adult coleopterans, *Aphelocheirus aestivalis* (Hemiptera: Naucoridae) (Fig. 14d) and *Acentropus niveus* (Lepidoptera). The plastron extends over much of the body surface and comprises a dense mat of very short hairs, each of which is bent over at the tip (e.g. Larsen, O., 1924; Szabo-Patay, J., 1924; Thorpe, W. and Crisp, D., 1947a). There is no subelytral bubble and hence the animal is heavier than water (Thorpe, W. and Crisp, D., 1947c).

Weevils typically possess plastron scales, which may be covered in hairs, as in *Eubrychius* (= *Phytobius*) *velatus* (Fig. 14e) which has a plastron belonging to group I (Thorpe, W. and Crisp, D., 1949), but often form more rigid, perforate structures (e.g. *Neochitina*) (Fig. 14f) (Hinton, H., 1976).

In the larvae and/or pupae of some coleopterans and dipterans the plastron occurs on a spiracular gill, which is an outgrowth of the spiracular atrium and/or of the region surrounding it (Hinton, H., 1966b, 1968a). The gill may be a single structure, as in simulids, tipulids and various coleopterans, or consist of a number of tubular outgrowths, as in the coleopteran *Psepheriodes* (Fig. 15) (Hinton, H., 1966b, 1967c). In the tipulid *Orimargula hintoni* the plastron consists of numerous evenly spaced tubercles (Fig. 16a) but in some other species of *Orimargula* and in *Antocha* and *Lipsothrix* the plastron comprises a series of grooves (plastron lines) roofed over by a flat (*Lipsothrix*) (Fig. 16b) or arched (*Orimargula* and *Antocha*) (Figs 17 and 18) network of cuticular struts (Hinton, H., 1966a, 1967a, 1968a). In all other dipterans and in the Torridinicolidae (Coleoptera) the plastron is formed from vertical struts, bearing slender side branches as in *Aphrosylus*, which apically are either fused with several other struts, e.g. *Geranomya* (Fig. 19a,b), or divided into more or less horizontal branches which are fused with the corresponding branches of adjacent struts, e.g. *Dicranomyia* and *Aphrosylus* (Fig. 18c–f) (Hinton, H., 1967b,c, 1968a).

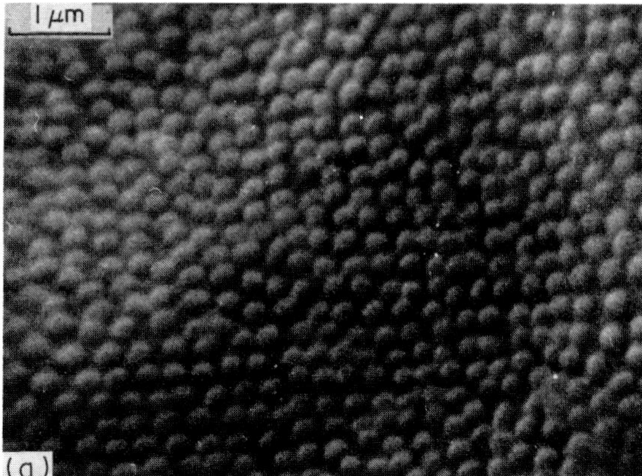

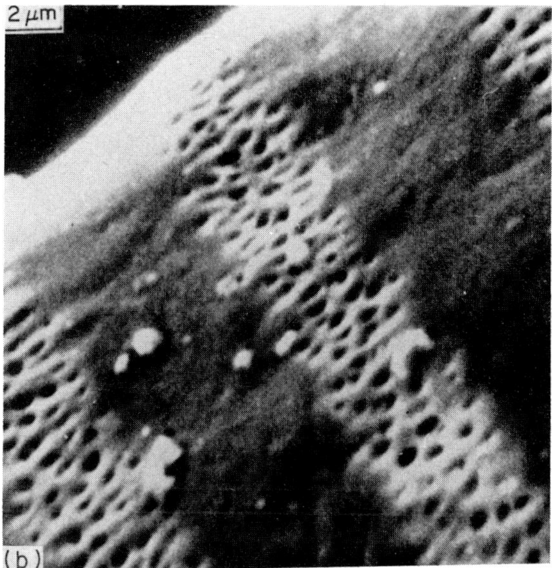

FIG. 16. Scanning electron micrographs of (**a**) the plastron on one of the spiracular gill branches of *Orimargula hintoni* and (**b**) the plastron lines on the spiracular gill of the tipulid *Lipsothrix remota*. (**a** from Hinton, H., 1968a; **b** from Hinton, H., 1967a.)

A number of insects have a plastron in their egg-shells, thereby enabling respiration to continue in both wet and dry situations; others simply have a system of openings (aeropyles) through which the respiratory gases enter and leave the egg. These may be scattered evenly over the surface or arranged in bands (e.g. some moths and beetles), clustered in a group at the front end (many hemipterans) or open

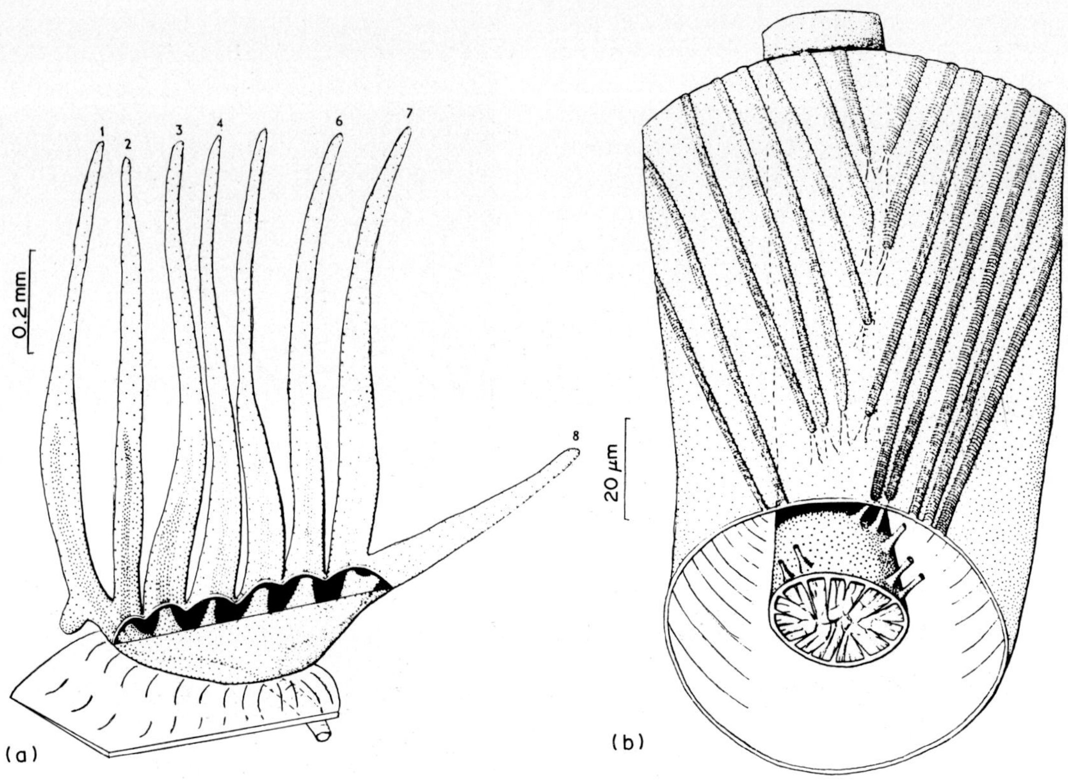

FIG. 17. (**a**) Diagram of the right spiracular gill of *Antocha bifida*. A section has been cut out of the base to show the branches of the spiracular atrium extending into the gill branches. (**b**) Diagram of part of the second branch of the spiracular gill of *Antocha bifida* near to its base. Note the plastron lines and their connections with the spiracular atrium. (From Hinton, H., 1966a.)

on the crest of longitudinal ridges (e.g. butterflies). In some cases the aeropyles are situated on the ends of stalks and in some hemipterans these stalks are more than half the length of the egg. The eggs of the lepidopteran *Nymphalis* are laid in masses, and when it rains a bubble of air is trapped over the front end of the egg mass and functions as a temporary gill (Hinton, H., 1970).

Where an egg plastron occurs it consists of modifications to the structure of the chorion (Margaritis, L., vol. 1). Essentially a layer of air is held by a system, of spaces formed by branching in the chorion, with openings (aeropyles) to the exterior (Hinton, H., 1960a,b, 1962, 1967d, 1968b; Telfer, W. and Smith, D., 1970; Smith, D. *et al.*, 1971). The spaces in the plastron network may fill with gas as the egg dries after being laid, but in some insects the liquid is removed from the spaces and they become gas-filled while the egg is still immersed in the fluid in the oviduct (Wigglesworth, V. and Beament, J., 1950; Hinton, H., 1970).

In muscids (e.g. *Calliphora*) the plastron is quite complex with an inner layer communicating via aeropyles with an outer layer, which in turn communicates with the exterior via pores in a superficial anastomosing plastron network (Fig. 20) (Hinton, H., 1963). The network may cover most or all of the surface of the egg, e.g. *Musca domestica*, in which case there is a serious problem of water loss in dry conditions. This is partly overcome in some such cases by the presence of a membrane under the shell, which only has small permeable areas and hence funnels the respiratory gases. Otherwise, as in *Musca*, the eggs cannot withstand being dried. In other cases the plastron may be restricted to a specific area, e.g. *Calliphora*, (Wigglesworth, V. and Beament, J., 1950) or may consist of discrete plastron craters scattered over the surface (Hinton, H., 1970). A similar level of complexity is found in the eggshell of the tettigonid *Homorocoryphus nitidulus vicinus*. However, the structure differs from that in dipterans, and Hartley, J. (1971) has

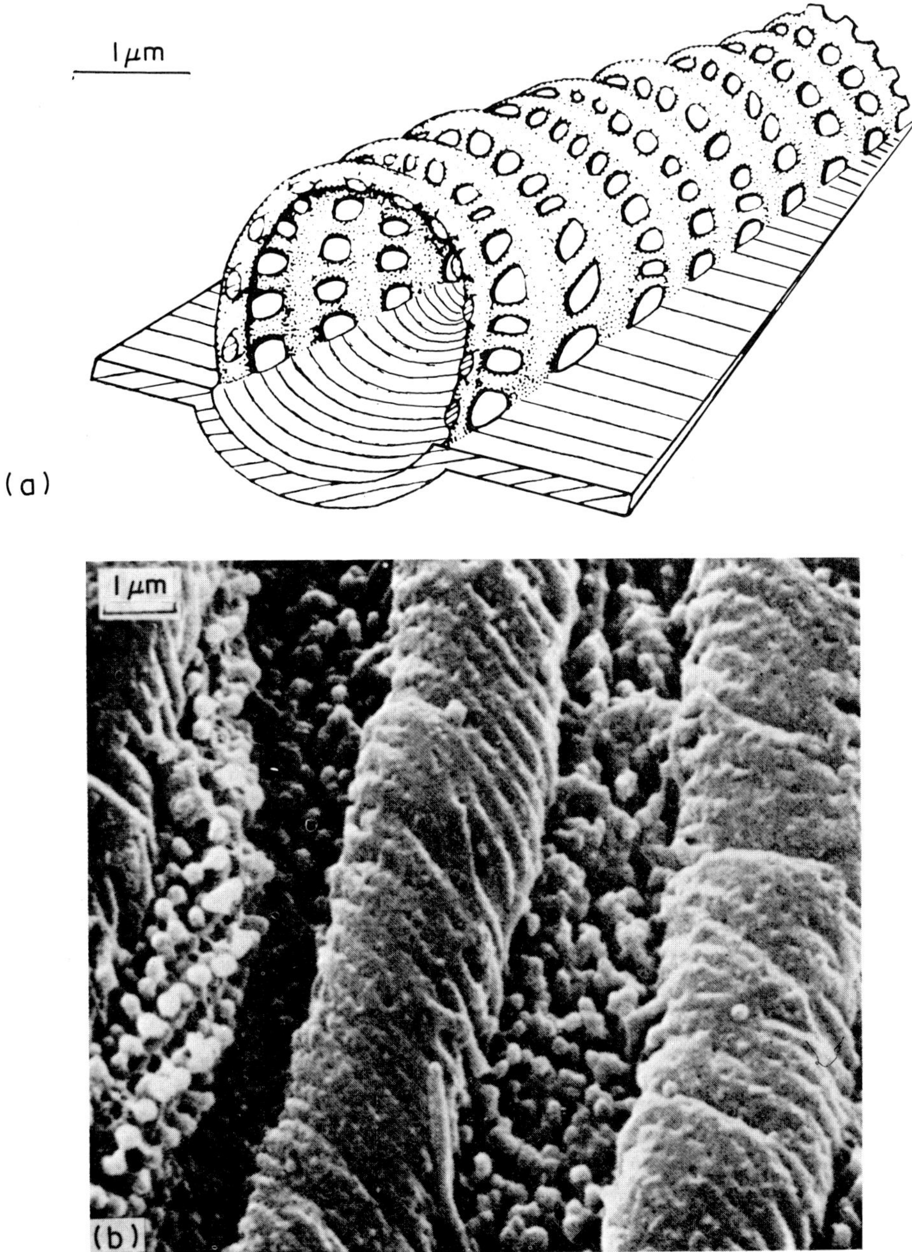

FIG. 18. (**a**) Diagram of a plastron line on the spiracular gill of *Antocha bifida*. (**b**) Scanning electron micrograph of plastron lines on the spiracular gill of *Antocha bifida*. (**a** from Hinton, H., 1966a; **b** from Hinton, H., 1968a, after Hinton, H., 1966a.)

suggested that the network in this species may be more important for water retention than for providing a high respiratory efficiency under water.

Alternatively, the eggs of some dipterans, hemipterans and hymenopterans have evolved one or more respiratory horns, the distal regions of which bear a plastron (Figs 21 and 22) (Wigglesworth, V. and Beament, J., 1950; Hinton, H., 1960c, 1961, 1969, 1970). Although in some insects, e.g. *Rhodnius*, the structure is probably insufficiently developed, in the staphylinid beetle *Oxypus* it has been calculated that, with an air–water interface of

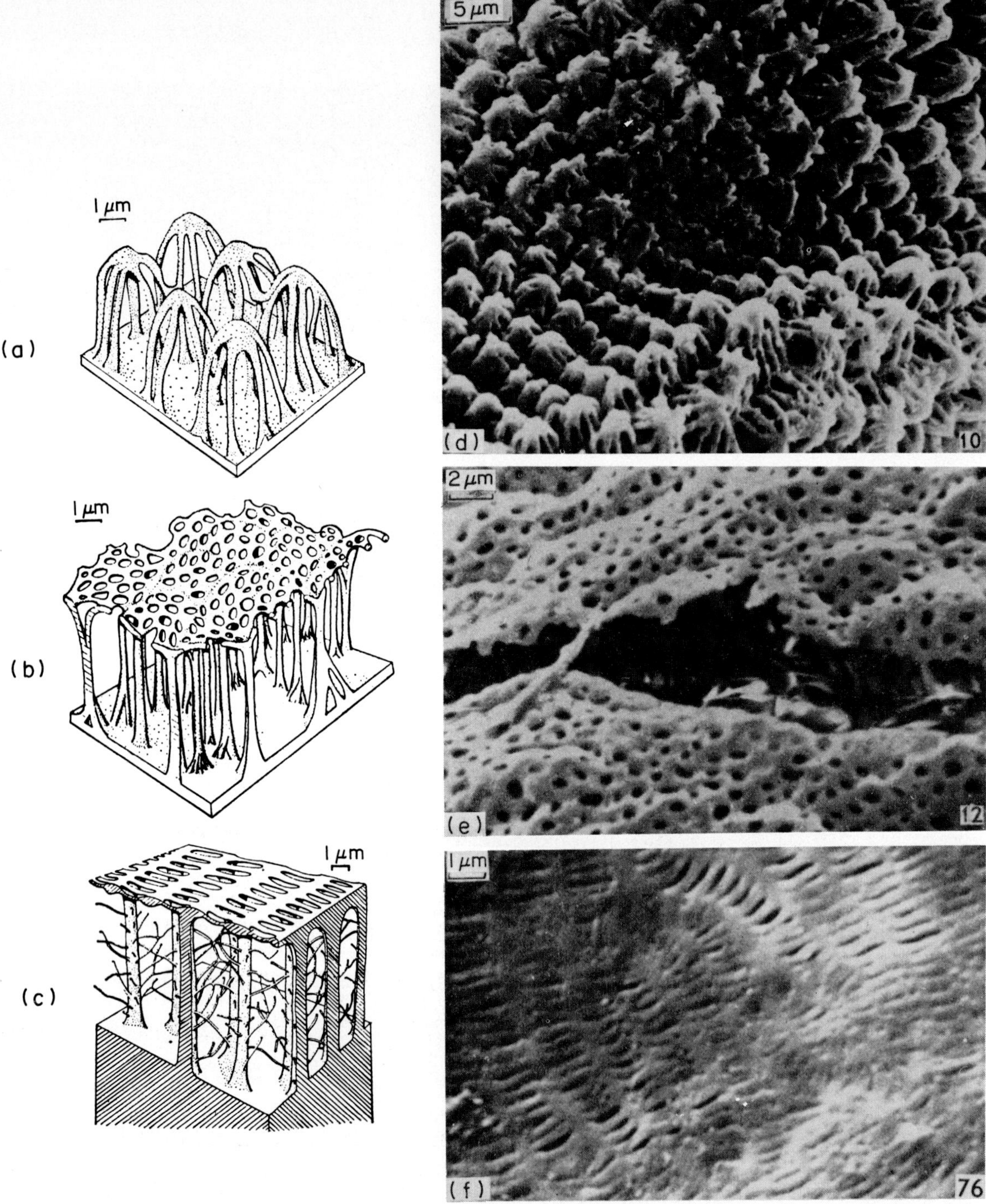

FIG. 19. Diagrams and scanning electron micrographs of the plastrons of (**a, d**) the tipulid *Geranomyia unicolor*, (**b, e**) the tipulid *Dicranomyia trifilamentosa* and (**c, f**) *Aphrosylus celtiber*. (**a, b** and **d–f**) and **c** from Hinton, H., 1968a; **e** from Hinton, H., 1967b.)

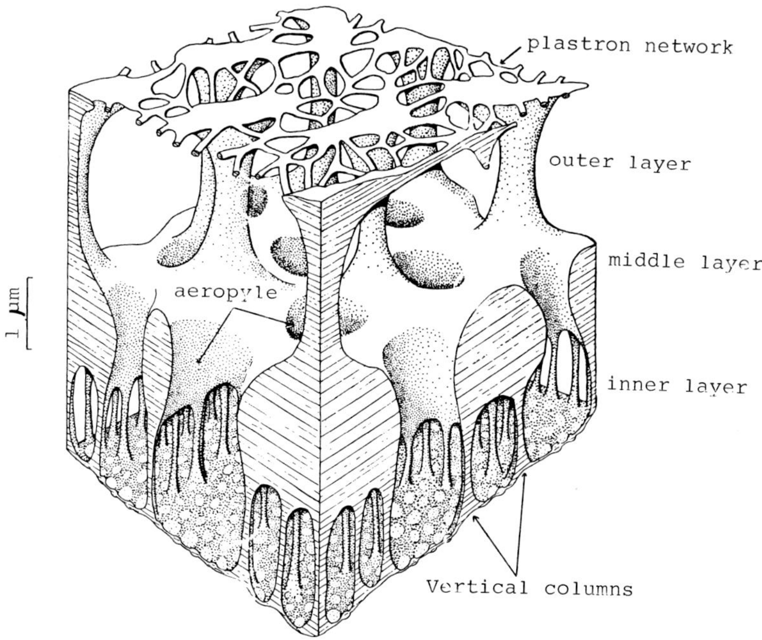

FIG. 20. Diagram of the median plastron of the egg of *Calliphora erythrocephala*. (From Hinton, H., 1963.)

3×10^3–4×10^3 $\mu m^2\,mg^{-1}$ tissue, it can support 25% of the normal metabolic requirements when the egg is submerged, and in the eggs of some other species, such as the lepidopteran *Antheraea* which has an interface of 4×10^5 $\mu m^2\,mg^{-1}$ tissue, it compares favourably with the most efficient plastrons of adult insects. Where a respiratory horn has developed the rest of the surface tends to be impermeable, and hence the amount of water loss under dry conditions depends on the cross-sectional area of the base of the horn (Hinton, H., 1970). Respiratory horns float and hence remain in contact with air if the egg is only submerged in shallow water.

As with temporary gas gills, when oxygen is utilized by the animal, gill pO_2 decreases and a ΔP_O is established. However, plastrons are incompressible and hence gill pN_2 must remain constant, thereby preventing nitrogen from diffusing out. Thus, in Fig. 13, the change in the proportions of nitrogen and oxygen is along the line AC, C being further to the left at higher metabolic rates and/or with gills of smaller area. Gill pO_2 can thus vary between pO_2 in air at 1 atm (i.e. 155 mmHg or 0.20 atm) and zero. At the latter level gill pN_2 (i.e. 595 mmHg or 0.78 atm) equals the total gas pressure of the gill (Rahn, H. and Paganelli, C., 1968), which in *Coxelmis novemnotata* is well within its safety limits (Davis, C., 1942).

A measure of the respiratory efficiency of a gas gill can be obtained from the respiratory index, $1/q_b R$, where q_b is the basal metabolic rate and R is the diffusion resistance of oxygen along a closed path and is defined as the ratio of ΔP_O to oxygen uptake (q) (Thorpe, W. and Crisp, D., 1947b), i.e.:

$$R = \frac{\Delta P_O}{q}$$

In adult *Aphelocheirus aestivalis* diffusion of oxygen through the cuticle is only just sufficient for its basal metabolic requirements and hence an active animal must utilize its plastron (Thorpe, W. and Crisp, D., 1947b). Values for average differences in oxygen tension $(\Delta P_O)_{av}$ are given in Table 3.

The plastrons are quite extensive and the question arises as to whether the whole of the plastron is functional, i.e. whether ΔP_O is significant at the extremities of the plastron as well as near the spiracles. The average difference in oxygen tension across the water–air interface $(\Delta P_O)_{av}$ is given by

$$(\Delta P_O)_{av} = \frac{q}{Ai_O}$$

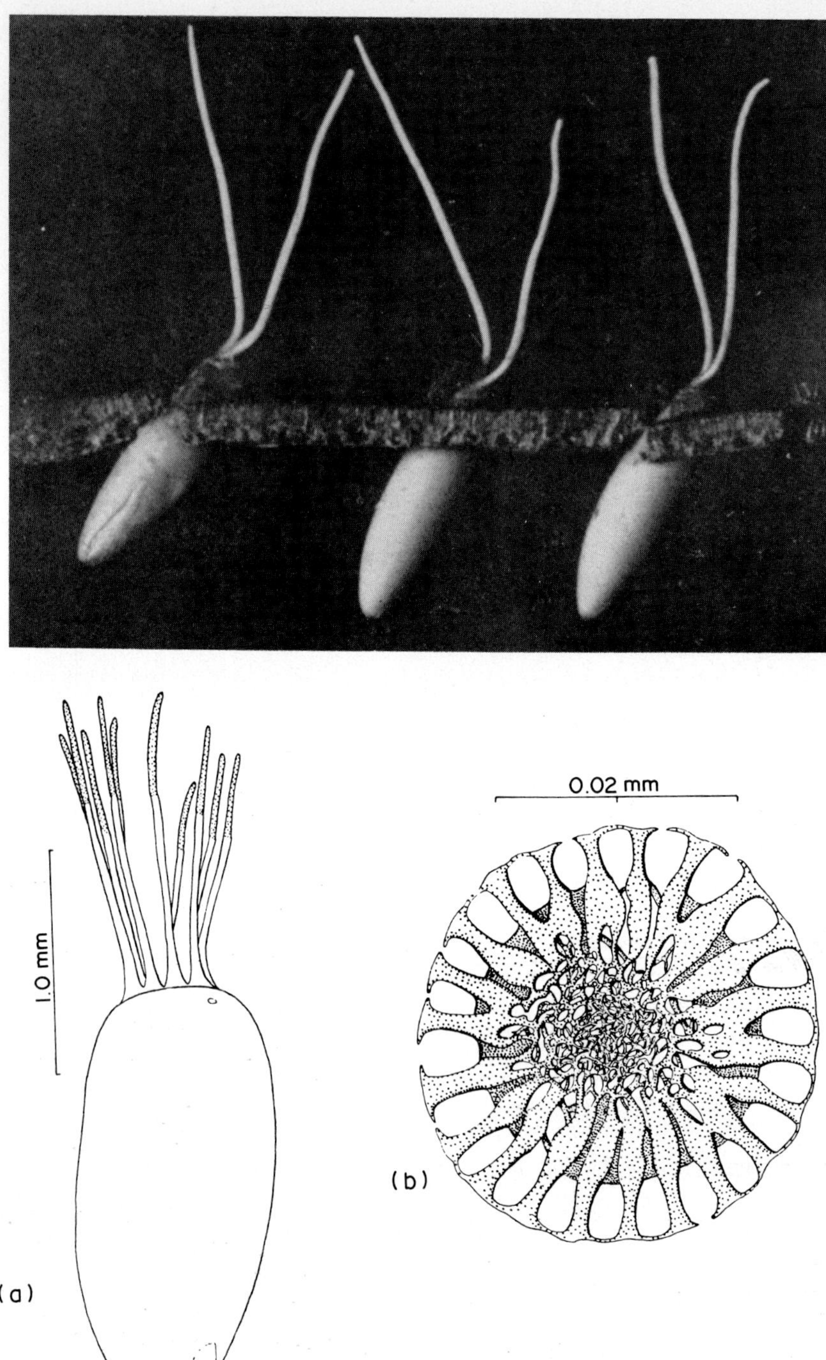

FIG. 21. Photograph of eggs of *Ranatra linearis* laid in a leaf of *Nymphaea alba*, viewed from the side and showing the respiratory horns. (From Hinton, H., 1961.) FIG. 22. Diagrams of (**a**) an egg of *Nepa cinerea* to show the respiratory horns and the extent of the plastron (stippled) and (**b**) a transverse section through the plastron-bearing distal region of a respiratory horn of the egg of *Nepa cinerea* to show the structure of the plastron. (**a** from Hinton, H., 1961; **b** from Hinton, H., 1966a.)

The actual drop in pressure at distance x from the spiracle $(\Delta P_O)_x$ is given by

$$(\Delta P_O)_x = (\Delta P_O)_{av} \frac{nx_1 \cosh n(x_1 - x)}{\sinh nx_1}$$

where x_1 is the maximum distance from the spiracle (Fig. 23) and $n = (i_O/Dh)^{\frac{1}{2}}$ (where D is the diffusion constant of oxygen within the plastron and h is the height of the plastron). $(\Delta P_O)_x$ is maximal at the spiracle, where $x = 0$, i.e.:

$$(\Delta P_O)_{x=0} = (\Delta P_O)_{av} \frac{nx_1}{\tanh nx_1}$$

and minimal at the outer limit of the plastron, where $x = x_1$, i.e.:

$$(\Delta P_O)_{x=x_1} = (\Delta P_O)_{av} \frac{nx_1}{\sinh nx_1}$$

Thus the efficiency of the plastron, in terms of the uniformity of oxygen uptake over its surface, is dependent on nx_1:

$$nx_1 = \left(\frac{i_O x_1^2}{Dh}\right)^{\frac{1}{2}}$$

(Thorpe, W. and Crisp, D., 1947b; Crisp, D. and Thorpe, W., 1948; Crisp, D., 1964) or, in the case of tubular spiracular gills, where h is replaced by $r/2$ (where r is the radius)

$$nx_1 = \left(\frac{2i_O x_1^2}{Dr}\right)^{\frac{1}{2}}$$

(Crisp, D., 1964; Hinton, H., 1968a).

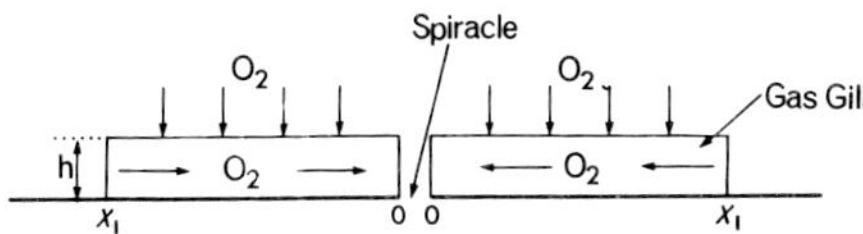

FIG. 23. Diagrammatic view of a flat plastron to illustrate the diffusion pathways of oxygen. h, height of plastron; x_1, maximum extent of plastron. (From Mill, P., 1974, after Crisp, D. and Thorpe, W., 1948.)

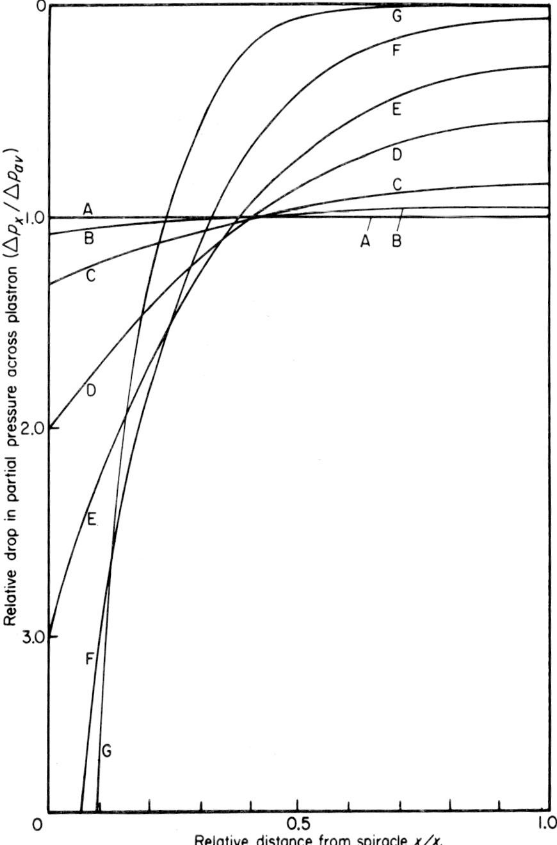

FIG. 24. The relationship between the relative drop in oxygen tension across the plastron interface and the relative distance from the spiracle. Each curve is derived from $(\Delta P_O)_x = (\Delta P_O)_{av} \cosh n(x_1 - x)/\sinh nx_1$. $(\Delta P_O)_x$, actual drop in oxygen tension at distance x from the spiracle; $(\Delta P_O)_{av}$, average drop in oxygen tension; x_1, maximum extent of plastron; n, $(i_O/Dh)^{\frac{1}{2}}$; i_O, the invasion coefficient of oxygen; D, the diffusion constant of oxygen within the plastron; h, the height of the plastron. nx_1 is a measure of the functional efficiency of the plastron. The curves are derived for various values of nx_1, i.e. 0.1 (A), 0.5 (B), 1.0 (C), 2.0 (D), 3.0 (E), 5.0 (F) and 10.0 (G). (From Crisp, D., 1964.)

Figure 24 shows the relative drop in pO_2 across the plastron $[(\Delta P_O)_x/(\Delta P_O)_{av}]$ plotted against relative distance from the spiracle (x/x_1) for different values of nx_1. The lower the value of nx_1, the greater is the efficiency of the plastron towards its periphery. Some values of nx_1 are given in Table 4 and all are less than 1. Hence, the principal diffusion barriers are at the water–air interface and between the tracheoles and the tissues (Thorpe, W. and Crisp, D., 1947b).

The hair plastron of *Haemonia* collapses at an excess pressure of 0.5–2.0 atm, which corresponds to a depth of 5–20 m, while *Hydrocyrius* can withstand an excess of 2.0 atm for at least 5 min, although some flattening of the hairs occurs (Miller, P., 1961), and the structurally more efficient one of *Aphelocheirus* requires an excess of at least 3 atm (Thorpe, W. and Crisp, D., 1947a; Thorpe, W.,

1950; Hinton, H., 1976). However, Hinton, H., (1968a, 1976) has suggested that wetting occurs before the plastron collapses in *Aphelocheirus*, and in spiracular gill plastrons wetting occurs well before any structural damage. He has also demonstrated that the length of exposure time to excess pressure is important. Thus, in *Geranomyia unicolor* wetting occurs immediately at 1.4 atm, takes 10 min at 1.0 atm and over 2 h at 0.3 atm. Wetting may also occur at low surface tensions, but again there is a high safety margin. Thus in *Aphelocheirus*, *Antocha* and *Simulium* wetting does not occur until the surface tension falls to 25–26 dynes cm^{-1}, whereas the surface tension in their normal environment generally exceeds 70 dynes cm^{-1} (Thorpe, W. and Crisp, C., 1947a).

Most insects which possess a plastron live in well-oxygenated water (e.g. fast currents, the edges of lakes and the intertidal zone). This type of environment is particularly suitable for such animals since at low environmental pO_2 the plastron will work in reverse and extract oxygen from the tissues. They do not need to leave the water to obtain oxygen and indeed plastron-bearing coleopterans which live in this habitat are unable to swim. A number of species experience large fluctuations in the water level of their environment and may be left exposed for considerable lengths of time but, since plastrons are rigid, they allow oxygen to diffuse inwards while helping to prevent water loss. However, those hydrophilid beetles which possess plastrons, and many of the plastron-bearing weevils, live in stagnant conditions and are able to swim. This is an advantage since they need to leave the water if the pO_2 falls too low (Hinton, H., 1968a, 1976).

3 VENTILATION

Ventilatory mechanisms serve to renew the respiratory medium, and in aquatic insects which possess gills the ventilatory current reduces the thickness of the boundary layer (section 2.2.1).

Table 4: Values of the function nx_1 *(see text and Figs 23, 24)*

Hair plastrons		
Macroplea (= Haemonia)	0.47	Thorpe, W. and Crisp, D. (1949)
Spiracular gill plastrons		
Antocha bifida	0.19	Hinton, H. (1968)
Simulium ornatum	0.97	Hinton, H. (1968)

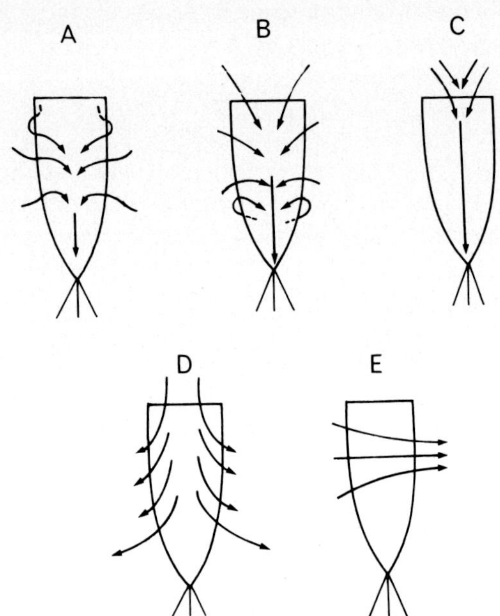

FIG. 25. Diagram showing the direction of the ventilatory currents flowing over the abdomen of (**A**) *Ecdyonurus dispar*, (**B**) *Leptophlebia marginata*, (**C**) *Ephemera vulgata*, (**D**) *Cloëon dipterum* and (**E**) *Caenis horaria*. (From Eastham, L., 1932.)

3.1 External gill ventilation

In ephemeropteran larvae the gills produce ventilatory currents (Fig. 25) and show an anterior–posterior metachronal rhythm with, in *Leptophlebia marginata* and *Ephemera vulgata*, an intersegmental phase lag of about one-eighth of an oscillation (i.e. 45°). In *Ecdyonurus dispar*, *Leptophlebia marginata* and *Ephemera danica* the flow of water is from the sides and/or below the animal towards the dorsal surface and then posteriorly along the dorsal midline (Fig. 25a–c); in each of these species the second to the sixth gills produce this current (Eastham, L., 1936, 1937, 1939). In *E. dispar* and *L. marginata* the gills move in an elliptical path to produce the current. In the former the lamella of each gill moves backward and upward with the convex posterior surface leading and gradually increasing its angle to the direction of movement, thus producing a backward and upward current (Fig. 26a). During the return stroke, which is forward and downward, the gill is feathered.

The current is augmented by the delay between the movement of successive pairs of lamellae, whereby the lower border of each lamella comes

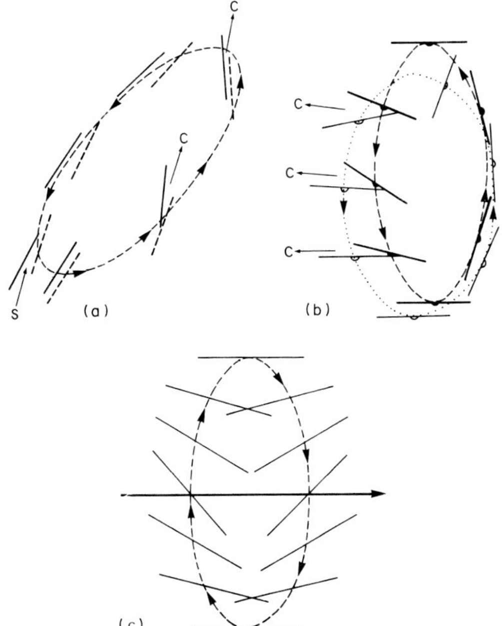

FIG. 26. Diagrams showing the paths of gills of various ephemeropterans. (**a**) The paths of two adjacent left gills of *Ecdyonurus dispar* viewed from behind. The solid line represents the anterior of the two. (**b**) The paths of the two lamellae of the second pair of gills of *Leptophlebia marginata* viewed from behind. The solid line represents the anterior of the two. (**c**) The path of any gill of *Caenis horaria* viewed from behind with the ventilatory current flowing from left to right. C, compression; S, suction. (From Mill, P., 1974; **a** after Eastham, L., 1937; **b** after Eastham, L., 1936; **c** after Eastham, L., 1934.)

into contact with the surface of the next posterior one near the start of the effective upward stroke. As this proceeds the angle between the lamellae steadily decreases and so water is forced out from between them in an upwards direction. Conversely, during the latter part of the downward movement water is drawn between them from below and the side (Eastham, L., 1937). In *L. marginata* the effective stroke is backward and inward, so producing a posteromedian current; the lamellae are again feathered on the return stroke. In this species each gill has two lamellae. The posterior one does not reach as far forward as the anterior one (Fig. 26b). Hence, during the effective stroke the anterior lamella meets the posterior one at an angle which is progressively reduced during the effective stroke, thus squeezing water inwards. Furthermore, successively more posterior gills are held at lower angles to the abdomen and hence water is drawn from dorsolaterally at the anterior end of the abdomen and from ventrolaterally at the posterior end (Eastham, L., 1936). The action in *E. danica* is rather different. The gills in this species are lanceolate and fringed and are held dorsally over the abdomen when at rest. During ventilation they move downward and inward and then upward and outward, but with the gills held at an appreciable angle to their direction of movement during both phases. The current is produced during the downward stroke by an undulating wave which passes from the base to the tip of each gill and another which passes posteriorly across the gill surface. The second pair of gills are twisted spirally so that water can be drawn inwards from the sides, but the degree of overlap between the gill processes otherwise prevents water from passing between them. The effect of this is that water is only drawn into the dorsal mid-line from the anterolateral region of the abdomen (Fig. 25c) (Eastham, L., 1939).

In *Cloëon dipterum* the water is drawn posteriorly along the dorsal mid-line but passes outwards and upwards between the gill lamellae (Fig. 25d) and, as in *E. dispar* and *E. vulgata*, the phase lag enhances the backwardly directed ventilatory current by alternately producing suction and compression (Eastham, L., 1932).

Caenis horaria lives partly buried in mud, and the second pair of gills are greatly enlarged to form a protective branchial chamber which partly encloses the third to sixth pairs. The second pair are raised to an angle of 30–40° during ventilation. The other gills produce a laterally directed current (Fig. 25). They move in an elliptical path, and during the effective downward and backward stroke the marginal filaments of each lamella lie straight and close together so that water cannot pass between them, but lag behind the lamella during the upward and forward stroke. The lamella presents an angle to its direction of movement during both phases, but swivels at the end of each stroke so that it faces in opposite directions, thereby producing a screw effect (Fig. 26c). The intersegmental phase lag in this species is about one-third of an oscillation (i.e. 120°). Also the members of each pair have a phase difference of similar magnitude. If the current is from left to right the left gill of each pair leads the right one, and the underside of each lamella faces the right on the downstroke and the left on the

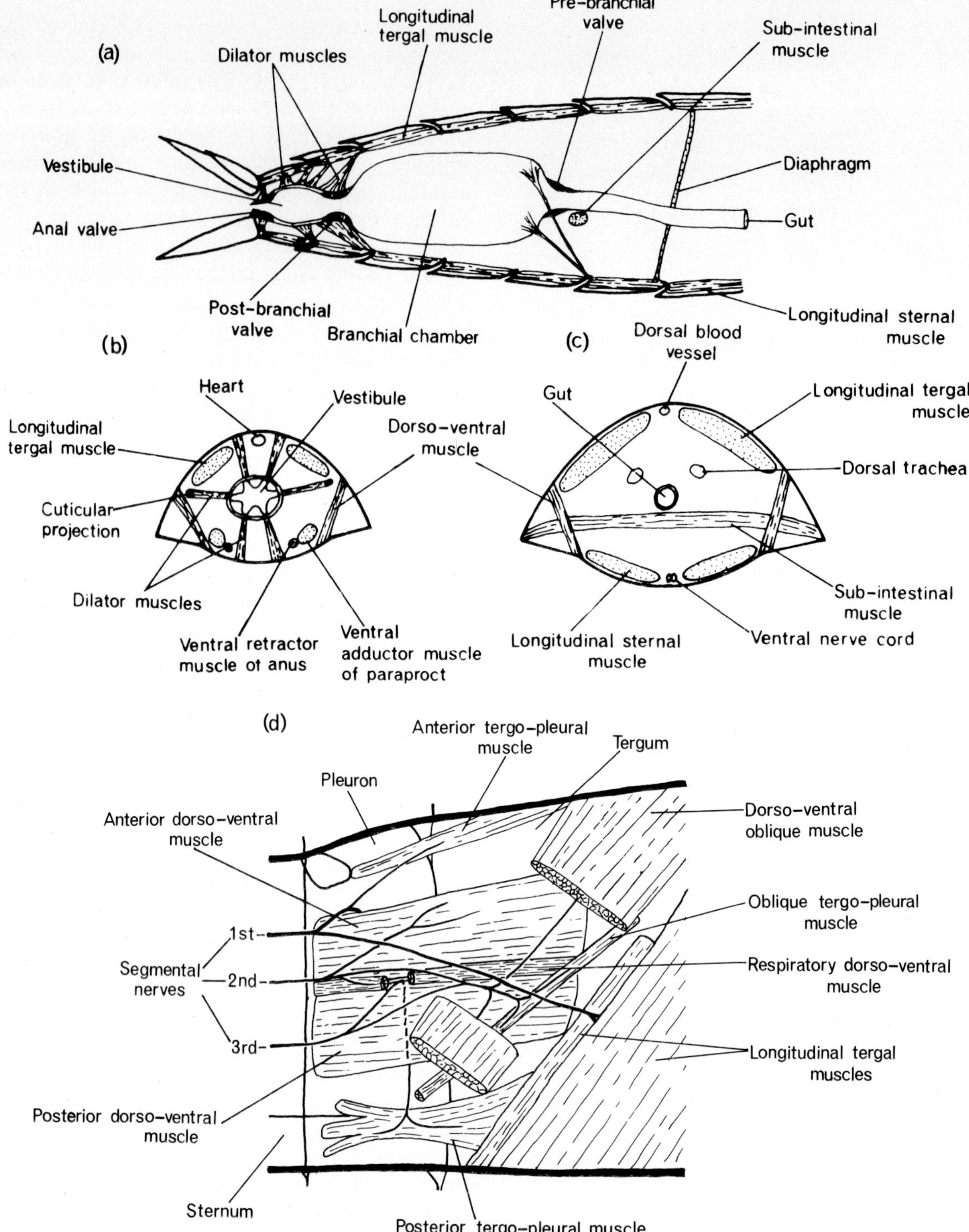

FIG. 27. Diagrams of the ventilatory system of an aeshnid dragonfly larva. (**a**) Longitudinal section through the abdomen. (**b**) Transverse section through the vestibule region. (**c**) Transverse section through the region of the subintestinal muscle. (**d**) Tergopleural region of the right side of the fifth abdominal segment to show the motor branches of the second and third lateral nerves. (**a** and **b** from Hughes, G. and Mill, P., 1966; **c** from Mill, P., 1974, after Mill, P. and Pickard, R., 1972b; **d** from Mill, P., 1974, after Mill, P., 1965.)

upstroke, and vice-versa. Thus the movement of each gill, the lateral phase difference and the metachronal rhythm all contribute to the current (Eastham, L., 1934).

Similarly, ventilatory currents are produced by the lateral gills of the larvae of the megalopteran *Corydalus cornutus*. In low pO_2 (8%) the frequency of gill movement increases (Kinnamon, S. and Kammer, A., 1983; Kinnamon, S., Kammer, A. and Kiorpes, A., 1984).

3.2 Branchial chamber ventilation

In anisopteran dragonfly larvae the gills lie within a modified region of the hindgut, the branchial chamber, and this has been described in detail in aeshnids (Fig. 27) (Tillyard, R., 1917). The wall of the branchial chamber is lined with a thin layer of circular muscle and six bands of longitudinal muscle fibres, and is attached anteriorly to the ventral body wall by a pair of very thin muscles. It is interesting to note that the gut contents do not foul the gills since they enter the hindgut enclosed in a peritrophic membrane formed in the midgut. Immediately posterior to the branchial chamber there is a small, highly muscular vestibule which, in addition to its intrinsic circular and longitudinal muscles, has a number of pairs of dilator muscles attached to it. The vestibule connects to the exterior via the anus. There are valves between the gut and the branchial chamber, between the branchial chamber and the vestibule, and guarding the anus (Fig. 27).

Ventilation (V) starts with expiration; this is effected by contraction of the segmental expiratory dorsoventral muscles, sometimes aided by the anterior dorsoventral muscles. This causes the sterna to be lifted and a consequent rise in the internal abdominal pressure. The abdomen is prevented from extending by a concurrent increase in tension in the longitudinal muscles (section 4) and hence the pressure increase is transmitted to the branchial chamber, where pressures in the order of 2–4 cmH_2O have been recorded, with a rate of increase of 6–12 $cm\,s^{-1}$ (Fig. 28a,b). At the beginning of the expiratory phase the anal valve is opened to about one-third of its maximum extent and the increase in pressure within the branchial chamber causes water to be jetted from the anus. Inspiration is effected by the contractions of two large intersegmental abdominal muscles, the diaphragm (between segments 4 and 5) and the sub-intestinal muscle (between segments 5 and 6) (Fig. 27). They augment the natural elasticity of the sclerites, and the sterna hence move downwards to their original resting position. At the same time the dilator muscles of the vestibule contract. The combined effect of the downwards-moving sterna and the contraction of the vestibular dilator muscles is to produce a slight negative pressure (about 0.5 cmH_2O) in the branchial chamber (Fig. 28a,b). The anal valves are fully opened and hence water is drawn into the branchial chamber. After a pause the ventilatory cycle is repeated. A summary diagram of these events is given in Fig. 29 (Amans, P., 1881; Matula, J., 1911; Wallengren, H., 1914a; Whedon, A., 1918; Steiner, L., 1929; Tonner, F., 1936; Snodgrass, R., 1954; Mill, P., 1965, 1970; Hughes, G. and Mill, P., 1966; Mill, P. and Hughes, G., 1966; Mill, P. and Pickard, R., 1972b). The expiratory dorsoventral muscles receive a rich supply of tracheoles and the fibres contain numerous mitochondria which are arranged regularly on either side of the Z-lines (Mill, P. and Lowe, D., 1971) (Fig. 30). This rhythmic ventilation usually occurs at a frequency of 12–57 cycles min^{-1} in aeshnids, but higher rates (40–114 cycles min^{-1}) have been seen in *Libellula* (Hughes, G. and Mill, P., 1966), and the frequency tends to increase with increase in temperature (Babák, E. and Foustka, O., 1907). At 20° a 4 cm long aeshnid larva changes about 83% of its branchial chamber and vestibule contents during one cycle, which is about 0.05 cm^3 (Wallengren, H., 1914b). Larvae of the anisopteran dragonfly *Cordulegaster boltonii* respond to visual stimuli by ceasing ventilation; they can detect prey objects at distances of 10–15 cm and a level of illumination of 5–10 lux, providing that the stimuli move at a minimum angular velocity of $10°\,s^{-1}$ (Weber, T. and Caillère, L., 1978).

Other forms of ventilation also occur. Thus, at low oxygen concentrations (pO_2 below about 2.5 $cm^3\,l^{-1}$ at 17–18°) the larva comes to the surface and protrudes the tip of its abdomen so that it can pump air into its branchial chamber (Wallengren, H., 1914b). Occasionally so-called "gulping" ventilation (Vg) occurs. The anal valve opens and water is taken into the vestibule. The anal valve then

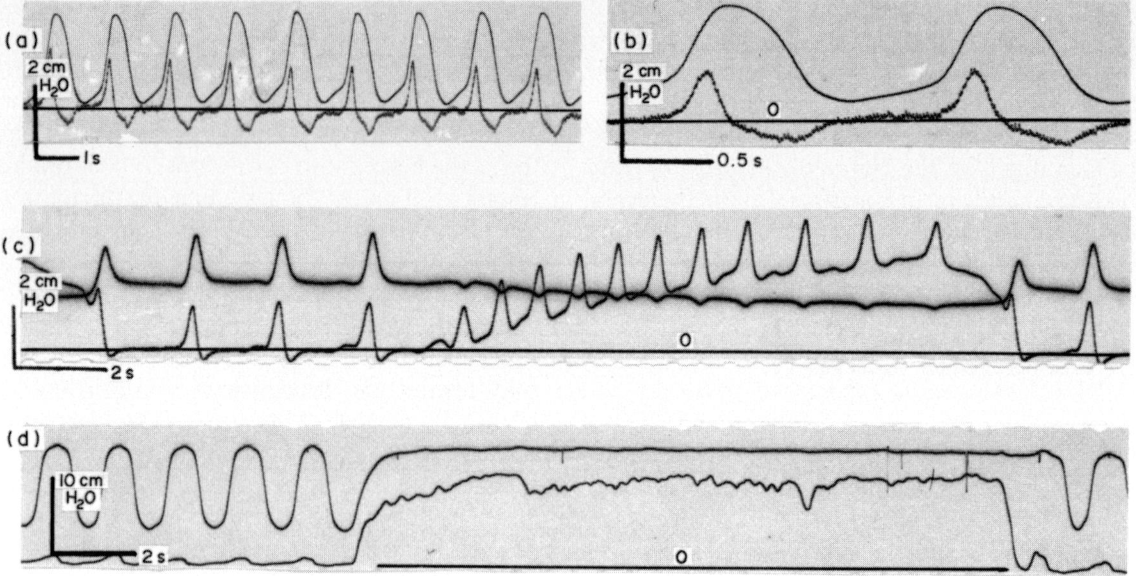

FIG. 28. Recordings of (**a,b**) normal ventilation in *Aeshna*, (**c**) gulping ventilation in *Aeshna* and (**d**) maintained abdominal compression in *Anax imperator*. The upper traces show the dorsoventral movements of the sterna (upwards indicates lifting, i.e. expiration). The lower traces show the pressure changes in the branchial chamber in relation to the outside pressure (**O**) (upwards indicates positive pressure). (From Hughes, G. and Mill, P., 1966.)

closes and the post-branchial valve opens, the vestibule contracts strongly and water is forced into the branchial chamber. This is repeated several times in succession, thus gradually pumping up the branchial chamber and causing dorsoventral expansion of the abdomen (Fig. 28b). The water in the branchial chamber is then ejected, either immediately the pumping stops or after a period during which the intrinsic musculature of the branchial chamber appears to be agitating its contents ("chewing" ventilation). Sometimes a maintained compression, with an associated increase in branchial chamber pressure, occurs periodically (Fig. 28c) (Tonner, F., 1936; Hughes, G. and Mill, P., 1966).

The dragonfly larva can jet-propel itself through the water and this is effected by a very rapid expiratory movement, in which the posterior dorsoventral muscles aid the expiratory and anterior dorsoventral muscles while, at the same time, the longitudinal muscles contract. The effect is to produce an increase in pressure of up to 40 cmH_2O at a rate of increase of about 1000 $cm\,s^{-1}$ within the branchial chamber. As the animal is propelled forwards, the legs are moved posteriorly to lie alongside the body, presumably to aid streamlining. This jet-propulsive activity must certainly subserve ventilation (section 3.3.2.4, autoventilation) (Wallengren, H., 1914a; Tonner, F., 1936; Hughes, G., 1958; Hughes, G. and Mill, P., 1966; Mill, P. and Pickard, R., 1972a).

3.3 Tracheal ventilation

While diffusion may be sufficient to meet the respiratory needs of small insects, ventilation of the tracheal system occurs in many species. It may result from passive suction or from active ventilatory movements, or be a consequence of other activities such as walking or flight (autoventilation) (Miller, P., 1974, 1981b; Kestler, P., 1978, 1980).

3.3.1 PASSIVE SUCTION VENTILATION

A number of insects rely on this mechanism. Effectively what happens is that a partial vacuum develops in the tracheae and air is drawn in through slightly open spiracles. Under such conditions carbon dioxide and water vapour cannot escape, although a small amount does pass directly through the cuticle (Buck, J., 1958). Thus, periodic opening of the spiracles is required to flush carbon dioxide out of the system. This method of ventilation is of advantage where water conservation is of import-

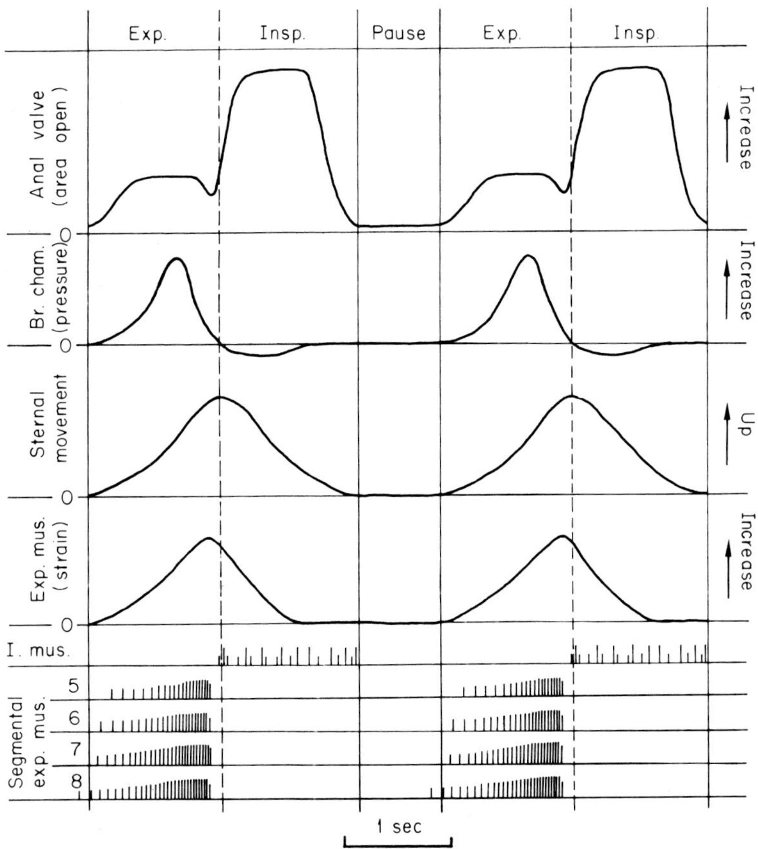

FIG. 29. Summary diagram of normal ventilation in aeshnid dragonfly larvae showing, from the bottom, expiratory and inspiratory muscle activity, the strain produced by a single expiratory dorsoventral muscle, sternal movement, branchial chamber pressure and opening of the anal valve. Exp., expiration; Exp. mus., expiratory muscle; Insp., inspiration; I. mus., inspiratory muscle. (From Mill, P., 1972, after Mill, P. and Pickard, R., 1972b.)

ance since the animal will only lose appreciable amounts of water when the spiracles are opened to remove the excess carbon dioxide. High levels of internal carbon dioxide are made possible by banking most of it as bicarbonate.

In fleas (Herford, G., 1938) and mosquito larvae (Miller, P., 1974) the tracheae have been observed to collapse and then reinflate when the spiracles open, and in fact in fleas the whole system can be reinflated by the first abdominal spiracles. Indeed there is evidence for passive suction ventilation occurring in a variety of small quiescent insects and also in the pupae of insects such as *Agapema* and *Hyalophora* (Buck, J. and Keister, M., 1955, 1958; Punt, A. *et al.*, 1957; Schneiderman, H. and Williams, C., 1953, 1955; Kestler, P., 1971). The interval between successive spiracular openings may be in the order of several hours, but is very variable and is dependent on temperature and the state of maturity of the animal, as well as showing individual differences. The cycle of events which occurs has been particularly well studied in the pupa of *Hyalophora*. Each cycle comprises three phases, the total duration depending on various factors such as temperature as well as exhibiting individual differences.

In the first "contraction" phase (C) the spiracles remain virtually closed, the tracheal pO_2 falls from 18% to 3.5%, the tracheal pCO_2 increases slightly from 3% to about 4% and the pressure in the tracheae decreases from atmospheric to -4 mmHg, thereby creating a partial vacuum (Fig. 31). This negative pressure may allow some air to be drawn in through the spiracles. However, carbon dioxide and nitrogen cannot escape. The second phase is characterized by very small, brief opening movements of the spiracular valves (fluttering) (F) and

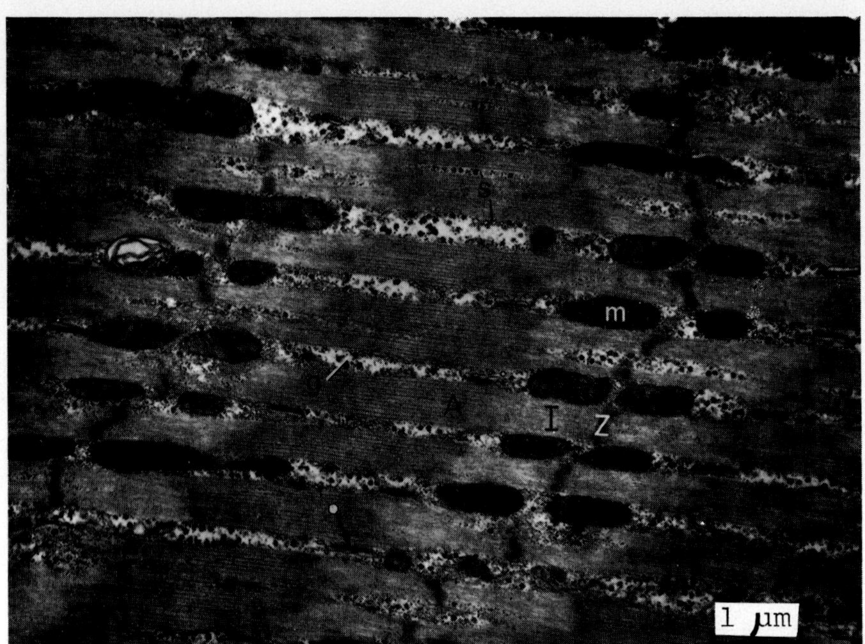

FIG. 30. Electron micrograph of a longitudinal section through a relaxed expiratory dorsoventral muscle of the larva of an aeshnid dragonfly to show the mitochondria (m) arranged on either side of the Z-lines (Z), the position of the dyads (♂) at the A–I band junctions, the sparse sarcoplasmic reticulum (s) and the abundant glycogen (g). A, A-band; I, I-band. (From Mill, P. and Lowe, D., 1971.)

the first few openings allow air to be sucked in by the negative tracheal pressure so restoring it to atmospheric. Subsequent openings produce oscillations of pressure of between −0.03 and −0.15 mmHg. These fluttering movements may be initiated by the direct effect of low pO_2 on the ganglia. They allow more air to enter the tracheae than during the second phase and some outward diffusion of nitrogen, but are insufficient to allow carbon dioxide (with a shallower partial pressure drop than nitrogen) to escape. Thus, while tracheal pO_2 remains at about 3.5%, tracheal pCO_2 steadily increases to about 6.5%.

In the third stage, which lasts 15 to 30 minutes, some or all of the spiracles open fully (O). This is initiated by the high level of tracheal pCO_2, (and possibly the level of bicarbonate also) which probably has a direct effect on the spiracular closer muscles (section 3.3.2.3). Indeed, in 14% CO_2 the spiracles open fully whatever the level of pO_2. The spiracular valves remain open for a while and then start to close. Towards the end of the phase they are closed, but occasionally flutter partly open. About 350 mm³ of carbon dioxide is released in this burst phase. Since the tracheal volume of a 5 g individual is only in the region of 400 mm³ only about 10% will be direct from the tracheae; the remainder is probably derived from bicarbonate formed in the haemolymph and tissues. In a dry atmosphere similar amounts of carbon dioxide and water are lost (Fig. 32) and indeed the water loss is about the same as is produced metabolically. During this phase the uptake of oxygen increases the tracheal pO_2 to 18% while the loss of carbon dioxide reduces the tracheal pCO_2 to 3% (Fig. 31) (Schneiderman, H., 1960; Kanwisher, J., 1966; Levy, R. and Schneiderman, H., 1966a–c; Schneiderman, H. and Schechter, A., 1966; Brockway, A. and Schneiderman, H., 1967; Kestler, P., 1971). This sequence of events has been termed the CFO cycle. (The numbering of the phases differs in different papers (see reviews by Miller, P., 1974, 1981b).) Pupae living under humid conditions do not show this sequence of events (Sláma, K., 1960).

The larvae and pupae of *Orthosoma brunnei* live in moist conditions in wood which contains a minimum of 2% oxygen and a maximum of 15% carbon dioxide. In conditions where the oxygen is

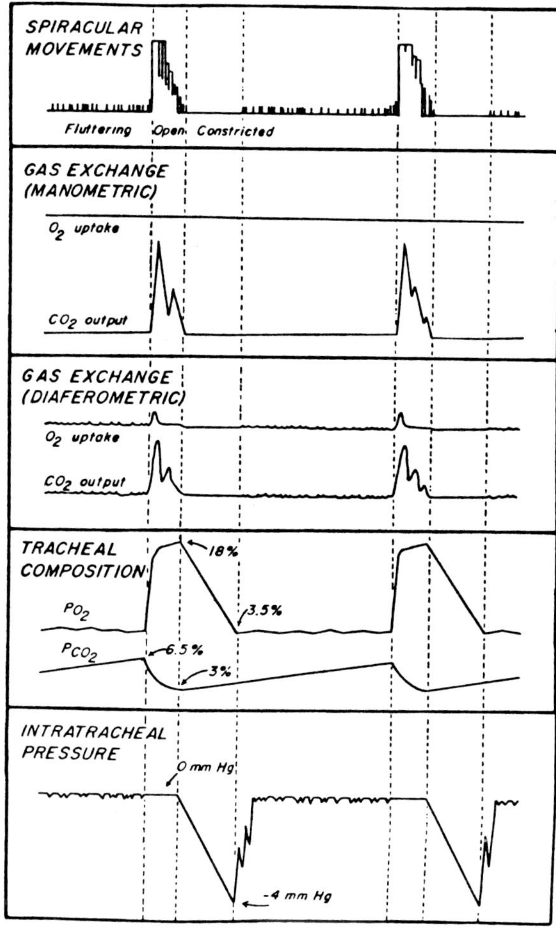

FIG. 31. Diagram to show the correlation between spiracular movements, oxygen uptake, carbon dioxide output, composition of the tracheal gases and intratracheal pressure in the three phases of passive suction ventilation in *Hyalophora*. (Diaferometric data are from Punt, A. *et al.*, 1957). (From Levy, R. and Schneiderman, H., 1966c.)

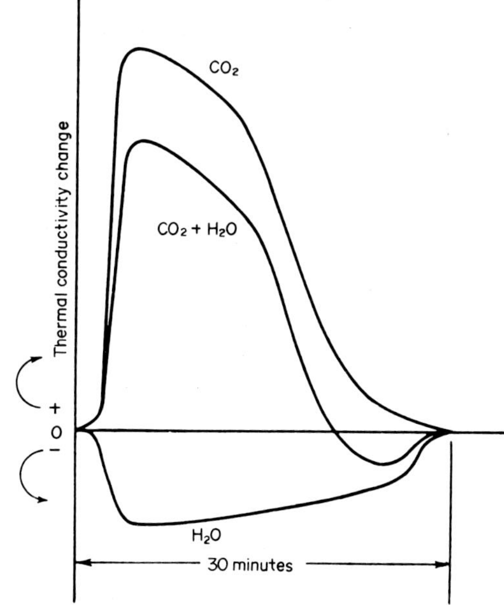

FIG. 32. Carbon dioxide and water vapour output (thermal conductivity recordings) during a single burst in a pupal *Hyalophora* in diapause. (From Kanwisher, J., 1966.)

below 4% or the carbon dioxide above 14% the spiracles remain open, as in *Hyalophora* (see above). In atmospheric air, CO_2 is released in bursts from moulting larvae, prepupae, pupae and resting adults. The spiracular valves of the prepupae and pupae, at least, show low-amplitude fluttering movements but periodically, at intervals corresponding to the release of the bursts of CO_2, the amplitude of the flutters increases. It is interesting that this mechanism is so highly developed in animals which spend much of their life in a moist environment where water conservation is of little importance (Paim, U. and Beckel, W., 1963a,b).

In quiescent larger insects, such as *Periplaneta* and *Schistocerca*, the discontinuous release of CO_2 (Wilkins, M., 1960; Hamilton, A., 1964) is associated with periodic ventilation. Thus the O phase is replaced by a period of abdominal pumping (V) which actively ventilates the tracheal system, and this has been called the CFV cycle (section 3.3.2.1).

3.3.2 Active Ventilation

Compression of any region of the body raises the internal body pressure, and this in turn acts upon any compressible region of the tracheal system; hence air is forced out through any spiracles which are open. Abdominal expansion has the reverse effect. It is thus extremely important that there should be coordination between spiracular opening and closing and the ventilatory pumping movements.

Many insects, especially, but not exclusively, the larger ones actively ventilate their tracheal system, alternately forcing air out of, and drawing air in through, the spiracles. However, there is considerable variation in the overall pattern. Thus some insects, such as dragonflies, locusts and small phorid flies, ventilate more or less continuously; others, such as wasps, ventilate only during and immediately after activity; yet others, such as some cockroaches, ventilate intermittently (e.g. Matula, J., 1911; Schreude, J. and de Wilde, J., 1952; Miller, P., 1960a, 1979; Myers, T. and Fisk, F., 1962; Paulpandian, A., 1964; Komatsu, A., 1977).

The normal effective movement involves compressing the abdomen dorsoventrally, and the resultant increase in internal pressure is sufficient to force air out of any spiracles which are open (expiration), the reverse movement producing a negative internal pressure and hence drawing air in through the spiracles (inspiration). This basic ventilatory rhythm may be augmented, especially under conditions of respiratory stress, by various mechanisms such as shortening the abdomen or by movement of other regions of the body. Furthermore, opening and closing of the spiracles are generally synchronized with the ventilatory movements, often producing a one-way air flow through the main tracheal trunks. This can be achieved by, for example, opening anterior spiracles only during the inspiratory phase and opening posterior spiracles only during the expiratory phase. Furthermore, there may be other, longer-term rhythms in addition to the basic expiratory–inspiratory cycle. Active ventilation will be considered under four headings: abdominal pumping, auxiliary mechanisms, spiracular synchronization and autoventilation.

3.3.2.1 *Abdominal pumping.* As mentioned above, rhythmic ventilation comprises alternately compressing and expanding the abdomen and always involves dorsoventral movements. In the Heteroptera, Coleoptera and phorid flies, the terga are lowered, but in most insects it is the sterna which are raised (sometimes accompanied by small tergal movements). In the Odonata, Acrididae, aculeate Hymenoptera and some Diptera the normal rhythm involves only dorsoventral movements, but in some other insects these dorsoventral movements are normally accompanied by ventrolateral movements of the membraneous pleura (Tettigonidae, Neuroptera, Lepidoptera and many Trichoptera) or by longitudinal movements of the sclerites (Phryganeidae (Trichoptera), Mantodea and some Hymenoptera) (e.g. Plateau, F., 1884; Packard, A., 1898; Babák, E., 1912; Snodgrass, R., 1935; Miller, P., 1981b).

The resting position, apnoea, is with the abdomen expanded and thus the first phase of ventilation is always expiration (Huber, F., 1960; Mill, P., 1972; Wigglesworth, V., 1972). Expiration is invariably an active process involving the contraction of segmental dorsoventral muscles, e.g. *Periplaneta americana*, the mantid *Hierodula membranacea*, *Schistocerca gregaria* and *Locusta migratoria migratorioides* (Snodgrass, R., 1935; Shankland, D., 1965; Lewis, G. *et al.*, 1973; Kerry and Mill, in preparation). As in dragonfly larvae (section 3.2) there is activity in the longitudinal muscles during expiration in order to prevent abdominal extension, and in *Blaberus craniifer* and *Hierodula membranacea* dorsal longitudinal muscles aid expiration by producing a longitudinal contraction during normal ventilation (Miller, P., 1981a; Kerry, C., 1982; Kerry and Mill, in preparation). The role of the individual muscles will be discussed further in section 4. In most cases the expiratory phase involves a smooth continuous compression, but in resting *Schistocerca* there is an expiratory pause with the sterna remaining only partly raised for a few hundred ms (Fig. 33) (Miller, P., 1965).

Inspiration generally follows immediately after expiration, but at lower frequencies in *P. americana* and *Locusta migratoria* there is a brief maintained compression (expiratory plateau) (Kestler, P., 1971; Hustert, R., 1974); maintained compressions also occur at certain stages in the ventilation of

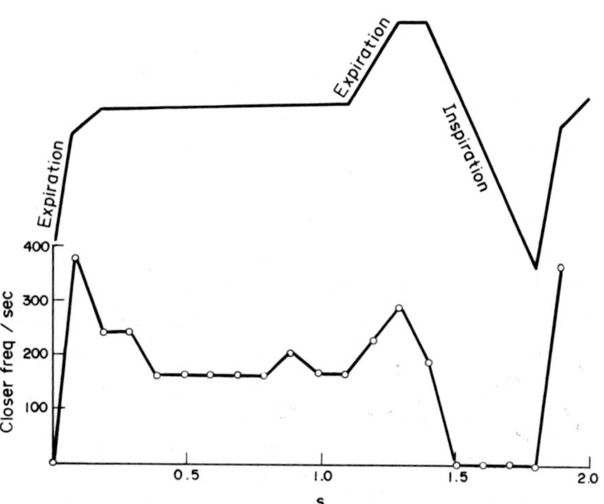

FIG. 33. The relationship between dorsoventral sternal movements (upper line; upwards indicates abdominal compression, i.e. expiration) and the frequency of motor impulses in the nerve to the closer muscle of spiracle 1 of a locust (lower line; overall frequency of both closer motor neurons). (From Miller, P., 1965.)

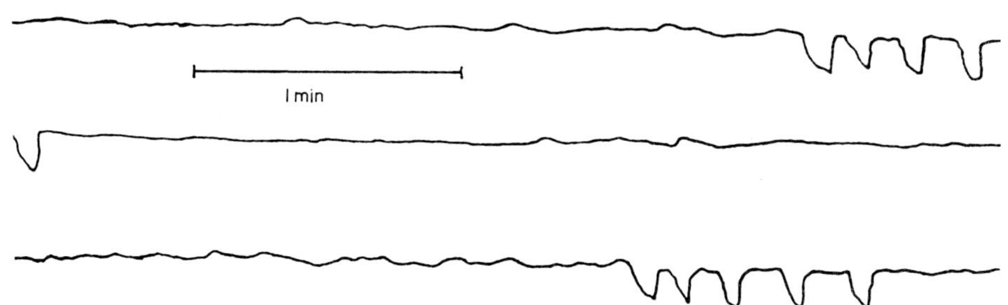

FIG. 34. Intermittent ventilation in the cockroach *Byrsotria fumigata*. Expiration is downwards. The records are continuous. (From Myers, T. and Retzlaff, E., 1963.)

H. membranacea (see below). In a number of insects, such as *Dytiscus* (Hughes, G., 1958) and *Periplaneta* (Farley, R. *et al.*, 1967), there are no inspiratory muscles, the sterna being returned to their resting position by elastic recoil which is primarily due to the resilience of the bowed sterna. However, this may be augmented by the presence of resilin, as in the beetles *Oryctes* and *Melolontha* (Andersen, S. and Weis-Fogh, T., 1964). In some insects inspiration is initiated or aided by active muscle contraction. Thus *S. gregaria* and *L. migratoria* have a pair of dorsoventral inspiratory muscles near the anterior border of most abdominal segments (muscle 192 in segment 4 and muscle 207 in segment 5; see section 4.1.2). These are arranged with one end of each muscle inserting onto a lateral projection of the sternum so that contraction causes the sternum to move away from the tergum (Snodgrass, R., 1935; Lewis, G. *et al.*, 1973; Hustert, R., 1975). Similar muscles occur in adult dragonflies, but are probably only used under conditions of respiratory stress (Miller, P., 1962). There is generally a pause following inspiration before the next expiratory phase.

In quiescent adults of *P. americana* periods of ventilation lasting 3–4 min and containing 30–40 ventilatory cycles occur about every 10 min at 29° (Paulpandian, A., 1964) and this CFV cycle (section 3.3.1) has been studied in detail by Kestler, P. (1971, 1978, 1980). According to Kestler, P. (1971) the pumping frequency is temperature-sensitive and is about 3.8 min^{-1} at 20° and 5.7 min^{-1} at 30°. Similar intermittent ventilation is also seen in the Cuban burrowing cockroach, *Byrsotria fumigata* (Myers, T. and Fisk, F., 1962; Myers, T. and Retzlaff, E., 1963), in *Blaberus giganteus* and *B. craniifer* (Miller, P., 1966c, 1981a,b) in locusts (Hamilton, A., 1964, 1972; Hustert, R., 1974, 1975) and in *Periplaneta australasiae* (Komatsu, A., 1977). Thus in *Byrsotria* groups of between 3 and 15 (average 5) ventilatory movements with a frequency of about 5 min^{-1} occur, separated by periods of apnoea which average about 7 min in duration (Fig. 34). In resting *B. craniifer* the periods of ventilation last 1–5 min and comprise 5–30 cycles, the frequency of which usually declines during the period. The intervening periods of apnoea last 5–10 min. In locusts 2 h after their last ecdysis, periods of ventilation were recorded every 1.7 min. In 8-day-old locusts ventilation was more continuous, but in animals more than 40 days old it was again intermittent. During the rest period in adult locusts, miniature inspirations occur which briefly expand the abdomen even further. A more complex cycle occurs in *Hierodula membranacea*. In a resting adult male of this species ventilation follows a period of apnoea in which the abdomen is in the expanded state (E). It consists of a ventilatory phase (V) in which normal abdominal ventilatory movements occur at a frequency of 30–45 min^{-1}. In the early stages these may be irregular, but become regular and of fairly low amplitude, but towards the end of this phase their amplitude increases. The start of the next, "compression" (Co), phase is initiated by the sterna being held in the fully raised position. If this phase is short the sterna tend to remain raised throughout it; otherwise occasional rhythmical inspiratory–expiratory movements occur. The compression phase is terminated by the sterna returning to their rest position either directly or after a few rhythmical movements, and is followed by a period of apnoea again (Fig. 35). The length of each cycle (including the period of apnoea) lasts

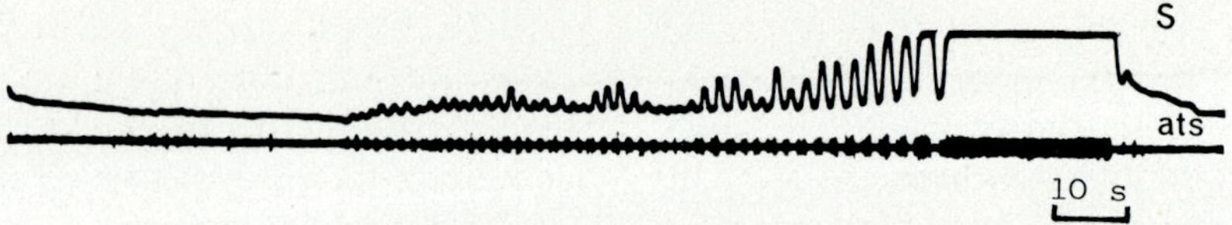

FIG. 35. Recording of the ventilatory cycle of *Hierodula membranacea*. Dorsoventral movements of the sterna (S) are shown on the upper trace (upwards indicates abdominal compression, i.e. expiration). The lower trace is a chronic recording from the "expiratory" anterior tergo-sternal muscle. (From Kerry, C., 1982.)

between 3 and 5 min (i.e. 20–12 cycles h^{-1}) (Kerry and Mill, in preparation). These are similar to the phases of the cycle which occurs during moulting in *S. gregaria* (section 3.4), except that in *H. membranacea* the sequence is EVCo rather than ECoV.

In the lepidopteran *Papilio machaon* there is evidence that pumping of the haemolymph may aid ventilation. The heart periodically reverses the direction of haemolymph flow. During normal heart-beating the haemolymph is pumped into the thorax and accumulates there because a valve reduces the anterior–posterior flow in the perineurial sinus. When the heart reverses the flow, haemolymph is pumped into the abdomen. During the period when haemolymph is pumped into the thorax, abdominal pumping occurs for a time at about 10 cycles min^{-1} (at 16°), presumably to aid ventilation of the abdominal tracheae. Furthermore, it has been suggested that air may be sucked in through the abdominal spiracles when haemolymph accumulates in the thorax, and through the thoracic spiracles when haemolymph accumulates in the abdomen (Wasserthal, L., 1976, 1980). In the giant silk moth *Attacus atlas*, expiration and inspiration via the thoracic spiracles occur slowly as a consequence of the haemolymph accumulating in the thorax and in the abdomen respectively. In contrast to *P. machaon*, active ventilation of the abdominal tracheal system in *A. atlas* occurs when most haemolymph is accumulated in the abdomen (Wasserthal, L., 1981).

The volume of air moved by the abdominal pump has been measured in *Schistocerca* and averages about 0.7 (range 0.2–1.2) ml g^{-1} min^{-1} in the resting animal (Weis-Fogh, T., 1967). This is equivalent to about 5% of the tracheal air being exchanged during each ventilatory cycle (Miller, P., 1974).

In insects which are active or under other respiratory stress (e.g. increased CO_2), ventilation becomes continuous and the frequency and depth of ventilation may both increase. In *Periplaneta americana* the expiratory plateau is abolished and in *Schistocerca* the respiratory pause similarly disappears (Miller, P., 1965, 1981a). In *P. americana* the ventilatory rate at 28° increases from 5 min^{-1} in 5% CO_2 to a maximum of 150–180 min^{-1} in 20–30% CO_2 (Hazelhoff, E., 1927), but further increase in CO_2 causes a decrease in the frequency (Schreude, J. and de Wilde, J., 1952). Similarly, in *Blaberus craniifer* the period of apnoea is reduced and the frequency of the ventilatory cycles within each ventilation period increases when the animal is exposed to increased CO_2 or reduced O_2. Continuous ventilation occurs in 5–10% CO_2 or less than 5% O_2 (Miller, P., 1981b). Stress caused by handling can also elicit an increase in ventilatory frequency (e.g. in *Blaberus discoidalis* and *B. craniifer*) (Kaars, C., 1979; Miller, P., 1981a). The overall effect is to increase the volume of air pumped. Thus, in the initial stages of flight the average abdominal pumping rate in *Schistocerca* increases to 3.1 (range 1.7–4.4) ml g^{-1} min^{-1} and in a resting *Schistocerca* exposed to 5% CO_2 even higher values are attained, i.e. 4.0–4.7 ml g^{-1} min^{-1} (Weis-Fogh, T., 1967; see also Altman, P. and Dittmer, D., 1971).

3.3.2.2 *Auxiliary mechanisms.* When insects are subjected to respiratory stress the dorsoventral abdominal movements may be augmented by in-phase longitudinal abdominal movements, which further increase the pressure within the abdomen during expiration and hence allow for a greater turnover of air in the tracheal system during each ventilatory cycle. This auxiliary mechanism occurs, for example, in *Schistocerca* (Miller, P., 1960a), *Periplaneta* (Shankland, D., 1965) and the phorid flies *Megaselia scalaris* and *Paliciphora borinquensis* (Miller, P., 1979, 1981b). In *Schistocerca* and *Periplaneta* both tergal and sternal longitudinal muscles are involved in shortening the abdomen (expiration). Lengthening the abdomen during inspiration is also achieved by longitudinal muscles in these two species, probably aided by some elastic recoil. In *Schistocerca* a pair of sternal longitudinal muscles at the posterior end of most abdominal segments are arranged so that each muscle inserts at one end onto an anterior projection of the sternum of the next posterior segment (muscle 189 in segment 4 and muscle 204 in segment 5; see section 4.1.2). Hence, when the muscle contracts that sternum is pushed posteriorly. A similar effect is achieved in *Periplaneta* by contraction of the dorsal and ventral lateral external muscles in each segment.

Two other forms of auxiliary ventilation are also seen in *Schistocerca* under conditions of respiratory stress. "Neck ventilation" occurs next after longitudinal telescoping of the abdomen and involves alternate retraction and protraction of the head with respect to the prothorax in phase with abdominal expiration and inspiration respectively. Both movements are produced by muscular activity (Miller, P., 1960a). Neck ventilation also occurs in various other acridids and in the beetle *Melolontha* (du Buisson, M., 1924a,b; Miller, P., 1960a). Subsequently, in *Schistocerca*, "prothoracic ventilation" occurs, in which the prothorax is alternately retracted and protracted, again in phase with abdominal expiration and inspiration respectively. However, whereas prothoracic retraction is brought about by muscular action, protraction occurs as a result of the elasticity of the system (Miller, P., 1960a). Neck and prothoracic pumping have also been described in the prionine beetles *Malladon downesi* and *Macrotoma palmata*, which also exhibit metathoracic pumping movements. This metathoracic pumping is elicited by the contraction of a pleurosternal muscle, the action of which is to close the pleurosternal hinge in phase with each abdominal expiratory stroke. Other thoracic muscles may also be coupled temporarily to the ventilatory rhythm (Miller, P., 1971c). Thoracic pumping has long been known in the stick insect *Carausius* and the embiopteran *Embia* (du Buisson, M., 1924a,b) and in the beetles *Hydrophilus* and *Dytiscus* (Brocher, F., 1931).

Vigorous abdominal pumping in *Schistocerca*, involving both dorsoventral and longitudinal movements, has been shown to provide up to 300 l air kg^{-1} hr^{-1}, which works out at 167 mm^3 for a single ventilatory stroke in a locust weighing 2 g (Weis-Fogh, in Miller, P., 1960a). This can be augmented by neck and prothoracic pumping movements, which can provide up to 11% and 3% respectively of the total volume of air pumped by a non-flying locust. Furthermore, spiracle 1, which serves the head, is primarily affected by these two methods of auxiliary ventilation (Miller, P., 1960a).

3.3.2.3 *Spiracular movements and synchronization.* As mentioned in section 2.3.1, spiracular closing is elicited by the contraction of a closer muscle; opening either by activity in an opener muscle or by elastic recoil. It has long been known that CO_2 has a local action on spiracles (e.g. Hazelhoff, E., 1927; Kitchell, R. and Hoskins, W., 1935; Schreude, J. and de Wilde, J., 1952; Case, J., 1956, 1957; Miller, P., 1962; Burkett, B. and Schneidermann, H., 1974). In spiracle 2 of *Schistocerca*, and in the spiracles of *Hyalophora*, *Sphodromantis* and *Blaberus*, CO_2 acts directly on the closer muscles, with increase in pCO_2 causing a decrease in muscle tension, despite continuing motor excitation, and hence opening of the valve. In *Schistocerca gregaria*, at least, the CO_2 acts on the process of neuromuscular transmission, thereby reducing the magnitude of the excitatory junction potentials. This effect is probably of particular importance in controlling those spiracles where an opener muscle is absent. Potassium contracture and regenerative muscle potentials have been suggested as the mechanism whereby closure is maintained (Hoyle, G., 1959, 1960; van der Kloot, W., 1963; Miller, P., 1974).

Oxygen, however, has an effect on the central nervous system, and in dragonflies (Miller, P., 1964a) and *Hyalophora* pupae (Burkett, B. and Schneiderman, H., 1974) lowered pO_2 effects fluttering of the spiracular valves. Indeed there is an interaction of the central response to hypoxia with the peripheral response of the spiracular muscles to increased pCO_2, which determines spiracular fluttering in various insects, including *Hyalophora* pupae (e.g. Schneiderman, H., 1960; Burkett, B. and Schneiderman, H., 1974) and resting cockroaches (e.g. Kestler, P., 1971). Furthermore, although CO_2 does not appear to have an effect on the central control of spiracular fluttering in *Hyalophora* (Burkett, B. and Schneidermann, H., 1974), there is evidence for it acting on the central nervous system in *Schistocerca* at least, and the combined central effect of CO_2 and O_2 alters the

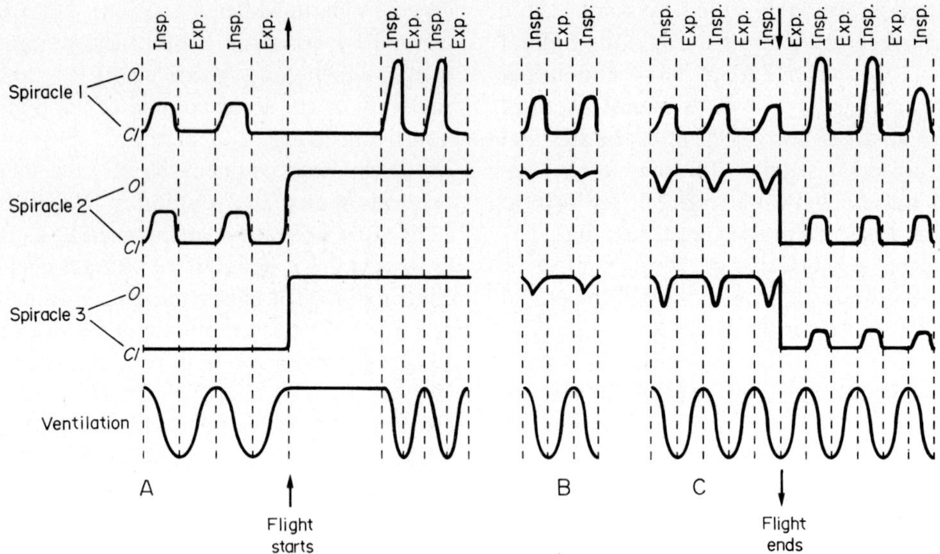

FIG. 36. Summary diagram of the behaviour of the spiracles of *Schistocerca gregaria* before, during and immediately after flight. (**A**) Before and the early part of flight; (**B**) about 30 min after the start of flight; (**C**) the end of flight and immediately after flight. Cl., closed; Exp., expiration; Insp., inspiration; O., open. (From Miller, P., 1960c.)

response in both spiracular and ventilatory motor neurons (Miller, P., 1974).

It is of paramount importance to the insect that the spiracles should not be open unnecessarily, since they are a major site of water loss. Thus, in *Locusta* normal water loss through the spiracles is about 2.9 mg g^{-1} h^{-1}, increasing to 6.2–8.0 mg g^{-1} h^{-1} during hyperventilation (Loveridge, J., 1968). It has been suggested that periodic ventilation in teneral *Schistocerca* is a water-saving mechanism (Hamilton, A., 1964). Indeed this may be the prime reason for the periodic ventilation which occurs in many insects.

In the resting insect comparatively few of the spiracles normally open. Thus fleas use spiracles 3 and 10 (Wigglesworth, V., 1935) and adult mosquitoes use spiracles 1 and 2 (Krasfur, E. *et al.*, 1970).

If spiracles remain open when ventilation is occurring then air will simply be pumped in and out through them, i.e. tidal ventilation. However, in many insects the spiracular movements are synchronized with the ventilatory movements, some spiracles only being open during inspiration, others only being open during expiration. In larger insects in particular there is often a unidirectional flow of air through the longitudinal tracheal trunks, air usually entering through anterior spiracles and leaving via posterior spiracles (e.g. Lee, M., 1925; McArthur, J., 1929; Fraenkel, G., 1932b; Bailey, L., 1954; Miller, P., 1960c; Weis-Fogh, T., 1967). Thus in resting *Schistocerca* air enters through spiracles 1, 2 and 4 and leaves through spiracle 10. However, when activity increases, spiracles 5–9 also become involved in expiration, the more posterior being recruited first, while spiracle 3 may open during inspiration. When the animal starts to fly spiracle 1 opens more fully during inspiration, while spiracles 2 and 3 open fully and remain open, thereby providing a marked tidal flow to the pterothorax. Later in flight the valves of spiracles 2 and 3 tend to show a slight and brief closing flutter during inspiration. The valves of spiracles 4 and 5–10 continue to show opening movements during inspiration and expiration respectively (Fig. 36). Auxiliary ventilating mechanisms do not occur during normal flight but appear briefly along with hyperventilation, when flight ceases.

In slow ventilation in *Periplaneta americana* spiracles 1 and 3 open during inspiration, while spiracle 2 tends to open during the maintained compression between expiration and inspiration. With increase in the ventilatory rate (i.e. induced by increase in pCO_2) spiracle 2 opens during inspiration, spiracles 1 and 3 may remain open throughout the cycle, and

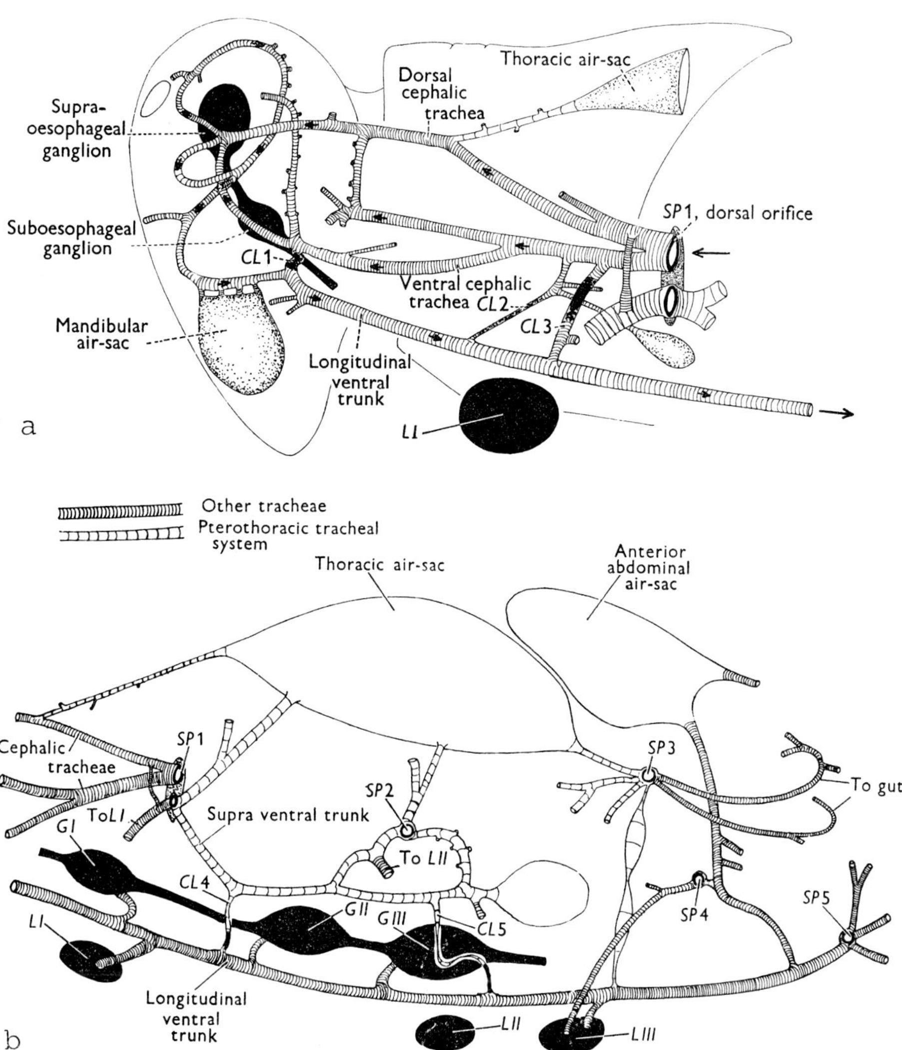

FIG. 37. Diagram of (**a**) the main tracheae to the head and their relationship to the dorsal orifice of spiracle 1, and (**b**) the pterothoracic tracheal system. CL1–CL5 degenerate cross-linking tracheae; G1–G3, thoracic ganglia; L1–LIII, bases of legs; SP1–SP3, spiracles 1–3. The arrows in (**a**) indicate the probable direction of the movement of air which results from abdominal ventilation. (From Miller, P., 1960c.)

spiracles 4–10 open during expiration (Kestler, P., 1971). Similarly in *Blaberus giganteus* some thoracic spiracles remain open (Miller, P., 1973). Hence in *P. americana* and *B. giganteus* part of the air flow is tidal.

As mentioned earlier, spiracle 1 in *Schistocerca* has two orifices, the dorsal one providing for the primary air supply to the central nervous system and the rest of the body, other than the pterothorax; the ventral orifice of spiracle 1 and spiracle 4 primarily supplying the prothoracic and metathoracic legs respectively (Fig. 37). Air entering the dorsal orifice in both *Schistocerca* and *Sphodromantis* passes to the head. During expiration spiracle 1 is closed and the air passes into the ventral longitudinal tracheae and is conveyed posteriorly to the abdomen (McArthur, J., 1929; Miller, P., 1960b,c, 1974).

Although the unidirectional air flow is normally anterior–posterior, temporary reversals can occur, at least in some insects. Such reversals have been observed in *Arphia*, *Nyctobora*, *Chortophaga*, *Dissosteira*, *Periplaneta*, *Blaberus*, *Sphodromantis* and *Schistocerca* (McArthur, J., 1929; McGovran, E.,

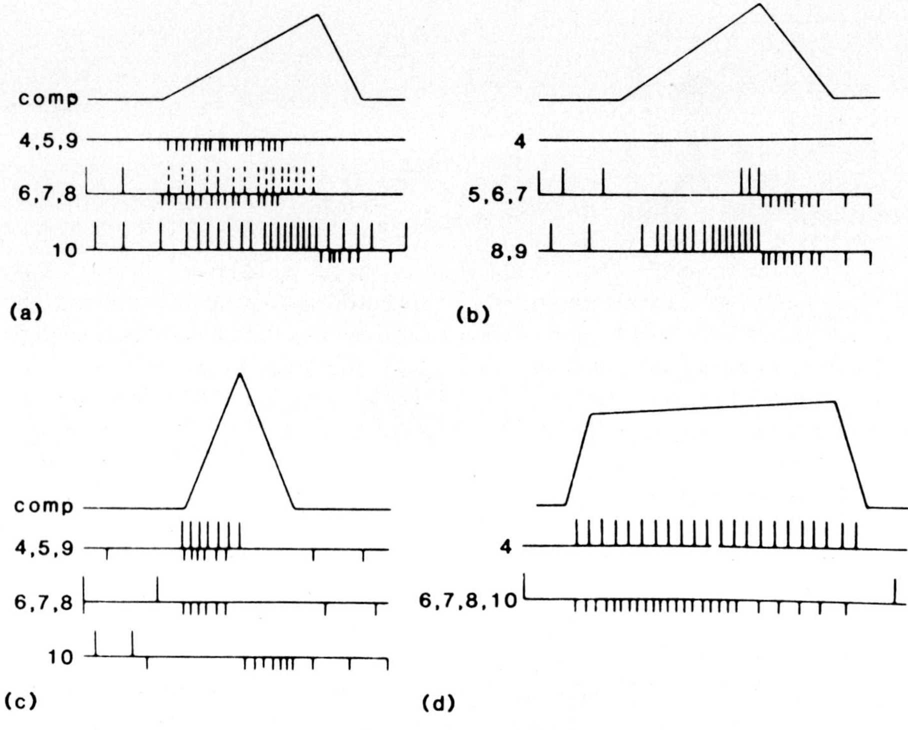

FIG. 38. Summary of sternal movements (upper lines; upwards indicates compression, i.e. expiration) and the associated activity pattern in the motor neurons of spiracles 4–10 (upward strokes, opener spikes; downward strokes, closer spikes) in *Blaberus discoidalis* and *Gromphadorhina portentosa*. Dashed opener spikes indicate variable activity. (**a**) Normal ventilation in *Blaberus discoidalis*; (**b**) normal ventilation in *Gromphadorhina portentosa*; (**c**) stress ventilation in *Blaberus discoidalis*; (**d**) Stress ventilation (hissing) in *Gromphadorhina portentosa*. (From Kaars, C., 1979.)

1932; Kitchell, R. and Hoskins, W., 1935; Miller, P., 1973, 1974). In the Hawaiian cockroach, *Nyctobora noctiviga*, and *Blaberus giganteus*, reversal can be elicited by hypoxia (Kitchell, R. and Hoskins, W., 1935; Miller, P., 1973). In *Sphodromantis* the reversal may only last for a single ventilatory cycle, with the abdominal spiracles opening during inspiration and spiracle 1 opening during expiration (Miller, P., 1974) (section 4.2.3). In *Gromphadorhina portentosa* reversal is complex and involves spiracles 8 and 9 becoming inspiratory while air may still enter spiracles 1 and 2 and leave from spiracles 5 and 6 (Kaars, C., 1979).

Of the air entering the tracheal system, the proportion which flows unidirectionally may vary in different insects and under different conditions. Thus in *Schistocerca*, while the amount flowing unidirectionally remains about the same under a variety of conditions (i.e. $0.55\,\mathrm{ml\,g^{-1}\,min^{-1}}$), the proportion of the total air flow decreases from 80% in a resting animal to 18% during the initial stages of flight because of the tremendous increase in volume being pumped tidally by the pterothorax (section 3.3.2.4) (Weis-Fogh, T., 1967). In contrast, experiments on the distribution of air entering spiracle 1 of *Sphodromantis* indicate that not only does the amount entering it increase markedly when ventilation is active (i.e. $1.0–1.2\,\mathrm{ml\,g^{-1}\,min^{-1}}$ compared to $0.25–0.33\,\mathrm{ml\,g^{-1}\,min^{-1}}$ when resting), but also the proportion flowing unidirectionally increases from 45% to 95% of the total amount (Miller, P., 1974). However, these two examples are not strictly comparable since in the former the increased oxygen demand is required primarily by the pterothorax, rather than being a response to an overall metabolic need.

The flow rate of the anterior–posterior longitudinal airstream has been measured in the Hawaiian cockroach, *Nyctobora noctivaga*. In young instars (0.4–0.8 g) it reaches $9.6\,\mathrm{ml\,g^{-1}\,h^{-1}}$ but in larger animals (1.6–2.0 g) is rather less ($7\,\mathrm{ml\,g^{-1}\,h^{-1}}$). In the presence of 15% CO_2 the rate increased by about six times (Kitchell, R. and Hoskins, W., 1935).

In blaberid cockroaches, which are very broad-bodied animals, transverse currents may be produced by only opening the expiratory spiracles on one side of the body (Kaars, C., 1979). A similar unilateral opening has also been observed in abdominal spiracles of *B. giganteus* (Miller, P., 1973). In *Blaberus discoidalis* and *Gromphadorhina portentosa* spiracles 1 and 2 are inspiratory. In resting *Blaberus craniifer* spiracles 10 open synchronously if the expiratory phase is short but if it is long there is a brief unilateral opening towards the end of abdominal compression (Miller, P., 1981b). In resting *B. discoidalis* only spiracle 10 may be open during expiration, but increase in respiratory demand results in the progressive recruitment of spiracles 8, 7 and 6, all unilaterally, but hyperventilation (in the presence of increased pCO_2) results in spiracles 4–10 being bilaterally coupled and opening during expiration (Fig. 38). In resting *G. portentosa* spiracles 5–9 are expiratory and function unilaterally with recruitment for increased metabolic demands working anteriorly from spiracle 9 (Fig. 38) (Kaars, C., 1979). The complexities in the pattern of spiracular coupling in the blaberids is no doubt partly because spiracles 4, 5 and 9 in *B. discoidalis* are involved in defensive behaviour (Kaars, C., 1979), and in *G. portentosa* spiracle 4 is used for sound production while spiracles 3 and 10 are reduced in size (Dumortier, B., 1965; Nelson, M., 1979).

The timing of spiracular opening is such that those spiracles involved in expiration tend to open towards the end of the expiratory phase of abdominal movement, and their period of opening may be very brief, sometimes as little as 40–60 ms in *Blaberus giganteus*. This means that there is a short period when all the spiracles are closed while the abdomen is being compressed and thus there is a resultant small build-up in pressure. This reaches about 10 mmHg in *Schistocerca*, increasing to 15 mmHg during hyperventilation. Prior to inspiration −4 mmHg have been recorded in a hyperventilating animal. The increase in pressure prior to expiration may increase the volume of air expelled (McCutcheon, F., 1940; Hrbracek, J., 1949; Watts, D., 1951; Weis-Fogh, T., 1967).

3.3.3 Autoventilation

Autoventilation is ventilation which arises directly as a result of other movements, notably those concerned with locomotion; one example, jet-propulsion in dragonfly larvae, has already been discussed (section 3.2).

Autoventilation has been particularly well studied during flight in *Schistocerca*, where the dorsoventral movements of the nota and the lateral movements of the pleura associated with the wing beat, effect a marked exchange of air in the pterothoracic tracheal system (Fig. 39). Indeed, during average horizontal flight 5.4 ml air g^{-1} min^{-1} ventilates the flight muscles. Of this about 4.2 ml g^{-1} min^{-1} are provided by the thoracic movements, while abdominal pumping movements contribute the other 1.2 ml g^{-1} min^{-1}. The abdominal pump also provides about 1.3 ml g^{-1} min^{-1} to the rest of the body. The unidirectional anterior–posterior air flow through the longitudinal tracheal trunks produced by abdominal pumping remains about the same as at rest, i.e. about 0.55 ml g^{-1} min^{-1}, and its main function is probably to keep the central nervous system well supplied with oxygen (Table 5). The potential capacity of the thoracic flight pump, based on the volume changes which occur during a single wing-beat cycle, is considerably higher than 4.2 ml air g^{-1} min^{-1}, and is about 12.7 ml g^{-1} min^{-1} in steady flight and as much as 15.8 ml g^{-1} min^{-1} in strong flight. The pressure changes caused by thoracic pumping are between 10 mmH_2O and 25 mmH_2O (Weis-Fogh, T., 1967). The pterothoracic spiracles (2 and 3) remain open during flight, although their valves do make slight closing movements during inspiration if flight is prolonged (Miller, P., 1960c).

Adult dragonflies also rely largely on thoracic autoventilation during flight. In *Aeshna cyanea* the potential ventilatory rate has been calculated as 13.3 and 26.7 ml air g^{-1} min^{-1} at wing-beat amplitudes of 57° and 90° respectively (Weis-Fogh, T., 1967). Although these values are of the same order of magnitude as for *Schistocerca*, the organization of the tracheal system in dragonflies (Miller, P., 1962; Weis-Fogh, T., 1964a) is such that it seems likely that most of their potential ventilation is utilized to renew the air in the central tracheae of the flight muscles. Furthermore, dragonflies have high

Table 5: Ventilatory rates (ml air g^{-1} min^{-1}) of the abdominal and thoracic pumps during rest and average horizontal flight in the locust, Schistocerca.

	Abdominal pump			Thoracic pump
	Unidirectional flow (CNS)	Abdomen	Thorax	Thorax
Rest	0.55	0.15		—
Average horizontal flight	0.55	1.3	1.2	4.2

(After Weis-Fogh, T., 1967).

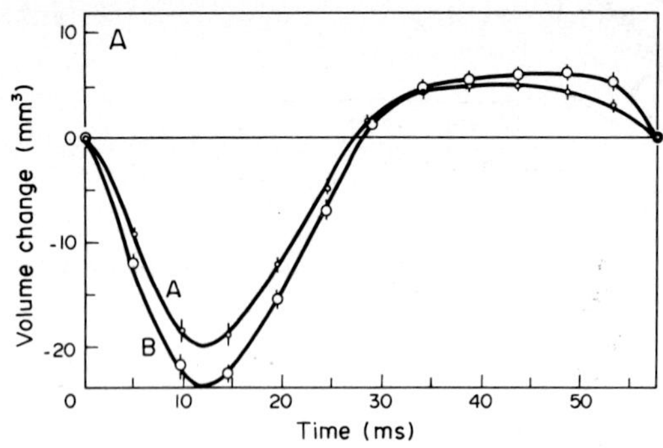

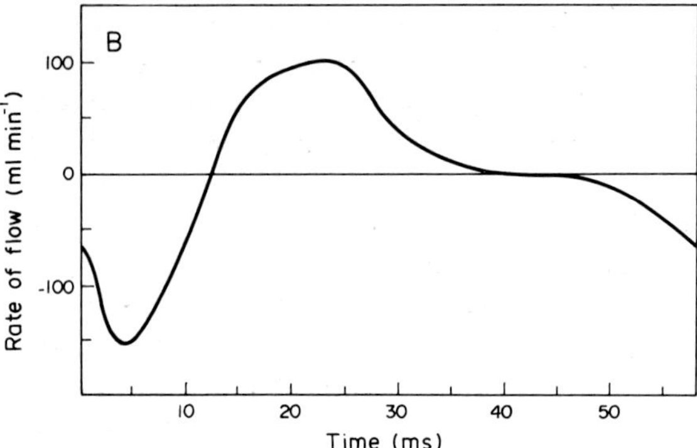

Fig. 39. (**A**) The change in thoracic volume caused by the wing movements of *Schistocerca gregaria* during a wing stroke. The two plots are for slightly different wingstroke amplitudes. (**B**) The potential rate of air flow through the thoracic spiracles of *Schistocerca gregaria* caused by the wing movements during a wingstroke. Negative values indicate that air is being sucked in through the spiracles. (From Weis-Fogh, T., 1967.)

metabolic rates and the tracheoles do not indent the muscle fibres (section 2.1). Indeed, at a rate of about 6.7 ml air g^{-1} min^{-1} the oxygen deficit will be about 5%, which is close to the critical limit for diffusion from the central tracheae to the muscle fibres. However, a safety factor is provided by the increase in ventilatory rate which can be achieved by increasing the wing-beat amplitude (Weis-Fogh, T., 1967). Abdominal pumping does not occur during flight in *Periplaneta americana* (Fraenkel, G., 1932a) and so autoventilation is probably important in this species also (Miller, P., 1981a).

Autoventilation occurs during flight in many other insects, but the details of the systems vary. In the Cerambycidae (Coleoptera), which have no air sacs, thoracic autoventilation is augmented in larger species (e.g. *Petrognatha gigas*) by draught or ram ventilation (due to a Bernoulli effect), whereby air is ducted through large, low-resistance primary tracheae. In *Petrognatha* there are two pairs of such tracheae, which run through the metathorax between spiracles 2 and 3, and from which numerous deformable secondary tracheae arise to invade the flight muscles (Fig. 40). In a wind speed of 5 m s^{-1}, 63 ml air g^{-1} min^{-1} are ducted through them, entering via the second spiracles and leaving via the third spiracles. During flight the thoracic pump exchanges 32.4 ml air g^{-1} min^{-1}, mainly by its effect on the secondary tracheae. In at least the larger scarabaeid (e.g. *Goliathus regius*) and buprestid coleopterans, which possess abundant air sacs in the pterothorax, strong abdominal pumping supplements autoventilation (Miller, P., 1966b).

In hymenopterans and dipterans autoventilation is probably insignificant, although it might effect the movement of gases and haemolymph within the pterothorax (Weis-Fogh, T., 1964b, 1967). Abdominal pumping, together with diffusion, apparently provides for the additional requirements of flying hymenopterans such as *Vespa* (Weis-Fogh, T., 1967), and in the flying honeybee there is a marked unidirectional flow of air in through the first and out through the third spiracles (Bailey, L., 1954). However, in *Bombus* the large third spiracles remain open during flight, and the mesophragma, to which the dorsal longitudinal flight muscles are attached, lies close to these spiracles and moves slightly during the wing-beat. Miller, P. (1981b) suggests that this may be of autoventilatory significance. Dipterans have an extensive system of pterothoracic air sacs and

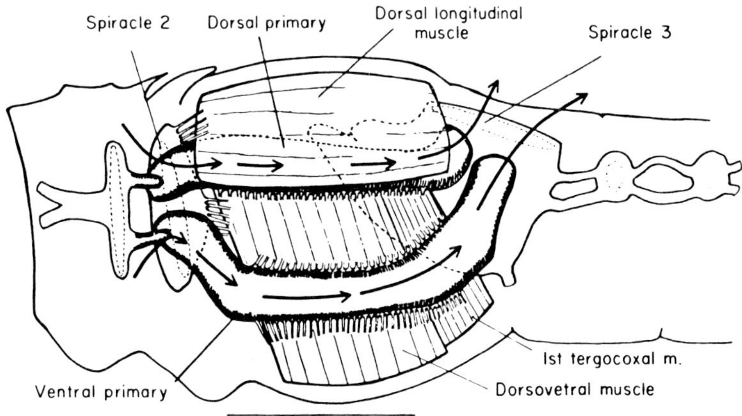

FIG. 40. Diagram of a medial section through the pterothorax of *Petrognatha gigas* to show the two pairs of giant primary tracheae which run between spiracles 2 and 3. Horizontal scale, 10 mm. (From Miller, P., 1974; after Miller, P., 1966b.)

apparently depend on diffusion alone (Weis-Fogh, T., 1964b; Whitten, J., 1968, 1972).

3.4 Ventilation during moulting

The changes in ventilation during moulting have received little attention (Reynolds, S., 1980). The spiracles are probably temporarily inactivated, and the presence of the old tracheal lining cannot be very helpful. Certainly moulting is a complex and energy-consuming behaviour and the process of shedding the tracheal linings must interfere severely with gaseous exchange. Thus, ventilatory movements may well be enhanced at this time in order to provide the animal with sufficient oxygen for its metabolic needs. Furthermore, after the moult is complete the newly expanded segments and appendages must harden properly, and the spiracles remain closed for a period (Miller, P., 1981b). Gill-bearing aquatic insect larvae must also undergo a period when gaseous exchange is impaired by the presence of the old cuticle over the new. The problem is exacerbated when the moult involves metamorphosis from a gill-breathing aquatic larva with a closed tracheal system to a terrestrial adult with an open tracheal system, as in many exopterygotes. In aeshnid dragonfly larvae, at least, this involves a considerable change in the relative proportions of the body, as well as in the musculature, notably in the abdomen (Mill and Whittles, in preparation).

The information available is virtually restricted to the moult from the fifth-instar larva to the adult in *Schistocerca gregaria*. Six ecdysial stages can be recognized (Hughes, T., 1980a–d). In stage 1 the insect stops feeding and selects a perch. In stage 2 it hangs down from its perch and inflates its gut. During this stage hyperventilation occurs, possibly as a result of a hypoxic effect on the ganglia brought about by the inefficiency of gaseous exchange mentioned above. This ventilation is characterized by marked dorsoventral abdominal pumping movements with normal synchronization of the spiracles. Towards the end of this stage the abdominal pumping movements start to occur in groups, with intervening periods of maintained abdominal compression. In the preparatory (pre-ecdysial) phase of the cricket *Teleogryllus oceanicus*, abdominal pumping movements occur. They involve longitudinal, synchronous contractions of the abdomen and resemble normal ventilatory pumping, but in this case they produce swelling of the head and thorax (Carlson, J., 1977).

Stage 3 of the locust ecdysis involves the adult extricating itself from the larval cuticle, including the removal of the old tracheal linings, and is dominated by peristaltic waves of abdominal contraction which effect this process. Following each wave a few ventilatory cycles occur, but as the stage proceeds each such period of ventilation is reduced to a single abdominal compression which lasts 1.0–1.5 s. The peristaltic waves may have an autoventilatory role. In stage 4 the adult hangs upside-down from the exuvium and then climbs up and selects a site for wing expansion. During this phase there is a strong expansion (E) of the abdomen (inspiration), followed by a maintained abdominal compression (Co) which occurs with all of the spiracles closed, producing intratracheal pressures of up to 40 mmHg. There is then a short period of dorsoventral abdominal pumping (V). This ECoV cycle lasts 0.5–1.5 min. It also occurs in stage 5, when the wings are expanded and folded, and for the first 2 h of stage 6 (Hughes, T., 1980a). Stage 6 is essentially a period of quiescence and lasts for about 24 h. Apart from the early period of ECoV, ventilation is intermittent in this stage and there are also occasional non-ventilatory pumping movements in which the spiracles open once for every four pumping movements (Elliott, C., 1980). In the moth *Manduca sexta* (Truman, J. and Endo, P., 1974) and in crickets (Carlson, J., 1977) this (expansional) stage consists of a long-maintained compression with no periods of ventilation.

Many muscles are coupled to the ECoV cycle at various times. Thus the abdominal spiracle openers are only coupled to the expiratory strokes in V, while the abdominal dorsoventral and longitudinal expiratory muscles, thoracic spiracle muscles, head retractor muscles and wing elevators are all coupled both to the expiratory strokes in V and to Co. The abdominal spiracle closers and the abdominal dorsoventral and longitudinal inspiratory muscles are coupled to the inspiratory strokes in V and to Co. The wing depressors, wing-folding muscles and mandible adductors are coupled to Co only. However, there is some variability in this coupling. Since the abdominal spiracle closers are inactive during Co, it has been suggested that the valves may be kept closed either elastically or myogenically (Elliott, C., 1980).

The changes which occur in the ECoV cycle in stages 4–6 of the locust moult are shown in Fig. 41. The duration changes steadily except for a marked decrease between stages 5 and 6 (Hughes, T., 1980c). The duration of Co, and the frequency and number of abdominal pumping movements in V, all increase with cycle duration, whereas E decreases in duration. The increase in frequency of the abdominal pumping movements is primarily a result of a decrease in the duration of each expiratory stroke as the number of pumping movements in the cycle increases.

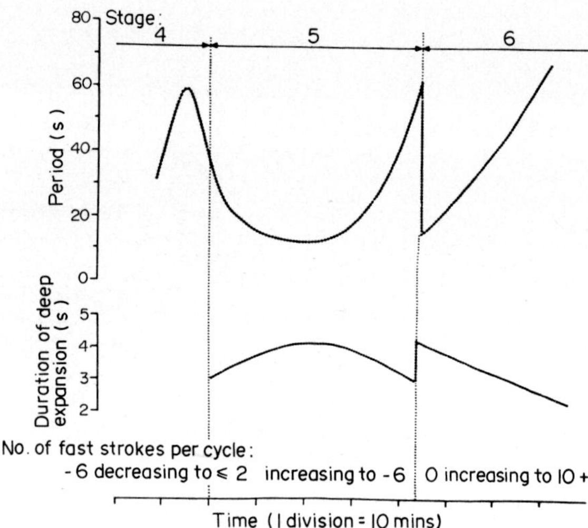

FIG. 41. Summary diagram showing the changes in ventilatory cycle time (period), the duration of deep expansion and the number of fast ventilatory strokes in the fourth (4), fifth (5) and sixth (6) phase of imaginal ecdysis in *Schistocerca gregaria*. (From Hughes, T., 1980c.)

In stage 4 an increase in pCO_2 elicited a slight increase in the ECoV cycle duration and an increase in the number and frequency of pumping movements in V (sometimes the first few pumping movements were rather long), but a decrease in the duration of Co. Hence ventilation starts earlier in the cycle. Conversely, hypoxia may cause a decrease in the ECoV cycle duration (Hughes, T., 1980c).

4 NEURAL CONTROL MECHANISMS

It is useful to summarize the main patterns of ventilation which have been described (Table 6) (section 3). The abbreviations which have been introduced fairly recently into the literature (e.g. Kestler, P., 1971, 1978, 1980; Hughes, T., 1980a–d; Miller, P., 1981b) help in handling the available data, but ultimately will almost certainly need some modification and rationalization as more is learnt of the nature of these cycles and as more systems are investigated thoroughly.

4.1 Ventilatory and spiracular muscles and their innervation

It is pertinent first to consider the various muscles

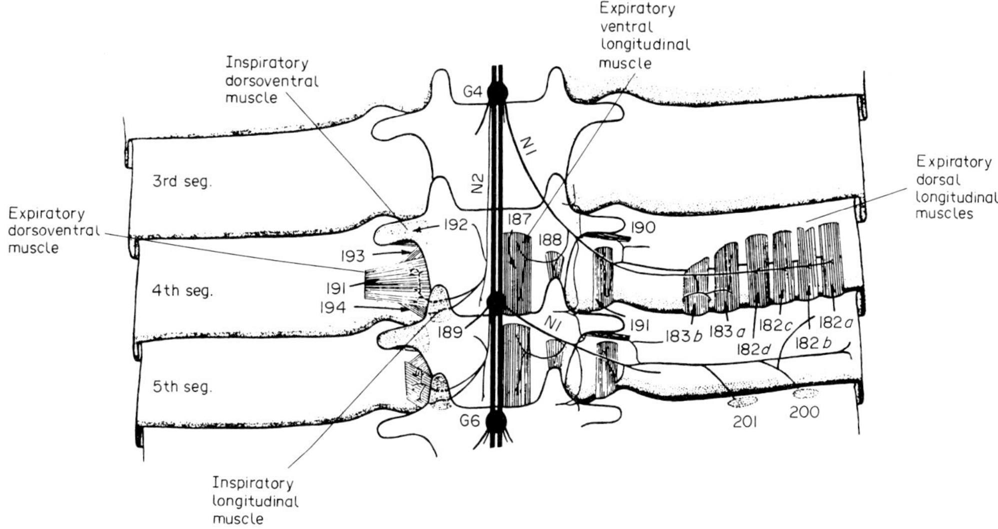

FIG. 42. Diagram of the muscles and nerves of abdominal segments 4 and 5 of *Schistocerca gregaria*. The muscles innervated by the first nerve have been removed on the left, while the main dorsal longitudinal muscles have been removed in segment 6. The median nerves are omitted. (From Lewis, G. *et al.*, 1973.)

associated with ventilation and spiracle control, and to describe their innervation. As seen earlier (section 3) the number of muscles involved varies in different species. However, there is some degree of conformity. Thus, in the abdomen there is at least one pair of segmental expiratory dorsoventral muscles and, where there is more than one, it is often possible to distinguish a primary expiratory muscle as opposed to various other muscles which may aid the process of expiration, either obligatorily or by being recruited when the respiratory demands of the insect require stronger ventilation.

4.1.1 ABDOMINAL EXPIRATORY MUSCLES

In aeshnid (*Anax imperator* and *Aeshna*) dragonfly larvae there is one pair of (primary) expiratory dorsoventral muscles (rdv) in most segments of the abdomen. They are aided by a second pair of dorsoventral muscles, the anterior dorsoventral muscles (adv). Similarly in *Hierodula membranacea* there are two pairs: the primary expiratory posterior tergo-sternals and the anterior tergo-sternals (Kerry, C., 1982; Kerry and Mill, in preparation).

In *Locusta migratoria* two of the dorsoventral muscles (206 and 209 in segment 5) appear to be primary expiratory muscles, 209 being more sensitive than 206. In *Schistocerca gregaria* certainly, muscle 206 is a primary expiratory muscle and 209 is also an expiratory muscle. In both species two other dorsoventral muscles (205 and 208 in segment 5) aid this process (Table 7 and Fig. 42) (Lewis, G. *et al.*, 1973; Hustert, R., 1975).

In aeshnid larvae and *L. migratoria* longitudinal contraction is not normally seen but, nevertheless, both dorsal and ventral longitudinal muscles receive an excitatory supply during expiration. In *L. migratoria* those involved are the dorsal longitudinal muscles 197, 198 and 203 (in segment 5) during shallow ventilation and, in addition, the ventral longitudinal muscles 199 and 202 (in segment 5) during stronger ventilation. In *S. gregaria* and cockroaches longitudinal contraction occurs during hyperventilation. In *S. gregaria* it appears to be only the dorsal longitudinal muscles which are involved (Fig. 42) but in *Periplaneta americana* there may be as many as five dorsal and two ventral longitudinal muscles eliciting this movement (Shankland, D., 1965). In *H. membranacea* there is some shortening of the abdomen during expiration, but it is uncertain which longitudinal muscles are involved. All of these expiratory dorsoventral and longitudinal muscles are innervated by lateral segmental nerves. The cell bodies of the neurons innervating the expiratory dorsoventral muscles in aeshnid larvae lie ipsilaterally, close to the bases of the

Table 6: Summary of the principal types and patterns of ventilation

Type	Sequence	Description	Examples
CFO	i Spiracles constricted (C) ii Spiracles fluttering (F) iii Spiracles fully open (O)	passive suction ventilation (section 3.3.1)	smaller insects and pupa of e.g. *Hyalophora*
CFV	i Spiracles constricted (C) ii Spiracles fluttering (F) iii Abdominal ventilatory pumping (V)	passive suction ventilation followed by active ventilation (section 3.3.1)	larger insects when quiescent, e.g. *Periplaneta*
V	Abdominal pumping (V)	continuous for long periods (sections 3.2 and 3.3.2.1)	generally in active insects, e.g. flying *Schistocerca*, also aeshnid larvae, or as part of a pattern
Co	Abdominal compression (Co)	sterna kept in the raised position (sections 3.2 and 3.3.2.1)	either occasionally e.g. aeshnid larvae, or as part of a pattern
E	Abdominal expansion (E)	a period of rest (probably apnoea) with the sterna kept in the lowered position (section 3.3.2.1)	either occasionally, e.g. aeshnid larvae, or as part of a pattern
V_g	Active abdominal expansion (V_g)	a series of cumulative small lowerings of the sterna (gulping ventilation); may involve a maintained expansion during which chewing ventilation occurs (section 3.2)	occasionally in aeshnid larvae pumping up the branchial chamber
EVCo	i Abdominal expansion (E) ii Abdominal ventilatory pumping (V) iii Abdominal compression (Co)	(section 3.3.2.1)	e.g. adult *Hierodula*
ECoV	i Abdominal expansion (E) ii Abdominal compression (Co) iii Abdominal ventilatory pumping (V)	(section 3.4)	e.g. moulting *Schistocerca*

References are given in the text

Table 7: Summary of muscles involved in abdominal ventilation

	Expiratory muscles			Inspiratory muscles		
	Primary dorsoventral	Secondary dorsoventral	Longitudinal	Dorsoventral	Longitudinal	Transverse
Aeshnid dragonfly larvae	expiratory dorsoventral (rdv)	anterior dorsoventral (adv)	dorsal and ventral (incl. lt_1 and lt_2)	—	—	diaphragm and sub-intestinal
Schistocerca gregaria and *Locusta migratoria* (segment 5*)	209 and 206	205 and 208	dorsal (197, 198) ventral (199, 202, 203)	207	204	ventral diaphragm
Hierodula membranacea	posterior tergo-sternal	anterior tergo-sternal	some longitudinal	—	—	—
Cockroaches	tergo-sternal (e.g. muscle 25 in *Periplaneta americana*	?	some dorsal and ventral	—	dorsal (18) and ventral (26) lateral externals (in *Periplaneta americana*)	

Data from Shankland, D. (1965); Mill, P. and Hughes, G. (1966); Lewes, G. *et al.* (1973); Hustert, R. (1975); Pickard, R. and Mill, P. (1975); Kerry, C. (1982); Kerry, C. and Mill, P. (in preparation).
* Corresponding muscle numbers in segment 4 can be obtained by subtracting 15.

first and second lateral nerves (Komatsu, A., 1980).

4.1.2 Abdominal inspiratory muscles

There are no segmental dorsoventral inspiratory muscles in cockroaches and larval dragonflies, but they do occur in *S. gregaria* and *L. migratoria* (e.g. muscle 207 in segment 5). These orthopterans also have paired longitudinal inspiratory muscles (e.g. muscle 204 in segment 5) (Table 7 and Fig. 42) (Lewis, G. *et al.*, 1973; Hustert, R., 1975), while in *P. americana* the paired dorsal (18) and ventral (26) lateral external muscles may aid inspiration (Shankland, D., 1965). In aeshnid larvae there are two large transverse abdominal muscles which aid inspiration. These are the diaphragm, which lies between segments 4 and 5, and the sub-intestinal muscle, which lies between segments 5 and 6 (Fig. 27) (Mill, P., 1965; Mill, P. and Hughes, G., 1966). The inspiratory longitudinal muscles in *S. gregaria* are innervated by the second lateral nerves, but the inspiratory dorsoventral muscles in this animal, and the transverse muscles in aeshnid larvae, are innervated by the median nerves. The cell bodies of the axons in the median nerves lie posterolaterally (e.g. Zawarzin, A., 1924; Komatsu, A., 1980).

4.1.3 Spiracular muscles

There is always a closer muscle present which is invariably innervated by a median nerve. In *S. gregaria* this is the only innervation (Fig. 43a) but in cockroaches the closer of spiracles 1 and 4–10 also receives innervation from a lateral nerve (Fig. 43b); in spiracle 1 of *Blaberus discoidalis* and spiracles 4 and 5 of *Gromphadorhina* this has been shown to be an inhibitory supply (Miller, P., 1969; Nelson, M., 1979). The axons in the median nerves have ventromedial cell bodies (e.g. Burrows, M., 1975b; Nelson, M., 1979), while the cell body of the inhibitory neuron is contralateral (Nelson, M., 1979).

In cockroaches, spiracles 4–10 possess an opener muscle (Miller, P., 1981a) and in most segments this receives its innervation from a lateral nerve. This is also the case for spiracles 5–10 of *S. gregaria* but the opener of spiracle 1 of this latter species is innervated by a median nerve from the prothoracic ganglion as well as by a lateral nerve from the mesothoracic ganglion, while the openers of spiracles 3 and 4 only receive innervation from median nerves (Fig. 43) (Lewis, G. *et al.*, 1973; Miller, P.,

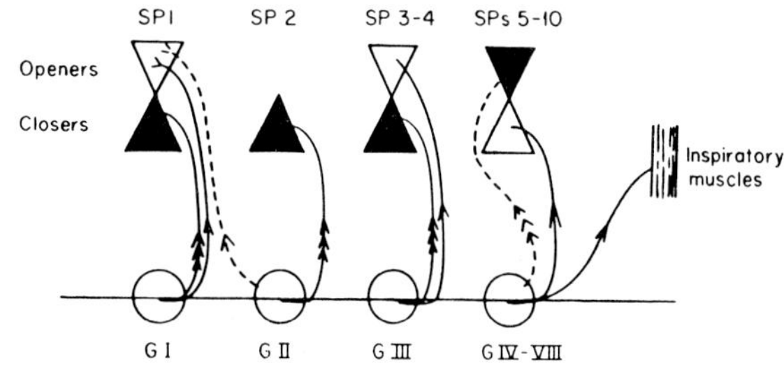

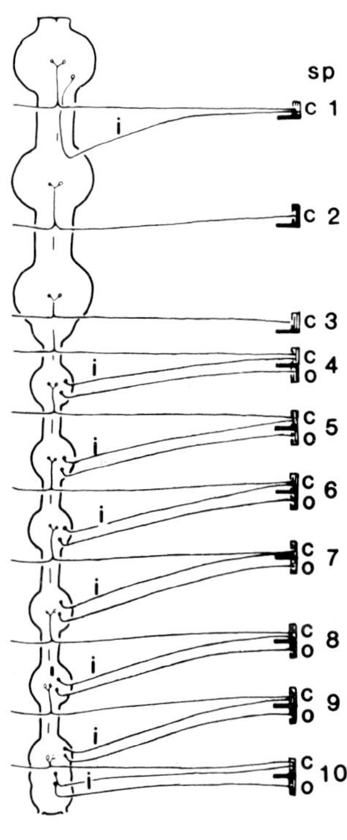

Fig. 43. (**a**) Summary diagram of the innervation of the spiracle muscles in *Schistocerca gregaria*. Black triangles indicate muscles active during expiration. Continuous lines, median nerves; dashed lines, lateral nerves. GI–III, thoracic ganglia; GIV–VIII, abdominal ganglia; sp, spiracle. (**b**) Summary diagram of the supposed excitatory and inhibitory (i) motor neurons innervating the opener (o) and closer (c) muscles of spiracles (sp) 1–10 in cockroaches. (See text for confirmatory evidence of inhibitory motor neurons). (**a** from Miller, P., 1974; **b** from Miller, P., 1981a.)

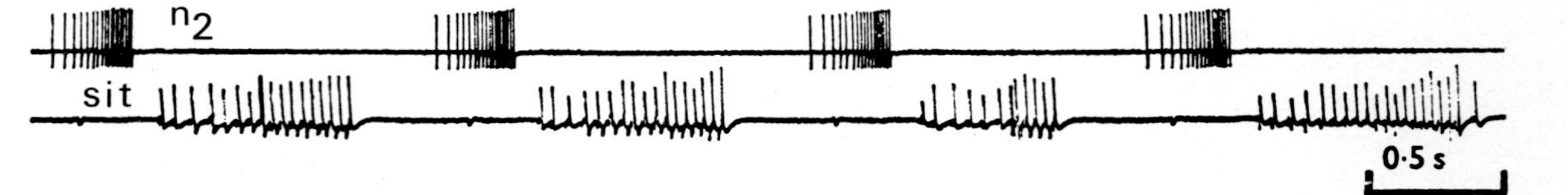

FIG. 44. Alternation of expiratory bursts in a second lateral nerve (n_2) and inspiratory bursts in the subintestinal muscle (sit) in an aeshnid larva. (From Mill, P., 1970.)

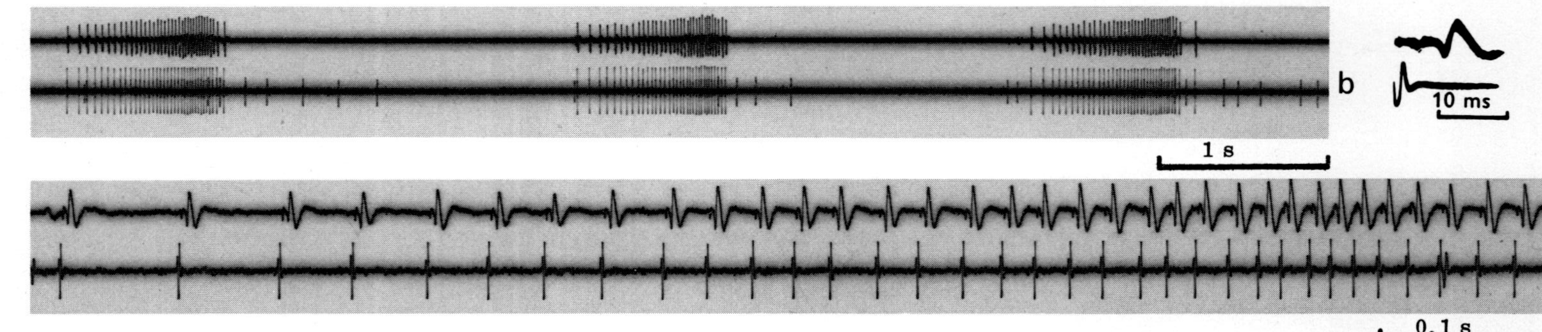

FIG. 45. Recording of expiratory bursts from a second lateral nerve (lower traces) and from the "primary" expiratory dorsoventral muscle which it innervates (upper traces) in *Aeshna*. (**a**) Consecutive bursts; (**b**) superimposed potentials in a single burst; (**c**) single burst. Note the facilitation of the muscle potentials. (From Mill, P. and Hughes, G., 1966.)

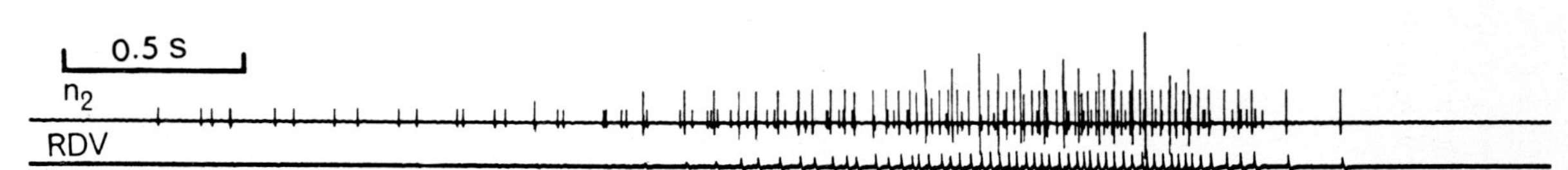

FIG. 46. Recording of an expiratory burst from a second lateral nerve (n_2) and from the "primary" expiratory dorsoventral muscle which it innervates (RDV) in *Anax imperator*. Note the 1:1 relationship between one of the units in the nerve and the muscle potentials. (From Mill, P., 1970.)

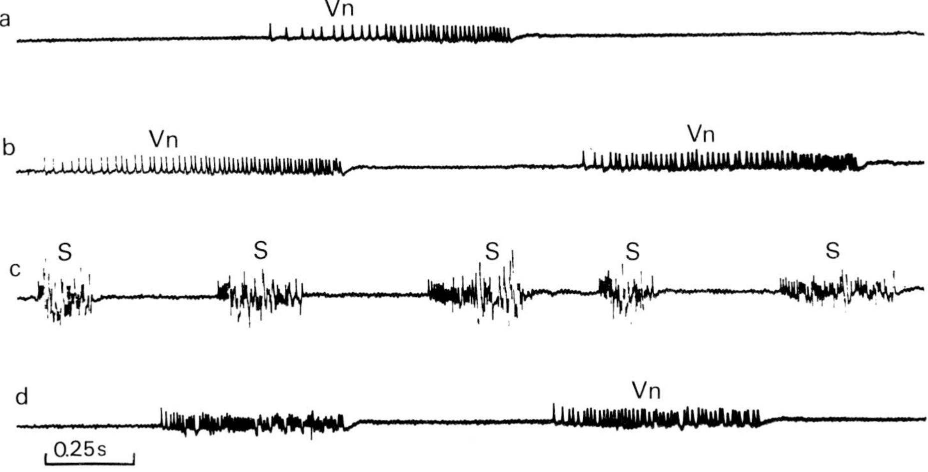

FIG. 47. Extracellular, chronic recordings from an expiratory dorsoventral muscle of an unrestrained aeshnid larva showing the transition from normal ventilation (Vn) to jet-propulsion (S) and back again; **a**–**d** are continuous records. (From Mill, P. and Pickard, R., 1975.)

1974). The cell bodies of the neurons innervating the latter two spiracles are located ventromedially (Burrows, M., 1975b).

4.2 The motor output

In a number of insects expiratory and inspiratory bursts alternate rhythmically during normal ventilation and the opening and closing of the spiracles (in those with an open tracheal system) is often synchronized with one or other of these phases. Similarly, in those insects which use their gills to produce a ventilatory current, retractor and protractor bursts show an alternating rhythm. It is convenient to deal with some of these aspects separately before proceeding with a discussion of intrasegmental and intersegmental coordination.

4.2.1 EXPIRATION AND INSPIRATION

Expiratory bursts containing one or more units are typical of many insects, including aeshnid dragonfly larvae, *S. gregaria*, *L. migratoria*, *H. membranacea* and *G. portentosa*. They have been recorded both from the segmental nerves and from individual muscles.

In aeshnid larvae the expiratory dorsoventral muscles are innervated by three motor neurons in the second segmental nerves. In abdominal ventilation in a quiescent animal only one of these is normally active and its firing pattern within each expiratory burst is rather unusual, comprising a steady increase in frequency to a maximum, which is maintained for a short period before firing stops, usually fairly abruptly (Figs 44 and 45; Table 8). The corresponding muscle potentials show facilitation, increasing in amplitude with increase in motor neuron frequency (Mill, P. and Hughes, G., 1966; Mill, P., 1970) (Figs 45 and 46). During stronger ventilation, as is observed for instance prior to and immediately following jet-propulsive swimming, this pattern is lost, the unit firing throughout the burst at a higher frequency (Fig. 47). Indeed, during jet-propulsive swimming the firing frequency is still higher, the muscle potentials are larger and one or two extra motor units are probably recruited; also the bursts are shorter (Fig. 47) (Mill, P. and Pickard, R., 1975). The effect of the expiratory burst is to increase the tension on the expiratory dorsoventral muscle; this contracts, raising the sterna. However, the sterna continue to lift after the tension has started to fall (Fig. 29); this is probably an inertia effect of the water leaving the branchial chamber (Mill, P. and Pickard, R., 1972b; Pickard, R. and Mill, P., 1972).

The axons of the motor neurons innervating the anterior dorsoventral muscle also run in the second lateral nerves. At low ventilatory frequencies they are often silent (apparently more so in *Aeshna* than in *Anax imperator*) (cf. Figs 44 and 45). When they do fire the pattern in the (expiratory) bursts is fairly regular (Table 8), but the frequency is lower than in

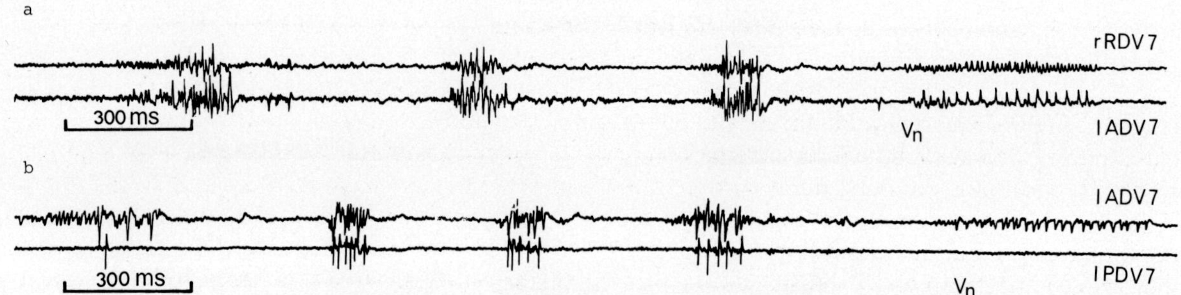

FIG. 48. Extracellular, chronic recordings from an unrestrained aeshnid larva during ventilation (V_n) and jet-propulsion. **(a)** Recordings from expiratory (RDV) and anterior (ADV) dorsoventral muscles. **(b)** Recordings from anterior (ADV) and posterior (PDV) dorsoventral muscles. l, left; r, right; numeral indicates abdominal segment. (From Mill, P. and Pickard, R., 1975.).

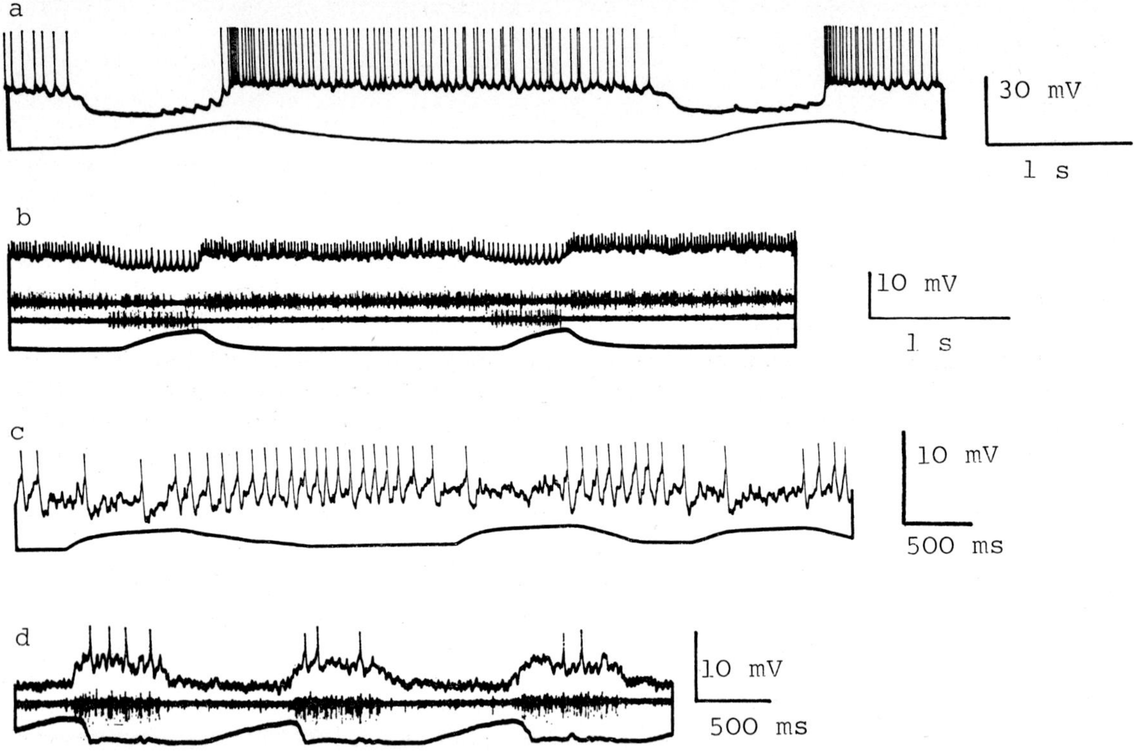

FIG. 49. Intracellular recordings from four different types of expiratory motor neurons in *Schistocerca gregaria.* **(a)** A bursting motor neuron showing a decrease in frequency during the burst. **(b)** A tonic motor neuron with higher-frequency expiratory bursts. **(c)** a motor neuron the activity of which waxes and wanes. **(d)** A phasic motor neuron which only fires at high ventilatory rates. Expiration is indicated by downward movement of the lower traces. (From Burrows, M., 1974.).

FIG. 50. The relationship between dorsoventral movements of the sterna (downwards on the upper trace indicates abdominal compression, i.e. expiration) and expiratory bursts recorded chronically from one of the expiratory muscles (anterior tergo-sternal) in the third (3) and sixth (6) abdominal segments of *Hierodula membranacea.* (From Kerry, C., 1982.)

expiratory bursts recorded at the same time from the expiratory dorsoventral muscle. The bursts tend to be fairly well synchronized with those to the latter; again the frequency within the bursts and the amplitude of the muscle potentials increase during jet propulsion, while the burst duration decreases (Fig. 48). The posterior dorsoventral muscle is not involved in normal expiration but receives low-frequency "expiratory" bursts during jet-propulsion (Mill, P. and Pickard, R., 1975).

The main expiratory dorsoventral muscles in *S. gregaria* and *L. migratoria* (206, 208 and 209 in segment 5) are also innervated by expiratory motor neurons in the second lateral nerves but muscle 205, which is brought into action only during hyperventilation, receives its innervation from the first lateral nerves (Lewis, G. *et al.*, 1973; Hustert, R., 1975).

In *Schistocerca gregaria* four types of expiratory motor neuron have been described, based on their firing pattern and on their interneuronal input (Table 8). One of these neurons starts firing with a fairly high frequency which decreases steadily throughout the burst (Fig. 49a). The others are a bursting tonic neuron, which is similar to the above except that it continues to fire (at a lower frequency) throughout inspiration (Fig. 49b); a neuron which tends to fire regularly at a fairly low overall frequency and which waxes and wanes over the cycle with its maximum frequency occurring during, but not necessarily at the beginning of, expiration (Fig. 49c); and a phasic neuron which fires just a few times during expiration (Fig. 49d) (Burrows, M., 1974). The last two are similar to those neurons which innervate the dorsal longitudinal muscles in aeshnid larvae (see below). At low ventilatory rates in *S. gregaria* the overall pattern tends to be somewhat variable. Expiratory neurons sometimes start to fire tonically as soon as inspiration ceases but do not cause abdominal compression until later, when they produce a high-frequency burst. Alternatively there is an initial marked increase in frequency, when the sterna are partly raised, followed by a lower-frequency tonic firing, producing a pause in the expiratory movement, and a second high-frequency phase which completes the expiratory compression (Lewis, G. *et al.*, 1973). The bursts are normally initiated by slow units; fast units (when they occur) firing later in the burst. At higher amplitudes of ventilation more units are recruited (motor recruitment) (Hinkle, M. and Camhi, J., 1972; Lewis, G. *et al.*, 1973). The situation in *Locusta migratoria* is rather similar. During the expansion (rest) phase low frequencies of firing can often be recorded from the primary expiratory muscle 209 (in segment 5). Muscles 206 and 208 are usually involved with expiration but are only activated by a slow unit during short and weak ventilatory strokes. Muscle 205 tends to be recruited only during stronger ventilation. There are two slow motor neurons innervating all four of these muscles, while 205 and 206 also receive innervation from fast motor neurons. The slow unit in muscle 206 fires fairly regularly throughout the burst (Table 8) (Hustert, R., 1975). Expiratory bursts have also been recorded from the posterior and anterior tergo-sternal muscles in *Hierodula membranacea* (Fig. 50) (Kerry, C., 1982; Kerry and Mill, in preparation) and from the tergo-sternal muscle in *Gromphadorhina portentosa* (Mill *et al.*, unpublished observations).

The longitudinal muscles involved in expiration are innervated by axons in the first lateral nerve in aeshnid larvae, *S. gregaria* and *L. migratoria*. In aeshnid (e.g. *Aeshna* and *Anax*) larvae the firing pattern during expiration is somewhat variable, consisting of anything from one or two action potentials to an increase in frequency in a continually firing neuron just before the start of the expiratory burst to the expiratory dorsoventral muscle (Fig. 51 and Table 8). In *L. migratoria* axons innervating the dorsal longitudinal muscles 197 and 198 and the ventral longitudinal muscle 203 (in segment 5) are active during shallow ventilation. With stronger, more rapid ventilation extra motor units are recruited and motor neurons innervating the ventral longitudinal muscles 199 and 202 fire weakly (Table 8). In aeshnid larvae and *L. migratoria* the effect on the longitudinal muscles is presumably just sufficient to prevent the increase in abdominal pressure from causing elongation; but is insufficient to elicit a contraction. A similar situation occurs in *S. gregaria* but well-formed expiratory bursts have been recorded from the first lateral nerves during hyperventilation (Lewis, G. *et al.*, 1973; Hustert, R., 1975; Pickard, R. and Mill, P., 1975).

When aeshnid larvae undergo jet-propulsive swimming, short bursts of activity synchronized with those to the expiratory dorsoventral muscles

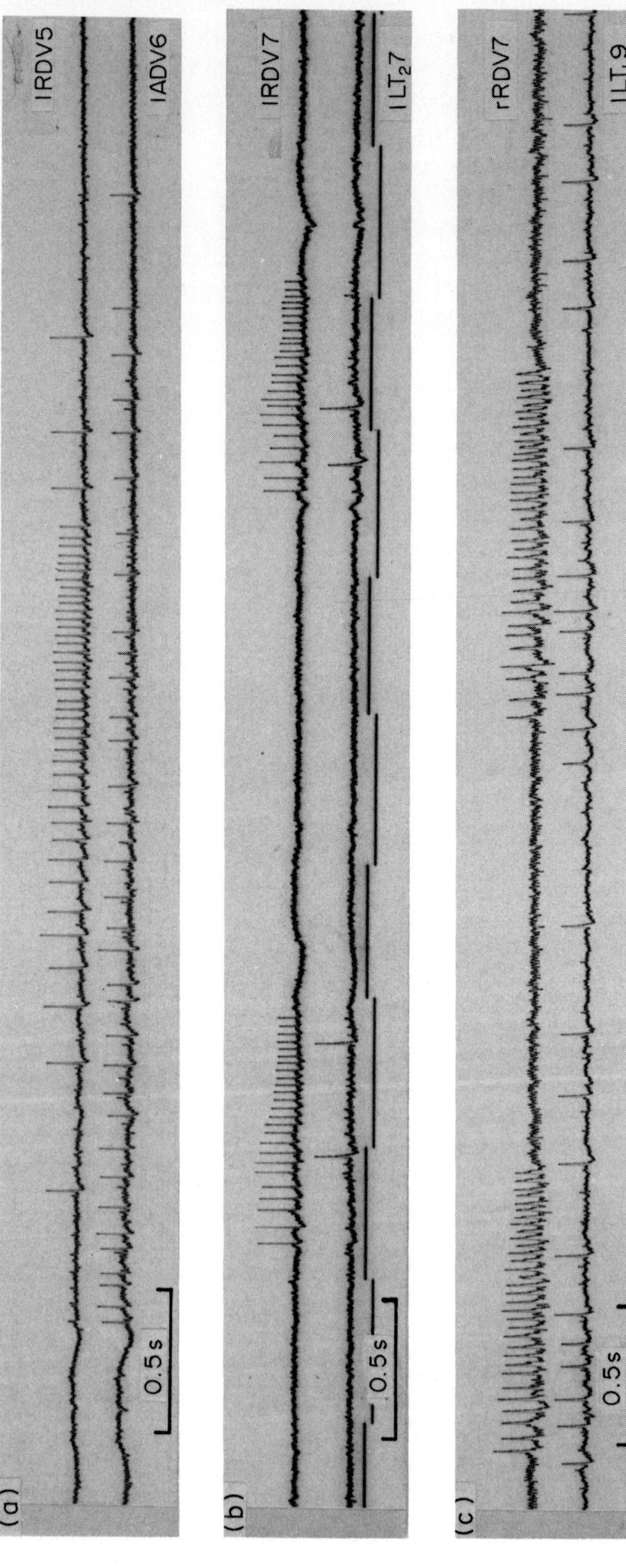

FIG. 51. Chronic recordings from muscles during normal ventilation in an aeshnid dragonfly larva. The top trace in each record is from the expiratory dorsoventral muscle (RDV). The lower traces are from **(a)** the anterior dorsoventral muscle (ADV), **(b)** a longitudinal tergal muscle (LT_2) and **(c)** a different longitudinal tergal muscle (LT_1); l, left; r, right; final numeral indicates abdominal segment. (From Pickard, R. and Mill, P., 1975.)

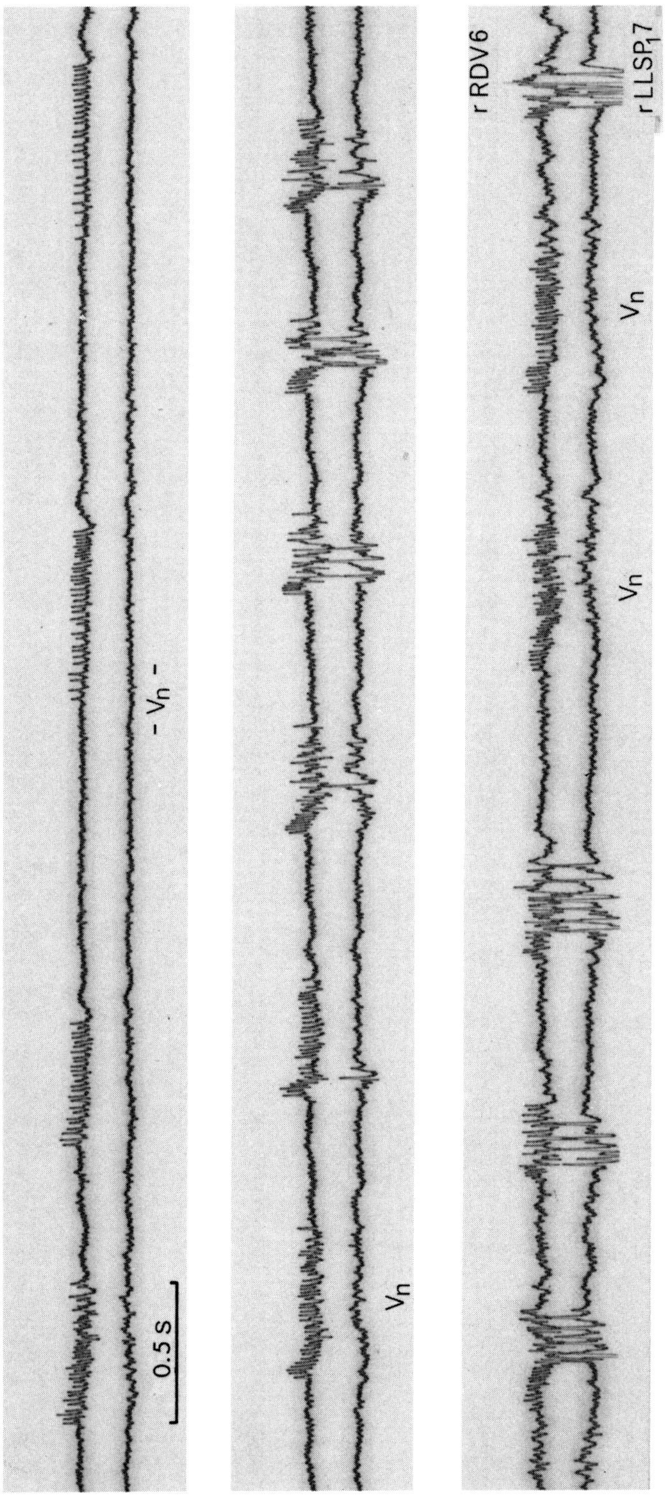

FIG. 52. Extracellular chronic recordings from an unrestrained aeshnid larva during ventilation (V_n) and jet-propulsion. Upper traces are from an expiratory dorsoventral muscle (RDV); lower traces are from one of the ventral, longitudinal muscles ($LLSP_1$). l, left; r, right; numeral indicates abdominal segment. (From Mill, P. and Pickard, R., 1975.)

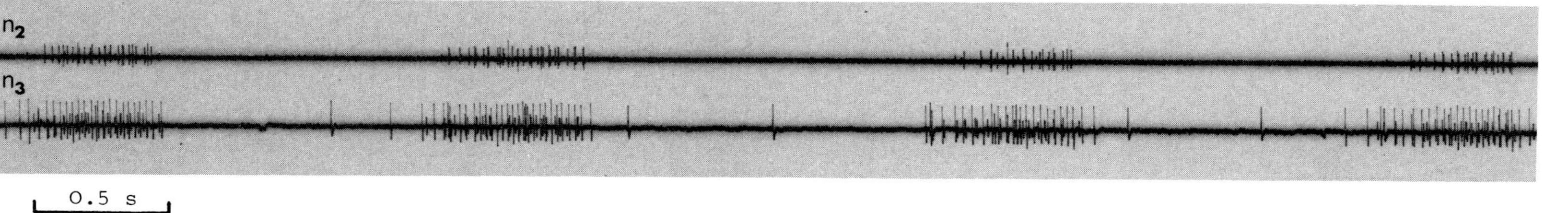

FIG. 53. Recordings of expiratory bursts in second (n_2) and third (n_3) lateral abdominal nerves of a teneral adult of *Aeshna*. (From Mill and Whittles, in preparation.)

occur in the neurons innervating the longitudinal muscles (Fig. 52) (Mill, P. and Pickard, R., 1975). In teneral adult aeshnids the primary expiratory dorsoventral muscle is probably the posterior tergopleural muscle, the expiratory dorsoventral muscle of the larva having disappeared at metamorphosis. The expiratory bursts, in the third segmental nerves, contain two or three motor units, the most active one firing fairly regularly throughout the burst (Table 8 and Fig. 53) (Mill and Whittles, in preparation).

In *S. gregaria* the duration of the expiratory burst increases with increase in cycle duration (Fig. 54) (Lewis, G. *et al.*, 1979). The output to each side of a ganglion is closely synchronized but there is not normally a 1:1 relationship between the action potentials in corresponding motor neurons (e.g. Mill, P. and Hughes, G., 1966). In *Blaberus craniifer* Miller, P. (1981a) noted that the expiratory bursts tended to be either short (about 200 ms) or long (2–4 s), with few of intervening duration.

In gulping breathing in aeshnid larvae, where the branchial chamber is pumped up (section 3.2), there is an increased frequency of expiratory bursting in the motor output to the expiratory dorsoventral muscles. During short pauses in ventilation there may be continual low-level activity in the expiratory dorsoventral muscle but longer periods of apnoea tend to be accompanied by a cessation of firing of

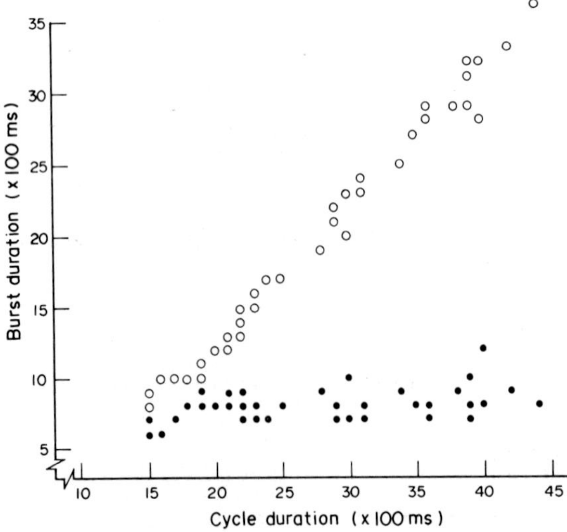

FIG. 54. The relationship between the duration of expiratory (○) and inspiratory (●) bursts and the duration of the ventilatory cycle in *Schistocerca gregaria*. (From Lewis, G. *et al.*, 1973.)

the expiratory motor neuron innervating this muscle, although there may still be activity in the (inspiratory) transverse sub-intestinal muscle (Pickard, R. and Mill, P., 1972, 1975). In *H. membranacea* the compression (Co) phase is elicited by high-frequency firing of the expiratory motor neurons (Fig. 35) (Kerry, C., 1982; Kerry and Mill, in preparation).

The expiratory and inspiratory bursts alternate and, since the fully expanded position of the abdomen is the rest position, inspiration always follows expiration in the normal ventilatory rhythm. (Of course, as mentioned in section 3.3.2.1, active inspiration does not occur in all insects which show active ventilatory movements.)

The segmental inspiratory dorsoventral muscles, where they occur, and the diaphragm and sub-intestinal muscle of anisopteran dragonfly larvae, are innervated by a pair of motor neurons in the appropriate median nerve. They lie on either side of the mid-line and both send a branch to each side; hence, unlike in the expiratory motor system, the output to each side of a ganglion is identical. Typically, each unit normally fires with a fairly constant frequency throughout the inspiratory burst; this is the case in aeshnid larvae, *S. gregaria* and *L. migratoria* (Figs 44, 55 and 56; Table 8) (Mill, P. and Hughes, G., 1966; Mill, P., 1970; Miller, P., 1973; Burrows, M., 1974; Hustert, R., 1975), although in the larva of the aeshnid *Anax parthenope* Komatsu, A. (1980, 1982) noted a slightly lower frequency at the start and end of each burst. In central recordings from inspiratory motor neurons in *S. gregaria*, Burrows, M. (1974) noted that they sometimes continue firing throughout most or all of inspiration, but at around a third of the frequency of the inspiratory burst. During the expansion (rest) phase the inspiratory dorsoventral muscle in *L. migratoria* (207 in segment 5) often continues to exhibit muscle potentials at a low frequency. Indeed, in this species occasional miniature inspiratory bursts occur during the expansion phase in both this muscle and the longitudinal inspiratory muscle (204 in segment 5), with concurrent inhibition of the primary expiratory dorsoventral muscle (209 in segment 5) (Hustert, R., 1975).

In *S. gregaria* at least the duration of the inspiratory bursts remains fairly constant over the cycle duration range of 1.5–4.4 s (i.e. 40 to 14 cycles

Table 8: Types of motor neurons in dragonfly larvae and locusts

Type	Example
Expiratory	
(a) Bursting (increase in frequency) 	abdominal expiratory dorsoventral muscle (rdv) in aeshnid larvae.
(b) Bursting (decrease in frequency) 	metathoracic ganglion, lateral nerves and abdominal dorsoventral muscle (206) in *Schistocerca gregaria.*
(c) Bursting (regular) 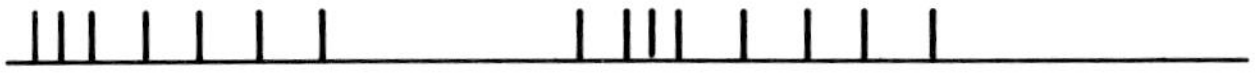	fifth abdominal ganglion and abdominal anterior dorsoventral muscle (adv) in aeshnid larvae; dorsoventral muscle (206) (slow unit) in *Locusta migratoria.*
(d) Bursting (tonic) 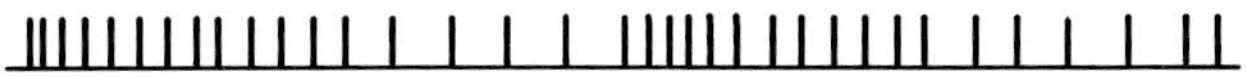	metathoracic ganglia of *Schistocerca gregaria.*
(e) Waxing and waning 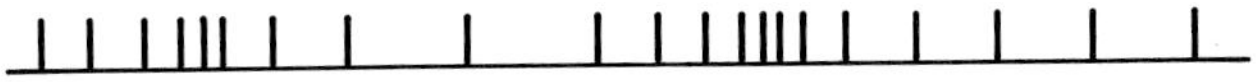	abdominal primary longitudinal tergal muscle (LT1) in aeshnid larvae; metathoracic ganglion in *Schistocerca gregaria.*
(f) Phasic 	abdominal secondary longitudinal tergal muscle (LT2) in aeshnid larvae; metathoracic ganglion (during stressed ventilation) in *Schistocerca gregaria*; abdominal longitudinal ventral muscle (199) in *Locusta migratoria.*
Inspiratory 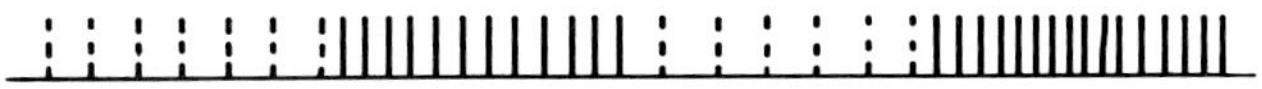	fifth abdominal ganglion, diaphragm and sub-intestinal muscle in aeshnid larvae; metathoracic ganglion and median nerves 3 and 4 in *Schistocerca gregaria*; abdominal dorsoventral muscle (207) in *Locusta migratoria*
Closer 	closers of spiracles 1 to 4 in *Schistocerca gregaria.*
Opener 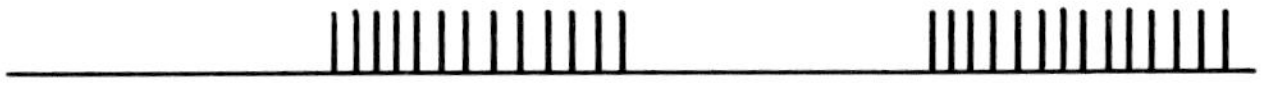	opener of spiracle 4 in *Schistocerca gregaria.*

Muscle numbers for *Schistocerca gregaria* and *Locusta migratoria* refer to segment 5. Corresponding muscle numbers in segment 4 can be obtained by subtracting 15.

Data from Mill, P. and Hughes, G. (1966); Lewis, G. *et al.* (1973); Burrows, M. (1974, 1975b); Hustert, R. (1975); Pickard, R. and Mill, P. (1975); Miller, P. and Mills, P. (1976); see also Mill, P. (1977).

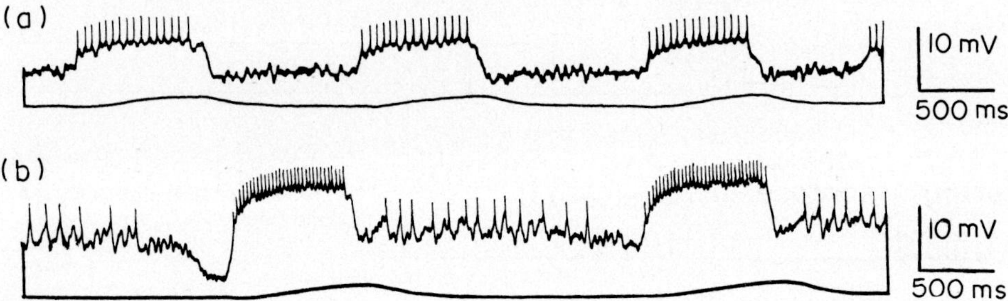

FIG. 55. Intracellular recordings from inspiratory motor neurons in *Schistocerca gregaria.* (**a**) Bursts of spikes only during inspiration. (**b**) Bursts of spikes during inspiration together with a lower firing frequency during expiration. Inspiration is indicated by upward movement of the lower traces. (From Burrows, M., 1974.).

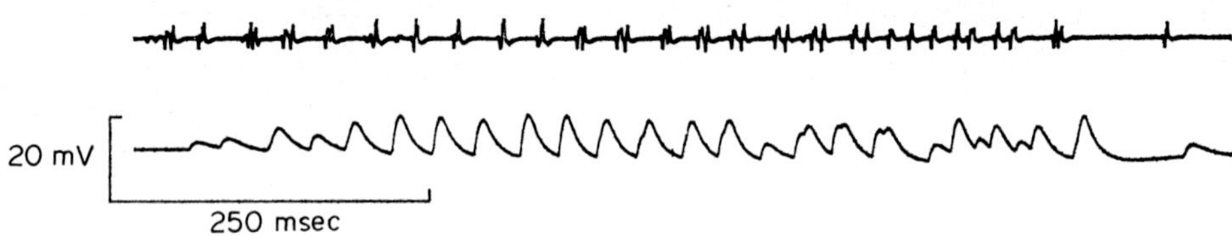

FIG. 56. Recordings of an inspiratory burst from a transverse nerve (upper trace; extracellular) and the inspiratory muscle (192 in segment 4) which it innervates (lower trace; intracellular) in *Schistocerca gregaria.* (From Lewis, G. *et al.*, 1973.)

min^{-1}) (Lewis, G. *et al.*, 1973) (Fig. 54). Hence, in this species cycle duration is primarily adjusted by change in the duration of the expiratory bursts.

In aeshnid larvae and in *Schistocerca gregaria* the delay between an expiratory burst and the subsequent inspiratory burst is generally shorter than that between inspiration and expiration (Fig. 44) (Mill, P., 1970; Lewis, G. *et al.*, 1973).

4.2.2 Gill Retraction and Protraction

The aquatic larva of the megalopteran *Corydalus cornutus* alternately retracts and protracts its gills (section 2.2.2.1) to produce a ventilatory current. Gill retraction is elicited by a single (power stroke) retractor muscle, whereas several muscles are involved in protraction. The gill retractor muscle is innervated by four motor neurons, bursts of firing in one of which alternate with bursts in one of the two gill protractor motor neurons (Fig. 57). Both types of neuron often fire at an initially high frequency, which then declines throughout the burst (Table 7), but the overall firing frequency of the retractor neurons is higher than that of the protractor neurons. The retractor bursts are shorter in duration than the protractor bursts but both decrease in duration while the intra-burst firing frequency increases as ventilation frequency increases above 54 beats min^{-1}. Below this ventilatory frequency retractor burst lengths are constant, but the protractor motor neurons may cease firing (Kinnamon, S. and Kammer, A., 1983). In low pO_2 (8%) the retractor bursts decrease in duration and the retractor motor neuron firing rate increases; gill protractor motor neurons are recruited (Kinnamon, S., Kammer, A. and Kiorpes, A., 1984).

4.2.3 Spiracular Movements and Their Coordination with Ventilation

All spiracles possess a closer muscle innervated by a pair of motor neurons, the cell bodies of which lie one on either side of the mid-line of the ganglion (section 4.1.3). In *S. gregaria* they are adjacent to the pair which innervate the inspiratory dorsoventral muscles. Like the latter, the axon arising from each cell body divides shortly after leaving the ganglion, one branch running to each side. In the absence of a ventilatory rhythm, which occurs regularly in resting cockroaches (e.g. *Blaberus* and *Periplaneta*) and is quite common in adult dragonflies, especially quiescent ones, but less frequent in *S. gregaria*, those innervating the anterior spiracles, at least, fire continuously at a fairly low frequency. This has been termed "free-running" and the pair of

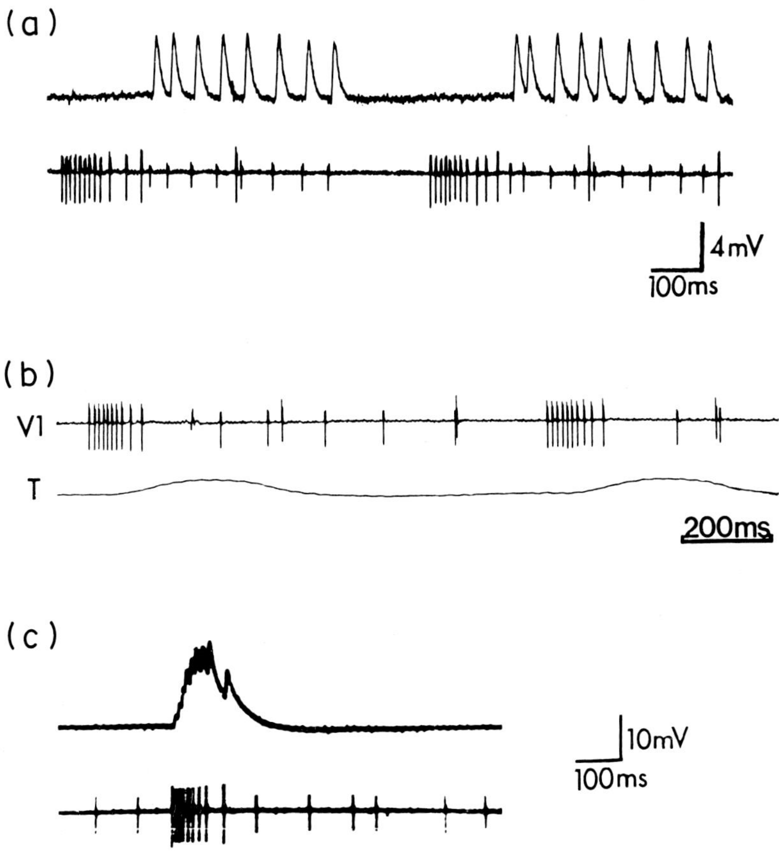

FIG. 57. Recordings from the gill retractor and protractor motor neurons and muscles of *Corydalus cornutus* during ventilation. (**a**) Intracellular recording from a gill protractor muscle (upper trace) and extracellular recording from the nerve which innervates both retractor and protractor muscles (lower trace). (**b**) Extracellular recording from the nerve innervating the gill muscles (V1) and movement of the gill (T). The burst in the upper trace is retractor activity. (**c**) Extracellular recording from the nerve innervating the gill muscles (lower trace) and intracellular recording from a fibre in the retractor muscle (upper trace). Calibration marks refer to intracellular traces. (**a**, **c** from Kinnamon, S. and Kammer, A., 1983; **b** by kind permission of Kinnamon, S. and Kammer, A.)

neurons in any one ganglion have approximately equal firing frequencies (Fig. 58). In *S. gregaria* those innervating spiracle 1 tend to fire at 17–19 s^{-1}, those innervating spiracle 2 at 8–10 s^{-1}. Since the frequencies of the members of each pair are not identical, they are sometimes in phase, which leaves slightly longer periods when the closer muscle is not receiving excitation; hence the latter relaxes briefly and slightly, allowing the spiracles to flutter open (Case, J., 1957; Miller, P., 1962, 1965, 1973).

The anterior spiracles tend to close during expiration and open during inspiration (section 3.3.2.3). For example, in slowly ventilating *S. gregaria*, where there is a pause in the expiratory phase, there is an initial high-frequency burst in the closer motor neurons (up to 250 s^{-1} in each axon) followed by a maintained firing at a lower frequency (about 40 s^{-1} per axon) during the pause, and culminating in another burst during the final stage of sternal lifting. Firing then ceases as the abdomen is expanded during inspiration. The two motor neurons are fairly closely coupled during this activity. A similar relationship occurs in spiracle 2 (Miller, P., 1965). With increase in ventilatory frequency the expiratory pause disappears and the burst in the closer motor neurons shows an initial high frequency which then declines throughout the burst (Fig. 59) and the action potentials tend to occur in groups of two or three (Table 7 and Fig. 60) (Miller, P., 1965; Miller, P. and Mills, P., 1966; Burrows, M., 1974, 1975b).

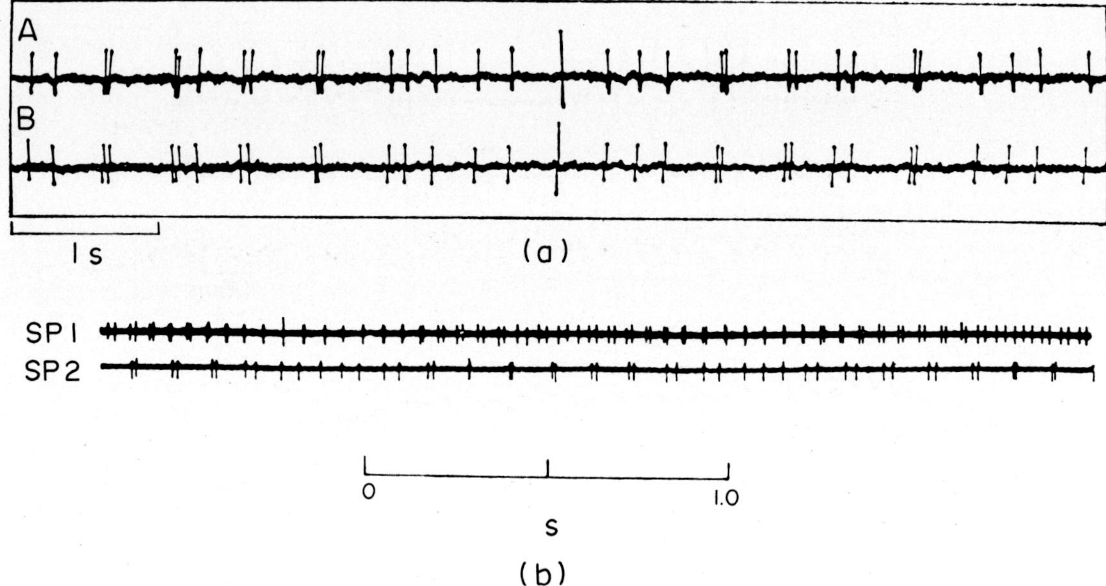

FIG. 58. "Free-running" activity in the closer motor neurons innervating (**a**) spiracles 2 (A, right; B, left) of *Periplaneta americana* and (**b**) spiracles 1 (SP1) and 2 (SP2) of *Schistocerca gregaria*. In (**b**) the connectives between the meso- and metathoracic ganglia had been severed. (**a** from Case, J., 1957; **b** from Miller, P., 1965.)

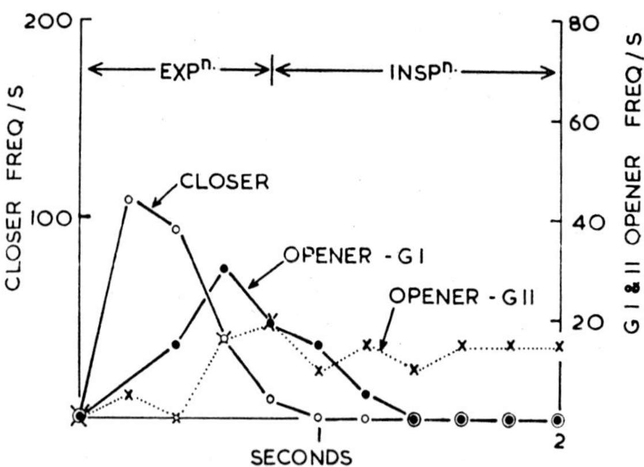

FIG. 59. The frequency of motor impulses in the nerves to the closer and opener muscles of spiracle 1 of *Schistocerca gregaria* during ventilation at a higher frequency than shown in Fig. 33. G1, activity in the mesothoracic opener neurons; G2, activity in the prothoracic motor neurons. (From Miller, P., 1974.)

The opener muscle of spiracle 1 in *S. gregaria* receives excitation from two axons in the median nerve which fire with a burst towards the end of expiration. Each neuron shows a fairly steady frequency throughout the burst. They nevertheless have a positive effect on opening since tension in the opener muscle decays more slowly than that in the closer muscle. In the higher ventilatory rates induced by CO_2 these opener motor neurons start to fire earlier in the expiratory phase and the burst extends into the inspiratory phase. Ultimately these neurons fire throughout the cycle. A third opener motor neuron to spiracle 1, in a lateral nerve, also tends to fire throughout the cycle, but is more active during inspiration (Miller, P., 1965) (Fig. 59). The bursts in the opener motor neurons in the median nerve innervating spiracle 4 in *S. gregaria* occur during inspiration. Again they show a fairly steady firing frequency, but perhaps rather higher at the start of the burst (Fig. 61 and Table 7) (Burrows, M., 1975b). There is no reason to believe that these are not the same type as those in the median nerve innervating spiracle 1. In *Blaberus giganteus* it has been shown that, of the two motor neurons innervating the opener muscle of spiracle 10, one is a "slow" axon eliciting 1–2 mV postsynaptic potentials while the other is a "fast" axon producing postsynaptic potentials in the region of 50 mV (Miller, P., 1973). However, in spiracles 4–10 of *Blaberus discoidalis* and *Gromphadorhina portentosa* the opener muscle is innervated by a common inhibitor neuron, activity in which causes rapid relaxation of the slowly contracting part of the opener muscle (but not of the fast-contracting part) (Kaars, C., 1983). Kaars mentions that it sometimes innervates

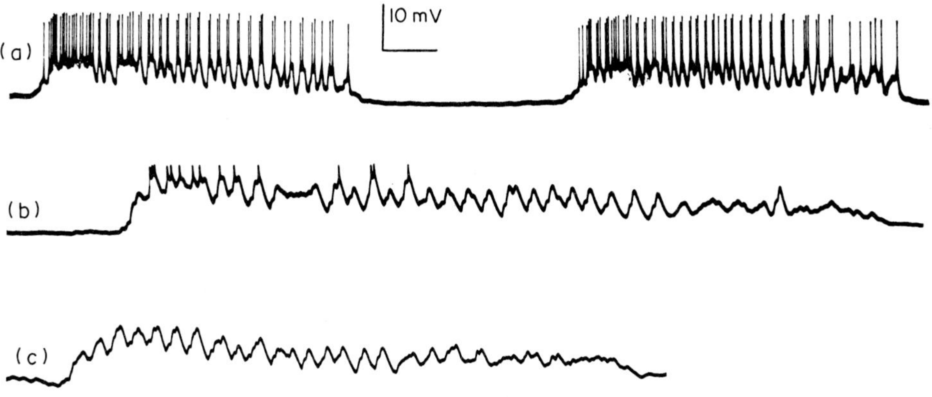

FIG. 60. Intracellular recordings from a spiracle closer motor neuron in *Schistocerca gregaria*. The neuron is kept hyperpolarized in (**b**) and more so in (**c**), so that the nature of the excitatory postsynaptic potentials is revealed. Horizontal scale: (**a**) 250 ms, (**b, c**) 125 ms. (From Burrows, M., 1974.).

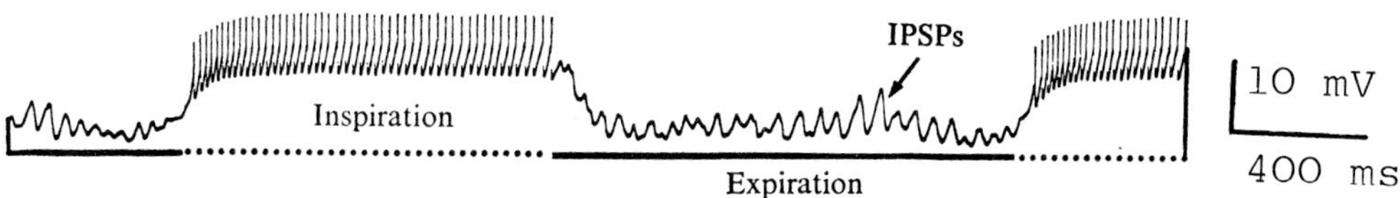

FIG. 61. Intracellular recording from an opener motor neuron of spiracle 4 in *Schistocerca gregaria*. IPSPs, inhibitory postsynaptic potentials. (From Burrows, M., 1975b.).

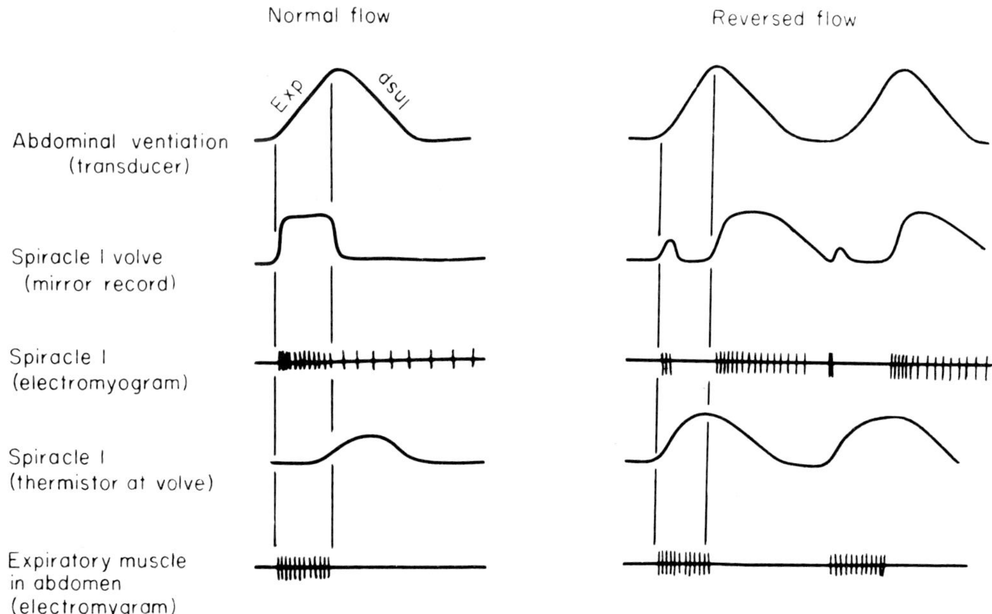

FIG. 62. Diagram showing the activity of spiracle 1 of *Sphodromantis* during normal unidirectional and reversed ventilation. Upwards on the spiracle valve record indicates closing. The thermister records of spiracle 1 show air movements through the valve (upwards on the trace) during inspiration with normal flow and during expiration with reversed flow. (From Miller, P., 1974.)

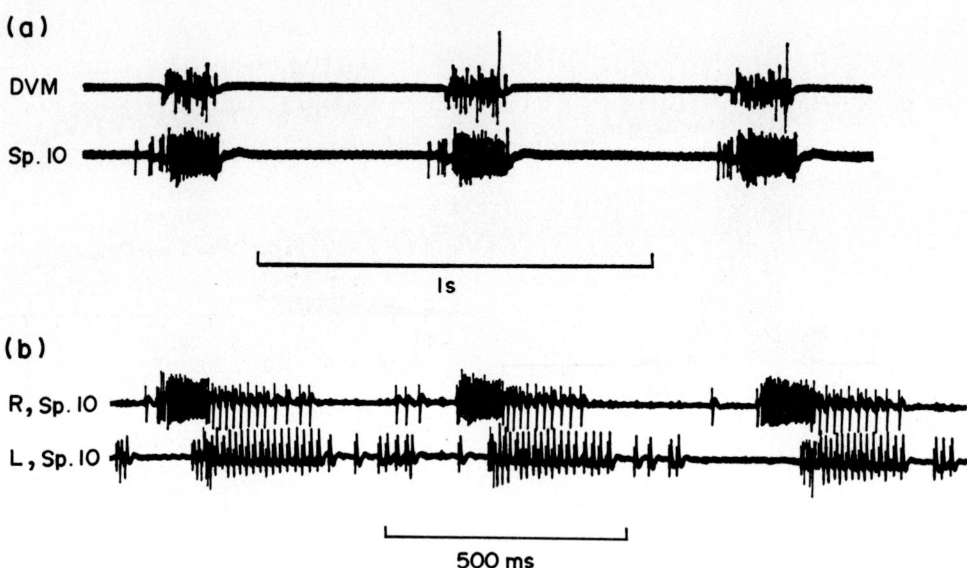

FIG. 63. (**a**) Muscle activity in an expiratory dorsoventral muscle (DVM) and the opener muscle of spiracle 10 (Sp. 10) in *Blaberus giganteus* during rapid ventilation. (**b**) Muscle activity in the opener muscle of spiracle 10 (L, left; R, right) during transitional coupling in *Blaberus giganteus*. The right spiracle is dominant and shows strong expiratory bursts and weak inspiratory bursts; the left spiracle is subordinate and shows slight expiratory activity and strong inspiratory bursts. (From Miller, P., 1973.)

closer muscle fibres. Thus it is possibly the same neuron as is described by Nelson, M. (1979) as a specific closer inhibitor of the abdominal spiracles.

The opening of spiracle 1 in the mantid *Sphodromantis*, as in *S. gregaria*, is normally coupled with inspiration, the closer motor neurons producing a high-frequency burst which coincides with the expiratory burst (Fig. 62). However, it occasionally switches over and its opening becomes coupled with expiration. This occurs when the "expiratory" burst in the closer motor neurons is attenuated to a few action potentials at the start of expiration or disappears altogether, while its firing frequency during inspiration increases (Miller, P., 1974). Similarly, the opening of spiracles 10 in *Blaberus giganteus* can be coupled with either expiration (Fig. 63a) or inspiration. When opening is coupled with inspiration in *B. giganteus*, bursts occur in the fast axon. While, in general, spiracles on opposite sides of one segment tend to be coupled, as indeed is the case here, the separate innervation of the spiracle opener muscles on opposite sides does allow the possibility of independent action. Indeed, when opening in spiracle 10 is coupled with expiration, only one spiracle is actively opened. The fast opener on one side shows high-frequency bursts during expiration (dominant) while that on the other side is either silent or produces a few spikes (subordinate), usually at the end of the dominant neuron's burst, but gives a stronger burst during inspiration than does the dominant neuron (Fig. 63b). Spiracles 3, 6, 7 and 8 behave in a similar fashion to spiracle 10 but their thresholds are higher (Miller, P., 1973).

4.2.4 INTERSEGMENTAL COORDINATION

In probably most insects, expiratory (and inspiratory) bursts occur in most abdominal segments almost synchronously, although detailed analysis has revealed a tendency for an anterior–posterior rhythm in *Schistocerca gregaria* (Fig. 64) (Lewis, G. *et al.*, 1973), and in shallow ventilation in *Locusta migratoria* the segmental delay may be quite marked (Hustert, R., 1975). There are, however, certain exceptions. Thus, in abdominal ventilation in aeshnid larvae the rhythm is posterior–anterior. The expiratory bursts first appear from the eighth abdominal ganglion and emanate from successively more anterior ganglia as far forward as the fourth with an intersegmental delay in the region of 100 ms (Fig. 65) (Pickard, R. and Mill, P., 1972). In *Corydalis*, gill retractor bursts start in the third abdominal segment and appear from successively more posterior ganglia and then from the second

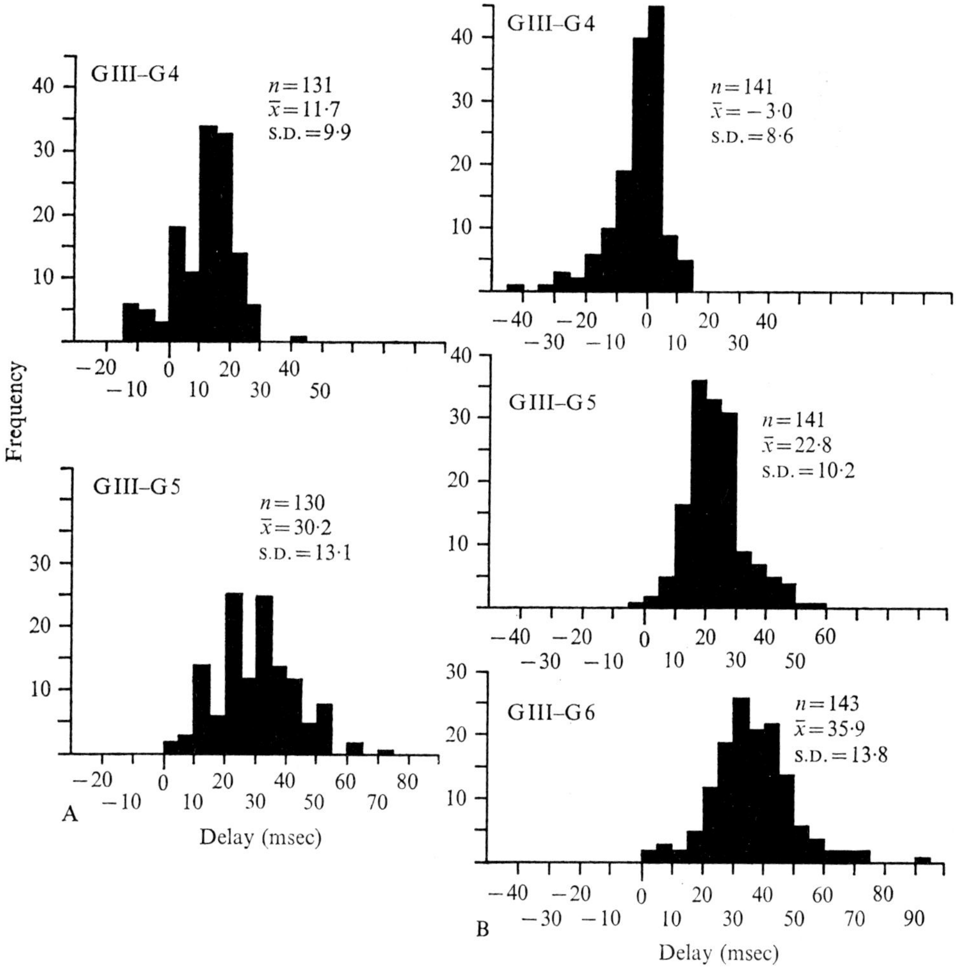

GIII–G4 $n=131$ $\bar{x}=11{\cdot}7$ S.D. $=9{\cdot}9$

GIII–G5 $n=130$ $\bar{x}=30{\cdot}2$ S.D. $=13{\cdot}1$

Frequency

A Delay (msec)

GIII–G4 $n=141$ $\bar{x}=-3{\cdot}0$ S.D. $=8{\cdot}6$

GIII–G5 $n=141$ $\bar{x}=22{\cdot}8$ S.D. $=10{\cdot}2$

GIII–G6 $n=143$ $\bar{x}=35{\cdot}9$ S.D. $=13{\cdot}8$

B Delay (msec)

FIG. 64. Histograms of the intervals between the production of (**A**) expiratory motor bursts in the lateral nerves of GIII and G4 (upper) and of GIII and G5 (lower), and (**B**) inspiratory motor bursts in the median nerves of GIII and G4 (top), GIII and G5 (middle) and GIII and G6 (bottom). Negative values indicate that bursts appear in a posterior ganglion earlier than in an anterior one. The results are all from one locust. (From Lewis, G. *et al.*, 1973.)

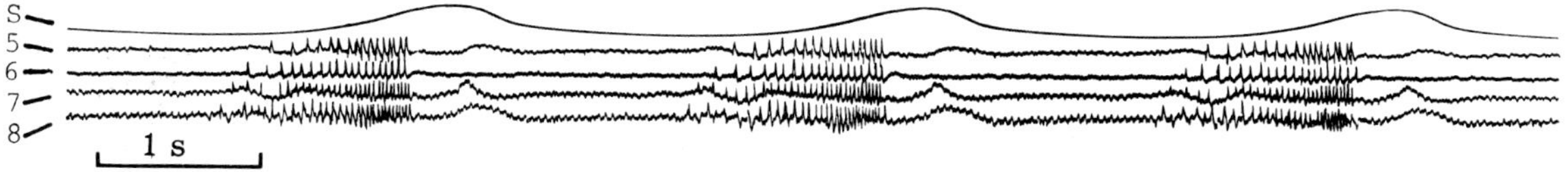

FIG. 65. Expiratory bursts recorded from the dorsoventral expiratory muscles of abdominal segments 5–8 of a larva of *Anax imperator*. S, Sternal movements (upwards indicates compression (i.e. expiration)). (From Pickard, R. and Mill, P., 1972.)

and first ganglia. The intersegmental delay is generally between 8 and 40 ms, except between the third and second segments, where it is noticeably longer (75–105 ms) (Fig. 66) (Kinnamon, S. and Kammer, A., 1983).

Free running does not show any signs of intersegmental coupling. However, during active ventilation spiracular activity is often coupled such that anterior spiracles are open during inspiration and posterior ones during expiration, and hence a unidirectional flow of air is produced in the longitudinal tracheal trunks (section 3.3.2.3).

In spite of this there is not necessarily any direct coupling between the motor neurons in adjacent

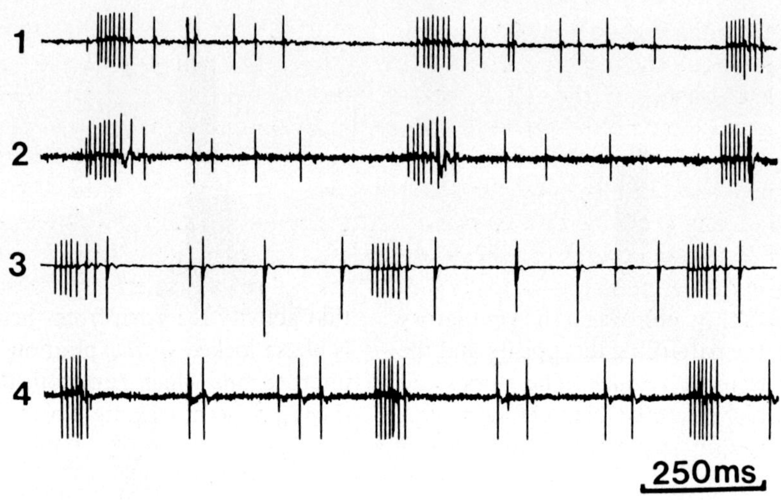

FIG. 66. Retractor bursts recorded from segments 1–4 of a larva of *Corydalus cornutus*. (From Kinnamon, S. and Kammer, A., 1983.)

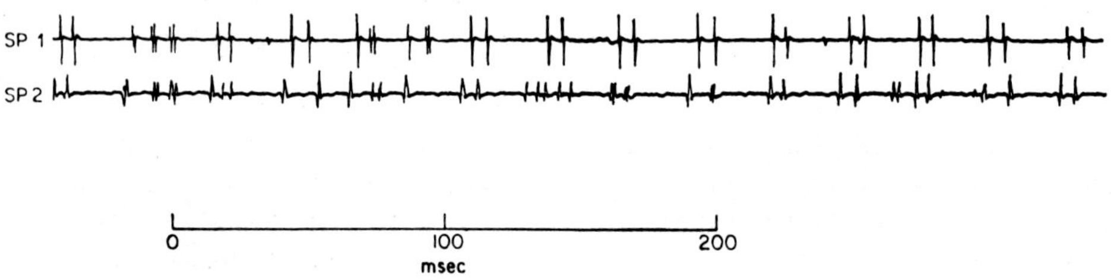

FIG. 67. Simultaneous recordings from the motor nerves to the closer muscles of spiracles 1 (SP1) and 2 (SP2) in a locust during an expiratory pause (see Fig. 33). (From Miller, P., 1965.)

segments. For example, the firing of the closer motor neurons of spiracles 1 and 2 in *S. gregaria* does not show any intersegmental coupling during the high-frequency burst at the start of expiration. However, they do show such coupling during the expiratory pause, with those to spiracle 2 leading those to spiracle 1, i.e. an anteriorly directed sequence in the thorax (Fig. 67) (Miller, P., 1965).

4.3 Central control

The available evidence indicates that the motor neurons are not themselves naturally oscillating; rather they receive a mixture of excitatory and/or inhibitory inputs and the type of input varies in the different neuron types. Thus, in *S. gregaria* the expiratory bursting neurons which show a decrease in frequency during the burst are triggered by EPSPs which continue throughout expiration and the burst is terminated by the cessation of excitation, there being no inhibitory input. As the ventilatory frequency increases the EPSPs occur earlier in the inspiratory phase. On the other hand, the continually firing motor neurons which burst during expiration are controlled by a mixture of excitation and inhibition. Like the bursting neurons, the expiratory bursts are initiated by EPSPs which continue throughout the inspiratory phase. However, the burst is terminated not only by cessation of excitation but by a burst of IPSPs. The expiratory motor neurons which wax and wane receive EPSPs and IPSPs during both phases, their relative strengths of excitation and inhibition determining the frequency. The phasic expiratory motor neurons, which only fire during rapid ventilation, probably receive EPSPs during expiration, but IPSPs throughout the whole cycle, although the latter occur at a higher frequency during inspiration.

Bursts in the inspiratory neurons are triggered by EPSPs, which persist throughout inspiration, as well as by rebound from inhibition. Although the start is rapid it is insufficient to produce an initial high frequency. The burst is ended by IPSPs which persist during expiration. These either suppress, or at least reduce the frequency of, the action potentials. The bursts in the closer motor neurons to spiracles 1–4 are usually active during expiration. The bursts are started by EPSPs, the resultant depolarization gradually declining throughout expiration. The EPSPs tend to occur in groups of two or three (which summate) and hence the spikes are produced in groups (Fig. 60). When the ventilatory frequency is high the patterning disappears and the spike frequency is more regular. The bursts are ended by IPSPs which continue throughout inspiration in the closer motor neurons of spiracles 1 and 2, but not in those of spiracles 3 and 4. In the latter two spiracles the membrane potential of the closer motor neurons gradually depolarizes during inspiration. In the opener motor neurons of spiracle 4 the bursts normally occur during inspiration. It is uncertain whether they result from excitation or from inhibitory rebound. However, the bursts are ended by IPSPs which continue throughout expiration (Burrows, M., 1974, 1975b). The inspiratory motor neurons of the larvae of the aeshnid *Anax parthenope* have also been observed to receive IPSPs during expiration (Komatsu, A., 1980). Even allowing that the same interneuron can excite some motor neurons and inhibit others, the above indicates an absolute minimum of six interneurons in *S. gregaria* (Mill, P., 1977). Neither in *S. gregaria* nor in larvae of *A. parthenope* do there appear to be any direct connections between motor neurons (Burrows, M., 1974; Komatsu, 1980).

In oscillating systems such as ventilation it is generally accepted that the pattern generator consists of a local control centre (oscillator) in each half-segment. The system is driven by one or more command interneurons and, while these may or may not affect all of the local control centres, the centres in one segment are activated, possibly because they have the lowest threshold, and serve as the pacemaker for the system. The pacemaker drives the other local control centres which, as just mentioned, possibly also receive input from the command interneuron(s), with coordinating interneurons ensuring the correct phase relations between them.

The ventilatory pacemaker in *S. gregaria* appears to reside in the metathoracic ganglion. Thus Miller, P. (1965) has shown that, during the expiratory pause, the action potentials in the meso- and prothoracic spiracles are coupled, the former leading with a segmental delay which indicates an interneuronal conduction velocity of 3–4 $m\,s^{-1}$ at 25°. Similarly Lewis, G. *et al.* (1973) indicated a tendency for expiratory bursts to emanate first from the metathoracic ganglion and then from successively more posterior ganglia, and they suggested that activity recorded from the connectives, which is phase-locked with expiration, occurs in a pair of interneurons which run from the metathoracic to the last abdominal ganglion and control the expiratory bursts.

There is good evidence in *S. gregaria* for a pair of interneurons, i.e. one on each side of the nerve cord, each member of the pair synapsing bilaterally with a number of ventilatory and spiracular motor neurons in at least the thoracic and first abdominal ganglia (Burrows, M., 1975b) and also providing excitatory input to many of the flight motor neurons (Burrows, M., 1975a). These interneurons have an excitatory effect on those motor neurons which are active during expiration and this has been confirmed for the closer motor neurons to the first four pairs of spiracles as well as for some of the motor neurons which control the expiratory phase of abdominal ventilation and neck pumping; all apparently receive matching EPSPs. Furthermore, the two inputs are responsible for the slow and fast rhythms observed in individual closer motor neurons. Similarly IPSPs matching the EPSPs have been recorded from the opener motor neurons of spiracles 1, 3 and 4 as well as from some of the motor neurons controlling the inspiratory phase of abdominal ventilation and neck pumping (Burrows, M., 1975b). These interneurons are thought to have their soma in the metathoracic ganglion, one branch of the axon ascending to the meso- and prothoracic ganglia, the other running posteriorly at least as far as the first unfused abdominal ganglion (Burrows, M., 1975b). Indeed, Miller, P. and Mills, P. (1976) and Elliott, C. (1980) have recorded from an interneuron in the abdomen which fires in accelerating bursts just before and during expiration; Pearson, K. (1980) has recorded similar bursts from a pair of interneurons in the metathoracic ganglion, depolarizing one of which is sufficient to reset ventilation.

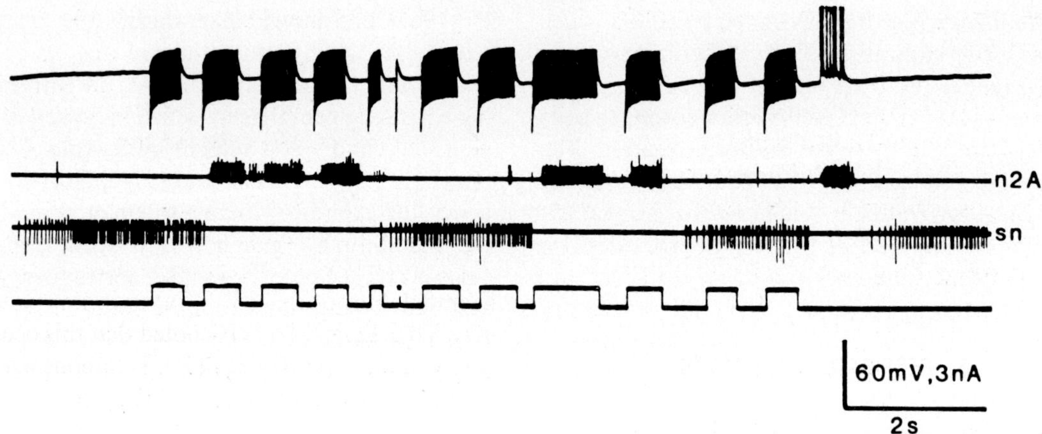

FIG. 68. Intracellular recording from the ascending excitatory (AE) interneuron (top trace) together with expiratory bursts in a second lateral nerve (n2A) and inspiratory bursts in a median nerve (sn) of the larva of *Anax parthenope*. Bottom trace indicates periods of stimulation (upwards) of AE. Vertical scale refers to top (mV) and bottom (nA) traces. (From Komatsu, A., 1984.)

In *Periplaneta americana* it has also been suggested that the pacemaker resides in the metathoracic ganglion (Farley, R. *et al.*, 1967). They observed rhythmic bursts of action potentials in the nerve cord which were conducted from the metathoracic to the last abdominal ganglion at a velocity of about $3\,m\,s^{-1}$ and which produced synchronous contractions of the abdominal expiratory muscles. In *Corydalis cornutis* the evidence indicates that the pacemaker is in the third abdominal ganglion (Fig. 66) (Kinnamon, S. and Kammer, A., 1983). Mill, P. and Hughes, G. (1966) and Pickard, R. and Mill, P. (1972) have provided evidence that the pacemaker is located in the last abdominal ganglion in larval aeshnids (*Aeshna* and *Anax imperator*) (Fig. 65). The posterior–anterior sequence of expiratory bursts which they described has been confirmed by Komatsu, A. (1984). Komatsu, A. and Kusachi, R. (1982) have described a pair of ascending excitatory (AE) interneurons in *Anax parthenope*, the cell bodies of which are located ventrally near the posterior end of the last abdominal ganglion. Rhythmic bursts of potentials occur in these neurons, there being a slight increase in frequency during the burst, as in the primary expiratory motor neurons, and the start of each burst slightly precedes an expiratory burst in the second lateral nerves. Bursts in the AE interneuron trigger the expiratory bursts, their effect apparently being on the two smaller units in these bursts. Direct stimulation of AE elicits an expiratory burst in the second lateral nerves, but does not inhibit the inspiratory motor neurons (Fig. 68). There is also evidence for a pair of ascending inspiratory (AI) interneurons (Komatsu, A., pers. comm.).

4.4 Effects of sensory input

The effect of carbon dioxide and oxygen on the ventilatory rhythm and on spiracular control, with excess carbon dioxide and low oxygen both having an effect on the central nervous system while the former also has a peripheral effect on the spiracles, has been discussed already.

It might be expected that proprioceptive stimulation would play a relatively insignificant role in the control of ventilation. However, there are indications that the dorsoventral movements in *P. americana* and aeshnid larvae, for example, are monitored by internal proprioceptors (Farley, R. and Case, J., 1968; Pill, C. and Mill, P., 1981). Furthermore, it has been shown that electrical stimulation of an abdominal first lateral nerve in an inter-expiratory period can initiate an expiratory burst in the second lateral nerves and reset the ventilatory rhythm in aeshnid larvae (Mill, P. and Hughes, G., 1966). This phenomenon has been confirmed by Komatsu, A. (1984) who also noted that this stimulation elicits bursts in the AE interneuron. Furthermore, Mill, P. (1970) showed that the expiratory bursts could be entrained to a frequency of stimulation slightly greater than the intrinsic ventilatory frequency, at least for a few cycles. A similar resetting effect has been observed as a result of repetitive abdominal compressions in *P. americana* (Farley, R. and Case, J., 1968). However, electrical stimulation of a first lateral nerve has the reverse effect, but this may be a result of antidromic stimulation of the expiratory motor neurons in this animal since there is only one pair of lateral nerves per segment.

5 RESPIRATORY PIGMENTS

Haemoglobin only occurs in a small number of insects, presumably because the body fluids are not involved to any great extent in the transport of the respiratory gases. It is found in the larvae of some chironomids (e.g. Fox, H., 1945) and in the larvae of *Gastrophilus* and in the adults and larvae of the notonectids *Anisops* and *Buenoa* (e.g. Hungerford, H., 1922; Poisson, R., 1926; Bare, C., 1929; Keilin, D., 1944; Miller, P., 1966a). In chironomids it is of low molecular weight (31,400) and indeed is the smallest known haemoglobin, consisting of only two unit molecules and having a sedimentation constant of 2.0. It occurs in the plasma; this is unusual for such a small molecule but is presumably possible because insects do not use ultrafiltration, which can remove low molecular weight proteins from plasma, in their excretory process. In notonectids, on the other hand, the haemoglobin is contained within richly tracheated groups of cells in the abdomen.

The function of the haemoglobin differs in the two groups and it is helpful first to consider the underlying basis for this difference. Thus, the maximum partial pressure (tension) of oxygen (pO_2) in the environment is generally no more than 159 mmHg (dry air at sea level). The oxygen tension at which the respiratory pigment is half-saturated (p_{50}) is used to give an indication of the affinity of the pigment for oxygen. Hence a low value for p_{50} indicates that the haemoglobin releases the bound oxygen only at low values of pO_2 and thus it has a high affinity for oxygen. Conversely the oxygen affinity is low when the p_{50} is high. In the chironomids the haemoglobin is a high-affinity pigment while in the notonectids it is a low-affinity pigment.

5.1 Haemoglobin as a high-affinity pigment

The haemoglobin of *C. riparius* has a p_{50} of 0.5 mmHg at 10° and 0.6 mmHg at 17°; hence the effect of temperature is very small. Furthermore there is no Bohr effect (Fox, H., 1945).

Chironomus larvae live in mud burrows in still water where the environmental pO_2 is probably often rather low, and they apparently only use their haemoglobin to transport oxygen at low values of pO_2. Thus Leitch, I. (1916) noted that this occurred below about 7 mmHg (at 17°), i.e. about 4.4% air saturation, while Ewer, R. (1942), who compared normal larvae with larvae in which the haemoglobin was rendered functionless by the use of carbon monoxide, showed that the pigment in *Chironomus plumosus* only functions below about 40% air saturation (i.e. $3\,ml\,O_2\,l^{-1}$) but down to 15% air saturation oxygen uptake is independent of the pO_2 of the environment. Larvae kept in U-tubes periodically show ventilatory movements, with pauses for filter feeding or resting, but below 10% air saturation ventilation is continuous. In larvae exposed to carbon monoxide, ventilation becomes continuous when their environment is reduced to 26% air saturation. Thus Walshe, B. (1950) suggested that the haemoglobin helps to keep the larvae aerobic and active at low oxygen tensions. Larvae kept for a period under anaerobic conditions (i.e. in nitrogenated water) can build up a small oxygen debt which is repaid as soon as the water is aerated and the larvae become active (Fig. 69); the haemoglobin is thought to be involved only during the initial increase in oxygen consumption, when the water is still not fully aerated, and indeed they can repay their oxygen debt in 7% saturated water (Walshe, B., 1947a). However, the chironomid *Tanytarsus brunnipes*, which also lives in mud tubes but in well-aerated water, is very intolerant of low environmental pO_2. Nevertheless the haemoglobin does not transport oxygen above 25% air saturation (Thienemann, A., 1923; Walshe, B., 1947b).

Although a storage function for chironomid haemoglobin has been proposed (e.g. Miall, L. and Hammond, A., 1900; Pause, J., 1918) it only provides enough O_2 for around 9–12 min aerobic respiration in a resting animal (Leitch, I., 1916; Walshe, B.,1950).

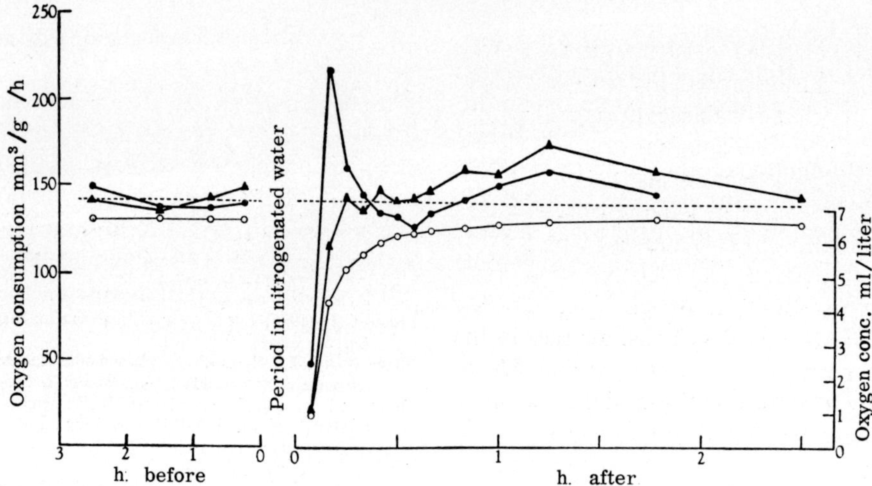

FIG. 69. Rate of oxygen consumption of *Chironomus plumosus* larvae at 17° before and after 16 h in anaerobic conditions. ●, normal animals; ▲, carbon monoxide-treated animals; ○, oxygen concentration of the water. The dotted line indicates the level of average oxygen uptake before the period of anaerobis. (From Walshe, B., 1947a.)

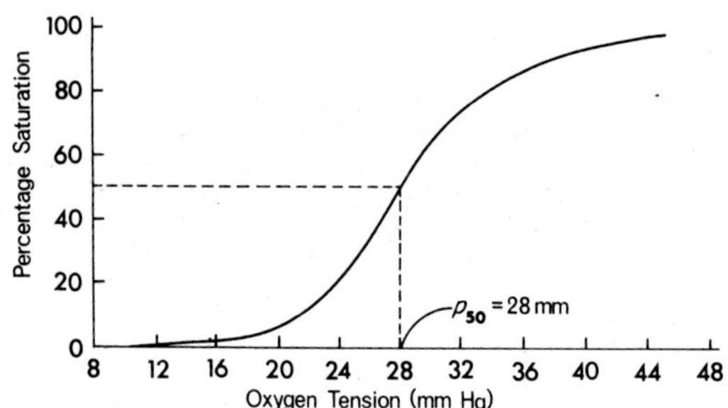

FIG. 70. Oxygen dissociation curve for the haemoglobin of intact adult *Anisops pellucens* at 24°. (From Mill, P., 1974, after Miller, P., 1966a.)

5.2 Haemoglobin as a low-affinity pigment

The p_{50} of the haemoglobin of *Anisops pellucens* is 28 mmHg at 24° (Fig. 70); there is a considerable temperature effect but no Bohr effect (Miller, P., 1966a).

Adults of *Anisops* and *Buenoa* have a compressible gas gill (section 2.2.3). During a dive they are able to maintain a period of neutral buoyancy due largely to the storage function of the haemoglobin (e.g. Bare, C., 1926, 1929; Jaczewski, T., 1936; Hungerford, H., 1958; Miller, P., 1964b,c, 1966a). An average dive of an adult *A. pellucens* lasts about 5 min, but is reduced to about 1 min after treatment with carbon monoxide. Thus the storage function of the haemoglobin allows the duration of the dive to be increased by about 4 min (Miller, P., 1966a). It extends the dive of larval *Gastrophilus* by about the same amount (Keilin, D. and Wang, Y., 1946) but only provides an additional 1–2 min in *Anisops debilis* (Miller, P., 1964c).

Miller, P. (1966) has suggested that at the start of the dive the animal has to actively overcome its buoyancy but, as the gas gill progressively decreases and it reaches a state approaching neutral buoyancy, the activity of the animal becomes minimal. At this stage the haemoglobin starts to give up its oxygen to the gas gill via the abdominal tracheae in order to maintain the phase of neutral buoyancy. The oxygen is then probably taken up through the thoracic spiracles. When the gill pO_2 reaches about 20 mmHg the store will have been fully used up, and so the gas gill will start to decrease in size again. This results in a decrease in buoyancy and hence an increase in activity, the animal ultimately returning to the surface to renew its oxygen stores.

6 CONCLUSIONS

There is considerable diversity within the respiratory system of insects and it is perhaps not surprising to find that parts of the system have come to serve a variety of other functions. Some of these have already been alluded to, such as the role which the system plays during growth in the locust, where the reduction in volume of air sacs during the course of an instar allows other internal organs to grow (Clarke, K., 1958). Air sacs are also involved in the detection of sound waves in some insects, acting as resonators (Pringle, J., 1954) and can also provide heat insulation (Church, N., 1960a,b; Weis-Fogh, T., 1964a). Airs sacs and/or gas gills aid the buoyancy of various insects. Thus, air sacs in the larvae of the nematocerans *Corethra* and *Mochlonyx* have a hydrostatic function and the buoyancy of the animal can be altered so that it can float at different levels (e.g. Damant, G., 1924; Wigglesworth, V., 1953). The decrease in buoyancy caused by the reduction in size of a gas gill during a dive probably provides the stimulus for *Corixa* and *Notonecta glauca* to return to the surface (Botjes, J., 1932; Both, M., 1934); *Nepa* and *Aphelocheirus* have sense organs which are sensitive to differential pressure changes, and also to absolute pressure in the latter species (e.g. Hamilton, M., 1931; Thorpe, W. and Crisp, D., 1947c).

Tracheoles are involved in firefly lanterns (Peterson, M. and Buck, J., 1968) and in water movement (Oschman, J. and Wall, B., 1969). The lateral gills of the mayfly *Hexagenia* probably aid balance (Morgan, A. and Grierson, M., 1932), while rapid ventilation of the branchial chamber in larval anisopteran dragonflies produces jet-propulsive swimming. In the cockroach *Gromphadorhina portentosa* the fourth spiracles and the tracheae leading to them are modified to produce a hissing sound by the expulsion of air through one or both spiracles (Nelson, M., 1979).

REFERENCES

AHEARN, G. A. (1970). The control of water loss in desert tenebrionid beetles. *J. Exp. Biol. 53*, 573–595.

ALTMAN, P. L. and DITTMER, D. S. (1971). Respiration and circulation. *Biological Handbooks*. Fed. Amer. Soc. Exp. Biol., Bethesda, Maryland.

AMANS, P. (1881). Recherches anatomiques et physiologiques sur la larve de l'*Aeschna grandis*. *Rev. Sci. Natur. Montpellier* (sér. 3) *1*, 63–74.

AMBÜHL, H. (1959). Die Bedeutung der Strömung als Ökoligischer Faktor. *Scheiz. Z. Hydrol. 21*, 133–264.

AMOS, W. B. and MILLER, P. L. (1965). The supply of oxygen to the active flight muscles of *Petrognatha gigas* (F.) (Cerambycidae). *Entomologist 98*, 88–94.

ANDERSEN, S. O. and WEIS-FOGH, T. (1964). Resilin. A rubber-like protein in arthropod cuticle. In *Adv. Insect Physiol.* Edited by J. W. L. Beament, J. E. Treherne and V. B. Wigglesworth. Vol. 2, pages 1–65. Academic Press, London.

BABÁK, E. (1912). Untersuchungen über die Atemzentrentätigkeit bei den Insekten. 1. Über die Physiologie der Atemzentren von *Dytiscus* mit Bemerkungen über die Ventilation des Tracheensystems. *Pflüg. Arch. ges. Physiol. 147*, 349–374.

BABÁK, E. and FOUSTKA, O. (1907). Untersuchungen uber den Auslösungsreiz der Atembewegungen bei Libellulidenlarven (und Arthropoden uberhaupt). *Pflüg. Arch. ges. Physiol. 119*, 530–548.

BAILEY, L. (1954). Respiratory currents in the tracheal system of the adult honeybee. *J. Exp. Biol. 31*, 589–593.

BARE, C. O. (1926). Life histories of some Kansas "Backswimmers". *Ann. Ent. Soc. Amer. 19*, 93–101.

BARE, C O. (1929). Haemoglobin cells and other studies of the genus *Buenoa* (Hemiptera, Notonectidae). *Univ. Kansas Sci. Bull. 18*, 265–349.

BAUDET, J. L. and SELLIER, E. (1975). Recherches sur l'appareil respiratoire des Blattes. II.–Les vésiculisations trachéennes et leur évolution dans le sous-ordre des Blattaria. *Ann. Soc. Ent. France (N.S.) 11*, 481–489.

BEAMENT, J. W. L. (1964). The active transport and passive movement of water in insects. *Adv. Insect Physiol. 2*, 67–129.

BECKEL, W. E. (1958). The morphology histology and physiology of the spiracular regulatory apparatus of *Hyalophora cecropia* (L.) *Proc. 10th Int. Congr. Ent.* (Montreal, Canada, 1956). *2*, 87–115.

BOTH, M. P. (1934). Die Regulation des Luftschopfen bei *Notonecta glauca* L. *Z. Vergl. Physiol. 21*, 167–175.

BOTJES, J. O. (1932). Die Atemregulierung bei *Corixa geoffroyi* Leach. *Z. Vergl. Physiol. 17*, 557–564.

BÖVING, A. G. (1910). The natural history of the larvae of the Donaciinae. *Int. Rev. Hydrobiol. 7*, (Biol. Suppl.) 1–108.

BROCHER, F. (1909). Recherches sur la respiration des insectes aquatiques adultes. La notonecte. *Ann. Biol. Lac. 4*, 89–138.

BROCHER, F. (1912a). Recherches sur la respiration des insectes aquatiques adultes. Les *Haemonia*. *Ann. Biol. Lac. 5*, 5–26.

BROCHER, F. (1912b). Recherches sur la respiration des insectes aquatiques adultes. Les elmides. *Ann. Biol. Lac. 5*, 136–179.

BROCHER, F. (1913). Recherches sur la respiration des insectes aquatiques adultes. La notonecte. *Zool. Jahrb. Physiol. 33*, 225–234.

BROCHER, F. (1931). Le méchanisme de la respiration et celui de la circulation du sang chez les insectes. Résultats de mes recherches pendant les vight dernières années. *Archiv. Zool. Exp. Gen. 74*, 25–32.

BROCKWAY, A. P. and SCHNEIDERMAN, H. A. (1967). Strain-gauge transducer studies on intratracheal pressure and pupal length during discontinuous respiration in diapausing silkworm pupae. *J. Insect. Physiol. 13*, 1413–1451.

BROSEMER, R. W., VOGELL, W. and BÜCHER, TH. (1963). Morphologische und enzymatische Muster bei der Entwicklung indirekter Flugsmuskeln von *Locusta migratoria*. *Biochem. Z. 338*, 854–910.

BUCK, J. B. (1958). Cyclic CO_2 release in insects. IV. A theory of mechanism. *Biol. Bull. 114*, 118–140.

BUCK, J. B. and KEISTER, M. (1955). CO_2 release in diapausing *Agapema* pupae. *Biol. Bull. 109*, 144–163.

BUCK, J. B. and KEISTER, M. L. (1956). Diffusion phenomena in the tracheae of *Phormia* larvae. *Physiol. Zool. 29*, 137–146.

BUCK, J. and KEISTER, M. (1958). Cyclic CO_2 release in diapausing pupae. II. Tracheal anatomy, volume and pCO_2; blood volume; interburst CO_2 release rate. *J. Insect Physiol. 1*, 327–340.

BUCKLER, W. (1875). On the larva and habits of *Paraponyx stratiotalis*. *Ent. Month. Mag. 12*, 160–163.

BUISSON, M. du (1924a). Observations sur la ventilation trachéene des insectes. I. La ventilation trachéene chez un acridien. *Bull. Acad. Roy. Belg. Cl. Sci.*, Sér. 5. *10*, 373–391.

BUISSON, M. du (1924b). Observations sur le mechanisme de la ventilation trachéene chez les insectes. II. *Bull. Acad. Roy. Belg. Cl. Sci.*, Sér. 5. *10*, 635–656.

BURKETT, B. N. and SCHNEIDERMAN, H. A. (1974). Roles of oxygen and carbon dioxide in the control of spiracular function in cecropia pupae. *Biol. Bull. 147*, 274–293.

BURROWS, M. (1974). Modes of activation of motoneurones controlling ventilatory movements in locust abdomen. *Phil. Trans. Roy. Soc. B. 269*, 29–48.

Burrows, M. (1975a). Co-ordinating interneurones of the locust which convey two patterns of motor commands: their connexions with flight motoneurons. *J. Exp. Biol. 63*, 713–733.

Burrows, M. (1975b). Co-ordinating interneurones of the locust which convey two patterns of motor commands: their connexions with ventilatory motoneurones. *J. Exp. Biol. 63*, 735–753.

Burrows, M. (1980). The tracheal supply to the central nervous system of the locust. *Proc. Roy. Soc. B. 207*, 63–78.

Bursell, E. ed. (1970). *An Introduction to Insect Physiology*. Academic Press, New York.

Burton, A. J. (1962). The flight mechanism of rhinoceros beetles. Thesis: University of Natal, Pietermaritzburg.

Byers, C. F. (1930). A contribution to the knowledge of Florida Odonata. *Univ. Fla. Path biol. Sci. 1*, 1–327.

Calvert, P. P. (1929). Different rates of growth among animals with special reference to the Odonata. *Proc. Amer. Phil. Soc. 68*, 227–274.

Carlson, J. R. (1977). The imaginal ecdysis of the cricket (*Teleogryllus oceanicus*) I. Organization of motor programs and roles of central and sensory control. *J. Comp. Physiol. 115*, 299–317.

Case, J. F. (1956). Carbon dioxide effects on the spiracles of flies. *Physiol. Zool. 29*, 163–171.

Case, J. F. (1957). The median nerves and cockroach spiracular function. *J. Insect Physiol. 1*, 85–94.

Chapman, R. F. (1969). *The Insects, Structure and Function*. Elsevier, New York.

Church, N. S. (1960a). Heat loss and the body temperature of flying insects. I. Heat loss by evaporation of water from the body. *J. Exp. Biol. 37*, 171–185.

Church, N. S. (1960b). Heat loss and the body temperature of flying insects. II. Heat conduction within the body and its loss by radiation and convection. *J. Exp. Biol. 37*, 186–212.

Clarke, K. U. (1957). The relationship of oxygen consumption to age and weight during the post-embryonic growth of *Locusta migratoria* L. *J. Exp. Biol. 34*, 29–41.

Clarke, K. U. (1958). On the role of the tracheal system in the post-embryonic growth of *Locusta migratoria* L. *Proc. Roy. Ent. Soc. Lond. A. 32*, 67–79.

Comstock, J. H. (1887). Note on the respiration of aquatic bugs. *Amer. Nat. 21*, 577–578.

Corbet, P. S. (1962). *A Biology of Dragonflies*. Witherby, London.

Crisp, D. J. (1964). Plastron respiration. *Recent Progr. Surface Sci. 2*, 377–425.

Crisp, D. J. and Thorpe, W. H. (1948). The water-protecting properties of insect hairs. *Discussions Farad. Soc. 3* (*Interaction of water and porous materials*), 210–220.

Cummins, K. W. (1964). Factors limiting the microdistribution of larvae of the caddisflies *Pycnopsyche lepida* (Hagen) and *Pycnopsyche guttifer* (Walker) in a Michigan stream (Trichoptera: Limnephilidae). *Ecol. Monogr. 34*, 271.

Damant, G. C. C. (1924). The adjustment of the buoyancy of the larva of *Corethra plumicornis*. *J. Physiol. 59*, 345–356.

Davis, C. (1942). Oxygen economy of *Coxelmis novemnotata* (King) (Coleoptera, Dryopidae). *Proc. Linn. Soc. New South Wales 67*, 1–8.

Dodds, G. S. and Hisaw, F. L. (1924). Ecological studies of aquatic insects. II. Size of respiratory organs in relation to environmental conditions. *Ecology 5*, 262–271.

Dolley, W. L. and Farris, E. J. (1929). Unicellular glands in the larvae of *Eristalis tenax*. *J. N.Y. Ent. Soc. 37*, 127–133.

Dufour, L. (1847). Description et anatomie d'une larvae a branchies externes d'Hydropsiche. *Ann. Sci. Nat. 8*, 341–355.

Dumortier, B. (1965). L'émission sonore dans le genre *Gromphadorhina brunner* (Blattodea, Perisphaeriidae) étude morphologique et biologique. *Bull. Soc. Zool. France 90*, 89–101.

Eastham, L. E. S. (1932). Currents produced by the gills of mayfly nymphs. *Nature, Lond. 130*, 58.

Eastham, L. E. S. (1934). Metachronal rhythms and gill movements of the nymph of *Caenis horaria* (Ephemeroptera) in relation to water flow. *Proc. Roy. Soc. Lond. B. 115*, 30–48.

Eastham, L. E. S. (1936). The rhythmical movements of the gills of nymphal *Leptophlebia marginata* (Ephemeroptera) and the currents produced by them in water. *J. Exp. Biol. 13*, 443–449.

Eastham, L. E. S. (1937). The gill movements of nymphal *Ecdyonurus venonus* (Ephemeroptera) and the currents produced by them in water. *J. Exp. Biol. 14*, 219–228.

Eastham, L E. S. (1939). Gill movements of nymphal *Ephemera danica* (Ephemeroptera) and the water currents caused by them. *J. Exp. Biol. 16*, 18–33.

Eaton, A. E. (1883–8). A revisional monograph of recent Ephemeridae or mayflies. *Trans. Linn. Soc. Lond.* (2nd ser.) *3* (Zoology) (in 6 parts), 1–352.

Edwards, F. W. (1919). The larva and pupa of *Taeniorhynchus richiardii* Fic. (Diptera, Culicidae). *Ent. Mon. Mag. 3* (5), 83–88.

Edwards, G. A., Ruska, H. and deHarven, E. (1958). The fine structure of insect tracheoblasts, tracheae and tracheoles. *Arch. Biol. 69*, 351–369.

Edwards, R. W. (1958). The relation of oxygen consumption to body size and to temperature in the larvae of *Chironomus riparius* Meigen. *J. Exp. Biol. 35*, 383–395.

Ege, R. (1915a). On the respiratory function of the air stores carried by some aquatic insects (Corixidae, Dytiscidae and *Notonecta*). *Z. Allg. Physiol. 17*, 81–124.

Ege, R. (1915b). On the respiratory conditions of the larvae and pupae of Donacine. *Vid. Medd. Dansk. Naturh. 66*, 183–196.

Ege, R. (1926). In *Physiological Papers Dedicated to August Krogh, London*.

Elliott, C. J. H. (1980). Neurophysiological analysis of locust behaviour during ecdysis. D.Phil. thesis, Oxford, U.K.

Eriksen, C. H. (1963a). The relation of oxygen consumption to substrate particle size in two burrowing mayflies. *J. Exp. Biol. 40*, 447–453.

Eriksen, C. H. (1963b). Respiratory regulation in *Ephemera simulans* Walker and *Hexagenia limbata* (Serville) (Ephemeroptera). *J. Exp. Biol. 40*, 455–467.

Eriksen, C. H. (1968). Ecological significance of respiration and substrate for burrowing Ephemeroptera. *Canad. J. Zool. 46*, 93–103.

Ewer, R. F. (1942). On the function of haemoglobin in *Chironomus*. *J. Exp. Biol. 18*, 197–205.

Facheu, M. J., and Sellier, E. (1971). L'ultrastructure de l'intima cuticulaire des sacs aeriens chez les insectes. *C. R. Acad. Sciences, Paris 272*, 2197–2200.

Farley, R. D. and Case, J. F. (1968). Sensory modulation of ventilative pacemaker output in the cockroach, *Periplaneta americana*. *J. Insect Physiol. 14*, 591–601.

Farley, R. D., Case, J. F. and Roeder, K. D. (1967). Pacemaker for tracheal ventilation in the cockroach *Periplaneta americana* L. *J. Insect Physiol. 13*, 1713–1728.

Feldmeth, C. R. (1970a). The respiratory energetics of two species of stream caddis fly larvae in relation to water flow. *Comp. Biochem. Physiol. 32*, 193–202.

Feldmeth, C. R. (1970b). The influence of acclimation to current velocity on the behaviour and respiratory physiology of two species of stream Trichoptera larvae. *Physiol. Zool. 43*, 185–193.

Fox, H. M. (1920). Methods of studying the respiratory exchange in small aquatic organisms, with particular reference to the use of flagellates as an indicator for oxygen consumption. *J. Gen. Physiol. 3*, 565–573.

Fox, H. M. (1945). The oxygen affinities of certain invertebrate haemoglobins. *J. Exp. Biol. 21*, 161–165.

Fox, H. M. and Sidney, J. (1953). The influence of dissolved oxygen on the respiratory movements of caddis larvae. *J. Exp. Biol. 30*, 235–237.

Fox, H. M., Simmonds, B. G. and Washbourn, R. (1935). Metabolic rates of ephemerid nymphs from swiftly flowing and from still waters. *J. Exp. Biol. 12*, 179–184.

Fox, H. M., Wingfield, C. A. and Simmonds, B. G. (1937). The oxygen consumption of ephemerid nymphs from flowing and from still waters in relation to the concentration of oxygen in the water. *J. Exp. Biol. 14*, 210–218.

Fraenkel, G. (1932a). Untersuchungen uber die Koordination von Reflexen und automatisch-nervosen Rhythmen bei Insekten. II. Die nervose Regulierung der Atmung wahrend des Fluges. *Z. Vergl. Physiol. 16*, 394–417.

Fraenkel, G. (1932b). Untersuchungen uber die Koordination von Reflexen und automatisch-nervosen Rhythmen bei Insekten. III. Des Problem des gerichteten Atemstromes in den Treacheen der Insekten. *Z. Vergl. Physiol. 16*, 418–443.

Fraenkel, G. and Herford, G. V. B. (1938). The respiration of insects through the skin. *J. Exp. Biol. 15*, 266–280.

GARDNER, A. E. (1956). The Biology of Dragonflies. *Proc. S. Lond. Ent. Nat. Hist. Soc., 1954–5*, 109–134.

GERTZ, V. K. H. and LOESCHCKE, H. H. (1954). Bestimmung der Diffusions-Koeffizienten von H_2, O_2, N_2, und He in Wasser und Blutserum bei konstant gehaltener Konvektion. *Z. Naturforsch. 9b*, 1–9.

GREVEN, H. and RUDOLPH, R. (1973). Histologie und Feinstruktur der larvalen Kiemenkammer von *Aeshna cyanea* Müller (Odonata: Anisoptera). *Z. Morph. Tiere 76*, 209–226.

GUTHRIE, D. M. and TINDALL, A. R. (1968). *The Biology of the Cockroach.* Edward Arnold, London.

HAMILTON, A. G. (1964). The occurrence of periodic or continuous discharge of carbon dioxide by male desert locusts (*Schistocerca gregaria* Forskal) measured by an infra-red gas analyser. *Proc. Roy. Soc. B 160*, 373–395.

HAMILTON, A. G. (1972). The combined use of a twin channel null-balance paramagnetic O_2 analyser and an infra-red CO_2 analyser for measuring respiration in insects. *Lab. Prac. 21*, 807–809.

HAMILTON, M. A. (1931). The morphology of the water scorpion *Nepa cinerea* Linn. (Rhynchota, Heteroptera). *Proc. Zool. Soc. Lond.*, pp. 1067–1136.

HARNISCH, O. (1958). Untersuchungen an den Analkiemen der Larve von *Agrion. Biol. Zbl. 77*, 300–310.

HARTLEY, J. C. (1971). The respiratory system of the egg-shell of *Homorocoryphus nitidulus vicinus* (Orthoptera, Tettigonidae). *J. Exp. Biol. 55*, 165–176.

HASSAN, A. A. G. (1944). The structure and mechanism of the spiracular regulatory apparatus in adult Diptera and certain other groups of insects. *Trans. Roy. Ent. Soc. Lond. 94*, 103–153.

HAZELHOFF, E. H. (1927). Die Regulierung der Atmung bei Insekten und Spinnen. *Z. Vergl. Physiol. 5*, 179–190.

HERFORD, G. M. (1938). Tracheal pulsation in the flea. *J. Exp. Biol. 15*, 327–338.

HINKLE, M. and CAMHI, J. M. (1972). Locust motoneurons: bursting activity correlated with axon diameter. *Science 175*, 553–556.

HINTON, H. E. (1960a). Plastron respiration in the eggs of blowflies. *J. Insect Physiol. 4*, 176–183.

HINTON, H. E. (1960b). The chorionic plastron and its role in the eggs of the Muscinae (Diptera). *Quart. J. Mic. Sci. 101*, 313–332.

HINTON, H. E. (1960c). The structure and function of the respiratory horns of the eggs of some flies. *Phil. Trans. Roy. Soc. B. 243*, 45–73.

HINTON, H. E. (1961). The structure and function of the egg-shell in the Nepidae (Hemiptera). *J. Insect Physiol. 7*, 224–257.

HINTON, H. E. (1962). The fine structure and biology of the eggshell of the wheat bulb fly, *Leptohylemyia coarctata. Quart. J. Mic. Sci. 103*, 243–251.

HINTON, H. E. (1963). The respiratory system of the egg-shell of the blowfly, *Calliphora erythrocephala* Meig., as seen with the electron microscope. *J. Insect Physiol. 9*, 121–129.

HINTON, H. E. (1966a). The spiracular gill of the fly, *Antocha bifida*, as seen with the scanning electron microscope. *Proc. Roy. Ent. Soc. Lond. (A). 41*, 107–115.

HINTON, H. E. (1966b). Respiratory adaptations of the pupae of beetles of the family Psephenidae. *Phil. Trans Roy. Soc. B. 251*, 211–245.

HINTON, H. E. (1967a). Structure of the plastron in *Lipsothrix* and the polyphyletic origin of plastron respiration in the Tipulidae. *Proc. Roy. Ent. Soc. Lond. (A). 42*, 35–38.

HINTON, H. E. (1967b). Spiracular gills in the marine fly *Aphrosylus* and their relation to the respiratory horns of other Dolichopodidae. *J. Mar. Biol. Assoc. U.K. 47*, 485–497.

HINTON, H. E. (1967c). On the spiracles of the larvae of the suborder *Myxophaga* (Coleoptera). *Aust. J. Zool. 15*, 955–959.

HINTON, H. E. (1967d). The respiratory system of the egg-shell of the common housefly. *J. Insect Physiol. 13*, 647–651.

HINTON, H. E. (1968a). Spiracular gills. *Adv. Insect Physiol. 5*, 65–162.

HINTON, H. E. (1968b). Structure and protective devices of the egg of the mosquito *Culex pipiens. J. Insect Physiol. 14*, 145–161.

HINTON, H. E. (1969). Respiratory systems of insect egg-shells. *Ann. Rev. Ent. 14*, 343–368.

HINTON, H. E. (1970). Insect eggshells. *Sci. Amer. 223* (Aug.), 84–91.

HINTON, H. E. (1976). Plastron respiration in bugs and beetles. *J. Insect Physiol. 22*, 1529–1550.

HODGMAN, C. D. (1958). In *Handbook of Chemistry and Physics*. 40th edition. Edited by C. D. Hodgeman. Chem. Rubber Publ. Co., Cleveland, Ohio.

HOUGHTON, W. (1868). Caddis worms and their metamorphosis. *Pop. Sci. Rev. 7*, 287–295.

HOULIHAN, D. F. (1969). Respiratory physiology of the larva of *Donacia simplex*, a root-piercing beetle. *J. Insect Physiol. 15*, 1517–1536.

HOULIHAN, D. F. (1970). Respiration in low oxygen partial pressures: the adults of *Donacia simplex* that respire from the roots of aquatic plants. *J. Insect Physiol. 16*, 1607–1622.

HOYLE, G. (1959). The neuromuscular mechanism of an insect spiracular muscle. *J. Insect Physiol. 3*, 378–394.

HOYLE, G. (1960). The action of carbon dioxide gas on an insect spiracular muscle. *J. Insect Physiol. 4*, 63–79.

HRBRACEK, J. (1949). Morphology and physiology of the spiracles of the family Hydrophilidae (Coleoptera). *Vest. Cesk. Zool. Spolecnosti 13*, 136–176.

HUBER, F. (1960). Experimentelle untersuchungen zur nervosen Atmungsregulation der Orthopteren (Saltatoria: Gryllidae). *Z. Vergl. Physiol. 43*, 359–391.

HUGHES, G. M. (1958). The co-ordination of insect movements. III. Swimming in *Dytiscus*, *Hydrophilus* and a dragonfly nymph. *J. Exp. Biol. 35*, 567–583.

HUGHES, G. M. and MILL, P. J. (1966). Patterns of ventilation in dragonfly larvae. *J. Exp. Biol. 44*, 317–333.

HUGHES, T. D. (1980a). The imaginal ecdysis of the desert locust, *Schistocerca gregaria*. I. A description of the behaviour. *Physiol. Ent. 5*, 47–54.

HUGHES, T. D. (1980b). The imaginal ecdysis of the desert locust, *Schistocerca gregaria*. II. Motor activity underlying the pre-emergence and emergence behaviour. *Physiol. Ent. 5*, 55–71.

HUGHES, T. D. (1980c). The imaginal ecdysis of the desert locust, *Schistocerca gregaria*. III. Motor activity underlying the expansional and post-expansional behaviour. *Physiol. Ent. 5*, 141–152.

HUGHES, T. D. (1980d). The imaginal ecdysis of the desert locust, *Schistocerca gregaria*. IV. The role of the gut. *Physiol. Ent. 5*, 153–164.

HUNGERFORD, H. B. (1922). Oxyhaemoglobin present in backswimmer, *Buenoa margaritacea* Buenoa. *Canad. Ent. 54*, 262–263.

HUNGERFORD, H. B. (1958). Some interesting aspects of the world distribution and classification of aquatic and semiaquatic Hemiptera. *Proc. 10th Int. Congr. Ent.* (Montreal) (1956), *1*, 337–348.

HUSTERT, R. (1974). Morphologie und Atmungsbewegungen des 5. Abdominalsegments von *Locusta migratoria migratoriodes. Zool. Zb. Physiol. 78*, 157–174.

HUSTERT, R. (1975). Neuromuscular coordination and proprioceptive control of rhythmical abdominal ventilation in intact *Locusta migratoria migratorioides. J. Comp. Physiol. 97*, 159–179.

IMMS, A. D. (1957). *A General Textbook of Entomology*. 9th edn (revised by O. W. Richards and R. G. Davies). Methuen, London.

JACZEWSKI, T. (1936). Contributions to the knowledge of aquatic Heteroptera of Egypt. *Ann. Mus. Zool. Polon. 11*, 171–211.

KAARS, C. (1979). Neural control of homologous behaviour patterns in two blaberid cockroaches. *J. Insect Physiol. 25*, 209–218.

KAARS, C. (1983). Innervation and control of tension in abdominal muscles of *Blaberus discoidalis* and *Gromphadorhina portentosa J. Insect Physiol. 29*, 371–376.

KANWISHER, J. W. (1966). Tracheal gas dynamics in pupae of the Cecropia silkworm. *Biol. Bull. 130*, 96–105.

KEILIN, D. (1944). Respiratory systems and respiratory adaptations in larvae and pupae of Diptera. *Parasitology 36*, 1–66.

KEILIN, D., TATE, P. and VINCENT, M. (1935). The perispiracular glands of mosquito larvae. *Parasitology 27*, 257–262.

KEILIN, D. and WANG, Y. L. (1946). Haemoglobin of *Gastrophilus* larvae. Purification and properties. *Biochem. J. 40*, 855–866.

KEISTER, M. and BUCK, J. (1961). Respiration of *Phormia regina* in relation to temperature and oxygen. *J. Insect Physiol. 7*, 51–72.

KEISTER, M. and BUCK, J. (1964). Respiration: some exogenous and endogenous effects on rate of respiration. In *The Physiology of Insecta*. Edited by M. Rockstein. Vol. 3, pages 617–658. Academic Press, New York.

KERRY, C. J. (1982). The Neural Control of ventilation in the Praying Mantid, *Hierodula membranacea*. Ph.D. thesis, Leeds, U.K.

KESTLER, P. (1971). Die Diskontinuierliche Ventilation bei *Periplaneta americana* L. und anderen Insekten. *Dissertation*, Wurtzburg.

KESTLER, P. (1978). Atembewegungen und Gasaustausch bei Ruheatmung adulter terristrischer Insekten. *Verh. Dtsch. Zool. Ges. 1978*, 269.

KESTLER, P. (1980). Saugventilation verhibdert bei Insekten die wasserabgabe aus dem Tracheensystem. *Verh. Dtsch. Zool. Ges. 1980*, 306.

KINNAMON, S. C. and KAMMER, A. E. (1983). Neural control of ventilatory movements in the aquatic insect *Corydalus cornutus*: the motor pattern. *J. Comp. Physiol. A. 153*, 543–555.

KINNAMON, S. C., KAMMER, A. E. and KIORPES, A. L. (1980). Control of ventilation in an aquatic insect larva: central effect of hypoxia. *Amer. Zool. 20*, 942.

KINNAMON, S. C., KAMMER, A. E. and KIORPES, A. L. (1984). Neural control of ventilatory movements in the aquatic insect *Corydalus cornutus*: Central effect of hypoxia. *Physiol. Ent. 9*, 19–28.

KITCHELL, R. L. and HOSKINS, W. M. (1935). Respiratory ventilation in the cockroach in air, in carbon dioxide and in nicotine atmospheres. *J. Econ. Ent. 28*, 924–933.

KLEKOWSKI, R. Z. and KAMLER, E. (1968). Flowing-water polarographic respirometer for aquatic animals. *Polskie Arch. Hydrobiol. 15*, 121–144.

KOCH, H. J. (1934). Aandeel van bepaalde organen aan de zeuerstofopname door het gesloten trachaeinsystem, bij de larven dur Odonata Zygoptera. *Natuurvet. Tijdsch. 16*, 75–80.

KOCH, H. J. (1936). Recherches sur la physiologie du système trachéen clos. *Mem. Acad. Roy. Belg. Cl. Sci. 16*, 3–98.

KOMATSU, A. (1977). Change of respiratory movement after emergence in the cockroach, *Periplaneta australasiae* (in Japanese). *Jap. J. Appl. Ent. Zool. 21*, 179–183.

KOMATSU, A. (1980). Synaptic input driving respiratory motoneurons in dragonfly larvae. *Brain Res. 201*, 215–219.

KOMATSU, A. (1982). Respiratory nervous activity in the isolated nerve cord of the larval dragonfly, and location of the respiratory oscillator. *Physiol. Ent. 7*, 183–191.

KOMATSU, A. (1984). Ascending interneurons that convey a respiratory signal in the central nervous system of the dragonfly larva. *J. Comp. Physiol. A. 154*, 331–340.

KOMATSU, A. and KUSACHI, R. (1979). Responses of respiratory motoneurons to segmental nerve stimulation in the dragonfly larvae, *Anax parthenope*. *J. Physiol. Soc. Japan 41*, 405.

KOMATSU, A. and KUSACHI, R. (1982). Ascending respiratory interneurons controlling motor discharges in dragonfly larvae. *J. Physiol. Soc. Japan 44*, 502.

KOMNICK, H. (1977). Chloride cells and chloride epithelia of aquatic insects. *Int. Rev. Cytol. 49*, 285–329.

KOMNICK, H. (1978). Osmoregulatory role and transport ATPases of the rectum of dragonfly larvae. *Odonatologica 7*, 247–262.

KRASFUR, E. S., WILLMAN, J. R., GRAHAM, C. L. and WILLIAMS, R. E. (1970). Observations on spiracular behaviour in *Aedes* mosquitoes. *Ann. Ent. Soc. Amer. 63*, 684–691.

KROGH, A. (1919). The rate of diffusion of gases through animal tissues, with some remarks on the coefficients of invasion. *J. Physiol. 52*, 391–408.

KROGH, A. (1920a). Studien über Tracheenrespiration. II. Über Gasdiffusion in den Tracheen. *Pflüg. Arch. Ges. Physiol. 179*, 95–112.

KROGH, A. (1920b). Studien über Tracheenrespiration. III. Die Kombination von mechanischer Ventilation mit Gasdiffusion nach Versuchen an Dytiscuslarven. *Pflüg. Arch. Ges. Physiol. 179*, 113–120.

KROGH, A. (1943). Some experiments on the osmoregulation and respiration of *Eristalis* larvae. *Ent. Medel. 23*, 49–65.

KROGH, A. (1948). Determination of temperature and heat production in insects. *Z. Vergl. Physiol. 31*, 274–280.

KROGH, A. and ZEUTHEN, E. (1941). Mechanism of flight preparation in some insects. *J. Exp. Biol. 18*, 1–10.

LARSEN, O. (1924). Zur Kenntnis von *Aphelocheirus aestivalis* Fabr. *Ark. Zool. 16*, (No. 16) 1–21.

LAWTON, J. H. (1971). Ecological energetics studies on larvae of the damselfly *Pyrrhosoma nymphula* (Sulzer) (Odonata: Zygoptera). *J. Anim. Ecol. 40*, 385–423.

LEADER, J. P. (1970). Effect of temperature, salinity, and dissolved oxygen concentration upon respiratory activity of the larva of *Philanisus plebeius* (Trichoptera). *J. Insect Physiol. 17*, 1917–1924.

LEE, M. O. (1925). On the mechanism of respiration in certain Orthoptera. *J. Exp. Zool. 41*, 125–154.

LEITCH, I. (1916). The function of haemoglobin in invertebrates with special reference to *Planorbis* and *Chironomus* larvae. *J. Physiol. 50*, 370–379.

LEVY, R. I. and SCHNEIDERMAN, H. A. (1966a). Discontinuous respiration in insects. II. The direct measurement and significance of changes in tracheal gas composition during the respiratory cycle of silkworm pupae. *J. Insect Physiol. 12*, 83–104.

LEVY, R. I. and SCHNEIDERMAN, H. A. (1966b). Discontinuous respiration in insects. III. The effect of temperature and ambient oxygen tension on the gaseous composition of the tracheal system of silkworm pupae. *J. Insect Physiol. 12*, 105–121.

LEVY, R. I. and SCHNEIDERMAN, H. A. (1966c). Discontinuous respiration in insects. IV. Changes in intratracheal pressure during the respiratory cycle of silkworm pupae. *J. Insect. Physiol. 12*, 465–492.

LEWIS, G. W., MILLER, P. L. and MILLS, P. S. (1973). Neuromuscular mechanisms of abdominal pumping in the locust. *J. Exp. Biol. 59*, 149–168.

LEYDIG, F. (1859). Zur Anatomie der Insecten. *Arch. Anat. Physiol.* 149–183.

LIEFTINCK, M. A. (1940). Revisional notes on some species of *Copera* Kirby. With notes on habits and larvae. (Odon., Platycneminidae). *Treubia 17*, 281–306.

LOCKE, M. (1957). The structure of insect tracheae. *Quart. J. Mic. Sci. 98*, 487–492.

LOCKE, M. (1958a). The formation of tracheae and tracheoles in *Rhodnius prolixus*. *Quart. J. Mic. Sci. 99*, 29–46.

LOCKE, M. (1958b). The co-ordination of growth in the tracheal system of insects. *Quart. J. Mic. Sci. 99*, 373–391.

LONGLEY, A. and EDWARDS, J. S. (1979). Tracheation of abdominal ganglia and cerci in the house cricket *Acheta domesticus* (Orthoptera, Gryllidae). *J. Morph. 159*, 233–244.

LOVERIDGE, J. P. (1968). The control of water loss in *Locusta migratoria migratorioides* R. & F. II. Water loss through the spiracles. *J. Exp. Biol. 49*, 15–29.

LUCAS, W. J. (1900). *British Dragonflies (Odonata)*. Upcott & Gill, London.

LUTZ, P. E. and JENNER, C. E. (1960). Relation between oxygen consumption and photoperiodic induction of the termination of diapause in nymphs of the dragonfly *Tetragoneuria cyanosura*. *J. Elisha Mitchell Sci. Soc., Chapel Hill, N.C. 76*, 192–193.

McARTHUR, J. M. (1929). Function of spiracles in Orthoptera. *J. Exp. Zool. 53*, 117–128.

McCUTCHEON, F. H. (1940). The respiratory mechanism in the grasshopper. *Ann. Ent. Soc. Amer. 33*, 35–55.

McGOVRAN, E. R. (1932). The effect of some gases on tracheal ventilation of grasshoppers. *J. Econ. Ent. 25*, 271–276.

MACNEILL, N. (1960). A study of the caudal gills of dragonfly larvae of the sub-order Zygoptera. *Proc. Roy. Irish Acad. B. 61*, 115–140.

MATSUDA, R. (1975). *Morphology and Evolution of the Insect Abdomen*. Pergamon Press, Oxford.

MATULA, J. (1911). Untersuchungen uber die Functionen des Zentralnervensystems bei Insekten. *Pflug Arch. ges. Physiol. 138*, 388–456.

MIALL, L. C. (1895). *The Natural History of Aquatic Insects*. Macmillan, London.

MIALL, L. C. and DENNY, A. (1886). *The Structure and Life-History of the Cockroach (Periplaneta orientalis). An Introduction to the Study of Insects*. Lovell Reeves, London.

MIALL, L. C. and HAMMOND, A. R. (1900). *The Structure and Life History of the Harlequin Fly*. Oxford University Press, London and New York.

MILL, P. J. (1963). Neural activity in the abdominal nervous system of aeschnid nymphs. *Comp. Biochem. Physiol. 8*, 83–98.

MILL, P. J. (1965). An anatomical study of the abdominal nervous and muscular systems of dragonfly (Aeshnidae) nymphs. *Proc. Zool. Soc. 145*, 57–73.

MILL, P. J. (1970). Neural patterns associated with ventilatory movements in dragonfly larvae. *J. Exp. Biol. 52*, 167–175.

MILL, P. J. (1972). *Respiration in the Invertebrates*. Macmillan, London.

MILL, P. J. (1974). Respiration: Aquatic Insects. In *The Physiology of Insecta*. Edited by M. Rockstein. 2nd edition. Vol. 6, pages 403–467. Academic Press, New York and London.

MILL, P. J. (1977). Ventilation motor mechanisms in the dragonfly and other insects. In *Identified Neurons and Behavior of Arthropods*. Edited by G. Hoyle. Pages 187–208. Plenum Press, New York.

MILL, P. J. (1982a). *Comparative Neurobiology*. (Contemporary Biology Series). Edward Arnold, London.

MILL, P. J. (1982b). A decade of dragonfly neurobiology. In *Advances in Odonatology*. Vol. I. *Proceedings of the Sixth International Symposium of Odonatology*. (Chur, Switzerland) (1981), pp. 151–173.

MILL, P. J. and HUGHES, G. M. (1966). The nervous control of ventilation in dragonfly larvae. *J. Exp. Biol. 44*, 297–316.

MILL, P. J. and LOWE, D. A. (1971). Ultrastructure of the respiratory and non-respiratory muscles of the larva of a dragonfly. *J. Insect Physiol. 17*, 1947–1960.

MILL, P. J. and PICKARD, R. S. (1972a). A review of the types of ventilation and their neural control in aeshnid larvae. *Odonatologica 1*, 41–50.

MILL, P. J. and PICKARD, R. S. (1972b). Anal valve movement and normal ventilation in aeshnid dragonfly larvae. *J. Exp. Biol. 56*, 537–543.

MILL, P. J. and PICKARD, R. S. (1975). Jet-propulsion in anisopteran dragonfly larvae. *J. Comp. Physiol. A. 97*, 329–338.

MILLER, P. L. (1960a). Respiration in the desert locust. I. The control of ventilation. *J. Exp. Biol. 37*, 224–236.

MILLER, P. L. (1960b). Respiration in the desert locust. II. The control of the spiracles. *J. Exp. Biol. 37*, 237–263.

MILLER, P. L. (1960c). Respiration in the desert locust. III. Ventilation and the spiracles during flight. *J. Exp. Biol. 37*, 264–278.

MILLER, P. L. (1961). Some features of the respiratory system of *Hydrocyrius columbiae* Spin. (Belostomatidae, Hemiptera). *J. Insect Physiol. 6*, 243–271.

MILLER, P. L. (1962). Spiracle control in adult dragonflies (Odonata). *J. Exp. Biol. 39*, 513–535.

MILLER, P. L. (1964a). Factors altering spiracle control in adult dragonflies: hypoxia and temperature. *J. Exp. Biol. 41*, 345–357.

MILLER, P. L. (1964b). Possible function of haemoglobin in *Anisops*. *Nature, Lond. 201*, 1052.

MILLER, P. L. (1964c). The possible role of haemoglobin in *Anisops* and *Buenoa* (Hemiptera: Notonectidae). *Proc. Roy. Ent. Soc. A. 39*, 166–175.

MILLER, P. L. (1965). The central nervous control of respiratory movements. In *The Physiology of the Insect Central Nervous System*. Edited by J. E. Treherne and J. W. L. Beament. *Proc. 12th Int. Congr. Ent.* (London, U.K., 1964). Pages 141–155. Academic Press, London.

MILLER, P. L. (1966a). The function of haemoglobin in relation to the maintenance of neutral buoyancy in *Anisops pellucens* (Notonectidae, Hemiptera). *J. Exp. Biol. 44*, 529–543.

MILLER, P. L. (1966b). The supply of oxygen to the active flight muscles of some large beetles. *J. Exp. Biol. 45*, 285–304.

MILLER, P. L. (1966c). The regulation of breathing in insects. *Adv. Insect. Physiol. 3*, 279–354.

MILLER, P. L. (1967). The derivation of the motor command to the spiracles of the locust. *J. Exp. Biol. 46*, 349–371.

MILLER, P. L. (1971a). Rhythmic activity in the insect nervous system. I. Ventilatory coupling of a mantid spiracle. *J. Exp. Biol. 54*, 587–597.

MILLER, P. L. (1971b). Rhythmic activity in the insect nervous system. II. Sensory and electrical stimulation of ventilation in a mantid. *J. Exp. Biol. 54*, 599–607.

MILLER, P. L. (1971c). Rhythmic activity in the insect nervous system: thoracic ventilation in non-flying beetles. *J. Insect Physiol. 17*, 395–405.

MILLER, P. L. (1973). Spatial and temporal changes in the coupling of cockroach spiracles to ventilation. *J. Exp. Biol. 59*, 137–148.

MILLER, P. L. (1974). Respiration — aerial gas transport. In *The Physiology of Insecta*. Edited by M. Rockstein. 2nd edition. Vol. 6, pages 345–402. Academic Press, New York.

MILLER, P. L. (1979). A possible sensory function for the stop-go patterns of locomotion in phorid flies. *Physiol. Ent. 4*, 361–370.

MILLER, P. L. (1980). Pacemaker neurones and rhythmic behavior. In *Insect Biology in the Future*. "VBW 80". Edited by M. Locke and D. S. Smith. Pages 819–846. Academic Press, London & New York.

MILLER, P. L. (1981a). Respiration. In *The American Cockroach*. Edited by W. J. Bell and K. G. Adiyodi. Pages 87–116. Chapman & Hall, London.

MILLER, P. L. (1981b). Ventilation in active and inactive insects. In *Locomotion and Energetics in Insects*. Edited by C. F. Herreid and C. R. Fourtner. Pages 367–390. Plenum Press, New York.

MILLER, P. L. and MILLS, P. S. (1976). Some aspects of the development of breathing in the locust. In *Perspectives in Experimental Biology* (*Proc. 50th Anniv. Meeting Soc. Exp. Biol.*) *1* (Zoology). Edited by P. Spencer Davies. Pages 199–208. Pergamon Press, Oxford.

MORGAN, A. H. and GRIERSON, M. C. (1932). The functions of the gills in burrowing mayflies (*Hexagania recurvata*). *Physiol. Zool. 5*, 230–245.

MORGAN, A. H. and O'NEIL, H. D. (1931). The function of the tracheal gills in larvae of the caddis fly, *Macronema zebratum* Hagen. *Physiol. Zool. 4*, 361–379.

MYERS, J. G. (1935). Aquatic 'wooly-bear' caterpillars from British Guiana. *Proc. Roy. Ent. Soc. Lond. 10*, 65–70.

MYERS, T. B. and FISK, F. W. (1962). Breathing movements of the Cuban burrowing cockroach. *Ohio J. Sci. 62*, 253–257.

MYERS, T. B. and RETZLAFF, E. (1963). Localization and action of the respiratory centre of the cuban burrowing cockroach. *J. Insect Physiol. 9*, 607–614.

NELSON, M. C. (1979). Sound production in the cockroach *Gromphadorhina portentosa*. The sound production apparatus. *J. Comp. Physiol. 132*, 27–38.

NUNOME, J. (1944). Studies on the respiration of the silkworm — I. Diffusion of oxygen in the respiratory system of the silkworm (in Japanese). *Bull. Seric. Exp. Stn. Japan 12*, 17–39.

OLESEN, J. (1972). The hydraulic mechanism of labial extension and jet propulsion in dragonfly nymphs. *J. Comp. Physiol. 81*, 53–55.

OSCHMAN, J. L. and WALL, B. J. (1969). The structure of the rectal pads of *Periplaneta americana* L. with regard to fluid transport. *J. Morph. 127*, 475–509.

PACKARD, A. S. (1898). *A Textbook of Entomology*. Macmillan, London.

PAGANELLI, C. V., BATEMAN, N. and RAHN, H. (1967). Artificial gills for gas exchange in water. *Proc. 3rd Symp. Underwater Physiology*. (Edited by C. J. Lambertsen). Pages 452–468. The Williams & Wilkins Co., Baltimore.

PAIM, U. and BECKEL, W. E. (1963a). Seasonal oxygen and carbon dioxide content of decaying wood as a component of the microenvironment of *Orthosoma brunneum* (Forster) (Coleoptera: Cerambycidae). *Canad. J. Zool. 41*, 1133–1147.

PAIM, U. and BECKEL, W. E. (1963b). The influence of oxygen and carbon dioxide on the spiracles of a wood-boring insect, *Orthosoma brunneum* (Forster) (Coleoptera: Cerambycidae). *Canad. J. Zool. 41*, 1149–1167.

PARSONS, M. C. and HEWSON, R. J. (1974). Plastral respiratory devices in adult *Cryphocricos* (Naucoridae: Heteroptera). *Psyche. Camb. 81*, 510–527.

PAULPANDIAN, A. (1964). Cyclic ventilation movement in the common cockroach, *Periplaneta americana*. *Curr. Sci.* (Bangalore) *33*, 404–405.

PAUSE, J. (1918). Beitrage zur Biologie und Physiologie der Larve von *Chironomus gregarius*. *Zool. Jahrb. 36*, 339–452.

PEARSON, K. G. (1980). Burst generation in coordinating interneurons of the ventilatory system of the locust. *J. Comp. Physiol. 137*, 305–313.

PENNAK, R. W. and MCCOLL, C. M. (1944). An experimental study of oxygen absorption in some dragonfly naiads. *J. Cell. Comp. Physiol. 23*, 1–10.

PETERSON, M. K. and BUCK, J. (1968). Light organ fine structure in certain asiatic fireflies. *Biol. Bull. 135*, 335–348.

PETITPREN, M. F. and KNIGHT, A. W. (1970). Oxygen consumption of the dragonfly, *Anax junius*. *J. Insect Physiol. 16*, 449–459.

PHILIPS, M. E. (1939). The anterior peristigmatic glands in trypetid larvae. *Ann. Ent. Soc. Amer. 32*, 325–328.

PHILIPSON, G. N. (1954). The effect of water flow and oxygen concentration on six species of caddis fly larvae. *Proc. Zool. Soc. Lond. 124*, 547–564.

PICKARD, R. S. and MILL, P. J. (1972). Ventilatory muscle activity in intact preparations of aeshnid dragonfly larvae. *J. Exp. Biol. 56*, 527–536.

PICKARD, R. S. and MILL, P. J. (1974a). Ventilatory movements of the abdomen and branchial apparatus in dragonfly larvae. *J. Zool. 174*, 23–40.

PICKARD, R. S. and MILL, P. J. (1974b). The effects of carbon dioxide and oxygen on respiratory dorso-ventral muscle activity during normal ventilation in *Anax imperator*, Leach. *Odonatologica 3*, 249–255.

PICKARD, R. S. and MILL, P. J. (1975). Ventilatory muscle activity in restrained and free-swimming dragonfly larvae (Odonata: Anisoptera). *J. Comp. Physiol. A. 96*, 37–52.

PILL, C. E. J. and MILL, P. J. (1981). The structure and physiology of abdominal proprioceptors in larval dragonflies (Anisoptera). *Odonatologica 10*, 117–130.

PLATEAU, F. (1884). Recherches expérimentalis sur les mouvements respiratoires des insectes. *Memoires de l'Académie royale des sciences, des lettres et des beaux-arts de Belgique. 45*, 1–219.

POISSON, R. (1926). L'*Anisops producta* Fieb. (Hémiptère, Notonectidae). Observations sur son anatomie et sa biologie. *Arch. Zool. Exp. Gén. 65*, 181–208.

POONAWALLA, Z. T. (1966). The respiratory system of adult Odonata. Part 1: The spiracles. *Ann. Ent. Soc. Amer. 59*, 807–809.

PRINGLE, J. W. S. (1954). A physiological analysis of cicada song. *J. Exp. Biol. 31*, 525–560.

PUNT, A., PARSLER, W. J. and KUCHLEIN, J. (1957). Oxygen uptake in insects with cyclic CO_2 release. *Biol. Bull. 112*, 108–119.

RAHN, H. and PAGANELLI, C. V. (1968). Gas exchange in gas gills of diving insects. *Resp. Physiol. 5*, 145–164.

RÉAUMUR, R. A. F. (1734–43). *Mémoires pour servir a l'histoire des insectes*. Paris, vol. *3*.

REYNOLDS, S. E. (1980). Integration of behaviour and physiology in ecdysis. *Adv. Insect Physiol. 15*, 475–595.

RICHARDS, A. G. and KORDA, F. H. (1950). Studies on arthropod cuticle. IV. An electron microscope survey of the intima of arthropod tracheae. *Ann. Ent. Soc. Amer. 43*, 49–71.

RICHTER, P. O. (1969a). Spiracles of adult Scarabaeoidea (Coleoptera) and their phylogenetic significance. I. The abdominal spiracles. *Ann. Ent. Soc. Amer. 62*, 869–880.

RICHTER, P. O. (1969b). Spiracles of adult Scarabaeoidea (Coleoptera) and their phylogenetic significance. II. Thoracic spiracles and adjacent sclerites. *Ann. Ent. Soc. Amer. 62*, 1388–1398.

ROBERT, P.-A. (1958). *Les Libellules (Odonates)*. Delachaux et Niestlé S.A., Neuchâtel.

SADONES, J. (1896). *La Cellule, Louvain. 11*, 273.

SAYLE, M. H. (1928). Factors influencing the rate of metabolism of *Aeshna umbrosa* nymphs. *Biol. Bull. 54*, 212–230.

SCHNEIDERMAN, H. A. (1960). Discontinuous respiration in insects: role of the spiracles. *Biol. Bull. 119*, 494–528.

SCHNEIDERMAN, H. A. and SCHECHTER, A. N. (1966). Discontinuous respiration in insects. V. Pressure and volume changes in the tracheal system of silkmoth pupae. *J. Insect Physiol. 12*, 1143–1170.

SCHNEIDERMAN, H. A. and WILLIAMS, C. M. (1953). The physiology of insect diapause — VII. The respiratory metabolism of the Cecropia silkworm during diapause and development. *Biol. Bull. 105*, 320–334.

SCHNEIDERMAN, H. A. and WILLIAMS, C. M. (1955). An experimental analysis of the discontinuous respiration of the Cecropia silkworm. *Biol. Bull. 109*, 123–143.

SCHREUDE, J. and DE WILDE, J. (1952). Analysis of the dyspnoeic action of carbon dioxide in the cockroach (*Periplaneta americana* L.). *Physiol. Comparata et Oecol. 2*, 335–361.

SCUDDER, G. G. E. (1963). Adult abdominal characters in the lygaeoid-coreoid complex of the Heteroptera and the classification of the group. *Canad. J. Zool. 41*, 1–14.

SHANKLAND, D. L. (1965). Nerves and muscles of the pregenital abdominal segments of the American cockroach, *Periplaneta americana* (L). *J. Morph. 117*, 353–385.

SLÁMA, K. (1960). Physiology of sawfly metamorphosis. I. Continuous respiration in diapausing prepupae and pupae. *J. Insect Physiol. 5*, 341–348.

SLEIGHT, C. S. (1913). Relations of Trichoptera to their environment. *J.N.Y. Ent. Soc. 21–22*, 4.

SMITH, D. S. (1961). The structure of insect fibrillar flight muscle. A study made with special reference to the membrane systems of the fiber. *J. Biophys. Biochem. Cytol. 10* (*suppl.*), 123–158.

SMITH, D. S. (1963). The structure of flight muscle sarcosomes in the blowfly *Calliphora erythrocephala* (Diptera). *J. Cell. Biol. 19*, 115–138.

SMITH, D. S. (1966). The organization of flight muscle fibers in the Odonata. *J. Cell. Biol. 28*, 109–126.

SMITH, D. S. (1968). *Insect Cells, Their Structure and Function*. Oliver & Boyd, Edinburgh.

SMITH, D. S., TELFER, W. H. and NEVILLE, A. C. (1971). Fine structure of the chorion of a moth, *Hyalophora cecropia. Tissue Cell 3*, 477–497.

SNODGRASS, R. E. (1935). *Principles of Insect Morphology*. McGraw-Hill, New York.

SNODGRASS, R. E. (1954). The Dragonfly larva. *Smithson. Misc. Coll. 123* (2), 1–38.

STEEN, J. B. (1971). *Comparative Physiology of Respiratory Mechanisms*. Academic Press, New York.

STEINER, L. F. (1929). Homologies of tracheal branches in the nymph of *Anax junius* based on their correlation with the muscles they supply. *Ann. Ent. Soc. Amer. 22*, 297–309.

STOCKNER, J. G. (1971). Ecological energetics and natural history of *Hedriodiscus truquii* (Diptera) in two thermal spring communities. *J. Fish. Res. Bd. Canada. 28*, 73–94.

STRIDE, G. O. (1955). On the respiration of an aquatic african beetle, *Potamodytes tuberosus* Hinton. *Ann. Ent. Soc. Amer. 48*, 344–351.

STRIDE, G. O. (1958). The application of a Bernoulli equation to problems of insect respiration. *Proc. 10th Int. Congr. Ent.* (*1956*) *2*, 335–336.

SZABÓ-PATAY, J. (1924). Sur la morphologie et la fonction de l'appareil respiratoire des *Aphelocheirus*. *Ann. Mus. Nat. Hungary 21*, 35–55.

TELFER, W. H. and SMITH, D. S. (1970). Aspects of egg formation. In *Insect Ultrastructure* (*Symp. Roy. Ent. Soc. 5*). Edited by A. C. Neville. Pages 117–134. Blackwell, Oxford and Edinburgh.

THIENEMANN, A. (1923). Die beiden Chironomusarten der Tiefenfauna der Norddeutschen Seen. *Verh. Int. Ver. Limnol. Kiel. Stuttgart. 1*, 108.

THORPE, W. H. (1950). Plastron respiration in aquatic insects. *Biol. Rev. 25*, 344–390.

THORPE, W. H. and CRISP, D. J. (1947a). Studies on plastron respiration. I. The biology of *Aphelocheirus* (Hemiptera, Aphelocheiridae (Naucoridae)) and the mechanism of plastron retention. *J. Exp. Biol. 24*, 227–269.

THORPE, W. H. and CRISP, D. J. (1947b). Studies on plastron respiration. II. The respiratory efficiency of the plastron in *Aphelocheirus*. *J. Exp. Biol. 24*, 270–303.

THORPE, W. H. and CRISP, D. J. (1947c). Studies on plastron respiration. III. The orientation responses of *Aphelocheirus* (Hemiptera, Aphelocheiridae (Naucoridae)) in relation to plastron respiration; together with an account of specialized pressure receptors in aquatic insects. *J. Exp. Biol. 24*, 310–328.

THORPE, W. H. and CRISP, D. J. (1949). Studies on plastron respiration. IV. Plastron respiration in the Coleoptera. *J. Exp. Biol. 26*, 219–260.

TILLYARD, R. J. (1916). Life-histories and descriptions of Australian Aeschninae. *J. Linn. Soc. Lond., Zool. 33*, 1–83.

TILLYARD, R. J. (1917). *The Biology of Dragonflies*. Cambridge University Press, Cambridge.

TONAPI, G. T. (1957). A comparative study of spiracular structure and mechanisms in some Hymenoptera. *Trans. Roy. Ent. Soc. A. 110*, 489–520.

TONNER, F. (1936). Mechanik und Koordination der Atem- und Schwimmbewegung bei Libellenlarven. *Z. wiss Zool. 147*, 433–454.

TRUMAN, J. W. and ENDO, P. T. (1974). Physiology of insect ecdysis: neural and hormonal factors involved in wing-spreading behaviour of moths. *J. Exp. Biol. 61*, 47–55.

VAN DAM, L. (1937). Uber die Atembewegungen und das Atemvolumen von *Phryganea*-larven, *Arenicola marina* und *Nereis virens*, sowie uber die Sauerstoffausnutzung bei *Anodonta cygnea*, *Arenicola marina* und *Nereis virens*. *Zool. Anz. 118*, 122–128.

VAN DER KLOOT, W. G. (1963). The electrophysiological and nervous control of the spiracular muscle of pupae of giant silkmoths. *Comp. Biochem. Physiol. 9*, 317–333.

VARLEY, G. C. (1937). Aquatic insect larvae which obtain oxygen from the roots of plants. *Proc. R. Ent. Soc. Lond. A. 12*, 55–60.

WALLENGREN, H. (1914a). Physiologisch-biologische Studien uber die Atmung bei den Arthropoden. II. Die Mechanik der Atembewegungen bei Aeschnalarven. *Lunds Univ. Arss. N.F. Afd. 2 10*, (4), 1–24.

WALLENGREN, H. (1914b). Physiologisch-biologische Studien uber die Atmung bei den Arthropoden. III. Die Atmung der Aeschnalarven. *Lunds Univ. Arss. N.F. Afd. 2 10*, (8), 1–28.

WALSHE, B. M. (1947a). On the function of haemoglobin in *Chironomus* after oxygen lack. *J. Exp. Biol. 24*, 329–342.

WALSHE, B. M. (1947b). The function of haemoglobin in *Tanytarsus* (Chironomidae). *J. Exp. Biol. 24*, 343–351.

WALSHE, B. M. (1950). The function of haemoglobin in *Chironomus plumosus* under natural conditions. *J. Exp. Biol. 27*, 73–95.

WASSERTHAL, L. T. (1976). Heartbeat reversal and its correlation with accessory pulsatile organs and abdominal movement in Lepidoptera. *Experientia 32*, 577–578.

WASSERTHAL, L. T. (1980). Oscillating haemolymph "circulation" in the butterfly *Papilio machaon* L. revealed by contact thermography and photocell measurements. *J. Comp. Physiol. 139*, 145–163.

WASSERTHAL, L. T. (1981). Oscillating haemolymph "circulation" and discontinuous tracheal ventilation in the giant silk moth *Attacus atlas* L. *J. Comp. Physiol. 145*, 1–15.

WATTS, D. T. (1951). Intratracheal pressure in insect respiration. *Ann. Ent. Soc. Amer. 44*, 527–538.

WAUTIER, J. and PATTÉE, E. (1955). Experience physiologique et experience ecologique. L'influence du substrat sur la consommation d'oxygene chez les larves d'ephemeropteres. *Bull. Mens. Soc. Linn. Lyon (N.S.), No. 7*, 178–183.

WEBER, T. and CAILLÈRE, L. (1978). Thermistor telemetry of ventilation during prey capture by dragonfly larvae (*Cordulegaster boltonii*, Odonata). *J. Comp. Physiol. A. 128*, 341–345.

WEIS-FOGH, T. (1964a). Functional design of the tracheal system of flying insects as compared with the avian lung. *J. Exp. Biol. 41*, 207–227.

WEIS-FOGH, T. (1964b). Diffusion in insect wing muscle, the most active tissue known. *J. Exp. Biol. 41*, 229–256.

WEIS-FOGH, T. (1967). Respiration and tracheal ventilation in locusts and other flying insects. *J. Exp. Biol. 47*, 561–587.

WELCH, P. S. (1916). Contribution to the biology of certain aquatic Lepidoptera. *Ann. Ent. Soc. Amer. 9*, 159–190.

WELCH, P. S. and SEHON, G. L. (1928). The periodic vibratory movements of the larva of *Nymphula maculalis* Clemens (Lepidoptera) and their respiratory significance. *Ann. Ent. Soc. Amer. 21*, 243–258.

WHEDON, A. D. (1918). The comparative physiology and possible adaptions of the abdomen in the Odonata. *Trans. Amer. Ent. Soc. 44*, 373–437.

WHITTEN, J. M. (1968). Metamorphic changes in insects. In *Metamorphosis: A Problem in Developmental Biology*. Edited by W. Etkin and L. I. Gilbert. Pages 43–105. Appleton-Century-Crofts, New York.

WHITTEN, J. M. (1972). Comparative anatomy of the tracheal system. *Ann. Rev. Ent. 17*, 373–402.

WICHARD, W. (1973). Zur Morphogenese des respiratorischen Epithels der Tracheenkiemen bei Larven der Limnephilini Kol. (Insecta, Trichoptera). *Z. Zellforsch. 144*, 585–592.

WICHARD, W. (1974a). Zur morphologischen Anpassung von Tracheenkiemen bei Larven der Limnephilini Kol. (Insecta, Trichoptera). I. Autökologische Untersuchungen im Eggstatter Seengebiet im Chiemgau. *Oecologia 15*, 159–167.

WICHARD, W. (1974b). Zur morphologischen Anpassung von Tracheenkiemen bei Larven der Limnephilini Kol. (Insecta, Trichoptera). II. Adaptationsversuche unter verschiedenen O_2-Bedingungen während der larvalen Entwicklung. *Oecologia 15*, 169–175.

WICHARD, W. (1977). Structure and function of the tracheal gills of *Molanna angustata* Curt. *Proc. 2nd Int. Symp. Trichoptera*. The Hague, The Netherlands. Pages 293–296.

WICHARD, W. (1979a). Structure and function of the respiratory epithelium in the tracheal gills of mayfly larvae. *Proc. 2nd Int. Conf. Ephemeroptera*. Krakow, Poland, 1975. Pages 307–309.

WICHARD, W. (1979b). Zur Feinstruktur der abdominalen Tracheenkiemen von Larven der Kleinlibellen-Art *Epallage fatime* (Odonata: Zygoptera: Euphaeidae). *Ent. Gen. 5*, 129–134.

WICHARD, W. and KOMNICK, H. (1971). Zur Feinstruktur der Tracheenkiemen von *Glyphotaelius pellucidus* Retz. (Insecta, Trichoptera). *Cytobiologie 3*, 106–110.

WICHARD, W. and KOMNICK, H. (1974a). Zur Feinstruktur der rektalen Tracheenkiemen von anisopteren Libellenlarven. I. Das respiratorische Epithel. *Odonatologica 3*, 121–127.

WICHARD, W. and KOMNICK, H. (1974b). Zur Feinstruktur der rektalen Trachenkiemen von anisopteren Libellenlarven. II. Das rektale Chloridepithel. *Odonatologica 3*, 129–135.

WICHARD, W. and KOMNICK, H. (1974c). Fine structure and function of the rectal chloride epithelia of damselfly larvae. *J. Insect Physiol. 20*, 1611–1621.

WIGGLESWORTH, V. B. (1930). A theory of tracheal respiration in insects. *Proc. Roy. Soc. B. 106*, 229–250.

WIGGLESWORTH, V. B. (1932). The extent of air in the tracheoles of some terrestrial insects. *Proc. Roy. Soc. B. 109*, 354–359.

WIGGLESWORTH, V. B. (1935). The regulation of respiration in the flea, *Xenopsylla cheopsis* Roths. (Pulicidae). *Proc. Roy. Soc. B. 118*, 397–419.

WIGGLESWORTH, V. B. (1953a). Surface forces in the tracheal system of insects. *Quart. J. Mic. Sci. 94*, 507–522.

WIGGLESWORTH, V. B. (1953b). *The Principles of Insect Physiology*. 5th edn. Methuen, London.

WIGGLESWORTH, V. B. (1959). The role of the epidermal cells in the migration of tracheoles in *Rhodnius prolixus*. *J. Exp. Biol. 36*, 632–640.

WIGGLESWORTH, V. B. (1970). Structural lipids in the insect cuticle and the function of the oenocytes. *Tissue Cell 2*, 155–179.

WIGGLESWORTH, V. B. (1972). *The Principles of Insect Physiology*. 7th edn. Chapman & Hall, London.

WIGGLESWORTH, V. B. and BEAMENT, J. W. L. (1950). The respiratory mechanisms of some insect eggs. *Quart. J. Mic. Sci. 91*, 429–452.

WILKINS, M. B. (1960). A temperature-dependent endogenous rhythm in the rate of carbon dioxide output of *Periplaneta americana*. *Nature, Lond. 185*, 481–482.

WINGFIELD, C. A. (1937). Function of the gills of the mayfly nymph, *Cloëon dipterum*. *Nature, Lond. 140*, 27.

WINGFIELD, C. A. (1939). The function of the gills of mayfly nymphs from different habitats. *J. Exp. Biol. 16*, 363–373.

ZAHNER, R. (1959). Uber die Bindung der mitteleuropaischen *Calopteryx* — Arten (Odonata, Zygoptera) an den Lebensraum des stromenden den Wassers. I. Der Anteil der Larven an der Biotopbindung. *Int. Rev. Hydrobiol. 44*, 51–130.

ZAWARZIN, A. (1924). Uber die histologische Beschaffenheit des unpaaren ventralen Nervs der Insekten. *Z. Wiss. Zool. 122*, 97–115.

Species Index

Author Index

Subject Index